THE FLETCHER JONES FOUNDATION

HUMANITIES IMPRINT

The Fletcher Jones Foundation has endowed this imprint to foster innovative and enduring scholarship in the humanities.

The publisher gratefully acknowledges the generous support of the Fletcher Jones Foundation Humanities Endowment Fund of the University of California Press Foundation.

THE COPERNICAN QUESTION

THE COPERNICAN QUESTION

Prognostication, Skepticism, and Celestial Order

Robert S. Westman

UNIVERSITY OF CALIFORNIA PRESS

Berkeley Los Angeles London

University of California Press, one of the most distinguished university presses in the United States, enriches lives around the world by advancing scholarship in the humanities, social sciences, and natural sciences. Its activities are supported by the UC Press Foundation and by philanthropic contributions from individuals and institutions. For more information, visit www.ucpress.edu.

University of California Press
Berkeley and Los Angeles, California

University of California Press, Ltd.
London, England

Library of Congress Cataloging-in-Publication Data

Westman, Robert S.
   The Copernican question : prognostication, skepticism, and celestial order / Robert S. Westman.
       p.   cm.
   Includes bibliographical references and index.
   ISBN 978-0-520-25481-7 (cloth : alk. paper)
     1. Astronomy, Renaissance—Europe—History—16th century.
2. Science—Philosophy—Europe—History—16th century.
3. Copernicus, Nicolaus, 1473–1543.   4. Galilei, Galileo, 1564–
1642.   5. Kepler, Johannes, 1571–1630.   I. Title.
QB29.W47 2011
   520.94'09031—dc22   2009039562

Manufactured in the United States of America

20   19   18   17   16   15   14   13   12
10   9   8   7   6   5   4   3   2

In keeping with a commitment to support environmentally responsible and sustainable printing practices, UC Press has printed this book on Rolland Enviro100, a 100% post-consumer fiber paper that is FSC certified, deinked, processed chlorine-free, and manufactured with renewable biogas energy. It is acid-free and EcoLogo certified.

For my wife, Rachel, my sons Aaron and Jonathan,
and to the memory of my brother, Walt

How many people [who] are immersed in a discipline are used to reducing everything to it, and not because of a desire to explain everything by it, but because things really seem like that to them. What happens to them is like someone who walks immersed in snow and to whom everything ends up appearing white . . . like someone who loves in vain and sees the face of his beloved in everything. . . . So he who is a theologian, and nothing but a theologian, takes everything back to divine causes; he who is a doctor takes everything back to corporal states, the physicist to the natural principles of things, the mathematician, like the Pythagoreans, to numbers and figures. In the same way, the Chaldeans were entirely occupied with the measurement of celestial movements and the observation of the positions of the stars . . . and all things were stars to them, and they willingly took everything back to the stars.

GIOVANNI PICO DELLA MIRANDOLA,
Disputationes contra Astrologiam Divinatricem, 1496

TEOFILO: Coming back to Nundinio, at this point he started to show his teeth, gape his jaws, squint his eyes, wrinkle his eyebrows, flare his nostrils and utter a capon's crow from his windpipe, in order to make the people present understand by that laughter that he understood well, that he was right while the other was saying ridiculous things.

FRULLA: Is not the truth of what the Nolan said proven by the fact that Nundinio laughed at it so much?

GIORDANO BRUNO, La cena de le ceneri, 1584

I see some who study and comment on their almanacs and cite their authority in current events. With all they say, they necessarily tell both truth and falsehood. *For who is there who, shooting all day, will not sometime hit the mark?* [Cicero] . . . Besides, no one keeps a record of their mistakes, inasmuch as these are ordinary and numberless; and their correct divinations are made much of because they are rare, incredible, and prodigious.

MICHEL DE MONTAIGNE, "Of Prognostications," in Essays, 1580

It's tough to make predictions, especially about the future.

Attributed to YOGI BERRA
(and, in a slightly different version, to Niels Bohr)

The only function of economic forecasting is to make astrology look respectable.

JOHN KENNETH GALBRAITH

CONTENTS

ILLUSTRATIONS

PREFACE AND ACKNOWLEDGMENTS

Under what conditions do people change or give up beliefs to which they are most deeply committed? This general concern lay behind my original interest in the topic of this book: why and how Copernicus changed his own thinking about the organization of the heavens and what made his discovery persuasive to others after its publication in 1543. When I was first drawn to these historical questions in the late 1960s, they were central to the prevailing postwar narrative of the historiography of science, a story cast in the language of "scientific revolution" and "the origins of modern science." Although it is surely uncontroversial to say that there were many profound conceptual changes between 1500 and 1700, it is dubious whether *revolution,* a term normally used to designate a short-term political upheaval or rupture, is the right governing metaphor. Yet a difficulty about origins still remains: what, after all, was the question that Copernicus was trying to answer? This book is as much about that question—and the problems it generated and left unresolved—as it is about the kinds of answers offered by Copernicus and those who followed him.

In the 1970s and '80s, what seemed to me missing from discussions of the Copernican episode was the strong connection between transformations of concepts and the communal dimension of scientific change that Thomas Kuhn's work had foregrounded theoretically in *The Structure of Scientific Revolutions* but left unaddressed in his *Copernican Revolution.* Thus initially I sought to probe how Copernicus's work was read and

used in different cultural-political settings and how I should write about those spaces without introducing anachronistic categories, such as *research, scientist,* or *the scientific community.* In a series of papers that I published between 1975 and 1990 I identified and probed three major arenas of controversy about the arrangement of the heavens: the university, the aristocratic court, and the church (Westman 1975b, 1980a, 1980b, 1983, 1986, 1990). Readers familiar with those papers may occasionally recognize in this study a few surviving remnants of passages or turns of phrase; but in all cases, I have critically reexamined my earlier writings and revisited the sources on which they were based.

Put somewhat differently, I rejected the sensible advice of friends to get on with my career and republish my earlier studies as a modestly amplified collection. My decision to take an alternative route can be explained in part by a passion to get to the bottom of the questions posed at the beginning of this preface, even while appreciating that historical narratives, like the meanings we make of our own lives, are always subject to revision and reinterpretation. Still, I confess that the prospect of writing a long-term history felt daunting, not only as a practical matter of integrating the unending stream of new studies but because there were precious few models to emulate. To add to the arguments against such an undertaking, some historians had concluded that the *longue durée*—the so-called "big picture" history of science—was nothing but a residue of a less sophisticated historiographical era that projected a preformed template of rationality and progress onto the past (Porter 1986; Cunningham and Williams 1993). Against such late-modernist confidence, a postmodernist, anthropologizing, and sociologizing sensibility had begun to take hold, producing an opposing image of scientific knowledge as contingent, embedded practice. This perspective made rationality situational rather than universal and eschewed progress in favor of temporally circumscribed, "thick" treatments of knowledge-making practices.

In early 1991, my work took a new turn when Luce Giard arrived from Paris for her annual stint of teaching at the University of California, San Diego and suggested that I might show her an outline of the project—an outline that she quickly judged capable of yielding two books rather than one. Thanks in part to her generous willingness to read still more, I turned my entire attention to this book, a project that I estimated (as usual, incorrectly) would take about five years to complete. Once under way, the writing took on a life of its own as I found myself unexpectedly moving from my intended starting date of 1543 back into the fifteenth century. That redirection resulted in large measure from reading a remarkable group of astrological prognostications acquired by the British Museum in 1898 from the estate of an Italian prince and early historian of mathematics, Baldassarre Boncompagni (1821–94). This collection, first described by the indefatigable Lynn Thorndike in volume 5 of his *History of Magic and Experimental Science* (1941), contains several prognostications by Domenico Maria Novara, a master of astronomy with whom the young Copernicus lived and worked during his student days in Bologna. Eventually, I was able to find more than a dozen forecasts by Novara dispersed among several European libraries. Studying these prognostications caused me to rethink my entire approach and to see a way forward to a new kind of big picture—although this narrative makes no claims to finality.

The Chicago Manual of Style instructs that a preface should normally include the author's "method of research (if this has some bearing on readers' understanding of the text)." Apart from my visits to libraries, duly noted below, I seem to work best in an atmosphere filled with people, classical music, good cappuccino, and biscotti (but not smoking). I wrote most of this book in

various independent coffee houses in San Diego, moving from one to the next as each in turn succumbed to the vagaries of the business cycle. By 2005 I had completed a good-enough draft of the project, writing at a modestly appointed but welcoming establishment called "The Urban Grind." I am most grateful to the University of California for supporting my idiosyncratic work habits with two short-term fellowships—a President's Humanities Research Fellowship and a UCSD Center for the Humanities Fellowship.

Luce Giard read and commented perceptively on every chapter. Floris Cohen provided a long and very helpful critique of the entire penultimate draft. Michael Shank offered invaluable critical comments on early versions of chapters 1–4. Several colleagues and students kindly read drafts of individual chapters, helped sharpen my historiographical and philosophical perspectives, and reassured me that I was not writing fiction: Anna Alexandrova, Margaret Garber, Andrew Hamilton, Marcus Hellyer, Sandra Logan, Ernan McMullin, John Marino, Naomi Oreskes, Pamela Smith, Jacob Stegenga, Mary Terrall, Steven Vanden Broecke, and Eric Van Young, as well as the late Amos Funkenstein and the late Richard H. Popkin. Numerous stimulating and enjoyable exchanges with Harun Küçük greatly encouraged my thinking, beyond which he also saved me considerable time and effort by creating a bibliography from my endnotes and by preparing several of the illustrations. I thank everyone for helping to make this a better book.

I owe much to the support of good friends who were there for me over many years, in weather fair and foul: Joseph Aguayo, Jeffrey Prager, and Debora Silverman, and, at a critical juncture, Michael Meranze and Helen Deutsch. David W. Jacobs helped immeasurably to keep my focus centered. Friends and colleagues whose support has been very meaningful to me, although they might not realize how much, include Frank Biess, Eric Blau, Michael Dennis, Julie Gollin, Virginia Gordon, Sarah Hutton, Judith Hughes, Stephanie Jed, Hasan Kayali, Edward Lee, Lynn Mally, Marjorie Milstein, Robert Moeller, Rebecca Plant, Ted Porter, Kathryn Shevelow, Julia Smith, Ulrike Strasser, Eric Van Young, and my wonderful neighbors on Cranbrook Court and Sugarman Drive. And finally, I will always be grateful for the warm receptions I received from Karel and Anka Bergman in Cardiff, Wales; from the family of Karel and Sylvia Fink on my first visit to Prague during the "Prague Spring" of 1968; and from Honza and Zuzka Rous on many subsequent research trips to that city, before and after 1989.

For decades, Owen Gingerich has worked in many of the same venues, and my debts to his work will be apparent throughout. I have also profited from the many fine studies of Miguel Angel Granada, a veritable cascade of whose articles appeared as I was composing this book. I received a copy of André Goddu's excellent book *Copernicus and the Aristotelian Tradition* (2010) as I was editing copy, and I have been able register my engagement with its most important insights.

My mentors in London, the late A. Rupert Hall and Marie Boas Hall, first suggested the general topic before I realized that it was a topic worth pursuing. I presented a much shorter version of chapter 1 at a conference in 2001 honoring David C. Lindberg, who sparked my interest in the history of science when it was a quite different field.

Invitations to present seminars and lectures allowed me to test my ideas before different audiences. I am grateful for the thoughtful reactions offered by those who attended my presentations at the British Society for the History of Science (Essex, 1993); the University of Ferrara (1993); Cornell University (1994), Harvard University (1994); the History of Science Society (Vancouver, 2000); the University of Wisconsin (2001); the University of California, Los Angeles (2001 and 2008); Indiana University (2003); the Joint Meeting of the British, Canadian, and U.S. History of Science Societies (Halifax, 2004 and Oxford, 2008); the University of California, San Diego (2000, 2005, and 2009); Roskilde University (2007); the University of British Columbia (the Stephen Straker Memorial Lecture, 2008); the University of Edinburgh (2009); the Henry E. Huntington Library (2009 and 2010); the American Historical Association (San Diego, 2010); and the Johns Hopkins University (2010).

The garnering of research materials for this book began in the predigital age, when consulting the British Museum's reference catalogue was a matter of multivolume weight lifting and when the fountain pen, the typewriter, the photocopier, and the microfilm reader were the main tools of historical scholarship. I am immensely grateful to the research committees of the UCLA and UCSD academic senates, whose support enabled me to travel to European libraries and to build up a substantial microfilm collection of rare books and manuscripts. Even now, when it is possible to check references in the British Library catalogue from a California café, there is still no substitute for consulting the original sources in their home repositories, and I am grateful to the directors and librarians of those many remarkable institutions who granted both access and support: the Archivio di Stato di Bologna; the Bayerische Staatsbibliothek, Munich; the Biblioteca Ambrosiana, Milan; the Biblioteca Apostolica Vaticana; the Biblioteca Colombina y Capitular, Seville; the Biblioteca Comunale Ariostea, Ferrara; the Biblioteca Comunale dell'Archiginnasio, Bologna; the Biblioteca Estense, Modena; the Biblioteca Medicea Laurenziana, Florence; the Biblioteca Nazionale Centrale di Firenze; the Biblioteca Nazionale Centrale di Roma; the Biblioteca Riccardiana, Florence; the Biblioteca Universitaria di Bologna; the Biblioteca Universitaria di Pisa; the Biblioteka Uniwersytecka we Wrocławiu; the Bibliothèque Mazarine, Paris; the Bibliothèque Nationale de France; the Bibliothèque de l'Observatoire de Paris; the British Library, London; the Bodleian Library, Oxford; the Cambridge University Library; the Columbia University Library; the Herzog August Bibliothek, Wolfenbüttel; the Huntington Library; the Istituto e Museo di Storia della Scienza, Florence; the Národní Knihovna České Republiky (Klementinum), Prague; the Österreichische Nationalbibliothek, Vienna; the Royal Observatory, Edinburgh (Crawford Collection); the San Diego State University Library (Zinner Collection); the Schaffhausen Stadtbibliothek; the Schweinfurt Stadtarchiv; the Stanford University Library; the Strahovská Knihovna, Prague; the Universitätsbibliothek Erlangen; the Universitätsbibliothek Tübingen; the University of California, Los Angeles (Charles E. Young Research Library); the Warburg Institute, London; the Württembergische Landesbibliothek, Stuttgart; and the Yale University Beinecke Library. At the University of California, San Diego, my home institution, I am especially grateful for the indispensable research facilities of the Geisel Library and for the support of Lynda Claassen (Special Collections), Sam Dunlap (European Studies and History of Science), and Brian Schottlaender (Audrey Geisel University Librarian).

At a time when intellectual property rights are

in flux, libraries are struggling to support services partly with reproduction and permission fees, and publishers are under massive pressure to control costs, the publication of images has become a major burden to authors. I thank all the institutions that kindly provided copies of images for this book. I wish to register special thanks to the University of Oklahoma History of Science Collections for its enlightened policy of making available free online use of visual material from its rich collections of rare primary sources. In Bologna, I cannot forget the generosity of Giuliano Pancaldi, who supported my lodging and also managed to open otherwise inaccessible doors to libraries, villas, and even the Vatican, and without whose assistance some crucial images would not have appeared in this book. To Marco Bresadola, Michele Camerota, and Paolo Galluzzi, my thanks as well for helping me to obtain several other images.

I am pleased to acknowledge permission to use materials and occasional phrasing derived from a few of my previously published articles: "The Melanchthon Circle, Rheticus, and the Wittenberg Interpretation of the Copernican Theory" (© 1975 by the University of Chicago Press); "Three Responses to the Copernican Theory: Johannes Praetorius, Tycho Brahe and Michael Maestlin" (© 1975 by the Regents of the University of California); "The Astronomer's Role in the Sixteenth Century: A Preliminary Study" (© 1980 by Science History Publications, Ltd.); "The Copernicans and the Churches" (© 1986 by the Regents of the University of California); "Proof, Poetics and Patronage: Copernicus's Preface to *De revolutionibus*" (© 1990 by Cambridge University Press); "Two Cultures or One? A Second Look at Kuhn's *The Copernican Revolution*" (© 1994 by the University of Chicago Press); and "Was Kepler a Secular Theologian?" (© 2008 by the University of Toronto Press).

I cannot say enough for the high quality and professionalism of the editorial staff of the University of California Press. Stan Holwitz's intellectual commitment and long experience in publishing convinced me to submit the manuscript. Erika Büky brilliantly edited a complex and demanding text, challenged me to improve my constructions, and meticulously corrected infelicities at all levels. Working within the narrow dictates of allowable space, Claudia Smelser brought her tasteful contributions to the book's design. In different ways, Chalon Emmons, Laura Harger, Sheila Levine, and Rose Vekony skillfully managed the overall process of production. To everyone, my deepest thanks. I alone am responsible for any remaining errors or misstatements.

My sons Aaron and Jonathan have never known me not to be working away on something or other about Copernicus—often late at night. Their presence in my life has brought me incomparable cheer. Now that they are grown, I hope that they will forgive their father's eccentricities and his failure to take them to more baseball games.

To my wife Rachel, I owe the most. She has always understood the joys and demons of writing. Her love, inspiration, and companionship made it possible for me to undertake this journey.

Introduction

THE HISTORICAL PROBLEMATIC

Is the Earth motionless at the center of a finite, star-studded sphere, or is it a planet moving in an annual circuit around the center? Medieval scholastic natural philosophers debated all sorts of imaginative questions of this kind: whether there are, or could be, more worlds; if there were several worlds, whether the earth of one could be moved naturally to the center of another; whether the spots appearing on the Moon arise from differences in parts of the Moon or from something external; whether the Earth is fixed in the middle of the world and has the same center of gravity; and whether the Earth rotates around its axis.[1]

There were two motivations for entertaining such alternative possibilities. The first arose from natural philosophers answering theological worries about threats to God's unlimited, absolute power: for example, could God not make several worlds, if he so wished? But the second source of alternatives was already built into Aristotle's argumentational and rhetorical practices. Aristotle frequently reported the claims of his predecessors only to reject them in favor of his own positions. One such view was that of the Pythagoreans, who "affirm[ed] that the center is occupied by fire, and that the earth is one of the stars, and creates night and day as it travels in a circle about the center."[2] From the thirteenth to the seventeenth centuries, Aristotle's description of the Pythagorean view became a standard part of the argument that students learned—and then learned to reject—in support of the Earth's centrality and immobility. It was only sometime in the last years of the fifteenth and the first decade of the sixteenth century that a Polish church canon and sometime astronomical practitioner named Nicolaus Copernicus posed the Pythagorean idea to himself in a new way. He did so not in the thirteenth-century-philosophical style, as an alternative to be rejected, but rather as a mathematical assumption in the style of Claudius Ptolemy, reinterpreting the old Pythagorean idea as an astronomical explanation for two perplexing problems: first, the Sun's apparent motion as mirrored in the planets' motions, and second, the disputed ordering of Venus and Mercury. Yet not until 1543 did Copernicus finally publish a full-dress defense of this explanation and mobilize it as a vehicle for persuading others.

The Copernican Question opens with a paradox of historical context. Why ever did Copernicus concern himself about the order of the planets when the burgeoning late-fifteenth and early-sixteenth-century heavenly print literature, directed to learned elites and ordinary people alike, was overwhelmingly preoccupied with astrologically driven anticipations of the future, sometimes coupled with powerful apocalyptic fantasies that the world would soon come to an end? For those who read Copernicus's book, *De Revolutionibus Orbium Coelestium* (On the Revolutions of the Heavenly Spheres), what did getting the structure of the heavens right have to do with more accurately predicting the future? And with printing tech-

Theologus. Aftronomus.

1. Theologian and astronomer-astrologer seeking concordance, [Petrus de] Alliaco 1490. Linda Hall Library of Science, Engineering & Technology.

nology making possible the production, circulation, and comparison of an increasing number of prophetic schemes, which prophecies or which combination of prophetic authorities—biblical, extrabiblical, astrological—were to be trusted?[3] Indeed, could heavenly knowledge support prophecy? During the Great Schism of 1378–1414, when the theologian Pierre d'Ailly worried that three men all claiming to be pope betokened the imminent arrival of the Antichrist, he turned for assistance to conjunctions of Saturn and Jupiter and slow, long-term motions of the sphere just beyond the fixed stars—seeking reassurance in a "concordance" of the Bible with astrology and ultimately concluding that the Antichrist's coming would not occur before 1789.[4]

Among those immersed in such categories and authorities was Copernicus's early contemporary, Christopher Columbus (1451–1506), who

regarded his "Enterprise of the Indies" as but a step toward the fulfillment of his own guiding fantasy in the service of the Spanish crown: the liberation and reconquest of Jerusalem. Steeped in the astrological and biblical prophecies of Pierre d'Ailly and following Saint Augustine's figure of seven thousand years for the world's duration, Columbus believed that the world had entered its last 155 years. He regarded himself (invoking the meaning of his name as "Christ-bearer" [Christoferens]) as a major participant in the enactment of this drama: "God made me the messenger of the new heaven and the new earth of which he spoke in the Apocalypse of St. John after having spoken of it through the mouth of Isaiah; and he showed me the spot where to find it."[5]

Columbus was by no means the last discoverer to represent himself as a divine messenger heralding a new world, and he was far from the only one of Copernicus's contemporaries to be preoccupied with prophetic knowledge. Andreas Osiander, the influential Lutheran pastor who shepherded Copernicus's book through the press at Nuremberg, published in 1527 a prophecy "not in words, but in pictures alone," from materials appropriated from a much earlier prophecy—all meant to show the papacy's decline into tyranny, moral decay, and secular power as a powerful symptom of the end times.[6] And indeed, even as Galileo, Kepler, and others began to move the Copernican arrangement into the modernizing currents of the seventeenth century's first decade, they and other heavenly practitioners retained an intense preoccupation with the future.

Who, then, could be trusted to speak about the future in an age when the heavens were a major theater of cultural and political anxieties? And who decided which methods of prognostication were acceptable? Those were the major questions of the Copernican moment. But if so granted, then why did *De Revolutionibus* not make explicit a connection between planetary order and the success of astrological prognostication? In this book, I argue that Copernicus himself *did* see them as related even as early as his student days in Krakow and Bologna during the 1490s. Claudius Ptolemy's *Tetrabiblos*, the fundamental astrological text of antiquity, assigned to the planets certain essential capacities and differential powers to produce specific physical effects on Earth that were directly tied to the order of the planets. Because astrology depended on astronomy to deliver reliable positions for the planets, if astronomy's principles were called into doubt, then the relationship with the companion discipline was also imperiled. And indeed, the iconoclastic Renaissance philosopher Giovanni Pico della Mirandola (1463–94) undermined these very relations in his sharp, far-reaching assault on astrology, posthumously published in 1496. Uncertainty about astral powers and planetary order would become one of the problems—perhaps even the crucial one—to which Copernicus's reordering of the planets was a proposed, if unannounced, solution.

Historians—including myself—have not generally regarded the new Sun-centered ordering of Copernicus and his followers to involve a response to contemporary concerns about astral powers.[7] With the exception of the prescient suggestions of John North and Richard Lemay,

scholars have granted to astrology no historiographically significant place in the Copernican literature.[8] This is to some extent understandable, as there is not a single word about celestial influences in any of Copernicus's extant writings. Conceptual revolution, the idiom in which the long-term Copernican narrative was most often cast in the twentieth century, effectively hid these kinds of questions from scrutiny because it foregrounded the physical problems raised by Copernicus's achievement as the first step on the road to the great breakthroughs of seventeenth-century natural philosophy. The narrative of the "Copernican Revolution" is organized around discovery, diffusion, reception, and assimilation. Theoretical illumination or breakthrough provides the narrative center; the subsequent epistemic history charts theoretical amplification, empirical verification, and sometimes obdurate resistance to truth, while exiling prediction to the thematic backcountry.

Thomas S. Kuhn's still-influential, philosophically informed historical study *The Copernican Revolution* is one variant of this kind of historical writing; elements of it may be found as early as William Whewell's *History of the Inductive Sciences* (1837).[9] When Kuhn called Copernicus's achievement "revolution-making" rather than "revolutionary," he meant to suggest that Copernicus's own contribution was incomplete; other things had to happen in order to consummate the change. Copernicus's contribution was to work out a series of detailed planetary models that fit together as a genuinely interconnected system rather than as a group of discrete calculational devices. He pursued this theory in the face of observed effects, like falling bodies, that he could not convincingly explain. Still, the new theoretical framework that he opened functioned as a heuristic that allowed and encouraged others to think differently and, over time, to accommodate more and more evidence coherently within the new ordering of the universe. Thus Kuhn's account is not "realist" in any straightforward sense; it was not so much the new theory's correspondence to reality, its "truth-to-nature," but rather its "fruitfulness," its heuristic power, that was noteworthy. Copernicus's original insight, on this account, both culminated an earlier tradition and initiated a new one, a crucial imbrication without which Kepler, Galileo, and Newton might not have imagined their worlds. A fundamental innovation in the relatively narrow technical specialty of astronomy "transformed neighboring sciences and, more slowly, the worlds of the philosopher and the educated layman." It was in this sense, rather than in the paradigm-changing rupture of his later work, that Kuhn regarded the entire development as a revolution that Copernicus initiated and of which he was a necessary and central part.[10]

Talk of "deep" revolution or long-term upheaval at the level of both scientific concepts and standards no longer comes as easily as it did at the time of Kuhn's original writing in those historiographically optimistic, if not quite innocent, years following the end of World War II.[11] Nostalgia for so-called big-picture history still exists, but for many it is strongly resisted by a sense that only something like an anthropological immersion in local sites of knowledge making—seeing things "from the native's point of view"—can yield real insight about the actual practices of science.[12] Yet, as revealing as such concentrated localist probings may be, the anthropological tool kit does not provide the methods needed to study change over long periods.[13] Quite the contrary: this approach leaves open the task of explaining how, across time, specific readings, meanings and evaluative judgments made in one cultural setting circulated, metamorphosed, persuaded, or dropped away in others. The present study takes seriously the elements of both sorts of projects—meanings formed at local sites as well as the long-term movement of standards, reasons, and theoretical commitments—seeking a treacherous middle course between the Scylla of internalist conceptualism and the Charybdis of the localist turn.

By way of introduction, consider the specific questions and difficulties that Copernicus's work raised for sixteenth-century readers. First, if the main problem faced by sixteenth-century heavenly practitioners was how to shut down, or at least limit, doubts about predicting the future—whether the occurrence of celestial events, human happenings in the near term, or the end of the world—after 1543, they had to consider whether reordering the planets would help in those efforts. Yet Copernicus's reordering was far from the only strategy that might be used to make astrology's predictions persuasive; indeed, as rapidly became clear, some saw his planetary models rather than his planetary arrangement as

having a bearing on the casting of new tables of motions. In any case, the sixteenth century witnessed many different approaches to these questions, all of them beset by difficulties. Moreover, the heliostatic ordering itself came at the price of introducing new kinds of objections, many of them quite serious.

One immediate and enduring problem lay in the preeminent astronomical text of antiquity, Claudius Ptolemy's *Almagest*. Ptolemy's work enjoyed a revival in the fifteenth century largely through the efforts of Georg Peurbach (1423–61) and his brilliant pupil Johannes Regiomontanus (1436–76), who both completed Peurbach's translation and added to it some of his own ideas. In Regiomontanus's *Epitome of the Almagest* (1496), late-fifteenth-century readers could learn that Ptolemy had anticipated the possibility of the Earth's daily motion and produced arguments to show its absurdity.[14] The other major problem for the rehabilitated Pythagorean view, conflict with the Bible's authority about what moves and what does not, was obviously not a concern of the pagan Alexandrian Ptolemy—nor even yet one which elicited comment from Regiomontanus—but it clearly was by the time that Copernicus's book appeared. Catholic theologians and Protestant reformers alike regarded Scripture as a criterion of the truth of heavenly knowledge. For astrologically inclined practitioners, however, the main question was whether you could extract prognosticatory benefits from Copernicus's proposal without taking on board the parts of it that undermined Aristotelian physical intuitions and ran up against those passages of the Bible that could be read literally as resisting the Earth's motion. In other words, could the theory's utility as an instrument of prediction be separated from its physical truth and scriptural compatibility?

The inclusion of scripture among the criteria considered essential to judge the adequacy of claims about the heavens became ever more urgent during the Protestant Reformation and the overlapping period of Catholic spiritual renewal and response in the sixteenth century. It was widely held that divine messages could be read both in the words of the Bible and in events of the natural and civil worlds. But what was the relation between the Bible and these events? Should the words and sentences of the Bible be taken always to mean literally what they said and, for that reason, to describe actual events and physical truths? Was the subject matter of the biblical text always conveyed by the literal or historical meaning of its words? And who had the ultimate authority to decide on the mode of interpretation appropriate to a given passage? Finally, when the subject matters of two different kinds of texts were seen to coincide—for example, astronomical-astrological with biblical, or astronomical with natural-philosophical—who had the authority to decide which standards of meaning, and truth should govern their assessment? Questions of this sort were inextricably interwoven into issues faced by Copernicus's sixteenth- and early-seventeenth-century followers.

Beyond scriptural and physical criteria, there were other, more strictly logical standards for judging claims about heavenly motions. But, again, who was taken to have the authority to decide which standard should prevail? For celestial prognosticators, tables of mean motions and observations were the principal standard. Yet from the observational evidence alone, long available since antiquity—daily risings and settings of heavenly bodies, retrograde motions, changes in speed, the occurrence of eclipses, and so forth—Copernicus could not deduce a theory uniquely founded on the Earth's motion.[15] Worse still, if Copernicus aspired to make even stronger claims about the nature of reality, then he would have to satisfy a logical ideal widely held among philosophers. Aristotle's standard of scientific demonstration—never itself a logic of discovery—raised the bar impossibly high: it demanded a syllogism called *apodictic*, in which from a true, necessary, and incontrovertible major premise, a true conclusion was inferred. Yet the logic of Copernicus's central claim did not fit that stringent argument-form because, like Ptolemy in his *Almagest*, Copernicus used the less robust conditional syllogism as his preferred pattern of reasoning—starting with the Earth's motion as an assumed, rather than incontrovertibly true, premise. In that sense, it was a supposition or hypothesis that might or might not be true, yet from which true consequences could be deduced.

To add to such logical considerations, Copernicus had opened a question that had been glimpsed in antiquity but which previously had not been seen to possess far-reaching consequences: how to choose between different models of heavenly motion supported indifferently by the same observational evidence. A simple version of this problem had appeared around the first century B.C., when the Alexandrian Greeks Apollonius and Hipparchus recognized the phenomenon of geometrical equivalence for the case of two different models of the Sun's motion: the simple eccentric and epicycle-cum-deferent.[16] Ptolemy's reference to this problem when discussing his own model for the Sun was a major source for sixteenth-century writers.[17] Those who read Ptolemy in Regiomontanus's *Epitome*, however, were shown an equivalence that went unrecognized in the *Almagest*: transformation of epicyclic into eccentric models for the inferior planets, Venus and Mercury.[18] Not until 1543 did geometrical equivalence show up as a question of much wider significance, the choice between arranging the entire heavens around a Ptolemaic-Aristotelian central Earth or a Copernican central Sun. As Copernicus himself put the matter: "It makes no difference that what they [the ancients] explain by a resting Earth and a universe whirling round, we take up in the opposite way so that together with them we might rush to the same goal. For in such matters, those things that are thus mutually related agree, in turn, one with the other."[19]

Visualizing these geometrical transformations was by no means straightforward. Early readers who focused only on the now-famous diagram of concentric circles in book 1, chapter 10, would have had trouble appreciating the passage's real significance. Work would be needed to bring out the equivalences. The same applies to the Earth's motion(s) as the source of visual illusions:

> Why should we not admit that the appearance of daily revolution is in the heavens but the truth [*veritatem*] in the [motion of the] earth? This situation closely resembles what Virgil's Aeneas says: "Forth from the harbor we sail, and the land and the cities slip backward." For when a ship is floating calmly along, the sailors see the image [*imago*] of its motion in everything outside, while on the other hand they suppose that they are stationary, together with everything on board. In the same way, the motion of the earth can unquestionably produce the impression that the entire universe goes around.[20]

Throughout his main argument, Copernicus played on such deceptions of the visual *imago*. But the boat analogy addresses only one motion. The Earth was not just "floating calmly along." It was describing a more complex motion, something like a carnival horse on an imaginary merry-go-round—able to rotate daily with respect to its own axial pole while simultaneously revolving, over a period of a year, in the opposite direction with respect to the platform's central axis. However, such planetary merry-go-rounds, or orreries, would not enjoy their heyday as visual assists until the eighteenth century (see figure 34).[21] At best, Copernicus could argue that if the Earth has an additional, annual motion, then that motion would be apparent in viewing the other planets, showing up as "a sort of parallax produced by the earth's motion."[22] Each of the planets mirrors the Earth's unfelt motion as a component of its own total motion.

In comparable language, known to Copernicus, Peurbach had already called attention to this same peculiar phenomenon with reference to the Sun: "It is evident that each of the six planets shares something with the sun in their motions and that the motion of the sun is like some common mirror and rule of measurement to their motions."[23] Had Copernicus drawn explicit attention to this passage in Peurbach's book, widely taught in the universities, it might have helped to highlight this problem, if not to persuade otherwise skeptical readers that he offered a viable solution to it. But Copernicus makes no references to Peurbach and, for that matter, few to other contemporaries. Good humanist that he was, Copernicus represented himself as though in an exclusive dialogue with the ancients. Meanwhile, other observational consequences, such as the variation of the Moon's apparent diameter, in which Copernicus claimed "greater certainty" than Ptolemy, in no way depended on the new ordering of the planets.[24] Modern reconstructions of the mutual advantages and disadvantages of the Copernican and Ptolemaic arrangements have made these considerations much easier to grasp; yet, as a consequence, they have unwittingly made the situation faced by contemporaries seem more obvious than it really was.

Hindsight may also intrude in another way. The heliostatic theory, taken as a timeless entity all of whose entailments are known, predicts certain effects that were not immediately observed. The historical question is, when did those effects become real questions for the agents? And further, when and how were those effects seen to be implications of the Sun-centered theory rather than of its alternative? For example, if the Earth

moves, you ought to be able to detect a slight parallactic effect in a distant star; over a period of six months or a year, the star should appear to shift its position. Also, Mars at opposition should have a diurnal parallax greater than the Sun and hence should be closer than the Sun to the Earth. Or yet again, if Venus is revolving around the Sun, then it ought to display a complete set of phases, like the Earth's moon. And finally, if the Earth is set in motion, the resulting distances create serious problems regarding the plenum of nesting eccentric spheres that many believed transported the planets themselves. In 1543, Copernicus himself recognized that the Earth's motion entailed the appearance of an annual parallactic effect in the fixed stars, and he acknowledged that the stars exhibited no such appearance; however, he did not allude to the possibility of Venusian phases; and, as he was hardly a systematic natural philosopher, he did not comment unambiguously on the ontology of the heavenly spheres.[25] Copernicus explained the absence of parallax as a consequence of the universe's large, hitherto-underappreciated size. However, his first disciple, Georg Joachim Rheticus (1514–74), stated quite bluntly that "Mars unquestionably admits a parallax sometimes greater than the sun's" and then proceeded to infer that "therefore, it seems impossible that the earth should occupy the center of the universe."[26] The question of measuring stellar—or planetary—parallax does not seem to have been grasped as approachable by anyone before Tycho Brahe in the 1580s and, yet more optimistically, by Galileo after 1610; and there was no stable consensus that the problem had been resolved until Wilhelm Gottfried Bessel produced measurements of stellar parallax in 1838.[27]

Another sort of entailment concerned physical effects inferred from observations unmediated by any sort of new technologies of magnification. If, contrary to ordinary sense experience, the Earth can be imagined to rotate in twenty-four hours from west to east [A], then, if you have Aristotelian intuitions, you will expect all kinds of calamitous terrestrial effects [B]. Ptolemy himself had already articulated just such objections:

[B:] All objects not actually standing on the earth would appear to have the same motion, opposite to that of the earth: neither clouds nor other flying or thrown objects would ever be seen moving towards the east, since the earth's motion towards the east would always outrun and overtake them, so that all other objects would seem to move in the direction of the west and the rear. . . . Yet [not-B:] we quite plainly see that they do undergo all these kinds of motion, in such a way that they are not even slowed down or speeded up at all by any motion of the earth.[28]

The argument pattern in this example, as in the argument from stellar parallax, was that known to logicians as *modus tollens*: if A, then B; but not-B, therefore, not-A. The real workhorse in logically valid reasoning of this sort is denial of the consequent, not-B—the premise standing for unobserved or unmeasured effects—just as it had been in the reasoning patterns of various sorts of Greek scientific and medical writings.[29]

Against Ptolemy and Aristotle, Copernicus sketched an alternative theory of gravity that retained the intelligibility of Aristotle's "natural," "simple," and "place" as the right categories in which to describe and explain motion. But then Copernicus reshuffled Aristotle's natural motions, assigning uniform circular motion to the Earth, the planets, and the elements and demoting all rectilinear motions to the status of temporary, nonuniform deviations from circularity.[30] In this new account, all the planets shared "a certain natural desire, which the divine providence of the Creator of all things has implanted in the parts, to gather as a unity and a whole by combining in the form of a globe."[31] Hence, if [A] the Earth rotates daily and/or revolves around the Sun annually, then, as in Ptolemy's implication [not-B] above, objects detached from the Earth would not appear to speed up or slow down.

That Copernicus even troubled to devise an alternative to the Ptolemaic-Aristotelian theory of gravity shows that he was working to recast his role as a traditional astronomer-astrologer principally concerned with prognostication—that he was actively looking for alternative arguments to block objections from traditional natural philosophy. This move, in turn, raises the specific historical question of how Copernicus convinced himself to pursue his own seemingly absurd hypothesis, at least as a conditional argument, sometime between 1497 and 1510. Certainly, a major consideration must have been his recognition that the assumption of a Sun-fixed ordering allows many observations to be intelligibly connected in a way that has no comparable explanation on the Ptolemaic account.[32] Perhaps Coper-

nicus intuited that such "explanatory loveliness" betokened the potential for inferring the best possible explanation of the planetary phenomena and their arrangement.[33] In 1543, *De Revolutionibus* foregrounded its most lovely entailment—the universe's well-proportioned orderliness or *symmetria:* from the assumption of the Earth's motions, "not only do [planetary] phenomena follow therefrom but also the order and size of all the planets and spheres, and heaven itself is so linked together that in no portion of it can anything be shifted without disrupting the remaining parts and the universe as a whole."[34]

This was surely a new claim. Ptolemy (and Regiomontanus) had failed to mention these consequences for planetary order that Copernicus detected on setting the Earth in annual motion, although they had noticed and rejected the physical consequences of the Earth's daily rotation. Yet whatever explanatory gain Copernicus had found, his reasoning, like Ptolemy's, also follows a conditional form. To many contemporaries, it was quickly obvious that the argument violated *modus tollens* because, strictly speaking, Copernicus would have been making an invalid inference called "affirming the consequent": If A, then B; B is affirmed, therefore A follows. If Copernicus was in possession of other arguments, he chose not to make such crucial evidence part of his public presentation in *De Revolutionibus,* nor did he claim that his theory yielded tables of motions superior to those of Ptolemy. Indeed, why would he have withheld his best evidence—after some four decades of considered reflection—if he really possessed it? Further, if the predictions yielded by the new arrangement were no better than the alternative, then how could it possibly be said to improve astronomical *or* astrological forecasting? These questions of evaluative judgment, not fully or clearly unpacked in the compact phrasing and limited visualizations of *De Revolutionibus,* greatly affected the considerations of later practitioners.

Finally, there are questions of how celestial practitioners (and nonpractitioners) responded in the face of negative or even potentially refuting instances as well as inferences that were drawn from confirmations. Astrology's predictions frequently failed or, at least, appeared to fail. Who could tell whether this was the fault of inaccuracies in the planetary tables, the principles on which those tables were based, or the astrologers' interpretations of the chart based on the planetary positions? And to reverse the question, if Copernicus's hypothesis was taken to be true, would that guarantee the accuracy of the astronomical (or astrological) predictions based upon it? Similarly, what inferences might be drawn about the truth of Copernicus's hypothesis from belief in the success of the ephemerides derived from the Copernican planetary tables? For, in following Aristotle's apodictic standard of demonstration, how could any of Copernicus's advocates be sure that they had met the demanding standard that ruled out all possible alternative arrangements?

The scientist, philosopher, and historian Pierre Duhem (1861–1916), who studied and commented on many of the original texts involved in the Copernican episode, was the first to call attention to this last question as a problem of broader epistemological interest. Among philosophers of science, it has since come to be known as the problem of *underdetermination*.[35] In an 1894 essay, Duhem maintained that a physical theory is not an isolated hypothesis analogous to the wheels and cogs of a watch that can be disassembled into its individual parts. Rather, a physical theory is like an organism that must be taken as a "whole theoretical group": "Presented with a sick person, the doctor cannot perform a dissection to establish a diagnosis. The doctor must decide the seat of the illness only by inspecting the effects produced on the whole body. The physicist charged with reforming a defective theory resembles the doctor, not the watchmaker."[36]

Dramatic and sobering consequences follow from this vestige of late-nineteenth-century holism. First, if a physical theory is holistic rather than atomistic—an interconnected network rather than a set of independently standing empirical propositions—it is uncertain at best which parts of the theory are refuted when a prediction (or an experiment) fails. In the 1950s, the philosopher W. V. O. Quine further radicalized Duhem's claim by arguing that when Nature pushes back at a theory, it is always possible to make pragmatic adjustments or additions to the beliefs that make up the theory so that, at least logically, one is never forced to give up the whole web of beliefs. For my purposes it is unnecessary to consider the different possible interpretations of Quine's views.[37] But it is worth noting an especially radical version of Quine's thesis that maintains that "*any* seemingly disconfirming observa-

tional evidence can *always* be accommodated to *any* theory."[38] Physical theories on this account thus have unusual staying power. Apparently endless adjustments can block the refutations of *modus tollens*. A further significant consequence is that both Duhem and Quine denied the possibility of crucial experiments in physics. In geometry, you can follow the method of exhaustion, reducing all contrary propositions to absurdity; but in physics you cannot because, as Duhem argued, you "are never certain that [you] have exhausted all the imaginable hypotheses concerning a group of phenomena."[39] How then could a theory's full web of background assumptions ever be shown to be refuted?

Sixteenth-century celestial practitioners, of course, were not aware of underdetermination as a general epistemological problem. At ground level, Copernicus's followers and his adversaries were simply cognizant of the problem of blocking the uncertainties and refutations produced by rival alternative accounts. All sides, indeed, shared considerable confidence that demonstrations, sometimes quite strong ones, could be delivered. And it is these historically situated efforts that will especially interest me in this book. Only from the perspective of long historical distance can the Copernican question be viewed, epistemologically, as the first full-scale "Duhemian situation" in the history of science—and even then, not in the sense that Duhem or Kuhn imagined.[40] Duhem, for his part, read the history of astronomical theory from the Greeks to the Renaissance as vindicating a powerful scientific antirealism, the view that the propositions of science predict but do not describe features of the world.[41] In his classic essay *To Save the Phenomena* (1908), Duhem famously read Copernicus, Kepler, and Galileo as misguided in pursuing a realist theory, one that they believed corresponded to the world. The difficulty for Duhem was that these thinkers had prematurely abandoned the well-founded tradition of treating astronomy's models as no more than convenient predictive instruments that "saved the phenomena" but with no claim to truth. Had they remained committed to the view that astronomical hypotheses are fictions, Duhem counseled, then the problem of ranking geometrically equivalent hypotheses would have been irrelevant. According to Duhem's provocative—and oversimplified—interpretation, the Church was scientifically warranted in maintaining a skeptical view of Galileo's Copernican claims. The Church was thus made to stand squarely in line with Duhem's reading of astronomical tradition from the time of the Greeks. Even before Maffeo Barberini became Pope Urban VIII in 1623, he warned Galileo that God, being all-powerful and omniscient, already knows all possible orderings of the universe; and yet, as the medievals had often argued, he chose to use his unfathomable power to build only a finite universe, a view long taught by the Church as a matter of tradition.[42] Humans, opined the pope, were not to fall prey to their own pride in believing that they could imagine all other possible worlds. "The man who was to become Urban VIII," wrote Duhem,

> had clearly reminded Galileo of this truth: No matter how numerous and precise are experimental confirmations, they can never render a hypothesis certain, for this would require, in addition, demonstration of the proposition that these same experimental facts would forcibly contradict all other imaginable hypotheses.
>
> Did these logical and prudent admonitions of [Cardinal Robert] Bellarmine and Urban VIII convince Galileo, sway him from his exaggerated confidence in the scope of experimental method, and in the value of astronomical hypotheses? We may well doubt it.[43]

Powerful words, perhaps not unexpected from a believer who wished to harmonize the truths of nature—or at least its methods of investigation—with those approved by the Church. But, although Duhem was a brilliant investigator, the first to establish the existence of a flourishing scientific culture in the medieval period, his own antirealist commitments led him to indulge in some historical attributions that are questionable at best, wrong at worst. For example, Duhem dubiously attributed to Galileo an "impenitent realism" that made him appear to hold that Copernicus's theory had been incontrovertibly demonstrated. It was this position that Duhem's Urban then corrected with his theologically-grounded skepticism. Thus, ultimately the blame for Galileo's condemnation was to be assigned to the Copernicans' excessive zeal for an "illogical realism."[44] Likewise, as Geoffrey Lloyd has shown, Duhem's reading of the core ancient writers Geminus, Proclus, Ptolemy, Simplicius, Theon, Hipparchus, and Aristotle do not support

the Duhemian interpretation of a robust astronomical instrumentalism.[45] Moreover, Peter Barker and Bernard Goldstein have unearthed passages in some sixteenth-century astronomical writings that further deflate confidence in a global application of the distinction between realism and instrumentalism.[46] And Maurice Clavelin has pointed out that Duhem's unmitigated continuist perspective, which regarded Galileo's theories of motion as nothing more than developments of fourteenth-century natural philosophy, had the effect of marginalizing the Copernican framework as an alternative approach for Galileo's science of motion—in effect, regarding it as "a detail without conceptual implications."[47]

The danger of imposing inappropriate analytic categories points again to the need for a more rigorous historicism, ruthlessly attentive to the pastness of the agents' own categories but also informed and balanced by a judicious cultivation of modern epistemic resources. Such an investigation, beginning with a careful excavation of the resources for classifying knowledge, finds Copernicus and his successors engaged in trying to answer a series of questions in which the premises of astronomy and astrology were somehow linked. In this sense, one might say that both Duhem and the early Kuhn, whose *Copernican Revolution* bears signs of Duhem's influence, were insufficiently holistic in their historical treatment of astronomy and astrology. Once we see the two as part of a shared complex, we can ask new questions. For example, how could an astronomy believed to be well grounded secure the foundations of astrology against criticism and refutation? What astronomical choices were open to practitioners in the face of astrology's often failed predictions? How did Copernicus and his followers seek to eliminate the traditional, alternative world ordering while advancing proofs for their own? What difference did it make that there were two alternative planetary arrangements when a comet and a nova appeared unexpectedly in the 1570s? And how were these choices made within the space of the logical, rhetorical, literary, and disciplinary possibilities available to the people of that long-ago time?

SUMMARY AND PLAN OF THIS WORK

This book is divided into six parts. Chronologically, it ranges from Copernicus's intellectual formation in the 1490s to Galileo's telescopic dis-coveries of 1610—something of a "long sixteenth century"—but it reserves the historian's right to reach backward (to the 1470s) and forward (into the seventeenth century).[48]

The language that I use in the part divisions is primarily analytic rather than actor-specific. Of particular importance is the word *space*—now something of a term of art—which I often use in the ordinary sense of a physical place or locus (like a city, princely court, university, or meeting site). In this case, *possibilities* might refer to some-thing like a domain of material transactions, such as finding (or not finding) a book, the likeli-hood of meeting (or not meeting) certain people at a specific location, or the movement of people and objects between locations; but it might also suggest a site associated with the identity of cer-tain interpretations and meanings.[49] *Space* can also be used to suggest a domain defined by tem-poral and cultural categories not uniquely speci-fied by a physical location—in this sense, the or-ganization of a book, the representation of an author's identity, the schemes of knowledge clas-sification, and the literary conventions in which authors chose to write. And finally, *space* is con-nected with time—and memory. Spaces of pos-sibility change as time passes and memories of earlier moments fade. A book or a discovery may be remembered or encountered even as the origi-nal circumstances of its use are forgotten or par-tially obscured. It is the continual juxtaposition of the residual traces of past spaces, elements of an emergent present, and representations of an unformed future that seriously complicate an "innovation-diffusion" model. And it is within such life spaces—at very close range—that I propose to look for the formation, meaning, and movement of concepts and practices.

Part 1 (chapters 1–3) lays out an unfamiliar set of categories. In Copernicus's lifetime, the resources of authorship were in considerable flux not only because of the increasing number of books produced but also because of the ex-panded range of choices of literary form and au-thorial self-definition attendant on the intro-duction of print. To avoid oversimplification, one needs overlapping schemes to map the coordi-nates within which Copernicus formed his proj-ect. I emphasize authorial self-representation, printers' interests, and knowledge domains: writ-ers chose what to call themselves on their title

pages, the genres in which they wrote, and the knowledge classes in which they located their subject matter.

Chapter 1 consciously joins in its title the ana-lytic neologism "Literature of the Heavens" to the actors' category "Science of the Stars." The first brings the advantage of circumventing the re-strictive categories *astronomy* and *cosmology*, with their presentist overtones, and points to all the genres in which an author could choose to write about the heavens in the late fifteenth and six-teenth centuries. "Science of the Stars" refers to a fourfold matrix of epistemic categories wherein astronomy and astrology were classified as parts of a common enterprise, but split into theoretical and practical parts, respectively. This simple divi-sion, whose historical roots I explore, accounts for various difficulties obscured by shoveling everything into the undifferentiated categories *astrology* and *astronomy*. For example, contempo-raries generally considered the actual making of astrological predictions to be part of the domain of astrological practice rather than of astronomi-cal theory. Yet I suggest that the entire set of categories making up the science of the stars was seen to be a single complex, combining both quantitative and hermeneutic elements that to-gether yielded predictions of meaningful social and atmospheric effects. Among these imbri-cated resources of classification, chapter 2, gives special attention to academic astrological prog-nostications and popular prophecies as new phe-nomena of print in the last quarter of the fifteenth century, and thereby prepares the background for a thick description of Copernicus's intellectual formation.

The crucial problem of chapter 3 is what I call Copernicus's exceptionalism. If by now it is a truism that astrology was widespread in the Renaissance, then why was Copernicus—of all people—exempt from involvement in its prac-tice? His surviving writings include not a word on that subject, and some historians have as-sumed that his great mind must have shielded him from such an unsavory discipline. My ap-proach is, as it were, through the back door, that is, through his intimate connection with Dome-nico Maria Novara. Novara was neither an iso-lated figure nor, as Kuhn and others have argued, a Neoplatonist, but rather one among a flour-ishing group of astrological practitioners in late fifteenth-century Bologna. Just a few months before Copernicus's arrival in Bologna in 1496, Giovanni Pico della Mirandola's massive attack on astrology's foundations was issued by a Bolognese publisher, and I argue that a major, hitherto unnoticed element of Copernicus's early problematic was precisely the effort to resolve Pico's skeptical objections concerning the ordering of the planets.

This theme of Piconian skepticism and the responses to it runs like a red thread through the entire narrative, providing at least one major element of thematic unity to the sixteenth-century scientific movement missing from current accounts of that period. Time and again, Pico's arguments show up as the singular historical fulcrum point around which theologians and natural philosophers rejected astrology; but it was also this same body of objections in response to which a small group of mathematical practitioners developed bold new projects of heavenly knowledge. A century after Copernicus wrestled with Pico's skepticism in Bologna, Kepler was still engaging with his arguments in Prague.

De Revolutionibus was a work incubated in the northern Italian university towns of the late fifteenth and early sixteenth centuries—in its language and ideals, in its imitation of Ptolemy, in its dialectical and rhetorical mode of introduction. By the time it appeared in 1543, however, Europe was already entering a new epoch marked by confessional divisions, the subject of part 2 (chapters 4–7). Between roughly the mid-sixteenth and the mid-seventeenth century, early modern intellectual productions concerning the natural world had to contend, to one degree or another, with the concerns of this increasingly confessionalized Christian Europe that was fragmenting along religious fault lines. Issues of denominational identity and loyalty were emerging. Some humanist friendships came under serious strain because of confessional pressures. In the process of early modern state building, as Heinz Schilling argues, the German territorial states used ecclesiastical institutions, including schools and universities, as a basis for social integration, for producing voluntary obedience and acceptance of the hierarchy of social classes. Baroque princes from the mid-sixteenth century began monopolizing the churches; the prince became *defensor fidei*. Sacralization of the ruler preceded and assisted the subsequent monopolization of military force and taxation. Many

political theorists subscribed to the axiom *Religio vinculum societatis* ("Religion is the bond of society").[50]

Whatever their social positions, heavenly practitioners, and men of letters more broadly, could not avoid encounters with concerns like the defense of divine authority and providence, the maintenance of scriptural authority and status of prophetic wisdom, anticipations of salvation or signs of the apocalypse, and demonic threats. Yet remarkably, even within these less-welcoming circumstances, important possibilities for interconfessional cooperation still existed. The production of Copernicus's work is such an instance. As Copernicus approached the end of his life, his work finally appeared in print through the indispensable assistance of the Lutheran Rheticus.

In chapter 4, I contend that Rheticus and Copernicus pursued a dual strategy of authorial presentation. This is not the usual view. It is commonly held that Rheticus's *Narratio Prima*, which is really the first published account of Copernicus's heliocentric proposal, was entirely his own work, even if it was a paraphrase of sections of Copernicus's still-unpublished manuscript. I argue that it was directed to an audience of heavenly practitioners in Wittenberg and Nuremberg. At Wittenberg, Philipp Melanchthon was alone among Protestant reformers in regarding astrology as an important and legitimate tool for contemplating and anticipating elements of the unfolding of the divine plan in the six thousand years of world history. Rheticus believed that Copernicus had successfully refuted Pico's arguments and that his theory itself supported a world-prophetic vision. Copernicus, however, explicitly dedicated *De Revolutionibus* to the pope and, more broadly, to Catholic circles; and, in that work, although there is no hint of astrology or millennial prophecies whatsoever, there are clues pointing to a suppressed reference to Pico (chapter 3).

Wittenberg figures prominently in this account both because of Melanchthon's influential views on the providential import of astrology and because of his institutional influence: he systematically built up a cadre of mathematically skilled practitioners who fanned out to a cluster of German universities informed by his principles of education. Here, the foremost figure was Erasmus Reinhold (1511–53). Neither Reinhold nor Melanchthon shared Rheticus's enthusiasm for

Copernicus's more radical proposals, but Reinhold quickly understood the possibility of cleaning up the calculational errors in *De Revolutionibus* and using its planetary models to improve the tables on which astronomical and astrological predictions were founded. Reinhold supported this long, costly, exceedingly tedious, but ultimately successful enterprise through complex negotiations with Albrecht, duke of Brandenburg. Chapter 5 shows just what was at stake for both patron and client. Reinhold's achievement, with its emphasis on utility rather than truth, constituted a partial answer to Pico's objections. It led many contemporaries to the conclusion that Copernicus's models could be used to improve the planetary tables without going so far as to upset the traditional place of the science of the stars in the hierarchy of the arts disciplines. And, partly as a result, Rheticus's presentation of Copernicus—with its more serious consequences for natural philosophy—was largely swept aside.

One may characterize the Wittenberg problematic as an absence, an institutionally shared silence about the heliocentric planetary ordering—as I did in a much earlier study.[51] However, in this book, I have tried to change the character and direction of the question from an absence to a presence. What did the Copernican world hypotheses mean to contemporaries obsessed with hopes, fears, and doubts about mastering the powerful astral forces that bathed the Earth in a stream of influences at the center of the theater of the world? One may ask, in a positive sense, when and how did practitioners connect Copernicus's planetary arrangement to the problematic of practical astronomy and astrology, or to eschatological predictions of the end of the world? If Reinhold's reconstitution of Copernicus succeeded, it was largely because it purported to have something new and useful to offer to the practicing astrologer.

But had this narrative exclusively foregrounded Copernicus, the reader might have been led to the conclusion that astrological credibility rested only on improving astronomical accuracy. This was not the case. In the middle decades of the sixteenth century, as I argue in chapter 6, quite different ideals of astrological knowledge and different sorts of practice were in play. This was the heyday of new astronomical textbooks (largely issued by Wittenberg authors), new editions of the *Tetrabiblos*, and new manuals of astrology. Astrol-ogy was practiced in different settings all over Europe. It was taught in many universities, its practitioners were solicited by rulers of all confessional stripes and, with the notable exception of Melanchthon, theologians regarded it as a threat to divine foreknowledge.

Such worries about theological authority foreshadowed the anxieties that would entrap Galileo more than sixty years later: if special knowledge of the stars could be used to predict the future (e.g., the death of a pope) or retrodict an important past event (e.g., the birth of Christ), then purveyors of the heavens might usurp the authority of theologians, and in that case, the devil must be at work. Here again Pico's skepticism proved to be an ally of theologians who could use his arguments to undermine the diabolical pretenses of the subject without throwing out the entire science of the stars. Much new and excellent scholarship has now shown how astrologers actually worked. In this chapter, I seek to integrate those researches into the larger question of the diverse ways in which astrologers sought to sustain their credibility to speak as authoritative intermediaries for the stars' messages. These efforts took place at both the high and the low ends of the science of the stars. At the low end, for example, Giuliano Ristori successfully predicted the death of Duke Alessandro de'Medici in 1537, with a prescience that assured his reputation into the seventeenth century. At Louvain, on the other hand, Reiner Gemma Frisius, Gerard Mercator, and Antonio Gogava engaged in high-end efforts to reform the theoretical foundations of astrology by reconstituting its optical principles. It was this exciting, top-down project that fired the imagination of the young John Dee on his first visit to the Continent in 1548.

The Catholic reaction to the surging tide of prognosticatory activity, efforts at astrological reform, and the emerging challenge of Copernicus's work was not uniformly coordinated; but, as chapter 7 shows, it gradually converged in the middle and later decades of the sixteenth century. A very early negative reaction to *De Revolutionibus* at the papal court (instigated by Giovanni Maria Tolosani) failed to gain political traction. Yet, although not adopting Copernicus's radical proposals, many Catholic as well as Protestant practitioners used the numbers of Reinhold's *Prutenic Tables,* which managed to avoid being seen as confessionally tainted. In the writings of the prominent Spanish theologian Miguel Medina, however, one sees how theological resistance to this-worldly divination was more pronounced among Catholics than among Protestants. Catholic heavenly practitioners thus faced a new difficulty: if astrology was off-limits, then how could the study of astronomy be justified? One major response was provided by the Jesuit Christopher Clavius, whose influential textbook of astronomy set the key problems for many late-sixteenth-century readers, including Galileo and Kepler. Where Melanchthon had argued the indispensability of astronomy for astrology, Clavius restricted the practical value of astronomy to matters of the calendar, the weather, and navigation.

The crucial development that produced engagement with the question of the reordering of the universe was the unexpected encounter with singular heavenly novelties in the 1570s. Unlike the "Italian nature philosophers" (e.g., Bernardino Telesio, Francesco Patrizi, and Tommaso Campanella) who challenged the Aristotelian corpus by replacing old physical qualities with new ones, this development triggered a problematic of disciplinary authority: the question was no longer merely whether it was legitimate for those who identified themselves with mathematical subjects to philosophize, but rather in what register it was acceptable to do so. That issue might show up as a matter of general explanatory prerogative: what sorts of explanations (or Aristotelian causes) might mathematizing practitioners legitimately introduce into a natural philosophy of the heavens? Or it might show up as a problem of reconciling different parts of the science of the stars with one or another natural philosophy: consider what physical foundations must be assigned to astrology if the Earth is classified as a planet. Or it might take the form of the question opened up by Tycho Brahe and later pursued by Kepler: if there are no hard, impenetrable spheres to carry the planets, then what sort of explanatory resources can physics supply to theoretical astronomy to explain why the planets move?

The three chapters that make up part 3 (chapters 8–10) move our discussion from confessional to disciplinary divisions and tensions, to unexpected appearances in the skies and their import for the ordering of the planets. *De Revolutionibus* had been studied, read, and sometimes heavily annotated for thirty years. But, apart from Rheticus, it had attracted no one who held its central propositions as true or even as a basis for further pursuit. This situation began to change in the 1570s, coinciding with a new generation of practitioners. During the last three decades of the century, two clusters of problems gradually became connected. The first concerned who had the right resources to speak for the future: official theologians, with their claim to exclusive interpretive authority over holy scripture and prophecy, or astrological prognosticators, with their planetary tables, lists of aphorisms, technical schemes, charts, and anxiety-producing judgments? The second problem concerned the authority to speak for the new orderings of the planets against the traditionalist alliance of natural philosophers, theologians, and mathematical practitioners whose teachings still dominated university curricula. The two are not usually seen as connected. If the possibility of the Earth's daily rotation had been entertained in the fourteenth century only as a speculation in natural philosophy, what now moved late-sixteenth-century mathematical prognosticators to stake out special authority to reorder the heavens and, through it, to lay claim to the domain of natural philosophy?

Chapter 8 shows that it was a small subset of mathematical prognosticators who shook the traditional structures of disciplinary authority. The unexpected nova of 1572 and the comet of 1577 complicated the status of prognosticators for a reason not previously emphasized by historians: because these events had eluded prediction. The prognosticators no longer faced the customary question of predicting where the planets would be at certain times so that their influences could be engaged or avoided, but that of how to accommodate—and hence explain—celestial events that had not been anticipated and which did not recur. Willy-nilly, those prognosticators who believed that these events were superlunary turned either to theology or natural philosophy (or both) to explain what they had failed to foresee. For this group, the problem of the unpredictables, therefore, was that one could only speak of them in retrospect: they were singular events that had occurred and then disappeared.

What did this have to do with the future? For many, if not most, commentators, the solution was to give meaning to such unanticipated heavenly occurrences by subsuming them under a long-term apocalyptic narrative. Apocalyptic prophecy essentially foresaw that there would be more, and more unusual, "breakdowns" in the natural and human realm as the end of the world grew closer. So even if one had not foreseen any particular occurrences, one could still say that, as a class, all preternatural events, that is, events that seemed to go beyond the ordinary course of nature, counted as evidence for the end-of-the-world narrative. By this reasoning, a heavenly prognosticator could maintain his credibility by speaking (retrospectively) for the Apocalypse without having successfully foreseen particular instances of its unfolding in time. And this was a game that was difficult to lose.

Yet there were a few dissenting voices among practitioners of theoretical astronomy, like Michael Maestlin, Tycho Brahe, Johannes Kepler, and (implicitly) Christopher Clavius, who questioned whether one could really read such an apocalyptic narrative in the heavens. They did not deny that the heavens contained divine meanings or that the Last Judgment was approaching, but they did deny that the heavens foretold specifically when the world would end. The heavens were simply not the right place to look for such definite anticipations: they were to be contemplated and worshipped, read for signs of changing weather and—at least for the Lutherans—changing personal and political fortune, but not plumbed for eschatological meanings, as Pierre D'Ailly and Johannes Lichtenberger had urged in the fifteenth century. And hence, for these more cautious starmen, the proclamations of late-sixteenth-century apocalyptic prognosticators were to be regarded as, at best, misguided.

Chapter 9 investigates how these unanticipated novelties of the 1570s were implicated in the appearance of a new generation of planetary reorganizers. Indeed, exactly why and how were Michael Maestlin and Thomas Digges attracted to Copernicus's views? Was it an accident that these first favorable followers of Copernicus since Rheticus both wrote about the new star and the comet? Or, as I discuss in chapter 10, that Tycho Brahe introduced his new system of the world in the middle of a treatise on the comet of 1577? As Copernicus's central proposals finally attracted a new generation of advocates, we see that the differences between the Copernicans Maestlin and Bruno were vastly greater than those between Kepler and the non-Copernican Tycho Brahe. This striking diversity of positions among the Copernicans is perhaps the most compelling reason for abandoning both isms and Kuhnian incommensurabilities. Moreover, for the first time since De Revolutionibus, the question of the ordering of the planets moved to the foreground in the 1580s and took on the character of a polemic. Why and how this happened is the main subject of chapter 10.

Johannes Kepler's intellectually formative moment, the subject of part 4 (chapters 11–12), came at the very end of this development and was shaped by it. Historians have long recognized his special place among Copernicus's new disciples. Earlier conceptualist historiography built on the contrast between Copernicus's "conservative" innovation (the shift of the reference frame from Earth to Sun) and Kepler's "revolutionary" break with what Alexandre Koyré famously called the "obsession with circularity." The tendency was to emphasize the admittedly radical character of this conceptual achievement rather than his proximate historical evolution out of the problems of the tumultuous 1580s. Chapter 11 claims that what set Kepler apart from both the mathematicians and the natural philosophers of his time was his shaping of a new role for the heavenly practitioner, moving beyond even that of the philosophizing prognosticators of the 1570s and '80s. Kepler's project, already evident from his earliest years, was nothing less than a wholesale revision of the principles of the science of the stars—not merely theoretical astronomy, on which much of the historiographical emphasis has been placed, but practical astronomy and theoretical and practical astrology. Kepler systematically modeled his investigations on Aristotle's full range of explanations: formal, material, efficient, and final causes. From the very beginning, there were telling logical and physical considerations. Unlike Copernicus and Maestlin, Kepler wanted to show that the celestial investigator could hope to entertain more than a mere weighing of alternatives. And in searching for a physical explanation of planetary motion, he turned to none other than Pico's treatise against astrology. Thus the religious fervor of his writings bespoke a conviction that the astronomer could have demonstrative and not just dialectical knowledge, physical as well as mathematical understanding of the divine plan.

Few were persuaded. As I describe in chapter 12, the resistance to Kepler's astonishing cosmic schemes and conjectures was rapid. But when books had strong affinities to personalized craft objects, and exchanges traveled through hand-carried letters subject to the vagaries of couriers, *resistance* had a meaning different from that in the emergent culture of journals of the late seventeenth century, the mass steam-print economy of the early nineteenth century, or Bruno Latour's technoscientific public sphere in the twentieth. This chapter underlines a theme already encountered in Copernicus's Bologna, Melanchthon's Wittenberg, and John Dee's Louvain: the small scale of social spaces and the possibilities for personal exchange. It focuses on the confessionalized response from the Tübingen theological faculty and the court of the Duke of Württemberg—the one within walking distance from the other—and the handful of astronomical adepts who bothered to write to Kepler (and whose letters survive).

Historical periods do not typically map neatly over calendrical divides. Yet the first decade of the new century comes as close as anything to marking both a culmination and an emergent moment in European representations of the order of the universe. What makes this development at once fascinating and difficult to grasp is precisely its protean character. Part 5 (chapters 13–15) shows that arguments over celestial order were conducted along uncoordinated fronts of engagement—sometimes making contact, sometimes not—across different institutional contexts and within diverse literary forms. Overall, the sixteenth-century problem of securing and improving astrological forecasts against skeptical objections remained the paramount concern among heavenly practitioners, but now a new space of questionings was emerging, linking celestial order and the causes of planetary motion to proliferating possibilities in natural philosophy. The astral practitioners who gradually forged this *via moderna* found themselves assuming unaccustomed disciplinary postures as philosophers of the heavens, as they shifted comets and novas into the realm of ordinary phenomena. Souls, aetherial fluids, angels, intelligences, and magnetic powers appeared as explanatory resources in the discursive space that was emerging as an alternative to the still-hegemonic planetary spheres of traditionalist heavenly natural philosophy.

The century's first decade is a crucial moment.

Yet, like the Scientific Revolution as a whole, it has often been portrayed as a two-sided battle of ancients and moderns. And, of course, there is more than a little truth in that contention, because it echoes categories widely used by the historical agents themselves. Galileo's *Dialogue concerning the Two Chief World Systems* is perhaps the best-known example. But that description fails to capture a salient, large-scale feature of the century's first decade: a proliferation of multi-sided struggles sometimes within, sometimes between traditionalists, modernizers, and the middling sorts who freely appropriated from each. The year 1600 is emblematic of such multiple-sided engagements. Giordano Bruno was burned at the stake in Rome by zealous defenders of tradition, thereby casting an extended pall over the Italian scene. Yet the middling modernizer Tycho Brahe, and soon Kepler, rejected Bruno as a threat to the science of the stars. Meanwhile, William Gilbert's daring philosophy of the magnet appeared in London but met with differing responses from Galileo and Kepler. And finally, Kepler arrived full of hope at Tycho Brahe's castle in Prague, where the latter quickly charged him with the task of defending a world system in which he did not believe and destroying the world scheme of Brahe's competitor Raimarus Ursus, along with his reputation.

Chapter 13 focuses on the tense and unconsummated relationship between two of the leading modernizers of the decade, Kepler and Galileo. Each has attracted a veritable scholarly industry, but little effort has been made to track their relationship as a problem in its own right. One difficulty is the assumption that there was no relationship because Galileo broke off correspondence with Kepler between 1597 and 1610. A related difficulty is the view that Galileo did not begin to pursue a "full" Copernican program until after he acquired a telescope in 1609 and moved to the Tuscan court the following year. Hence, it is claimed, unlike Kepler, Galileo simply showed little interest in Copernicus or astronomy.

I dispute both propositions. Already in 1597 the two were pursuing quite different types of Copernican projects. Within a few years, Kepler was openly seeking both a Copernican astrology and a heliostatic astronomy. The conundrum lay as much in their personal styles as their philosophical predilections: Galileo—not unlike Tycho Brahe—was most comfortable with disciples, whereas Kepler regarded Galileo, like his teacher Maestlin, as someone to be converted from a position of private to public advocacy. Although Galileo maintained his wall of silence for thirteen years, he and Kepler continued to monitor one another's positions through an obscure English intermediary named Edmund Bruce. Moreover, Galileo did not drop his interest in Copernican issues. But Bruno's public execution sharply constrained his space of philosophizing even as he and Kepler developed strikingly different views on what to value in the new magnetic philosophy of the modernizer William Gilbert.

These themes continued to play out, crisscrossing and interacting in unexpected ways across the decade. Evidence of a new temper among celestial practitioners—not just modernizers—came with the appearance of another nova in 1604. As chapter 14 shows, unlike the earlier nova, this one persuaded even traditionalists to begin to move these unpredicted events out of the extraordinary realm of divine, miraculous meanings and causes and into the ordinary domain of strictly natural causes. This move did not push divine purpose completely out of range, but it put a premium on the search for natural causes. This was an important development. Effectively, God should not be the explanation of first and sole resort. And with that reclassification, a cluster of critical questions about the unpredicted events began to merge with the problem of the constitution of the universe. Kepler fully integrated the nova into his wide-ranging celestial philosophizing and into his struggles with other modernizers at the Rudolfine court in Prague. In contrast, although Galileo had cultivated a support network of sympathetic Venetian nobility and church modernizers, powerful traditionalist resistance in the universities framed his approach to the nova.

Celestial theory traveled in all the modes typical of communication in preindustrial European society: letters, manuscripts, books and their margins, and, occasionally, books of collected letters. Books showed up for sale at fairs and stalls or were sent with couriers and itinerant scholars as personal gifts. Unlike Galileo, Kepler made extensive use of the printing press. And chapter 15 shows that Kepler's arguments with Pico, Bruno, Gilbert, and Tycho soon arrived in England through the *Stella Nova* (1606), that is, three years before the public account of the elliptical astronomy for which he is much better known today. The *Stella Nova* proved to be a multipurpose vehicle, carrying arguments for Kepler's Copernican astrology as well as against Bruno's infinite universe. Kepler accompanied the book with letters that gained mixed support from constituencies as varied as King James I and the modernizers Christopher Heydon and Thomas Harriot.

The first decade of the seventeenth century also poses another problem. The period prior to Galileo's telescopic announcements has often been organized with a focus on a dominating figure but without adequate attention to the overall scene of engagement. As shown in the chapters 13–15, continuing conflicts over the legitimacy of astrology, the execution of Bruno, and the accommodation of unexpected celestial events already complicated this overly tidy picture. Chapter 16 adds another layer of complexity with the three-way struggles over planetary order. Just a year before his death, Tycho Brahe was aggressively seeking to win over Galileo and Clavius. His untimely death upset the balance of power in Prague and, as James Voelkel has shown, led to a drawn-out struggle between Kepler and Tycho's heirs and remaining followers that also affected both Kepler's claims and the way in which he presented them.[52] Had Tycho Brahe lived for another ten years, Kepler would almost certainly not have risen to prominence as the most forceful spokesman for a Copernican-based philosophy of the heavens. For even with Kepler's visibility at the Rudolfine court, the Copernicans had little philosophical authority in the universities. I explore the question of why there was no alliance among the Copernicans of this period, what this absence shows about the three-way conflict to establish the Copernican ordering as natural philosophy, and why Kepler's variant of the Copernican system failed to shut down skeptical objections to it.

The contingencies of Tycho's early death and the appearance of another nova in 1604 were followed in 1609 by still another historical accident of sorts: Galileo's unanticipated acquisition of magnifying lenses and his application of them to detect unexpected heavenly novelties. The juxtaposition of these contingent events with the three-sided conflicts of the decade form the background to part 6 (chapters 16–18). Although "research," the discovery of new knowledge, is the guiding presupposition of all areas of knowledge in our own age, in the early seventeenth century novelty was not part of the traditionalist ethos: it was neither sought nor expected. In fact, in the tradition-laden world of the university, it was seen as dangerous. Heavenly practitioners of that period attempting to advance claims to new knowledge had to seek out other grounds on which to build their positions. And the question of the final two chapters is just how Galileo managed to make his newly grounded claims stick.

Chapter 17 turns to the question of the kinds of sociabilities that characterized the world in which Galileo made and advanced his claims for recurrent celestial novelties. Certain princely courts, like those of Landgrave Wilhelm IV of Hesse-Kassel, Emperor Rudolf II in Prague, and King James I in London, were clearly open to modernizing currents. But were these courts and others, like that of the Medici in Florence, more than just alternative spaces for nontraditional philosophizing? An ambitious case has been made that patron-client relations were the dominant form of early modern scientific sociability, and further, that the instrumentalities of patronage provided the key structures of court society of which Galileo's rhetoric of motives and authorial strategies were a function. In this chapter, I argue against patron-centered accounts both in their mitigated Mertonian version and in their strong structural-functionalist version. In place of the central role accorded gift-giving, patronpleasing, and status-seeking strategies, I propose that Galileo adapted and transformed his pedagogical practices to the learned sociability of aristocratic culture while using his new platform at the Medici court to continue by other means his war with the traditionalists.

Chapter 18 considers in detail the question of why and how Galileo's startling telescopic claims so rapidly overcame heavy resistance among traditionalist practitioners, as compared with reports of the slightly earlier nova and Kepler's Copernico-elliptical astronomy. Here I examine the relative importance of print in reporting and establishing the credibility of his initial performances and discoveries, the diffusion of telescopes, and Galileo's failure to win support through direct demonstrations with the instrument in front of live witnesses. I also revisit the question of Galileo's motives for returning to Florence and what he was really seeking from the Medici. And I consider in detail the interesting question of why Kepler, although personally ambivalent about Galileo, both trusted and brilliantly defended the claims in the *Sidereus Nuncius* when he had never once looked through a telescope and when Galileo had still made no effort to reestablish any sort of contact or collegial relations. Print had a much more powerful role than patronage in both establishing and undermining credit. Yet even after Galileo finally broke his thirteen-year silence, he was less than forthcoming to Kepler and never sent him a telescope. The ambivalent and tenuous collaboration between these two towering Copernican modernizers ended with a failure to meet, argue, and explore points of commonality. Toward Kepler, Galileo maintained a distantly respectful demeanor while continuing to withhold meaningful credit and publicly suppressing his own debts. And that division was not only personal: it would continue to mark the paths taken by their followers for much of the seventeenth century until the arrival of Isaac Newton.

CATEGORIES OF DESCRIPTION AND EXPLANATION

Some prior linguistic and conceptual clarification may assist readers in following the terms of this study. To begin with, there will be no place here for nineteenth-twentieth century role designators like *scientist* or *men of science* to characterize early modern heavenly practitioners—a usage that one ought to resist as strenuously as dressing up seventeenth-century kings in World War I uniforms.[53] In recent years, the recommended prescription for avoiding this lingering practice has been to substitute the term *natural philosopher;* but although undoubtedly salutary, this substitution does not resolve the objection that most of the characters in *The Copernican Question* used a fluid range of self-descriptors, at least as varied as *astrologer, astronomer, mathematician, prognosticator,* and *lover of the stars.* Lumping all such agents under categories like *natural philosopher, real astronomer,* or *mathematical astronomer* clearly directs the representation before the question has been settled.

There is a similar problem with knowledge domains. A good deal of erasure is achieved by combining the noun *science* with the verb *to be:* the construction "Science is . . . " essentializes our problem, overriding the fact that classifications of knowledge change over time and thus should not be presumed in advance. Thus, I treat knowledge classification itself as a kind of practice, and I present historical agents as actively arranging, rearranging, renaming, and sometimes suppressing or merely gesturing at their own categories of heavenly knowledge.

A different sort of narrative from this one might have emplotted, say, the theme of progress and the advancement of knowledge. But "progress" had neither the well-articulated status nor the authority of a standard of heavenly knowledge that it acquired well after this study's period of concern.[54] Standards other than progress were used by Copernicus, Rheticus, Galileo, and Kepler to argue for the preferability of the Copernican arrangement. In fact, disagreements over what standards *should* prevail are themselves part of this story, as indeed they are of some other, much later episodes in the history of science, like the twentieth-century problem of continental drift.[55] Hence, if we want to understand the *historical* problem—how practitioners actually decided to pursue, adopt, and abandon different knowledge claims—then we must be careful not to fall into a usage widespread among philosophers and scientists, that eventually theory X—in a cleaned-up, "mature" form—came to be seen and accepted as true. For that is a different kind of story, involving a form of erasure that, as Kuhn rightly observed in *The Structure of Scientific Revolutions,* simply suits the needs of a particular way of doing and presenting current science.[56]

Of course, *eventualism,* as I shall call it, has uses that should not be overlooked. For example, I suspect that no one familiar with Newton's formidable mathematical approach to natural philosophy still needs to be persuaded that Newton both explained and predicted a great deal more than the natural philosophies of his predecessors.[57] But, taking the question differently, when was it that "Newtonian mechanics" rather than "Sir Isaac Newton's mechanics" came to be taught in a version that is still taught to physics students and is accepted by all physicists as absolutely true—at least at a certain level of description? In other words, how did Newtonian knowledge achieve such high authority as it circulated through time and place? For the elementary particle physicist Steven Weinberg, the answer is to be found not in Newton's lifetime nor even in the Enlightenment but with the much later formulations of Pierre Simon de Laplace and Joseph Louis Lagrange in the early nineteenth century: after them, there was a once-and-for-all consensus that the truth had been established. "Newton," as Weinberg likes to quip, "is pre-Newtonian."[58] A more historical way of tackling the problem—one that does not so readily make investigation of the intervening period seem epistemically unproblematic and dispensable—is to think of the question in Nicholas Jardine's important formulation as one of "calibrating" both the reality and the truth value of assertions as one moves from an earlier to a later "scene of inquiry."[59] Instead of eventualistic judgments, one might ask, in what sorts of calibrations did natural philosophers actually engage once the notion of progress began to be deployed as a standard? How useful were very long-term notions of progress to investigators of Newton's own time? In the late seventeenth century, for example, Christiaan Huygens and Gottfried Leibniz ascribed truth to Copernicus's planetary arrangement but judged Newton's force of universal gravitation to be an unintelligible foundation. It was a "perpetual miracle"—at best, a nice piece of mathematics, but a natural philosophy without intelligible, physical causes and hence with no explanatory value at all.[60] The critical judgment that Huygens and Leibniz made was not a long-term comparison between the theories of Newton and those formulated by Ptolemy and Aristotle but between those of Newton and the immediately rival theories of René Descartes. The same sort of evaluative practice can be seen in 1692 when Newton's ally David Gregory intoned in his inaugural lecture at Oxford that, with Newton, "*at length* there dawned that most desirable day in which to the immortal glory of this age and people the physical forces of natural bodies were assimilated to a genuine pattern, that is, to geometry." By contrast, Descartes merely tried to "investigate the causes of things logically, or rather, sophistically" because he was "intoxicated by easier and less composite laws, and, not applying his geometric ability in the slightest, fell into errors from which we were at length liberated by the aid of geometers."[61] In such a case, history of science has something to say about the contested reality of questions and explanations as well as changes in standards of progress and truth and how historical agents actually applied them.[62]

The permanent truths of nature of which Steven Weinberg speaks are another matter, for they would seem to require the excision of all cultural residues once taken to be real: Kepler's and Newton's theologies and standards of judgment as well as Maxwell's and Faraday's. For Weinberg, considerations of this sort may still have the capacity to entertain us but certainly possess no enduring truth value, and thus, as far as science itself is concerned, "have been refined away, like slag from ore."[63] From that perspective, the present study may seem to contain, at best, a great deal of unrefined ore and, at worst, an overabundance of slag.[64] This is not entirely my fault. For even statements of eternal truths that claim to correspond to an "out there" must be intelligible to practitioners within a consensus of linguistic conventions and categories of knowledge.[65] And, in that respect, the conventions of the cultural world into which Copernicus introduced his new heavenly representation look very strange to us. This was a world in which the Bible counted as a criterion of proof in matters of natural knowledge. It was also a world in which divine or demonic presence endowed objects, events, and their relations with reality, because both God and the devil, so it was believed, could manifest their intentions through all sorts of natural signs and, indeed, through the events of history itself. Could geometrical theorems and tabulations of angular positions really yield insight into the divine plan? Or was that objective to be reserved for those disciplines concerned with the nature of matter, the causes and goals of motion (physics), or the sacred scriptures and their commentators (theology)?

Attention to such temporally bound epistemic conventions has a value that eventualistic judgments simply ignore. Such consideration can tell us something about the tortuous formation of new conceptual proposals and how they could ever have seemed intelligible to anyone at the time. Of equal importance, I want to recover what might be called the context of doubt: how practitioners made epistemic choices in the face of often daunting—and sometimes quite structured—uncertainties. That problem was especially intractable for the historical agents in this study: predicting the future and explaining the occurrence of objects that had not been predicted.

An investigation of this sort need imply no full-blown skepticism about an external reality. Unlike some social constructivists, in this study Nature—in the sense of "things themselves" that push back at human senses and conceptions—does have more than an incidental role. But Nature shows up only through mediating conceptions and representations, and it is about these surviving residues that history really has something to say. The considerable attention that contemporaries devoted to the meaning and location of unanticipated comets and novas between 1572 and 1604 serves to illustrate the point. If Copernicus had never lived or proposed his new principle of planetary order, what effect would the reports of these objects have had on the traditional classification of natural realms? A counterfactual question of this sort is not as far-fetched as might first appear, as this question simply did not constitute a problem for the great majority of writers. Surely we can imagine that a few heavenly practitioners (among them Michael Maestlin, Thomas Digges, and Tycho Brahe) and a few natural philosophers resistant to the traditionalist Aristotelian mainstream would have used the unanticipated appearances to knock down or at least modify the long-lived Aristotelian distinction between incorruptible celestial heavens composed of solid, planet-bearing spheres and a changeable terrestrial realm. Yet the proposition that the heavens were made of an incorruptible fluid stuff was already prevalent before the recovery of the full corpus of Aristotle's writings in the thirteenth-century Latin West, and it never quite died out.[66] What distinguished the period 1572–1604 was that the initial challenge originated from a nontextual, nonhuman source: natural events in the "out there"—believed by contemporaries to be divinely caused—actually impinged on the perceptual apparatus of those who claimed to have observed something new.

We might want to say that there is some resemblance between the earlier and the later representations. But it is unnecessary to say anything much more fancy about this statement than that no one in the past 430 years has disputed the occurrence. That is not the case with some other sorts of entities taken to be credible in the sixteenth century, like the report of a monk-calf replete with human head, tonsure, friar's habit, pig's feet and a vivid resemblance to "Martino Utero" (Luther).[67] But scholars of the heavens for centuries have continued to believe that something new indeed occurred in a position in the heavens defined by the same configuration of "fixed stars" as that first detected by naked-eye observers of the novas of 1572 and 1604. *Positionality, sudden change in brightness,* and *novelty* remain as continuous properties, even if estimates of distance and the meanings and agencies attributed to the events themselves clearly do not. In 1934, the Caltech astronomers Fritz Zwicky and Walter Baade introduced the idiom of representation that continues to generate consensus today. They redesignated those past events as "Tycho's star" and "Kepler's star" and attributed their occurrence to explosions that produced an unusually large flare-up in brightness and left all sorts of "remnants," like the Crab Nebula and black holes.[68] Such occurrences, they said, were exceedingly rare: in a given galaxy, they happened approximately once in a thousand years. Although miraculist explanations were out of style at Caltech in the 1930s, Zwicky and Baade believed that such awesome and huge events clearly merited a Latinate word that connoted "really big and powerful," like "supermarket" or "superman," words that were just achieving currency at that time in the United States; hence, *supernova.*[69]

One ought to be comfortable with the view that from both standpoints—that of the 1930s as well as that of the 1570s—these celestial occurrences possessed local reality as well as partial overlap or family resemblance in the ways that they were represented across hundreds of years. Whether such celestial representations are true or false to nature, however, is not sufficient to the historical task of explaining their credibility and tenacity between 1572 and 1604. The mere occurrence of the events, that is, did not determine the historical character of their representations; nor did it determine the cultural uses to which European, rather than, say, Chinese or Indian, witnesses put them. Indeed, the question here is why and how Europeans paid attention—why there were debates and conversations about how novas and comets fit into contemporary classifications of heavenly knowledge. In this study, such historically specific knowledge categories constitute the main locus of actors' questions—the "space of possibilities" or "problematic."

Not the least reason for studying such spaces of possibility is that they carry their own intellectual and cultural integrity. Once one recovers a local community's questions, then one is in a position to appreciate that their spaces of possibility were no less real or meaningful to them than ours are to us—even if theirs now seem remote and dissociated. On this formulation of historical relativism, the history of science is not merely about concepts achieved and still currently valued but rather about the interrogatory conditions that made such knowledge possible, real, and meaningful in particular places and circumstances, and about what enabled or blocked movement from one community of practitioners to another. And how theories and claims traveled from one locus to another—always carrying along bits of cultural "slag"—is a major theme in this work.[70]

In that regard, the more dramatic discontinuist gestures of Thomas Kuhn and Ludwik Fleck look too extreme.[71] I reject many of Kuhn's ambitious claims in his earliest and most interesting formulation of *The Structure of Scientific Revolutions:* the existence of a paradigm as demarcating "mature science" from other forms of knowledge, a universal set of problem-solving values transcending paradigmatic divides, and "paradigm shifts" producing incommensurable, revolutionary domains of scientific knowledge.[72] The Copernican question does not warrant the application of incommensurability in any sense of the term of which I am aware.[73] But rejection of the part does not authorize dismissal of the whole. Individual elements of Kuhn and Fleck's work have proved quite sturdy and unfailingly rich and suggestive.

My rejection of incommensurabilities is part of a larger theme in this study. For this is also a history that is suspicious of disjunctions—not so much among actors' as historians' categories. It is not, for example, a story of "for or against," as an undue emphasis on the Galileo episode often encourages. Moreover, in the sixteenth century, there was no sharp division between pro- and anti-Copernicans. One does not even notice the category *Copernicans* until the early seventeenth century, and certainly no debates among different adherents of world systems emerged until the 1580s.

I also avoid the widely favored analytic usages of *Copernicanism* and *heliocentrism*—even as terms designating family resemblance—because they unnecessarily homogenize the considerable diversity of sixteenth-century representations and usages to which the new theory was put. But *geostatic* and *heliostatic* are sometimes useful in denoting the "near-centrality" or "off-centeredness" of the central, resting body in the universe.

I have also developed an unease with the word *cosmology*, which, although coined in the early seventeenth century, did not then possess the same domain of meanings later ascribed to it.[74] Such terms, however useful they may be to us, were no part of the historical agents' descriptive taxa. In approaching the writings of Copernicus and his followers, it is more helpful to follow Michel de Certeau in speaking of ways in which historical agents used, adapted, and transformed.[75] The expression *world system* is an example of an actors' term of such interpretive malleability.[76]

A term like *Copernicanism* looks like a representation of Copernicus's ideas that he could have been brought to recognize but about which he would have been puzzled because, even more than *cosmology*, it is a noun of retrospective historical analysis. Indeed, to the best of my knowledge, no one appears to have used it in the sixteenth or seventeenth centuries. *Ism*-talk can be found occasionally in the sixteenth century to designate a philosophical school (e.g., *Platonism* [1571]) or in the seventeenth century as a designator of adherence to a particular religious doctrine (e.g., *Catholicism, Arminianism*), but "-ism" suffixes seem first to have become widespread in the nomenclature of political and economic doctrines around the 1830s, with the rise of a mass print economy. Consider, for example, *Chartism, scientific socialism, capitalism;* thence, soon after, the usage was applied to bodies of scientific theory (most notably *Darwinism*) and religious outlook (*agnosticism*).[77] So far as I can determine, *Copernicanism* made its first appearance in 1855, in Augustus De Morgan's "The Progress of the Doctrine of the Earth's Motion, Between the Times of Copernicus and Galileo; Being Notes on the AnteGalilean Copernicans."[78] De Morgan's main distinction was "between a *physical* and a *mathematical* use of the Copernican or any other theory," which he represented as follows:

A *mathematical* Copernican was one who saw that, come how it might, the heavenly appearances are such as *would* take place *if* the earth *did* move about the sun, and also about its own axis: and that, consequently, the supposition of such motions, true or false, would be a convenient and efficacious mode of explaining and predicting celestial phenomena. A *physical* Copernican added to the above the belief that the reason why things appear as they would if the earth had these motions, is that it really *has* them. The first said that the hypothesis *explains* or *demonstrates* phenomena; the second said that the hypothesis is a true statement of the causes which *produce* phenomena.[79]

De Morgan spoke of those who used Copernicus's data as "numerical Copernicans" (Erasmus Reinhold) and later drifted into the designation "physical Copernicanism" as a synonym for "physical Copernican."[80] Just a few years later, in his discussion of John Milton's *Paradise Lost*, David Masson reached without apology for terms like "Ptolemaism," "the pre-Copernican mode of thinking," and "pre-Copernicanism," and left his Victorian readers with an "-ismed" but unnuanced frame of meaning: "Before he [Milton] began to write it in its epic form, his Ptolemaism had greatly abated, if it had not been wholly exchanged for Copernicanism."[81]

The great latitude permitted by *Copernicanism* as an analytic term of classification is what needs to be questioned, as it came to have considerable currency in twentieth-century historiography. At best, it looks like a term of family resemblance—implying multiple similarities but also dissimilarities—among different elements constituting a shared domain of meaning. But how much needs to be shared to constitute a functional conceptual family? This study shows that the resemblances are often weaker than expected: indeed, the more rigorous and fine-grained the focus, the more the differences push against the similarities. Minimally, might one not expect agreement about visual representations of the planetary arrangement? Or acceptance of one (or more) of the Earth's motions? Rejection of some or all of Ptolemy's propositions about the arrangement of the universe? Inclusion of propositions, such as the infinitude of the universe, to which Copernicus himself did not explicitly commit himself? Or, similarly, inclusion of propositions regarding astral influence, about which he had nothing to say? Similarities easily slide into totalities. In the twentieth century, holistic categories like *worldview* and political ideologies like *fascism* (1920s) or *communism* (a word dating to the 1840s) simply became part of the standard repertoires of political and intellectual discourse. And from there, it is easy to see how Kuhn's *paradigm* or Michel Foucault's *episteme* could comfortably resonate with the same tendencies to lump together many diverse elements and, once granted, could then encourage deeply compelling discontinuist metaphors of change (*revolution, rupture, incommensurability,* and so forth). Affectively powerful and analytically useful though they may be, such terms of first approximation ultimately must be used with caution, if for no other reason than that the evidence does not cooperate with the hopeless complexity of the historical process—unless perhaps one is willing to admit qualifications without number. Some current historiographical misgivings about the term *Scientific Revolution* may derive from concerns about residual meanings associated with its earlier attachments to such totalizing categories. The solution to this problem is not to abandon all present-day analytic categories but only those that encourage further totalizing tendencies and imperil a discerning balance of analytic and agents' categories.[82]

PART I

Copernicus's Space of Possibilities

The Literature of the Heavens and the Science of the Stars

In the fifteenth century, a vast and complex literature described, explained, and invoked the motions of the heavens and their influences on the Earth. From the 1470s onward, the learning of the heavens, much of it inherited from the ancient and medieval worlds, began to acquire a new sort of accessibility as it was reproduced in the medium of print. This chapter describes the broad contours of that literature and its various classifications. It shows how those categories evolved, how it worked as a body of knowledge, and the peculiar forms that it took in the sixteenth century. This corpus of writings—rather than an exclusive and autonomous stream of planetary theory—constituted the foundational categories of the intellectual world in which Copernicus was educated at Krakow and Bologna in the 1490s and in which his work took form and was later evaluated.

Interest in astrological prognosticating had begun to catch on in the Latin West as far back as the twelfth century, with the arrival of sophisticated Arabic astrological writings. Among the most influential of such works was the *Great Introduction to Astrology* of Albumasar (Abu'Mashar), which emphasized the preeminent effects of great planetary conjunctions.[1] Soon, a good many medieval practitioners were attracted by the prospect of using the heavens in medical prognosis as well as retrospective diagnosis. The popular "zodiac man," representations of which abounded by the fourteenth century, mapped signs of the zodiac onto the body parts that they ruled: it assisted surgeons in deciding when to bleed the patient and guided physicians in prescribing a diet that would counteract a specific disease.[2] The Black Death (or bubonic plague) of 1347–51, which killed one-quarter to one-third of Europe's inhabitants, greatly accelerated a sense of loss of social control and, with it, augmented the special credibility of Albumasarian causal explanations grounded in the power of planetary conjunctions.[3] In the last decade of the fifteenth century, another new and frightening disease entity appeared, accompanying the massive movement of French armies into Italy. It too killed, but first by attacking the genitals. Was this "French disease," as many non-Frenchmen called it, caused by a conjunction of Saturn and Jupiter on 25 November 1484? Was it, soon afterward, augmented by a "horrible" solar eclipse on 25 March 1485? Or did God act directly, without need of celestial influence, to punish men for their sins? Whatever the preferred explanation, "astrology had come to stay," as Olaf Pedersen has aptly observed, "and many scholars came to regard astronomy principally as a theoretical introduction to astrological practice."[4]

It is difficult to generalize with confidence about the full range of astrological works that were composed before the era of print. The extant remains of the considerable library of Simon de Phares, astrologer to the French king Charles VIII,

2. Astrological prognostications compared with works of theoretical astronomy published in the German lands, 1470–1630. Based on Zinner 1941, 73.

may be a useful indicator; it was principally a collection devoted to the destinies of individuals.[5] Insofar as medical astrology concerned individual patients, that would partly account for such a focus. However, the arrival of syphilis with Charles's marauding armies spawned a genre of writing about the new plague that applied not just to individuals but to groups. Ptolemy had already classified prognostications into two kinds—those concerning "whole races, countries and cities" (general) and those relating to individuals (specific).[6] Print technology made possible the first kind in a way that had not previously existed. Just over twenty years after Gutenberg published the first book in the West, an almanac for the year 1448, the urban or regional forecast became

COPERNICUS'S SPACE OF POSSIBILITIES

3. Astronomia on her throne, flanked by Ptolemaeus, "prince of astronomers" (right), and Urania, "the Heavenly Muse" (left). Sacrobosco 1490. Courtesy History of Science Collections, University of Oklahoma Libraries.

a standard part of the literature of the heavens and soon dwarfed all other types. Although these annual prognostications occasionally circulated in manuscript, by the 1470s they appeared regularly in print and gradually began to displace hand-produced predictions.

Annual astrological prognostications were part of a larger pattern. Overwhelmingly, the celestial productions that the early printers chose to put on their trade lists were short works intended for practical use: single-leaf wall calendars, almanacs, ephemerides (tables of daily planetary positions), lunar tables, and eclipse forecasts. Ernst Zinner's bibliography of "astronomical literature" published in "Deutschland" over the period 1448–1630, comprising more than five thousand

items, illustrates this contention by enabling a gross count of different sorts of writings produced by publishers in the domains of the Holy Roman Empire. One can only guess at bibliometric patterns for the rest of Europe, and it is impossible to determine absolute numbers of copies.[7]

Gradually the emerging culture of print dressed up its products. It used a variety of new techniques to encode already existing literatures of heavenly representation, such as visually compelling title pages; epistolary dedications to a patron or general dedications to the general reader; and didactic woodcuts displaying spheres, circles, angles, and movable planetary discs, or volvelles.[8] Regiomontanus, the earliest printer of celestial works, pioneered techniques of setting type for astronomical woodcuts, including those that he used to illustrate the models for Peurbach's *New Theorics of the Planets*.[9] Print technology also had undeniable consequences for the conditions of prognostication that expanded the limited possibilities previously open to the hand copyists. First, it made possible the rapid replication and distribution of forecasts. Second, as the annual prognostication became a unique feature of print, it helped to make the astrologer into a more public figure. It also fostered demand for astrology's theoretical-foundational texts and further promoted the authority of the works of theoretical astronomy on which they depended. And third, because such works were public rather than private, it changed the possibilities for offering advice to rulers and hence the conditions of prognosticatory authorship. How this shift in the social and literary conditions of forecast and advice occurred and how it was implicated in Copernicus's astronomical project is an important concern of this and the two following chapters.

Some preliminary chronological parameters will assist. Between roughly the 1480s and the 1550s, the fundamental texts of Greek and Arabic astrology were published in one edition after another. By 1524, the date for which an enormous quantity of prognostications predicted a flood of biblical proportions, the forecasting literature itself reached a scale unimaginable prior to the invention of printing. This surge in heavenly writings issued from those regions where printing had initially taken hold in various cities of the Holy Roman Empire—notably Nuremberg, Leipzig, Augsburg, and Wittenberg—and the great northern Italian city-states (particularly Venice): soon these sites were joined by another great port, Antwerp, and its neighboring university town, Louvain.[10]

Zinner's bibliography need not be regarded as a definitive count of editions of the literature of the heavens so much as a heuristic for questioning the meaning that planetary theory held for contemporaries.[11] Much of the received historiography makes planetary theory the core of a narrative that leads, willy-nilly, to the undeniably major achievements of Newton. Such narratives generally take their endpoints as justification for the inquiry into what precedes. What they do not explain is the vast quantity of prognosticatory literature that contemporaries viewed as significant and the relation of the genre of planetary

theory to it. The primary reason that the calculation of planetary positions held such great importance was that it was necessary for the production of quantitatively based knowledge of the future of the human realm. This book, while also ending with Newton, arrives there by a route that makes prognostication central rather than peripheral.

COPERNICUS'S EXCEPTIONALISM

Why should this matter of prediction be of any concern to a study about Copernicus and the subsequent meanings that contemporaries ascribed to his achievements? Copernicus's formidable position in the history of astronomy and in the historiography of the Scientific Revolution is hardly open to dispute. Yet both among his biographers and among many historiographers of the Scientific Revolution, Copernicus appears as something of a pristine figure in relation to astrology, let alone the bibliometrics to which I have referred. Despite some suspicion that this view is not quite viable, no one has yet seriously challenged the strong position articulated by Edward Rosen.[12] "Did Copernicus believe in astrology?" asked Rosen, and he answered his own question as follows: "This is an extraordinary aspect of Copernicus's mentality. He lived in an age when many of those in power as well as of those on the lower rungs of the social ladder believed in astrology. [Copernicus] did not."[13] And, in his comments on the single instance of the term *astrology* in Copernicus's extant writings, he stated forcefully: "Fortune-telling astrology received absolutely no support from Copernicus. In this respect he differed markedly from Brahe, Galileo, and Kepler, to mention only a few of the celebrated astronomers who believed in astrology and practiced it for one reason or another. In particular, the contrast between Copernicus and his disciple Rheticus in this regard is complete. Nowhere in the *Revolutions* nor anywhere else in the unquestionably authentic writings of Copernicus can the slightest trace of belief in astrology be found. On the other hand, Rheticus's addiction to astrology is notorious."[14]

Even authors quite willing to admit into account considerations rejected by Rosen have not found Copernicus easy to integrate into their narratives. "The Copernican Revolution provides the blueprint for the Scientific Revolution as a whole," Charles Webster declared in his influential 1980 Eddington Memorial Lectures. But Webster began with Paracelsus because he could find no evidence to link Copernicus himself with prophecy and eschatology, let alone astrology.[15] And in one of an excellent collection of essays devoted to assessing astrology in early modern science, Keith Hutchison presented a great quantity of convincing illustrations from churches, town halls, instruments, and frontispieces showing that the sun was frequently placed symbolically at the center or associated with the figure of the king, but he did not find any direct evidence linking Copernicus to astrology.[16]

The closest that anyone has come to drawing a plausible connection was J. L. E. Dreyer in his 1905 classic *History of Astronomy from Thales to Kepler.* Dreyer called attention to a political prophecy inserted near the middle of Rheticus's *Narratio Prima* (1540), the first work to describe Copernicus's claims in print. This cyclical prophecy predicted that as the Earth's eccentricity slowly changed, different kingdoms would rise and fall. Dreyer admitted that "nothing of this theory of monarchies is mentioned by Copernicus himself" but then suggested that Rheticus would not have inserted such a prophecy without Copernicus's permission. Rosen effectively dismissed Dreyer's speculation. He wrote off the forecast as superstitious nonsense, ascribing it to Rheticus's exuberance and youth while absolving Copernicus of any association with it whatsoever.[17]

The Rosen-Dreyer disagreement still divides scholars. Yet in my opinion Dreyer was on the right track: indeed, his observation can be taken a good deal further. Here I briefly anticipate a preliminary argument against Copernicus's supposed immunity from astrological concerns. Planetary order became problematic for Copernicus within a shared structure of literary and epistemic possibilities that included both the domains of planetary theory and the prognostication of earthly effects. One reason that the link to astrological prognostication is not so obvious is that Copernicus, like other authors of his time, followed conventions of compositional form that included and excluded certain subjects. The prevailing view, widespread among the humanists, held that ancient works represented the ideal stylistic models for the organization and presentation of knowledge. Stylistic models, however, were not merely the subject of high literary theory; stylized conventions were communicated by repetition through the curriculum of Renaissance grammar schools. This practice is well documented for Italy, where students were given Cicero's writings—especially his letters—as examples to be closely emulated for vocabulary, content, and form.[18] Early modern authors thus had a well-schooled sense of rhetorical boundaries and decorum.[19] Copernicus's major work was thus not exceptional in closely following the organization of Ptolemy's treatise of theoretical astronomy, the *Mathematical Syntaxis* (commonly latinized as the *Almagest,* after the Arabic). The *Almagest* provided the models and parameters from which one could make specific predictions for the planets' angular positions but said nothing about its effects on particular persons or geographical regions. For Ptolemy, the prediction of specific effects fell into the separate domain of astrology, and to that subject he devoted a separate work, the *Tetrabiblos* (or *Quadripartitum*). Later I will show that, for sixteenth-century readers, the *Tetrabiblos* was effectively more than a single work.

PRACTICES OF CLASSIFYING HEAVENLY KNOWLEDGE AND KNOWLEDGE MAKERS

The question of Copernicus's exceptionalism is entangled in a dense thicket of knowledge categories and forms of presentation that are anything but obvious. If we want to make sense of his stated intentions as well as his silences (which are many), our account should try to mirror the thickness of these representational resources. To begin, however, it is useful to remind ourselves of what they were not. Copernicus did not present his work in a culture of emerging specialization and professionalization like that of, say, nineteenth-century Germany or England. There were no self-conscious specialty groups with their own journals, no characteristic research techniques and professional ideals of academic advancement, let alone a concern with common standards of measurement.[20] The late-fifteenth- or early-sixteenth-century academic practitioner of Copernicus's time had little resemblance to his counterpart in the bureaucratic university of the late twentieth century, which one historian has called a "factory system"—"the student . . . a 'pair of hands' working for the greater glory of his supervisor, the department as a conveyor belt for the production of Ph.D.s, the publication of papers as a sort of dividend."[21]

The sixteenth-century sense of the learned professions and disciplines was hierarchical. Some writers imagined the organization of the professions as a mirror of the aristocratic hierarchy of social ranks or the order of the natural world. But a variety of different criteria were employed for organizing the ranks of knowledge. They might include the subject matter's moral dignity, nobility, historical ancestry, or degree of abstraction; its degree of certitude; its practical value; and the order in which the disciplines were best taught—or some combination thereof.[22] The Renaissance rhetorical fashion for praising or satirizing the professions depended on which of these criteria were favored and in which combination.[23] Regiomontanus, for example, praised Euclid's theorems for possessing the same certitude as they had a thousand years earlier, while opposing them to the uncertainties betokened by the many branches of scholastic philosophy.[24] Copernicus praised the heavenly art ("which is labeled astronomy by some, astrology by others") for the perfection of its subject matter and for its pleasures in contemplation prior to describing the disagreements of its practitioners about principles and assumptions.[25] For Francesco Capuano de Manfredonia, a prolific commentator on John of Sacrobosco's *Sphere*—the standard, elementary introduction—astronomy's subject matter was physical in its concern for bodies in motion, celestial spheres, and influences, and, in that sense, it fell under natural philosophy; but its methods were also mathematical and, in that sense, were capable of secure demonstrations.[26] Yet, ultimately, Capuano decided that astronomy's demonstrations were "more physical than mathematical."[27] A century after Regiomontanus, Tommaso Garzoni imagined a "universal piazza of all the professions in the world," a survey that ranged from university professors and theologians to cooks, chimney sweeps, prostitutes, and latrine cleaners. Even as Garzoni used comic inversion to rebuke and undermine, he assumed the hierarchical pretensions of the higher professions. Yet pedagogically, the early modern academic could have competences in quite different subjects and was capable of teaching in quite different disciplines while respecting and never challenging the boundaries separating them. Although some prominent early seventeenth-century voices favored the discovery of new knowledge, research as an ideal that embodied originality did not emerge until at least the German philology seminars of the late eighteenth century.[28]

Consequently, not much can—or should—be taken for granted, not even epistemic categories as crucial as *science, theory, practice,* or *truth;* disciplinary designators as seemingly apparent as *astronomy, astrology,* and *cosmology;* or genres of writing, authors' print identities, and titles assigned to works. None of these notions carried quite the meanings that we would now ascribe to them. But why assume that they should? At best we can try to work out some stable points of signification. Consider, for example, the exchangeable Latin terms *scientia stellarum, scientia astrorum,* and *syderalis scientiae,* which may all be translated as "the science of the stars." Whatever their earlier medieval usages, in the sixteenth century these words—rather than astronomy or mathematical astronomy—actually covered the entire subject matter of the study of the heavens.[29] Although arguably the term *scientia* can be rendered as "knowledge," that translation has the disadvantage of being too vague and general; it fails, for example, to target distinctions that historical agents sometimes made between explanation and description, cause and effect. On the other hand, *scientia* clearly did not carry the later connotation of *science* as a special type of *method* for gaining knowledge (e.g., a specialized or singularly rigorous form of self-correcting knowledge) or vocational training in highly specialized skill sets.[30] Thus we can admit the term *science* into our descriptive tool kit as long as we are careful not to conflate earlier with later usages. Further along, we shall see that for contemporaries, the phrase "science of the stars" actually encompassed the subject matters of *both* astronomy and astrology, and each might be further subdivided into theoretical and practical parts.

Terms denoting social roles or identities are another matter. The word *scientist,* as is now well known, did not exist before the 1830s.[31] It was not an operative category in preindustrial Europe. Its appearance coincided with the emergence of a professionalizing impulse in the scientific move-

COPERNICUS'S SPACE OF POSSIBILITIES

ment that evolved in early Victorian Britain. Historians of twentieth-century science do not worry about the strangeness of this term; but historians of our period of concern are now more comfortable designating Newton as a *natural philosopher* or a *physico-mathematician* than a *scientist*.[32] The diversity of early modern usages—especially in those areas of knowledge that mixed together mathematical and physical elements—is a question of empirical investigation.

One approach to sustaining the historical integrity of past social agency is to make a virtue of our limited knowledge and to focus on how authors represented their identities or those of others in the works that they published. Copernicus, for example, wrote that "mathematics is written for mathematicians." Edward Rosen chose to translate this famous passage as "Astronomy is written for astronomers," straining to get the text's language into agreement with his own assumptions about how Copernicus conceived his role.[33] But neither my rendering nor Rosen's is quite satisfactory without further qualification. For the historian to call Copernicus a mathematician evokes confusing associations with the current domain of meaning, in which mathematicians may or may not test hypotheses against the physical world;[34] and to call him an astronomer overrides the meaning that *mathematicus* had in the sixteenth century, that is, someone skilled in any subject that involved mathematics—for example, optics, music, statics, or astrology. An author's *self*-representation or print identity—the way he presented himself on the title page and within his writings—thus carries considerable methodological utility. The title-page identity is how that author was often known to readers. Similarly, the language that one author used to characterize another can often provide clues to the larger field of representation—how, for example, he understood (or misunderstood) the authors he had read and how he classified their aims and historical circumstances.

Authorial self-designations were at least as strangely varied as the often-interchangeable singulars *astronomus, astrologus,* and *mathematicus* or the conjunctions *medicus et astronomus, iatromathematicus, medicus et mathematicus,* preferred by academics who held more than one chair in those subjects; the pompous *theologastrosophus* (wise theologian of the stars); the nonacademic prognosticator who likened himself to a *cosmo-graphus* (cosmographer or geographer) by the designation *astrographus* (astrographer); the late fifteenth-century author of an astrological judgment who said that "not for nothing" was he called *phisicus et astrologus;*[35] the slightly ambiguous "lovers-of-wisdom" forms that may have been used to conceal the absence of other credentials, such as *astrophilus, philomathus, Mathematik Liebhaber,* or simply *astronomiae studens;* and the quite specific identity of an author as the "student of" someone famous (*discipulo del*).[36]

Apart from what authors called themselves on their title pages, it is also useful to note how authors coded others of their kind. Such contemporary practices of authorial classification were anything but standardized. A useful example is that of the Florentine Francesco Giuntini (Junctinus, 1523–90), who, at the end of a massive two-tome work on astrology, included a "Catalogue of Learned Men whose notes and productions benefited us in completing the Mirror of Astrology." Effectively, it was an index of names distributed throughout the work's more than 2,500 pages. The censors who granted permission for the work to be published described Giuntini himself as a "Doctor of Sacred Theology" ("sacrae Theologiae Doctoris") and the royal privilege ("Extraict du Priuilege du Roy") described him with the title "Doctor of Theology and Chaplain to Our Very Dear and Much-Loved Brother, the Duke of Anjou."[37] Never mind that the work of this "theologian" was a massive portable library for defending "good astrology" that contained an edition of Ptolemy's *Tetrabiblos,* spherics and theorics, lengthy treatises on all aspects of theoretical astrology, and an abundance of horoscopes of famous men.[38]

The ninety-nine names in Giuntini's catalogue illustrate—although they do not exhaust—some typical features of the language and assumptions that he and others used in constructing authorial classifications. The overall structure does not lack coherence, but its anachronisms, inconsistencies, and omissions, easily spotted with hindsight, provide clues for illuminating the unregulated and arbitrary space of authorial representation. Anaximander and Thales, for example, have been readily assigned sixteenth-century professions ("astrologus"; "astronomus"). Likewise, Hipparchus is an "astrologus" but notably not an "astronomus." In addition, the list reflects the contingencies of what Giuntini himself took from his own reading—especially authors' print identities. Marsilio Ficino is an "astrologus" associated with a school ("Platonicus") but is not called a philosopher,[39] whereas Pico della Mirandola is a "poet, orator and philosopher" but is not associated with a school.[40] Vitelo, who wrote on optics in the thirteenth century, is called a "mathematicus," whereas Ptolemy was an "Egyptian astrologer" but not a "mathematicus," and Hermes was an "Egyptian" but not an "astrologus." Messahala (Masha'allah ibn Athari), a Jew who wrote about astrology, is an "Arab," while al-Battani, an Arab, is an "Egyptian astrologer."

Coming closer to Giuntini's own time, Regiomontanus is "a man famous in all kinds of mathematics" but is not presented as having any relation to astrology. Leovitius and Stadius are both represented as "astrologers," but although both were well known at the time for their ephemerides, the former is called an "astronomus," whereas the latter is designated a "mathematicus." In much the same way, Christopher Clavius, author of a highly regarded commentary on Sacrobosco's *Sphere*, is called a "mathematicus" but not an "astronomus" or a "Jesuit." By contrast, the thirteenth-century Spanish king Alfonso X, who lent his name to the important planetary tables still being used in Giuntini's time, curiously merits the title "astronomus." The list excludes Protestant authors and is unusually attentive to authors' membership in different orders of the Catholic Church; but Giuntini does not hesitate to analyze (negatively) the horoscopes of numerous Protestant authorities, such as Osiander and Melanchthon.[41]

Finally, Giuntini used the adjectives *famous* (*insignis*) and *excellent* (*eximius*) to designate authors of whom he was especially approving: his own teacher, Brother Giuliano Ristori, the "Famous Carmelite Mathematician"; Lucio Bellanti, "Famous Astrologer and Physician of Siena"; Georg Peurbach, "Famous Astronomer"; and Roger Bacon, "Astrologer and Excellent Philosopher." These examples suffice to illustrate the great variability in attributions of authorial identities and the dangers of too hastily imposing our own role designators and categories. Thus we cannot be altogether surprised by an entry in Giuntini's astrological calendar: "Nicolaus Copernicus of Torun, Varmian Canon, born 19 February 1472, 4h 48m P.M."[42] More interesting than Giuntini's error on the year of Copernicus's birth is his decision to call him a canon rather than a famous astronomer.

Besides authorial representation, the language and syntax of a book's title were important means by which an author signaled the category to which the work belonged—a quick guide to a book's rhetorical location. The sixteenth and seventeenth centuries were a period of great innovation, warfare, and religious polemic. Heavenly works announcing controversy or difference sometimes borrowed or echoed such rhetoric in their titles with terms like *new, great, defense, against, mystery,* and *reformed.* Conversely, works announcing new entities like comets often resorted to the more neutral-sounding *observation, description,* or *method.* By far the most common terms were the plethora of designations learned in the schools: *commentaries, principles, elements, rudiments, questions, disputations, doctrines, dissertations, assertions, propositions, problems, demonstrations, epitomes, instructions, uses and exercises, introductions* (*isagoge*), *forerunners* (*prodromus*), and *visual sketches* (*hypotyposes*).[43] And finally, there were works following ancient models that simply named their subject with the prepositions *of, on,* or *about. De Revolutionibus* was such a work, a general exposition of a subject.

Other, more general epistemic criteria could also be deployed as resources for organizing one kind of knowledge with respect to another by rank or precedence. This practice is apparent where an author defines one work as the necessary prelude to another. Ptolemy, for example, defined the *Tetrabiblos* in relation to astronomy, and sixteenth-century readers took "astronomy" to refer to the *Almagest,* his major work on the planets. In contrast, the original relationship of the *Almagest* to the *Planetary Hypotheses,* the work that contained Ptolemy's physicalized representations of the *Almagest*'s models, was unknown because no Latin work explicitly attributed the latter work to Ptolemy's authorship.[44]

The criteria of precedence underlying the relationships of such works were ultimately beholden to Aristotle's own disciplinary classifications. In the *Physics,* Aristotle classified astronomy, optics, and harmonics as sciences that combined mathematics and physics. These mixed subjects were to be regarded as in some sense "rather more physical than mathematical" or "the more physical part of mathematics."[45] The difficulty of establishing the meaning of Aristotle's imprecise language reflects, in turn, his own struggle with Plato's view that unchanging reality resided not in matter but in universal Forms. For Aristotle, the student of optics as well as the physicist would be interested in the object's mathematical form—say, the apparent diameter of the Moon—as mentally abstracted and hence separable in thought from matter, but then adding to that form a reality that was physical. Richard McKirahan discusses an example of this kind of subordination from Euclid's *Optics,* a treatise on illusions of linear perspective: as different angles are formed with the eye E at the vertex (AEB, BEC, CED), the segments representing the Moon's diameter (AB = BC = CD) appear to change, such that AB will appear greater than BC and BC greater than CD. One inferred physical content effectively by invoking the translation rule that straight lines represent light rays. The geometer is only concerned with the lines and angles as a matter of reasoning and mental abstraction, whereas the optician is interested in the apparent changes in the Moon's size and apparent brightness as produced by the angles and the physical substrate of light rays entering the eye.[46]

In this example, geometry treats properties of lines, angles and their relative magnitudes; but physics concerns the nature of the ray itself. They consider and describe different aspects of the same phenomena. In his work on logic, the *Posterior Analytics,* Aristotle says mathematics is the subject that provides the "cause" or "the reasons why." Yet the geometer need not know the subject matter of optics (the "fact" about the nature of light rays) in order to demonstrate various properties of lines, angles, and triangles; the geometer "explains" or "demonstrates" that abstracted visual lines trace out straight lines that form angles with the eye.[47] This did not mean, however, that mathematics had epistemic precedence over physics, as Aristotle maintained that the forms with which mathematics are concerned are "not said of any underlying subject."[48] Although mathematics could account for the shape of a physical object, it did not thereby determine the object's nature. This ontological tension between form and matter in Aristotle's diverse writings reflects his encounter with Plato's doctrine of forms, a position toward which Aristotle may have moved over his lifetime. In any case, Aristotle's scattered statements left plenty of work for his later commentators and thereby allowed differential uses of his authority.

Ptolemy is an example. He was explicitly in-debted to Aristotle's divisions of knowledge, but he leaned toward Plato in granting to mathematics a secure capacity for knowing, rather than to physics, whose subject was sensory and variable, or to theology, whose subject was eternal and could only be guessed at. For if matter is ever-changing and corruptible, the mathematician can still determine whether motion "from place to place" is circular, or whether a body moves in straight lines toward or away from the center. As for the main physical propositions needed to ground the astronomy of the *Almagest*—the finite sphere and the Earth's immobility at its center—Ptolemy used a counterfactual argument based on what we would observe about falling (or flying) bodies were the Earth to rotate daily on its axis, but he made no reference to the Pythagoreans' annual motion.[49] Thus essentially he endorsed the same conclusion about the Earth's immobility that Aristotle had reached when he appealed for explanation to the physical natures of the things moved. But Ptolemy reserved for the *Planetary Hypotheses* an account of the unobservable ether's parts as fine, small, rarefied, and also "more homoeomerous," that is, more similar in shape than elemental bodies.[50] And the solid orbs in which he wrapped his geometrical models were essentially built to fit; one could simply translate the geometry of the motions into a geometry of solid, convex, and concave shapes which the perfectly similar aetherial bodies could be said to constitute.[51] By thus carefully segregating his discussion of the eternal aether from the *Almagest,* Ptolemy reinforced the separation of the specific mathematical techniques and routines of calculation from considerations of celestial physics.[52]

In the sixteenth century, an important instance of how the geometrical models were wedded to the physical spheres occurs in the case of Peurbach's "theoric of orbs." Interpreting Peurbach's diagrams for his readers, the Wittenberg astronomer Erasmus Reinhold wrote: "The eccentric orb having been set up, they [astronomers] then gather together physical reasons whereby they attach to it two other orbs of unequal thickness—one above, the other below—so that the total sphere is made concentric with the center of the world, lest it be necessary to assume a vacuum or that celestial bodies are mutually torn apart."[53] On Reinhold's reading—which not all commentators shared—the astronomer first constructed the geometrical model and then

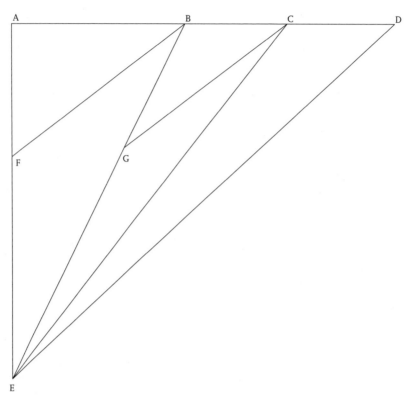

4. Subalternation. Euclid, *Optics*, proposition 4. For full geometrical proof, see McKirahan 1978, 200.

added (or subalternated) the physical substrate, thereby filling up the gaps to preserve a plenum. The off-center "partial orb" (white region), the epicycle diameter of which defined its width, was surrounded by a composite of two, dark-shaded, lunule-like shapes—one concave, the other convex. Reinhold had nothing further to say about the precise physical status of this inner, shaded region.

Reinhold's contemporary Andreas Osiander provides another important instance of how these uncertain relations could be managed. He took it upon himself to defend Copernicus in his famous, unsolicited "Ad Lectorem" ("Letter to the Reader"), which he added without the author's permission to the front matter of *De Revolutionibus*. Osiander emphasized that the mathematical part of astronomy can produce hypotheses but that these hypotheses can never be known to be true or even probably true. Because the premises of the mathematical part could never be known with certainty, this part could never be a demon-strative science. However, the physical part could be demonstrative in the sense that its premises are "divinely revealed." Thus, without specifying any such divinely revealed physical premises, Osiander implied that natural philosophy must be the superior science within the disciplinary couple constituting astronomy, thereby providing it with whatever secure proofs it possesses. To think otherwise would "throw the liberal arts into confusion"—a charge from which Osiander thought to protect Copernicus.[54]

THE SCIENCE OF THE STARS

The question of how to classify heavenly knowledge in relation to the mixed sciences—essentially its place within what Aristotle and Ptolemy called "theoretical philosophy"—does not exhaust the problem of classification. There was another, complementary structure of classification and, at least from the thirteenth century, a relatively stable organization. It consisted of another disci-

clinat ut patebit. Sʒ epicyclus ei⁹ motu duplici mouet́ scʒ i lon/
gũ & in latũ. In longitudine quidẽ sicut epicycli supiorũ semp
tñ in decenouem mensib⁹ solarib⁹ fere semel reuoluit. unde so/
lem in hoc sicut supiores nõ respicit. Terminorũ expositiones
p oĩa sũt hic sicut in trib⁹ supiorib⁹. DE MERCVRIO.

Ercuri⁹ habet orbes qñcʒ & epicyclũ. quoⱬ extre/
mi duo sũt eccétrici ḱm qd. supficies nãcʒ cõuexa
supremi & cõcaua infimi mũdo cõcentricę sũt. cõ/
caua aũt supremi & cõuexa infimi eccétricę mũdo
sibiipsis tñ cõcentrice. & centrũ earũ tñ a centro
ęquantis quantũ centrũ ęquantis a centro mundi distat. Et ipsũ
THEORICA ORBIVM MERCVRII.

5. The theoric of Mercury's orbs.
From Peurbach 1485. By permission
of San Diego State University Library, Special Collections, Historic
Astronomy Collection.

plinary couple, astronomy and astrology, the former sometimes divided into theoretical and practical parts. But at the outset, this pair poses various interpretive difficulties. First, many contemporaries did not label the complex in readily consistent ways. Sometimes practices involved labeling subject matter interchangeably (and, to us, confusingly) as either "astrology" or "astronomy." The editor of the 1493 edition of Ptolemy's *Tetrabiblos*—one of the earliest to be printed— described astrology as "judicial astronomy" and astronomy as "quadrivial astrology."[55] At other times, the term "science of the stars" designated either the whole complex *or* only a part of it. In other words, the term could be used as a shorthand designator, one that presumed but did not specify its entire domain of reference, thereby leaving for us a trail of ambiguities and disturb-

ing questions. Did it refer to astronomy or astrology alone or together? Did it refer to the theoretical principles of astrology or only to their practical use? The difficulty is that individual writers did not commonly present all categories of the subject at once in a given writing. Consequently, the most secure general rule about contemporary usage can be obtained only from inspection of specific cases. In the period of concern here, I have found it useful to proceed as if all parts of the science of the stars were present, even if not explicitly articulated—something like codes on a computer screen that are present and essential to the functioning of the software yet remain hidden to the eye. With this approach, one can observe which elements an author has chosen to treat and which are disregarded. Figures 6 and 7 summarize the principal elements.

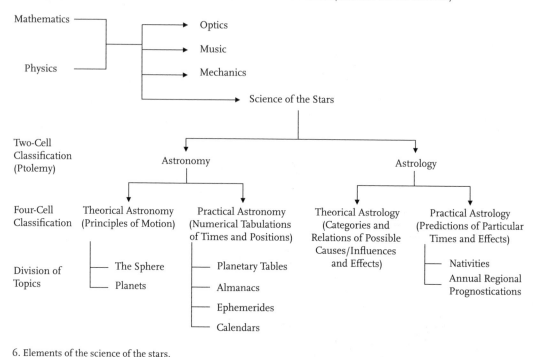

6. Elements of the science of the stars.

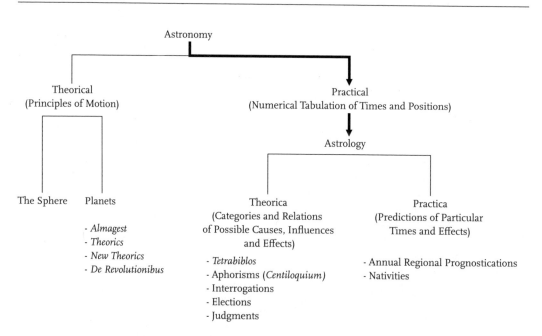

7. Typical structure of prognosticatory practice. The heavy arrow indicates that to the astrological practitioner the practical part of astronomy (tables and ephemerides) was the most crucial part of his work.

Occasionally, an author drops a remark en passant that reveals the existence of the underlying codes. A sixteenth-century astronomical textbook writer with the marvelous name of Erasmus Oswald Schreckenfuchs defined his subject as follows:

The principles of the astronomical discipline are to be sought. The science of the stars is divided into two parts. This science encompasses astronomy and astrology. Astronomy is the subject [*doctrina*] which, by means of geometry and arithmetic, inquires into and also demonstrates the various motions, sizes, and distances of the heavenly bodies such that, to put it simply, all variations and changes of the appearances may be saved—as much with the planets as with the remaining stars.

Astrology, however, is the subject [*doctrina*] that predicts from the stars' motion and virtue and also from the star's position and its nature the varieties of qualities and quantities of motion in bodies. *There will be nothing concerning astrology in this little book, since that requires its own proper and special treatise,* and it is evident that it is more complex and also broader than can be related in brief detail.

Schreckenfuchs's division of astronomy and astrology was very common and an example of what I shall call a *two-cell* classification, one without further division into theoretical and practical parts. The example provides a clue to the more general practice underlying the classification: the author situates his subject explicitly with respect to a part that he will not treat—in this case, astrology. Thus, although the subject is divided into two parts, the book treats only one. Furthermore, within the division with which Schreckenfuchs does concern himself—the elements of the sphere—he does not differentiate his subject by reference to the parameters of planetary models or the use of tables. Without specification, his work is thus introductory to the second part of theoretical astronomy: "The title of this book is 'On the Sphere' because it contains the treatise about the sphere, that is, about the round or spherical body constituted of different circles which, through the imagination, the student ought to transfer from the material sphere to the celestial sphere."[56] In short, he does not intend to burden students with topics concerning the planets that follow the sphere.[57]

At the heart of the separation between astronomy and astrology lay Ptolemy's division of the study of the heavens into a book of principles for predicting planetary positions and a separate book of principles for making sense of planetary influences and meanings. At the outset, that distinction is made in the *Tetrabiblos,* where he merely gestures at astronomy without mentioning geometrical devices or planetary spheres; but in the *Almagest,* he ignores all reference to astral influences—observing exactly the same division (and silence) as Copernicus. As noted earlier, Ptolemy states in the *Tetrabiblos* the relationship between two kinds of prediction, the one celestial, the other terrestrial—a distinction that marks degrees not only of abstraction but of certitude:

Of the means of prediction through astronomy, O Syrus, two are the most important and valid. One, which is first both in order and effectiveness, is that whereby we apprehend the aspects of the movements of sun, moon, and stars in relation to each other and to the Earth, as they occur from time to time; the second is that in which by means of the natural character of these aspects themselves we investigate the changes which they bring about in that which they surround. The first of these, which has its own science, desirable in itself even though it does not attain the result given by its combination with the second, has been expounded to you as best we could in its own treatise [the *Almagest*] by the method of demonstration. We shall now give an account of the second and less self-sufficient method in a properly philosophical way, so that one whose aim is the truth might never compare its perceptions with the sureness of the first, unvarying science.[58]

Ptolemy, followed by his later commentators, constructed the definition of his subject matter in relation to a companion domain. Astronomy pertained to the heavenly motions themselves, had its own treatise (the *Almagest*), operated by Euclidean-style geometrical demonstrations, and relied on no principles other than its own. Astrology concerned the changes brought about on Earth by the heavenly motions, had its own treatise (the *Tetrabiblos*), and yet operated by what Ptolemy clearly asserted to be a "less self-sufficient method."[59] In other words, this was a relationship of subordination, or subalternation, between a higher, autonomous discipline and a lower one that was dependent on it. Astrology depended on astronomy to provide the positions of the planets; without astronomy, there could be no astrology. But, unlike astronomy's claims, those of astrology were conjectural and concerned the transient, unstable material world. Thus, astrology was subalternated to astronomy and added to it astral causes and physical effects.

In the thirteenth century, Ptolemy's distinction was further subdivided to yield a three-cell classification. Campanus of Novara expressed this division with great lucidity in the prologue to his widely used *Theorica Planetarum*: "This science of great nobility was divided by those who professed it in ancient times into two headings: for we may both consider the celestial motions by themselves and also relate them to earthly things, according as they influence [*influunt*] the latter and as they cast their rays on them [*irradiant*]; the first study belongs to a science of proof [*scientie demonstrantis*], but the second to a science of judgment [*scientie judicantis*]."[60] The ancient and medieval distinction between "judgments" about celestial rays and their influences and the works of "demonstration" that they presupposed clearly provided the epistemic basis for the generic separation observed by later authors. But Campanus then further distinguished the demonstrative component as follows:

Again, that part which relies on proofs is divided in its turn into theorical and practical parts; its theorical part is the one that deduces [*sillogizat*] the quantities of the individual celestial motions and the relative sizes of the orbs and the distances of their centers, as well as the dimensions of the bodies and other such things, by most certain methods just as from the first principles of geometry. Its practical part is that which applies [*applicat*] the above conclusions, which have been demonstrated by suitable geometrical figures, to use by clothing them in the numbers which are peculiar to arithmetic; for that reason no one can be an apt student of this art unless he has first been instructed in the theorems of geometry and arithmetic.

However, since this practical part is very useful [*ualde perutilis*], being the immediate end of the science of proof and the *necessary antecedent of the science of judgment* it has been reduced by philosophers to a handy form of tables to make it easier for the inquirer. By means of these tables even those with only a modest education, provided that they are handy in dealing with numbers, can easily find the exact positions of all the planets for any given time.[61]

In this important passage, Campanus admitted that when he divided astronomy into "theorical" and "practical" parts, it was the latter portion—practical astronomy—that really provided the most useful point of contact for those who turned to astronomy for assistance in other endeavors, such as astrology and the calendar. For, as he said quite candidly, "very many are deterred" from the complicated operations of calculation even if they were "lovers of this noble science." Consequently, the geometrical must be separated from the arithmetical, the "theorical" from the "practical," for the sake of those who, "either through occupation with practical affairs or lack of experience or weakness of understanding, are unable to deal with the above-mentioned difficulties [and in order that they] may have the means to find out the exact positions of the planets at all times, while avoiding the detailed numerical complications which I mentioned. . . . [For such people] beg from others the true place computed for a year, which they call an 'almanac,' and thus make up for the deficiency arising from their preoccupation or ignorance."[62]

Campanus resolved the gap between astronomical theoric and practice with an instrument, the planetary equatorium. The equatorium broke down the planetary motions into individual circular components, each made of some material—brass, cardboard, or thick parchment.[63] The device then saved laborious calculation by manual manipulations of threads and deferent and epicycle disks through which mean longitudes could be "equated" (converted) into true ones. The equatorium is an example of what Jim Bennett, in another context, has called a mimetic instrument, one in which the plane figures of a geometrical representation were constructed from a material substance. Campanus described it as "an instrument which is perceptible to the senses and which brings about a motion similar to the rotation of the heavenly bodies."[64] Campanus's planetary *theorica* wedded theoric to practice, form to matter: it was both a material instrument (*materiale instrumentum*) and a written text that explicated the (Ptolemaico-geometrical) principles and parameters from which the instrument could be built.[65] Emmanuel Poulle has shown further that the equatorium was both a didactic instrument and an instrument of calculation. And Campanus's equatorium became the paradigmatic instrument bridging astronomy and astrology from

COPERNICUS'S SPACE OF POSSIBILITIES

Geoffrey Chaucer down to sixteenth-century authors such as Johannes Schöner, Oronce Finé, and Petrus Apianus.[66]

Theorica, then, carried two related, but distinct, meanings. First and foremost, it referred to a classificatory division: that part of the science of the stars that contained the (primarily) geometrical tool kit for modeling the motions (astronomy) and defining the numerous possible configurations thought to produce physical effects (astrology). The practical part—the numbers in the planetary tables, ephemerides, and so on—connected the angles in the geometrical theorics or models to tabulated and observed positions. But practical astronomy also linked the tables to rules for making astrological forecasts, which is why in the sixteenth century one frequently finds quite extensive treatment of astrological theoric in an ephemerides rather than in a commentary on Peurbach's *theorica*.[67]

Second, the theoric was also a demonstration apparatus, a material instrument, an *equatorium* useful for teaching concepts and perhaps for calculations.[68] The word *theoric* thus became a kind of shorthand term—school jargon—for a classificatory principle and a teaching instrument. It is tempting to declare further that the planetary theorics themselves were physical, that they actually referred to solid, material, and even impenetrable orbs—a move that some historians recommend with little reservation, but which one should be cautious in making from the pictures alone.[69] If the term *theoric* still retained the sense of concrete, representational instrument in the sixteenth century, however, it obviously lacked the connotations that instruments for discovering novelties, like the telescope and microscope, would acquire in the seventeenth century.[70] Later, we shall see how sixteenth-century commentators tried to organize and define an explicitly deductive relationship between the plane line drawings of the *Almagest* and the thickened orb diagrams of the theorics.

It is now clear why the fifteenth- and early-sixteenth-century prognosticators or composers of "practices," as well as the calendar and almanac makers, could maintain a common silence about "theorical" principles. The principles that they needed were actually embedded in the instrument used to prepare their forecasts or simply presupposed by other texts designed to provide "the reason why." As long as the instrument served

their predictive needs, there was no motivation to include discussion of its principles.

In sum, from the time of Campanus of Novara, the ground was laid for the science of the stars to be constructed as what one may think of as a three-cell matrix. Following Campanus, astronomy, but not astrology, was subdivided into "theorical" and "practical" parts, that is, according to the sort of knowledge that each was thought to be capable of achieving. The distinction between *theorica* and *practica*, which is treated in detail below, was modeled after Aristotle's division between the theoretical part of philosophy (metaphysics, mathematics, and physics) and the practical part (for example, ethics and politics). The first provided proofs, demonstrations, causes, or general principles; the second described how to use such principles either to make or to do things.[71] Not surprisingly, Campanus used Ptolemy—rather than Aristotle—because Ptolemy's entire book devoted to astrology presupposed an astronomy that, unlike Aristotle's, could deliver viable predictions. Moreover, Ptolemy—again, unlike Aristotle—grouped astronomy and astrology as predictive disciplines; he organized these not by their social prestige but by the criterion of which provided greater certitude and hence could stand on its own. Thus the principal text for establishing astrology's authority remained Ptolemy's *Tetrabiblos,* especially as works of Arabic astrology began to enter the Latin West in the twelfth century.[72] Later commentators often replaced Aristotle's predicate *theoretical* (philosophy) with such terms as *speculative, contemplative,* or *theorical;* and in place of *practice,* they used various terms like *active, report,* or *observation.* By the time that Copernicus arrived at university in the late fifteenth century, the science of the stars was evolving into a fourfold matrix that included practical astrology.

Although Aristotle did not provide a place for astrology in his division of the sciences,[73] Renaissance writers readily incorporated his authority into their accounts. The gloss of Haly Abenrodan (Haly Heben Rodan or Ali ibn-Ridwan), the earliest published commentary on the *Tetrabiblos,* nicely facilitated this move. A crucial chapter of that work (book 1, chapter 4)—to which we shall return below—concerns the strengths of the planetary influences, or "virtues," of which Haly Abenrodan remarked that "Ptolemy takes the same path as Aristotle" and that "it is apparent that he was in agreement with those philosophers called walkers or peddlers, that is, 'peripatetics,' because of their moving schools."[74] Another important passage that linked Aristotle's authority to astrology came from the *Meteorologica*, where Aristotle referred to the effect of the celestial on the sublunary region.[75]

Viewed from the perspective of the astrological *practicas*, however, we meet again the classificatory practice of excluding reference to other parts of the four-cell matrix of the science of the stars. Astrological practices or prognostications tended to make use of but only occasionally to refer to theoretical principles. Again, if the astrological *practicas* were largely silent about astronomical theoretical principles, the astronomical or planetary theorics, for their part, provided only a mathematical description of the planetary mechanisms. Commonly, authors left the job of fully explicating or demonstrating the principles of that subject to Ptolemy's *Almagest*. The original reason for this generic division seems to be that the *Almagest* was a difficult technical work both to read and to translate. In the twelfth century, Spanish and Italian translators (John of Seville, Gerard of Cremona) wrote *theoricae planetarum* as elementary manuals or abridgements of the *Almagest* in order to provide an introduction to the latter.[76] But if the *theorica* was preparatory to the *Almagest*, then its relationship to the natural philosophy of Aristotle and to the Arabic version of Ptolemy's *Planetary Hypotheses* is harder to determine with confidence.[77]

The theoric of the planets as a genre embodied the mathematical principles underlying astronomical calculation and established for it a stable descriptive vocabulary.[78] Later, the term *theoric* would take on other meanings. As Olaf Pedersen has amply demonstrated, the "theoric of the planets" was a somewhat late thirteenth-century addition to the more or less standard compendium of works that formed the university curriculum. This collection typically included Sacrobosco's elementary *Sphere,* a treatise on how to build and operate the astrolabe, a set of lunar tables, and one or another treatise on the calendar.[79] By the thirteenth century, however, Islamic astrological texts had arrived in the Latin West. To describe this assemblage, Pedersen coined the term *corpus astronomicum,* but a better term might be that frequently used by the historical agents: *the science of the stars.*[80]

THE CAREER OF THE *THEORICA*/ *PRACTICA* DISTINCTION

Three further points need to be made about the distinction between *theorica* and *practica*. First, although this book is concerned with the subject matter of the heavens, such divisions were not restricted to astronomy and astrology. As early as the twelfth century, Domingo Gundisalvo (fl. 1140) wrote that "there is one part of philosophy which makes us know what ought to be done and this is called 'practical' (*practica*); and there is another which makes us know what ought to be understood and this is 'theoretical' (*theorica*). Therefore, one is intellect, the other is in effect. One consists in the cognition of the mind alone, the other in the execution of work." As an example, theoretical geometry considers immobile magnitude (line, surface, solid) that the intellect abstracts from matter while practical geometry considers these things "as they are mixed in matter with other things [color, sound, and so on]."[81]

This scholastic distinction had a long career, although not without important shifts in its meaning. Late in the seventeenth century, for example, Newton was still engaged with it when he felt it necessary to point out to his intended readers that the subject of the *Principia* was "rational mechanics" as distinct from "practical" or "manual" mechanics. The former, whose principles he famously proposed to reconstitute as "the [mathematical] science of the motions that result from any forces whatever," he described as part of philosophy—natural philosophy; the latter he regarded as an artisanal subject developed by the ancients and concerned with "making bodies move."[82]

Second, and most important, theoretical and practical divisions of knowledge did not map consistently over modern classifications of scientific labor, as one tends to expect from the typical later pairings of theoretician and practitioner or theorist and experimenter.[83] A literary division certainly existed for early modern authors, but rules of authorship did not project consistently and orthogonally into the space of discrete social roles. The presentist assumption here concerns specialization. Someone who wrote a work of theoric was not, thereby, a theoretician or a specialist in theory. Authors could write freely in both theoretical and practical genres. Moreover, there was no obligation to examine theoretical and practical functions of the same subject matter.[84] As a result, the division between theoric and practice had but a distant family resemblance to nineteenth-century disputes about the authority of "pure" scientists over "practical" engineers.[85] William Whewell's neologism *scientist,* analogous to *artist,* suggests that he was responding to a felt need for a term to describe a practitioner engaged in specialized disciplinary practices.[86] Even so, the term *scientist* did not really take hold more widely until the end of the nineteenth century, especially in the United States. In Britain, where Whewell's coinage went largely unrecognized, it was taken to be a vulgar term of American provenance.[87]

In the fifteenth century, if the distinction between theoric and practice represented any division of labor at all, it was as a mode of pedagogical organization. The 1405 statutes at the University of Bologna, for example, organized medicine hierarchically. Throughout the fifteenth century, lecturers in theoric read principally from Galen, with stress on fevers, critical days, and the general principles of physiology; Hippocrates' *Prognostica, Aphorismata,* and *De morbis acutis;* and the all-important first section of Avicenna's *Canon,* which presented the principles or causes underlying all of academic medical learning and medicine's relation to cognate disciplines.[88] Parts of Aristotelian natural philosophy were also included. In fact, as Charles Schmitt has emphasized, the strong tendency in sixteenth-century Italian universities was to regard natural philosophy, on the authority of Aristotle's *De sensu et sensato,* as preparatory to medical studies: "It is further the duty of the natural philosopher to study the first principles of disease and health; for neither health nor disease can be properties of things deprived of life. Hence, one may say that most natural philosophers, and those physicians who take a scientific interest in their art, have this in common: the former end by studying medicine, and the latter base their medical theories on the principles of natural science."[89] In the sixteenth century, this meant that for physicians, natural philosophy functioned, in large measure, as a preparation for theoretical medicine: that is, it assisted in the formation of a rational, cause-based medicine. Hence, the (ideal) "theorical" physician was one who possessed both philosophical and philological skills—the latter especially, under the impact of the humanist program

of restoring the original Greek texts.[90] The lectures on theoric, moreover, took place during the day and carried higher salaries and disciplinary status.

In contrast, different lecturers read practical medicine from Avicenna's third *Canon* in the evenings. This concerned the visible signs of diseases: symptoms and treatment of specific organs "from head to toe" (*a capite ad calcem*).[91] By the 1480s, the "practicing" physician could carry with him a printed "medical practice." The number of such vademecums had increased considerably by the turn of the century, and many were reprinted or reedited later in the sixteenth century. Three-quarters of a century later, the Tübingen physician Leonhard Fuchs illustrated the tendency to try to unify the causal, theorical account within the frame of a practice. His *Institutionum Medicinae* (1555) begins with theoretical definitions and divisions but ends with a short, practical head-to-toe section.[92] But titles can also mislead. Giovanni Marquardi offered a promising (unpunctuated) conjunction in the title of his work on diseases, *Practica Theorica Empirica Morborum Interiorum,* which might be translated as "A Practice or Empirical Theoric" or "A Practical, Empirical Theoric"; but in fact the work was strictly a practical work that moved from headaches to hernias without any philosophizing.[93]

Further evidence of authorial divisions carrying over into pedagogical roles may be found in the specifications of the Bologna statutes for the astronomy lecturer: in addition to teaching the sphere, "he shall prepare the judgment and the almanac."[94] Johannes Paulus de Fundis provides an apt example: he composed a "New Theoric of the Planets," a "New Sphere," a commentary on Sacrobosco's *Sphere,* a defense of astrology against Nicole Oresme, a *Question concerning the Duration of This Age of the World,* and "a little judgment" (*meo iudiciolo*) for the year 1435.[95]

During the sixteenth century, the categories of theoric and practice became a *universal* convention for organizing knowledge. Beyond medicine, astronomy and astrology, we find it carried over into subjects at least as varied as geometry, arithmetic, cosmography, law, music, writing, palmistry, painting, war, mechanics, navigation, and dance.[96] But one must be careful not to assume that authors of such works divide into neat social groupings that correspond directly to the genres in which they wrote. For example, Franchino

Gaffurio (1451–1522), like De Fundis, wrote in both modes, as evinced by his *Theorica Musice* (1492) and *Practica Musice* (1496). The first followed a Boethian organization, beginning with a humanist praise of music (*laus musice*) and then covering the theory of proportionality, the nature of sound and number, the formation of consonances, and species of intervals. The second dealt with the practice of rhythmics, the use of the theory of proportion in polyphonic or contrapuntal composition, and so forth.[97] Gaffurio was a prolific composer and choirmaster at the Ambrosian Cathedral in Milan. Should one designate him a musician? A theoretician? A practitioner? Restrictive decisions of this sort often leave the impression of a degree of specialization or a notion of experthood that simply did not exist.

In the same period, the distinction between theoric and practice was clearly operative in dividing arithmetic along lines similar to music. Some twenty-six Boethian-style Latin arithmetics appeared in the late fifteenth century. These were concerned with the theory of number, proportions, figurative numbers, and "number mysticism" (number having its own being).[98] In the same period, there appeared some thirty practical arithmetics concerned with commercial reckoning and business affairs. Some were *algorisms,* which explained how to count and manipulate Hindu-Arabic numerals.[99] Others used Roman numerals and the abacus, often employing a "counter" or computing table divided into place-value columns in which physical objects were moved around, "carried over" across the columns, and so forth. The earliest printed arithmetic was published at Treviso, written in Venetian dialect, and is a fine exemplar of the principle that an arithmetical practice involved learning by doing.[100] "Here beginneth a Practica, very helpful to all who have to do with that commercial art commonly known as the abacus."[101] In 1568, the Florentine philosopher Francesco Vieri summarized the same twofold divisions of arithmetic and geometry in his lectures at Pisa: "Plato in the *Philebus* calls the 'speculative' the arithmetic of philosophers and the other that of the common people [i.e., of merchants]."[102]

The same commercial demands for reckoning that made the Italian arithmetical and medical practices so important may also account for the appearance of legal practices. Possibly the earliest printed work of its type—is the *Practica Nova Iudi-* *cialis* (New Judicial Practice) of Petrus de Papia (1473; Nuremberg: Anton Koburger, 1482). Like some medical practices, this legal work began with an alphabetized glossary that enabled rapid reference according to specific legal problems.[103]

In the sixteenth century, the distinction continued to operate, even if the terms *theorica* and *practica* were not overtly deployed in book titles. For example, the Louvain globe maker and mathematics teacher Gemma Frisius published a work titled *On the Principles of Astronomy and Cosmography and concerning the Use of the Cosmographic Globe* (Antwerp, 1548). The first part briefly outlined the aetherial and the elemental worlds, defined the principal geometrical divisions of the sphere, and then showed how the cosmographer's sphere could be divided into zones, climates, parallels, habitats, and so forth. The *and* in the title now moved the discussion into the domain of practice: "Having touched upon the theorical principles of this art, let us provide for practice and how we can teach about various and many instruments."[104]

The tendency to join theoric and practice became more explicit and mutually integrated in the last quarter of the sixteenth century. In some instances, this movement was driven by the practitioners of subjects like painting, who wished to be associated with the authority of the university and hence with theoric. For example, Giovanni Paolo Lomazzo's *Treatise on the Art of Painting* advertised in the subtitle that it contained "all the Theoric and Practice to make a Painting" ("Ne' quali si contiene tutta la Theorica, & la prattica d'essa pittura"). His definition of painting as an art was dynamic. It involved not only the use of light and perspective to "imitate" material bodies using the right proportions of lines but also the representation of motion so as to visually produce specific affects and passions of the soul. Put logically, painting was subalternated to the arts that were "liberal" (geometry, arithmetic, perspective, and natural philosophy). To move from the principles of geometry to the representation of figures in the world meant that something—in this case, the hands—needed to be added to the liberal art in question. Yet the fact that a geometer used his hands to draw circles and squares did not thereby put geometry into a "servile condition."[105] Hence, to "conjoin" theoric with practice was a "virtue."[106] Another "handy" example of such conjoining appears in Johannes Rothmann's little work of chiromancy or palmistry. Chiromantic theoric taught that the universe is alive, linked together by the spirit of the world soul. Rothmann's book, however, was mostly devoted to practice. With knowledge of a person's birth date and the positions of lines on the palms, one might draw up a horoscope, inscribe one or both palms within, and proceed to interpret.[107] And finally, use of the hand to link theoric with practice may be seen in Sigismund Fandi's attractive work on handwriting. Bringing together his skills as a geometer and arithmetician, Fandi showed that elegant lettering required the careful application of the tangencies of lines and circles.[108]

When fifteenth-century printers and authors began to construct the architecture of a heavenly literature for emerging markets, *theoric* and *practice* were thus easily carried over as naturalized categories of classification. The new print economy subtly, but decisively, initiated a change in the meaning of *practice* as printers began to issue hundreds of copies of "practical" treatises across the entire panoply of knowledge. The theoric and the practice as literary forms thus began to acquire a life of their own.

THEORETICAL ASTROLOGY
FROM THE ARABIC TO THE REFORMED, HUMANIST *TETRABIBLOS*

Until the mid-1530s, 10 to 20 percent of the astrological publications in Zinner's bibliography were concerned with basic theoretical principles; between 1535 and 1560, there was a spike. Works of this sort did not make predictions for specific times and places. That was the aim of practical astrology. Instead, the literature of theory did two things: it defined the many categories needed to make the forecast; and, more practically, it specified the considerable number of combinations and permutations of planetary locations that the astrologer should regard as capable of delivering meaningful effects.[109] The logic of this domain was deductive and its consequences normative. If anyone in group A (e.g., male or female) has planets X and Q in sign H, then they should beware of effects Z or, on certain days, should avoid journeys or elect to be bled.

The *Tetrabiblos*, that singularly important and widely consulted work, was the preeminent resource for such theorizing. Its popularity in the sixteenth century may also be regarded as one manifestation of the humanist campaign to restore and interpret the original texts of classical antiquity.[110] Unlike modern editions, however, the *Tetrabiblos* was by no means a single, stable, uniformly agreed-upon text: printing in effect transformed it into a collection of texts, not unlike the corpus of medieval celestial works earlier described.[111] The conjoined publication of astronomical and astrological works helps further to explain the interchangeability of the terms *astronomy* and *astrology*. Thus someone wishing to prepare an astrological judgment at the end of the fifteenth century would often turn to a corpus of ancient and medieval works published as a conglomerate with the *Tetrabiblos*. This practice has not been adequately recognized, and a brief, though incomplete, excursus will repay attention.[112]

There were effectively two principal constructions of Ptolemy's *Tetrabiblos*: one Arabic, the other Greek. The first offered Aegidius de Tebaldi's Latin translation from the Arabic, typically with the commentary of Haly Abenrodan; it was the creature of late-fifteenth-century Italian presses. The second presented the Greek text with fresh translations, but without the Arabic commentary. The 1484 Latin edition fell into the first category. Issued by the German Erhard Ratdolt at his new establishment in Venice, it divided the text into sixty chapters, added to which was the all-important *One Hundred Aphorisms, or the Fruit of the Four Books* (the *Centiloquium*) attributed to Ptolemy, together with the commentary of Haly Abenrodan.[113] This "fruit" gave flesh to Ptolemy's classification of astrology as a kind of sophisticated guesswork, unable to predict particulars with absolute exactitude, while offering a vademecum of midrange guidelines in the form of aphorisms useful for medical counsel, preparation of horoscopes, and annual prognostications.[114] Medical aphoristics helpfully connected conjunctions, eclipses, aspects, houses of the zodiac, and the appearance of comets with symptoms (e.g., fever and nausea) as well as preferred therapeutics (e.g., when to purge, bleed, or amputate).[115] Different versions attributed authorship variously to Ptolemy and to the mythical Egyptian prophet Hermes Trismegistus. *Iatromathematica*, another work appearing under the authority of Hermes, became an important source for linking astrology with medicine, a point that has re-

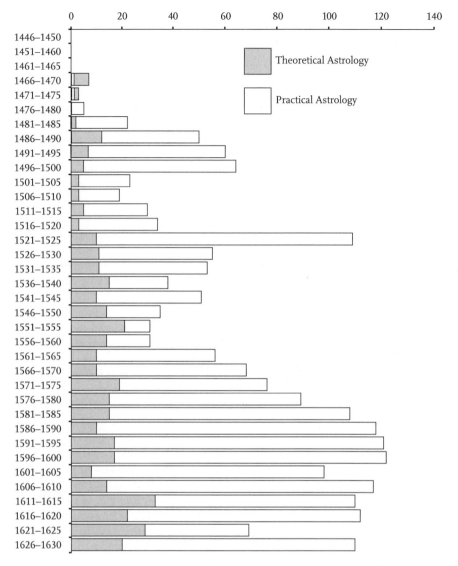

8. Works of theoretical and practical astrology published in the German lands, 1466–1630. Based on Zinner 1941, 73.

ceived little attention from those who stress the importance of the Hermetic writings.[116]

The 1493 edition was, for all practical purposes, a little astrological library. It was produced in a dense, double-columned folio volume, with sixty-six lines of small type to the page, by Boneto Locatelli, a priest from Bergamo and a major Venetian printer of academic books in the 1490s, and underwritten by the publisher, Ottaviano Scotto.[117] In the Locatelli-Scotto edition, the highly compacted text itself was typeset anew; it differed from the Ratdolt edition in that its chapters were broken up by the lengthy commentaries of Haly Abenro-

dan. It was only with special effort, therefore, that one could read through the entire intercalated text without pausing to study the concomitant glosses. Individual editors and publishers added their own interpretive stamp. For example, the fifteenth-century editor Girolamo Salio of Faventino appended his own introduction, a detailed table of chapter headings, and thirteen auxiliary works by different authors.[118] To this compilation, individual owners might bind in still other works of their own preference, thereby creating an even more personalized object. For example, one copy held by the Herzog August Bibliothek in Wolfen-

COPERNICUS'S SPACE OF POSSIBILITIES

büttel is bound with Regiomontanus's *Epitome of the Almagest,* another with Sacrobosco's *Sphere of the World* (together with three commentaries) and Capuano of Manfredonia's gloss on Peurbach's *New Theorics of the Planets,* and still another with Haly Abenragel's *Libri de iudiciis astrorum* (Books concerning Judgments of the Stars).[119] The 1493 Salio edition was apparently successful, because it was issued again in February 1519 by the "friends and heirs of Ottaviano Scotto."[120] A few months later, still another variant appeared at Paris, edited by Jean Sievre and based again on the same text, but without Haly Abenrodan's commentary.[121]

The word *revolutions* that Copernicus chose to emphasize in the title of his major work was undoubtedly inspired by language that was standard within this complex of works packaged with the *Tetrabiblos*—for example, Messahalah's *De Revolutionibus Annorum* (included in the 1493 Salio edition) or the *Alfonsine Tables* (which Copernicus could have owned already in Krakow). *Revolution* thus shared both astronomical and astrological meanings, referring to any point in the zodiac to which the Sun or any of the planets started and returned (the Sun's "ingress" into the sign Aries bearing special importance). Astrologically, these starting and returning points were believed to be among a multitude of indicators of significance in a person's nativity—for example, the place of the Sun at the exact moment or "radix" of birth. In addition to returning to its own "place" at the birth moment, each planet eventually laps the revolution points occupied by the other planets, and such "returns" (*reversiones*) could strengthen or weaken the other planet's qualities and tendencies. For example, Jupiter's returning to its starting point in the birth scheme could signify the birth of a child in addition to wealth, good health, and high esteem; but Jupiter's returning to the point occupied by Saturn in an individual's nativity would augment Saturn's already existing inclinations.[122] Copernicus's central astronomical argument was that the sidereal return times of the planets with respect to a fixed, central Sun held the key to their true ordering.

By the 1530s and 1540s, Northern humanist printers were issuing the *Tetrabiblos* in three major publishing centers: Nuremberg, Basel, and Louvain. New attention was now paid to the Greek text, and a quite different corpus of auxiliary works was appended. Elements of these changes, but not yet the full range, are already evident in the 1533 Strasbourg edition from the Basel printer Johannes Herwagen (Hervagius). Nicolaus Pruckner (Prügner), a Strasbourg physician and maker of calendars and prognostications, introduced this *corpus astrologicum* by extolling the value of astrology as "above all necessary to the art of healing."[123] Pruckner included the typical panoply of supplementary works, such as the friendly *Little Book of Definitions and Terms in Astrology* of the work's dedicatee, Otto Brunfels, the Strasbourg humanist, herbalist, and Grecophile.[124] The edition opened with Julius Firmicus Maternus's *Astronomicon* (or *Liber Matheseos*), a work of general value for interpreting nativities.[125] The text of the *Tetrabiblos* then appeared (in Latin) as a single, continuous unit—but without commentary.[126] Once the birth chart had been constructed, the additional works assisted interpretation of the prognostications' meanings.

Two years later, Johannes Petreius produced a Greek edition of the *Tetrabiblos* at Nuremberg, together with Joachim Camerarius's full Latin translation of the first two books and selections from the last two. This "humanist's" *Tetrabiblos* marked a break with the earlier editions. It contained none of the accompanying Arabic works to be found in the Venetian or recent Basel productions.[127] Nevertheless, as Petreius conceded in an unsigned "Letter to the Reader," the new translation was incomplete, and therefore he had decided to include the old (Arabic) version of the last two books:

> Since Camerarius undertook to translate only the first [two] books, we wish, Reader, to add here these two last books from the old translation: first, on account of those who have not yet learned to walk on their own legs in Greek, lest they be conquered by an imperfect work; secondly, on account of those who might be pleased as well to see the difference between a learned interpreter and author translating into his own language and a foreign [author and interpreter] translating into Arabic, a barbaric language. In view of this [contrast], we have noted in the margins those places where Camerarius has translated from these [languages] so that they [the differences] may be more easily collected. Finally, the old translation, such as it is, is concerned with the same material; and since it may instruct where nothing more elegant exists, there is no reason for us to abandon it. Farewell.[128]

The Nuremberg collection was soon followed by still another from the office of Heinrich Petri in Basel in 1541. It announced the complete works of Ptolemy but notably "excluding the *Geography.*" This corpus, introduced briefly by its editor, Hieronymus Gemusaeus, advertised the advantage of "bringing together in one volume for the first time the works of Ptolemy so divided and beautifully ordered that whatever that one author professed may be [easily] cited."[129] This was to be something like an "astronomer's companion" to the *Tetrabiblos,* because George of Trebizond's fifteenth-century translation of the *Almagest* was now allegedly improved by the addition of new illustrative material: "We place the engravings at the very beginning of this work since, by their force, they ought to be able to contribute great light to the more naturally obscure matters. Likewise, we portray very exactly the figures of all the constellations for the special convenience of learned men." As further inducement, the publisher included in his edition the annotations of the Neapolitan astrologer and prognosticator Luca Gaurico as well as Proclus Diadochus's *Hypotyposes Astronomicarum* (translated by Giorgio Valla), the latter work being "an epitome and compendium of everything in the *Almagest,* all of which is of assistance in recalling most topics to memory."[130]

One way to justify new translations was to complain about previous translators. In 1551, for example, Heinrich Petri's office in Basel added to its rich lists of astrological texts Haly Abenragel's *Books concerning Judgments of the Stars.* The translator and editor, Antonio Stupa of Verona, took pains to announce its improvements and differences from both the Arabic and the earlier 1525 edition. He complained loudly about the "barbaric Latin expressions," the "need to cleanse the text," and the "translator's ignorance of the Latin language." According to Stupa, Yehuda, the son of Moses, originally translated the work from Arabic into Spanish; then Aegidius de Tebaldi and Petrus Regius translated it into Latin, but in such a way that Spanish, French, and Italian phrases were preserved. A purge (*purgatio*) of the vernacular phrases was required, and the Latin needed to be "changed and made more pure."[131] In the end, surprisingly, the completion of the last two books of Camerarius's 1533 translation came neither from highly productive Nuremberg nor from Basel, but from Louvain. The center of a vigorous and long-lived prognosticatory tradition run by

the De Laet family, the university in Louvain was also the home of an important group of mathematical practitioners; and it was there that Antonio Gogava published a complete Latin translation in 1548, to which he appended two treatises on the burning mirror and conics.[132]

As printers competed for market share, newness was claimed for virtually any set of changes that could differentiate a work from available alternatives. By 1553, Johann Oporinus (Herbst), the Basel publisher of Andreas Vesalius's *De Humani Corporis Fabrica* (1543), offered readers new resources for astrological study. Like the Nuremberg edition, this edition of the *Tetrabiblos* provided the Greek and Latin texts, Camerarius's comments, and Pontano's *Centiloquium;* but now the reader was offered a Latin translation of the entire four books as well as an introduction and commentaries by Camerarius's good friend Philipp Melanchthon, the Wittenberg Hellenist, pedagogue, and humanist. Melanchthon politely made little adjustments to Camerarius's edition.[133] In the very next year (1554), however, Petri issued still another Latin edition at Basel, this one by Girolamo Cardano. It sought to differentiate itself from the profusion of other versions by combining Haly Abenragel's title with that of Ptolemy, which yielded "Commentaries concerning the Stars or, as they are commonly called, the Books of Four Parts Joined Together."[134] On the authority of Galen, Cardano explained in the prologue that the translator should not put his words in place of the author's but should stay close to the author's own words. As with the earlier editions of the *Tetrabiblos,* however, Cardan inflated the value of his own edition by denigrating the Arabisms of Haly Abenrodan. "Haly Abenrodan . . . published an Arab [version] worthy of such an author. Yet had this man made his translation true to the intent of Ptolemy's words, perhaps he would have freed us from this labor. Now, however, since this book, on account of its brevity, is not by itself clear nor is it useful for the explanation of other matters that have not yet been published nor is what Haly produced complete, I am compelled to lower myself down to this new labor for the sake of public utility and Ptolemy's glory."[135]

Cardan exaggerated. He did not translate the *Tetrabiblos* afresh but simply pirated Antonio Gogava's 1548 Louvain edition, to which he added a new preface and extensive commentaries. In

1578, Petri warranted the publication of still another "new" edition on the grounds that it had been "corrected by the author" (*ab autore castigata*); it also included a posthumous work by Cardano ("Concerning the Seven Qualities and Forces of the Moving Stars") and, to add to its appeal, the "scholia and resolutions or tables" on Ptolemy of the Strasbourg clockmaker and astronomer Conrad Dasypodius.[136]

As the sixteenth century proceeded, then, what it meant to read or refer to Ptolemy changed considerably. Ptolemy, "prince of astrologers" or "prince of astronomers," as he was often known, was actually a cramped Latin text buried in a pile of (mostly) Arabic commentaries.[137] Beginning in the 1530s, the northern humanists had started to reconstruct the text in their own image, so that anyone wishing to study the influences of the stars needed either to know Greek or to be sure which translation was reliable. This development exactly parallels the proliferation of Greek editions of Aristotle, which began in the late fifteenth century and soon brought into question the existence of the "true Aristotle."[138] Similarly, in 1514, the first Greek text of the New Testament for the Complutensian Polyglot Bible, completed under the direction of Cardinal Francisco Ximénez de Cisneros, marked the beginning of Renaissance humanist New Testament scholarship.[139] The evolution of the text of the *Tetrabiblos* resembled this development of sophisticated biblical and Aristotelian textual criticism and commentary. And the question was the same: What did Ptolemy or Aristotle or the biblical authors *really* say and mean? Although textual improvements were sometimes exaggerated, publishers and authors collaborated in producing a constantly amplified apparatus of interpretation that required the skills of the mathematician as well as the philologist, the physician, or the theologian. Advertisement of these disciplinary skills helps, in part, to account for the curious proliferation of singular and compound print identities that authors of this expanding heavenly literature assigned to themselves.

Prior to this expanded availability of the *Tetrabiblos*, other considerations had already contributed to the growth of astrological practice. In the fifteenth century, the humanists' use of print to "restore" the purity of the original Greek text of the *Tetrabiblos* simply added another powerful motive to the already problematic tendencies of late-medieval Arabic conjunctionist astrology. Krzysztow Pomian and Paola Zambelli have suggested that this astrology, which called attention to the importance of the long-term effects of great planetary conjunctions rather than eclipses, constituted a kind of antitheology of history. Long in advance of their occurrence, it claimed to be able to forecast wars, successions of kingdoms, the fall of empires, natural catastrophes, and, most dangerous of all, alternations in religious confessions.[140]

Already in the fourteenth century, there was a pronounced Scholastic reaction in response to this development. The Parisian theologian and university chancellor Jean Gerson worried about the unacceptable fatalism of a conjunctionist periodization of history in which different planets governed the dominance and fall of different religions.[141] Major critiques of astrology issued from the great commentators on Aristotle: Nicole Oresme at Paris in the fourteenth century, and Henry of Langenstein at the University of Vienna and Blasius of Parma in the fifteenth.[142] This tradition of criticism came to a head, however, in 1496 when Giovanni Pico, count of Mirandola, unleashed a massive, devastating, and exceptionally learned attack against all forms of astrology. Contrary to the pagan, conjunctionist astrology of catastrophic cycles and restorations favored by the Stoics and the Arabs, Pico argued for a Christian, linear view of history extending from the Creation to the Last Judgment. Human fate was not bound by Albumasar's cycles. Men could freely overcome the constraints of the stars, broadly assisted by the Bible and even the pre-Revelation ancient philosophies from other cultures. In the debate that followed, a more limited Ptolemaic view of astral influence gradually evolved: some believed that only certain conjunctions were efficacious, namely those alignments of the Sun and Moon that produced eclipses. This eclipsist view was championed by the Neapolitan humanist Giovanni Pontano and his disciple Agostino Nifo, and it was later supported by Philipp Melanchthon and his important followers, such as Johannes Schöner.[143] But, in spite of this trend, the credibility of Arabic conjunctions did not disappear. The printers and almanac makers went right on providing materials on which alarming forecasts could be based, perhaps the most influential example being Cyprian Leovitius's forecast of the end of the world in 1584.

9. Urania as Astrologia, instructing a student to follow Ptolemy. Sacrobosco 1527, title page. By permission of San Diego State University Library, Special Collections, Historic Astronomy Collection.

THE ORDER OF THE PLANETS AND COPERNICUS'S EARLY FORMATION

During Copernicus's formative period as a student, planetary order was a subtopic in Sacrobosco's *Sphere* but not an autonomous genre of writing within the science of the stars. In the late fifteenth century, the topic was a somewhat minor theme inhabiting dispersed loci within the emergent print literature of the heavens. The earliest printed editions of Sacrobosco's *Sphere* visually fixed the Ptolemaic ordering as a series of concen-

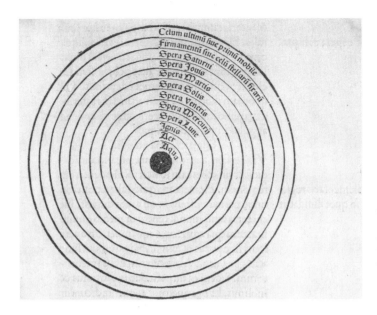

Celum ultimū siue p̄imū mobile
Firmamentū siue celū stellarū firaū
Spera Saturni
Spera Iouis
Spera Martis
Spera Solis
Spera Veneris
Spera Mercurij
Spera Lune
Ignis
Aer
Aqua

10. Planetary spheres concentrically surrounding the four elements, earth, air, fire, and water. Sacrobosco 1478, title page. Image courtesy History of Science Collections, University of Oklahoma Libraries.

tric spheres that constituted the "clear, aetherial region . . . (which the philosophers call the fifth essence)," moving circularly and immune to all change.[144] In describing the periods of revolution, however, Sacrobosco left no uncertainty about the order of Venus and Mercury: "The sun [completes its motion] in three hundred and sixty-five days and six hours, Venus and Mercury about the same."[145] Neither Campanus's *Theorics* nor Georg Peurbach's *New Theorics* devoted any attention to the subject of order. It was left to the commentators—or to printers and readers—to decide how to supplement the *theoricas*. An important early example came from the Augsburg printer Erhard Ratdolt, who set up a new establishment in Venice in the early 1480s. In 1482, he bundled into a slim, continuously paginated quarto volume Sacrobosco's *Sphere*, Peurbach's *New Theorics*, and Regiomontanus's *Disputations against the Nonsense of Gerard of Cremona's Theorics of the Planets*—an "omnibus edition" or compendium of works thereafter frequently conjoined. Cast as a dialogue between two interlocutors—one from Krakow, the other from Vienna—Regiomontanus's epideictic rhetoric of praise and blame made the disputations into something of an early promotional tool, blasting the technical errors of the old theorics as fables and old wives' tales while associating Peurbach's work with the certitude of mathematics.[146]

This practice of juxtaposing different works

was quite common among early publishers, and I have listed such omnibus collections in my bibliography.[147] It created a new sort of product, much like the little libraries of auxiliary works that were packaged with the *Tetrabiblos*. For readers, it made it seem natural that certain books belonged together. Individual readers were thus free to construct their own connections between different elements of the works that the printer or editor chose to bundle; or, as books were usually sold without bindings, purchasers might also bind together their own preferred collections.[148] Thus different commentaries on the *Sphere*—some containing more visual assists—were often conjoined with Peurbach's *New Theorics*.

Taken as a work in its own right, however, Peurbach's text still treated each planet independently. He represented the planet's motions as a composite set—an assembly of off-center circles constituting what he called the "partial sphere," a grouping that nested contiguously within a "total sphere." The two touched at only one point, and the larger of the two, in turn, was concentric to the center of the universe. Considered geometrically, the partial spheres were shells composed of concave and convex surfaces eccentric to the universe's center. The 1485 Ratdolt edition tried to assist visualization by coloring in the planet's total sphere, a concentric, rubricated donut shape containing an off-center, circular channel within which the planet performed its revolutions. This

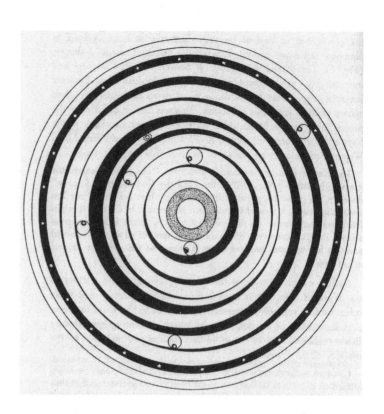

11. Reconstruction of an alleged fifteenth-century cosmological scheme showing nested, eccentric, solid spheres. From De Santillana 1955. Drawing by William D. Stahlman.

copy's reader has annotated the concave and convex surfaces (see figure 5). But Peurbach himself did not speak of a "cosmological arrangement" of total and partial orbs within a single, all-embracing sphere of the sort illustrated so suggestively—and somewhat misleadingly—in Giorgio De Santillana's *The Crime of Galileo* (1955) (see figure 11). The diagram is a modern construction, prepared by William D. Stahlman, not a reproduction of an early modern original. Neither Regiomontanus's original edition of Peurbach nor any later edition displayed any cosmic arrangement of this sort among its illustrations.[149] The title page of the 1543 Paris edition of the *New Theorics* (figure 12) is the only example of which I am aware that illustrates an alternative strategy—evidently a publisher's decision. It displays a standard, uncomplicated, Sacrobosco-like representation of concentric spheres on its title page while depicting the familiar individual theorics separately—but with no further comment.

Although lacking a separate treatment of planetary order, Peurbach's *New Theorics* was not altogether devoid of comment on the interrelations of the respective orb sets. In his discussion of Mer-

cury's notoriously complex motions, Peurbach noted that the period of revolution of the orb in which Mercury's epicycle moves eastward is one year—exactly the same time that the Sun takes to make one complete revolution in mean longitude. He further observed that "Mercury in this respect behaves like Venus toward the sun. For it always happens that the mean longitude of the sun is also the mean longitude of these two planets."[150] Likewise, "the mean longitude of any of the three [superior] planets, added to its motion in the epicycle, becomes equal to the mean longitude of the sun in degrees and minutes."[151] Hence, although there was no single principle that united the planetary motions, Peurbach pointed out that the three superior planets shared the Sun's mean motion in longitude. Following the comment about Mercury and Venus, Peurbach inserted the aside that I noted in the introduction (here more fully parsed): "Therefore, from these things and from what has been said above, it is evident that each of the six planets shares something [*aliquid*] with the Sun in their motions, and that the Sun is like some common mirror and rule of measurement to their motions."[152]

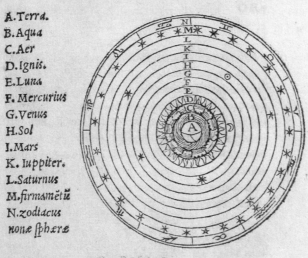

THEORI-
CAE NOVAE PLANE-
TARVM GEORGII PVR-
bachij fœliter incipiunt.

FIGVRA NOVEM SPHAERA-
rum & elementorum ordinem defignans.

A. Terra.
B. Aqua
C. Aer
D. Ignis.
E. Luna
F. Mercurius
G. Venus
H. Sol
I. Mars
K. Iuppiter.
L. Saturnus
M. firmamētū
N. zodiacus
nonæ fphæræ

PARISIIS.
Apud Chriftianum Wechelum fub fcuto Bafilienfi in
uico Iacobæo: & fub Pegafo in uico Bellouacenfi.
An. M. D. XLIII.

12. Georg Peurbach's title-page *figura*, "showing the order of the nine spheres and the elements." Peurbach 1543. Image courtesy History of Science Collections, University of Oklahoma Libraries.

Because this passage falls directly within the discussion of mean longitudes, the shared "something" was clearly the Sun's annual motion (or second anomaly) rather than its daily rising and setting, or even its light and influence. But, most important, the passage goes strangely beyond what was required at this point in the discussion of Mercury's theoric—as though Peurbach intended to develop the insight further but for some reason never returned to it. Regardless, the *New Theorics* also went beyond Sacrobosco's *Sphere* and the *Almagest* in calling attention to the Sun's contribution of an annual component to the motions of Mercury, Venus, and the superior planets. The passage was thus available to any reader or commentator who might be moved to speculate further about an alternative explanation. Copernicus, I believe, was one of those attentive readers.

There were very few others. An early, anonymous annotator noticed the otherwise buried comment, remarking in the margin of his copy: "Without doubt, all the planets have a certain relation and proportion to the Sun; therefore, it is as if the Sun were at once leader, prince and moderator of all."[153] In 1495, the Paduan commentator Francesco Capuano of Manfredonia echoed the original text, referring to the mean motions of Venus and Mercury "as if a mirror or measure" of the Sun's motion.[154] Yet the 1535 Wittenberg edition, with Philipp Melanchthon's preface to Simon Grynaeus, took no special notice of this passage. In contrast, Erasmus Reinhold appended a comparatively lengthy scholium in 1542, together with an illuminating diagram, where he clearly stated that the three superior planets "take notice of" the Sun while the two lower planets, Venus and Mercury, maintain its "friendship." "For all the planets," he concluded, "it is therefore necessary to know the Sun's mean motion."[155] Just a year earlier, Reinhold's Wittenberg colleague Rheticus had used language very similar to that of Peurbach in making central the claim that the (stationary) Sun provides a "common measure of the planetary orbs with respect to one another."[156] In modern times, Peurbach's comment seems to have been noticed first by Ernst Zinner, who owned the annotated copy mentioned above, and more recently by Eric Aiton, who underscored the passage in his translation of the *New Theorics*.[157]

Neither the old nor the new *theorica*s, apart from Peurbach's important comment, put special emphasis on planetary arrangement; evidently they reserved that topic for Sacrobosco's *Sphere*, which preceded Peurbach's *New Theorics* in the order of teaching, or for the *Almagest*, the text to which they were explicitly meant to serve as an introduction. Having dealt with the Sun and Moon (book 9, chapter 1), Ptolemy did take up the problem of order very briefly as a prelude to a detailed analysis of the five remaining planets. His discussion described a consensus among "almost all the foremost astronomers": after the fixed stars, Saturn, Jupiter, and Mars follow in sequence. But with respect to Venus and Mercury he reported disagreement: "We see that they are placed below the sun's [sphere] by the more ancient astronomers, but by some of their successors these too are placed above [the Sun's], for the reason that the sun has never been obscured by them [Venus and Mercury] either."[158] No one in antiquity ever claimed to have observed Venus or Mercury transit the Sun.[159]

The abridgment of the *Almagest* begun by Peurbach and completed by Regiomontanus, but not published until twenty years after the latter's death, followed Ptolemy's organization, at times adorning or abridging his language or developing his treatment in new ways.[160] In book 9, chapter 1—again following Ptolemy's organization—Regiomontanus explicitly used the word *controversy* to describe the lack of consensus about the ordering of Mercury and Venus. Moreover, he reported

that al-Bitruji (Alpetragius) located Venus above the Sun and Mercury below it. Thus, the *Epitome* sharpened the sense that, even after Ptolemy, there was disagreement about the order of Venus and Mercury among the ancients or their successors.[161] This sense of controversy about the order of Mercury and Venus was further reinforced in the more probing commentaries on Sacrobosco's *Sphere*, of which, after 1499, Capuano de Manfredonia's was the most extensive.[162] Yet, ultimately, the *Sphere*'s Ptolemaically ordered concentric woodcuts left a strong impression of celestial stability.

This general Ptolemaic consensus, in turn, was powerfully reinforced by the astrology presented in the *Tetrabiblos*. Ptolemy simply brought over the ordering that he had established in the *Almagest* (Earth, Moon, Mercury, Venus, Sun, Mars, Jupiter, Saturn). In the *Tetrabiblos,* book 1, chapter 4, he described the planets as possessing active, natural powers (heating, humidifying, cooling, drying) derived from their proximity or distance from the Sun (associated with heat), the Moon (associated with humidity), and Earth (associated with moist exhalations), or from their placement with respect to their neighbors. Jupiter, for example, derives his powers from his location between the "cooling influence of Saturn and the burning power of Mars." Venus shared in Jupiter's powers, but "because of her nearness to the sun," she does the opposite: she "warms moderately" and also humidifies like the Moon "because of the amount of her own light and because she appropriates the exhalations from the moist atmosphere surrounding the earth." Meanwhile Mercury, like Venus, "never is far removed in longitude from the heat of the sun" and for that reason tends to have the power to dry and absorb moisture; but because Mercury is also "next above the sphere of the moon, which is closest to the earth," it has the power to humidify. Thus, Mercury can "change quickly from one to the other, inspired as it were by the speed of his motion in the neighborhood of the sun itself."[163]

To be clear, because the planets themselves were made of an unchanging substance, their own constitution was not elemental; rather, they were associated with the particular *effects* that they were capable of causing in the sublunar region. Ptolemy's entire theoretical astrology was built on these fixed, elemental qualities, which were thus cemented firmly into his planetary ordering. All

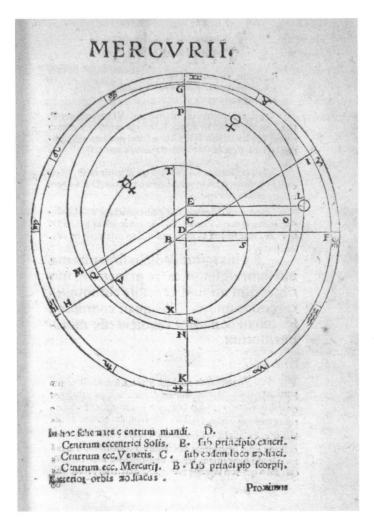

MERCVRII.

In hoc schemate c centrum mundi. D.

Centrum eccentrici Solis. E. fub principio cancri.
Centrum ecc.Veneris. C. fub eodem loco zodiaci.
Centrum ecc. Mercurij. B. fub principio fcorpij.
Exterior orbis zodiacus .

Proximus

13. Erasmus Reinhold's interpretation of Peurbach's shared-motions passage, showing eccentric circles with centers B (Mercury), C (Venus), E (Sun) and center of the world (D). Reinhold 1542. Image courtesy History of Science Collections, University of Oklahoma Libraries.

subsequent combinations and mixings—of which human temperament was only one of many—proceeded from these initial assignments. For example, beneficent Jupiter and Venus belong together because they share the hot and the moist; Saturn and Mars are maleficent because they join excessive cold with excessive dryness; and the Sun and Mercury swing both ways because they partake of both sets of qualities from the opposing groups and thus have "a common nature" that joins them to all the other planets. Similarly, the Moon and Venus are feminine "because they share more largely in the moist"; but masculine dryness belongs with the Sun, Saturn, Jupiter, and Mars, while Mercury, "common to both genders," can produce both moist and dry.[164] Further groupings, associated with multifarious divisions of zodiacal or equatorial planes rather than linear

contiguity, generated distributions of elemental qualities in the signs, houses, triangles, exaltations, and so forth.[165]

In 1491, when Copernicus entered Krakow as a student, Peurbach's *New Theorics* was still a relatively new book, just twenty years old. Issuing from Regiomontanus's short-lived printing operation in Nuremberg in 1472–73, it was the first work of theoretical astronomy ever to appear in print.[166] Within ten years of its initial publication, the *New Theorics* had acquired its first commentator, Albertus de Brudzewo (ca. 1445–ca. 1497) at Krakow.[167] Brudzewo's detailed 1482 commentary was published at Venice only in 1494 by Ottaviano Scotto and the next year by Ulrich Scinzenzaler at Milan, immediately prior to Copernicus's arrival in Italy to further his studies.[168] Although Brudzewo taught at Krakow from

1474 until 1495, there is unfortunately no direct evidence confirming that Copernicus attended his lectures on Peurbach. Yet, given Copernicus's inarguable interests in studying the heavens and the small scale of universities like Krakow, it seems entirely reasonable to assume that he received his earliest introduction to planetary theorics through one of Brudzewo's students who used the master's unique commentary. After his arrival in Bologna, Copernicus could easily have acquired one or the other copies of the published version.[169]

Brudzewo's commentary is of unusual interest for what it reveals of Copernicus's intellectual formation. It puts Peurbach into Copernicus's hands at an early date in his studies—perhaps at Krakow, at the very latest in Bologna—and it ties planetary theorics directly and explicitly to Ptolemy's astrology of elemental qualities. Brudzewo followed Ptolemy's ordering without argument while engaging in some minor adjustments of the astral attributes. In addition, he invoked Albertus Magnus as an astrological authority to finesse Ptolemy's distinctions between different kinds of qualities.[170] In this account, the Earth was the locus where the primary physical qualities—heat, cold, moisture, and dryness—were in a constant state of mingling and recombination. Saturn, the planetary sphere most distant from the Earth, was cold and dry, and the slowest of planets; it was associated with the least mixing of qualities. At the other end of the heavens, the Moon was closest to the sphere of mixing, so that it was reasonable that it shared in the Earth's moisture and returned the favor by causing the motion of the tides. Jupiter, after Saturn the next speediest, was hot and moist, but Brudzewo differentiated its moisture from that of the Earth by calling it a spiritual quality, the "carrier of life virtue." Jupiter's sphere could not initiate the mixing of matter, but it could influence matter that was already moved and mixed. Mars, next after Jupiter, was "moderately distant"; like the Sun, to which it was adjacent, it was hot and dry, but because of its distance it could not burn like the Sun. The Sun, on the other hand, had a type of heat and dryness that Mars did not possess. Located "in the middle of the planets, like a heart," the Sun's heat "distributes and ripens the seed that gives life." Next came Venus. Because it was the Sun's neighbor, it too could "give life," but it was also moist and hence capable of combining

with the Sun. Finally, although all planets were capable of mixing, only Mercury had the virtue of mixing with both of its adjacent neighbors; it derived this capability from its place between the cold, moist Moon and hot, moist Venus.

Between 1482 and 1495, Krakow students were thus presented with a version of Peurbach that included an explicit astrological dimension. I know of no other printed commentary that directly framed the *New Theorics* in this way.[171] To justify his treatment of the planets in connection with astral influences, Brudzewo located the topic of planetary order using classificatory resources from both Haly Abenrodan's *Tetrabiblos* commentary and Campanus's *Theorics*. Although his treatment has a certain logic, it must be extracted from the details: a two-cell division of astronomy, the first part ("theorical or speculative") dealing with planetary motions and positions, and the second part ("practical") treating the effects of those motions and bodies on different places.[172] Brudzewo then introduced two further qualifications. First, looking over his shoulder at Campanus, he acknowledged that "the practical part is treated in different ways by different authors, some including in it various instruments, others different tables."[173] Next, he subdivided the theorical into what he called the "narrative or introductory" part (under which he included Peurbach and Campanus) and the demonstrative (Ptolemy's *Almagest*). By his reasoning, Peurbach's *New Theorics* belonged under the descriptive part of theoretical astronomy because it presented the models without showing how they were derived. Planetary order thus belonged, by elimination, to the *demonstrative* part of theoretical astronomy (the *Almagest*) because it received no treatment in either Campanus's *Theorics* or Peurbach's *New Theorics*. Because Brudzewo clearly modeled planetary order on the *Tetrabiblos* (in turn derived from the *Almagest*), essentially he was locating planetary order within both astronomical theoric and theoretical astrology, the latter part subalternated to the former.[174]

Although Peurbach's and Regiomontanus's works were the major texts of theoretical astronomy at the end of the fifteenth century, they were not the only resources of planetary ordering. From the 1480s, editions and compilations of ancient works began to appear. Among the latter was the massive, encyclopedic compendium of

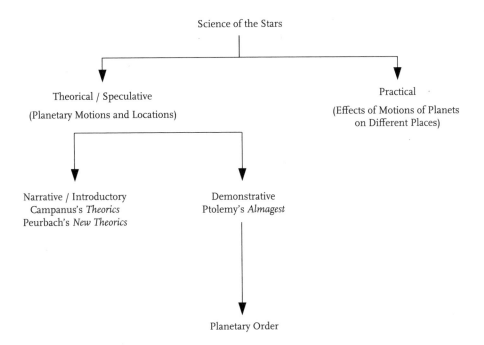

Science of the Stars

Theorical / Speculative
(Planetary Motions and Locations)

Practical
(Effects of Motions of Planets
on Different Places)

Narrative / Introductory
Campanus's *Theorics*
Peurbach's *New Theorics*

Demonstrative
Ptolemy's *Almagest*

Planetary Order

14. Albertus de Brudzewo's classification of the science of the stars (1495).

literally all branches of knowledge produced by the humanist prodigy Giorgio Valla (1447–1500), professor of Greek at the University of Pavia from 1466 to 1500. Valla's *Concerning What to Seek and What to Shun* appeared just as Copernicus returned for a second round of studies in Italy.[175] "Copernicus," Edward Rosen quipped, "both sought and shunned it."[176]

For Valla, anything worth knowing had already been found by the Greeks. Consequently, his discussions were not merely commentaries on the famous Aristotle and Plato but also examined lesser-known figures like Xenophanes, Cleomedes, Philolaus, and Aristarchus. Copernicus, who shared in the conceit for antiquity, helped himself to Valla's learning not merely for its use as rhetorical adornment but because, as a budding Hellenist in his own right, he had some degree of trust in it. Indeed, he would cast his masterwork as an encounter with ancient learning. Thus it was from Valla that he learned that Aristarchus held that the "earth moves around the solar circle" and that "Aristarchus makes the sun stationary"; but he clearly shunned Valla's statement that the stationary Sun was "beyond the fixed stars."[177]

COPERNICUS'S PROBLEMATIC
THE UNRESOLVED ISSUES

The chronology of Copernicus's encounter with the theme of planetary order may be summarized as follows:

Krakow, 1491–95. Because the *Almagest* was not available in print until 1515, Copernicus encountered the principles of theoretical astronomy through Sacrobosco's *Sphere* and Peurbach's *New Theorics*—perhaps in the *Sphaera Mundi* compendium, available from Ratdolt as early as 1482, but more than likely through Albert of Brudzewo's commentaries. Because he owned early editions of the *Alfonsine Tables* (Venice, 1492) and Regiomontanus's *Tabulae Directionum* (Augsburg, 1490), it is possible—even reasonable—that he had already acquired some advanced calculatory skill before arriving in Italy.[178] Next he would have had Brudzewo's astrological gloss on Peurbach. We also know that at some point—perhaps this early—he owned Haly Abenragel's *In Iudiciis Astrorum*. And finally, in *De caelo* (chapter 2.13–14), where Aristotle himself developed his own argument for the Earth's motionlessness and centrality, Copernicus would have found rejected the Pythagorean belief that "the center is occupied by fire, and that the Earth is one of the stars, and creates night and day as it travels in a circle about the center."[179] Familiarity with this passage shows up later quite clearly in the *Commentariolus,* Copernicus's earliest sketch of his new ideas, where he explicitly attributes the Earth's motion to the Pythagoreans, but where he makes no mention of the Pythagorean Central Fire or Counter-Earth (*antichthon*). Copernicus was thus reading the Pythagoreans in the setting of the science of the stars.[180]

Bologna, 1496–1500. Sometime after arriving in Bologna, Copernicus could have had occasion to find Regiomontanus's *Epitome of the Almagest;* and, as chapter 3 shows, it is certain that he had access to the Salio edition of the *Tetrabiblos,* together with a group of related publications that appeared in rapid succession.

Padua, 1501–3. Sometime after 1501, Copernicus had before him Valla's compost heap of Greek wisdom and still another reference to "Philolaus the Pythagorean," the Central Fire, and the Counter-Earth.[181] How soon he had seen Martianus Capella's *The Marriage of Philology and Mercury* after its publication in 1499 is harder to say.[182]

To these considerations one must add several others. In Regiomontanus's presentation—as in the *Almagest* itself (book 9, chapter 1)—planetary order received scant discussion. Although we now know that Regiomontanus privately entertained doubts about Ptolemy's inconsistent criteria of planetary ordering, all that Copernicus knew from the posthumously published *Epitome of the Almagest* was that there was a "controversy" about the ordering of Venus and Mercury.[183] And, insofar as that arrangement could be taken as established, it required that no gaps exist between the individual orbs, because "Nature does not permit an empty place."[184] As with Peurbach, calculations with the individual models were not dependent on planetary order. Moreover, in Brudzewo's commentary as much as in the *Tetrabiblos* itself, the order of the planets was associated directly with the order in which the astral properties were fixed. Because all the many combinations and permutations of planets and places

that followed therefrom were based on these primary assignations, one stumbles on the interesting conclusion that Ptolemy made planetary order *physically* central to his astrology in a way that it was not for the *Almagest* or for Regiomontanus's *Epitome.* Therefore, to shuffle the order of the planets in the domain of theoretical astrology would be to require a justification for changing the supposedly fixed planetary attributes responsible for producing specific earthly effects—or to ignore that question altogether as simply belonging to a different category of writing.

What then was the problem for which Copernicus's heliostatic reordering of the planets was the solution? This question, well posed recently by Bernard Goldstein, takes us to the heart of Copernicus's problematic.[185] It raises in turn the question of whether the solution was initially motivated by uncertainties within the domain of planetary order or that of planetary modeling. When the solution was achieved—when Copernicus reached his original insight—is a matter of ongoing debate; ultimately, it may be less interesting than what the problematic was and when its elements were in place.[186]

In general, historians have tended to discount planetary order as the originating ground for Copernicus's idea of the heliostatic arrangement. Since the reconstructions of Curtis A. Wilson and especially Noel Swerdlow in the 1970s, the conceptual fulcrum has been sought in Copernicus's preliminary sketch, usually called the *Commentariolus,* and in the mature masterwork, *De Revolutionibus,* where evidence points to Copernicus's manifest dislike of the motion of points (or the bodies occupying them) as nonuniform with respect to their own proper centers.[187] Wilson glosses Copernicus's inclination against such nonuniformities as consistent with a wider Renaissance sensibility of order. For Swerdlow, by contrast, the fundamental motive for the same phenomenon is unambiguously physical, grounded in the assumption that Copernicus regarded the planets as embedded in solid, impenetrable spheres whose uniform rotations could only be achieved mechanically around their diametral axes.[188] From these motives, Copernicus is said to have worked his way to transformations of the models that removed the offending inequalities—for Swerdlow, eccentric constructions that Copernicus would have encountered uniquely in Regiomontanus's modeling of Venus and Mercury.[189]

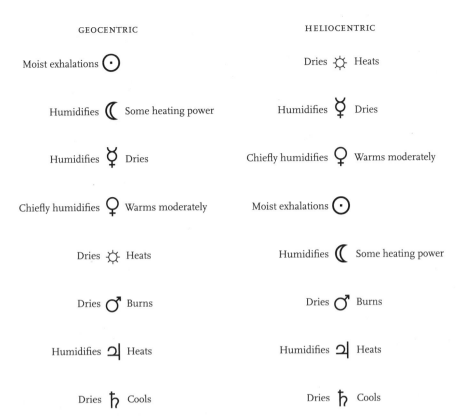

GEOCENTRIC	HELIOCENTRIC
Moist exhalations ☉	Dries ☼ Heats
Humidifies ☾ Some heating power	Humidifies ☿ Dries
Humidifies ☿ Dries	Chiefly humidifies ♀ Warms moderately
Chiefly humidifies ♀ Warms moderately	Moist exhalations ☉
Dries ☼ Heats	Humidifies ☾ Some heating power
Dries ♂ Burns	Dries ♂ Burns
Humidifies ♃ Heats	Humidifies ♃ Heats
Dries ♄ Cools	Dries ♄ Cools

15. Order of astral-elemental qualities according to Ptolemy (left); reshuffled order of the same astral qualities according to a Copernican arrangement (right).

On the other hand, Wilson speculates that the same search for uniform motions around proper centers may have led Copernicus to experiment with the complex option of mounting the annual solar epicycle onto the biepicyclic scheme already introduced as a replacement for the equant associated with the superior planets.

If Copernicus read Peurbach's shared-motions passage as posing a problem, then Wilson's plausible reconstruction sheds light on what Copernicus might have done to provide an answer to it, at least when applied to the superior planets. On that reading, the large annual epicycle in the geostatic framing can be transformed directly, without loss of equivalence, into the Earth's yearly circuit. But for Mercury and Venus, Copernicus would have needed the propositions in Regiomontanus's models to which Swerdlow first called attention.[190] Either transformation, whether from the superior planets or from Venus and Mercury, could have led to circumsolar constructions that might have suggested the possibility—even the likelihood—of the Earth's motion without di-

rectly requiring the inference. Indeed, it was this inference that Tycho Brahe would refuse.[191]

The final step in the narrative that privileges planetary modeling as the starting point is where the difficulties (and the interpretive differences) occur. What convinced Copernicus to make the Earth's annual motion the critical assumption? Was he compelled to do so by the potential intersection of hard, impenetrable orbs, as Swerdlow speculates? Or was it a metaphysical commitment to an orderly *machina mundi,* as Wilson proposes? In either case, if Regiomontanus was the critical resource for such heliostatic framings, one wonders why he himself did not see (and act on) either possibility. For no one was more familiar with the details of Peurbach than his own remarkable student and publisher.

Bernard Goldstein's proposal offers a different, most promising direction from which to attack the early problem situation. He reminds us of Copernicus's well-known appeal to the principle—set forth on the authority of Euclid's *Optics* in *De Revolutionibus,* book 1, chapter 10—according to

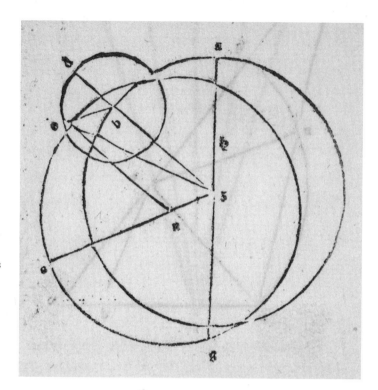

16. Parallelogram transformation of Venus's epicycle (OB) into eccentric circle with center R. Regiomontanus 1496, book 12, proposition 1. Image courtesy History of Science Collections, University of Oklahoma Libraries. (See also Dennis Duke's animation of Regiomontanus's model at http://people.sc.fsu.edu/~dduke/venhelio2.html.)

which the farther the orbs are from the immobile center of motion, the longer are their sidereal periods.[192] In the *Commentariolus,* Copernicus wrote: "One planet exceeds another in rapidity of revolution in the same order in which they traverse the larger or smaller perimeters of their circles."[193] Was this Copernicus's guiding assumption? Or was it a consequence of some other? Goldstein believes that it was the first.[194] Moreover, because this period-distance principle already applies to Mars, Jupiter, and Saturn in a geostatic arrangement, Copernicus's problem would have concerned the annual motion shared by the Sun, Venus, and Mercury. Hence the main difficulty would have been one of determining the heliocentric periods of Venus and Mercury. But although Copernicus cites values for the periods, he offers no personal calculations, instead using Martianus Capella's *The Marriage of Philology and Mercury* for the ordering rather than, say, Vitruvius's *Ten Books on Architecture* to lend to his contention the credibility of ancient authority.[195]

In Goldstein's scenario, however, it is ultimately the problem of planetary distances that drives the account: first, Copernicus's dissatisfaction with the gaps yielded by the method of "nesting," whereby the Ptolemaic ordering was given and the distances were established additively: the maximum distance of a lower planet was assumed to be equal to the minimum distance of the next highest planet. Unacceptable gaps occurred, however, when these distances were tested by a different method, that of measuring solar and lunar parallax, for the space so obtained failed to correspond exactly to the spheres nested therein. The period-distance principle provided an alternative approach to finding the distances; effectively, the periods became independent variables and the distances a function of them. But how could one be sure of the values for the periods—especially those of the problematic Venus and Mercury? Goldstein's scenario is motivated by an assumption (the periods are proportional to the distances from the center) and an empirical demand (the periods of Mercury and Venus). Somehow—was it from reading Martianus Capella?—Copernicus convinced himself that Mercury's period was shorter than that of Venus, and hence the ordering

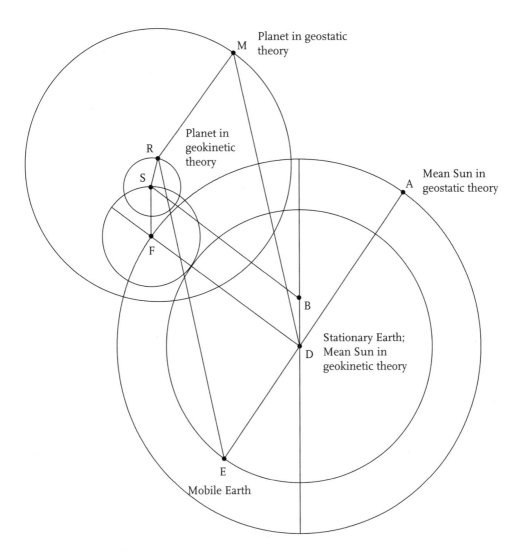

Labels in figure:
M — Planet in geostatic theory
R — Planet in geokinetic theory
S
A — Mean Sun in geostatic theory
F
B
D — Stationary Earth; Mean Sun in geokinetic theory
E — Mobile Earth

17. Parallelogram transformation of the annual epicycle of a superior planet (MR) into the orbit of the Earth (DE). Adapted from Curtis Wilson 1975. By permission of the University of California Press.

of those two bodies would fall nicely into conformity with the general principle that the planets with shorter periods are closer to the center of the universe. From there, Copernicus would have had to see that exchanging the Sun's position with the Earth would best fit the period-distance relation. The ensuing arrangement would have presented him with all sorts of interesting new consequences: for example, a simpler explanation of retrograde motions and a new method for computing the planets' relative distances from the Sun.[196] In Goldstein's reconstruction, then, planetary order solves the problem of the distances but

precedes the question of planetary modeling. Copernicus makes his initial commitment to the heliocentric system after noticing at least some of its advantages in explaining the appearances; from there, he proceeds to work out the details of the respective planetary models along the lines to which Swerdlow called attention in Regiomontanus's *Epitome*, book 12, chapters 1 and 2.[197]

This reconstruction has at least three attractions. First, it provides a possible explanation for why Copernicus would not have been content to rest with a Tychonic or geoheliostatic arrangement—if he had ever countenanced it:

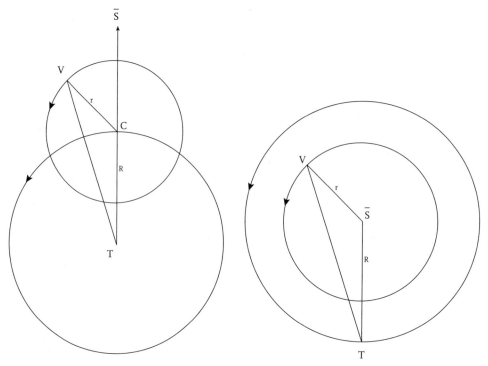

(a) A geocentric model for an inner planet.   (b) A heliocentric model for an inner planet.

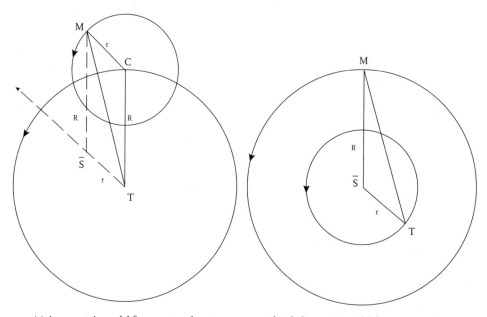

(c) A geocentric model for an outer planet.   (d) A heliocentric model for an outer planet.

18. Transformation of geocentric into heliocentric models. (a) A geocentric model for an inner planet. (b) A helio-centric model for an inner planet. (c) A geocentric model for an outer planet. (d) A heliocentric model for an outer planet. Goldstein 2002, 227. By permission of Bernard R. Goldstein and *Journal for the History of Astronomy*.

such a system would not satisfy a strict application of the period-distance requirement. Second, a version of the period-distance relationship was readily available in Aristotle's *De caelo,* book 2, chapter 10, which Copernicus would have known from his studies at Krakow.[198] Third, this reconstruction does not require any complicated defenses of Copernicus's (much-debated) position on the solidity or hardness of the spheres.

On the other hand, Goldstein's account pushes back the explanatory problem another step: If the period-distance relation was so readily available to astronomical practitioners, then what could have motivated Copernicus's early attention to it? When Regiomontanus pointed to a "controversy" about the ordering of Mercury and Venus, for example, he focused on an old, empirical uncertainty, not a principled incoherence in the ordering. When the Bologna Averroist Alessandro Achillini (1463–1512) commented on *De caelo,* book 2, chapter 10, he endorsed the Aristotelian relation: "It is also evident that the rank of these movers relative to the first mover must follow the order of the spheres in space because their precedence in position and magnitude determines their hierarchy in nobility. But with regard to their velocity of motion, we find them in the opposite situation, I mean the closest to the Earth is the fastest."[199] Although Achillini seriously disputed the reality of Ptolemaic epicycles and eccentrics, he did not dispute the ordering of Mercury and Venus or gaps between the spheres.[200] Evidently, Aristotle's ordering principle was itself subject to different interpretations.

If, as seems all but certain, Copernicus first came upon Peurbach's shared-motions reference accompanied by Albert of Brudzewo's commentary, he would have encountered it embedded in a layer of astrological interpretation: "The Master adds a corollary," wrote Brudzewo, "such that *all* the planets have some communication with the Sun's motion in their own motions. Now this is because they have a natural connection, as with a luminous body, just as Ptolemy says in Book One of the *Tetrabiblos;* and therefore they [the planets] participate in its motion, influence and operation."[201] Here Brudzewo went beyond Peurbach and connected his corollary directly to the Sun as the source of an elemental power (heat) necessary to the integrity of the whole balance of astral influences as well as to the luminosity and mo-

tions that the planets share with the Sun. One is reminded of Marsilio Ficino's encomium to the Sun in *Three Books on Life* (1489), although only in the context of drawing down the Sun's life-giving force rather than in establishing its position in relation to the other planets.[202] The Brudzewo corollary does not, of course, compel the Sun to be placed at rest at the center of the universe; but it adds an astrological inflection to Peurbach's otherwise-buried shared-motions proposition, namely, that in addition to the Sun's annual motion, the planets also share something of the Sun's power as a source of heat and astrological influence. And precisely here is a place where Copernicus *could have* made the "Pythagorean move," concluding that the source of heat, luminosity, and influence for all the planets should be at the center and at rest.[203]

To summarize, nothing in Copernicus's early problem situation absolutely compelled him to move to a heliocentric hypothesis.[204] The elements that we have been reviewing were underdeterminative with respect to adoption; but taken together, they explain how Copernicus could have recognized the Earth's annual motion as a possible solution to Peurbach's shared-motions conundrum. In that understanding he was clearly alone. Regiomontanus, who could well have discussed the shared-motions question personally with Peurbach, (and who, of course, published the book containing the passage in question), appears not to have associated the period-distance rule with the issue of the annual solar component or second anomaly.[205] Indeed, Regiomontanus remained publicly invested in the ordering set forth in the *Almagest* and hence in the *Tetrabiblos.*[206] Moreover, he had explicitly rejected the claim that the Earth rotates daily on its pole (*localiter, circulariter*), as if on a spit—the better for the Sun to roast it nicely and evenly. Regiomontanus's counterarguments followed the usual Aristotelian-Ptolemaic form *modus tollens* in appealing, for example, to the properties of air.[207]

The unanswered questions are thus two: what historical circumstances could have made the ordering of Venus and Mercury into a matter of such urgency that it motivated a new look at the application of Euclid's or Aristotle's version of the period-distance principle? And why was Copernicus motivated to adopt the Sun rather than the Earth as the center of all, rather than some, celestial motions?

2

Constructing the Future

THE ANNUAL PROGNOSTICATION

Although divinatory practices were well-known from the time of the Babylonians, print, chronic warfare, the growth of towns and cities, and the arrival of syphilis converged in the late fifteenth century to create a new horizon for astrological forecasting.[1] Copernicus's arrival in Bologna to continue his studies in 1496 coincided with these emergent conditions. Violence and insecurity were almost continuous from the French invasion of Italy in 1494 to the Peace of Westphalia in 1648. The military historian J. R. Hale wrote of the sixteenth century:

> Wars between individual states and between coalitions of states, civil wars, wars of intervention, wars against deliberate unbelievers: these familiar forms of conflict came to be waged on a larger scale than ever before, and they were joined by two others, novel and perturbing. Catholic against Protestant[:] wars were fought in the name of religion within Christendom itself and against people so far beyond its limits that the greatest ingenuity was required to show that they were occupying lands to which any European could advance any demonstrable claim. On the battlefield new tactics were introduced, a hitherto scarce-proven class of weapon—pistol, musket, cannon—became ubiquitous, the appearance of towns changed dramatically as new methods of fortification replaced those which had served the middle ages. Wars became more expensive, touched more pockets, induced more radical changes in administrative institutions and in constitutional practice.[2]

War in the sixteenth century also became, for the first time, an object of widespread study and public comment—largely the province of lawyers, political theorists, historians, and theologians. Luther famously remarked that it was "as necessary as eating, drinking, or any other business."[3] To contemporary thinkers, it seemed to be a state of affairs that was natural, inevitable, and even beneficial—part of the divine scheme. "Warre," wrote one Elizabethan wise head, "is the remedy for a State surfeited with peace, it is a medicine for Commonwealths sicke of too much ease and tranquilitie."[4]

The literature of annual prognostication—a literature of cities, towns, and regions—both exacerbated and lived off the anxiety, fear, and consolation that coincided with this development. Both kinds of writings—those of warfare and those of astrological forecast—owed the possibility of their widespread presence to print. Although references to warfare were common enough in practical astrology, only rarely did such allusions appear in the genre of astronomical theory. One exception was Kepler's extended wordplay on his conquest of Mars, "restrained by the bonds of Calculation, who, so often escaping the hands and eyes [of ordinary astrologers], was accustomed to deliver vain prophecies of the greatest moment, concerning War, Victory, Empire, Military Greatness, Civil Authority, Sport, and even the cutting off or calling forth of Life itself."[5]

The arrival of print technology in the mid-fifteenth century also fed, accentuated, and ultimately transformed a confluence of conflicting political and cultural impulses. One of these movements was the papacy's attempt to reestablish its cultural leadership throughout Europe after the debilitating period of the Great Schism. During the pontificate of Nicholas V (1447–55), humanism made major inroads into the Roman bureaucracy. Clerical humanists who staffed the papal and cardinalate households made superb ambassadors to secular courts that were leaning toward Rome.[6] Cardinal Piccolomini's mission to Vienna is a notable example of the success of this new papal policy; another instance is Cardinal Bessarion's missions as papal legate to the Commune of Bologna and later to Vienna. Regiomontanus's *Tradelist* of works that he planned to publish—with its reformist and elitist classicism, its exaltation of Greek mathematical ideals, its editing and translating projects, its promotion of *theorica* as a necessary prelude to *practica*—embodied the humanist ideals propagating within the papal and imperial network. Regiomontanus's dedication to his *Epitome of the Almagest* represented the ancient Ptolemy as "Prince of Astronomers," Peurbach as "My teacher and the Emperor's Astronomer," Cardinal Bessarion as "Prince or Patron of the Good Arts," and the *Epitome* itself as a "monument" to be placed and guarded amongst the other books in Bessarion's library.[7]

Historiographical emphasis on Regiomontanus and planetary theory is hardly misplaced; but it has tended to make invisible numerous thickets of prognosticatory culture. These cultures mimicked, as it were, the complex political and territorial arrangements of Europe, so that virtually every major city and every prince in the fifteenth and sixteenth centuries supported a forecasting presence. Indeed, the great demand for nativities and regional prognostications—not to mention the associated theoretical literature—arose, first and foremost, from the political, religious, and financial anxieties of the dominant European ruling families: the Hapsburgs and the Jagellons, the electors of Brandenburg, the house of Aragon in Spain and (by the end of the century) in Naples and Sicily; the numerous Italian princely courts, including the Sforza of Milan, the Gonzaga of Mantua, the Este dukes of Ferrara, the Bentivoglio lords of Bologna, and the Medici of Florence; and from the papal court in Rome dominated by the Medicis (Leo X and Clement VII) and the Farnese (Paul III) for much of the sixteenth century.

Astrological forecasting effectively represented the state's scientific arm, and the astrologers who composed the judgments for its rulers constituted a sort of mirror-image principiate. The dominant images of prognosticatory authority played on the double meaning of Ptolemy as "king/astrologer" or "king/astronomer" and Regiomontanus as "prince of astronomers" or "second Ptolemy."[8] And the arrival of print rapidly made possible a new economy for the prognosticators. Fourteenth- and fifteenth-century prognostications, as far as I can determine, were composed in Latin; from the 1470s onward, the printed forecast opened a vernacular space as well, and established a tone that was more often alarmist than consolationist. Prophecies, weather forecasts, or other sorts of appended predictions helped to sell the ubiquitous single-page calendars or almanacs. Academic and court forecasts dedicated to the local ruling class were always eponymous and typically appeared on an annual basis. The discursive styles of these genres of forecast varied from the academic to the popular, from the prophetic to the naturalistic. Learned prognosticators quoted, misquoted, and recycled a variegated library of Greek, Arabic, and Jewish authorities; but from the 1480s onward, Latin editions of Arabic and Jewish works appeared in rapid succession, creating a wealth of theoretical resources for the prognosticators.[9] The library of the fifteenth-century French astrologer Simon de Phares gives an idea of the diversity of theoretical works available to a major practitioner, both in print and in manuscript. Ordinary prognosticators were most likely to own the printed *Tetrabiblos* and the *Centiloquium,* Aristotle's *Physics* and *Meteorology,* perhaps Avicenna's *Canon,* a table of new and full moons, and the works of medieval Arabic and Jewish writers. Modernizers, such as Regiomontanus, might be used but were not often cited. And, as already noted, the production of the forecast presupposed but did not require a display of astronomical *theoric:* none of the prognostications that I have seen mention a single epicycle or eccentric.[10] Something else was needed to make the prognostication work. The judgment required knowledge that was sufficiently well-informed and local as to endow it with a measure of immediate relevance.

Recent scholarship has begun to help us to understand some dimensions of this phenomenon.[11] But the local cultures of the prognosticators have not yet been adequately investigated. How did they construct their internal social boundaries, structure their forecasts, employ astrological theory, defend their territorial practices, and generally sustain their authority, credibility, and personal safety from year to year? Answers to questions such as these—which deserve a separate study in their own right—would go far toward explaining the continuing fascination that the heavens held for early moderns. In this chapter and in the next, I shall begin to address this problem by looking more closely at the prognosticatory cultures of two of the three cities in which Copernicus pursued his Italian studies: Bologna and Ferrara.

All the annual prognostications manipulated their readership along a spectrum between alarm and consolation. To do so, they needed to operate at a level of specificity sufficient to name something particular, yet general enough not to imperil the practice of prognostication. For example, as part of their annual forecasts for the year 1500, two of the university's prognosticators at Bologna offered judgments about the kinds of sicknesses to be expected that year. Giacomo de Pietramellara wrote:

> Concerning illness that will disturb men's bodies: of this, truly, there will be plenty. And those who will become ill—whether they be men or women—shall not die but later will be restored to health.

> Doubtless (*senza dubio*), there will be smallpox . . . inside and outside of Italy and some places will be infected by plague, some by poisonous fevers or other unknown illnesses.

> Many women will die in pregnancy and many shall abort or, in truth, shall experience unaccustomed pain during childbirth and will give birth with great difficulty.

> There will be many illnesses of the eyes and ears and many of these through catarrhs and throat abscesses and aching ribs; these people shall not die. But many old and young men shall return to health only with difficulty or will perish of pleurisy.

> Not only shall we behold the aforesaid illnesses, but also many tertian and other long fevers, such as the quartans, phlegmatics, stomach aches and intestinal fluxes.[12]

The level of specificity in this forecast was typical for this period: the prognosticator named and linked particular illnesses and groups but, not surprisingly, offered no quantification beyond *some, many,* and *plenty.*

The other Bologna forecast for the year 1500 was issued by Domenico Maria Novara (1454–1504), Pietramellara's older colleague at the university and also a man whose influence on young Copernicus was substantial. Novara's language was equally alarmist in tone, though less optimistic about the prognosis for recovery and even more strikingly divergent in the specific illnesses forecast and etiologies adduced.

> This year will bring difficult and evil illnesses, and there will appear a swelling illness in the back, in the hand, and in the feet, and in part of the throat, as well as many rheums and catarrhs; and in women, ulcers will multiply on the breasts and in the throat and in men on the testicles. . . . But the place that will most feel the plague is Asia Minor, which now is called Turkey. And the influence will fall most of all on the lands of Turkey, the sign of the conjunction of their sect being damned. Then, [the influence] will fall on Constantinople, which is now the primary land of the Turks. This same influence will fall on a great part of that great [land] and many places in Italy, such as Milan, Pavia, and almost all of Romagna and Rome, and many places in the Kingdom of Puglia. But the places that are situated near the water will be the more damned because the putrid vapor rising from the water will be the main cause of poisoned air. On account of this, the mineral water that they administer for health to the sick is polluted and will bring little health; and those who drink it will be harmed by the bad quality that multiplies in it. And the waters in the deepest, stinking abysses are the kind that is the most poisonous. And the illness's greatest force will occur when Saturn enters into the sign of Gemini, and then a grievous illness will appear, in which there will always be an admixture of melancholic humor and into which the sick ones shall fall again.[13]

Here the prognosticator is more general in pointing to affected social groups (men, women), but more specific in naming body parts (back, hands, feet, throat, breasts, testicles) and geographic regions that will be struck by the planetary influences. The worst-hit region will be infidel Turkey, but certain Italian cities will share in the misfortune, especially those that drink polluted mineral water. Unlike other types of healers, the prognosticators appear not to have been involved in contracting with clients to deliver specific cures, although the kinds of agreements that they might have made is a subject still to be investigated.[14]

The practice of judging the future was not a strictly algorithmic exercise in which the prognosticator unproblematically followed or "applied" rules found in the astrology manuals. Apart from the association of specific planets with geographic regions, not all of the above information derives from sources like the *Tetrabiblos* or Albumasar's *Concerning the Great Conjunctions*. The prognosticators inevitably read and used the *Tetrabiblos* and other astrological masterworks in quite locally situated ways.[15]

And yet, for all that, the prognosticator did not operate without complex theoretical resources. Astrological theoric essentially supplied a toolkit of general rules, classifications, and taxonomies of (celestial) causes and (terrestrial) effects. Astronomical theorics and tables provided the quantities and relations that enabled the forecaster to predict when, say, the Moon would be new or full, when the Sun would enter a particular sign, when and by how much these great luminaries would be eclipsed, when two or more planets would line up with the Earth, and so forth.

Moreover, Ptolemy himself had anticipated the need for some kind of guidance in the prediction of specific events. In book 2 of the *Tetrabiblos*, Ptolemy addressed himself to the principles for making particular predictions for countries and cities—just the kind of forecasts that began to flourish in the late fifteenth century. All events would occur in a particular place, defined by geographical climes; thus forecasts used the elemental qualities to differentiate effects among peoples by geographical "parallels" (north-south) and "angles" (east-west). These events would have a beginning, a middle and an end, and would cause good or bad effects on certain classes of beings, human or animal. Unlike that of a country, a city's fate resembled that of an individual nativity. Ptolemy said that there were "affinities" or "familiarities" between the time of a city's founding or a founding ruler's birth date, locations of the Sun and Moon, and those bodies' subsequent influence.

Ptolemy assigned special causal importance to eclipses, based not only on when they would occur but in what zodiacal sign. To describe the duration of the eclipse's effect, he used a language of "intensification" and "relaxation" by analogy with the tightening and loosening of the strings of a musical instrument.[16] The event predicted from a solar eclipse would last as many years as the number of equinoctial hours of the eclipse's duration; for a lunar eclipse, the effect would be measured in months. Intensification and abatement of the eclipse's force, as specified by their effects at the coordinates of a particular terrestrial horizon, allowed the prognosticator to foresee (time-lagged) effects at that location many years into the future. However, many other significant planetary configurations, such as conjunctions and retrogressions, could act together with the eclipse to affect the strength and timing of these effects.

Because there might be a great many possible influences acting in consort, astrological theory faced the prognosticator with the problem of adding up the strengths of the influences in order to calculate the net or dominant effect. Ptolemy described this singularly important issue as follows: "Such are the effects produced by the several planets, each by itself and in command of its own nature. Associated, however, now with one and now with another, in the different aspects, by the exchange of signs, and by their phases with reference to the sun, and experiencing a corresponding tempering of their powers, each produces a character, in its effect, which is the result of the mixture of the natures that have participated, and is complicated." He then added, most candidly: "It is of course a hopeless and impossible task to mention the proper outcome of every combination and to enumerate absolutely all the aspects of whatever kind, since we can conceive of such a variety of them. Consequently questions of this kind would reasonably be left to the enterprise and ingenuity of the mathematician, in order to make the particular distinctions."[17]

Here, then, Ptolemy marked quite precisely the space within which the prognosticator might operate. The prognosticator constructed a judgment about the meaning of particular mixtures of elemental qualities whose influences jointly brought about particular effects. Physically, the influence was "a certain power emanating from the eternal ethereal substance . . . dispersed through and permeat[ing] the whole region about the earth, which throughout is subject to change."[18] Astrological theory thus left open the possibility that, even if the forecaster got the astronomical part right, he might still misjudge the outcome of particular combinations of influences. In short, Ptolemy had left the door open for the prognosticator to make legitimate misjudgments without ever imperiling the theory itself.

A good or reliable astrologer might rightly conjecture the effects of multiple combinations of influences on a particular city, region, or individual. But there was an additional reason for the compelling nature of astrology as a system of thought. Astrological discourse was eminently suited to addressing the anxieties of rulers, and the unstable early modern political system created an ideal climate in which such anxieties flourished. It was an authoritative resource of policy making that offered a malleable discourse of domination, gender, and hierarchy, a symbolic stage onto which an abundance of social and political relations might be projected. In contrast to other genres of early modern advice literature—the mirror for a prince and the courtiers' manuals— it combined a thoroughgoing naturalistic causal apparatus with this language of domination. And finally, it concerned itself with explaining phenomena in all those areas that in the twentieth century fell under the disciplines of political science, economics, psychology, sociology, weather forecasting, and medicine. A common mnemonic for remembering the meanings associated with each of the twelve astrological houses exhibits the full range of categories that were covered:

Vita, lucrum, fratres, genitor, nati, valetudo:
Uxor, mors, pietas, regnum, benefactaque carcer.

(Life, money, siblings, parents, children, health;
Spouse, death, piety, offices, honors, and prison.)[19]

From the 1470s onward, publishers put into print a canon of ancient and medieval astrological classics (some of them already on Regiomontanus's tradelist) and enabled prognosticators to issue their own forecasts independently of the university stationer, the place where manuscripts were copied. For some, a considerable range of theoretical material was still obtainable from manuscripts, as shown by the extensive contents of the library of Simon de Phares and from Pico della Mirandola's numerous references.[20] But, as shown in the previous chapter, printing made it possible for more would-be astrologers to own such basic tools of their trade, just as the new medium encouraged authors to advertise themselves as more than just "Master So-and-So."[21]

THE POPULAR VERSE PROPHECIES

Printing also multiplied and further empowered another extraordinary genre: popular verse prophecies. Verse prophecies were usually produced as cheap, anonymous, poor-quality octavos and quartos of two to four (and rarely as many as twelve) leaves. They successfully created a new readership, ranging from workers and artisans to more "refined bibliophiles," like Christopher Columbus's son Ferdinand, who built a substantial collection of such prophecies.[22] Their authorial identity was fluid: some were falsely attributed to academic or theological writers. One pamphlet described Pietro d'Abano, the thirteenth-century Paduan professor of medicine and astrology, as "the most reverend necromancer" (el reverendissimo negromante); another item was ascribed to "glorioso santo Anselmo," inaccurate attributions meant to augment their prestige and credibility. The authenticity of these works was further bolstered by the coding of their titles with reference to their antiquity (e.g., *Profetie antiche*), to archaeological objects, or to more official and legitimate works of prognostication (e.g., "Prophecy or Prognostication Found in a Pyramid in Rome in Latin Verses, Translated into Common Language").[23] Hawked by traveling peddlers, the print prophecies were a profitable source for the early publishers.

Unlike the academic prognostications, the prophecies were composed in a versatile and figural language of signs that easily permitted readings framing an almost limitless range of current events. Moreover, the absence of any mechanisms of regulation easily enabled the same tracts to be reissued with new titles or new authors. Most such tracts were devoid of any astrological apparatus altogether. They relied, in fact, much more on visual imagery, on discursive resources of deformity, unexpected or sudden apparitions of portents or pseudoprophets, alarming predictions of floods (for which there was, as it turned out, some basis in fact),[24] and inflated expectations of the arrival of an imperial figure who would bring peace and spiritual re-

ⓒ Zu wiſſen das diſs monſtrum geboren worden iſt in diſem iar ſo man zelt M. D. vnd
vi. vmb ſant Jacobs tag zu Florentz võ ainer frawen. vnd ſo es kund gethon iſt vnſerm
hailigen vatter dem babſt . hat ſein hailigkait geſchaffen man ſolt ym kain ſpeyſung gebē
beſunder on ſpeyß ſterben laſſen.

19. Representing disorder: the Florentine monster, 1506. German broadside. This image circulated in different but similar versions and was taken to represent an unnatural being, born a hermaphrodite with avian characteristics. The pope, according to the caption, ordered that the child-monster be allowed to die of starvation. Courtesy Bayerische Staatsbibliothek, Munich.

newal. Not surprisingly, this literature constituted a major resource of political, sometimes
anti-Lutheran, propaganda. There are also indications that some of these tracts were meant to be
recited out loud.[25] Ottavia Niccoli, who has studied this literature extensively, argues that its development is certainly, if not uniquely, associated
with "the political and religious disintegration
that occurred during the first decades of the sixteenth century—with the *guerre horrende de Italia*
and the continual passage of troops that brought
with it the scourge of pestilence, famine, and dev-
astation that gave credibility to the catastrophes
predicted by the preachers and the tale singers."[26]
As an example, Niccoli cites the anonymous author of *El se movera un gato,* who laments: "O Italy,

weep, he said—sob, sob, sob, gasp, gasp; Italy!—
for your land will see foreigners and disasters,
and these will be enterprises that your lords will
involve you in. And great and small will be troubled out of measure and with great bloodshed."[27]
Prophecies of this sort, or reports of unnatural
events, like the Florentine monster of 1506, were
far more flexible resources of alarm than the annual academic predictions because they were easily adapted to new circumstances.

If Ptolemy's *Tetrabiblos* was the paradigmatic
text for the academic tradition of prognostication,
popular print prophecy drew many of its most
powerful figures, tropes, and typologies from the
extraordinarily influential *Prognosticatio in Latino*
(1488) of Johannes Lichtenberger, né Grümbach

20. Divine inspiration of the prophets: Ptolemy, Aristotle, Sibyl, Saint Bridget and Brother Reinhard. Lichtenberger 1488. Reproduced by permission of The Huntington Library, San Marino, California.

(d. 1503). This work marks an interesting development in the prognostication literature. Lichtenberger tried to fuse apocalyptic and eschatological prophecies with the authority of high academic prognostication. Identifying himself as the "holy empire's judge of the stars" (*astrorum iudex sacri imperii*), he caused the peasants in his district to keep a respectable distance, fearing that the black raven on his desk was an evil demon.[28] Lichtenberger's *Prognosticatio* went through more than fifty unabridged and twenty-nine excerpted editions (in Latin, German, and Italian), its claims supported by a wealth of authorities ranging from the Bible, the Church fathers, ancient Greek philosophers and astrologers, and the Sibylline prophecies to the medieval prophecies of Joachim of Fiore. The title page showed in a striking woodcut the rays of inspiration that illuminated all levels of prophets from the "learned" (Ptolemy and Aristotle) to the "popular" (the Sibyl, Saint Bridget,

and Brother Reinhard). Forty-five such woodcuts informed the text—all far below the level of artistic craftsmanship of the many intricate, single-leaf cuts that were produced at this time for the court of Maximilian. Nonetheless, such visual resources made the book's general message accessible to a wide audience without requiring literacy to study the accompanying text.

The text itself was not without dramatic effect: it is filled with eschatological images and populated by agents who enact prophetic dramas. Lichtenberger believed that in his own time, various apocalyptic prophecies would be fulfilled, and these upheavals would then be followed by an age of peace. For example, when the emperor-savior defeats the Turks, there follows a "new order" (*nova ordo*) and a "new reform of the church" (*nova restauratio in ecclesia*).[29] However, because he attempted to synthesize quite different prophetic plots, Lichtenberger sometimes had the

German eagle (*Aquila*) play the role of the chastising emperor who renewed the Church; at other times, the wicked tyrant-eagle persecuted the Church, leaving *renovatio* to the angelic popes (symbolized by the lily). In an attempt to resolve the contradictions of coexisting pro- and anti-German prophecies, Lichtenberger introduced the good agent of God in the form of the oracle Saint Bridget, who led the Germans on their chastising mission.[30] His problem was then to identify specific agents with the generic dramas. Here Lichtenberger was called on to make a political judgment. For reasons that are not clear, he ruled out his immediate patron, Frederick III, and another obvious candidate, King Matthias of Hungary, settling instead on Frederick's son Maximilian as the *rex pudicus facie*.[31] Lichtenberger's bevy of prophecies was so flexible— Luther's apt term was "amphibolic"—that it permitted quite authentically opposite readings of its meanings and ensured thereby its phenomenal longevity, even until World War I.[32]

Lichtenberger's work, however, was important in a new respect: to the compilation of earlier prophecies, he now introduced the element of astrological causality. "A conjunction of Saturn and Jupiter announcing threatening calamities occurred on the 25th of November in the year 1484 at six hours and four minutes after noon and measured at one degree of Cancer above the horizon." The effect of this conjunction was augmented by a "horrible" solar eclipse in 1485 and a conjunction of the bad planets Saturn and Mars on 30 November 1485 at 23°43' Scorpio; but still another conjunction—"benign" Jupiter and "choleric" Mars—will "ameliorate" the effects of the 1484 alignment.[33] Ironically, it was the *Prognosticatio*'s own "conjunction" of these elements that gave the legitimacy of high astrological authority to what was already an extraordinarily flexible resource of prophecy. If some of Lichtenberger's sources derived effortlessly from medieval prophecies, such as Saint Bridget's *Liber coelestis revelationum*, Lichtenberger borrowed the authorship of cause and timing of the predicted event, without attribution, from someone who was unquestionably alive—the Duke of Urbino's personal physician and astrologer, Paul of Middelburg (1445–1533).[34] The practice of redeploying an earlier, successful *practica* to reinforce the authority of a recent one seems to have been common enough at the time not to excite any untoward reaction.[35] But in this case, Lichtenberger's appropriation became the occasion for the stinging *Invective of the Most Celebrated Master Paul of Middelburg, Surely a Prophet, against a Certain Superstitious Astrologer and Sorcerer*.[36]

The polemical title alone makes it clear that the boundary between legitimate and illegitimate prognostication was at stake. This theme persisted into the seventeenth century. The self-described legitimate prophets, such as Paul himself, used "divine, judicial astrology" (*mathesis*) and "astrological theory" (*astrologica theoria*), whereas the object of his attack indulged in "the secret ring of his sacred discipline, contaminated by superstitious sorcery and vain divination."[37] In sarcastic and personal tones, Paul said that he had come across this judgment (*judicium*) only in the last few days and that its author (Lichtenberger) sometimes wished to be known by name, but sometimes concealed his identity: "He shouted as much as Ruth while collecting grain and hiding in the woods so that I could guess that a certain German, Johannes Lytheberger, had published this prophecy [*vaticinium*]." To add to the shadowy authorship, the "prognostication" (*prognosticum*) used "various, worthless pictures" to foretell the imminent arrival of the Antichrist. Some of his pictures showed "religious women in labor"; others displayed pigs and hens singing together and the Antichrist teaching and thereby destroying the (Holy) Roman Emperor. Still elsewhere, Paul observed, the author confused different kings and princes. Lichtenberger represented himself as a minor prophet who predicted the coming of a new law (*novamque legem*). This, said Paul, is "certainly ridiculous no less than absurd and superstitious," because the Antichrist can only be predicted from the greatest conjunction of the most important planets, and Lichtenberger had failed to demonstrate either. Furthermore, the works of Abbot Joachim on which this "Lytheberger" relied so heavily were condemned by the Lateran Council in 1215. Yet, in spite of all these offenses, publishers had put this dreadful book up for sale in both Italy and Germany. Even worse, Paul revealed: "When I had read through it, I discovered that a great part is culled and woven together from my prophecy [*vaticinio*] of the great conjunction . . . but, like a deaf man having borrowed from the labor of another man, he mixes things up when he ascribes them to himself."[38]

Because Lichtenberger wove together things

that he did not understand, he had corrupted the proper "office of the astrologer" (*astrologi officium*): "For it appears that he has written not so much a prognostication [*prognosticum*] as a destructive exhortation which urges the most civil and peace-loving German princes to enter into a confederation so that they may become the most awful Christian neighbors, that they may make war on their neighbors; and, finally, not least, he [Lytheberger] behaves indecently when he speaks ill of the highest popes and detracts not a little from the honor and dignity of the Roman Church."[39]

Within the growing culture of print prophecy, this important controversy marked the emergence of an increasingly contested space of power. Regiomontanus had used print technology and the standard of correct translation (castigation) in the 1470s to contest a well-entrenched academic genre.[40] When in 1462 he raised the Greek mathematicians to the rank of pagan demigods in a lecture delivered at Padua, he elevated mathematics and the mixed sciences above all other forms of knowledge in a rhetoric of certitude that would reverberate down into the seventeenth century.[41] By the 1490s, the "nonsense" (*deliramenta*) to be exposed and condemned included not only "barbaric" translations but also a serious, rival genre of prophecy with threatening popular images, strategies of visual representation, and propagandistic power. Paul's *Invective* thus defined a new moment: the beginning of a struggle by Ptolemaic-style prognosticators to appropriate and even to monopolize the right to make prophecies.

SITES OF PROGNOSTICATION

This late-fifteenth-century state of contestation had its roots in the burgeoning economy of prognostication, which found sustenance in cities during the earliest decades of print. A group of well-established printers and mathematical prognosticators produced a steady stream of annual prognostications in Germany and Italy from at least the 1470s.[42] The prognosticators were, in fact, among the earliest groups to capitalize on the availability of print technology. Many of these forecasters were officially associated with universities, such as the Bologna prognosticators Pietramellara and Novara; others, like Paul of Middelburg and Johannes Stabius, were affiliated with princely and imperial courts. In the north, the academics used a variety of print identities, most commonly "Magister" or "Magister Artium Liberalium," occasionally, "Astrologus," and more rarely "Philosophus ac mathematicus."[43]

In Italy, Bologna and Ferrara appear to have been the leading sites of prognosticatory activity. Most of the Bologna prognosticators were members of both the medical and the arts faculty. Hence their print identities typically produced conjunctions such as "Doctore delle Arte e Medicina," to which might be added "Mathematicus." At Ferrara, the court astrologer Antonio Torquato sometimes identified himself as "Medico e Astrologo," sometimes as "Cultore de le Matesie" or "Medicine Cultore." Pietro Buono Avogario (Advogarius), the ducal prognosticator for nearly thirty years, used at one time "Eximium Artium et Medicine Doctorem," at another, "Phisico astronomo maistro." Someone outside these official institutions, however, was expected to announce his status, as did the Jew Boneto de Latis: "Hebreo Provenzal Doctor de Medesina Maestro infra li Zudei."[44] The quantity of editions that these prognosticators produced was considerable, the number of authors relatively small.

The most active sites of academic prognostication in the transalpine north during the last part of the fifteenth century were the university cities of Krakow and Vienna.[45] The university in Krakow had been founded in 1364 on the Bologna model, with students electing the masters, but after a period of decline, it was reformed, like the university in Vienna, on the Parisian model, with the masters securely in control.[46] Regiomontanus's *Deliramenta* was cast as a dialogue between a Krakovian and a Viennese. Two of the most prolific academic prognosticators of the period, Johannes Virdung and Johannes Stabius, had studied or lectured at Krakow at one time or another.[47] The former carefully associated himself in the titles of his prognostications with both university and town identity: *Judicium Baccalarij Johannis Cracoviensis de Hasfurt* (The judgment of John of Hasfurt, bachelor of Krakow).[48] In 1487, Virdung paid Albert of Brudzewo four Hungarian gold pieces to review a series of horoscopes, and the following year he attended Brudzewo's lectures on Peurbach.[49]

One of the most important indigenous Krakovian prognosticators was John of Glogau (Jan of Głogów). Virdung had attended his lectures in 1486.[50] John's predictions span at least from 1479 to 1508 and, as was typical of the genre of northern forecasting, they were as much dominated by the weather as by their region's social and political fortunes. Furthermore, unlike the Italian prognosticators, John typically invoked the honor of the city and the university rather than a particular prince: "I, Master John of Glogovia, undertake to write in a high and gentle style for God's honor and the fame and glory of the Studium of Krakow. I have divided my prognostication into three *differencia* in which may be considered, first of all, the daily condition of the elements and changes occurring in the air. In the next part there will appear the disposition of the elements. And finally, for human utility and according to the stars' testimony, you will find added certain days chosen for bathing and traveling and directing human actions."[51]

As Gustav Hellmann's figures in table 1 suggest, the empire led the way with an enormous production of such annual forecasts. Many of these forecasts were composed by university masters like John of Glogau and Johann Virdung, but a great many others, issued by city physicians and Lutheran pastors, mirrored the spread of the Reformation as it moved from the pulpit along the network of southwest German publishing cities.[52] That many of the northern prognosticators addressed their predictions to a city rather than to a prince may confirm, to some degree, Machiavelli's famous observation in his *Report on the Affairs of Germany*: "The power of Germany certainly resides more in her cities than in her princes . . . [they are] the real nerve of the Empire."[53]

The rise of the annual prognostication as an independent genre seems to be tied as much to the technology of print as to the rise of the power of cities in the fifteenth century. In Italy, however, the highly personalized political structure of the princely despotisms connected the prognosticators' own fortunes closely to those of their princely patrons even when the forecasters held university or municipal appointments. Ferrara, with its absolutist dukes of the house of Este, its famous *studium* and library, and the remarkable astrological frescoes of the Palazzo Schifanoia, had long been a congenial site for prognosticatory activities.[54] Peurbach and Regiomontanus had lectured in Ferrara at different times. Domenico Maria Novara identified his family lineage as "Ferrariensis." Giovanni Bianchini, retained in 1427 by the

TABLE 1
Gustav Hellmann's estimates of total number of sixteenth-century prognostications

Countries	Authors	Prognostications	Extant
Germany	213	1,719	849
Low Countries	38	295	93
Denmark	3	3	3
England	28	89	53
France	17	70	30
Italy	55	380	147
Poland	14	113	32
Spain	10	17	12
Czech lands	3	10	9
TOTAL	381	2,686	1,228

SOURCE: Hellmann 1924, 35.

duke of Ferrara as "maestro generale del conto della sua Camera," corresponded with Regiomontanus about observational and computational problems.[55] In 1452, Bianchini presented his *Tabulae Caelestium Motuum Novae* to Peurbach's patron, Emperor Frederick III—or, rather, as vividly portrayed in the astrologically freighted manuscript that dramatically represented the event, the duke commemorated his own vassalage by offering Bianchini and his tables to the emperor. In addition, Pellegrino Prisciani, who wrote a history of Ferrara, was a prognosticator widely read in the astrological classics. The highly influential prognosticator Luca Gaurico delivered a stirring defense of astrology at Ferrara in 1508.[56] In 1503, Copernicus received a degree in canon law from the university.

The career of the court and university prognosticator Pietro Buono Avogario offers some tantalizing and unusual hints about the actual practice of publishing a forecast. Avogario lectured on the science of the stars at the *studium* of Ferrara between 1467 and 1506.[57] A medal struck in his honor reads: "Petrus Bonus Advogarius of Ferrara. Remarkable Doctor. More Remarkable Astrologer."[58] Between 1475 and 1501, Avogario issued a series of annual prognostications dedicated to Ercole, or Hercules d'Este, duke of Ferrara. The first was published in 1479. Typically, Avogario's forecasts assessed not only the duke's fortunes

21. Dedication scene: Giovanni Bianchini presenting his *Tabulae Caelestium Motuum Novae* to Emperor Frederick III (1452). The figures' intertwined poses symbolize relations of power through hierarchy and reciprocity. The emperor, at far left, presents his coat of arms to his vassal, Duke Borgo d'Este, who reciprocates by offering the kneeling astrologer-astronomer Bianchini and his tables. The duke links the two groups of figures, appearing to step out of the circle of three courtiers at right. Courtesy Biblioteca Comunale Ariostea, Ferrara. MS. Cl. I, n. 147, f. 1r.

but also the astrological revolutions of courts in Spain, France, Rome, Hungary, Milan, Mantua, Urbino, Rimini, and Pesaro. The narrative structure of the predictions seemed designed as much to protect the astrologer's fortunes as those of his patron. In general, the ruler could expect good tidings; but he was also warned to avoid specific dangers in specific months. For example, in 1496, Avogario offered his patron the following prediction: "Most illustrious and excellent duke of Ferrara, my singular lord: you will enjoy very good fortune in the present year because the sign of life in the revolution owns the highest part of the heaven—the Moon and Venus. But Your Excellency shall not be without danger. All these things are confirmed by the revolution of your nativity: Your Excellency should still be careful of the month of August that he not be moved to vexation by any infirmity or by any other cause; in such a manner, this most excellent prince can have the greatest stability of his reign."[59]

The next year's forecast was even more sanguine: "My most singular lord, most illustrious and excellent duke of Ferrara: Through the influence of benevolent stars, this year will be made cheerful by happy tidings. Nevertheless, nothing will come to pass without great expense. Many will be set on acquiring things; but not in order to contribute to the stability of the Empire. Thus, Your Excellency should be careful about voyages by water even more so than by land. The place of your empire in the tenth celestial house promises you happiness and, undoubtedly, stability of the Empire."[60] Apart from the generality of the "cheer-

ful" forthcoming year, Avogario was quite specific that the duke should be wary of all voyages.

Prior to the publication of his first judgment, at the end of February 1479, Avogario wrote an extraordinarily revealing letter to the duke in a mixture of Latin and Italian:

Most illustrious and unconquerable lord duke, my most extraordinary lord. I send eternal greetings and wish you triumph and victory over [your] enemies, etc. I have just completed the judgment for the coming year, and because it is time to publish it, I first send it to Your Excellency, as is customary [*como e usanza*], so that you, and no one else, might be the first to see it, as is the custom [(*ut moris est*]. The judgment is really terrible, as Your Excellency will see. Nevertheless, if it pleases your will, blessed and praised in all ways, you, who are the great king by whose direction the whole *machina mundi* is governed, can arrange all this to be changed and varied. I send the judgment to Your Excellency bound with that of the present year—the better to attack enemies when needed and to win victories. Also I send you a forecast of the coming year's unfortunate events, in which it indicates that one should neither fight battles nor assault the enemy because the greatest dangers will befall him who commences them.[61]

There is evidence that Avogario always checked his predictions directly with the duke in advance, and that the mutual protection of the prognosticator and his patron was at stake. For example, on February 14, 1490, he wrote to the duke: "I think that Your Excellency will be pleased with what I

22. Rare representation of a prognosticator at work. Bookstand, armillary sphere, and compasses lie close at hand. From Pietro Buono Avogario of Ferrara, *Pronosticum*, 1497. © British Library Board. All Rights Reserved. I.A.27832. (Nineteenth-century library stamp of British Museum visible at center.)

have to publish, as I undertook to do in previous years. I willingly send it [the forecast] to Your Ducal Excellency because there is no sad influence that I have seen in Your Excellency's revolution. This judgment will have to be sent both throughout Italy and outside Italy and will likewise give renown to our happy country; but first read the present judgment, which is [dedicated] to the glory and praise of Your Most Illustrious Nobility."[62]

It would be interesting to know just how widespread was this practice of advance consultation with the patron and whether the "custom" both preceded and followed the introduction of print. Clearly, for weather forecasts and recommendations for bleeding of the sort that we find in many of the northern European judgments, advance consideration was not particularly useful. For politically sensitive predictions about the duke and his enemies, however, the distributional and propagandistic potential of print would have made

such prior consultation almost mandatory.[63] The problem was to judge whether some part of the prediction ought to be withheld because of damage that might rebound onto the ruler. What the language of the forecast says, therefore, about the ruler's relation to the natural world may well represent quite accurately the patron's power to edit the forecast: "He can arrange all this to be changed and varied." This practice of propagating the desired image of the ruler could assure the reputation of both the patron and the prognosticator and might help, at least in part, to explain the durability of the latter's reputation.

At the same time, there is no evidence to suggest that the duke was involved in any kind of crude dictation of the contents of the forecast. There is no reason, in fact, to doubt that both the ruler and the prognosticator believed the prediction, as much of the advice was personal and was given year after year. For example, in a letter of June 20, 1484, Avogario sent Ercole the follow-

23. Planimetric view of Ferrara (1499), a major site of prognostication in the fifteenth and early sixteenth centuries, where Copernicus also obtained his degree in canon law (1503). From Alessandro Sardi, *Annotazioni istoriche*. MS. Italiano 408, alfa F.3.17. Biblioteca Estense, Modena. By permission Ministero per i Beni e le Attività Culturali. Further reproduction by any means is prohibited.

ing uncontroversially sound advice, aphoristic in its style: "I say that he who would strike the enemy first, knows through the marvelous celestial influence, that if the number of armed men is more than tenfold greater, then he will lose his force and will be destroyed with all his men."[64] This advice, sufficiently cautious in balancing generality and specificity, was followed by a list of quite specific days for journeys and engaging in battle.

> Your Excellency received from me the other day the selection for journeys on two days, that is, the 21st and 22nd. If possible, Your Excellency ought to go on the 26th of June, that is, next Saturday. [On that day] Your Excellency will have the best election for vanquishing your enemies and for attaining every victory. And Your Excellency will meet the best outcome in your undertakings because at that time the Moon will embrace Jupiter and Venus in beautiful aspects and the Moon itself will be at crescent light. And so, God willing, seize the aforesaid day of the 26th freed from all [worries] and, with God's help, it will be good for you in all ways.[65]

The relationship between the rulers and prognosticators of Bologna, as the next chapter shows, was somewhat less close than at Ferrara.

By the last decade of the fifteenth century, prognosticatory authority was an expanding, increasingly self-confident, yet still chronically uncertain business—a contested zone. It was grounded in such a vast array of considerations that failure in any part of the science of the stars could easily explain the failure of a specific prediction while leaving the rest intact—or unexamined. Astrological prognostication thus offered the perfect Quinean situation: any seemingly disconfirming observational evidence could always be accommodated to theoretical astrology. Moreover, different combinations of local considerations not mentioned by Ptolemy might establish or enhance a forecast's credibility with a given audience: the author's social location and political acumen; the language of the text, Latin or vernacular; the precision of the forecast's language; the perceived fulfillment of the prediction; the explanation of failed predictions; the borrowing from, or plagiarism of, earlier authorities; the joining of predictions with biblical prophecy; association with official, theological authority; the nature of the political message; and, finally, trust in the planetary tables and ephemerides used in the forecast. On such foundations an author could bolster or endanger his reputation. In 1496, however, a new threat appeared that could not so easily be dismissed: Giovanni Pico della Mirandola's attack on the foundations of divinatory astrology. This attack appeared from a Bologna publisher just months before Copernicus arrived in that venerable university town.

3

Copernicus and the Crisis of the Bologna Prognosticators, 1496–1500

Copernicus was involved in a culture of astrological prognosticators during his student years in Bologna. Although not a single word about astrology has survived in his writings, a great deal can be said about the specific circumstances that framed his involvement with that subject as a local practice. Indeed, much can be learned about various elements that shaped his early problematic and that pertain to questions unresolved in chapter 1: his map of knowledge domains, the cluster of major questions that preoccupied him for the rest of his life, why the ordering of Venus and Mercury became a matter demanding of solution, and his concern with the period-distance rule. The four years that Copernicus spent in Bologna were a critical phase of his formative intellectual period. During this time, the prognosticators were under serious pressure both to justify their forecasts and to defend the theoretical foundations of their practice against the massive criticisms of Pico della Mirandola. All the astronomical and physical considerations that historians have emphasized are relevant to this account; but Copernicus's problematic makes more sense when one incorporates astrological practice into the story. More broadly, these conclusions suggest how we might make sense of the subsequent evolution of the Copernican question into the seventeenth century.

THE BOLOGNA PERIOD, 1496–1500
AN UNDISTURBED VIEW

The usual story is that Copernicus went to Italy in 1496 to study law. He came with some knowl-

edge of astronomical theory, acquired at the feet of his Krakovian masters. During his four years of study at Bologna, he made the acquaintance of the astronomer Domenico Maria Novara (1454–1504).[1] According to some historians, Copernicus studied with Novara and helped him (in some unknown way) to make celestial observations. Most significantly, Novara first acquainted the young Polish astronomer with difficulties in Ptolemy's theories, notably an apparent shift in the direction of the terrestrial pole. This anomaly is alleged to have stimulated his own ideas about moving the Earth.[2] According to others, Novara's critique of Ptolemy arose from Neoplatonic and Neopythagorean sources to which he had been exposed through the Platonic Academy of Florence.[3] Copernicus's complaints about the old world system—both the Ptolemaic equant mechanism and the ordering of the planets—were then thought to have arisen from the same intellectual ground.[4]

Given how little is known about Copernicus's short time in Bologna, it seems legitimate to ask how one can justify an entire chapter devoted to this period of his life. My first contention is that there is more to Copernicus's relationship with Novara than hitherto has been appreciated. A great deal of very admirable work has been done by positivist historians, such as Carlo Malagola, Leopold Prowe, and Ludwik Birkenmajer. But earlier historiographical presuppositions about what was useful and "scientific" or dismissible and "superstitious" in Novara's writings effec-

24. Domenico Maria Novara. Unknown eighteenth-century artist and cherub. *Raccolta iconografica*, vol. 12, fascicle 13, no. 58. Courtesy Biblioteca Communale Ariostea, Ferrara.

tively hid a great deal of information about the Bolognese master. These assumptions also seem to have inhibited the study of Novara's extant writings, which were all astrological in character. My second contention, which is far more ambitious, is that Domenico Maria Novara—and through him, Copernicus—were part of a flourishing community of prognosticators in Bologna.[5] A better appreciation of that culture is needed to understand the circumstances that framed the motivation of Copernicus's astronomical project.

FROM THE KRAKOW COLLEGIUM MAIUS TO THE BOLOGNA STUDIUM GENERALE

Nicolaus Copernicus arrived in the fall of 1496 at the old Bologna *studium generale*—the medieval term by which the university was still known—to receive instruction in "both laws," civil and canon. With three years of arts study at Krakow between 1491 and 1494 (but no degree), he probably arrived with some competence in Peurbach's *New Theorics*. The main Krakovian teachers were intimately familiar with theoric. John of Glogau had composed a commentary on the old *Theorics* of Gerard of Cremona; his pupil Albert of Brudzewo, as noted in chapter 1, was the first to comment on Peurbach.[6] Moreover, both Glogau and Brudzewo are known to have issued annual prognostications; thus it would be very surprising if Copernicus had not been acquainted with one or more of their forecasts.[7] Whether Copernicus himself engaged in any prognosticatory activities at Krakow, however, is unknown. The only extant book from his library that could date to the Krakow period and that suggests direct familiarity with astrological theory is Haly Abenragel's *In judiciis astrorum*. Copernicus's copy of this work is bound with Euclid's *Elements*.[8] Finally, just as Copernicus arrived in Bologna, Regiomontanus's *Epitome of the Almagest* was published in Venice.

Although both Krakow and Bologna were major sites of annual prognostication, Copernicus's formal choice of curriculum at Bologna was associated directly with his decision to follow in his uncle's footsteps as a church administrator or canon. The legal statutes of the Varmia chapter of the Church clearly decreed that any newly admitted canon should receive at least three years' training in one of the higher faculties of "some preeminent *studium*," unless he be already a "Master or Bachelor of Sacred Letters or a Doctor of Medicine or Licentiate in Physic or the Decretals [i.e., Canon] or Civil law."[9] These stipulations make sense when it is recalled that theology, law, and medicine were the three superior faculties beyond the bachelor of arts degrees in the medieval university structure.

In the pre-Reformation Italian city-states, however, there were no theology faculties of the sort to be found in the transalpine universities. Theology instruction occurred inside the houses of religious orders, usually mendicant orders (Dominicans and Franciscans), and sometimes at episcopal palaces or secular colleges. Although the corporations or *universitates* of law and medicine elected student rectors and therefore had a significant role in university governance, the *collegia theologorum* were, at best, examining boards run by mendicant friars, in which secular clergy had no part.[10]

Copernicus and his uncle followed a curricular path that involved no formal degree in theology. Indeed, their education was no different from that of many Polish students in the fifteenth century who attended the Italian universities to pursue legal or medical studies, although there was ample precedent for Krakow professors' having studied astronomy at Bologna.[11] Copernicus's uncle, Lukas Watzenrode (1447–1512), had himself studied canon law at Bologna between 1469 and 1473, eventually receiving a doctorate in that subject, and had then proceeded to amass a large number of wealth-yielding prebends and sinecures.[12] Copernicus, however, showed no inclination to follow the same ambitious path even after he returned to Poland. For the rest of his life, he enjoyed the modest economic security of a canonry and an absentee teaching post or scholastry at Vratislavia (Wrocław) arranged by his uncle.

Soon after coming to the Bologna *studium*, however, Copernicus became associated with the senior master of astronomy in a capacity that presupposed astronomical competences in the sphere and theorics acquired during his liberal arts training at Krakow. The most reliable statement that we possess occurs in Rheticus's *Narratio Prima*, written during 1539 while he was staying with Copernicus in Frombork (Frauenberg). "My teacher made observations with the utmost care at Bologna, where he was not so much the pupil as the assistant and witness of the learned Dominicus Maria."[13] Precisely because it was made in passing, Rheticus's allusion should be taken very seriously. If Copernicus was an "assistant and witness" rather than a "pupil" of Novara, then his primary relationship to the latter was not mainly an academic association in which texts were read and commented on but rather some kind of joint practice. That association is the main subject of this chapter.

BOLOGNA AND THE "HORRIBLE WARS OF ITALY"

The city of Bologna at the end of the fifteenth century was not a sleepy university town. Copernicus arrived at a moment of extraordinary political crisis for the delicately balanced Italian city-state system, of which the Church territories were a significant part. The city had a mixed government, ruled both by the Bentivoglio family and by the pope through his legate. The Bentivoglio, like the rulers of other Italian city-states, such as Ferrara, Mantua, or Urbino, were a family of mercenary captain-princes or *condotierre*.[14]

Only Giovanni I (d. 1402) had ruled legally. Sante Bentivoglio (d. 1463) ruled de facto through a small group of powerful patrician families known as the Sedici Riformatori dello Stato di Libertà or Regimento. This senatorial body, however, had been forced to share power with papal suzerainty.[15] It was Giovanni II (1443–1508), the last of the Bentivoglio despots, who maintained for most of his reign a workable balance between the commune's republican institutions—which he controlled—and papal interests, represented by its legate on the Sedici. This body met every day in the Palazzo del Commune, with the legate and Giovanni present. Its broad functions included appointing and communicating with diplomatic envoys, hiring mercenaries, sentencing exiles, balancing the budget, and regulating all matters associated with morals, public health, commerce, building, and food supply. It also had the task of wooing professors to the university from other institutions, making appointments, and establishing salaries.[16] The ruler himself, moreover, might intervene to settle intramural political disputes at the university, rule on the granting of free degrees, or play a role in bringing scholars to Bologna from other places.

Just prior to Copernicus's arrival, these arrangements were already under the shadow of threatening clouds. In September 1494, the French Valois king, Charles VIII (r. 1483–98), passed into Italy, marching down the western coastal region at the head of a massive, tightly unified army of thirty thousand men. It was the largest such military force to appear in centuries, and it assisted in promoting Charles's imperial, messianic image. Within a few months he had taken Milan and Pisa, and when he entered Florence, the reigning Medici family fled. By January 1495, Charles was advancing on Rome; by late February, he was in Naples. These French successes allowed other Italian states to take advantage of the imbalance of power. Venice, already more powerful than the other states, profited the most. It was involved in the formation of several military coalitions. The first of these leagues joined the duke of Milan, the Venetian Republic, and pope Alexander VI, together with the great foreign potentates Ferdinand, king of Aragon and Maximilian I, the Hapsburg emperor. Reinforced by Spanish and Venetian naval power, this so-called Holy League forced Charles to retreat north to Milan, where he suddenly died of a stroke in 1498. The alliance broke up as soon as the king retreated, showing how weak and opportune were the ties, and Venice now joined the new French king, Louis XII (r. 1498–1515), in the same year.

A few years after Copernicus left Bologna, the governo misto collapsed. Pope Julius II (1503–13) exploited the French presence to undermine the independent states of Romagna and Ancona and to assert papal hegemony. Bologna's mixed governance and Bentivoglio despotic rule came under intense pressure.[17] In October 1506, Julius exercised a prerogative already established by the medieval popes. He issued a bull putting Bologna under interdict, an action designed to foment popular opposition by forbidding priests to administer the sacraments. The senate protested, but as the interdict took effect, the churches began to shut down, and the religious orders departed. The handwriting was on the wall. The French now entered Bolognese territory at Castelfranco. Giovanni Bentivoglio was isolated; he received no support from either the dukes of Ferrara or the Florentines. Julius II told Niccolò Machiavelli that Giovanni Bentivoglio could either surrender directly into papal hands or leave the city. When the French king offered him asylum in Milan, he accepted. On the morning of November 2, the Bentivoglio family and their supporters gathered in the main piazza and, with some five hundred horses (and, as a chronicler wrote, amid much weeping and sighing), rode out of Bologna, with safe passage through the French lines.[18]

On November 11, 1506, the pope made his triumphal entry. The Bentivoglio palace was razed, and the pope immediately began construction of a new fortress, for which Michelangelo was called upon to make a large bronze statue of Julius. The effect on the university was almost as severe as on the palace. The university closed for two months. Many students left. A substantial number of rhetoricians and poets, whose fate and status had been tied to the Bentivoglio court, left for other universities.[19] Shortly after the invasion, a German student in Bologna wrote to a friend: "In Bologna I myself saw pestilence, an earthquake, high food prices, and every kind of distressing condition; and fortune spared me so that I might also see war, internal dissension, and three changes of government of Bologna in as many days. I hope that all of these things will at some time be of no little use to me."[20]

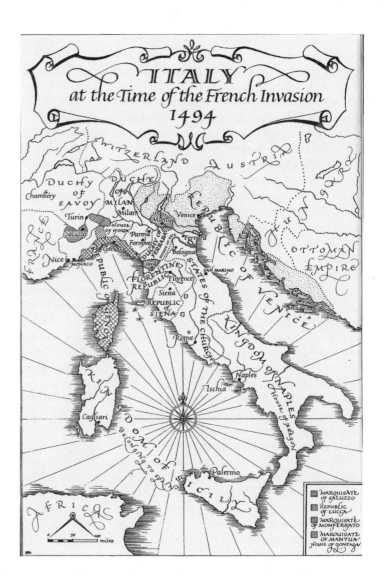

25. Italy at the time of the French invasion, 1494. From Guicciardini 1969. Courtesy of Princeton University Press.

Copernicus had missed the worst of these events. A humanistic culture centered on the Bentivoglio palace flourished throughout his stay in Bologna. But he would certainly have been aware of the threat of disorder during the entire period of his studies. The time he spent in Bologna was a time of considerable troop movements in the duchies of Milan and Ferrara and especially in the Republic of Florence. It was also a period when the Bentivoglio were engaged in almost constant secret negotiations with the French, and Girolamo Savonarola was burned at the stake in Florence. The year 1500 was a jubilee or holy year, and Rome had throngs of visitors who came to buy indulgences, some of the proceeds from which were used to finance future papal military expeditions.[21] Copernicus left Bologna in the spring of 1500 and, according to Rheticus, appeared at Rome as a "public lecturer on mathematical subjects before a large audience of students and also to a circle of great men and to craftsmen skilled in this kind of learning."[22] This is, unfortunately, all that we know of his visit. When Copernicus returned to pursue medical studies at Padua between 1501 and 1503, the French were still consolidating their control of the north and the Spaniards claiming the south, and the situation in Bologna was already worsening. The final collapse occurred only after he had returned to Varmia and had finally obtained a doctoral degree in canon law at the University of Ferrara in May 1503.

　　COPERNICUS'S SPACE OF POSSIBILITIES

The long-term consequences of the developments that were just beginning during Copernicus's student days were profound. Although class, family, and status rivalries already characterized the medieval Italian cities, within a few short years the French invasions significantly changed the balance of power not only in Bologna but in Italy and in Europe at large.[23] The real victor in the thirty-six-year "Italian wars" would be the Holy Roman Empire, which emerged in 1530 under Charles V (1519–56) with a dynastic union with Aragon and vastly larger resources of soldiers, equipment, and money than the French had.[24]

Francesco Guicciardini (1483–1540), the skilled Florentine diplomat, vividly captured these changes in his famed *Storia d'Italia*, composed in the late 1530s. The French passage into Italy, he said, "not only gave rise to changes of dominions, subversions of kingdoms, desolation of countries, destruction of cities and the cruelest massacres, but also [to] new fashions, new customs, new and bloody ways of waging warfare, and diseases [notably syphilis] which had been unknown up to that time. Furthermore, his [Charles VIII's] incursion introduced so much disorder into Italian ways of governing and maintaining harmony, that we have never since been able to re-establish order, thus opening the possibility to other foreign nations and barbarous armies to trample upon our institutions and miserably oppress us."[25]

Guicciardini attributed the great disorder that Charles wrought to his monstrously unnatural body and personal character: "Charles, from boyhood on, was of very feeble constitution and unhealthy body, short in stature, very ugly (aside from the vigor and dignity of his eyes), and his limbs so ill-proportioned that he seemed more like a monster than a man. Not only was he without any learning and skill but he hardly knew the letters of the alphabet; a mind yearning greedily to rule but capable of doing anything but that, since he was always surrounded by courtiers over whom he maintained neither majesty nor authority."[26] This trope of disorder served Guicciardini's purposes as a historian by exposing the irony of Italian weakness at the hands of an unworthy conqueror.[27] It was also a great exaggeration, as Charles was quite skilled at both war and diplomacy.[28] Yet the image of order and disorder (eventually deployed by Copernicus as a resource in his proof of a new planetary order) was also significant for enhancing the prognosticators' legiti-

macy. Writing retrospectively, Guicciardini said that the prognosticators' correct predictions of the invasion agreed with the unnatural, monstrous effects associated with the disorderliness of the king's body:

> The very heavens and mankind concurred in predicting the future woes of Italy. For those whose profession it is to foretell the future by means of skilled knowledge or divine inspiration [*o per scientia, o per afflato divino*][29] unanimously affirmed that more frequent and greater changes were in store, and stranger and more horrible events were about to occur than had been seen in any part of the world for many centuries. With no less terror did rumors spread everywhere that in various parts of Italy there had appeared things alien to the usual course of nature and the heavens. In Puglia one night, three suns appeared in the sky, surrounded by clouds, and with frightful thunder and lightning. In the territory of Arezzo an infinite number of armed men on enormous horses were seen for many days passing through the air with a terrible clamor of drums and trumpets. In many places in Italy the sacred images and statues had openly sweated. Everywhere monsters of men and other animals were being born. Many other things beyond the usual course of nature had happened in various sections: whence the people were filled with unbelievable dread, frightened as they already were by the fame of French power and by the ferocity of that nation which had already marched across all of Italy. . . . People were only surprised that amid so many prodigies there did not appear a comet which the ancients reputed to be an unfailing messenger of the mutation of kingdoms and states.[30]

Guicciardini's representation of the invasions made use of and elevated the importance of the prognosticators and their profession. But in addition, Guicciardini, narrating his account many years after the events, took for granted a distinction between two classes of prognosticators: those who based their forecasts on learned disciplines (*per scientia*) and those who claimed to foretell the future by means of some special prophetic gift (*per afflato divino*). Domenico Maria Novara's forecasts belonged to the first genre. Guicciardini effortlessly appropriated the images and explanatory resources of the second genre, the popular verse prophecies that made forecasts without any quantitative apparatus and which were profusely available at the time of the initial invasions.[31]

The heavenly forecasting literature functioned as a powerful resource of warning and consolation. As opposed to the promiscuous prophecies often read aloud in the streets, the learned genre advocated engaging in or avoiding certain kinds of actions based on a technical assessment of the convergence of celestial influences at specific moments. It was a form of advice literature that invoked natural causes and effects; because it was temporally grounded, it complemented the genres of so-called mirrors, or advice to princes and advice to courtiers. Niccolò Machiavelli's *The Prince* (1532) and Baldassare Castiglione's *The Book of the Courtier* (*Il cortegiano*, 1528) are only the most famous and influential examples of such aphoristic advice-giving practices.[32]

If the prognosticators profited from the anxiety and uncertainty created by the invasions, they were not without their own difficulties. By the last decade of the fifteenth century, the many producers of annual prognostications confronted a dangerously changing situation that threatened to undermine the authority of their forecasts. Paul of Middelburg, physician and astrologer to the duke of Urbino, already felt it necessary in 1492 to defend the legitimacy of "the astrologer's office" against the "superstitious" prophecies of Emperor Maximilian's astrologer, Johannes Lichtenberger (see chapter 2). His fears were not baseless. The verse prophets readily used carnivalesque images to mock openly the learned astrological culture by transforming their instruments into common kitchenware: "These astrolabes of yours are frying pans, your spheres are juggling balls, the quadrant is a pot, a jar; your tables are a [dining] tables set, where you put good things to eat. *Cuius, cuia, coioni,* you are part prophet and part diviner when you have drunk well of wine. Go with your Almanac into the kitchens, to the stoves in the back alley, where there is always a flood of grease and fat."[33] Apart from this sort of popular belittling satire, which drew liberally from the resources of carnival literature, Cicero's *On Divination* provided the learned source that was used most widely in resisting astrologers' claims.[34]

At the end of the fifteenth century, however, a new assault on astrology appeared that coincided with the ever-burgeoning literature of astrological theory and prognostication. In July 1496, Bene-

dictus Hectoris (Benedetto Ettore Faelli), one of the five leading Bolognese publishers of the late fifteenth century,[35] published the crabbed, voluminous, unfinished manuscript of Giovanni Pico della Mirandola's disputations on astrology. It was Pico's nephew, Gian Francesco Pico, who had assembled the manuscript for publication soon after the untimely death of his uncle on November 17, 1494, the very same day on which Charles VIII entered Florence. According to Gian Francesco Pico, his uncle's aim was to safeguard the authority of biblical prophecy and God's preeminent agency in ruling the cosmos apart from the heavens. Whatever other personal motives may have driven him, Pico took aim precisely at the vast corpus of ancient Arabic and Jewish conjunctionist astrology on which many of the prognosticators relied to make their alarmist predictions. Hectoris's reputation as a publisher of the celebrated Pico was quickly established, because he now added the work against the astrologers to the collection of Pico's other works and letters. As further evidence of his prestige, soon thereafter, in 1498, a pirated edition of Pico-Hectoris was issued by the Venetian publisher Bernardino Vitali—this in spite of a formal testimonial granted by the duke of Milan, Ludovico Sforza, giving Hectoris exclusive privileges to publish Pico's works.

Meanwhile, in 1497, the opponents of astrology opened a second front: Girolamo Savonarola, the fiery, charismatic Dominican preacher, reconstituted and radically abridged the learned, philosophical diatribes against astrological authority by his close friend Pico della Mirandola. Savonarola came from Ferrara, where Pico had studied briefly at the university (1479–80) and where the Este dukes sustained and promoted a rich astrological culture at both court and university.[36] Savonarola had encouraged Pico's attack on the astrologers, and he now produced a vernacular work that condensed Pico's arguments for a popular audience. He directed this work to all levels of Florentine society, encouraging them to "castigate and punish" greedy astrologers, satirizing the astrologers' principles, and affirming the superior power of religious prophecy. As for the zodiac, wrote Savonarola, it is a human construction: "There is no man who, in such a multitude of stars, coupling them in various manners, cannot imagine whatever figures he wants. . . . Just as men have imagined animal figures, they could have imagined houses, or castles, or trees, or

IOAN·PICVS·MIRANDVLA

26. Portrait of Giovanni Pico della Mirandola by Christofano dell'Altissimo. Date unknown. Courtesy Soprintendenza Speciale per il Polo Museale Fiorentino. (For full-color version, see http://en.wikipedia.org/wiki/File:Pico1.jpg.)

other similar things . . . but to believe that God and nature have drawn in the sky lions, dragons, dogs, scorpions, vases, archers, and monsters is a ridiculous thing."[37] On May 23, 1498, a year after the appearance of Savonarola's volume, at the urging of the pope, its author was hanged and burned for preaching a reform of the Church, for prophesying that both Florence and the Church would be scourged and reformed, and for allegedly plotting in secret against the ruling council or Signoria.[38]

Just as the Savonarola affair was reaching its climax, on May 8, 1498, the publisher Gherardus de Haarlem put up for sale the trenchant *Book of Questions concerning Astrological Truth* and *Replies to Giovanni Pico's Disputations against the Astrologers* by Lucio Bellanti (d. 1499), who identified himself only as a "physicus et astrologus" from Siena. The denomination *physicus* reflects that of an author trained in scholastic natural

philosophy.[39] Bellanti modestly declared in his "Letter to the Reader" that he had not written either of these two books "out of enmity or anger" for the "venerated Pico," although his tone was biting, if not sarcastic.[40] In the dedication, Bellanti cast his justification for astrology on the grounds of its necessity for medicine (*medicina scientia*), especially the importance to the physician of "practical astrology (which authors call judiciary)," but also on the basis of the higher authority of the theologians John Duns Scotus and the (frequently cited) "divine Thomas [Aquinas]." At this moment, the appeal to Rome and Catholic orthodoxy was as necessary for a proponent as for an opponent. Conscious of longstanding Church opposition to prediction based on the stars, Bellanti had to navigate a careful course between a good astrology of spiritual forces and a bad one controlled by demons.[41] Astrology had nothing to add to revealed knowledge, but through "nature's light" it could still be "useful" in "avoiding errors" and "coming to know God." His chapter-by-chapter rebuttal of Pico's arguments, which makes up the second work, quickly became a rich resource for the astrological prognosticators. One testimony to its continuing appeal is the fact that at least two further editions (each of two issues) appeared in the sixteenth century.[42]

Among Bellanti's considerable range of topics, organized as for-and-against *quaestiones* (always followed by the author's *resolutio*), is the question of where to locate astrology in the division of knowledge: "Whether astrology can be prognosticatory"; "Whether all astrology is useful for knowing the divine and for avoiding errors"; "Whether heaven is the universal or particular cause of events in the lower world"; "Whether astrology is *theorica* or *practica*." To the last question, Bellanti responded that astrology is partly practical, partly theorical. For if theoric is a kind of knowledge based on actions that have no object other than themselves ("intrinsic" or "speculative"), as for example, the part of astrology that concerns the planetary distances and motions themselves (*theorica planetarum*) or the nature of planetary influence, and if practice is a kind of knowledge that considers a mode of operation (*modus operandi*) that has a transitive effect (*actio transiens*), such as promoting the good and minimizing the bad (for example, medicine, military advice, ethics, and theology), then astrology is both practical and theoretical.

Besides its critique of Pico, Bellanti's treatment went beyond the *Tetrabiblos* to include questions dealing with the physical foundations of theoretical astronomy. Among these were "Whether the heaven is a fluid (*fluxibilis*) or solid substance" and "Whether there are eight, nine, or ten moving spheres." Yet, for all of his detailed arguments, Bellanti was often cited in the sixteenth century for claiming that astrologers had correctly predicted Pico's death at the age of thirty-three—a squib that, although false, became a convenient way to disparage Pico's objections.[43]

Bellanti's negative representation of Pico leaves open a question (to which I shall only allude) of how Pico's position evolved from an earlier espousal of astrology to a condemnation of it. D. P. Walker has argued that Pico, like everyone else at this time (including Pico's friend Marsilio Ficino), distinguished between a good astrology that maintained human and divine free will and a bad astrology that denied it.[44] The good kind invoked the Thomist formula *Astra inclinant, sed non necessitant* (The stars dispose but do not determine). Walker is probably right that Pico privately held the Thomist position. But one would be hard pressed to find this mitigated view maintained in the highly polemical *Disputations against Divinatory Astrology*. In the last two years of his life, Pico's position on astrology shifted dramatically from that of his earlier writings. And he made ample use of earlier critics, including Nicole Oresme's well-tempered fourteenth-century arguments.[45]

PICO AGAINST THE ASTROLOGERS

Pico's attack radically questioned astrology's theoretical and practical foundations and anticipated the broader revival of Ciceronian and Pyrrhonian skepticism in the sixteenth century, promoted by, among others, his nephew Gian Francesco Pico.[46] There is no need to rehearse here the entire battery of arguments and replies generated by the Piconian controversies, but it is important to appreciate the work's overall scope, tenor, and ferocity as well as aspects of Pico's critique that have been largely misunderstood or simply overlooked. The *Disputations* brought Pico's new objections together with earlier ones into a single massive synthesis. "Our Pico," wrote his nephew, "utterly burned up and reduced to ashes that unfortunate tree [of divinatory astrology]—from the root to the trunk, from the trunk to the branches, and from the branches to the leaves. He did this by means of his own incomparable natural talent and by means of the most scorching fire of the true philosophy and the true theology."[47]

In his table of chapter summaries, Pico said that all major authorities "damn" astrology, for it weakens religion, encourages superstition, promotes idolatry, makes people unhappy and miserable, and transforms free beings into slaves.[48] There is nothing weighty to be found in the books of the astrologers. In fact, "among their authors, there is no authority; among their reasons, nothing reasonable; in their experiences, nothing is established, nothing is constant, nothing true, nothing credible, nothing solid. There is only contradiction, falsehood, absurdity, and empty conceit so that you can scarcely believe what they write." In his own books, then, he claims that he will expose the "uselessness, falsehood, and ig-

COPERNICUS'S SPACE OF POSSIBILITIES

norance of the whole profession and the carelessness of the professors [of astrology] of our time. Likewise, why they sometimes predict the truth when the profession itself is false." And in the sixth book, astrology's main boundary designators will be destroyed: "The house, the sign, the aspect, the obsessions and combustions, the antisia, the retrogradations, the dragon's head and tail, the exaltations, trigons and triplicities, the faces, the termini, the dodecathemoria, the degrees, the parts, and the climacteric year."[49] In other words, none of astrology's categories are reliable. And furthermore, most damagingly, neither astronomers who calculate planetary positions nor astrologers who use these positions to establish the play of planetary influences on the lower terrestrial region agree among themselves about what is reliable. Pico constantly took the absence of consensus as an indication of the fallibility of both astronomy and astrology.

In natural philosophy it was otherwise for Pico; there, different authorities were made to reign harmoniously. Pico did not hesitate to bring into concordance the ancients (Aristotle, Plato, Plotinus) and their many commentators (Averroës, Avicenna, Aquinas). Material and efficient causes received preference. The zodiac, he contended, was a human construction. It was merely an arrangement of the stars that was useful for mathematicians. But, in themselves, the shapes of the constellations had no material essence, no independent capacity to induce effects. If the heavens did influence the terrestrial realm, then it was because the celestial contained the "most perfect of all natural bodies." Commensurate with heavenly perfection was the most perfect kind of motion (round or circular, *orbicularis*), and to the senses no quality was more perfectly perceptible than light. Through circular motion and light, the heavens acted as a "universal, efficient cause" to generate change. Here light worked physically as a vital principle and logically as a necessary condition—although not itself alive, it prepared or disposed all bodies capable of life to receive it. Like the Stoic *pneuma* and Ficino's *spiritus,* heat functioned as light's emissary, a mediating continuum between the soul and the body. According to Pico, heat came from light, as though a property: "A heat that is not fire nor even air, but rather a celestial heat . . . the most efficacious and most salutary, which penetrates, warms, and orders all things."[50]

In this theory of the heavens as the general condition of motion and life in the terrestrial world, Pico altogether ignored both Ptolemy's and the *Centiloquium*'s standard qualifications about predicting particulars. For Pico, these were mere lip service: the problem was much more serious. The astrologers could neither explain nor predict because change induced by the heavenly heat did not produce specific differences: "For who does not see that the heavens generate the horse with the horse and the lion with the lion and that there is no position of the stars under which the lion is not born of the lion, and the horse from the horse?"[51] But particulars were produced by different proximate (noncelestial) causes that varied as much as their effects. For this reason, the same cause (the heavens) could not explain the particularity of Aristotle's natural talent as a philosopher any more than it could explain divine miracles, such as the virgin birth of Christ.[52] There are glimmerings here of the central insight of the Duhem-Quine thesis: Pico's theory of celestial influence underdetermined the production of specific differences and thus became an important basis for his rejection of the astrologers' authority.

In books 9 and 10 of the *Disputations,* which have been either ignored or not well understood, Pico brought together a mixture of astronomical and astrological considerations that he took to be further evidence of the uncertainty that divided the astrologers.[53] For example, in book 9, chapter 7, he claimed that the astrologers were unsure how to divide up the houses. Should they use the method of Campanus of Novara or of Regiomontanus?[54] And, if they were unsure about the house divisions, how could they be sure exactly where the planets were in the houses? If they were unsure about the planets' locations, then how could they know the planets' influences?

And more. The astronomers—on whom the astrologers rely—did not agree on the length of the tropical year (the time between two vernal equinoxes, marked by the moment when the Sun crosses the equator northward). Hipparchus believed that the Sun completed its annual revolution in 365¼ days. Ptolemy, 285 years later, determined the extra fraction to be $\frac{1}{300}$ of a day. Al-Battani, 743 years after Ptolemy, found the Sun to be moving more slowly, because the fractional part was now $\frac{1}{106}$. Then Thabit ibn-Qurrah (d. 901) said that the length of the year is 365 days, 6 hours, 9 minutes, and (mistakenly) 12 de grees.[55] Some, like Ptolemy and King Alfonso, thought all years to be of equal length, whereas others, like the medievals al-Zarqali (Arzachel), Henry Bate, and Isaac Israeli, believed them to be unequal. They attributed this variation to the nonuniform motions of the eighth sphere.

To add insult to injury, Abraham ibn Ezra, in the sixth book of his *De revolutionibus nativitatum*, doubted that horoscopes could be cast with accuracy because no instrument could determine with sufficient precision the moment when the Sun enters the first degree of Aries (the conventional starting point for measuring longitude).[56] Likewise, Abraham Judaeus narrated a sobering episode in his book on the composition of astronomical tables: "Two astrolabes were constructed with the greatest care and of such size that the diameter of each measured nine palms. Then the constructors of the instruments, the two brothers Bersechit, observed together the Sun's altitude and its entry into Aries. The two instruments, however, did not give the same results but differed between them by two minutes."[57] "The instruments are imprecise" was not quite an established medieval trope, but it was certainly a well-established worry well before Thomas Hobbes used it against Robert Boyle's air pump in the seventeenth century.[58]

Pico now pointed to the ripple of implications that followed from an error of even one degree. The planet found to be at the extreme limit of one zodiacal sign would move to the next, so that the Moon, happy in Taurus, would be unhappy in Gemini. Further, a masculine quality would become feminine, a lucid one opaque, an opaque one shadowy, and so forth. Pico asked rhetorically: "In whatever way the planet's position varies— whether by one degree or one minute—will not the stellar virtue or *influxus* be profoundly changed by such an error?"[59] In fact, Pico argued, the best mathematicians acknowledged the uncertainties of their numbers. One of his examples was Paul of Middelburg. Pico described him as "a famous *mathematicus*, who now lives with Guido, Duke of Urbino, the most cultivated prince," and who denied that the tables of planetary latitudes were trustworthy (*fideles*).[60] No trust in the numbers, no trust in the prediction of effects.

In book 10, chapter 4, Pico came to a matter of crucial interest here: the order of the planets and the assignment of elemental qualities. The occasion for this discussion concerned five different

forms of arguments by astrologers that he found to be weak and ineffective. One of these he called the argument "from the order of the numbers;" it concerned the arbitrariness of the "natural affinity" between planetary order and the order of the elements. Pico did not ascribe this position to any particular authorities. If Saturn, for example, is the first in the order of the planets, it should have an affinity with fire, the first in the order of the elements. "And yet," said Pico, "according to the astrologers, what could be more different from Saturn than fire?"[61] Similarly, Mars is the third planet and water the third element, yet "does not Mars differ from water as much as water from fire?" The same was true of the signs: Aries is the first sign, yet the astrologers deny that Saturn has any correspondence with it. The problem was not so much the order in which the planets should be numbered—whether beginning with Saturn and counting down to the Moon or beginning with the Moon and counting up to Saturn—but that "the place and order of the intermediate planets is entirely uncertain."[62]

Pico, writing just before the appearance of Regiomontanus's *Epitome*, was nevertheless well aware of the fact that both the ancients and the moderns disagreed about the order of the planets with respect to the Sun. He used language anticipating that of Regiomontanus, calling it an "ancient controversy" whether the Sun comes right after the Moon or whether the Sun is in the middle of the other planets. The Egyptians thought that the Sun was close to the Moon; the Chaldeans placed it among the planets, as did Ptolemy and the moderns. Others dissented: Geber, a most intelligent *mathematicus*, and Theon, a Greek commentator on Ptolemy, as well as Plato and Aristotle, placed the Sun immediately above the Moon. Not a surprise, said Pico, for many arguments were borrowed and none could be thought certain, as calculation could not establish which planet was above another.

The same uncertainty held also for Venus and Mercury. According to Ptolemy, the Sun was placed aptly (*convenienter*) in the midst of the planets: Saturn, Jupiter, and Mars digress from it in longitude, while Venus and Mercury closely follow it. Pico dismissed this as a "frivolous and inconsistent conjecture," however, because the Moon, like the superior planets, also digresses from the Sun, and yet it is not thereby situated between those planets that digress and those that

COPERNICUS'S SPACE OF POSSIBILITIES

do not. An "equally weak" argument was that of al-Bitruji (Alpetragius), who believed that the inferior planets should move the fastest of all the planets. But then he placed the allegedly inferior planet Venus above the Sun and Mercury, thereby locating a supposedly fast-moving body where it ought to have been moving more slowly. Consequently, al-Bitruji adopted a view that no one accepted. Even the ancient view that placed the Sun close to the Moon was better than al-Bitruji's, because the Sun was thereby never lost to us by the interposition of Venus and Mercury. But perhaps there was another explanation why the Sun and Moon appeared to be contiguous while Mercury and Venus remained invisible: "Either Mercury is very small or Venus, being close to the Sun, is enveloped and cut off by its rays, whence it could not obstruct the path by which the rays descend toward us whereas the more distant Moon could do so. Thus, being very tenuous and not as earthly as the lunar thickness, they do not resist the rays; moreover, they have their own light which they send forth in the vicinity of the solar [light] They do not [so much] lose this light as change it and, as a result, an eclipse [of the Sun] is not observed."[63]

After criticizing these inadequate arrangements, Pico then cited a passage from Averroës' *Paraphrase on Ptolemy's Syntaxis* as an apparent exception to his argument that Mercury and Venus are usually lost in the Sun's light. This work was available only in Hebrew, but Pico was something of a Hebraist; he had studied Hebrew at Florence with Elia del Medigo, who was also quite familiar with Averroës' writings.[64] According to Pico, Averroës said that "he had once observed two dark spots on the Sun, and having made a calculation, he found that Mercury was opposite [or in line with] the Sun's rays."[65] Here Pico introduced a consideration that even Regiomontanus did not mention: the question of transits of the Sun. He then concluded from this passage, as well as on the authority of Moses the Egyptian and unnamed "others," that "the order of the Sun, Mercury, and Venus remains uncertain."[66]

A serious problem followed from this discussion: if there was disagreement on the order of the planets, then the principles governing the various associations of elemental qualities to the planets would be gravely undermined. Beholden to the superior science of astronomy, astrology could no longer be certain of its core association

of celestial causes and corresponding effects. Pico's questioning of Ptolemy's ordering of Mercury and Venus was itself not unprecedented—as we have seen, it was already in the *Almagest*—but the context was strikingly new. Now, for the first time, an uncertainty about planetary order was situated in the context of the assignment of qualities and powers to the individual planets. As a consequence, an uncertainty about the order would put the whole scheme of astrological influences at risk—including what young Copernicus had learned just recently from Albert of Brudzewo's commentary on Peurbach's *New Theorics of the Planets.*

DOMENICO MARIA NOVARA AND COPERNICUS IN THE BOLOGNA CULTURE OF PROGNOSTICATION

Copernicus must have encountered Pico's *Disputations* sometime after arriving in Bologna with his brother Andreas. In fact, he encountered more than a book: he entered a cultural space where, because of the frequent and uncertain disruptions provoked by the movements of troops, the prognosticators' warnings were valued more than ever. Copernicus moved into the house of Domenico Maria, known both by his city of ancestral origin (Novara) and by the city where he was born and studied (Ferrara). We have it on the reliable authority of Copernicus's disciple Rheticus that "he lived with Domenico Maria of Bologna, with whose ideas he was plainly acquainted and with whose observations he assisted."[67] Unfortunately, the house no longer exists today. A plaque at no. 65 via Galliera in the parish of San Giuseppe, erected in 1973, marks the spot where it is believed to have stood before the Risorgimento—"next to a bakery and along the public road"—and where subsequent urban renewal has produced an unassuming apartment building that now abuts, in an ironic postmodern juxtaposition, onto an international hotel and car park.[68] Exactly how long Copernicus lived at this location and how he found it is not known. There is some evidence that, either before or after his residence in the house of Novara, he also lived near the Porta Nova in the parish of San Salvatore, not far away. It may have been as a young law student seeking lodging that he first encountered Novara. Such a connection would explain why Rheticus took care to dissociate his teacher

27. Sala di Bologna, Vatican Apostolic Palace. Detail of mural map of Bologna. Encircled area indicates location of Domenico Maria Novara's residence, today via Galliera 65. Photo Vatican Museums.

from a strictly pedagogical relationship with the "learned Dominicus Maria."

But we have still another important clue to their relationship. The house in which Novara lived was owned by the notary Francesco Callegari. At the behest of Novara's beneficiaries, Callegari had prepared an inventory of Novara's possessions at the time of his death on September 5, 1504. In 1920 Lino Sighinolfi, the Bologna city librarian, published a small section of this inventory, bemoaning along the way the fact that the twenty-six titles in Novara's library had been omitted because the beneficiaries valued the furnishings more highly.[69] Callegari noted in a mar-

gin that the will had been drawn up just one month earlier by another notary, Ser Lorenzo de Benazzi. Novara, in fact, had probably lived for a time in Benazzi's house, because the inventory states: "He paid Lorenzo de Benazzi one hundred pounds for two years' rent of the house."[70]

The Benazzi, like many of the Bolognese notaries, were a noble family. Lorenzo's son, Giacomo (1471–1548), was one of the three masters of astronomy at the university from 1501 onward.[71] Giacomo announced in a prognostication published in 1502 that he was a pupil of Novara.[72] Lattanzio Benazzi (1499–1572), no doubt a younger relative of the family and possibly the son of Giacomo, held the astronomy professorship from 1537 to 1572 and issued numerous prognostications.[73]

Taken together, these details are highly suggestive of unsuspected cultural connections: at the time of Copernicus's stay in Bologna, the families of notaries and prognosticators were closely bound together. In fact, as the notarial archives reveal, most of the native Bologna prognosticators—Manfredi, Benazzi, Pietramellara, Scribanario, Vitali—came from prominent notarial families. The name Scribanario may derive from "someone who writes." Probably Novara was something of an exception to this rule. Novara's grandfather Bartolino (or Bartolomeo) Ploti di Novara was an engineer and builder of fortifications who designed the Castello di San Michele, the castle of the dukes of Ferrara.[74] At any rate, our evidence permits the speculation that the notarial profession—and perhaps even the profession of prognosticator—was passed down through generations.

The notaries were probably among the best-informed people in the city.[75] Notaries were concerned with securing and managing personal capital and its transmission between individuals and generations. They made up wills and household inventories. They wrote contracts and composed letters. They had contacts with people at all social levels. For example, they kept records of artisans' meetings, recorded oaths of office, and kept auditing records. They were also consummate deal makers. Furthermore, in Bologna, the Palace of the Notaries (Palazzo dei Notai) was located just across the main square from the palace where the senate held its weekly meetings. Some notaries held seats in that body. Lorenzo de Benazzi himself was a member of the Sedici in 1460.[76] The Church, the senate, and various university structures were no more than two to three minutes' walking distance apart. Effectively, the relatively small geographical and social scale of the city at the time of Copernicus's student days made much more probable all kinds of links among people that might not occur quite so readily in the city at its current size.

Although it is difficult to document precisely what the prognosticators might have learned from the notaries, the prognosticators were aware that an effective forecast required not only knowledge of how and when the planetary causes operated but also how their effects might be received and distributed in the always variable and complex terrestrial world. Put otherwise, because the same celestial cause could produce different effects in different places, effective interpreters of celestial forces and effects required good access to local knowledge.[77] The credibility of their predictions required them to be politically and socially well-informed. We have seen evidence of this kind of privileged knowledge in Avogario's candid letters to the duke of Ferrara. Behind the generalized language of Novara's 1492 prognostication we may detect a muffled echo:

> Wishing to prognosticate the nature of the celestial effects, it is necessary not only to contemplate the forces of the celestial bodies, but it is also necessary to understand how the subjects' passive dispositions adapt to the natural agents, because the same celestial nature produces different effects in different places. Therefore, ought not the rational prognosticator to have also an understanding of that part of philosophy which concerns the [diverse] locations of the entire world? That is to say, of all terrestrial and maritime things and the diversity of human customs, animals, plants, and also fruits which have such diversity throughout the universe.[78]

Concerned as the prognosticators were with the dispositions and future fate of members of different social groups, notarial contacts might have given them information about the immediate distribution of social and political capital and hence improved the quality of their local predictions. Little is known about astrological considerations in notarial practice, although Ottavia Niccoli has located direct evidence that notaries in Piacenza, Cesena, and Udine were among the transcribers and circulators of nonastrological prophecies.[79]

When Copernicus lived with Domenico Maria Novara and acted as his "assistant and witness," we can speculate that he somehow participated in this network of prognosticators and notaries. That he never issued any prognostications of his own at Bologna indicates one meaning of *assistant*. Copernicus, for all his well-known talent, was still something of an outsider. After all, he was formally obligated to the Varmia chapter. His uncle had already lined up a church sinecure for him, which he accepted at Bologna by proxy and through a notary. Finally, he was neither a native of Bologna nor, like Novara, a member of the faculty of arts and medicine. He came from neither a noble family nor a family of notaries nor a family of prognosticators. He had no degree of any kind, let alone a doctorate, with which to announce his *officium* or public identity: thus he could not (yet) invoke the social authority of the university as a warrant for his credibility. He could not be a full-fledged academic prognosticator, but he could still assist in the production of the forecast and thereby contribute to the practice of the prognosticators.

The formal authority to prognosticate at Bologna fell directly under academic jurisdiction. Since 1404, it had been a requirement of the university statutes that the lecturer on spherics and theorics issue an annual prognostication. The Bologna statutes specified a set of precise obligations that the astrologer was supposed to meet. He must provide free to the university an annual judgment (*judicium*) and, in addition, he had the obligation to dispute (*disputare*).

> Let the doctor elected to the salary of astrologer dispute two questions (*questiones*) in astrology and determine these not later than eight days before the said day of the disputation. And also, let him dispute about any subject (*quodlibet*) in astrology at least once. . . . And let the said *questiones* and the said *quodlibet* be written down and sent to the Stationer within fifteen days after the "determinations," and let it be done in good letters and on good, unshaved sheets of parchment. . . . And let the said *questiones* remain continuously at the Stationer so that copies may be made from them.[80]

The statutes make clear that these scribal prognostications enjoyed only limited distribution in Bologna because access to the documents was restricted to the university's stationer, where they would have been copied. The earliest Bologna prognostication that I have been able to locate is that of Johannes Paulus de Fundis (who lectured there from 1428 to 1473), written on February 7, 1435. It is evidently a copy, as it has been written not on unshaved parchment but on relatively thin, loosely sewn paper folio sheets (21.5 by 31 cm). It is also considerably longer than the later printed prognostications, which were typically eight folios folded in quarto.[81]

The arrival of print, as the previous two chapters show, changed the possibilities of readership. At Bologna, active publication of the yearly prognostications began at least as early as 1475 with Girolamo Manfredi (1455–93). Manfredi's

forecasts appeared either from his own press or at his own expense and mostly in the vernacular.[82] They were generally dedicated to the ruling family, the Bentivoglio, but in 1489, he addressed the legate and the senate as well. Therefore, by 1482, when Novara received his appointment, publication of the annual judgment had replaced the parchment requirements of the statutes as something of a stable practice.

Novara's surviving prognostications were all printed and dedicated to the Bentivoglio. They have consistently the same rhetorical structure. Typically, they begin with a general exordium offering some display of learning or general argument about heavenly influence. This brief section, amounting to one or two pages, might have been read publicly at the university. Perhaps it represented the condensed summary of a disputation. Whether the judgment that followed was delivered orally is also difficult to determine. It was commonly divided into subsections that treated major planetary conjunctions and eclipses, the time of the Sun's entry into Aries, the prospects for war and disease, and the specific fates of Bologna, Venice, Florence, Pisa, the Turks, and various rulers of other domains. Remarkably, I have found not one reference to the Bible in any of the Bologna prognostications—an important difference, so far as I can tell, between Italy and the transalpine north. The most commonly used authorities were Greek, Jewish, and Arabic: Messahalah, Albumasar, Ptolemy (*Tetrabiblos; Centiloquium*), Aristotle, and Haly Abenrodan's commentary on the *Tetrabiblos*. At the end of each prognostication appeared a table of new and full moons—effectively the basic materials of an almanac, the elements of eclipse prediction. Occasionally, as for the year 1500, Novara included a list of lucky and unlucky days.[83]

Domenico Maria Novara's prognostications, issued between 1483 and 1504, were composed in Latin, although occasionally the same forecast appeared in Italian as well. The Latin versions were, in all probability, the original texts, prepared by the prognosticator himself, whereas the vernacular version—which often dropped some technical references and expressions—might have been translated by someone else afterward, perhaps an assistant.[84] Leading Bologna publishers produced all but the forecasts for 1484 and 1497: Ugo Ruggieri (1492), Caligola Bazilieri (1496), Giustiniano da Rubiera (1500), and Benedetto di Ettore Faelli

(1501–4).[85] The Bologna prognosticators' annual predictions appeared simultaneously with those of other members of the university: Manfredi until 1493, and then his successors, Antonio Arquato (1493–94), Francesco de Papia (1493–97), Scipio de Mantua (1484–98), and Giacomo Pietramellara (1496–1536); Novara's pupil Giacomo Benazzi (1500–1528), Luca Gaurico (1506–7), Marco Scribanario (1513–30), Ludovico de Vitali (1504–54), Lattanzio de Benazzi (1537–72), and so forth.[86] Interestingly, Scribanario issued his judgments mostly in Italian and addressed them all to the papal legate, Cardinal Ascanio Sforza. Pietramellara made separate forecasts for the pope, the emperor, and the kings of France and Spain.[87] These important differences suggest some kind of political division of intended audiences among the university's prognosticators. Perhaps the existence of different predictions for the same year assisted in protecting their general credibility: someone was bound to hit the mark. More likely, however, they simply reflected the division of the *governo misto*: one set was directed to the secular princes, the other to the church.

The themes of Novara's individual exordia reveal something of the changing concerns and strategies of the forecasts. Prognosticators might choose a scholarly text through which to display their learning or begin with an aphorism drawn from the *Centiloquium*. In the 1489 prognostication, for example, Novara delivered a long commentary on a passage from Ptolemy's *Geography*, in which he conjectured that the Earth's pole had shifted slightly since the time of Ptolemy. The entire passage was quoted by Giovanni Antonio Magini in 1585 from what he called "a certain old prophecy made at Bologna in the year 1489" (*quodam antiquo vaticinio anni 1489 Bononiae*) that he claimed was difficult to find even at the time of his own writing![88] In the introduction to his 1496 forecast, Novara used an aphorism about love and hate from the *Centiloquium* as an occasion for a didactic lecture on the moral value of seeking out the mean between extremes.

At least from 1499 onward, a more combative and defensive tone is evident.[89] Signs of Pico's critique and the French invasion break into the preambulatory material. The 1499 prognostication is especially important because it appeared at exactly the time that Copernicus was in Bologna and because it has never been translated or analyzed.

Novara began the Latin prognostication by contrasting the wise and the ignorant. Constructing an epideictic rhetoric of praise and blame, he avoided mentioning specific names or parties, yet his account was undoubtedly aimed at an ongoing debate within the university faculty about the scope and legitimacy of what he called the "science of the stars" (*scientia astrorum*): How should such knowledge be ranked among the liberal arts? Of what practical or civil use is it? Who is competent to discourse about it? His preliminary answer was that there are few really wise men who understand the science of the stars and many who criticize and clumsily imitate its language. A correct use of (technical) language separates true astronomers from their imitators.[90]

In the second, somewhat less rhetorical part of the prefatory section, Novara emphatically developed the position that astronomers do not make claims of necessity in human affairs. Here, he followed the customary two-cell Ptolemaic division between the mathematical-astronomical part of the prognostication that he regarded as certain (and whose subject matter and methods are described in Ptolemy's *Almagest*) and the judgmental part that was not as firm and strong.[91] On this occasion, he did not make explicit use of the *Centiloquium*, which said that "the 'judgment' is in the middle between the necessary and the possible."[92] Instead, Novara shaped his appeal explicitly on the authority of Galen, perhaps because he was directing his remarks to critics of astrological medicine in Bologna: "Galen distinguishes in several places two kinds of knowledge in medical science. One kind is certain; but the other is cunningly conjectural, acceptable only insofar as its judgment is in the vicinity of the truth. Now, of these two kinds of knowledge, the astronomer's is the first—both certain and scientific[93]—because he deduces natural inclinations in human affairs from celestial causes. But again, the other [kind of knowledge] is conjectural because what it foretells by considering the natural inclination can vary with the free will."[94]

In this passage, Novara seems to be invoking a version of the conventional formula that the stars causally determine human dispositions but that knowledge of these dispositions does not guarantee that the astronomer can predict exactly how a human agent or a nation will behave in a particular circumstance. Moreover, Novara did not use this occasion to deepen astronomy's purchase on demonstrative knowledge. For example, he did not claim that such knowledge conforms to Aristotle's stringent ideal of the scientific demonstration, in which the major premises are "true, primary, immediate, better known than, prior to, and causative of the conclusion."[95] Nor did he take up any of Pico's pointed astronomical criticisms about the uncertainty of the Sun's motion or the order of the planets. Because this was the introduction to a prognostication, its aim was limited, and the focus was appropriately on the conjectural and conditional basis of the astronomer's forecast. Novara revealed, in the process, how he thought about the general logical structure of predictions in human affairs.

His main example shows that military events were a central preoccupation, in this case the possible movements of the French armies. "The astronomer predicts that, on account of the celestial influence, this year the French are naturally inclined to make war against the Italians. But precisely here, [the astronomer] is not content with this as a proper and certain prediction and so places it second, as a *conjectural inquiry*." Now, if the problem was "conjectural," then what guesses could be made about French "inclinations" in 1499? Novara did not say directly, but instead offered a decidedly philosophical disquisition about different kinds of inclinations:

> And he [the astronomer] says: "He who is most naturally inclined to pursue some goal pursues it by the greatest sensual appetite and natural inclination. For it appears that all desire the good." However, to us, it appears better [to say] that "the natural disposition inclines us," since, especially in choosing one of two possibilities, the mind does not [always] choose the more certain [even if all desire the good]. And so it appears to be more evident that it is [the mind] whose natural disposition is pleased. Hence, the astronomer says thus: . . . "If the French attack Italy this year, then it will happen for the most part that the effect follows because the natural inclination was strong." The astrologer thus banishes necessity in human actions. Indeed, [the astrologer] always infers the necessary as a conditional, just as if I were to say that "the French will surely fight against the Italians if they follow their natural inclinations." You, magnanimous prince, will not be surprised, however, if we say that the astronomer's judgments are of this kind. If the naysayers speak abusively, they do what they ought to do, since they do not know how to speak well.[96]

This passage appears to be more than a conventional defense of astrology. The science of the stars (like that of genetic predispositions today) was believed to produce knowledge of human dispositions and inclinations, and this knowledge was conditional on secure knowledge about the positions of the stars. It also allowed room for human volitional behavior. Once the astrologer determined which celestial configurations governed particular classes of individuals, then statements of possible effects of a particular character might be specified and annual prognostications drawn up. Such a pointed justification of the status of astrological knowledge is not to be found in any previous forecast by Novara of which I am aware.[97] Its timing, some three years after the publication of the *Disputations against Divinatory Astrology* and just a few months after the appearance of Lucio Bellanti's book, strongly suggests that it represents Domenico Maria Novara's reply to Pico della Mirandola.

Novara's defense of the logical status of prognostication shows that the fight for the status of prognostication was a two-front war against both Piconian skepticism and other claimants to forecasting competence. This is clear again from a prognostication by Giacomo Benazzi, a confirmed student of Novara. Benazzi used his very first prognostication to set up a classification of opponents. He spelled out the distribution of authority between the astronomical "school" of his teacher Novara and several other genera of forecast. First, "men inspired by divine will or revelation" make "prophecies;" such works are based on faith alone. Second, physicians "imperfectly" prognosticate critical days and the course of disease from individual bodily signs. Third, in a likely reference to Marsilio Ficino, he remarked that "therapeutic physicians" speak profusely (but without evident success) about "melancholics." Benazzi used these barely articulated genera to dismiss what would appear to be a certain faction in the medical faculty. Against this group, he praised the "astronomers" and their *methodus pronosticandi*—obviously the group with which he and his teacher were identified. The astronomers' method is "the more perfect" because it moves deductively and swiftly to "harmonic effects" from celestial motions and virtues. Also, in a barely veiled reference to Pico, Benazzi stated that astronomy "does not reckon to be an inquiry of the demons, as certain people unwittingly bear witness to in their publications." Astronomy provides us with powerful foreknowledge of good influences that we might follow and bad ones that we might avoid. "As Ptolemy says, in the sagacious sayings of the *Almagest: Sapiens dominabitur astris* [The wise man shall rule the stars]. For astronomy shall be able to protect us from infinite misfortunes that can occur while managing and governing commonwealths. And, in addition, says Ptolemy, the astronomer shall protect against many evils because the extraordinary and special man knows very well beforehand the nature of these things from the stars."[98] *Sapiens dominabitur astris* was a topos among prognosticators, but the phrase comes from pseudo-Ptolemy's *Centiloquium* rather than from the *Almagest*. As this casual citation shows, Ptolemy's roles as astrologer and astronomer were easily blurred. Benazzi here represented Ptolemy not as the author of the *Almagest* but as an astrologer whose writings could help men to free themselves from the baleful influences of the stars.

Rheticus's report that Copernicus was "plainly acquainted with Domenico's ideas" increases confidence that no later than 1499, and probably earlier, Copernicus was familiar with Novara's retort to Pico's charges against the hopeless uncertainty and allegedly deterministic dangers of astrological knowledge. Moreover, the 1499 prognostication elevates the likelihood that Copernicus was already reading Pico's book in Bologna and that it pushed him at that early date to think about a strategy for defending the astronomical foundations of astrology. In other words, even if judiciary astrology could strive only for conjectural knowledge in making the judgment, astronomy could hope to achieve knowledge that is securely grounded, even if not fully demonstrative in the strict Aristotelian sense. A defense of practical astrology grounded in the discipline of theoretical astronomy would differ considerably from the sort offered both by Novara's pupil Giacomo Benazzi and by Lucio Bellanti because it would be concerned with fixing the mathematical principles on which the planetary tables were constructed. And Copernicus's skill at *theorica* must have been obvious to Novara when he took him on as an assistant.

In the remaining sections of this chapter, then, I shall take up three questions. First, what could Novara have communicated to the young Copernicus about the Piconian critique as political news

circulated through the Bologna senate, the Bentivoglio palace, and the university? Next, how exactly might Copernicus have assisted Novara in the preparation of his annual judgments? And finally, how might Pico's critique have entered into Copernicus's hypothesis about the new ordering of the planets?

PROGNOSTICATORS, HUMANISTS, AND THE SEDICI

The Bologna prognosticators occupied an important social position in the city, a status that traded on their familial connections to the notaries and their rank in both the medical and arts faculty at the university. Girolamo Manfredi and Domenico Maria Novara, for example, were not only teachers in the university but also members of both the medical and the arts faculties.[99] The polymath Giovanni Garzoni (1419–1505)—a philologist, moralist, historian, and one-time teacher of Savonarola—lectured on Avicenna's *Canon 3* in the chair of practical medicine for thirty years.[100] During this time, he also composed *De Eruditione Principum Libri Tres*, a work in the evolving advice-to-princes genre that grew out of personal discussions with Giovanni Bentivoglio.[101] "I believe," wrote Garzoni, "that anyone who does not know astrology cannot succeed in being a good philosopher, physician, or poet."[102]

Garzoni's statement points to another source of the prognosticators' influence: their discursive and explanatory resources were widely valued and circulated beyond their local sphere of activity. The prognosticatory genre structured a representation of the immediate future for all social groups. Members of the highest cultural and political circles found their energizing tropes in the astrologers' vocabulary of necessity and freedom to act. An expression of this view appears, for example, in one of the short stories that make up Sabadino degli Arienti's *Le porretane*. A Milanese nobleman, Gabriele Rusconi, receives the following advice: "Gabriel, I am persuaded that since you have arrived at the age where you thirst for written books, you ought to agree and understand clearly—both through practice and theoric [*per pratica e per teorica*]—that the stars, through their influences, entirely dispose and govern our active life."[103] The prognosticators' language, drawing its resources from learned treatises like the *Tetrabiblos*, was not quite as malleable as the language of verse prophecy; nonetheless, it helped to contain ruling-class anxieties about imminent destruction from external forces. Such high, astrologized discourse, in other words, bound together the linguistic space of the poet, the physician, the philosopher, and the astronomer with men in the world of political action.

Yet it seems that the Bologna prognosticators themselves did not enjoy the same intimate relationship with the Bentivoglio as did their confreres at the Este court in Ferrara. It was the humanists, rather, with positions as rhetoricians, poets, or grammarians, who had special ties to the Bentivoglio court, sometimes as tutors of humanist subjects or as panegyrists. Francesco del Pozzo of Parma (known as Puteolano), for example, lived at the Bentivoglio palazzo near the university for many years; he lectured on rhetoric and poetry from 1467 to 1477, published an edition of Ovid's works in 1471, and tutored the Bentivoglio sons. Puteolano was also involved in setting up the first printing establishments at Bologna and Parma.[104] Giovanni Bentivoglio II described him as "a person most learned in poetry, oratory, and the liberal arts."[105]

The dominating Bolognese humanists of the end of the fifteenth century also had strong ties to the court: Filippo Beroaldo the Elder (Philippus Beroaldus; 1453–1505) and his colleague and rival Antonio Codro Urceo, who held the chair of grammar, rhetoric, and poetry (1482–1500).[106] Beroaldo had been a pupil of Puteolano and lectured on rhetoric and poetry at the university from 1479 to 1503; he also tutored Annibale and Alessandro Bentivoglio and dedicated a commentary on Suetonius to the former.[107] His political and intellectual connections were substantial. Giovanni Bentivoglio described him as "persona virtuosa e degna."[108] He was also a magnet for foreign humanists. The Portuguese poet Hermico Cayado, later praised by Erasmus, was in Bologna between early 1495 and May 1497 in order to study with Beroaldo. The publisher Hectoris produced Cayado's *Aeclogae Epigrammata Sylvae*.[109] Cayado and Beroaldo were linked in another way: both sang the praises of a Polish nobleman, Pawel Szdłowiecki (Paulus Sidlovitius), who was probably one of their patrons. It is possible that Copernicus came into contact with Beroaldo through Cayado and Szdłowiecki.[110] Beroaldo was also a close friend of Pico della Mirandola; some of his letters to Pico were published by Hectoris in the

1496 edition of Pico's works.[111] In fact, Hectoris published most of Beroaldo's writings, and they stand as exemplars of Bolognese humanism.[112] One of Codro's most important pupils, Antongaleazzo Bentivoglio, became chancellor of the College of Jurists in 1491.[113]

The major patron of the Bolognese humanists was Mino di Bartolomeo Rossi or Minus Roscius, the son of Bartolomeo (1455–1503). Beroaldo referred to him frequently in his writings with flattering phrases: "To me, no man was friendlier, dearer or more close"; he was "the refuge of all lettered men" ("asilo di tutti i letterati").[114] Cayado also referred to him as "Dictator amplissimus."[115] And of Rossi's writings Garzoni exclaimed: "You would have said that Cicero had been restored to life."[116] In fact, Rossi was a figure central to the city's political life both before and after the beginning of the French invasions. He was made a member of the Sedici in 1482. He accompanied Giovanni II on two trips in 1485 and 1488. In 1492, he was part of the embassy sent to Rome on the election of the new pope, Alexander VI. Two years later, he was accompanied by Antonio Urceo Codro and Alessandro Bentivoglio on a mission to the duke of Milan. Again in 1499 he spent six to seven weeks with the king of France in Milan negotiating the fate of Bologna. Bentivoglio confidence in Rossi was sufficiently high that he was again sent in April 1500 to represent the commune to the duke of Milan. And finally, sometime in 1502, he departed on a diplomatic mission to France, where he remained for nearly six months.[117]

Rossi was a man of political affairs, but his humanist sensibilities were no mere adornment: both he and Beroaldo had studied with Francesco Puteolano. Throughout his writings, Beroaldo mixed laudatory with biographical references to Rossi. In his commentary on Apuleius's *Golden Ass*, published by Hectoris on August 1, 1500, he engages in an extended digression praising the country villa of his friend Mino Rossi, with its spacious gardens, fountains and staircases, which lay about seven miles from the city in the valley of the river Reno. Every year—no doubt in the summer, as the palazzo was not well heated—Giovanni Bentivoglio came to this villa with his sons to be royally entertained by Rossi.[118] Cayado recited poetry there on at least one occasion.[119] Beroaldo was clearly part of this company, and it is evident that being entertained at this level was

28. Mino Rossi (1455–1503), senator of Bologna, patron of Bologna humanists, friend and correspondent of Pico della Mirandola, and dedicatee of at least two prognostications by Domenico Maria Novara. Courtesy of Marchese Ippolito Bevilacqua Ariosti.

not unusual for him; of course, mention of it was also a way of publicizing his own status by representing his social proximity to a patron. An earlier, informative reference can be found in a letter to Pico della Mirandola (April 10, 1486), where Beroaldo referred to having been entertained by Pico together with Mino Rossi at the former's residence. Hectoris published this letter as part of Pico's collected works in 1496.[120] Again, as late as July 31, 1502, Beroaldo referred to a supper party hosted by Rossi for Giovanni Bentivoglio.[121]

There is no independent evidence that Domenico Maria Novara was ever invited to such a dinner, but he had been well received at one of Rossi's residences. In 1501, he began the disputational part of his forecast with the report of an important conversation: "A few days ago, when I was at the home of Mino Rossi, Senator of Bologna, a man of course not ignorant of the Latin language, he brought forth for the public good (interrupting, as we were discussing some things) a certain doubt about astronomical matters, a difficulty, in fact, not unworthy of philosophical inquiry." This opening strategy is revealing and unusual because such references to personal meetings with senators are exceptional in the

prognostication literature, and there are no such references in any of Novara's earlier (extant) prognostications. Note what it achieved. First, the prognosticator signals that his own status is sufficiently important that it allowed him to engage in a "conversation" with a member of the Sedici at his home (perhaps the Reno Valley villa). Second, he takes care both to indicate his own lower status (Rossi interrupts him) and to reveal that the senator was the one who introduced the question for debate (a *dubium*).[122] Third, he lets it be known not only that Rossi was knowledgeable enough to make a significant objection ("worthy of philosophical inquiry") but also that he knew how to do it in the right way, that is, in good Latin (the language of the university). Even if Novara did not partake of dinner with the senator, they shared Latinity and serious philosophical argumentation. Together, these aspects of the meeting signal Novara's access to the center of political power and afford legitimacy to the encounter.[123]

The 1502 prognostication posed an interesting conundrum. Again, the prognostication begins with a scene of conversation, although this time without reference to "the home of Mino Rossi":

"There is a common proverb among speakers: 'The imagination often causes an event,' that is, frequently something happens just as we have imagined. From this puzzling proposition, the magnanimous prince Mino Rossi, a man (like us) not ignorant of the Latin language, brought forth for the public good a single, certain difficulty while he discoursed about many and various matters, such as befits senatorial dignity." Again it is the senator, now referred to as a "prince," who initiates discussion of a difficulty (*arduum certe*) that he knows the prognosticator will use as an occasion, as in the previous year, for a published academic discourse. Again the arrows of mutual social legitimation fly in both directions: Mino Rossi allows himself to be associated with the prognosticator and the university; Novara acquires a public association with a leading member of the senate. Perhaps there is also a hint that something else is being exchanged: on the one hand, senatorial political gossip of value to the prognosticator in constructing his forecast; on the other hand, authoritative arguments designed to quell doubts about the forecast's causal and epistemic foundations.

The strands of this investigation now come together to form a new space of probable understanding and hypothesis. The prognosticators were part of system in which privileged information and political intelligence in Bologna circulated through the notaries, the professoriate, and certain members of the senate. The credibility of the advice offered in the annual forecasts must have relied, at least in part, on the quality of this information—it certainly could not have come entirely from the astrology manuals. Proper calculation of eclipses and important conjunctions then furnished material for the causal, explanatory part of the judgment as well as the grounds for specific advice on the timing of when to act or not. Although Cicero's *On Divination* was a common authority for questioning astrology, there developed a deeper uneasiness, if not outright skepticism, about the dominion of the stars after the appearance of Pico's *Disputations*. The exacting placement of a *dubium*, attributed to Rossi, at the head of the 1501 and 1502 prognostications suggests the possibility that Novara was responding directly to these patronal concerns.[124]

The printer and bookseller Benedictus Hectoris did not hesitate to profit from both the believers and the doubters. Hectoris was the publisher of all of Novara's later prognostications even as he published the prestigious works of Pico. For Hectoris, the major threat came from piracy by other printers. Two Lyonnais publishers, Jacobinus Suigus and Nicolaus de Benedictis, pirated the first volume of his edition of Pico's *Works*, and the *Disputations against the Astrologers* was appropriated by the Venetian printer-pirate Bernardinus de Vitalibus on August 14, 1498.[125] Prestige accrued to printers by virtue of their association with authors of noble birth and sometimes from authors' provocative claims. With the publication of Pico, Hectoris had the advantages of both.

COPERNICUS, ASSISTANT AND WITNESS

A remarkable conjuncture of publications intersected Copernicus's arrival in Bologna in the fall of 1496: at Milan, in 1495, a new issue of Albert of Brudzewo's *Most Useful Commentary on the Theorics of the Planets* from a different publisher than the first issue; in August, from Simon Bevilacqua in Venice, another—indeed, the only other—commentary on Peurbach, by the public lecturer on astronomy at Padua, Francesco Capuano da Manfredonia; and also in Venice, at the end of August 1496, Regiomontanus's *Epitome of the Almagest*. Finally, in July 1496, Pico's *Disputations* appeared at Bologna. In short, within one year, publishers in key north-central Italian venues had created a kind of mini-revival of Regiomontanus's short-lived Nuremberg printing project, some of it framed by Italian and Polish interpreters. To these significant resources of the science of the stars must also be added the slightly earlier 1493 edition of the *Tetrabiblos*.[126] Girolamo Salio of Faventino, a physician with significant Bologna connections, explicitly dedicated that work to Novara.[127] Although the text reads "Domenico Maria of Anuaria, Doctor of Arts and Medicine and Most Excellent Astrologer," "Anuaria" is not a known place name, and the designation was unquestionably a typographical error for "Nouaria."[128] Hence the 1493 *Tetrabiblos*, the principal resource of theoretical astrology of the late fifteenth century, must be the edition used by Domenico Maria Novara and Nicolaus Copernicus, the young man who assisted him in his work. Indeed, the copy of the 1493 *Tetrabiblos* in the University of Bologna Li-

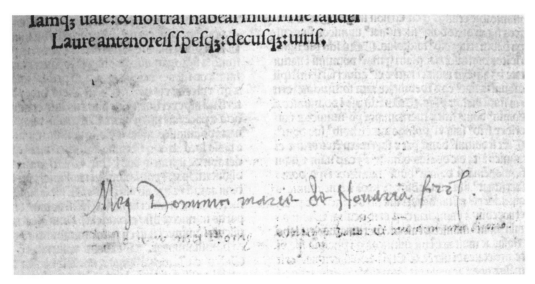

brary, which is bound together with the *De Nativitatibus* of Albubather Alhasan (Abū Bakr) and the *Summa Anglicana* of John of Eschenden (Johannes Eschuid), strikingly supports this hypothesis. The last page of the bundled collection contains Novara's signature and is the only book from his library currently known to survive.[129] Copernicus had thus arrived in Bologna and was living in the home of one of its major astrological practitioners just as Pico's critique appeared amid a rapid-fire sequence of publications concerning the science of the stars.

Novara was also the likely source of Copernicus's first knowledge of Regiomontanus's *Epitome*. Given everything now known, it would be surprising if the Novaran had not quickly acquired a copy of that work. Novara's motive would have been more than a concern with theoretical astronomy: he referred to "Magister Joannes de Monte Regio Germanus: Praeceptor Meus" (Master John Regiomontanus the German, my teacher) in a manuscript concerning the Moon's influence at the moment of human conception.[130] It seems clear that Novara, like Paul of Middelburg, regarded Regiomontanus as the right kind of astrologer.[131] And Novara may have owned some of Regiomontanus's manuscripts.

Apart from these key intellectual referents, all Copernicus's known observations from his Bologna period involve the Moon. For example, in the famous observation of *De Revolutionibus*, book 4,

chapter 27, Copernicus reports that he has seen the Moon eclipse "the brightest star in the eye of the Bull," namely Aldebaran (in the constellation Taurus). In 1543, he uses this observation to advance a theoretical claim: that the Bologna observation confirmed exactly the size of the apparent lunar diameter. Copernicus wrote: "These values agree so thoroughly with the observation that nobody need doubt the correctness of my hypotheses and the statements based on them."[132] It is curious that Copernicus chose to report an observation represented as witnessed only by himself and made forty-six years earlier when he might have used another, later observation. Noel Swerdlow and Otto Neugebauer comment that although Copernicus used this observation to test Ptolemy's lunar parallax, "one cannot be certain about what Copernicus did with [these observations] at so early a date."[133]

Was Copernicus acting then on March 9, 1497, as an assistant and witness to Novara? It would seem more likely that in this instance Copernicus was acting alone as an observer, as mention of Novara's presence as a witness surely would have enhanced the authority of the observation. But certainly one reason why Copernicus might have been interested in new and full moons was to assist in checking the lunar tables for Novara's 1498 prognostication. This concern with the reliability of the tables is entirely consistent with the usual account that he was disturbed by Ptolemy's theory

30. Table of contents showing full list of works included in the omnibus edition containing Ptolemy's *Tetrabiblos*, 1493. Note additional titles added by hand. By permission of Biblioteca Universitaria di Bologna. A.V.KK.VI.26.1. Further reproduction by any means is prohibited.

of the Moon because he had read Regiomontanus's *Epitome*, book 5, proposition 22. But, as Jerzy Dobrzycki and Richard Kremer have shown in the case of the little-known Viennese physician and ephemeridist Johann Angelus (Johann Engel; d. 1512), there is evidence for a more widespread interest in improving the common Alfonsine-based almanacs by introducing innovations into planetary theory.[134] And the examples of Angelus and Copernicus begin to provide a different sort of reason for the turn to theory, a reason situated in the space of practical astrology.

Consider the other Copernican observations about which there is some knowledge. Copernicus observed conjunctions of Saturn and the Moon in Taurus on January 9 and March 4, 1500. For the year 1500, both Novara and Marco Scribanario predicted a conjunction of Mars and Saturn in Taurus in February, an ill omen. Novara's forecast for the year 1500 went to press on January 20, eleven days after Copernicus's first observation. We can only speculate that the second observation, two months later, might have been made as a check on the first.

Perhaps the most interesting observation is that of a partial lunar eclipse that Copernicus made at Rome on November 6, 1500. Copernicus reported that he observed the eclipse at two hours after midnight. Novara's prediction of that event, made on January 20, 1500, affords a rare opportunity to correlate a specific prognostication with an observation by Copernicus:

This year an eclipse of the moon will occur on the fifth of November at night. And the moon will be eclipsed around the northern node at 24 degrees of Taurus and almost all of the moon's body will fall into the shadow. The beginning of the eclipse will be at 7:30 P.M.; the middle of the eclipse will be a little later at 9 P.M., although some astronomers place the beginning of the eclipse at 7 P.M. or 8 P.M. And so there is some error in the calculation: indeed, whoever makes a mistake in even one sixth part of an hour [10'] in the equations will fall into error. An even more serious error concerns the motion of the eighth sphere because it is necessary to acknowledge that the equinoxes are continually moving. . . . And whoever fails to notice this will suppose one thing, but, in the conclusion, will reach the opposite result.

Novara says no more about the eighth sphere but concludes that "the effects of this eclipse will not appear this year." A wise conclusion indeed! His prediction was off by at least five hours, depending on how one interprets Copernicus's description of his Rome observation: "I observed the . . . eclipse, also with great care, at Rome on 6 November 1500, two hours after the midnight which initiated 6 November . . . ten digits in the north were darkened."[135] Novara made no reference to this discrepancy in the next year's forecast; but it is clear that short-term inaccuracies of this magnitude offered no special difficulty for the credibility of the forecast. The discrepancy could either be ignored (as in this case) or rolled over into a long-term explanation of terrestrial effects.

It now seems reasonable to conclude that the late-fifteenth-century Bologna culture of prognostication furnished an important context for Copernicus's initial concerns about the lunar theory. As I have argued, Copernicus likely encountered Regiomontanus's *Epitome* at Bologna through Novara. Both Copernicus and Novara would have had a keen interest in Regiomontanus's account of the Moon because the Moon had a crucial role in the computation of eclipses; and eclipses, because of their long-term effects, were of great significance in the annual prognostications.[136] Rheticus's remark in the *Narratio Prima* contains more than an echo of Copernicus's early problematic: "The theory of eclipses all by itself seems to maintain respect for astronomy among uneducated people; yet we see daily how much it differs nowadays from common calculation in the prediction of both the duration and extent of eclipses."[137]

THE AVERROISTS AND THE ORDER OF MERCURY AND VENUS

As we have seen, Copernicus was not alone in regarding planetary order as disputed knowledge. Ptolemy's *Almagest* had already expressed doubts about the order of the inferior planets with respect to the Sun; in 1496, Regiomontanus renewed the doubt by calling it a controversy. And Pico's critique seriously aggravated the problem by connecting these doubts to the general viability of astrological theory and practice. Together, these developments must have put some pressure on Sacrobosco and *Tetrabiblos* commentators as well as natural philosophers for whom the order of Venus and Mercury was unproblematic.

Among Copernicus's contemporaries at Bolo-

gna, the Averroist Alessandro Achillini (1463–1512) was the most authoritative philosophical voice on this subject. He was well connected to the Bentivoglio court and eventually became one of the Sedici's highest-paid professors at the university. In argument he was said to be so sharp that he merited the squib "It is either the Devil or Achillini."[138] Benedictus Hectoris published his enormously dense *On the Orbs* in 1498. As a defender of the Averroist-Aristotelian system of concentric orbs, Achillini regarded Ptolemy's eccentrics and epicycles as absurd fictions; he was also critical of Ptolemy's handling of Mercury and Venus. According to Achillini, Ptolemy contradicted himself. In the *Almagest*, book 9, chapter 1, he said that he could not detect whether the inferior planets ever eclipsed the Sun; hence he concluded that they must be above the Sun. Subsequently, however, Ptolemy conceded that Mercury and Venus fell onto the same line that connected the Sun and the observer—a position with which Geber agreed.[139] Achillini believed that this second or "lower" ordering was correct. In its defense, he inferred the presence of Mercury and Venus while explaining away their invisibility: the Sun's brightness blots out the tiny planets when they are close; their transparency may not block the Sun's rays; moreover, their great distance from Earth causes the cone of the eclipse's shadow to be "lost in the air" or terminated before it reaches the Earth. These were the reasons why no one had observed solar transits. Although Achillini's commentary picked out for exegesis exactly the same passage cited by Pico just three years earlier, Achillini failed to mention Averroes' claim that he had observed "dark spots" or to draw the Piconian moral that dissensus betokened uncertainty. Rather, in the end, he heralded agreement: "It is unnecessary to say that Aristotle supposed the opposite from Ptolemy, as Averroes said." In fact: "Experience conforms to this order of the planets . . . it can be rationally conjectured through exact instruments."[140]

Yet there was an Averroist answer to Pico's complaint about the astronomers' disagreement concerning the order of the inferior planets and the arbitrary assignment of physical qualities. It came from Lucio Bellanti, the Sienese philosopher and astrologer who had taken the trouble to answer—or at least to say something about—each and every one of Pico's objections, including the arguments in *Disputations*, book 10, chapter 4:

He [Pico] does not know that there is a clear demonstration concerning the place of Mercury and Venus beneath the Sun, whether such [a demonstration] be elicited from the size[s] of the epicycles or from the greater apparent size of these bodies; [consequently] those who follow after him are unfamiliar with these [phenomena]. But Averroës says that these things are of no moment, since he asserts in *Metaphysics* XII that the appearances of these planets can be saved by different poles in the same orb, while denying those very foolish epicycles, which even now deceive philosophy. Nevertheless, he suffered from old age and could not learn astrology and thus did not understand either that of the ancients or of his own time.[141]

Bellanti was as imprecise on this matter as he was emphatic. He did not bother to supply his "clear demonstration," nor did he describe Averroës' solution to the uncertain ordering. Did he mean that there are two planets within the same orb? To Ptolemaic-style astronomers like Copernicus and Novara, Bellanti's denial of the value of epicycles could only have seemed superficial and objectionable—perhaps one further source of Copernicus's notorious mistrust of natural philosophers. The proposed solution to the problem of order was no solution at all: even if one tinkered with the poles around which the (concentric) orbs of Mercury and Venus were said to revolve, such an approach did not address the controversy over their relative order. And to anyone who hoped to use planetary models to make astrological predictions, Bellanti's reply would have appeared entirely useless.

Yet Bellanti's argument with Pico provided a striking textual locus that again called attention to one of Pico's most important astronomical objections to the claims of astrology. Furthermore, the Averroistic reference could have directed Copernicus's attention to Achillini's just-published *De Orbibus* and to its unsatisfying treatment of the ordering of Mercury and Venus. As Hectoris published (and sold) not only Achillini's work but also Pico's *Disputations* and Domenico Maria Novara's prognostications, there is every likelihood that these volumes would have been found on the same shelf in Hectoris's shop, or perhaps in Novara's home library.[142] Copernicus and Novara, introduced to the Viennese initiative of Peurbach and Regiomontanus in its Italian incarnation, surely would have seen Bellanti's Averroistic reply to Pico's objections about planetary ordering as awkward and ineffective. Perhaps this confluence of ideas, appearing in such rapid succession, set the twenty-five-year-old Copernicus to thinking about a different kind of approach to the ordering of the inferior planets.

COPERNICUS'S *COMMENTARIOLUS*, OR, PERHAPS, THE *THEORIC OF SEVEN POSTULATES*

Copernicus was not a systematic philosopher either in the traditional sense of a scholastic commentator like Achillini or in the later sense of a Descartes.[143] The *Commentariolus* more closely resembles an astronomical theoric, a laying down of principles preparatory to a later work. It is, as Albert of Brudzewo characterized Peurbach's work, descriptive or narrative rather than demonstrative, like the *Almagest*. It is compact and tightly argued, but it is also spare and unforthcoming about what is prior or presupposed. By contrast, *De Revolutionibus* and, even more, Rheticus's *Narratio Prima* are more readily open in stating their assumptions, as one might expect, because the former had many decades to ripen on the vine. It is thus often tempting to use the later works to interpret assumptions made in the earlier one.

Sometime before 1514—perhaps in 1510, perhaps much earlier—Copernicus composed this characteristically terse work, one whose list of basic postulates shows notable correspondences to the *conclusiones* of book 1 in Regiomontanus's *Epitome of the Almagest*.[144] Historians commonly call it the *Commentariolus*, although Copernicus's own title is unknown.[145] For convenience, I follow the traditional designation. The *Commentariolus* presents a new "order of the orbs" in the form of a derivation: "If there be granted to us some postulates [*petitiones*], which some call axioms."[146] There are seven of these postulates, which together lay out the elements of a new celestial order. Copernicus relied here, it seems, on Aristotle's sense of a principle or *petitio* in the *Posterior Analytics*: "any provable proposition that is assumed and used without being proved" rather than "that which is in itself necessarily true and must be thought to be so."[147] It is in the first sense that Copernicus may have intended his remark that "anybody familiar with the art of mathematics" (*e quibus mathematicae artis non ignarus*) ought to understand what he is doing.

Noel Swerdlow has aptly observed of the seven postulates that only two of them (numbered 3 and 6) function as logically prior, whereas four (numbered 2, 4, 5, and 7) actually serve as consequences of the aforementioned two. Consider first the assumptions, then the consequences.

> 3. All orbs encircle the Sun, which is, so to speak, in the middle of all of them, and so that the center of the universe is near the Sun.[148]

I suggest that this postulate, and even more explicitly number 6, can be read as Copernicus's resolution of Peurbach's conundrum: the Earth's annual motion explains the Sun's apparent motions as a mirrored effect.

> 6. Whatever motions appear to us to belong to the Sun are not due to [the motion of] the Sun but [to the motions of] our orb, [the orb of] the Earth, with which we revolve around the Sun just as any other [moving] star. The Earth is thus possessed of several motions.[149]

In a revealing comment at the very end of the postulates, Copernicus contended that both his own hypotheses and those of the natural philosophers should be regarded as inferences from the appearances—from effects to causes—rather than established by authority or by apodictic demonstration.[150] Copernicus's assumptions followed the Pythagoreans, but, rather than being arbitrary, as theirs were, his were to be seen as reasoned: "Lest anybody suppose that, with the Pythagoreans, we have asserted the Earth's motion rashly, he will find here strong evidence in [our] explanation of the circles. For, the arguments by which the natural philosophers try above all to establish the Earth's immobility rest for the most part upon appearances. But all their arguments are the first to collapse here, since we overturn the Earth's immobility also by means of an appearance."[151] Rheticus and Copernicus would later expand the claim: "This one movement of the Earth satisfies an almost infinite number of appearances."[152] And they would reinvent Plato and the Pythagoreans as "the greatest mathematicians of that divine age," whom Copernicus followed "in judging that circular motions ought to be attributed to the Earth's spherical body in order to explain the appearances."[153]

At this point, a counterfactual speculation may assist. What if Copernicus had written a different kind of book from the one that he ultimately produced, a book much closer to the spirit of Osiander, in which he argued simply that the Earth's daily and annual motions, although unproven assumptions, actually save more appearances than had hitherto been imagined by the Pythagoreans? Had he done so, he might have spared himself a good deal of trouble. He could have dedicated such a book to an earlier pope, published it much sooner than 1543, and avoided the anonymous letter to the reader, the defense of a seemingly absurd premise, and the search for any revisions of natural philosophy and theology. Yet already in the *Commentariolus* it is clear that he had decided to advance the Earth's motions as physical postulates. Of course, whether the Earth is a point or a physical body, when the Earth and the Sun are mutually transposed, then the period-distance relation nicely follows: "One [planet] exceeds another in rapidity of revolution in the same order in which they traverse the larger or smaller perimeters of [their] circles."[154] Further, once the sidereal periods of Venus and Mercury were determined separately from the Sun's annual motion, then the period-distance relationship definitively resolved their ordering.[155]

Thus, Copernicus's insight left a dilemma: the period-distance relation *underdetermines* the Earth's status as either a mathematical point or a physical body. Although the period-distance entailment might have satisfied his psychological intuition that he was right to make the mathematical assumption, it did not strictly authorize the move to make the Earth's annual motion a physical premise. So, why did Copernicus choose the second, more difficult, alternative? I can think of only one reason: If the Earth did not *really* move, then one could not claim that the planets—and their associated powers—really were ordered by their periods of revolution. To uphold the conclusion, Copernicus somehow needed to persuade himself of the truth—*or at least the likelihood*—of the fundamental premise and its corollaries. Without such a demonstration, he was in no position to answer Pico's critique concerning the astronomical foundations of theoretical astrology.

Already in the *Commentariolus*, this dilemma compelled Copernicus to address the physical implications of the Earth's motion. For example, even if the Earth's center is not the center of the universe, it is still the center of gravity toward which heavy things move (postulate 2). If the Earth rotates daily on its axis, explaining the apparent risings and settings of the stars, then the three other contiguous elements must move together with the Earth (postulate 5). Further, the size of the universe must be much larger than usually conceived, as the distance between the Earth and the Sun is "imperceptible" compared with the distance to the fixed stars (postulate 4). By starting from a mathematical rather than a physical premise, Copernicus opened up problems of physical justification and proof of a kind that his predecessors had never faced.

The problems were daunting, and it is perhaps no surprise that Copernicus remained silent about various other physical implications of his main assumption. For example, if he was following Sacrobosco's Sphere in the Commentariolus— and it is by no means clear that he was—then the Earth and its elements would need to be embedded in and continuous with a circularly moving celestial sphere that possessed qualities of an eternal, unchanging substance.[156] In that case, how could it also be a "center of gravity"? Indeed, how could a perfect, revolving sphere also contain diverse rectilinear motions, violent weather events, and complex human behaviors? And, in addition, by reordering the planets, how could one justify upsetting the planets' fixed elemental qualities? Further, how could one have a traditional—or indeed any—astrology? Copernicus's characteristic silence on such matters— his refusal to philosophize beyond the bare minimum—both allows and encourages the view that he must have struggled with the physical status of his primary assumption.

These considerations may also help to explain his choice of literary form. The organization of the postulates, as Swerdlow plausibly suggests, may simply indicate that this was a preliminary draft. But if Jerzy Dobrzycki's hypothesis is correct, then the postulates were consciously modeled after corresponding topics in Regiomontanus's Epitome.[157] This illuminating comparison opens the possibility of viewing Copernicus as both imitating and rewriting Regiomontanus's Ptolemy rather than remaining within the framework of authority delimited by the scholastic practices of Sacrobosco and his commentators.[158] Later, in De Revolutionibus, Copernicus developed a further textual transformation, dropping the austere strategy of axioms and postulates and introducing humanist logical and persuasive resources to defend the equivalents of the main postulates from his earlier treatise. A further consideration is that the Epitome also suited Copernicus's general approach. Although Copernicus combined geometrical and physical reasoning, as one would obviously expect in a mixed science such as optics, harmonics or astronomy, his main practices were geometrical. His major claims about planetary order rested on geometrical constructions to which was added time: hence the periods of circular revolutions of points. In this sense, the weight of his stronger physical claims rested initially on the way that he put together models of this sort. He was, as sixteenth-century writers often remarked, a "second Ptolemy" (and not a second Aristotle, a second Averroës, or even a second Peurbach). If Copernicus intended to deliver true conclusions about the world from a group of assumed premises, this was surely neither the practice followed by the philosopher Pico nor the typical method of philosophical disputations in which, like Achillini, Bellanti, and Capuano, the disputant was expected to argue an exercise in utramque partem, that is, from both sides of a proposition, before rejecting the affirmative arguments and providing his own solution or reply to the question.[159]

The abbreviated form, the succinct style, and the absence of prefatory material make it difficult to say from explicit references which audience Copernicus hoped to reach in the Commentariolus. It was certainly not some hypothetical— and anachronistic—"scientific community."[160] Our reconstruction of the Bologna period makes credible a more exacting guess about the kind of audience that Copernicus intended: academic prognosticators (exemplified by Domenico Maria Novara, Giacomo Benazzi, and Marco Scribanario) and court astrologers (Paul of Middelburg, Luca Gaurico); sympathetic fellow canons at Varmia (Tiedemann Giese); former friends and teachers at Krakow (John of Glogau, Albert of Brudzewo); ephemeridists (like Johannes Stöffler and Johannes Angelus); and opponents of Pico— in short, those later referred to under the generic designation "mathematicians" in the preface to De Revolutionibus. Copernicus was consistently dismissive of those whom he called "philosophers," even as he himself engaged in various kinds of philosophical bricolage.

When, on May 1, 1514, Matthew of Miechów (1457–

1523), a Krakow University physician, geographer, and historian, composed an entry in his personal library for a manuscript or a gathering of six leaves, he gave it the following title: "A Six-Folio Theoric Asserting That the Earth Moves while the Sun Remains at Rest" (*Sexternus Theorice Asserentis Terram Moveri, Solem vero Quiescere*).[161] For Matthew, and perhaps also for Copernicus, the *Commentariolus* was a theoric—in the sense of a group of seven postulates or assumed propositions.[162] The *Commentariolus* left full "mathematical demonstrations" of the individual planetary models explicitly to "another volume."[163] It is difficult to escape the impression that in assuming these postulates, Copernicus hoped they would resolve disagreements among astronomers. Should the planets revolve around the center of the universe on concentric orbs, as the Averroists wished? Should they revolve around off-center axes, as Ptolemy allowed? And was there a secure common principle by which to order the planets against the criticisms of Pico? The hope for consensus on these questions lay not in the premises but in the power of the derived consequences. For Copernicus could not appeal to the force of demonstrative reasoning in which, as Aristotle held in the *Topics*, "things are true and primary which command belief through themselves and not through anything else; for regarding the first principles of science it is unnecessary to ask any further question as to 'why,' but each principle should of itself command belief."[164] At best, Copernicus could resort to dialectical strategies of argumentation, as André Goddu suggests; and this meant that he could show a probable world but not a necessary one.[165]

Pico della Mirandola's name did not show up either in Copernicus's characteristically terse *Commentariolus* or in *De Revolutionibus*. But there is important evidence in the *Narratio Prima* that increases our confidence that Copernicus was trying to answer Pico's complaint. Rheticus, whose sympathetic testimony provides an unusually reliable firsthand acquaintance with Copernicus's thoughts, had composed his book in 1539 while living with Copernicus in Frombork. Rheticus had consistently presented Copernicus as the equal of Ptolemy and Regiomontanus, the true successor to Ptolemy. After reviewing Copernicus's solar theory, Rheticus noted with an air of triumph that "if [my teacher's] account of the celestial phenomena had existed a little before our

time, Pico would have had no opportunity, in his eighth and ninth books, of impugning not merely astrology but also astronomy. For we see daily how markedly common calculation departs from the truth."[166] This telling remark—the sole explicit reference to Pico in the Copernican corpus—urges a more ambitious question: Was Pico's objection about planetary order really a concern of the earlier work?

COPERNICUS, PICO, AND *DE REVOLUTIONIBUS*

The *Commentariolus* is much less revealing of its origins than either the *Narratio Prima* or the mature *De Revolutionibus*. Copernicus never published the former, and it is not clear that he ever intended to do so. His full vision of a Sun-centered cosmos evolved over more than three decades before it finally appeared in 1543. Rheticus's *Narratio Prima*, the first published statement of Copernicus's theory, was richly interspersed with biographical references, citations of ancient and modern authorities, and the author's own enthusiastic outbursts of praise for the man to whom he always referred with undisguised adulation as "my teacher." Not surprisingly, *De Revolutionibus*, the mature work, is also much closer to the *Narratio Prima* than to the *Commentariolus* in its full engagement with supporting and opposing authorities. Traces and residues from Copernicus's early formative period show up more clearly in the later works than in the earlier. It is to these later footprints that one must turn in order to illuminate the earlier tracks.

This brings us back to the famous book 1, chapter 10, of *De Revolutionibus*, where Copernicus reviewed the ancient philosophers' claims about the order of the planets. Some of his language bears a strong resemblance to Pico's account. For example, Copernicus, like Pico, grouped together Plato, Ptolemy, "a good number of the moderns" (*bona pars recentiorum*), and al-Bitruji. Of the last he wrote, in his characteristically compressed way, that "al-Bitruji places Venus above the sun, and Mercury below it." Copernicus's language does not carry quite the detail provided by Pico, but the enumeration of authorities is strikingly similar.[167] Unlike Pico, however, Copernicus consistently framed his problem and his criticism in the discourse of theoretical astronomy. He did not, like Pico, merely list authorities and their positions, nor did he simply dismiss the inconsistencies between them. As would be expected, he introduced a different sort of reasoning when he said that "those who locate Venus and Mercury below the Sun base their reasoning on the wide space which they notice between the Sun and the Moon." Copernicus then provided values for the absolute distances that define the space between the Moon and Mercury, Mercury and Venus, and Venus and the Sun.

Immediately following this discussion, however, Copernicus took up a question that Pico dealt with directly and about which he could have had no knowledge independently of Pico.[168] In his critical editorial notes to De Revolutionibus, Edward Rosen showed that Copernicus could have known the previously cited Averroës passage only from the Disputations, because the Arabic text did not exist and the work was available only in a Hebrew translation.[169] "In his Paraphrase of Ptolemy," wrote Copernicus, "Averroës reports having seen something blackish when he found a conjunction of the Sun and Mercury indicated in the tables. And thus these two planets are judged to be moving below the Sun's sphere."[170] Not surprisingly, Copernicus judged the Averroistic claim to be "weak and unreliable" (infirma sit et incerta). His argument was based on the space that would be required to fill up the region between the Earth and the Sun, a space that could not accommodate Venus's epicycle, "which carries it 45 degrees more or less to either side of the sun [and] must be six times longer than the line drawn from the Earth's center to Venus' perigee, as will be demonstrated in the proper place."[171]

From Averroës, Copernicus moved to reject Ptolemy's use of the criterion of elongations (limited versus unlimited) for determining the planetary order. Once again, both the language and the conclusion of book 1, chapter 10, strongly resembled those of Pico: "Ptolemy [Almagest, book 9, chapter 1] argues also that the Sun must move in the middle between the planets [medium ferri Solem] which show every elongation from it and those which do not. This argument carries no conviction because its error is revealed by the fact that the Moon too shows every elongation from the Sun."

Finally, Copernicus moved away from the Piconian text altogether. He used the Roman Martianus Capella as an authority for what he took to be the correct ordering of Venus and Mercury but again encoded his discussion in the unmistakable idiom of the theorics. As Swerdlow has pointed out, when Copernicus said that all the planets are "related to a single center," he spoke of the "convex orb" of Venus's sphere and the "concave orb" of Mars.[172] This part of the account ended with the famous dithyramb to the Sun, with its mixture of pagan and Christian images, and the suggestively poetic passage that echoes Giovanni Pontano's De Rebus Coelestis (1512): "Meanwhile, the Earth has intercourse with the Sun, and is impregnated for its yearly parturition."[173] This humanist admixture of theorics and poetics was prelude to a final restatement of Copernicus's own aesthetic, already presaged in the preface to his work: "We discover a marvelous fitting together of the parts [symmetriam] of the universe, and an established harmonious linkage between the motion of the spheres and their size, such as can be found in no other way." For the ordering of Mercury and Venus, then, Copernicus would not make an ad hoc appeal to authority but invoked a general structural principle that applied uniformly to the entire planetary order.

That Copernicus's new representation of the machina mundi avoided explicit reference to astrological matters shows at least the persistent force of ancient genres of writing in the sixteenth century. "My teacher," Rheticus wrote, "has written a work of six books in which, in imitation of Ptolemy, he has embraced the whole of astronomy, stating and proving individual propositions mathematically and by the geometrical method."[174] Ptolemy, as we have seen, did not treat astrological matters in the Almagest—and neither did Regiomontanus in his Epitome—but had carefully reserved their consideration to a separate work, the Tetrabiblos. It seems that Copernicus had in mind something comparable when, in a suppressed passage of the autograph manuscript of De Revolutionibus, he wrote: "I have also assumed that the Earth moves in certain revolutions, on which, as the cornerstone, I strive to erect the entire science of the stars."[175]

If Copernicus ever intended to compose a companion astrological work based on heliocentric principles—no mean feat in its own right—one might expect that, "in imitation of Ptolemy," he would have followed the same classificatory practice. Yet, in spite of his assisting Domenico Maria Novara in his prognosticatory undertakings, his later medical training at Padua, and his own medical practice in Varmia, we know of no horoscopes that he cast, no prognostications that he issued, and no orations in praise of astrology such as were commonplace in his lifetime. And, of course, these absences have only nurtured the image of Copernicus as immune from any interest or engagement with astrological practice.

Copernicus might have believed that if astronomy's foundations were reformed as he envisioned, then that change alone would be sufficient to sustain the traditional astrology found in the *Tetrabiblos*. But following Pico's critique of the arbitrary association of elemental qualities and planetary order, it seems far more likely that Copernicus would have recognized that a radical revision of the prevailing celestial arrangement would require a corresponding reform of astrology's principles. Indeed, besides the reassignment of physical qualities made necessary by the planetary reordering, Pico's other objections would need to be answered in a manner superior to that of Bellanti—for example, the house-division problem and the uncertainty of the instruments and tables. We may well wonder whether he would have turned to Regiomontanus's astrology, as he did with his planetary theory. In any case, just as a circularly moving Earth was incoherent with Aristotle's theory of the elements, so too the physics and meteorology underlying traditional astrology would need to be rethought.[176] Reformulating Ptolemy's *Almagest* was evidently more than enough for one man. Perhaps Copernicus believed that he could leave to the young and astrologically skilled Rheticus the reconfiguring of the *Tetrabiblos*, just as he had explicitly left debate about the world's infinitude to the philosophers.

Still, these absences ought not to obscure the scope of Copernicus's explanatory ambitions. At one stroke, by assuming the Earth's annual revolution, he had provided a single explanation for Peurbach's puzzle concerning the Sun's recurrent appearance in the planets' motions and a solution to the old controversy about the ordering of Venus and Mercury, now forcefully energized by Pico's assault on the planetary ordering that underlay theoretical astrology. Yet, for that explanation to carry weight, the Earth's motion had to be shown to be real. Because *De Revolutionibus*'s manner of presentation hid much of its origins, leaving only traces of the initial problem situation, excavation of the original problematic from the surviving residues shows what the heliocentric hypothesis was initially intended to explain.[177] Once having posed the hypothesis, however, Copernicus gradually noticed other entailments—new explanations of phenomena for which he was probably not even looking (such as why Mars, Jupiter, and Saturn appear brightest at opposition but Mercury and Venus do so at inferior conjunction) and, at the same time, new problems for which he did not have ready-made solutions (including the disordering of the planets' elemental attributes, the fall of heavy bodies to earth, the precessional motion, and the size of the universe). The first group must have given him the confidence that he had good reason to persevere as much as the second must have given him serious pause about publishing before he could produce credible solutions. The subsequent history of the Copernican question would continue to mirror these early, unresolved tensions.

PART II

Confessional and Interconfessional Spaces of Prophecy and Prognostication

4

Between Wittenberg and Rome

THE NEW SYSTEM, ASTROLOGY, AND THE END OF THE WORLD

Copernicus first formulated his new arrangement of the heavens amid the intellectual skepticism and political insecurity of the late fifteenth- and early-sixteenth-century prognosticatory culture of the northern Italian university towns. When his mature hypotheses of celestial order finally appeared between 1540 and 1543, however, it was at a time of historic upheaval no less conflicted about the legitimacy of knowledge of astral forces and their effects. Both the Roman church and the German Protestant reform movement were obsessed with world-historical biblical prophecies; but for the Lutherans there was, as Robin Barnes has argued, a uniquely urgent sense of imminent crisis and belief in an apocalyptic "End of Time."[1] The world was going to end soon. But when? And what natural "signs" of the divine plan were reliable indicators of this end? Neither the questions nor the apocalyptic resources were entirely new: they were all appropriated from well-established medieval sources.[2] But now the apocalyptic sensibility was heightened by Martin Luther's break with the Church. For Luther, Rome was the seat of the Antichrist, and the "last days" were rapidly approaching. In the dedication to his translation of the Book of Daniel (1530), he told his protector, John Frederick of Saxony, that "the world is running faster and faster, hastening towards its end, so that I often have the strong impression that the Last Day may break before we have turned the Holy Scriptures into German."[3]

On the eve of the Council of Trent (1545–63), Copernicus's hypotheses quickly became the oc-

casion for discussion and engagement among students of the heavens at Lutheran Wittenberg. The question was no longer merely whether prognostication of natural events could be accommodated to a Bible-governed narrative, but rather what relevance the Bible had for conflicting hypotheses of celestial order in theoretical astronomy. What implications did the new hypothesis of heavenly order hold for various sorts of theoretical and practical divination? And was this order really a manifestation of God's plan for the world?

The agents most immediately involved in transforming Copernicus's manuscript into printed texts were all preoccupied, in one way or another, with prognosticatory and apocalyptic considerations. They were also Lutherans who were located either in Wittenberg or its main outpost in southern Germany, the powerful city of Nuremberg: Georg Joachim Rheticus (1514–74), a protégé of Philipp Melanchthon (1497–1560) at Wittenberg; Johannes Schöner (1477–1547), the dedicatee of Rheticus's *Narratio Prima* (1540); Andreas Osiander (1497–1552), the influential Nuremberg preacher; Achilles Pirmin Gasser (1505–77), a pupil of Schöner and Melanchthon, later town physician of Feldkirch, and the author of a prefatory letter to the second edition of the *Narratio;* and Johannes Petreius (1497–1550), the Nuremberg publisher who had also studied at Wittenberg.

By 1543 there were three representations of Copernicus's new celestial scheme. The *Commentariolus* was known in a limited way in Catholic circles in Varmia and in Rome.[4] The two published accounts—the *Narratio Prima* (First Narration) and *De Revolutionibus*—were carefully crafted to appeal to different audiences. The first was implicitly directed to a Lutheran audience; the second was formally dedicated to the Pope.[5] Rheticus, undoubtedly with Copernicus's approval, addressed his *Narratio Prima* to Schöner, a widely reputed Nuremberg astrologer, prognosticator, and geographer who taught mathematical subjects at the city's *Gymnasium* from 1526 onward. Schöner was at first a Catholic and (unlike Copernicus) a priest and a chaplain (to the Bishop of Bamberg). He soon, however, moved easily into the politically moderate intellectual orbit of the Wittenberg reformer Philipp Melanchthon and developed a friendship with Andreas Hosemann, or Osiander.[6] He sided with the Reformation in Nuremberg, married, and had a son.[7]

Copernicus, meanwhile, dedicated *De Revolutionibus* to Paul III, a pope renowned for his wide learning and patronage of astrologers (such as Luca Gaurico), but also, like Melanchthon, well schooled in Greek. Among other accomplishments, he had called into session a reforming council at Trent, and under his reign both the Roman Inquisition and the new order of the Society of Jesus were founded. Just as Rheticus's name was excluded from any mention in *De Revolutionibus*, so the pope's name was not used in the *Narratio*.[8] Evidently, Copernicus's and Rheticus's dedicatory decisions were part of a dual strategy to shape a favorable reception for the new world hypotheses in a Europe that was just beginning to show evidence of serious splits along confessional lines.

MELANCHTHON, PICO, AND NATURALISTIC DIVINATION

University courses built around Aristotle's physical teachings and disputed by scholastic philosophers and theologians constituted the main arena for debating questions about the nature of the heavens in the Middle Ages; and, as Edward Grant has shown, such discussions, posed in the question-answer format, persisted well into the seventeenth century. But the school philosophers, perhaps affected by the Church's serious injunctions concerning the stars' threat to human free will, gave little or no space to astrological matters.[9] Resistance to the inclusion of astrology in natural philosophy began to change significantly during the Reformation. The crucial figure in this development was Melanchthon, rector of the university where Luther taught and known famously as the *Praeceptor Germaniae* (Teacher of Germany).

The Lutheran reformers were by no means united in their assessment of the value of natural knowledge. Martin Luther himself undoubtedly encouraged his followers in the work of prophetic interpretation—he even wrote a preface to Johannes Lichtenberger's prophecies—but he was distinctly ambivalent about naturalistic prophecy compared with Melanchthon, his close associate.[10] Throughout his life, Melanchthon advocated a strongly naturalistic theology. Commentators have variously characterized it—Stefano Caroti, for example, has called it a "theophanic view of reality" and Sachiko Kusukawa a "providential natural philosophy."[11] In Melanchthon's view, the Creator disclosed his providential plan through natural signs and great historical events; the Word was revealed as much through nature as through scripture and history. The point was to make systematic theology hegemonic in all naturalistic investigation. Harmony, design, order, and intent were visible in the created works. Also, certain persons, according to Melanchthon, had special gifts of prophecy that permitted "secret insight or otherwise hidden sense." Sometimes prophetic insight came in dreams that were subsequently fulfilled. Even here, it was stellar influence that caused "the inborn and natural prophetic power hidden in men to be awakened and excited to such an extent as to announce future things."[12]

Divinatory practice was thus not only a legitimate expression of the natural desire to know the Creator's works and to achieve divine grace, but it was also ethically desirable: it made one a better Christian.[13] Melanchthon gave the widest latitude and authority to all kinds of well-established naturalistic divination, ranging from medical astrology to dream interpretation and the interpretation of monstrous births, portentous comets, and other *mirabilia naturae*.[14] Also, having studied at Tübingen with Johannes Stöffler, Melanchthon had been deeply impressed by the claims of the prognosticators. Even the failure of the 1524 flood forecast did not, so to speak, dampen his enthusiasm.[15] On the other hand, Luther regarded Melanchthon's views with skepticism:

It pains me that Philipp Melanchthon is so strongly devoted to astrology, because most of the time he is

31. Lucas Cranach, *Philipp Melanch-thon*, 1532. Courtesy National Gallery of Victoria, Melbourne.

deceived. For he is easily impressed by heavenly signs and fooled by his conceptions. He has often failed, but he cannot be convinced otherwise. Once when I arrived from Torgau, quite exhausted, he said that my death was imminent. I have never wanted to believe that it was so serious. I do not fear the heavenly signs because man is greater than all the stars and cannot be subjected to them. Were our bodies to be subjected to them, I [still] would not fear the heavenly signs. That I shall leave to the clever wise men.

And, in his *Table Talks,* Luther exclaimed: "Nobody will ever persuade me, for I can easily overturn their flimsy evidence. They take note of everything that supports their case; whatever does

not, they pass over in silence. If a man throws a dice for long enough, he will throw Venus, but that happens by chance. That art of theirs is so much manure [*dreck*]."[16] His final word on astrology was: "Whoever fears the influences of the stars should know that prayer is stronger than stargazing."[17]

Luther's views would prove to be typical of theologians in the sixteenth century. Melanchthon, on the other hand, was keen to wrap a protective belt around astrology: he regarded some divinatory practice as illegitimate or, more to the point, superstitious and diabolical. The critical issue was the maintenance of the authority of scripture and divine providence. Wherever God and his Word were endangered by errors and excesses, there lay the work of the devil.[18] Interpreting biblical miracles astrologically in the manner advocated by Pierre d'Ailly, for example, was considered to be dangerous.[19] Also, Melanchthon rejected forecasting for its own sake as "vain curiosity" and "superstitious divinations."[20] He regarded prognosticatory questions like "Who will be victorious, France or Burgundy?" as undesirable because they were devoid of providential import.[21]

In 1553, Melanchthon's son-in-law Caspar Peucer (pronounced *Beucker*) produced a massive work of classification. Peucer's aim was to demarcate Christian from diabolical divination, and his work covered not only astrology but also many other kinds of divination, such as from the parts of the body (chiromancy) and from animal entrails.[22] Most natural divination is good, so he maintained, because it is based on natural or physical causes, but, in practice, things are not always easy, because matter is unstable: mixtures of primary qualities keep changing, and thus so do predicted outcomes. The devil is a trickster. Demons delude people's imaginations, causing them to believe that they can do things that they cannot do. For example, demons can simulate legitimate activities such as the making of predictions or the production of cures. The Catholic use of relics and invocation of saints were good examples of the devil's activities. But, toward all forms of astrology—with the exception of astrological images created by human artifice—Peucer was quite favorably disposed. Although astrology could be abused, there was a true and legitimate astrology deriving its justification from the "force of light" created at the beginning of the world—as described in Genesis—and, of course, from that part of the science of the stars that describes the celestial motions and measures distances and intervals between bodies and the sizes of bodies and orbs.[23]

For Melanchthon, anyone opposed to the sciences of the natural order was seen to have endorsed an Epicurean theology, a world of matter devoid of meaning and divine purpose. And to Melanchthon, the principal opponent of the divinatory sciences was none other than Giovanni Pico della Mirandola. Pico's views were not merely wrong; they could seriously mislead the young. Melanchthon regarded Schöner as an ally in this endeavor to protect students against Pico's pernicious claims. Schöner said that he had seen a handwritten marginal note in a copy of the 1504 Strasbourg edition of Pico's *Disputationes* owned

by the bishop of Bamberg. This note accused Pico of plagiarizing all of his ideas from unidentified authors.[24] Rheticus knew about this comment directly through Schöner, and his knowledge of it then passed, probably by word of mouth, to Melanchthon and Copernicus.[25] In a world where large private libraries were still rare, knowledge that people believed to be trustworthy could be discovered not only in the printed word but in comments written in the margins.

Melanchthon's reputation as a pedagogue was no accident. His books were extraordinarily influential models of pedagogy. They offered clear definitions of terms and effectively chosen examples and drew on a comprehensive range of ancient, medieval, and modern authorities. He organized his books in scholastic form as questions, with extensive answers ordered in the form of arguments. Among other topics, he wrote textbooks of dialectic, rhetoric, and physics as well as extensive commentaries on the Psalms, the Book of Daniel, and Genesis. Many of his writings were also cast as prefaces to student texts or to the writings of authors whose views he wished to promote. When he announced his intention to write a full defense of astrology against Pico, he selected as his venue a preface to Johannes Schöner's *Tabulae Astronomicae Resolutae* (1536).

From the stars' positions many things may be revealed about bodily health, about talents and temperaments, about many misfortunes in life, stormy weather, and changes in republics. But most of all, contemplation and attention to such matters is conducive to prudent behavior. The Christian religion neither objects to this opinion, nor do sacred writings damn such predictions, for they occupy the same part of Physics as do the predictions of the medical doctors; and, in fact, they presume natural causes. Some heavenly influence is imparted by the Sun, some by the Moon, as though some is like the force of pepper, the other like the force of a purgative; therefore, it is both pious to understand God's works and to observe the forces imparted to them. However, this entire argument is longer than can be treated here, and there are many books, written most eruditely, which answer the dishonest accusations of Pico and others.[26]

The continuing need to defend astrology against Pico's arguments in the 1530s and '40s shows that the force of Piconian skepticism had by no means dissipated.[27]

Melanchthon's and Camerarius's "purified" humanist translations of the *Tetrabiblos*, alluded to in chapter 1, became the principal texts for avoiding excessive reliance on Arabic conjunctionist astrology and for reaffirming Christian authority. The *Tetrabiblos* constituted the center of the natural philosophical curriculum at Wittenberg. Systematic justification for astrological knowledge was a prelude to its inclusion in teaching about the natural world. Melanchthon developed his views in several places, of which two are especially important: the preface to Schöner's *De Iudiciis Nativitatum* (1545) and his textbook of natural philosophy, the *Initia Doctrinae Physicae* (1549). In both places, Melanchthon kept the traditional two-cell distinction between the principal parts of the science of the stars, "of which the one shows the most certain laws of motion, the other, *mantike* or divination, shows the effects or meanings of the stars."[28] Now, it is clear that Melanchthon considered Pico's main threat to astrology to be the attack on the divinatory part, theoretical and practical astrology, rather than the attack on theoretical astronomy that had worried Copernicus. The *Praeceptor Germaniae* believed that Pico's arguments—which, following Rheticus, he believed to have been plagiarized—had been refuted by "learned men, [Lucio] Bellanti and certain other people."[29]

Melanchthon reached two important conclusions in his defense of theoretical and practical astrology. First, in response to the criticism that astrological judgments could be wrong, it was only necessary to acknowledge that theoretical astrology, like theoretical medicine, was a fallible, human art that could predict some events with probability, but not all. This was no different from what Ptolemy had claimed about the prediction of particulars in the *Tetrabiblos*.[30] It was also consistent with Melanchthon's Stoic definition of an "art" as a teaching or collection of certain propositions that offer a certain utility—but not absolute certainty—in life.[31]

When Melanchthon returned explicitly to Pico's main arguments against divinatory astrology in the *Initia Doctrinae Physicae*, he reached a second, important conclusion concerning the question of whether and how astrology can explain particulars. Specifically, Aristotle had not adequately justified the connection between universal cause and specific effects: "Aristotle says: 'The astrologers seek out particular effects—some

many, others fewer—how one or another motion of the stars affect various qualities [of matter].' But Aristotelian physics passes over this doctrine concerning the particular effects of the stars, remaining content with a general forewarning, that the heavenly bodies are the universal, efficient cause that incites and tempers matter by means of motion and light."[32]

In a later section on physical fate, Melanchthon confronted Pico's important objection that for astrology to be a science would require replication of identical cases. Pico had argued that even if the astrologer knows the exact configuration of the heavens at the moment of a man's birth, the same groupings or alignments never return or do so only after thousands of centuries. If the astrologer limited his observations only to the most frequently recurring configurations, Pico objected, his observations would be imperfect, because he did not consider the same part in relation to the same group of entities.[33] Against this, Melanchthon argued that because universal causes determine all particulars in nature, astrology, like medicine, needs only a few verified cases to establish that heavenly arrangement A is causally connected with singular terrestrial experience B.[34] However, by "singular experience," Melanchthon actually meant any experience that was a member of a particular class. For example, all children born when the Moon is joined with Mars and Saturn in the sixth house are potentially sickly; eclipses generally announce sad events. In other words, Melanchthon defended the prediction of singular events of a sort to be found in the annual prognostications and the *Centiloquium*, where certain arrangements of celestial bodies caused certain classes of terrestrial events.

That Melanchthon ignored Pico's attack on the uncertain order of the heavens should probably evoke no surprise. The main worry came from the threat to the causal nexus between heavenly motions and earthly events. This had also been the principal concern of Lucio Bellanti and other opponents of Pico. For Melanchthon, as for these earlier writers, the "precepts of the heavenly motions" were not called into question. Thus, in the *Initia Doctrinae Physicae*, his most systematic statement, Melanchthon treated the science of the heavens unproblematically on the basis of ancient authority: "according to the usual teaching of Ptolemy."[35] In short, he took for granted that there was a consensus among the astronomers.

RHETICUS'S *NARRATIO PRIMA* IN THE WITTENBERG-NUREMBERG CULTURAL ORBIT

G. J. Rheticus, a member of Melanchthon's circle at Wittenberg, wrote the *Narratio Prima* during the first few months of his stay with Copernicus in Frombork between May 1539 and the end of September 1541. Their relationship was undoubtedly a close one: Rheticus had a unique opportunity to become well acquainted with Copernicus in his last years. During Rheticus's stay, he prepared a map of Prussia, a biography of Copernicus, and a treatise arguing that the Earth's motion does not contradict holy scripture. Neither the map nor the biography are extant, but Reijer Hooykaas has recently found and published the important work on scripture.[36] A *Narratio Secunda* or *Altera* is frequently mentioned, although it never appeared. Because of the special opportunity for the older and the younger man to develop a trusting and familiar relationship, the question of authorial responsibility for the *Narratio Prima* remains an important consideration. How much of it reflected Copernicus's own views, and how much those of Rheticus? Indeed, what conventions of joint authorship were operative? Who was the intended audience?

The work did not dissimulate. It was cast in the form of a letter to Johannes Schöner, a real, rather than a fictional, person.[37] Nonetheless, in various places, Rheticus made skillful rhetorical use of Schöner's character as a literary resource to present Copernicus's claims and arguments. As I have already remarked, Schöner was a prominent member of an influential network of humanists and astrological practitioners whose focus was fixed in Melanchthon's Wittenberg. Schöner had studied astronomy with Bernhard Walther at Nuremberg; Walther had acquired the papers of Regiomontanus, and many of these subsequently came into Schöner's possession. From 1526 until the end of his life, Schöner taught mathematical subjects at the Nuremberg *Gymnasium*. Melanchthon had revamped the curriculum at Nuremberg much as he later did at many other *Gymnasia* and academies of Germany. Schöner also had a printing press at his own home in Kircheherenbach, like Peter Apianus's original press in Landschut.[38] Later, he became well-known for having published the bulk of Re-

giomontanus's literary remains (from 1531 onward), much of it at the Petreius presses in Nuremberg.[39] In fact, Schöner was deeply involved in the revival and consolidation of Regiomontanus's reputation as a great mathematician and astrologer. He mentioned using Regiomontanus's *Tabulae Directionum* in his own astrological calculations. And it was probably Schöner who first informed Rheticus about Copernicus[40] and who undoubtedly represented him as a practitioner worthy of a place in the Regiomontanus pantheon.

Schöner was also a prolific author in his own right, a major contributor to the German literature on the heavens of the 1520s and '30s. From 1515 onward, he published something nearly every year, a veritable torrent of *practicas*, ephemerides, instrument treatises, wall calendars, reports of comets, and general astrological works. Some of these were *canones* or how-to books: rules for constructing and using clocks and astronomical globes, the kinds of instruments for which Nuremberg was beginning to build a reputation.[41] Along with the prognosticator Johann Virdung (at Heidelberg), Stöffler's student Sebastian Münster (at Nuremberg) and Petrus Apianus, Schöner enjoyed a considerable reputation within the imperial territories. He had also amassed a rich library of astral literature. Among its holdings was the copy, mentioned in chapter 3, of Domenico Maria Novara's treatise "De Mora Nati" (On determining the moment of natal conception), in which the Bologna prognosticator referred to Regiomontanus as "my teacher."

Rheticus arrived in Nuremberg in October 1538, where he spent at least one month with Schöner. From there, Rheticus moved northwest to Ingolstadt, where he visited Peter Apianus, and thence to Tübingen, where he met Joachim Camerarius. There is little doubt that Rheticus's tour of the Nuremberg orbit was motivated by Melanchthon. Melanchthon, Camerarius, and Sebastian Münster had all studied at Tübingen with the flood prognosticator and calendar reformer Johannes Stöffler. Melanchthon probably arranged the trip, perhaps with the hope that the twenty-four-year-old Rheticus could improve his competence as a prognosticator by visiting Schöner. At any rate, as Rheticus narrated in 1542, it was on this trip that "I heard of the fame of Master Nicolaus Copernicus in the northern lands, and although the University of Wittenberg had made me a Public Professor in those [mathematical] arts, nonetheless, I did not think that I should be content until I had learned something more through the instruction of that man. And I also say that I regret neither the financial expenses nor the long journey nor the remaining hardships. Yet, it seems to me that there came a great reward for these troubles, namely, that I, a rather daring young man [*iuvenili quadam audacia*], compelled [*perpuli*] this venerable man to share his ideas sooner in this discipline with the whole world."[42] The reference to expenses, the long journey, and so forth suggests that Frombork was not on the original itinerary, and that the decision to visit Copernicus was made only after Rheticus's southwest journey had begun. This inference suggests that it was not Melanchthon who had referred Rheticus to Copernicus. The decision to dedicate the *Narratio Prima* to Schöner, therefore, was evidently a way of directing that work to Melanchthon and his famous circle of students and followers at Wittenberg.

Astrological interests were undoubtedly foremost in the one-month encounter between Rheticus and Schöner at Nuremberg. We can well imagine that Schöner's *Little Astrological Work, Collected from Different Books* (*Opusculum Astrologicum, ex diversorum libris . . . collectum*), in press with Petreius and due to appear the following year, would have been on the agenda of discussions. This work was typical of the sorts of "collections for the use of the studious" that Petreius was increasingly interested in publishing. It also fitted well into his program of weeding out the "superstitious" Arabic elements from astrological practice. Schöner's *Opusculum* bundled together various works of theoretical astrology. It included his own instructions for reading ephemerides, conveniently tabulated columns correlating planets with relevant terrestrial effects, an introduction to judiciary astrology, "succinct rules of nativities," and "common elections." To these he added the treatise on elections of Lorenzo Buonincontro, an author in some demand,[43] and Eberhard Schleusinger's *Declaration against the Slanderers of Astrology*. Schöner also published a vernacular prognostication for 1539 in Nuremberg, and that too might have been part of his discussions with Rheticus.

It follows that Rheticus and Schöner were inter-

ested in Copernicus's work because of its potential value for astrological prognostication, and the same was true for the publisher, Petreius. Moreover, Copernicus already had something of a reputation in Nuremberg for interpreting nativities.[44] Both Rheticus and Schöner had spoken with Petreius about publishing some works by Copernicus in Nuremberg.[45] Immediately after the appearance of the *Narratio Prima* around March, 1540, Petreius wrote a public letter to Rheticus. Significantly, this letter appeared at the head of the text of a fourteenth-century treatise by Antonius de Montulmo titled *De iudiciis nativitatum* (Concerning the judgments of nativities). Petreius surrounded Montulmo's work with symbolic evocations of high cultural authority: the Montulmo manuscript came from the library of a prominent Nuremberger (Schöner); it had a Regiomontanus association (having appeared on Regiomontanus's *Tradelist* of works to be published); the Petreius edition contained annotations attributed to Regiomontanus; and the work was published together with the treatise of a prominent Italian astrologer (Luca Gaurico).[46]

For Petreius, who had himself studied at Wittenberg, this publication was clearly another element in the Melanchthonian program to promote a legitimate Christianized astrology. As he phrased it: "This part of philosophy concerning nativities has sure and great advantages for conducting the course of life properly without superstition."[47] Nonetheless, the businessman Petreius was not entirely averse to publishing works of Arabic astrology if he thought they possessed some utility in promoting the casting of nativities.[48] Likewise, Petreius believed that even though Copernicus's theory departed from "the common explanations by which these arts are taught in the schools," it could still be of great use to "this part of philosophy concerned with nativities."[49] Indeed, it may have been Schöner who cast a horoscope of Copernicus based on information supplied by Rheticus. The horoscope agrees better with Schöner's *Tabulae Resolutae* than with Copernicus's own numbers.[50] Undoubtedly Schöner also believed that Copernicus's work could be of value to various branches of astrology beyond that of casting horoscopes.[51]

The dedication in the first edition of the *Narratio Prima* foregrounded Schöner's reputation and authority: "To that Most Famous Man Johann

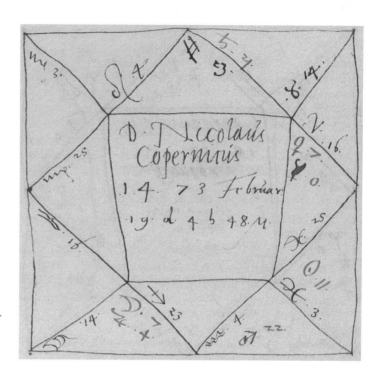

32. Copernicus's horoscope, ca. 1540.
Courtesy Bayerische Staatsbibliothek, Munich.

Schöner." Apart from the appeal to Melanchthon, the decision to dedicate the volume to Schöner suggests that both Rheticus and Copernicus believed that the association had value in legitimating the new enterprise. Through his numerous publications, Schöner was known to a wide readership for his mathematical and astrological skills and for his association with Regiomontanus. Moreover, at Nuremberg (and hence within the imperial territories) the frontispiece and dedication were of considerable value in publicizing the new hypotheses—perhaps of greater value than having Schöner act, like Andreas Osiander, as an editor or publication facilitator. The title continued: *Concerning the Books of Revolutions of that Most Learned Man and Excellent Mathematician, the Venerable Doctor Nicolaus Copernicus of Toruń, Canon of Varmia.* Throughout, Rheticus addressed Schöner with paternal deference, "as to his own revered father." This deference carried over to Copernicus, who, for reasons that were more than rhetorical, was constantly represented as "my teacher."[52]

The second issue (1541) was also a product of palpable Wittenberg-Nuremberg associations. It contained a new foreword by Achilles Pirmin Gasser, who knew Rheticus well from their hometown of Feldkirch. Gasser had typically wide humanist interests and skills. Like Rheticus and Petreius, he had studied at Wittenberg; he also held a medical degree from Montpellier and had followed Rheticus's father as city physician of Feldkirch.[53] He wrote five brief reports (*Unterrichten*) on the plague and an equal number of short "descriptions" (*Beschrybungen*) and "reports" on comets that appeared in 1531, 1532, 1533, and 1538.[54] In 1538, Melanchthon dedicated to Gasser an edition of John of Sacrobosco's *Libellus de Anni Ratione,* in which he also praised Rheticus. The following year Gasser published *Elementale Cosmographicum* at Strasbourg, a short work treating the "rudimenta" of astronomy and geography.[55] Between 1543 and 1545, Johannes Petreius published four of Gasser's prognostications; the prognostication for the year 1546 is dedicated to Rheticus.[56] Finally, in September 1543, Petreius inscribed as a gift to Gasser a copy of *De Revolutionibus.*[57]

Gasser's reputation as *medicus* and *astrologus,* as well as his prominence in Wittenberg-Nuremberg friendship circles, helps to explain why Rheticus recruited him to add a dedicatory letter to the second edition. Gasser called attention to the book's potential interest for astrological physicians or "iatromathematicians" by addressing a former schoolmate and fellow physician, Georg Vögeli of Konstanz (d. 1542): "So, dear Georg, we see that we are liberated from the majority of difficulties in astronomy and that other more obscure matters are cleared up for us, thus I beg you to read this little book that I am sending to you fully and with care; and after you have read it, criticize it rigorously and then recommend it especially to all those who love mathematics, in particular those who are close to you."[58] Gasser left no doubt that this was an unusual book—not merely "new" and "useful," but daring: it went against common sense and against the "theorics" usually taught in the schools. And it had reformist overtones: monks might even declare it to be "heretical." Nevertheless, this was a book that Gasser praised in vivid and unprecedented terms: "It genuinely appears to offer the restoration and even rebirth of a new astronomy that is completely in agreement with the truth; for, with the utmost vigor, it presents propositions most clearly upon those kinds of subjects which, as you know, have been a matter of controversy everywhere on the earth both among the most learned mathematicians and the greatest philosophers as well." According to Gasser, these controversial topics included "the number of celestial spheres, the distance of the stars, the sun's governance [in the universe], the planetary circles and their places, the constant length of the year, knowledge of the equinoctial and solstitial points, and finally the motion of the earth itself [nowhere mentioned in the title] and other difficult topics." Two audiences would find this book of especial value: "learned men of our time" (*ab nostri saeculi eruditis*) and "men moderately trained in mathematics" (*mediocriter mathesi imbutos*), and, of these, especially the "makers of ephemerides" (*ephemeridistas*).[59] Both theoricists and ephemeridists would like this book because astronomy—by virtue of its infallible precision, the surest of the sciences—was troubled by disagreements between observations, times, and what was promised by the models.

The rhetoric of Gasser's appeal to Vögeli mimicked Rheticus's strategy in dedicating his work to Johannes Schöner. Effectively, Schöner functioned as a stand-in for the audience of general readers. The stated purpose of the *Narratio* was both to "explain" and to "convince" Schöner that the ideas of Copernicus were worthy of comparison with the best thinking of the ancients (Ptolemy) and the moderns (Regiomontanus). Rheticus, in turn, presented himself as an earnest and admiring student rather than invoking his official position as a *magister*, a *mathematicus* on leave from the University of Wittenberg. He made a point of saying that he had had but a short time (ten weeks) to master the essentials of "a work of six books in which, in imitation of Ptolemy, he [Copernicus] has embraced the whole of astronomy, stating and proving individual propositions mathematically and by the geometrical method."[60] This passage clearly referred to the work promised by Copernicus in the *Commentariolus*. Rheticus stressed his own intellectual limitations. In addition to the short time available to him, he mentioned a "slight illness" and a "restful" side trip to Lubawa (Löbau) with Copernicus "on the honorable invitation of the Most Reverend Tiedemann Giese, bishop of Kulm." One gains the impression that they spent much time together and that Rheticus was integrated into Copernicus's circle of acquaintances.

The inclusion of so much biographical information suggests a strategic consideration: the creation of a verisimilar representation of the author. Responsibility for any misrepresentations was to be attributed to the fallible young student Rheticus rather than to the ideas of the master Copernicus. And it is possible that this way of structuring authorial responsibility had a conscious objective, namely, to permit stronger, more enthusiastic—perhaps even more controversial—claims to be made on behalf of the heliocentric hypothesis.

The interweaving of autobiography and the order of topical presentation and omission also served a further strategic function. Rheticus said that he had "mastered the first three books, grasped the general idea of the fourth and begun to conceive the hypotheses of the rest." However, he claimed that it would be "unnecessary to write anything to you" about books 1 and 2, "partly because my teacher's doctrine of the first motion does not differ from the common and received opinion."[61] In other words, Rheticus used the excuse of his own limited time to avoid beginning his presentation with the controversial questions of book 1.[62] Later, in spite of these earlier disavowals, he would return to this section of the manuscript. Consciously or not, the reader of the *Narratio Prima* was urged to identify with the author as someone working his way through the master's own work. As a result, the book opened with no hint of the "new hypotheses" to be introduced.

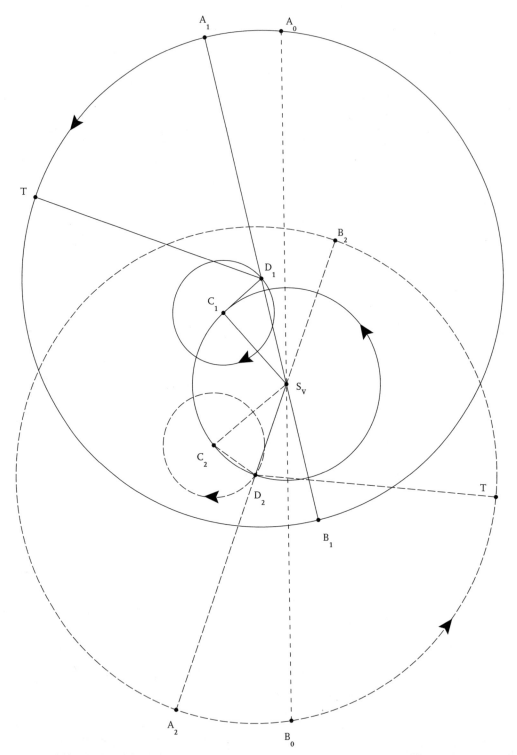

33. Complex modern reconstruction of Rheticus's Wheel of Fortune. The Earth (T) revolves counterclockwise in one sidereal year around D, a point off-center or eccentric to the Sun (S). But eccentric point D, shown at two positions (D₁ and D₂), in turn, revolves clockwise in 3,434 years around the small circle (or epicycle) with center C, shown at two positions (C₁ and C₂) as C revolves about fifteen times more slowly than D counterclockwise around the true Sun (S). Rheticus's Wheel of Fortune is designated as the circle centered on point C because its motion controls the Earth's maximum (A₁, A₂), minimum (B₁, B₂), or mean (not marked) distances from the Sun, and Rheticus believed that such changes in the Earth's eccentricity governed the times when great empires (Rome) and religions (Islam) would rise and fall. He also believed that the motion of D "did not differ much" from Elijah's prophecy of six thousand years for the second coming of Christ (i.e., 1¾ revolutions: 3,434y+2,575y = 6,009y). From Rheticus 1982, 153–55. Courtesy of the editors.

Rheticus then devoted the first seven chapters of the *Narratio* to the sorts of astronomical issues that would be of direct interest to astrological prognosticators like Schöner: topics underlying the stability and accuracy of the calendar, such as the motions of the fixed stars, the problem of the lengths of the tropical and the sidereal year, changes in the obliquity of the ecliptic, variations in the eccentricity of the solar apogee, the lunar theory, and eclipses. These are precisely the highly technical subjects that Copernicus addressed at great length in books 3 and 4 of *De Revolutionibus*.

WORLD-HISTORICAL PROPHECY AND CELESTIAL REVOLUTIONS

Rheticus then broke with generic convention. He departed from his autobiographical asides, from descriptions of the mechanisms of theoretical astronomy, and, unbeknownst to his readers, from Copernicus's manuscript. He introduced chapter 5 very simply: "I shall add a prophecy: That the Kingdoms of the World Change with the Motion of the Center of the [Earth's] Eccentric." He did not say, "*My teacher* adds a prophecy." So we can presume that Rheticus and Copernicus had decided to keep prophecy making in Rheticus's domain. But even if we grant that the idea did not originate with Copernicus himself, it cannot be the case that Rheticus "added" the prophecy without Copernicus's permission.[63] Rheticus stayed with his mentor for over a year after the publication of the book. And there is no evidence that Copernicus in any way objected to its contents, as the second edition appeared with some minor changes to the title but none to the text.[64] Moreover, although the prophecy has worried some modern commentators by appearing to interrupt the discussion of the eccentric's motion,[65] the marginal chapter designations added by Rheticus's secretary, the geographer Heinrich Zell (1518–68), allay this concern: Zell's side notes did not mark this section as a *digressio*. Hence there is no doubt that Rheticus meant the previous discussion to be continuous and hence to establish the astronomical basis for the prophecy.[66]

The kind of prophecy under discussion is also important. Rheticus used the word *vaticinium*. Significantly, he did not choose the terms that he deployed in a later vernacular forecast: *Prognosticon oder Practica Deutsch*.[67] Nor did he use the

occasion to write a general oration in praise of astrology—the topic of his master's disputation in 1535.[68] The reason for this choice of language is obvious once we look more closely at what he was doing in the *Narratio*: "We see that all kingdoms have had their beginning when the center of the eccentric was at some special point on the small circle . . . it appears that this small circle is in very truth the Wheel of Fortune, by whose turning the kingdoms of the world have their beginnings and vicissitudes. For in this manner are the most significant changes in the entire history of the world revealed, as though inscribed upon this circle." This was an apocalyptic, world-historical prophecy rather than a prognostication for the coming year. It began with the Roman Empire (at the Earth's maximum eccentricity) and then, as the eccentricity diminished, Rome declined "as though aging, and then fell." When the eccentricity reached the quadrant of mean value, the "Mohammedan faith" came into being and with it another great empire. Rheticus prophesied that in one hundred years, at minimum eccentricity, "it will fall with a mighty crash." The return of the eccentric's center to the other boundary of mean value, where it was at the world's creation, would herald the return of Jesus Christ. "This calculation," Rheticus added, "does not differ much from the saying of Elijah, who prophesied [*vaticinatus est*] under divine inspiration that the world would endure only six thousand years, during which time nearly two revolutions are completed."[69] In other words, knowing exactly the revolutions of the Wheel of Fortune eccentric allowed one to interpret the celestial cause properly and, hence, the meaning of the prophecy of Elijah.

The Elijah prophecy was very well known and much commented on when Rheticus composed these words.[70] One of the key eschatological texts of the thirteenth-century Joachimite prophecies was undoubtedly operative: "Helias cum veniet restituet omnia" (When the Messiah arrives, all will be restored).[71] But when in world history would the Messiah come? At Wittenberg, the *Chronicle* of Johannes Carion became the principal text for interpreting the meaning of the four monarchies and the Elijah prophecy. The entire work was organized into three books, following the three periods of world history allegedly prophesied in the saying of Elijah. In 1550, it was rendered into English:

The worlde shall stande syxe thousand yeares
and after shall it falle.
Two thousand yeares wythout the Lawe.
Two thousande yeares in the lawe.
Two thousand yeares the tyme of Christ.
And yf these yeares be not accomplyshed, oure
synnes shall be the cause, whyche are great
and many.[72]

Many other chronologies, such as those of Achilles Pirmin Gasser, were modeled on this tripartite structure.[73] The *Chronicle* claimed to reveal God's plan as the master narrative of world history, the key to understanding the biblical prophecies. The heavenly revolutions were part of this divine plan and helped to explain particular fortunes and misfortunes in the larger scheme of sacred and profane history. With good reason, Robin Barnes contends that the *Chronicle* was "the main vehicle for the entry of the latter [Elijah] scheme into Reformation thought."[74]

Melanchthon and Johann Carion had both studied with Stöffler at Tübingen; later, Carion became court astrologer to the elector of Brandenburg. Melanchthon rewrote Carion's manuscript, with assistance from Caspar Peucer, who prepared for Wittenberg students a large fold-out table of the book's topics. The two detailed volumes of the rise and fall of kingdoms and rulers appeared in at least one English and fifteen German editions before 1564 and, after 1558, in various Latin editions.[75] Melanchthon also connected the Elijah prophecy explicitly to the study of astronomy in his *Oratio de Orione:* "The opinion attributed to Elijah should not be condemned: the world will last 6,000 years, and then, after that, there will be a conflagration. 2,000 years of idleness; 2,000 years of Law; 2,000 years until the days of the Messiah."[76] And Luther gave prominence to Elijah's prophecy in his *Supputatio Annorum Mundi* (Wittenberg, 1541), a great, biblically derived chart of world history, modeled on analogy with the six days of creation. Luther estimated that Christ had been born when the world was 3,960 years old—not quite in agreement with Elijah's figure of 4,000—and that by his own time in A.D. 1540, some 5,500 years had elapsed.[77]

Reformation prophesiers freely appropriated exegetical resources from pre-Lutheran authors. Ironically, the Wittenbergers owed much to Pico della Mirandola before his skeptical period: he discussed the Elijah prophecy at length in his Genesis treatise, the *Heptaplus.*[78] As Rheticus certainly knew the *Heptaplus* directly, it is possible that Copernicus was also familiar with it.[79] According to Pico, Elijah's prophecy pertained to the fourth day of Genesis, that is, the fourth millennium after the Creation. Quoting the Hebrew text, Pico translated as follows: "The sons or disciples of Elijah said: six thousand years for the world; two thousand empty, two thousand for the law, and two thousand for the day of the Messiah, and because of our sins, which are many, there have passed those which have passed."[80]

These numbers left some exegetical difficulties for Pico. Against the "Hebrew interpreters," he argued that the period from Adam to Abraham was only 1,848 years: "Thus it came about that the fullness of the law succeeded the emptiness not after the second millennium but within its limits."[81] Likewise, Christ appeared 3,508 years after the beginning of the world—hence, within the limits of the fourth millennium rather than after its limits had passed.[82] Pico thus represented a pre-Reformation, Catholic alternative to Carion's estimate for the time of Christ's arrival. He concluded that the Catholic Church appeared just as does the "light of the moon" and that it shines on the world with a "countless multitude of martyrs, apostles, and doctors who all became famous within 500 years after the death of Christ."[83]

These learned reckonings about Christ's return help us to appreciate a further reason why Rheticus and Copernicus chose to call their work the *First Narration concerning the Books of Revolutions,* and why the masterwork itself was called *Six Books concerning the* Revolutions *of the Celestial Orbs.*[84] Knowledge of the proper periods of heavenly revolution permitted one to understand and forecast not merely annual and local terrestrial effects but also *longue durée,* biblical, and world-historical consequences. Indeed, at the end of his brief interpretation of the Elijah prophecy, Rheticus addressed Schöner: "God willing, I shall soon hear from your own lips how it may be inferred from great conjunctions and other learned conjectures of what nature these empires were destined to be, whether governed by just or oppressive laws."[85] The book's audience, in other words, was invited to draw further connections between the new solar model, the qualities of rising and falling empires, and the second coming of Christ. Yet, in view of the sensitive authorial arrangement—a Lutheran author narrating the astronomical hypotheses of a Varmian canon—it is also understandable why there could be no mention of the Melanchthon-Carion, antipapist reading of world history.

After Copernicus's death, however, Rheticus continued to develop further entailments of astrology and scripture, as for example in his preface to Johannes Werner's *On Spherical Triangles* and *On Meteoroscopy*:

> We know that the stars govern things below according to the order of nature, but the Creator of the heavens, who calls the stars by their name and who prescribes their measure and limit, who causes them to stop in their paths whenever he wishes, governs the effects of the stars as he wishes. Equally, through Joshua, he stopped the Sun in the sky, and through Ezekiel he caused it to reverse its path. But, as far as the stars are concerned I have no doubt that for the Turkish empire there is impending disaster, momentous, sudden, and unforeseen, since the influence of the Fiery Triangle is approaching, and the strength of the Watery Triangle is declining. Moreover, the anomaly of the sphere of the fixed stars is nearing its third boundary. Whenever it reaches any such boundary, there always occur the most significant changes in the world and in the empires, according to historical reports. And it is at this moment that God has exercised his judgment and that he has deposed the powerful from their thrones and lifted up the lowly—which happened to Xerxes when he invaded Greece with his large army.[86]

Immediately following this passage, Rheticus added a revealing remark: "Nicolaus Copernicus, the never sufficiently praised Hipparchus of our century, was the first to discover the law of the anomaly of the orb of the fixed stars, as we explained long ago. For when I had traveled to Prussia, for about three years, just as I was about to depart, this remarkable old man exhorted me to try to bring to perfection that which he himself had the means to bring to perfection, but which he had been prevented from doing by virtue of his age and destiny."[87]

From this passage, it seems quite clear that Copernicus saw in Rheticus a person who could complete the great reform of the science of the stars initiated by Regiomontanus and carried forward by himself. It was Copernicus's "destiny" to reform astronomy, the theoretical part of the heavenly disciplines that concerned the calculation of

the revolutions. Now that Copernicus was an old man, Rheticus saw it as his fate to carry on with a reform of the practical parts of astronomy, trigonometry, and the tables of motions, and perhaps also astrological theoric, the equivalent of the *Tetrabiblos*. Here, one should recall again Rheticus's chastisement of Pico for impugning both astronomy and astrology—a mistake that allegedly he would not have committed had he lived to witness Copernicus's achievement. Rheticus placed this triumphant comment just at the end of the "prophecy" passages in the *Narratio*.[88]

CELESTIAL ORDER AND NECESSITY

The first published statement of Copernicus's heliocentric theory occurred, almost inconspicuously, one-third of the way through Rheticus's *Narratio* (at the end of chapter 7).[89] "It is assuredly a divine matter," wrote Rheticus, "that the sure explanation of the celestial phenomena should depend on the regular and uniform motions of the terrestrial globe alone."[90] The placement of this announcement so late in the text differs considerably from the structure of *De Revolutionibus*, where Copernicus introduced the principal claim in the preface and built his case systematically in the first ten chapters of book 1. The difference may be accounted for by considering that Rheticus's rhetorical strategy was directed to the Wittenberg network rather than to a papal audience: first he recommended Copernicus's astronomical improvements, then he associated these with a world-historical prophecy, and only then did he begin to develop the case for connecting those benefits of the new astronomy with a new celestial order.

The new hypotheses were finally introduced and defended (in chapters 8–10). The strategy now drew the reader directly into the scene of persuasion. Again, Rheticus used biographical narrative to create an atmosphere of verisimilitude. He presented to Schöner the reasons alleged to have persuaded Copernicus himself to depart from the hypotheses of the ancients (chapter 8 was titled "The Principal Reasons Why the Hypotheses of the Ancient Astronomers Must be Abandoned"). He then asked Schöner whether the reasons that persuaded Copernicus might be taken to be good reasons (titling chapter 9 "Transition to the Enumeration of the New Hypotheses of Astronomy as a Whole").

Chapter 8 is especially interesting because Rheticus there engaged in some of his most aggressive claims on behalf of Copernicus's hypotheses. Some of these involved the appearance of necessity: things could not be otherwise. This impression was enhanced because Rheticus did not consider alternative explanations. He mixed such allegedly necessitarian arguments together in a list of other reasons that are congruent with the more dialectical or probabilist passages of *De Revolutionibus*. One is tempted to follow the judgment of historians who say that in this case Rheticus had acted alone. But again, I believe that Copernicus would not have allowed such claims to be published had he not agreed with them.[91]

Separate consideration of three arguments helps to show where and how Rheticus tried to bring readers to see persuasive connections between the Earth's motion and other celestial phenomena. First, Rheticus asserted that "the indisputable precession of the equinoxes and the change in the obliquity of the ecliptic persuaded my teacher to assume that the motion of the earth could produce most of the appearances in the heavens, or at any rate, save them satisfactorily."[92] Disagreements about the length of the tropical year were the result not of defective instruments ("as was heretofore believed") but rather of a "completely self-consistent law." If the containing stellar sphere is at rest, then what causes the equinoxes to precess? Rheticus answered that Copernicus discovered that a motion of the Earth was responsible ("or, at any rate, could save the heavenly appearances satisfactorily") and, further, that it was this discovery that "persuaded my teacher to assume" the Earth's motion. Once again, Rheticus used a biographical reference rather than a formal demonstration to promote assent: he wanted Schöner to take notice that Copernicus began to think about, and then to assume, the motion of the Earth in association with the precessional effects. At least one historian, Jerry Ravetz, has even tried to link this passage to Copernicus's work on the calendar and to his discovery.[93]

The second claim was more audacious: "My teacher saw that only on this theory [*hac unica ratione*] could all the circles in the universe be satisfactorily made to revolve uniformly and regularly about their own centers, and not about other centers—an essential property of circular motion."[94] Uniformly revolving circles are an al-

legedly unique consequence, in other words, of assigning an annual motion to the Earth.[95] This argument immediately evokes the first *petitio* in the *Commentariolus* (although it is noteworthy that Rheticus speaks here of circles rather than of spheres and orbs).[96] Yet, if this claim were logically true—which it is not—then it would make the (equantless) planetary mechanisms a direct result of the assumption of a moving Earth. Again, we may have here a biographical residue, or an eager young man's exaggerated misunderstanding of a theory whose author he admired.[97] In either case, the feature of necessity was allowed to stand in the text.

The third reason also involved the assertion of an unwarranted inference. Here, Rheticus used the authority of Pliny to assert that "the planets have the centers of their deferents in the vicinity of [*circa*] the sun, taken as the middle of the universe [*medium universi*]." Still on Pliny's authority, he continued to the placement of Mars: "Mars unquestionably shows a parallax sometimes greater than the sun's, and therefore it seems impossible that the earth should occupy the center of the universe." It is interesting that Rheticus made this weighty assertion without employing the persuasive appeal of a biographical reference to a specific observation made by Copernicus. This is one of the few places in the text where an alleged calculation without reference to an observation plays a prominent role in the discussion of planetary order. Rheticus evidently thought it sufficient to state the matter as a conclusion that Mars's variations in distance "surely cannot in any way occur on the theory of an epicycle" and, thus, "a different place must be assigned to the earth." Over a half century later, Michael Maestlin would endorse Rheticus's statement in his edition of the *Narratio Prima* (1596) on the authority of an observation reported in a letter from Tycho Brahe to Caspar Peucer.[98]

NECESSITY IN THE CONSEQUENT

Apart from these necessitarian gestures, Rheticus followed a tack closer to the strategy employed by Copernicus three years later and immediately available in the copy of Copernicus's manuscript sitting on the desk before him. He presented the new world hypothesis as an assumption that his teacher felt compelled to adopt "as a mathematician" because, in comparison with the hypotheses of Ptolemy, it led to a multitude of true and harmonious consequences. Rheticus now used for the first time a powerful phrase that never appeared in *De Revolutionibus*: "A most absolute system of the motions of the celestial orbs" (*motus orbium coelestium absolutissimo systemate*). Here, Rheticus repeatedly drove home the systematicity of the new celestial hypotheses on the basis of what looks like inference to the simplest explanation. He tried to augment the uniqueness of the case with a variety of rich tropes and dialectical commonplaces that are absent or only hinted at in *De Revolutionibus*.

The first sort of simplicity is the kind obtained when, as with a clock, many effects are derived from a single cause. This is the sort of image that one usually associates with the seventeenth-century trope of God as clockmaker, creator of an economical system of geared wheels. Rheticus, however, did not yet speak with quite the uninhibited confidence of the following century. Like Copernicus, he used humanist-style, rhetorical-dialectical questions to put the reader in a position where the answers to his questions would seem obvious and irrevisable: "Should we not attribute to God that skill . . . ?" "What could dissuade my teacher, as a mathematician, from adopting a theory suitable to the motion of the terrestrial globe?"[99]

Rheticus then tried to make it seem that astronomy's uncertainty and the existence of a rule of celestial order were generally established matters. His teacher realized that the main "reason for all of the uncertainty in astronomy" ("omnis incertitudinis in astronomia causam") was that a certain "rule" (*regulam*) had been ignored. This rule held that "the order and motions of the heavenly spheres agree in an absolute system."[100] For the remainder of this section, Rheticus continually contrasted Copernicus's adherence to this rule and its disregard or violation by the ancients and their successors.

I think that the purpose of this emphasis on order and its rhetorical amplification was to draw attention away from the inability of Copernicus and Rheticus to satisfy Aristotle's standard of necessary demonstration: reasoning that starts from true premises that require no prior justification, rather than dialectical reasoning, which begins from probable premises or commonplace topics ordinarily held to be true.[101] Both Rheticus and Copernicus had opportunity to become acquainted

with Aristotle's standard, as it was quite well established in the curricula at Krakow, Bologna, Padua, and Wittenberg.[102] And their (understandable) difficulty in meeting its demands could well explain the *Narratio*'s earlier strained emphasis on presumed features of necessity. Notable Copernicans later in the century, such as Kepler and Galileo, would grapple, sometimes quite publicly, with the matter of furnishing an apodictic proof for the new hypotheses.

Strikingly, Rheticus altogether avoided references to discussions of apodictic proof in the *Posterior Analytics*. He returned again and again to commonly accepted tropes of harmony and order: the phenomena appear to be linked together "as by a golden chain";[103] "the remarkable symmetry and interconnection of the motions and spheres . . . are not unworthy of God's workmanship"; such "relations . . . can be conceived by the mind (on account of its affinity with the heavens) more quickly than they can be explained by any human utterance."[104] And he applied, quite vividly, the metaphor of musical harmony itself: "We should have wished them [the masters of this science of astronomy], in establishing the harmony of the motions, to imitate the musicians who, when one string has either tightened or loosened, with great care and skill regulate and adjust the tones of all the other strings, until all together produce the desired harmony, and no dissonance is heard in any."[105] Rheticus immediately invoked Arabic astronomers to exemplify the consequences of failing to follow this rule. If only al-Battani had followed this common precept, he lamented, "we should doubtless have today a surer understanding of all the motions." In fact, because the widely used Alfonsine tables built on al-Battani, eventually astronomy would collapse altogether. These were the "principal reasons" for abandoning the ancients' hypotheses.

And with this assessment, Rheticus came at last to the Sun's place in the universe. With no explicit reference to Peurbach, he stated that even under the usual principles of astronomy, the celestial phenomena are connected to the Sun's mean motion. Further, the ancients already regarded the Sun as possessed of an important metaphoric status as "leader, governor of nature, and king." But these ancient solar encomia were later ignored. At this point, Rheticus adduced Aristotle as an authority on the Sun—not the *Narratio*'s last favorable reference to him: "How does the Sun accomplish its task [of governing nature like a king]? Is it in the same manner as God governs the entire universe (as Aristotle has magnificently described in his *De mundo*)? Or, does the Sun, in traversing the entire universe so often and resting nowhere, act as God's administrator in nature? This question appears not yet altogether explained or resolved."[106] And who better to decide this question than "geometers and philosophers (on condition that they have a smattering of mathematics)"?

With clear echoes of Regiomontanus and Domenico Maria Novara, this is the first suggestion that the resolution of the problem of celestial order was going to involve a new representation of the conditions of disciplinary authority. Yet there was no claim that the old approach must be entirely rejected—just the role of the Sun: "My teacher is convinced . . . that the rejected method of the sun's rule in the realm of nature must be revived, but in such a way that the received and accepted method retains its place. For he is aware that in human affairs the emperor need not himself hurry from city to city in order to perform the duty imposed on him by God; and that the heart does not move to the head or feet or other parts of the body to sustain a living creature, but fulfills its function through other organs designed by God for that purpose."[107] In this remarkable and inspired image, Rheticus called attention to the very phenomenon noticed by Peurbach: the presence of the Sun's motion as a component of each of the planetary models. Copernicus had shown the efficient cause of this perceived effect in the planets' apparent motions. Equally important, the same cause produced another major consequence: "A sure doctrine of celestial phenomena in which no change should be made without at the same time reestablishing the entire system" ("certam rerum coelestium doctrinam, in qua nihil mutandum, quin simul totum systema . . . restitueretur").[108] In other words, the necessity that failed to be found in the major premise now turned up as a consequent.

Were the reasons that allegedly persuaded Copernicus now sufficient to persuade Schöner and other readers like him? "I interrupt your thoughts, distinguished sir," Rheticus wrote, "for I am aware that while you listen to the reasons [*causas*] investigated by my teacher with remarkable learning and great devotion, for renewing the hypotheses of astronomy, you thoughtfully

consider what foundation [*ratio*] may finally prove to be suitable for the hypotheses of the astronomy reborn."[109] This transition seems to hint at the possibility that Copernicus's reasons were not enough, that different authorities and different, perhaps more general sorts of reasons, must be brought to bear. In fact, in chapter 9 ("Transition to the Enumeration of the New Hypotheses for the Whole of Astronomy"), Rheticus began to shift the burden of proof away from Copernicus alone. His main theme was that both astronomy and natural philosophy, as disciplines in their own right, proceed inductively: "In physics as in astronomy, one proceeds as much as possible from effects and observations to principles." Rheticus worked here with a double-edged sword: if astronomy and natural philosophy begin with a study of consequences and then work their way back to first principles, then this procedure should apply to all its individual practitioners, as much to Aristotle and Ptolemy as to Copernicus. Notably absent was the more stringent standard of necessary demonstration from the *Posterior Analytics*. In fact, Rheticus carefully avoided citing any of Aristotle's logical treatises. He presented instead a quite modest and epistemologically restrained Aristotle, as in this passage, where he cited *De caelo:* "Anyone who declares that he must be mindful of the highest and principal end of astronomy will be grateful with us to my teacher and will consider as applicable to himself Aristotle's remark: 'When anyone shall succeed in finding proofs of greater precision, gratitude will be due to him for the discovery.' "[110] Or, again: "If he could hear the reasons for the new hypotheses, he would recognize what he had proved in [his physics disputations] what he had assumed as principle without proof."[111]

Et tu quoque, Ptolemy: "In my opinion, Ptolemy was not so bound and sworn to his own hypotheses that, were he permitted to return to life, upon seeing the royal road blocked and made impassable by the ruins of so many centuries, he would not seek another road over land and sea to the construction of a sound science of celestial phenomena."[112] Here was the familiar Renaissance topos put in the service of a nascent, inductivist image: the smart and reasonable ancients, capable of changing their minds, versus unreasonable, hidebound contemporaries. If both Aristotle and Ptolemy regarded astronomy as *revisable* knowledge, then someone who came along with better explanations of the phenomena should prevail.

In the opening of chapter 10 ("The Arrangement of the Universe"), it is again Aristotle's authority that grounds the relationship in a weaker logic of relevance rather than a stronger one of causal necessity between hypothesis and results: "Aristotle says: 'That which is the cause of truth in the derived effects is the most true.'" This dialectical approach, echoed later in *De Revolutionibus*, would do much work for both Copernicus and Rheticus. In later sixteenth-century discussions, it turned out to be the point of greatest logical vulnerability, because many writers easily recognized that it was logically valid for a true conclusion to be derived from false premises. Yet Rheticus was quite firm about Copernicus's pattern of reasoning: he has "assumed such hypotheses as would contain causes capable of confirming the truth of the observations of previous centuries" as well as "all future astronomical predictions of the phenomena."[113]

In chapter 10, however, the phenomena were all qualitative. Rheticus engaged in brief, but close, paraphrases of *De Revolutionibus* book 1, chapters 6, 8, and 10, which covered the fixed, outermost sphere, the "common measure of the planetary orbs" ("communis orbium planetarum inter se dimensio"), the magnitude of the universe as "truly similar to the infinite," and "the remarkable symmetry and interconnection of the motions and spheres" ("admiranda . . . motuum et orbium symmetria ac nexus"), as compared with the arbitrary ordering of the "common hypotheses."

At times, the explications were clearer—or, at least, fuller—than the treatment of comparable topics in *De Revolutionibus*. For example, Rheticus wrote that "the orb of each planet advances uniformly with the motion assigned to it by nature and completes its period without being forced into any inequality by the power of a higher orb."[114] This passage makes it evident that there was another account with which Rheticus was arguing—perhaps Achillini's version of the Eudoxan-Aristotelian celestial physics—in which the outermost spheres communicate motion to the lower.[115] In *De Revolutionibus* book 1, chapter 4, however, Copernicus wrote more tersely and was less inclined to display alternative possibilities. At other times, Rheticus introduced suggestive speculations that find no comparable treatment in *De Revolutionibus*: for example, "The

larger orbs revolve more slowly, as is proper, whereas the orbs that are closer to the sun, which may be said to be the source of motion and light, revolve more swiftly."[116]

Most prominently, at the end of chapter 10, Rheticus speculated on the cause of there being only six planets. The concern may well have been his alone, as it is not attributed to "my teacher." The early Lutherans, as we have seen, were deeply preoccupied with both the beginning and the end of the world. They were obsessed with the unfolding of prophecy through world history, with all kinds of natural signs of the end of time. If Rheticus regarded the motion of the Earth's eccentric as governing the Elijah prophecy, then he must have seen in the Copernican celestial order the effects of divine planning. However, Rheticus considered the celestial harmony (*harmonia coelestis*), that is, the "entire system" (*totum systema*), to be the consequence but not the cause of there being six planets. Because, for Copernicus, the Moon no longer counted as one of the planets, it was the only remaining body whose center of revolution was the Earth. Rheticus proposed that the number six "is honored beyond all others in the sacred oracles of God and by the Pythagoreans and the other philosophers. What is more agreeable to God's handiwork than that this first and most perfect work should be summed up in this first and most perfect number?"[117] Evidently Rheticus could not find a biblical prophecy to accommodate to Copernican celestial order that was comparable to the Elijah prophecy earlier associated with the Earth's eccentric. He settled, instead, for a pre-Christian/Pythagorean revelation that further amplified the Pythagorean authority invoked in *De Revolutionibus*.

Here and elsewhere, Rheticus's enthusiasm was palpable—more so than that of anyone else until Kepler. Yet he did not match the "heat" generated by his metaphors and necessitarian claims with an effective didactic apparatus. Considering that he was arguing for a new picture of the world, it is strange that there are no woodcut representations of the "world system"; in fact, Rheticus provided detailed descriptions of quite technical matters but no diagrams at all, let alone moving ones.[118] Although the text was scattered with parameters, Rheticus made no effort to organize and tabulate the distance parameters and eccentricities in a way that could prove immediately useful for computation or pedagogy. Michael

Maestlin later felt the need to add diagrams to his edition of the *Narratio Prima* and to augment Rheticus's text with a separate treatment of the planetary theories to make it more useful for teaching purposes.[119] The work, then, was, as its title indicated, a first exposition of something new. In spite of its clarity and its skillful use of humanist rhetorical resources, it seems unlikely that the *Narratio Prima* was ever intended as a pedagogic text.

THE ASTRONOMY WITHOUT EQUANTS

Theoretical astronomy must describe the world accurately, and it must provide arguments for its claims. But what kinds of arguments? Rheticus wrote rather ambitiously: "The hypotheses of my teacher agree so well with the phenomena that they can be mutually interchanged, like a good definition and the thing defined"—as though the principles that things are few rather than many (simplicity) entailed the world's being this way and no other (necessity).[120] From the start, Copernicus seems to have been driven by a wish for necessity, based on a conviction in the economy of the assumptions. Rheticus strongly articulated the voices of both sides of the explanatory coin— the beauty of the assumptions, the necessity of the consequences. Orbs revolving uniformly about their own centers and propelled by "their own nature"—rather than by contiguous orbs—lay at the heart of the new, "simple" astronomy. Ptolemy's equant circles, on the contrary, involved equalizing motions that produced uniform motion about a point that was neither the center of the universe nor the center of the circle on which the planet revolved.[121] For Copernicus, the equant model was incompatible with a physical principle: all celestial motions are uniform or compounded of uniform, circular motions. This is simply what orbs do. There is no property of impenetrability required to achieve these motions; or, at least, Rheticus and Copernicus mention none.[122] Rheticus repeatedly trumpeted Copernicus's replacement of the equant. Oscillatory, rectilinear motions, such as deviations in latitude and slow changes in the equinoctial points, would be cleverly explained using a combination of two uniformly moving circles.[123] Much of the last half of the *Narratio* was a summary of these matters— what Copernicus later called in *De Revolutionibus* the "demonstrations" of the planetary mecha-

nisms, the geometrical demonstrations long since promised in the *Commentariolus*. What it took Copernicus 133 folios to do in books 3 and 4, however, Rheticus compressed into a few pages.

The Earth, itself a globe, now took over the duties of the "first motion" generated in the old hypotheses by the outermost sphere and the Sun (daily risings and settings). But once a single motion had been ascribed to the Earth—"like a ball on a lathe"—then, Rheticus said, other motions might be ascribed to it.[124] The second motion consisted of the "center of the earth, together with its adjacent elements and the lunar sphere carried uniformly in the plane of the ecliptic by the great circle." Rheticus here described the second motion of the Earth in the manner of the mathematicians, that is, as an assumption. He gave no hint that there were any physical difficulties to be solved until he reached the very last section of the *Narratio*—the rich and suggestive "Praise of Prussia" (*Encomium Prussiae*).

Here Rheticus displayed his humanist credentials, once again adroitly representing Aristotle as a fallible, time-bound human, author of a provisional physics. Setting the scene locally, Rheticus invoked the authority (and the words) of Copernicus's lifelong friend, the Varmia canon Tiedemann Giese. He presented Aristotle not as a scholastic logician of the universities, laying down rules of proper procedure, but as a natural philosopher who had followed the astronomers of his own time. This time- and culture-bound "humanist's Aristotle" was one whose judgment was not fixed but rather subject to criticism and reversal. Aristotle said that he had followed the mathematicians in assuming that the Earth is at the center of the universe. By the same token, Giese believed that now contemporaries too would be compelled to take another look at the "true basis of astronomy": "By returning to the principles with greater care and equal assiduity, we must determine whether it has been proved that the center of the earth is also the center of the universe."[125] Giese then raised a series of dialectical questions to suggest the direction of new answers: "If the Earth were raised to the lunar sphere, would loose fragments of Earth seek, not the center of the Earth's globe, but the center of the universe, inasmuch as they all fall at right angles to the surface of the Earth's globe? Again, since we see that the magnet by its natural motion turns north, would the motion of the daily rotation or the circular motions attributed to the Earth necessarily be violent motions? Further, can the three motions, away from the center, toward the center, and about the center, be in fact separated?"[126]

Were these Copernicus's questions as well? Rheticus did not explicitly ascribe them to "my teacher," and the *Encomium* itself ended a few lines later. But it is difficult to believe that such criticisms of Aristotle were those of Rheticus and Giese alone. Rheticus wrote his treatise in the space of ten weeks; Copernicus had been thinking about these problems for at least three decades. Using the humanist strategy of posing rhetorical questions, the questions Giese posed were ideal for student academic disputations and, after the republication of the *Narratio Prima* in 1566, they undoubtedly provided an important heuristic for second- and third-generation Copernicans.

PRINCIPLES VERSUS TABLES WITHOUT DEMONSTRATIONS

Rheticus's *Encomium Prussiae* contains what appear to be the residues of a debate about how the new hypotheses ought to be presented and to which audiences they ought to be directed. Those considerations alone raise the suspicion that much conscious strategizing preceded the book's appearance. Indeed, it helps to explain why there were two quite different presentations of Copernicus's views.

The *Narratio* was explicitly directed to a Nuremberg astrologer; it made no reference to Melanchthon and Wittenberg. Yet Rheticus suffused his *Encomium* with classical and astrological images of a sort that clearly echoed Melanchthon's own. Using Pindar's Olympian ode, he painted a lofty analogy. Once, the Sun-god Apollo brought forth riches from the isle of Rhodes, previously hidden from the Sun's rays beneath the sea; now, "by an act of the gods, Prussia passed into the hands of Apollo, who cherishes it as once he cherished Rhodes, his spouse." The progeny of Prussia and Apollo are its great cities, its great laws, councils, and literature, and its great men: Königsberg (which produced Albrecht, duke of Prussia, margrave of Brandenburg); Toruń (Copernicus); Gdańsk (its council); Frombork (Bishop Johann Dantiscus, the head of Copernicus's order); Malbork (the king of Poland's "treasury"); Elbląg (an "ancient settlement where the sacred pursuit of literature is undertaken"); Chełmno (formerly

Kulm; "famous for its literature" and the Law of Chełmno, and also the seat of Bishop Tiedemann Giese). The *Encomium* is a poetically veiled praise of all that has preceded, presenting the Prussian Copernicus rising like the once-hidden isle of Rhodes to receive the rays of the Sun, whose true principles he now exposes.[127]

Yet Rheticus presented a moderate and cautious Copernicus, worried about the likely effects that his views would have, especially among natural philosophers. Even though he realized that the observations required new hypotheses that would "overturn" (*eversurae essent*) the old ideas of celestial order and "do violence to the senses" (*sensibus nostris pugnaturae*), Copernicus

decided that he should imitate the Alfonsine astronomers rather than Ptolemy and compose tables with accurate rules but no proofs. In that way he would provoke no dispute among philosophers; ordinary mathematicians would have a correct calculus of the motions; but true scholarly men trained in the arts, upon whom Jupiter had looked with unusually favorable eyes, would easily arrive, from the numbers set forth at the principles and sources from which everything was deduced. . . . And the Pythagorean principle would be observed, according to which one ought to philosophize in such a way that philosophy's inner secrets are reserved for learned men, trained in mathematics, etc.[128]

Giese, Copernicus's close friend and sympathizer, was represented as urging the full revelation of these "inner secrets." Contrary to the Pythagorean injunction, the new hypotheses should appear in print. Now, of course, "My friends urged me to publish" was a well-known early modern topos. But Rheticus's naming of Giese makes it likely that this was a genuine reference to the discussions at Lubawa in 1539. In Rheticus's text, Giese pushed strongly for the view that, although desirable, Copernicus must present to the world more than an improved calendar for the church and better tables of planetary motion. The language here is emphatic: "His Reverence pointed out that such a work would be an incomplete gift to the world [*imperfectum id munus reipublicae*] unless my teacher set forth the reasons for his tables [*causas suarum tabularum*] and also included, in imitation of Ptolemy, the system or theory and the foundations and proofs upon which he relied to investigate the mean motions and prosthaphaereses and to establish epochs as initial points in the computation of time." The planetary tables must not appear without an account of their underlying assumptions. This must be both a theoretical and a practical astronomy. But Giese thought that this situation was all the more serious because "the required principles and hypotheses are diametrically opposed to the hypotheses of the ancients." In other words, it was a matter not just of giving reasons but of giving reasons more persuasive than the alternative. Giese continued: "Among men capable of speculation [artifices][129] there would be scarcely anyone who would hereafter examine the principles of the tables [tabularum principia] and publish them after the tables had gained recognition as being in agreement with the truth." There was no place here, he asserted, for the practice frequently adopted in kingdoms, public affairs, and deliberations, "where decisions are kept secret until the subjects see the fruitful results and remove from doubt the hope that they will come to approve the plans."[130]

Giese's voice in the narrative now vigorously took on Aristotle and the philosophers: "After convincing himself that he had established the immobility of the earth by many proofs, Aristotle finally takes refuge in the argument" that the Earth's placement at the center of the universe is an assumption that saves the phenomena.[131] The more learned philosophers (prudentiores et doctiores) would recognize that Aristotle had made a contestable assumption (in De caelo, book 2, chaps. 13–14)—precisely in opposition to the Pythagoreans—and they would then need to investigate whether Aristotle had really demonstrated that the center of the Earth was also the center of the universe.

THE PUBLICATION OF
DE REVOLUTIONIBUS
OSIANDER'S "AD LECTOREM"

Andreas Osiander was a prominent leader of the Reformation movement in Nuremberg, theologically headstrong but extremely influential and effective in spreading his views.[132] Various political leaders sought his counsel, among them Albrecht, margrave of Brandenburg-Ansbach (later duke of Prussia), an important prince whom Osiander succeeded in converting to the Reformed view and whose interest in the stars was considerable. Both Rheticus and Erasmus Reinhold dedicated

works to him (respectively, Chorographia tewsch, 1541, and Prutenic Tables, 1551). In addition, Thomas Cranmer, the future archbishop of Canterbury, lived in Osiander's Nuremberg house during a long visit to the Continent for the purpose of soliciting advice about King Henry VIII's pending suit for annulment. The relationship was a warm one: Cranmer eventually married Osiander's niece, Osiander dedicated his Harmony of the Gospels (1538) to Cranmer, and the king eventually found a satisfactory legal resolution to his marital difficulties.[133]

Osiander also gave advice on the publication of De Revolutionibus. His involvement in that process was no accident, as his authority on civil and religious questions was considerable: Could the children of Anabaptists be forcefully baptized? No, said Osiander, but their parents could be exiled and the children reared and baptized by a Lutheran family. Could one swear an oath "by all saints"? Yes, replied Osiander, because the word saints does not refer exclusively to the saints of the Roman Church. Regarding books that could be printed and sold, Nuremberg had a censorship board, and it was said of Nuremberg's citizens that "what Osiander holds and believes, they must also believe."[134]

Osiander was also a respected member of the Nuremberg-Wittenberg friendship circle centered on Melanchthon. He had been involved in Schöner's appointment to the Nuremberg Gymnasium in 1526. Schöner named his son Andreas after Osiander. Melanchthon invited him to contribute an "Ornamentum" to Schöner's Tabulae Resolutae in 1536, although Osiander did not comply with the request.[135] In March 1540, when Andreas Aurifaber (1512–59) sent a copy of the Narratio Prima to Gasser, he sent another copy to Osiander, his future father-in-law.[136] Between 1543 and 1546, Petreius published five works by Osiander.[137] All of these contacts suggest that Rheticus's decision to entrust the manuscript of De Revolutionibus to Osiander must have had something to do with the esteem in which Osiander was held by Melanchthon, Schöner, Petreius, and even Copernicus himself.

Most important, when Osiander received the manuscript at Nuremberg—some time after Rheticus left for a new post at the University of Leipzig in October 1542—his knowledge and perceptions of the new hypotheses had already been shaped principally by reading the Narratio Prima.

Hence Osiander would have been familiar with the earlier strategic discussions (reported in the *Encomium Prussiae*) among Giese, Rheticus, and Copernicus concerning theoretical principles versus practical tables without demonstrations. He would have known, as well, that Giese had pushed for the stronger of the two positions. If he knew the second edition, then he would have been familiar also with Gasser's judgments about the "restoration of the most true astronomy."

Just one month after the second edition appeared, Osiander wrote to both Rheticus and Copernicus regarding the presentation of *De Revolutionibus*. Fragments of these letters, written on the same day (April 20, 1541), later came into Kepler's possession and are known only through his excerpts.[138] They show that Osiander was already pushing privately for the skeptical view of astronomical knowledge that he would later articulate in his anonymous "Letter to the Reader." The discussion was, in part, strategic, intended to forestall criticisms. Osiander wrote to Copernicus that something could be done to placate the "peripatetics and theologians whose future opposition you fear." Because Osiander himself was a theologian, perhaps Copernicus and Rheticus had initially sought his personal advice on this matter. But Osiander's counsel was not motivated so much by his own fear of opposition as by what he himself believed about hypotheses and about the proper organization of the domains of knowledge. "I have always felt," wrote Osiander to Copernicus, "that [hypotheses] are not articles of faith but rather foundations of calculation, so that it matters not at all whether they be false so long as they display exactly the phenomena of motion. . . . For this reason, it would be desirable if you would touch upon something about this matter in the preface."[139]

Osiander the preacher believed that theology concerns itself with "articles of faith," astronomy with "foundations of calculation." Hence astronomy can operate quite well from false premises. Moreover, as Osiander told Rheticus even more fully in the second letter, "The Peripatetics and Theologians will easily be placated if they hear that there can be different hypotheses for the same apparent motion and that these [of Copernicus] are not presented because they are certain but, rather, because they permit the most convenient way to calculate the apparent and compounded motions; and, it is possible that some-

one else may contrive other hypotheses so that to explain the same apparent motion one person may present suitable mental images (*imagines*), another even more suitable; and, each one is free—even better: each should be thanked—if he contrives even more convenient hypotheses." Osiander then added his opinion about how such a form of presentation could induce gradual assent: "In this manner, induced to leave behind their severe critique in order to pass over to the pleasures of investigation, first they will become more reasonable; then, after they have sought in vain, they will come over to the author's opinion."[140]

When *De Revolutionibus* appeared two years later, it contained Osiander's views, placed anonymously in the form of a polemical "Ad Lectorem" ("Letter to the Reader") immediately after the frontispiece, and without the permission of either Rheticus or Copernicus. Osiander had shown the same independence in this matter as he had in the theological controversies in which he was embroiled and which eventually caused him to fall into disfavor in Nuremberg.[141] The "Ad Lectorem" made no direct reference to the *Narratio Prima*, but Osiander clearly presumed the existence of that work when he began it as follows: "Since the novelty of the hypotheses of this work has *already* been widely reported."[142] Immediately, the letter took on a sustained, argumentative tone that echoed what is known from the prior correspondence. The overriding theme was the reassurance that the work would not disrupt the presumed hierarchy of the disciplines: "learned men" need not fear that "the liberal arts established long ago upon a correct basis" will be "thrown into confusion." The higher disciplines of theology and philosophy seek to know the causes of things; in fact, they seek to know true causes, although "neither of them will understand or state anything certain, unless it has been divinely revealed to him." Astronomy, on the other hand, is incapable of finding true explanations: "For these hypotheses need not be true nor even probable; if they provide a calculus consistent with the observations, that alone is sufficient." Osiander offered here his only example of astronomy's limited epistemic capacity: the perplexing relationship between the size of Venus's epicycle and the planet's apparent diameter.[143] Rheticus had mentioned Venus's epicycle as an example of the "vast commotion" stirred up by the opponents of astronomy—a problem now solved by Copernicus's hypotheses![144] And hence Osiander's reference was probably not the least cause of his anger when he read the illicitly attached letter.[145]

The more profound source of Rheticus's ire, however, was Osiander's view of astronomy as a discipline fundamentally incapable of knowing anything with certainty. For Rheticus, this extreme position surely must have resonated uncomfortably with Pico della Mirandola's attack on the foundations of divinatory astrology. And, in fact, Osiander was as deeply familiar with Pico as were most of his learned contemporaries—indeed, not merely familiar, but sympathetic. Unlike the naturalistic reformer Melanchthon, his protégé Rheticus, and the canon Copernicus, it would seem that Osiander now offered new grounds for endorsing Pico's conclusions: not merely was the disagreement among astronomers grounds for mistrusting the sort of knowledge that they produced, but now Osiander proclaimed that astronomers might construct a world deduced from (possibly) false premises. Thus the conflict between Piconian skepticism and secure principles for the science of the stars was built right into the complex dedicatory apparatus of De Revolutionibus itself.

Osiander's view of astronomy's limited epistemic capabilities was not at all inconsistent with his attitude toward prophetic speculation about the Last Things. It was desirable to try to reckon the time of Christ's coming for the spiritual comfort it might afford, but such reckoning was ultimately conjectural and speculative. In 1544, the publisher Petreius issued Osiander's Conjectures on the Last Days and the End of the World just one year after De Revolutionibus and four years after Osiander had read the Narratio Prima. The Conjectures was supposed to carry forward Melanchthon's commentary on the Book of Daniel by offering a more exact calculation of the prophesied epochs. Of the four conjectures, the first dealt with the prophecy of Elijah, the second calculated that 1,656 years had elapsed between Adam and the Flood, the third connected Christ's age on Earth (33 years) with the end of the Church, and the fourth predicted from Daniel that Rome would twice achieve world dominance.[146]

Osiander's conjectures employed neither astronomical nor astrological methods.[147] Indeed, there was no mention at all of Rheticus's interpretation of the Elijah prophecy linking the motion of the Earth's eccentric to the rise and fall of monarchies. If Osiander was willing to entertain any hermeneutic method auxiliary to scriptural exegesis, it was the earlier Christian Kabbalah of Pico della Mirandola: "These conjectures also use Joan picus merandulane in ye yere of our Lord M.CCCC.lxxxvi. & did put up this one among his disputable 90[o] conclusions saying: if there be any humane conjecture of the last time, we may serche & finde it by the most secret way of Cabbalist, the end of the world to come hence of 514 yeres."[148] Pico's willingness to use the Kabbalah while later preserving a strong skepticism about naturalistic divination was coherent with the view, emphasized by Bruce Wrightsman, that Osiander regarded scripture as the only unerring source of truth.[149] Astronomy was useful only insofar as it assisted in improving the accuracy of the calendar or reckoning more precise biblical chronologies.[150] Other than that, it could lay no claim to the truth of statements about the order of the heavens.

HOLY SCRIPTURE AND CELESTIAL ORDER

The role of Scripture at this point is, nonetheless, curious. Osiander was not a literalist with respect to the Bible's language.[151] But if the Bible was not literal in every respect, how could it be said to be a secure, apodictic resource for knowing what moves and what does not? At stake was the relevance—and hence the authority—of a small group of biblical passages that used nouns (Sun, Moon, stars) and verbs (rise, set, move) allegedly referring to the heavens. Scripture certainly contained no vocabulary drawn from spheric or theorics (e.g., ecliptic, equinoctial points, right ascension, orb). But perhaps such categories could be used to make sense of obscure passages. Here again was the central point of contention for the defenders of Christian doctrine: at what point was it appropriate for natural knowledge to be deployed in assisting faith?

Hooykaas's publication in 1984 of Rheticus's lost work reconciling holy scripture with the motion of the Earth now allows important progress to be made in understanding this problem. To begin with, the exact title of the work is, unfortunately, unknown. However, an important letter from Giese to Rheticus refers helpfully to "the little work by which you have skillfully protected the motion of the Earth from disagreement with the Holy Scriptures."[152] The designation *Opusculum quo a Sacrarum Scripturarum dissidentia Telluris Motus vindicatur* is a more plausible title for the work than either the inscription on the title page ("Epistola de Terrae Motu") or the heading ("Dissertatio de Hypoth. Astron. Copernicanae") chosen by the seventeenth-century Utrecht publisher Johannes van Waesberge. For one thing, the work is not cast in epistolary form; for another, Rheticus never refers to "Astronomia Copernicana," a decidedly seventeenth-century rendering with overtones of Keplerian and Galilean language. The dating of the work also affects what we make of it. If, as Hooykaas plausibly speculates, Rheticus wrote the treatise before September 1541—that is, while still with Copernicus in Frombork—then his arguments and interpretations would have been known to Copernicus and could have been communicated easily enough by either Copernicus or Rheticus to Osiander and Melanchthon.[153] As late as July 26, 1543, when Tiedemann Giese received his copy of *De Revolutionibus*, he expressed the hope that Rheticus would attach both his biography of Copernicus and the *Opusculum* (as I shall call it) to all the remaining printed copies. This shows that Giese regarded it as normal to join together genera of a related nature—much like the conventional practice of forming collections of astronomical and astrological works. It would also suggest that Giese saw it as appropriate to join the *Opusculum* to the finished *De Revolutionibus* rather than to the preliminary *Narratio Prima*. Also, the fact that there is a separate work devoted to the question of scripture, the Earth's motion having been argued for as a desirable "assumption" in another place, helps to explain why Copernicus referred to the matter only glancingly—if not arrogantly—in the preface to *De Revolutionibus*. Finally, if Osiander was acquainted with this treatise or knew of its contents through conversation or correspondence with Rheticus, then one would have expected at least some response to it in the "Ad Lectorem." But there is no direct evidence that either Osiander or Melanchthon was aware of the *Opusculum*.[154]

The *Opusculum* is remarkable not least because it shows that Rheticus and Copernicus had worked out the basic elements of a systematic defense of the compatibility of scripture with the new hypotheses. They knew that they had a problem. Theologically, the work strives for a moderate stance—separating scripture from natural philosophy and using Augustine as the guiding authority with frequent protestations of catholicity and multiple appeals to traditional authorities.[155] This approach was certainly plausible for a Lutheran such as Rheticus, but in practice it might not have satisfied Melanchthon's providential and strongly scripture-driven natural philosophy. For this reason, with an eye on Melanchthon, Rheticus may have had justifiable reasons for hesitating to publish it. Giese's role in urging its publication suggests that the approach was more acceptable to the moderate middle ground of Varmian Catholicism than to the polemical Lutheranism of Osiander or even the theophanic naturalism of the *Praeceptor Germaniae*. Had Copernicus lived, therefore, he might have encouraged the publication of Rheticus's work—as did his fellow canon, Giese. With Copernicus's death on the eve of the Council of Trent (1545–63), this brief gesture of philosophical and exegetical openness would go unheeded until second- and third-generation Copernicans independently revived Saint Augustine's principle of accommodation more than a half century later.

The essence of Rheticus's argument was its appeal to this more flexible Augustinian standard. Among other things, accommodation allowed a separation of the requirements of confessional allegiance from the freedom to philosophize.[156] This method permitted the interpreter to say that in those (few) places where the Bible speaks of natural things, it does so according to common speech. In Rheticus's apt terms: "It borrows a kind of discourse, a habit of speech, and a method of teaching from popular usage."[157] The Bible's purpose determined its discourse—salvation and moral lessons, not philosophical or natural-philosophical teaching. Hence Rheticus urged what amounted to an intentional discursive boundary between the Bible and natural philosophy. The Bible may speak in accord with the senses even if what it says is erroneous with respect to what is held in natural philosophy. On certain matters, however, the Church had declared its position long ago and without ambiguity—for example, with the doctrine of the creation. In such cases, it was fair to regard scripture as having a direct bearing on philosophical beliefs not merely because the Bible said so but because the biblical meaning had the endorsement of ancient patristic authority. But in most other cases—for example, the rising and setting of the Sun—passages that appear to teach about Nature are not to be read in the discursive frame of such technical disciplines as *astronomia theorica* and *practica*.

Both the learned and the unlearned, then, may benefit from the Bible's moral lessons, while the philosophically inclined may construct their beliefs on independent natural foundations. The meaning of difficult passages should be sought by textual comparison rather than by introducing separate technical vocabularies, assumptions, methods, categories, and so forth. From this reasoning followed an important *prohibition*. According to Augustine—echoed by Rheticus—it is sacrilegious to overinterpret by trying to "extract" one's own philosophical views from holy scripture. "For Saint Augustine desires that we should never let ourselves be so happy with our own opinion on nature, which we believe to have extracted from the sacred writings, that, when truth has taught us otherwise, we are ashamed to retract, and fight for our own view, as if it were the teaching of Scripture."[158]

For Rheticus, some commentators exemplified Augustinian exegetical caution (esp. Nicholas of Lyra), whereas others violated the prohibition against "rooting out" (*eruendi*) philosophical views from the scriptures held on independent and prior grounds. One of the offenders was the Roman writer Lactantius, "otherwise a man of great learning and eloquence, [but who] ridicules those who claim that the earth is round."[159] The prime offender, however, was Pico della Mirandola:

> Many passages of Scripture could be collected by way of showing that Scripture often accommodates itself to popular understanding, and does not seek exactness in the manner of Philosophers. So, on the authority of Nicholas of Lyra, it was because of the uncultivated state of the people that, in the beginning of Genesis, no mention is made of the air, much less of the element fire, as being beyond the perception of the uneducated. It is clear that for the same reason, except sun and moon, nothing is said in that place of the other planets,—however much Pico in his *Heptaplus* tries to extract them therefrom—not to speak of still other things that are left out in the same place.[160]

Pico's offense in the *Heptaplus* (1489) was consistent with his offense in attacking the foundations of astrology and astronomy in the *Disputationes*, although Rheticus did not explicitly draw the comparison. Pico's desire was to put scripture ahead of natural divination. But while defending the primacy of scripture, Pico painted an esoteric and secluded image of knowledge: he claimed that its meaning was not on the surface but veiled in the depths of its words. This deeper meaning was accessible only to "the few disciples who were permitted to understand the mysteries of the kingdom of heaven, openly and without figures."[161] Otherwise, Christ proclaimed the gospel to "the crowds" in the form of parables. When Moses spoke at the summit of the mountain, the Sun would illuminate his face "wondrously bright," but "since the people with their owl-like and unseeing eyes could not endure the light, he used to speak to them with his face veiled."[162] By what method could one gain access to the "buried treasures" and "hidden mysteries" of the first chapter of Genesis?

The answer for Pico was that one needed the assistance of an independent hermeneutic that could make sense of the symbolic and often highly condensed manifest language of scripture. In short, one needed a fully developed theory of the Creation itself in order to make sense of the Mosaic account. This is just what Pico provided in the *Heptaplus*. The "Second Exposition," in particular, is relevant to Rheticus, as it deals with the celestial world. Here Pico outlined a ten-sphere heaven—seven planets, the sphere of the fixed stars, the ninth sphere "apprehended by reason, not by sense" and the tenth "fixed, quiet and at rest, which does not participate in motion." Pico offered in support a mix of medieval authorities rather than arguments and empirical evidence: Walafrid Strabo and Bede ("Christians") and the Hebrews Abraham the Spaniard ("a great astrologer," also a favorite source in the *Disputationes*) and Isaac ben Solomon Israeli ("the philosopher").[163] Pico then declared that the eight lowest spheres correspond to what Genesis calls "earth." Following this, Pico then "found" the specific terrestrial elements in the sky in two quite different orders: the Moon corresponded to earth, Mercury to water, Venus to air, and the Sun to fire. Then, "in inverse order": Mars corresponded to fire, Jupiter to air, Saturn to water, and the eighth, "unwandering" sphere to earth.[164] This explicitly figural interpretation of the meaning of *earth* created a notable silence that Pico then addressed: "See how he [Moses] has shown us the nature of the moon and the sun figuratively and in a few words. But why is he silent about the rest, when we promised in our proems that he would treat sufficiently and learnedly of all? Why, I say, when he has made mention of the tenth, ninth, and eighth spheres, and also of Saturn, the sun, and the moon, is there not even a word of the four that are left, Venus and Mercury, Jupiter and Mars?"[165]

Here Pico rejected any appeal to the principle of accommodation as an excessively facile refuge: "I cannot without blushing betake myself to it, since I swore that Moses omitted nothing which might make for a perfect understanding of all the worlds." As described in chapter 3, he later rejected any appeal to the "astronomers" and the "astrologers" because of their longstanding internal disagreements. Thus, Pico's solution lay neither within the Bible itself nor in the domain of the natural philosophers and mathematicians: "I believe that yet more deeply hidden here lies a mystery of the ancient wisdom of the Hebrews, among whose dogmas on the heavens this is important: that Jupiter and Mars are included by the sun, and Venus and Mercury by the moon. If we weigh the natures of these planets, the reason for this belief is not obscure, although the Hebrews themselves offer no reason for the doctrine."[166] Because this arrangement does not occur in Genesis, and the Hebrews "offered no reason" for it, the future author of the *Disputations against Divinatory Astrology* supplied his own, astrologically pregnant interpretation:

> Jupiter is hot, Mars is hot, and the sun is hot, but the heat of Mars is angry and violent, that of Jupiter beneficent, and in the sun we see both the angry violence of Mars and the beneficent quality of Jupiter, that is, a certain tempered and intermediate nature blended of these. Jupiter is propitious, Mars of ill omen, the sun partly good and partly bad, good in its radiation, bad in conjunction. Aries is the house of Mars, Cancer the dignity of Jupiter: the sun, reaching its greatest height in Cancer and its greatest power in Aries, makes clear the kinship with both planets. . . . The moon . . . clearly shares in the waters of Mercury, and shows how great an affinity it has with Venus by the fact that in Taurus, the house of Venus, it is so exalted that it is judged to be nowhere more propitious or beneficent.[167]

Pico concluded his association of the elemental qualities and the ordering of the planets with

the confident judgment that "Moses has spoken sufficiently so far of the empyrean heaven, the ninth sphere, the firmament, the planet Saturn, and the sun and the moon which represent the rest, suggesting their inclusion to us by his very silence."[168]

These informative passages show us how far Pico had moved between the *Heptaplus* and the *Disputations*. Equally important, however, is the light shed on the position of Rheticus and Copernicus. They shared with Pico the trope of uncovering deep mysteries. Their quest, however, was a mathematical one. And, as such, Copernicus and Rheticus rejected Pico's view that the order of the planets could be found by reading Genesis either *sensu literalis* or *sensu allegorico*.

DE REVOLUTIONIBUS
TITLE AND PREFATORY MATERIAL

If the *Narratio Prima* was directed to an audience of Nuremberg and Wittenberg preachers, prognosticators, natural philosophers, and theologians, Copernicus's preface addressed *De Revolutionibus* explicitly to a Roman ecclesiastical audience. However, although the preface was cast in the idiom of church patronage and reform, it is not the language of office seeking. This is the argument of a man drawing on his richest intellectual resources and hoping to gain support, near the end of his life, for what he thinks to be the intelligibility of the heavens and, by implication, what the Church ought to teach about it. When Copernicus composed the preface in June 1542, two issues of the *Narratio Prima* were in circulation; Rheticus had already left the manuscript with Petreius; and at sixty-nine, the old canon must have sensed that he did not have long to live. A few months later, he lay paralyzed from a stroke, and on May 24, 1543, he died just as the book was placed into his hands, publishing as he perished.

Even before Rheticus arrived, however, Copernicus's ideas already had supporters in Rome at the level of both the papal Curia and the cardinalate court. Paul III's predecessor, Clement VII, had heard Copernicus's new hypotheses described verbally before him. His young secretary, the Bavarian Johann Albrecht Widmanstetter (1506–77), was a brilliant biblical scholar who, in 1555, published the first Syriac edition of the New Testament.[169] It was Widmanstetter who explained the new theory to the pope in the Vatican gardens in 1533 before two cardinals, a bishop, and the pope's physician. In return, Clement presented his secretary with the gift of a Greek manuscript containing several philosophical treatises.[170] Two years later Widmanstetter moved into the service of a recently promoted Dominican cardinal, Nicholas Schönberg (1472–1537), and after Schönberg's death Widmanstetter became a secretary (*Secretarius Domesticus et Familiaris*) to the succeeding pope, Paul III.

In November 1536, Schönberg wrote to Copernicus, urging him to send a copy of his manuscript to Rome and even offering to provide as amanuensis the representative of the Varmian chapter in Rome, Theodoric of Reden. No mention is made of support for publication. Nonetheless, Copernicus understood the proper signs of epistolary display for seeking approval and protection in Rome, and he could easily have interpreted Schönberg's letter as a sign of eventual papal approbation; at the very least, Widmanstetter's continued presence indicated support in the highest curial circles.[171] Copernicus placed Schönberg's letter in *De Revolutionibus* immediately after the title page and just before his own preface to Paul. In this way, he allowed the Dominican Nicholas Cardinal Schönberg to provide the first description of his new "account of the World": "In it you teach that the earth moves; that the sun occupies the lowest, and thus the middle, place in the universe."[172] Thus, when Osiander placed his anonymous "Letter to the Reader" just after the title page and ahead of the cardinal's favorable letter, he was knowingly interfering as much with the author's methodological aims as with his intended strategy for seeking the pope's protection. Rheticus was so incensed by this unauthorized interference that he tried to seek legal redress against both Osiander and the publisher Petreius from the Nuremberg city council—but without success.[173] However, in the insecure days before the existence of laws of copyright, authors lacked legal recourse concerning the integrity of their works.[174]

By the time that Copernicus drafted his preface, however, his curial supporters were no longer around: Pope Clement and Cardinal Schönberg had died. Nonetheless, in deciding to address Paul III (1534–49), Copernicus could not have been unaware of his reputation. The new pope had, like Copernicus himself, a strong humanist

training: he had studied at the University of Pisa, he was a poet, he knew Greek, and he was respected for his wide learning.[175] As the former Cardinal Alessandro Farnese, he also came from a wealthy, noble family. He could afford to pay his servants with his own money rather than strictly from ecclesiastical revenues. In 1526–27, his household contained 306 persons.[176] Although we cannot be sure what Copernicus knew about the papal finances, it is likely that he was quite familiar with the idiom in which specific appeals for patronal protection had been made to the cardinal before he became pope. For example, Pomponio, the brother of the prognosticator Luca Gaurico, dedicated a commentary on Horace's *Ars poetica* to Cardinal Farnese that was published in 1504.[177] The text of this commentary would have been written while both Pomponio and his brother were together in Padua with Copernicus. Much later, Girolamo Fracastoro, also known to Copernicus in Padua, dedicated his Averroist *Homocentricorum Siue de Stellis Liber Unus* (Venice, 1538) to Paul III.[178]

In addition, there were appeals to Paul of a more direct, practical nature: Luca Gaurico issued prognostications in 1529 and in 1532 forecasting that Alessandro Farnese would become pope. These successful predictions did eventually result in the desired papal favor. Gaurico found himself a regular dinner companion of the cardinal. Then, in 1543, he presided at an astrological ceremony for the laying of the cornerstone in the Farnese wing of the Vatican Palace. Gaurico calculated the exact hour and zodiacal sign for the event, assisted by the Bolognese prognosticator Vincenzo Campanacci, who "found the proper time on the astrolabe and announced it in a loud voice." Three years later, Gaurico was rewarded with a bishopric.[179] As one might expect, Copernicus's strategy was much closer to that of Pomponio than that of Luca Gaurico. He kept a distinct silence on prognosticatory matters, making no predictions about the pope's health, longevity, or political future, or the best times to make important journeys. Nor were there echoes of Rheticus's millennial prophecy.

The only hint of astrological connotations is the word *Revolutions* in the book's title. Selecting a title entailed deciding among recognizable genres of writing, genres that readers could recognize, that publishers could use to market works, and for which royal or imperial privileges might be granted. (It is understandable that Copernicus could not appropriate the title of Peurbach's well-established school book, the *New Theoric*. It is less clear why he did not think to choose the more Ptolemaic-sounding *New Almagest,* used by G. B. Riccioli in 1651.) The title on which Copernicus finally settled resonated with the conventional medieval association between revolutions and nativities.[180] There was, as far as I know, no generic precedent for a title that linked revolutions and celestial orbs.[181] And, although Copernicus might have connected revolutions and prognostications, he had no new alternative to the *Tetrabiblos.* His immediate disagreement was with the first principles of the *Almagest* on which his work was modeled.

Yet both in his preface and in the suppressed introduction to book 1, Copernicus stressed again and again not the standard Ptolemaic theme (from the *Tetrabiblos*) of astronomy's capacity for certitude at the general level of its mathematical models (compared with astrology's fallibility in making judgments about the unstable physical world) but rather what we now know to be Pico della Mirandola's complaint: the *uncertainty* among traditional astronomers—including the disagreement about the ordering of Venus and Mercury. Copernicus constantly contrasted the beauty and purity of astronomy's subject matter with the "perplexities" and "disagreements" that encumber its hypotheses. Besides disagreements about hypotheses, there was the additional reason that "the motion of the planets and the revolution of the stars could not be measured with numerical precision . . . except with the passage of time." As a consequence, "very many things do not agree with the conclusions which ought to follow from his [Ptolemy's] system." An example of one such entailment is the uncertain length of the tropical year, about which Copernicus cited Plutarch's view that this value had so far eluded the skill of the astronomers. And further: "It is well known, I think, how different the opinions concerning it have always been, so that many have abandoned all hope that an exact determination of it could be found." He then added to this skepticism the further comment that "The situation is the same with regard to other heavenly bodies."[182]

In the preface, Copernicus approached astronomy's uncertainty by means of ironic contrasts.

He presented himself as someone worthy of laughter and derision, someone who goes against tradition and whose theories will surely be repudiated. The tone recalls the diffident, yet righteous Saint Socrates of Erasmus's *Godly Feast.*[183] Two ecclesiastic friends, the cardinal of Capua (Nicholas Schönberg) and the bishop of Chełmno (Tiedemann Giese), had repeatedly urged him to publish. They argued that even if his theory appeared to be crazy, "so much the more admiration and thanks would it gain after they saw the publication of my writings dispel the fog of absurdity by the most transparent proofs [*liquidissimis demonstrationibus*]." Finally he had acceded to their entreaties and would "permit it to appear after being buried among my papers and lying concealed not merely until the ninth year but by now the fourth period of nine years."[184] The allusion to a fourfold Horatian waiting period of thirty-six years probably has some truth, as historians have noticed; but, at this point in the text, it is difficult not to take Copernicus's failure to mention the *Narratio Prima* as anything but a conscious omission—part of a deliberate strategy not to mix the dedications to the two audiences.

The first of the traditional "disagreements" to which Copernicus alludes had been anticipated in the *Commentariolus:* it concerned the preferred foundations of theoretical astronomy. Those who use "homocentric circles" cannot get their theories to fit the phenomena absolutely; those who use eccentrics can deduce the phenomena from the arrangements of spheres but violate "first principles."[185] Worst of all, neither tradition can deduce what Copernicus calls "the arrangement of the universe and the sure commensurability [*symmetria*] of its parts."[186] In short, we have an ironic reversal: It is tradition itself that is full of monstrous incoherence and absurdity. And here follows the famous trope to which Kuhn and others have attached so much significance and which allegedly ties Copernicus to Florentine Neoplatonism through Domenico Maria Novara: "With them it is just as though someone were to join together hands, feet, a head, and other members from different places, each part well drawn, but not proportioned to one and the same body, and not in the least matching each other, so that from these [fragments] a monster rather than a man would be put together."[187]

Copernicus's *symmetria* bears some resemblance to the images amplified so enthusiastically by Rheticus, but his source is the unmistakable and forceful opening lines of Horace's *Ars Poetica*—an allusion that Copernicus probably knew to be pleasing to the pope:

> If a painter chose to join a human head to the neck of a horse, and to spread feathers of many a hue over limbs picked up now here now there, so that what at the top is a lovely woman ends below in a black and ugly fish, could you, my friends, if favoured with a private view, refrain from laughing? Believe me, dear Pisos, quite like such pictures would be a book whose idle fancies shall be shaped like a sick man's dreams, so that neither head nor foot can be assigned to a single shape. "Painters and poets," you say, "have always had an equal right in hazarding anything." We know it: this licence we poets claim, and in our turn we grant the like; but not so far that savage should mate with tame, or serpents couple with birds, lambs with tigers.[188]

The central theme emphasized by Horace and noticed by his Renaissance commentators was the principle of "fittingness" or "belongingness." Style must fit its subject, diction its characters; characters must preserve decorum and appropriateness; the beginning must fit the end.[189] Significantly, the audience is the custodian of "appropriateness" and rejects through laughter what it perceives not to agree with nature. What moves or delights or persuades the audience is what makes for good poetry. And it was this rhetorical view of poetry that many Renaissance commentators so appreciated in Horace.

The uses to which Copernicus put the Horatian text were important and unprecedented. First, he tacitly transferred the literary aesthetic ideal of good poetry into the domain of astronomy. Just as one prefers a coherent to an incoherent literary work, so a theory of the planets possessing mathematical coherence (*symmetria, armoniae nexus*) is to be preferred over one that does not. The implication is that such a world picture is not arbitrary, for art imitates nature; hence, a decorous audience will judge such a theory to be true, while shunning as absurd one lacking in *symmetria*. If such an argument did violence to the *Posterior Analytics* by illicitly mixing the subject matters of poetry and astronomy and by rejecting strictly demonstrative knowledge, it was entirely in keeping with humanist commentators on Horace. For example, in 1482

Christoforo Landino, a well-known rhetoric teacher at the Studio of Florence, commented: "Since all art imitates nature, the poet will be laughed at just as the painter will be scorned if he portrays the monstrous, viz. if he places a human head on the neck of a horse and to this horse's neck he paints in the body from the various parts of birds and makes the lowest members those of the fishes."[190]

Copernicus found Horace's image helpful for another reason. It offered a reply to the view of astronomical hypothesis articulated by the likes of Osiander and Fracastoro.[191] Astronomers, like painters and poets, might possess "an equal right in hazarding anything," but even the latter did not have unlimited license to join lambs with tigers and serpents with birds. Copernicus acknowledged "the freedom to imagine any circles whatever"—including his own supposedly absurd hypothesis of assigning an imaginary circle to the Earth "for the purpose of explaining the heavenly phenomena." The beauty and irony of the "absurd" assumption was that, comparatively speaking, it led to "sounder demonstrations" (*firmiores demonstrationes*). Thus, beyond its well-known astrological connotations, the choice of the word *revolutions* in the book's title sought to focus attention on the new astronomical meaning in the multiplicity of desirable entailments that flowed from the putatively absurd conditional premise. And, among those entailments, Copernicus stressed that his hypothesis alone yielded the *symmetria* lacking in rival alternatives: "The order and size of all the planets and spheres, and heaven itself is so linked together that nothing can be moved from its place without causing confusion in the remaining parts of the universe as a whole."[192]

The logic of this claim was relative rather than absolute, and it was the best that Copernicus could offer. It built on the image of weighing and balancing alternative hypotheses against a commonly accepted standard rather than against the stringent Aristotelian ideal of a *cognitio certa per causas,* in which true conclusions, deduced from true, proper, and necessary premises, rule out all possible alternatives.[193] Copernicus thus brought forward humanist rhetoric and dialectic as an antidote to Piconian skepticism. There is a dialectical topos of whole to part operative in the Horatian trope.[194] As in the contemporary Renaissance view of Horace's *Ars Poetica,* the audience was sup-

CONFESSIONAL AND INTERCONFESSIONAL SPACES

posed to play a crucial role in deciding what was a good work of poetry, or a well-crafted *machina mundi*. Copernicus, like Landino, held that the audience is the final arbiter—but not just any audience. In the most famous line from the Preface—"Mathemata mathematicis scribuntur"—Copernicus emphasized that only a certain kind of community has the special competence to make judgments: those with mathematical skill and training. Most immediately, he meant those within the Church who were mathematically literate and who would not only understand and approve of his theory but would also accept the new standard for judging it; those without such disciplinary credentials would misunderstand and deplore it.

The preface, however, was at best reticent about the names of recent theoretical astronomer-astrologers. Neither Regiomontanus nor Peurbach, neither Girolamo Fracastoro nor Domenico Maria Novara was mentioned. And, most significant, it omitted Rheticus and the *Narratio Prima* altogether. The audience that Copernicus constructed in the preface divided the Church into two parts: those who were enlightened in mathematical subjects and those who were not. In the first group Copernicus included popes Leo X and Paul III, Cardinal Schönberg, Bishop Tiedemann Giese, and Paul of Middelburg. In the second, he placed untutored theologians—Copernicus called them "idle talkers"—who know nothing about mathematics and who he imagines will deride him by distorting scripture for their own purposes. The sole example of the latter was Lactantius. This contrast is quite interesting in light of our earlier discussion. Paul of Middelburg was clearly supposed to evoke orthodoxy with ecclesiastical associations to theoretical astronomy, legitimate astrological prognostication, and calendar reform; but the casual reference to Lactantius's mathematical incompetence makes sense only in the context of Rheticus's treatise on scripture, to which the preface makes not a single reference.[195]

Other elements of the Horatian topos resonated well with a variety of Copernicus's aims and in a wider semantic field.[196] First, Copernicus found what rhetoricians, artists, and poetic and visual theorists had long found in the *ut pictura poesis* formula: a discourse of bodily and literary coherence and an evocative aesthetic, connecting the poetic and the visual. The likely source for this ideal would have been available to Copernicus already in his student days at Padua. There, not only did Copernicus study medicine, but also it is very likely that he was involved with an active culture of artists. It has been suggested that Copernicus both knew Pomponio Gaurico and moved in the ambience of the Venetian artistic world, which included the revolutionary painter Giorgione and the Paduan artists Tullio Lombardo, Andrea Riccio, and Giulio Campagnola.[197] In 1504, just one year after Copernicus left Padua, Pomponio published his treatise *De Sculptura.* "A constant characteristic of the treatise," writes Robert Klein, "is the adaptation of rhetorical and poetical categories to the plastic arts."[198] The ideal property of the proportionally sculpted body, according to Pomponio, was its *symmetria:* "In every way our body is fitted together most precisely from measured parts, so that plainly one may regard it as nothing but a kind of perfect, harmonious instrument set in good order according to all the numbers."[199] It was but a short step from here to Copernicus's aesthetic in which "nothing can be moved from its place without causing confusion in the remaining parts and the universe as a whole."

Second, the Horatian image converged remarkably well with the political vocabularies of humanist curial reformers in the early sixteenth century as well as with familiar visual images of Reformation popular propaganda. At that historical moment, one of its connotations was reconciliation and reform. Within subtly overlapping arenas of language and image, one can see Copernicus treading a delicate path, recommending both that the Church pursue a reform of practical astronomy (alluding to the scandalous state of the calendar) and that it reconsider its association with theoretical teachings about the order of the heavens.

Copernicus represented the pope not as a granter of offices but as a protector. The pope "holds dominion over the Ecclesiastical Commonwealth" and answers not merely to an omnipotent deity but to a God of order—"The best and most orderly Artisan of all." As an advocate of new theoretical knowledge, Copernicus was striking forth into uncharted territory. He associated himself with the pope as protector of truth-seeking philosophers and the church's view of the heavens. The holy father's authority, moreover, came not merely from God but also from his specific human qualities: "Even in this remote corner of the earth where I live, you are considered the highest authority by virtue of the dignity of the office and love of the mathematical arts and all learning." Copernicus urged the pope to protect him against the hostility, uncertainty, and disagreement engendered by certain astronomers and philosophers: "By your authority and judgment, you can easily suppress calumnious attacks."[200] The language of the preface resonates effectively with Catholic reform literature of the same period. The writings of Raffaele Maffei (1451–1522), an influential curial figure, well exemplify this sort of political text. After reciting a litany of abuses that he hopes the pope will correct, Maffei uses the topos of the head and the body to urge a purging of the unnaturally greedy parts. "Your city, O father, must be cared for and renewed that it might not be ruled over by others who (as the Apostle says) would neglect their own home. Above all, [your city] must be restored to its primitive *libertas* and purged of the greed that goes contrary to your morals, since nature seeks that in which the members can be brought into conformity with the head, citizens with the prince, and, in a similar way, the flock with its shepherd."[201] The pope at the head of his flock is not corrupt; it is he who must expurgate those who are, and thereby protect Rome from abuse.[202]

Catholic church reformers used the image of the head to symbolize order and authority, but the many-headed monster also functioned as an image of moral disorder at the popular level. In German broadsheets of the 1520s, Lutherans employed the beast of the Apocalypse, with its seven unequal heads, as visual propaganda to attack papal indulgences, and Catholics later portrayed Luther as a many-headed, fanatical wild man.[203] Luther himself recognized the immense instructive power of visual signs, "above all for the sake of children and the simple folk, who are more easily moved by pictures and images to recall divine history than through mere words or doctrines."[204]

These "high" and "low" reformist points of reference help to situate Copernicus's moral imagery associating head and body, papal authority and celestial reform. Joined with his use of the Horatian aesthetic is a language of natural order and ethical conviction, underwriting a belief that astronomy's first principles are at once true and untainted.[205] By addressing Rome in this idiom of order and reform—resonant with both elite and popular connotations—he also evoked moral and political associations to which fellow clerics from his own local region of Varmia could respond. Indeed, Varmian religious politics was strongly humanist, irenic, and Erasmian in spirit, amid growing anti-Catholic sentiment and conversions to Luther's doctrines. Copernicus's closest Varmian friend, Tiedemann Giese, corresponded with Erasmus.[206] In the preface, Copernicus credited Giese with encouraging the publication of *De Revolutionibus*. In the *Encomium Prussiae*, Rheticus portrayed Giese as a radical reformer who pushed the cautious Copernicus not to conceal the theoretical principles from which he deduced his new system of planetary motions:

> The Most Reverend Tiedemann Giese, bishop of Chełmno . . . realized that it would be of no small importance to the glory of Christ if there existed an exact calendar of feasts in the Church as well as a sure theory and explanation of the motions [*certa motuum ratio, ac doctrina*]. . . . His Reverence pointed out that such a work would be an incomplete gift to the world [*munus Reipublicae futurum*], unless my teacher also set forth the sources [*causas*] of his tables and also included, in imitation of Ptolemy, the plan or method and the foundations and proofs [*quo consilio, quaue ratione, quibusque nixus fundamentis, ac demonstrationibus*] upon which he relied.[207]

Giese also wrote a treatise reconciling Copernicus's theory with the Bible (*Hyperaspisticon*, now lost), in which he borrowed the first word (*hyperaspistes*, or "shield bearer") from a polemic written by Erasmus against Luther.[208] And Giese shared the characteristic Erasmian view that gentle persuasion could achieve more than sharp criticism and satire. Differences of opinion could be resolved through love and toleration; Christian unity must come from within the church.[209] It was on this vulnerable but tolerant middle ground that the Lutheran Rheticus met the canons Giese and Copernicus.

The strategy of persuasion that Copernicus followed in his preface of 1543 undoubtedly reflects the outcome of earlier discussions with Giese and Rheticus, echoes of which are heard in the

Encomium Prussiae. Copernicus attempted to side-step conservative elements in Rome while carefully omitting all references to the Lutherans Rheticus, Melanchthon, Gasser, and Schöner. His reformist rhetoric was neither stridently polemical nor satirical, but gently Horatian and Erasmian. It sought an end to controversy among astronomers, and, by implication, astrologers. It proposed an internal cadre of humanist *mathematici* to reform church teaching on the heavens by providing theoretical principles from which planetary order and calendrical accuracy could be restored. The entire enterprise was legitimated by papal authority and by appeal to a range of ancient pagan sources. The approach evokes Erasmus's broad reconciliation of Christian and pagan letters and also echoes the Beroaldo circle of the Bologna period in expounding a *philosophia Christi,* a life of lay piety modeled on the true life of Christ and the earliest sources of Christianity rather than on empty ceremonial practices and overly subtle Scholastic definitions.

The iconographical representation of Copernicus above the epitaph in his hometown parish church, Saint John of Toruń, provides evidence that his successors, at least, regarded his life in an Erasmian spirit, as a kind of *astronomia Christi.*[210] J. J. Vogel, the seventeenth-century artist, adapted his woodcut from an anonymous devotional painting of the astronomer (c. 1583). Melchior Pyrnesius (d. 1589), a younger fellow townsman and physician, commissioned the portrait and the epitaph. That Pyrnesius was carrying out Copernicus's wishes in the choice of wording for the epitaph cannot entirely be ruled out. The Latin, in sapphic meter, was one of thirty-four odes on Christ's suffering written in 1444 by Aeneas Sylvius Piccolomini, who later reigned as Pope Pius II (1458–64):

> Not grace the equal of Paul's do I ask,
> Nor Peter's pardon seek, but what
> To a thief you granted on the wood of the cross,
> This I do earnestly pray for.[211]

THE "PRINCIPAL CONSIDERATION"

The wider semantic field of the preface—its convergence with political and visual vocabularies of order and disorder—extended toward a broader audience that included the pope, and the hope of winning his public endorsement.[212] Regardless of its status as a dedicatory epistle, however, the preface went beyond standard epideictic gestures of praise and blame. It embodied seriously and accurately the logical structure of Copernicus's principal claim about the world. The essence of that claim was the same as the one stated in the *Narratio Prima* and the tenth chapter of book 1. In short, the assumption that the Earth is a planet may seem absurd, but the consequences of making that assumption make it more desirable than any other alternatives. And the consequence that Copernicus and Rheticus stressed most emphatically was the determinate ordering of the planets by their periods of revolution—about which there had been disagreement since the time of the ancients. The Copernican solution to that problem was presented by Rheticus as putting an end to the controversy initiated by Pico over the deeper grounds of natural divination.

The arrival in Frombork of the youthful, impassioned, and talented Rheticus seems to have energized the canon of Varmia and awakened his determination to move his ideas into print. The wealth of dialectical topoi that Rheticus wove into the *Narratio Prima* did not change the structure of Copernicus's logic; the gestures at necessity did not measurably strengthen the causal connection between the conclusion and the assumed premise. But the *Narratio* rhetorically amplified the more restrained presentation found in *De Revolutionibus* and left no doubt that astronomy aspired to true explanations, even if it could deliver only probable ones. By contrast, the position that astronomy's premises need not be true or even probable, incorporated by Osiander into his unauthorized "Ad Lectorem," represented the alternative that I believe Copernicus had already rejected more than thirty years earlier when he composed the *Commentariolus.*

Thus, baked into the prefatory structure of *De Revolutionibus* were the traces of Copernicus's early problem situation as well as the residual tension that defined the Copernican question: mutually contrary premises that logically entail the same conclusion. Osiander, to be clear, did not say explicitly that the Earth is a fictional point, but his position distinctly allowed and encouraged that view. Whether the Earth is a point or a real body revolving with the planets around a stationary Sun, the planets will be ordered such that the shorter the period of revolution, the closer it will be to the Sun, and the longer the period, the farther the planet's distance from the central body. Neither Copernicus nor Rheticus nor Osiander understood this situation as the problem of underdetermination as it was gener- alized some four centuries later by Duhem and Quine "for any finite body of evidence."[213] To para- phrase Peter Dear's well-conceived observation, we must learn to ascribe meanings in correct sixteenth-century ways to what appears to us now as self-evident.[214]

5

The Wittenberg Interpretation of Copernicus's Theory

Copernicus's reputation as a learned astronomer was established very quickly in the two decades after the appearance of the *Narratio* and *De Revolutionibus*. Although he was highly regarded in Catholic circles before the appearance of his main work, the publication of *De Revolutionibus* now spread his ideas all over Europe. His name became known, and his main work, in spite of its terse Latin style and its forbidding technical material, was widely read.[1]

How easy it was to grasp is another matter. The remarks of the learned Jesuit astronomer Christopher Clavius were not atypical. Commenting on Copernicus's discussion of the precession theory, he wrote: "He speaks confusingly, and he explicates and describes with extreme difficulty, so that soon it appears to me to be written so that everything is in conflict with everything else."[2] The Kassel *mathematicus* Christopher Rothmann acknowledged that Copernicus had not explained well how the Earth's axis maintains its orientation: "I know that in this regard Copernicus is quite obscure and is not easily comprehensible."[3] To the readers of his *Mysterium Cosmographicum* (1596), Kepler, following the suggestion of his teacher Maestlin, recommended the persuasive explications of Rheticus's *Narratio Prima* and acknowledged that "not everyone has the time to read Copernicus's books *On the Revolutions*."[4] In 1615, Galileo remarked that the book was not "absurd" but was "difficult to understand."[5] More than a century after its first publication, the Dutch Copernican Martin Hortensius complained that

Copernicus's major work was "too obscure in his writings to be understood by everyone"; moreover, his theory would not have been condemned so readily had astronomers presented it "popularly" in the form of visualizable material globes.[6] The point is significant because even globe makers, who were well disposed to Copernican astronomy for its improvements in such areas as lunar theory, eclipse prediction, and the securing of stellar longitudes (among them Gemma Frisius and Gerhard Mercator) did not make Copernican orreries of the sort that became commonplace in the seventeenth and eighteenth centuries.[7] Hortensius's comment attests to ambiguities in Copernicus's style that afflicted even some contemporaries—and not without reason.[8]

But *De Revolutionibus* was not the only resource for disseminating Copernicus's views. After Reinhold's *Prutenic Tables* appeared in 1551, Copernicus's renown within the literature of the heavens became firmly anchored to the domain of practical astronomy, even among constituencies unfamiliar directly with *De Revolutionibus* itself.[9]

From the perspective of the historical agents, there is a simple explanation for this state of affairs: the dominant preoccupation of those who possessed techniques of celestial investigation was the making of knowledge about the future. And in the mid-sixteenth century, those concerns and competences were most powerfully located in the circle of students and scholars gathered around Philipp Melanchthon at Wittenberg.

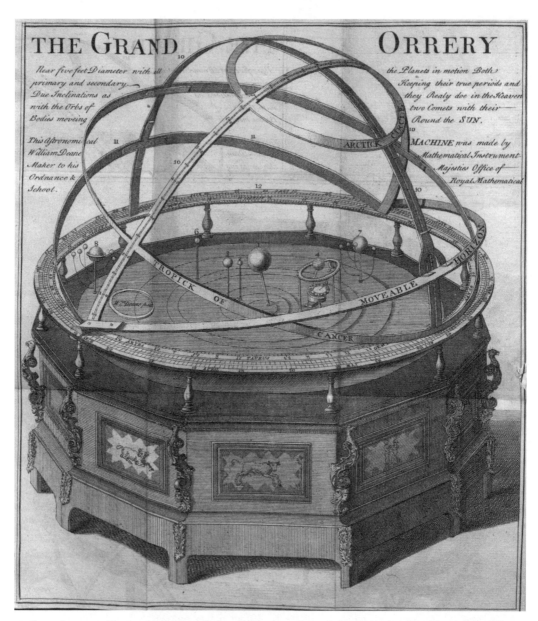

34. Copernican orrery illustrated in Deane 1738, described by the author as "a true Representation of the Motions of the Heavenly Bodies themselves." The orrery contributed to the naturalization of the Copernican system in the eighteenth century. Witnessing the geared motions of the gilded ivory balls on stiff wires in the five-foot-diameter fine ebony frame, under a canopy of silvered celestial arcs and circles supported by brass pillars, was supposed to convince observers of the model's validity in a way that tables, calculations, and diagrams could not: "Those Gentlemen and Ladies who delight in the Study of Astronomy and Geography, will, by seeing this *Grand Machine*, comprehend at one View the Reason of the several *Phaenomena*, or Appearances, in the Heavens, resulting from the various Motions of the Bodies which compose this Solar System; and will edify more from a few Lectures, than by a Year's close Application to Study" (90–91). For further discussion, see Westman 1994, 110. By permission of San Diego State University Library, Special Collections, Historic Astronomy Collection.

MELANCHTHON AND THE SCIENCE OF THE STARS AT WITTENBERG

Soon after arriving at Wittenberg in the early 1520s, Melanchthon set up a small *schola privata* in his own house, a preparatory or "trivial" school where the main emphasis was on reading and speaking Latin and the subsequent study of logic, rhetoric, and grammar.[10] Such use of professors' houses for private instruction was common in Wittenberg as well as in Tübingen, where Melanchthon had previously studied and taught. The school was intended not merely to benefit the students but also to supplement the typically meager professorial salaries (sometimes provided in the form of grain or wood)—especially in the arts and philosophy faculties. In Melanchthon's house, advanced students were taught Greek and really good students Hebrew. Language preparation of this sort was the main requirement for advancement to the subjects of the trivium and quadrivium. On Sundays the headmaster read and explained the Gospel before the whole school, and the boys were required to memorize the Lord's Prayer, the Apostles' Creed, and the Commandments.[11] The school was meant to inculcate civil, intellectual, and ecclesiastical discipline, and the university merely continued this goal. It was clearly a desirable means of inculcating loyalty to the state.

In October 1524, Melanchthon was invited by the Nuremberg city council to set up a *Gymnasium*, the very school where Schöner would later teach. Although Melanchthon declined to become the head of this school, he did offer an inaugural declamation that laid down the basic lines of its organization. The joining of evangelical principles with a program of classical texts—such as Pliny, Galen, Ptolemy, and Aristotle—became the hallmark of Melanchthon's curricular reforms. The curriculum had to serve religious requirements as well as moral and political purposes.[12] Such elements formed the basis of his own works of natural philosophy as well. Later he would involve himself extensively in establishing and reforming the principles of organization for the Protestant universities which, beyond Wittenberg, included Tübingen, Leipzig, Frankfurt an der Oder, Greifswald, Rostock, and Heidelberg. All the newly founded universities reflected the Melanchthonian humanist spirit of education:

Marburg (established in 1527), Königsberg (1544), Jena (1548), and Helmstedt (1576).[13]

Melanchthon's vision of the Reformation was thus firmly spread through the schools and through men who carried with them Wittenberg models of teaching and scholarship. Forty years later that spirit was still very much alive: "To Philip and the school of Wittenberg for a long time by God's grace this praise was peculiar, that he both instructed the minds of the students in varied knowledge and especially formed the judgment of youth as to true opinions concerning things and eminently prepared them for public service. Hence it came about that all who . . . were true disciples of Melanchthon employed a very similar style and form of oration in speaking and writing, moulded and turned out in imitation of their most erudite preceptor."[14]

Through his numerous prefaces and by means of his own example, Melanchthon actively lent his authority as much to the evangelical curriculum as to the science of the stars. His prefaces resembled little humanist orations with moral lessons. Euclid's *Elements,* for example, taught students moderation, discipline, and justice. "Indeed these very words, 'Let no one ignorant of geometry enter here,' teach that the opinion of Plato, which he engraved on the portals of his school, is not inappropriate as applied to general conduct. It excludes from schools those who do violence to geometric proportion, who do not understand or regard the organized system of honorable offices, who rush headlong without control, wherever their impulse carries them."[15] Arithmetic and geometry, in turn, prepared the way for the science of the stars. And again, Melanchthon regarded classic texts as the best for inculcating pedagogical lessons. But he also used a variety of other works to promote the heavenly disciplines. For example, he wrote an "Oration on the Dignity of Astrology" (1535),[16] an encomium to astrology and the mathematical disciplines (1536), and a foreword to a bundled collection of astronomical works by Alfraganus and al-Battani, which also included Regiomontanus's 1472 *Paduan Oration* in praise of mathematics (1537).[17] He authored prologues or added previously published letters to editions of the *Sphere* of Sacrobosco (1538), to Schöner's *Tabulae Resolutae* (1536), and to two editions of Peurbach's *New Theorics of the Planets* (1535, 1542).[18] He included Erasmus Reinhold's oration to Regiomontanus in a book of his Wittenberg declamations.[19] He added a preface to Cyprian Leovitius's edition of Regiomontanus's *Tables of Directions and Profections.*[20] It is clearly an oversimplification to characterize these writings collectively as merely pertaining to "astronomy."[21]

Strongly consistent with Melanchthon's general principles, these prefatory benedictions, covering all categories of the science of the stars, urged readers to regard the main texts not as ends in themselves but as useful in strengthening reverence for the divine plan. Thus Melanchthon did not so much change the fundamental content of the traditional curriculum for his schools as offer a new sort of legitimacy and organization for a group of well-established texts. Good divination based on the natural order was to be seen as a sacred activity.

> To recognize God the Creator from the order of the heavenly motions and of His entire work, *that* is true and useful divination, for which reason God wanted us also to behold His works. Let us therefore cherish the subject which demonstrates the order of the motions and the description of the year, and let us not be deterred by harmful opinions, since there are some who—rightly or wrongly—always hate the pursuit of knowledge . . . [I]n the sky, God has represented the likeness of certain things in the Church. Just as the moon receives its light from the sun, so light and fire are transfused to the Church by the Son of God.[22]

Natural order was to be considered a mirror of civil and ecclesiastical order within a scriptural master narrative. Put succinctly, the world was created by God (Genesis), and it would end when God decided to bring an end to it (the prophecies of Daniel). For its duration, the world is an orderly realm, visible especially in the heavens (according to Plato and Ptolemy) and unfolding in historical time. Humans are part of this order and can and should make it their business to know and revere it. Knowledge of both expected and unexpected heavenly events could thus help to give a degree of control—although not absolute control—over events that appeared to defy the natural orderliness of life (Ptolemy, *Tetrabiblos*). Here is how Melanchthon put the case in his 1535 "Oration on Astrology":

> What are eclipses, conjunctions, portents, shooting stars, and comets but God's oracles which

warn of great calamities and changes in the life of man? Anyone who scorns these [signs] despises God's warnings. Yet the more that these [events] are to be feared, generally the more that ought to be attributed to the heavenly signs; for the great man may learn the same from sacred letters, lest desperation or impudence or impiety creep up into his soul. Just as it is not irreligious in farming and in navigation to attend to the forecasts of storms, so too [it is not irreligious to] consider natural signs in the management of [everyday] affairs, which God displays so that he may render us more watchful and he may provoke in us care; and thus for the faithful, it is useful.[23]

Melanchthon's evangelical precepts, style, and personal example inspired generations of pupils and teachers to promote what they regarded as legitimate divination. The entire enterprise had a remarkable coherence that bore a distinctive and easily recognizable systematizing impulse. Between about 1540 and 1580, Wittenberg-trained authors were among the leaders in producing great quantities of texts in all categories of the literature of the heavens "for the use of the studious": the sphere, planetary tables and planetary theory, astrological theory, and prognostication.

THE MELANCHTHON CIRCLE, RHETICUS, AND ALBERTINE PATRONAGE

The dominant figures in this Wittenberg movement—Erasmus Reinhold (1511–53), Caspar Peucer (1525–1602), and Georg Joachim Rheticus (1514–74)—have all been mentioned, but not yet as a group. In fact, they were all connected with one another, under Melanchthon's aegis, in a circle whose social and affective core was structured by strong paternal-filial bonds. Two of the three, Reinhold and Peucer, eventually held the rectorship of the university. Peucer lived in Melanchthon's house as a student and eventually married one of Melanchthon's daughters, Magdalena. A poem at the beginning of Peucer's *Elements of the Doctrine of the Celestial Circles* (1551) endearingly, and not surprisingly, dubs Melanchthon "Father," and a chronological list of "Astrologi," starting with the Creation and proceeding up to 1550, ends with Erasmus Reinhold, "Praeceptor mihi carissimus."[24] This was a connection of no small importance at a time when kinship relations functioned as a major resource in acquiring professorships in the German universities.[25] And it was these three men who ultimately had the greatest role in shaping and legitimating enduring readings of Copernicus's work.

There was, however, another significant force shaping the political space of the Melanchthon group, and this was the patronage of the territorial prince, Albrecht Hohenzollern, margrave of Brandenburg-Ansbach and duke of Prussia (1490–1568). Ducal Prussia was the first Protestant territory in Europe. Albrecht had been grand master of the Teutonic Order, but in April 1525 he became a secular hereditary duke holding the duchy as a royal vassal of the king of Poland, Sigismund I. Politically, both Luther and Andreas Osiander had played important roles in this transition. It was Luther who successfully promoted the idea of changing the Teutonic lands in Prussia into a secular duchy, and it was Osiander's preaching in Nuremberg that secured the conversion of Albrecht to the Lutheran cause.[26] Indeed, Albrecht became the first avowedly Lutheran prince, an enthusiastic supporter in the spread of the Lutheran school system, and a strong promoter of theological and mathematical studies in his patronage of the new university of Königsberg, the Albertina, which was founded on Melanchthonian principles in 1544.

Albrecht was, in fact, part of a wider process of early modern state building, whereby princes of territorial states used ecclesiastical institutions, including schools and universities, as a basis for social integration, for producing voluntary obedience and acceptance of the hierarchy of social classes.[27] His frequent and close correspondence with Melanchthon testifies to how self-consciously the Wittenberg model was exported to the ducal seat in Königsberg. And finally, ducal Prussia directly impinged on Copernicus's geographical and political space: during Copernicus's lifetime, Varmia (also Warmia; in German, Ermland) was a small, triangular Catholic enclave of about four thousand square kilometers, ruled by a prince-bishop and almost entirely surrounded by the newly Lutheranized territory of ducal Prussia but for a short western border with royal Prussia.[28]

We should not mistake these boundaries, however, for the ideological ones that prevailed in cold war Poland and Germany. Even though Albrecht actively sought to bring evangelical clergy into Prussia, he was not an authoritarian ruler; G. H. Williams has even called him "an early proponent of religious freedom."[29] Ducal Prussia became something of a haven for religious dissenters—Dutch Sacramentists, Spiritualists, and Anabaptists. Let us recall that Rheticus ended the *Narratio Prima* with his *Encomium Prussiae,* and it was in this section of the book that he painted a confessionally pluralistic picture: "Königsberg, seat of the illustrious prince, Albrecht, duke of Prussia, margrave of Brandenburg, etc., patron of all the learned and renowned men of our time."[30] But he also listed among the lights of Prussia the bishop of Frombork, John Dantiscus (r. 1530–48), a poet and diplomat of some note, who was an important adviser to Albrecht's liege lord, Sigismund I, but certainly no friend of the Lutheran cause.[31] In this section Rheticus also praised Tiedemann Giese, the bishop of Chełmno, for having advocated a reformed calendar and "a correct theory and explanation of the motions."

Albrecht was thus early regarded as friendly to religious dissenters and a supporter of scholars, humanists, and skilled artisans in the Nuremberg-Wittenberg orbit. He retained Johann Carion (1499–1537) as his court astrologer until his death. Georg Hartmann (1489–1564), an instrument maker and the vicar of Saint Sebald's Church in Nuremberg, was an important correspondent of the duke between 1541 and 1544. During this time, he supplied him with clocks, political information, and reports of his demonstrations of the special properties of magnets in the presence of Emperor Ferdinand I.[32] Albrecht also supported Melanchthon's close friend Joachim Camerarius, first rector of the Nuremberg *Gymnasium,* who translated the first two books of the *Tetrabiblos* and cast genitures for the duke.[33]

It is not surprising, therefore, that Duke Albrecht was also regarded as a prospective protector by religious moderates like Copernicus, Giese, and Rheticus. Contacts with Albrecht had begun while Rheticus was still with Copernicus at Frombork. Most significant, on April 23, 1540, one month after the *Narratio Prima*'s publication at Gdańsk, Giese sent Albrecht a copy of the book. In the accompanying letter, Giese told the duke that "Nicolaen Cüpernic," a doctor and canon of Frombork, had introduced an *astronomische speculation,* a novelty (*newigkeit*) which, to the unskilled, appeared "strange" (*seltzam*). But now, this Cüpernic had inspired and roused a "deeply learned mathematician from the University of Wittenberg" who wished the opportunity to study the foundations of this "opinion," and he had come to Prussia to do just that. This (obviously Lutheran) mathematician had now written a short report that had been published in the form of a little book in which he also praised the land of Prussia, and "he has not passed over the name of Your Highness in silence." Giese ended by asking the duke to offer his "protection" (schutz) to the author of this work.[34] Perhaps there was some hope—although there is no direct evidence here—that the reissue of the Narratio might bear the direct marks of ducal authority and legitimacy in the form of a title such as Astronomia Prutenica or even Astronomia Albertina.

On August 28, 1541, still in residence with Copernicus in Frombork, Rheticus now gestured with a practical production: he sent Albrecht a short German work called Chorographia, a map of Prussia that he had made, and a small instrument about which nothing is known. Chorography differs from geography, according to Ptolemy, in that the latter is a picture of the whole known habitable world "as a unit in itself," whereas the former, "selecting certain places from the whole, treats more fully the particulars of each by themselves—even dealing with the smallest conceivable localities, such as harbors, farms, villages, river courses, and such like." Thus, the chorographer was like a painter who depicts only the eye or the ear by itself.[35] Chorography was also connected with astrology, as it delineated the regions subject to specific planetary and zodiacal influences.[36] In the letter dedicating his Chorography to Albrecht, Rheticus was firmer than Giese: he called Copernicus "my lord teacher," and he did not lessen the status of Copernicus's ideas by referring to them either as a "speculation" or an "opinion": "Thus, we shall have in the praiseworthy work of the truly most learned Doctor Nicolaus Copernicus, my lord teacher, an exact reckoning of the time and the years [i.e., the calendar] as well as the paths of the Sun, the Moon and the other stars; and in what fashion and order they must be and were created. For, as is well known, hitherto there have been great deficiencies and obstacles on this matter."[37] In the same dedication, Rheticus praised astrology and its link to geography: without an accurate knowledge of terrestrial longitudes and latitudes, one would not

be able to predict the areas that eclipses would affect on Earth or to make "learned conjectures" concerning the effects on the inhabitants of those regions.[38] Rheticus thus tied the appeal for patronal support of a work of astronomical theoric to its value for chorographical and astrological practice.

Albrecht immediately replied with the gift of a handsome Portuguese gold coin, a lisbonino worth about ten ducats, intended for honorific display rather than use as practical currency. The very next day Rheticus, capitalizing on the positive response, sent a new request: could the duke use his influence with the elector of Saxony to request further "leave time" on Rheticus's behalf from his teaching obligations at Wittenberg so that he could attend to the publication of De Revolutionibus? Again the duke obliged. Near-identical letters were sent at once to the elector and to the officials at Wittenberg.[39] Thus, although Albrecht said nothing about Copernicus's theory, his appreciation of Rheticus's practical work was clear and his endorsement of further time off unambiguous.

In these negotiations, we have a hint of a more general consideration that motivated much Renaissance princely patronage of the sciences of the heavens. Patrons were greatly interested in practical results—material and visual resources that could be used in some specific way. But courtly scientific artifacts carried more than merely functional meaning. Just as books carried singular (printed) dedications to specific patrons, and just as the patron encoded his books in unique bindings, so court craftsmen strove to make their products bear the unique signs of their patrons. Thus it is no surprise that the craftsmen who made maps, terrestrial and celestial globes for navigation, timekeeping devices, measuring or observing instruments, triumphal arches, or great palace murals were also adept at attaching symbols of the status of their patrons to their productions. The goal was display and symbolic control.[40] The same was true for the astrologers who provided in their nativities a functional instrument to guide rulers in their future decisions and, at the same time, used the stars to mark the patron's uniqueness. Insofar as natural philosophy was seen to be a precondition for this sort of practical knowledge (for example, medical or political prognostication) and insofar as it was theologically coherent, it would be seen to be worthy of praise and support but not capable of being exhibited, like a globe, in a court setting. In fact, philosophy was the province of the universities, and the category of the middle sciences was an academic construction. Astronomical theory without connection to other parts of the science of the stars possessed no special cultural niche in the courts.

Melanchthon, as will be recalled, regarded astrology as part of physics, and physics concerned itself with the full spectrum of explanatory resources, from the material and purposive to the efficient and formal. Mathematics was one part of the mixed coupling that constituted astronomy; but mathematics alone had no epistemic capacity to generate true statements about the Earth's nature. The best that the mathematical part of astronomy could do was to take its initial physical premises from scripture or from Aristotle and Ptolemy. In spite of Melanchthon's praises of the science of the stars, the discipline of mathematics was still an arts subject in the curricular hierarchy. At Wittenberg, as in other universities at this time, there was no provision for the doctorate in mathematics or astronomy, no formal licensing that recognized those disciplines as autonomous professions. The goal of the universities was to teach students the disciplines prerequisite to the professions of teaching, preaching, medicine, and law. Someone who wished to make an academic career as a *mathematicus* with competence in *astronomia practica*, therefore, had a choice of seeking a higher degree in one of the higher faculties of law, theology, or medicine— most typically the last—or becoming a professional teacher of the mathematical arts, after which one could practice astrology.[41]

Rheticus was a teacher. Thus, after he returned to Wittenberg at the end of September 1541, he was expected to take up the pedagogical duties from which he had been exempted for some two and a half years while he lived with Copernicus. He resumed the teaching of arts students who were preparing for careers in medicine or theology. But now, returning laden with a stack of new books as well as Copernicus's manuscript, an unanticipated development occurred that, for a teacher of mathematical subjects, was a signal honor: he was elected dean of the faculty, a position of considerable status for the twenty-seven-year-old—and one that he could not turn down without insulting his friends and supporters.[42]

Because the academic term lasted from October 18, 1541, to April 30, 1542, Rheticus could not personally oversee the printing of *De Revolutionibus* at Nuremberg. During this time, Rheticus gave numerous lectures on the "precepts of the heavenly motions" (*doctrina de motibus coelestibus*), including, of course, many on astrological topics, but there is no evidence that he actively used the *Narratio Prima* to teach *theorica*, nor that he promoted the heliocentric theory in either public disputations or private groups.[43]

RHETICUS, MELANCHTHON, AND COPERNICUS
A PSYCHODYNAMIC HYPOTHESIS

It is unfortunate that there is no fuller or more secure explanation for what transpired on Rheticus's return to Wittenberg. Burmeister suggests that Rheticus simply did not stay in Wittenberg long enough for a "Copernicus school" to get off the ground.[44] Although there is of course no way to know what might have happened had Rheticus stayed longer, it is relevant that no such group of followers formed about him either in Leipzig or in Krakow; nor did either the *Narratio Prima* or *De Revolutionibus* have the effect of producing assent to Rheticus's reading of Copernicus among those who studied at Wittenberg or the other Melanchthonian universities in the century's middle decades. This was not for lack of familiarity with *De Revolutionibus*. Among those whom Burmeister has been able to identify as being present at Rheticus's lectures in 1541–42 (Caspar Peucer, Hieronymus Schreiber, Matthias Stoius, Johannes Homelius, Joachim Heller, Matthias Lauterwalt, Friedrich Staphylus, Joachim Acontius, Johannes Stigelius, and Hans Crato von Krafftheim), I have found none who endorsed the representation of celestial order and its prophetic import found in the *Narratio Prima*, or the dialectical, Horatian-Erasmian defense found in the masterwork itself. Surviving copies of *De Revolutionibus* belonging to the first five individuals on Burmeister's list, as well as to Erasmus Reinhold, show that in general books 3 and 4 were annotated with great care, whereas book 1 was barely marked at all.[45] The annotations generally follow the same pattern that one finds in Reinhold's notes: they appropriate the new planetary models while ignoring the hypothesis of planetary order. The main exception was Rheticus's close friend and public supporter, Achilles Pirmin Gasser, whose copy of *De Revolutionibus* does not reflect the Reinhold annotation choices.[46]

From the fervor of Rheticus's writings and, by contrast, from the apparent pallidness of their reception at Wittenberg, it is apparent that others did not appreciate the meaning of the intense personal experience that Rheticus had undergone during his long stay with the canon of Frombork. When he returned, he not only reported his new beliefs to Melanchthon, Reinhold, and others, but, as Hans Blumenberg has written, he also returned full of fire, a believer and a convert.[47] He was, as we might say, overheated.

Although Melanchthon could easily see from the *Narratio* that Rheticus regarded Copernicus as an ally in the struggle against Pico, the *Praeceptor Germaniae* did not share the young man's enthusiasm for the new astronomy. Indeed, he evidently could not understand the intensity and meaning of Rheticus's experience with Copernicus; nor was he predisposed to upset the order of the curriculum or drastically revise the manuscript of his lectures on natural philosophy to accommodate Rheticus's passion. As Melanchthon wrote to his friend Camerarius on July 25, 1542, after Rheticus had been back for some ten months: "I have been indulgent toward the age of our Rheticus in order that his disposition, which has been incited by a certain enthusiasm (*quodam Enthusiasmo*), as it were, might be moved toward that part of philosophy in which he is conversant. But, at various times, I have said to myself that I desire in him a little more of the Socratic philosophy, which he is likely to acquire when he is the father of a family."[48]

In an earlier study, I suggested that the personal meaning that Copernicus's theory had for Rheticus might well lie in his unconscious relationship with the older man and in the death of his own father. In brief, Rheticus's father, Georg Iserin, a physician from Feldkirch, was convicted in 1528 on charges of swindling and stealing from his patients and was beheaded.[49] One painful result for Rheticus's family was that his father's name could no longer be employed legally after his execution, and hence the unfortunate widow had to revert to her Italian maiden name, De Porris; her son, Georg, later took the scholarly name of "Rheticus," having been born within the old boundaries of the Roman province of Rhaetia.[50]

At the age of fourteen, and just over a decade before he met Copernicus, Georg Joachim Iserin Jr. had lost not only his father but also part of his legal family identity. We do not possess Rheticus's direct testimony about the father's death; nor, indeed, do we know whether he was a direct witness to it. Moreover, we have virtually no information on the general patterns of his early familial relationships with which to mount a viable psychodynamic explanation. Perhaps, given this paucity of evidence, it would be easiest to abandon further speculation, let alone any search for explanations that are currently unfashionable and unlikely to command any agreement.[51] Yet we cannot avoid the simple fact that Rheticus was the only figure who could possibly be called a "disciple" of Copernicus, or even a "follower," prior to Michael Maestlin and Thomas Digges in the 1570s. It is not merely that Rheticus was convinced by Copernicus's (dialectical) arguments where others were not, but that he so clearly identified with Copernicus; he became, in effect, his collaborator in presenting the new theory of celestial order in the Wittenberg cultural orbit. Copernicus's act of unifying previously diverse fragments seems to have had a liberating, almost intoxicating effect on Rheticus. In sharing in the presentation of the theory of this man whom he so admired, he wrote as though a veil that had covered an inner personal turmoil was lifted.[52]

A psychodynamic explanation based on early childhood experiences and perceptions, of fantasies, unconscious wishes, and dreams must be able to draw, at the very least, on sufficient information about significant patterns in an individual's early relationships with others in order to establish accurate parallels with later behavior. Conflicts that remain unresolved in early years tend to reappear later on in different guises, and the historian's skillful identification of repetitive patterns, can help to explain the cause of later actions that seem inappropriate, exaggerated, unexpected, and even destructive in a particular individual. In addition to the general models of relating to mother, father, brothers, and sisters, specific traumatic events can sometimes intensify a low-lying conflict or perhaps initiate a new one. An individual's response to the devastating experiences of rape, warfare, or the witnessing of a violent death, for example, depends on factors such as age, basic ego strength, and external support, but these events will be upsetting even for a normal individual. And yet, even with good evidence, the historian must be content with probable and intuitive conjectures, because the clinical procedure of testing against free associations in the present is not available.[53] Still, if historians are at a disadvantage in lacking immediate clinical evidence, they do possess one source of considerable value, as Frank Manuel has pointed out: "Historians . . . have a completed life before them, and the end always tells much about the beginnings."[54] In the case of Rheticus, the key pieces of evidence are three: the violent death of his father, the personal rhetoric of the *Narratio,* and a recollection about Copernicus at the end of his life.

First, the father's death. What must it have been like? We do not possess Rheticus's direct testimony, but we know that Georg Joachim Iserin Jr. had lost not only his father but also part of his old identity. In the paternalistic German family culture of this period, we can imagine what a profoundly painful experience it must have been for the young adolescent to have lost not only the man whom all young boys so greatly admire and fear but also the social identity associated with his father's name and profession. Psychologically, we would predict that his later relationships with men, particularly older men, would be marked by a deepened and intensified ambivalence. On the one hand, the manifest horror and grief over the power of aggression that led to the violent death of his father might later appear in his need to atone for even the vaguest of hostile feelings. Together with this fear of his own aggression, we should expect to find determined efforts—in the search for wholeness, strength, and harmony—to unconsciously repair the damage earlier wrought on his father. On the other hand, he must also have felt an unconscious sense of liberation—of being freed at last from the tyranny of the old man—a feeling which was fully consonant with his later identification with intellectual rebellion in the persons of Copernicus and Paracelsus.[55]

Let us now see if we can find some support or refutation for these hypotheses in Rheticus's linguistic and stylistic choices in the *Narratio.* To begin with, we find evidence of ambivalence in his representation of Johannes Schöner. The *Narratio Prima* starts with the following dedication: "To the Illustrious John Schöner, as to his own revered father, G. Joachim Rheticus sends his greetings."[56] The motto on the title page is, however, of an entirely different tenor. It is really a sort of revolutionary manifesto embodied in a Greek quotation from Alcinous, which does not appear in *De Revolutionibus:* "Free in mind must be he who desires to have understanding."[57]

Reverence and rejection of authority: the ambivalence is announced at the beginning of the work and the themes reappear clearly in the text itself. Thus he writes of Schöner:

> Most illustrious and most learned Schöner, whom I shall always revere like a father, it now remains for you to receive this work of mine, such as it is, kindly and favorably. . . . If I have said anything with youthful enthusiasm (we young men are always endowed, as he says, with high rather than useful spirit) or inadvertently let fall any remark which may seem directed against venerable and sacred antiquity more boldly than perhaps the importance and dignity of the subject demanded, you surely, I have no doubt, will put a kindly construction upon the matter and will bear in mind my feeling toward you rather than my fault.

The tone is more than polite etiquette. It is cautious, apologetic, and deferential, as if to implore his father unconsciously not to beat him for his aggressive sentiments. In the next sentence, we are reassured that Copernicus was also the kind of man who was very respectful of ancient authority: "Concerning my learned teacher I should like you to hold the opinion and be fully convinced that for him there is nothing better or more important than walking in the footsteps of Ptolemy and following, as Ptolemy did, the ancients and those who were much earlier than himself."[58]

Rheticus then describes the rebellious, tradition-breaking act of Copernicus almost as though it were beyond his control but ends with the quotation from Alcinous that had already appeared on the title page. "However, when he [Copernicus] became aware that the phenomena, which control the astronomer, and mathematics compelled him to make certain assumptions even against his wishes, it was enough, he thought, if he aimed his arrows by the same method to the same target as Ptolemy, even though he employed a bow and arrow of a far different type of material from Ptolemy's. At this point, we should recall the saying: 'Free in mind must he be who desires to have understanding.' "[59] The imagery of the bow and arrow, symbolizing the new and different theoretical assumptions used by Copernicus in his attack on the phenomena, suggests an important part of the appeal that Copernicus's work held for the young Rheticus, for he himself had shared in the attack by influencing Copernicus to publish his new theory after so many years. In fact, no one else—including Giese—had been able to extract the treasure from him. An interesting event near the end of Rheticus's life reinforces this interpretation. A young mathematician named Valentine Otho (b. 1550), who had studied with the successors of Rheticus at Wittenberg—Caspar Peucer and Johannes Praetorius—came to visit Rheticus in the last year of his life in 1574 in the town of Cassovia (Kaschau or Kosice) in the Tatra Mountains.[60] In Otho, the aging Rheticus saw himself as he had once been, almost as though he were looking into a mirror of the past. Otho reports: "We had hardly exchanged a few word on this and that when, on learning the cause of my visit, he burst forth with these words: 'You come to see me at the same age as I was myself when I visited Copernicus. If I had not visited him, none of his works would have seen the light.' "[61]

In Copernicus, Rheticus had found a kind and strong father who, unlike Melanchthon, had a streak of youthful rebellion in him: a man who was different, as Rheticus's father had been; a father who could attack ancient authority with his intellectual weapons without himself being destroyed; a father who, *like the system he created*, had a head and a heart that were connected to the same body. Such a man Rheticus was eager not only to identify with, but to idolize and deify.[62]

Rheticus's personal relationship with Copernicus, therefore, was singular; and it had a closeness and intensity that no one at Wittenberg would ever fully comprehend. This may explain why Rheticus's linking of the Elijah prophecy with the new celestial ordering failed to register support at Wittenberg. There, Copernicus was regarded neither as a prophet nor as a god. The Socrates whom Rheticus portrayed in the *Narratio*, however, was the godlike Copernicus, the man who had lifted the fog for him and shot his arrows true to the target. His theory could predict the rise and fall of kingdoms in world history. The Socrates of Melanchthon, on the other hand, was a hard-working, contemplative, married man with children. Peucer and Reinhold easily fit this description, Rheticus not at all. In 1550, he was forced to leave Leipzig on a charge of sodomy (*Sodomitica et Italica peccata*).[63]

Among Rheticus's few supporters was Achilles Pirmin Gasser. Gasser had endorsed both Rheticus and Copernicus in his preface to the *Narratio* and thereby also supported the connection between prophecy and the new celestial order. Moreover, in his *Chronicle*, Gasser had reinforced the Elijah prophecy. As the Feldkirch city physician, Gasser was one of the immediate successors to Rheticus's father. And it is evident that Rheticus held Gasser in the highest regard, as he presented to him one of his father's personally signed medical books in October 1542, soon after returning from his stay with Copernicus.[64]

By contrast, the failure to win Melanchthon's endorsement proved to be a decisive moment both in Rheticus's life and in the evolution of the Copernican question. It meant not only that the *Narratio* did not appear in the Wittenberg curriculum, but also that Rheticus could not rely on the *Praeceptor Germaniae* for critical support in seeking ducal protection and hence patronal legitimation. There would be no *Astronomia Prutenica*.

ERASMUS REINHOLD, ALBRECHT, AND THE FORMATION OF THE WITTENBERG INTERPRETATION

Perhaps as early as 1540 Reinhold had first become directly acquainted with the theory of Copernicus through (unrecorded) communications with Rheticus and by reading the *Narratio*.[65] In his *Commentary on Peurbach's New Theorics of the Planets*, published in 1542 after Rheticus's return, he expressed a note of dissatisfaction with the state of planetary theory that resonates with the preface to *De Revolutionibus*: "In order to reveal the causes of the various appearances, some learned astronomers have supposed or decreed eccentric deferent circles, others a multitude of orbs and motions.... The great number of celestial orbs thus amassed must be attributed to the [astronomer's] art, or rather, to the weakness of our understanding."[66] Somewhat later he referred to a new hope for the reform of astronomy: "I know of a recent author who is exceptionally skillful and who has roused in everyone a great expectation that astronomy may be restored. And he now has ready the publication of his labors. Indeed, in explaining the phases of the Moon he abandons the Ptolemaic model where he assigns to the moon an epicyclic epicycle."[67]

Thus Rheticus's presentation of Copernicus's

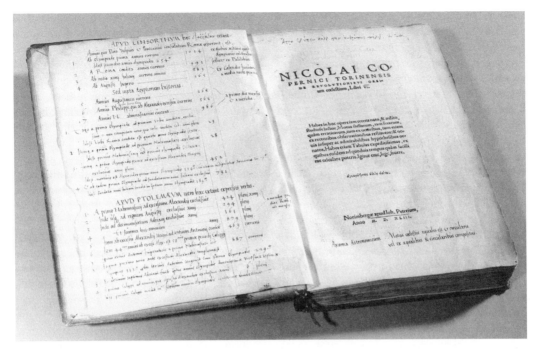

35. Reinhold's copy of *De Revolutionibus* (1543), title page and recto of front leaf. Courtesy Crawford Library, Royal Observatory, Edinburgh. Cr.8.43.

equantless hypotheses, immediately prior to the publication of *De Revolutionibus*, had already raised in Reinhold "a great expectation." Of Copernicus's treatment of the Moon's second inequality, Rheticus wrote: "He assumes that the moon moves on an epicycle of an epicycle of a concentric."[68] And, on the following page, Rheticus took the point one step further: "Furthermore, most learned Schöner, you see that here in the case of the moon we are liberated from an equant by the assumption of this theory, which, moreover, corresponds to experience and all observations. My teacher dispenses with equants for the other planets as well."[69]

It is interesting that Reinhold made his appreciative remarks about the "epicyclic epicycle" in his preface to Peurbach, a work built securely on equant theorics. For as soon as he acquired his own copy of *De Revolutionibus*—which must have been soon after its publication—it was clearly Copernicus's *substitution* of alternative models for the equant that attracted Reinhold's admiration. At the bottom of the title page of his personal copy of *De Revolutionibus*, in beautiful and carefully formed red letters, Reinhold foregrounded the guiding principle of Copernicus's planetary mod-

eling: "Axioma Astronomicum: Motus coelestis aequalis est et circularis vel ex aequalibus et circularibus compositus" (The Astronomical Axiom: Celestial Motion is both uniform and circular or composed of uniform and circular motions).[70]

It was precisely this principle that Copernicus had tried to satisfy, if not always successfully, in his construction of the planetary models and which Reinhold consistently singled out in his extensive, private annotations. Thus, in book 5, having completed his long discussions of the Sun and Moon, Copernicus initiated his descriptions of the remaining planetary theories by attacking the ancients for conceding that "a circular motion can be uniform with respect to an extraneous center not its own, a concept of which Scipio in Cicero would hardly have dreamed. And now in the case of Mercury the same thing is permitted, and even more."[71] Copernicus articulated what he called the "principles of the art" in terms of three particular variants: "either an eccentreccentric or an epicyclepicycle or also a mixed eccentrepicycle, which can produce the same nonuniformity, as I proved above in connection with the sun and moon."[72] Once again, Reinhold's pen boldly annotated the point, with a great flourish reserved for Mercury's orb: "Orbis Mercurij Eccentrj Eccentrus Eccentrepicyclus" (The orb of Mercury is an eccentric on an eccentric on an eccentric circle carrying an epicycle).[73] And, in commenting on Copernicus's scheme for the precession of the equinoxes—an oscillatory motion of the poles that Copernicus colorfully likened to "hanging bodies swinging over the same course between two limits"—Reinhold observed that "from two equal and regular circular motions, there is produced: I. motion in a straight line; II. reciprocal motion lest it be infinite; and, III. unequal or differing motion which is slower at the extremes than near the middle."[74] Such was the power of equantless astronomy, which the Maragha astronomers had developed in the fourteenth century and which now Reinhold, and later Tycho Brahe, cherished as a major development in the restoration of astronomy to its original foundations.

Equally significant is that Reinhold so pointedly ignored the Earth's motion, which he failed to classify as either an axiom or a postulate of Copernicus's astronomy. In only one of his private annotations did he explicitly argue that the Earth's motion was absurd.[75] Reinhold designated the grounding principle of astronomy as neither physical nor metaphysical. For him, the phrase astronomical axiom was meant to refer to the mathematical part of astronomical theoric, that part of the "precepts of the celestial motions" upon which the practical tables were based. Later, in the Prutenic Tables, Reinhold stated in his twenty-first precept of calculation that Copernicus and Thabit ibn-Qurrah (contrary to Ptolemy) believed the sidereal year to be more secure than the tropical year because "observations of the sun around the solstices are not sufficiently firm and constant."[76] On this account, says Reinhold, "All posterity will gratefully celebrate the name of Copernicus. The tenets of the celestial motions [doctrina coelestium motuum] were almost in ruins; the studies and works of this author have restored them. God in His Goodness kindled a great light in him so that he discovered and explained a host of things which, until our day, had not been known or had been veiled in darkness."[77]

The timing and circumstances of Reinhold's decision to begin working on new planetary tables mark another significant juncture in the evolution of the Copernican question. The use of Copernicus's ideas embodied in the Prutenics would become the dominant one even as new readings of De Revolutionibus emerged in the 1580s. Reinhold had already initiated a connection with the duke of Prussia in May 1542, just as Rheticus had concluded his term as dean and just one month before Copernicus composed his preface to De Revolutionibus. The patronage strategies in this initial approach usefully illustrate the specific nexus of local practices and negotiations from which the Prutenics emerged. These developments also reveal much about how an influential gloss on the meaning of Copernicus's work came to be established.

Reinhold approached the duke directly by correspondence rather than through an intermediary, and such contacts continued for some years (although, later on, Melanchthon and some of his students also acted on Reinhold's behalf). Reinhold introduced himself as someone unknown to the prospective patron.[78] He represented Albrecht as a preeminent supporter of "Astronomy and Cosmography" and himself as moved by the duke to undertake such studies.[79] He sent along as a gift a copy of a "little book" that he had just published and which he hoped would be seen as fitting and appropriate. This book was none other than his commentary on Peurbach's New Theorics, dedicated to Duke Albrecht. He confessed to the duke that the little book was intended for the schools and, significantly, that it was not a showy thing like an instrument.[80] He defended his little book as providing the foundations and the key to the proper practice of the discipline—without which one could neither understand nor use Ptolemy, tables of motions, or instruments. He begged the duke to accept his book "in humble submission from a Professor of Mathematics since, as Your Grace knows, our art is usually cared for by princes but rather little by other people."[81]

Three months later, the duke sent an encouraging reply, which is interesting not least because of its direct and specific engagement with Reinhold's concerns. He had not only received the book but also read and learned much from it.[82] He explicitly confirmed his wish to "honor and promote" the arts of astronomy and cosmography (echoing Reinhold's own language), but he also expressed his support for "all other free and praiseworthy arts." The duke then acknowledged the association of his name with the Peurbach commentary: "In the book that you have written you have done honor to our name when you have dedicated to us the foundations of this praiseworthy art."[83] And now, in exchange for what the duke called Reinhold's "gracious kindness," he did not wish to fail to send something in return "as an expression of our gracious goodwill": a small goblet (*Becherlein*), no doubt for drinking beer.

Peurbach for a beer mug. It was a promising start, an exchange marked by various reciprocal gestures of civility in which a rather learned—but hardly a new—schoolbook of theoric, written in Latin but with some erudite Greek references, was accepted as an honorable tribute by a prospective patron. In the course of the exchange, the duke signaled that he had taken the trouble to read the book (or, at least, he represented himself as having read it); he also conferred his esteem on Reinhold as a professor of mathematics and on the disciplines of astronomy and cosmography, which he located on the same level as the other arts subjects. And, as with the works of Rheticus, the duke confirmed that he was pleased to have his name associated with an astronomical work, even a book for young students.

In the fall, another exchange further consolidated the favorable climate of mutual benefit. Reinhold used historical precedent and analogy to amplify his gratitude, reemphasize his deference, and create new linguistic resources for further requests: the "distinguished mathematicians" Aratus and Diocles had dedicated works to King Antigonus of Macedonia, and Eratosthenes was the first and best to write about Europe, Asia, and Africa and to dedicate his tables to King Ptolemy. Reinhold continued: "Although I myself am not the equal of these excellent workmen, still Your Estimable Grace has allowed your favor to fall in such a way that we mathematicians may adorn this art with the name of Your Grace and thereby establish a model for young students." And he now let drop his further intention to produce unspecified "great works" which, at this "very unsettled time in Christendom," were more necessary than ever—particularly because "our [mathematical] arts are so little cared for."[84]

The duke replied very favorably in November within a customary language of social hierarchy and deferential gesture: "Your highest thanks for our small, well-meaning honorarium for the work you provided were hardly necessary because in being able to show all the more our merciful support and will to you or to other skillful and learned people, we want to be seen as the merciful lord in all appropriate matters."[85] The duke wished himself to be represented as patron of learning. More important, Albrecht signaled clearly that he could be counted on for further support. As an example, he mentioned that occasionally he would be happy to have "your Judgment based on the stars." Reinhold had thus gained the duke's confidence together with positive signs for the future.[86]

In January 1544, Reinhold put this confidence to the test. He was now well acquainted with Copernicus's work. Perhaps some of the annotations with which we are familiar were already in place. Rheticus had left more than a year and a half earlier to take up a post in Leipzig. Reinhold now reported to the duke that he had a new project that had been underway "for at least a year," that is, at least since the appearance of *De Revolutionibus* in May 1543. This project, as he did not tire of stressing, was a most "useful" one.

A year ago I received a letter from your grace, in which you requested that every now and again I write to you and send some of the works of my art. It was my obligation to respond to this favorable disposition to me and the praiseworthy fervor of your grace with my hard work; thus I began already to draft at that time a new and useful work, namely "New Tables of Heavenly Motions," which would have the title *Prutenic Tables*—a work that I have decided to dedicate to Your Grace and in which I follow the observations of Copernicus, comparing them with the observations of other older and newer scholars. Moreover, there will also be new tables of previous and future solar eclipses especially useful for illuminating the history of all times and for improving the chronicles. Also, I shall add ephemerides to these for many future years. I did not wish to write again to Your Grace before I was able to send a part of this work or a specimen of it. Work proceeds slowly for the sole reason that I am kept busy by a great number of affairs in this school. Nevertheless, to provide evidence of my continuing gratefulness to Your Grace, I am sending to you this letter, and I enclose with it a small work, namely a newly published calendar that I hope will be useful to persons who study these [mathematical] disciplines. It was not possible to have these loose pages bound before I sent them off to you. However, shortly, with God's help, I shall send over something about observations of solar eclipses and the revolutions of a few princes.[87]

In this frank and revealing document, we see clearly how the quest for princely support was couched and managed in a rhetoric of utility rather than of truth. This language corresponded to expectations at the level of practical use. The duke wanted to see results, and he expected to see them in the form of written evidence or the products of material craft. Reinhold understood what was expected of him. As for Copernicus—already well known to the duke through Rheticus's and Giese's earlier approaches—his authority as a (new) source of observations was assumed rather than argued for by Reinhold. There was also a shared organization of knowledge in which Reinhold promised that the numbers in the new tables would be useful in composing improved chronicles of world history as well as calendars and genitures. The duke did not question these connections in subsequent correspondence, as they were natural and obvious.

The letter is also interesting in that it was not directed to the official patron of the University of Wittenberg, the elector of Saxony. Evidently the elector was not a reliable supporter of the subjects deemed important by Melanchthon. Thus, in writing to Albrecht, Melanchthon showed a keen sense of how to manage the appeal for material support. As we might say, he knew which buttons to press: utility joined with praise of aristocratic honor.

> Only very few people now learn mathematics, and likewise only very few of the mighty support these studies. Here [at Wittenberg] there is a learned man who devotes himself singularly to this study, and he has begun some works that will be useful for spreading this knowledge. *However, our court is little concerned with these studies.* If your Grace could give him something annually—be it a donation or a scholarship—then I hope that this generosity would also be useful for these same arts. And I would also remind you that he would bring to light useful works and ephemerides. *Otherwise, if the prince does not support these kinds of works, we are not going to have ephemerides.* But please, Your Grace, excuse this perhaps inappropriate petition.[88]

These several appeals—Melanchthon's in particular—had the desired effect. Albrecht quickly replied in August 1544 that although he had his own scholars to worry about, "nonetheless, in order to make you happy so that you can see that your petition was not for naught, I am going to give him [Reinhold] 100 Rhenish Gulden for two years for his support, and he will receive this in four parts [i.e., every six months]."[89]

In their replies, both Reinhold and Melanchthon reframed the meaning of the duke's favorable action. His response was to be seen as an endorsement not only of Reinhold the impoverished mathematician but also of the more general proposition that the mathematical study of the stars was a subject especially dependent on princely support. And further, this dependency on the state was consistent with Melanchthon's familiar emphasis on the orderliness of nature as reflected in the Creator's plan: "Human life must have arithmetic, the art of surveying, the calendar and cosmography. The beautiful order of the yearly course of the sun during the year is a clear testimony from God that Nature was created by a wise, orderly master artisan, and thus we should all love these arts in praise of God because, through them, this order is explained. Now, to be sure, private individuals without the help of rulers cannot make instruments for observing the sun, stars, eclipses, equinoxes, and so forth because it costs something."[90] Similarly, Reinhold linked the duke's financial grant to the status of the subject matter of mathematics and to church teaching. The duke's support of these subjects was essential "because there are two things that are true: (1) that these arts are necessary for an orderly life and these arts adorn Christian teaching in the church, as God himself says, that the Sun is there to make the year for us. Now, without the help of rulers, there would be no understanding of the year, no observation, no art would be learned, and . . . (2) these arts could not survive. Thus, it is praiseworthy that you, Your Grace, extend your hand to us poor mathematicians, and I accept the gracious promise of this stipendium with my humble thanks."[91]

In December 1545, the epistolary connection between court and university was transformed into a directly personal one when the duke visited Wittenberg and met with Reinhold. They must have discussed the subject of astrological genitures on that occasion, because afterward Reinhold sent the duke a list of genitures of various rulers, from which he was to pick out the ones that he desired. The register was impressive: it included, among others, the nativities of Luther and Copernicus. If Zinner is correct, Reinhold's birth charts later fell into the hands of Luca Gaurico.[92] The duke subsequently asked for the genitures of his own family—his wife, his daughter Anna Sophia, his cousin Margrave Albrecht (Alcibiades), and the son of Margrave Casimir of Brandenburg—as well as Emperor Charles V and the young King Sigismund I of Poland. He asked Reinhold to prepare these as quickly as possible and promised that he would send Reinhold a gift in return.[93] Reinhold said that he was inclined to prepare the genitures, although, perhaps as part of the negotiation, he remarked pointedly that the responsibilities of his position at the university greatly occupied him and hinted that it might take considerable time to calculate and explain them.[94]

For the next two years, however, free time proved to be the least of Reinhold's worries. Luther died in February 1546. The loss of the reform movement's powerful and charismatic figure affected not only the university but also the fragile religious and political alliances that Luther had managed to hold together. Now a period of complex and bitter religious wars began, lasting until the Peace of Augsburg in 1555. In April 1547, the imperial troops of Charles V crossed the Elbe at Mühlberg, between Dresden and Wittenberg. There they virtually annihilated the Protestant armies of the Schmalkaldic League and took prisoner the elector Johann Friedrich of Saxony (1532–47; d. 1554), patron of the university.[95] University life was disrupted. Many students and professors, including Reinhold, fled from Wittenberg. The *mathematicus* wandered for a whole year with his family, unable to continue work on the tables and able to send but few nativities to the duke. Sometime during this period, it appears that his wife died and left him to care for his three children—a burden both emotional and financial. Yet throughout, he continued to receive strong indications of support from both Melanchthon and Albrecht.[96] To the duke, Melanchthon continued to reiterate his contention that princely support was essential for mathematical studies because such studies brought almost no profit or income and were generally neglected and derided. The duke made it clear that he would look on it very favorably if Reinhold were to decide to prepare for a career in theology.[97]

This gesture, surprising in our terms, was evidently a signal of ducal willingness to promote Reinhold to a higher, more secure status from which he could continue with his astronomical work. Consider what such a position might mean. The theology faculty at Wittenberg had been given the power, from 1535, to formally examine candidates for the ministry. And, as Germany began to split into ever more particularistic territorial jurisdictions, individual princes took over ecclesiastical functions that had once belonged to the bishops. The medieval bishop's consistorial court, for example, now became a civil court to which the elector appointed theologians and lawyers. This court controlled all ecclesiastical discipline, tried matrimonial cases, maintained uniformity of worship, passed judgment on popular morals, and had the right of excommunication. After Luther's death, the consistories became central institutions by which territorial rulers controlled daily social and religious life.[98] Academic theologians, in addition to their scholarly work, could also serve the state in such civil and juridical roles.

Reinhold left open the door to such a shift in position, but he did not push it. In April 1548, his academic protector Melanchthon wrote to Albrecht that Reinhold would be done with the tables within a couple of years. He could then devote himself more fully to a life of Christian reading and prayer, although he was already a God-fearing man who lived morally and was very learned in philosophy and theology. Even now, said Melanchthon, he knew enough to be promoted to a doctor of theology. Perhaps at a later date he might enter the theology faculty at the duke's own university at Königsberg.[99]

The tables were completed in November 1548, earlier than Melanchthon had anticipated. Reinhold soon learned through Melchior Isinder, professor of Greek and theology at Königsberg, that the duke planned to send him fifty thalers as a reward for completing the work.[100] Soon after, Reinhold remarked to the duke that "we who pursue these mathematical studies for the benefit of the State and the Church, find it indispensable to have the support and generosity of princes." Moreover, he promised that "in a short time we shall have in hand the best tables of heavenly motions and, in the future, Your Grace, I shall turn to the *other* part of Astrology, that which *interprets* the stars' influences."[101]

In spite of constant expressions of support, however, the publication of the technically arduous *Tables* was going to be costly and would require a very good printer with experience in the typographical composition of mathematical works. This task was going to impose even further financial demands on the duke, whose reliability in fulfilling his obligations was becoming questionable. Petreius was the obvious choice for a publisher. As we have seen, he had an extremely rich list that included Schöner, Osiander, Gasser, Cardano, and Copernicus. As early as November 1544, Melanchthon had asked Schöner to approach Petreius with the suggestion of publishing Reinhold's tables.[102] By September 1549, Reinhold had in hand an imperial privilege that granted him the right to publish the tables as well as a number of other works, including a commentary on De Revolutionibus and a new work of theoric.[103]

Reinhold now wrote a long and revealing letter to another former Melanchthon student, the Königsberg theologian Friedrich Staphylus. His intention was to gain further support through Staphylus's mediation. This letter shows how a mediator might facilitate a negotiation, permitting certain thoughts to be expressed more openly and allowing the intermediary to shape the request more forcefully and effectively. To Staphylus, Reinhold expressed his rather desperate financial condition: he was "the poor father of a very large family"; he spoke as well of "the poor students who help me with my calculations." He acknowledged the duke's many years of past support and his own unwavering intention to dedicate the tables to this "generous Maecenas." He emphasized the uniqueness and superiority of his work, which he referred to as "New Astronomical Tables" (Novae Tabulae Astronomicae): "From these one could calculate backward almost three thousand years or surely to the times of Ezekiel."[104]

Reinhold now frankly spilled out his concerns. He complained of the physical and financial toll that the work of calculation had taken on his health and on his family. He needed a Maecenas who would take care of his children. He had shelled out approximately 500 gulden from his own salary; of the 250 thalers provided by the duke, only some loose change ("klingenden Gelt") now remained. It was seven years since he had stopped teaching in a private school in order to devote himself to his mathematical studies. For some two hundred years, scholars had had the astronomical tables of King Alfonso of Spain. These tables had required the work of twenty-four men, and, in return, the king had paid them with a ton

of gold ("einige Tonnen Goldes zur Belohnung geschenkt haben soll"). But now Alfonso's tables no longer agreed with Ptolemy, and Reinhold hoped that his own tables, greatly corrected, would carry the fame of the patron after whom the work was named for the next five hundred years. Reinhold associated this fame with that of Copernicus: "I have many reasons why I call these tables Prutenic Tables and why I am pleased to dedicate them to the noble prince Duke Albrecht of Prussia; and this is the principal reason: I have borrowed most of the observations from the most renowned Nicolaus Copernicus, a Prussian, and I have conceived and executed these tables from [these observations] as principles and foundations."[105] Having emphasized the duke's social gain, Reinhold now clarified specifically what he wanted in return: "I am not so shameless that I should demand a ton of gold for my great and lengthy work. I wish only proportionate compensation for my costs and my pains so that, as the fruit of my great work, I might bequeath not only poverty to my children."[106]

Staphylus communicated Reinhold's message succinctly and effectively. In addition to emphasizing the uniqueness of the work, he stressed Reinhold's poverty and skillfully added a further twist of his own: "Someone has advised him," he told the duke, "that he should dedicate the work to the Emperor Charles V and give it the title Tabulae Carolinae, just as the Alfonsines were named after King Alfonso of Spain, who brought them to fruition with a great sum of money. Master Erasmus merely says that he would rather dedicate it to Your Grace and name it the Prutenic Tables in praise and honor of Your Grace."[107] Staphylus's veiled threat to name the tables for the emperor rather than the duke was evidently made without Reinhold's knowledge. Staphylus also added that he could not advise the duke on how to respond.

The Reinhold-Staphylus strategies achieved only partial success. The duke once again reiterated his love of the arts, his wish to help, and so forth, but he also alluded, for the first time, to his own great expenses. As a consequence, he was disposed to offer only five hundred gulden "at an agreed date and place," and that was to be the end of it: "He will be satisfied with that," wrote Albrecht.[108]

Reinhold was not satisfied, and perhaps was a little disappointed at Staphylus's handling of the matter. Even as he scrambled to get his tables into production with Petreius—might the work be printed on "regal" (size) rather than "common crown" paper?—he grumbled to Staphylus: "Certainly I do not doubt that if you had dealt with my matter verbally with the duke, everything would have turned out much better; for I can also strengthen with an oath what I have told you about my costs and many difficulties."[109] He wanted the full amount paid within a year. He and his children had suffered enough. Perhaps the duke could also add to the promised monetary sum a cloak or "some other gift of honor." Moreover, the duke should also be reminded of the ongoing account. According to Reinhold, he had been paid a total of about 300 gulden—232 gulden in cash and "two gilded drinking mugs"—and, in return, he had dedicated to the duke his commentary on Peurbach.[110]

From a client's point of view, the distinction between gifts of honor and gifts of money could be a thin one, as objects made of precious metals might be converted into currency. And when a patron was slow in producing monetary payment, there might be a temptation to convert the Maecenas's drinking vessels into gulden rather than to fill them with some inebriating liquid. From a patron's perspective, it might sometimes be cheaper to pull a valued object off a (possibly full) shelf than to dip into the castle treasury. Furthermore, judged against currency values available to us from early sixteenth-century Nuremberg, the *Prutenic Tables* were costly. Albrecht Dürer, for example, received 85 gulden for making portraits of the emperors Charles IV and Sigismund in 1512. The prognosticator Jörg Nöttelein received 21 gulden for an annual forecast. Veit Stoss, a Nuremberg sculptor, paid 800 gulden for a house in 1499. The wealthy burger Willibald Imhof entered into his expense book for annual costs, respectively, 156 gulden for the household, 134 for wine, 45 for beer, 12 for medical, 70 for his wife and 140 for his children. In 1565, Imhof paid 400 gulden in taxes.[111] Judged by these standards, Reinhold was asking for more than three times Imhof's annual household expenses to cover publication of the *Prutenic Tables*. This comparison may well help to explain why the duke continued to drag out the payments.

Meanwhile, a new obstacle appeared that exposed the fragility of the relationship between the author, printer, and patron. In the summer of 1550, Petreius unexpectedly died. At that time, he already had in his possession some part of Reinhold's manuscript. But his death now meant that the *Tables*, the natural sequel to *De Revolutionibus*, would not appear in Nuremberg. It also meant the demise of whatever resources the printer Petreius had agreed to put up for the publication.[112] In fact, authors often received no compensation from the sales of their books but relied heavily on gifts of honor and money from patrons such as Albrecht.[113] To add to the difficulties of losing his printer, Reinhold himself was unusually overburdened with teaching and administrative responsibilities. He had been elected dean of the faculty for 1549–50; now he assumed the office of rector, which required him to write and oversee the printing of all official university announcements and proclamations. To complete the *Tables*, he needed to retrieve his manuscript from Nuremberg, obtain a new printer, and win the duke's approval for his dedicatory letter.

This last task proved to be trivial compared with collecting on the full promised payments. The dedication contained themes already articulated and tested in previous correspondence. These were now amplified and integrated in the ducal dedication. For example, Reinhold introduced his dedication with Melanchthon's familiar providential justification for the science of heavenly motions: "I affirm what is most true: that all this wisdom, the science of number, measure, and the celestial motions is a spark of light from heaven in the minds of men, placed there that it may show that this world was not put together by chance combinations of Democritus's atoms but that the eternal, just and good Mind is a work of architecture; and [that these sciences] bring the greatest utility to life."[114]

To have the duke's name associated with the eternal heavens was "much more splendid" than to have his name or the memory of his virtues inscribed in various writings, histories, trophies and buildings. Reinhold continued by explaining why he was going to use the work of Copernicus on which to construct his tables: "Copernicus, the most learned man whom we are able to name other than Atlantus or Ptolemy, even though he taught in a most learned manner the demonstrations and causes of motion based on observation, nevertheless fled from the job of constructing tables, so that if anyone computes from his tables, the computation is not even in agreement with his observations on which the foundation of the work rests."[115] There followed a reference to the seven anguished years he had spent on this project, costly to himself and his family and a sad time for the country. He might have spent his time on "more lucrative works or divinations," but he had preferred this computational endeavor, as now one might compute the celestial motions back to the beginning of the world and correctly compute eclipses and great planetary conjunctions. Finally, he invoked the example of previous patrons. It had (allegedly) cost King Alfonso 400,000 pieces of gold to have his tables constructed; and Alexander had (allegedly) given Aristotle 80,000 talents to investigate nature—a grand total of 480,000 pieces of royal money. The moral did not need to be spelled out: the present work was a bargain.

The duke read the dedication and made his own calculations.

> We have read the *epistola dedicatoria* diligently and let it please us well, above all because we appreciate and believe that there will be people who, in this matter, understand more than we do, since we are bad in Latin [*ein schlechter Latinus*]. And, so that you may more conveniently finish the work that you have begun as soon as possible, we have agreed to a payment of five hundred gulden for the realization of our favorable will, calculating the gulden at twenty-one Meissen groschen or thirty groschen of our Prussian coinage, and such [monies] shall be paid to you over five years. Thus, when you have finished the work, the *Prutenic Tables,* the first one hundred gulden will come to you immediately and, thereafter, [another installment of] one hundred gulden—so that you should dedicate to us all the works that you have described in the catalogue that you sent us.[116]

The duke was clearly determined to squeeze as many dedications out of Reinhold as he could manage. And he knew exactly from the publication privilege granted by the emperor the numerous titles that Reinhold had in mind. Among these was a work to be titled *Hypotyposes orbium coelestium, quas vulgò vocant Theoricas Planetarum, congruentes cum tabulis Astronomicis suprà dictis,* another titled *Eruditus Commentarius in totum opus Reuolutionum Nicolai Copernici,* a work on "the doctrine of triangles," a commentary on all the books of Euclid, a commentary on Ptolemy's *Geography* with a new Latin translation, a treatise on the quadrant, and a previously unpublished work on optics by Alhazen. All these, and more, were supposed to carry dedicatory praises of the duke—and, perhaps, the dedication was all that he would actually read.

In the end, however, only the *Tables* bore the duke's name. And, when he finally received the book in October 1551, fresh from the Tübingen presses of Ulrich Morhard, he received as well a letter from Reinhold urging him to verify their value by showing them to learned men. In particular: "Your Princely Grace might seek the judgment not only of other experts but also that of the honorable Dr. [Andreas] Osiander who, in our time, because of his erudition in mathematics, has a great reputation."[117] Reinhold's recommendation strongly suggests that he knew of Osiander's role in affixing the anonymous letter to the reader to *De Revolutionibus*—probably by hearsay from Rheticus—and also that he believed that Osiander had the duke's trust. In this he was not mistaken. On March 21, 1552, the duke replied: "We have not failed to show Andreas Osiander your work which, after he quickly skimmed it, pleased him very much—just as now we are concerned that everyone praise and honor this same book. And Mr. Osiander has asked to read and review the book carefully when he has found some time."[118]

We do not know if Osiander ever completed the reading. He died in the same year. Moreover, although the duke continued to press Reinhold for more nativities, the plague soon forced the mathematician to flee Wittenberg, and he died in his hometown of Saalfeld on February 19, 1553.[119] "I have lived, and you, Christ, have shown me the way; now, I pass on," he is said to have uttered at the last.[120] But Reinhold never found his way to the duke's treasury. To his son Erasmus (who eventually became court physician to Ludwig, count of the Palatinate) and his two daughters (about whose lives we know nothing), he bequeathed a long and ultimately futile struggle to collect the overdue balance.[121]

THE *PRUTENIC TABLES*, PATRONAGE, AND THE ORGANIZATION OF HEAVENLY LITERATURE

It is now clear why the *Prutenic Tables* had immediate import for the science of the stars but none for the organization of the middle sciences. The *Prutenics* did not tamper with the conventional

    CONFESSIONAL AND INTERCONFESSIONAL SPACES

relationship between the mathematical and the physical parts of astronomy any more than the Alfonsine tables did. Reinhold's patronage negotiations, his place within a certain configuration of power, instead reinforced the prevailing disciplinary organization of the heavenly sciences. Patron and client shared a common representation of knowledge within a socioeconomic relationship of dominance and dependency. Driving the cognitive aspect of the enterprise was the desire for prognosticatory knowledge—knowledge that was not merely symbolic but useful. Both Reinhold and Albrecht were negotiating the association of ducal honor with control of this new resource of celestial prediction at a moment when the Melanchthonian wing of the Lutheran movement was placing special ideological importance on the sacred, revelatory power of nature.

The role of Copernicus's work in this development is fairly straightforward. Because *De Revolutionibus* was a work of general principles and models, it did not offer practical precepts for carrying out specific operations. In contrast, the *Prutenics* had direct utility for anyone wishing, for whatever purpose, to make planetary calculations. Effectively, Reinhold rewrote *De Revolutionibus*, transforming it from a work of "principles" and "theories" into a work of "rules" and "precepts" modeled after the thirteenth-century Alfonsine tables.[122] Anyone familiar with the practices of calculation based upon the Alfonsines would find the transition to the *Prutenics* relatively easy. Reinhold clearly saw his work as a treatise for the practitioner, a ready tool for the prognosticator and the maker of ephemerides.

The astronomical problem that the *Prutenics* solved was the gap between Copernicus's theoretical achievement and its direct utility for the practitioner. Unlike Ptolemy, Copernicus had provided no *Handy Tables* to accompany his theoretical work. He had "fled from the job of constructing tables."[123] Owen Gingerich has shown how Reinhold provided this missing element: he engaged in a massive recalculation of Copernicus's original work. Among the benefits was that the *Prutenics* tabulated the correction angles or prosthaphaereses for every degree rather than for every third degree and carried the values to seconds, rather than minutes, of arc. In addition, the starting positions or radices were neatly tabulated instead of being distributed throughout the text, as in *De Revolutionibus*. Furthermore, although Co-

pernicus had shown that each planet's line of apsides underwent slow, long-term change, he offered no method for finding the apsidal line for a specific time. Reinhold provided such a method and also greatly improved the values for computing the orbital parallax of Jupiter. Finally, Gingerich's computerized recalculation of the *Prutenics'* numbers shows that Reinhold closely followed Copernicus's eccentric-cum-epicycle model rather than Ptolemy's equant.[124] However, Reinhold did not advertise to his readers the theoretical basis of the *Prutenics* but instead promised to treat these in a separate work: "In our commentaries which I wrote on Copernicus's *Work on the Revolutions*, I have laid bare the causes and explanation of the individual constructions."[125] Only the small circle of learned readers at Wittenberg who had studied with Reinhold or who had had local access to his well-annotated copy of *De Revolutionibus* would have been directly aware of the *Prutenic Tables'* theoretical basis.[126]

As Reinhold was careful to explain, however, the *Prutenic Tables* was not itself a work of astrology. Reinhold states in his preface to the reader that it is only linguistic custom to equate the terms *astrology* and *astronomy*:

> "Astrology" is the old name by which [the Ancients] at one time understood not only the precepts of forces and effects but also [the precepts] of the motions of the stars and of the celestial bodies. However, in a later age, it became customary to call by the term *Astronomy* that *doctrina* which contemplates the explanation of the stars' motions and chases them down with numbers; and, to adapt the word *Astrology* solely to the predictions of events caused by the stars' motions and positions or what those events might mean in this lowest nature. However, this divinatory part will be spoken of at another time.[127]

Thus, Reinhold did not regard himself or Copernicus as writing directly about the "divinatory part" of the "precepts of the celestial motions"—a clear example of the self-conscious separation of the epistemic domains of the heavenly literature.

Yet throughout the *Prutenics*, Reinhold missed no opportunity to teach the use of his tables by the specific example of Albrecht's birth (May 17, 1490). The ruler's astrological destiny could assist the reader to calculate the Sun's mean motion, adjust calculation to a specific meridian (Onolsbach, the duke's birthplace), compare the

Alfonsine and the Copernican methods for calculating the mean motion, show how Copernicus incorporated the precessional motion, and so forth.[128] The duke's birth date functioned in effect as a pedagogical and a political resource for bridging the gap between Copernican theoric and Reinholdian astronomical practice. Such use of a ruler's horoscope was not uncommon: Peter Apianus had honored Charles V by using the emperor's nativity to illustrate the use of the hand-colored planetary volvelles in *Astronomicum Caesareum* (1540).

In sum, Reinhold presented his *Tables* under the patronal authority of the Lutheran convert Duke Albrecht and the intellectual authority of the East Prussian canon Copernicus, for the purpose of undergirding the providential and apocalyptic Melanchthonian vision of the world and its history. Reinhold made abundantly evident that his *Tables* were important for the church. The Lutheran church, he said, counts the years from Christ's birth for good reasons: first because God wished to show his great mercy for man's sinfulness through his Son, and second because "the Church expects that the Messiah is about to return soon and thus, so it may prepare the pious for eternal life and glory."[129] If, as he concluded, the world was 3,962 years old at the time of Christ's birth, the assignment that Reinhold left to his successors was to determine how soon the world would end.

THE CONSOLIDATION OF THE WITTENBERG INTERPRETATION

At Wittenberg, after Rheticus's departure, Copernicus's achievement was no longer contested knowledge. Reinhold's readings of Copernicus as a practical astronomer became authoritative at many different levels. While keeping the Earth at rest, Wittenberg authors deployed Copernican resources in all genres of the heavenly sciences, from elementary treatments of the sphere to planetary theorics and from table making to prognostications of the weather and individual fates. It may well be that among the students who helped Reinhold calculate the *Prutenics* in the 1540s were some who were taught from his notes on *De Revolutionibus*. Joachim Heller, a student of Reinhold and Rheticus and successor to Schöner at the Nuremberg *Gymnasium*, referred quite favorably to Copernicus in his prognostica-

tion for 1549.[130] This may well be the first reference to Copernicus in a *practica*, although there is no explicit trace of Reinhold's influence. In 1551, the year in which the *Prutenics* appeared,[131] Heller began his forecast as follows: "The basis of my calculations in my prognostications follows the new and well-founded astronomy which Nicolaus Copernicus demonstrated in his work and has brought to light on the firm basis of heavenly observations; but I have had to take care over the difference in our meridian [and his]."[132]

Heller made no further use of Copernicus's name to enhance the public authority of his *practica,* and I have been unable to find any further references in his later forecasts. Yet it is clear that by 1551, he believed that even such a vague reference would enhance the credibility of his predictions. *De Revolutionibus* was obviously a resource of significant personal value to him as a prognosticator. In fact, we cannot rule out the possibility that he owned more than one copy of that work, as he gave an unannotated copy as a gift to his friend Christopher Stathmion, one of the most prolific German prognosticators of the mid-sixteenth century.[133]

We do not know the fate of Reinhold's copy at his death. Did it go with him to Saalfeld and end up in the hands of his son Erasmus? Did it go directly to his student Peucer? Or might Peucer have purchased the book from Reinhold's impoverished family? Whatever the case, because it was well known in Wittenberg that Reinhold intended to publish a commentary on *De Revolutionibus,* there is likely to have been immediate interest in the fate of his library.[134]

The man largely responsible for consolidating and institutionalizing the Wittenberg interpretation within the Melanchthonian universities was Caspar Peucer. The son of an artisan from Bautzen (Czech Budyšin), then part of Bohemia, he showed such remarkable talents as a young man that he began to attend the University of Wittenberg at the age of fifteen.[135] Besides studying the heavenly sciences with Reinhold and Rheticus, he was taught arithmetic by Michael Stifel (1487–1567). Upon Reinhold's untimely death in 1553, he became his successor. In 1559, he was named to the chair of medicine, and in 1560 he succeeded to the rectorate of the university on the death of his father-in-law.

The Wittenberg interpretation, as reflected in Peucer's work and the writings of other Wittenberg schoolmasters, firmly echoed the views of Peucer's two mentors, Melanchthon and Reinhold. Melanchthon's natural-philosophy lectures, delivered orally for many years, appeared in 1549 under the title *Initia Doctrinae Physicae*.[136] The first edition made strong statements against Aristarchus's old and absurd "paradox" and warned students that it conflicted with Holy Scripture and the Aristotelian doctrine of simple motion.[137] These passages might be taken as evidence of Melanchthon's rejection of Rheticus's views. In contrast, there is a more sympathetic treatment of such topics as Copernicus's lunar theory, because it is "so beautifully put together" (*admodum concinna*), and in a few places Melanchthon uses Copernican data for the solar apogee and for the apogees of the superior planets.[138] For a textbook of natural philosophy it pays greater than usual attention to astronomical concerns. The second edition of the *Initia*, which appeared the following year, significantly toned down the negative allusions to Copernicus by deleting phrases about those who argue, "either from love of novelty or from the desire to appear clever," that the Earth moves.[139] If Melanchthon had seen the manuscript of Reinhold's *Prutenic Tables* only after the first edition of his lectures appeared, it might help to account for this moderation of tone.

Peucer's introductory textbook followed Melanchthon's practices in the first edition of the *Initia*. For example, he cited Copernicus on the absolute distances of the Sun and Moon from the Earth in his discussion of eclipses and on the length of the day.[140] And he portrayed Copernicus as the reviver of the theory of Aristarchus and his ideas as contrary to holy scripture and Aristotelian simple motion.[141]

Piecemeal references of this sort were entirely typical in Wittenberg astronomical textbooks and in classroom presentations from the 1550s onward because of the authoritative precedents established by Melanchthon and Reinhold. Within ten years of the appearance of *De Revolutionibus*, references to Copernicus were becoming common; within twenty-five years, they were entirely normal. Many occurred in elementary presentations of the *Sphere*, others in new editions of the *Theorica* and Reinhold-based ephemerides. In 1566, just over two decades after the first edition of *De Revolutionibus* and fifteen years after the appearance of the *Prutenics*, a new edition of *De Revolutionibus* appeared. This time it was bun-

dled with the *Narratio Prima*. The frequent textbook citations of Copernicus and recommendation for its use by students may well have stimulated the new edition, published by the Basel printer Heinrich Petri. Whatever the explanation, the joint 1566 edition greatly increased opportunities for owning—and reading—both the "Lutheran" and "Catholic" presentations of the heliocentric texts. Owen Gingerich has calculated a printing of 400 to 500 copies for each edition, which means that, in principle, some 800 to 1,000 copies were in circulation by the late 1560s.[142]

What gave many of the Wittenberg textbooks fresh pedagogical value was neither the novelty of their subject matter nor the references to Copernicus but their friendly, accessible and deductive forms of organization and their provision of clear definitions. As a work of natural philosophy, Melanchthon's *Beginning Precepts of Physics* established a precedent for integrating arguments from physics with the teaching of the sphere. This lent a somewhat more Aristotelian and less Ptolemaic cast to the teaching of the rudiments of the sphere. The *Beginning Precepts* thus made it easier for students to appreciate the part of astronomy that was physical and the part that was mathematical. In turn, it was easy to see why Melanchthon subordinated astrology to physics, as its domain concerned stellar causes and terrestrial effects. Wittenberg textbooks that focused on the mathematics of the sphere also carried over Melanchthon's emphasis on clearly integrating theological and physical propositions.

One example of the way a typical Wittenberg "question" was handled comes from Sebastian Theodoric of Winsheim. He organized his *New Questions on the Sphere* in the popular question-answer format, structuring the answers as numbered arguments. Theodoric gave students easily memorized answers that could be committed to memory. One of these questions was "Is the Earth moved or not?"

The Earth is not moved but remains at rest in the middle of the world, and this is confirmed by the following reasons:

I. Whatever Holy Scripture affirms is certain without any doubt.

II. Holy Scripture affirms that the Earth is fixed and immobile.

Therefore, the Earth remains at rest in the middle of the world.

I. These sentences confirm the minor.

Psalms 104. "He who created the Earth on firmness will not move it ever and for eternity."

Ecclesiastes chap. 1. "The Earth remains in position for eternity, the Sun rises and sets, returning to the place from whence it rises, etc."

II. If the Earth were to move forward with local motion, absurdities would occur. . . . For it would move from the middle either to the equator or to the world's axis or in between both.

But these absurdities do not occur.

Therefore the Earth does not move with local motion.

III. It is necessary that the surface toward which the heavy body falls along a straight line be immobile, for otherwise, a heavy body could not fall in a straight line since . . . immediately afterward, the point vertically above would change. For nothing falls in a straight line that does not fall from the vertical.

All heavy bodies fall to rest on the Earth's surface in a straight line, and, were they not prevented by the Earth's solidity, they would fall to its center.

Therefore the Earth is fixed and immobile at the center of the world.

IV. If the Earth were to move, it would move either with straight or circular motion.

But it is moved neither with straight nor with circular motion.

Therefore it is not moved.

PROOF THAT THE EARTH IS NOT MOVED WITH A STRAIGHT MOTION.

For if the Earth were moved with straight motion, on account of its rapid motion, all heavy bodies would regress and remain hanging in the air.

PROOF THAT THE EARTH CANNOT BE MOVED WITH CIRCULAR MOTION.

I. If it were moved circularly, it would hinder the growth of living beings. For rest is required for growth and growth occurs through the adhesion, likening and bordering of like parts; and motion impedes it.

II. Living creatures could not stand firmly on its surface.

III. Objects thrown up would fall neither at the place from which they were thrown nor fall to Earth at right angles.

IV. If the Earth were moved toward the East, meteors and all other things in the air would appear to be carried toward the West—and the opposite. If, indeed, the air and all those things which exist in the air were to turn around together with the Earth, all things would appear to rest.

But experience testifies [*experientia testatur*] that none of these events occur.

Therefore the Earth is not moved but remains at rest at the center of the world.[143]

With the exception of the argument from scripture, the arguments from physics conformed to typical *modus tollens* constructions (If X were to happen, then consequences Y would occur; but not-Y, ergo not-X).

"For the use of students": this was the motto of the Wittenberg textbooks on spherics and theorics. And it helps to explain why so many of these elementary, didactic productions issued within the Melanchthonian university orbit were written by men who were in no sense specialists in astronomy or mathematics but professional teachers with multiple pedagogical competences. Some rose to become theologians; others taught the humanistic languages, Greek and Hebrew; many held positions in both mathematics and medicine, thereby increasing their income and status. A few parlayed these last competences into positions as municipal or court physicians in which they were also expected to cast genitures for their patrons. All reflect the influence of Reinhold's *Prutenics*. We also know that quite a few had their own personal copies of *De Revolutionibus*.

Under Melanchthon's influence, a generational cadre of students produced a "reformed" literature of the heavens beyond Wittenberg. Leipzig was a particularly important site. There could be found Melanchthon's close friend and biographer Joachim Camerarius, joined by Rheticus in 1542; and Victorinus Strigelius (Strigel; 1524–69), a prolific theologian who published a brief work on the sphere (*Epitome of the Doctrine Concerning the First Motion Illustrated by Some Demonstrations* [Leipzig, 1564]) amid a profusion of biblical commentaries, theological orations, and disputations on original sin and free will. Strigelius possessed a copy of *De Revolutionibus*, which, at his death, would pass into the library of Michael Maestlin at Tübingen.[144] Strigelius's colleague Johannes Homelius (Hommel; 1528–62) had been a pupil of Rheticus and Reinhold. He published nothing and seems only to have taught mathematical subjects at Leipzig, but he also held the title of *mathematicus* to the elector August of Saxony and to the emperor Charles V. He too owned Copernicus's main work and may have been the first to men-

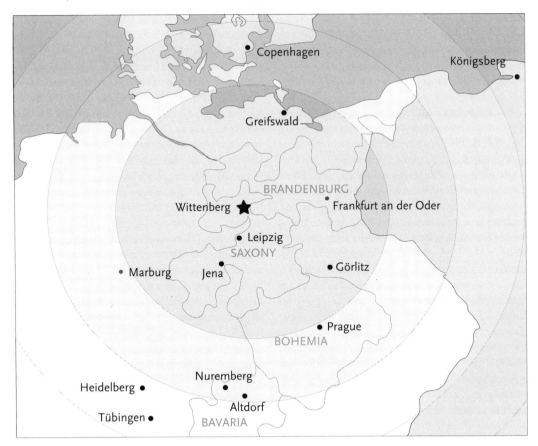

36. The Wittenberg orbit, showing key locations of Melanchthon's influence: Copenhagen (kingdom of Denmark); Frankfurt an der Oder (margraviate of Brandenburg); Görlitz (margraviate of Upper Lusatia); Greifswald (duchy of Pomerania); Heidelberg (electorate of the Palatinate); Königsberg (duchy of Prussia); Leipzig, Jena, and Wittenberg (electorate of Saxony); Marburg (landgraviate of Hesse); Nuremberg, Altdorf (duchy of Bavaria); and Tübingen (duchy of Württemberg).

tion it to the young Tycho Brahe.[145] Bartholomeus Scultetus of Görlitz (Schulz; 1540–1614; M.A. Wittenberg, 1564), Homelius's successor, is known to have taught Tycho and to have had direct access to the annotations of his predecessor.[146] Michael Neander (1529–81) was appointed to the faculty of arts at Wittenberg in 1551 but left shortly thereafter to teach Greek and mathematics at Jena, and in 1561 he published a short work on the *Sphere*.[147] His successor at Jena, Jacob Flach (1537–1611), had also studied at Wittenberg and became professor of mathematics (1572–81) and then of medicine (1582–1611).[148]

Another former Melanchthon pupil, Hermann Witekind (1524–1603), held the chair of Greek at Heidelberg but became professor of mathematical subjects at Neustadt in 1581. Like Neander, he too published a *Sphere* but, as it appeared in 1574,

it contained, in addition to (by now) common references to Copernican parameters, a brief discussion of the new star that had appeared in 1572.[149] Theodorico Edo Hildericus, or Von Varel (1533–99), studied theology and mathematics at Wittenberg (receiving his M.A. in 1566) and Tübingen. Later, he taught mathematical subjects at Jena (1564–67) and Wittenberg (1567–73), Hebrew and history at Frankfurt an der Oder, theology and Hebrew at Heidelberg (from 1578), and finally theology at the new academy of Altdorf.[150] Erasmus Flock (1514–68) taught mathematics for a short period at Wittenberg (1543–45) and served as dean (1544), overlapping with Reinhold's tenure, but later became municipal physician at Nuremberg.[151] Meanwhile, at Marburg, Victorinus Schönfeld (1525–91) taught mathematical subjects from 1557 to 1591 after having been trained at Wittenberg during the Peucer years (1554–57).[152] And Johannes Garcaeus Jr. (Gartze; 1530–75), an important student of Peucer, began teaching astrology and astronomy at Greifswald from 1561, following Erasmus Reinhold's brother Johannes by a few years.

Not long afterward, the Dane Georgius Christophorus Dybvadius (Jørgen Christoffersen Dybvad; d. 1612), a onetime student of Peucer and Sebastian Theodoric, published a curious work in the year he received his M.A., titled *Short Comments on Copernicus's Second Book, which by unmistakable arguments prove the truth of the doctrine of the first motion and show the composition of the tables*.[153] Later, he assumed the extraordinary chair of theology, natural philosophy, and mathematics at Copenhagen (1575–1607). At Tübingen, one could find the Wittenberg-trained Samuel Siderocrates (Eisenmenger; 1534–85), who published a prognostication in 1561 and wrote *Oration on the Iatromathematical Method of Conjunction* (Strasbourg, 1563).[154] After Siderocrates became physician to the margrave of Baden, he was succeeded at Tübingen by Philipp Apianus (Bienewitz; 1531–89), the son of the famed and prolific astronomer and printer Peter Apianus of Ingolstadt and one of the few stargazers in the Melanchthonian orbit not to have come from Wittenberg.[155] And finally, Johannes Praetorius (Richter; 1537–1616; MA 1571), who lectured at Wittenberg in the early 1570s, became the first holder of the mathematics chair at Altdorf. All told, the Wittenbergers were largely responsible for the significant spike in astronomical textbook production at midcentury.

THE ADVANCED CURRICULUM AT WITTENBERG

Advanced students who progressed to the study of theoric had access to Reinhold's syllogistically organized commentary on Peurbach. The 1553 edition contained a few more references to Copernicus, but teachers found that its pedagogical clarity and the acuity of Reinhold's notes made it a highly serviceable classroom textbook. More than twenty years later, former students of Peucer at Wittenberg, such as Johannes Praetorius, Andreas Schadt, and Caspar Straub, were still using Peurbach to introduce students to planetary theory. As late as 1594, Kepler was probably using Reinhold's Peurbach commentary to pre-

pare his classes at Graz.[156] Peurbach was an introduction to Ptolemy's *Almagest*, as Melanchthon emphasized and Reinhold underscored in his commentary.[157] Learning to describe and visualize the models thus preceded their geometric and trigonometic derivation. In 1542, Reinhold glossed this classification with Aristotle's distinction between "knowledge of the fact" or effects available to the senses (*tou hoti = quia*) and "knowledge of the reasoned fact" or causes of particular effects (*dioti = propter quid*). The application of this division to theoretical astronomy became canonical in the 1540s.

After 1543, however, a new pedagogical challenge arose. If one wished to teach the more complex topics of planetary theory using Peurbach and the *Almagest*, then, as *De Revolutionibus* had been modeled after the *Almagest*, the obvious thing to do would be to use Copernicus's work to illuminate corresponding problems. This explains why Peucer, Schadt, Straub, and Praetorius advised their students to consult *De Revolutionibus* directly and why they themselves were familiar with Copernicus's work.[158]

What no commentator on Peurbach provided was a representation of Copernicus's planetary hypotheses in relation to a stationary Earth. Reinhold had obviously intended to do this and may well have been presenting such models to his students in the 1540s. In the absence of a text that worked through the corresponding topics in each book, the best that one could do would be to assign both the *Almagest* and *De Revolutionibus*— for students, clearly an expensive and cumbersome solution. Eventually, a new work on astronomical hypotheses appeared under suspicious circumstances at Strasbourg in 1568 with exactly the same title granted to Reinhold by the Imperial privilege of 1549. This work represented an explicit transformation of Copernicus's title: *Hypotyposes Orbium Coelestium*, rather than *De Revolutionibus Orbium Coelestium*.[159] The edition that I have used contains no authorial attribution;[160] however, another issue of the same work contains a preface by the Strasbourg mathematician and clockmaker Conrad Dasypodius (Hasenfratz; 1531–1601), who claimed that it was a work by Reinhold. And still another issue—this one under the title *Hypotheses astronomicae*—appeared at Wittenberg in 1571 with Caspar Peucer's name on the title page, dedicated to Wilhelm, landgrave of Hesse-Kassel, and claiming to be Peucer's own lectures.[161] Exactly how these authorial reattributions occurred is not clear unless we suppose that the manuscript was pirated from Wittenberg by one of Peucer's students or, as seems most likely, by Dasypodius himself. In any event, although there is little reason to doubt that the *Hypotyposes* were, in fact, based on lectures given by Peucer, the contents also bear the clear impress of Reinhold.[162]

The *Hypotyposes Planetarum* was the first textbook that, like the *Prutenics*, purported to be "in agreement with both the Copernican and Alfonsine Tables." It is a long book, some 534 pages of exceedingly prolix descriptions and demonstrations. In its organization, it follows neither *De Revolutionibus* nor the *Almagest*. It begins by presenting its subject as "astronomy" but immediately defines the mathematical part as "prior." Although there is mention of rectilinear motion, there is no separate discussion of physical topics, such as gravity, planetary intelligences, or celestial matter. It is not, like Melanchthon's *Initia*, a work of natural philosophy; nor does it seek any systematic integration of that discipline. In this sense, it is un-Melanchthonian. Peucer refers the reader to Ptolemy's *Tetrabiblos* for the physical part, which teaches "the forces and effects of the stars" (pp. 1–3). He adds that the terms *astronomy* and *astrology* interchangeably denote the book's subject matter. Throughout, he uses the word *hypothesis* (rather than the title word *hypotyposis*, or "vivid sketch") in the sense of "authors' inventions or feigned stories . . . supposed or assumed." Thus the revised title of 1571, *Hypotheses Astronomicae*, easily fitted the original text. Chief among the astronomical hypotheses are the ones that Reinhold called "axioms" and which Peucer defined as follows: "That celestial motions are circular and eternal or composed from many circles; the other, that they are uniform and regular." There are further evidences of a Reinholdian or Copernican vocabulary in such words as *homocentrepicycle* and *theoria* rather than the Peurbachian *theorica*—although the running head throughout the book reads "Theoricae planetarum." Peucer is also comfortable using the expression *systema circulorum* ("system of circles") or *systema coelestium orbium* ("system of celestial orbs,"), but he made no effort to show the sort of necessity insisted upon by Rheticus. Somewhat later, however, he assigned a special role to the Sun that clearly reflected Peurbach's image of the

sun as a "common mirror and rule of measurement": "Each of the planets is connected by certain rules to the Sun's motion, so that the Sun appears to be the moderator and governor of all the planets and celestial motions or, as it were, to prescribe and order the rules of motion upon which it is not permitted to infringe."[163] This is a striking remark: it was precisely on this point that Peucer ignored Rheticus's inspired passages in which he described the mean motion of the Sun as not only "established by the imagination" but as "self-caused," "both choral dancer and choral leader."[164]

Notably absent from the *Hypotyposes* is any extended treatment—diagrammatic or otherwise—of Copernicus's planetary models, in the absence of which it would be hard for any reader to see exactly how the relevant "accommodations" or transformations might be made between heliocentric and geocentric reference frames. For the planets, the closest that Peucer came was in a section titled "Accommodation of these Hypotheses to the Copernican and Prutenic Canons" which followed his description of the Ptolemaic lunar theory. Here he gave a verbal summary of Copernicus's biepicyclic hypothesis and then proceeded to compare it, angle by angle, with the Ptolemaic model. However, Peucer ended this section by advising students that if they wished to study further the "demonstrations" of these hypotheses, they ought to consult "Copernicus himself."[165]

Yet throughout, there was the persistent assumption—never made fully explicit, let alone generalized—that no problem impeded Copernican parameters being carried across to a geocentric ordering scheme. For example, in discussing absolute distances in Earth radii, Peucer readily cited Copernican values (from the Earth to the Moon: $55\frac{1}{8}$ to $65\frac{1}{2}$; from the Earth to the Sun: 1179).[166] And, most suggestively, Peucer claimed that Copernicus's precessional model could be "transferred" to a geostatic arrangement if ninth and tenth spheres were added:

> If these hypotheses were to be transferred [*hypotheses si transferantur*] to the eighth orb by means of two other spheres, the ninth and tenth; and if, in the same manner, there were to be established in the heavens a moving equator with moving poles and axes as well as those same points where the equator intersects the ecliptic and those [points]

which are most distant from it; and, assuming that the ecliptic with its poles always remains immobile with respect to the eighth sphere, then I believe that the same [effects] would be achieved without having to change the ancient hypotheses.[167]

Not merely were the hypotheses "transferred," but so were the diagrams and some crucial phrasing. It was a *textual* transformation. Peucer produced two diagrams that he had lifted directly from *De Revolutionibus*, including the libration mechanism and the "intorta corolla," only replacing Roman with Greek lettering.[168] Shortly thereafter, the book ended abruptly without generalizing the results.

A group of notes found by Owen Gingerich and based on Wittenberg lectures from the period 1564–70 (now at Gonville and Caius College, Cambridge) shows that the printed textbooks discussed above provided a skeleton of questions within the Wittenberg classroom, much as Sacrobosco's *Sphere* had allowed considerable freedom to manage the content of the material covered in the thirteenth century.[169] In 1297, Bartholomew of Parma wrote a commentary in which he proposed to "say many things about the sphere and about those matters pertinent to comprehension of the sphere which John of Sacrobosco did not say in his treatise."[170] It was just such freedom to work within established categories that allowed commentators to assimilate materials and to structure topics from both natural philosophical and astrological texts. In the fourteenth century, Pierre d'Ailly added questions concerning whether astrology is part of the mathematical or the natural sciences and whether mathematical entities are wholly abstracted from motion, matter, and sensible qualities.[171] The Wittenberg commentator Jacobus Praetorius was working within this sort of commentatorial tradition in 1564 when he said that in the sixth chapter of the *Sphere* readers should refer to the part of philosophy called "physics," that is, "the tenets of the body." However, one also needs to know "all of philosophy" whose subject matter encompasses "the art of good speech," the "precepts of natural things," the "rules of motions," the "precepts for living," and the "rules of morality."[172] Another lecturer, Johannes Balduinus, followed Theodoric's *New Questions*, while Bartholomeus Schönborn, in 1567, followed Peucer's *Elements* and *On the Size of the Earth* as well as Strigelius's *Epitome* with

a few references to *De Revolutionibus*.[173] The last major section of the manuscript ends with Peucer's *Hypotyposes Astronomicae*, which shows that a natural progression of the curriculum was followed from the sphere to theorics. These topics were succeeded by Sebastian Theodoric's lectures of April 1569 on how to compute solar and lunar eclipses using the *Prutenics* and an actual calculation for an eclipse on August 15, 1570. Finally, at the end of this revealing collection of lectures on the heavenly sciences there appeared a nativity, albeit without explicit instructions on how to make it, which was evidently the culmination of Sebastian Theodoric's lectures.

The absence of lectures on the construction of genitures in the Gonville and Caius manuscript is, however, an artifact of its survival, since a complementary set of Wittenberg lecture notes in the Wrocław University Library contains a *Treatise on Genitures* bound immediately after Schönborn's summer lectures on Peucer's *Sphere* (June 19–August 22, 1570).[174] Thus more advanced students could proceed to the different sorts of astrology easily enough once they had mastered the *Sphere* or, better yet, Peurbach's *New Theorics* and the rules for using the *Prutenics*. Several different kinds of printed resources were available for teaching astrological theoric, all published at Nuremberg or Tübingen in the 1530s and '40s: Camerarius's edition of the *Tetrabiblos* (1535), Antonio of Montulmo's *Concerning the Judgments of Nativities* (1539), Schöner's *Little Astrological Work* (1539) or *Three Books concerning the Judgments of Nativities* (1545), and Joachim Heller's edition of Messahalah 's *Three Books concerning the Revolution of the Years of the World, concerning the Meaning of the Planets' Nativities, concerning Reception* (1549).[175] But it is unclear how many copies of these works were in circulation and affordable to Wittenberg students in the early 1550s. And none of them taught how to construct and calculate a nativity using the *Prutenics*, the *Hypotyposes*, or Copernicus's *Revolutions*. The need for a work that would show how to use the *Prutenics* for casting nativities must have been more urgent after Reinhold's death in 1553. Peucer himself was busy writing astronomical textbooks and a work on divination.

The task of adding the final link in the rebuilding of a reformed heavenly science fell to Johannes Garcaeus (Gartze), a valued student of Peucer and Reinhold. His first work, discussed below, was called *A Brief and Useful Treatise on Erecting Heavenly Figures, Verifications, Revolutions, and Directions.*[176] But astrological prognostication was never far from the master narrative of world history. In 1563, Garcaeus published an eschatological tract with thirty questions and answers about the end of the world and the return of Christ. Robin Barnes has well paraphrased the main questions: "Why has the last Judgment been announced beforehand? What proof do we have it will come? Why has God waited so long? What do the Epicureans think about this belief? Why did God keep the time of the end hidden from us? What signs do we have that the day is now approaching? What will follow the Judgment? How should we ready ourselves?" Natural signs pointing to the end were much in evidence: "How often have we beheld haunting images in the heavens—terrifying comets, blazing chasms and wondrous portents. Likewise, [we behold] terrifying winds, earth tremors, great floods, spiraling prices, wars, disease—astonishing signs that have already been fulfilled and those that daily are still being fulfilled."[177] In 1576 Garcaeus relaxed his apocalyptic gaze long enough to produce an ambitious presentation of 400 birth charts of famous men, a feat that easily dwarfed Luca Gaurico's 1552 collection of 160.[178]

The *Brief and Useful Treatise* illustrates the range of kinds of works that constituted the science of the stars. It is a handbook for calculating nativities, not a book about how to interpret judgments based on the calculations. It appeared in 1556 from a major Wittenberg publishing house, the heirs of Georg Rhau, with a dedicatory poem from Caspar Peucer. The epilogue says that "the true foundations of astronomy" and "astrology, as it is now called," lie at the basis of this work. Garcaeus evidently meant to recommend the *Tetrabiblos* and the *Centiloquies* rather than any Arabic astrological works: "It now remains for students to make use of these examples in order to exercise this teaching and to pass on to the reading of Ptolemy, who, as the most eminent author, writes such judgments that rarely fail, as those of us who follow Ptolemy experience in making judgments." Garcaeus dedicated his work to the elector of Saxony rather than to the duke of Brandenburg, and the elector's nativity appears on the back of the title page (b. July 30, 1526, 5:30 A.M.).

Garcaeus offered a typical Melanchthonian justification that authorized both the contemplative and the active value of the science of the stars. First, he asserted that there is no Democritean necessity in the world. Yet there are direct signs of divine providence manifest in the beauty of the stars and their orderly motions, and these witnesses of divinity alone would be sufficient to justify a contemplative approach to the subject of "sidereal science" even if the stars' positions signified nothing about the weather and human temperament.[179] But, of course, Garcaeus did believe that the stars' positions possessed meanings and produced effects, and his book would explain the categories and resources needed to make such predictions, though not to interpret their meanings. What was needed?

A principal requirement was to find the Sun's true position on the meridian for a particular latitude. To do this, one needed to know how to use an ephemerides and thus to distinguish between the tropical and sidereal years, how to make intercalations between different calendars, and how to take account of the precessional motion (on which Garcaeus followed Copernicus). Garcaeus made particular use of various chapters in *De Revolutionibus*, book 3, and precept 21 from the *Prutenics* and sometimes referred to "the new ephemerides of Copernicus," by which he evidently meant Reinhold's work of 1550. This casual interchanging of Copernicus for Reinhold shows how rapidly Reinhold's work was taken to be a conventional marker of authorship for Copernicus. Garcaeus then proceeded to methods of domification (proposition 10), where he mentioned four different ways to divide the zodiac, including the systems of twelve equal divisions ascribed to the ancients; the system of Regiomontanus and Abraham ibn Ezra, with its twelve unequal divisions; Campanus's system; and finally, that of Alcabitius and John of Saxony, rejected by the "learned Regiomontanus." In the end, Garcaeus recommended the first of these as the simplest for students to grasp and, for those well versed in astronomy, the Regiomontanus system.

Throughout, the only example Garcaeus gave was the elector's birth date and place, the meridian of Freiburg im Breisgau. And finally he turned to the problem of actually constructing the horoscope or, as he called it, the "revolutions of the geniture," that is, the interval of time needed for the Sun to return to the same place in the dodecatamories that it occupied at the moment of the individual's birth.[180]

Garcaeus's otherwise unremarkable little work thus took its place in this pedagogical reformation of the science of the stars. In the second half of the sixteenth century, these Wittenberg astronomical-astrological textbooks became the most influential models for instruction in safe predictive knowledge based on natural causes. They assisted in both the standardization and the replication of certain sorts of definitions, arguments, and generic modes, and they served consciously to complement strictly biblically based prophecy.[181]

GERMANY AS THE "NURSERY OF MATHEMATICS"

Sometime in the 1560s, the French humanist and educational reformer Peter Ramus (1515–72) conducted a survey of the state of mathematical studies in Europe. Although he personally visited several cities of the Empire in the late 1560s—including Nuremberg, Augsburg, and Basel—he mostly relied for sources on his friends, students, and epistolary correspondents in various countries, including the Englishman John Dee.[182] His account is interesting not least because it offers a contemporary representation from the end of the decade I have just described.

Ramus concluded with envy and special admiration that Germany (the Empire) was "the nursery of mathematics."[183] Using the classical division of knowledge into contemplative and active functions, Ramus found reason to praise the active, practical parts of all areas of knowledge—as opposed to their contemplative, theoretical value—and the special importance of mathematics to the diverse disciplines. Germany's history, as he (often erroneously) constructed it, was replete with great discoveries and discoverers: "German mathematicians, I maintain, discovered the three extraordinary arts of artillery [bombardica], printing and navigation." These discoveries all had practical value and consequences. For example, the Venetians made use of artillery machines in their war against the Genoese around 1400; the benefits of printing appeared in the publication of Peurbach's tables by Regiomontanus; and the first book ever printed (he claimed), Cicero's Office, was published by Johann Fust in Mainz in 1466 (a copy of which Ramus professed to own).[184] Ramus offered several discriminating explanations for Germany's preeminence in the realm of knowledge. It had

an abundance of gold and silver mines as well as a great military capacity. It had great artisans and painters, like Albrecht Dürer. Moreover, only in Germany were there such learned princes as Wilhelm IV, landgrave of Hesse-Kassel (1532–92), who were interested in patronizing mathematical studies. He reserved the greatest praise, however, for Philipp Melanchthon:

> Just as Plato revived the study of mathematics in Greece through the great power of his eloquence and erudition, so Melanchthon found [mathematical studies] already greatly encouraged in most academies of Germany, with the exception of Wittenberg. Whereupon, through the force of much and varied instruction and through the example of a pious and upright life, which, at least in my opinion, no doctor or professor in that country had ever attained, he wondrously ignited [those studies]—with the result that Wittenberg became superior not only in theology and eloquence, in which fame it especially excels, but also in the studies of the mathematical disciplines.[185]

Ramus valued the Wittenbergers's special combination of linguistic eloquence and mathematical acuity: he referred to Reinhold as "the eternal publicist [praeconium] of mathematical studies"; to Peucer as "another Melanchthon, and the future publicist of mathematical studies"; and to Rheticus as "another Copernicus."[186] Elsewhere in the Empire he listed more than a dozen other professors of mathematics (including many not trained at Wittenberg) as examples of German mathematical prowess and rich patronage that he earnestly hoped would be emulated by France and other nations. The list included both Catholics and Protestants and did not seek to distinguish between present and past occupants of university chairs. A good number of these writers were probably known to Ramus and his informants because they represented the critical mass created by the surge of midcentury textbook authors: Gemma Frisius (Louvain); Philipp Apianus (Ingolstadt); Johann Stöffler, Johann Scheubel, and Samuel Siderocrates (Tübingen); Sebastian Münster and Christian Wursteisen (Basel); Erasmus Oswaldus Schreckenfuchs (Freiburg); Valentine Naibod (Cologne); Christian Herlin and Conrad Dasypodius (Strasbourg); Valentine Engelhardt (Erfurt); Georg Rheticus and Johann Hommel (Leipzig); and Johann Virdung (Heidelberg).

For all his praise of the German mathematicians, however, Ramus projected his own peculiar vision of the heavenly sciences: an astronomy devoid of all theoric—hence, effectively, a practical astronomy of tabulated numbers, or, as he called it, "an astrology without hypotheses."[187] This was certainly not the selective reading of *De Revolutionibus* that I have been describing as the Wittenberg interpretation, which sought to improve Ptolemy's hypotheses by adapting Copernicus's to them. Ramus objected to Copernicus not merely because he had argued for a false conclusion (the Earth's motion) but because he had inferred it from false causes (planetary hypotheses).[188] In fact, not only Copernicus but all "astrologers" were to be criticized for employing any theorics at all—"fictions" (*commenta*), that is, "the most absurd fable in order to demonstrate the true facts of nature from false causes."[189] Ramus expressly hoped that Rheticus would be the one to "liberate astrology from [all] hypotheses" and invited him to discuss the matter with him in Paris.[190]

Other stargazers, however, were understandably chagrined by Ramus's radical proposal. The Scottish physician and mathematician Duncan Liddell, professor at Helmstedt, the founder of a chair of mathematics in Aberdeen and an acquaintance of the Tycho Brahe–Wilhelm IV circles, wrote in the margin of his copy of Ramus's *Scholarum Mathematicarum*: "He demands absurd and impossible things."[191] Tycho Brahe, who met Ramus in Augsburg, objected years later that without hypotheses one simply cannot understand celestial phenomena.[192] In 1609, Kepler charitably (and cleverly) appropriated Ramus's call for an "astrology without hypotheses" for his own "astronomy of true physical causes" and laid claim in his *Astronomia Nova* to the Royal Chair of Mathematics in Paris that Ramus had promised as a reward to anyone who could meet his ideal of heavenly reform.[193]

CONCLUSION

By way of summary, let us return to Copernicus's problematic in the 1490s. There were three difficulties that plagued the science of the stars: (1) Pico della Mirandola's claim that astronomers' disagreement concerning planetary order undermined the epistemic basis of divinatory astrology; (2) a widespread belief that there were inaccuracies in the planetary tables on which calendars, almanacs, and prognostications were based; and (3) a disagreement about the right sort of planetary models to use in deriving the tables: homocentrics, eccentrics, or eccentrepicyclics. To these considerations one must add Peurbach's "shared motions," which, as far as can be determined, no one but Copernicus regarded as a difficulty, much less one worthy of transformative explanation. The hypothesis of the Earth's annual motion, however Copernicus arrived at it, was essentially a response to the first and fourth worries of his Krakow and Bologna periods. Copernicus's resolutions to the second and third problems were then built into the articulation of the new ordering but were in no way necessitated by it.

The Wittenbergers recognized this disjunct between planetary order and planetary modeling. Consequently they were able to reject or ignore the hypothesis of the Earth's motion as a basis for further theoretical work both because it threatened the order of the middle sciences and because they recognized (rightly) that it was unnecessary to resolving the inaccuracy of the tables. Thus the Wittenbergers prized Copernicus's solution to the second problem most highly. His sidereal frame of reference and modified Ptolemaic eccentric were seen as better solutions for the tables than the theory of Averroist homocentrics (whether in the representations of Achillini, Fracastoro, or Amico).[194] Because they believed that Copernicus's biepicyclic models really did improve the accuracy of the planetary tables (as Reinhold's *Prutenics* seemed to confirm), it was believed that his improvements would resolve the uncertainties of astrological forecast in its many forms.

However, it is also striking that Copernicus's anomalous solution to the problem of planetary order received no comment. Reinhold and Peucer clearly acknowledged the solar component shared by the planetary motions, but they did not see Copernicus's explanation of it as in any way justifying Rheticus's exuberant endorsement. Furthermore, if the impenetrability of the celestial spheres had played any role in convincing Copernicus to set the Earth in motion, he certainly did not use that premise to argue his central case in 1543. Indeed, Copernicus lacked an apodictic demonstration, and no one—not even Rheticus—claimed that *De Revolutionibus* had met such conditions of proof. As for the defenders of theoretical and practical astrology, the question of planetary order simply does not appear to have posed a serious problem—from Bellanti's early polemics against Pico down to the many midcentury editions of the *Tetrabiblos*. And these responses again underscore Copernicus's highly idiosyncratic solution to the astrologers' dilemma.

Melanchthon's reforms meanwhile increased the number and concentration of practitioners of "legitimate divination" in the reformed evangelical universities between the 1540s and the 1570s. Reinhold's reading of *De Revolutionibus* promoted a planetary theoric of compounded circular motions within a close-knit circle of students and colleagues at Wittenberg. Complemented by Melanchthon's *Initia Doctrinae Physicae* and numerous Wittenberg textbooks, Copernicus's physical claims for a moving Earth were unambiguously rejected on both physical and scriptural grounds. Meanwhile, Albertine patronage reinforced the utility of connections between the new theoretical principles of Copernican astronomy and astrological practice. For the remainder of the century, Reinhold's *Prutenics* was the most important resource for making astronomical and astrological predictions. In this sense, Reinhold's achievement did constitute a Copernican answer to Pico's skepticism about the possibility of astrological divination. But in the end, it was only a partial answer because it ignored Pico's charges concerning the order of the planets and therefore astronomy's status as a middle science. Although Reinhold's efforts encouraged more widespread study of *De Revolutionibus*, a major outcome of Melanchthon's support was that Rheticus's contested reading of the heliocentric hypothesis was successfully ignored. It was not the *Narratio* that assisted in the construction of nativities and annual prognostications, but rather *De Revolutionibus* and the *Prutenics*. The Wittenberg interpretation thus cut the bond that the *Narratio* had tried to establish between prophecy and the new celestial order.

6

Varieties of Astrological Credibility

MARKING THE DANGERS
OF HUMAN FOREKNOWLEDGE

In the middle decades of the sixteenth century, the surging tide of prognosticatory activity exacerbated tensions among different claimants to foreknowledge. Although early modern popes and cardinals were notorious consumers of astrological advice, leading theologians, both Catholic and Protestant, were united in the belief that their god alone had secure knowledge of the future. Yet the stars and planets were a constant reminder of the residual presence and power of the pagan gods and their secular virtues as well as the threat of purely natural determination.[1] The worry about such pagan residues involved theologians in demarcationist practices. They strove to maintain their own authority to determine what counted as legitimate ways of producing short-term foreknowledge in relation to their understanding of the master narrative of holy writ and the sacred prophets. The rest was diabolical superstition. The devil was the concrete embodiment of mistrust: he had the power to confuse the senses and the intellect, to substitute falsehood for truth, to fool people through dreams and fantasies, to distract people from proper worship by dwelling in statues, and by making them believe that they knew the future when, in fact, they were the victims of his deceptions. Also worrisome was the possibility—and the uncertainty—that bad demons rather than good spirits were the cause of planetary effects.[2] Yet although

there was agreement that the devil must be fought, there was no consensus about where and how he was operating. Even a churchman or a prince could be infected by the devil's tricks and machinations.[3] Defending astrology's credibility thus carried with it the worry that, even if successful, the sort of knowledge it might attain would be tainted.

Copernicus's name became associated with an optimistic and safe view of prognosticatory practice, especially through the *Prutenic Tables*. After the *Prutenics* appeared in 1551, Copernicus became the essential friend, the *vademecum* of all legitimate ephemeridists and nativity casters regardless of confessional persuasion, national affiliation, or theoretical allegiance. As a resource of astrological forecast, the Wittenberg articulation of Copernican-based tables and mechanisms marked the special confidence of the Melanchthonian wing of the Protestant movement in decoding the divine plan through its manifestations in nature. Melanchthon and his son-in-law Peucer, as we have seen, allowed the greatest latitude for different kinds of divination; but Luther was much more wary than Melanchthon about any sort of prophecy that was not exclusively based on the Bible and other sacred texts. John Calvin was more cautious than the Melanchthonians but more moderate than Luther. In 1549, four years before the appearance of Peucer's classification of good and bad kinds of divination, Calvin wrote a vernacular treatise ostensibly directed to the "unlettered" and which unambiguously ruled out all but "natural astrology"—which meant foretelling the weather, the tides, and vicissitudes of the human body. The grounds for his attack on "judiciary astrology," which encompassed for him all human activities and interactions, was based largely on familiar Piconian and Augustinian doubts.[4] Astrologers, for example, relied too much on the moment of birth rather than that of conception, although the latter was admittedly hard to determine. Errors of a few seconds in calculating the time of birth could make enormous differences in predicting the "complexion" or combination of qualities in a personality. Proximate causes, such as the parents' seed, were "one hundred times more forceful than all the stars."[5] An omnipotent God simply did not need the stars in order to give man special grace or to decide who would be eternally saved. Clearly, the end-of-century exchange between Pico della Mirandola and Lucio Bellanti remained the principal axis around which epistemological positions were organized. Official Calvinist and Catholic theological opinions on the domain of acceptable kinds of astrology were thus very close.

By undermining Pico's authority, many prognosticators hoped to weaken theological resistance. One strategy was to cast (astrological) aspersions on the skepticism to which the theologians appealed. In a published collection of the genitures of famous men, the polymath Girolamo Cardano repeated Bellanti's charge that "an astrologer" had predicted Pico's death at age thirty-three. Likewise, Savonarola's geniture with the Moon and Mars at the midheaven in Capricorn meant "without doubt" a public death by fire: "And so," wrote Cardano, "he was burned at Florence."[6] Another approach was to fight one theological authority with the arguments of another, as for example with Christ's horoscope. Could the stars' configuration have exerted effects on the son of God? No, said Francesco Giuntini in 1573, on the authority of Albertus Magnus: the stars did not cause Christ's destiny, but they announced its meaning.[7]

Compared to such defensive gestures, Copernicus and Rheticus's project to bring an end to disagreement among astronomical theorizers sat at the high end of the science of the stars. It was presented as a revision of astronomical principles rather than directly as a contribution to astrological practice. Yet in the 1540s, before the appearance of the *Prutenics,* both Rheticus's preliminary formulation and the detailed models of *De Revolutionibus* opened new questions with a possible bearing on astrology's credibility: Did the Copernican planetary order improve the accuracy of astral predictions? Could it offer a more robust explanation for variations in the strengths of the planetary influences? Did it resolve lingering questions about the occurrence of eclipses, the positions of the stars, or the slow movement of the equinoxes? The 1540s and '50s would prove to be a period noteworthy for different attempts to reground astrology as credible heavenly knowledge.

BECOMING A SUCCESSFUL PROGNOSTICATOR

Nothing was quite so telling for astrology's reputation as a forecast for a prince that appeared to come true. In 1528, a Carmelite monk named Giuliano Ristori of Prato (1492–1556) issued a prognostication, no longer extant, in which he reportedly foretold the early death of Alessandro, illegitimate son of Lorenzo de'Medici the Younger, soon to become ruler of the Florentine Republic (1530). In January 1537, the hapless Duke Alessandro came to an unfortunate end, just as his astrological adviser had foretold. He was strangled in his bed, and rule quickly passed to his eighteen-year-old cousin Cosimo I.[8] The Senate elevated Cosimo to the rank of duke, and thus began the consolidation of Medici rule . . . and also Ristori's reputation.

Ristori's prediction became a noted exemplar for the rest of the century and beyond. Its wider credibility derived not least from the fact that it perfectly fitted a familiar anxiety of the ruling classes: the early, untimely death of a prince. The absence of a published forecast did nothing to prevent Ristori from acquiring tremendous fame beyond Pisa, Siena, and Florence, the region where he taught and prognosticated. Luca Gaurico, by the 1540s a well-established astrological writer, used the example of Ristori to imply the success of his discipline: "Brother Julian of the Order of Carmelites calculated this heavenly scheme and established it for the rule of Duke Alessandro de'Medici, who, in 1537, was strangled in his bed by his cousin."[9]

Decades later, the skeptic Sicke van Hemminga (Sixtus ab Hemminga, 1533–86), who knew of the case from Gaurico's rendering, gave it expansive, if bruising, treatment.[10] Although further tinkering with the prediction later proved necessary, it was in no way problematical for Ristori's reputation. In 1571, Francesco Giuntini wrote that as a student, he had heard Ristori "rectify" Alessandro's nativity in his lectures on Ptolemy's *Tetrabiblos* delivered at Pisa in 1548. Rectification was a standard practice whereby the astrologer used a known outcome retrospectively to recalculate the forecast so as to bring planetary angles and house cusps into agreement with what had actually occurred.[11] Astrology in this respect resembled Kuhn's account of "normal science"—the carpenter, not his tools, was always at fault.[12] As late as 1618, the philosopher Rudolf Goclenius the Younger (1572–1621), following Giuntini, reported that Ristori's prediction had astonished almost all of Italy.[13]

On June 28, 1537, six months after Cosimo I took power, Ristori presented to the new grand duke a detailed and lengthy work on his patron's future prospects, a private nativity adorned with learned marginal references.[14] Ristori situated his forecast within a two-cell classification scheme:

> Among the human sciences, both parts of astrology, that is, the speculative and the practical, are most noble subjects because the one holds most worthy all the bodies in the heavens, while the other has [as its subject] man, the most excellent of the animals. One part proceeds by means of the certainty of geometrical reasoning, the other by means of continuous experience so that they are worthy, I say, of holding dominion over all things—and of the two, most of all the practical, generally called "judiciary" because it is intended for governance and reasoning in human affairs. Indeed, what kingdom, what state, what republic or family would not want the best, well-ordered advice to follow in its affairs upon seeing how much the heaven inclines or disposes?[15]

Ristori's categories provide a good example of the flexible applications of ancient and medieval resources of classification. Unlike Reinhold, who foregrounded the convenient tabulation of the numbers for practical astronomy, Ristori privileged the interpretive side of astrological practice. Thus, he called his subject "the human sciences" rather than "the science of the stars." Furthermore, he subsumed Ptolemy's division between astronomy and astrology in the *Tetrabiblos* under a twofold division of "astrology": "speculative" and

"judiciary." The former corresponded to the subject that Campanus of Novara classified as "theorical astronomy" and which Ptolemy called simply "astronomy." The latter term corresponded to both theoretical and practical astrology, that is, to the *Tetrabiblos* as well as to the specific advice allegedly based on it.

Designing his prognostic as an extensive advice treatise, Ristori combined an image of Cosimo as the perfect prince with a good deal of practical advice on whom to watch out for (Pope Paul III) and whom to befriend (Emperor Charles V and the dukes of Urbino, Mantua, and Ferrara).[16] The inexperienced Cosimo certainly needed all the counsel that he could get. Although perhaps accustomed by his upbringing to palace intrigue, he inherited a world filled with major political tensions: between the papacy and the Medici, France and the Empire, the Empire and Florence. A group of disgruntled Florentine exiles would soon challenge the new grand duke's authority on the battlefield.

In 1543, Cosimo had many good reasons to reward the famed Ristori, who had earlier taught theology at Siena, with an appointment to a chair of astrology at Pisa.[17] And when Cosimo celebrated his family's consolidation of power by ordering an elaborate series of murals to adorn his main palace, the Palazzo Vecchio, Ristori's nativity became an authoritative resource, reaffirmed by later astrologers, in the palace's iconographical design.[18] Cosimo's rising sign (ascendant or *horoscopus*), the definer of life prospects, was in Capricorn (Saturn was the planetary ruler of that sign), and Ristori—like earlier Medici astrologers—placed Saturn in Capricorn at the moment of Cosimo's birth. To add to the wonderful convergences, all of which seemed to point to the inevitability of Cosimo's assumption of power and the heralding of a Golden Age, it was known that Capricorn had also been the ascendant sign in the nativity of the emperor Augustus, regarded as the founder of the city of Florence, and also in that of the emperor Charles V.[19] And when Cosimo won the great battle of Montemurlo on August 1, 1537, soon after receiving Ristori's June prognostic, his victory coincided with the date of Augustus's victory at Actium.[20] No wonder that Saturn-in-Capricorn themes were decidedly prominent in the artistic choices made for the astrologically meaningful placement of figures in Medici palaces and villas.[21] Indeed, auspicious features of Cosimo I's nativity were brought into Medici iconography with what the art historian Janet Cox-Rearick calls "a frequency unprecedented among Renaissance princes."[22]

If Ristori's authority as an astrological consultant to the new grand duke made him into a highly credible source for the design of the central artistic motifs in the Palazzo Vecchio, his case also illustrates one way in which astrological judgment making functioned as a bridge between academic and court culture in an important mid-sixteenth-century Italian court. As in the case of Reinhold and Duke Albrecht, the prince was seriously concerned with reliable knowledge about astral influences that would constrain or direct the possibilities of his political actions.

Another example illustrates again just how a prognosticator's advice figured in making crucial decisions that built his reputation.[23] During the spring of 1554, Cosimo was preoccupied with strategies for a siege that he had laid on Siena. His first commander, Giangiacomo de'Medici, the Marchese di Marignano, headed a good-sized division of 4,500 infantry, 400 cavalry, 20 pieces of artillery, and 1200 *guastatori*, experts in the destruction of fortifications. In January Marignano had twice failed to take the city, and at the end of the month, Cosimo, still hoping for a winter success, temporized by dividing the army into three parts to save time and money. On March 27, 1554, Cosimo received a prognostic from the Roman astrologer Formiconi that covered the period from the late spring of 1553 to the late summer of 1554. It focused especially on the duke's problems of February and March 1554. Although the spring was usually regarded as an optimal time for fighting, by May the Florentines had relaxed the blockade to the point of ineffectiveness. The Sienese, under Filippo Strozzi, recognized the weakening of the blockade, and Strozzi marched out of Siena with eight thousand infantry and one thousand cavalry on June 11—much to the consternation of Marignano, who regarded the actions of the younger Cosimo as completely misdirected. Cosimo, following Formiconi's prediction about his fortunes, maintained his restraint until July 12, when, contrary to the advice of Marignano in the field, he directed his armies to take up the offensive in earnest. In early August, Piero Strozzi was mortally wounded, and the Florentine and imperial armies won a great victory at Marciano, a "cosa fatale," exactly on the

anniversary of Cosimo's victory at Montemurlo and the Roman Augustus at Actium.[24]

Apparently court astrologers could win credibility just by making a single successful prediction. One success betokened the probability of more; and in the wildly unregulated world of astrological forecasting, there were different ways to blur, rectify, or simply ignore failed predictions. Astrologers like Ristori and Formiconi were thus more akin to diplomats in the military and political culture of the court advisory structure than to the ideal courtier famously described in Baldassare Castiglione's *Il cortegiano*. Their credibility rested on a combination of calculational skills with the capacity to render advice that was both prudent for themselves and their patrons. Yet, surprisingly, Castiglione did not even mention the stargazer as a social type in his detailed work.[25]

There seems to have been a Florentine tradition of astrologers in holy orders who cast genitures for the Medici and other clients. Ristori and Formiconi were not the only astrologers to serve the grand dukes of Florence. Nor was Ristori the only cleric: another was Ristori's enormously prolific student Francesco Giuntini, who in the two massive tomes of his *Mirror of Astrology* went so far as to claim that astrology actually derived from theology.[26] Another Medici astrologer was the Dominican Egnazio Danti. Danti tutored his patron in the elements of astrology and was later awarded the title of cosmographer to the grand duke.[27] Little is known of the lives of Giovanni da Savoia, from whom we have a forecast for Cosimo I, and Giovanni Battista Guidi, who cast thick annual horoscopes for Cosimo I's son and successor, Francesco, between 1567 and 1583.[28] Galileo Galilei, as will be seen in chapter 13, belonged to this Florentine astrological tradition, but he seems to have been one of the only nonclerics in this lineage.

MULTIPLYING GENITURES

If an astrologer's reputation could be established by a single dramatic prediction for a famous person, like Ristori's prognostic for Alexander de'Medici, why not collect and publish many genitures of famous people together? In fact, just as Ristori was issuing his prognostic for Cosimo, Girolamo Cardano initiated just such an approach in his *Libelli duo* (1538 and 1543), and its influence lasted well through the middle decades.[29] The idea, however, was by no means new. The practice of assembling and comparing genitures had already occurred in the ancient world.[30] Thorndike found it in the recording of medical histories (*consilia medica*) of the fourteenth and fifteenth centuries.[31] And Hilary Carey has shown that English royal astrologers in the thirteenth through the fifteenth centuries were keeping records of horoscopes.[32] By the middle decades of the sixteenth century, however, astrological compilations became just one more expression of a much larger development: printers developed resources, hitherto unimaginable, for the multiplication and comparison of information. Especially notable was the use of dramatic visual resources in productions as various as Andreas Vesalius's *On the Fabric of the Human Body* (1543), Leonhard Fuchs's *Book of Plants* (1545), Conrad Gesner's *History of Animals* (1555), Aeneas Vico's *Images of Emperors from Antique Coins* (1553), Vincenzo Cartari's *Images of the Gods* (1556), and Joachim Camerarius's *Collection of Symbols and Emblems* (1593–1604).[33] As will be seen shortly, the new astrological compendia participated in this emergent climate of inductive display, at least in their aspirations.

Here a distinction should be drawn between the activity of collecting, the mode of representing the accumulated items, and the logic of the claims made on the basis of such information. Unlike the works mentioned above, the new astrological compilations did not depend for their rhetorical authority on any special visual allure. Their power rested on considerations as various as the sheer quantity of genitures, the fame of the individuals whose charts were drawn and the possibility of attracting them as clients or patrons, allegations about the quality of the biographical evidence and frequent flattery of those genitured, the use of personal anecdote to highlight the astrologer's special forte, claims to superiority over other astrologers, and—wherever personal knowledge allowed—astute character assessment.[34]

As with the annual prognostications, therefore, local knowledge crucially enhanced the credibility of the astrologer's forecast. At the same time, the compilations lent themselves to virtually endless opportunities for disagreement over the interpretation of the occurrence and meaning of particulars. But because the compendia also included the names of important rulers, scholars, and cities, it is difficult to see how they could have lacked for popularity, political value, and scholarly interest.[35] Moreover, the large format often used for such collections allowed ample room for readers to add annotations.[36] Thorndike found fifty-two horoscopes copied into a 1546 edition of Albohali on the judgments of nativities.[37]

Cardano's increasingly large geniture aggregations of 1538 and 1543 (*Libelli duo*) and 1547 (*Libelli quinque*), as Anthony Grafton has shown, were undoubtedly an important inspiration for the emergent fashion of multiples. And, interestingly, it was Petreius who undertook to publish Cardano's works at just the time that he was augmenting his list of other works on the science of the stars. Although Cardano assured the reader that his collections of astrological biographies of great men were based on solid information and that his technical methods were beyond dispute, they were more often than not filled with idiosyncratic decisions and arbitrary claims.[38] But compared to the Bologna annual prognostications, which predicted rulers' fates only for the short term and with which Cardano would have been familiar, his geniture collection displayed brief details pertaining to the complete lives of rulers—ranging from the dukes of Milan, the emperor Charles V, King Henry VIII, and popes Leo X and Paul III to ecclesiastics such as Luther, Osiander and Cardinal Bembo. And some of these genitures came to Cardano from the notebooks of other practitioners. A striking example, pointed out by Grafton, is the four genitures that Rheticus sent to Cardano, including those of Pico and Savonarola, noted above, as well as those of Peurbach and Albrecht Dürer.[39]

But there was more. In March 1546, Rheticus visited Cardano in Milan.[40] Cardano's self-serving description of the encounter, dropped in among the lists of his *Astronomical Aphorisms* (1547), reveals still another strategy for promoting both his own credibility and that of the sort of astrology that he practiced. Cardano did not represent Rheticus as an author—let alone one with a significant relationship to Reinhold, Melanchthon, or the late Copernicus—but rather more vaguely as someone "most skilled in the motion of the stars" and "a cultured man and an expert in mathematics . . . [a] gentleman who is honorable and meticulous in carrying out all of his duties."[41] Worse yet, he created an image of himself as the master who knows all and of Rheticus as his fawning and incredulous pupil: "He [Rheticus] heard me say more than once that I had invented and taught an art, by virtue of which, once given a horoscope, I could predict many extraordinary things about the body, the character, and the major experiences of the subject without knowing whose horoscope it was. He tried this out twice, and it worked."[42] Cardano continued with an interpretation that portrayed Rheticus in the role of an inquisitive but awestruck and venerating pupil:

Finally on 21 March 1546, he came to me with the following horoscope, not informing me of the name or the subject, since even he did not at that time know the name. He asked me to say something about it, saying that a great event had happened in it. But he had set the third degree of Aquarius as the ascendant, since he had made it earlier, fixing the ascendant not on the basis of the given time but by his own computation. Looking at it, I said: "This man is Saturnine and melancholic." He replied: "Where do you get that?" I answered, "Because Saturn rules over the ascendant, and holds the degree opposed to it, and looks at it. And Saturn is in Leo, which adds to the sorrow." Then I added: "But he is capable of smooth and easy speech, and seems gentle and calm." He asked, "Where do you get that?" "Because," I answered, "Aquarius is a human sign, and Saturn produces men who are smooth of speech, and the head of the dragon—which is very important—is in the ascendant. It makes men who are gentle of speech and demeanor." As I examined it, he added, "You captured the man perfectly, it couldn't be done better, but it's not all that surprising. You always do this, and you yourself admit it's fairly easy. But please, go through the rest." To which I replied: "He will certainly die a bad death." "But how do you know that?" he asked, to which I replied, "He holds Saturn condemned with the dragon's tail, in the seventh house. Therefore, my method says that he will die badly." "How?" he asked. "By hanging," I said. "How do you know that?" "Since Saturn and the dragon's tail, in the seventh house, show that he will be hanged." "But," I added, "after he's hanged, he will be burned." Looking at me in amazement, he asked, "How do you know that?"[43]

As I have suggested earlier, Rheticus was drawn to older, famous men—especially those who seemed to have deep secrets to reveal—and whom he idealized and had an inordinate need

to please; and, once again, such considerations might have attracted him to Cardano. Rheticus had arrived in Milan with offerings in the form of genitures to feed Cardano's insatiable appetite for horoscopes—of Vesalius, Regiomontanus, Agrippa, Poliziano, Osiander, and so forth.[44] It is possible that, in looking to fill the emotional void left by the death of Copernicus three years earlier, Rheticus had desperately and worshipfully prostrated himself before Cardano. Furthermore, if Cardano was acquainted with Rheticus's literary persona from the *Narratio Prima*, he might have believed that he could represent himself publicly in a role comparable to that of the revered *praeceptor meus* of that earlier work. In any case, as scattered references suggest, Rheticus felt mistreated by Cardano, and he used every opportunity to deride him to others.[45]

Five years later, Luca Gaurico outdid Cardano with 160 genitures, published at Venice.[46] Gaurico had been collecting for at least two decades, some of his charts gathered on a pass through Wittenberg in the early 1530s. Gaurico's effort, which was to be his last publication, included genitures of popes (7), cardinals and prelates (29), lettered men (41), musicians (9), artists (5), individuals who died a violent death (46), and mutilated or monstrous people (9). There was no love lost between him and Cardano. Gaurico, ever promoting himself, used his volume not only to display his personal successes as a prognosticator but also, evoking the narcissism of small differences, to undermine Cardano's earlier interpretations by introducing subtle changes into the fine details of the horoscopes.[47]

The collecting trend continued when the Wittenberger Johannes Garcaeus unleashed some four hundred genitures in 1576, dedicated, like his earlier work, to the elector of Saxony.[48] Of these, some one hundred were genitures of men of learning, including many Wittenberg authors, such as Peucer, Rheticus, Homelius, and Reinhold. And still another work in this small but influential group was Heinrich Rantzov's *Exempla quibus Astrologicae Scientiae Certitudo Astruitur* (Examples on which the Certitude of Astrological Science Is Built), first printed at Antwerp in 1580 but reaching a third edition within five years.[49] Rantzov, a good friend of Tycho Brahe, sorted rulers by how long their reigns lasted, how long they lived, and in which months they died.

But what did this collecting add up to? Or

rather, what did practitioners believe they could say about the subject's logical status on the basis of their new hunting and gathering practices? Ptolemy's classification of astrology as "less self-sufficient" than astronomy (*Tetrabiblos*, book 1, chapter 1) provided a cardinal opportunity for later editors and translators to offer comment. In his expansive commentary on the *Tetrabiblos* (Basel: Henricus Petri, 1554), Cardano looked especially to three ancient authorities: Hippocrates, Galen, and Aristotle. From Galen and Hippocrates he appreciated that astrology, like medicine, builds knowledge from singular examples. Unlike his many efforts to enhance the credibility of his geniture collections against the claims of rival prognosticators, in the *Tetrabiblos* commentary he reached for a higher justification that might be persuasive to philosophers. And here he did not conclude that astrology was merely "less certain." Astrology was to be seen as a "conjectural art" in an epistemically positive sense.

Cardano cast his ideal of astrological scientificity according to the method of resolution and composition that he had learned in his student days at Padua.[50] In moving from the fact to the reasoned fact (from the *tou hoti* to the *tou dioti*, from describing to explaining, logicians spoke of "resolving" the objects of sense perception into their principles or elements or causes. A proper scientific explanation was obtained when one could "recompose" the abstracted factors to show their causal connection with the observed facts. The entire loop, according to this account, was called the "method of resolution and composition."

For Cardano, astrology was the best of a group of predictive sciences that used this way of proceeding.

The method [*via*] of this teaching, just like most of the liberal arts, is achieved through resolution and composition. . . . Causes are known from the effects. [Ptolemy proceeds] by deducing causes from the similarities and from the composition of many things. This art, therefore, is conjectural [*ars coniecturalis*] rather than an exact science [*exquisita scientia*]. One then proceeds from causes to effects; but this procedure is not as sure in natural things as in mathematical ones, where the causes are known per se and not from the effects. The arts that teach or know the future in this manner are agriculture, nautics, medicine, physiognomy and its parts, the interpretation of dreams, natural magic, and astrology. Of all these, astrology is the most noble because it concerns everything whereas other [arts] concern only a certain kind.[51]

Cardano here acknowledged that astrology, like any physical knowledge of the natural world, derived from variable and insecure sensory information. That was just the sort of knowledge that physics was capable of delivering. Unlike geometry, where invariant axioms and postulates were given beforehand, astrology could not claim exactitude; in fact, it was prey to many weaknesses. For example, Cardano was presciently aware that a highly negative prediction, if made public, could have social consequences that might affect both the prognosticator and his subject—not unlike the U.S. Pentagon's short-lived proposal in July 2003 to create a futures market for betting on the probability of terrorist attacks.[52] But rather than try to fix astronomy from above, as it were, Cardano believed that astrology could work up to some sorts of general statements based on individual judgments even if it could not push its inferences all the way up to the categories of astronomy itself. Such general statements might seem to be exemplified by Cardano's representation of the sixty-seven genitures that Petreius published in the same year that *De Revolutionibus* came off his presses at Nuremberg:

Here are expressed all the different forms of death, by poison, by lightning, by water, by public condemnation, by iron, by accident, by disease; and after long, short, or middling periods; also the various forms of birth that yield twins, monsters, posthumous children, bastards, and those in the course of whose birth the mother dies; and then the forms of character, timid, bold, prudent, stupid, possessed, deceptive, simple, heretics, thieves, robbers, pederasts, sodomites, whores, adulterers: and also, with regard to the disciplines of the rhetoricians, the jurisconsults, and the philosophers, and those who will become the greatest physicians and diviners, and famous craftsmen, and also those who will become despisers of the virtues. I have also followed out the different incidents of life, explaining what sort of men kill their wives, suffer exile, prison, and continual ill health, convert from one religion to another, or pass from the highest position to a low status, or, on the other hand, from a low fortune to kingship or power.[53]

Of course, neither this opprobrious survey nor the sixty-seven genitures of 1543, nor the one hundred that followed in 1547, demonstrated astrology to be strictly following the resolution-composition ideal. With its blend of character assessment, local political knowledge, and number crunching, the various parts of astrology inevitably amounted to a kind of interpretive practice or judgment making. For Cardan and later horoscope compliers, aphorisms were the crucial mediating link that allowed singulars to be mobilized more generally.[54] It would not be the last time that authoritative statements of logical procedure were used rhetorically to support the notion that assembling specific instances was the way to arrive at reliable knowledge, let alone to secure astrology's status as chief among the divinatory disciplines.

Johannes Garcaeus was perhaps the most ambitious of the nativity gatherers. Moving beyond Cardano and Gaurico—whose groundwork laid the precedent—Garcaeus's detailed genitures were systematically embellished with aphorisms, occasional comments on the relevance of dreams (whether caused by God or the devil), and frequent "cautions" in which he introduced qualifications or compared his interpretations with those of other astrologers. Yet he did not try to justify astrological knowledge along the lines of Cardano's commentary. Working with a two-cell classification, his justification was decidedly Melanchthonian. He presented astronomy as that part of the science of the stars that was governed by well-grounded, "certain" rules of motion, the planets causally producing effects in the lower region of the universe. These effects were known to astrology by what Garcaeus could not describe much better than "experience in continual agreement with itself" (*ex perpetua sibi consentiente experientia*).[55] Hence, astrology could not be ranked higher than an "art." He admitted further that "very few demonstrations" of planetary effects were known, although, in spite of such weakness of human understanding, God still wished that there might remain some light in men's minds. Astrology was to be seen as a "useful art." It encouraged piety by showing that God had created an orderly world rather than one of chance; it could explain differences of human temperament; it could help the physician to mitigate bodily ills; and so forth. In sum, Garcaeus exhibited pages and pages of genitures, but no more than Cardano could he strictly connect these to the principles from which they were putatively

derived. In other words, he was not able to frame the connections as demonstrative knowledge; furthermore, neither he nor Cardano appealed to Copernicus's theorizing as a basis for inspiring confidence in their geniture collections. Fortunately or not, it would not be the only enterprise of this period directed to winning for astrology an improvement in its epistemic status.

FROM WITTENBERG TO LOUVAIN
ASTROLOGICAL CREDIBILITY AND THE
COPERNICAN QUESTION

The 1540s and '50s were also notable for high-end efforts to block Piconian skepticism using claims about the epistemic security of the mixed sciences, astronomy and optics. This development coincided with the earliest circulation of the Copernican proposals. *De Revolutionibus*, it will be recalled, had advertised its main thesis as a response to the disagreements among mathematicians; yet for whatever reasons, that same work had failed to name Pico, who had made such disagreements the heart of his critique of astrological knowledge. Here Rheticus's local role is important. At Wittenberg, no one besides Rheticus had direct, personal knowledge of the connection that Copernicus sought to establish. Yet Rheticus himself failed to be an effective advocate. After his abrupt departure for Leipzig, Reinhold's private reading of *De Revolutionibus* circulated relatively rapidly at Wittenberg and quickly overshadowed the local authority of the *Narratio Prima*. Even Cardano, who had no access to the Reinhold annotation group but did have personal contacts with Petreius and Rheticus, made no gestures to use *De Revolutionibus* to reinforce astrology's theoretical foundations.[56] As is evident from his long comment on *Tetrabiblos*, book 1, chapter 4, Cardano followed Ptolemy's planetary ordering and made no mention of any controversy or disagreement about the order of Venus and Mercury with respect to the Sun.[57]

Louvain represents a variant node in these developments. Piconian worries were readily in play during the aftermath of the failed 1524 flood prediction. For example, Cornelius de Scepper (d. 1555), a learned humanist, sometime astrological practitioner, and adviser to the exiled Danish king Christian II, used the *Disputationes* not entirely to exclude but to limit astrology's reach through double determination. God's actions were preeminent. He could act through natural causes to achieve natural effects; he did not have to wait for a planetary conjunction to occur. He could make floods happen "by a fire emitted from the heavens, or by releasing waterfalls, or by imposing a tempest."[58] It was in this environment that the first edition of the *Narratio* (March 1540) arrived in the hands of the physician and mathematical practitioner Reiner Gemma Frisius (1508–55) sometime before July 1541. It had circulated through a Danzig merchant (Jacobus à Barthen) and the aforementioned De Scepper.[59] The 1524 abortive flood prediction was still a living memory in Louvain; *De Revolutionibus* was almost two years from publication and the *Prutenics* still a decade away. Furthermore, no one in Louvain knew Rheticus personally. Yet the *Narratio*—the dissident reading of *De Revolutionibus*—was now available, well before the formation of the Wittenberg interpretation.

Gemma Frisius exemplifies a common social type of this period, whose engagement with mathematically based subjects was combined, at least for a time, with a position in the medical faculty.[60] Holder of a public chair in medicine at the university in Louvain (1537–39), he was also connected to the craft culture of map, globe, and instrument making centered on Antwerp, and he privately tutored students in mathematical subjects, including astronomy and probably astrology.[61] Some of these students—Gerard Mercator (1512–94), Johannes Stadius (1517–79), Sicke van Hemminga (1533–86), Antonio Gogava (1529–69), and John Dee (1527–1608)—eventually established considerable reputations in their own right. Later, his son Cornelius (1535–ca. 1578) became a medical professor at Louvain and issued ephemerides and prognostications.[62] What is much harder to establish with confidence is the character of the grouping. Given the spacing of their birth dates, for example, did they study with Gemma sequentially or simultaneously? Did they cohabit or see themselves self-consciously as a group, with Gemma as leader, patron, or paterfamilias? Did they share a common set of theoretical or practical concerns, problems, or positions?[63]

The picture is filled with hopeful evidence and just as frequent qualification. Consider the retrospective testimony of John Dee. In his famous *Mathematical Preface* to Billingsley's edition of Euclid (1570), Dee referred briefly to his experience at Louvain in 1548–49: "I was, (for *21.yeares ago) by certaine earnest disputations, of the Learned *Gerardus Mercator,* and *Antonius Gogaua,* (and other,) thereto so prouoked: and (by my constant and inuincible zeale to the veritie) in obseruations of Heauenly Influences (to the Minute of time,) than, so diligent:"[64] It is noteworthy that Dee credited his early formation in astrology to "Louayn." Antonio Gogava was roughly the same age. Yet given that Dee certainly knew Gemma, it is surprising that he failed to name him or, indeed, his important, but much older, student Johannes Stadius, the margins of whose ephemerides Dee used for many years as a kind of astrological diary.[65] Likewise, Dee dedicated his *Propaedeumata aphoristica* (1558) to Mercator but again failed to mention Gemma in that work. Another piece of evidence involves the circulation of the text of *De Revolutionibus*. Gemma's copy of that work is one of the most extensively and heavily annotated now known to exist.[66] Consequently, it is likely that most, if not all, students who came into contact with Gemma could have acquired some familiarity with that book and his reading of it. Yet among extant copies we find no indication of shared or overlapping marginal comment comparable to the Reinhold annotation group at Wittenberg. Finally, there are Gemma's public remarks and allusions in dedicatory letters praising the works of two students: Antonio Gogava's Latin translation of the *Tetrabiblos* (1548) and Johannes Stadius's *Prutenic*-based *Ephemerides* (1556). These last will be considered in the context of the development of Gemma's engagement with the Copernican writings.

Gemma had first heard about Copernicus's theory almost ten years before the publication of the *Narratio Prima* through Johannes Dantiscus, sometime ambassador to the king of Poland and successor to Copernicus's uncle as bishop of Varmia.[67] A short time before July 20, 1541, he had Rheticus's book itself, about which he exuded to Dantiscus: "If that author of yours will have demonstrated and proved these matters—which the "introduction" (*proemium*) that he sent beforehand strongly presaged—then, does this not mean that he is giving us a new earth, a new heaven and a new world?"[68]

Such enthusiastic rhetoric raises the prospect of a full-dress endorsement until one notices that the details of which Gemma approved in the *Narratio* did not include the ordering claims that

Rheticus had so emphatically trumpeted. What instead attracted consideration were parameters like Mars's position in longitude, the Moon's apparent size, the length of the tropical year, and the equinoctial precession—all significant features of Rheticus's presentation but none tied uniquely to the Earth's diurnal and annual motions. And, indeed, immediately after his ejaculation about a "new world," Gemma clarified: "Here I shall not talk about the hypotheses, which he [Copernicus] uses to support his demonstration, of whatever kind they are, or whatever amount of truth they contain. To me it does not matter whether he claims that the earth moves around, or whether it stands still; as long as we have the stars' motions and time intervals very precisely determined and reduced to extremely accurate calculations."[69] One need not wave the Duhemian instrumentalist flag over this passage: after all, Gemma did not go so far as to say that we *cannot* know anything about the nature of the heavens or that all astronomical theories are merely instruments of calculation. He simply allowed that this was not going to be the place where he would address matters of truth; nor, moreover, was one required to do so. Perhaps one reason that Gemma held off such a determination is that he had still not seen the masterwork which he knew to be forthcoming.

He did not have to wait long. Sometime before 1545, Gemma acquired *De Revolutionibus*, and the extensive traces of his annotations show that he studied it with great care and acumen. Cindy Lammens's exhaustive and exemplary study of the notes shows that Gemma marked, underlined, or postillated almost all aspects of the text.[70] But most of the comments are brief paraphrases of Copernicus's language and rarely betray Gemma's judgments on the arguments and claims themselves. Ultimately, Gemma seemed to show the same primary interest in the theory's utility for calculation that he had tagged in his 1541 letter to Dantiscus.[71] Strikingly absent are any notes on the preface and Osiander's "Ad Lectorem"; consequently, Gemma paid not the slightest attention to Copernicus's jabs at the theologians or to Osiander's skepticism about astronomical knowledge. In thus ignoring the bearing of holy scripture on the central Copernican claims, his private comments, at least, were silently at odds with Wittenberg judgments.

In 1548, after several years of studying *De Revolutionibus*, an excellent opportunity to connect astrological reform with Copernican principles arose. In that year, Gemma's prodigy Gogava produced his new Latin translation of the *Tetrabiblos*, based, as Gemma noted, "partly upon that of [Joachim] Camerarius, partly that of Antonio Gogava of Graven."[72] Gogava admitted that Ptolemy's text was difficult—especially the last two books, which even the most learned Camerarius had not translated. And it was Camerarius's translation that Gogava now completed, making it the first full Greek and Latin edition of the *Tetrabiblos*, preceding even that of Melanchthon. Gogava's edition, as shown in chapter 1, was just the latest in a long-term trend to secure the Ur-text. Gemma's "Letter to the Reader" explained that Ptolemy's ancient text had become choked with "dense weeds, slugs, and mushrooms"; Ptolemy's own language was so ponderous that it had deterred many people from reading the whole. The new Louvain edition built directly on the Wittenberg renewal of ancient astrology, claiming the usual improvements over the Arabic text, but also leaning toward the moderns: "Let the ancients truly give way, provided that better things follow."[73]

And Gogava did, indeed, go further with his auxiliary apparatus, adding to his edition two treatises that he presented as "uncommon" (*non uulgares*): the first a treatise on the parabolic conic section ascribed to Apollonios; the second on the burning mirror, "from which studious and learned mathematicians will doubtless take great delight since Apollonios's *Conic Elements* is necessary but not publicly available at this time."[74] Never mind that these works were really of Arabic origin, authored in the tradition of Alhazen and his followers in the tenth and eleventh centuries and transmitted through Roger Bacon.[75] The purpose of the appendixes seems to have been associated with an effort to propose a physical and optical explanation for the intensity of the rays, the astral rays to be seen as acting in some sense analogously to the burning mirror. But neither Gemma's letter nor Gogava himself sought to assemble such an optical physics or to tie it to the Copernican planetary ordering. Conceivably, Gemma had some such project in mind, but, obeying convention, he followed the rhetorically appropriate path, confining himself to comments on the subject matter of the work to which praise was directed, in this case a work of theoretical astrology rather than theoretical astronomy.

Because of Melanchthon's great devotion to improving astrology, the question of the rays' intensities was also of interest at Wittenberg. When Melanchthon's *Initia Doctrinae Physicae* appeared in 1549, just a year after Gogava's *Tetrabiblos*, Melanchthon addressed a question on which Gogava's texts might be seen to have a bearing: How did variations in the planetary distances affect the strengths of the astral influences? Here was an occasion for discussing the relevance of the Copernican arrangement. Yet, as we have seen in chapters 4–5, one reason that Melanchthon regarded Rheticus's arguments for the Copernican ordering as inflated and unconvincing was his excessive and seemingly misplaced enthusiasm; another was that he could see no reason to relinquish his trust in the alternative accounting afforded by ancient knowledge. That position can here be further specified. Melanchthon believed that it was enough to consider how the planetary forces vary with the planet's position on its epicycle. He stated this understanding clearly: "Experience teaches which forces the planets have to influence those lower bodies. It is necessary that the forces exercised on those [lower bodies] be weaker when these [planets] are located at the greatest apsides of their epicycles and, thus, when they are most distant from Earth. They are made much more efficacious and powerful, however, when situated some thousands of earth diameters closer to us in the lowest parts of their epicycles." Following Ptolemy's assignment of fixed qualities, as detailed in *Tetrabiblos*, book 1, chapter 4, Melanchthon proceeded to specify the character of the forces issuing from the superior planets: "Saturn has the force to make things cold and to gently dry them out. Mars, on the other hand, vigorously dries and burns. But Jupiter is the mean between these and has a tempered nature. Simultaneously, it warms and humidifies, it excites and sustains spirits most suited for fertility and generation."[76]

Thus, the intensity—but not the character—of the planet's innate qualities would vary as it moved closest to the Earth and became brightest at the perigee of its epicycle; but, for the inferior planets, the centers of the epicycles were a matter of uncertainty and disagreement. Lack of consensus on this issue was, of course, one of Pico's complaints. Melanchthon stated explicitly that the epicycle centers of the superior and inferior planets divided into two classes: the inferiors always lie along a line of sight connecting the Earth to the Sun, whereas the superiors are not bound at all.[77] Hence, the mean periods of the two inferior planets were necessarily equal (one year) and the longitudinal digressions of Mercury and Venus with respect to the Sun were bounded (Mercury: 27°37'; Venus: 46°); those planets were thus said to be in a "perpetual conjunction" with the Sun. As Melanchthon remarked, they were "like satellites which minister and care for the king's body."[78] But nothing was said about the *order* of the attendant bodies "accompanying" the Sun. On this point, Melanchthon, unlike Cardano, openly admitted that the ancients disagreed about whether Mercury and Venus were above the Sun (as Plato believed) or below it (as Cicero and Ptolemy maintained), and he asserted that "we shall retain the opinion of the most ancient astrologers, as did also Cicero, Ptolemy, and other recent mathematicians with great unanimity." And he added: "Therefore, we say that Venus is located nearest to and below the Sun; and under it is Mercury, above the Moon's sphere."[79] For Melanchthon, the solution was to trust the authority of Ptolemy and modern commentators like Erasmus Reinhold.

In 1556, five years after the *Prutenics* appeared, Johannes Stadius and John Feild, respectively, published the first ephemerides based on the new tables. Recent commentators have found in Gemma's dedicatory letter to Stadius persuasive evidence that he was more favorably inclined to Copernicus's theory than previous evidence would suggest. Less clear is the strength of his conviction and whether he had shifted his earlier views or was now simply prepared to make earlier-held views public. The passage in question—amounting to about a half page out of three and a half—will repay close scrutiny. I shall organize my gloss around a division of the text into three parts.

I. There remains the final difficulty concerning the motion of the Earth and the paradox of the Sun at rest in the center of the universe. Those, however, who lack [training in] philosophy and the method of demonstration do not understand the causes or use of hypotheses. For, in fact, authors do not set up these [hypotheses] as if things must necessarily be so and could not be established in some other way. But in order that we may have a sure reckoning [*certa ratio*] of the motions corresponding to the apparent places of the stars in the heaven, for the future and the past as well as for the present time, we have made assumptions agreeable to nature's principles rather than to ones that are utterly absurd.[80]

This passage seems appropriate to the kind of work to which it is attached. Gemma distinguishes between the heavens' apparent motions and the assumptions or hypotheses (geometrical models) needed to account for those appearances. Unlike Osiander, whose text was by now available to Gemma, he does not take a skeptical position: he does not say that astronomers are incapable of knowing which among their hypotheses are true or even probable. Rather, he invokes a comparative standard for enabling choice. Some of the possible hypotheses are preferable to others even if astronomers do not claim that the hypotheses so preferred are unique and necessary. In fact, at first blush, Gemma's standard sounds very much like the one adumbrated by Copernicus and Rheticus—as is borne out by the succeeding passage:

II. While at first sight Ptolemy's hypotheses may seem more plausible than Copernicus's, the former nevertheless commit rather many absurdities, not only because the stars are understood to move nonuniformly on their circles, but also because they do not offer reasons [*causae*] for the phenomena as clearly as [*tam euidentes*] Copernicus's hypotheses. For Ptolemy assumes that the three superior planets (by way of example), when they are achronic or diametrically opposite the Sun, are always in the perigees of their epicycles, and that is [also] a fact [*tou hoti*]. The Copernican hypotheses, however, insert this same fact as a necessity and give the reason for it [*dioti*]. And [the Copernican hypotheses] attribute hardly anything absurd to the natural motions, from which a greater knowledge of the planetary distances is in this instance deduced than from the other [hypotheses].[81]

The second passage introduces the familiar Aristotelian logical distinction between "the fact" (*tou hoti; quia*) and the "reason for the fact" (*tou dioti; propter quid*), a distinction that, although undoubtedly known to Gemma from his university training, he could easily have found applied specifically in Reinhold's commentary on Peur-

bach's *Theorics* (1542), the predecessor to the *Prutenics*. Reinhold, it will be recalled, had claimed that Peurbach's models provided the *tou hoti*, whereas full-dress models of Ptolemy's *Almagest* offered the *dioti*.[82] This was a relationship of complementarity. But Gemma used this distinction differently, to exemplify the application of a comparative or dialectical standard: hypotheses that provide reasons for the described motions (which he called natural or necessary) were to be preferred over those that did not (described as absurd). Although he came close to saying that *all* of Copernicus's hypotheses succeed in this respect where Ptolemy's do not, he stopped well short of such a claim. His sole example, developed from *De Revolutionibus,* book 1, chapter 10, was followed by the gesture that Copernicus provides "greater knowledge" of the planetary distances than do the hypotheses of Ptolemy.[83]

I suggest that this passage, if set next to the sections earlier cited from the *Initia Doctrinae Physicae,* may be read as Gemma's reply to Melanchthon's position. Gemma was interested in just that arrangement where the three superior planets appear brightest and where their rays are most intense. For Melanchthon and Ptolemy, this configuration was simply a *tou hoti* coincidence accounted for by their simultaneous alignment at the perigees (or six-o'clock position) on their respective epicycles, when they are allegedly closest to the Sun; but for Copernicus, there was an explanation *dioti*: the planets are brightest when closest to the Earth and the Sun lies on the other side of the Earth, at opposition.[84]

Although promisingly Copernican, Gemma's curiously incomplete position has led commentators to mixed judgments: Cindy Lammens, for example, speaks of a "qualified appraisal," Fernand Hallyn of a "prudent realism."[85] These characterizations can be further developed. For one thing, Gemma might have added a good many other "absurdities" of a kind similar to the one that he mentions, such as the explanation of retrograde motion, the ordering of the inferior planets, and the more "natural" ordering of the planets according to their periods of revolution.[86] At best, he was merely hinting at a class of arguments. And where he referred to the "greater knowledge" of planetary distances, he mentioned nothing about "more plausible" measurements and values or a new method for calculating the relative distances from the Sun. Moreover, he blatantly ignored the physical and scriptural consequences that Melanchthon spelled out in his lectures. In other words, Gemma, limited not least by literary convention, was in no position to make a full-blown argument from eliminative necessity.

> III. If, indeed, anyone so wishes, he may also transfer to the heavens those motions of the earth that he posits, over and above [*praeter*] the first two, and still use the same rules of calculation. Yet, because of his invincible natural talent, it did not seem pleasing to a most learned and most prudent man to turn upside down the entire order of the hypotheses but to rest content to have posited what would suffice for the true discovery of the phenomena.[87]

The third part of the crucial passage ends on a note with recognizably Wittenbergian resonance, entirely in keeping with the fact that Gemma was praising an ephemerides rather than a work of natural philosophy, and hence was staying within the rhetorical expectations of that kind of book. Gemma, echoing Osiander, acknowledged that, for purposes of calculation, there was no reason to upset traditional assumptions. Yet, balancing this view, he also held the opposite: for purposes of calculation, there was no reason not to assume some or all of the Earth's motions. Here, Gemma seemed to align himself thoroughly with a practical, calculational agenda, with the inflection on practical astronomy: the Earth's several motions could be used as assumptions in calculation without accepting the reality of those motions— possibly the position that Copernicus held some years before he composed the *Commentariolus*. Gemma did not deny thereby that astronomers are incapable of knowing such a reality; he was merely silent on the matter. His strategy, in this respect, departed from the Wittenberg consensus, which was always careful in its pronouncements to declare its teachings in natural philosophy and theology. In sum, we may locate Gemma somewhere between the Wittenberg Interpretation and Copernicus's mature position in *De Revolutionibus*. And we may find further confidence in this judgment by the determinations of contemporaries: no one in the sixteenth or seventeenth centuries classified Gemma as an adherent of the Copernican ordering. Nor is Gemma known even to have made a Copernican globe. Nonetheless, his views, succinct as they were, would still prove influential.

JOHN DEE AND LOUVAIN
TOWARD AN OPTICAL REFORMATION OF ASTROLOGY

John Dee appears to have been the first Englishman of this period to develop serious contacts with Gemma Frisius and other Louvain mathematical practitioners of the 1540s. Although his stay was brief (1548–50), it shaped the space of possibilities for his earliest intellectual projects. It would be nice to know why and how he chose Louvain for his studies and who funded him, but we can only speculate about the circumstances.[88] Dee had already pursued mathematical disciplines at the one college at Cambridge known for its strength in those subjects, St. John's.[89] It is not known why he had not chosen to study in Italy, especially Padua, the more common path traveled by English students of the Tudor era. It is possible that Dee had made preliminary contact with Flemish émigrés in London, such as the well-established printer Thomas Gemini (b. ca. 1510).[90] Gemini had published Andreas Vesalius's *Fabrica* at London in 1545; he later published Leonard Digges's *Prognostication of Right Good Effect* (1555). Gerard L'E Turner speculates that Mercator and Gemini might have received engraver's training in the workshop of Gaspar van der Heyden in Louvain.[91]

Once in Louvain, Dee quickly befriended key members of the Louvain scene skilled in mixed mathematical subjects. However he publicly attributed his early inspiration to Gogava and Mercator rather than to Gemma Frisius or Johannes Stadius. Whatever the meaning of these attributions and omissions, Dee was clearly working to make himself impressively skilled in all dimensions of astrological practice. Part of this formation was a budding taste for book acquisition, well suited to the accelerating production of high-quality mathematical and astrological works then pouring out of the major Continental presses. Many of the books that Dee appears to have purchased during his brief Louvain period include items from Petreius's list.[92] Over the course of his bibliophilic life, he acquired all the current tools of the science of the stars, among which were a substantial number of editions of the *Tetrabiblos*.[93] In August 1551, he also purchased in Paris (and annotated) the earlier 1519 Locatelli edition (no. 37 in Dee's library), with the admired gloss of Haly Abenrodan. These acquisitions illustrate a voracious collecting appetite that eventually led him to accumulate more than 2,292 printed works and 199 manuscripts, including weather observations for Louvain dated 1548.[94]

In 1558, some eight years after leaving Louvain, Dee published in London a curious farrago of some 120 apothegms—some pithy, others prolix—titled *An Aphoristic Introduction to Certain Especially Important Natural Powers (Propaedeumata Aphoristica)*. Rather than a group of handy aphorisms of middling generality to guide the interpretation of birth charts—an auxiliary tool like the *One Hundred Aphorisms* attributed to Ptolemy or Hermes—Dee's list constituted, as it were, a raft of theoretical aphorisms, an ancillary to the *Tetrabiblos* itself: an astrological *theorica*. Later, in his *Mathematical Preface*, Dee explicitly connected his definition of "astrologie" in that work with the apodictic ideal of the *Propadeumes*. Astrology, he said, is founded, "not onely (by Apotelesmes) *tou hoti*, but by Naturall and Mathematicall demonstration *dioti*. Whereunto, what Sciences are requisite (without exception) I partly haue here warned: And in my *Propadeumes* (besides other matter there disclosed) I have Mathematically furnished up the whole Method."[95]

Dee's loosely organized string of numbered aphorisms was a literary form that enabled him to avoid, say, the question-and-answer form of the popular academic *epitome* found in the Wittenberg orbit and instead suggests a resemblance to student *exercitationes* of the kind deployed by Cardano and Julius Caesar Scaliger.[96] No matter, he then claimed to present his subject in the form of a demonstration: "Not only can a means be found of proceeding demonstratively [*Apodictiè procedendi*] in the art with regard to an infinite number of particular situations, but also, besides the main principles of the art have been laid down and established here."[97] This was to be a work that gave astronomical and optical reasons for the effects and their varying strengths. Recent commentators—unlike most contemporaries—have since exercised considerable ingenuity in bringing a sense of order to Dee's list, visualizing its proposals with illustrations that the original utterly failed to offer and shining a strong light on its otherwise concealed intellectual debts.[98]

Dee's work was also implicitly a reply to Pico's (and perhaps Calvin's) objections against connecting astral powers with earthly effects.[99] Here it leaned toward Louvain in its mobilization of the optical resources accompanying Gogava's *Tetrabiblos*. The *Propadeumes*, as the neologizing Dee Englished his work, was a project ambitiously designed to offer a new kind of explanation—a *propter quid* or explanation—for the variation in the powers of the stars.[100] The planets could be imagined rather like burning mirrors concentrating the rays of light. Dee gave his attention to a spread of variables: planetary diameters, surface illumination, ray angles incident with the horizon, angular and linear distances, and duration of effect.[101] Predicting the moment of greatest effect involved all sorts of questions about the functional dependencies of different causes and effects—how long the effect would last, how much surface area was covered, how close or far away was the illuminating source, how great was the density of different rays bunched together, and the net effects of different combinations and permutations of planets acting conjointly. All told, Dee's fulsome account provided a bevy of new ways to judge changes in the quantity of the qualities preassigned by Ptolemy in *Tetrabiblos*, book 1, chapter 4. But Dee leaned toward Wittenberg in not adjusting the ordering of the planets. As late as 1582, he still expressly associated himself with Reinhold and avoided Copernicus's theoretical claims: "Yt notable Mathematician Erasmus Reinoldus Salueldensis, in his Prutenicall tables Astronomicall, did reduce and make perfect Copernicus his most diligent labour and excellent observations of ye heavenly motions. . . . Ye said Copernicus his Calculation and Phaenomenies: excepting his Hypotheses Theoricall, not here to be brought into question."[102]

The Louvain practitioners must have held views with some family resemblance to Dee's vision of a reformed Ptolemaic astrology—Gemma Frisius as evinced by his association with Gogava, and Mercator by virtue of being the dedicatee of the *Propadeumes*.[103] Whether or not Dee's "catoptric art" actually enacted that reform in quite the way that the Louvain group would have done is harder to say; the only solution that we have is, after all, that of Dee.[104] And what is more, it seems apparent that Dee implicitly grounded the astronomy on a modified version of the Melanchthon-Ptolemy model.[105] Angular distances and the analogy of heat were also important for

Dee (think of the altitude of the noonday sun at the summer solstice). Perpendicularity was the angle of privileged strength: "The more nearly the radiant axis of any star approaches perpendicularity over any elemental surface, the more strongly the star will impress its forces upon the place exposed to it: directly, to be sure, because of the greater nearness of the agent, but also by reflection, because such reflected rays are joined more closely with the incident ones."[106] Of course, in principle, the planet's rays could strike Earth perpendicularly whether the planet was at its furthest distance (apogee) or its nearest (perigee). For Melanchthon, such perpendicularity would yield an increase in strength only at the point of closest approach. But on this question, Dee introduced an opposing amendment. The powers of celestial bodies in relations of special significance to other bodies in the zodiac ("proper signification") are strongest at apogee, when the base of the radiant cone is largest and the rays converge angularly on the middle, axial ray.[107] In this case, optics trumped planetary ordering because propinquity to the effect was not a relevant variable; De Revolutionibus, therefore, was irrelevant to the solution.

JOFRANCUS OFFUSIUS'S SEMI-PTOLEMAIC SOLUTION TO THE VARIATION IN ASTRAL POWERS

Sometime before 1556, in a curiously polemical work that appeared only posthumously in Paris in 1570, Jofrancus Offusius launched a novel effort to contain the Piconian challenge: *Concerning the Divine Power of the Stars, against the Deceptive Astrology (De Divina Astrorum Facultate, in Laruatam Astrologia.).*[108] Like Gemma Frisius, Offusius was intimately familiar with the text of *De Revolutionibus;* unlike John Dee, he was inclined to use it as a resource for addressing the variation in the astral powers. And unlike either of the aforementioned—not to mention the rest of his contemporaries—he was prepared to criticize Ptolemy's designation of elemental qualities in the *Tetrabiblos,* book 1, chapter 4, and to incorporate some features of Copernicus's planetary ordering.

Given these distinctive characteristics of his work, it is unfortunate that we cannot place him more precisely. The print identity on his title page ("a German, lover of knowledge") points to a general regional association while also confirming the absence of either a university or a court affiliation. Other evidence places Offusius's birthplace at Geldern in the Lower Rhine region of Westphalia, but he does not appear on the matriculation lists of several leading German universities.[109] The slim published evidence, however, puts him in Paris. For example, the author's dedicatory letter, dated at Paris in January 1556, names the Hapsburg emperor Maximilian II as a patronal target, and the dating raises at least the possibility that he was living and working in that city during the preceding years.[110] Furthermore, by the time that his widow arranged for the book to be published by the royal printer Jean Royer, who had issued Offusius's ephemerides in 1557, the primary dedication was to the French crown.[111]

A related issue is Offusius's association to John Dee. After Dee left Louvain, he was in Paris in 1550 and back in England by 1551; whether or not he and Offusius met on that occasion, it is certain that they did meet later in London, at Southwark, in 1552 or 1553. And finally, we know that after Offusius's book was published, it moved in (and out of) Parisian circles. We do not have enough evidence to claim that it made a splash, but we know that it quickly sparked considerable interest. For example, in 1577, Friedrich Risner, a close collaborator of Peter Ramus and the editor of the optical works of Alhazen and Witelo, sent a copy to Lazarus Schoener (1543–1607), professor of philosophy and mathematics at Marburg and later the editor of Ramus's mathematical works.[112] The king's surgeon, Franciscus Rassius de Noens, owned copies of both Offusius and *De Revolutionibus.*[113] Tycho Brahe commented seriously on Offusius's use of "mystical numbers" for deriving the planetary distances, but we do not know when he first became acquainted with his work.[114] The learned Oxonian Henry Savile also appears to have owned a copy.[115] Evidently *The Divine Power of the Stars* was the kind of book that attracted attention from the same group of readers who were interested in and capable of reading Copernicus.[116] With the exception of one reader, a comparable level of interest has yet to be documented for Dee's *Propadeumes.* That exception, as will be seen shortly, is Jofrancus Offusius.

Offusius located his work in distinctions familiar at least from Pico's *Disputationes* and Marsilio Ficino's *De Triplici Vita* and developed at

JOFRANCI

OFFVSII GERMANI

PHILOMATIS, DE DIVI-

na Aſtrorum facultate, In laruatam

Aſtrologiam.

AD.

Sereniſſimam Chriſtianiſſimamq̧ Galliæ
Reginam.

Quod pauci intelligent, multi reprehendent.

SOLA DEI MENS.
IVSTITIÆ NORMA.

PARISIIS,

Ex Typographia Iohannis Royerij, in Mathematicis
Typographi Regij.

1570.

Cum priuilegio.

37. Offusius's *De Divina Astrorum Facultate* (1570): Friedrich Risner's gift to Lazarus Schoener in Marburg, March 10, 1577. Courtesy Rare Book and Manuscript Library, Columbia University.

length by Caspar Peucer in his 1553 treatise on divination.[117] Offusius was at pains to mark the difference between a bad astrology, based on illusions, and one that was good or "divine." The bad kind was motivated by the devil's deceptions and the machinations and fallacies of evil men, and it was said to be condemned in the Bible. The derisory word that Offusius used in his title is *larvatum*, which has connotations of an entity lacking in substance, like a ghostly or deceptive apparition, a mad, demented, crazed or bewitched being. However, although some opponents had rightly attacked bad astrology, they had mistakenly thrown out the wheat with the chaff. Chief among astrology's detractors who had gone too far in discarding the entire subject was Pico: "Pico [della] Mirandola fought against it [the deceptive astrology], but, in my opinion, he did no more than to make it nicely available to the ears of the unlearned; and, having set forth everything without good reason, teaching or advice, the unlearned—as if assuming all to be true—delightedly invented things from it about future events with some injury to Philosophy."[118]

Offusius's objections to Pico in this passage were so general and ambiguous that a reader who did not know Pico's text would have found it difficult to know anything about the contents of his specific objections. Yet Offusius shared in the widespread sixteenth-century practice of attack by proxy. He conducted his argument largely through and in the name of earlier authorities while giving minimal reference to his own contemporaries, thereby concealing engagement with them. He seemed to take for granted that his readers were familiar with Pico's criticisms but otherwise afforded him no further authorial credit. It seems reasonable to assume that anyone who chose to own a book like *The Divine Faculty of the Stars*—readers like Risner, Schoener, Rassius Noens, and Tycho Brahe—would have been familiar with the Piconian arguments against which Offusius inveighed.

The polemical phrase *deceptive* [or *masked*] *astrology* functioned as the repository of difficulties concerning the elemental qualities left unaccounted for in the *Tetrabiblos*, and the latter are what initially invite comparisons between Offusius's book and the Dee-Gogava project. For although he confessed great admiration for Ptolemy's works (and also thought that few really understood the *Almagest*), Offusius was thoroughly dismissive of the account in *Tetrabiblos*, book 1, chapter 4, which was still accepted by contemporary commentators like Cardano, Gaurico, and Melanchthon and which connected the planets' capacities to heat and humidify with their mutual interrelations and to their propinquity to Earth:

> In book I, chapter 3 [*sic: Tetrabiblos*, book 1, chapter 4] he [Ptolemy] asserts that the Moon humidifies because he says that it comes near to the Earth, from whence humid exhalations issue forth. Likewise, he says that Saturn dries things because it is furthest from the Earth's humidity. Now surely it is not a wise saying to claim that the eternal is self-evident by comparison with the corruptible. Yet, in the same way, his reasons imprison us: for it would follow that Jupiter is drying up because it falls between two drying [planets], since he says that stars carried between a cooling and a heating [planet] are tempered.[119] And there are other things from him that are unworthy of a man who is a philosopher.[120]

Offusius was the first of the midcentury Ptolemaic commentators actually to criticize this crucial chapter publicly and directly. Evidently, his hope for ridding astrology of Ptolemy's difficulties was not to be achieved solely by collecting genitures in the manner of Cardano and Gaurico: rather, as he explained in his dedication to the emperor, "We shall proceed partly physically, partly mathematically, surrounding doubts with likelihoods, and we hope that for posterity it [our art] might become worthy of the name science . . . Our own art is based upon nothing but an examination of the divine work and upon rules and patterns collected from long experience."[121] Offusius referred grandiosely to having collected some 2,700 observations in his travels.[122]

But where was it all leading? Offusius presented a novel proposal for handling the variation of planetary intensities. He thought that this could be done neither by inferring the distances from the planets' physical "influencing" properties nor by using Ptolemaic values derived from ratios of the planets' eccentricities, as in the traditional Melanchthon-Ptolemy procedure for stacking the orbs.[123] Offusius's approach was a priori and deductive: his idea was that the distances were specified by special numbers with harmonious significance, and the language that he applied was that of Copernican *symmetria*.

TABLE 2

Planets' mean, least, and greatest distances from Earth, in terrestrial diameters, according to Offusius

Planet	Mean	Minimum	Maximum
Moon	30 3/8	25 7/8	34 7/8
Venus	81	49	113
Mercury	216	144	288
Sun	576	551 1/2	600 1/2
Mars	1,536	851 1/2	2,220 1/2
Jupiter	4,096	3,448	4,744
Saturn	10,922 2/3	10,082 1/6	11,763 1/6

SOURCE: Offusius 1570, fol. 7v.

TABLE 3

Derivation of the "Quantity of Qualities" (576) according to Offusius

Regular Solid	Number of Faces	Number of triangles in each face	Product
Tetrahedron	4	6	24
Cube (hexahedron)	6	4	24
Octahedron	8	6	48
Dodecahedron	12	30	360
Icosahedron	20	6	120
TOTAL			576

SOURCE: Offusius 1570, fols. 3–4.

He also looked to Plato's *Timaeus,* finding inspiration in the five Platonic solids that would later attract Kepler so profoundly, and, remarkably, he found assurance in much the same metaphysic: the Creator used ideas of harmony when he built the world; he introduced nothing superfluous; and the world was a work of divine architecture made for man.[124]

Unlike Copernican harmony, however, Offusian *symmetria* did not use the sidereal periods as an independent criterion of planetary order from which to calculate the distances; it did not begin, that is, with the assumption of the Earth's motion, but rather with the traditional arrangement, grounded in Aristotelian and Ptolemaic arguments for the Earth's centrality and rest. This point is crucial: Offusius's astronomical intuitions, like those of other midcentury would-be heavenly reformers, were still geostatic. This was not for lack of detailed appreciation of Copernicus's proposal. Offusius proposed essentially a modified version of the traditional Ptolemaic stacking principle. But instead of using the eccentricities of the models, each taken individually, he postulated a single, unifying principle: that a planet's mean distance from Earth (expressed in Earth diameters rather than radii) was exactly 8/3 that of the next closest planet. For example, if one assumed that Mercury follows Venus with respect to the Earth, then its distance was exactly 2 2/3 times farther from Earth than Venus. Offusius considered the number 8/3 to be a pleasing, if not "divine," number for determining the intervals between the planets. But what made it so?

Offusius chose the "divine number" of 576 Earth diameters (1,152 Earth radii) for the absolute solar distance and as the starting point for stacking the other planets. He claimed that the desirability of this value rested on a new argument to be found in Plato's five regular geometric solids. According to Offusius, each face of these solids can be divided into a finite number of triangles—either isosceles (with two equal sides) or scalene (with no two sides of equal length). Multiplying the number of triangles in each face by the number of faces in each solid yields the values seen in table 3. Offusius was aware that other values had been proposed for the solar distance: he mentioned those of Ptolemy (580), Aratus (555), "the moderns" (Copernicus had used 571), and even "I myself writing against Cardano's *De subtilitate*" (where he posited the number as 579). But, there were other reasons for thinking that 576 ought to be accepted as a candidate for special numberhood. Dividing the five polyhedra into triangles, the ancients already recognized the number 576 as "the soul of the world," and now Offusius thought that it could be seen to contain a "harmony" with earthly consequences. He called this harmony "the quantity of the qualities . . . by virtue of which the heavens cause [effects] in us."[125]

The qualities whose strengths Offusius sought were the familiar elemental pairs hot and cold, moist and dry—each quality being distributed among the planets according to the original prescriptions of *Tetrabiblos* book 1, chapter 4. If guided by *symmetria,* Offusius thought, one could make reasonable guesses about the strengths of these qualities—unmodified by any resistance from the lower world that they affect. Offusius's *symmetria* cashed out as a set of special quantities. He associated cubed numbers with the strengths of the qualities hot and cold; those associated with the moist and dry planets were squared—with the exception of Venus's moisture and Mercury's dryness. Each elemental quality was assigned to one of four regular polyhedra—pyramid (heat), icosahedron (moisture), octahedron (cold), cube (dryness)—and each planet was assigned numbers corresponding to its qualities. For example, the Sun is hot (27, or 3 cubed) and dry (49, or 7 squared), while Saturn is cold ($107\frac{11}{64}$, or $\frac{19}{4}$ cubed) and dry ($12\frac{1}{4}$, or $\frac{7}{2}$ squared). These "quantified qualities" totaled 360, which, conveniently, was the number of triangles in the subdivided dodecahedron and hence the signifier of the divine, heavenly container. And, to add to its numerological sex appeal, Offusius pointed out that 360 lies almost exactly between the number of days in the solar year ($365\frac{1}{4}$) and the number in the lunar one ($354\frac{1}{4}$).[126]

As an exercise in Pythagorico-Platonic reasoning, so far, so good. Offusius now came to the question of the uncertain order of the inner planets, Venus and Mercury, the very problem that Gemma Frisius had ignored in his letter to Stadius. Here Offusius's discussion was clearly indebted to *De Revolutionibus*, book 1, chapter 10, in a way that was, thus far, without precedent. Although lengthy, the passage is worth our attention:

> Here, I place Venus below Mercury, lest the fitting together [*symmetria*] may thus be harmed. Martianus Capella, who published the *Encyclopaedia*, believed that these stars traveled together around the Sun. For a long time, thinking that the only explanation of the revolution was in each [planet] separately, I subscribed to that credulity. [Meanwhile,] Plato in the *Timaeus* believed that both [planets] were situated above the Sun. Alpetragius placed Mercury below the sun, but Venus above it; and Ptolemy, followed by nearly everyone from the school of modern mathematicians, places Venus between the Sun and Mercury. However, next to him, at last came Copernicus, a man not inferior to any of the remaining [mathematicians]—in fact, better than those whose fortune was to be born at this time (he alone among the moderns has helped [the quality of] everyone's observations). Coperni-

cus demonstrates that the errors of both these stars (and thus also of all the others) may be saved if they travel together around the resting Sun and from his theory (each planet being considered), it follows that Venus is always much closer to the Earth than Mercury. Now, lest this argument might constrain my own liberty, wherever anyone might wish these stars to be located, who will not concede that I might prescribe a theory altogether agreeable with the observed excursions and apparent diameters?[127]

Offusius's statement was the boldest of any on the question of order since the appearance of *De Revolutionibus* in 1543. He did not merely state his agreement with Martianus Capella and Copernicus, but he also justified the arrangement of Mercury and Venus by appealing explicitly to the Copernican principle of the proper fitting together of the parts. And yet in this passage he also resisted endorsing the Earth's motion. In fact, although he adopted the Capellan-Copernican ordering of Venus and Mercury, his tables of planetary distances always show the Sun and Moon as planets. Like Gemma Frisius and Erasmus Reinhold, Offusius maintained a loud silence on the issue of the Earth's motion.

Offusius's engagement with the Copernican text was hardly accidental. He was actually more deeply acquainted with Copernicus's *De Revolutionibus* than he revealed in his book. Owen Gingerich and Jerzy Dobrzycki have shown that no less than eight surviving copies of *De Revolutionibus* contain his marginal jottings or those of others connected to him—an annotation group whose size puts it in the same league as that of Reinhold's network at Wittenberg. Moreover, Gingerich and Dobrzycki provide compelling evidence that Offusius was active in Paris roughly between 1552 and 1558 (after which the trail goes cold).[128]

The remarkable, detailed handwritten summary of book 1 bound into the back of the National Library of Scotland (Edinburgh) copy of *De Revolutionibus* is especially valuable for the light that it sheds on a crucial passage of Copernicus's preface that remained unannotated by Gemma Frisius and Erasmus Reinhold. Copernicus had written: "Therefore, I also, having found the occasion, began to consider the mobility of the Earth. And although the idea seemed absurd, nevertheless because I knew that others before me had been granted the liberty of devising whatever circles they pleased in order to explain the heavenly phenomena, I thought that I, too, would be readily permitted to test whether, by positing some motion for the Earth, more reliable demonstrations than theirs could be found for the revolutions of the heavenly spheres." It will be recalled that the conditional formulation in which Copernicus frames his hypothesis leaves him open to the objection that the assumed "idea" might actually be false, rather than merely absurd. But this is not the position that Offusius takes in his comment: "Copernicus does not altogether [omnino] assert the motion of the Earth (as it appears to many uninformed persons) but, from the hypothesis of the Earth's motion and from other suppositions, he infers and explicates what can be observed in the heavenly bodies and in their orbs. He also presents the method and sound rules for mathematical reasoning as well as the means of judging the phenomena or appearances and of calculating the motions of the heavenly bodies."[129]

Offusius's comment shows his admiration for both Copernicus's assumption and his reasoning, but it does not carry him so far as to endorse the truth of the assumed premise itself. Indeed, the gloss suggests that Offusius read Copernicus in much the same way that Gemma Frisius encouraged in his "Letter" of 1556: assuming the Earth's motion, for purposes of calculation, would lead to better (geometrical) demonstrations of the apparent motions.

Elsewhere in his unpublished commentary, Offusius clearly maintained much the same scriptural and physical objections to the Earth's motion that had been advanced by Melanchthon and his disciples. There is every likelihood that he knew Melanchthon's *Initiae Doctrinae Physicae* and/or Peucer's *Elementa Doctrinae de Circulis Coelestibus et Primo Motu* (Wittenberg: J. Crato, 1551).[130] He was certainly well acquainted with Reinhold's *Prutenic Tables*, as his own *Ephemerides* of 1557 drew, albeit without credit, on Reinhold; but he was eager to mark his differences from the 1556 *Prutenic*-based ephemerides of Johannes Stadius of Antwerp.[131] Most important, for all of his numerological enthusiasms, he did not mention Rheticus's allusion to the putatively sacred number six; nor did he refer to Rheticus's reasons for preferring a new planetary ordering. Evidently some divine numbers were better than others.

In sum, within a decade or so of its publication, Copernicus's main hypothesis was apparently being taken seriously as a calculational assumption by those looking for a way to improve astrology's credibility from above. Offusius had produced a theory that purported to quantify astral radiations and to connect the variations in the strengths of the planetary qualities with terrestrial effects. Unlike John Dee's more traditional project, which followed the Ptolemaic ordering scheme still recommended by Melanchthon, Offusius's main variables were the linear distances built around what Giovanni Battista Riccioli in 1651 called the "semi-Tychonic system." Although Offusius recommended Platonically derived numbers as a new way to derive the distances, he did not feel compelled to apply the principle of *symmetria* in the same way that Copernicus had done. In fact, he applied it inconsistently because, if the Moon is considered a planet, then its mean period of 30 days around the Earth and the periodicities that follow it are the Capellan Venus's 225 days and Mercury's 88 days. Not until Kepler was there an attempt to offer an interpretation of *symmetria* as a robust principle of planetary order.

SKIRTING THE MARGINS OF DANGEROUS DIVINATION

This chapter began by exploring the boundary between divine and demonic divination—the perilous divide on which our hardy midcentury defenders of astrology's credibility balanced their goals. Negotiation of the boundary between legitimate and illegitimate divination was, as we have seen, a matter of major concern, especially at Wittenberg, in just the period when Dee, Offusius, and the Louvain practitioners were advancing new proposals for a safe and effectual astrology. The resources of classical scholarship, recent astronomical theorizing, and medieval optics might be seen to assist understanding of the strengths of the planetary powers, but such study was no guarantee of the moral or religious significance of the influences thought to be known. If one could, in fact, extract power from the pagan planetary gods, could the effects be kept within the bounds of Christian orthodoxy?

That was a question already broached in Marsilio Ficino's ebulliently optimistic project of Neoplatonized medical astrology, a work purporting to exercise control over natural forces to improve life and health. The highly influential *Three Books on Life* (1489) held out the promise that the physician, obligingly informed by Ficino and the stars, could prescribe remedies for defeating melancholy, the scholar's nemesis. Scholarly melancholy, in the Ficinian version, was brought on by an imbalance between bodily inactivity and mental overactivity. Sluggishness of the body produced phlegm that "dulls and suffocates the intelligence," while the brain's excesses caused an overproduction of black bile, which in turn caused melancholy. "Hence," wrote Ficino, "it can justly be said that learned people would even be unusually healthy, were they not burdened by phlegm, and the happiest and wisest of mortals, were they not driven by the bad effects of black bile to depression and even sometimes to folly."[132]

Ficino's *Three Books on Life* hit a major chord: it was an attractive work that no learned scholar could do without. There were only two problems of concern: one religious, the other, as it were, scientific. As to the former, the depressed scholar might prefer to forgo attending mass in favor of burning incense, eating vegetables (rather than "the more fatty or harsh foods"), playing soothing solarian music on his lute, and working out regularly (for "with too little physical exercise, superfluities are not carried off and the thick, dense, clinging, dusky vapors do not exhale").[133] Naturally based magic could be seen to threaten the Church's monopoly over the care of souls. "The Church," as D. P. Walker aptly noted, "has her own magic in the mass; there is no room for any other."[134] The "scientific" problem, the problem of explaining the strength of the effects delivered by the planetary forces, was less worrisome.

Among the books that John Dee collected was the 1516 edition of Ficino's *De vita*, although whether he acquired it before or after his visit to Louvain is uncertain.[135] In a tantalizing marginal annotation, first noticed by Nicholas Clulee, Dee mentioned that he had met both Girolamo Cardano and Jofrancus Offusius "in 1552 or 1553" at the home of the French consul in Southwark.[136] The date is plausible, as we know from independent evidence that Cardano was in England around 1552 and that he cast nativities for Edward VI and his tutor, John Cheke.[137] Moreover, Dee's note is virtually indistinguishable stylistically from the entries in the margins of the various ephemerides that he used as a kind of diary to record meetings with visitors as well as births, ailments, celestial observations—and meetings with spiritual beings.[138] Both the date of the Dee-Cardano-Offusius meeting and the location of the marginal comment strongly suggest, therefore, that at least one topic of conversation at Southwark was the specific passage in Ficino's work against which the note was written.

Ficino's theme in book 3, chapter 15, was not melancholy but rather the harnessing of the mimetic power of images and identifying terrestrial objects (such as lodestones and gems) that could draw down the power of specific planets or constellations—a practice not altogether without risk because, as Ficino remarked, "I learned from the theologians and Iamblichus that makers of images are often possessed by evil daemons and deceived." But he continued: "I personally have seen a gem at Florence imported from India, where it was dug out of the head of a dragon, round in the shape of a coin, inscribed by nature with very many points in a row like stars, which when doused with vinegar moved a little in a straight line, then at a slant, and soon began going around, until the vapor of the vinegar disappeared."[139]

For Dee, Ficino's putatively Indian gem was no mere textual reality. While visiting the French ambassador's residence in the company of Cardano and Offusius, Dee believed that he had laid eyes on just the sort of object that Ficino described: "I have seen a similar stone of the same quality."[140] But Ficino's text situated the object within a larger web of meanings—specifically, rules for drawing down astral influences and for picking out the most receptive entities in a world mediated by an all-pervasive cosmic spirit (*spiritus mundi*). It would appear that Dee himself understood the stone within Ficino's frame of meaning and that the other witnesses also agreed with Ficino's description of the Indian gem as possessed by an evil demon. This was a note about a witnessed event rather than merely a reader's paraphrase. The absence of objections in the marginal observation suggests that Dee associated himself with the full meaning of Ficino's constitution of the object. In fact, Dee's marginal entry has much the same quality as one of his later diary entries. It is noteworthy that he composed his diary largely in English but reserved for Latin any natural-philosophical observations.

Was Ficino then the topic of further conversation? The evidence does not permit a definitive answer. But shortly after the Southwark encounter in the fall of 1553, Dee recorded that Offusius had requested of him that he share his "hypotheses for the confirmation of astrology" concerning the "causes of atmospheric changes." Nicholas Clulee conjectured that these hypotheses "refer to the '300 Astrological Aphorisms' that Dee claims to have written in 1553" and that this work was very likely a draft of Dee's first published work, the *Propaedeumata Aphoristica* (1558).[141] If this conjecture is correct, then one may add an inference: that at least one objective of the *Propaedeumata* was to provide secure demonstrative foundations for the "good" astrology and talismanic practices described in Ficino's work. In fact, although Ficino's physical account explained that the World Soul spread its power by means of a *spiritus*—"a very tenuous body, as if it were now soul and not body, and now body and not soul," he did not say by how much the strength of the *spiritus* varied in causing effects within each being. He simply asserted that it did so because of likenesses between causes and effects. This claim was not so very different from Ptolemy's qualitative account claiming that the planets' elemental natures depended on their proximity to the Sun or the Moon. Yet without a quantitative account, one could have only a very vague sort of talismanic practice and a mere list of recipes (without dosage) for avoiding scholarly melancholy. If Ficino had given the *Tetrabiblos* a theoretical basis in Stoic-Neoplatonic terms, he had (in Dee's view) failed to provide more than a good theory of how to match the right stellar influences with the right terrestrial entities: he had not provided a good explanation for the varying strengths of the influences. When, several years later, Dee accused Offusius of plagiarizing his aphorisms, he thereby confirmed the deep importance of his earlier association.[142]

The *Propadeumes* of 1558, however, did not appear in conjunction with Ficino's *Three Books on Life*, itself a work of astrological theory. On the other hand, Dee's work could have been joined together with any number of recent practical works, such as those by Schöner, Gaurico, Cardano or Heller. If Petreius had still been alive, the enterprise might have been produced at Nuremberg, but Dee opted for the English publisher Henry Sutton. Either Dee or Sutton (or both together) decided that the first edition of the *Propadeumes* should appear with its own frontispiece

but be bound together with a work concerned with the preparation of astrological judgments. The work chosen for this purpose was Cyprian Leovitius's recently published *Brief and Clear Method for Judging Genitures, Erected upon True Experience and Physical Causes* (1557). Like Schöner's work, from which it clearly benefited, Leovitius's manual was filled with the kinds of conveniently compact charts and matrices that would enable a practicing astrologer to sort out, conventionally, the net strength of a planet's influence.[143] To complete the package, the publisher prefixed to Leovitius's work Hieronymus Wolf's short dialogue in defense of astrology: *A Warning concerning the True and Lawful Use of Astrology*.[144] Perhaps this *defensio astrologiae* was intended to protect Dee against the perception of Ficinian excesses. Wolf was secretary to the wealthy banker Jacob Fugger in Augsburg; he was also a sometime acquaintance of Tycho Brahe and, like his friend Melanchthon, a famed Hellenist. Because Leovitius (Lvovický de Lvoviče) himself was the Fugger's *mathematicus* for a time, the annexation of Wolf's dialogue to this work added a veneer of humanist legitimacy. The publisher described Wolf as "a man superior in his knowledge of all human letters, languages, arts and mathematics."[145]

In the dialogue, Wolf constructed his "disciple" as a defender of astrology against the Piconian charge of astral determinism:

> *Disciple:* What think you of the saying of Picus, who commandeth the same common and civil prudence to be Judged by the propensions of the mind, the temperature of the body, and the success of things?
>
> *Astrologer:* I praise him, and obey him: nevertheless Astrology is not to be accompted useless.

This was followed by a reassuring statement of the astrologer's proper domain of operation:

> *D:* And what doth the Astrologer Counsel?
>
> *A:* He doth not affirm his good or evil predictions, but only pronounces them; neither doth he Counsel any thing else, but only perswades.
>
> There is neither good nor bad to any man, but what comes from the Divine providence, therefore they are to be born with a contented mind.
>
> That with piety, prudence, and diligence, good things may be increased and evil ones abated: and this is the chief fruit of predictions: neither do we

affirm an inevitable necessity of all events; but we Judge of the events of all things, according to the strength of the significators.[146]

The world described in Dee's *Propadeumes* contained within it many Ficinian elements that have attracted scholarly notice. Attraction and repulsion: a natural universe bound together by similitudes; the parts of the universe resonating like a lyre, other parts producing "harsh dissonances"; the world like a great "glass" reflecting the divine plan, active power flowing from the stars "like seals whose characters are imprinted differently by reason of differences in elemental matter" (aphorism 26). Also, like Ficino, Dee believed that whoever knew how to play the world lyre could control men's minds and bodies (aphorism 23). Beginning with aphorism 22, Dee introduced a Piconian dimension to his discussion, the power of light (what he called in the *Mathematical Preface* "light Iudgement" or "perspective").[147] The allusion to Genesis, though not explicit, was clearly present: " It is the faculty of the first and chief sensory form, namely, light, that without it all the other forms could do nothing" (aphorism 22). Besides Pico, a major source of Dee's interest in light was the thirteenth-century Franciscan Roger Bacon, many of whose manuscripts he collected. Bacon showed how the effects of light could be studied by means of lenses and mirrors that could concentrate the sun's rays to produce heat and combustion. In Bacon, Dee also found a distinction between bad, demonic magic (condemned by Bacon) and "the legitimate performance of marvellous feats by human artifice using the secrets of nature as instruments."[148]

The Baconian distinction—reinforced by Wolf's appended treatise—effectively provided Dee a safe rubric for associating himself with activist, Ficinian, magical, and talismanic technologies and separated him from the more cautious sort of divination approved by Peucer and Melanchthon at Wittenberg. It also allowed Dee to separate himself from the ordinary prognosticator: "The common and vulgare *Astrologien*, or Practiser who . . . like a simple dolt, & blinde Bayard, both in matter and maner, erreth: to the discredit of the *Wary*,

and modest *Astrologien*."[149] When Dee was jailed for a time in 1555 on charges of "calculating," "conjuring," and "witchcraft," was he already suspected of putting Ficinian principles into practice?[150] As for Rheticus, whose father had been condemned and executed twenty years earlier, or Luca Gaurico, who was jailed for making a prediction that came true, or Caspar Peucer, who later spent many years incarcerated for his heterodox theological views, the main danger faced by sixteenth- and seventeenth-century men of learning was being seen as engaging in practices that the state or Church classified as unorthodox or "superstitious." This was a major reason why they needed protectors or patrons.

Whatever the immediate effects of his own brief incarceration, Dee continued to probe the effective boundaries of orthodoxy. Sometime after 1570, he turned from the use of strictly planetary forces for combating scholarly melancholy. Instead, he tried a new method, perhaps more Protestant in its private style. Closeted in a room of his house, Dee created his own sort of chapel in which, together with an assistant or "scryer" and with the help of a light-gathering crystal, he engaged in rigorous and frequent prayer. In reply, so he represented it, angelic beings carried back divine messages. On December 22, 1581, an angel told Dee to exercise a level of discipline more suited to a monk than to a married man with children. It included fasting, abstention from sex and overeating, attentive bodily grooming, and lots of prayer.[151] A few months later, Dee enlisted the scryer Edward Kelly, who promoted himself as highly skilled in "spirituall practice" and with whom Dee worked for several years.[152] Dee's wife, Jane, did not like Kelly. Yet constant work with the angels at close quarters evidently fostered an intense bonding between the two men, and in 1587 an angel prescribed that they should share their wives. As Deborah Harkness has shown, the wish seems to have been fulfilled. In May 1587, Dee entered in his diary, "Pactu[m] factu[m]."[153] Soon afterward, Kelly left the household—evidently to Dee's chagrin, because as late as 1591, he was reporting troubling dreams about his former scryer.

7

Foreknowledge, Skepticism, and Celestial Order in Rome

Paul III, the pope to whom Copernicus carefully and elegantly crafted the dedication to *De Revolutionibus,* knew where to seek astrological advice on the most propitious moment to lay the cornerstone of the wing of the Vatican palace named after his family; but there is no evidence that his astrologers advised him beforehand about the arrival of a technical work on planetary theory. Paul was deeply preoccupied with other matters. Voices urging spiritual and institutional renewal and reform had been growing louder for more than a century.[1] By the mid-sixteenth century, the Roman Church was deeply preoccupied with affirming its traditional authority against the culturally and politically fragmenting effects of the "German Schism." Also hanging like a great shadow over the reigns of Clement VII (1523–34) and his successor, Paul III (1534–49), was the memory of the humiliating and devastating Sack of Rome in 1527 by Hapsburg armies.[2] Beginning in the last years of Paul's reign, the Catholic response assumed many forms, some defensive, some quite innovatory. The Society of Jesus, formed in 1542 through the initiative of Ignatius de Loyola (1491–1556), a Spanish nobleman, brought new and remarkable cultural energies and initiatives into the Church. A great reforming council was held at the Italian city of Trento between 1545 and 1563 (with gaps in 1547–51 and 1552–62). The overriding issue confronting the Council of Trent was the need to give the faithful some feeling of security by restoring strict clerical discipline and renewing theological doctrine. It was reasonably successful in achieving

the latter but largely failed in the former. Whatever real reforms were eventually made, however, some of the new initiatives created an atmosphere of obsessive control over detail, endless doctrinal clarifications by councils, synods, and theologians, suspicion of deviancy, and a proclivity for inflexible, legalistic remedies in areas of social conflict—a climate to which the Protestants also contributed in their own way.[3]

Copernicus's book was received and studied at the papal court in Rome just after the start of the Council of Trent, but it did not figure in the council's formal decrees. Even calendar reform, which had been a major concern at the Fifth Lateran Council (1512–17), was not in any sense a primary issue of consideration.[4] There were other arenas where the Church saw more urgent need for drawing unambiguous boundaries between the legitimate and the illegitimate: (1) divinatory practices that claimed foreknowledge of future events, (2) the authoritative edition of the Scriptures, and (3) the correct exegetical standards for interpreting the Bible. By the end of the Council of Trent, an attitude toward these matters had emerged that was markedly different from that of the Melanchthonians.

DE REVOLUTIONIBUS
AT THE PAPAL COURT
A STILLBORN (NEGATIVE) REACTION

Copernicus's theory was not discussed at the Council of Trent.[5] Matters of natural philosophy or even calendar reform were not in any sense primary issues of discussion.[6] Thanks, however, to the discovery by Eugenio Garin of a new document, we can now say that at about the time the council was beginning, there was a considered reaction within the highest circles of the papal court. A year after the appearance of *De Revolutionibus*, an almost exact contemporary of Copernicus, the Florentine Dominican Giovanni Maria Tolosani (1470/1–1549), finished a large apologetic work titled *On the Truth of Sacred Scripture*.[7] Tolosani's treatise, never published, concerned itself precisely with certain issues that were about to be debated at Trent. Between 1546 and 1547 Tolosani added a cluster of "little works" dealing with a variety of topics: the power of the pope and the authority of church councils, the emendation of the calendar, conflict between Catholics and heretics, justification by faith and by works, the dignity and office of cardinals, and the structure of the Church. Tolosani was thus an adviser on matters of doctrine, and he was also well placed in the curial hierarchy. In addition, he was an astronomer of no mean ability. He had written a treatise on the reform of the calendar, and, unlike Copernicus, he had actually attended the Fifth Lateran Council, no doubt at the behest of the head of the calendar commission, Paul of Middelburg.[8]

As early as 1544, Tolosani had obtained a copy of *De Revolutionibus*. It may have been the very copy sent to the pope on the authority of Copernicus (although we cannot be certain). No direct evidence exists that Copernicus had arranged the wording of his dedication with Rome beforehand; had he done so, it is likely that he would have made some mention of it. Surely the Lutheran Rheticus, by virtue of his blatant Wittenberg association, could not have acted as an intermediary in this regard. The absence of a dedication to a patron would have been taken as a sign of authorial illegitimacy and vulnerability; beyond that, the relationship between a book and a patron's approval of its content was indeterminate. As one English writer expressed it in 1620, the absence of a dedication and a patron "calleth it into suspicion that either the author hath no friends of worth, or that the worke is not worthie of patronage."[9] Ideally, patrons should offer their protection by dint of reading and agreeing with the book's contentions;[10] but in cases where books were offered to patrons not personally known to their authors,

the author could hope to win over the prospective patron and other readers only through the persuasiveness of his arguments. Such appears to be the most likely explanation for Copernicus's strategy in dedicating his volume to the pope.

However, the possibility cannot be ruled out that Copernicus knew of Tolosani from the time of the Fifth Lateran Council and perhaps had requested that a copy of the work be sent directly to him by virtue of his status as a *mathematicus* who, like Paul of Middelburg, possessed the right kind of expertise to evaluate his proposals.[11] It seems clear that *De Revolutionibus* had arrived while Tolosani was preparing his much longer work on the Church and holy scripture, because he added an extensive opinion about *De Revolutionibus*, book 1, as the fourth of the twelve additional works.[12]

The contrast with Copernicus's reception in Wittenberg—at precisely the same moment— could not have been more striking. At the papal court, there was no Rheticus to promote the theoretical value of the work, no Reinhold to translate the new models into tables, and no Melanchthon to urge its utility for prognostication to a prospective patron. Whatever Copernicus himself may have known about the pope and his advisers skilled in mathematical subjects, we do not know whether the pope read the carefully crafted preface directed to him.[13] And Tolosani himself refers neither to the pope nor to the Copernican preface. What we can be sure of is that Copernicus did not receive a sympathetic reading from the learned Dominican.

Tolosani, certainly skilled in astronomical matters, shaped his reading of *De Revolutionibus* using Aristotelian natural philosophy and scripture and his own favored commentator, Thomas Aquinas. Tolosani's lack of comment on any topic beyond book 1 in *De Revolutionibus* is clearly a matter of choice rather than a failure of competence. His critical strategy focused on physical and theological difficulties, and he did not hesitate to belittle the author's failures as a reader and thinker: "Nicolaus Copernicus," he remarked dismissively, "neither read nor understood the arguments of Aristotle the philosopher and Ptolemy the astronomer."[14] Also, it seems apparent that Tolosani had no hint of Osiander's role in adding the "Ad Lectorem," as he was not part of the local Nuremberg-Wittenberg network where this information was in circulation. However, his reading was at least sufficiently astute and careful that he recognized the "Ad Lectorem" to have been written by someone other than Copernicus, an "unknown author."[15] But Tolosani did not interpret the comments of the "unknown author" in the moderate sense favored at Wittenberg. He quoted a passage directly from the "Ad Lectorem,"[16] but, unlike Osiander, he produced an interpretation that was highly critical of Copernicus: "By means of these words [of the "Ad Lectorem"], the foolishness of this book's author is rebuked. For by a foolish effort he [Copernicus] tried to revive the weak Pythagorean opinion, long ago deservedly destroyed, since it is expressly contrary to human reason and also opposes holy writ. From this situation, there could easily arise disagreements between Catholic expositors of holy scripture and those who might wish to adhere obstinately to this false opinion."[17]

Tolosani's readings served his own apologetic needs. In characterizing *De Revolutionibus* as a revival of "the Pythagorean opinion," he explicitly situated that work in an Aristotelian-Thomist framework rather than, as Copernicus did, in a Ptolemaic one. As Aristotle had refuted the Pythagoreans, it was but a short step from there to Tolosani's position as a defender of the faith—completely in line with his work on the truth of scripture: "We have written this little work for the purpose of avoiding this scandal." As a consequence, although Tolosani was pleased to use the "Ad Lectorem" as a weapon against *De Revolutionibus,* he could not accept the letter's skeptical contention without at the same time giving up altogether the possibility of certitude in astronomy.

Yet, although Tolosani rightly read *De Revolutionibus* as making natural philosophical claims, he avoided engagement with key arguments offered in the preface. There were no allusions to the Horatian metaphor, and there is no evidence that Tolosani knew the tropology and logic of the *Narratio Prima.* Nevertheless, he was not content only to demolish Copernicus's physical claims on the basis of Thomist-Aristotelian physics: he also chose to locate *De Revolutionibus* explicitly within the scheme of the Thomist-Scholastic hierarchy of the disciplines and to present the theory as a violation of its principles of classification.

He [Copernicus] is expert indeed in the sciences of mathematics and astronomy, but he is very deficient in the sciences of physics and logic. More-over, it appears that he is unskilled with regard to [the interpretation of] holy scripture, since he contradicts several of its principles, not without the danger of infidelity to himself and to the readers of his book. . . . The lower science receives principles proved by the superior. Indeed, all the sciences are connected mutually with one another in such a way that the inferior needs the superior and they assist one another. No astronomer can be complete, in fact, unless first he has studied the physical sciences, since astrology presupposes natural heavenly bodies and the motions of these natural [bodies]. A man cannot be a complete astronomer and philosopher unless through logic he knows how to distinguish in disputes between the true and the false and has knowledge of the modes of argumentation, [skills] that are required in the medicinal art, in philosophy, theology and the other sciences. Therefore, since the aforesaid Copernicus does not understand the sciences of physics and logic, it is not surprising that he should be mistaken in this [Pythagorean] opinion and accepts the false as true through ignorance of these sciences. Call together men well read in the sciences, and let them read Copernicus's first book on the motion of the Earth and the immobility of the sidereal heaven. Certainly they will find that his arguments have no force and can very easily be taken apart. For it is stupid to contradict an opinion accepted by everyone over a very long time for the strongest reasons, unless the impugner uses more powerful and insoluble demonstrations and completely dissolves the opposed reasons. But he [Copernicus] does not do this in the least.[18]

Tolosani's appraisal was the first polemic against Copernicus. It exhibits the characteristic Catholic emphasis on the weight of tradition, an emphasis that grew with the decrees of the Council of Trent. Just as one does not lightly abandon the opinions of the Church fathers, it argued, so one does not easily give up long-established opinions in physics and astronomy. Unlike Osiander, who tried to protect Copernicus's work by stressing the separation between the mathematical and the physical parts of astronomy, Tolosani brought out the dependency of astronomy on the higher disciplines of physics and theology for the truth of its conclusions. Physics and theology are disciplines superior to mathematics by virtue of their nobler subject matters and the strength of long tradition.[19] According to Tolosani, Copernicus mistakenly "imitated the Pythagoreans." He advanced a false physical claim—that the Sun (an incorruptible body) was located at the center of the universe (a place subject to change)—just as the Pythagoreans erroneously located an element (fire) at the center from which it naturally moves away.

Tolosani ended his little treatise with the following important revelation: "The Master of the Sacred and Apostolic Palace had planned to condemn this book, but, prevented first by illness and then by death, he could not fulfill this intention. However, I have taken care to accomplish it in this little work for the purpose of preserving the truth to the common advantage of the Holy Church."[20] The Master of the Sacred Palace was Tolosani's powerful friend Bartolomeo Spina, who attended the opening sessions of the Council of Trent but died in early 1547.[21] As trenchant as was Tolosani's critique of Copernicus, there is simply no evidence that it received any serious consideration either from the new master or from the pope himself. Meanwhile, the unpublished manuscript, written in the spirit of Trent, was probably shelved in the library of the Dominican order at San Marco in Florence, awaiting its use by some new prosecutor. There is some evidence that it was known to a later Dominican, Tommaso Caccini, the same person who in December 1613 delivered a sermon in Florence highly critical of Galileo.[22] However, until Caccini, Tolosani's position became a kind of "dormant" viewpoint that had no audience in the Catholic world; no Catholic astronomer or philosopher worked under any formal prohibitions from the Index or the Inquisition for the rest of the sixteenth century.

THE HOLY INDEX AND
THE SCIENCE OF THE STARS

Under the short pontificate of Paul IV (1555–59), the powers of the Inquisition increased, the Roman Jews were physically confined to a new ghetto, and all works of the major reformers (including Erasmus and Melanchthon) were prohibited in Rome.[23] The Rome Index of 1559 and the Trent Index of 1564 marked a movement to centralize the powers of the Index in Rome itself rather than to distribute its authority among diocesan inquisitors.[24] In two of the last sessions at Trent (February 25, 1562 and December 3–4, 1563), the council formalized a series of longstanding local censorship practices by recommending the establishment of a published list of books that

Catholics were forbidden to read without authoritative corrections or without special permission (*Index librorum prohibitorum*).[25] The Congregation of the Index was finally established in March 1571.[26]

By the time that the Index was formally constituted, however, the major works of the reformers were flooding the markets, and the literature of the heavens was sweeping like an unstoppable wave across Europe. What could be done to control this flood? In practice, there was a certain degree of variation and overlap in the Index lists and a continual evolution of classificatory strategies. As a rule, local indexes—issued in specific cities—tended overwhelmingly to list the titles of theological works; but already during the 1540s it was increasingly common to condemn specific authors without naming their works. The practice of classification by author rather than title was obviously connected to the author's religious identity and, at least at first, it made the inquisitors' task much easier, as it was no longer necessary to decide whether specific books—or even specific passages—required prohibition. Thus the Index lists included many authors whose works fell into the category of the science of the stars but who actually were being condemned simply for having written an objectionable theological work or even for being associated with a prominent Protestant reformer or ruler.[27] For example, the 1559 Index of Paul IV lists, among others, the names (but not the writings) of many authors from the Wittenberg orbit. Besides Melanchthon himself, we find Achilles Pirmin Gasser, Andreas Osiander, Caspar Peucer, Cyprian Leovitius, Erasmus Oswaldus Schreckenfuchs, Hartmann Beyer, Jacobus Mylichius, Joachim Camerarius, Johann Carion, Johann Schöner, Victorinus Strigelius, and, for the first time, Georg Joachim Rheticus.[28] Noteworthy by his absence is Erasmus Reinhold—all the more surprising because, as we have seen, Reinhold pointed out the specific value of his *Tables* for the Lutheran church, and it was named explicitly for the (heretical) duke of Brandenburg.[29] Like Reinhold's *Tables*, Schöner's *Three Books concerning the Judgments of Nativities* (*De Iudiciis Nativitatum Libri Tres*) was dedicated to Duke Albrecht, but it contained a preface by Melanchthon that may have flagged it for the Index: the copy in the Bibliothèque Nationale, for example, was once shelved in the restricted library of a religious order.[30]

In addition to wholesale exclusions by authorial identity, the 1559 Index also defined other dangerous categories of divinatory writings which in 1586 were to receive the full sanction of a papal bull:

All books and writings [that concerned divining from] chiromancy [palm lines], physiognomy [facial features], aeromancy [airy phenomena], geomancy [scattering of pebbles, grain or sand], hydromancy [water color, ripples and waves] onomancy [letters of a name], pyromancy, [fire] or negromancy [black or evil magic] or books about soothsaying, sorcery, omens, haruspication [divination from the entrails of animals], incantations, divinations from the magical arts *or* judiciary astrology concerning future contingent events or the results of events or cases of chance—with the exception of those books and writings composed from natural observations for the purpose of navigation, farming or for use in the art of medicine.[31]

Hereafter, this same language recurred in the form of a standardized rule for prohibiting all books in these categories, with the important exceptions noted above.[32] And, interestingly, these legitimate areas of prediction were just the ones that Pico had singled out.[33] Thus, while the Catholic Church did not altogether rule out certain kinds of astrology, its sweeping damnation of magic and divination effectively positioned it against Melanchthon and Peucer, who declared that there were some kinds of magic, prophecy, and foreknowledge not owned by the devil. Yet where Peucer left the door open to individual readers—as well as to university visitation or oversight committees and to the secular courts—to determine where the devil might be operating, the Inquisition, by naming authors, printers, and cities, left rather less to the imagination.[34]

This technology of prohibition developed gradually and imperfectly. As for the success of the lists, Bujanda has observed that, precisely because the censors never came to master the material, they consisted of "misrepresented names, distorted titles, repetitious or overlapping prohibitions. All these errors, these inaccuracies, these failures of understanding show, initially, and in an obvious way, the cultural and intellectual limits . . . of the officials who, at different levels, gathered and communicated this information."[35]

It will suffice here to show something of what the officials were attempting to do, however awkwardly, in constructing the lists. Prior to the Church's effort to rule out entire classes of books, officials followed the practice of naming specific works. For example, in 1549, the Venetian inquisitor Giovanni della Casa included Osiander's *Conjecturae de Ultimis Temporibus;* in 1550, the Sorbonne explicitly listed Cardan's *De Subtilitate;*[36] in 1559, although Paul IV's Index had constructed the general category of magic and divination, his inquisitors still thought it necessary to name Luca Gaurico's *Tractatus Astrologicus* and *De Astrologia Judiciaria.*[37] Still other classificatory maneuvers evolved. At Antwerp (1571), particular authors were named within their discipline: theology, jurisprudence, medicine, philosophy, mathematics, and the humane disciplines. Among the mathematicians mentioned, all the names, with the exception of Peter Ramus, came from well-known Protestant courts, universities, and cities in the empire: Cyprian Leovitius (court of Ott-Heinrich of the Palatinate), Erasmus Oswald Schreckenfuchs (Basel, Freiburg), Caspar Peucer (Wittenberg), William Xylander (Heidelberg), Johannes Schöner (Nuremberg), Sebastian Theodoricus (Wittenberg), and Sebastian Münster (Nuremberg).[38] The Parma Index of 1580 caught Johannes Garcaeus's *Astrologiae Methodus* and Marsilio Ficino's *De Vita Coelitus Comparanda* but not the items listed at Antwerp.[39] The Portuguese Index of 1581 broadened the net still further. It warned against any book published in Frankfurt, Zurich, Basel, Schaffhausen, Geneva, Tübingen, Marburg, Nuremberg, Strasburg, Magdeburg, or Wittenberg. It also conceived a virtual "Who's Who" of especially dangerous publishers: Andreas Cratander, Bartholomeus Ousethemerus, Johannes Hervagius, Johannes Opporinus, Robertus Stephanus, Christophorum Froscoverus, Christanus Egenolphus [of Marburg], Henricus Petreius, Thomas Wolfius, and Crato Milius. And in 1583, the Spanish Index tried another tactic. It prohibited all books by leaders of heresies ("heresiarchs") published after the year 1515; but it carefully noted that "the books of Catholics are not prohibited unless works of the aforesaid heresiarchs are inserted in them."[40] Had Osiander's identity been associated publicly with the "Ad Lectorem," it might have led to an earlier condemnation of *De Revolutionibus.*

MAKING ORTHODOXY
LEARNED ADVICE FROM TRENT

The Index's generalized rules guiding the prohibition of dangerous books was one factor that promoted a need for learned comment to describe and explain the right conditions of faith and to guide its inquisitorial enforcement. Trent would have a serious effect on the entire world of Catholic scholarship: an early and particularly influential expression of Tridentine positions was Michael (Miguel de) Medina's *A Christian Exhortation, or Concerning the Right Faith in God (Christianae Paraenesis siue de Recta in Deum Fidei)* in seven books (Venice, 1564). Medina (1489–1578) was a member of the Franciscan order; in 1550 he was elected to the chair of holy scripture at the University of Alcalá, and in 1560 Philip II sent him to the Council of Trent. Although later doubts about Medina's orthodoxy led to his incarceration at Toledo, he was Spain's leading representative at the Council.[41]

The purpose of Medina's dense and exhaustive tome, comprising some 289 folios, was nothing less than to define the grounds of orthodox faith. Book 2 alone is nearly 80 folios. It covers all forms of divination, prophecy, and magic, with prolific citation of both ancient and modern authorities. Medina wrote in a clear, if typically prolix, scholastic vocabulary and made unambiguous distinctions to enable the faithful to steer away from the diabolical.

The central concern of book 2, chap. 1, is what Medina called "prophetic foreknowledge."[42] Its object was to define all the possible categories of human knowledge of future events and to single out the dangerous kinds. The problem had important theological precedents, especially the Scotist discussion of future contingents. But, as we shall see, Medina's authorities were by no means restricted to theologians. Indeed, he grounded his intellectual authority widely by associating himself with "the ancient Church fathers and many pagan philosophers"—most notably Cicero's *On Divination*. But his major intellectual sources in this section were Giovanni Pico della Mirandola and his nephew Gian Francesco Pico.[43]

Of the two, Medina's debt to the elder Pico was the more substantial. In the ten folios where he concentrated his attack on the astrologers, Giovanni Pico's name appears not less than three times, and the source of Medina's positions is obvious even where Pico's name is not explicitly mentioned: for example, where he begins his discussion acknowledging the existence of a domain of legitimate predictions based on natural knowledge.[44] For Medina, such predictions may be made a posteriori by assembling different experiences (*tum ad artem comparandam experimentis*) or by collecting both proximate and remote causes. It turns out that this kind of predictive knowledge falls into the very areas that Pico, like Nicole Oresme before him, regarded as acceptable: agriculture, seamanship, and medicine. Medina also added to these Piconian disciplines several other classes of divination—from parts of the body (physiognomy), including metoposcopy (using the features of the face), chiromancy (the hands), and podomancy (the feet); from the elements, including aeromancy (air), hydromancy (water), geomancy (earth), and pyromancy (fire); and from dreams (somnispicy).

Each of these areas had astrological correlates (for example, "astrological physiognomy").[45] Medina's central concern was to define the boundary that determined which forms of divination were "superstitious" and which were "divine". The bible, for example, contains reports of divinely inspired dreams; but astrology is the source of much superstition.

Medina attacked from many angles. He brought together ancient authorities to show the force of learned precedent for his own position. Following Cicero, of whom he was particularly fond and whom he quoted at length, he condemned "judiciary astrology" as superstitious and called it the "vain invention" of the Chaldeans, Babylonians, and Egyptians. He separated this astrology of judgments from "theorical astrology," by which he obviously meant theoretical astronomy rather than the principles of astrology to be found in the *Tetrabiblos*.[46] And he cited Tacitus's view that "mathematicians (for thus they were commonly called) . . . are a kind of men who are treacherous to princes, deceitful to those who trust them, always prohibited but never expelled from our city."[47]

Medina also hammered away with skeptical questions, such as: If astrology is a legitimate subject, then why are there so many errors in the astrologers' prognostications? And why are the prognostications written in such ambiguous language? How can astrology call itself a *scientia* if it lacks an adequate physics and metaphysics?

Medina then criticized the use of Aristotle's *Meteorology* as a physical basis for justifying the presence of planetary influences.[48] As we have seen in chapter 4, Melanchthon had called attention to the inadequacy of Aristotle's (limited) statements concerning the causal influence of the celestial on the terrestrial region. But where Melanchthon had seen this as an occasion for improving on Aristotle's account, Medina emphasized the contrary: that Aristotle never intended to say that the celestial bodies produce invisible "influxes." Rather, he contended that it is the Sun that principally affects the earthly region because celestial light generates (*progignitur*) heat, and light rays reflect or refract on striking solid bodies. From the activity of light, the sensible qualities—hot, cold, wet, and dry—are generated. In this sense, the celestial bodies themselves possess "virtually but not formally" the properties that they produce in terrestrial objects.[49] On the other hand, defenders of astrology (Medina cited Lucio Bellanti, among others) make up stories and fictions about planetary influences' causing the generation of gold, silver, copper, and so on. The real cause was the Sun's natural heat, which, like the father's sperm, is a "formative virtue" that warms the Earth's maternal uterus and brings about a mixing together of the elements to conceive gold, silver, iron, and so forth.

Medina now brought up a major Piconian theme: the endless controversy over the rules and authorities of astrology. There were allegedly many astrologies stretched across many centuries and civilizations. Which among these often contradictory teachings was to be taken as a reliable basis of astral foreknowledge? Take the properties of the different zodiacal houses: "Immortal God, how much diversity there is to grasp: the Egyptians teach one thing, the Arabs another, the Greeks still another, the Latins another, the ancients another, the moderns another, Ptolemy another, Paul another, Heliodorus another, Manilius another, [Firmicus] Maternus another, [Haly] Abenragel another, and Johannes Regiomontanus another. Which of these, then, are we to believe to be the true astrology? For they all say that their astrologies agree with experiences."[50]

To add to such uncertainty, there were other areas of controversy that plagued astrology. Astrologers did not know for sure how many spheres exist in the universe and—in an echo of Copernicus's preface—they disagreed over assigning concentric, eccentric, and epicyclic orbs. Medina's claim invoked lists of authorities arranged in historical sequence. Up to the time of King Alfonso of Castile, for example, many believed that there were just eight spheres (he cites Plato in the *Timaeus;* Aristotle; "Plato's pupil" Eudoxus; Proclus the Platonist; Averroës; the "father of astrology," Claudius Ptolemy; and Albertus [Magnus]). On the other hand, the "followers of the Babylonian Hermes" pressed the view that there are nine spheres (he cites Rabbi Isaac, Alpetragius, Thabit ibn-Qurra, Alfonso, Rabbi Abraham ibn Ezra, Rabbi Abraham Zacuto, and Rabbi Levi ben Gerson). Against the rabbis, the common view was now that of Albertus Magnus, that there are ten spheres. Finally, there were those who thought that there was no celestial substance and, following Pliny, that the planets move around in "the space of air" (*in aëris spatio*), which extends above the Earth and the waters.[51]

There was then further disagreement about how many motions should be attributed to the eighth sphere to account for the precessional motion—one or several? But again, there were different ways of defining the sphere's parameters. Regiomontanus and Arzachel, following Thabit, put two little circles in the "heads," that is, in opposite signs (Aries and Libra) of the equinoxes of the ninth sphere. The slow motions of these circles then produced a resultant oscillatory motion of the equinoxes of the eighth sphere (*accessus et recessus*). But the quantities themselves posed further difficulties. Arzachel thought that the "head" was ten parts away from the "fixed point" (i.e., the first point of Aries); Thabit placed it four parts, nine to ten minutes distant, and Regiomontanus located it not more than eight parts away. From these differences, Medina drew the predictable Piconian conclusion: there was not merely disagreement about which representation was correct, but even among those who agreed on the model, there was discord regarding its exact parameters.[52]

In the totally unregulated world of sixteenth-century citation practices, an author—especially someone associated with the Spanish crown—was free to bring to bear on his subject all kinds of authorities in any ways that suited his purposes. Medina often cobbled together clusters of diverse authorities, ancient and modern, as a device for creating convincing testimony. Resources for undermining certitude thus lay close at hand. Medina produced a relentless list of what in his Scholastic vocabulary he stiffly and academically called "difform" (i.e., different) uncertainties and categories of authorities. Medina's authorities were also difformly lumped: "famous astrologers," the "gravest" or "sharpest philosophers," "skilled mathematicians," "rabbis," and so on.[53] This cresting flood of authorities now raised up the specific uncertainties that imperiled knowledge of planetary motion. Consider, for example, the differing views of these authorities on the time required for the eighth sphere to move just one degree: Ptolemy (100 years); al-Battani (60 years); and, following him, Rabbis Levi and Abraham Zacuto and King Alfonso in his corrected tables; Arzachel (75 years); Hipparchus (78 years); Rabbis Joshua, Moses Maimonides, Abraham ibn Ezra of Toledo, and Haly Abenrodan (70 years); Regiomontanus (80 years); and finally Agostino Ricci the Neapolitan (66–70 years). Add to these very different values the belief of Rabbi Zacuto, as Ricci reported from Indic tradition, that there are two diametrically opposite stars that complete their revolution in 144 years.[54]

It was not only the eighth sphere that was troublesome. Many critical observations were alleged not to agree with the positions of the planets: the location of Mars, the moment of the Sun's entry into the vernal and autumnal equinoxes, and the Sun's relation to the other planets. Medina enumerated these doubts as follows:

> Favorinus [of Arles], in his *Oration against the Genethlialogues* [as reported] in [Aulus] Gellius agrees that Mars's motion is unknown to the entire astrological discipline; Johannes Regiomontanus, eminent among astrologers, frankly bemoans it in his *Letter to Blanchinus*; Guillaume de St. Cloud, a famous astrologer, recounted the error of its [Mars's] motion in his observations written down two hundred years ago; Rabbi Levi [ben Gerson] confirms that it is impossible to determine the Sun's entry into the equinoctial points; and, as for the shape and form of the heavens and the fixed stars, the Indians relate them one way, the Chaldeans another, the Egyptians another, the Arabs otherwise, the Hebrews still otherwise, Timothy one way, Hipparchus another, Ptolemy and the moderns still another.

Dissension among authorities was exactly the inverse of what Catholic theology employed as a central criterion of church teaching: consensus among the Church fathers. In his last major claim about heavenly knowledge, Medina again used a litany of authorities to accentuate disagreement:

> Nevertheless, that which was brought to view more than all other [claims] was the following: the Egyptians, Pythagoras, Timaeus Locrus, Plato, Aristotle, Eudoxus, Ptolemy's commentator Theon, and almost all the Greek philosophers situated the sun in the second sphere. On the other hand, Anaximander, Metrodorus Chius, and Crates Thebanus placed the same body as the highest of all the planets. The Chaldeans, Archimedes and Claudius Ptolemy located it in the fourth sphere; but Xenocrates, differing from all these [philosophers], believed that it was contrary neither to the celestial appearances nor to correct philosophy, that all the planets and other stars were moved on the surface of the heavens.

With this welter of dissenting authorities as background, Medina finally reached his own epoch:

> Nicolaus Copernicus advanced an opinion with great force of reason—as it appears to many learned men—an opinion not heard of for many centuries and crushed by all ancient astrology. He contends with impunity that the Sun is located at the center of the world while the Earth is moved and, on the other hand, the heavens remain at rest (like the opinion about the Earth's motion among certain ancients and also Celio Calcagnini, whom it pleased us to remember), and the Earth departs from the middle and hurtles around the heavens as if it were a star. All of these things show—even if not true (for I cannot anymore summon faith in this view than to others)—how little even the most solid minds who study the nature of the celestial bodies had the ability to advance, even after long study.[55]

In one respect, Medina's representation of Copernicus recalls Tolosani's portrayal a decade earlier: Copernicus was to be seen as a philosopher reviving an ancient "opinion," a proposition about which authorities could agree or disagree, not an astronomer presenting a hypothesis for further investigation. Although negative, the Franciscan Medina's representation was much less harsh than the Dominican Tolosani's. He did not attack Copernicus's competence; nor did he deride the disciplinary location of *De Revolutioni-*

bus; nor did he confront it with scriptural passages and propositions from Aristotle's *Physics*. Medina used his brief summary of Copernicus's "opinion" to add more weight to the claim that the astronomical foundations of human foreknowledge are uncertain.

ASTROLOGY, ASTRONOMY, AND THE CERTITUDE OF MATHEMATICS IN POST-TRIDENTINE HEAVENLY SCIENCE

Michael Medina's defense of the faith would not be the last effort by a Catholic theologian to make use of Pico's withering critique of the foundations of the science of the stars. But his revived Piconian skepticism occurred explicitly in a new context: the post-Tridentine defense of Church tradition and prerogative and the demarcation of legitimate prophecy from illegitimate naturalistic foreknowledge. In its wake, Medinian skepticism left open serious questions: If astronomy was incapacitated by controversy, what sort of heavenly knowledge could the Church endorse? And which voices within the Church should be taken as authoritative on this matter? What epistemic grounds could be regarded as theologically safe, legitimate, and viable? Quite clearly, both scripture and tradition demanded that the Church say something about the meaning of the creation as described in Genesis. Furthermore, for centuries the Church had claimed its prerogatives over the calculation of holy feast days, such as Easter. Calendrical computation did not require any ontological commitments in natural philosophy, but if the Bible was read literally, then engagement with the physical part of astronomy became unavoidable. And, as Tolosani acknowledged, one should not hastily overturn a natural philosophy with which the Church had long associated itself.

Within a few years after the appearance of Medina's tome, new approaches to the heavens appeared in the Catholic world. The "hard line" on astrological forecasting had taken hold in many places, but most evidently in Catholic Italy. One significant outcome was a shift toward justifying the foundations of astronomy independently of those of astrology. This development took different forms among different writers, but the whole process was already well under way at least thirty years before Sixtus V's 1586 "Bull against Divination" codified and formalized the well-entrenched inquisitorial practice of enumerating illegitimate areas of foreknowledge.

Like John Dee, many of these writers stressed the certitude of mathematical demonstrations in ways similar to that earlier heralded in Regiomontanus's Paduan *Oration* of 1462. Yet, ironically, while the trumpeting of mathematics' foundational superiority became a rallying cry from the 1560s onward, it also served once again to underscore the persistent dilemma of underdetermination. For mathematical demonstrations could be used either to argue for the viability of astrological practice—by appeal to theoretical astrology's secure foundations in theoretical astronomy—or to justify the independence of astronomy from all astrological concerns. That is, if astronomy's theorems were rooted in geometry and geometry's proofs were epistemically secure as deductive knowledge, then astronomy would have no need to justify itself by appeal to astrological practice. The pivotal intellectual resource in this discussion derived from Proclus Diadochus (411–85), head of the Platonic Academy of Athens.

Francesco Barozzi published the first Latin edition of Proclus's *Commentary on the First Book of Euclid's Elements* at Padua in 1560.[56] But as Eckhard Kessler has shown, Proclus's ideas had already entered scholarly discourse as early as 1501, through Giorgio Valla's encyclopedic grab bag *Concerning What to Seek and What to Shun*. Valla had translated and incorporated parts of Proclus's *Commentary* into his own work without attribution to its ancient author.[57] As we have seen, Copernicus was among Valla's attentive readers, although he clearly avoided endorsing any of the metamathematical elements in Proclus's work. Proclus's influence might have been felt earlier in the century had Valla's pupil Bartolomeo Zamberti published his full Latin translation in 1505.[58] In 1533, the Basel publisher Hervagius issued a Greek edition of Proclus by Melanchthon's friend Simon Grynaeus (1493–1541). Grynaeus argued in his little preface that geometry was important not only as basis for studying the sensible world but, more generally, like logic, as a "rule for all arts." The only difference was that logic teaches by general laws of reasoning while geometry teaches by example.[59] Melanchthon, as we know, did not adopt such a strong view of mathematics, but he was clearly sympathetic to the general defense of that subject in the context of his own pedagogical reforms and his strong commitment to astrology.[60]

After 1560, the availability of Barozzi's Latinized Proclus stimulated a debate about the status of mathematics in Italy.[61] Thus, it is clearly no accident that an efflorescence of works extolling the value and scope of mathematics—meaning Euclid's *Elements*—began to appear after 1560. Besides Dee's famous preface to Billingsley's Euclid (1570), there were Henry Savile's Oxford lectures (1570–72), François Foix de Candale's edition of Euclid's *Elements* (1566), and Peter Ramus's *Scholarum Mathematicarum* (1569). Less well known is a stunningly detailed chart of the mathematical sciences published in 1577 by the Dominican Fra Egnazio Danti (1536–86). Christopher Clavius clearly belongs to this notable development—his own *Geometry* first appeared in 1574—and Descartes appears to represent its culmination.[62] But, as I argue later, Kepler must surely count as one of its preeminent figures.

The new enthusiasm for mathematics did not prevent the end of the long tradition of annual prognostications at Bologna. Lattanzio Benazzi, at his death in 1572, became the last of the annual Bologna prognosticators. Astrological pedagogy did not completely die out in that university, but a shift in the attitude toward astrology is already apparent in the work of Benazzi's successor, Egnazio Danti. Danti became public professor of mathematics at Bologna only after having been forced to leave the Medici court at Florence in the fall of 1575, after twelve years as cosmographer to the grand duke of Tuscany under Cosimo I (1519–74).[63] During that time he had worked closely with Cosimo on a variety of astronomical, geographical, and engineering projects, including the (successful) great quadrant on the facade of the Dominican church of Santa Maria Novella, the (unsuccessful) building of a canal from the Tyrrhenian to the Adriatic Sea, and the design of two globes, one terrestrial, the other celestial, for the Sala di Geografia in the Palazzo Vecchio.[64] It is difficult to improve on Thomas Settle's compelling description:

> He was a mathematician, in the sixteenth-century sense of the word, a geographer, possibly a goldsmith, certainly an instrument maker, architect and engineer, an astronomer of some merit, inquisitive and interested in a general reform of astronomy, a translator and annotator of mathematical texts, a professor of mathematics at the University of Bologna, eventually a member of the Papal Commission for the Reform of the Calendar, and finally a Bishop.

While in Florence, in the twelve years from 1563 to 1575, he had the title of court cosmographer; his duties were a mixture of most elements on the above list and included the teaching of all the mathematical subjects.[65]

The grand duke was so interested in Danti's projects that he sometimes visited Danti's workshop at the *convento* of Santa Maria Novella. Eventually, thanks to Cosimo's efforts, Danti was permitted to move out of the *convento* into the palace.[66] It comes as no surprise, then, that Danti instructed the scientifically inquisitive Cosimo in the rudiments of astronomy and in the casting of horoscopes. One of his earliest publications was a summary of the sphere condensed into four pages; perhaps Danti used this material when teaching the grand duke and members of his family.[67]

Five years later, in 1577, after he had moved to Bologna and perhaps as a sign of his new pedagogical status, he published an expanded version of his earlier five tables in the form of a magnificent folio volume, *The Mathematical Sciences Reduced to Tables*.[68] This work is comparable to and even exceeds in detail (although not in scope) John Dee's much better-known "Groundplat of Mathematics" (1570). Its purpose was not, like Dee's, to reform the mathematical sciences but rather to lay open what was already known, under the authority of Proclus, for contemplation of mathematical concepts as objects "midway between the natural sciences and metaphysics."[69]

In his classification of mathematical subjects, Danti divided geometry into three parts: *pratica, speculativa,* and *mista*.[70] Unlike Dee, he used an explicitly scholastic language of classification ("speaking as do the Philosophers") to place astronomy as a "science *subalternated* to geometry" just as music is to arithmetic.[71] In other words, geometry furnishes the principles that serve as the foundation of astronomy; one analyzes the heavens using lines and circles and then adds the physical part, the motion of physical bodies. For the study of the sphere, he explicitly recommended Proclus,[72] but for the theorics of the planets, he admitted that the subject was difficult for those who did not already know it and who thus did not retain the diagrams in their memories. Thus, with apologies for not providing visual aids, he recommended that the Paris edition of Reinhold's *Commentary on Peurbach* be read in conjunction with his outline.[73] Danti's occasional references to Copernicus were respectful and were clearly indebted to his exposure to Reinhold.[74] Danti then displayed outlines of other sciences subalternated to geometry, including perspective, catoptrics, gnomonics, meteoroscopy (measurement of stellar and planetary positions), dioptrics (the science of astronomical instruments), mechanics (the study of machines), architecture (including fortification, painting, and sculpture), the measurement of time, geography (the description of the Earth), and hydrography (descriptions of the sea, shores, and winds). Conspicuous by its absence from the detailed outlines of the sphere and of the theorics was the discipline of astrology.

With the appointment of Giovanni Antonio Magini in 1588 as the afternoon lecturer in mathematics (the morning lecturer was Pietro Antonio Cataldi), Bologna brought to the faculty a highly talented and prolific ephemeridist. Magini was not a member of any religious order. He published substantial books in all areas of the science of the stars, including theoretical astrology and astronomy. Yet the practice of issuing the annual prognostications did not revive with his appointment.[75]

THE JESUITS' "WAY OF PROCEEDING"
THE TEACHING MINISTRY, THE MIDDLE SCIENCES, ASTROLOGY, AND CELESTIAL ORDER

Unlike the Melanchthonians, whose pedagogical reforms involved taking over and reestablishing the charters of existing universities under the patronage of converted Lutheran princes, the Jesuit teaching ministry established independent colleges adjacent to existing universities. Initially, the colleges were just lodging houses in which the Jesuits stayed while studying at the neighboring university. The teaching ministry was central to the early formation of the Jesuit order. By 1544, there were seven such places of lodging near established universities in Paris, Louvain, Cologne, Padua, Alcalá, Valencia, and Coimbra. Not fully satisfied with the manner of teaching at those institutions, however, Ignatius of Loyola decided to allow and encourage the Jesuits themselves to conduct drills and exercises for other Jesuits. In 1551, the order opened its main college at Rome, the Collegio Romano, with a telling inscription over the door: "School of Grammar, Humanities, and Christian Doctrine, Free."[76] Juan Alfonso de Polanco, Loyola's secretary, expressed this policy well: "First of all, we accept for classes and literary studies everybody, poor and rich, free of charge and for charity's sake, without accepting any remuneration." It was a remarkable policy, and it underlined the Jesuits' activist and positive view of human nature. The refusal to charge tuition was one among several factors that led to the rapid expansion of Jesuit schools; and the policy of not charging for services extended, as well, to the confession and to other ministries. As John O'Malley has shown, "[Jerónimo] Nadal [the close friend of Loyola and highly influential interpreter of his ideas] insisted that a Jesuit accept no gifts of any kind from anybody whose confessions he had heard, even gifts that were later sent to him when no direct relationship to the sacrament was intended."[77] Moreover, although the elites patronized the Jesuits wherever they sought to establish their schools, the order did not, in principle, favor the sons of the rich, and their schools generally had a mix of social classes.[78] Still, this social policy was not always easy to maintain and to reconcile with the demands of the humanist, as opposed to vernacular, curriculum, as the humanistic curriculum presupposed skills in reading and writing that boys from lower social classes could not easily acquire.[79] By 1560, Polanco had formulated a clear vision of the importance of teaching in relation to other Jesuit ministries: "Generally speaking, there are [in the Society of Jesus] two ways of helping our neighbors: one in the colleges through the education of youth in letters, learning, and Christian life, and the second in every place to help every kind of person through sermons, confessions, and the other means that accord with our customary way of proceeding."[80] As Luce Giard has argued, the entire development could not have been foreseen in 1540 when the order was created: it evolved gradually and, however coherent Polanco's formulation, without an initially clear design.[81]

Important disagreements arose within the Society about the character of the teaching ministry as expressed in the general organization of the educational curriculum. What should be taught, in what form, and by whom? With regard to astronomy, the argument eventually shifted Jesuit discussions toward the question of the kind of theoretical foundations that ought best inform that discipline as a mixed science—should its subalternating principles be mathematical, physical, or scriptural? This shift did not erase the division between the theoretical and the practical, but it called into question the principles on which that division should be based. This contested framework of the mixed sciences was ultimately the informing context within which planetary motion, influence, and order were filtered.

The Jesuits played a significant role in resituating astrology within the science of the stars.[82] Not only did they object to its perceived dangers, as various Church authorities had done for centuries, but ultimately they called into question the basic justification for studying the heavens. A decade before Medina's *De Recta Fidei* (1563), a hint of early Jesuit sentiment appeared in 1552 in a remark of Nadal (1507–80): "The *mathematicus* can expound nothing of judiciary astronomy; his entire business consists of speculative mathematics."[83] Nadal's remark, unfortunately terse, suggests that he favored the theoretical or contemplative part of a two-cell division of the science of the stars as well as the Aristotelian division between physics and mathematics that was later widely favored in the Thomistically inspired Jesuit manuals of natural philosophy. Among the early and most original of these philosophical syntheses was the systematic exposition of Aristotle's *Physics* in *On the Common Principles and Dispositions of All Natural Things* by the Spanish Jesuit Benito Pereira (1535–1610).[84]

Pereira maintained that mathematical objects have a different (and lower) kind of reality from physical entities because they are mere mental abstractions from matter. As Thomas Aquinas said: "Mathematical things do not subsist separated according to being, because if they subsisted there would be good in them, the good namely of their very being." Pereira, however, had relatively little to say about the heavens compared with Melanchthon in his *Initia Doctrinae Physicae*, and nothing at all about the contemporary proponents of homocentric astronomy. Generalized references to astronomy and mathematics were common devices among writers on natural philosophy who wished to maintain disciplinary distance from mathematical subjects while sustaining the lower position of these subjects. And when Pereira did speak of the heavens,

he was consistent in maintaining that the domain of the mathematician's understanding is limited. He acknowledged that the physicist and the astrologer both study the heavens but argued that their modes of consideration differ: in essence, the astrologer is restricted in his understanding to quantitative relations such as distances, times, and angular motions.[85] Pereira said that the astrologer "does not care [*non curat*] to search for and to lay claim to true causes and to those [causes] which agree with the nature of things." The physicist, on the other hand, concerns himself with such matters as the reasons why the planets move, what the celestial stuff is made of (Pereira followed the customary view that celestial matter was fundamentally different from terrestrial), and why heavy bodies fall—in other words, he deals with "proper and natural causes." Because astrologers make a posteriori inferences from what they see in the heavens, they cannot be sure that their explanatory devices are correct; and Pereira explicitly followed Averroës in claiming that such devices as epicycles and eccentrics "conflict with nature and with right reason."[86] Theoretical astronomy's models may be useful for prediction, but they have no physical reality. So if the physicist can have reliable knowledge about the true causes of everything in the heavens, whereas the (mathematical) astrologer cannot, can there be a judicial astrology grounded in acceptable physical principles?

Pereira's answer was negative. For all of his praise of the physicist's epistemic capabilities, Pereira turned out to be a skeptic with respect to both celestial causality and its possible effects. In 1591, Pereira published an extensive and frequently republished critique of astrology as part of a broader work, *Against the Fallacious and Superstitious Arts, That Is, About Magic, the Heeding of Dreams and Astrological Divination.*[87] All of the ideas about astrology contained in that work could easily have been expressed thirty years earlier, as none were premised on writings that appeared after that date. And because we are dealing with a small, close-knit group of men, I shall proceed on the assumption that Pereira's ideas were well known in the Collegio Romano before 1591. In that work, Pereira developed several categories of objections to judicial astrology—for example, its inconsistency with holy scripture, its weak or defective explanatory structure, and the devil's use of it to manipulate men's fates. Many of these arguments had already been adumbrated by Pico and Medina.[88] Pereira used the excuse of Pico's "prolixity" to help himself to the Mirandolan's ideas, although not without his own original contributions.[89]

To begin with, Pereira criticized the notable difficulty of using Aristotle as an astrological authority: "*Aristotle*, whom the world acknowledgeth the Prince of Philosophers, candidly and ingenuously confessed that concerning many Celestial things he had no certain or exquisite knowledge, but onely an imaginary and conjectural skill, and being destitute of real and manifest reasons, he was compelled to make use of probable arguments and conjectures."[90] To add to these deficiencies, Pereira noted that Aristotle "maketh no mention or speaketh the least word concerning this kinde of Astrology."[91]

On the matter of celestial causality, he followed Pico in asserting that the heavens produce only general causes (motion and light) rather than proximate ones, and hence, we cannot be certain of particular future contingent effects.[92] But even if astrologers could have a perfect knowledge of heavenly causes, they would still fail to predict particular events. Take the classic Augustinian example of the unequal fortunes of twins born under the same planetary configuration and extend it to thousands of nativities: "When *Homer, Hippocrates*, Aristotle, and *Alexander the Great* were born, were not also many others born, in the very same times and moments; yet in the several excellencies proved [un]equal to them."[93] Or again, consider the vanity of presuming to predict the next pope's election: "The promotion of any one to this high degree of honour depends not upon the will and power of the party himself, or any other particular person, but upon the decree and suffrages of the whole Conclave, whose charge it is to elect the chief Bishop; and therefore it is not onely necessary (suppose I affirme *Peter* to be the man) to know the constellation of his *Nativity;* but I must also know exactly the Constellations and Position of the Stars in the several births of all those whose joynt suffrages vote and elect *Peter* to that high preferment."[94] Grant further that we could know all causes and all effects and that there is no human free will: "What good will it be to know things so long before-hand if they cannot be declined and avoided? . . . [F]or what can be more grievous, then not onely to be oppressed with present calamities, but also to be rackt and tormented with an inevitable expectation of miseries to fall upon us hereafter?"[95] Even if such predictions occasionally do come to pass, Pereira claimed (without example) that astrologers' predictions are rarely fulfilled: when they do occur as predicted, "they unknowingly stumble upon a truth," and their results are thus "at randome." Indeed, "Why should we give credit to an Astrologer, for one true Prediction, who tells so many which are quite contrary to verity?"[96]

Pereira also condemned both the conjunctionist astrology of history and elective astrology based on the moment of laying the foundation stone for a castle or town. And, like Melanchthon, he targeted Pierre d'Ailly as a worrisome example ("hugging himself in his own conceit concerning the concord of History and Astrology"). According to Pereira, the credulous Cardinal d'Ailly believed the astrologers when they foretold that the Council of Constance would end in a disaster for the Christian religion; that no such result occurred should have been enough to put him off the subject altogether, but it did not.[97]

Finally, Pereira objected to the theory of celestial influence. The main difficulty concerned the attribution of contrary properties to the planets. For instance, Saturn shares in common with all the stars the general property of illumination, but unlike the other planets, it is supposed to possess the special property of producing cold. It did not matter to Pereira whether Saturn itself was cold or whether it merely had the virtual capacity to produce cold as an effect of its inherent power. However, if celestial, as opposed to terrestrial, bodies were homogeneous in their constitution, then by their nature they should have one and only one natural property—light. Yet light produces not cold but the contrary quality of heat; ergo, the doctrine of celestial influence must be incoherent and false.

It is consistent with Pereira's views about the lower status of mathematics that his style of argumentation entirely avoided the astronomical basis of the forecasts and the accuracy of the planetary tables. This general silence about mathematicians' views makes the lack of reference to Copernicus unsurprising. Compared with Medina, who stressed disagreement among the astronomers, for Pereira it was astrology's false physical principles and its conflict with general "experience" that ultimately convicted that subject. Thus, although the timing of Pereira's work may have been influenced by Sixtus's 1586 bull against the astrologers, the philosophical foundations of his objections still aligned him squarely against those who had long believed that the Jesuit curriculum should give mathematics a central place.

The mathematicians' arguments eventually left their mark on the Jesuit curriculum of studies (*Ratio Studiorum*, in versions of 1586, 1591, and 1599), but only after deep and protracted debate with philosophers like Pereira, whose views were widely shared within the order. By the time that the final version appeared in 1599, the *Ratio* had integrated mathematics and related subjects into a larger ideal of intellectual life.[98]

The leading figure in this development was Christopher Clavius of Bamberg (1538–1612). Mathematics was central to Clavius's calling as a Jesuit: he came to be known as "the Euclid of our times."[99] He entered the novitiate in 1555 and was a member of the important first generation who personally knew Ignatius Loyola, the founder of the order. Between 1557 and 1560, Clavius studied philosophy and logic at Coimbra with Pedro de Fonseca. His first (known) interest in mathematics appeared in Fonseca's course on the *Posterior Analytics,* a text in which Aristotle presents the example of the sum of the three angles of a triangle as equal to two right angles.[100] This example was evidently formative for Clavius; the relationship of mathematics to Aristotelian philosophy lay at the heart of later curricular conflicts in the Jesuit order. By the academic year 1563–64, just as the Council of Trent had drawn to a close, Clavius was teaching mathematics at the Collegio Romano. The following year, he offered a course on the sphere. In an unpublished manuscript of his own on the same subject, he included some astrological elements.[101] However, by the time that Clavius's commentary on Sacrobosco's *Sphere* appeared in 1570, these astrological references had completely vanished.

Athens—or, in this case, Alexandria—was acceptable to the Church only so long as it served Jerusalem. But given that astrological practice was so widespread, how exactly was a balance to be struck? Likewise, given the utility of mathematics, where should its limits be drawn? Reference to the name Averroës (or to "the Averroists") was often made to evoke fears of an excessive reliance on reason and philosophy to the detriment of revelation and mystery.[102] In choosing the geocentric Sacrobosco as a literary and organizational model, Clavius signaled an affirmation of tradition.[103] Regardless of confessional loyalty, it was still the schoolmaster's ideal format. At Wittenberg, as in Rome, the study of astronomy was placed in the service of a reforming church. The Jesuit and the Melanchthonian Sacroboscos—as well as their variants—thus became tools for teaching students how to use astronomy to support the faith. But in the post-Tridentine climate, threats to the faith manifested themselves in increased suspicion about astral forces and the literature that taught them. Protestant printers and authors, as seen earlier, were represented in the Index as the main producers of that literature. The strong prohibitions contained in Paul IV's 1559 Index suggest that Clavius's *Commentary on the Sphere* was knowingly sensitive to strictures against foreknowledge based on the stars.[104] And this is exactly what one finds in Clavius's scheme of classification at the beginning of his *Commentary.* Clavius used the familiar distinction between *theorica* and *practica,* but he constructed it somewhat differently from earlier authors. First, he designated the entire subject matter of his book as "astronomy" rather than as the "science of the stars." Second, he restricted his subject to only two categories—theoretical and practical astronomy. Unlike Schreckenfuchs, he did not promise to cover astrology in a separate work.

> Astronomy is divided into *theorica,* that is, the contemplative part, and *practica,* that is, the making and doing part. *Theorica* considers the whole fabric of the world [*machina mundi*] as it is in itself, describing the arrangement of the world and dividing the world's entire embrace into the aetherial and elementary regions. Then, it [*theorica*] investigates the number, size and motion of all the celestial bodies and also it considers the rising and setting of the planets and stars. Likewise, it considers the images and figures of the signs and all the constellations and it teaches how to reckon by calculation the true places of both the fixed stars and the moving ones, called planets. Similarly, it [*theorica*] inquires diligently about the planets' forward and backward motions and stationary positions, their conjunctions and oppositions, together with eclipses of the luminaries, the Sun as well as the Moon, and an almost infinite number of other relations of this kind.

Clavius then provided a sampling of works that he recommended in this category. His list remained constant from the *Sphere*'s first edition of 1570 until the last in 1611. "This [theorical] astronomy is explained partly in Ptolemy's *Almagest* or *Great Construction* or, if you like, in Johannes Regiomontanus's *Epitome,* in Albategnius's astronomical work, in Alfraganus's little work, in Georg Peurbach's *Theorics of the Planets,* in Nicolaus Copernicus's *Celestial Revolutions,* and in the nearly innumerable volumes of other authors." The early inclusion of *De Revolutionibus,* without qualification and without objection, shows that he regarded it as an obvious and legitimate resource of theoretical astronomy, much as the Protestant schools of this period did. This easy inclusion would also seem to confirm his unfamiliarity with Tolosani's critique of Copernicus. However, Clavius now shoveled into *theorica* the subject matter usually assigned to practical astronomy.

> Theorical astronomy partly concerns the many instruments invented by astronomers with the greatest diligence in order to put the celestial motions before our eyes. Commonly, this includes Ptolemy's astrolabe or planisphere, Gemma Frisius's catholic or universal astrolabe, also Juan de Rojas's universal planisphere, the astronomical ring, quadrant, torquetum, astronomical radius, and other instruments of this kind. Finally, theorical astronomy is taught in that part which is usually called *tabular,* whereby astronomers probe the celestial motions through numbers arranged in tables, such as the tables of King Alfonso of Spain, Johannes Regiomontanus, Johannes Blanchinus of Ferrara, and Nicholas Copernicus—usually called the *Prutenic Tables*—and many others.[105]

Clavius did not argue for his regrouping of the common elements of *theorica;* he simply put forward his classification as though it was natural. The "theorical" thus immediately eliminated astrology from any possible claim to having met the standard of demonstrative—rather than conjectural—knowledge and brought Clavius into line with post-Tridentine positions on astral foreknowledge. His scheme also consigned astrology to the practical part of astronomy, usually reserved for tables, almanacs and instrumentation: "Practical astronomy, or what some call judiciary, prognostic, or divinatory, accommodates all those things useful in human life. But many use this part rashly and so wish to broaden out this prognostic part so that the subject then becomes superstitious and, hence, quite

rightly, the Church is suspicious. St. Augustine damns it in the books concerning the Christian doctrine. . . . So, in the meanwhile, I think that we ought to say nothing at all about this [prognostic part]."[106]

In the 1585 edition of the *Commentary,* Clavius went one step further. He changed the period to a comma at the end of the final sentence in the above quotation and introduced the following revealing addition: "if only because the following authors utterly destroy it [*funditùs evertunt*]: Giovanni Pico Mirandola in *Twelve Books Written against the Astrologers;* his nephew Gianfrancesco Pico in the books *Concerning Preconceived Ideas;* Antonio Bernardo Mirandola, the Bishop of Casertanus, in books 22, 23, and 24 of his *Monomachia;*[107] Michael Medina, book 2, chapter 1, *Concerning the Right Faith in God;* and Julius Syrenius in the books *On Fate.*"[108] One can only speculate about Clavius's decision to strengthen the exclusion of astrology fifteen years after the first edition,[109] or his further decisions to add an exacting list of mostly modern authorities (as contrasted with Pereira's deployment of mostly classical and ancient ecclesiastical authorities) and to choose the strong verb *everto*—which can also mean "overthrow," "subvert," "confute," or "confound." Among Jesuit astronomical practitioners, Clavius's formulation would become the standard posture on astrology.[110]

In this way, Clavius recommended the Piconian objections to astrology but also favored tradition against Copernicus. This left him with a textbook that could easily have been taken as an advanced complement or preparation for Reinhold's *Commentary on Peurbach:* a geocentric arrangement that favored Peurbachian orbs and spheres and, hence, Ptolemaic calculating practices (including the equant). However, the *Sphaera* was not a full-blown *theorica,* and although Clavius promised to publish such a book, he never did.[111]

What options were left between Medina's skeptical undermining of astronomy and the Fracastorian-homocentrist rejection of eccentric devices? Evidently, when the first edition of the *Sphere* appeared in 1570, these extremes were the positions that framed Clavius's problematic, although Clavius continued to add arguments against Fracastoro in the 1581 edition. Perhaps as a way of minimizing personal antagonisms, Clavius constructed many of the polemics in his commentaries as arguments against generic rather than eponymic designators, such as "the skeptics," "the adversaries," and "the Averroists." Pereira, for example, was not mentioned by name, although Fracastoro, who died in 1553, was explicitly included. And although Copernicus was mentioned, he was not represented as a figure with any special claim to novelty but rather as a "follower of Aristarchus of Samos who lived four hundred years before Ptolemy."[112] One consequence of representing Copernicus and his universe in this manner was that it obscured the important difference between the simple Aristarchan picture of concentric circles and the system of complex off-center Copernican mechanisms. Moreover, the *Sphaera* contained no implication of the existence of an academic "school" or an organized social movement, no talk of "Copernicans" or "followers"—not even a heliocentric visual. In this respect, Clavius's (over)simplified representation of the Copernican arrangement entirely resembled other textbooks of this period.

CLAVIUS ON THE ORDER OF THE PLANETS

Clavius's *Sphere* contains one of the most extensive discussions of planetary order in the sixteenth-century tradition of Sacrobosco commentary. Unlike Sacrobosco, Clavius classified his account under the heading of the "order of the planets," a designation that carried with it the meaning of *sequence* but not necessarily *systematicity*, for nowhere did Clavius use the label *cosmology*. Nonetheless, this use of language shows that even in a traditionalist work, Copernicus's treatment of the "revolutions" as an autonomous topic affected Clavius's organization. The notion of "order" then made it easy for Clavius to constitute his discussion as a tradition of "opinions" (*opiniones, sententiae*) of which Copernicus's (and his own) were instances. "Among the ancients there were some, of whom the leader was Aristarchus of Samos, four hundred years before Ptolemy—whom Nicolaus Copernicus, among the moderns, follows in his work on the celestial revolutions—who fashioned this order among the bodies making up the whole universe: the Sun located immobile in the center or middle of the world, around which is Mercury's orb, then Venus's orb, and around that is the great orb containing the Earth with the elements and the

Moon, and around that is the orb of Mars; then Jupiter's heaven; after that, the globe of Saturn; and, at last, there follows the sphere of the fixed stars."[113] As was typical of this kind of treatment in the teaching manuals, Clavius omitted any account of Copernicus's or Rheticus's ordering principles. Even the few private annotations on his own copy of *De Revolutionibus* focused on errors in the short section on geometry rather than on the standard of proof articulated in the preface and book 1, chapter 10.[114] Like the majority of his contemporaries, Clavius did not know that Copernicus intended his proposed arrangement as a solution to the Piconian critique of celestial order. If he was familiar with Rheticus's *Narratio,* he omitted any reference, even to denounce its author as a heretic.

Less important than the fact that Clavius favored Ptolemy's planetary ordering, however, is that his approach to the problem was, like Copernicus's, in the tradition of Ptolemy and Regiomontanus. Unlike the Aristotelian commentators of his era, Clavius marshaled argumentational resources that mainly involved claims about astronomical effects (such as eclipses, occultations, elongations, angular speeds, relative visual diameters of the planets, and diurnal parallax). In this sense his practice and his authorial identity as a mathematician resembled that of the author of *De Revolutionibus.*

Ancient authority was still powerful and immediate in Clavius's intellectual world. Clavius well knew from Regiomontanus and Copernicus—and also Pico—that the order of Mercury and Venus with respect to the Sun was a matter of controversy among the ancients. Yet it is noteworthy that he did not represent his own discussion as a matter of controversy among the moderns. Thus, after the opinion of Aristarchus, Clavius presented the following alternatives: "The most ancient Egyptians, Plato in the *Timaeus,* Aristotle in book 2, chapter 12 of *On the Heavens* and book 1, chapter 4 of the *Meteorologica* thought that this was the order of the celestial spheres: the Moon occupied the lowest place; immediately following is the Sun; then Mercury; then Venus; fifth Mars, sixth Jupiter, seventh Saturn, and finally, eighth, the starry heaven or firmament. Only Aristotle in his little book to Alexander (if indeed it is to him) *On the World* locates Venus immediately above the Sun and Mercury below it." Note that, for Clavius, this

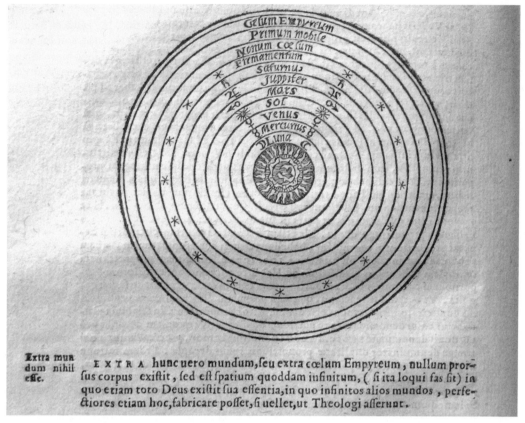

Celum Empyreum
Primum mobile
Nonum coelum
Firmamentum
Saturnus
Jupiter
Mars
Sol
Venus
Mercurius
Luna

Extra mundum nihil esse. EXTRA hunc uero mundum, seu extra coelum Empyreum, nullum prorsus corpus exiftit, sed eft fpatium quoddam infinitum, (fi ita loqui fas fit) in quo etiam toto Deus exiftit fua effentia, in quo infinitos alios mundos, perfectiores etiam hoc, fabricare poffet, fi uellet, ut Theologi afferunt.

38. Clavius's woodcut of spheres displaying "The Number and Order of All the Bodies in the Entire Universe." Clavius 1591, 72. Author's collection.

opinion was no better than that of Aristarchus and Copernicus: "Such an order of the planets and heavens is also undoubtedly refuted by the astrologers."[115]

Between the 1570 and the 1581 editions of the *Sphaera*, Clavius adjusted his designation of the Ptolemaic heavenly ordering from the "true" (*verus*) to the "truer" (*verior*) opinion. The change of this one term occurs in a postil.[116] What significance should be attached to this modification? It has been suggested that the mitigated language may reflect a peculiarly Jesuit view of knowledge that favored "probabilities" (in the sense of "more likely opinions"). This observation is important, although it is not immediately clear why Clavius's view would have become more "Jesuitical" between the earlier and the later editions.[117] His original arguments for the Ptolemaic ordering remained unchanged. In fact, he held the view that no single argument alone sufficiently (*sufficienter*) confirmed that arrangement.

By collecting together Clavius's scattered references, it is possible to summarize the main assumptions guiding his arguments.[118]

1. The closer a planet is to the first mover, the slower it moves; the farther from the first mover, the faster it moves.

2. The closer a celestial body is to Earth, all things being equal, the greater is its parallactic angle.

3. Any star that occults another is closer to Earth.

4. Planets closer to the Sun are more intensely illuminated than those more distant.

5. The Sun's motion is the "rule and measure" of the other planets, and the Sun is therefore the "mean" between the three superiors and three inferiors.[119]

6. Between two adjacent, concentric orbs there can exist neither a body nor a vacuum.

7. Planets with irregular motions, that is, those involving more orbs, belong closer to other planets with irregular motions.

Individually, these propositions require some comment. The first proposition is the same one that Goldstein regarded as motivating Copernicus's move to reorder the planets; yet Clavius remained silent and unreceptive even to the Capellan proposal. Clavius's discussion of daily parallax was geometrical but not observational. The theory of daily parallax involves triangulating the position of a planet, using the observer's position on the Earth's surface and the center of the Earth to make up the necessary points. Clavius offered a diagram that showed different ways in which the triangles can be put together. A difference is wanted between the apparent angle of the object as seen by the observer and the true angle as calculated from the center of the Earth—often called the parallactic angle but labeled by Clavius the "diversity of aspects." This method was said to work for all bodies from the Moon to the Sun, with smaller parallactic angles obtained for more distant bodies. Presumably, this sort of investigation would solve the order of Mercury and Venus with respect to the Sun. Yet if Clavius attempted measurements of any sort, his passive grammatical constructions ("it was discovered that") concealed the observational evidence from his readers.[120] Moreover, the technique, so Clavius acknowledged, did not work for bodies above the Sun, such as Mars—the very planet on which Tycho Brahe would later conduct an extensive observational campaign.

Next, Clavius argued from his first assumption above, ordering by speeds ("ex velocitate & tarditate motus"). Here he was quickly forced to admit that "by this method, nothing certain can be established about the Sun, Venus and Mercury among themselves . . . since their proper motion from West to East is performed in the same time."[121] Exactly at this point, Clavius could have mentioned Peurbach or Copernicus's arguments or even those of Martianus Capella; but, as with many other such opportunities to draw contrasts with his position, he was silent. The postil, however, confidently announced that "the order of the heavens is proved from the swiftness and slowness of motion."

The next postil, using the third proposition (above) as foundation, told the reader that "the

order of the heavens is confirmed by eclipses." The Moon was, of course, a good place to start because Clavius believed that it was capable of occulting any star—which put it in the lowest position in the universe. But for Venus and Mercury, he could only assert that "by equal reason, Mercury will be beneath Venus and Venus below Mars and thus to the end." Again, Clavius had no new observations to produce; nor did he refer to earlier ones. He repeatedly returned to the problem of these two planets.

The overall force that Clavius ascribed to these three arguments is not altogether obvious, as he referred to them as established *infallibiliter* (parallax), *convenienter* (inverse speeds), and *firmiter* (eclipses). Terms like *infallibly, suitably* and *firmly* might conceivably resemble conditions of demonstration that were much later formalized as *necessary* and *sufficient*. But here the usage seems to possess largely rhetorical force. "None of these reasons sufficiently (*sufficienter*) establishes this order, but all of them together greatly affirm (*sumptae confirmant*) that the heavens are arranged in that order."[122]

Clavius next marked off a second cluster of arguments, obviously important but evidently not intended to carry the same force as the first group. He introduced this new group as follows: "To have a greater understanding of this order, I think that I shall not digress from the subject if I make public other arguments of astronomers from which it will be evident that the suitability of this order is the greatest of all."[123] These arguments are:

1. That the Sun is in the "middle" (*medius*) of the planets, that is, between the upper and the lower three. The orbs of Mercury and Venus fit exactly into the space (1,006 Earth radii) between the Moon's greatest distance from Earth (64 Earth radii) and the Sun's closest approach (1,070 Earth radii). This space must be filled, as "Nature abhors a vacuum." This claim was presented directly on the authority of Regiomontanus, Ptolemy, and al-Battani, but not Aristotle—an example of how Clavius was willing to cite or ignore authorities explicitly to suit his purposes.

2. That the Sun is the "measure and rule of motion" (*regula et mensura motuum*). Clavius's account starkly recalls the shared-

motions passage in Peurbach; but, like that of Albert of Brudzewo, his interpretation conserved Peurbach's position: The upper three planets "agree with the Sun's motion" because their epicycles all have periods of one year; in contrast, the Moon, Mercury, and Venus agree with the Sun in their deferent motions, "as is displayed in the *Theorics of the planets*."[124] The choice of language also bears distinct echoes of Copernicus's and Rheticus's phrasing referring to the fixed distance between the Earth and the Sun, the "common measure of the planetary orbs" (*communis orbium planetarum inter se dimensio*).[125] This issue is, of course, precisely the fulcrum on which Copernicus had made the Earth-Sun transposition. Yet, thirty years after the appearance of *De Revolutionibus*, Clavius's account again underscores that practitioners did not feel logically or observationally compelled by the heliocentric ordering.

3. The Sun will be positioned in the middle as "king" and "just as if the heart of all the planets." Having established the sense in which the Sun is in the "middle" of the planets, Clavius proceeded to detail the qualities of the planets: "Saturn, the adviser, on account of his old age; Jupiter, the judge of all things, because of his generosity; Mars, leader of soldiers; Venus, like the mother of a family, the distributor of all good things; Mercury, her scribe and chancellor; finally, the Moon discharges the office of messenger."[126]

It is notable that Clavius's description did not assign overtly Christian virtues to the planets; they were still like pagan gods, but now deprived of their astral powers.[127] Given his exclusion of astrology from the science of the stars, this crumbling of astral potency is, of course, exactly what one would expect. Clavius thus effectively positioned his account in sharp contrast to the views of the astrologers of the secular courts and universities, who still claimed the authority to conjecture the powers and meanings conveyed by the astral gods and thus to speak for them.

4. Planets closer to the Sun are more vigorously illuminated. This claim follows from the third principle above, a property of the Sun's middle position. Clavius said that the Sun's light (*lumen*) illuminates all the planets "equally" (*aequabiliter*), but the closest ones—Mars and Venus—are illuminated more brightly than the others. This claim, although it is nothing more than an assertion, counted for him as further confirmation—at least by virtue of coherence—of the correct order of Venus and Mercury.

5. Another argument for the Sun's position then followed, drawing on the authority of Albumasar, from the Sun's active capacity to produce heat. If the Sun were located in Saturn's place, then things below would be too cold; if it were too close to Earth, things on Earth would burn up; hence, it belongs in the middle, where its "actions" may be "tempered," and it can accommodate itself better to lower things.

6. With the Sun's position established, Clavius returned to the question of Mercury and Venus. Here, he invoked something like a criterion of simplicity. Mercury, he said, is "fittingly" (*convenienter*) located above the Moon and below Venus because its motions on its five orbs and one epicycle are "more irregular" (*magis irregularis*) than Venus's, which require only three orbs and an epicycle. This sort of argument has the appearance of an ad hoc gesture, as Clavius did not apply it consistently to the other planets. It was not uncommon to invoke regulative ideals of simplicity, as we have seen from the example of Rheticus, but no one had yet formulated general rules for judging the relative simplicity of two theories in the same domain of application.

7. The ancients used the planets to name the days of the week and the hours of the day. Clavius's planetary ordering does not map neatly over the sequence of weekdays; but with some nimble, ad hoc juggling he managed to produce a fit. That he thought it plausible to include such an argument may perhaps be explained by the need to add spice to a student textbook, perhaps a mnemonic for remembering the planets and their ordering.[128]

Clavius then inserted a short paragraph in which he asserted that the aforementioned arguments agree with Sacrobosco's ordering and provide grounds for rejecting the diverse "opinions" of Metrodorus and Cratus (that the Sun and Moon are the highest planets); Democritus (that Mercury is above the Sun); al-Bitruji (that Venus is above the Sun); Plato and Aristotle (that the Sun and Moon are a "low place"). This pause seems curiously out of place. It does not, as one might have expected, conclude the commentary. Moreover, it allows no place for the "moderns." In fact, it precedes an objection that might have come earlier and constitutes one of the longest subsections in the entire set of arguments.

8. The postil reads: "Why the Sun is not eclipsed even though Mercury and Venus are below it." Clavius admitted that "some people" have made this objection but that, on the authority of Ptolemy and Regiomontanus, it has no force because the very small apparent diameters of Mercury and Venus prevent their observation during a transit. Clavius did not ignore "the very small" (as he did the parallax argument); he trustingly reported that "according to al-Battani and Thabit [ibn-Qurra] and other astronomers, the ratio of the sun's visual diameter to Venus's visual diameter . . . is tenfold."[129] The squares of the respective diameters yield a 100:1 ratio of the areas; and thus "Venus will occult only one hundredth part of [the Sun], which is of no importance . . . a fortiori, therefore, neither can Mercury [occult the Sun], since its visual diameter is much smaller than the visual diameter of Venus."[130]

Clavius's description completely ignored the reference, singled out by Copernicus, to Averroës' report of two "blackish objects" on the sun. Had he overlooked it in his reading of *De Revolutionibus*? Did he feel that it was an opinion unworthy of consideration, less important than the order of the planets and the names of the weekdays? Or did he simply have no reply? We do not know.

Finally, Clavius summarized his conclusion with a woodcut titled "The Number and Order of All the Bodies in the Entire Universe." This picture appeared in all editions of the *Sphere* and also included an explanation for what exists outside this universe: "No body exists, but it is rather a certain infinite space (if the divine may be spoken of in this way) in which God's entire essence exists and in which God, if he so wished, could build an infinite number of other worlds, even more perfect than this one—as the Theologians claim."[131] Clavius's formulation was quite in line with thirteenth- and fourteenth-century voluntarist views about the possibilities of divine omnipotence, and, like that of Copernicus, exemplified the sort of targeted—rather than systematic—natural philosophizing and discipline-bounding in which astronomers were generally prepared to engage.

DISCIPLINARY TENSIONS

In the second half of the sixteenth century, the teaching manuals of Clavius and Melanchthon (and his disciples) were two of the most influential resources for introducing students to the topic of planetary order. Not only were they well written and well structured, but they were also reproduced as a part of efforts at large-scale pedagogical reform. As reformers within vigorously expanding religious movements, both authors represented astronomy as useful to their respective churches and the astronomer as someone who taught the actual structure of the heavens; and thus, in different ways, both promoted a view of astronomy as part of traditional natural philosophy.

The local situations that the two faced, however, were different, and these differences help to explain both the parameters of their presentations and also the ways in which they approached and used Copernicus. Clavius triangulated his case for strengthening the authority of astronomy with respect to several different ideals of heavenly knowledge: (1) the Melanchthonian science of the stars, grounded in the control of astral forces; (2) the attempted unification of astronomy with Averroist-Aristotelian physical principles; (3) the post-Tridentine skeptical demand for a justification of astronomy apart from the powers of the stars; and (4) Robert Bellarmine's view that only scripture and the consensus of the Church fathers could serve as reliable foundation of astronomy.

As to the first consideration, Clavius was forced to fight for the legitimacy of mathematical studies against traditional philosophers without the Melanchthonian appeal to astrological utility.

With the Jesuits agreed on which parts of the science of the stars to reject, justificatory resources for the study of astronomy, like Melanchthonian astrological physics and John Dee's *Propadeumes,* were obviously unavailable. Melanchthon, with the backing of Albrecht of Brandenburg, was at first far more successful than Clavius in quickly building a cadre of students and faculty with the mathematical expertise necessary for practical astronomy and astrology. In the careful discussions about the curriculum that the Jesuits conducted over many years, Clavius's arguments were not specifically about astronomy but more broadly about raising the status of the professors of all mathematical subjects in relation to the teachers of natural philosophy. According to the 1586 Jesuit curriculum of studies, known as the *Ratio Studiorum,* the mathematics professor must be given the right to dispute publicly with the other professors during formal ceremonies: "By this it will easily come about that the pupils, seeing the professor of mathematics together with the other teachers taking part in such acts and sometimes also disputing, will be convinced that philosophy and the mathematical sciences are connected, as they truly are; especially because pupils up to now seem almost to have despised these sciences for the simple reason that they think that they are not considered of value and are even useless, since the person who teaches them is never summoned to public acts with the other professors." Next, the mathematics professors should not be encumbered with "other occupations." And finally, students must be made to see that mathematics is the best preparation for philosophy and the other sciences: "Because of their ignorance of the mathematical sciences, some professors of philosophy have very often committed many errors, and those most grave; and what is worse, they have even committed them to writing, some of which it would not be difficult to bring forward."[132]

By 1586, the successful inclusion of such bold and critical language signaled the victory of Clavius's arguments and measures for the indispensability of mathematics not only to philosophy but also to all other disciplines. It is apparent from other passages and phrases in the *Ratio* that the target was specifically Averroist readings of Aristotle (at one point in the *Sphere* Clavius refers, with ill-concealed sarcasm, to *Erroistas*). Clavius here intended to signal an unhealthy dependency on philosophy at the expense of Ptolemy's mathematical astronomy, although there is no evidence that, as some of his contemporaries were doing, he was moving to replace Aristotle with Plato.[133] The *Ratio* pushed for a different Aristotle, stating that philosophy professors must know mathematics in order not to distort "the passages in Aristotle and other philosophers which concern the mathematical disciplines." These professors should desist from raising problems that denigrated the authority of mathematics in the eyes of students, "such as those in which they teach that the mathematical sciences are not science, do not have demonstrations, abstract from being and the good, etc.; for experience teaches that these questions are a great hindrance to pupils and of no service to them; especially since teachers can hardly teach them without bringing these sciences into ridicule (which I do not know just from hearsay)." And finally: " To our disgrace, we lack professors who can give the teaching of mathematics that is needed for so many and excellent uses. At Rome too, if you except one or two, scarcely anyone will be left who is qualified either to profess these disciplines or to be at hand at the Apostolic Seat when there is discussion of ecclesiastical times."[134]

At Wittenberg, there were no comparable negotiations between the professors of mathematics and philosophy. In the 1530s and '40s, the curriculum had evolved a smooth consensus on the proper texts, ranging from the various Melanchthon-endorsed elementary spherics to Reinhold's Peurbach commentary to the *Prutenics* to the *Almagest.* Ptolemy was the preeminent authority in theoretical astronomy. In his 1549 physics lectures, Melanchthon readily dismissed the Averroist criticism of planetary mechanisms without extensive argument: "One should detest the perversity and petulance of Averroës and many others who deride this teaching [of eccentrics] built upon [Ptolemy's] great composition [*Almagest*]; consequently, it cannot be asserted that such devices [*machinas*] are really to be found everywhere in the heavens. Let not the studious allow themselves to be deterred by this sophistry, for then they shall fail to understand the [heavenly] motions. . . . Indeed, it is unnecessary that such orbs be carved in the heavens."[135]

Averroist homocentric astronomy, deriving its authority from Achillini, enjoyed a rather different fate in Italy. In quick succession, Girolamo Fracastoro and Giovanni Battista Amico published somewhat different versions of concentrically organized sphere arrangements at Padua in the mid- to late 1530s.[136] How quickly these rehabilitations became known north of the Alps is harder to say.[137] Paul Wittich, who annotated several copies of *De Revolutionibus* in the late 1570s and early 1580s, guessed that Copernicus knew the work of Fracastoro.[138] To some degree, such regional availabilities may explain different uses of both Copernicus's work and the new homocentric renditions. Whereas *De Revolutionibus* at Wittenberg was immediately linked to Melanchthon's apocalyptically motivated astrological concerns, in Rome, Clavius explicitly enlisted Copernicus in his own struggle with Fracastorian astronomy.[139]

Copernicus himself had devoted precious little printed space to homocentrist claims.[140] He brusquely dismissed "homocentric circles" in his 1542 preface as failing to account for the phenomena and for having contributed to the confusion that had created of the universe a disjointed "monster." Had Copernicus continued thence with a defense of the utility of eccentrics, epicycles, *and* equants—as though writing like Clavius principally for an audience of teachers and students—he might well have written a quasi-Peurbachian defense of traditional, theoretical astronomy. But, as we know, Copernicus replaced the equant and freely used eccentrics and epicycles for purposes of calculation. Why he objected to the equant has occasioned considerable debate.[141] Whether, as has been suggested, he believed in the existence of hard, impenetrable orbs is certainly difficult to determine from his language, which is at best ambiguous. Neither Copernicus nor Rheticus made the old controversy about the physical status of planetary spheres central to their works.

In the 1581 edition of the *Sphere,* however, Clavius added such a discussion in the form of a long disputation. This expanded commentary, as recent interpreters have remarked, reveals much about Clavius's own representations of astronomy in relation to natural philosophy. It is also highly revealing of his appropriations of Copernicus as an ally in this controversy.[142] Clavius argued that both astronomy and natural philosophy work inductively—a posteriori, from sensory effects back to causes. He took no note of the fact that Rheticus and Copernicus had also deployed this position in the *Narratio Prima* to loosen up

the relations between astronomy and physics.[143] In a typical humanist gesture, they reminded their readers that Aristotle too was human, that he had borrowed his astronomical principles from his own time-bound contemporaries Eudoxus and Callipus, and thus, were he still alive, he would be capable of changing his mind about the arrangement of the heavens. The idea was to take away from the natural philosophers the possibility of arguing deductively from physical or metaphysical first principles and to lean on the truth of an argument's consequences. Clavius also made use of this trope when he said that Aristotle maintained that one should always look to the astronomers.[144] But unlike Copernicus, Clavius used this logic to argue that, contrary to Fracastoro's homocentric spheres, the Ptolemaic eccentric orbs found in the common *Theorics* lead to no false or absurd consequences.[145]

The other side of this logical coin was the objection *ex falso sequitur verum*. If eccentrics and epicycles are fictions, then the truth of the causes is uncertain, "since it is permitted to draw from false premises a true conclusion, as Aristotle maintains in his *Logic*."[146] The argument against *ex falso verum* lay at the crux of Clavius's controversy with the Fracastorians. Here Clavius made an interesting distinction. He contended that the rule invoked by "the adversaries" pertained to the logical form of the syllogism but not to the form for making astronomical predictions. In the former case, from a conclusion already known to be true, many false premises can be conceived. For example, one can make a conclusion of the proposition that all animals are sensitive and then construct the syllogism "All plants are sensitive; all animals are plants; therefore, all animals are sensitive." Or: "All stones are round; all stars are stones; therefore, all stars are round." Clavius argued that the example of epicycles and eccentrics belonged in a different category because, unlike such deductions that yield no new information, the conclusion drawn from a prediction is not known in advance. It is noteworthy that Clavius invoked as his examples astronomical predictions but not astrological prognostications or nativities.

From eccentric orbs and epicycles, not only the appearances of past things already known are defended, but also future things are predicted, the time of which is completely unknown. So, if I were to doubt whether, for example, at the full moon of

September 1587 there will be a lunar eclipse, I can be sure from the motions of the eccentric orbs and epicycles that there will be an eclipse, so I would doubt no more. Indeed, from these motions I know at which hour the eclipse will begin and how much of the moon will be obscured. And in the same way, all eclipses, both solar and lunar, can be predicted and also their times and magnitudes; even though they have no particular order among themselves, such that a determinate time interval lies between two successive eclipses, but sometimes in a year two occur, sometimes one, sometimes none.[147]

This example warrants the conclusion that eccentrics and epicycles can be used to make predictions but not that they exist in the heavens.

Clavius regarded Copernicus as a confederate of sorts in this argument. "He [Copernicus] does not reject eccentrics and epicycles as fictitious and repugnant to philosophy. In fact, he supposes the Earth itself to be like an epicycle, and he puts the Moon on an epicycle on an epicycle." In this temporary alliance, Clavius did not share Copernicus's major premise but only the part that furthered his case against the homocentrists: that the planets vary in their distances from Earth.

Copernicus freely acknowledges . . . [that] the planets always have unequal distances from the Earth, as is clear from his teaching that the Earth's position is in the third heaven away from the center of the world. Yet it may only be concluded from his position that Ptolemy's arrangement of the eccentrics and epicycles is not completely certain, since there is an alternative way in which many phenomena can be defended. In this question, we do not try to persuade the reader of anything other than that the planets are not borne always at equal distances from Earth; indeed, either there are eccentric orbs and epicycles in the heavens in the order that Ptolemy established, or surely we ought to posit some cause of these effects that is equivalent to eccentrics and epicycles.[148]

Clavius here left open the possibility that a different explanation for the variation in distances might someday be found. But how much we should make of his ecumenical gesture is not entirely obvious, as he failed to discuss Copernicus's alternatives to the Peurbachian models and entirely neglected Rheticus.[149]

Opposing Fracastorian homocentrics was as far as Clavius would bring himself to agree with

Copernicus and as far as he would distance himself from traditional natural philosophy. He soon turned the other edge of his sword, wielding the objection *ex falso sequitur verum* from which he had earlier defended his own views.[150] The false premise was now the Earth's motion rather than the homocentrics:

If the position of Copernicus involved no falsities or absurdities there would be great doubt as to which of the two opinions—whether the Ptolemaic or Copernican—should better be followed as appropriate for defending phenomena of this kind. But in fact many absurdities and errors are contained in the Copernican position—as that the Earth is not in the center of the firmament and is moved by a threefold motion (which I can hardly understand, because according to the philosophers one simple body ought to have one motion), and moreover that the sun stands at the center of the world and lacks any motion. All of which [assertions] conflict with the common teaching of the philosophers and astronomers and also seem to contradict what the Scriptures teach.[151]

So Copernicus now stood refuted, rejected on the basis of (by now) traditional physical arguments and on the authority of the very philosophers whom Clavius had previously attacked. The argument was something rather less than comprehensive. Clavius made no effort to refer to Copernicus's main contentions in support of his new planetary arrangement. That was not a controversy to which young Jesuit students were to be exposed. The disputation would end with both Copernicus and the Fracastorians rejected yet still credited with possibly having some kind of reliable knowledge about the heavens: "On that account Ptolemy's opinion is to be preferred to Copernicus's invention. From all of which it is clear that it is just as probable that the existence of eccentrics and epicycles ought to be granted as it is probable that eight or ten moving heavens be granted; since like the number of heavens, astronomers discovered the [eccentric and epicyclic] orbs from the phenomena and the motions."[152] This conclusion put Clavius in opposition as much to the far-reaching (although not total) skepticism of Medina and Pico as to the philosophical "adversaries" against whom he argued. Astronomy was not to be thrown out with astrology; it was a credible discipline even if the claims of the astrologers were superstitious.

Ultimately, then, the value of astronomy was to rest on arguments both from utility and from its status as a form of contemplation of God through the visible world.[153] This astronomy had no need of appeal to astrological prognostication. "Not unjustly, Ptolemy says at the beginning of the *Almagest,* which later Arab tradition preserved, that this one science is the right way and path to knowledge of the highest God. [The letter of] Saint Paul to the Romans 1 does not depart from this judgment where he says 'God's invisibles are perceived by the intellect through those created things made by him.'" A kind of aesthetic justification was emerging that Clavius explicitly called "natural theology" and which modestly anticipated Kepler's more fully developed arguments from divine design.[154] Clavius glossed Paul as referring especially to the celestial bodies, the most beautiful and noblest of created things. He further amplified his point with two lines from Psalms, the first found already in Sacrobosco and cited by most authors thereafter: "The heavens declare the glory of God, and the firmament announces the work of his hand." The second quotation, less frequently used, later served Riccioli in the famous frontispiece to his *Almagestum Novum,* a work whose title was clearly indebted to Clavius: "I shall behold your heavens and the works of your fingers, the moon and the stars, which you have founded."[155]

An important although unexceptional practice underlay Clavius's attempt to wed astronomy to scripture: he presumed that the astronomer's method worked independently but never at odds with the meaning of holy writ. This view was much in line with the position found in traditional and innovatory manuals of astronomy and natural philosophy, a position echoed as much by Tolosani in 1546 as by Riccioli in 1651. Clavius had a reply to the skeptics' assertions that the astronomers' claims concerning the heavens are fallible. Against Medina (as usual, unnamed), he claimed on astronomical grounds that some opinions about the heavens were more probable— better established—than others. Medina's style of argument was typical of many theologians' hermeneutical practices: lining up authorities and seeking to reconcile them by consensus. Clavius's work still left open the question of whether theologians might have more to say about the heavens. Perhaps the answer to Medinian skepticism was to assign a more active role to scripture, using

the consensus of the Church fathers as a secure source of truth about the heavens. At the fourth session at the Council of Trent in April 1546, appeal to patristic consensus was officially sanctioned as one criterion for interpreting uncertain words and passages of Scripture. The larger issue was defense of the Church's interpretive authority. The Church as a whole, not the individual, was the final arbiter of the meaning of holy scripture.[156]

ASTRONOMY IN A HEXAMERAL GENRE
ROBERT BELLARMINE

In a series of lectures delivered at the Catholic college at Louvain, Robert Bellarmine (1542–1621) introduced patristic consensus as a criterion for interpreting scriptural passages pertaining to the heavens. His conclusions were original and unprecedented. In 1616, Bellarmine's scripturalism would undergird the position that he adopted on the status of the Copernican question in his engagement with Galileo and the Carmelite P. A. Foscarini. Bellarmine's extraordinary lectures were delivered between October 1570 and Easter 1572, that is, in the period shortly after the first edition of Clavius's *Sphere* and just prior to the dramatic appearance of a nova in November 1572.[157] Although never published, Bellarmine's lectures had a long-term underground reputation among the Jesuits and their friends, and it is a testament to the lectures' importance that Clavius took pains to develop a battery of counterarguments in the *Sphere* (although he never named Bellarmine).[158] Recent commentators have not missed the unorthodox character of many of the ideas contained in the Louvain lectures—the vision of the fluid, fiery, mutable heavens, the rising and setting phenomenon explained as a simple, westward rotation of the Sun, Moon, and other planets, the spiral planetary trajectories in which the celestial bodies move themselves "like the birds of the air and the fish of the water."[159] Because the author of these sentiments was later a central player in the Galileo affair, the historical interest of these ideas is indisputable.

The overall character of Bellarmine's discussion, however, is not so easily captured. Even some thirty years after Copernicus's *De Revolutionibus,* the term *cosmology* was still not in general use, and Bellarmine himself was no exception to this practice. The Louvain lectures fell into the hexameral genre; they comprised a series of discrete comments concerning the supposed creation of the world in six days. Thus its authority already derived from the kind of text on which it was a commentary: the last ten questions of Thomas Aquinas's "Treatise on the Work of the Six Days." Among these questions were "Whether heaven by its nature is corruptible" and "Whether heaven may in fact be corrupted."

Although ostensibly about the heavens, Bellarmine's hexameral topics involved different interpretive practices from those of Sacrobosco's *Sphere*. Genesis commentators might use all sorts of different materials to assist in making sense of their text's moral and physical language; but unlike the Sacrobosco commentators, they had no need of astronomy or geometry as interpretive tools.[160] In the sixteenth century, Genesis commentators increasingly applied to their text the tools of natural philosophy. Further, as Edward Grant has shown, medieval *quaestiones* and *sententiae* were typically composed as a series of discretely ordered propositions. Such compartmentalization of contents, which encouraged treating topics separately, would help to explain why and how Bellarmine took up certain topics and not others.[161]

Attention to the genre in which Bellarmine developed his ideas immediately calls attention to other striking differences from Clavius's *Sphere*. Because Bellarmine discussed astronomical issues within the framework of Aquinas's theology and natural philosophy, he did not follow the *theorica-practica* classificatory conventions of the science of the stars. Consequently, neither the order of the planets nor the matter of astral influence arose as natural topics in the main text—still another indication that Bellarmine did not understand his discussion as a cosmology, a spherics, or a theorics of the planets. Instead, because the lectures were about what was supposed to have happened during the Creation, they made a series of existence claims concerning the constitution of the heavens: that because the heavens are made of fire, not only are they subject to change, but it is also "most certain" that in the future, profound changes will occur in that region. His argument was thus cast in the terms of natural philosophy, clearly constructed in opposition to Aristotle's belief in the world's eternity.

There is another possible motive as well: the widespread apocalyptic preoccupation during this period with the decline of nature, the expectation that the world would soon end. Bellarmine appeared to endorse this familiar sensibility in citing 2 Peter 3:10–13:

> The Day of the Lord will come like a thief, and then with a roar the sky will vanish, the elements will catch fire and fall apart, the earth and all that it contains will be burnt up. Since everything is coming to an end like this, you should be living holy and saintly lives while you wait and long for the Day of God to come, when the heaven will dissolve in flames and the elements will melt in heat. What we are waiting for is what he promised: the new heavens and new earth, the place where righteousness will be at home.

However, unlike Lutheran prognosticators of this period (shortly to be discussed), Bellarmine did not prophesy about precisely when the "new heavens and new earth" would arrive. Rather, his motive was to mobilize scripture in rebuttal of Aristotle's view of the world's eternity: "The followers of Aristotle interpret texts such as 'the heavens will pass on' in the sense that the heavens will stop and no longer move; but certainly 'they will perish' does not have this meaning."[162] Throughout, Bellarmine's interpretive style and method was analytic and scholastic; it drew on the writings of the Church fathers, making subtle distinctions rather than gnostically unveiling hidden truths: "It is uncertain whether the dissolution of the heaven is to be taken in a substantial or accidental sense. Saint Gregory in Book 18 of his *Moralia*, chapter 5 says: . . . 'there will be a new heaven and a new earth' not in the sense that others will be formed but in the sense that the actual ones will change their aspect."[163] Thus, on the eve of the appearance of the nova in Cassiopeia, Bellarmine was cautious in interpreting the character of the change that would occur at the end of the world.[164]

Bellarmine's lectures also illustrate diversity and interpretive flexibility within the Jesuit order during the formative period of its educational mission. Clavius became the defender not only of the traditional Ptolemaic planetary arrangement but also of Ptolemaic astronomical practices. Clavius, unlike Bellarmine, held that self-moving bodies would simply not know where to go. Only geometrical forms tell us how the planets move and return with regularity. In this sense, whatever his views about what moves and what is at rest, Clavius as a practitioner was closer to Copernicus and Reinhold. And it is in this sense that Copernicus was widely represented in the sixteenth century as a "second Ptolemy." Clavius's arguments for the necessity of mathematics in the curriculum were thus fully consonant with Francesco Barozzi's revival of Proclus's *Commentary on Euclid* in 1560, as well as the earlier mathematizing approaches of Regiomontanus and Copernicus.

Bellarmine pointed in another direction: if the astrologers and astronomers agreed on the explanations of the phenomena, then, because there is only one truth, scripture must be brought into agreement with those explanations; but where heavenly observers disagreed, then, said Bellarmine, "it is possible for us to select among them the one which best corresponds to Sacred Scriptures."[165] And it was here that the theologian, relying on the consensus of the Church fathers, the literal meaning of scripture, and his own interpretive intuitions, could feel confident in advancing positions independent of astronomical tradition.

PART III

Accommodating Unanticipated, Singular Novelties

8

Planetary Order, Astronomical Reform,
and the Extraordinary Course of Nature

ASTRONOMICAL REFORM AND THE INTERPRETATION OF CELESTIAL SIGNS

Attention to the science of astronomy, already so well sustained in the Wittenberg cultural sphere, received an unexpected boost with the dramatic and unheralded arrival of two apparitions in the skies of the 1570s. One of these was a brilliant entity—represented variously as a meteor, a comet, or a new star—that appeared in 1572 and remained until May 1574; the other—represented almost universally as a "bearded star" or comet—could be seen for just over two months between November 1577 and January 1578. These unforeseen appearances, taken to be evidence of God's extraordinary capacity to intervene in the orderly course of nature,[1] attracted interest across Europe at a time when German popular *practica* and *prognostica* were filled with the most dire warnings about the anticipated closure of all closures: the arrival of the Antichrist (embodied by the Turks and the Roman Church), desertion of the true faith (again, represented by the Catholic Church), the rise of deviant sects (Calvinists, Anabaptists, Enthusiasts, Antitrinitarians), and the conversion of the Jews. Not surprisingly, for many Lutheran preachers, physicians, and theologians, the new apparitions became occasions not just for astronomical study and astrological judgment but also for gleaning eschatological meanings from key apocalyptic passages in the New Testament, prophetic writings, and the works of the Church fathers.[2] And there was no shortage of biblical pas-

sages to feed the pessimism believed to be written in the heavens. Around 1570, for example, Simon Pauli (1534–91) helped himself to the prophetic material in Luke 21:25–26 for a sermon preached at Rostock: "There shall be signs in the sun, and in the moon, and in the stars; and upon the earth distress of nations, with perplexity; the sea and the waves roaring; Men's hearts failing them for fear and for looking after those things which are coming on the earth: for the powers of heaven shall be shaken." He also foresaw "many comets and blazing lights."[3] Interpretations of the recent heavenly messengers easily fitted this wide frame of meaning, heralding the impermanence of the natural world and the imminent breakup of the creation.

Coincident with this intensification of apocalyptic attitudes and hermeneutical practices, a new sort of social triangulation began to take shape. A small group of celestial practitioners began to forge a new space of possibility. They shifted the space of discussion of these "unpredictables" exclusively to the heavens, writing about the new happenings as heavenly events in their own right rather than as meteorological events below the Moon that were responding to astral influences. These practitioners defined themselves as superior in mathematical skill to those whom they regarded as run-of-the-mill prognosticators; and they confidently began to map out a new set of discursive rights for themselves, asserting claims about the nature of the heavens—the customary and exclusive domain of traditional commentators on *Gen-*

esis and Aristotle's *Physics* and *On the Heavens,* and of nonmathematical, anti-Aristotelian philosophers of a Stoic or Platonic persuasion.⁴ These new voices were diverse, intellectually, socially, and geographically. Some were academic mathematicians; others were practitioners of noble station. The academic position of *mathematicus* might exist within different disciplinary and institutional arrangements. Involvement of nobles and gentlemen as active practitioners of astronomy and astrology, however, was unprecedented and marks an important development in this period. A few examples will illustrate the main patterns of differentiation.

Because there was no higher faculty for the study of the science of the stars, most academic practitioners of mathematical subjects tended to hold higher degrees in medicine, the only subject in the upper university faculties that dealt broadly with the natural world. Nonetheless, this was not the only chair-holding configuration. At Tübingen, for example, the upper faculty was theology, because the higher educational institution in the duchy of Württemberg was, like the lower one, a theological seminary.⁵ Michael Maestlin (1550–1631) is a prime instance: he was strongly identified with Tübingen, the territorial evangelical university where he had studied. Between 1570 and 1576, the period that included the appearance of the new star, he was a mathematical *Repetent,* a teacher who supervised afternoon repetitions of exercises introduced in the morning lectures at the Tübingen *Stift.* When he wrote about the comet, he was a parish pastor in the town of Backnang (1576–80), but by the time a new comet appeared in 1580, he had moved to the University of Heidelberg as *mathematicus,* and he returned to Tübingen in 1583 or 1584.

Whereas Maestlin exemplified the common Tübingen pattern of taking a higher degree in theology, Helisaeus Roeslin, another Tübinger, exemplified the more common configuration that linked skills in understanding the motions of the planets to medicine. Roeslin was a physician who had studied the science of the stars; his teacher at Tübingen was the onetime-Wittenberger Samuel Eisenmenger (Siderocrates) and the predecessor of Maestlin's teacher, Philip Apianus. He became a court physician (to Count Palatine [Pfalzgraf] Georg Johann I of Veldenz-Lützelstein). Cornelius Gemma (1535–79), like Roeslin, was a physician, but one who stayed in the university. In this

regard, he had a special advantage: he was the son of the ephemeridist Gemma Frisius, and this connection may have helped to prepare him to become Ordinary and Royal Professor of Medicine at Louvain. Another filial prognosticator was Elias Camerarius, the son of the prolific Nuremberg astrological humanist Joachim, who became professor of mathematics at Frankfurt an der Oder. Less common was the concatenation of chairs held by Jerónimo Muñoz (1517?–91) at Valencia: Hebrew (1563) and mathematics (1565).⁶

Although rulers had long supported the "bridge" figures who taught mathematical subjects in the university and prepared nativities or annual prognostications for the ruler, the emergence of the noble or exclusively court-based practitioner marks a critical moment in the evolution of the astronomer's role. The social privileges of nobility and court practitioners without pedagogical responsibility opened new spaces of rhetorical, authorial, and material possibility. Because these practitioners' primary commitment was not to pedagogy, their position fostered a new freedom of literary form. Each court had its own character and traditions; no one was expected to write school manuals. Hapsburg imperial patronage in the sixteenth century, especially during the Rudolfine era (1576–1612), was, beside the Danish court of Frederick II, perhaps the outstanding source of support for celestial studies in Europe. In Prague, this was not simply a matter of financial support but what R. J. W. Evans has aptly called the "cosmopolitan freedom" among the intellectual elite and the diversity of philosophical attitudes that populated the Hradčany Palace.⁷ At the same time, it would be a mistake to confuse such noble or court practitioners with the idealized literary characters of the prescriptive manuals of proper conduct, among whom, in any case, the figure of the heavenly practitioner found no place.⁸

The important examples from this period immediately call attention to the singularity of this phenomenon: Thomas Digges (1546–95), the English landed gentleman-prognosticator and sometime student of John Dee; Thaddeus Hagecius ab Hayck (Tadeáš Hájek z Hájku; 1525–1600) and Paul Fabricius (1519/29?–89), physicians to three Hapsburg emperors; Landgrave Wilhelm IV (1532–92), ruler of the lands of Hesse-Kassel, where he had built an observatory at his castle and retained a small group of skilled mechanicians; and Tycho

Brahe (1546–1601), scion of an old and important Danish family, who eventually acquired a small island on which he built a unique castle devoted exclusively to the science of the stars. By the end of the 1570s, the terms *astronomer* and *mathematician* had begun to acquire new meanings from these authors' pens, as the use of parallactic calculation—not in itself an entirely novel practice—gradually made it possible to appropriate the explanatory resources of disciplinary domains hitherto restricted to theologians or natural philosophers. Yet Aristotle's aether and the planetary spheres were not immediately undermined by the appearance of the nova of 1572, because even the most daring parallax wielders believed that God had created it outside the ordinary course of nature.[9]

When God exercised his power in this manner, natural philosophers spoke of the event as extraordinary or "hyperphysical"—beyond the bounds of natural explanation. God's creation of the world, for example, was an example of a hyperphysical act: he created the world from nothing. By contrast, ordinary events in the universe were the result of natural causes, usually "efficient" in Aristotle's sense, like a falling rock causing ripples in a lake, although God could also act within nature through immanent causes, in which case he would be said to do so "physically."[10] Whether making physical or hyperphysical explanations, a small group of parallax-wielding mathematicians were able to display their emerging confidence by voicing claims across the traditional disciplinary boundaries between the science of the stars, natural philosophy, and theology.

To highlight the *historicist* character of this thesis, consider an earlier formulation. In 1957, Thomas Kuhn struggled to accommodate the heavenly novelties of the 1570s within a narrative organized around the gradual ascendancy of Copernicus's theory. Although Kuhn had not studied the primary sources, he appreciated that the new appearances did not immediately confirm Copernicus's planetary ordering. Then, in an effort to save "Copernicanism" as the centerpiece of his analysis, Kuhn thought that he could claim such confirmation by expanding his time frame from a decade to a century.

Somehow, in the century after Copernicus's death, all novelties of astronomical observation and theory, whether or not provided by Copernicans, turned themselves into evidence for the Copernican theory. That theory was proving its fruitfulness. But, at least in the case of comets and novas, the proof is very strange, for the observations of comets and novas have nothing whatsoever to do with the earth's motion. They could have been made and interpreted by a Ptolemaic astronomer just as readily as by a Copernican. . . . But neither can they be quite independent of the *On the Revolutions* or at least of the climate of opinion within which it was created.[11]

Kuhn's "somehow" and "climate of opinion"—open-ended phrasing for which he was famous—point to the narrational and logical difficulties of making "Copernicanism" the driving explanation of his story.

Meanwhile, without further ado, Kuhn put on his philosopher's cap and used the comet of 1577 in *The Structure of Scientific Revolutions* to emphasize a theme of epistemic rupture that had been absent from the earlier work. "Using traditional instruments, some as simple as a piece of thread [referring to Maestlin], late sixteenth-century astronomers repeatedly discovered that comets wandered through the space previously reserved for the immutable planets and stars. The very ease and rapidity with which astronomers saw new things when looking at old objects with old instruments may make us wish to say that, after Copernicus, astronomers lived in a different world. In any case, their research responded as though that were the case."[12] The evidence I consider in this chapter no more sustains this sort of reading than it does the totalizing account of social, logical, and linguistic rupture to which it was supposed to lend support. Two generations after the publication of Copernicus's work—let us say thirty years—there was no "Copernican revolution" in the Kuhnian sense of a radical transformation of incommensurable paradigms or even a pronounced struggle between two opposing "cosmologies." Why not?

Here again, the notion of revolution—in either its earlier or its later Kuhnian formulation—obscures the space of sixteenth-century categories of knowledge and discursive possibilities. In the decade or so after 1572, instead of a widespread Kuhnian-style debate between Copernican and Ptolemaic "cosmologies" in a disciplinary environment of "specialist astronomers" and "nonastronomers," the divination-driven science of the stars continued to organize the central classificatory matrix within which authors of heavenly tracts advanced claims, tested proposals, set up new projects, and argued about the structure of the whole enterprise. And it was within this prognosticatory framework that the emergent group of authors began to put forward proposals to improve the technologies for interpreting heavenly signs and making annual predictions.

At stake for them was the very intelligibility of the divinely authored "book of nature." Did God actually speak directly to men through natural causes and signs, as the Melanchthonians argued, or only through scripture, revelation, and ancient prophecy, as Luther and his more orthodox followers maintained? And, if he spoke through nature, then did he speak in the same way through nature's regularities as through contingent, miraculous irregularities? And finally, which professional groups were best authorized to interpret the meaning of these natural signs: theologians, natural philosophers, or mathematicians? Who could be sure that the mathematicians were reliable in their judgments if their instruments, tables, and theories were so uncertain? The republication of both Pico's attack on astrology (in 1557 and 1572) and Bellanti's counterattack (in 1553, 1554, 1578, and 1580) shows that publishers still believed there to be a willing readership for these issues; it suggested that the debate about the epistemological status of the divinatory part of the heavenly sciences had not been resolved. And within just this emergent discursive space, Maestlin and Digges introduced the Copernican proposal for a new planetary order—but, as chapters 9 and 10 show, it would be no more than one among several resources for accommodating heavenly novelties.

THE NEW PICONIANS

Inspired by the late-fifteenth-century controversies over divinatory astrology, a new round of critiques began to appear after midcentury. One may read these later productions as building on and reconstituting the earlier ones while applying them to new ends. Two themes were prominent in these discussions: either genuine foreknowledge was reserved for God alone, or human agency was advanced as a preferable kind of explanation for the events in question. For the prognosticators, politics was always subject to the constrain-

ing effects of astral influences, but there were signs that political theory was beginning to declare its independence from the hegemony of the stars. Machiavelli's theory of the state as a civil body located the constraining influences in the whims of unpredictable Fortuna. As usual, Shakespeare had his finger on the pulse of contemporary debate: "The fault, dear Brutus, is not in our stars, but in ourselves."[13]

An important contributor to this emerging debate about the interpretation of natural signs was Thomas Erastus (1523–83), a Swiss Protestant physician, theologian, and political theorist. Erastus studied at Bologna and Padua in the 1550s, just when the new editions of Pico and Bellanti were appearing and just prior to the imposition of heavy restrictions on the science of the stars laid down by the 1559 Index. When he returned to Germany to take up a bridge position (as court physician to the count of Hennenberg and professor of medicine at Heidelberg [1558–80]), he published a German translation of Savonarola's *Treatise against the Astrologers* (1497) under the title *Astrologia Confutata* (1557).[14] He may have chosen Savonarola's work because it was considerably shorter than Pico's *Disputations* and was directed toward a vernacular audience: not until 1581 was Savonarola's treatise translated into Latin by Tommaso Buoninsegni, a Dominican professor of theology at the University of Florence.[15] Erastus followed his own translation with an extensive polemic in 1569 against the Wittenberg prognosticator and municipal physician Christopher Stathmion in *A Defense of the Book of Jerome Savonarola concerning Divinatory Astrology against Christopher Stathmion, a Physician of Coburg.*[16] Stathmion, it will be recalled, was the same prognosticator to whom Joachim Heller had made a gift of *De Revolutionibus*. Because Erastus's work was organized as a defense of Savonarola, and because Savonarola had built his arguments on the scaffolding of Pico's critique, the work includes many Piconian themes; but, in general, this treatise—and its sequel—tended to privilege Savonarola's emphases. Erastus continued his attack on astrology with an assault on Paracelsian medical and astrological doctrines.[17]

Yet Erastus was not the only neo-Piconian in this period. In 1560, John Dee's London publisher Henry Sutton issued William Fulke's indignant, but diffuse and ineffective, *Antiprognosticon*.[18] In

1583, Fulke (Fulco) was easily outmatched by Sicke van Hemminga of Frisia (Sixtus ab Hemminga) in his *Astrology Refuted by Reason and Experience* (Antwerp). Van Hemminga had known Gemma Frisius, although he was too young to have been part of the Louvain circle of the 1540s. He had studied at the University of Groningen and seems also to have taught medicine and astrology at Cologne. But by the end of his life, he had undergone a radical reversal in his views. Thoroughly acquainted with the Piconian debates, he now believed that there was need for a new sort of approach: "Previously, Giovanni Pico Mirandola wrote quite learnedly about foreknowledge against the astrologers. Lucio Bellanti of Siena replied to him; and then Cornelius Scepper, a man of much erudition, and a great many others, wrote for and against them.[19] But all these arguments, opposing arguments, disputations against other disputations, and words against words brought about a situation where no one that I know descended to practice and experience and grabbed hold of the rod for the purpose of refuting the particular predictions of the nativities."[20]

Van Hemminga himself did not hesitate to take hold of this schoolmasterly image of the rod, aiming his strokes of worldly experience explicitly at three of the most prolific astrological prognosticators of the century: Cyprian Leovitius, Girolamo Cardano, and Luca Gaurico.[21] Like these midcentury horoscope compilers, Van Hemminga amassed a group of genitures, half of which were for prominent rulers. He examined some thirty of these in unusually sharp and unflattering detail, reviewing the declinations, right and oblique ascensions, oppositions, and whatnot. In each case, he showed that actual outcomes produced disturbing incongruities with specific forecasts and that human volition could explain these outcomes better than the arrangement of the planets. For example, Cardano had predicted a long and healthy life for King Edward VI of England, but unfortunately the king had died soon after the sanguine forecast. And what of King Henry VIII's serial marriages? After examining the fates of Henry's wives, he noted that the king's nativity forecast sterility, impotence, and indifference to women. "Tell us, astrologer, why did he repudiate his wife? Why did he want to have illegitimate children?" he asks. "Was it because of the position of Venus and the Moon? Not at all; it was because he wished

it."[22] And then pushing his point into the most vulnerable territory: "Why did he [King Henry] change the religion [of his realm] after the Roman pope shut him out of his kingdom? Was it, as Cardano wishes, on account of the position of Venus? Not at all; he wanted to do it rather than suffer the indignity brought about by the pope's negative approbation. But still, why did he wish this? Because he thought that he could thus conduct his affairs."[23]

Van Hemminga also devoted a full section to the infamous case of Alessandro de'Medici (see chapter 6 above). He concluded that the young man had in fact survived the most dangerous years predicted by his nativity. For example, for his eighth, forty-second, and sixty-fourth years, the nativity predicted drowning, suffocation, strangulation, ruin, and poisoning. But the year in which he was actually killed was not among those or any other fatal years. "Thus, the uncertainty of the teachings of this profession as much as the useless labors of its professors are shown, for they try and also promise to predict the future, but soon the outcome of things proves with the greatest certainty the falsehood [of their predictions]."[24] Finally, Van Hemminga devoted nearly 20 percent of the book to his own nativity.[25] He compared configurations found in his scheme to those of famous people who shared the same configurations in their birth schemes but whose destinies were entirely different from his own.

In 1586, the same year in which the pope issued his bull against the astrologers, the prolific Tübingen poet and historian Nicodemus Frischlin (1547–90) appended an untechnical but *Solid Refutation of Astrological Divination Repeated from the Best Modern and Ancient Authors, Whose Names You Will Find after the Preface* to a treatise that lauded astronomy (*Five Books on the Astronomical Art in Agreement with the Teachings of Celestial and Natural Philosophy and Collected from the Best Greek and Latin Writers, Theologians, Doctors, Mathematicians, Philosophers and Poets*).[26] Frischlin praised selected Wittenberg mathematicians of the 1540–50 era (Schöner, Milich, Peucer, Winsheim) but then castigated Melanchthon and his circle for defending astrology.[27] As an instance of the excesses of astrologers, he cited Cardano's alleged sacrilege in drawing up a horoscope of Christ. Not surprisingly, Frischlin drew liberally on Piconian and neo-Piconian resources.

MISTRUSTING NUMBERS

Besides this revitalized tradition of astrological skepticism, a new doubt about the modernizers emerged: Were the *Prutenic Tables* more reliable than the *Alfonsines* in predicting astrologically significant conjunctions and eclipses? Cyprian Leovitius, whose work on nativities Van Hemminga attacked, had based his *New Ephemerides for 1556–1606* (Augsburg, 1557) on the *Alfonsines*.[28] However, when Tycho Brahe passed through Lauingen (Bavaria) in 1568 on his way to Augsburg, he became personally acquainted with Leovitius and questioned him about his observational practices—indeed, he asked why Leovitius's Alfonsine-based numbers never agreed with actual observations. Leovitius, who was for a time *mathematicus* to the famous Fugger banking family (and later to Ott-Heinrich of the Palatinate), replied that he did not have any suitable instruments but that the solar eclipses he had witnessed "by means of the Fuggers' clocks" were more in accord with the *Prutenics,* whereas lunar eclipses were in greater agreement with the *Alfonsines.* Similarly, the positions of the three superior planets were in greater accord with the *Prutenics,* whereas those of the two inner planets agreed better with the *Alfonsines.*[29]

These matters were of considerable moment for his prophetic interests. Leovitius had recently amassed a large quantity of astro-historical information in which he reviewed the (largely negative) effects of past conjunctions, eclipses, and comets on the fates of important historical figures going back to the time of Christ and Julius Caesar. Indeed, any famous ruler who had lost a battle, fallen from power, or died was fair game for Leovitius's calculus, as the lingering effects of recent (or not so recent) eclipses and conjunctions could be (and were) easily associated with the demise of anyone, including recent figures ranging from Luther and Melanchthon to Pope Paul IV and Emperor Charles V.[30]

With the conclusion of his lugubrious survey in 1563, Leovitius was confirmed in his detection of a larger biblical narrative in the heavens: the fulfillment of the prophecies of Daniel and the Fourth Monarchy. This bit of hermeneutical work emboldened him to add a separate forecast of Lichtenbergian proportions.[31] Leovitius's forecast virtually left the ground—almost as if he

were conducting the celestial choir himself in its own finale. Among various heavenly events, he predicted for May 1583 a major conjunction of the superior planets in Pisces; for the end of March and the beginning of April 1584, an even more powerful gathering of all the planets in Aries; a little later, a solar eclipse in 20° Taurus. Were these arrangements signs of the fulfillment of the Elijah prophecy? Evidently this must be the prophesied moment, because an earlier summit conference of the planets at the time of Charlemagne, when the world was closer to five thousand years old, had not eventuated in the end of the world. Moreover, the next conjunction of similar magnitude, in eighty years, would exceed Elijah's six-thousand-year prophecy by some forty years.[32]

The authority of Leovitius's drastic forecasts easily impressed apocalyptic writers who used the science of the stars even if they could claim little expertise in its various parts. Two works appeared in England that relied heavily on Leovitius while also invoking the Elijah prophecy and the anti-Piconian literature, from Bellanti and Melanchthon to Giuntini: the Dutchman Sheltco à Geveren's *Of the End of this World, and second comming of Christ, a comfortable and most necessarie discourse, for these miserable and daungerous daies* (London, 1577) and Richard Harvey's *An Astrological Discourse upon the great and notable Coniunction of the two superiour Planets, Saturne and Iupiter, which shall happen the 28. day of April 1583* (London, 1583).[33]

Unlike these derivative apocalypticians, Tycho Brahe was not so easily swayed: his earlier encounter with Leovitius was one among several experiences that convinced him of the need for a radical change in the quality of celestial observations and the establishment of a new, rigorous standard of precision. Long after this meeting, he remarked of Leovitius that "if he had only cultivated astronomical calculation, in which he was altogether quite capable and exact and if he had either left astrological judgments alone or if he had at least touched upon them with more moderation and circumspection, perhaps this would have been more worthy of honor not only for him but for the whole art, and he would not have been exposed, here and there, to slanders."[34] Brahe's critical assessment of Leovitius mirrored his retrospective account of his own life. In 1598, he described his decision to turn to astronomy at age sixteen as tied to his recognition of the uncertainty of both the new and the old tables as well as the (new) ephemerides.

> Soon my attention was drawn toward the motions of the planets. But when I noted their positions among the fixed stars with the help of lines drawn between them, I noticed already at that time, using only the small celestial globe, that their positions in the sky agreed neither with the Alfonsine nor with the Copernican tables, although the agreement with the latter was better than with the former. After that I therefore noticed their positions with ever increasing attention, and I frequently made comparisons with the numbers in the *Prutenic Tables* (for I had made myself acquainted with these also without any help). I no longer trusted the ephemerides, because I had realized that the ephemerides of Stadius, at that time the only ones that were founded on these numbers [i.e., of Reinhold], were in many respects inaccurate and erroneous.[35]

The situation described by Tycho occurred in 1562–63, several years before his meeting with Leovitius, while he was a university student in Leipzig under the care (and very close observation) of his tutor, Anders Vedel, and where he had close contact with the Wittenberg-trained *mathematici* Johannes Hommel (Homelius) and Bartholomew Schulz (Scultetus). Tycho's response to this situation anticipated in some respects Descartes's parable of trusting his personal experience more than the books of the schools.[36] Tycho, in this instance, represented himself as trusting his own capabilities more than those of either the ancient or the modern table makers. But in making his own observations, he needed to violate the prohibitions against manual labor imposed by his family and his social class. Even Baldassare Castiglione's famed *Il cortegiano* included no mention of the celestial arts as worthy of a nobleman.[37] And here the astronomical instrument was the crucial consideration. Tycho's well-known preoccupation with instrumentation had, thus, a twofold meaning. It was at once a practical means for improving the stellar map against which all celestial motions were gauged and a symbolic resource for establishing an exceptional identity: a heavenly practitioner among nobles, a nobleman among (ordinary) heavenly practitioners. Thus, in a narrative typical for its self-praise, he lauded the success of his elementary, youthful observational endeavors against the established compilations of numbers.

> Since, however, I had no instruments at my disposal, my governor having refused to let me get any, I first made use of a rather large pair of compasses as well as I could, placing the vertex close to my eye and directing one of the legs toward the planet to be observed and the other toward some fixed star near it. Sometimes I measured in the same way the mutual distances of two planets and determined (by a simple calculation) the ratio of their angular distance to the whole periphery of the circle. Although this method of observation was not very accurate, yet with its help I made so much progress that it became quite clear to me that both tables suffered from intolerable errors. This was amply apparent from the great conjunction of Saturn and Jupiter in the year 1563, which I mentioned in the beginning [of the whole narrative], and this was precisely the reason why it became my starting point. For the discrepancy was a whole month when comparison was made with the Alfonsine numbers, and even some days, if only a very few, on comparison with those of Copernicus.[38]

To Tycho, the accuracy of the tables was clearly a greater concern than the arrangement of the planets. The numbers were the principal resource underlying the entire practice of forecasting, whether astronomical or astrological. And one wonders why such a concern had not arisen earlier in the century, around the time of the failed forecasts of a massive flood in 1524. One answer is that the sort of precision that Tycho valued had arisen as a problematic within a new space of possibility. The publication of the *Prutenic Tables* in 1551 had created expectations of higher accuracy; it had also opened up a field of comparison with the *Alfonsines* and had quickly spawned a proliferation of ephemerides at different locations: London (J. Field, 1557), Cologne (J. Stadius, 1560, 1570), Venice (J. Carelli, 1557; G. Moleti, 1563; G. A. Magini, 1580), Tübingen (M. Maestlin, 1580).[39] It was in this context that an astronomical "reform"—with its obvious resonance with religious renewal—would become Tycho's governing trope for changing and improving the whole enterprise of the science of the stars in a decade of apocalyptic expectation and the unanticipated appearance of two celestial novelties.

THE RISE OF THE THEORETICAL ASTRONOMER AND THE "SCIENCE" OF THE NEW STAR OF 1572

Quite apart from the work of Copernicus and Rheticus, it was the unexpected appearance of a celestial novelty in 1572 that occasioned a significant shift in what it meant to be an astronomer. Lynn Thorndike was on the mark when he wrote that the "new star of 1572 came as a greater shock than the publication of the Copernican theory in 1543."[40] Unlike *De Revolutionibus*, which was a difficult, learned work—and one whose central claim was not directly accessible to the senses—the unusual event in the skies in November 1572 required no literacy to discern. Earlier academic prognosticators and street prophets had prepared the way for the occurrence of unusual events such as this by creating a space of meaning in which reports of monsters and unforeseen portents *were to be expected*.[41] The Melanchthonians had fueled this environment by linking the appearance of monstrous entities with signs of the impending end of the world. Such signs of the degeneration of nature were taken as optimistic indexes that the sacred narrative was unfolding as it should— even if the prognosticators could not agree on an exact timetable.[42] But this event immediately raised the question of who had the resources to manage its classification and meaning.

To answer this question, recall that a comparable event in A.D. 1006, described recently as "the brightest new star on record," was narrated only in scattered (handwritten) records, such as the personal diary of the Japanese poet and courtier Fujiwara Sadaie, Haly Abenrodan's commentary on the *Tetrabiblos*, and various European monastic chronicles.[43] Print, as shown in chapter 1, began to transform the space of prognostication in the 1470s and was intimately connected with the practice of annual forecasting. A century later, the association between printing and prognostication had, if anything, grown stronger.[44] For example, Joachim Heller in Nuremberg described the comet of 1556 and its future influences in a *practica* issued from his own press; the emperor's *mathematicus* Paul Fabricius issued a description of the same comet at Vienna, in which he publicly regretted that in his *practica* for that year he had predicted that no comets would appear.[45] Now, as several short accounts of a *stella nova* rapidly appeared throughout 1573 and into 1574, print

transformed that entity into a public event such as would not have been conceivable in 1006. And, unlike the recent comet, it was never, to my knowledge, framed in the conventions of a *practica*. Only learned, mathematically skilled prognosticators could provide publishers with accounts of this entity as a "new star," because although the object's novelty was never in question, its status as a *star*, as opposed merely to a *new appearance*, was inextricably bound up with the techniques and skills for measuring its distance, namely determining the small angle of parallax. And those who claimed that they had failed to measure an angle larger than the moon's parallax used their observations to make claims about the star's distance and its physical or theological meanings. Almost imperceptibly, the new star was incorporated into the space of theoretical astronomy.

When Tycho Brahe assembled book 1 of his *Astronomiae Instauratae Progymnasmata* (Exercises Preliminary to the Reform of Astronomy, 1602), a work that he was writing in the early 1580s and intended as part of a trilogy aimed at a full astronomical reform, he used parallax as a criterion for organizing a twofold division of authors' views on the new appearance of 1572. Of course, the very deployment of this criterion profited Tycho's self-representation as *the* reformer of astronomy, as it allowed him to position himself at the center of a mere handful of heavenly practitioners who had proclaimed a null result. By contrast to these null-parallax authors (or Nullists, as I will call them), the members of the (sizeable) remaining group claimed some value for the angle of parallax larger than the Moon's. Among other things, then, the Nullists, as a literary entity, worked to support Tycho's representation of "astronomers" as those practitioners in possession of superior skills and knowledge. Although the Nullists were generally cautious in interpreting the entity as a portent with astrological significance, they were unrestrained in considering it as susceptible to interpretation with the tools of natural philosophy and theology. A nova could be a portent, but only a *stella nova* could become a resource for contesting the Aristotelian proposition of celestial immutability. Such a claim to starhood rested on different calculational resources from those used in the 1524 flood forecast. The astronomical authority of the 1524 deluge derived from predictions of planetary conjunctions based on the 1499 Stöffler-Pflaum ephemerides. It was

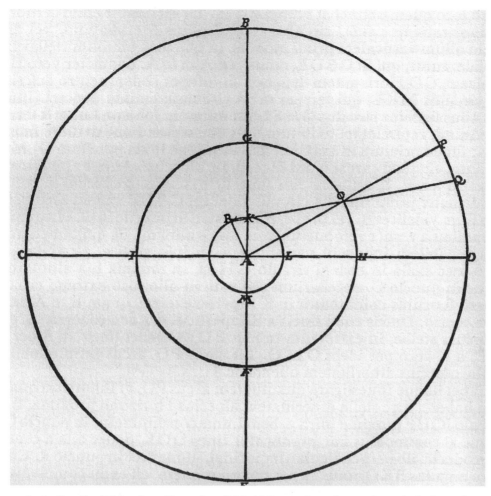

39. Angle of parallax (BOA) as formed by two lines (BOQ, AOP)—one drawn from a point on the Earth's surface (B) through the Moon (O) to a point on the sphere of the fixed stars (Q), the other from the center of Earth (A) through the Moon (O) to a different point on the stellar sphere (P). The greater the parallactic angle, the closer the object is to the observer; the smaller the angle, the farther away the object. Tycho Brahe, *De Stella Nova* (1573), in Brahe 1913–29, 1:26.

an almanac-based prediction, long believed to be an occurrence certain to happen (until it didn't), and its central meaning issued from the anticipation of catastrophe.[46]

The 1572 nova was a different sort of event. It had not been anticipated, of course, and its astronomical credibility rested on techniques for computing *cometary* parallax—ironic, in that those who claimed the nova to be a *stella* used the method to show that it was *not* a comet—or, at least, not a *typical* comet. The *longue durée* of this calculational practice with respect to the planets bears mention. In the third century B.C., Aristarchus had measured the parallactic angle of the Sun and Moon during eclipse, and his values

gained authority and long-term stability when Ptolemy incorporated them into the *Almagest*.[47] The other heavenly bodies were too distant for an eclipse (or a transit) to be visible. However, in 1472, Regiomontanus composed *Sixteen Problems on the Magnitude, Longitude, and True Location of Comets,* a group of general parallactic theorems for determining the positions of comets. The work is less original than was once believed: as Bernard Goldstein and Peter Barker contend, Regiomontanus generalized Ptolemy's parallax method for the moon, and his techniques are the same as those employed in the *Astronomy* of Levi ben Gerson (1288–1344). The significant point here is that the *Sixteen Problems* did not become an active resource for prognosticators until Schöner, Regiomontanus's great promoter, published it in 1531.[48] Thereafter, although the technique was widely used, Regiomontanus's geometry alone did not lead to the presumption (or the conclusion) that comets were superlunary. In fact, his purpose may have been to calculate the dimensions of the region of air and fire in which comets were generally believed to move.[49] Thus, a century after Regiomontanus, the prognosticators needed to establish that a technique with widespread credibility for measuring the distances of sublunary objects (comets) could be applied to an object in a hitherto-restricted region. As a celestial entity, the nova was unlike a comet in that it did not appear to have any proper motion; but, like a comet, it was an unanticipated, transient occurrence that could be made to function as an event presaging future effects.

Two crucial issues thus need to be underlined. First, the prognosticators who located the object within the classificatory grid of heavenly—as opposed to meteorological—objects and meanings established themselves as possessing a special interpretive authority about its future meanings and effects as well as its reasons for being. There were, of course, already some independent natural philosophical tendencies (notably Paracelsian) toward accepting the possibility of generation and corruption in the heavens.[50] The point is that some of those prognosticators who concluded that the object was a new star were also conscious of using their measurements to challenge the authority of the Aristotelian doctrine of celestial immutability. Such criticism did not immediately undermine the entire hierarchical division between celestial and terrestrial stuff, but it did give special weight to the technique of distance determination on which the critics' epistemic authority relied. At the same time, it opened up the question of what sort of credibility might be available to astronomical practitioners who brandished parallactic claims.[51] It will come as no surprise that the Nullists devoted considerable effort to describing and defending their technical methods. Equally interesting, they did not stop with such measurements; a few used their conclusions about the nova to draw new implications for the order of the heavens. This had not occurred with either of the other two major comets that appeared earlier in the sixteenth century (1532 and 1556). Using the generic and disputational resources at their disposal, these writers now began to test conventional boundaries between astronomical, theological, and natural philosophical practice.

Now, contrast the Nullists with the Wittenberg network of prognosticators that had coalesced in the immediate aftermath of the publication of *De Revolutionibus* and which sought to use the formidable technical resources of that text for astrological purposes but with no commitment to Copernicus's representation of planetary order. Rheticus had failed to convince Melanchthon or anyone else at Wittenberg that the long-term motion of the Copernican solar eccentric counted as good evidence in favor of the Elijah six-thousand-year prophecy or the new planetary arrangement. With the exception of his old friend Achilles Pirmin Gasser, Rheticus had also failed to persuade anyone that Copernicus's theory might be accommodated to holy scripture or that it could overcome the well-entrenched objections of the Peripatetic natural philosophers.

Precisely the opposite reaction occurred among the small group of geographically scattered men who convinced themselves that the new apparition of 1572 was a *stella*. These men believed that the nova must be some kind of divine message, that it was an entity of either natural or preternatural origin; most believed the latter. Moreover, whatever else it might have been, the star functioned as a resource in forging an unprecedented authorial network, a kind of Fleckian "thought collective" of heavenly practitioners that successfully transcended traditions of local reference and practice.[52]

This development is well worth describing, even if it is not especially mysterious. At first, its members had no other connection than the fact of having produced an "inquiry" or "observation" or "contemplation" (these were among the rhetorical categories chosen) about what they all took to be the same celestial phenomenon. Printers issued their little reports so quickly that references to writings published less than a year earlier were not unusual, and already there were signs of comparison and elements of critical engagement. Such comparisons were used differently, sometimes by proclaiming agreement, sometimes by accentuating disagreement. In both cases, the intended effect was to bolster the authority of the later author's claims.

The nova, moreover, had other historical meanings that would have a deeper impact on the science of the stars. It was Tycho Brahe who did more than anyone else to construct the new star as a crucial element—a precursor or preparation—to a more general astronomical reform. His massive *Astronomiae Instauratae Progymnasmata* was intended to do just that.[53] In the 1570s and '80s, aided by friends in rich centers of book production, he worked hard to collect treatises written about the new star.[54] The project of knowing by collecting bore some resemblance to the horoscope compilations of Cardano, Gaurico, Giuntini, and Garcaeus. However, the *Progymnasmata* was a far more substantial undertaking—a veritable sourcebook of long excerpts and sometimes full, verbatim transcriptions of writings on the new appearance—dividing its authors into two groups—those who got it right and those who did not. Rhetorically, the first part of the title evoked school exercises (the *gymnasium*), intended perhaps to understate the great program of reform to follow. Logically, however, the book's organization laid the grounds for a proof of the existence of the new star by showing the relative inadequacy of the observations of the sublunarists. Inadvertently, it also constituted a kind of social survey of late-sixteenth-century heavenly practitioners. However, by the time that the *Progymnasmata* appeared in 1602, Tycho had just died, and the publication was handled by his son-in-law Franz Tengnagel and his Prague successor, Johannes Kepler.

When Tycho Brahe first observed a bright object on the evening of November 11, 1572, he was at his uncle's fief holding, Herrevad Abbey, in the company of his own servants. We know this from the report he made public only many years after the event. Although the episode is well known, Tycho's exact language has not been subjected to detailed scrutiny.

> When I had satisfied myself that no star of that kind had ever shone forth before, I was led into such perplexity by the incredibility of the thing that I was not ashamed to admit to doubting the faithfulness of my own eyes; and so, turning to the servants who were accompanying me, I asked them whether they too could see a certain brilliant star when I pointed out the place directly overhead. They immediately replied with one voice that they certainly saw it and that it was extremely bright. But in spite of their affirmation, still being doubtful on account of the novelty of the thing, I enquired of some rustic country people who by chance were traveling past in carts, whether they could see a certain star on

high [*in sublimi*]. Indeed, these people shouted out that they saw that huge star, which had never been noticed so high up. And at length, having confirmed that my vision was not deceiving me, but in fact that an unusual star existed there, and astonished at such a new phenomenon beyond anything that had ever occurred in the heavens, compared with the other stars, I undertook immediately to measure with my instrument.[55]

This description is all the more interesting because it underlines the fact that Tycho turned to people of no special skill or learning and of clearly lower social station to assist in making credible to himself the existence of a new entity just after he had noticed it. The experience was so profound that, many years later, when composing his *Progymnasmata*, he felt that it would enhance the persuasiveness of his account if he used his memory of the event and its witnesses—in a way that he had not done in 1573—to introduce his own account of how he came to regard the entity as a new star.

Tycho's initial strategies for establishing the apparition as a *stella* thus differed from those in the later work just described. In both the earliest report (1573) and the later account (1602), Tycho's aim was the same: to transform his observation into a *learned* claim, one based not merely on rough, unaided observation but on theoretical argumentation.[56] And again, not merely learned, but noble: it was Tycho's alone and bore the stamp of his social station. In the little book that he published in 1573, he employed several methods for persuading his readers. The novel appearance had coincided with his work on a "meteorological diary," but he represented that latter project to be incomplete and unworthy of being read by the learned. Thus he had decided to publish only the first part of it, the part that deals with the star, and then only at the urging of his close friend Johannes Pratensis (Jean de Près), a Frenchman who taught medicine at the university in Copenhagen.

The story of being urged to publish was a trope, but it also possessed some degree of truth. Tycho had initially convinced de Près and the French diplomat Charles de Dançay of the existence of the new star by describing it to his incredulous friends at a dinner and then taking them outside to observe it. The instrument that he used to make his observations was a quadrant that he had made three years earlier for his friend Paul Hainzel, a nobleman and member of the Augsburg municipal council. Yet if Tycho's first audience consisted of noble friends and academics, the subsequent discussion was not about whether a new entity existed but rather where it was located. Was it a *stella* or a sublunary object? And, as it was also without precedent, where did it belong on the map of knowledge?

THE GENERIC LOCATION
OF THE NEW STAR

There was no established genre for describing new stars. There were plenty of works about comets and prodigies, but all such entities were sublunar. What, then, was the meaning of the title that Brahe assigned his small book in 1573—*A Mathematical Contemplation concerning the New Star, Never Seen before This Time but First Observed for the First Time Not Long Ago in November of the Year 1572*? The adjective *mathematical* told readers that this treatment would use methods drawn from the mathematical part of astronomy, as opposed to meteorology. The words *contemplative* and *speculative* were cognate words for *theorical* and thus signaled that this study fell into astronomy's theoretical part. In other words, it would show the geometrical principles needed to understand this object as a "star," a *lux* far enough distant from Earth to be above the Moon and within the eighth sphere: hence, a *stella* that fell within the domain of astronomical rather than meteorological application. The work then followed common prognosticatory practice in explicitly separating the "astrological judgment" from the astronomical prelude, but it is clear that astrological considerations were accorded great significance, as Tycho joined to his treatise two other works of that character.

According to J. L. E. Dreyer, Tycho was already writing the first of these at the time that he observed the star: *A New and Learned Method for Composing a Meteorological Diary.*[57] This work contains important clues about Tycho's early thinking. Meteorology, which treated of "the fertile region of the air," was clearly an important subject, as it dealt with heavenly influences. However, Tycho believed that "today, it is dishonored in many places since it does not escape the notice of unskilled and ordinary men, *the vain and worthless authors of annual prognostications*. We shall discuss the abuse and depravity of these men more extensively on another occasion in a little book which we have called *For Astrology, against the Astrologers*."[58] Notice that Tycho did not criticize astrology per se, but rather its "common and unskilled practitioners" who misunderstood what he called "the Astronomy of the Microcosm." Likewise, in a remark perhaps intended for Leovitius, he was dismayed that observations of eclipses based upon the "usual tables of celestial motions, either the *Alfonsines* or the *Copernian* . . . do not correspond exactly to their calculations." As a result, he announced that at some point he would issue a "*Catalogue of Celestial Observations* for each ten-year period . . . and not only for the motion of the luminaries [Sun and Moon] but also for the remaining moving and fixed stars, especially for Mars and Mercury." Here, in a nutshell, were nascent elements of Tycho's life program: an observational reform immediately associated with a change in the foundations of prognostication. All this was published in the company of his presentation of the new star and a forecast of a lunar eclipse for December 1573 which displayed a detailed calculation of the eclipse using the *Prutenic* and then the *Alfonsine Tables*, followed by an *Astrological Judgment concerning the Effects of This Lunar Eclipse.*

These three works were published together as a continuous volume (Copenhagen: Lawrence Benedict, 1573), and Tycho clearly meant the works to be read and considered together.[59] Like Thaddeus Hagecius, considered below, Tycho constituted the unforeseen object within the established categories of the science of the stars. In the "mathematical contemplation," Tycho sought to establish the credibility of his observational and parallactic claims. Tycho next described the *Meteorology* as "a certain astrological and meteorological diary."[60] That work outlined a philosophy of nature that presented the "world machine" as a "theater," an interconnected harmony of higher and lower entities comprising the heavens and the terrestrial region, with man residing at the center like a mirror of the whole world. This representation had familiar Neoplatonic and Paracelsian overtones, although he carefully eschewed any use of Paracelsian celestial forces to explain the nova's generation.[61]

In the third of these works, Tycho presented himself as a practical astronomer showing how a specified alignment of the Sun, Earth, and Moon

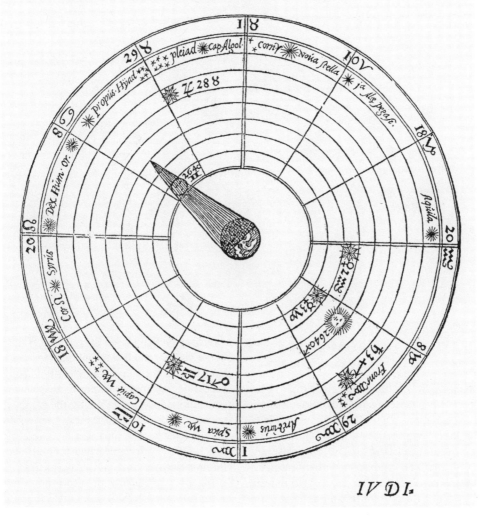

40. New star's position in the zodiac in relation to the planets, according to Tycho Brahe, *De Stella Nova* (1573), in Brahe 1913–29, 1:57.

would yield an eclipse on December 8, 1573, at 8:10 P.M. His first diagram followed conventional practices for representing lunar eclipses. But the illustration at the beginning of the judgment unconventionally located the eclipse within a full, Ptolemaic, unscaled, planetary ordering scheme. For the first time, a single representation sought to connect a predicted event (the eclipse) in relation to an unforeseen event (the new star) within the frame of planetary order. In a companion illustration, Tycho took a further step. If the explanation for the nova's generation was miraculist, nonetheless it had a beginning, like a castle or a city, and thus could be linked to astrological meanings and effects within a theater of astral sympathies. Tycho Brahe's nova thus organized a new problematic—reconciling the extraordinary and the ordinary course of nature.

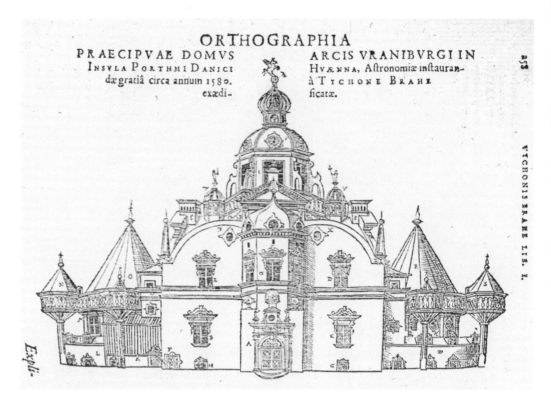

ORTHOGRAPHIA

PRAECIPVAE DOMVS
INSVLA PORTHMI DANICI
dægratiâ circa annum 1580.
exædi-

ARCIS VRANIBVRGI IN
HVÆNNA, Astronomiæ instauran-
à TYCHONE BRAHE
ficatæ.

Expli-

41. Tycho Brahe's Uraniborg castle. From *Epistolarum Astronomicarum Liber Unus* (1596). Image courtesy History of Science Collections, University of Oklahoma Libraries.

COURT SPACES AND NETWORKS
URANIBORG, HAPSBURG VIENNA AND PRAGUE

Tycho Brahe was a hereditary nobleman rather than one who had been recently ennobled. His social position was not unlike that of a traditional feudal lord. King Frederick II of Denmark had granted him a tiny island "to have, enjoy, use, and hold quit and free," said the document in its unmistakably feudal legal language, "without any rent, all the days of his life, and as long as he lives and likes to continue and follow his *studia mathematices*."[62] The position was exceptional. On August 8, 1576, the foundation stone of the main building of Uraniborg was laid by a familiar friend—the French ambassador Dançay, at whose residence Tycho had defended the science of the stars in an oration almost two years earlier. The stone's brief inscription registered that Tycho Brahe, nobleman of Knudstrup, had commenced building "in accordance with the King's will . . . for the contemplation of philosophy, and especially of the stars" and that the stone was laid "as a memorial and a happy omen." Tycho clearly

construed a link between the purpose of this new sort of castle, intended specifically for the "contemplation of the stars," and the act of laying the cornerstone at an astrologically auspicious moment. Immediately after giving the words of the inscription, he described Dançay's arrival on the "destined day": "Accompanied by several noblemen besides by some learned men among our common friends to attend this performance, and on the 8th of August in the morning, when the rising sun together with Jupiter was in the heart of Leo, while the moon was in the western heavens in Aquarius, he laid this stone in the presence of all of us, having first consecrated it with wine of various kinds and praying for good fortune in every respect, in which he was joined by the surrounding friends."[63]

The moment was almost hopelessly rich in astral imagery. The stone was laid in the eastern corner of the castle's foundation, in the direction of Jupiter and the rising Sun, literally establishing a "new foundation" for a more secure (practical) astronomy to underlie the study of stellar influences. The architectural plan for the palace em-

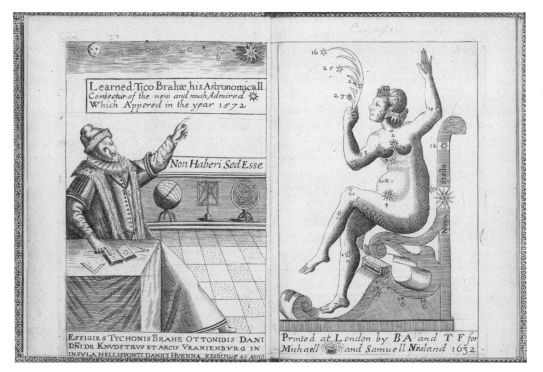

42. Left: Tycho Brahe pointing to the new star of 1572. Seventeenth-century English representation. Tycho's motto: "Non haberi, sed esse" ("Better to be than to appear to be"). Right: Cassiopeia personified with her *stella nova*. Brahe 1632, facing front plates. Courtesy Linda Hall Library of Science, Engineering & Technology.

bodied the Vitruvian principle of *symmetria* in the immediate sense of the proper fitting together of the parts into a harmonious whole, the adjacent towers falling into 2:1 Pythagorean ratios and the central tower serving as an axis of symmetry.[64] Tycho also construed the heavens and Earth according to an architectonic of vertical correspondences. Heavenly influences rained down on the Earth, while animals, vegetables and minerals were "influenced" to grow on and in the earth. But, as Pseudo-Ptolemy's *Centiloquy* had it, the wise man, unlike lower beings, was not dominated by the stars; and, in this case, the wise ruler of the whole system of harmonious correspondences was Tycho himself. Embodying that wisdom were the instruments of heavenly rule, the great measuring devices, lavishly depicted and hand colored in the *Astronomiae Instauratae Mechanica* (1598)—the volume itself carefully heralded by emblems symbolizing the vertical correspondences. "In looking up, I look down," says a heaven-gazing figure leaning on a stellar globe; "In looking down, I look up," says a companion figure who sits astride a mound that half

drapes underground furnaces and (al)chemical apparatus. This image of heavenly knowledge cohered strikingly with the passages just noted in the *Stella Nova* and with passages examined below, in the Copenhagen Oration (1574) and the *Astronomiae Instauratae Mechanica*.

Although Tycho had developed strong connections with celestial practitioners in the German universities, he eventually attained a degree of political, social, and economic independence unmatched among heavenly practitioners of this period. His social position was such that he could confer the economic and status privileges of nobility on the activity of astronomical investigation. Even among noble practitioners, the result was a new role model and new prestige for the science of the stars. As is well known, by the mid-1580s, he had built up a stock of observing instruments capable of observations averaging one minute of arc—unique in its time even compared with the remarkable instruments built for Landgrave Wilhelm.[65] From 1584 onward, he also had his own printing press.[66] This enabled him to exercise a striking degree of control over the presentation of his ideas. He could publish what he liked, when he liked (except when there were shortages at his paper mill), and, more important, for whom he liked. The first publication to appear from his presses under his own name appeared in 1588; it contained his account of the comet of 1577 and a brief description of his new system of the world, and it was not for sale. This was not true of everything that he published during his lifetime. For example, the *Epistolae Astronomicae* (1596) was sold at the Frankfurt Book Fair in the spring of 1597; after his death in 1601, Tycho's heirs immediately moved to publish various works.[67] Nevertheless, whatever practical problems he faced, as a feudal lord Tycho could avoid the free capitalist market and limit the distribution of copies of his books to constituencies whom he thought would praise and understand them. Tycho's self-representations as a nobleman and as a heavenly practitioner were thus firmly bound together through the technology of print. And Tycho's position as lord of a castle devoted entirely to study of the heavens was unique in its own time.

The medical physician or instrument maker was a more common role for court astral practitioners. Thaddeus Hagecius was a physician at the imperial court in Vienna and later in Prague. He appears on first impression to have been another bridge figure, one who moved easily between university and court and was fluent in the discursive practices of both. In this regard, he resembles celestial practitioners at the Italian princely courts—Avogario, Ristori, Danti—as well as his fellow Bohemian Cyprian Leovitius at the court of the Palatinate. But, on closer inspection, he seems to have been more like Tycho Brahe in his close connections with academic mathematicians and in his eventual rejection of all teaching connections. He enjoyed great favor from the emperor and was ennobled in 1595, near the end of his life.[68] His duties included the issue of annual prognostications, although if there were nativities for the emperor, as there must have been, I am unaware of their existence. He was also one of the most learned court figures in the last quarter of the century.[69] His house, like Tycho's Uraniborg palace, was sometimes used for alchemical work. John Dee stayed there for a few months (August 1584–January 1585) and, in his *True and Faithful Relation*, described Hagecius's study as filled with books "and very many *Hieroglyphical* Notes *Philosophical* in Birds, Fishes, Flowers, Fruits, Leaves and Six Vessels, as for the Philosopher's works."[70]

The Maximilian and Rudolfine courts also attracted a coterie of brilliant German and Swiss instrument makers, such as the Wittenberg-trained Johannes Praetorius (who moved to the new academy at Altdorf in 1575), Jost Bürgi (once a prized mechanician for the landgrave), the highly prolific Augsburger Erasmus Habermel (ca. 1538–1606), and Heinrich Stolle.[71] One of the ways in which aristocratic egos were gratified and inter-aristocratic status competition intensified was in the building of a concrete display of "scientific" capital—beautiful observing and timekeeping instruments, and the retaining of practical mechanicians and mathematically trained men who, in sharing their learning with the ruler, gave prestige to him as he, in turn, rendered higher status to their occupation.

Yet, although they shared many common concerns and competences, the court figures who contributed under the emperor's aegis to the *theory* of the heavens were not the emperor's mechanicians but, like Hagecius, his physicians. This is not surprising given that medicine was the most common academic preparation for studying the science of the stars beyond the baccalaureate. And all the sixteenth-century Hapsburg emperors had more than one personal physician, or *Leibarzt*. Even while the imperial Hofburg was still located in Vienna, from the reign of Maximilian I onward the physicians were clearly the group with the greatest interest in theorizing about the heavens. These were the men who possessed the wide-ranging philosophical commitments that have lent historical definition to the "intellectual universe" of the Hapsburg court—variously characterized as Neoplatonic, Hermetic, Paracelsian, magical, occultist, alchemical, and Mannerist. Although there is some value in each of these characterizations, the most remarkable feature of the late-sixteenth- and early-seventeenth-century Hapsburg court seems to me to have been its capacity to contain such a diversity of perspectives and emphases rather than a single, uniform outlook.[72] Equally interesting was the wide range of social and intellectual associations in which the court functioned as a node linked by travel and correspondence—not so much a medical academy as a loosely bound humanist culture, richly cosmopolitan, confessionally tolerant and philosophically variegated.[73]

43. View of Prague and Hradčany Castle in 1591, during the reign of Rudolf II. By Joris Hoefnagel (1542–ca. 1600), Flemish court artist to Rudolf II. Author's collection.

Although there is no simple explanation for this openness to philosophical difference, a prominent factor must have been the emperor's proclivity to accommodate religious heterodoxy— not for altruistic reasons, of course, but because, as R. J. W. Evans claims, Rudolf was as uncomfortable with Protestant sectarians as with the Catholic Church of which he was a member.[74] The Hapsburg court during the Maximilian and (especially) Rudolfine periods thus welcomed Melanchthonian, moderate Calvinist, and even Jewish intellectual currents within an irenic Catholic culture in which the Jesuits were extremely prominent.[75] This relatively relaxed and unusual confessional environment was not so much a reflection as a refraction and accommodation of tensely arrayed sectarian tendencies. It seems to have translated into a general willingness to entertain philosophical differences, if only for a brief historical moment, before the whole enterprise collapsed with Rudolf's death in 1612 and the beginning of the Thirty Years' War in 1620. The court physicians, exemplified by

Hagecius, Paul Fabricius, and Johannes Crato von Krafftheim were not so much doctrinally opposed to academic learning as they were sympathetic to criticisms of Aristotle, open to Paracelsian ideas and willing to look broad-mindedly at new schemes of planetary order, such as the proposals of Copernicus, Brahe, and, later, Nicolaus Reimarus Ursus and Kepler. In this respect, again, there were some important parallels with the situation in Denmark, where there was a considerable receptivity to Paracelsian ideas. Paracelsus and Copernicus exemplified the sorts of new tendencies to which court epistemological diversity was open. Rheticus, for example, who had earlier gravitated to the Albertine court, also maintained epistolary contact with Hagecius and his former student Crato as well as with Emperor Ferdinand himself.[76] In a confessionally rigid world, not all courts were philosophically open, but those intellectuals most expressive of new tendencies knew where to look for the most tolerant arenas. And in their theoretical productions, they too contributed to the reputation of the court.

As the foremost court practitioner of the early Rudolfine period, Hagecius illustrates both conventional and novel tendencies within this milieu. Born into a well-to-do and well-educated Prague family, Hagecius exemplified erudite Prague humanism. He had studied at both Prague and Vienna in the 1540s. He had lectured at Charles University from 1554 to 1557 and maintained connections with the university. It is not known exactly when he established a formal relationship with the imperial court, which was then residing at the Hofburg in Vienna, but it was probably as early as the reign of Ferdinand I (b. 1503; r. 1556–64).[77] He served Maximilian II (b. 1527; r. 1564–76) and then followed Rudolf II (b. 1552; r. 1576–1612) to Prague in 1576, the same year in which Tycho began to construct Uraniborg.[78] He was there during the great building period of the Hradčany Castle, which extended throughout the 1580s, as well as the elaborate decoration of the palace over two decades and the construction of the royal gardens.[79] After completing his studies at Prague and Vienna in the 1550s, Hagecius gave up teaching. His substantial contacts with learned men in courts, universities, and humanistic circles outside Prague and Vienna made him a pivotal facilitator, introducing new ideas into the imperial court through the circulation of people, books, letters, and manuscripts.

As was typical of academic physicians, Hagecius was well versed in the science of the stars. Like Tycho Brahe, he owned and annotated a copy of Carelli's user-friendly *Ephemerides,* one of the first such productions of the post-*Prutenics* era.[80] His diverse writings included a treatise on eclipses (1550), an oration in praise of geometry (1557), a treatise on reading character from facial lines (1562), an astrological fragment from an unknown author (1564), and several prognostications written in Czech (1554, 1557, 1560, 1564, 1565, 1567, 1568, 1570, 1571).[81] Hagecius's *Book of Aphorisms on Metoposcopy* offered readers various resources, some of them astrological, for interpreting personality based on the geometry of furrows in the brow.[82] His interests extended from stars to medicinal plants and their astrological signatures—a reminder that the vegetable and mineral kingdoms also resided under astral influence. When the Bolognese naturalist Pier Andrea Mattioli (b. 1500) became one of the court physicians (from 1554 to 1577), Hagecius prepared a Czech translation of Mattioli's sump-

tuous *Herbarium.*[83] Mattioli, who was a close friend of Ulisse Aldrovandi, helped to establish important intellectual contacts between Bologna and the imperial court.[84] And Hagecius's decision to translate this work into the vernacular exemplifies one sort of intellectual production favored at court.

HAGECIUS'S POLEMIC ON THE NEW STAR

Hagecius's strategy in presenting the nova differed from Tycho's solo production: where Tycho had introduced the nova in the context of his noble friendships, Hagecius bundled his book with a small corpus of previous writings by court and university mathematician-physicians, aiming to show that some phenomena were celestial, others meteorological and sublunary. The new *stella* belonged in the first of these classes. Hagecius's 127-page *Inquiry concerning the Appearance of a New and Formerly Unknown Star* (*Dialexis de Novae et Prius Incognitae Stellae Apparitione*) represented the dominant voice in this collection.[85] The entire book only vaguely resembled the structure of a humanist dialogue, in which different voices engage in mutual conversation and in which no definitive conclusion is reached.[86]

Hagecius's dedicatory apparatus sheds light on the sorts of appeals that such an author believed would enhance his credibility—in this case, a quasi-autobiographical letter to the emperor justifying his mathematical credentials and testimonials from other physicians. Unlike Tycho, for whom mathematical studies had been forbidden fruit, Hagecius recounted a story meant to display his own academic preparation. Before his strictly medical studies pulled him away, he was the only student in the greatly under-enrolled class of Andreas Perlach at the Vienna Archgymnasium in 1549–50. Hagecius helped out, so he says, by teaching mathematics to younger students so that Perlach would have a few more auditors with adequate preparation. Thus it was that Hagecius happened to be so skilled mathematically when the works of Gemma and Muñoz appeared, and he decided to enter the controversy about the existence of a new star.

This letter advertising Hagecius's credentials was followed by two testimonial letters further underscoring his credibility. The first was from his medical colleague Johannes (Hans) Crato, en-

THADDAEI
HAGECII AB HAGEK
DOCTORIS MEDICI,
Aphorifmorum Metopofcopi-
corum libellus vnus.

Editio fecunda.

IN FACIE PRVDENTIS RELVCET
SAPIENTIA: PROVERB. XVII.

FRANCOFVRTI
Apud hæredes Andreæ Wecheli,
MDLXXXIIII.

44. Astrology based on facial lines. Title page, Hagecius 1584. By permission of Strahov Library, Prague.

dorsing Hagecius's skill in parallactic measurement. The second was an excerpt of a letter from Muñoz to Hagecius's Viennese friend Bartholomew Reisacher begging for the appearance of Hagecius's work "on parallaxes." After his own work, Hagecius adjoined works and laudatory letters by "the Emperor's Physician and Mathematician" (Paul Fabricius) and the "Royal Professor of Medicine at Louvain" (Cornelius Gemma).[87] Hagecius augmented this collection with a brief excerpt from a work attributed to Regiomontanus on the comet of 1475 and Johann Vögelin's *On the Meaning of the Comet of 1532*. These writers further marked the local association between university and court at Vienna: Vögelin, like Hage-

cius, had studied with Perlach at Vienna and had used Regiomontanus's parallactic methods in his own study.

A quarter of a century before Francis Bacon's *Advancement of Learning* (1605), therefore, Hagecius represented his inquiry as collective and comparative. As the main author, he could associate his own arguments, methods, and conclusions with the authority and conceptual resources of the mathematical physicians and prognosticators who supported his views (Cornelius Gemma, Muñoz, Fabricius). Because the observations and conclusions of these writers were so close to his own, it seemed to Hagecius that "they came from the same source of truth because we applied ourselves to exploring the same thing, although separated by different times and a great interval of place."[88] This union of conclusions, made at geographically dispersed locations, led him to the idea that he should publish his observations, "joined to the writings of other learned men who did not differ from our cogitations" because "I judged that in building the truth, a great many impulses will carry forward to posterity." The idea was consonant with the coterminous multiplication of genitures. But in Hagecius's formulation lay hints of a more explicit departure from the Aristotelian proof standard of demonstrative necessity: "In the teaching of the truth, agreement is [based on] a probable but not a necessary argument."[89] Tycho Brahe seems to have brought this kind of standard to its greatest consummation in his *Progymnasmata* of 1602.

If agreement allegedly led to probable knowledge, however, then presumably those who denied the existence of a nova and upheld the Aristotelian doctrine of the impossibility of celestial blemishes could also claim some degree of probability.[90] Hagecius conceded none of this. First, he invoked the trope that the master is more open to changing his mind than his followers: Aristotle himself knew nothing about parallax, because the astronomy of his time was very "coarse and rough"; yet had he been alive in our time, he would have changed his mind. Hagecius then cast his argument in deductive form (*syllogimus scientificus*):

Major premise: Any body that has no parallax or a parallax less than the moon belongs to the aetherial and in no manner to the elementary region.

Minor: Our star is discovered to have no parallax. [Evidence for this from Gemma, Muñoz, and Hagecius himself.]

Therefore, it belongs to the aetherial and not the elementary region.[91]

Now, it followed that if the null-parallax syllogism was correct, then, unlike hypothetical arguments for the Earth's motion or the existence of an infinite void, which were merely *possible,* the argument from God's omnipotence would not be sufficient (had he so wished, God could have made a nova). Thus Hagecius emphasized God's *ordained* power. And he made God, rather than the heavens, the efficient cause: God could and did choose to create this prodigy ex nihilo, by his word, just as, at the time of Christ's birth, he cre-

ated a divine meteor.[92] Moreover, although Hagecius himself could well have chosen many authorial targets to represent the rejection of his minor premise, his work ended with an appendix attacking the Veronese physician Annibale Raimondo, who had denied the existence of the nova.

Rhetorically, the attack on Raimondo marked a departure from the otherwise consciously moderate tone of the *Inquiry.* But compared with the controversy in which Hagecius became deeply embroiled two years later—and which continued over the comet—it was the epitome of politeness. Hagecius had received Raimondo's latest irritation in Regensburg on October 9, 1575—where he met Tycho Brahe for the first time—and by January 10, 1576, he had signed off on his indelicate *Response to the Writing of the Virulent and Abusive Anibale Raimondo of Verona, born under Montebaldo, who again endeavors to confirm that the Star which shone . . . in the year 1572 and '73 was not a nova but an old star.*[93]

Raimondo's treatise served Hagecius's evolving purposes. It allowed him to mark off an emergent, antitraditionalist consensus whose members not only agreed in their claims and reasoning but also held the virtues of skill, academic association, and class standing.[94] Besides Gemma and Muñoz, Hagecius included Thomas Digges and Tycho Brahe: "All men highly skilled in mathematical subjects . . . the last two of whom, besides being of singular learning and industry in the mathematical arts, are also famous noblemen."[95] Noble birth enhanced the mathematical worthiness of Digges and Brahe rather than endowed them with that quality. Raimondo functioned as the Other; and, in choice Rabelaisian language worthy of the political and religious polemics of his age, Hagecius liberally portrayed him in various insulting terms: a barking dog, a blind man, an impudent slanderer and blasphemer, an atheist, a sycophant, a sophist, a liar, a holder of foolish and idiotic opinions, an unskilled and crude degrader of astronomy who hides out in his castle, an uncivilized and unpolished counterfeit stargazer—in a word, an "astrologaster." This rhetorical inversion recalls the late-fifteenth-century boundary struggle between the mathematically based astrological prognosticators and the popular prophesiers; it also sets in relief Copernicus's far more tempered appeal to mathematicians as the only group able to understand his book. This would not be the last

45. University of Copenhagen, from Pontoppidan 1760. Courtesy the Research Library, Getty Research Institute, Los Angeles, California.

instance in which claimants to heavenly knowledge sought to promote their disciplinary standing by turning attributes of skill upside down to dramatize a valence of sensory and cognitive deficit, moral and religious disapproval.

AN EMERGENT ROLE
FOR A NOBLE ASTRONOMER
TYCHO BRAHE AND THE COPENHAGEN ORATION

In September 1574, the year after the appearance of his tract on the nova, Tycho delivered a long oration at the French Embassy in Copenhagen on the antiquity and value of the mathematical sciences. This oration was prelude to a series of academic lectures on astronomy. The audience consisted of university students, faculty—including a personal friend, the physician Johannes Pratensis—and the French legate to the Danish court, Charles de Dançay. The possibility that the last two had arranged the event cannot be excluded.[96] The whole performance carried the weight of royal legitimacy, as Tycho made clear in both his title and his introduction, that he spoke at the king's command (*ex Regis Voluntate*).[97] This oration has long been recognized as extremely im-

portant for what it reveals about Tycho's early thinking.[98] The circumstances of its delivery also testify to Tycho's active representation of himself as a new sort of astronomical practitioner, as he addressed the university in the best humanist manner, yet outside its own walls.

The event was typical of Tycho's resistance to conventional social roles. Already at Leipzig, he had pursued his university astronomical studies under the shadow of familial disapproval. A decade later, he again resisted conventional possibilities, this time at castle and court: "I did not want to take possession of any of the castles our benevolent king so graciously offered me. . . . I am displeased with society here, customary forms and the whole rubbish. . . . Among people of my own class . . . I waste much time."[99] And, to the king's new physician in ordinary, Petrus Severinus, he showed his awareness of the uncertainties of court life: "For the court accepts one and all with flattery and benevolence, but sends them away unsatisfied and against their will."[100] It was no accident that, when Tycho did agree to accept a royal benefice on the isle of Hven in 1576, it was one of his own design and suited exactly to his own purposes.

In the Copenhagen Oration, Tycho did not hesitate to declare his independence from what he represented as considerable faculty opposition to astrology, especially by the professors of theology and philosophy. As an orator in the academic style, Tycho made use of common resources of mathematical-humanist propaganda to exhort the university to grant high status to all disciplines that involved mathematics. Thematically at least, Tycho's work belongs in the same genre as other paeans to the value of mathematically predicated disciplines: for example, Regimontanus's Paduan oration, Melanchthon's many prefaces, Peter Ramus's *Scholarum Mathematicarum* (1569), Henry Savile's Proemium Mathematicum at Oxford (ca. 1570), John Dee's preface to Billingsley's *Euclid* (1570), and Clavius's introduction to the *Sphere* commentary (1570). The oration stressed the now-familiar themes of the certitude and utility of mathematics, and these became the grounds for arguing for its preeminence among other disciplines. If philosophy had anything to offer, it was on account of mathematics: "I believe that the ancient philosophers mounted to such heights of erudition because they were taught geometry from early childhood, while most of us waste the best years of adolescence in the study of grammar and languages."[101] Unlike the works just mentioned, however, Tycho's oration placed special emphasis on astronomy and the value of astrology for medicine. If one approached the subject with circumspection, avoided superstition, and protected free will, it was a valuable discipline.[102] This was a moderately conservative astrology. There is also a strong likelihood that Ramus's *Scholarum Mathematicarum* had been the immediate inspiration for the oration because, as Dreyer has shown, Tycho and Ramus had met at Augsburg in 1570. This meeting may also have added to Tycho's sense of the need for a renewal of astronomy, but Tycho did not agree with Ramus that it should be an "astrology without hypotheses," a discipline somehow grounded in numbers alone. Any improvements had to be based on the traditional view of astronomy as a geometrically based science.[103]

The Copenhagen Oration offered an amplified Melanchthonian justification for the study of the heavens that contrasted strikingly with those made by Clavius and Calvin. It made use of the Paracelsian microcosm-macrocosm analogy as the link between the heavens and Earth and was grounded in the Old rather than the New Testament: not the Apostle Paul cited by Clavius (Romans 1:10) but Adam, Seth, and the patriarch Abraham were the sacred figures who, according to Tycho (following Flavius Josephus), demonstrated the earliest knowledge of the stars (*cognitionem astrorum*). However, Tycho declared that the science of the stars (*scientia astrorum*) came down through the Greeks (Timocharis, Hipparchus, Ptolemy), the Arabs (al-Battani), the Latins (Alfonso, king of Aragon), and, "in our age," Nicolaus Copernicus, "another Ptolemy." Everything we now know came from Ptolemy and Copernicus.[104]

Key elements of Tycho's views on Copernicus were displayed in the oration, but not because of any strong academic opposition to the heliocentric theory. Importantly, Tycho did not represent Copernicus as the reviver of an ancient doctrine but as a "modern." He was an authority to be spoken of in the same breath with Ptolemy; he was to be seen as a Ptolemaian who was critical of the ancient master. This view was consonant with the images of Copernicus presented by Rheticus, and by Paul Fabricius in the triumphal entry arch that he designed for Rudolf II's entry into Wrocław in 1577.[105] According to Tycho, Copernicus judged that Ptolemy's observations and hypotheses "sinned against the mathematical axioms" and did not agree with the Alfonsine calculations. And so, by the remarkable skill of his genius, he invented new hypotheses; he constructed some elements contrary to physical principles. Here Tycho explicitly represented Copernicus's hypotheses not in the school-manual style, as a set of concentric circles, but as deriving from two founts of theoretical principles, mathematical and physical.[106] He followed the Wittenberg interpretation, praising Copernicus for the principles of his planetary theory:

In our time, Nicolaus Copernicus, who has justly been called a second Ptolemy, from his own observations found out something was missing in Ptolemy. He judged that the hypotheses established by Ptolemy admitted something unsuitable and offensive to mathematical axioms; nor did he find the Alfonsine calculations in agreement with the heavenly motions. He therefore arranged his own hypotheses in another manner, by the admirable subtlety of his erudition, and thus restored the science of the celestial motions and considered the course of the heavenly bodies more accurately than anyone else before him. For although he holds certain [theses] contrary to physical principles, for example, that the Sun rests at the center of the universe, that the Earth, the elements associated with it, and the Moon, move around the Sun with a threefold motion, and that the eighth sphere remains unmoved, he does not, for all that, admit anything absurd as far as mathematical axioms are concerned. If we inspect the Ptolemaic hypotheses in this regard, however, we notice many such absurdities. For it is absurd that they should dispose the motions of the heavenly bodies on their epicycles and eccentrics in an irregular manner with respect to the centers of these very circles and that, by means of an irregularity, they should save unfittingly the regular motions of the heavenly bodies. Everything, therefore, which we today consider to be evident and well-known concerning the revolutions of the stars has been established and taught by these two masters, Ptolemy and Copernicus.[107]

At the end of the oration, Tycho hinted at a new possibility, as yet only vaguely described, "according to Copernicus's opinion and numbers but referring everything to a stable Earth, rather than as he [Copernicus] supposed, by three motions [of the Earth]." Further, were it not for the fact that he could not stay long ("I was inclined to make a certain trip to Germany"), he intended to show how this sort of analysis for the Sun and Moon could be applied to the remaining planets. Such an exposition, he promised, would surpass "the hypotheses set down in vain in a certain book published recently by Peucer and Dasypodius. For these men applied the calculations of Copernicus unsuitably to the Ptolemaic and Alfonsine hypotheses."[108] These passages show that Tycho was already quite familiar with *De Revolutionibus* at this early date and that he was clearly cognizant of the published Reinhold-Peucer gloss on that work through familiarity with the *Hypotheses Astronomicae* (1571); but his knowledge of the Wittenberg interpretation probably derived only from conversations with Peucer in 1566 rather than from direct knowledge of the trail of annotated copies circulating in Wittenberg from the early 1550s.

Although the Copernican thematic helps to locate Tycho's early thinking about planetary theory, it is something of a subsidiary theme in the Copenhagen Oration. Indeed, much of the remainder of that work is taken up, as would befit such an oration, with a defense of the utility of astronomy based upon its value for astrology. Tycho's defense shows the strikingly divergent paths on which justifications of astronomy were moving in the 1570s: Melanchthonian, Jesuit, Maestlinian, Diggesian. Tycho made only a cursory bow to astronomy's utility for timekeeping before he announced that among its greatest utilities was its link to a "separate doctrine" which "they" call astrology and whose governing, self-evident premise was that "there is no doubt that this lower world is ruled and impregnated by the higher."[109] Astrology studied sidereal effects and influences in the lower world (which are "more occult and external to the senses") and made judgments based upon them. Without naming Ptolemy, Tycho then advanced the traditional Ptolemaic distinction: "many" assign to astrology a "conjectural rather than a demonstrative" sort of understanding. But Tycho immediately took a Melanchthonian view of astrology: he insisted that astrology's conjectural character derived from the physical rather than the mathematical parts of astrology.[110] Astrology, then, was like medicine: it was part of physics but gained whatever reliability it had from its dependence on the mathematical part of astronomy.

TYCHO AND PICO, GENERIC AND NAMED ADVERSARIES

The longest and most substantive part of Tycho's oration was a defense of astrology against its "adversaries." Copernicus, Clavius, Bellarmine, and Offusius all followed the stylistic practice of generically naming opponents, and Tycho fully participated in this practice. Typically, he designated astrology's opponents as philosophers or theologians but never as mathematicians or astronomers. Named opponents do not appear until the end of Tycho's oration, and when they do, they help to unveil the meaning of the intellectual authorities underlying the positions against whom he was explicitly arguing. The immediate local adversary was Niels Hemmingsen (1513–1600), senior theologian at the University of Copenhagen, although the legate Dançay shared the view that evangelical teachings contradicted astrological predictions, especially nativities.[111] As John Christianson has shown, Hemmingsen was a leading spokesman in Denmark for Melanchthonian theology (or "Danish Philippism"), a conviction that Tycho shared; but on astrological matters, Hemmingsen did not follow Melanch-thon.[112] Instead, his arguments against astrology seem to have come for the most part from "Calvin's little book against the astrologers."[113] But, as Tycho noted, although Calvin was otherwise a clear and clever author, he was simply ignorant on this subject.[114] And Calvin, like Erastus (also noted) and all sixteenth-century opponents of astrology, drew their arguments mostly from Pico.[115]

It is likely that Tycho had studied Pico's *Disputations* by the early 1570s, but just how carefully is hard to document. There is, however, no doubt that he had read Lucio Bellanti's book, because he referred to it in the Copenhagen Oration as that "learned volume against his [Pico's] objections." The copy of Bellanti's work—still extant amid the remnants of Tycho's library in the Prague Klementinum—confirms the Dane's acquaintance with the fin-de-siècle controversy concerning the science of the stars. Moreover, following Bellanti's rhetorical techniques, Tycho both praised and deplored "the miraculously learned Pico, count of Mirandola, who from a young age was gifted in the mathematical sciences, as also in the precepts of astrology and its influences on human destiny, as it was then commonly treated, and in which he was not superficially experienced." With this bow to Pico's noble birth and to his extraordinary intellect, Tycho now submitted a large qualification: "Either because of the superstitions and falsehoods of certain pseudoastrologers or on account of the hatred of several contemporaries (as it is thought) who did not approve of its [astrology's] delights, he [Pico] brandished a great volume against astrology, digested into thirteen books, in which he produced the unnecessary, frivolous, silly, and trifling writings of unskilled astrologers rather than shattering the more solid labors of this art—which no one beholden to the truer or more secret astrology would approve."[116]

Tycho's distinction between a "truer" and a "frivolous" astrology immediately recalls the title of the work proposed in the *Meteorological Diary* a year earlier, where he railed against the "vain and futile authors of annual prognostications": "Against the Astrologers, in Favor of Astrology."[117] Tycho admitted that there was an intrinsic weakness in astrology, but crucially, his concern was not so much with astrology per se but with Pico's criticism of the entire science of the stars. "The aforesaid count of Mirandola took the freedom

to call into doubt astrological truth (which may easily become an object of controversy, being physical and conjectural, and also because the variable flow of matter may be altered) *and also astronomical truth*—for example, the change in the maximum obliquity of the ecliptic from ancient times."[118]

The obliquity or slanting of the ecliptic refers to the angle created by the plane in which the Sun appears to move eastward with respect to the stars and the plane of the celestial equator, with respect to whose poles the stars and planets appear to rise and set daily in a westward direction. The two points where these planes intersect are called the equinoxes (vernal and autumnal) and are crucial for defining the calendar, the seasons, and related feast days.[119] Characteristically, the very feature that Tycho singled out from among Pico's criticisms was the accuracy of the values for the obliquity, not the order of the planets. According to Tycho's compressed presentation of Pico's critique of astronomy, the greatest change in the angle of obliquity that Hipparchus and Ptolemy found was one-sixtieth of a degree (1'), whereas "in our time," it was generally agreed that the obliquity was a third of a degree (20') less than in the time of the ancients. However, Tycho did not use this occasion to make a sustained and substantive answer to Pico's charges. In the style of an oration, he explicitly invoked Bellanti's "learned" rebuttal, and he associated himself with the famous and rhetorically flippant story of the forecast of Pico's death:

Certain Italian writers—among whom is Luca Gaurico, bishop of Giffoni, famed in the profession of astrology—say for certain that three famous Italian astrologers [predicted Pico's life span] as the fatal year of thirty-three from the "direction" of Pico's horoscope . . . , which he draws well in the themata of the horoscope. And although the same Pico took pains to refute this prediction and to render it vain, insofar as it was that, it is said that he hid in some cloister at around this time of his life; and nevertheless, in the same year for which the prediction was made, he gave way sufficiently so that he tried to put the certitude of astrology to the test as much with his own body and his own life as he attempted to undermine it with his natural talent and his pen.[120]

So, here again was the delicious sixteenth-century irony of Pico (and his arguments against astrol-

46. Tycho Brahe's copy of Bellanti's *Liber de Astrologica Veritate* (1554), showing embossed initials (top: TBO) and date of his binding (bottom: 1576). Courtesy National Library of the Czech Republic, Klementinum. M 34 (14 A 66).

ogy) dying just at the moment allegedly prognosticated by the "famous astrologers." Yet, significantly, Tycho did not leave his rebuttal at the somewhat rhetorical level of the 1574 oration. Glancing ahead to 1598, we find the reflective Tycho of the *Mechanica* looking back and specifying more exactly the same path of an astrological reform that he saw as the direct sequel to his accomplishments in astronomy.

> Our purpose was to rid astrological studies of mistakes and superstition, and to obtain the best possible agreement with the experiences on which they are based. For I reckon that it will hardly be possible to find in these studies the equal of such exact reasoning as [is to be found] in geometrical and astronomical truth. In my youth I was excessively enslaved to the prognostic part of astronomy that deals with divination and builds on conjectures; but later

on, feeling that the motions of the stars upon which it builds were insufficiently known, I put it aside until I should have remedied this misfortune. At length, after I learned more exactly about the paths of the stars, I again took up [the prognostic part of astrology] and I concluded that this inquiry—although it is considered idle and meaningless not only by ordinary people but also by very many learned men, among whom there are even several *mathematici*—is really more certain than one would readily think; and this is as much true with regard to meteorological influences and [weather] predictions as with regard to nativities. [121]

THE TYCHONIAN PROBLEMATIC, 1574

The Copenhagen Oration marks a significant moment in the formation of Tycho's problematic. The theological and philosophical objections about free will were relatively easy to dismiss: the stars dispose, but do not determine, and one who knows the placement of the stars can avoid their malevolent influences. But the objections from astronomy were of a different order. As early as 1574, Tycho had arrived at a juncture bearing residues from the Piconian problematic of Copernicus's Bologna period. Like Copernicus, he had convinced himself that the best defense of astrology, a conjectural discipline, would be through some kind of a reform of astronomy. And in that regard, he agreed strongly with Copernicus's introduction of the equantless, biepicyclic mechanisms—perhaps even more emphatically than did Peucer, who maintained some lingering affection for the equant. Like Dee and Offusius, he was still unsympathetic to a full Copernican solution to the Piconian objection about the ordering of Mercury and Venus. In fact, even his profound investment in the Vitruvian architectural principle of *symmetria* was not sufficient to persuade him to commit himself to the Copernican astronomical principle of *symmetria*. And it had not yet occurred to him, so far as can be discerned, that Mercury and Venus could be reordered in circumsolar paths without positing the motion of the Earth. Rectification of the faults and imprecisions in the annual prognostications, so he believed, would require a program of new stellar observations along the lines glimpsed in the title mentioned in his *Meteorological Diary: A Catalogue of Heavenly Observations*. As a collective undertaking, the scale of that project was unprecedented, dwarfing even the tedious recalculations involved in Reinhold's *Prutenic Tables*.

A TYCHONIC SOLUTION
TO PICO'S CRITICISM?
NAIBOD'S CIRCUMSOLAR ORDERING
OF MERCURY AND VENUS

Tycho's early biographer, J. L. E. Dreyer, wrote: "The idea of the Tychonic system was so obvious a corollary to the Copernican system that it almost of necessity must have occurred independently to several people."[122] Although this corollary may seem obvious in retrospect, the idea was not quite so evident in the 1570s; nor is the motivation for introducing it any clearer today. Actually, what one would have expected to be "obvious" was a principal component of the theory, well known in the Middle Ages and available in

print since 1499: Martianus Capella's *On the Marriage of Philosophy and Mercury* ("The stars of Mercury and Venus . . . do not go around the Earth at all, but around the Sun in freer motion").[123] It was Copernicus himself who, in *De Revolutionibus*, book 1, chapter 10, made Capella's representation into a more exacting astronomical hypothesis and may well have played some role in moving him toward the new planetary ordering after 1499.[124] Although Tycho could easily have been familiar with the Copernican gloss at the time of his Copenhagen lecture, he did not use the occasion to mention it then, even though, as he says, he lectured to the students "according to the framework and numbers of Copernicus, albeit reducing everything to the resting earth."[125]

However, in 1575, during an extended trip to various cities in the Empire, the bibliophile Tycho acquired some important new textual resources. In 1575, he met Hagecius for the first time at Regensburg, just before the coronation ceremonies for Rudolf, and was presented by the emperor's physician with a gift of the *Commentariolus*, a valuable manuscript that evidently had been in Hagecius's family for many years.[126] The gift set the right tone for a longstanding bond of intellectual friendship and mutual respect that was consummated many years later in Tycho's appointment as imperial astronomer. Tycho now had immediately at hand the earliest statement of Copernicus's ideas, cast in propositional form and without the expansive argumentative structure that it later acquired. Thus, three decades after Copernicus's death, he was in a good position to appreciate the evolution of this beautiful yet not altogether persuasive theory. He also bought more than thirty books for his library (although we have no reason to believe that his library ever approached the size and breadth of Dee's).[127] It was in the course of this book buying that Tycho purchased Valentine Naibod's *Three Books of Primary Instruction concerning the Heavens and Earth and the Daily Revolutions of the World*.[128]

Little is known about Naibod except that he taught at the Catholic universities in Cologne and Erfurt and was murdered in Venice in March 1593. He was part of the new breed of textbook writers who had some familiarity with Copernicus's work and took it for granted as part of the corpus of writings in astronomical theory. This acquaintance may have impelled him to include the chapter titled "Of Various Opinions concern-

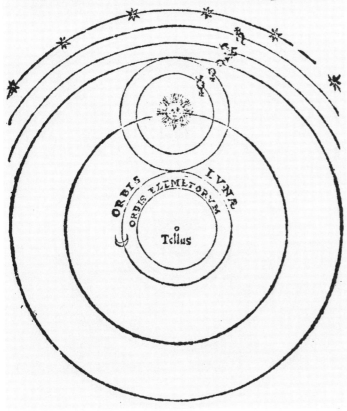

Syftema maximarum vniuerfitatis partium ex fententia Martiani Capellæ.

47. Capellan ordering of Venus and Mercury. Naibod 1573, fol. 41. Courtesy National Library of the Czech Republic, Klementinum. 5.J.2.

ing the Order of the Celestial Orbs," in which he rehearsed the views of familiar ancient authorities (Plato, Cicero, Pliny, Ptolemy) on the old question of whether Venus and Mercury are above or below the Sun.[129] We cannot rule out the possibility that it was Offusius's recently published *Divine Power of the Stars* that had called his attention to this question.

There followed an unusual diagram captioned "The System of the Principal Parts of the Universe according to the Opinion of Martianus Capella." Naibod's representation of Capella's figure employed the widespread convention of concentric circles, and the artist made Venus's orb, at its greatest distance, tangent to Mars's orb at its least distance from the Earth. It is obvious that the illustration's textual locus was *De Revolutionibus* because it was followed immediately by another diagram, captioned "The System of the Universe According to the Opinion of that Great Man Nicolaus Copernicus of Toruń." However, a glance at

this heliocentric representation shows that it was not guided so much by a close analysis of *De Revolutionibus,* in which the spheres were no longer tangent, as by a slight modification of the artistic convention of depicting planetary order using contiguous concentric circles. This led the illustrator into a quarry of scaling errors, as when he represented the Moon's orb as equal in diameter to that of Venus and tangent to that of Mars.

The basic Capellan ordering is significant. In Naibod's version, Venus and Mercury appear to circulate like epicycles around the Sun and within the limits of its total orb, at least as defined by the orbs of the Moon and Mars.[130] For all the arbitrary symmetries of Naibod's diagram, it contained the nugget of a resolution to the Piconian rebuke concerning the ordering of Mercury and Venus—and it did so without moving the Earth. Moreover, it is certain that Naibod's book attracted Tycho Brahe's notice no later than the date stamped on the vellum binding ("TBDO/1576"),

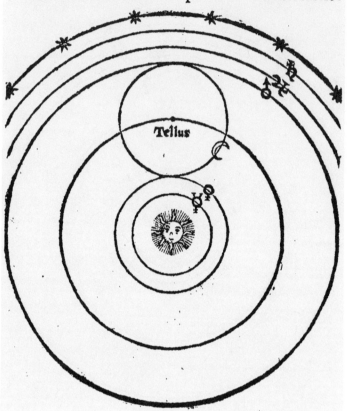

Syſtema vniuerſitatis de ſententia ſummi viri Nicolai Copernici Torinenſis.

48. Copernican ordering of the
planets. Naibod 1573, fol. 41v. Courtesy
of The National Library of the Czech
Republic. Klementinum. 5.J.2.

an identifying mark identical to that of several
other volumes in his collection.[131] Furthermore,
the Capellan diagrams appeared in one of the
few sections with any owner's markings. Exactly
what meaning Tycho ascribed to the new visual
resource is not known, but it is difficult to believe
that the picture could have failed to stimulate
new ideas about circumsolarity in conjunction
with a resting Earth, as it was not yet common
for astronomical manuals to represent comparative orderings of the planets. At the very least,
the illustration riveted attention on Copernicus's
reference to Capella in *De Revolutionibus,* book 1,
chapter 10, at around the time that Tycho would
have been familiarizing himself with that book.
Can it be any coincidence that soon thereafter
he placed the comet of 1577 not merely above the
Moon but in a circumsolar trajectory outside that
of Venus?

THE COMET OF 1577
AND ITS DISCURSIVE SPACE

The comet of 1577, like the nova, was as much a
phenomenon of the culture of print and prognostication as an object of natural investigation.
In 1578 alone, well over one hundred authors described and pronounced judgments on the latest
apparition to flash across the skies of Europe.[132]
To judge by the frequency of explicit, published
cross-referencing among authors, at least by the
Nullists, these writings produced a heightened
sense of communal engagement, mutual comparison, and borrowing that had not previously
existed among heavenly practitioners. Both the
comet and the nova authors constructed themselves in this literature as a class of learned men.
In general, *learned* carried the meaning of someone well versed in the liberal arts, but in this case

UNANTICIPATED, SINGULAR NOVELTIES

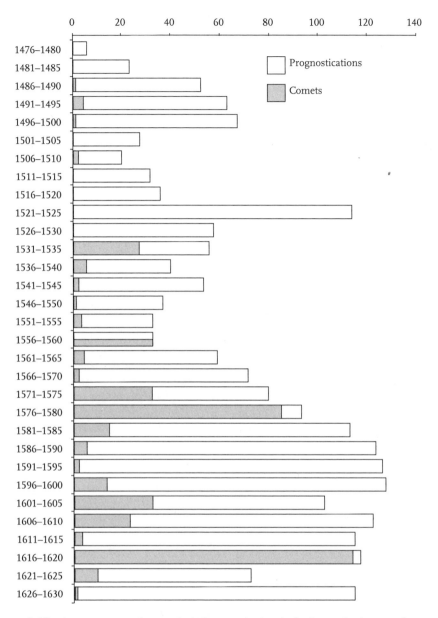

49. Publications on comets and on astrological prognostications in the German lands, 1470–1630. Based on Zinner 1941, 73.

the primary meaning was one especially skilled in mathematical subjects. This notion of a common group of proficient practitioners extended further: it included an implicit right to differ among themselves, as though anyone regarded as possessing this kind of knowledge might even commit errors without loss of honor. For example, Maestlin, who disagreed with both Hagecius and

Andreas Nolthius, described Hagecius as "that most learned man" and Andreas Nolthius as "a learned man and famous mathematician." Hagecius spoke of "many learned and pious men who in our age appear to hold the same opinion."[133]

This respectful rhetoric of social reference produced a discourse of inclusivity within which differing observations, calculations, and hypotheses could be reported and compared. Although such men opposed themselves to those described as "unskilled" (*imperiti*), they also further differentiated themselves using the kind of disciplinary designators that we have already met in this study. Generic terms like *astronomer* or *physicist* were often used in the comet and nova literature to signify generally known authorities. For example, when Maestlin summarized opinion about the material constitution of comets, he wrote: "All physicists believe that it is a mass of hot and dry vapors drawn up from the Earth by the force of the stars, etc. Not a few astronomers are of the same opinion."[134] As the next chapter shows, this process was transformed when, in 1588, Tycho published a new sort of work—a lengthy and systematic summary of his own observations and calculations of the comet, tabulated and compared with those of a large number of other authors.

ASTROLOGICAL AND ESCHATOLOGICAL MEANINGS OF COMETS

As was typical of predictions based on such representations of these entities as transient and fearsomely blazing, the social and political prognostications described in many of the astrological judgments tended to be more alarmist than consoling. To make an astrological judgment based on a comet, the technical problem for the prognosticator was to associate the comet's visible properties (e.g., color and longitude) with the familiar character of corresponding planets and then to link these with local political concerns. The force of such effects could be augmented when combined with an anticipated celestial event derived from the prognosticators' normal repertoire, such as a major eclipse, conjunction, or, in some cases, an eschatological prophecy. But there were some surprising failures to use these apparitions as messengers of the end. The prolific astrologer and table maker Cyprian Leovitius prognosticated that the world would end sometime after 1584, in association with the great conjunction of Saturn and Mars in 1583. The appearance of the 1572 nova, unanticipated though it was, fell nicely into the build-up of dramatic natural events. Leovitius, however, missed a fine opportunity to write a suitably dramatic and rhetorically commensurate work; instead, the 1573 London edition of his *Great Conjunctions* contained only a brief, almost negligible judgment, written as a poem, that proclaimed a comet linked to the star of Bethlehem, which had announced the coming of Christ; by implication, rather than explicit announcement, the new comet was to be associated with the Second Coming.[135]

If sixteenth-century theological opinion was divided on the question of linking astronomy with astrology because it appeared to threaten divine freedom, there was also a split between prognosticators over whether comets bore eschatological significance. Tycho Brahe was quite prepared to attribute astrological meaning to the nova and the comet as "harbingers" of the great earthly changes to be brought about by the anticipated great conjunction of 1583. But among the null-parallax authors, both he and Maestlin stand out as unusually restrained in making eschatological interpretations of the comet's meaning.[136] The astrological connection drawn by Tycho had Leovitian resonances but not Leovitian conclusions. Even Tycho drew a sharp distinction between what it was possible to know from ordinary, predictable concatenations of heavenly events and extraordinary, unforeseen occurrences whose explanation was known only to God: "There are actually no reliable grounds for predicting the end of the world from the heavenly constellations," he wrote,

> for this knowledge does not come from the light of nature and its understanding but from divine prophecy and the predestination of God's will, which is known to no man nor even to the angels of heaven . . . it should be understood that this comet, even though it has also been born unnaturally in the heavens, cannot signify the end of the world, for comets were seen before the birth of Christ, and they have been seen at various times ever since the beginning of the world. . . . Therefore, although the end of the world, according to the signs of Christ and the prophets, cannot be too distant, yet the termination itself cannot reliably be foreseen, either from a natural eclipse of the sun and moon, or from other heavenly constellations, or comets, and I thus assert that the end of the world is known only to God the Almighty and to no creature.[137]

Tycho's significant dissenting position here allowed him to pursue the sort of Melanchthonian astrology that he had defended in his Copenhagen oration while keeping clear of the more extreme apocalyptic astrology engaged in by many of his contemporaries.[138] Astronomy, in this regard, did not need to be "accommodated" to the messages of the prophets.

THE LANGUAGE, SYNTAX, AND CREDIBILITY OF COMETARY OBSERVATION

Commentators agree that this comet, unlike any previous ones, had profound consequences for astronomy, natural philosophy, and planetary order. But these consequences were quite separate from its astrological or apocalyptical uses. Like the nova, the comet became an occasion for a small group of heavenly practitioners to undermine—although not to destroy—the ontological spell that marked the boundary between heavenly and terrestrial stuff. And once again, the claim to its existence as a superlunary entity was bound up with the credibility of the observations and calculations of the individual null-parallax authors. Just five years after the nova, the practitioners who placed the brilliant comet of 1577–78 in the heavens emerged as a strong subset of the same group that had opened the new discursive space for arguing the existence of a new star. The overlap cannot have been accidental: once breached, the once-improbable null parallax had rapidly become a measurement that one might expect to make—although most practitioners did not share that conclusion. Of the original participants in the nova group, a majority (Brahe, Maestlin, Gemma, Postel, Muñoz, and Landgrave Wilhelm) entered the discussion about the comet, joined now by a new and outspoken figure, the Tübingen-educated (1561–69) municipal physician of Hagenau, Helisaeus Roeslin (1544/5–1616).[139] Yet the original bonds of agreement tying the nova group together did not automatically carry over to the new group. Unexpectedly, and for quite different reasons, the active and prolific Hagecius and the Melanchthonian Caspar Peucer dissented, while the fervent Thomas Digges and John Dee published nothing at all. Nor at this early date did Tycho try to use his authority as a nobleman to influence other heavenly practitioners to accept his claims. The same was true of the widely respected court physician Hagecius. How, then, did the superlunarists make credible their descriptions?

For these writers, some of the classificatory problems bearing on the astronomical placement of the comet were the same ones that had been raised by the nova. The comet, like the nova, raised new questions about the reliability of heavenly observation. How was one to evaluate individual personal descriptions? Indeed, what was one to do with the multiplicity of individual, separately made observation reports that rolled off the presses throughout 1578? Here one must be careful to distinguish between the standardized or idealized notion of an "observation report" (the modern philosophers' beloved O) and the unregulated late-sixteenth-century practice of describing an observation or even the seventeenth-century practice of appealing to a second or third witness as a warrant for the credibility of an observation.[140] To make matters worse, without long-term tables of mean motions, there was no obvious way to predict the comet's course. Calculating a path was sheer guesswork. And further, there was now the new problem of explaining where the apparition came from. Was it a natural or a preternatural event? If the latter, could one make standard astrological judgments about its effects?

Astronomically, the biggest differences between the nova and the comet were that the latter exhibited a proper motion with respect to the fixed stars and it had a visible tail (hence the common name *bearded star*). Observers, regardless of whether they ultimately classified the comet as a sub- or superlunary event, constructed their reports on a few daily rather than many long-term observations. As a result, although they relied on the ancients for knowledge of stellar positions, evidence of its motion with respect to the stars depended mainly on their own sightings. Thus, unlike the sort of generalized notion of experience that counted as evidence in natural philosophy—nature's ordinary course—these reports were often quite historically specific as to time and place.[141] This kind of specificity was as true of private as of published accounts, as can be seen in Tycho's description of the comet prepared for King Frederick. In that account, meant only for the king's eyes, Tycho distinguished between the comet's first appearance (November 11, "in the evening soon after sunset"), its "true beginning" ("on 10 November, about an hour past midnight, though it is true that various seafaring people have reported that they saw it on the evening of 9 November in the Baltic Sea, but I cannot vouch for that"), and the moment when he himself first viewed it ("It was first observed by me through my instrument on 13 November, because until then, the heavens were not clear enough for such observations").[142] What Tycho presented as reliable in this account, as elsewhere, did not depend on his own social status or that of the seafarers or the "passing peasants" who had noticed a bright apparition five years earlier. He attributed the reliability in his reports to his own observational practices. In 1588, he would transform the grounds of persuasion by systematically comparing his specific observations with those of other claimants.

PLACE AND ORDER, THE COMET AND THE COSMOS

GEMMA, ROESLIN, MAESTLIN, AND BRAHE

Even the most anti-Aristotelian writers could not let go of the Aristotelian premise of a heterogeneous universe made up of two ontologically different regions.[143] Although the nature of each realm became a matter for debate (was the upper region fluid or solid?), few wished to relinquish the idea that a corporeal object must have a "place," that it must "belong" to some region that was proper to it. But in this case, because a comet, unlike a nova, possessed its own motion, it must be more like a planet—some kind of a wandering star that could not belong in the eighth sphere. And this fact raised questions that had not occurred with the nova: either the comet was attached to one of the already-existing, uniformly moving planetary orbs, or it had an orb of its own. And, if it had its own orb, where did that carrying vehicle come from? For example, had it been there since the Creation, or had it been created de novo?[144] Still again, the comet was visible for only three months; hence, if it was a periodic phenomenon located in an orb, then there was insufficient information to confirm even one complete circuit. Furthermore, even the best estimates of its distance from Earth were educated guesswork. Even guesses, however, had to be guided by some theoretical scheme, as the function of theory was to bring predictable order to a handful of discrete observations. Once again, the practice of theorizing proved to be full of surprises.

If the comet was attached to a planetary orb, then prior convictions about planetary order had to come into consideration, because the few discrete observations alone could not determine its path. Cornelius Gemma, the son of the Louvain ephemeridist Gemma Frisius, was one of the earliest to publish on the comet. Like Tycho, with whose little treatise on the nova he was acquainted, he accepted the traditional Ptolemaic planetary ordering. As with the nova, he located the comet just above the Moon, and because he believed that Mercury was the next highest orb, he decided that the newest apparition must be contained within it.[145] But Gemma bestowed such scant attention on this matter, compared to the comet's longitudinal position and its astrological and apocalyptic meaning, that one must conclude that it had relatively little importance for him.[146]

Roeslin, like Gemma, might easily have followed the path of many German municipal physicians by devoting himself to his patron, composing nativities, and issuing annual prognostications.[147] Yet Roeslin was a physician of a more philosophical and apocalyptic bent: in his exegetical practices, using biblical passages and prophetic writings to interpret the meanings of the astronomical claims of other writers, he was clearly indebted to the tradition of Lichtenberger, although unlike the latter he was more candid in his borrowings from the work of his contemporaries. He completed his medical degree in the same year that Maestlin began his general studies at Tübingen, and the two corresponded for many years. Yet in his substantial treatise on the comet, he did not mention Maestlin at all: instead he relied heavily on Gemma's treatise, which he greatly admired. He also drew on Hagecius's *Inquiry*. Gemma's treatise, including its illustrations, functioned for him virtually as a paratext. His admiration also extended to Gemma's cometo-planetary hypothesis, although in representing the comet's motion he had his own ideas. Roeslin depicted it not in relation to the planets but in a strictly stellar frame of reference.[148] He agreed that the comet's motion was similar to that of the planets insofar as it was regular. Similarly, Hagecius had argued that the new star must be aetherial because its daily motion, unlike that of objects in the sublunar region, was uniform.[149] For Roeslin, this regularity alone, rather than Gemma's parallactic measurements, was a sufficient reason for placing it above the Moon, as all ordinary Aristotelian comets moved nonuniformly.[150] But, for

Roeslin, regularity meant any two or three intervals in the zodiac between which one could pick out a Pythagorean ratio. This search for harmonies strongly resembled the practice of Offusius, which he did not cite but with which he might easily have been familiar. For example, according to Roeslin, the comet's average daily motion in Capricorn was $2°\frac{4}{13}'$, but in Aquarius, the next sign, its motion diminished by exactly half to $1°\frac{2}{13}''$, and in Pisces, he alleged that it diminished by half again, to $34\frac{8}{13}''$ [sic]—in all, a fourfold diminution between Capricorn and Pisces. Roeslin persuaded himself that the motions displayed Pythagorean regularities: "If the comet's motion be considered either with respect to the previous [zodiacal] sign or three of its motions with respect to one another, we discover clearly that its motion follows sure proportions—especially the double [2:1], the sesquialteral [3:2] and the sesquitertiary [4:3] which, in music, clearly make the octave, the fifth and the fourth, the most joyous and perfect of all consonances." Ignoring the inexactitude of his figures (drawn from his own observations and those of Gemma), Roeslin then hastened to assure his readers that he did not wish to defend the old Pythagorean teachings of a (general) celestial harmony but merely wanted to show that the comet moved along according to the aesthetically pleasing consonances of the Pythagorean tuning system: "the most simple, the most constant, the most perfect and uniform harmony."[151] It was neither the first nor the last time that an ideal of knowledge would underdetermine its domain of possible applications.

Roeslin now proposed against Aristotle the fine idea that God had created a special zone in the heavens within which certain comets and new stars were generated. His novel, though complicated, scheme appears to have been an amalgam of Gemma's two separately published figures for the nova and the comet. Compared with ordinary comets, which allegedly affected the weather, he believed that the apparition of 1577 was so rare and so unusually placed that it carried metaphysical and prophetic meanings: it had its own pole, placed symmetrically along the axis of the solstitial colure as much to the east of the zodiac pole as the equinoctial pole lay to the west—a kind of bisected eccentricity. Roeslin thought that this "symmetry" merited a "new sphere of comets and God's miracles." The idea

of distinguishing between two sorts of comets, sub- and superlunary, was, like much else in Roeslin, appropriated directly from Gemma and Hagecius, but the attempt to organize a link between the comet and the nova within the same region explains why Roeslin had to speak of the comet's motion only "by analogy" with that of Mercury. Because Roeslin had put aside the parallactic argument in favor of Pythagorean musicology, the analogy, such as it was, had to be with Mercury's longitudinal motions rather than with Mercury's distance from the other planets.[152] Here, Roeslin's guesswork was more consonant with Clavius's hodgepodge for ordering the planets than with Pythagorean criteria.

In his final chapter, Roeslin's astronomical speculations became the occasion for an unsparingly detailed eschatological exegesis of the celestial messengers. The method resembled that of Reformation theological interpreters looking for a single narrative linking the Old and New Testaments. According to Roeslin, God, from the time of the creation, had providentially created "miracles, prodigies and portentous signs," but now this unusual comet and new star had appeared "against the usual order of nature" and pointed explicitly toward the last times, the end of the world as foretold by Elijah's prophecy. Along the way, Roeslin offered up twenty-five propositions in which he made further connections (although in this case not naming his sources) between the novelties and Leovitius's prediction of a conjunction in the fiery trigon and Rheticus's predicted diminution of the Earth's eccentricity.[153] As if this "concordance" were not sufficiently convincing, the Pfalzgraf's physician ended his treatise with a medical "prognosis" of the "critical years" of the last years of the world, using the microcosm-macrocosm analogy to warrant the forecast that the world was entering the last four phases of its illness, the "macrocosmic crisis" beginning with the new star's demise in 1574 and then consisting of critical forty-year periods (ending in 1614, 1654, and 1694). He concluded: "These critical years ought to persuade us all the more because they aptly correspond to the analogy between the Old and New Testaments. All prophecies and oracles are ended and closed; and in addition, there will occur testimony from nature drawn from that great Book of the World and created beings."[154]

Although Roeslin and Maestlin had come out

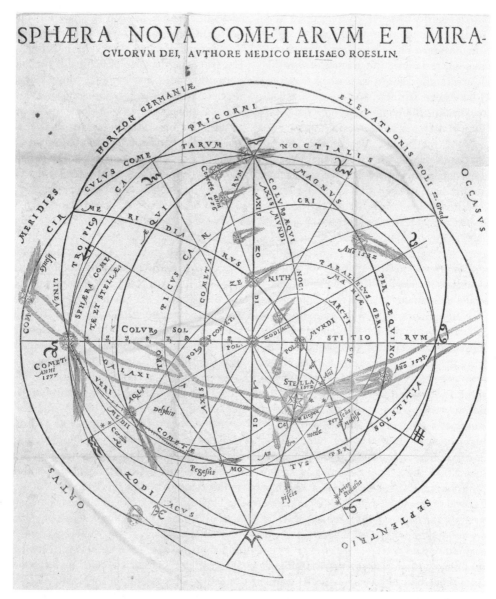

50. Roeslin's "New Sphere of Comets and God's Miracles," including comets of 1532, 1533, 1556, and 1577, as well as the new star of 1572. Roeslin 1578. Courtesy Bibliothèque nationale de France.

of the increasingly restrictive orthodoxy of the Tübingen confessional world, Roeslin seems to have drifted toward the Schwenckfeldian views of his mentor Siderocrates.[155] The contrast between their astronomical competences could not have been more pronounced. Maestlin, already persuaded by the Copernican claims, was the only practitioner to use fully Copernican astronomical resources to construct a theory of the comet's path. Equally significant, where Roeslin

had reveled in biblical and apocalyptic interpretation, Maestlin carefully disavowed even minimal astrological judgments. And this caution was matched by the modesty with which Maestlin spoke of his astronomical "conjecture" and Tycho referred to Maestlin's account as a "speculation" or a "hypothesis."

Maestlin's theory was also the most astronomically detailed and complex of all the accounts published in the immediate aftermath of the comet's appearance.[156] Tycho was so impressed by Maestlin's treatise that he gave it primary place and scope in his summary of 1588: "It is clear that Maestlin's discovery shows great wisdom and industry, and it smacks of the remarkable quality of genius."[157] What was at stake? Unlike the nova problematic, the difficulty involved finding an orb that would fit a handful of observations for an entity that resembled none of the planets in its steep latitudes. Unlike Roeslin, who had relied largely on Gemma for his observations and did not describe how he made his own, Maestlin explained how he made his own sightings with the help of a thread. Again, where Roeslin relied on the authority of Gemma and Hagecius, Maestlin presented his account as a narrative of struggle in which, having tried one conjecture after another, he finally hit on a Copernican circumsolar orbit; and, adapting the libration device that Copernicus had used for Venus, he applied it to the unusual latitudes through which the comet appeared to move.

The Maestlinian solution was novel in its appropriation of Copernican modeling techniques, but it was not without its difficulties: the comet's motion was retrograde with respect to the planets, and this would have presented a serious difficulty were it really located in Venus's sphere, as the comet and the planet would then be moving in opposite directions. Maestlin might have solved this problem by assigning the comet to a sphere just outside that of Venus, "as if it were an adventitious and extraordinary planet," as Tycho Brahe proposed in 1588.[158]

However, Tycho had not yet arrived at this interpretation in 1578, when he penned his account for the king. As John Christianson shows, the manuscript of that never-published work shows three variants of the passage in which he describes the comet as located "in the sphere of Venus." The language that follows shows Tycho taking another step beyond his Copenhagen Oration, adopting the Copernican-Capellan placement of Mercury and Venus while keeping the Earth at rest. Thus he writes: "I conclude that it [the comet] was in the sphere of Venus, if one wants to follow the usual distribution of the celestial orbs. But if one prefers to accept as valid the opinion of various ancient philosophers and of Copernicus in our own time, that Mercury *is next to* the sun, and Venus around Mercury, both with the sun at the approximate center of their orbs, which reasoning is not entirely out of harmony with truth, even if the sun is not put to rest in the center of the universe as the hypotheses of Copernicus would have it."[159]

The case of Roeslin is more explicit and confirms his heavy reliance on other writers for his astronomical information. He sent Maestlin a letter in October 1578 in which he proposed to assign Maestlin's Venusian comet to his new sphere of comets and stars. Maestlin was only too pleased to reprint the letter at the front of his *New Ephemerides . . . Calculated from the Prutenic Tables*, but without endorsing Roeslin's "new sphere."[160]

CONCLUSION

For the vast majority of authors, the two dramatic apparitions of the 1570s were divine signs or portents that carried apocalyptic and astrological meanings but no new consequences for natural philosophy or the science of the stars. It was only a relatively small group of writers who used the comet and the nova to open up a breach in the traditional Aristotelian ontology of the world; and it was not primarily their Copernican sympathies that drove them to this view. The widely shared conclusion was that God, if he wished, could miraculously create singular celestial mutations. This view of the conditional possibilities of divine power was already common among natural philosophers in the thirteenth and fourteenth centuries. The new and significant conclusion was really the emphasis placed on God's use of his ordained power at a great moment of expectation—in 1572 and then again in 1577—to send prophetic messages to pious men, messages that seemed palpably to embody all sorts of end-of-time plots in the heavens.[161]

Significantly, the men who believed that they were interpreting God's providential signs—the bearers and interpreters of these heavenly meanings—were mathematical prognosticators. For these same authors, the nova of 1572 and the comet of 1577 stimulated a new sense of disciplinary identity both as theoretical astronomers and as practitioners whose mathematical skill enabled them to speak as privileged interpreters not merely in the political domain, to which they were already long accustomed, but also in the socially and epistemically circumscribed domain of natural philosophy and theology. Simultaneously, these events provided occasions to theorize about the order of the planets that were unprecedented in the comet literature. The effects of this stimulus, however, did nothing to create a new consensus. If anything, by the end of the decade there was even less agreement among heavenly practitioners about planetary order than before.

In fact, with all the wildly confident eschatological declarations and the attacks on Aristotle's immutable heavens, a counterimpression of epistemic uncertainty lingered along the trail left by the relentless explosion of heavenly literature of the 1570s. "I see some who study and comment on their almanacs and cite their authority in current events," wrote the ever-astute, Pyrrhonizing Michel de Montaigne. "With all they say, they necessarily tell both truth and falsehood. *For who is there who, shooting all day, will not sometime hit the mark? . . .* Besides, no one keeps a record of their mistakes, inasmuch as these are ordinary and numberless; and their correct divinations are made much of because they are rare, incredible, and prodigious."[162] Montaigne's skepticism was not narrowly Piconian but much indebted to Cicero and the influential writings of Sextus Empiricus.[163] Ever appealing to reason in order to undermine its claims, Montaigne's glancing reference to Copernicus underlined an emergent fallibilist temper in that otherwise sanguine moment: "When some new doctrine is offered to us, we have great occasion to distrust it, and to consider that before it was produced its opposite was in vogue; and as it was overthrown by this one, there may arise in the future a third invention that will likewise smash the second. . . . What letters-patent have these, what special privilege, that the course of our invention stops at them, and that to them belongs possession of our belief for all time to come? They are no more exempt from being thrown out than were their predecessors."[164] If the 1570s offered fertile ground for such fallibilist sentiments, the next two decades would do little to diminish the possibilities for further doubt.

The Second-Generation Copernicans

MAESTLIN AND DIGGES

A generation can be thought of as defining for a group of practitioners a space of temporally bound experiences and conceptual possibilities. The generation informed by the Wittenberg consensus, born largely in the 1540s, was the cohort that, by the 1570s—and most dramatically in the 1580s—began actively to engage the full text of *De Revolutionibus*. Because of the expectation, shaped by recent science, that revolutionary change ought to occur relatively rapidly, this delayed development is of a sort that one might have expected to occur—but which did not—in the first decade or two after 1543. Indeed, it is just this slowly evolving pace that led Kuhn to caution that the effects of Copernicus's work were "revolution-*making*" and that earlier writers, unconcerned with revolution as the governing trope, simply designated as "gradual." Both the chronology and the categories of this shift will concern us in this chapter and the next.

The second generation after Copernicus was an activist generation that no longer read *De Revolutionibus* solely as a tool of astrological prognostication but now took the text seriously as a resource for theorizing about planetary order and planetary models.[1] Such theorizing itself represented a break with conventional disciplinary practices. Most of these practitioners were astronomers with mathematical skills strong enough to be noticed, admired, or feared by other contemporaries; with the notable exceptions of Giordano Bruno and Diego de Zuñiga, most respected the boundaries that defined the typical interpretive practices of philosophers and theologians.

Among this generational aggregate, Michael Maestlin (1550–1631) and Thomas Digges (1546–96) were, after Rheticus, the earliest followers of Copernicus. Their work contains none of the qualifications, hedgings, and silences to be found in the otherwise technically well-versed prognosticators Gemma Frisius and Jofrancus Offusius. Yet neither had known Copernicus personally, and neither knew the other, either personally or through their respective networks. Maestlin was a German academic, Digges a landed gentleman and protégé of John Dee. Both were skilled in the mathematics of their day and were as capable of appreciating the technicalities of *De Revolutionibus* as were Gemma Frisius and Offusius. Yet they held opposing positions on the status of astrological prognostication. Digges was a Melanchthonian in astrology whereas Maestlin was guarded, critical and sympathetic to much of the Piconian critique. Although their divergences were not extreme, they were entirely typical of differences among practitioners who regarded themselves as adherents of Copernicus's central ordering claims.

MICHAEL MAESTLIN
PASTOR, ACADEMIC, *MATHEMATICUS*, COPERNICAN

Michael Maestlin, more than ten years younger than Christopher Clavius, was a university teacher of mathematical subjects for much of his life, and like Clavius he devoted a certain amount of his efforts to writing student textbooks. His *Epit-*

ome *Astronomiae,* first published at Heidelberg in 1582, went through many editions.[2] He also wrote many astronomical works that were never published but which exist today in manuscript. His writings show him to have been humanistically inclined and well versed in the literary and astronomical classics of antiquity, but also, like Brahe and Clavius, extremely knowledgeable in a variety of mathematical disciplines, of which astronomy was to him the most important.

As an influential textbook writer, Maestlin was, in a sense, a Lutheran counterpart to Clavius, and although he was never referred to in that way during his lifetime, some of his astronomical work was marked by explicit confessional import. In 1586 he wrote a detailed and strident attack on Clavius.[3] In 1588, he polemicized about the calendar with another Jesuit, Antonio Possevino.[4] Using the *Prutenic Tables,* Clavius had engineered a solution to the civil calendar; it was promulgated by Pope Gregory XIII in 1582 as the official calendar for all Christendom. Clavius's adjustment meant that, as a result of Rome's calculations, everyone was supposed to set their calendars ahead by ten days. This change was obviously a political issue with serious practical consequences beyond even the concerns of heavenly practitioners: it effectively made prognostications and almanacs dependent on a decree from Rome for the organization of time. Maestlin advised against such dependency, and as a result, the German Protestant territories resisted Rome by staying on the old-style calendar until Napoleon changed it centuries later.

At about the same time, but quite independently, Maestlin and Thomas Digges formed commitments to Copernicus's representation of planetary order. A great deal more is known about Maestlin's intellectual formation as a Copernican simply because of the fortunate survival of many of his books and manuscripts—which included a first edition of *De Revolutionibus,* now in the city library of Schaffhausen, Switzerland. He acquired this copy while still a young student at Tübingen. He wrote on a back inner flyleaf that he had purchased the book on 6 July 1570 from the widow of Victorinus Strigelius. The latter, as we recall, had been a member of the Melanchthon circle but had later become a member of the Tübingen theology faculty. Strigelius's elementary astronomical textbook, written in the Melanchthonian question-and-answer style, shows no Co-

pernican sympathies, and the copy of *De Revolutionibus* that passed over to Maestlin arrived with neither an original ownership provenance nor reader's notes. Yet by the time a Swiss schoolmaster named Stefan Spleiss carried this book from Tübingen to Schaffhausen after Maestlin's death in 1631, together with many other volumes from Maestlin's library, it had more annotations by Maestlin in its margins than almost any other copy that has survived from this period.

Maestlin's very early acquaintance with Copernicus's major work must be set against his early intellectual formation as an arts student of unusual mathematical talent in the orthodox Lutheran religious seminary (*Stift*) and university at Tübingen (1568–71). Just one month after receiving a master's degree in 1571, Maestlin published an edition of Reinhold's *Prutenic Tables* which showed that he had swiftly surpassed the usual elementary requirements in the sphere and *theorics.*[5] How his interests and competence had expanded so far beyond the normal curriculum we do not know. We know only that he studied the science of the stars with Philip Apianus (1531–1589), the successor to Samuel Eisenmenger (Siderocrates) and son of the more famous Peter, who arrived in 1569 after having been dismissed on grounds of heresy from the university in Catholic Ingolstadt.[6] (Apianus would later suffer a comparable fate at Tübingen when he was dismissed as a crypto-Calvinist.)

We may also infer that Maestlin had an unusually good relationship with the controversial poet, dramatist, and historian Nicodemus Frischlin, who had arrived at the university in 1568, the same year that Maestlin began his studies. When Frischlin composed a long poem, celebrating the apparition of 1572 as a new star, Maestlin's *Astronomical Demonstration* was appended as a sequel to the poem. Although astral poetry was a well-known genre, this somewhat unusual juxtaposition of writings may be connected with the fact that Frischlin recognized the student Maestlin's unusual astronomical ability and capacity for meticulous work.[7] They must have become acquainted with one another in 1571–72 while Maestlin was studying theology (he completed his studies in January 1573) and when, during Apianus's absence, Frischlin gave the astronomy lectures. Evidently Maestlin must have caused Frischlin to change his original views, as Frischlin acknowledged in his patronal preface that he had followed

Aristotle "until I was taught something truer and better by other, more learned men."[8]

After 1577, Maestlin assumed pastoral duties as deacon of the parish church at Backnang. In 1580, he was appointed *ordinarius* in *mathesis* or mathematical subjects at the University of Heidelberg.[9] When Apianus departed Tübingen in 1584 under suspicion of unorthodox religious beliefs, Maestlin was asked to take over his position and held the post until his death in 1631.

As is apparent from his profuse annotations in the margins of *De Revolutionibus*, Maestlin had an early and altogether more favorable reaction to Copernicus's central representation of the planetary arrangement than did Tycho Brahe. There is some reason to think that Apianus stimulated this initial interest, much as Maestlin later ignited Kepler's; but Apianus's annotations (on a different copy) do not lead one to suspect that he shared his student's reaction.[10] In his religious views, Maestlin was orthodox and reliable; yet in celestial affairs, he cautiously but persistently followed the *via moderna*. At the same time, unlike Tycho Brahe, Maestlin expressed no reservations about the *Prutenics*. In 1576, while replacing Apianus as lecturer, he published a *Prutenic*-based ephemerides that extended Stadius's tables from 1577 to 1590.[11] One may suspect that the critical formative circumstance here was his early reading of Rheticus's *Narratio Prima*. Although Maestlin was strongly drawn to the astronomical arguments in Copernicus's preface, the placement of Paul III's name at the head of that freighted letter was politically incompatible with Maestlin's strongly antipapal sentiments.[12] Also, as he published a new edition of the *Narratio Prima* in 1596, it is possible that his acquaintance with that work dates back to the time that he acquired *De Revolutionibus*, although this dating must remain a matter of speculation.

Maestlin's few folios on the new star already make specific references to several different passages in *De Revolutionibus* that reveal extensive and early familiarity with the entire scope of the work. Most crucially, Maestlin appreciated that the universe must be large enough to contain the new star; and in this respect he found an explanatory advantage in the huge size of the Copernican world (*quasi infinitum*). He referred to the immense and unknowable distance of Copernicus's outermost sphere (*De Revolutionibus*, book 1, chapter 6), to the "motion in commutation" that the object would have were it affixed to one of the planetary orbs (book 5, chapter 3),[13] and to Copernicus's demonstration of the "certain distances of the planetary orbs from the center of the world" (preface; book 1, chapter 10).[14] The last reference follows a passage where Maestlin openly broke with Aristotle's doctrine of celestial immutability. He declared that the appearance was not a comet but a new star, the cause of whose appearance was not natural. Even comets, he thought, could be generated either in the region of elements or in the starry orb "which, according to Copernicus, is the highest heaven that contains all things, and for this reason, we wish to say against Aristotle and all the physicists and astronomers that the heaven is not lacking in generation and corruption."[15] We can take Maestlin to intend this remark to mean that God chooses to use his power to cause mutations in the heavens. Maestlin admitted that it might be "absurd" to describe the heavens themselves as mutable, but no less so than to deny that the universe is large enough to contain this enormous star.

> Since the altitude [radius] of the orb that carries the stars is immense and extends so far up, it is not certain that the distance between the Sun and the Earth can be compared to it (thus did testify Copernicus, after Ptolemy the prince of astronomers, who, when demonstrating the planetary orbs' sure distances from the center of the world, stops short of the stellar orb).[16] For this reason, it is impossible to measure either this star's true magnitude or its altitude from the center of the world [to the stellar orb]. Nevertheless, it is certain that [this star] surpasses in apparent size all the [fixed] stars of first magnitude and that it is immeasurably larger than the Earth.[17]

Maestlin's abbreviated statement shows that as early as 1573 he had accepted Copernicus's argument for the distances of the planets.

But as to the cause of the apparition, Copernicus had nothing to offer, and Maestlin was left to speculate.[18] "I do not see an explanation, unless perhaps that its creation depends upon a hyperphysical cause. What, therefore, would prevent us from saying that all this is hyperphysical? That this New Star was created by the greatest Creator in these newest times, that it began and will cease miraculously, and the cause of both would escape all human capacity."[19] Like Brahe, Hagecius and others, Maestlin followed the logic of denying the consequent (*modus tollens*), that is, if the apparition were an object *x* (a comet, a planet, or a fixed star), then it would have certain attributes *y* (a tail, not twinkling, twinkling). Those attributes were not observed (not-*y*); ergo, it was not any of those objects (not-*x*). Thus, having excluded the new star from the category of any known natural entity, he used Copernicus's authority to warrant the credibility of the star's insensible distance from Earth and to justify a miraculist or "hyperphysical" explanation for its origin. Maestlin hinted that "in these newest times" all sorts of miracles (exceeding human comprehension) were to be expected.

Maestlin's explanation of the new star as an extraordinary divine intervention in the normal course of nature raises the related question of its prophetic import. Had its appearance been foreseen in any sacred writings? More relevant to our theme, if Maestlin knew the *Narratio Prima* at this early date, why did he not take the opportunity to connect the Elijah prophecy to the nova or to the revolutions of the solar eccentric? The explanation may be found in his later work against the Gregorian calendar reform, in which he took a more restrained position on the authority of that prophecy than that accorded to it by the Wittenbergers Rheticus, Melanchthon, Peucer, Luther, and others. For Maestlin, the Bible took precedence over the Elijah prophecy, just as biblical authority on the closeness of the end obviated any need for the "false calendar."[20] Even in 1596 Maestlin offered no special annotation on the Elijah passage in his commentary on the *Narratio Prima*.[21] Thus, although Maestlin shared with his Lutheran contemporaries a belief in the biblical prophecies that forecast an end to the world, he hesitated publicly to connect those prophecies with the appearance of the nova.[22] It was left to Frischlin to interpret the nova's meaning.

MAESTLIN'S HESITATIONS ABOUT ASTROLOGY

Maestlin's caution in assigning specific prophetic significance to the new star echoes his general circumspection in astrological matters—quite unusual for a Lutheran celestial practitioner of this period. More difficult to determine is whether this caution was principled, grounded in his conservative temperament, or somehow associated with the political environment. Consider that

Maestlin restricted his account to the new star's astronomical coordinates and did not include an astrological judgment in the manner of commentators such as Brahe, Hagecius, and Cornelius Gemma. In later works, including those on the comets of 1577–78 and 1580, he presented conjectures on the meanings of those apparitions, but at the same time he showed an explicit reluctance to engage in astrological interpretations. In the 1578 book on the comet, he used the dedication to the duke of Württemberg to fix limits:

> I ought to confess here that I will not generally satisfy many people's expectations into whose hands my book will fall; for, although I have put together what an astronomer can say about this comet, I have noted down conjectures taken not so much from the foundations of astrology as derived from elsewhere. However, I hope that I may not be unworthy of pardon in the cause of this matter. For, although I am not a little familiar with both abstract and concrete mathematics, in concrete [mathematics], under which the heavenly motions are placed, I jealously guard considerations of astronomy rather than astrology.

Even so, one of his worries concerned the reliability of the tables of motions. Many quarrels among learned men devolved on the numbers; Maestlin hoped that he would be able to restore "absolute integrity" to the tables by bringing together the observations of the ancients (Hipparchus, Ptolemy, al-Battani) and the moderns (Regiomontanus, Peurbach, Copernicus). Yet again, "I always preferred astronomy to astrology. And, for this reason, I do not wish nor can I here claim the astrological judgment for myself but instead I relinquish it to others, many of whom I see are quite apprehensive so that they prognosticate boldly (inasmuch as this is easy). So, for this reason that I have explained, the comet is placed under astronomical science in order that the wisdom and omnipotence of the Best and Greatest God may be seen in this and other ways."[23]

These passages suggest personal rather than foundational objections. Other compatible evidence reinforces the impression of self-doubts about engaging in astrological hermeneutics. For example, in an undated letter found by Richard Jarrell, Maestlin responded to a request for a child's nativity: "The astrological opinion that is asked of me I am unable to write, nor do I have the skill, because both publicly and privately I have often protested [against it]. . . . I have never undertaken astrology."[24] Further, there is the revealing list of works for which Maestlin was granted an imperial publishing privilege in April 1580. This considerable register contained neither proposed astrological prognostications nor any work of astrological theory.

> A compendium of astronomy. A richer explanation or commentary on the spherical doctrine or the first part of astronomy, concerning the first motion. Theories of the planets or a commentary on the second part of astronomy. A clear, common arithmetic. A most complete doctrine of plane and spherical triangles. A commentary on Cleomedes. Likewise, a learned commentary and demonstrations of Theodosius's propositions in his books on the sphere. Various sundials and new hanging instruments by which the daily or nightly hour may be investigated through the sun's shadow or its altitude or that of the stars. Likewise, other instruments useful for observations of heavenly phenomena and for [measuring] planimetric and stereometric dimensions. Revolutions of the celestial orbs in imitation of Ptolemy's *Almagest* and Nicolaus Copernicus's *Revolutions*. Tables of the motions of celestial orbs derived from these new *Revolutions* in imitation of the *Alfonsine* and *Prutenic* tables. Likewise, resolved tables in imitation of Bianchini's tables. And new ephemerides calculated from these new tables.[25]

Maestlin's proposed agenda of publications was a virtual update of Reinhold's earlier one in the *Prutenics,* which Maestlin clearly knew from his edition of that work. Yet Maestlin's ephemerides contained nothing faintly resembling the extensive treatment of astrology found in the ephemerides of Leovitius and Magini. Moreover, although Hieronymus Wolf contributed a testimonial to the dedication, he made not a single allusion to the stars' influences.[26] Likewise, Maestlin's extremely clear and comprehensive textbook of astronomy gave not the least serious attention to astrological theory or practice.[27] And finally, as late as 1619— well after Kepler had developed his own astrological reform—Maestlin did not budge from his opposition to making astrological judgments.[28]

This reluctance to engage in practical astrology seems to have turned on a further distinction— between judgments about the meanings of events that had been foreseen (eclipses, conjunctions) and those unforeseen (comets, novas). Maestlin could easily accommodate interpretations about the meaning of unpredicted celestial events because, theologically, these were specific, singular acts of God, flowing from his absolute power to act in the world while communicating a message (of love or anger) to humankind. This interesting position was in step with Tübingen theological opinion. For example, Jacob Heerbrand (1521–1600), one of Maestlin's theology professors, preached a vigorous and frightening sermon about the comet of 1577 that judged it to be the result of a direct and unforeseen act by God, a punishing sign demanding human repentance.[29] Heerbrand, a onetime student of Melanchthon— but not unambiguously a Melanchthonian in his views about astrology—believed that divine purpose could be known through the "book of nature." But, although Heerbrand warranted mathematics as a legitimate tool for interpreting the book of nature, he did not specify how it was to be used. For example, it did not follow that the distance of a heavenly object was actually knowable through parallactic measurements or that a comet's "place of birth" in the zodiac required an astrological judgment—none of which Heerbrand attempted to provide.[30]

If apocalyptic prophecy could be linked with astrological prediction—as was done famously by Johannes Lichtenberger in 1488—the connection was by no means a necessary one. Evidently a separation between the two was well developed at Tübingen even before Maestlin's arrival. Explicit rejection of astrological prognostication showed up in a sermon by the chancellor of the university and the senior professor of theology, Jacob Andreae (1528–90), who preached against it in 1567. Once again, theology made common cause with Pico's arguments: the uncertainty of prognosticating particulars and the absolute preeminence of God's foreknowledge. Human curiosity drove men to want predictions about future weather and disease from the "annual practices," but to quell such anxieties, Andreae quoted the prophet Jeremiah: "Do not be afraid of the signs of the heavens."[31]

Maestlin's hesitations about practical astrology thus far appear to be in harmony with the principled skepticism of the powerful Andreae, even if they were not motivated by it. Moreover, Maestlin certainly must have had firsthand familiarity with the main skeptical objections from classical authorities as well as from Piconian or neo-Piconian sources. Many were mentioned in a polemic—familiar to Maestlin—that Nicodemus Frischlin inserted into his astronomical textbook of 1586.[32] Frischlin's list of authorities shows the kinds of works that were circulating in Tübingen during the 1570s and '80s.[33] His appeal to Luther and Calvin, while excluding all mention of Melanchthon, clearly betokened the existence of some sort of anti-Melanchthonian current in Tübingen.[34] When the duke asked Maestlin to render a formal opinion on Frischlin's book, Maestlin carefully refrained from launching a counterpolemic: he said nothing in his report about Frischlin's attack on astrology but subtly undermined the book by voicing general doubts about Frischlin's mathematical competence.[35]

Maestlin's trimming, together with his reluctance to attack astrology's foundations, leaves open the final possibility that he held out the hope for some sort of adjustments to astrological theory. More than a decade later, when Kepler was required to issue annual prognostications at Graz, Maestlin objected to Kepler's introduction of arguments about theoretical astrology into his forecast for 1598: "It appeared to me that [this matter] might be reserved for another [sort of] treatise in which it could be disputed only with the learned . . . [rather than with] country bumpkins and idiots . . . a prognostication lasts only one year, whereas there is another kind of writing that is permanent and in which the matter can be treated more suitably."[36] Maestlin's worry was not about Kepler's theorizing, which he clearly supported, but about pursuing it in the proper genre and with the right kind of audience. And meanwhile, Maestlin's silences on astrology in his published works allowed the appearance of a position closer to that held by Christopher Clavius, his famous opponent in calendrical matters. Whatever their differences, both Maestlin and Clavius found themselves justifying the study of astronomy on the basis of scriptural passages that legitimized contemplation of the heavens as a sacred activity in its own right.

THE PRACTICE OF THEORIZING
MAESTLIN'S GLOSSES ON COPERNICUS

The "Copernican" passages of the nova tract show how premises from the first and fifth books of De Revolutionibus were brought to bear for the first time on an independent and unforeseen event: the accommodation of a new, superlunary entity. The example reveals how an Aristotelian assumption (that everything in the universe has a proper place) was still at work in a universe where the meaning of up and down had been radically changed; in addition, it illustrates the wider tendency among sixteenth-century astronomers to avoid deploying Copernicus's theory as a totality. Moreover, Maestlin did not use the occasion to present himself as a disciple, in the manner of Rheticus, or to argue for more extensive justification of the necessity of a heliocentric planetary arrangement. Much the same was true in his treatment of the comet of 1577, where he claimed to have found a "place" for the bearded star within the circumsolar sphere of Venus and where, for the first time, a practitioner used Copernicus's libration mechanism to account for radical changes in the latitude of a comet.[37] Eventually this limited posture would prove vulnerable. Tycho Brahe later criticized him for aligning himself with a proposition based on unproven premises: Maestlin could not persuade anyone that there was such a vast interval between Saturn's sphere and the outermost sphere unless he could first demonstrate that the Sun is at rest and the Earth revolves around it.[38]

However, Maestlin's glosses on those passages of De Revolutionibus corresponding to the propositions that he used in the nova tract show that he held stronger views in private than he was prepared to espouse in public. There is some coherence to these notes. Where Copernicus made his crucial claim in the preface that planetary order cannot be rearranged at will, Maestlin jotted down the following remark: "This argument is wholly in accord with reason. Such is the arrangement of this entire, immense machina that it permits firmer demonstrations; indeed the entire universe revolves in such a way that nothing can be transposed without confusion of its [parts] and, hence, by means of these [more solid demonstrations], all the phenomena of motion can be demonstrated most exactly, for nothing unfitting occurs in the course of their forward motion."[39] This passage anticipates the arguments for Copernicus that Maestlin developed in his subsequent notes and show the extent to which his annotations were more explicitly favorable than those of Gemma Frisius and Offusius. It also displays many typical characteristics of his glossing practices.

The annotations imitate, clarify, amplify, and sometimes rephrase Copernicus's language from the accompanying text; typically, Maestlin did not replace Copernicus's reasoning but added his own judgment. The presence of so many notes and the occasional use of the first person singular suggests that he intended to publish a new edition with commentary, perhaps something more ambitious than his edition of the *Narratio Prima*. Occasionally, as in a parenthetical aside at the end of the preceding passage, he inserted a revealing comment: "As for astronomy, Copernicus wrote this entire book as an astronomer, not as a physicist." Copernicus worked like a geometer rather than a natural philosopher: he structured his proofs in the manner of Ptolemy (and Euclid) rather than Aristotle. He proved his conclusions by the use of geometry, optics, and observations rather than by seeking out Aristotelian final, material, and efficient causes. Yet Maestlin did not believe physical propositions to be entirely out of bounds for astronomers.

The annotations also show that Maestlin believed that astronomy as a discipline was capable of making true claims about the world. One indication is the clear rejection of Osiander's position. Maestlin entertained strong suspicions even before he learned directly from Apianus of the true identity of its author:

This preface was fastened on by someone, whoever its author may be (for indeed, the words and thinness of style reveal them not to be those of Copernicus) . . . and no one will approve of these [hypotheses] at first glance from concern for novelty, in spite of his opinion, but rather one first reads and rereads, and then one judges. Hereafter, indeed [if one accepts this letter], it will be the case that a proof cannot be contradicted. For the author of this letter, whoever he may be, while he wishes to entice the reader, neither boldly casts aside these hypotheses nor approves them, but rather he imprudently squanders away something which might better have been kept silent. For the disciplines are not made stronger by shattering their foundations. Wherefore, unless this [hypothesis of Copernicus] is defended better than by this person, he labors in vain, because there is much weakness in his meaning and reasoning. And therefore, I cannot approve of all the simple and confused parts of this letter, much less defend it.[40]

But how was astronomy supposed to reach truth? Against the author of the "Ad Lectorem," Maestlin enthusiastically endorsed Copernicus's foundational argument in which *symmetria* was a consequence of assuming the Earth's annual motion. This argument was for him an example of how the truth is always consistent with itself: "Certainly this is the great argument," he notes, "that all the phenomena as well as the order and distances of the orbs are bound together in the motion of the earth."[41] In a remarkable and extensive comment in the opposite margin, he glossed Copernicus's reasoning in setting the Earth in motion and his own grounds for acceptance:

Moved by this argument, I approve of the opinion and hypotheses of Copernicus, which, as I see it, many would do if they did not fear the displeasure of others who know the hypothesis of the Earth's immobility to have been confirmed for a long time, as can be seen in the commentary that [Erasmus] Oswald Schreckenfuchs wrote on the *Sphere*.[42] Now Copernicus did not play about, showing off his cleverness like some skilled master, but rather he set about to restore the motions which were nearly in ruins, and for this purpose he judged internally consistent hypotheses [*hypotheses conuenientibus*] to be necessary. However, since he noticed in the demonstrations that he collected together that the ordinary hypotheses were insufficient, that most demonstrations failed, and that many absurdities were admitted, he approved at length this very same opinion [*opinionem*] of the Earth's mobility, since indeed, it not only amply satisfied the phenomena but also admitted nothing absurd into the whole of astronomy, or rather [it admitted only] what followed as a logical consequence [*consequens*].

In fact, should anyone purge the ordinary hypotheses so that they would agree with the phenomena and admit of no inconsistencies, then, to be sure, I should regard that person seriously, and clearly he would win over most others. But I see that in fact some men, most outstanding in mathematical disciplines [*mathematicis*], have worked on these [hypotheses], yet, in the end, without success. Therefore, it appears to me that unless the usual hypotheses are reformed (a task for which I am inadequately prepared because of my rash nature), I shall endorse [*approbabo*] the hypotheses according to the opinion of Copernicus, after Ptolemy, prince of all astronomers.[43]

Maestlin clearly understood this passage in the sense intended by Copernicus, as a dialectical argument involving the comparative weighing and balancing of claims by an audience. The idea was that the side with the greater coherence and fewer faults was the one that ought to be chosen. He presented the situation as involving two possible hypotheses, both within the domain of astronomy: the "usual" and the new. One side had (undetailed) observational faults and inconsistencies, the other possessed logical coherence and no (mentioned) incompatibilities. This clear imbalance weighed in favor of Copernicus's position.

But when Maestlin wrote this note in the early 1570s, he ran into a major difficulty involving the practical use of such a dialectical standard: there simply was no larger community to whom he could gesture specifically in support of his judgment. An inductive counting up of supporters, a strategy similar to that employed in the two previous decades by the horoscope compilers Cardano, Garcaeus, and Gaurico, was not available to him. Neither was a moral or political reckoning. The Horatian aesthetic standard underlying Copernicus's head-and-body language in the preface, associating papal authority with celestial reform, obviously did not evoke in him any positive moral imagery.[44] For the Lutheran Maestlin, astronomy's first principles surely could not be grounded in papal authority!

Maestlin was simply *unaware* of the local political meanings associated with Copernicus's ideal of knowledge as expressed in the *De Revolutionibus*. In fact, the only thing that Maestlin knew for sure about the immediate political context of the publication of that work was that someone had appended an offensive (anonymous) letter which all but disemboweled the book's most forceful premises. Thus, Rheticus's presentation of the heliocentric arrangement was more apt because it was not directed to the pope; and, besides the work's relative brevity, this could help to explain why Maestlin later chose to publish a new edition in conjunction with Kepler's *Mysterium Cosmographicum*.

The early 1570s was a critical moment, therefore, when the Copernican arrangement as a claim about the real world might well have failed to arouse any advocacy at all. Who else besides Maestlin and Rheticus (with whom there is no evidence of contact) weighed the core arguments in a manner favorable to the author of *De Revolutionibus*? There were simply *no* academic commentators who had presented themselves as advocates. Much to the contrary, the *via antiqua*, as represented in the school manual literature of the heavens, had developed objections against

the Earth's motion whose wide authority was only amplified by frequent repetition. At best, there were a few favorable remarks in Gemma Frisius's letter attached to an ephemerides tabulated for Antwerp.

Under the circumstances, in his key annotation, Maestlin had no choice but to make a passing reference to an example of someone who appeared to be evaluating the alternative hypotheses in a way that he himself was doing. That is the significance of his fleeting mention of the *Sphere* commentator Erasmus Oswaldus Schreckenfuchs, like himself a prolific professor of mathematical subjects teaching at a German university (Freiburg im Breisgau), who admired Copernicus's hypotheses enough to insert a short reference in his comment on the Earth's central place. Schreckenfuchs mentioned that he had written his own "Commentaries on Copernicus," in which he spoke "more fully and more clearly." But—and here was a crucial remark of value to Maestlin— Schreckenfuchs said that he "feared that certain men who are most obstinate in their veneration of the ancient philosophers will be offended."[45] The meaning was clear. Even if Copernicus's arguments were compelling, breaking with tradition would arouse dangerous criticism; it could provoke political "displeasure" and resistance. Because Maestlin himself well understood this kind of resistance, the remark could easily be taken as self-referential. It is significant, therefore, that he kept his own comments strictly within the framework of astronomy and was cautious in expressing any conviction in the Copernican ordering.

In sum, Maestlin approved of the Copernican hypotheses not because of the personal authority of Copernicus, the dedication to the pope, or Osiander's confessional stance. Indeed, on the basis of what can be inferred from the annotations, it appears that Maestlin disengaged Copernicus's dialectical ideal of knowledge and its associated argumentational forms from their original, controversial, political context.

At the same time, another standard was in operation: Maestlin read Copernicus's principle to mean that truth is an adequate standard in rendering a choice where both the premises and the conclusions of an argument are true. A version of this epistemic ideal sometimes appeared in the form of a favorite Maestlinian aphorism, both in his annotations and in his published works:

"Socrates is a friend; Plato is a friend; but a greater friend is Truth" ("Amicus Socrates, amicus Plato, sed magis amicum veritas").[46]

But how was the practicing astronomer supposed to translate the logic of this maxim? How was one supposed to infer that the premises of an argument were true when one had only visible phenomena and effects from which to draw inferences? In his preface to the pope, as André Goddu argues, Copernicus relied heavily on a topical logic that fell well short of apodictic certainty: "If the hypotheses which they [the ancients] assumed were not false, then everything which followed from the hypotheses would have been verified without doubt."[47] Early annotators of De Revolutionibus passed over this brief passage but Maestlin glossed Copernicus approvingly:

> The truth is consonant with the truth, and from the truth nothing but the truth follows. And if, in the reasoning process, something false and impossible follows from the received opinion or hypotheses, then necessarily a fault must be concealed in the hypothesis. If, therefore, the hypothesis of the Earth's immobility were true, then the true would also follow from it. But, in [the traditional] astronomy, there follow a great many inconsistent consequences and absurdities—as much in the arrangement of the orbs as in the motion of the planet's orbs. Therefore, there will be a fault in the hypothesis itself. At the least, the problem appears in the Sun's motion with regard to the length of the tropical year; likewise in the case of the motion of the three superior planets; most of all, however, in the case of Venus and that of the orb of the fixed stars.[48]

In this revealing, but still limited gloss, Maestlin did not take up the specific details of the mentioned astronomical difficulties or how they might be regarded as critical problems that one and only one system could resolve. Likewise, Maestlin took no notice of the logical possibility that true conclusions can follow from false premises—the very objection raised by Clavius in the 1581 edition of his Sphere commentary.[49] He also excluded nonastronomical criteria from his purview, including the common Wittenberg textbook objection that the Copernican hypothesis did involve "absurdities" (in natural philosophy) and "incompatibilities" (in the interpretation of certain passages of holy scripture); and he makes no mention of the astronomical "fault" trumpeted by Tycho Brahe in 1588 (the huge, unused space between the sphere of Saturn and the outermost sphere). Indeed, the absence of such later allusions strongly supports my view that the note was composed in the 1570s.

When Maestlin said that Copernicus wrote his book "as an astronomer and not as a physicist," I read his viewpoint as self-referential. For Maestlin, the practice of astronomical theorizing meant reading the text of De Revolutionibus within the space of the text's own interpretive practices and possibilities, coupled with a consequent refusal to cross the boundaries with interpretive practices from outside traditional astronomical theory.[50] In fact, all his examples of advantages for Copernicus involved claims that concerned regularities or irregularities in the motions and thus, in his terms, were "mathematical" in character. For example, in De Revolutionibus, book 5, chapter 3 ("General Demonstration of the Apparent Inequality on Account of the Earth's Motion"), where Copernicus showed that the intrusion of the Sun's mean motion into the models of inferior planets could be eliminated by setting the Earth in motion, Maestlin noted: "Another argument that confirms the mobility of the Earth. In the hypothesis of the Earth's immobility, if one clearly disregards what is called the [daily] revolution, the orbs of the Sun, Venus, and Mercury are distinct from one another. They are contiguous but not continuous with one another. Not only are they moved by one and the same mean motion, but also nowhere else does it occur that the orbs of two or more planets have one mean motion. The true revolution [of these orbs] appears admirably well in the hypothesis of the Earth's mobility, as can be seen in the text."[51] On the following page, Maestlin elegantly glossed Copernicus's explanation for what I have referred to as Peurbach's conundrum: "That motion, therefore, which the ancients called [the motion of] the epicycle is nothing other than the difference by which the speed of the Earth exceeds the planet's motion, as with the three superior planets, or the speed by which the Earth is exceeded—as with the two inferior planets."[52]

For Maestlin, truth had what might be called an *intrageneric* or intradisciplinary character. Insofar as he engaged with natural philosophy, he stayed close to the canonical practices of Ptolemy's and Copernicus's main works, where physical propositions had a limited auxiliary role in supporting astronomical practice. As a commentator, he felt free to judge their physical speculations, occasionally adding or substituting something of his own.[53] Maestlin's refusal to engage in new kinds of practices of physical theorizing later showed up in the form of paternal disapproval of his prized student's freewheeling speculations concerning efficient causes.[54] Or rather, as chapter 11 shows, this unaccustomed sort of theoretical activity--which would involve organizing a new kind of rhetorical presentation—was one of the important ways in which Kepler broke with Maestlin and his contemporaries.

THOMAS DIGGES
GENTLEMAN, MATHEMATICAL PRACTITIONER,
PLATONIST, COPERNICAN

In the same year that Maestlin published his unostentatious *Astronomical Demonstration* using elements of Copernicus's *De Revolutionibus* to argue for the place of the nova, Thomas Digges (1546–95) published an elegant and comparatively much longer work about the new appearance. To the first part of the title he assigned a soaring Platonic trope of ascent: *Mathematical Wings or Ladders, by which the Remotest Theaters of the Visible Heavens are Ascended and All the Planetary Paths are Explored by New and Unheard of Methods*. His print identity—"Thomas Digges of Kent, an author of noble lineage"—emphasized his class position above other considerations.[55] It modestly represented the fact that the Diggeses were an old Kentish family. Indeed, as we know from other sources, they had been long established in the vicinity of Canterbury and were related by ties of marriage and friendship to most of the prominent families of that region: the Sidneys, Sackvilles, Wyatts, Brooks, Clintons, Fyneux, and others.[56] A full-page engraving of the Digges family coat of arms at the end of *Wings or Ladders* visually represented the author's social station, playing on the association between the nobility of his birth and the wisdom attained by the Platonic soul through contemplation of mathematicals. Borne aloft by parallactic calculations beyond the sense-bound world of mortal humanity, the intellect ascends unimaginable distances into the "pure aether": Digges alluded to writers whose mathematical wings were, by contrast, like those of Icarus, which melted in the sublunary region.[57] *And so*, continues the title, *the Immense Distance and Magnitude of this Portentous Star, this Unexpected Tremulous Fire in the Northern World, and forthwith its Awe-Inspiring Place, may be found; and so also may God's Astonishing and Frightening Presence be displayed and known most clearly.*

Thomas Digges's father, Leonard (1520–63?), besides having authored several practical mathematical treatises, was politically controversial: he had been a leader in Wyatt's rebellion and was condemned for high treason under Queen Mary.[58] It was fortunate for the family that the estates were restored under Elizabeth with only a fine and pardon.[59] A close contemporary of John Dee, he was also the authorial model for his son, Thomas. Unlike the Danish nobleman Tycho Brahe, however, Thomas Digges did not face class strictures against engaging in mathematical practice. The son faithfully completed several of his father's publications posthumously and, in a perpetuation of the authorial lineage, he also transformed the paternal text by complementing the initial presentation with theoretical demonstrations and insertions of his own conception. As in the case of Rheticus, a deferential discourse of "following the father" was juxtaposed with displays of independence and difference. Plato was the first resource by which the son marked his difference, Copernicus the second. Leonard was especially interested in "geometrical mensurations," problems that involved the application of geometry to practical measurement of all sorts. Although he described himself as a "gentleman," he constructed his audience as one of "practitioners." He also wrote his works in the vernacular, contributing to a trend of this period in England.[60] For example, his *Tectonicon* offered practical advice for land surveying and building.[61] In 1571, Thomas published and augmented his father's *A Geometrical Practical Treatize, Named Pantometria*.[62] This work concerned fortification, the digging of mines, surveying of fields, military problems, the properties of mirrors, among other topics, all described and crisply illustrated in a book about measuring everything terrestrial. To this work Thomas added a separate work of theoretical geometry. The title informs that Thomas "thereunto adioyned a *Mathematical* treatise of the fiue regulare *Platonicall* bodies, and their *Metamorphosis* or transformation into fiue other equilater uniforme solides *Geometricall*, of his owne inuention, hitherto not mentioned of by any *Geometricians.*" Digges wrote that to "reache aboue the common sorte, I haue thought good to adioyne this Treatise of the 5. *Platonicall* bodies, meaning not to discourse of their secrete or mysticall applicances to the Elementall regions and frame of Celestial Spheres, as things remote and farre distant from the Methode, nature and certaintie of Geometrical demonstration."[63]

Digges's appended "mathematical discourse" was not the first sixteenth-century appreciation of these geometrical objects and their properties. Commentators and admirers of Euclid, such as François de Foix de Candale, Piero della Francesca, Luca Pacioli, and Albrecht Dürer had all found reason to praise the five regular solids as representing the consummation of Euclid's *Elements,* showing how the earlier theorems of plane figures led up to the demonstrations of five and only five solid objects constituted exclusively from the square, the equilateral triangle, and the pentagon.[64] This was a well-established theme in Proclus's *Commentary* and was in complete accord with Dee's preface to the Billingsley Euclid, although it added nothing to the geometry of practical mensuration.

Thomas's attraction to the Platonic polyhedra may have been stimulated by an early personal association with John Dee after the death of his father. Not enough has been made of this relationship. In *Wings or Ladders,* Digges referred to Dee as his "second parent" in mathematics and lauded Dee as having "sown many seeds of those most sweet sciences" during his "most tender years" and as having "nurtured and increased others which previously had been sown in a most loving and faithful manner" by his own father.[65] Dee, in turn, addressed Digges in his *Parallactic Device* as "my dearest young man and my worthiest mathematicus and heir."[66] With good reason, Francis Johnson read these passages as referring to Digges's life, suggesting that Digges may have been Dee's "ward and pupil" in the years after Leonard's death.[67] Although there is no independent evidence that he was a "ward," there was clearly a close enough relationship between the older and the younger man to believe that Digges had a firsthand acquaintance with the fundamental ideas of Dee's early period, that is, prior to his "conferences" with angels: the Platonizing preface to Euclid, the astrological reform in the *Propadeumes,* and Offusius's *De Divina Astrorum*

Facultate (as well as Dee's anger over Offusius's plagiarism).[68] We now know that the relationship began when Digges was about thirteen, as the Roberts-Watson study of Dee's library shows that Dee gave a copy of Archimedes' *Collected Works* to Digges in 1559.[69] And what else would be implied by the notion of "worthiest heir" than that Digges had a direct acquaintance with Dee's library, or that Dee was fully familiar with the arguments contained in *Wings or Ladders,* copies of which Digges had sent to close acquaintances?[70] But there is no evidence that such acquaintanceship included access to the private diary that Dee's biographers have found so valuable.

Another Digges father-son production, involving the juxtaposition of theory with practice, showed up most poignantly in Leonard Digges's *A Generall Prognostication.* Like the *One Hundred Aphorisms* of Ptolemy and Hermes, Digges offered "plaine, briefe, pleasant, chosen rules"— combinations and permutations of planets and signs *always* linked to meaningful terrestrial effects, from inclement weather to rainbows and earthquakes. It was reminiscent of Schöner's handy *Three Books on the Judgments of Nativities* (1545). Originally published in 1553, the title was changed after 1555 to *A Prognostication Euerlastinge of Righte Good Effecte,* a so-called perpetual almanac (a book of general rules rather than specific predictions), and it continued in print for at least fifty years thereafter. However, in 1576, Thomas unexpectedly added to this a translation of a few chapters from *De Revolutionibus,* book 1, which he called *A Perfit Description of the Caelestiall Orbes according to the most aunciente doctrine of the Pythagoreans, latelye reuiued by Copernicus and by Geometricall Demonstrations approued.*

The Platonism of *Wings or Ladders* is concerned with contemplating order rather than plotting the activity of celestial forces.[71] This does not mean that Digges eschewed astrology. We know that William Cecil, Lord Burghley, had requested of Digges an astrological judgment, although Digges included nothing in his work comparable to Tycho's nativity of the star's meaning. Nor does he seem to have taken up Dee's ambitious proposals for a reform of astrology's theoretical principles. Yet Digges did take pains to communicate privately an astrological prognostication to Burghley, in which he speculated cautiously on the new star's "unknowen influence":

I have waded as farre as auncient groundes of Astrologie and Aucthors preceptes of approved credytt will beare mee, to sifte out the unknowen influence of this newe starre or Comet: whiche is lyke to be no lesse vehement then rare, as by the first and seconde of the 7 notes heerin enclosed maye partly appeere. The third sheweth from what quarter the calamitye is to be expected. The fourth on what kinde of creature the influence is like to take effect, The 5th and 6th what Regions and Provinces heere on earth are menaced. The 7th betrayeth the efficient infernall cause, or sutche as are like to place the cheefest parte in this fatall Tragedie. More particularities by arte cannot bee gathered. [72]

The decision to write *Wings or Ladders* in an "astronomical" mode and to exclude the astrological judgment allowed Digges to postpone the question of how the two parts might be connected. *Wings or Ladders* referred forward to the Copernicus translation project of the *Perfit Description* but not backward to the *Pantometria*'s Platonic solids, or to what Dee called "number numbrying" in the Billingsley Preface, or again to Offusius's mystical Pythagorean numbers; its wings were the mathematical propositions of practical astronomy necessary for computing the position of the nova.[73]

Dee and Digges must have engaged in discussions about parallactic methods, because just a few days after Digges's work was published, Dee's *Parallacticae Commentationis Praxeosque Nucleus quidam* [*The Kernel and Practice of Parallactic Inquiry*] appeared (dated Mortlake, March 5, 1573). Curiously, Dee's work nowhere mentions the nova (not even in his private diary), although Digges says that "that honorable man John Dee . . . himself took over the task of treating this matter [of the new star], which I have no doubt will be resolved shortly to the glory of God and the pleasure, utility, and highest admiration of students of the mathematical arts."[74] Many extant copies of *Wings or Ladders* are bound with the *Parallactic Inquiry,* suggesting that contemporary readers had read Digges's comment as meaning that the two works belonged together.[75] *Wings or Ladders,* although it displayed a table of star positions in Cassiopeia at the very beginning, was, like the *Parallactic Inquiry,* largely a treatise of parallactic theorems rather than a "history" or compilation of observational reports. Tycho Brahe, who studied Digges's work very

carefully, remarked that the promise of the "brilliant and magnificent title" was not borne out by the very small number of observations on which the substantive claims were built. Instead of appending Regimontanus's treatise on comets, in the manner of Hagecius, Digges provided a series of geometrical "definitions" and "problems" devoted to the improvement of Regimontanus's methods for situating comets. And, as he remarked near the end: "I have discussed this [method] so profusely not because I truly wish to take anything away from Regiomontanus, a most skilled mathematician, but lest others in our time, more accustomed to the theoretical than to the practical part of astronomy, might lose first place in this Olympic battle for Truth. For there is no doubt that were Regiomontanus alive today, having repudiated his old method, he would have wished to invent new ones, that he might be able to search for the truth of this mystery lying amid the thick shadows."[76] Digges thus shared Regiomontanus's valuation of mathematical certitude while differing with him about how to set up the parallactic triangles.

DIGGES ON COPERNICUS
IN *WINGS OR LADDERS*

Wings or Ladders can also be read from the perspective of its Copernican content. The first thing to be noticed, however, is that the frontispiece foregrounded the existence of a new star and the methods for establishing its position in the heavens. As a divine miracle, the nova also heralded the second coming of Christ, with its eschatological overtones of change in the natural world.[77] Like Maestlin, Digges did not advertise any association with Copernicus in the title. He put all the Copernican references into the prefatory and concluding material—the brief preface to Lord Burghley; the second, much lengthier, and even more substantive one to the general reader; and the proemium to the work. All these remarks went considerably beyond the brief allusions to Copernicus provided by Maestlin in his *Astronomical Demonstration* or the praises of the Copernican *Prutenics* so typical in this period. Although abbreviated, *Wings or Ladders* was still the most substantive presentation of Copernicus's central hypotheses to appear in print since the foundational works of the early 1540s.

Digges, like Rheticus, built his position directly on the Copernican epistemic ideal of the dialectic weighing of probables—the opposition between the "monstrous" assemblage of homocentrics and eccentrics and the "symmetria" of the Copernican "hypotheses." However, for Digges, Copernicus's reformist imagery did not have the power to evoke moral or political associations with papal Rome and Erasmian Varmia—or, of course, Rheticus's private associations with his decapitated father. Digges's imagery was more explicitly and vividly Platonic, more than likely reflecting his strong connections with John Dee and the London circles in which he moved. Likewise, Digges's accounting of the incoherence of the received astronomy bore his own imprint: he declared that the ancients mistakenly proceeded to set up their "theorics" from false parallaxes and from these to "hunt for true distances"; instead, they should have established their theorics in "inverse order" (i.e., a posteriori), beginning with "observed and known parallaxes." This was not an objection made by Copernicus himself. Because *Wings or Ladders* was a book grounded in parallactic techniques, however, Digges evidently believed that his point would carry more than rhetorical force, although he must have known that neither Copernicus nor the ancients could have begun with measurements of planetary parallax.[78] Digges, obviously echoing Copernican concerns, offered an imprecise description of the "other" universe, the commonly received one, without any visual illustration: "rather crippled and mutilated, made up of mutually colliding and resisting eccentric orbs and epicycles moving irregularly around their own centers."[79] No such "collisions" appeared in Leonard Digges's illustration of the concentric planetary orbs. On the other side of the balance, Digges produced a paraphrase of the Horatian passage from the preface to *De Revolutionibus*. He followed this immediately by Copernicus's dialectical topos (the same lines inflected by Rheticus and Maestlin) that if the premises were not falsely assumed, then the conclusions would not have turned out to be false.[80] Digges did not say that he himself found these reasons persuasive. His approach was indirect and attributional: "These were the particular reasons why Copernicus, that man of singular industry and admirable skill, made use of other hypotheses and tried to erect a new anatomy of the heavenly machine." But without then dwelling further on the details of Copernican

"anatomy," Digges declared that the diminishing of the new star's apparent magnitude now furnished an "especially opportune occasion for making a trial." By a "trial" he meant an observational test to see whether the "sole cause" of future changes in magnitude was, "as supposed in the Copernican theorics," the Earth's motion.[81]

Here is the first instance of an author's using the hypothesis of the Earth's motion to predict consequences that he believed would rule out all other possible hypotheses. In other words, there was evidently a hope for an apodictic demonstration rather than a probable or likely argument of the sort found in the *Narratio Prima* and in *De Revolutionibus*: "I promise to demonstrate in the future . . . that Copernicus's paradox concerning the Earth's motion, thus far rejected, is actually the truest."[82] Digges's hypothesis, in this respect, looks like a stronger claim than that of Maestlin. Logically it resembled Galileo's later conviction that the phenomenon of the tides could only be explained by means of the hypothesis of the Earth's daily and annual motions. Propositionally, Digges's argument may be recast as follows:

1. There can be no alteration of any kind in the substance of the heavens.

2. Because the new star is part of the heavens, it cannot actually increase or diminish but can only appear to do so.

3. The cause of those apparent changes is the Earth's annual motion.
   More specifically: if and only if the Earth is the (sole) cause of the appearance of change in the star's brightness, then the star will appear to diminish around the vernal equinox, slowly increase in size as it approaches the summer solstice, reach a point of "unusual magnitude and splendor" at the autumnal equinox, and then diminish again as it moves toward the winter solstice.[83]

Digges's hypothesis built on analogy between the seasonal variation in the Sun's intensity and the predicted (seasonal) variation in the new star's apparent magnitude. But, whatever the underlying analogy, he offered no evidence that he had tried to make the required observations and calculation of stellar parallax—nor, given the great distances involved, could he have produced anything but a null result. Like those of Icarus, the wings of Digges's hypothesis were destined to melt as he got closer to the Sun.

(RE)CLASSIFYING THE STAR

Sometime between 1573 and 1576, Digges must have realized that he was not going to be able to produce an apodictic demonstration based on parallactic measurements. According to the hypothesis in *Wings or Ladders,* only one year of observations should have been needed to visually detect the effects of the Earth's motion in the star. We have no evidence that Digges attempted to make such observations. By 1574, however, the star again did something unexpected: after having been visible for sixteen months, it disappeared altogether. Various explanatory options were now open that involved making judgments of disciplinary classification. To make such judgments entailed a preference for certain sorts of tools to be applied. First, it could be classified strictly as a hyperphysical event, an ad hoc divine miracle, and thereby excluded from all physical accountability—a testimony to the ultimate impotence of human reason. Such a move would shift explanatory authority from astronomy and natural philosophy to theology. Second, the premise of celestial immutability could be rejected, thereby opening the possibility of a physical explanation for the generation and disappearance of the object. Third, Digges could remain within the traditional astronomical framework by rejecting the Copernican planetary arrangement and, with it, his hypothesis linking the Earth's motion to the variable appearances of the star. Fourth, he could retain the Copernican ordering and adopt something like Maestlin's solution: locate the star in the outermost, motionless Copernican sphere, a place so distant that no stellar parallax could ever be measured. But, of course, without invoking a Maestlinian "hyperphysical" explanation, Digges could not account for the star's disappearance.

That Digges chose none of these possibilities, that he did not give up his adherence to the Copernican ordering after 1574, shows that he was still searching for stronger grounds on which to build his commitment. Put otherwise: Digges continued to accept the "probability" of the Copernican celestial ordering on the basis of the *symmetria* claims in *De Revolutionibus,* but, guided by the Aristotelian apodictic ideal of knowledge, he then continued to look for new arguments and resources that would permit him to adopt a position that would exclude all other alternatives. As a venue for this project, he chose his father's quite successful, twenty-year-old *A Prognostication of Right Good Effect.* And, as is well known, he added to it his own appendix, with an English translation of sections from book 1 of *De Revolutionibus.*[84]

THE MATHEMATICIANS' COURT

Thomas Digges's choice of authorial form and expression raises interesting questions about his theoretical practices. We see him moving in a variety of ways to enhance the authority of Copernicus's central claims. Making the translation was itself a declaration about the value of the work and the kind of broader audience he hoped to reach. This "addition," as he labeled it, was not meant only for the learned. In choosing to adjoin his translation of Copernicus's astronomical theoric to his father's *practica* of astrological rules for making weather predictions and assigning appropriate times for bleeding, he consciously chose a conjoint format rather than a separate treatise like *Wings or Ladders.* Also, it is striking that he chose not to join his translation to the astrological theoric of Dee's *Propadeumes.* The rarely reproduced title page is a vivid reminder that this was first and foremost a type of prognostication and that the Copernican section was the part "lately corrected and augmented by Thomas Digges his sonne." It continued to be reissued in that form until the last edition of 1605.[85] Evidently, Thomas hoped to reach the same kind of readership in England already attracted to Leonard's works and hence already concerned with practical matters of prognostication. But in his foreword, titled "To the Reader," he made it clear that he aimed to bring notice to a neglected part of philosophy: "I thought it conuenient together with the olde Theorick also to publish this, to the ende such noble English minds (as delight to reache aboue the baser sort of men) might not be altogether defrauded of so noble a part of Philosophy."[86] Thus "noble English minds" would have an opportunity to contemplate both the new and old theorics within a single work of astrologically based weather prognostication.

The precedent for such a joint publication of theoric and practice was obviously the *Tectonicon* (1571). Yet it was one thing to juxtapose the *Tectonicon*'s Platonic solids in a work of practical measurement, but quite another to join a "Perfit Description" of a heliocentric scheme in an infinite stellar surround ("this most excellent and dyffycile parte of Philosophye")[87] together with a "Prognostication Euerlastinge" that was supposed to facilitate astrological predictions "for ever."

This juxtaposition left open the important question of how the two works were supposed to be connected to one another. Was Thomas Digges here proposing a Copernican solution to the problem of the distribution of celestial forces raised in John Dee's *Propadeumes*?[88] If so, he provided no specifics and would admit in his dedication only that "I have waded farder then the vulgar sorte Demonstratiuè & Practicè." Moreover, he carefully called his Copernican addition a "Perfit Description" rather than, say, a "Perfit Philosophical Demonstration." And then, following the strategy of Copernicus and Horace, he invested in his intended audience the custodianship of appropriate judgment. In his title, he used the phrase "by Geometricall Demonstrations *approued*." But Digges's trope was neither a Horatian audience moved by good poetry nor an Aristotelian one moved by necessary demonstration. Instead, he transformed the image into one of a legislative assembly before which he himself would argue the case like a barrister, one who appears to leave the rendering of judgment to his mathematically skilled peers: "God sparing life I meane though not as a Iudge to decide, yet at the Mathematicall barre in this case to pleade in such sorte, as it shall manifestly appeare to the World whether it be possible upon the Earthes stabilitie to deliuer any true or probable Theorick, & then referre the pronouncing of sentence to the graue Senate of indifferent discreete Mathematicall Readers."[89] Here Thomas Digges's rhetorical "pleading" for Copernicus's solution to the order of the planets thinly veiled the conflict with Leonard's geocentric diagram of the orbs ("The nature, course, coulour, and placing of these seuen Planets, according to Ptolomei") where the Sun, Mercury and Venus all shared the traditional period of one year.[90] Furthermore, the absence of any explicit links in *De Revolutionibus*

between the new planetary ordering and the science of astral forces occasioned no obvious place for direct commentary on that subject.

REORGANIZING COPERNICUS

A related question is why Digges chose to select for translation only four chapters from book 1 (reorganizing the chapters in the sequence 10, 7, 8, 9). And why did he place with his translation an eye-catching representation of a heliocentric "description of the caelestiall orbes" embedded within a starry orb "fixed infinitely up" and attributed to "the most auncient doctrine of the Pythagoreans, etc."? The choice and order of chapters might reveal something about where Digges located the fulcrum of the Copernican argument. Rather than follow the Ptolemaic organization in the *Almagest*, beginning with the sphere and building up through the principle of the individual motions to the order of the planets (chapters 1–10), Digges instead foregrounded planetary order (chapter 10) and then considered the inadequacy of the Aristotelian counterclaims (chapters 7, 8, and 9). In other words, he restructured Copernicus's own presentation so that it conformed to the model of a demonstration, in which the major premise is set forth and then objections (against Aristotle and Ptolemy) are proposed and eliminated. To manage Osiander's "Ad Lectorem," Digges simply eliminated it and substituted his own assertive "To the Reader": "*Copernicus* mente not as some have fondly excused him to deliuer these grounds of the Earthes mobility onely as Mathematicall principles, fayned & not as Philosophicall truly auerred."[91] Digges then explained why he had included sections of Copernicus dealing with Aristotle's physics: "I haue also from him [Copernicus] deliuered both the Philosophicall reasons by *Aristotle* and others produced to maintaine the Earthes stability, and also their solutions and insufficiency."[92] The "insufficiency" was to be judged against Copernicus's own epistemic principle: "There is no doubte but of a true grounde truer effects may be produced then of principles that are false, and of true principles falshod or absurditie cannot be inferred."[93] Here is the juncture where Digges (and Maestlin) chose to follow Copernicus's dialectical standard rather than the skeptical (and pessimistic) syllogism (*ex falsa verum*) that Cla-

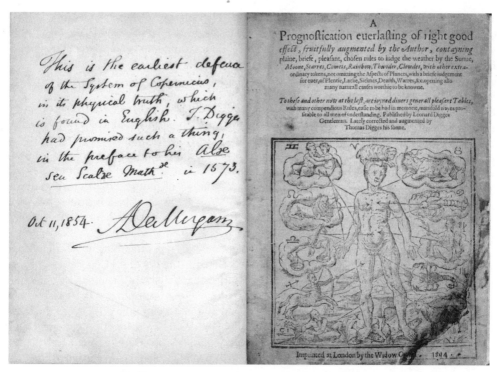

51. Augustus De Morgan's copy of Leonard Digges, *Prognostication Euerlastinge*. 1596 (but incorrectly showing 1594) with the "augmentation" by Thomas Digges but lacking the Pythagorean-Copernican diagram. Right: Title page displaying zodiac man (repaired). Left: Note by Augustus De Morgan: "This is the earliest defence of Copernicus, in its physical truth, which is found in English. T. Digges had promised such a thing, in the preface to his *Alae seu Scalae Math[ematic]ae* in 1573." Courtesy Senate House Library, University of London.

vius would shortly propose. In other words, Digges and Maestlin focused attention on what they regarded to be false principles (e.g., the Ptolemaic equant) and false effects (e.g., colliding orbs) considered to betoken false premises.[94] In turn, both managed the uncertainty by studiously ignoring the case in which false premises lead to true conclusions.

In *A Perfit Description*, Digges was most concerned to bring his "new theorick" as close as possible to demonstrative knowledge, effectively to compensate for the failure of *Wings or Ladders*. That this was his ultimate objective can be inferred from the title of a never-completed work that he mentioned in his *An Arithmeticall Militare Treatise, named Stratioticos* (1579): "*Commentaries* upon the *Reuolutions of Copernicus*, by euidente Demonstrations grounded upon late *Obseruations*, to ratifye and confirme hys *Theorikes* and *Hypothesis*, wherein also Demonstratiuelie shall be discussed, whether it bee possible upon the vulgare *Thesis* of the Earthes *Stabilitie*,

to delyuer any true *Theorike* voyde of such irregular Motions, and other absurdities, as repugne the whole *Principles* of *Philosophie* naturall, and apparant grounds of common *Reason*." This choice of wording makes evident that he was using "demonstration" in both the sense of astronomical proofs that employ geometrical theorems and are confirmable by observations, and the much stronger Aristotelian demonstration "voyde" of physical "absurdities." He was conscious of doing something different. More than once, he represented his own project in opposition to that of the learning of the universities: "Behold a noble Question to be of the Philosophers and Mathematicians of our Uniuersities argued not with childish Inuectives but with graue reasons Philosophicall and irreproueable Demonstrations Mathematicall."[95] That the "commentary" was still unfinished in 1579 suggests that Digges was never able to complete the full details of his "graue reasons Philosophicall." After that date, we now know that he turned to military and political concerns.[96] This interpretation makes logical sense of Digges's own "augmentations" to the translation where he intervened, effectively as a natural philosopher, in chapters 10 and 8 of book 1 of *De Revolutionibus*. The first of these "grave reasons Philosophicall" addressed the question of whether the universe is infinite, which Copernicus famously left "to be discussed by the natural philosophers."[97] The second insertion tried to explain the physical effects that would occur were the Earth to move.

Digges was first "discovered" in 1839 as a figure of special relevance to the Copernican question by the mathematician and historian Augustus De Morgan (1806–71). In the early 1930s, a Stanford English professor named Francis R. Johnson rediscovered Digges through a copy of the *Prognostication Euerlastinge* at the Henry E. Huntington Library (founded in 1919). What made the Huntington copy especially important is that, unlike some other extant copies, it contains intact Digges's visual representation of an infinitized heliocentric system. Neither of De Morgan's two copies of the work (1574, 1596) contains the diagram that Johnson found, and hence De Morgan made no mention of the "infinity" problem in his otherwise still-insightful historical writings.[98] Since Johnson's discovery, however, the Digges visual has become easily one of the most popular and recognizable illustrations in textbooks of the history of science for the early modern period.[99] It has served as a sort of icon of Copernicanism, a placeholder between Copernicus and Galileo. But we must now remind ourselves that that diagram appeared in the relatively traditional work of his father, Leonard Digges.

THOMAS DIGGES'S INFINITE UNIVERSE "AUGMENTATION" IN LEONARD DIGGES'S *PROGNOSTICATION EUERLASTINGE*

Much has been made of the "Englishness" of Leonard and Thomas Digges, but Leonard's book, like many such works of this period, was openly (and heavily) indebted to midcentury Continental sources. The work opens with what appears to be a strange polemic: "Agaynst the reprouers of Astronomie, and sciences Mathematicall." This is less a defense of "mathematics" or even of "mathematical practitioners" than of the science of the stars against (unnamed) skeptics. A panoply of writers favorable to astrology constituted his main authorities in the foreword: Guido Bonatti, Philip Melanchthon, Johannes Schöner, and Jerome Cardano.[100] "I appoint all nyce diuines, or (as *Melanchton* termeth them) *Epicurei Theologi,* to his hye commendations touching *Astronomie,* uttered in hys epistles to *Simon Grineus,* to *Schonerus* and at the peroration of *Cardanus* fyue bokes, where he sheweth how far wyde they alledge the scriptures agaynste the *Astronomer.*"[101] Leonard's references fully mobilized key bits of midcentury authorization for the kind of astrology presented in his book: Melanchthon's oft-published letters to Schöner and Grynaeus, with their declared intention to write a full defense of astrology against Pico; Cardano's presentation of astrology as a compilation of genitures; Guido Bonatti's claim that astrology was a science in Aristotle's sense.[102] The book appeared slightly too early to include reference to John Dee's *Propadeumes* (1558). We may take it that young Thomas knew these writings firsthand (they were probably in Leonard's library), and by republishing his father's work he gave them his own endorsement.

But Thomas did not approve of everything. Leonard Digges had included in his work an illustrated general scheme of the planets—both in the traditional concentric arrangement and in the quite unusual form of individual, free-floating globes representing the different volumes of the planets. The outermost sphere in the geocentric scheme was labeled: "Here the Learned do Appoyncte the Abitacle of God: and all the Elect."[103] It was not typical for prognosticators to include general schemes of the universe in their works, but Leonard's illustration afforded Thomas an opportunity to correct the earlier edition and to ascribe the errant condition to the printer: "Hauing of late (gentle Reader) corrected and reformed sondry faultes that by negligence in printing have crept into my fathers General Prognostication: Amonge other thinges I founde a description or Modill of the world and situation of Spheres Coelestiall and Elementare according to the doctrine of Ptolome, whereunto all Universities (ledde thereto chiefly by the auctority of Aristotle) sithens haue consented."[104] The alleged correction was doubtless a polite stratagem for undermining the authority of his father's diagram, as both illustrations appeared in the same volume from the first to the last edition!

Thomas Digges's picture is literally a spatial in-

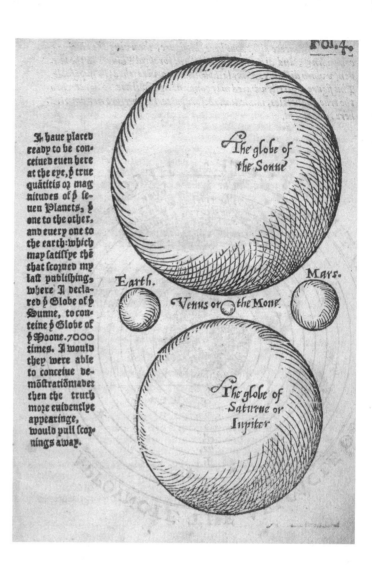

The globe of the Sonne

Earth.

Venus or the Mone.

Mars.

The globe of Saturne or Iupiter

J haue placed ready to be conceiued euen here at the eye, þ true quãtitis oʒ magnitudes of þ seuen Planets, þ one to the other, and euery one to the earth: which may satisfye thẽ that scoʒned my laſt publiſhing, where J declared þ Globe of þ Sunne, to conteine þ Globe of þ Moone. 7000 times. J would they were able to conceiue demõſtratiõmadeʒ then the truth moʒe euidentlye appearinge, would pull scoʒnings away.

52. Relative volumes of the bodies of the seven planets. L. Digges 1576. Reproduced by permission of The Huntington Library, San Marino, California.

version of his father's traditional ordering of the planets. Yet the differences in spatial organization are not matched by a comparable ontological inversion. The same moral opposition between saved and sinner is sustained by the soteriological theme encoded in both schemata, an opposition between the outermost, eternal "habitacle of the elect" (with its strong suggestion of Calvinist predestination) and the sinful "Globe of Mortalitie" or "the peculiar Empire of death" (the terrestrial region, composed of the four elements) on which the astral influences played. Because the two works were published as one, how could it be otherwise? Thomas still represented the outer sphere as an "orbe," but now an "unlimited" one and yet "filled" (if one can "fill" that which lacks limit) with "innumerable" lights so great in magnitude

that they exceed the sun "both in quantitye and qualitye." Commentators from Francis Johnson to Alexandre Koyré to Miguel Angel Granada have rightly pointed to the importance of Marcellus Palingenius's *The Zodiac of Life* as a conceptual resource for Digges's infinite scheme. This long, eclectically Stoic, pre-Socratic, and often explicitly anti-Aristotelian cosmic work was composed as a poem and was also widely used in England as a school text.[105] Palingenius (also known as "the Stellified Poet") hailed from Ferrara, that great site of fifteenth-century prognostication, and not surprisingly, the twelve-book zodiacal organization of his work does not fail to contain astrological themes.[106] But Digges cited only three passages (in Latin rather than the easily available English) to emphasize the contrast between celestial *bonum*

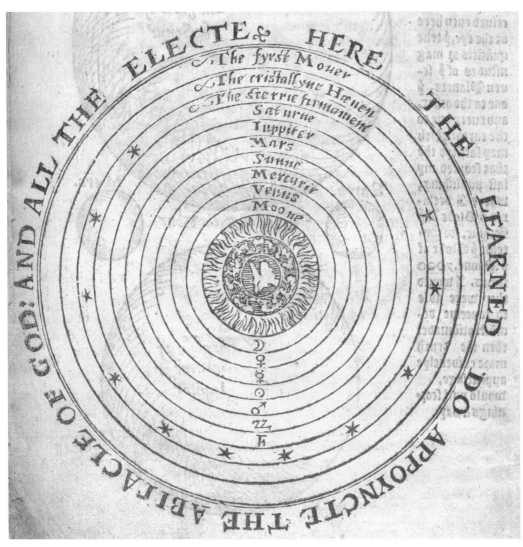

The fyrst Mouer
The cristallyne Heuen
The sterrie firmament
Saturne
Iuppiter
Mars
Sunne
Mercuri
Venus
Moone

ELECTE & HERE THE LEARNED DO APROYNCTE THE ABITACLE OF GOD: AND ALL THE

53. Geocentric woodcut. L. Digges 1576. Reproduced by permission of The Huntington Library, San Marino, California.

and terrestrial *malum,* and he gave no hint of the poem's breadth.

Might it be that Digges was looking in an Aristotelian sense for a "place" in which to locate the nova of 1572? Without signaling his own intervention to the reader, he inserted into his translation a passage that corresponds to Copernicus's heliocentric illustration (book 1, chapter 10), the language of which differs somewhat from the labeling of the diagram:

> That fixed Orbe garnished with lightes innumerable and reaching up in *Sphaericall altitude* with-

out ende. Of whiche lightes Celestiall it is to bee thoughte that we onely beholde sutch as are in the inferioure partes of the same Orbe, and as they are hygher, so seeme they of lesse and lesser quantity, euen tyll our sighte beinge not able farder to reache or conceyue, the greatest part rest by reason of their wonderfull distance inuisible unto us. And this may wel be thought of us to be the gloriouse court of ye great visible god, whose unsercheable worcks inuisible we may partly by these his visible coniecture, to whose infinit power and maiesty such an infinite place surmountinge all other both in quantity and quality only is conueniente.[107]

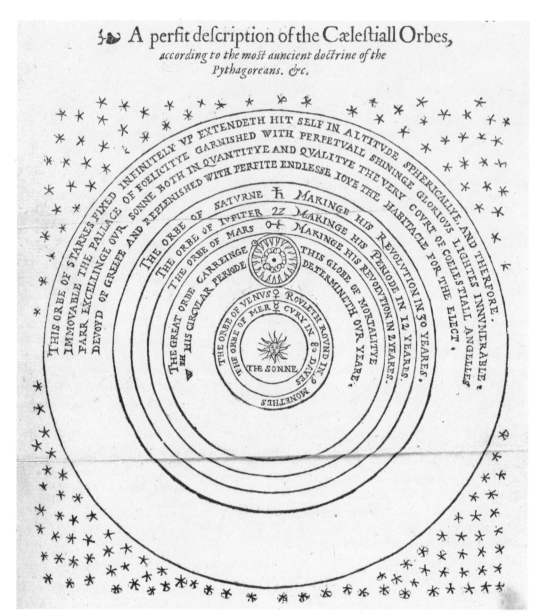

54. Infinite universe. T. Digges 1576. Reproduced by permission of The Huntington Library, San Marino, California.

Digges thus permitted Copernicus's voice, rather than his own, to conjecture the existence of an "infinite place" (called the "palace of foelicitye" in the diagram) whose "perpetual shinynge lightes" were mostly invisible. This place was also large enough to accommodate the nova, a star that greatly exceeded the Sun in magnitude. Yet, for all its possibilities, Digges did not park the nova in his celestial palace. His silence on the subject in 1576 suggests that he still could not explain

how a light that was supposedly eternal could appear and disappear in the space of sixteen months.

THE PLUMMET PASSAGE

With Digges's second original (and concealed) intervention in the translation, he again used resources already present in the Copernican text as an occasion to resolve a difficulty that he found in the argument. The theoretical practice involved in this instance, was, effectively, an imaginative extension of the problem of falling bodies—just as earlier Digges took Copernicus's refusal to speculate with "the philosophers" on the infinitude of the universe as an invitation to move across the boundary into natural philosophy. In chapter 8, Copernicus introduced the analogy of a ship to the Earth's daily motion, using a passage from Virgil's *Aeneid* (book 3, l. 72): "Forth from the harbor we sail, and the land and the cities slip backward." Characteristically, the humanist Copernicus used a literary reference to create a space for his philosophical voice: "For when a ship is floating calmly along, the sailors see its motion mirrored in everything outside, while on the other hand they suppose that they are stationary, together with everything on board. In the same way the motion of the earth can unquestionably produce the impression that the entire universe is rotating."[108] Copernicus then searched for new language, moving from the paradox of these appearances to speculate on possible explanations for why clouds and "other things that hang in the air" or float in the water share in ("link," "mingle," "accompany," "conform to," "are unaffected by") the Earth's motion. And finally, he touched on the case of falling and rising bodies, which, turning Aristotle's doctrine of simple motion against itself, he analyzed, still in Aristotelian categories, as a "compound of straight and circular."[109]

Digges evidently believed that a full demonstration of Copernicus's larger claims would need to improve on this account of the mixing of circular and rectilinear motions. And so, at this point in the translation, Digges silently augmented the discussion, making Copernicus say: "No otherwise then if in a shippe under sayle a man should softly let a plummet downe from the toppe alonge by the maste euen to the decke: This plummet passing always by the streight maste, seemeth also too fall in a righte line, but beinge by discourse of reason wayed his Motion is found mixt of right and circulare."[110]

Francis Johnson offered an empiricist gloss on this passage: "Digges brings forward the very pertinent experiment, which he had probably made himself, of dropping an object from the mast of a moving ship and noting that it appeared to fall to the deck in a straight line parallel to the mast."[111] Here Johnson overlooked two salient facts. First, Digges already had available from Copernicus the key elements needed to write this passage: the ship and the concept of "a mixt motion of right & circulare." As with the (now) well-known fourteenth-century discussions of Nicole Oresme and Jean Buridan, in the context of commentaries on Aristotle's *Physics,* it is at least as likely that Digges's description was a textual event as it was a "historical" one. And had Digges himself actually performed—or observed someone else perform—the ship experiment, he might have remarked on the place and time of its occurrence rather than presenting it in a conditional mood.[112] But he presented no such evidence; nor did he take any authorial credit for inserting the plummet passage into the translation. By contrast, in his introductory letter to the reader, following the three cited passages from Palingenius, he introduced in his own voice a quite novel (and suggestive) description of the globe of mortality: "In the midst of this Globe of Mortalitie hangeth this darck starre or ball of earth and water balanced and sustained in the midst of the thinne ayer onely with that propriety which the wonderfull workman hath giuen at the Creation to the Center of this Globe with his magnetical force vehemently to draw and hale unto it self all such other Elementall thinges as retained the like nature."[113]

In sum, although Digges evidently hoped for an apodictic demonstration of Copernicus's hypothesis, his conception involved no experiential performances, either making parallactic observations or conducting shipboard experiments with plummets. In this regard at least, he seems to have been a Platonist who never left his comfortable chair.

CONCLUSION

Maestlin and Digges became the first adherents of Copernicus in the thirty or so years after the appearance of *De Revolutionibus.* As we have seen, they shared a judgment about the betterness of the heliocentric planetary ordering. Such comparison of probables represented new territory, an uncharted realm in which there were no precedents for the formation of epistemic commitments. Ultimately, their judgments were warranted by dialectical reasoning which, as they both understood, fell short of the kind of strictly demonstrative knowledge prized by Aristotelian apodictic logic. This failure to secure an eliminative proof was the very conundrum with which Copernicus himself must have wrestled and that again came to the fore in the 1590s with the far more powerful arguments of Kepler and Galileo.

Maestlin and Digges were also both well aware of Wittenberg publications. But, as far as I can tell, neither of them worked inside the Wittenberg glossation network: neither had access to the Reinhold annotations, with their emphasis on books 3–6 of *De Revolutionibus*. And, ironically, their *ignorance* of this more articulated interpretive tradition could well have assisted in opening up a new reading, more consonant with Copernicus's original intent. Their understandings developed within a different network of interpreters. One would like to know more about those personal networks than I have been able to provide here, and that is clearly an area for further research.

What is clear from the available evidence is that, for the first time since Rheticus, two readers of Copernicus, with no knowledge of Copernicus's and Rheticus's local context (other than perhaps Maestlin's knowledge of Osiander's identity), granted a positive valuation to the Horatian aesthetic standard. Digges explicitly mingled his own valuation with Platonic mathematical images and ontology that strongly suggests the influence of John Dee. No such explicit connections are evident for Maestlin, and whatever Platonic preferences existed in his inventory may have come through Rheticus.

Against this Copernican aesthetic and its corresponding dialectical commonplaces, with their mitigated claim to probable knowledge, stood the traditional Aristotelian standard of strict demonstrative proof that Rheticus and Copernicus had publicly sidestepped by their silence. The new question for Copernicus's second-generation followers was whether the heliocentric ordering could explain appearances that neither it nor the Ptolemaic arrangement had predicted. Here, the nova and the comet together mark a critical moment. Both Maestlin and Digges tried—largely unsuccessfully—to capture these unanticipated entities for the Copernican representation. Both tried to deduce the appearances of the nova, either from the size of the Copernican universe or from the Earth's motion. And Maestlin tried to establish that only Copernicus's ordering of Mercury and Venus could accommodate the comet of 1577. One might regard these efforts as early attempts to connect the extraordinary and unexpected to the ordinary and recurrent. Within the domain of theoretical astronomy, Maestlin and Digges, like many of their contemporaries in this decade, were beginning to alter the role of the traditional astronomer as prognosticator. But their theorizing occurred in a commentatorial mode: Maestlin's annotations more closely hugging the text, Digges's interpolations to his translation more boldly amplifying and exploring its physical entailments. By juxtaposing his infinitist scheme with his father's *Prognostication*, Digges visually dramatized an opposition between the ancient and the modern ways of proceeding, while also creating the impression that the Copernican "modill" was somehow connected to the making of forecasts. Within a few years, this emergent space of possibility would evolve in a way that he could not have foreseen.

A Proliferation of Readings

In the 1580s, second-generation interpreters of *De Revolutionibus,* mostly Nullists, rapidly produced a spate of new readings. These readings opened up issues that Copernicus himself had tersely bounded off (such as the universe's infinitude), merely used as a piece of his main argument (the Capellan arrangement of Venus and Mercury), altogether neglected to develop (heliocentric and geocentric transformations), or treated ambiguously (the ontology of the spheres). Planetary order, left out of consideration by the Wittenbergers, now moved from liminal to central consideration, making it at times a matter of aggressive advocacy and defense of priority. That a planetary arrangement—a hypothesis, after all—might be regarded as a matter of novelty, discovery, and credit rather than, as Copernicus had it, the rediscovery of something already known to the ancients, was unprecedented. Yet although some agents began to use *systema mundi* as an alternative term for designating or describing an arrangement of the planets, such practitioners still did not constitute the topic itself either as an autonomous literary genre or as a separate epistemic category.[1] Significantly, Copernicus and Rheticus's strong sense of systematicity as an interdependency of the parts tended to be ignored.

Meanwhile, the topic of planetary order continued to be dispersed among different kinds of writings. For example, the daring and unorthodox Giordano Bruno, the first philosopher to align himself with Copernicus, chose to avoid rhetorical formats conventional within the literature of the science of the stars. Some of his writings were works of natural philosophy composed in the form of long poems; others were divided more conventionally into propositions; still others were philosophical dialogues marked by elements of satirical parrying reminiscent of Rabelais and Erasmus and flamboyant self-fashioning worthy of the Shakespearean stage. It was in these quite diverse kinds of writings that Bruno famously represented Copernicus's proposal as but the first step toward his own bold vision of innumerable heliocentered worlds distributed endlessly throughout an infinite homogeneous space.

During the same period a fertile middle ground was quickly expanding—a space that, like the Wittenberg consensus, lay somewhere between the traditional and the Copernican arrangement. By 1588, this *via media* had become the site of a highly contentious priority struggle within which Tycho Brahe's geoheliocentric scheme emerged as the most visible and, ultimately, the most influential alternative. These are the sorts of developments that in some form might have been expected, but which did not occur, in the decade or so after 1543. In this chapter, I shall investigate why and how these positionings developed in the 1580s.

THE EMERGENCE OF A *VIA MEDIA*

At least from 1570, the possibility of a *via media* is already apparent in scattered printed references to planetary ordering. Jofrancus Offusius had explicitly discussed the Capellan passage in *De Revolutionibus;* three years later, Valentine Naibod's circumsolar diagram of Mercury and Venus presented the first published illustration of that arrangement. The appearance of the 1566 Basel Petri edition of *De Revolutionibus* clearly stimulated Naibod's attention. This edition roughly doubled the number of copies in circulation, thereby disseminating the crucial presentation of Capella in the context of the central heliocentric arguments. Simultaneously, by an equal number, Petri thereby increased the availability of Rheticus's work. Many readers who knew *De Revolutionibus* had never laid eyes on the *Narratio Prima*. Its convenient bundling with the main work now yielded significant consequences. The new packaging meant that the two works traveled together, with Rheticus's gloss intensifying the focus on the *symmetria* claim and the inductivist, provisionalist image of astronomy and natural philosophy.

Independently, however, other sources occasioned attention to the Capellan scheme. In 1573, a Leiden scholar named Jacob Susius acquired a handsome manuscript of the Greek poet Aratea's *Phaenomena* (in Latin translation). It contained many attractive illustrations, among which was a striking representation of the Capellan arrangement. The manuscript circulated locally among Leiden humanists but seems not to have been known outside those circles. Among the Leiden group was Janus Dousa, the son of the head of the university. In 1591, Dousa published a long poem accompanied by what was effectively a Capellan scheme that he labeled "Delineation of the Orbs of Venus and Mercury according to the Opinion of the Egyptians and Pythagoras." In this strictly classical context, it seems to have had little or no impact among mathematical practitioners; nor was it seen to have any consequences for astrological prognostication. Knowledge of the Capellan ordering had thus clearly traveled without carrying along the original problem situation in which Copernicus had used it.[2]

Unlike the Dutch classicists, however, Tycho Brahe encountered the Capellan arrangement in the immediate context of contemporary astronomical-astrological writings, notably those of Offusius and Naibod. Although we cannot say exactly when Tycho came into possession of *De Revolutionibus* or even when he first studied its arguments, he surely would have heard mention of Copernican claims as early as his student days with

Johannes Homelius in Leipzig, in the early 1560s.[3] Parts of his 1574 Copenhagen Oration could not have been written without a close familiarity with the theory's details. Hagecius's gift of the *Commentariolus* manuscript in 1575 gave Tycho privileged access to Copernicus's very earliest, barebones formulation, without the derivations of the models later presented. Still another consideration was his 1575 book-buying trips, which took him through the key publishing areas of southern Germany, Basel, and as far south as Venice.[4] During these travels, he bought and embossed his seal on a substantial number of books. Naibod's little elementary work is among those that survive; *De Revolutionibus* is unfortunately not. The extant evidence thus points clearly to a moment of intensified consideration in the two years prior to Tycho's move to the isle of Hven in August 1576, just as the building of the infrastructure of Uraniborg commenced and the assembling of a staff got under way. In 1578, Tycho continued to edge toward a Capellan scheme. He wrote that the comet of the previous year was "in the sphere of Venus"; moreover, "if one does not want to follow the usual distribution of the celestial orbs but would rather accept as valid the opinion of various ancient philosophers and of Copernicus in our own time, that Mercury has its orb around the sun, and Venus around Mercury, with the sun at the approximate center of their orbs, which reasoning is not entirely out of harmony with the truth, even if the sun is not put to rest in the center of the universe as with the hypotheses of Copernicus."[5]

A month after Tycho arrived at Uraniborg, he made a passing remark to a correspondent that indicated he had been contemplating building a mechanical model to help him visualize Copernicus's threefold motion of the Earth.[6] The problem of visualizing the claims and models of *De Revolutionibus* would soon change. Just as Uraniborg was nearing completion in July, 1580, a Silesian mathematician named Paul Wittich arrived on the little island. Wittich circulated easily within a network of Hapsburg celestial practitioners that linked Wittenberg, Prague, Wrocław, Görlitz, and, in due course, Uraniborg. In 1579, he became associated at Wrocław with the circle of Andreas Dudith (1533–89), a Hungarian humanist, diplomat, theologian, and would-be student of the science of the stars. During his four-month stay at Uraniborg, Wittich left his mark. He carried with him several copies of *De Revolu-*

tionibus—perhaps as many as the five that still survive today.[7] Four contained extensive annotations, some of them copied directly from Erasmus Reinhold; still other notes in other copies overlapped the earlier ones. Two of these earlier copies contained additional blank pages bound into the back. Wittich used these sections as a sort of workbook in which to visualize and study the details of Copernicus's models. Although the copies once owned by Gemma Frisius, Michael Maestlin, and Jofrancus Offusius rival the Wittich copies in their profusion of annotation, Wittich's were the only ones to carry forward Reinhold's approach to reading *De Revolutionibus*.[8] Dudith aptly referred to him as "Witticho-Copernicus noster Regius" and "Neo-Copernicus."[9]

Wittich's stopover at Uraniborg seemed to make him a prime candidate for joining Tycho's unusual and growing household. That novel aggregation of skilled observers and artisans was always in need of further scholarly and technical labor to build and maintain the instruments, manufacture paper, print books, observe the stars and planets, record times and angles, and serve as couriers for Tycho's books and letters.[10] Wittich, however, could offer something unique and useful to both Tycho's practical and his theoretical concerns. He had devised the so-called method of *prosthaphaeresis*, a technique for mitigating the tedium of multiplying long numbers by reducing these calculations to operations of addition and subtraction (the essential insight behind logarithms).[11] But, beyond this valuable tool of calculation, conversations between Wittich and Tycho must have deepened the latter's understanding of Copernicus's models and brought fresh focus to the question of planetary order. The annotations can be read as residual traces of the content of those exchanges. They provide a privileged glimpse into the interpretive practices of a highly skilled mathematical practitioner trained in the Wittenberg style, with special access to the Reinhold glosses and with numerous personal contacts throughout the far-flung network of Melanchthonian universities and related court circles. Tycho surely appreciated what Wittich had to offer because, as far as we know, this was his first exposure to the Reinholdian glosses. As he left, Tycho gifted him with a sumptuous copy of Peter Apianus's *Astronomicum Caesareum*. Yet Wittich must have had some independent means—or prospect thereof—because the gift was not sufficient to hold him: after his four-month visit, he never returned to the island.

Wittich's reading of *De Revolutionibus* was important for Tycho in several ways. Following Reinhold's astronomical axiom ("Celestial Motion is both uniform and circular or composed of uniform and circular motions"), he went far toward visualizing the transformations of Copernicus's planetary models into corresponding geocentric versions. Wittich showed exactly how geocentric transformations of Copernicus's individual models would retain their parameters—and hence, by implication, their full predictive capacities. The undertaking was at times tricky. Copernicus had wrestled mightily—and not always successfully—with the problem of Mercury and, between the *Commentariolus* and *De Revolutionibus*, there are no less than four related, yet different models.[12] Wittich, who knew the *Commentariolus*, confronted three different heliocentric modelings—or "modes": an epicycle on an eccentric, an eccentric on an eccentric, and an epicycle on an epicycle. Wittich's diagrams showed the three heliocentric modes, but his geocentric transformations labored in the first two. Ultimately, the diagrams rendered *De Revolutionibus*, for anyone so inclined, into a geostatic modeling tool.[13]

Wittich's reading also provided Tycho with a more favorable gloss on aspects of the first book of *De Revolutionibus* that Reinhold had studiously ignored. For example, on Copernicus's account of gravity, Wittich commented: "Just as like seeks like, the elements do not desire the center of the universe but rather the center of their globe; and the only reason why the stars and the Earth are globes is because they are moved by their form."[14] Most strikingly, Wittich accentuated central features of what today might be considered the theory's explanatory likelihood.[15] He underlined Rheticus's statement that "there is something divine in the circumstance that a sure understanding of celestial phenomena must depend on the regular and uniform motions of the terrestrial globe alone."[16] His gloss reads: "The reason for the revival and establishment of the Earth's motion." Where Copernicus heralded the major consequence that the planets' ordering cannot be changed without upsetting the planets' *symmetria*, Wittich wrote: "Most of all, the evidence from the planets agrees precisely with the Earth's motion, and thus are strengthened [*confirmantur*] the hypotheses that Copernicus assumed."[17]

55. Copernicus's planetary arrangement, with Paul Wittich's annotations. Copernicus 1566. © Biblioteca Apostolica Vaticana (Vatican), MS. Ottob. Lat. 1902, fol. 9v.

Finally, Wittich carefully marked up (and corrected) the page where Copernicus, in a woodcut now famous, exhibited the order and periods of the planets.[18] Carrying the numbers out to three places, Wittich showed that, according to "a more exact reckoning of Copernicus," the mean sidereal periods of revolution increase continuously from the Sun to Saturn. In a second tabulation, he showed that the mean daily motions decrease continuously from the Sun; and in a third, he gave values for the synodic periods. At the top right of the same page, there is a revealing hint of

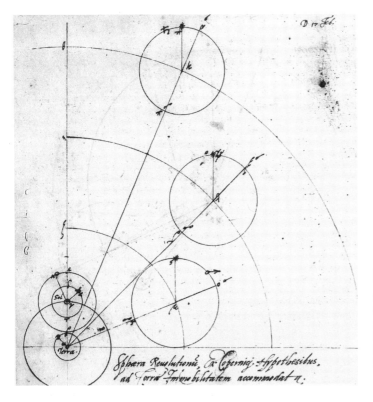

Sphæra Revolutionū, ex Copernici Hypothesibus, ad Terræ Immobilitatem accommodat η:

56. Paul Wittich's sketch, titled "Sphere of Revolutions Accommodated from the Hypotheses of Copernicus to the Immobility of the Earth," in Copernicus 1566. © Biblioteca Apostolica Vaticana (Vatican), MS. Ottob. Lat. 1902, fol. 210v.

his thinking: he located the Earth and Moon between the three superior and the two inferior planets ("the motions agree"); but, in a bracket that encompasses the entire group, he indicated that "the Sun participates." It is evident that Wittich understood and acknowledged Copernicus's *symmetria* argument, but he did not consider it decisive. He well appreciated what he had been taught in Reinhold's commentary on Peurbach— that without the Earth's moving, the Sun appears to participate in the motion of every planet.

Wittich thus left Tycho with both a clear sense of the Copernican theory's most significant explanatory gains and its calculational utility. Of course, calculational advantage was what all astronomer-astrologers wanted. And thus, on finishing his last drawings, Wittich could easily have followed Reinhold in ignoring the question of planetary order. This is exactly what one would expect had Wittich been engaged in the sort of pedagogical clarification typified by the diagramming and glossing strategies that Reinhold, Wursteisen, and Schreckenfuchs applied to such good effect in their editions of Peurbach's *New Theorics*. In other words, nothing in the conventional genres of the science of the stars would

have compelled him to engage in any further work. Yet Wittich's thorough immersion in the texts of *De Revolutionibus* and the *Narratio Prima* evidently left him unable to forget the compelling consequences that followed from setting the Earth in motion—the explanatory simplicity and celestial order that he himself had flagged as Copernicus's chief arguments.

In the very last sketch, Wittich tried his hand at creating his own *ordo*, knitting together the entire assemblage of planets into what he called the "Sphere of Revolutions." Effectively, the diagram presented a modified Capellan arrangement. In its use of concentric circles (rather than orbs or spheres) for the Sun, Venus, and Mercury, the sketch followed the artistic convention found in Naibod; in its descriptive language, it borrowed one word from Copernicus's title ("revolutions") but not the remainder ("of the celestial orbs"). However, in displaying the second anomaly for the superior planets, it broke precedent. Most strikingly, Wittich drew the epicyclic radii for the superior planets exactly parallel to the Earth-Sun radius, a visualization that evokes Peurbach's sun as "some common mirror and rule of measurement."

Beyond this suggestive move, however, there is evidence that the sketch was, in two respects, incomplete. First, although the diagram explicitly referred Venus's and Mercury's motions to the Sun, the circuits for the remaining three planets appear to be referred only to the Earth. Because Wittich's earlier, far more complex illustrations consistently show his correct understanding of how to make transformations using parallelograms (and trapezoids), it is surprising that he did not readily carry over the same practice by drawing parallel lines from the points representing the superior planets to the Sun. For if the radii always remain parallel to the Earth-Sun radius, then the planets are, in fact, revolving in exact coordination with the Sun. And if the parallelograms are completed, then the scheme becomes consistent with both a central Earth and with all the planets organized around a moving Sun. Second, sixteenth-century representations of planetary order were conventionally not calibrated; they were meant to display order but not distances. Thus, Wittich's unusual decision to include the solar epicycles for the superior planets invited further, proper sizing of the epicycle/deferent ratios. However, the diagram is not consistent with these ratios and hence depicts the epicycle diameters as equal to one another. Consequently, Wittich did not allow that Mars's epicycle, when properly calibrated, would cut the circle representing the Sun's circuit. Still another possibility is that Wittich was committed to a model of solid, material spheres and hence deliberately constructed the drawing so that Mars's epicycle would lie exactly tangent to the Sun's circle.[19]

Even as Wittich's sphere-of-revolutions diagram relaxed any lingering urgency about the traditional uncertainty concerning the ordering of Mercury and Venus, then, it posed a new uncertainty about Mars. Thanks to Wittich's fortunate dating of this diagram (February 17, 1578), we know that it fell at the end of a sequence of previous sketches made over a two-week period and, hence, a year and a half before he arrived on Tycho's island. Therefore, in his few months at Uraniborg, Wittich might well have raised Mars as a topic of conversation—perhaps in the castle's sumptuous Winter Room, where Tycho was accustomed to dine with his family and where such conversations often took place with visitors and more learned members of his staff.[20] This level of familiarity with the talented Silesian explains

why, after Wittich's early death in 1586, Tycho engaged in vigorous, unrelenting—and ultimately successful—efforts to procure from his widow his richly annotated library of Copernicana.[21]

The Wittich episode also highlights the persistent indeterminacy about celestial order that, time and again, resonates with Copernicus's early problem situation. Technical familiarity with the complex models of De Revolutionibus was certainly a prerequisite for taking advantage of the theory's imagined calculational benefits for astrological prognostication. But awareness of some of the theory's many virtues in explaining celestial phenomena did not automatically lead to the inference that this was a better—or even more likely—explanation than the traditional one.[22] Well into the seventeenth century, practitioners gave different weight to the importance of Copernicus's elimination of the annual motions. Astronomer-astrologers also had different tolerances for the premises of traditional natural philosophy. How, for example, could one decide between the long-held view that heavy bodies fall to the center of the universe and Copernicus's proposal that elements "desire" the centers of their own globes? And finally, how should biblical testimony and God's intentions be used to adjudicate between different hypotheses of celestial order?

In 1580, "Neo-Copernicus" left Tycho Brahe with much to think about: detailed demonstrations for adapting Copernican planetary models to a stationary Earth; an enhanced view of the Capellan circumsolar arrangement; the possibility of Copernicus's alternative to traditional terrestrial physics; and, overall, a much-heightened appreciation for Copernican aesthetics and explanatory allure.

ALONG THE VIA MEDIA
TYCHO'S PROGRESS

Tycho Brahe's encounter with Wittich nicely frames the problematic of the via media. Celestial practitioners of the second generation worked from mixed inclinations and ensuing uncertainties arising from a confluence of traditionalist and modernizing impulses. Time and again, the indeterminacy of that convergence recapitulated elements of Copernicus's original problem situation. Which way to lean: with the explanatory power (and physical uncertainty) yielded by a Sun-fixed arrangement toward the capture of singular novelties and possible further explanatory gain, or toward the conquest of singular novelties with just enough Copernican explanatory virtue not to imperil traditional physical and scriptural fundamentals?

After Wittich's departure, there is no evidence that Tycho ever wavered in his adherence to the Capellan arrangement of Mercury and Venus or to the Earth's central position. But whence to proceed? If two planets revolve around—and with respect to—the Sun, why only two and not others? And if more, then why not Mars, Jupiter, and Saturn (as Wittich's diagram all but declared)? Not least because of its proximity to Earth, the question of Mars crystallized this new issue of planetary order. Does Mars lie always beyond the Sun, as Wittich and Ptolemy argued? Or, as Copernicus believed, does Mars approach closer to Earth than the Sun does? Both in private correspondence and in later publications, Tycho consistently structured this conundrum as a choice between what he called the hypotheses of Ptolemy and Copernicus. Throughout, he noticeably excluded all mention of Wittich. And characteristically, Tycho turned for a solution to his great observing instruments—resources that, by 1581, were vastly more powerful and precise than any ever before constructed. It seems appropriate to ask what Tycho Brahe was looking for: What did he expect to find, and how did he represent his ventures?

The Copernican ordering, unlike the Ptolemaic—or, for that matter, its Wittichian variant—predicts that at opposition, the distance separating Mars from the Earth will be as little as two-thirds of the Sun's distance to the Earth. This close approach of Mars, was, at least, the entailment that Tycho himself circulated in 1584 and 1587.[23] As a matter of observation, when Mars is at its closest, very small differences in its apparent daily position should be detectable at the extremes of the arc joining the zenith point to the horizon, observations taken in the morning and at night. The heliocentric ordering also entails that Mars's daily parallax should yield a value greater than that of the solar parallax. However, although the now-accepted values for the solar (9") and Martian (<27") parallaxes would have been utterly out of reach for Tycho, he had accepted from the Greeks a number for the solar parallax that was twenty times greater (3'). Hence, the standard that he was using to guide his

expectations—although still subject to the notorious vagaries of atmospheric refraction—appeared to him to fall within the acceptable range of his instruments (1'). A value of less than three minutes would put Mars beyond the Sun; a value greater than three minutes would place it closer to the Earth than the Sun.[24]

Rheticus, as will be recalled, had already announced a further inference from the Copernican ordering—unwarranted, as it turns out—that Mars's approach justified the necessary conclusion that the Earth cannot be at the center of the universe. It is entirely conceivable that just this formulation of the Mars question had come up in discussion between Tycho and Wittich in late 1580, as one of the latter's several copies of De Revolutionibus was the bundled 1566 edition; and Wittich's annotations, although undated, show that he had read the very page on which Rheticus's remark occurs.[25] Furthermore, we cannot exclude the possibility that Tycho himself was familiar with Rheticus's little book even prior to Wittich's visit. Hence the choice that Tycho posed lay between, on the one hand, Wittich's expanded Ptolemaic-Capellan arrangement and, on the other hand, the full Copernicus-Rheticus ordering—or some third alternative. In January 1587, Tycho laid out the alternatives in these words: "Either the Earth is whirled about in an annual motion whereby all the planet's epicycles are eliminated, or another, hitherto-unconstructed arrangement of the heavenly revolutions must be sought."[26] This was a choice—it should be emphasized—made prior to any observations, from which the motions of Mars would support either Wittich's sketch or *that part of Copernicus's ordering that Tycho could incorporate into a new Earth-centered arrangement.*[27] At least as early as 1582, Tycho can thus be read plausibly as seeking a way to refute both Wittich and Rheticus. For Tycho, the issue between the two orderings hinged on a difference in Martian parallax of about two minutes of arc.

Two years after Wittich's departure, in late 1582 or early 1583, Tycho's observation logs show that he had begun to orient his large-scale instruments toward determining the diurnal parallax of Mars.[28] A letter to Brucaeus in 1584 indicates that he was explicitly seeking—and failing to find—evidence that Mars indeed approached Earth more closely than did the Sun. In Tycho's subsequent references, historians have noticed a curious lack of specificity to these earliest paral-lax observations.[29] The situation is further com-plicated by Tycho's claim in a letter of January 18, 1587, that he had determined that Mars did in-deed approach more closely to the Earth than did the Sun.[30] Much historiographical attention has focused on how Tycho could have claimed to have made an observation beyond his instru-ments' capabilities and how he could credibly have maintained that the paths of Mars and the Sun intersect. The matter held direct conse-quences for the ordering of the planets.

NEGOTIATING THE SPHERES' ONTOLOGY

The logic of this situation calls for some com-ment. If Tycho could persuade himself (and oth-ers) of a Martian observation of greater than three minutes, it would imply that, if the Earth remained at rest, there would be an interpene-tration of the spheres of Mars and the Sun. And if, in addition, the spheres were impenetrable, three-dimensional objects, then either there would be no motion at all, or there would be a fatal collision of Mars and the Sun. Because there was obviously no such celestial mishap, then solid, impenetrable spheres must not exist. But rejection of the spheres, in turn, led to a new dis-junction with new entailments: *either* the Earth must move, as the Copernicans Rheticus, Maest-lin, and Digges held, *or* the Earth stays at the cen-ter, and Mars, like Venus and Mercury, revolves with respect to the Sun; but Jupiter and Saturn split the difference, revolving with respect to ei-ther the Sun or the Earth. (In 1651, the prolific Jesuit G. B. Riccioli designated and endorsed the latter arrangement as the "semi-Tychonic.") In addition, an Earth rotating daily on its axis while remaining at the center of the universe is compatible with either of these orderings. These entailments are worth pondering, not least be-cause they show the breadth of conceptual possi-bilities at this moment; but they also forcefully highlight that the Copernican elements in Ty-cho's new arrangement were even more pro-nounced than is usually acknowledged.

The precipitating consideration came, as it had in the 1570s, from the realm of the extraordinary and, hence, the nonrecurrent. No later than June 1586, Tycho received an unfinished treatise on the fourth and most recent comet to appear in the

short period of eight years. This *Scriptum de Co-meta* was composed by Christopher Rothmann (1550?–1608?), *mathematicus* at the court of Hesse-Kassel. Rothmann, like Wittich, had studied at Wittenberg and was thoroughly familiar with *De Revolutionibus;* but, unlike Wittich, he was com-pletely persuaded by Copernicus's arguments. In the treatise that he sent to Tycho, he did not men-tion his views on planetary ordering, but he held back nothing in launching a robust interpreta-tion that altogether dispensed with the spheres:

Most philosophers heretofore have stated, and it is commonly believed, that the planetary spheres are solidly compact bodies, which by their own motion securely transport the planets attached to them. . . . The solid bodies permit no penetration of their region . . . This belief in celestial spheres is pro-mulgated by the greatest writers, and commonly possesses the authority of a general axiom. Never-theless, [impelled] by my love of truth, I shall prove that it is absolutely false. . . . I shall also show that between the sphere of the fixed stars and the earth there is nothing but this airy element, and that the seven planets are suspended only in air. . . . The planetary spheres are nothing but air, and they are marked off not in reality but only by reason, and are assigned to the planets so that each does not transgress the space allotted to it. . . . Thus it is now clear how a comet can be or move in the sphere of Saturn. . . . The very motion of the com-ets is the strongest argument that the planetary spheres cannot be solid bodies. For it cannot hap-pen that a solid body permits a penetration of its region. So you cannot pass with your body through a wall. For, two bodies cannot be in the same phys-ical space at the same time.[31]

Nothing as bold and decisive on this topic is found in any of Tycho's earlier writings. If any-thing, Tycho seems to have been uncertain about how to treat comets as physical entities. On the one hand, using the Capellan arrangement, he had assigned a special sphere to the comet of 1577–78, in the region between the Moon and Venus.[32] In the same treatise, however, relying on Paracelsian authority, he wrote: "The Paracel-sians hold and recognize the heavens to be the fourth element of fire, in which generation and corruption may also occur, and thus according to their philosophy it is not impossible for comets to be born in the heavens, just as occasional fabu-lous excrescences are sometimes found in the earth and in metals and monsters among animals."[33] This was something of a dilemma: if comets were like monsters—occasional, one-time abrogations of the natural order—then how could they appear and reappear, carried around by a planetary sphere? Indeed, if the comet of 1577 was just outside the sphere of Venus, then it ought to have reappeared as early as the fall of 1578. If it did not (and it did not), then it was not like a planet. It was thus an event outside the ordinary course of nature, and its disappearance required some new kind of explanation.

Yet within a few months of receiving Rothmann's still-incomplete treatise on the comet of 1585, Tycho replied with a crucial amendment: "I would readily agree with you, as you assert in the part you sent me, that the heaven is all air and does not consist of solid matter; it is indeed nothing but air, provided you understand that the air which is above the Moon is far more subtle than this element, to the extent that it deserves the name of most fluid and subtle aether rather than that of elemental air."[34] Tycho thus agreed with Rothmann on the issue of heavenly fluidity but judged that the heavenly air was aetherial rather than elemental. In other words, Tycho was not so ready to abandon the traditional ontological boundary between terrestrial and celestial stuff. Thus he left open the question of whether the heavens could generate anything new.

In the same letter, he added a statement that has since raised questions about his sincerity: "Truly, many years ago, I was *not* of the opinion that any orbs really exist in heaven, and that the heavenly material is hard and impenetrable."[35] In his reply, Rothmann acknowledged Tycho's position: "You correctly agree with me that the celestial spheres' matter is not hard or impenetrable, but fluid and fine, easily yielding to the planets' motion."[36] Tycho continued to repeat this view, although without reference to Rothmann, as when he famously wrote to Caspar Peucer in 1588: "I was still imbued with the opinion, accepted for a long time and approved by almost everybody, that Heaven is crammed full of certain real orbs carrying the stars."[37] These statements make it appear that Rothmann's treatise had merely reinforced a view that Tycho had already reached on his own.

Yet matters cannot have been quite so simple. If Tycho had struggled with the question of how the regularities of the heavens could accommodate such ad hoc monsters as comets, the parallax investigations of Mars, begun in 1582, represented a new and different problem: Can the paths of two regularly moving celestial bodies cross one another? That question obviously raised different issues about the nature of the heavenly substance: not whether the heavens can generate "fabulous excrescences" or whether such extraordinaries can execute a one-time crossing through the planets' spherical zones, but rather whether the eternal celestial substance is solid or fluidlike and airy. If there are no spheres, then what kind of stuff, if any, exists between the Earth and the planets? What causes the planets to move? Most critically, what explains the orderliness of that motion? What medium allows the planets and stars to cause effects on Earth? And what implications are there for reclassifying the Earth itself as a moving body?

Tycho Brahe had available to him a useful source for organizing his thinking on such questions—Lucio Bellanti's *Book concerning Astrological Truth*—a book in his possession by 1574, well before his Mars campaign and long before he received Christopher Rothmann's treatise. Bellanti held that the heavens were made of incorruptible, solid matter and that they moved circularly by their own nature.[38] Although he associated both Aristotle's and Ptolemy's authority with this view, Ptolemy had said nothing in the *Almagest* or the *Tetrabiblos* about spheres or their physical constitution; and, as is now well known, Ptolemy's authorship of the *Planetary Hypotheses* was not established until the mid-twentieth century. Bellanti took the task of defending astrology as *scientia* to mean that astronomy ("the other part of astrology") must be shown to have demonstrable physical foundations. In conventional scholastic fashion, he organized his conclusion for the solid constitution of the heavens as the outcome of the resolution of two groups of opposing arguments. This dialectical approach, characteristic of Aristotle's own practice, meant that he went to some length to lay out the very position that he then rejected. The refuted position, in this case, held that the heavens are made of a fluid or flowing substance (*fluxibilis*).

The affirmative arguments may be summarized. First, the lighter or thinner a body, the more fluid it must be; but the heavens are the lightest of all. And yet the heavens are also dense—or, at least, possess denser parts. According to Bellanti, whoever denies that the heavens are not dense should listen to Aristotle, who said that a star is the denser part of its orb. Solidity is that part of a body where there is matter; and porosity is that part where there is a scarcity or absence of matter. Furthermore, over a long distance, a great quantity of matter impedes vision, multiplies light rays, and causes the rays to be reflected and refracted. But because this does not happen in the heavens, it follows that the heavenly substance must be fluid. Next, the Sun and its parts are composed of a fluid substance. This is known because it possesses the property of fire, a fluid substance. Fire lights and heats and fosters life in animals. To be sure, the heavens turn around, and the elements, of which fire is one, move rectilinearly; but within its own sphere, fire, like air, moves circularly around the center, as is known from the philosophers and the weather.

Bellanti's reply to these arguments stemmed largely from worries about the coherence of the heavens. Spheres provide form, and contiguous spacing (solidity) sustains shape. If the heavens are fluid, then the parts will be spaced unevenly relative to one another—hence allowing a vacuum—and therefore, with no resistance a body would move in no time at all. Moreover, if the parts were unevenly distributed, then there would be nonuniform resistances from different parts, and therefore the moving intelligence, in order to move each part uniformly, would have to apply force not in one simple way but in an infinite number of ways—surely an "unphilosophical" idea. Moreover, different parts moving differently in various directions would dissolve all coherence: parts would separate from the whole, and heavenly bodies would fall apart. In short, the heavens would be corruptible.[39] Bellanti's objections arose largely from physical rather than from strictly optical considerations.

Bellanti's *quaestiones* might help to explain Tycho's otherwise bewildering claims—why, in the first place, he did not so readily give up a commitment to solid spheres after the appearance of the nova and the first comet. To abandon the spheres implied the need for an alternative physical account that would involve explaining how the heavens cohered rather than, as Reinhold had commented, losing the form of their movement and instead moving about freely "like birds in the air or fish in the sea."[40] Evidently Tycho did not

have such an explanation. But by 1585, no less than four comets had appeared, and the heavens still remained intact. Moreover, thanks to his new instrumentation, between 1582 and 1585 Tycho had become more sensitive to the presence of both planetary and stellar refraction. Indeed, if planets display greater refraction than stars, this might be crucial to explaining the failure to detect Mars's parallax.[41] In turn, the indisputable presence of refraction might have raised questions about a physical explanation. According to Bellanti, solid heavenly matter and distance did not impede the visual rays.[42] But in that case, what caused light from Mars to bend and for its apparent position to be distorted? Tycho's direct experience with comets and parallactic observations might well have caused him to think about the air below the Moon as the crucial medium of such distortion and predisposed him to look favorably—if not decisively—on Rothmann's bold inference that elemental air extends well beyond the Moon and throughout the heavens. But there is no evidence that he had made such an inference on his own.

ROTHMANN'S TRANSFORMATION AND THE FIRST COPERNICAN CONTROVERSY

It is unfortunate that more is not known about Rothmann's early years—even information so basic as when he was born or exactly when he moved to the court at Hesse-Kassel.[43] We know that he matriculated at Wittenberg in August 1575, a moment of transition in the faculty responsible for the curriculum of the science of the stars.[44] Melanchthon, Reinhold, and Rheticus had passed from the scene more than a decade earlier. Peucer, whose strong influence on the curriculum persisted, was incarcerated in 1574 on charges of crypto-Calvinism; and in the same year, Rheticus died in relative obscurity. News of these recent developments—especially Peucer's imprisonment—would have been fresh just as Rothmann was commencing his studies. There is no reason to believe that the lectures that Rothmann attended as a student would have differed much from those delivered just a few years earlier by Sebastian Theodoric and Bartholomeus Schönborn. The curriculum still placed heavy emphasis on Peucer's introduction to the sphere and Reinhold's Peurbach commentary. Caspar Straub and

Andreas Schadt each prepared their own set of annotations on Peurbach and, evidently used them as the basis of their lectures in 1575 and 1577.[45] Garcaeus's treatise on calculating nativities using the *Prutenic Tables* is another likely source of study; and for advanced students there was Reinhold and Peucer's *Hypotheses Astronomicae*, the 1571 edition of which was dedicated to Rothmann's future patron, Wilhelm IV.[46]

Rothmann might also have heard rumors about the authorship of the "Ad Lectorem." Johannes Praetorius had studied with Homelius and lectured at Wittenberg just before Rothmann's arrival (from 1572 to 1575). Praetorius had also learned some details of the early printing history of *De Revolutionibus,* and that information could easily have become part of Wittenberg oral culture. A note in his hand on one of his two known copies of *De Revolutionibus* specifically identified Osiander as the person to whom some unnamed sources attributed the authorship of the "Ad Lectorem."[47] On the second copy, earlier owned by Homelius (who had studied with both Rheticus and Reinhold), Praetorius composed a brief exposé of Osiander's role: "Rheticus affirmed that this preface was added by Andreas Osiander. But Copernicus did not approve of it. Also, the same person changed the title contrary to the author's will. For it ought to be 'De Revolutionibus.' Osiander added 'orbium coelestium.' "[48] Because these annotations are undated, we cannot say for sure exactly when Praetorius wrote them; but as Homelius died in 1562, it is certainly possible that Praetorius acquired the narrative—even if not the book in which he wrote it—while still at Wittenberg. If he did, then the story of Osiander's skullduggery and Rheticus's dismay about it could have spread at Wittenberg as early as the 1560s, thereby casting a shadow over that mitigated reading of *De Revolutionibus.*

At some point, Rothmann broke with the Wittenberg consensus and became a follower of Copernicus and of Rheticus, the lone dissenter at Wittenberg. Why and how this important shift occurred is still a matter of uncertainty. It is extremely unlikely that the transformation occurred in his student years, even if that is surely when the groundwork for it was laid. More likely Rothmann followed a path somewhat like that of Wittich and the early Tycho. An unpublished treatise that he seems to have begun writing around 1583 contains a Capellan diagram together with typical geocentric placements for the three superior planets, but without the annual epicycles found in Wittich's scheme.[49] Was this arrangement inspired—or perhaps reinforced—by Wittich? Again, the chronology is uncertain. We know only that in November 1584, Wittich was in Kassel and observed an eclipse with Rothmann.[50] When he arrived and how long he stayed is unknown; but in October 1585, the landgrave told Brahe that Wittich had helped to improve his instruments.[51] This reference was clearly meant as praise because at Kassel, expertise in the construction of mechanical celestial globes and astronomical clocks was highly valued.[52] Reports about these contacts suggest that once again Wittich could have used the occasion to share his readings of *De Revolutionibus.* Certainly experimentation with Wittich-like constructions was occurring at Kassel prior to Tycho Brahe's publication of his geoheliocentric scheme. In 1587, the landgrave requested that his Swiss mechanician Jost Bürgi (1552–1632) build a bronze model of a geoheliocentered arrangement.[53]

The crucial point is that at Kassel in the years 1584 to 1586, there was nothing like the Mars campaign then under way at Uraniborg. Rothmann's attention was focused on improving stellar observations and observing comets. It was in the comet treatise that he brought to bear his defense of the view that there is nothing but air stretching from the Earth to the heavens. It has been suggested—convincingly, I believe—that Rothmann had decided to associate himself with this claim on reading Jean Pena's preface to Euclid's *Optics* (1557).[54] Essentially, Rothmann set up a new argument in the familiar form *modus tollens*—but in a way that exactly reversed Bellanti's assumption about the optical properties of the spheres. If celestial spheres and orbs exist, so his argument went, then they entail refraction at high altitudes; because refraction does not occur at such heights, but only near the Earth, where it is produced by thick vapors no higher than fifteen to twenty degrees above the horizon, then there must be no celestial spheres.[55] On the other hand, comets do exhibit refraction, because they are the product of compressed terrestrial vapors that rise from the Earth high above the Moon where they catch (and bend) the Sun's rays.

The entire argument thus depended on what one was willing to infer from the evidence of refraction. Rothmann's elemental-air hypothesis

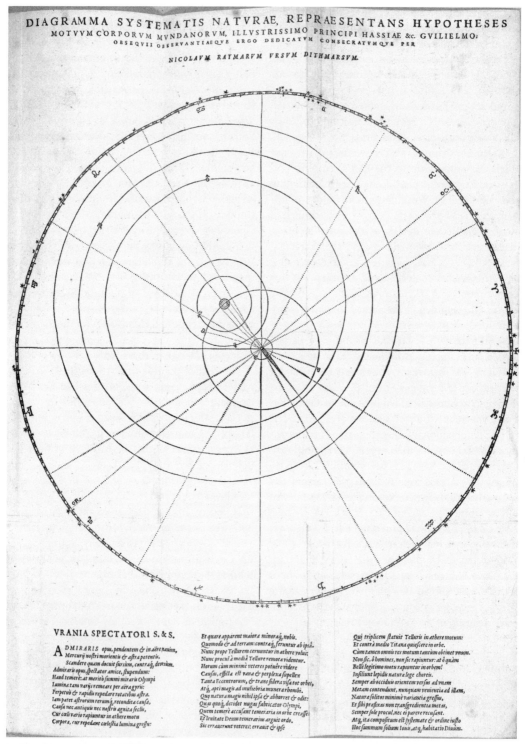

57. "Diagram of the System of Nature, Hypotheses Representing the Motions of the Bodies of the Universe." Ursus 1588.

provided Tycho with an alternative explanation for the nature of the celestial stuff, which, in the form that he had qualified it, at once freed him to push his own alternative to Wittich's scheme. Ironically, although Tycho's claim to have found a large Martian parallax gave him just the leverage for which he had been looking, it immediately opened up for Rothmann the possibility of endorsing an alternative that Tycho would not tolerate. Both claims—the first to the spheres' nonexistence, the second to a large Mars parallax value—underdetermined the choice between Tycho's new planetary arrangement and the Copernican hypothesis to which Rothmann had surprisingly turned. Once again, the logic of the situation guaranteed that the evidence would allow no finality.

A few months after Brahe issued *De Mundi Recentioribus Phaenomenis* from his own presses in the spring of 1588, he became embroiled in a two-front controversy. On one front, he commenced with Rothmann a two-year exchange of long letters—serially topical, rambling expositions—about the comparative value of their planetary schemes, the constitution of comets, and the heavens; on the other, he descended into a bitter priority dispute with Nicolaus Reimarus Ursus, the client of one of his noble friends, who had published a planetary scheme at Strasbourg at the very end of July 1588 that, at least in its basic geometry, was equivalent to Wittich's arrangement. Where Tycho had adroitly ignored Wittich, he could not as readily ignore the publication of a diagram that had appeared so soon after his *De Recentioribus Phaenomenis* and so closely resembled it. Thus, he could not quite so easily use the word *absurd* to dismiss Ursus in the way that he had dismissed the orderings of Ptolemy and Copernicus in the title to his new hypothesis.[56] The Ursus affair upset Tycho more than the controversy with Rothmann, very likely because it called into question the sensitive issue of the Mars parallax, on which his system and his reputation as an observer rested. He would manage the two matters rather differently, using the resources of his social position—mediated by his letter-writing contacts—to attack and defame rather than to debate the low-born Ursus.

Quite unlike Tycho's handling of the dispute with Ursus, his exchanges with Rothmann represented a well-articulated controversy—to be exact, the first such debate involving Copernicus's theory. It was organically interwoven both with the earlier discussion about the comet and the ontology of the spheres, and also with the evolving relations between the courts of Uraniborg and Kassel. The Uraniborg and Kassel courts, both ruled by noble celestial practitioners, together had perhaps the strongest concentration of skilled celestial practitioners in Europe.[57] Together they constituted a space unhindered by the constrictive disciplinary genres and pedagogical demands of the universities. Out of this environment came the book in which Tycho embedded the brief sketch of his world system—a work principally devoted to a full-scale treatment of the comet of 1577, directed to an audience of scholars (many of them university teachers) but never intended as a tool for the classroom. Although Tycho had sent Maestlin one of the earliest copies of his book (dated May 14, 1588, but not received until August), he did not engage with the Tübingen *mathematicus* over the Copernican question, let alone in any debate comparable to the extensive interchange with Rothmann. By the end of the year, Maestlin had lent his copy to Helisaeus Roeslin in Strasbourg. In the two weeks that Roeslin had the book in his possession, he managed to construct a narrative in which Ursus had been Tycho's pupil and had also plagiarized the Dane's arrangement while accenting it with his own addition, the Earth's daily motion.[58] Still another copy made its way to Thomas Digges in 1590 through an intermediary, but if it arrived at its destination, it is not known to have elicited a response.[59] Giordano Bruno did not receive a copy at all; he was neither a celestial practitioner nor part of Tycho's correspondence network. Unlike such initiatives—in which books moved as gifts or loans to a small cluster of scholars (among whom were the majority of second-generation Copernicans)—the Brahe-Rothmann exchange was deeply grounded in the mutual epistemic benefits that had motivated the initial interchange. Unlike the difficulties that Galileo would face in Rome less than twenty years later, this first extended encounter opposed the Copernican hypothesis not to the traditional ordering of the *via antiqua* but to Tycho's version of the *via media*.

The intercourtly letters between Brahe and Rothmann also show the sort of frank and flexible development of epistemic ideals, committed positions, arguments, disagreements, and appeals to experience made possible by shared understandings of disciplinary ordering and precedence.

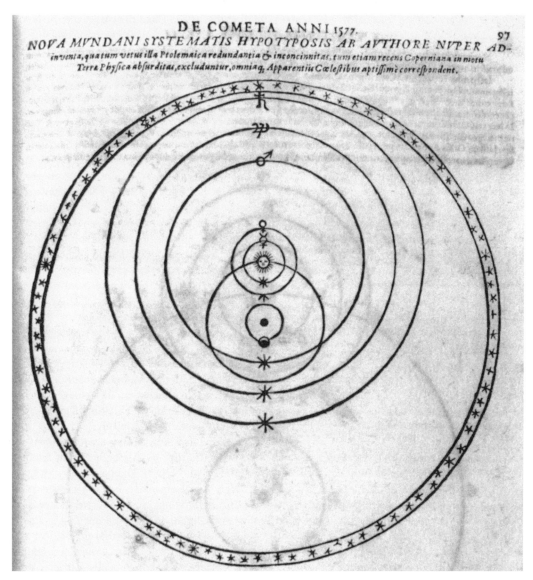

DE COMETA ANNI 1577.

97

NOVA MVNDANI SYSTEMATIS HYPOTYPOSIS AB AVTHORE NVPER AD-
*inventa, quatum vetus illa Ptolemaica redundantia & inconcinnitas, tum etiam recens Coperniana in motu
Terra Physica absurditas, excluduntur, omniaq, Apparentiis Cœlestibus aptissimè correspondent.*

58. Tycho Brahe's "New Hypotyposis of the World System," from Brahe 1610 (reissue of Brahe 1588). Image courtesy History of Science Collections, University of Oklahoma Libraries.

Common disciplinary identity provided a point of social and epistemic reference that could mitigate ultimately profound disagreements. Brahe and Rothmann represented themselves to one another as astronomers working through observations and measurements to conclusions within the domain of celestial physics. Following Rheticus, they saw astronomy as a subject that worked inductively from the appearances to conclusions about physical reality. Rothmann insisted that optical demonstrations "compelled" him to conclude that the constitution of the heavens is the same kind of air found below the Moon. The difference between sublunary and superlunary air was to be found in their relative thickness and in the ratios of the elemental mixture of air to earth and air to water: superlunary air was less thick, less mixed, hence "most pure" and transparent. Below the Moon, big storms showed how readily the air mixes with water, rattles windows, and sways trees. Up in the heavens, the sun's rays are extremely fine and rare; as is also apparent (down here) closest to the horizon, at dusk and twilight, they penetrate and travel unimpeded through the

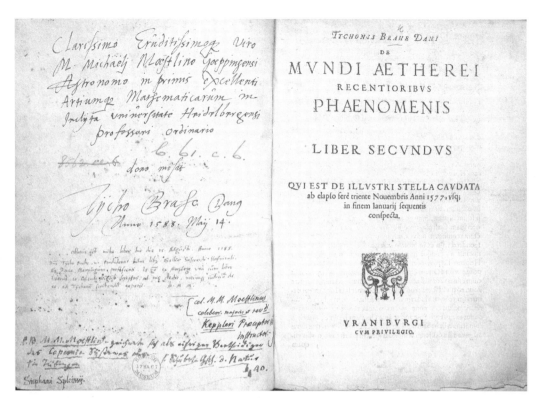

59. Maestlin's copy of Tycho Brahe's *De Mundi Aetherei Recentioribus Phaenomenis* (1588), with Tycho's dedication, May 14, 1588. © British Library Board. All Rights Reserved. C.61.c.6.

pure heavenly air without bending.[60] For Rothmann, these premises about the air-water mixture allowed measurable entailments: vapors around the horizon—but not exactly *at* the horizon—were six times denser than at thirty degrees of altitude; hence, greater refraction would be expected.[61] But above thirty degrees, Rothmann and Brahe found their point of disagreement: a difference in the refractive "transparency" would decide whether the heavens were made of Rothmann's "pure air" or Brahe's "pure aether." Neither conclusion would affect their common rejection of solid, material spheres.

Obviously, there was more than enough here to signal another point of commonality: their break with the traditional Aristotelian account of the elements and the constitution of the heavens. Rothmann invoked a rhetoric of understatement while inviting common cause with Tycho: "I am David, not Oedipus."[62] If Aristotle inferred the stability and permanence of the heavens from the unchanging sizes and clarity of the stars, then for the Earth the same inference should follow: its magnitude does not change. Yet, although it appears not to change, sensory experience testified that it is involved in all sorts of physical changes.[63] Following Melanchthon, astronomy also looked to holy scripture for guidance in matters of the creation. After receiving Tycho's book, Rothmann wrote that in the scriptures God has not revealed much about the question of celestial matter; thus how could physicists know anything certain about this subject? Whatever we do know we learn through mathematical and optical demonstrations.[64]

The immediate problem for Tycho was not in Rothmann's gestures at common approaches but in what was presumed in advance: if both aether and air are transparent, then what could be inferred reliably about the existence of one or the other heavenly substance from the measurement of optical effects? Despite much talk of such observations and measurements, Tycho did not really believe that the problem could be resolved in this way. His conclusion was assumed prior to any attempted measurements: he was already committed to a fundamental divide between heavenly and sublunary ontology. His belief in the existence of correspondences between upper "astral powers" that revolve and lower elemental things that "imitate" was premised on this distinction.[65] Lower things were, at best, useful as analogues to the nature of the upper realm. For example, rainwater, snow, and wine all contain impurities. These impurities cause measurable refractive effects. However, when the impurities are removed, these substances become transparent and are not distinguishable. If different substances, when purified, produce the same optical effects, then Rothmann's heavens could not be made of elemental air. "Here," wrote Tycho, "it appears to me that you infer more than this [your] assumption can sustain."[66] Better to follow Paracelsus in saying that comets have a fiery rather than an airy nature, while the heavens themselves are made of a permeable, fluid, incorruptible aether—heavenly, but not elemental, air.[67] In a subsequent letter, Tycho associated Rothmann's kind of air with the substance that animals breathe, "just as Jean Pena the Frenchman, you, and certain others dare to proclaim publicly."[68] By contrast, Tycho's heavenly air—called variously "aether," "the spirit of philosophizing wine," "the philosophical heaven," and "the universal essence of the heaven"—was analogous to the spirit in terrestrial wine.[69] The architecture of Tycho's castle was famously organized to reflect these correspondences and analogies—the celestial work pursued on the roof decks, the "terrestrial astronomy," or *labor pyronomicus*, in the alchemical ovens of the basement.[70]

The social and epistemic latitude for the Brahe-Rothmann disagreements bears comparison with its undeniable constraints.[71] Rothmann characterized the epistolary exchanges as "the most profound and subtle mathematical disputations."[72] Conflict about planetary order took place within the same matrix of disciplinary categories as disagreement concerning the ontology of the heavens: theoretical astronomy, physics, and theology. Rothmann's lower social rank did nothing to diminish the candor of his sharp criticisms. He addressed Tycho respectfully not only by his inherited titles of nobility (lord of Knudstrup) but also by the acquired status that he imputed to Tycho as "Most Distinguished among All Mathematicians of Our Age."[73] Rothmann held that controversies should be judged neither by patrons nor by the disputants themselves: "We cannot play the character of the plaintiff, the judge, and the

king; a third person is required to pronounce the sentence, a *philalethes* [lover of truth] absolutely free of any prejudice."[74]

But matters were not so simple. Who got to choose the lover of truth? Tycho said that he gladly accepted the challenge to find such a *philalethes* among "competent philosophers in Germany or elsewhere." And he soon proposed as his candidate the doyen of Wittenberg, Caspar Peucer: "I defer the whole matter to him, as he is knowledgeable foremost in the disciplines of Philosophy and Mathematics as well as the study of Theology." Peucer also possessed the advantage of age and independence; indeed, Rothmann should harbor no suspicions, as "he [Peucer] has never been acquainted with me."[75] And, as Tycho told Rothmann earlier in the same letter, it was the *philalethes* Peucer who had endorsed "my opinion on the basis of holy writings in a letter to me that the superelementary heavens are made of the thinnest, most clear matter."[76] On the other hand, while Tycho complained openly about Rothmann's failure to appreciate his achievements, he also sought to reassure that he would not publicly judge Rothmann's unpublished writing on the comet. He might pass on copies of his letters to various interested parties, but he declared his respect for Rothmann's freedom to discuss, prove, or disprove, both privately and publicly. Only just and proper reasons mattered.[77] Of course, these affirmations of Rothmann's freedom, even if sincere, occurred in the context of letters that Tycho had the power to select, edit and publish at his Uraniborg presses and to distribute as he liked.[78]

Rothmann's criticisms reveal the range of his possibilities in this intercourtly space. To begin, Rothmann questioned whether Tycho should be calling his arrangement "new." Rothmann took credit for having described much the same arrangement in an unpublished work titled *Elementa Astronomica*. Even so, he did not claim to be doing anything other than transposing Copernicus's hypotheses into the framework of a moving Sun and a stationary Earth, a transformation whose anticipation he attributed not to Peucer or Wittich but to Rheticus and Reinhold.[79] At this point Rothmann described the flat bronze model that the landgrave had commissioned from Bürgi, which exhibited on one side "not only the Sun and the theorics of all the other planets with their true longitudes, as well as their anomalies and centers, but also the latitudes of the three superior planets."[80] Rothmann called this model an "inversion" of Copernicus's, and although he was always careful to ask whether this was really "the same" as Tycho's "new hypotheses," it was this arrangement—essentially Wittich's—whose entailments he proceeded to outline and reject.

How, for example, was one to think about the geoheliocentric scheme as a physical representation? Rothmann contended that it was difficult to imagine an alternative physical account without spheres: "Who would ever believe that the center of the Sun's greater epicycle is provided with such power that it can drag all the planets after it—in fact drawing them from their spheres and returning them again—when they are connected by no clinging, corporeal matter?"[81] The visualizations that Bürgi's model allowed seem to have played some role—perhaps a critical one—in moving Rothmann against the Wittichian geoheliocentric organization and all the way to the Copernican model. When the landgrave actually saw the diagram of Tycho's hypothesis, Rothmann said that it reminded him of Bürgi's model: "Good God, he joked, the Sun's circle must be stronger than brass, since it can drag all the planets with it."[82] The issue at stake had been anticipated by Bellanti, Reinhold, and others: if there were no spheres, what would constrain the planets to follow orderly paths? Rothmann acknowledged that the planetary bodies themselves would not collide in the inverse Copernican ordering, but "confusion" would persist nonetheless because "there would remain no true or determinate distinction among the spheres."

The solution was to be found in the very position that Tycho rejected as absurd. Rothmann argued his point forcefully: "I can find nothing contrary to the singular truth of Copernicus's hypothesis. . . . In fact, Copernicus has sufficiently refuted the physical absurdities and more so."[83]

These arguments warrant careful summary. First, Copernicus's arrangement was consonant with a deity who was the "author of order, not confusion," one who assigned to each planet not a sphere but a "bounded space" (determinatum spatium).[84]

Second, although Tycho called his hypothesis "new," it seemed to be merely an "inversion" that could not better satisfy the appearances than those of Copernicus.[85] Yet this inverted Copernican model had its uses: Rothmann found it helpful in teaching Copernicus.[86]

Third, Copernicus was right in his proposal that gravity is a natural desire (appetitum quandam naturalem) that God implants in the parts of celestial bodies, a disposition (affectio) manifesting itself as a power (efficacia) that tends to gather and hold together the parts that form globes. The Earth is such a body. It has a round shape like the other planets and hangs freely in the air; why, then, should it be deprived of motion? Rothmann expanded the Copernican argument and his physical intuition about it through an analogy from general experience, ending with a suggestive counterfactual gesture: "If we raise up and suspend a globe most exactly and delicately on its pole axis and, while still freely hanging, we push it [so that it moves] in a circle, we see that it retains this motion for a sufficiently long time that its [motion] cannot suddenly stop. Now, if this artificially contrived but hindered motion is possible, how much more natural would it be were it not so impeded?"[87] At Uraniborg as at Kassel, there were plenty of globes to spin—hence the experience to which Rothmann appealed was at least locally familiar. But globes turning around unimpeded? Permanently impelled? We must resist the temptation to find in Rothmann a full-blown Salviati playing "What if?" to Tycho's Simplicio; but the expression of such hints of a budding physical intuition are exactly the kinds of possibilities enabled by this evolving genre of the scholarly courtly letter.

Fourth, Rothmann endorsed an accommodationist exegesis of sacred scripture.[88] Scripture was written for everyone, not just Tycho Brahe and himself; its true purpose was redemptive and salvational.[89] Many passages made no sense unless read in this modality, as for example the Genesis passage (1:16) that says that the Moon is greater than the other stars. Rothmann also used Romans 1:10 to argue, rather more aggressively, that God revealed more of his wisdom about nature through the natural world than through holy scripture.[90] Clavius had used the same text (as discussed in chapter 7) simply to urge astronomy's value as a form of contemplation apart from astrology.

Finally, in a passing reference, Rothmann elaborated a physical objection to the prophetic entailment adumbrated in the Narratio Prima. Rheticus's cyclical explanation for the rise and fall of empires "ought not to be accepted when, with al-Battani, he writes too freely and misuses the mysteries of astrology. . . . For how can the change of the Sun's [and hence Earth's] eccentricity cause the change of empires?"[91] Astronomers could legitimately infer conclusions about the heavens that were otherwise veiled in the Bible's ordinary language, but they should not use the heavens to make prophecy. Rothmann effectively aligned himself with Tycho and Maestlin in dissociating astronomical order and predictive practice from the sort of ambitious prophesying promoted by Cyprian Leovitius.

In February 1589, Tycho replied that his hypotheses precisely agreed with the appearances, that they far surpassed those of Ptolemy and Copernicus and corresponded far more to the truth.[92] The comparative syntax is notable in its mirroring of the structure of Copernicus's own argument, its avoidance of Aristotle's standard of demonstration, and its anticipation of the seventeenth-century phase of the controversy over celestial order. As a real exchange got under way, it is apparent that the shared criterion of order underdetermined its applications. Rothmann and Brahe could point to meanings of order, simplicity, or economy that the one arrangement possessed but that the other did not. Whose scheme, then, was more orderly? less disorderly? Aphorisms of simplicity were readily available, but there was no historical precedent for their application to this sort of problem; and even holy scripture, to which both would resort, carried its own uncertainties about which exegetical standard to use and when.

Against Rothmann's charge that Tycho's hypothesis would bring confusion to the planetary zones, Tycho returned fire in kind. If the Earth, the oceans, and the Moon revolved together annually, as if they were one body with a threefold motion, then the elemental air, the Earth, and the ocean waters would all mix together with the celestial revolutions. In that case, lower beings would become confused with higher ones, thereby turning upside down the whole natural order.[93] Tycho's hypotheses would not allow that sort of disorder. His hypotheses were neither inverted nor confused: the heavens were uniform from the Moon to the eighth sphere; the planets freely rose and set; the Sun was in the middle, around which the planets paraded "in the most beautiful harmony." Reintroduction of "real celestial orbs" would obviously destroy the harmony.

The linchpin of Tycho's argument against the inverted Copernican scheme—just as later in his conflict with Ursus—was the 1582 Mars observa-tions. With Rothmann, he grounded the authority of his claim for Mars's approaches in a general rhetoric of exactitude and accuracy.[94] Rothmann was evidently expected to take this rhetoric as a sign of the observations' authenticity—and there is every reason to think that he did, not least because logically his own position was not threatened. The Mars observations—whether authentic or not—failed to limit the choice between the Copernican and Tychonic hypotheses: in both scenarios, Mars approaches the Earth more closely than does the Sun. Thus, while affirming that Mars's close approaches to Earth refuted the ordering of "the Ptolemaics," Tycho simply ignored the fact that their schemes did not disagree on that score.[95]

No such disregard occurred over the criteria for reading holy scripture.[96] But who got to determine those criteria? The argument turned on the prophets' capabilities. Against Rothmann's accommodationist standard and his view that the prophets knew no more about the natural world than the common people, Tycho argued that the prophets indeed possessed more than ordinary skill in astronomy and physics, and he was thus willing to read the scriptures as a reliable basis of physical knowledge. As evidence, Tycho invoked a letter from his *philalethes*, Caspar Peucer, purporting to confirm that scripture denied the existence of solid spheres and supported their liquidity.[97] Where Rothmann had cited Augustine's "much freer" standard for reading scripture, Tycho replied that, as far as he knew, that church father had nowhere supported either the diurnal or the annual motion of the Earth.[98] With no help from Catholic juridico-theological commissions, scriptural authority was proving to be a difficult resource for these earnest Lutherans to apply unambiguously to the resolution of the uncertainties engendered by Reason's confrontation with Experience.

Finally, at the end of his November 1589 letter, Tycho challenged Rothmann to respond to the physical entailments of the Copernican motions. The first was essentially a version of Ptolemy's *modus tollens* objection in the *Almagest*. On a diurnally rotating Earth, a lead ball dropped perpendicularly from a tower passes violently through the air and thus fails to move circularly.[99] More objections of this form followed. If there were an annual motion, then the eighth sphere would be pushed back so far that it would appear to vanish.

The space between the Sun and the fixed stars would be so large that a third-magnitude star of 1' apparent diameter would be the size of the Earth's annual orb, or 2,284 Earth radii—absurdly larger than the Sun! Moreover, if there were an annual motion contrary to the daily motion, then nothing would ever appear to be at rest. And if the Earth possessed these two motions, it would upset the body's singular and simple nature. Finally, what of the "intricate librations" further imposed by Copernicus's motion of the Earth's axis?

Rothmann excused his delayed reply by an account of an illness for which he had been unable to find a cure—whether through baths, herbs, or the ministrations of the landgrave's Galenic physician, Dr. Butter.[100] Miserable though he was, the illness had evidently not affected the clarity of his mind or the firmness of his views. Rothmann continued to hold that Copernicus's account of the compounding of rectilinear and circular motions had "sufficiently made clear" why a lead ball would land at the foot of a tower even if the Earth underwent a daily rotation. Like Tycho, he argued comparatively. He invoked Tycho's Stoic-Paracelsian philosophy of sympathies and correspondences: Tycho ought to have "much less doubt" about a part sticking to its nature because "you know from your better philosophy that Nature is attracted unto Nature and Nature retains Nature."[101] A piece of gold still retains gold's nature, just as the lead ball retains its motion in the one second that the Earth moves hundreds of German miles. Tycho should also remember that nature always chooses to act with the fewest and most economical principles. Accordingly, is it intelligible that all the planets and stars rather than the Earth alone should rotate daily? That there should be two rather than one center of motion?

Against Tycho's objections that the annual motion would leave an incomprehensibly huge and useless vacuum between Saturn and the fixed stars, Rothmann turned seamlessly from simplicity criteria to medieval scholastic arguments from divine omnipotence and magnitude. No talk here of refraction, measurements, and elemental air. Rothmann countered Tycho's charge of "absurdity" with God's absolute power to create any logical or physical possibility that he so wished. However great the universe or the spaces between bodies, these gaps were as nothing when compared to the power of the infinite Creator.[102] The difficulty with such a move was that it could

cut both ways: just as the deity could choose to create an extra-big gap in the heavens, he could also have chosen to create a much smaller, tidier universe. Curiously, Rothmann did not think to counter Tycho's charge of *asymmetria* with the *symmetria* of Copernicus's period-distance relation, the very criterion on which Kepler would shortly build his own case.

Indeed, this was the last letter of Rothmann's to be published in the *Epistolae Astronomicae* of 1596; but it was not Tycho's last word. Immediately following, Tycho added a five-page summary section titled "The Author to His Reader concerning the Preceding Letters from Rothmann and the [Author's] Response to Them." At stake was Tycho's entire project of astronomical reform and, by implication, the very structure on which his Uraniborg castle was organized. Against Rothmann's emphasis on God's absolute power, Tycho accentuated the orderliness of the universe, closely (and ironically) appropriating the Copernican language and imagery of *symmetria*. The troublesome empty space was an *asymmetria*; Rothmann should recall that the painter Albrecht Dürer had depicted the *symmetria* of the human body as a "microcosm" just like the parts of the universe, "proportionately ordered and disposed to one another so that any of these parts would have as much a sure relation to the whole as to the parts."[103] There was such a Düreresque harmony in the universe, but Tycho's readers should look for it in the duality of motion and rest rather than in Copernicus's reordering. The heavens were active, alive ("provided with a vital spirit"), and moved eternally, and they sent down influences to the unmoving center. To displace the center would be to dissolve the "Theater of the World" and its receptacle Earth at the center:

> The Earth more aptly receives the influences directed to the center because it is passive and resting, just as the heavenly forces of revolution are active; for there is a reason why this second part of the universe exists, however lowly. Besides living things, this [lower] world contains so great a number of things analogous to the heavens. For this reason, it is written that God created heaven and Earth, where Earth is the second [in precedence] as if it is decreed and foretold that this part of the world is to be united with the heavens. Nor is the lowly image of this humble and indeed insignificant star (as the Copernican scheme considers it) disregarded or abandoned.[104]

Tycho's castle was organized to receive, study, and manipulate the effects of those influences. In short, to concede Rothmann's arguments about falling bodies would be to imperil all these applications of order and hence the astronomical reform and its astrological and alchemical sequelae. And thus, at the very end, Tycho added the following comment: "To this point, there has been no response to these refutations of Rothmann's preceding arguments about the motion of the Earth, which he used to uphold the Copernican assumption—even supposing that he has not seen them. Wherever he has gone, he could not stay with me, nor has he returned to his prince."[105] This is how Tycho ended the argument.

GIORDANO BRUNO
"ACCADEMICO DI NULLA ACCADEMIA
DETTO IL FASTIDITO"

"Academic of no Academy, called the Troublemaker": the subtitle of Giordano Bruno's comedy *Il Candelaio* (The Candlebearer) can well be taken as a self-description.[106] Compared to other second-generation advocates of Copernicus's theory, Bruno was a most unusual figure—and was thus seen by contemporaries. Like most Copernicans, he was deeply enmeshed in aristocratic circles but held no court position. In fact, he was chronically itinerant—beginning his studies in a Neapolitan Dominican monastery in 1565, being ordained a priest in 1572, receiving a doctorate of theology in 1575, yet soon after being declared a heretic and breaking with the monks in 1576, wandering for many years (in Geneva, Lyons, Toulouse, and Paris), often publicly criticizing academic and ecclesiastical authority, residing between 1583 and 1585 with the French ambassador in London (Michel de Castelnau), mixing in Elizabethan court circles, and evidently engaging in a disputation at Oxford.[107] From London, he returned to Paris (in 1585) but soon began to travel through the Empire, offering academic lectures and producing a steady stream of publications in Marburg and Wittenberg (1586–88), Tübingen, Prague, and Helmstedt (1588–1590), and then Frankfurt, Zurich, and again Frankfurt (1590–91). In 1592—the same year in which Galileo moved from Pisa to Padua—he made the ill-fated decision to return to Venice, and not long after, his Venetian host turned him over to the Inquisition.

Bruno's philosophical work was unorthodox and frankly contestatory. He exploited and experimented with the persuasive and subversive possibilities of the dialogue as a form of philosophical activity, consciously constructed in radical opposition to pedagogical genres. His writing was often at once playful, ironic, conversational, serious, critical, and comical, at times ambiguous, and typically genealogical rather than axiomatic or demonstrative.[108] Six brilliant Italian dialogues were published in London in 1584, all with Venetian imprints. Lacking secure internal time references and surviving correspondence, we do not know exactly when these works were composed; but given their large number and proximal outpouring, it is likely that composition was already under way perhaps three or four years before Bruno arrived in London.[109]

These works do not fit the typical dialogical format of the universities, with the figure of the master expounding dogma to a passive pupil as, for example, in Robert Recorde's *Castle of Knowledge* or the question-and-answer format of Maestlin's *Epitome* or Melanchthon's *Initiae Doctrinae Physicae*. Instead, Bruno's writings often present their subject as a gestational process of illumination somewhat in the manner of Rheticus or Kepler.[110] But unlike Rheticus, who was representing the work of his *praeceptor* Copernicus, Bruno fashioned his Copernicus with a literary self that, while indulging the seriocomic, used his characters to present views that Bruno himself could have once held and rejected. At the same time—such is our necessary caution in identifying Bruno's authentic voice—one cannot be entirely sure that the figure of the Nolan in the dialogues always represents Bruno's views. Indeed, one may sometimes question when the serious is meant to be comic and the comic serious, and whether a philosophical dialogue can ever be a comedy—a problem already recognized by Renaissance theorists of the dialogue.[111]

About Bruno's view of God there is no such uncertainty. This was not the God of potentialities such as surrounded Clavius's universe and withheld his full power. Bruno's deity was rather an all-powerful being ceaselessly enacting his infinite potential in a universe of incomparable magnitude, fittingly evoking the image that Bruno drew of himself as a breaker of boundaries, genres and teachings, both modern and ancient. Added to this self-grandiosity there was also, as Miguel Granada observes, an evangelical Bruno, announcing the cyclical return of a golden age of truths buried under the corruptions of Aristotle and his followers.[112] Tycho Brahe, who had his own reformation of the heavens to manage, dismissed him on the flyleaf of his copy of Bruno's *Camoeracensis Acrotismus* with the cruel pun "Nullanus, nullus et nihil. Convenjiunt rebus nomjna saepe sujs" (The Nullan [the Nolan], the Nobody, the Nothing. Names often agree well with their objects).[113] Although Kepler was always looking out for Copernican allies, Bruno would never be one of them. The more cautious Galileo kept a politically discreet silence and never once mentioned the Nolan. It will thus come as no surprise that other aspects of Bruno seemed strange, irritating, and objectionable to modernizers and traditionalists alike, not least his reading of *De Revolutionibus*.

BRUNO'S VISUAL, PYTHAGOREAN READING OF COPERNICUS

Bruno's reading of Copernicus in the Italian dialogues is notoriously difficult and elusive, again reminding us of the uncertain analytic utility of the category "Copernicanism." At least three considerations may assist. First, although Bruno was an immediate contemporary of Wittich, Rothmann, and Brahe, we know of no contacts that he had with any prognosticators in the Kassel-Uraniborg network until the late 1580s, after he had worked out the fundamentals of his own position. Second, although Bruno referred specifically to Ptolemy and Copernicus and generically to "mathematicians," he made no specific mention of Regiomontanus, Peurbach and his commentators, Clavius, Rheticus, or Reinhold.[114] Third, Bruno barely hinted at the arguments from harmony and order so heavily emphasized by Copernicus, Rheticus, Rothmann, Digges, Maestlin, and even Tycho Brahe.[115] He made clear that his own epistemic fulcrum lay above all in the discovery of physical explanations and that his main argument—his real battle—was with Aristotle. Given that his education was that of a Thomistic seminarian, his emphasis on theology, metaphysics, and natural philosophy hardly surprises.[116] In the *Cena delle Ceneri* (The Ash Wednesday Supper), Teofilo reports that the Nolan "had come neither to lecture nor to teach, but to answer; that the symmetry, order, and measure [*simmetria, ordine e misura*] of the celestial motions are assumed as they are and had been understood both by the ancients and modern men; that he did not dispute them on this, and had no case against the mathematicians, with which to deprive them of their measurements and theories, which he endorsed and believed; but that his interest was directed toward the nature and verification of the cause of these motions."[117]

Bruno's first explicit representation of planetary arrangements was a diagram published in the *Cena*. In only one respect does the illustration's crude orthography resemble the familiar woodcuts to be found in the sphere and *theorica* commentaries: it is a series of concentric circles. But a line divides these circles in half, with the top portion labeled *Ptolemaeus* and the bottom *Copernicus*. I will refer to these two parts using Bruno's designations. In its comparative intent, the only possible precedent was the Naibod diagrams of a decade earlier (see figures 47 and 48). Most important, the labeling of the image is confusing and incomplete. For example, in the *Ptolemaeus* portion, the symbols for the two outermost planets, Saturn and Jupiter, are reversed, mirror images of their conventional symbols. Likewise, the arrow in Mars's symbol points toward the left, or ten o'clock, rather than the conventional two o'clock. (If Bruno's diagram is held before a mirror, Saturn, Jupiter, and Mars immediately snap into their conventional orientations). Venus and Mercury, lying between the Sun and Moon, are upside down relative to the viewer's standard orientation, and their reflections remain so in a mirror. The Moon encircles the center of the diagram, but it is unclear whether the Earth is represented by the center point or by the circle that surrounds it.

Turning to the *Copernicus* half of the diagram, the problems multiply. The Sun in *Copernicus* is the most prominent symbol; unlike the Sun symbol in *Ptolemaeus*, it is an active, illuminating body whose rays occupy the equivalent of the entire lunar circle in *Ptolemaeus*. The radius of the lunar circle in *Copernicus* is about half that in *Ptolemaeus*, but even more strangely, the Moon occupies what appears to be the opposite side of a common epicycle with a point (or is it a body?) that appears to be the Earth. No other symbols

PTOLEMAEVS.

COPERNICVS,

60. Planetary orderings of Copernicus and Ptolemy, according to Bruno. From Giordano Bruno, *La cena de le ceneri* (1584), fol. 98v. Courtesy Bibliothèque nationale de France.

assist the labeling of the circles in *Copernicus*. Because of the way the diagram has been constructed, the reader is tempted to read *Copernicus* by following around the symbols in the upper half. This strategy yields predictable results for Saturn, Jupiter, and Mercury but fails for the others: Mars's circle in *Ptolemaeus* (at noon) passes through the position occupied by the Moon in *Copernicus* (7 o'clock); the Sun (1 o'clock) passes through the center of the lunar-terrestrial epicycle; Venus (11 o'clock) in *Ptolemaeus* passes through the terrestrial point; and the Moon encircles the Sun.

Neither here nor elsewhere did Bruno utilize the terms *cosmology* or *system*, let alone *Copernican theory* or *Copernican hypothesis*.[118] Instead he organized his representations around the figure of Copernicus as a heroic man (standing for truth against "the stupid mob"), a mathematician (in contrast to someone who searches out natural causes), and a kind of prophet ("ordained by the gods to be the dawn which must precede the rising of the sun of the ancient and true philosophy").[119] Bruno's linguistic choices thus did not readily fit into a genre with predictable rules and expectations, such as the sphere or the theorics. Making sense of the *Cena* diagram is a bit like peeling an onion. Bruno's interlocutors constitute the first layer, the characters whom they relate the second, the books and diagrams that they describe a third, and Bruno's own views yet another layer. Teofilo and Smitho describe an argument about Ptolemy and Copernicus between two characters: the Oxford don Doctor Torquato and the Nolan. The argument seems, at first, trivial enough: whether Copernicus's Earth is correctly represented by a point on the epicycle opposite the symbol of the Moon or by the epicycle's center—surely a matter that could be resolved by consulting *De Revolutionibus* itself. However, in his authorial capacity Bruno does not describe the diagram straightforwardly but rather has the character Torquato make the drawing. The whole episode seems like a performative moment, an occasion for stage play (with the reader as spectator), rather than a static, conventional representation: "Then they put some paper and an inkpot on the table. Doctor Torquato laid out a sheet of paper which was both wide and long, took pen in hand, and drew a straight line through the middle of it, from one side to the other. In the center he drew a circle, of which the aforementioned line, passing through the center, was the diameter. Inside one semicircle he wrote *Terra* and within the other *Sol*. On the earth side he drew eight semicircles, where the signs of the seven planets were placed in order, and around the last semicircle he wrote: *Octava Sphaera Mobilis,* and at the top: *Ptolemaeus.*"[120]

The important themes of the performance—as throughout the entire work—are those of manners, competence and judgment. The Nolan's voice regularly indulges in insulting language and unflattering imagery to contrast his own impeccable courtesy and superior intelligence with the ignorance and incivility of ordinary street folk (and Oxford academics). Finally, a copy of *De Revolutionibus* is fetched and the problem apparently resolved: "The reason for the error," says Smitho, "was that Torquato had looked at the pictures in the book without reading the chapters, or, even if he had read them, he did not understand them." So, the scene purports to represent an argument about the meaning of a diagram in *De Revolutionibus* (book 1, chapter 10), even though the picture found in the *Cena* is surely not that diagram. The Nolan then bursts into laughter and triumphantly announces that the epicycle's center is merely the mark of the compass point. "If you really want to know where the Earth is according to Copernicus's meaning, read his own words. They read, and saw that he said that the Earth and the Moon were as if contained in the same epicycle, etc."[121]

The reading—or the misreading—served the function of promoting the coherence of the dialogue and driving home the parallel themes of academic pedantry and incompetence. Smitho then ends the fourth dialogue with a revealing comment that firmly anchors the Nolan's own teachings: "The doctrine of Copernicus, though it is useful for calculations, is not always sure and specific as to the natural causes which are the most important."[122]

It appears that Bruno was doing at least two things simultaneously, although at different levels: he was using the disagreement over planetary ordering as a device to undermine the authority of traditionalist, academic learning (crude, Latinate, and pedantic) while playing to the fashions of an Elizabethan courtly audience (polite, Italianate, and learned); and he was using that same disagreement to align himself with a highly idiosyncratic reading of Copernicus while advocating his own, singular views about the universe. For modern commentators, quite different interpretations turn on the Nolan's so-called error in reading Copernicus's text. Attention has focused on the search for coherent intentions and meanings, as no one believes that Bruno completely misunderstood *De Revolutionibus.* Was he working unawares with a variant or corrupt version of Copernicus?[123] Was the diagram really meant as a magical or eucharistic hieroglyph and hence devoid of astronomical import?[124] Might the problem really be found in the obscure wording of Copernicus's own diagram in *De Revolutionibus*?[125] Or, like Ficino, was Bruno convinced that he had found an ancient wisdom, a *prisca philosophia,* that was veiled in Aristotle's writings?

There is much to recommend this last reading. Dario Tessicini has recently suggested that this "hidden wisdom" was really the original Pythagorean conception of the Earth and the invisible Counter-Earth revolving around the Central Fire.[126] On this account, Bruno read Copernicus through the lens of the Pythagorean Earth and Counter-Earth, making the relevant substitutions: the Moon for the Counter-Earth and the Sun for the Central Fire. Tessicini's reading lends plausibility to Bruno's placement of the Earth and Moon on opposite ends of the same epicycle-diameter. Bruno's exegesis of the Pythagoreans (via Aristotle, of course) was essentially the opposite of Copernicus's "astronomical" reading, which simply ignored (or erased) the counter-Earth. This read-

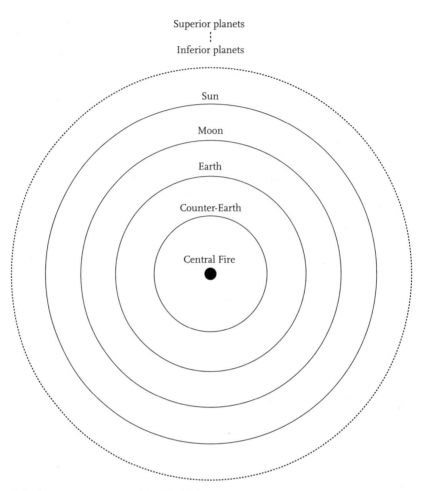

61. Pythagorean arrangement. After Tessicini 2007.

ing also has the virtue of removing an apparent absurdity from the *Cena* while also making it consistent with Bruno's discussion in the *De Immenso* of 1591, where Venus and Mercury inhabit opposite sides of the same deferent. In short, this "Pythagorean" reading of Bruno is consistent with all of the Nolan's other writings about the universe. Moreover, the Pythagorean framing had an additional thrust: because the universities built their curricula around Aristotle and Ptolemy, it gave Bruno a stance from which to point out the missed treasures already contained in *De caelo*. It also allowed him to separate himself from Copernicus, a "mere mathematician"—that is, an astronomer and astrologer. Copernicus had hit on only one part of the Pythagorean insight but then failed to appreciate the true meaning of his own discovery—the much deeper natural philosophy, ultimately rooted in the innumerable worlds and infinitist, homogeneous space heralded by Bruno. This truth could now be said to be lying hidden right under the noses of the narrow-minded pedants.

Although Bruno's earlier readings of Copernicus appear to have some coherence with later, more radical elements of what is sometimes conveniently called his "cosmology," the broader thematic unities that scholars have so patiently established were unavailable to the likes of Tycho Brahe, Rothmann, Galileo, and Kepler. What little we know of their access to Bruno's writings suggests that it was local and fragmentary. Because Bruno had no known entrée to the Tychonic correspondence network, we cannot even be sure how the *Acrotismus* reached Uraniborg or whether it found its way from Prague (via Hagecius?) via some Wittenberg intermediary or through one of Tycho's itinerant assistants.[127] The *Cena* and its

UNANTICIPATED, SINGULAR NOVELTIES

visualizations, on the other hand, were directed toward a local, courtly, Elizabethan audience and never again appeared in Bruno's writings; in fact, we have at present no direct evidence of how widely these images circulated.[128] Therefore, insofar as German celestial practitioners of the late 1580s and '90s were conversant with Bruno, it was through his nondialogic, Latin works.

BRUNO AND THE SCIENCE OF THE STARS

In 1586, Bruno began to translate his unorthodox ideas into the language and genres of the universities. Like Kepler in his *Epitome Copernicanae Astronomiae* (1618–21), Bruno clearly understood that philosophical, ecclesiastical, and secular authority was firmly embodied and interwoven in the academies. The political shift toward this audience may explain his direct engagement with the Aristotelians' arguments and the exclusion of satirization of their alleged pedantry. *One Hundred Twenty Articles concerning Nature and the World against the Peripatetics,* the basis of a disputation at the Collège de Cambrai, appeared at Paris in 1586. Two years later, Bruno published an expanded version at Wittenberg, which was preceded in March by a valedictory oration to the Wittenberg faculty.[129] These works rehearsed key elements of the radical vision already adumbrated in the *Cena* and *De l'infinito* while presuming but not explicitly building the account around Copernicus. Passing praise of Landgrave Wilhelm and the observational project at Hesse-Kassel—about which he exhibited only limited and indirect knowledge—shows that Bruno actively hoped for some kind of support in Protestant Germany for his anti-Aristotelian vision.[130] Not unlike the aggressive modernizer Descartes in the following century, Bruno appropriated anything from the astronomers of his own moment that seemed coherent with his own theorizing; however, the limitations of his connections with the Tychonic network probably explain his failure to mention Tycho's *systema*.

Insofar as Bruno entered into dialogue with the astronomers, it concerned the ontology of the spheres. He referenced several names from the comet literature—Cardano, Gemma, Roeslin, Brahe (but not Maestlin). There are also hints that might be taken (charitably) as evidence of familiarity with Rothmann's (unpublished) treatise

on the comet of 1585, but the absence of any detailed engagement with that text—comparable to his attention to the works of Copernicus and Aristotle—suggests that, during his stay in Marburg, he may have known of it only by hearsay.[131] These tantalizing and scattered references raise the further question of the kind of evidence that Bruno took the astronomers to be offering and what he considered to be their logical relation to his own claims.

It seems fairly clear that Bruno used the authority of the astronomers' names and status as a vehicle for presenting his own, already-well-formed views about comets as hidden entities and "kinds of stars." For example, in the *De Immenso* (1591), the final work of Bruno's German period, there is a revealing prose passage amid the poetry: "The astronomers of our time (of whom the prince and the most noble is Thyco the Dane) report and bear witness most surely to such things about comets (which we understand to be hidden Earths or stars because they show up infrequently as if in a mirror and make an angle with sun that is visible to us) so that they [the astronomers] cannot remain faithful any longer to the common view by which it is proclaimed that matter rising is kindled between the region of fire and the highest air."[132] This long, somewhat awkward, parenthetical statement is the heart of the matter. A second reference to Brahe, concerning the proper motion of the stars, follows the same pattern.[133]

These references to Tycho as a preeminent authority who rejected Aristotelian doctrine were no more than generalized gestures in support of Bruno's own propositions about the homogeneity of space and the absence of spheres and orbs. However, somewhat later, there comes a more specific allusion to observations by "Uraniborg astronomers": "Before the time of those Italian astronomers, Albumasar, one of the Peripatetic Arabs, declared there to be a comet above Venus. Moreover, there was a comet in 1585 that was said to be a round comet. Others saw a new star in the months of October and November which they declare to be in the heavens above Saturn, the observation of which, I have read, was made by Uraniborg astronomers."[134] Leaving aside the question of whether Bruno obtained this information from Kassel or Uraniborg, one is again struck by the deployment of generalized, often vague syntax to represent observational reports.

While arguing theses contrary to traditional philosophy, Bruno's practices of using experience to invoke (or evoke) "confirmation" were completely in keeping with those of other philosophers of his time.[135]

Apart from these efforts to appropriate the authority of Uraniborg and Kassel for his cause, it is hardly surprising that Bruno stood completely at odds with respect to the standard tools of astrological prognostication. In rejecting the entire machinery of epicycles, deferents, orbs and spheres, he was separating himself from the calculational resources on which practical astrology was based. It is no wonder, then, that his appropriations from the astronomers were both vague and selectively configured to fit his epistemic posture. Similarly, while rejecting the science of the stars, Bruno did not entirely forgo the celestial *influxus:* he spoke of "fragments of truth" that were "mingled with numerous vanities" in astrological writings.[136] Thus, while rejecting the widespread view that comets and eclipses could have astrological significance, Bruno did not relinquish the potential utility of medical prognostication.[137] Unlike Digges, whose infinitist Copernican scheme came tightly wrapped with a "prognostication everlastinge," Bruno's conclusions about astrological divination were largely in accord with Pico's skepticism. Bruno's celestial influences could be captured and manipulated but not calculated and predicted.[138] In that sense, there would be no need for Tycho's instruments: control of celestial influences could be "magical" without being astrological.

UNANTICIPATED, SINGULAR NOVELTIES

PART IV

Securing the Divine Plan

11

The Emergence of Kepler's Copernican Representation

THE COPERNICAN SITUATION AT THE END OF THE 1580s

At the end of the 1580s, Copernicus's theory was one alternative amid a proliferating field of representations of celestial order. Copernicus's proponents were distributed among different networks—and also largely separated by them. Yet the Wittenberg interpretation had made certain parts of Copernicus's work both familiar and credible. References to Copernican parameters in academic textbooks were common from the 1550s onward. Heavenly practitioners of all stripes were using Reinhold's Copernican planetary tables. Copernican planetary modeling practices had made serious inroads among a small group of unusually capable students of *De Revolutionibus*. Still, nearly fifty years after the appearance of *De Revolutionibus*, the Copernicus-Rheticus proposals to reorder the planets had attracted precious few adherents: in Salamanca, the biblical commentator Diego de Zuñiga;[1] in London, the gentleman–harbor engineer and sometime member of Parliament Thomas Digges; at Kassel, Landgrave Wilhelm's court mechanician, Christoph Rothmann; the peripatetic natural philosopher and sometime university lecturer Giordano Bruno; and at Tübingen, the professor of astronomy and mathematical subjects Michael Maestlin. These five followers were so diverse in their commitments, so different in the uses to which they put Copernicus's theory—and, as we have seen, even in their representations of Copernicus's arrangement—

that they did not form anything like a movement or a school. One still hesitates to use a term like *Copernicanism,* which leaves the misleading impression of an integrated network and a unified perspective.

With the sixteenth century nearing a close and the expectation of the end of the world ever powerful among the Lutherans, the attraction of Copernicus's theory remained, at best, tenuous. Copernicus's preface to the pope had failed to catalyze the envisioned Catholic following. The alternative Rheticus-Copernicus strategy in the *Narratio Prima* seems to have assisted the cause somewhat more successfully, especially after it reappeared in 1566 joined with the master work; but such success as it did enjoy cannot be explained by Copernicus's disciples' embrace of the revolutions of the Wheel of Fortune as evidence of the Elijah prophecy. The attempt to endow the new planetary arrangement with prophetic meaning had failed to stick. After Rheticus, no adherent of Copernicus's theory heralded the import of the Elijah prophecy, although Bruno, for his own reasons, proclaimed the dawn of a new age. Moreover, the many astronomical textbook writers, whatever their religious confessions, still gave no attention to the explanatory power of the Copernican ordering in their teaching manuals, in striking contrast to the frequently repeated battery of arguments about the absurdities of the Earth's motion. Some of the system's important implications thus went unacknowledged. For example, Rheticus had alluded in the *Narratio* (without vi-

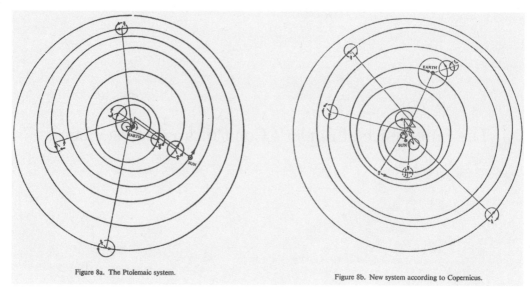

Figure 8a. The Ptolemaic system.

Figure 8b. New system according to Copernicus.

62. Comparison of Ptolemaic and Copernican systems. From Alexandre Koyré, *The Astronomical Revolution* (1992), 60–61. Drawing by William Stahlman. By permission of Dover Publications.

sual assistance) to unoccupied spaces between the spheres. In fact, the actual parameters of the distances in Copernicus's scheme imply the existence of sometimes immense spaces between the total spheres. Yet no one in the sixteenth century until Kepler took into account this radical departure from the Ptolemaic *plenum* of contiguous spheres.[2] Present-day readers familiar with the many illuminating diagrams that help us to visualize the central features of a "Copernican system" should ask why no sixteenth-century practitioners drew them.[3]

Other than common physical and scriptural objections, there is another important explanation for the neglect of the Copernican ordering scheme. No one had found in that arrangement a ready resolution to the problem of the variation in the strengths of the astral powers. If natural philosophers and astronomers alike were loath to give up the Aristotelian account of falling bodies, it was arguably at least as calamitous to relinquish Ptolemy's ordering of astral properties in the *Tetrabiblos*. Reassigning the order of planetary forces—inserting the Earth and removing the Sun and Moon from among the planets— amounted to overthrowing one of the traditional theoretical foundations of astrology. Those who had the capacity to appreciate the new heliocentric ordering—from the would-be astrological reformers Dee and Offusius to the astronomical

reformers Wittich and Tycho—apparently had good reasons to look for an astrological reform that would retain the resting Earth. Those, like Clavius and Pereira, who followed the Piconian rejection of astrology, argued that physical and logical reasons decisively undermined the Copernican arrangement. The great diversity in the character of the representations among Copernicus's most vociferous adherents did nothing to resolve the issue.

Tycho Brahe, as we have seen, appeared to have won the first phase of what was beginning to look like a full-fledged controversy, at least against those whom he regarded as opponents: Ursus, Wittich, Liddel, Roeslin, Rothmann, and Bruno.[4] It remained now for Tycho to persuade those who had not been part of the immediate scene of controversy. The great attraction of Tycho's geoheliocentric arrangement rested, of course, on its particular wedding of ancient and modern elements; the trade-off was that the observational claim about Mars's parallax had to be taken on trust in Tycho's observational skills and the technological infrastructure that only he possessed. After 1588, Brahe's planetary ordering sketch (for that is all that it was) moved to the center of discussions about what he and others were now calling the true *systema mundi*—not least through his own efforts at self-promotion. He sent copies of his 1588 work to influential noblemen and astrono-

SECURING THE DIVINE PLAN

mers all over Europe. He chased down the remnants of Wittich's library. Using a collection of his letters with the landgrave and Rothmann as a way to promote his own views, Tycho's *Epistolae Astronomicae* (Astronomical Letter Book) ignored Wittich's "sphere of revolutions," slandered Ursus's reputation, and debated Rothmann's arguments for the Copernican arrangement before declaring him to have given up.[5] Inadvertently, the book also created a space of debate about the true system of the world.

The two leading Copernicans of the 1570s were publicly silent in the following decade. From 1580 Thomas Digges had turned his attention away from theoric to practical engineering and active political life. Maestlin was the only one of the handful of Copernicans who was actually teaching astronomy.[6] But if Maestlin's introductory textbook, the *Epitome Astronomiae*, is taken as our sole guide, it is not certain that Maestlin was actually teaching anything about the Copernican arguments for planetary order in his regular introductory astronomy lectures—arguments that we know him to have been entertaining privately from the early 1570s—although he may have pointed out some difficulties with Ptolemy's account.[7] His oft-published instructional manual contains numerous references to Copernicus, but his remarks are entirely in line with the standard Wittenberg primers of the period. By comparison with such practices at Tübingen, the Salamanca statutes formally recommended that students read *De Revolutionibus*.[8] Again, however, we have no secure evidence of what that meant in the classroom. Was the capable Gerónimo Muñoz teaching anything different from what would have been presented in Wittenberg? When Diego de Zuñiga used Copernicus to render an interpretation of Job 9:6, he could probably assume that students and scholars at that university would understand the reference. The more significant questions, therefore, are not whether there were any adherents of Copernicus teaching in the universities, but what form their adherence took and in what context anyone taught the Copernican ordering claims.

COUNTERFACTUAL KEPLER

The problem can be made more precise by posing a counterfactual question.[9] Imagine that, in 1589, the year that he entered Tübingen as a young student, Kepler had chosen to attend a different university, whether Catholic or Protestant. Among his courses, suppose that he was introduced to the elements of the sphere. But because he was quickly recognized as a superior student, he was also taught Peurbach's *New Theorics* and Copernicus's *De Revolutionibus*. What would he have been taught about the explanatory advantages of the Copernican ordering? I believe that the short answer is *very little, if anything at all*.

It is possible to construct a sort of snapshot of mathematical practitioners teaching in European universities in Kepler's student years at Tübingen (1589–94), although our evidence about actual classroom instruction is extremely fragmentary. Let us think of it as a mouse-eaten picture. Consider Italy. At Padua or Bologna, Kepler easily becomes proficient in astronomical practice and theoric. From the ephemerides makers Giuseppe Moletti and Giovanni Antonio Magini, he learns how to use the *Prutenic Tables*, to adapt Copernican epicycles to a geocentric scheme, and to make astrological prognostications. At the Collegio Romano, he receives no comparable encouragement toward astrological practice from Clavius. But Kepler is well prepared to debate with "adversaries," championing the virtues of eccentric-epicycle astronomy (although not the Copernican version) against Fracastorian homocentrists and raising the customary physical arguments against the followers of Copernicus.

The case at Pisa raises somewhat different questions. Here is the young, newly appointed mathematician Galileo Galilei, required to teach the rudiments of the sphere as part of his duties. Presumably he can prepare his lectures by turning to any one of a number of good introductory German astronomical manuals. But because the names Melanchthon, Peucer, Strigelius, Schreckenfuchs, and Rheticus appear on the Index of prohibited books, he prudently avoids all German astronomy books and instead turns to Clavius. By good fortune, Galileo has met him personally in Rome in 1587, and he already knows of Clavius's reputation in relation to the calendar reform. He has composed a series of elementary astronomical lectures that draw heavily upon Clavius's *Sphere*.[10] If, as seems likely, he has used these notes at Pisa (and, after 1592, at Padua) to introduce his students to the fundamentals of astronomy, then presumably he does not teach his students any of the advantages of the Copernican

TABLE 4
Holders of university chairs in mathematics during Kepler's student years, 1589–94

University	Chairholder	Teaching Dates
Holy Roman Empire		
Altdorf	Johannes Praetorius	1576–1616
Dillingen (Jesuit)	Christopher Silberhorn	1594–96
Freiburg im Breisgau	Laurentius Schreckenfuchs	1575–1611
Graz (Jesuit)	Petrus Bastius	1589–91
	Laurentius Lupius	1591–94
Greifswald	David Herlicius	1584–97
Heidelberg	Valentin Otho	1588–1601
Helmstedt	Duncan Liddel	1590–96
Ingolstadt (Jesuit)	Christopher Silberhorn	1586–92
	Cornelius Adriansen	1592–93
	Johannes Appenzeller	1593–1601
Jena	Georg Limnaeus	1588–1611
Leipzig	Christoph Meuer	1585–1616
Marburg	Victorinus Schönfeld	1557–91
	Johannes Hartmann	1592–1609
Olomouc (Jesuit)	Thomas Williams	1590–97
Prague (Jesuit)	Christophorus Stephetius	1593–95
Rostock	Henricus Brucaeus	15??–1593
Tübingen	Michael Maestlin	1584–1631
Vienna (Jesuit)	Henricus Zittardus	1589–90
	Christoph Grienberger	1590–91
	Petrus Phrearius	1591–94
Wittenberg	Petrus Otto	1583–94
Würzburg (Jesuit)	Petrus Roestius	1590–91
	Jacobus Nivelius	1592–93

planetary arrangement.[11] It is hard to know what Galileo thinks privately about Copernicus in 1589, as we do not even know whether he was familiar with Rheticus's *Narratio* or Copernicus's *De Revolutionibus* at that time. Because Galileo is already able to cast nativities, however, he must know how to use an ephemerides and has undoubtedly learned *something* about Copernicus through Magini or Moletti.[12]

Beyond Tübingen, imagine that Kepler finds the former Wittenberger Johannes Praetorius at the new university in Altdorf (founded in 1575), near Nuremberg. Unlike the Italians, Praetorius is tied into the web of personal connections linking together Wittenberg, Nuremberg, the imperial court, and Tycho's Uraniborg. He has studied with Peucer and, like Valentine Otho (at Heidelberg), met Rheticus in Krakow (about 1570).[13] Like Maestlin, Praetorius has a substantial library of heavenly literature—including copies of Rheticus's *Narratio Prima* and Copernicus's *De Revolutionibus*. He has even read his copies, as testified by the many underlinings.[14] Tycho has referred to Praetorius and Maestlin as the two leading lights of astronomy in the German lands—and, in this judgment, even hindsight must agree.[15] For the study of mixed mathematical subjects in the German lands, Kepler cannot do much better.

TABLE 4 *(continued)*

University	Chairholder	Teaching Dates
Italy		
Bologna	Giovanni Antonio Magini	1588–1617
Pisa	Galileo Galilei	1589–92
Padua	Vacant	1588–92
	Galileo Galilei	1592–1610
Rome (Collegio Romano, Jesuit)	Christopher Clavius	1587–1612
Swiss Federation		
Basel	Christianus Wursteisen	1564–85
	Peter Ryff	1586–1629
Denmark		
Copenhagen	Jørgen Christoffersen Dybvad	1575–89
	Anders Krag	1590–1600
France		
Paris (Collège de France)	Maurice Bressieu	1581–1608
England		
Cambridge	Oliver Green	Gonville and Caius College, 1590–91
	Joseph Jessop	Kings College, 1582–91
Oxford	Frances Mason	Merton College, 1592
Spain		
Salamanca	Jerónimo Muñoz	1578–92

Praetorius's surviving lecture notes from the early 1590s allow us more than the usual hint of what Kepler might have heard. Praetorius does present certain advantages of Copernicus's hypotheses, but these concern the familiar Wittenberg-style substitution of homocentrepicyclic models for the equant device while keeping the Earth at rest. They have nothing to do with the premises of cosmic order.[16] In a 1594 lecture, however, Praetorius presents a diagram that lays out the Copernican ordering. One of the few examples we know of from this period, it is noteworthy for two reasons. First, Praetorius attaches values to each of the planets for the relative (maximum, minimum, and mean) distances with respect to the Sun. Second, he presents the Copernicus-Rheticus claim to *symmetria* as follows: "This symmetry of all the orbs appears to fit together with the greatest consonance [agreement], so that nothing can be inserted between them, and no space remains to be filled. Thus, the distance from Venus's convex orb to Mars's concave orb takes up 730 semidiameters, in which space the great orb contains the Moon and Earth and moving epicycles."[17]

Praetorius's rendering leaves an air of ambiguity. He fails to mention that the Copernican orbs fit together because the sidereal periods increase as the planet is farther from the Sun and the distances have a common measure in the Earth-Sun distance. That was the foundation of the *symme-*

tria emphasized by Rheticus and Copernicus. The phrase "no space remains to be filled" hints at a Ptolemaic nesting procedure whereby the maximum distance of a sphere is made equal to the minimum distance of the next highest. Copernicus says nothing about nested spheres, and in his arrangement there were gaps between the different planetary spheres. Perhaps Praetorius did not understand this Copernican entailment; perhaps he did and yet chose to present it "Ptolemaically" to his students. I do not know. In any event, it is significant that he assumed that students *ought* to consider the new ordering.

Moreover, since receiving a copy of Tycho's *De Mundi* in the fall of 1588, he had also been thinking about Tycho's new system.[18] In the 1594 lecture, he tried to present a heliocentric ordering for Mercury and Venus, a central resting Earth, and a Martian orb that does not intersect the Sun's. The diagram could have been drawn by Wittich or Ursus. The diagram had another difficulty: the outer epicycle was labeled "Mercurij Veneris" and carried the symbol for Venus, while the inner epicycle read "Veneris Mercurij" and bore the planetary symbol for Mercury. He had evidently fallen back into an earlier, Ptolemaic frame of mind and decided, finally, to slash an "X" through the scheme. But, the real difficulty was, once again, Mars: "It is certain that if we were to transfer the Sun with the surrounding orbs of Venus and Mercury to the Earth's place [on the Copernican theory] and, in like manner, the Earth, with the moon following at its feet, to the place of the Sun, then . . . it would be necessary to add epicycles of the size of the great orb, with the result that there would occur a great confusion of the orbs (especially with Mars)." Praetorius had appreciated the Wittichian point about the need to "add epicycles of the size of the great orb." Evidently, Tycho's alternative scheme had not persuaded him to allow Mars to invade the solar orb. "According to Copernicus, Mars's maximum distance from Earth ought to be 3,044 [Earth radii] and its minimum 427 [Earth radii]; yet, this simply cannot be allowed because it would then occupy not only the Sun's orb but also the great part of Venus."[19]

Praetorius's presentation of this Martian "confusion" in his lectures shows that he was prepared to introduce controversial material to his students, even issues on which he himself was still in some doubt. But his own general position

was clear: neither Tycho's nor Copernicus's world scheme was bound by necessary premises. The astronomer was free to pick and choose elements. Praetorius chose to stay with a central, resting Earth but to appropriate the enlarged Copernican distance to the fixed stars. "Copernicus has shown that the sphere of the fixed stars has an immense distance not only with respect to the Earth's semidiameter but also to the great orb; hence the sphere of Saturn is an almost immense distance from the region of the fixed stars." And if Saturn's orb could be so greatly expanded, then "nothing prohibits us, having changed the boundaries of revolution, from making Mars's orb greater so that it will not invade the territory of the Sun."[20] In other words, it was a simple matter of assuming the nesting principle: that Mars's minimum distance is the same as that of the Sun's (1,180 Earth radii) and Mars's maximum distance could be obtained by taking over the Copernican ratio for Mars's closest and furthest approaches. In the end, Praetorius would have presented our counterfactual Kepler with a Wittichian scheme.

Whatever the shortcomings of this brief and fragmented picture, its scope allows some escape from a focus on isolated, individual cases of representations of Copernicus's arrangement in the classroom. Perhaps the most interesting conclusion to emerge is the disjunction between public endorsement of the Copernican arrangement and the critical details that were actually taught in pedagogical settings. A student of the Copernican adherents Zuñiga and Galileo in 1589 was likely to learn little about the central Copernican arguments compared to a student of the progressive Wittenberger Praetorius. In the same year, Maestlin was the *only* professor of mathematical subjects in Europe who was communicating the explanatory advantages of the Copernican planetary ordering. Certainly he imparted his views outside the lecture hall; how much he was prepared to discuss inside the classroom is more difficult to say.[21]

KEPLER'S COPERNICAN FORMATION AT TÜBINGEN, 1590–1594

Kepler's formation as an active adherent of Copernicus's central theory was both singular and rapid. Unlike the great majority of Maestlin's students, Kepler was a willing, unusually capable beneficiary of his teacher's Copernican convictions; and so far as we know, he was the only one. His *Mysterium Cosmographicum* (Cosmographic Mystery), with an imprint of 1596, two years after he left Tübingen, contains many frank passages that refer back to his earliest student days. Thus, if used critically, it may assist in reconstructing the evolution of his views. In this section, I use the *Mysterium* as a resource for that purpose; in a later section of this chapter, I shall consider it as a work in its own right.

In a passage much quoted by all recent commentators, Kepler recalls that "six years ago [in 1590] when I worked under the direction of the very famous Master Michael Maestlin at Tübingen, I was disturbed by the many disadvantages in the usual opinion about the universe; also I was delighted by Copernicus, whom my Master often mentioned in his lectures." Here Kepler was already displaying the humanist stylistic practices typical of the entire corpus of his writings. Characteristically, he represented his views as the outcome of a gestational process—a consequence of human toil and effort rather than an assemblage of bare propositions. He also used his recollection as an occasion—one among many—to signal deference to his teacher. Kepler was curious enough about what he heard from Maestlin's frequent references to Copernicus to make it his business to learn more. And, in this manner, he developed a close and respectful relationship with the older man, although he did not fuse his identity with that of his teacher, as Rheticus appears to have done with respect to Copernicus. As Kepler said in the continuation of the above passage: "I collected together little by little, partly from Maestlin's words, *partly by my own efforts*, the advantages over Ptolemy."[22]

Maestlin's words, I suggest, were both oral and written. Because Kepler did not acquire his own copy of *De Revolutionibus* until the fall of 1595, he must have been allowed to consult Maestlin's personal copy.[23] Who else in Tübingen but Maestlin owned a copy of that book? How else would it have been possible in 1593 for Kepler to "defend the opinions [of Copernicus] at the disputations of the candidates in physics" or to compose "a thorough disputation on the first motion, arguing that it comes about by the Earth's revolution" if he did not have direct access to Copernicus's text?[24] And because the particular copy of the book used by Kepler was accompanied by Maestlin's profuse annotations, it seems fair to

say that Maestlin's glosses framed Kepler's initiation into the main Copernican arguments. Here were not only Copernicus's central arguments, but also, in many cases, Maestlin's direct reasons for approving them.

Maestlin was also involved in Kepler's intellectual development in other ways. In 1596, when Kepler was told by the Tübingen academic senate, on Maestlin's advice, to add further clarification of Copernicus's theory to the text of his *Cosmographic Mystery*, Kepler made ample use of what he remembered from Maestlin's reading of Copernicus. And Maestlin obliged further by adding visual illustrations to assist in understanding Copernicus's explanation for the variation in the sizes of the annual epicycles. He also prepared a well-annotated edition of Rheticus's *Narratio Prima* and a separate treatise of his own on the Copernican distances. That Maestlin prepared all these diagrams and notes so expeditiously suggests that they could have been part of his earlier teaching repertoire and hence part of the battery of considerations that also constituted the earliest reasons for drawing Kepler toward the Copernican planetary order.[25]

This development was neither a mysterious process nor a religio-scientific conversion, as Kuhn seems to regard it, but an appreciation and extension of the logic of the original Copernican problem situation. Copernicus's ordering scheme, as Koyré and others have well explicated, provided a simplified economy of explanations for astronomical phenomena that the Ptolemaic alternative could not.[26] Assuming the Earth's annual revolution allowed one to explain planetary inequalities—retrograde motion as an optical illusion, the annual component in the planetary motions as a projection of the Earth's motion—it also enabled a single criterion (sidereal periods) rather than a package of diverse criteria for ordering the planets, and ordering by periods rather than the stacking principle as the basis for calculating the relative distances from the center. And these purely astronomical considerations were among those that initially moved Kepler, as they had Maestlin: "What has *more* convinced me than [Copernicus's ability to predict] is that Copernicus alone gives, most elegantly, *the reason for things* where all other astronomers remain surprised and he alone removes the cause of this surprise, which resides in an ignorance of causes."[27]

I do not think that removing the "ignorance of

causes" was in itself enough to convince Kepler, as Koyré held, that Copernicus's system was "the true one."[28] Had such mathematical arguments been sufficient, then why would Kepler have launched an ambitious search for other kinds of support? Effectively, Kepler judged that Copernicus's system held an explanatory advantage. That was the point of the Horatian topos, as Copernicus used it: the audience was supposed to judge that his theory retained an aesthetic coherence that was lacking in the alternative. And it was this feature that would have made Kepler agree that Copernicus's system was "vastly superior to the traditional teachings" (Koyré) and "fruitful" (Kuhn); but it was not yet enough to convince him that it corresponded to the actual structure of the world.[29] The force of the Copernican entailments was, however, sufficient to encourage Kepler to search further, beyond the arguments from geometrical and astronomical harmony offered by Copernicus, Rheticus, and (had he known them) Rothmann.

I suggest that this weighting of the explanatory advantages was enough for Kepler—where it had not been for others, such as Dee, Offusius, Brahe and Wittich—because it was also backed by Maestlin's judgment concerning the character of Copernicus's arguments and by Maestlin's (alleged) success in accounting for an unusual and unforeseen occurrence, the comet of 1577. We can read in Maestlin's annotations what Kepler read and must have discussed with his teacher: that Copernicus's theory "admitted nothing absurd into the whole of astronomy, or rather [it admitted only] logical coherence." This view of Copernicus's theory also colored Kepler's estimate of Maestlin's hypothesis concerning the resemblance of the comet's attributes to those of Venus and hence its motion in the circumsolar sphere of that planet. Maestlin's "guess" (coniectura) about the comet's path was for Kepler a striking confirmation of the Copernican criterion of relevance between hypothesis and results—an instance of "how reliably truth is consistent with truth." Effectively, Maestlin proposed to reduce to a heavenly regularity what many writers of the 1570s took to be a transitory meteorological event. And Kepler even went so far as to declare in the Mysterium that "from this [Maestlin's] reason alone one draws a very strong argument for the arrangement of the Copernican orbs."[30] Beyond the importance of the comet in building Kepler's

confidence in Copernicus, he had also Maestlin's enthusiasm (in his annotations) for the "great argument": "All the phenomena as well as the order and distances of the orbs are bound together in the motion of the Earth."[31]

Kepler clearly regarded the Copernican claims as holding the potential for something stronger than Horatian coherence. "As in Virgil," he wrote, " 'the report grows by traveling and gains strength as it goes,' so for me the careful contemplation of these topics was the cause of further contemplation."[32] There were many things to contemplate, some of them political. Kepler had to consider that Copernicus presented his standard of proof in a preface directed to the pope; and obviously such an association could not be tolerated in orthodox Tübingen. Also, like Copernicus, he had to engage with the obvious divergences, which the theologians were sure to notice, between the Earth's motion and the literal words of holy scripture. And finally, there was the delicate matter that he already saw ways to defend Copernicus that went beyond Maestlin's astronomical arguments: "While Copernicus [and Maestlin] had attributed motion to the Earth rather than to the Sun on the basis of mathematical reasons, I did it on the basis of physical, or better still, metaphysical reasons."[33]

KEPLER'S SHIFT IN THE ASTRONOMER'S ROLE

Kepler called his new approach "cosmographical," a word that later caused him some regret, as it resulted in some booksellers classifying the Mysterium among geographical writings.[34] Yet he was looking for a word that would carry connotations of a full philosophical investigation of the heavens without associating himself with the via antiqua, the traditional philosophy of the schools. His approach thus continued the shift in the astronomer's role begun already among the null-parallax group of the 1570s.[35] But, unlike those writings, Kepler's project was not instigated by the appearance of exceptional events (comets and novas) occurring outside the ordinary course of nature, which could be explained as expressions of God's absolute power.[36] Kepler's starting point was the ordinary course of nature, the regularities of the heavens, the order that God freely chose to make with his potentia ordinata (ordained power). This preoccupation with the grounds of order—and es-

63. Academic disputation scene. Inset from frontispiece of Stierius 1671. The *Praeses* stands behind and on a higher platform and puts questions to the candidate or *Respondens*. Author's collection.

pecially the sort of systemic orderliness that Copernicus found among the planets—thus preceded and informed all of his work. And, unlike Bruno's infinitist philosophy, in which a traditional defense of astrology was impossible, Kepler's approach built within the frame of a finite universe that would (always) allow the possibility of a science of the stars. Eventually, he would extend his physical approach from planetary order to planetary theory, and, indeed, he would try to make Copernicus's insight the basis of a new and far-reaching philosophy of the heavens. This more ambitious search was inseparable from the creation of a new language with which to characterize the difference between his enterprise and the traditional one.

How did this phase of Kepler's enterprise begin? A fragment survives of an early disputation in which Kepler defended Copernicus's arrangement before the physics candidates in 1593. Georg Liebler was almost certainly the presiding master or *praeses*. Liebler had been teaching physics at Tübingen since 1552, and his *Epitome of Natural Philosophy* was the natural equivalent of Maestlin's *Epitome of Astronomy*.[37] Unfortunately we know nothing of how this event came about, why the topic was proposed and permitted, or how Kepler was chosen to dispute any such propositions. However, the very fact that such a disputation was held means that Kepler had enough confidence in Copernicus's geometrical arguments to defend the heliocentric arrangement against *Aristotle* rather than Ptolemy. The distinction is crucial. Just three years after being introduced to the Copernican theory by the math-

ematician Maestlin, Kepler had shifted his own approach from arguments of geometrical-astronomical advantage to arguments based on the system's claim to some sort of physical reality and metaphysical grounding. The later account of this transition in the *Mysterium* thus has a viable chronological dimension. If Kepler (as he said) was defending Maestlin's views before the physics candidates, he was also beginning to break away from Maestlin's conception of astronomy as a discipline limited to the study of formal causes.[38]

One passage from the Tübingen fragment has already been employed to make some fairly serious and wide-reaching claims about Kepler's early development. In his classic 1924 study, E. A. Burtt argued for the importance of metaphysical, religious, and especially Neoplatonic presuppositions in the development of early modern science. As shown in chapter 2, Burtt (mis)represented Domenico Maria Novara as a florid Neoplatonist whose ideas were highly consequential for Copernicus. In his parallel treatment of Kepler, he sustained the Neoplatonic theme. He used the Tübingen disputation to support his view that Kepler's motivation for becoming a Copernican amounted to "sun worship": "the exalted position of the sun in the new system appears as the main and sufficient reason for its adoption."[39] Echoes of Burtt's reading followed in an influential source. Thomas Kuhn cited Burtt's interpretation in *The Structure of Scientific Revolutions* as one piece of evidence in support of his provocative claim that, when considering "persuasion rather than proof, the question of the nature of scientific argument has no single or uniform answer. Individual scientists embrace a new paradigm for all sorts of reasons and usually for several at once. Some of these reasons—for example, the sun worship that helped make Kepler a Copernican—lie outside the apparent sphere of science entirely."[40] Here is Burtt's translation (with my emendations):

> In the first place, lest perchance a blind man might deny it to you, of all the bodies in the universe the most excellent is the sun, whose whole essence is nothing else than the purest light, than which there is no greater star; which singly and alone is the producer, conserver, and warmer of all things; it is a fountain of light, rich in fruitful heat, most fair, limpid, and pure to the sight, the source of vision, portrayer of all colors, though himself empty of color, called king of the planets for his motion, heart of the world for his power, its eye for his beauty, and which alone we should judge worthy of the Most High God, should he be pleased with a material domicile and choose a place in which to dwell with the blessed angels. . . . For if the Germans elect him as Caesar who has most power in the whole empire, who would hesitate to confer the votes of the celestial motions on him who already has been administering all other movements and changes by the benefit of the light which is entirely his possession? . . . Since, therefore, it does not befit the first mover to be spread out orbicularly,[41] but rather to proceed from one certain principle, and as it were point, no part of the world, and no star, accounts itself worthy of such a great honor; hence by the highest right we return to the sun, who alone appears, by virtue of his dignity and power, suited for this motive duty and worthy to become the home of God himself, not to say the first mover.[42]

Burtt correctly characterized this passage as an instance of Kepler's Neoplatonic inspiration. But this praise of the Sun's specialness cannot explain why Kepler was *initially* drawn to the Copernican ordering any more than it explains why Copernicus was drawn to the heliocentric system in the first place. Eulogies to the Sun's splendor were always premised on the Sun's spatial location in the heavens, and that was usually regarded as the mean between the three upper and the three lower planets in the Ptolemaic scheme.[43] So the Burtt-Kuhn passage attests rather to the "physical, or better yet, metaphysical" reasons that Kepler sought *after* having been persuaded by Maestlin's reading of Copernicus. Kepler was

grappling here with a connection between a central Sun and some sort of motive power issuing from that body, which would propel the planets. The connection had to be drafted as an explanation, and for the physics candidates the model of a good explanation was Aristotle's four causes: it would include not only the form of a thing or its spatial location, but also what it was made of, for what purpose it existed, and what effects it could bring about. Already in the Tübingen disputation Kepler was looking to transfer the role of primary mover from the outermost sphere to the Sun. In so doing, he was breaking with the view that astronomy could concern itself only with formal causes. He was beginning to expand his scope to the other Aristotelian causes, reasoning that the prime mover of the planets (efficient cause), by virtue of its degree of perfection (final cause) is the prime source of light and heat, and instigator of color (material cause) and should, therefore, proceed from the center (formal cause) rather than be distributed throughout the universe. All these attributes—giving light and heat and having the capacity to move—could be found in the Sun, the principal body in the universe. And when joined to the heliocentric periodicities, the result was a single power of nearly infinite motion at the center, steadily weakening as it issued forth from the smallest to the largest orbs and dissipating entirely in the spheres of the fixed stars—the latter immobile because they are almost infinitely far from the primary power.

Kepler presented these arguments in the context of an academic physics disputation, where Aristotelian texts defined the space of commentatorial practice. One might ask, Why did he not follow Copernicus in keeping the Sun passively at rest, while retaining some version of individual movers for each of the planets? And where indeed did he find the notion that the Sun functions as an efficient cause? Although Copernicus was clearly his dominant source of inspiration in matters astronomical, *De Revolutionibus* has little to say about why the planets move. A single passage in book 1, chapter 4, asserts that whatever force moves the individual planet must itself be constant. Although Copernicus's work exuded Platonic inspiration, it did not attribute any motive power to the Sun.[44] When eulogizing the Sun, Copernicus represented it as a stationary lamp or a visible god seated on a royal throne that "governs the family of planets revolving around it."[45] Kepler would need to look elsewhere.

An explicit source of information on planetary movers was the neo-Aristotelian Julius Caesar Scaliger's *Exercitationes Exotericae,* with which Kepler was very familiar from the beginning of his studies.[46] Scaliger is not mentioned in the extant 1593 fragment, but because Kepler had begun to study Scaliger's work in 1589, it is not unreasonable to assume that in the missing part he discussed Scaliger's views in something like the way that he brought up the matter much later in the *Epitome Astronomiae Copernicanae:*

> As a matter of fact Scaliger, who professed Christianity, and other followers of Aristotle dispute as to whether this movement of the spheres is voluntary and as to whether the beginning of will in the movers is understanding and desire. . . . Furthermore, motor souls were added, tightly bound to the spheres and informing them, in order that they might assist the intelligences somewhat; or because it seemed necessary for the first mover and the movable to unite in some third thing; or because the power of movement was finite with respect to the space to be traversed and the movement was not of an infinite speed but was described in a time measured out according to space: and that argued that the ratio of the motor power to the movable body and to the spaces was fixed and measured.[47]

Among other things, this passage shows us the kind of direction in which the student Kepler could have moved when philosophizing about the celestial movers. There was a rich tradition of Aristotelian commentary of which he could avail himself, and Scaliger represented something of the avant-garde of that moment.[48] Scaliger offered up the distinction between the moving power (efficient cause) and the intelligence (formal and final cause) that guided the form of the motion in time. One might think of a ship, a locomotive source, a pilot, and a map needed to yield directed motion.

Besides Scaliger, Kepler was also confronted with other, converging and countervailing developments. He accepted as a student that Tycho Brahe had refuted the existence of impenetrable celestial spheres. As will be recalled, Maestlin had received a copy of *De Mundi* directly from Tycho by the summer of 1588.[49] And without spheres, what work remained for the moving soul?

For Aristotle will readily grant that a body cannot be transported by its soul from place to place, if the sphere lacks the organ which reaches out through the whole circuit to be traversed, and if there is no immobile body upon which the sphere may rest. Moreover, even if we grant solid spheres, nevertheless there are vast intervals between the spheres. Either these intervals will be filled by useless spheres which contribute nothing to the state of movement; or else, if there are not solid spheres throughout these intervals, then the spheres will not touch one another or carry one another.[50]

If spheres and their movers were eliminated, then what of planetary intelligences? The intellective soul, unlike an animal soul, had no capacity to function as an efficient cause because it was a disembodied "mind." As such, it could make sounds and it had a will to act, but even so it had no capacity to move the planetary globe; and the planet, being an inert body, had no capacity to obey or to move itself.

Of course, we cannot say with certainty that Kepler made all of these arguments in the missing part of his disputation. But it is hard to believe that he would not have taken the trouble to refute Scaliger at the very time when he was most deeply engaged in chewing on Scaliger's exoteric exercises and when his conviction in the Copernican ordering motivated him to seek in the Sun a place for a single motive force.[51] Kepler's anti-Scaligerian emphasis on an active, central Sun as efficient cause—trading on the analogy with light and heat emanation—still appears to lack the theological underpinning that it would acquire by 1595, in his well-known conception of the trinitarian sphere. At least the representation of the universe as the visible image of the threefold deity, with the procreative Sun-center as the image of God, did not appear explicitly in the surviving Tübingen fragment—perhaps because he did not dare present such a risky notion before the physics candidates, perhaps because it has simply been lost to us.[52] But, with or without the theological metaphysics, Kepler clearly had by 1593 the crucial, if not yet honed, notion of a moving power that fit the Copernican order of periods, diminishing in strength as it spread out from the center.[53]

The exact nature of this power was another matter. Once into the new disciplinary territory of cosmography, the teachings of the best sixteenth-century astronomers—Reinhold, Peucer, Clavius, Praetorius, Brahe, Maestlin—were of little help, because their tools were geometrical and Kepler's physical questions were new ones.[54] Was light itself the actual efficient cause of planetary motion? Heat alone? Light and heat together? A separate moving force analogous to light? Or was light a vehicle for the moving force? Perhaps there was a separate Scaligerian intelligence, relocated from the planets to the Sun? And, if the Sun was a power source, then what of the planets? Were they passive recipients or themselves capable of producing their own influences, as held by traditional theoretical astrology? These questions and many others would soon form the basis of a new problematic of heavenly physics: the physical nature of light, the quantity of the power, and how it varied over distance. And again, all these new questions and conceptualizations occurred with remarkable alacrity, considering that by 1605, not much more than a decade after the Tübingen disputation, Kepler had arrived at the elliptical orbit for Mars. How many of these new kinds of questions Kepler was asking himself as a student is harder to say. If Scaliger's dense mélange of propositions and arguments was stimulating to Kepler and his classmates—perhaps helping by its chaotic organization to break up the icepack of orderly Aristotelian physical propositions—there is no evidence that any of the Tübingen physics faculty or their students were turning with equal enthusiasm to works of Neoplatonic inspiration.[55]

KEPLER'S PHYSICAL-ASTROLOGICAL PROBLEMATIC AND PICO

Where else could Kepler search for a new sort of physical cause other than in the conventional resources of academic natural philosophy—or the unconventional ones? Already, such tendencies are evident among practitioners of the *via media*. For example, Tycho Brahe and Helisaeus Roeslin comfortably, albeit in different ways, plumbed the resources of Paracelsian natural philosophy. Yet recent commentators unanimously agree that Kepler's early thinking was shaped by Neoplatonic and Stoic rather than Paracelsian thought (even if his early reading preferences lie hidden behind the obscuring glaze of late-sixteenth-century citation practices).[56] If Kepler was reading ancient or modern writers in those traditions,

such as Cicero and Nicolaus Cusanus (whom he cites), presumably he must also have read others, such as Plotinus, Marsilio Ficino, and John Dee (whom he does not).[57] That he had read Offusius's *Against the Deceptive Astrology* cannot be entirely ruled out. Yet apart from such writers, where could Kepler have turned for a new conception of a moving force if not to theoretical astrology?

The specific question that I want to address is whether Kepler had found such an idea of a solar moving power in Pico's critique of astrology. That important suggestion has been raised and explored by Louis Valcke in a recent study.[58] Valcke's Piconian texts recall our earlier discussion of Pico and Copernicus in chapter 3. Valcke calls special attention to those passages in the *Disputations* where Pico claimed that the heavens act on Earth only by means of light and motion rather than through individual planetary influences:

> We have said that heaven acts on us by motion and light. It is believed that heaven achieves three effects through motion: it moves, it warms, it carries light.[59]

> As for light, and by that I mean always this heat that emanates from light—the origin of all its influence resides in the Sun, eye of the world.[60]

> Light is another prerogative that suits the heaven.... It follows necessarily that light ought to possess some actualizing virtue, exercising itself on bodies and equally some vital principle—not that light itself is alive, nor that it gives life but that it most excellently prepares and disposes the body for life that is already alive; for heat proceeds as if it is a property of light, this heat that is neither fiery nor airy but celestial, as all light is a quality proper to heaven; it is a matter, I say, of a very efficacious heat, most salutary, penetrating, warming, and harmonizing all things.[61]

These are vivid passages. And it is both striking and revealing that Copernicus himself made no use of them—including the last, which is clearly indebted to Ficino's Stoically inspired pneuma-*spiritus*.[62] Copernicus's decision in that regard underlines, once again, his caution—and also Maestlin's—in moving no further into natural philosophy than he believed to be absolutely necessary. This hesitation usefully marks Kepler's difference. Valcke's study shows that there are clear resonances between these passages and various texts from Kepler's Prague (1600–1612) and Linz periods (1612–26). These passages all concern the similarities and differences that Kepler was trying to work out between light, heat, and the solar moving power. For example:

> Heat is proper to light.[63]

> The source of the world's life (which is visible in the motion of the heavens) is the same as the source of light which forms the adornment of the entire machine, and which is also the source of the heat by which everything grows.[64]

Besides Valcke's suggestive connections between Kepler and Pico, based on scattered textual resemblances from his post-Graz period, Sheila Rabin has shown that Kepler's engagement with Pico was by no means casual. In several works, Kepler engaged in an open, running debate with Pico. There are also many references in the surviving correspondence, beginning as early as 1599.[65] But it seems not to have been noticed that Kepler's first explicit reference to Pico occurs even earlier in an unpublished commentary on John Sleidan's *Three Books on the Four Great Empires* (1596), a reference strongly suggesting that he was in possession of the book at least as early as that date.[66] Even twenty years later, in the *Harmonics*, Kepler still believed that Pico's *Disputationes* was the major text with which he had to engage to establish the legitimacy of any kind of astrology, let alone a Copernican one: "I shall publish the books of Giovanni Pico della Mirandola with a commentary, if I am aware that I should be doing what would please students of philosophy, and if I am not deprived of the necessary means."[67]

The position that Kepler sustained fairly consistently throughout these writings was that truth and falsehood coexist both in astrology and in Pico's critique of it. Kepler then took it as his assignment to separate the wheat from the chaff—or, as he put it in his more vividly earthy way: "No one should consider it unbelievable that out of astrological foolishness and godlessness a useful sense and holiness could [not] also be found, that in unclean slime could not also be scraped out a snail, mussel, oyster, or eel useful for eating, that a silk spinner could not be discovered in a big heap of caterpillar egg droppings, and finally, that a good granule from a busy hen or a peach or a gold nugget might be found in an evil-smelling dung heap."[68] This conception of sorting out the holy from the godless, the edible from the inedible, strongly resembles Kepler's characteristic attitude toward other subjects—planetary arrangements and planetary theories. He was not going to throw out the baby with the bathwater. He would often self-consciously place himself as "the man in the middle."

How much of this view of Pico had Kepler worked out ten to fifteen years earlier, in his Tübingen period? If the encounter occurred only after he had worked out his initial ideas, then Pico presumably had comparatively little impact on Kepler's astrological-physical problematic. The earlier we can place these connections, the more problems are resolved: the initial sources of Kepler's Neoplatonic and Stoic commitments, the inspiration for the central moving power, the space of possibilities that framed his effort to construct a Copernican astrology, and the physical basis for a Copernican planetary arrangement.

DATING KEPLER'S ENCOUNTER WITH PICO
A TÜBINGEN SCENARIO?

For several reasons, Maestlin would seem to be the most likely source of a solution to the dating problem. We have established Maestlin's disinclination toward astrology, his ample library, and his willingness to let Kepler read his books. To these considerations may be added some others. Kepler's decision to align himself with Maestlin's views on Copernicus put him at odds with different constituencies among the higher faculties of the university: not only with the physics faculty but also, as will be seen shortly, with the theologians. And further, if, contrary to Maestlin, Kepler wished to retain the practice of astrology, he had to face the question of what it would mean to have a Copernican astrology. There were different issues to juggle within such an environment, and different solutions were possible. These difficulties centered on astrology as a physical theory. For example, no theory of celestial influence had been accommodated to a moving Earth. Then again, the Copernican planetary ordering demanded a revision of the sequence of elemental qualities from that described in the *Tetrabiblos*. In some respects, it would have been easier for Kepler to avoid the middle way, to follow either

Maestlin's public caution or Pico's full-throated disavowal of the divinatory arts.

Kepler's 1593 physics disputation is, once again, a central resource for investigating this issue because it is simply the earliest extant material that we have in which he presented his germinating ideas. The disputation exhibits a two-pronged development. On the one hand, Kepler openly defended Maestlin's position on Copernicus in the arena of physics. On the other hand, in his nascent probings of celestial natural philosophy or cosmography, Kepler was already boldly departing from Maestlin's guarded stance. It was not the only way in which he staked out a unique identity, nor was it the only agonistic arena to reveal important differences between teacher and student.

Kepler was drawn to astrology quite early in his student days. As early as the summer of 1592—a year after receiving a master's degree at Tübingen—he was in contact with Helisaeus Roeslin. Kepler requested help in prognosing the dangers of a fever and assistance in interpreting the exact details of a horoscope, whose identity as his own he did not reveal.[69] "Regard this as certain," wrote Kepler to Roeslin, "that Mars never crosses my path without involving me in a quarrelsome mood."[70] Without knowing that he was offering a judgment on Kepler's own nativity, Roeslin gave practical advice on how to judge the lag time between the astral cause and its effect, the problem of inexact observations, and the appropriate level of generality at which to cast the interpretation:

Anybody would be thoroughly mistaken who wants to restrict the effects emanating from the configurations to a particular year, let alone month and day. It is certain that the stars exert their effect. . . . But the matter is not so certain that we can assign it a definite time. For many details occur which conflict with such general rules of the heavens, so that the effect is either advanced or postponed. In addition, the motions of the heavenly bodies are not understood well enough, so that whole degrees, not to mention minutes, will be missing. But one degree [in the nativity] corresponds to a whole year in the configurations. In like manner, a quarter of an hour in the nativity corresponds to four whole years. It is therefore safest for the astrologer making predictions to stick to generalities. Let him say: around this age a burning fever would come, and this person would be in danger of losing his life, that is to say, around these or those years, and this may well happen earlier or later.[71]

A fascinating letter, to be sure, but why would Kepler need to write to someone like Roeslin for assistance of this sort? There were surely those in Tübingen who believed in astrology; was there no one in that city who could have assisted him in learning how to practice it? I suggest that Kepler was more than a little interested in horoscope astrology as a young man because he was anxious about his own fate, especially his chronically sickly body.[72] In his remarkable nativity of 1597, Kepler wrote a detailed description of his character (by coincidence, exactly three hundred years before Freud's self-analysis). Maestlin, whom he clearly revered, but who had long expressed reservations about astrology's fundamentals, had offered him no help on this score. In sympathy with both Pico and Tycho, Maestlin often complained about the poor state of astronomical observations. But as Maestlin had known Roeslin at Tübingen since at least the 1570s, it is reasonable that he might have mentioned Roeslin to Kepler as an experienced practitioner.[73]

This evidence of Kepler's budding interest in astrology raises related and equally serious questions. Could Maestlin have acquainted Kepler with Pico's critique of astrology in the same way that he had introduced him to De Revolutionibus? Discovery of a lost copy of the Disputations belonging to Maestlin would probably go a long way toward resolving that question. At a minimum, one would like to know which parts of Pico's critique were especially convincing to Maestlin. As will be recalled, Pico's attack on divinatory astrology was multifaceted, and the reception of Piconian astrological skepticism was at least as varied as the diversity of usages to which the Copernican proposals were put.

Writers emphasized or made different uses of various elements in Pico's critique; some were quite broad, others pointed or restricted. Miguel de Medina, for example, used Pico's contentions to undermine the entire science of the stars; Clavius, in contrast, excluded only the prognostication of celestial influences from his robust defense of astronomy. Pico argued against the possibility of predicting particulars, claiming that the only "influence" from the heavens was the general one of light—a claim that well suited the anti-mathematical Pereira. In 1583, Sicke van Hemminga accepted Pico's general objection against predicting individual differences, but, believing that the Piconian critique was insufficiently focused, he took the critique deep into the entrails of specific nativities in a way that Pico had not done. Together with Savonarola, Pico contested the physical reality of the zodiacal signs and houses, but surprisingly, apart from Erastus, few, if any, writers championed this objection in the sixteenth century. Against the astronomical foundations of astrology, Pico complained of the uncertainty of astronomers' numbers and instruments—an objection that Tycho Brahe, in his 1574 Copenhagen oration, regarded as deserving the strongest kind of reply. At midcentury, Offusius and Dee took up the question of the strength of astral forces in relation to the planets' distances from Earth, but, although fully conscious of Pico's objections, nowhere did they engage the objections raised by Pico and Savonarola. And, most significant for this study, in highlighting all possible disagreements among astronomers, Pico included the longstanding controversy about the order of the inferior planets that was centrally implicated in the formation of Copernicus's (and Rheticus's) problematic.

Maestlin's position, in this regard, was exceptional. He plainly believed with Rheticus that Copernicus's solution to the order of the inferior planets had answered Pico's objections to astronomy per se.[74] At the same time, we may reasonably surmise that Maestlin shared Pico's powerful doubts concerning the zodiac's reality. The evidence may be inferred from chapter 12 of the *Mysterium Cosmographicum*, where Kepler begins by saying, "There are many who have considered the division of the zodiac into twelve equal signs to be a human invention, that is, of a kind that has no basis in natural reality."[75] The presence of this assertion in a work in which Maestlin played a major editorial role—together with what we already know about his attitude toward astrology—increases our confidence that Kepler was echoing both Maestlin's and Pico's criticisms.[76] But, although Kepler continued to maintain (and augment) his adherence to the purely instrumental status of the zodiac in his later writings, he did not address this part of the critique in his physics disputation.

THE GOLD NUGGET

Perhaps Maestlin was not the proximate source of the 1593 disputation. Specific materials for such student exercises were usually provided by the pro-

fessor who oversaw the disputational ritual.[77] If Georg Liebler was the presiding master for Kepler's performance, as I have speculated, then his *Epitome of Natural Philosophy* might offer some clues.[78] Liebler, as Charlotte Methuen has shown, was Kepler's teacher of natural philosophy, and he was also a Melanchthonian with respect to astrology.[79] As such, one would expect him to have been an opponent of Pico. According to Liebler, astrology, like optics and music, was to be seen as a mixed discipline lying between mathematics and physics. Although Liebler as a philosopher did not concern himself with the mathematical part of astrology—casting horoscopes or writing annual forecasts—he did accept Melanchthon's view that the heavenly bodies act through physical forces on the terrestrial realm. The young astrologizing Kepler might have been drawn to Liebler's position on this matter, seeking a Melanchthonian physical justification for a practice in which he wished to engage as a mathematician. Yet, as it turned out, Liebler did not oblige; for, ever faithful to his Melanchthonian roots, he also rejected Pico's theory of celestial force! And Liebler's objection to Pico, I believe, became a significant topic in Kepler's 1593 disputation:

> The heaven is moved not so that it may heat but so that the stars may communicate their forces to the parts annexed to them, from one to the other, in accord with the capacity of matter to receive. I do not judge to be true what Mirandola contends in *Against the Astrologers*, book III: that the heavens have no particular force beyond the universal influence of motion and light; but certainly this celestial heat enlivens and stimulates these lower things to growth, while cold and dryness come about by accident. For we see that some days of winter are excessively hot and, on the other hand, some summer days are exceedingly cold. But I do not think that this could occur unless there were some specific force [*pecularia aliqua vis*] in the stars by whose qualities these lower bodies are moved. And I think that these forces [*vires*] of the stars come forth from their own particular forms. [80]

This passage is highly suggestive of important connections and theoretical positionings. A Copernicanizing and astrologizing Kepler evidently found in Liebler's rejection of Pico's light physics a key resource that he needed to construct the primary element of both a heliocentric dynamics and a heliocentric astrology: Rather than individual planetary movers, the Sun's light and capacity to cause motion are the source of both planetary motion and crucial celestial effects in the terrestrial realm. Kepler thus aligned himself with just that part of Pico's conception of heavenly influence that, for quite different reasons, both his teachers had disavowed: Liebler because he supported a traditional astrology of planetary forces, and Maestlin because he rejected all forms of astrology. Kepler had found a gold nugget in Pico's dung heap.

PROGNOSTICATING (AND THEORIZING) IN GRAZ

In 1594, Kepler was faced with an unexpected development that caused him to confront the practice of astrology in a new and even more serious way. The Tübingen senate recommended him for a teaching post at the evangelical school in the city of Graz, in the district of Styria; it was something like the *Stift* that Kepler had attended in Tübingen. However, the district also required the holder of the Graz position to issue annual calendars or almanacs, together with prognostications. Alexandre Koyré speculated that the senate sent Kepler to Graz to rid itself of a Copernican troublemaker: "It is probable that he was already regarded with suspicion by orthodox Lutheranism on account of his avowed and enthusiastic partisanship of Copernican cosmology, and that the university authorities sent him to Graz to a post socially inferior to that of a pastor in order to get rid of him."[81] I suggest, on the contrary, that Kepler's general learning and mathematical skills were well recognized and highly regarded in Tübingen. Moreover, within two years, the same senate that sent him to Graz would approve publication of the *Mysterium Cosmographicum*. The real issue at stake, therefore, was that Kepler was the only student of Maestlin's who had any kind of serious astrological competence. He had experience in casting nativities; somehow he needed to learn how to make annual practices for an entire region, without a Domenico Maria Novara or a Maestlin to help him. The problem was obviously not one of competence in theoretical astronomy but one of skill and confidence in making public forecasts for an entire region. In addition, Kepler might have been wondering how he could engage in such an activity while sustaining his conviction in Copernicus's theory.

Unlike Bologna a century earlier, Graz presented Kepler not with a scene of multiple, possibly competing prognosticators but rather with a well-established tradition in which a single astrologer generally covered the whole region. Furthermore, because the position was associated with the Lutheran *Stift*, the predictions foregrounded religious themes and were composed in the vernacular. This is particularly evident in the calendars and *practica*s of Kepler's immediate predecessor, the Wittenberg-trained Georg Stadius (1550–93).[82]

Kepler's three extant prognostications from Graz (1597–99) show that he had rapidly learned the essentials of the genre, at least partly by studying other prognosticators.[83] For example, in his *practica* of 1597, Kepler began the section on war and the state by disputing the exact birth hour of the Turkish Sultan Mehmet III as reported by the Nuremberg calendarist and church deacon Johann Paul Sutorius. Later, in the section on crops, he referred to Cardano as "our teacher" and to the weather predictions of the Rothenburg calendarist Georg Caesius.[84] As always, Kepler put his unique stamp on every topic: his preliminaries were four times as long as those of his predecessors, and his study of eclipse observations, associated with the preparation of calendars, led him eventually to develop a new theory of the Moon's motion.[85] Kepler's 1597 *Practica* also broke with earlier Graz prognosticatory practice by using the introductory section to introduce new physical arguments: "That no star, great or small, has in itself any particular force [*Krafft*] to act on earth other than universal light [*allgemaine Liecht*]. Flowing forth spherically from the Sun into all the stars, this light weakens unevenly; so, if the Sun's light were to be extinguished, soon all effects mediated by the heavens would cease, and mortal creatures would live only so long as they were carried along by the particular forces."[86]

Kepler made no explicit attempt to introduce the Copernican planetary arrangement into the preliminaries of the 1597 *practica*, but he made uncustomary use of the *practica* genre to continue the transgressive disciplinary work of the 1593 Copernican cosmographical disputation. He worked by analogy, finding resources in the physical part of astrological theory that would later show up in the main physical propositions of the *New Astronomy*. In the passage above, we see that from the Sun's life-giving force Kepler intuited an inertia in which just as living creatures would die without the Sun's light, so the materially passive planet naturally inclines toward rest and moves only when acted upon by the Sun.[87] Shortly thereafter, however, Kepler broke with traditional astrological—physical theory by proposing a reciprocal relation between Earth and Moon: "The Moon has all things in common with the Earth—not only are its effects mediated by the heavens but also its fate is affected and changed and the Earth rules the Moon no less than the Moon the Earth. In this regard, no practitioner, optician or astronomer, can doubt that, because of the sun's shining, our dark Earth is much brighter and stronger than the Moon, just as the Moon shines because of the Earth."[88] And in the *practica* for 1599, Kepler included reference to the theory of harmoniously effective aspects that he would develop formally in his *More Certain Foundations of Astrology.*[89]

KEPLER'S COPERNICAN COSMOGRAPHY AND PROGNOSTICATION

Even as he was issuing his earliest *practicas*, Kepler was drafting the *Mysterium* as a tool of persuasion. In this section I shall be concerned with some of his rhetorical decisions and joint strategies with Maestlin for presenting the work. The lengthy title bears witness to the fact that this was a book enmeshed in academic categories. Even some contemporaries would find it abstruse: *Forerunner of Cosmographic Dissertations containing the Cosmographic Mystery*—so the first part of the title was announced.[90] *Cosmography* ordinarily referred to the study of the terrestrial globe (geography), but it also bore a resemblance to astronomy in that both supposed the geometry of the sphere; more broadly, the term was sometimes used to refer to the study of the parts of the whole universe.[91] Here Kepler's *cosmographicum* created a neologism that combined the mathematical part of astronomy with a new sort of physics of the heavens. The unusual adjective *mysterium* bears the clear religious connotation of a secret (and sacred) reality intentionally hidden by God.[92] This view may have been inspired by Clavius's representation of astronomy as a form of contemplation of the invisible divine mirrored in the visible world.[93] The whole work was a "forerunner" in the sense that Kepler planned a series of further dissertations using the method of deducing God's

intentions from a priori metaphysical, theological, and physical axioms. These future disputations were also to be cosmographical—rather than strictly astronomical—because, like Aristotle's *On the Heavens,* they continued to make use of the full range of explanatory resources. Thus, *cosmography* firmly linked astronomy to natural philosophy; *mystery* connected both to theology. Kepler regarded himself as studying the Creator's plan not primarily through the Bible but cosmographically, through the Creation.

The second part of the title told how this was to be done. It addressed the structure of the universe, but without naming Copernicus. Although Kepler would argue pointedly for Copernicus in the body of the text, his title announced that his book offered a demonstration of why the heavens are ordered as they are: *Concerning the Admirable Proportion of the Heavenly Orbs and concerning the Genuine and Proper Causes of the Number, Size, and Periodic Motions of the Heavens, Demonstrated by Means of the Five Regular Geometric Solids.* Only at the bottom of the frontispiece did the reader learn that *There is added the learned* Narration *of Master Georg Joachim Rheticus, concerning the books of the Revolutions and the admirable hypotheses on the number, order and distances of the Spheres of the World by the most excellent Mathematician and Restorer of the Whole of Astronomy, the learned Nicolaus Copernicus.* The moving-power hypothesis—so critical in Kepler's work leading to the elliptical theory—received no explicit mention.

Maestlin's role in the entire production was considerable. It was Maestlin who decided (on his own) that the *Narratio* be adjoined to the main work because, revealingly, he thought that many of Kepler's readers—especially the members of the Tübingen academic senate—would not know or understand the basic elements of Copernicus's theory. Effectively, the *Mysterium* was to be bundled by analogy with the 1566 edition of *De Revolutionibus.* And it was Maestlin, not Kepler, who did the actual work of preparing the new, strongly Copernican introduction, postillating Rheticus's text, and adding illustrations that had not appeared in the earlier editions. It was Maestlin who carefully edited Kepler's text and caught a crucial error in Kepler's labeling of the equant, even while he expressed caution concerning his student's hypothesizing about a moving soul or virtue in the Sun.[94] Finally, it was Maestlin who added a treatise of his own that explained in great detail

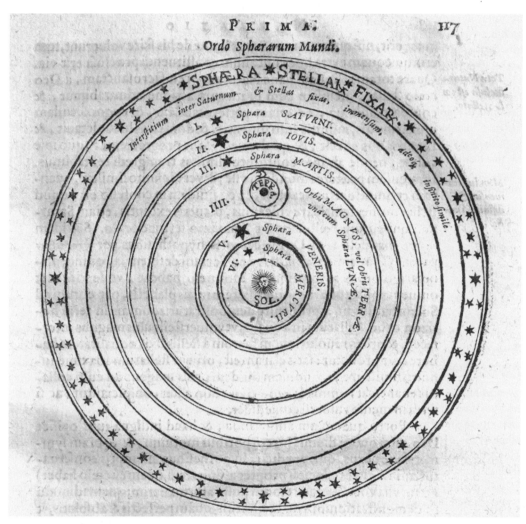

64. "Order of the Spheres of the World." From Maestlin's edition of Rheticus (1596). Image courtesy History of Science Collections, University of Oklahoma Libraries.

the dimensions of the planet's models. Maestlin's conception was that the ordinary reader would get the theoretical rudiments of Copernicus's system from Rheticus (rather than from *De Revolutionibus*); the parameters of the individual planetary models from his own *Dimensions of the Orbs and Celestial Spheres;* and the metaphysical, physical, and astronomical reasons for the planetary arrangement from Kepler's *Mysterium Cosmographicum*. It would be a complete, rather than a partial explanation of the divine plan of the heavens.

Publishers' compendia, as we have seen time and again, were a common enough sort of production in the literature of the heavens. If Ratdolt's 1491 omnibus edition is a typical example of a pub-

lisher's project, by contrast, the Kepler-Rheticus-Maestlin compendium directly involved the authors—especially Maestlin, who, with considerable effort, oversaw its production. To do so he had had to postpone a major polemical work attacking the Gregorian calendar, but he clearly thought it was worth the trouble. In return, the framework of the joint production provided Maestlin an opportunity to express his own views publicly for the first time. Somewhat as Rheticus had done for Copernicus, Kepler enabled Maestlin to associate himself in print with a robust version of the heliocentric planetary arrangement.

Yet, if cosmography concerned the philosophy of the heavens, what did it have to do with the remaining elements of the science of the stars? How could Kepler justify writing a book about astronomical theory when his livelihood was grounded in writing astrological practices and teaching the rudiments of the sphere? Already he had used his 1595 prognostication to announce that he had discovered a "secret" and would publish a full account soon.[95] And when Kepler eventually drafted the patronal dedication to the *Cosmographic Mystery*, he precisely defined his "mystery" in direct contrast to the genre of ordinary prognostications. Kepler opened the dedication by immediately recalling his promise: "Seven months ago, I promised you a work that would be judged by the learned to be beautiful and fascinating and *far preferable to annual prognostications*."[96] The strategy of the dedication shows that Kepler believed that he first needed to argue for the superiority of theoretical knowledge—knowledge that foregrounded explanations and justifications over practical utility, which was concerned only with effects. The dedication emphasized the opposition between the act of contemplating the heavens and the supposed profit to be had from the forecasts. "We accept painters, who delight our eyes, musicians who delight our ears, though they bring no profit to our affairs."[97] Such contemplative activity was its own reward— fit for nobles and princes, but not for ordinary people. "It is sufficient for astronomers to write for philosophers rather than those who babble and shout, for kings rather than shepherds." In short, although Kepler himself was a local prognosticator, it was the pursuit of theory that he regarded as ennobling; it was to be seen as a princely or even an imperial ideal.[98] That was the justification that Kepler commended for his sort of study. The divine plan required theoretical knowledge, knowledge of God's reasons for putting together the universe as he did. Such wisdom about the *mysterium* could not be achieved by the lowly and debased practices of ordinary prognostications. Put otherwise, the *Mysterium* was not directed to the more lucrative market of annual forecast.

The distinction between legitimate and illegitimate prognostication was not a new one. As we have seen throughout this study, the tension was not between two social groups—theoreticians and practitioners—but between the claims of different modes of knowledge and those who subscribed to them. Thus Paul of Middelburg, a century earlier, trumpeted his higher astrology over Johannes Lichtenberger's apocalyptics; Tycho Brahe elevated astronomers of the microcosm above common and unskilled practitioners; and Clavius denounced nativities and annual prognostications while permitting weather forecasting. For Kepler, the legitimacy of astrological practice rested on getting the principles of theoretical astrology right; but those principles, in turn, were to be grounded on the correct principles of theoretical astronomy—exactly the motivation that drove Copernicus's own project. The kernel of this connection between astrology and astronomy existed already in the 1593 physics disputation and continued in the *Cosmographic Mystery*; parts of it were developed more fully in his 1602 treatise, purporting to provide astrology with sounder foundations. It was then further refined in the *Stella Nova* (1606), in polemics with Helisaeus Röslin and Philip Feselius (1608–9), in the brilliant physics of the true sun of the *Astronomia Nova* (1609), and finally in the larger structure of the *Harmonice Mundi* (1619).

Glancing forward five years will assist our reading. As in his dedication to the *Mysterium Cosmographicum*, so in his treatise on the foundations of astrology Kepler began by defining the legitimacy of his subject against annual prognostications: "The public expects a mathematician to write annual prognostications. Since at the approach of this year, 1602[,] . . . I decided not to pander to the public's craving for marvels but rather to do my duty as a philosopher, namely to limit the scope of such prognostications. I shall begin with the safest assertion of all: that this year the crop of prognostications will be abundant, since, as the crowd's craving for marvels increases, each day will bring an increase in the number of authors."[99] Kepler's disdain for ordinary prognostication was palpably Piconian, although Pico's name was not invoked:

Some of what these pamphlets say will turn out to be true, but most of it time and experience will expose as empty and worthless. The latter part will be forgotten while the former will be carefully entered in people's memories, as is usual with the crowd. . . . Astrologers have regard to causes that are partly physical, partly political (for the greater part, indeed, inadequate, and mostly imaginary, vain and false) and, for the rest, causes that are completely null (when they allow enthusiasm to guide their pens). When they are carried away by this, if what they say is true the fact must be attributed to chance—unless we are to believe that more often and for the most part it is brought about by some higher occult instinct.[100]

Kepler's "sounder foundations" were indebted to the central proposition of Pico's light physics, as in thesis 5: "The most universal, powerful, and surest cause—which all men know about—is the increase and decrease of the height of the Sun at noon." This Keplerian-Piconian proposition actually inverted the central analogy underlying the ancient astrological notion of "influence." That analogy was between the Sun's intensifying and weakening effects on Earth and a comparable force in the other planets. Pico effectively negated the original analogy between solar and planetary influence and restricted all influences to the Sun. But whereas Pico denied that anything could be explained by the heavenly force other than the warming and cooling effects of the Sun, Kepler also retained the Moon as a source of power: "For experience shows that all things that consist of humours swell as the Moon waxes and shrink as it wanes. This single effect determines many of the predictions and choices of auspicious times that are to be made in matters of Economy, Agriculture, Medicine and Seafaring."[101]

In the *Mysterium Cosmographicum*, Kepler had continued to develop a physical astronomy. Most significant, he now centered it in the actual body, the true Sun, rather than in its mean position.[102] He proposed that there was no way to explain why the planets move other than by a single force issuing from the Sun's material body. And in his physical speculations about the Moon, he continued to appropriate conceptual resources from Pico. What caused the Moon to follow the Earth, "just as some steward moves about the master of the house, or just as those who walk on a ship do not tire during the journey unless the great force of the waters sways the unsteady or perhaps shifts those at rest?" Maestlin had conjectured that the Moon had continents, oceans, mountains, and seas just like the Earth.[103] Rejecting material agency, such as "bars, chains and hard orbs," Kepler found an explanation for the cohering of these similar bodies in a pervasive, immaterial, Stoic "heavenly influence" (*influxus coelestium*), carried by the surrounding air and taken into the body with respiration.[104] Yet he had not yet worked out exactly how the Sun and the Moon together could account for astrological effects on the earthly globe. Instead, having denied the physical efficacy of the zodiac, Kepler tried to offer a new sort of astrological theory based on the distinction between perfect and imperfect harmonies. The reasons why the Creator had used some kind of proportionality to space the planets must have had something to do with *which* configurations of the planets in the zodiac ("aspects") produced terrestrial effects.[105]

THE DIVINE PLAN, ARCHETYPAL CAUSES, AND THE BEGINNING OF THE WORLD

Throughout his many writings, Kepler made the Copernican planetary order his hermeneutical fulcrum.[106] On it, he provided first principles of justification adapted from Pythagorean-Platonic metaphysics and epistemology and tied them together in a theology of the created world. The same iconic patterns were given special value as structuring principles throughout. Of particular importance was what Kepler called the resemblance (*similitudo*) between those entities at rest in the visible world (the Sun, fixed stars, and intermediary space) and the invisible Christian trinity (God the Father, the Son, and the Holy Spirit) that was held only on faith. This was how God's threefold being showed up in the world, and Kepler was confident that much more could be known: "I had no doubt that since the immovables presented this harmony that the moving things would reveal themselves as well."[107]

That he already classified the Sun as a resting body shows how this theological metaphysics worked to naturalize Kepler's astronomical convictions. It also helped to distance him from traditional theological practices: the archetypal principles provided the proper tools for reading *Genesis*. And finally Kepler distanced himself from traditional natural philosophy by taking on board Pico's light physics; from this he evolved a theory that the cause of motion could arise only from a material entity and that the prime cause of planetary motion issued from the body of the Sun.

With each move in this inquiry, Kepler achieved new levels of integration and justification that linked different domains: metaphysics, theology, physics, astronomy, and astrology. As early as 1599, he seems to have believed that he could reach a unified set of principles from which all domains of knowledge derived. In that sense, Kepler was a Platonist not only in the usual sense of believing that the structure of the universe reflected a divine archetypal pattern, but also in the sense of believing that archetypal principles underwrote all domains of knowledge: theology, natural philosophy and the science of the stars. From such general foundations, it could be argued that accepting archetypal principles in the domain of theoretical astrology implied accepting them in theoretical astronomy.[108] Sometimes the philosophizing astronomer, sometimes the theologizing hermeneut, Kepler's is a unique voice in the emergent third generation after *De Revolutionibus*. Here was the early hope of justifying the Copernican planetary ordering using metaphysical and physical principles that would facilitate a once-and-for-all elimination of the alternative arrangements that Maestlin and Rothmann had failed to block in the previous decade.

Kepler's ostensibly Platonic conviction was that the Creator had used certain patterns or structures of consummate beauty to make the world of three-dimensional objects. Hence, it is no accident that this archetypal theme figured crucially in the *Mysterium Cosmographicum*. Kepler wanted it that way, as is evident from the arrangement of the title page, from the story he narrated about how he discovered the proper causes of heavenly order, from the dramatic foldout of the polyhedral scheme, and from the little poem that he placed strategically on the back of the title page. This poem evoked the first chapters of *Genesis* as well as Plato's *Timaeus*:

What is the world? What motivated God to create it and according to what plan? From whence has God drawn the numbers? Which rule governs such an enormous mass? Why has God created six circuits? Why has he created these spaces between each orb? Why are Jupiter and Mars separated by such a vast space when they are not the first two orbs? Here, indeed, Pythagoras teaches all this to you by means of five figures. He has shown us by his example that we can be reborn after two thousand years of error, provided that there is a Copernicus.[109]

Kepler added further authority to his enterprise—and its opposition to Aristotle—by locating it within this ancient filiation of wisdom or *prisca philosophia,* going back as far as Pythagoras.[110] In the *Harmonice Mundi* (1619), Kepler made explicit the connection to Plato's *Timaeus*

and pushed the filiation back to the beginning of the world. The *Timaeus* "is beyond all hazard of a doubt a kind of commentary on the first chapter of Genesis, or the first book of Moses, converting it to the Pythagorean philosophy, as is readily apparent to the attentive reader, who compares the actual words of Moses in detail."[111] However, unlike the numerous sixteenth-century commentators on Genesis, Kepler was already constructing a new, more secular, audience: he did not explicitly write this work as a theologian for other theologians, although he knew that the Tübingen authorities would read it.[112] He did not use the genre of the commentary, and he did not use the writings of the Church fathers to elucidate the meaning of the created world. Nor, like thirteenth- and fourteenth-century natural-philosophical commentators, did he speak hypothetically of a world that God *could* have built. He represented himself as offering the reasons for the arrangement of the planets, as first conjectured by Copernicus "from the phenomena" (*ex "phainomenois"*) and as God had actually made it to be (*a Creationis Ideâ*).[113] Following baroque humanistic practice, his authorities were a mixture of the ancients (Pythagoras, Plato, Euclid, Proclus, Ptolemy) and the moderns (Nicolaus Cusanus, Copernicus and his disciple Rheticus, Maestlin, Tycho, and the Euclid commentator François Foix de Candale)—with a decided leaning toward the authority of the moderns.[114]

If Kepler's Copernican-Platonic reading of Genesis set him apart from traditional Genesis commentators, it also marked off his enterprise quite sharply from another stream of writers—the apocalyptic astrologers and humdrum prognosticators, who, like Lichtenberger, Leovitius, and many others believed that signs of the end of the world could be read from heavenly phenomena. Kepler's stand represented an important and even a courageous break from the brooding anticipations of his coreligionists. Along with Tycho and Maestlin, Kepler was convinced that apocalyptics should be left to traditional biblical commentators.[115] Even the Platonic year—the moment when the planets were supposed to return to the alignment from which they started at the beginning of the world—could not, in his opinion, be known.[116] But, if knowledge of the end belonged to traditional theologians, knowledge of the *beginning* belonged to the astronomers. That was the mystery that Kepler believed only Copernican cosmography could unravel, and he now positioned himself as its preeminent spokesman.

That the five regular solids can, in fact, be inscribed more or less successfully in the spaces that separate the six Copernican planets seemed to Kepler to be the sort of argument that could move the Copernican hypothesis from the status of "relatively better than" to an assertion of apodictic necessity, that "God chose to make it that way and no other." To achieve that sort of claim would be to move astronomy into the domain of theological and physical cosmography and thereby to lay claim to astronomy as a form of demonstrative knowledge.

Initially motivating this move was Kepler's (and Maestlin's) revival of Rheticus's reading of Copernicus against what had become the dominant interpretation of Copernicus among the Wittenbergers and their followers throughout Europe from the middle of the sixteenth century. From Rheticus's presentation and from the emphases in Maestlin's annotations on *De Revolutionibus,* Kepler immediately grasped the implications of the true proportions of the distances between the orbs in Copernicus's scheme. Now, for the first time in the sixteenth century, Kepler produced in print a scaled representation of the Copernican orbs which showed clearly (and rather startlingly) that the universe was more empty than full.[117] The polyhedral hypothesis replaced the traditional gap-filling function of the contiguous Ptolemaic orbs; equally significant, it also filled the discontinuous interorb Copernican gaps with beautiful Platonic forms, accessible only to the eye of the mind, rather than with impenetrable and invisible matter—also accessible only to the intellect. And there were further significant advantages to the archetypal argument. For Kepler, the appeal to Pythagorean-Euclidean geometry as an explanation for why there are six planets improved considerably upon Rheticus's unlikely appeal to the sacred number six in Pythagorean arithmetic.[118] This was so, he explained, because of the priority of quantity in the creation of the universe: in the beginning, so Kepler claimed, God created "body"—in the sense of three-dimensional geometrical objects.[119]

Commentators have now well described the technical astronomical details of Kepler's polyhedral "discovery" (*inventum*).[120] One important part of this alleged demonstration was his conviction that the arrangement of the polyhedra provided a different route to the numbers representing the distances separating the planetary orbs. One could derive the numbers in the way that Copernicus had supposedly done, starting from the values already known from observation, or one could get rather close to them by starting with the metaphysically loaded geometry of the Platonic-Pythagorean solids. Maestlin no doubt approved Kepler's new method from prior causes because it seemed to provide a deeper justification for his own adherence to the criterion of mathematical harmony. Because the arguments were mathematical in character, they appeared to keep Kepler within the conventional bounds of astronomy while offering unexpected support for Copernicus from theology.[121] But Kepler, of course, was by no means the first sixteenth-century Platonizing astronomer to pay attention to the polyhedra, and their application was not restricted to cosmic fashioning.

The Platonist-Copernican Digges had devoted a whole treatise to the regular solids, while "meaning not to discourse of their secrete or mysticall applicances to the Elementall regions and frame of Celestial Spheres."[122] By contrast, the non-Copernican Offusius had vigorously asserted a cosmic role for the polyhedra in divining the correct planetary distances from Earth by counting up the equal triangles into which the polyhedral faces could be divided. One must wonder if it is entirely coincidental that all the elements that Kepler used in his polyhedral scheme were already present in Offusius's *Against the Deceptive Astrology:* the *Timaeus* and the polyhedra, the preoccupation with planetary distances, the idea of *symmetria* from Copernicus's *De Revolutionibus,* and the concern to answer Pico's objections. But, whereas Offusius was interested in the special numbers yielded by this procedure, as Rheticus had speculated on the significance of the number six, Kepler avoided numerology for this purpose.[123] The trials that he reports in trying to find a "reason" for the spacing of the planets are all geometrical, suggesting that he was already looking for answers restricted to that domain. Eventually, when he hit on the polyhedra, it was the procedure of inscribing and circumscribing that drew his attention; and that idea may have come to him because he had been able to obtain new books in Graz. One of these was François de Foix de Candale's derivation of the perfect solids presented in book 16 of his expanded edition of Euclid (1567).[124] Was Offusius another? One is

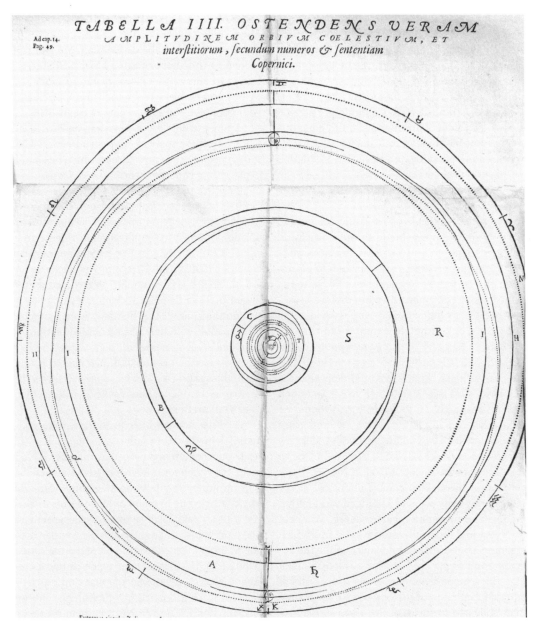

TABELLA IIII. OSTENDENS VERAM
Ad cap.14.
Pag. 49.
AMPLITVDINEM ORBIVM COELESTIVM, ET
interſtitiorum, ſecundum numeros & ſententiam
Copernici.

65. "True thicknesses of the celestial orbs and the gaps between them, according to the numbers of Copernicus." Kepler 1596, tabella 4. Image courtesy History of Science Collections, University of Oklahoma Libraries.

tempted to speculate that Kepler had encountered Offusius's book in Graz at just the moment that his concerns with astrology intensified.[125] Whether or not that was the case, Offusius's example pointedly highlights the priority of Kepler's Copernican commitment. Without Maestlin's Copernicus, Kepler might well have developed a representation along the lines of the geocentrist Offusius.[126]

FROM KEPLER'S POLYHEDRAL HYPOTHESIS TO THE LOGICAL AND ASTRONOMICAL DEFENSE OF COPERNICUS

Kepler narrated his polyhedral discovery as both a discovery and a divine inspiration.[127] What made it count as a "discovery" was that, in the manner of a Tübingen philosophizing theologian, he believed himself able to deduce it from first principles. He then had the new idea that, in the manner of a philosophizing astronomer, metaphysics should be tested empirically against the numbers yielded by the *Prutenic Tables,* much like ordinary planetary hypotheses. No one had done this sort of thing before. Copernicus had presented his aesthetic harmonies to the pope in a literary and political idiom, but he had hardly ventured to provide an irrevisable deduction purporting to show that the Holy Trinity was an archetype that manifested in the very structure of the world itself!

Like an evangelical reformer, Kepler wanted not only to contemplate but to spread God's word. He resisted the circumspect, private manner that had characterized Maestlin's approach to his annotations on the *De Revolutionibus.*[128] He represented the act of publishing his discovery as a way actively to honor a god who wished to be known not only through scripture but through the book of Nature.[129] Or, as Kepler put it charmingly in 1621: "Certainly God has a tongue, but he also has a finger. And who would deny that the tongue of God is adjusted both to his intention and, on that account, to the common tongue of men? Therefore in matters which are quite plain everyone with strong religious scruples will take the greatest care not to twist the tongue of God so that it refutes the finger of God in Nature."[130] To get such a book about nature published at the official university press in Tübingen was another matter. The university printer Georg Gruppenbach required the faculty senate's assent—which meant, of course, the acquiescence of the theologians—and to get that, Kepler needed to enlist Maestlin's support. In spite of his many years of outward caution, Maestlin proved himself more than adequate to the task. In a sense, his student's boldness licensed him to come forth in defense of views that he himself had long held. Moreover, both his confessional stance and his credibility as a mathematician were uncontested, and the manner in which he taught Copernicus in the classroom had thus far created no grounds for objection.[131]

Experienced in the politics of the small Tübingen faculty, Maestlin mounted an effective strategy. He emphasized to the senate the astronomical novelty and value of Kepler's achievement without actually displaying its details. In Maestlin's language, Kepler had "dared to think" and had then "dared to try to prove" that the number, order, and distances of the planets could be derived a priori from the divine plan. He had succeeded in this demonstration, Maestlin averred, but in so doing, he had not made his work sufficiently accessible to the public. There were many who knew nothing about Copernicus and who did not know Euclid's proofs. Master Kepler needed to explain these matters "less obscurely."[132] That is, he needed to add a preface in which he should explain Copernicus's theories more clearly.

Maestlin's gambit worked: there were no difficulties with Kepler's discovery, only the manner in which he had communicated it. The senate unanimously approved publication of this unprecedented presentation of Copernicus's theory on condition that Maestlin's recommendations be followed. Kepler was extremely pleased, as he had feared that the theologians would reject publication on the grounds that his views were somehow incompatible with holy scripture. He was right to have had such worries. For, even though he had an answer to their concerns, he decided to leave it out of this earliest work in response to the friendly yet firm recommendations of the theologian Matthias Hafenreffer.[133]

Just three months after formulating the polyhedral hypothesis, in July 1595, Kepler had already outlined an approach that would make up much of chapter 1, and which, by October, he communicated in a letter to Maestlin.[134] First he would affirm the agreement of Copernican theory with all past observations and its capacity to make predictions about the future. Neither here nor in the published text did he argue the much stronger claim that he later made in the *Astronomia Nova* (*New Astronomy*), that Copernicus's theories were geometrically equivalent to those of Tycho and Ptolemy.[135] Next, without mentioning Clavius or Osiander, he referred to the problem of the skeptical objection *ex falso sequitur verum.*[136] This was a stronger objection to astronomy than Pico's because it was based on the alleged deductive structure of all astronomical knowledge rather than the inductive claim that astronomers purportedly disagreed about everything on which astrology was built. Kepler's lengthy reply to this objection amounted to an extension of an argument made by Aristotle whose consequences Clavius had well understood and which Maestlin and Copernicus had affirmed rather too compactly: that true conclusions must ultimately

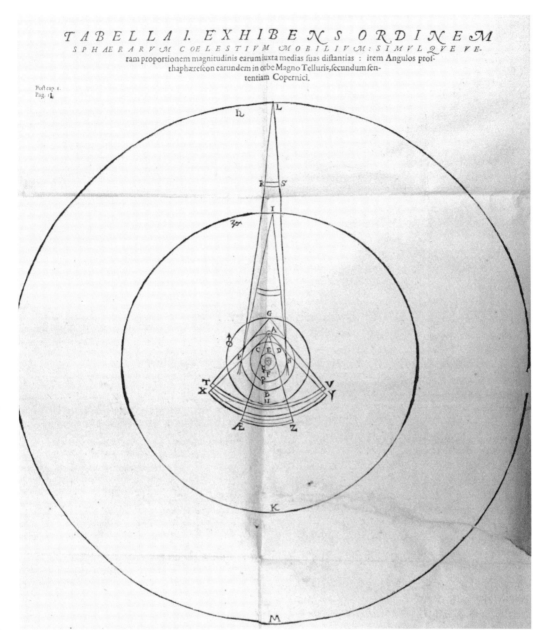

66. "The order of the celestial spheres and the true proportionality of their sizes according to their mean distances." Kepler 1596, tabella 1. Image courtesy History of Science Collections, University of Oklahoma Libraries.

follow from true premises. Aristotle had shown that although one may propose any old false ad hoc propositions in order to deduce a given conclusion (*quid*), one cannot then return to the original principles from which the conclusion was first derived (*propter quid*). Nor, because they are unconnected, can one deduce new conclusions that are coherent with the original premises. One simply has agreement about the existence of the phenomena and a series of unrelated propositions from which they may be derived following rules of logical validity. Consequently, one can never know the true "reason why."

In the remaining part of the chapter, the part that the senate had urged him to add, Kepler showed how Copernicus, unlike Ptolemy, never

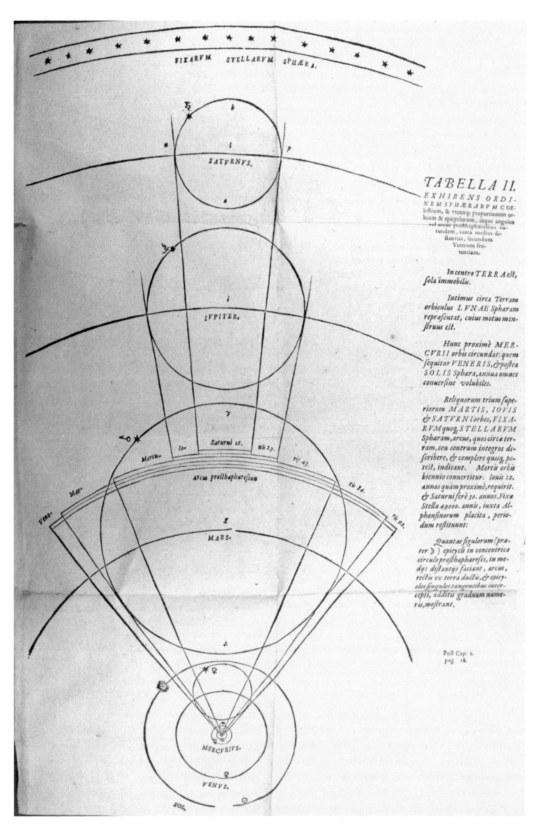

FIXARVM STELLARVM SPHÆRA.

SATVRNVS.

TABELLA II.

EXHIBENS ORDI-
NEM SPHÆRARVM COE-
lestium, & vtcunq; proportionem orbium & epicyclorum, atque angulos
vel arcus prosthaphæresion ro-
tundem, iuxta medias di-
stantias, secundum
Veterum sen-
tentiam.

In centro TERRA est,
sola immobilis.

Intimus circa Terram
orbiculus LVNAE Sphæram
repræsentat, cuius motus men-
struus est.

Hunc proximè MER-
CVRII orbis circundat: quem
sequitur VENERIS, &postea
SOLIS Sphæra, annua omnes
conuersione volubiles.

Reliquorum trium supe-
riorum MARTIS, IOVIS
& SATVRNI orbes, FIXA-
RVM quoq; STELLARVM
Sphæram, arcus, quos circa terram, ceu centrum integros describere, & complere quisq; potest, indicant. Martis orbis
biennio conuertitur. Iouis 12.
annos quàm proximè, requirit.
& Saturni ferè 30. annos. Fixæ
Stellæ 49000. annis, iuxta Al-
phonsinorum placita, periodum restituunt:

IVPITER.

γ

Mercu-   Io.   Saturni 12.   ris 22.   rij 43.

Arcus prosthaphæreseon

ris 34.

Mars·

rij 92.

MARS.

Quantas figulorum (præ-
ter ☽) epicycli in concentrico
circulo prosthaphæreses, in medijs distantijs faciant, arcus,
rectis ex terra ductu, & epicyclos singendos tangentibus inter-
cepti, additis graduum numeris, mostrant.

Vene-

λ

♀

♀

Post Cap. 1.
pag. 11.

MERCVRIVS.

♀

VENVS.

SOL.

67. "The ancient opinion about the order of the celestial spheres according to their mean distances, whatever the proportionality of their orbs and epicycles and also their arcs and angles of correction [prosthaphaeresis]." Kepler 1596, tabella 2.
Image courtesy History of Science Collections, University of Oklahoma Libraries. The epicycle corrects the planet's mean
motion and distance by adding to or subtracting from it. Compare the sizing of the radii of the epicycles in Kepler's calibrated diagram with those in Wittich's "Sphere of Revolutions" (figure 56).

had recourse to ad hoc hypotheses.[137] The arguments were thoroughly focused on planetary order and implicitly constituted a direct answer to Peurbach.[138] First, without explanation, Ptolemy's epicycles for the superior planets have the same period of revolution, while the deferents of Venus and Mercury share with the Sun the same period of revolution. Moreover, Ptolemy has no explanation for why all the planets, except for the Sun and the Moon, demonstrate retrograde motion. Third, the ancients could not explain why the epicycles for the superior planets increase as one moves from Saturn to Mars whereas, after the Sun, the deferents of Venus and Mercury are smaller. And finally, why at opposition to the Sun are the three superior planets always closest to the Earth, whereas at conjunction they are at their highest points? In all cases, Kepler showed that the same reason—the Earth's annual revolution—could explain the otherwise inexplicably observed effects.

The diagrams that Maestlin provided to illustrate these arguments were the first in the sixteenth century to depart from the conventional practice of representing the cosmic patterns without regard for relative scaling and proportionality. Even so, except in Kepler's annotated second edition of the *Mysterium* (1621), these illustrations were not further reproduced in the seventeenth century. Yet without these diagrams it would have been especially difficult to follow the third argument above. Thus, one began with the "opinion of the Ancients" in tabella 2, where it was easy to see gross, relative differences in the diameters of the three superior epicycles (which measured the correction due to the Sun's motion). But then one had to return to tabella 1 for the Copernican explanation: a Martian with eye at point G would observe the Earth's annual motion through angle TGV while, just above, a Saturninan situated at L would observe exactly the same motion through the smaller angle RLS. Kepler added to the geometry an imaginative argument that he had developed in a student disputation on the Moon: an observer on Saturn looking at Earth would observe the same kind of retrograde motions that an Earthling sees in the motions of Venus and Mercury.[139]

Kepler's final astronomical argument for Copernicus explicitly appealed to Maestlin's model for the comet of 1577, which he represented as a striking confirmation. This representation of Maestlin shows just how strongly indebted Kepler was to his teacher: it took no note of the possibility that Tycho's theory could account equally well for the few observed positions of the comet.[140] Chapter 1 thus became all the more explicitly a joint product of Kepler's and Maestlin's efforts. Among other things, it visually energized and improved on Rheticus's arguments, bringing them to bear directly and publicly against Ptolemy's claims for the first time since the 1540s. It made an unspoken alliance with Clavius's defense of the possibility of an astronomy that corresponded to the world. It made the comparative explanatory features of Ptolemaic and Copernican astronomy clearly visible for the first time in the sixteenth century. In 1905, J. L. E. Dreyer was so impressed that he found it "difficult to see how anyone could read this chapter and still remain an adherent of the Ptolemaic system."[141] Of course, like so much else that seems "obvious" in retrospect, things looked different in 1597, when Kepler's book finally went up for sale at the Frankfurt Book Fair, catalogued as a geography, its author ignominiously and mistakenly listed as Repleus.[142] That error was emblematic, a harbinger of the great resistance and misunderstanding that the *Mysterium*—and, indeed, most of Kepler's writings—would soon face.

12

Kepler's Early Audiences, 1596–1600

THE *MYSTERIUM COSMOGRAPHICUM*
THE SPACE OF RECEPTION

Kepler's early representation of the heavens embodied an unprecedented convergence of elements in the political space of the Tübingen theological orthodoxy. The *Mysterium Cosmographicum* aimed for a rigorous justification of the loose aesthetic standard that Copernicus had used to warrant a strong sense of *world system,* one that involved an interdependency of elements. This mathematical aesthetic never quite shook off all traces of its origins in classical literary theory (Horace), architecture (Vitruvius), music (Boethius), and art (Alberti), but Kepler tried the new tack of wedding it to a theological physics and a metaphysics of geometrical archetypes. His arguing of a rich and assertive Neopythagorean Platonism against Liebler's Melanchthonian natural philosophy and his pushing of a Platonic reading of Genesis in the house of the theologians underlines the radical boldness of Kepler's vision. He was as yet uncertain how to construct the physics (or metaphysics) of a Copernican astrology.[1] But he had managed to join a somewhat robust, even if idiosyncratic, physics to the Copernican astronomical premises that pushed not only the planets but also the limits of what Maestlin, Kepler's strongest advocate, regarded as the domain of the thinkable. The fertility of Kepler's conception, joined to its transgressive audacity, made it very hot to handle. Without Maestlin's support, Kepler's book would almost certainly have failed to navigate intact past the shoals of the Tübingen academic senate, for Maestlin well understood how to manage the theologians' sensitivities.

Yet, ironically, the very strategies that Kepler and Maestlin had used to establish the credibility of a Copernican defense also proved to be their greatest liability. Conceived within the knowledge structures and constraints of the local Tübingen constituency, a problematic that Kepler carried with him to Graz, the polyhedral representation looked strange even to many contemporaries.

The oddness of Kepler's new proposals raises the question of their discursive possibilities in other forums of reception. It hardly needs to be said that there were none of the institutional mechanisms of evaluation that would emerge later in the seventeenth century: neither societies devoted to natural philosophy nor anything resembling journals. Furthermore, as is already evident from our consideration of the quite varied usages and readings of *De Revolutionibus* in the 1580s, this was no age of modesty in public controversy. Of course there was an important literature of courtesy, which articulated idealized rules of mannered behavior for the upper classes, including the conduct of decorous conversation.[2] But, in the same period, conventions of invective in religious-political disputes had reached unprecedented levels of ingenuity and aggression. The same Bellarmine who in the early 1570s lectured soberly on Genesis and fluid heavens at Louvain published a series of lectures on virtually every topic of religious and political controversy after he moved to the Collegio Romano in 1576.[3] Manuals of how to conduct religious controversy proliferated.[4] Replies and counterreplies among controversialists raised vilification to a high art.[5] In contrast to these heavily personalized titles of religious polemic—let alone Bruno's trenchant satire—Kepler presented the Copernican scheme as a cosmography in the form of a carefully argued academic disputation. Forensic and dissertative, the *Mysterium Cosmographicum* was still a model of humanist restraint and of prudent practices of academic persuasion.

How did Kepler and Maestlin imagine that they would gain a positive reception for their revival of Copernicus? Kepler would receive an allotment of copies that he could distribute as gifts. The printer, seeking his own profit, would send a large allotment to the Frankfurt Book Fair. Beyond these measures, Kepler could write to sympathetic third parties, asking them to solicit or pass along any evaluations that came their way. Yet Kepler and Maestlin lacked the substantial printing resources and contacts of a Tycho Brahe: their immediate audience was still at the university in Tübingen and at the ducal court. How important was princely patronage for such a cause? What difference would the duke's authority make to a mathematician in the university of another German territory? To a mathematician in a Catholic land? Or even to a local Tübingen theologian?

THE TÜBINGEN THEOLOGIANS AND THE DUKE

The response from the theologians was not long in coming. The published text of the *Mysterium Cosmographicum* contained provocative materials not previously seen by the senate—bold Copernican diagrams and Rheticus's *Narratio* with Maestlin's notes and introduction.[6] Maestlin used his introduction to the *Narratio* both to praise Kepler's polyhedral discovery and, for the first time, to endorse Copernicus's notion that bodies approach or flee the Earth because of an "affinity" (*affectio*) for its roundness. He even went further, suggesting that this affinity might also reside in the Sun, the Moon, and the planets.[7] Worse yet, it was clearly irritating and absurd that the correct interpretation of Genesis should proceed through the ancient pagans Pythagoras and Proclus and the modern Copernicus. Suddenly, the earlier senate request for "less

obscurity" took on a new and unexpected meaning. In October Maestlin reported an increasingly tense atmosphere:

> Time and again Dr. Hafenreffer has assailed me (jokingly, to be sure, although serious tones too seem to be intermingled with the jests). He wants to debate with me, while defending his Bible, etc. By the same token, not long ago in a public evening sermon he expounded Genesis, chapter 1: "God did not hang the sun up in the middle of the universe, like a lantern in the middle of a room," etc. However, I usually reply humorously to those jokes as long as they remain jokes. If the matter were to be treated seriously, I too would respond differently. Dr. Hafenreffer acknowledges your discovery to be wonderfully imaginative and learned, but he regards it as completely and unqualifiedly in conflict with Holy Writ and truth itself. Yet with these men, who are otherwise fine and very scholarly, but have no adequate grasp of the fundamentals of these subjects, in like manner, it is better to act jokingly while they accept jokes.[8]

Maestlin and Kepler had crossed a political and disciplinary boundary with the Tübingen theologians. Hafenreffer gave "brotherly advice" to Kepler: he should remain a "pure mathematician," stay away from making true claims about the universe, and avoid provoking schisms in the Church. To Hafenreffer, who was by no means unfriendly to Kepler, the threat of schism reflected the sorts of anxieties that were all too prevalent at that moment in the reforming movement. But Hafenreffer and his colleagues did not move beyond this admonition—unlike the Roman church authorities in the more serious warning they gave to Galileo some twenty years later. The crucial difference lay in the process of confessionalization in the Protestant lands: the real head of the church, the *defensor ecclesiae*, was not a cardinal or pope but the duke himself. And although the theologians had the duke's ear regarding all sorts of important legal and moral matters that lay within their perceived area of competence, he turned naturally to his mathematicians for advice that normally fell into their domain.

In approaching the duke, Maestlin and Kepler avoided traditional academic arguments, saying nothing at all about holy scripture or Aristotle's physics. Instead they constructed an exaggerated appeal to reputation and renown, advising that "all the famous astronomers of our time fol-

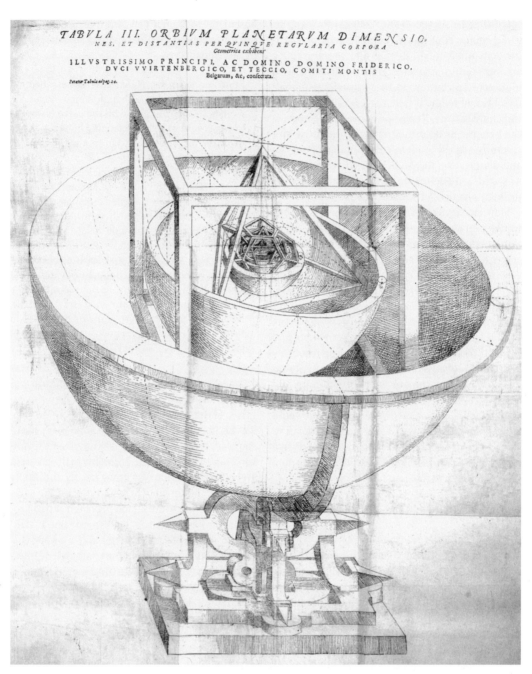

68. "The five regular, geometrical bodies produce the dimensions and distances of the planetary orbs." Kepler 1596, tabella 3. Image courtesy History of Science Collections, University of Oklahoma Libraries.

low Copernicus instead of Ptolemy and Alfonso" (after whom the thirteenth-century planetary tables were named).[9] At the same time, they stressed novelty. Maestlin emphasized that Kepler had found a new method, appealing to the metaphysical significance of the polyhedra rather than to observations alone, by which the planetary distances could be established and ephemerides improved.[10] And finally, Kepler requested financial support ("under a hundred gulden") to build a copper model of his polyhedral Copernican arrangement, with the planets represented as precious jewels. Like a Trojan horse in the ducal palace, this princely display would focus immediate visual attention on the divine a priori causes of the Copernican arrangement while undermining the authority of the traditional ordering. Moreover, the cosmographic mystery would become more readily apparent—and pleasurable to contemplate—as the duke turned the spigots that released the beers, wines, and liquors filling the copper hemispheres (which functioned as cups) representing the spaces between the planets. Like Kepler's physical solar moving power that pushed the planets around the Sun but weakened as it moved further from the center, the design called for "a costly aqua vitae" to flow from the Sun and a "very bad wine or beer" to be put into the last cup formed by Saturn and Jupiter. Wines and beers of diminishing strength and quality would fill the remaining spaces.[11]

Had it gained ducal support, Kepler's Copernican tavern and a proposed mechanical, heliocentric planetarium might perhaps have furnished the elements of a Kuhnian conversion scene. But although these projects failed to materialize, Kepler did not cease his efforts to link the duke publicly to what he called his *inventum*.[12] Ultimately, he achieved this goal in the heading of the famous, dramatic visual fold-out that appears as the third plate in the *Mysterium Cosmographicum*.[13] The polyhedral model is shown resting suggestively on an ornate supporting base—clearly intended for a patron to put on display—rather than as an abstract, two-dimensional representation, as was common for diagrams of planetary arrangements.[14] It was this diagram that embodied and conveyed patronal protection against theological criticism. As Maestlin remarked to Kepler: "They [the theologians] make no overt move because they are deterred by the authority of our duke, to whom the principal diagram is dedicated."[15]

What was at stake here, of course, was protection of a piece of academically precarious theoretical knowledge. Kepler had failed to smuggle his polyhedral liquor cabinet into the duke's court, but, in his typically ingenious way, he had found a different method, which exploited the resources of the printing press to attach his discovery to the duke's authority. This interesting episode highlights the special difficulties that Kepler faced in defending a consequential and disputed set of theoretical propositions about the heavens. Unlike the typical patronage productions of court mechanicians and mathematicians, the polyhedral model did not serve any practical function. It was not a timepiece, a terrestrial globe, a cannon, a mural, or a celebratory arch, although it traded on ornamental display; it promised no commercial return or alchemical gold, and it was explicitly not an astrological *practica*. Unlike Rheticus in his approach to Albrecht a half century earlier, Kepler did not try to argue that Copernicus's theory would improve the prospects of astrological forecast; and, like Maestlin, he chose not to avail himself of the opportunity to link his discovery to the Elijah six-thousand-year prophecy. Kepler understood that his defense of the Copernican arrangement would require new sorts of arguments about the legitimacy of theoretical knowledge. What would be the response outside of Tübingen and Graz?

THE GERMAN ACADEMIC MATHEMATICIANS
LIMNAEUS AND PRAETORIUS

Like bees to honey, three central participants from the world systems controversy of the 1580s were immediately attracted to the *Mysterium Cosmographicum*: Roeslin, Brahe, and Ursus. All had a publicly vested interest in their own representations of celestial order. Three others were established professors of mathematics whose convictions were held privately and without any effort at public advocacy: Georg Limnaeus (Jena), Johannes Praetorius (Altdorf), and Galileo Galilei (Padua). Their responses have much to teach about how university mathematicians engaged a theoretical novelty; in this section, I deal first with the traditionalists, Limnaeus and Praetorius.

Especially striking is the pattern of selective response that had also characterized the Wittenberg interpretation. Limnaeus (1554–1611), like Praetorius, had come out of the Melanchthonian tradition. He had studied with the Wittenberger Jacob Flach at Jena; he is known to have issued a prognostication in 1585 and was appointed to the mathematics chair in 1588 as Flach moved up to the medical faculty.[16] Even on the basis of such slim evidence, we may think of him as displaying the typical profile of a late-sixteenth-century German academic prognosticator. He had found Kepler's book at the Frankfurt Book Fair and wrote to identify himself as a philosophical ally, admitting to Kepler that he was something of a closet Platonist: "Most Illustrious Sir, never was I estranged from that most ancient philosophy of the Platonists—nor have I thought, as have several petty philosophers in our time, that it ought to be shunted outside the borders of the territory of the republic of letters."[17] This passage suggests that Limnaeus was covert in his Platonic sympathies at Jena, and it is possible that such suppressed philosophical convictions were common among university mathematicians of this period. Thus his joy to find a sympathizer in Kepler, who had dared to express his views publicly. Limnaeus's Plato, as he was eager to tell, was part of the filiation of ancient philosophy, inheritor of a deep wisdom descending to the Greeks through the Egyptians, the Jews (as he claimed to learn from Josephus), and Berosus the Babylonian. Limnaeus regarded Kepler's book as evidence of the revival of this sort of secret but powerful philosophizing. Learned men, and especially those who study astronomy, would now have "a new path to knowledge of the stars." Was Limnaeus hinting that he had been induced by the *Mysterium Cosmographicum* to shift his beliefs about the order of the planets? If so, he concealed his beliefs well; together with Kepler, he readily celebrated the ancient wisdom yet, unlike Bruno, without declaring the slightest adherence to its Copernican core.

Limnaeus had sent his opinions directly to Kepler. Would he have expressed himself in a more forthright manner had he written to a third person? Much of what Kepler himself learned about the reactions to his book came through the mediation of the Bavarian nobleman and chancellor Johann (Hans) Georg Herwart von Hohenburg (1553–1622). Herwart was one of the most important of a series of learned baroque noblemen with whom Kepler developed special ties of intellectual trust and friendship. Herwart had broad and open philosophical and scientific interests. Most important for Kepler, he was deeply sympathetic to the exciting views of the young Graz prognosticator—indeed, much more so than was Maestlin, as Herwart began to fathom the implications of Kepler's physical theory, or, later on, Tycho Brahe, as he realized that Kepler was harder to dominate than his other assistants. Kepler trusted both Herwart's intellectual openness and his capacity to serve as a sounding board for his new and partially formed ideas. Many of the

central elements of Kepler's *Harmonice Mundi* were worked out in early letters to Herwart.[18]

Johannes Praetorius was one of the university mathematicians who expressed himself to Herwart. If Limnaeus's response was favorable to Kepler's invocation of the ancient philosophy while utterly vague about his astronomical claims, Praetorius showed that he understood Kepler's cosmographical intentions quite well. He was as conversant as Maestlin was with the original Copernican texts, had personally known Rheticus, and was thus singularly well qualified to judge Kepler's polyhedral *inventum*. As will be recalled, he also had private knowledge from Rheticus about the conflict generated by Andreas Osiander's role in changing the title and adding the "Ad Lectorem."[19] However, none of this intimate knowledge helped to effect a more positive view of Kepler's book. Writing to Herwart, Praetorius offered an evaluation of the *Mysterium Cosmographicum* that invoked a version of the sharp distinction between astronomy and physics familiar to him from many sources: "I do not understand the least part of these [physical or metaphysical] matters, and I judge them to disagree quite a bit with the aim of astronomy; they rather pertain to physics and can scarcely assist the astronomer in any way."[20] Praetorius nowhere classified Kepler's project as "cosmological" or mentioned Kepler's physical theory of the Sun's moving power; but he clearly rejected the idea that the planetary distances could be deduced a priori from metaphysical principles. Unlike Limnaeus, he did not seem concerned that Kepler identified these principles as Pythagorean or Platonic (which he does not mention at all). The threat perceived by Praetorius was rather that Kepler was recommending a break in the theoretical astronomer's normal practices of building up (or adjusting) planetary hypotheses a posteriori from sensory observations alone.

In this assessment Praetorius was correct. Kepler really was proposing a shift in the theorizing astronomer's disciplinary role; but that change, as we have seen, involved license to make new explanations of a sort typically reserved for the natural philosopher. Maestlin had been willing to go part way toward Kepler's position, allowing—indeed, even celebrating—the introduction of archetypal (formal and final) principles into astronomy. That was the basis of his argument before the Tübingen senate and his laudatory remarks in the introduction to his edition of the *Narratio Prima*. Outside the pedagogical practices of the classroom, Maestlin believed that astronomy was capable of operating from true premises. But even here agreements were not uniquely determined. As the Rothmann-Brahe exchange shows, shared disciplinary identity and common metaphysical ideals underdetermined particular theoretical stances. Thus, too, in the context of Kepler's new cosmography, Maestlin and Praetorius disagreed about planetary order but shared common disciplinary skills and explanatory resources. In this case, the astronomer should not be permitted to introduce efficient and material causes into astronomy.

Of the two, Praetorius's position was the more exclusionary. As one would expect, Praetorius invoked Aristotle's authority (rather than that of Plato or even Ptolemy) to reinforce the boundaries that he wanted to maintain: "According to Aristotle, because astronomy defers to physics and cannot be counted as a science in itself and without relation to any other thing, it is proper (I believe) that the astronomer should put his doctrines to use as follows: the phenomena perceived by the eyes and sense should be consistent with his hypotheses, as though the different motions were brought about by such a definite cause."[21] Even in citing Aristotle, Praetorius did not explicitly invoke the logic of subalternation: he did not say that astronomy needed to borrow physical principles from the higher discipline in order to operate. Like Osiander, he emphasized the unbridgeable gulf between astronomy and physics, as in one of his (unpublished) lectures given at Altdorf in 1605: "The astronomer is free to devise or imagine circles, epicycles, and similar devices, although they might not exist in nature. . . . The astronomer who endeavors to discuss the truth of the positions of these or those bodies acts as a physicist and not as an astronomer—and, in my opinion, he arrives at nothing with certainty."[22] In the absence of transcendent criteria fixed by some authoritative social body, the matter of who could speak for the proper order of knowledge was a contingent one. Setting up the relationship among disciplines, as Praetorius had done, allowed him to dismiss Kepler's cosmographic project as a violation of proper astronomical practice.

Although this sort of boundary defense invoked the criterion of ranking the disciplinary differences according to their order of certainty, it seems that what was really at stake here were entrenched explanatory prerogatives: one should

not transgress onto the territory of the professors of natural philosophy—exactly the starting point of Kepler's cosmographic arguments in the disputations of the Tübingen physics candidates. Thus, Praetorius wrote: "That speculation [*speculatio*] of the regular [solid] bodies, what, I beg to know, does it offer to astronomy? It can (he says) be useful for defining the order or magnitude of the celestial orbs; yet clearly the distances of the orbs are derived from another source, that is, a posteriori from the observations. And with these things [the order and distances] defined, what does it matter whether or not they agree with the regular bodies"?[23] Praetorius ended his letter by charging Kepler with wanting to "correct" the observations of Reinhold and Copernicus so that they would agree with the polyhedral hypothesis.[24] This sharply negative reaction shows the tenacity with which the Wittenberg interpretation persisted at Altdorf and, indeed, in most of the German universities until the end of the century.

KEPLER'S *MYSTERIUM* AND THE *VIA MEDIA* GROUP

The year 1597 was one of those curious moments of unexpected convergence and culmination when tensions and discussions that had percolated during the 1580s suddenly burst into print. The 1597 spring catalogue of the Frankfurt Book Fair announced Kepler's *Mysterium*, Brahe's *Epistolae Astronomicae*, and Roeslin's *De Opere Dei Seu de Mundo Hypotheses*.[25] Tycho's Uraniborg anthology of intercourtly astronomical letters had already began to circulate, bound as a volume of substantial authority and girth, copies of which moved as gifts along distribution channels that he and his assistants managed. Within the boundaries of that circulation, the once-private Brahe-Rothmann exchange could more easily frame a controversy for a wider audience than a slim academic production like the *Mysterium*.[26] A few months after their appearance, Brahe's and Roeslin's works had reached Ursus, who in 1591 had risen above his once-lowly social status to become imperial mathematician in Prague. Together these two works would provide more than enough material to enable Ursus to publish a slashing attack on Rothmann, Brahe, and Roeslin while selectively and deviously citing for his own cause a laudatory letter sent to him by the young and relatively inexperienced Kepler.[27]

In these freighted circumstances, the prospects for Kepler's *Mysterium* were already dim, either as a promising inaugural defense of Copernicus by a young, unknown prognosticator or, more ambitiously, as the nucleus of some kind of nascent Copernican network. Unbeknownst to Kepler, the outlook for such coalition building had already darkened in 1592. Sometime after the landgrave's death in August, Rothmann departed from the richly supportive Kassel scientific environment, never to return; and Bruno disappeared into a Roman dungeon.[28] In the same year, another Copernican sympathizer arrived in Padua. Galileo's entry appeared to augur favorable possibilities for a fruitful, mutually supportive, Maestlin-style alliance. But, as the following chapter shows, the prospects for that relationship soon encountered their own serious obstacles.

It is now clear that for Maestlin by 1588, it was not only the Tübingen theologians and natural philosophers who constituted the main point of reference and contention but also the *via media* group. Yet, for the latter, Maestlin appears to have been anything but a threatening presence. His evident skill in handling the comet of 1577 and his caution in identifying himself as a Copernican seem to have protected him from attack. The copy of *De Mundi* that Tycho sent Maestlin was one of the earliest presentation copies—indeed, perhaps the very first—and within it, he thoroughly and respectfully discussed Maestlin's modeling of the comet of 1577 without once challenging his views on planetary ordering.[29] From Strasbourg, Roeslin deferred, time and again, to Maestlin's judgment and superior mathematical skill, as when he voiced misgivings about Ursus's system, with whose details he had sought to familiarize himself as early as the summer of 1588.[30]

Principal among Roeslin's worries in his correspondence with Maestlin were asymmetries in the gaps separating Mars and the Sun and Saturn and the eighth sphere. But after reading the *Fundamentum Astronomicum*, he went further, erroneously describing Ursus's arrangement as involving the interpenetration of the orbs of Saturn, Jupiter, and Mars—a scheme made possible only by the physical assumption that the heavens consisted of an airy rather than an aetherial medium. According to Roeslin, Ursus's system was no better than that of Copernicus or Ptolemy, and thus a new arrangement must be sought: "So far," he wrote to Maestlin, "I judge clearly that a

way that agrees with physics and astronomy remains unknown. However, such a system would have to be found in the first chapter of Genesis and, in the end, I have uncovered the matter hidden in this chapter and in the entire Holy Scripture."[31] At this point, Roeslin had no knowledge of the Brahe-Rothmann exchanges or the uncertainties of deploying scripture as a criterion of heavenly knowledge. The question of which exegetical tools should be applied to Genesis were for him entirely unproblematic.

In his reply (now lost), Maestlin sidestepped scripture and went directly to Roeslin's astronomical representations, pointing to his evident confusion in failing to realize that Ursus had assigned a daily motion to the Earth and that Tycho had not. After Roeslin received Maestlin's copy of Tycho's book, Roeslin then formed the impression that Ursus had plagiarized Tycho's arrangement! Roeslin also concluded that Tycho had failed to demonstrate that Mars displays a parallax greater than that of the Sun; in Roeslin's opinion, no astronomer had ever witnessed such an event.[32] Further, he objected to Tycho's dismissal of solid spheres because one could not otherwise account for the orderly motion of the heavens or, with Maestlin, for the regular motions of comets. Yet, with Tycho, he needed to retain a centrally resting Earth as the receptacle of astral influences. Roeslin's reflections on the organization of the heavens were, as Miguel Granada rightly concludes, the result of a thoroughgoing engagement with Ursus's *Fundamentum*.[33] Indeed, in this case, as also in his earlier appropriations of Cornelius Gemma's work on the comet of 1577, Roeslin's interpretive practices would prove to be less those of a mathematician than of a philosophizing Paracelsian physician.[34] Although Roeslin decidedly lacked the striking originality and skill of a Bruno or a Descartes, his awareness of the developing controversy and his struggle to enter it as an active contributor show that the controversy was beginning to draw in figures on the periphery of the central networks.

Maestlin's frequent exchanges with Roeslin suggest that by the time that Kepler enrolled at Tübingen in 1589, Maestlin was well enough aware of Roeslin's approach to Genesis that he could easily have shared it with the young Kepler.[35] Roeslin's main questions were little different from the ones that had motivated Kepler's cosmographic *dissertations*: What was the world's

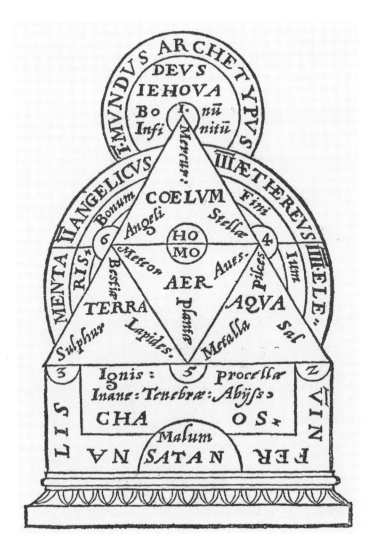

Figure labels (within illustration): ARCHETYPVS · I · MVNDVS · DEVS · IEHOVA · Bo · I · nū · Infi · nitū · Mercur: · III ÆTHEREVS · II ANGELICVS · COELVM · Bonum · Angeli · Stella · Fini · HO MO · Meteor · Aues · 6 · 4 · Pisces · ium · Bestiæ · AER · AQVA · III ELE· · TERRA · Planta · Metalla · Sal · Sulphur · Lapides · MENTA · RIS · ANGELI · 3 · Ignis: · 5 · Procella · z · Inane: Tenebræ: Abyss: · CHA · O S: · LIS · Malum · VN SATAN FER · VIN

69. Roeslin's Signaculum Mundi or "The World's Seal," showing man (*homo*, center) as a two-sided being who partakes of the Good and the Bad, the angels and the elements, God and Satan. Roeslin 2000 [1597]. Courtesy Bibliothèque nationale de France.

mysterium, the hidden meaning of Genesis? How did God's triune being show up in the created world? If the questions were the same, however, the exegetical resources were decidedly not. Kepler found the Trinity embodied in the sphere and looked further to the geometrical archetypes of the *Timaeus,* the period-distance relation of Copernico-Maestlinian *symmetria,* and the solar-force hypothesis. Roeslin, in contrast, helped himself to the trinity of Paracelsian principles (salt, sulfur, and mercury), Neopythagorean numerology, and the analogy between the macrocosm of Nature and the human microcosm that reflected it. He reduced this bit of metaphysical and theological physics to a single, dualistic image, which he called a *signaculum,* organized around an equilateral triangle with Jehovah and infinite goodness at the vertex and Satan and chaos at the base. And where Rothmann, Bruno, and (later) Kepler appealed to the principle of accommodation to explain why certain biblical passages appeared to affirm literally a motionless Earth, Roeslin's scriptural practices typically involved unexplicated gestures to books and verses.[36] Furthermore, unlike Kepler in the *Mysterium,* Roeslin presented planetary ordering not as a deduction from first principles but as the best choice among alternatives (from Ptolemy, Copernicus, Ursus, Tycho, and himself). He called these alternatives "hypotheses concerning the system of the world," and he arranged each hypothesis as a series of five or six numbered propositions and three consequences. He also provided in an appendix what was the first comparative visual summary

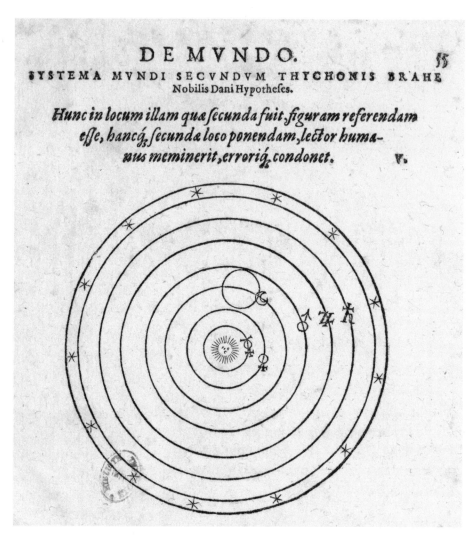

70. Copernican system, erroneously labeled "The World System according to Thyco Brahe," but with correction noted below caption. Roeslin 2000 [1597], image 5. Courtesy Bibliothèque nationale de France.

of all the competing planetary arrangements of the 1580s.

But what sorts of choices were enabled by Roeslin's representational practices? The woodcut images of "world systems" were egregiously mislabeled, and several did not tally properly with Roeslin's carefully numbered propositions. Most confusingly, the diagram of Copernicus's system sits in the fifth position and is labeled "The World System according to the Hypotheses of the Noble Dane, Thyco Brahe," beneath which is the correction: "The humane reader will keep in mind, as he pardons the error, that the diagram in this place should have been the one referred to in the second place, and the one in the second place

should be put here." However, while the legend for the second figure properly refers to Copernicus's "hypotheses," the figure that it actually contains is that of Ursus's hypotheses; and, to complete the mayhem, the caption that describes Ursus's world system heads a diagram that shows Tycho's ordering. These errors—and the obvious uncertainty about how to correct them—must have occurred with the Frankfurt publisher, where the manuscript had been taken sometime after 1595 by Roeslin's promoter, the French diplomat and jurist Jacques Bongars.[37]

However, other difficulties arose because the author himself followed the widespread convention of displaying planetary order without scaled

SYSTEMA MVNDI SECVNDVM PTOLOMÆI ET VE-
terum Philosophorum Hypotheses.

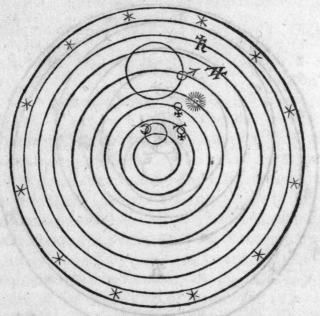

SYSTEM'A MVNDI SECVNDVM NICOLAI COPERNICI II.
Hypotheses.

71. Top: "The world system according to the hypotheses of Ptolemy and the ancient philosophers." Bottom: System of Ursus, erroneously labeled "The world system according to the hypotheses of Nicolaus Copernicus." Roeslin 2000 [1597], images 1 and 2. Courtesy Bibliothèque nationale de France.

HYPOTHESES

SYSTEMA MVNDI SECVNDVM HELISÆI
Roslin Medici hypotheses.   IV.

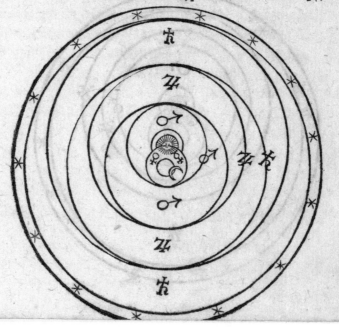

72. Top: System of Tycho Brahe, erroneously labeled "The world system according to the hypotheses of Raymarus Ursus Dictmarsus [*sic*]." Bottom: "The world system according to the hypotheses of the physician Helisaeus Roeslin." Roeslin 2000 [1597], images 3 and 4. Courtesy Bibliothèque nationale de France.

distances. This was a problem because Roeslin systematically mixed optical claims based on variable distances into his symmetrically calibrated diagrams. Because most of Roeslin's criticisms of the other arrangements depended on the distances between spheres—too great, too small—this matter was central. For example, in representing Ptolemy's *systema* with equal spacing between the respective circles, he objected that if Mars were given an epicycle that allowed it to come closer to Earth than the Sun and closer to Saturn than Jupiter approached, then it would lead to an unacceptable intersection of the orbs of Jupiter and the Sun.[38] In the *systema* for Copernicus—mislabeled as the system of Tycho Brahe!—Roeslin objected that there would be "a huge space and an immense vacuum" between the sphere of Saturn and the fixed stars (although no such gap is shown in the diagram).[39] Roeslin's representation also ignored the strong sense of planetary *symmetria* that Copernicus himself had emphasized; and he had nothing to say about the accountings of Digges, Bruno, or Rothmann. Meanwhile, Roeslin's most serious objections against Ursus were that the space allowed for the three superior planets was "too confined and narrow" and that Ursus allowed a daily motion for the Earth (the fall of heavy bodies explained by a magnetic force); in addition, Ursus rejected the solidity of the spheres and the aether in favor of an airy region of infinite extent.[40] And finally, against the "uncertain space" between the Sun and Mars posited by Tycho and Ursus and between Saturn and the fixed stars posited by Copernicus, Roeslin claimed that no other distribution of the spheres than his own contiguously packed, eccentric spheres could be known "with certain reasoning."[41]

Having just published a book of such firm conviction in its own understanding of the divine plan, Roeslin must have been quite surprised to come across a work by the same student of Maestlin's whom he had so recently instructed in the casting of nativities and who now purported to offer nothing less than his own grand reading of God's intentions. Yet we have no evidence that he expressed his reactions directly to Maestlin or to Kepler. Instead, between May and July 1597 Roeslin expressed his views twice, at some length, to a court intermediary, Herwart von Hohenburg, whom he must have known harbored deep interests in the science of the stars.[42] Since 1588 the

views of Ursus had claimed Roeslin's primary attention; but now, suddenly, he was confronted with a full-blown defense of Copernicus under the authority of Pythagoras.

Following his previous practices, Roeslin accepted the cards that were dealt and did not hesitate to reshuffle them without budging from his prior convictions. Kepler's discovery, so he alleged, was more in agreement with his own arrangement. The first letter, however, was vague on details, and Roeslin even admitted that although "the cube gives the distance of the spheres of Saturn and Mars [sic]," whether "another of the five regular solids could not also give such [a result], I do not know. I leave this judgment to those skilled in geometry. Such a thing is beyond me, nor can I pursue it. And although there may be five such distances of five planets, with every body especially arranged for specific planets, still I will not be of Copernicus's opinion for that reason."[43] A problem that Roeslin did not mention—and clearly had not yet solved—was that his own arrangement would require an extra solid for the Earth-Moon space. Had God really left only five gaps, or was it six? Roeslin changed the subject, pointing out that Kepler had also overlooked a gap: "The almost infinite distances between the sphere of Saturn and the fixed stars. . . . With which geometrical figure does he want to account for such an infinite empty space? As I consider it, such an invention with the five regular geometrical bodies will thus just suit any business without risk." And Tycho had done no better. He had allowed "too narrow a space to the three higher spheres of Saturn, Jupiter and Mars."[44]

By July 1597, Roeslin had a solution. Without showing any calculations, he asserted that the cube would give the size of Saturn's sphere, the tetrahedron would fit Jupiter, and the dodecahedron would fill the space between Mars and the Sun. And now came the critical move: "The fourth regular body, the icosahedron, gives the space from the Sun to the Moon, in which space Mercury and Venus traverse their course." In other words, the twenty-sided icosahedron would subsume three gaps, and that left only the octahedron to cover the gap between the Moon and the uppermost part of the air. If this gap filling was not decisive, Roeslin believed that Herwart would agree that his arrangement far better accommodated the polyhedra because of the last gap: "I do not require a further geometrical demonstration, because I put the uppermost part of the sphere of Saturn to be contiguous to the eighth sphere. Thus Your Grace sees how his invention confirms my system far better than his. And thus it is an argument applicable to either side. . . . They are not demonstrative arguments."[45]

It is revealing that neither Roeslin nor Ursus ever questioned the metaphysical relevance of the polyhedra as a new criterion for comparatively evaluating (and eliminating) multiple hypotheses of cosmic order. It is also telling that they turned to a comparative standard of this sort, indecisively supplemental to holy scripture and traditional natural philosophy. The legitimacy of that standard rested in the first instance on ancient authority, still the preferred idiom for discussing celestial order. Both Roeslin and Ursus accepted that the five solids betokened some sort of reality and hence Kepler's assertion that the spaces between the orbs "are revealed by the five figures of Pythagoras." But there was no agreement on how the standard should be applied. Roeslin's reply to Herwart, as shown above, simply asserted that the solids, without being determinative, best fit the spacings in his arrangement.

Unlike Roeslin, Ursus came by his knowledge of the *Mysterium* much sooner than the *Mysterium* itself, in a letter from Kepler of November 1595. The reader of Ursus's *De Hypothesibus Astronomicis* thus found the description of the polyhedral arrangement (without visual assistance) wrapped together with Kepler's effusive praises of Ursus.[46] In addition, the reader encountered Kepler's hint of his underlying intuitions: "The greater the amplitudes [of the mean motions], the more slowly they are moved round; and then there is the weakening [of impulsion] in the outer ones of the kind which happens by extenuation of rays of light."[47] Full citation of Kepler's letter served Ursus's interests, putting Kepler's laudatory testimonial before the reader with little comment while allowing deployment of Kepler's speculation as an unendorsed, friendly weapon against Roeslin's hypotheses.[48] Yet Ursus was not really holding Roeslin to Kepler's own standard, the a priori derivation of the relative distances from the ratios given by the inscribed and circumscribed polyhedra. Ursus's criticism was limited to Roeslin's arbitrary distance claims or, as he indelicately called them, his "gappy hypotheses" (especially the distance to the fixed

stars and the intersection of Mars with the Sun's path).[49]

Ursus structured much of his opprobrious tract explicitly around passages from Tycho's *Epistolae Astronomicae*. His strategy was to undermine any claims to recent credit and to prove that Tycho had found his central idea from Apollonius of Perga in an obscure passage of *De Revolutionibus,* that Copernicus's notion came from Aristarchus, and that Roeslin had awkwardly filched his system from Tycho.[50] The extremely meticulous, postillated commentaries also allowed him to deliver pinpoint rhetorical strikes at the originality of Tycho's and Rothmann's claims, mixed with colorful and demeaning jabs at their personal motives, character, alleged bodily disorders, general competence, and social status.[51] Although Ursus was replying in kind to names that he himself had been called in the *Epistolae* (e.g. "dirty scoundrel"), class tensions are readily discernible. Having risen to the exalted status of imperial mathematician, the former north German shepherd and talented mathematical boy could now call the noble Tycho Brahe a "mere mechanic" who "made much of discerning double stars through the triple holes in his nose"; the court mathematician Rothmann merited the name Sniveller (Rotzmann); and the Strasbourg physician Roeslin earned the names "absurd manikin" and "fibbing little man."[52] For a brief historical moment, the *via media* group's hybrid, spin-off planetary systems—the offspring of *De Revolutionibus*—were charged with high cultural value.

The last point was not lost on Tycho Brahe, who had largely constructed the social space in which these struggles were launched. But in 1597 Tycho's own fortunes had changed. He had lost the support of the new king, Christian IV, and been forced to move from his storied island. He was now engaged on different fronts with efforts to conserve his reputation and to secure a new patron.[53] In 1598, while in residence with his friend Heinrich Rantzau, Tycho managed to publish a sumptuous book—some copies impressively hand-colored and silk-bound—showing in magnificent detail his large instruments as well as the elaborate front view of his Uraniborg castle. This volume, which ignored the ongoing world-systems controversy, featured no diagram of his own planetary arrangement; it became a vehicle acknowledging the emperor's favor for the very post occupied by the threatening upstart Ursus.[54]

At the same time, Tycho continued his efforts to obtain the remnants of Wittich's library and, in particular, the several copies of *De Revolutionibus* that he knew it contained.[55] During this period he also obtained a copy of the *Mysterium,* and in March 1598 a supplicating letter from Kepler reached Tycho at Rantzau's residence.

Kepler's situation was dire. The religious and political climate in Styria had become extremely difficult for Lutherans. In Graz, Kepler was ordered to pay a fine for evading the Catholic ritual on the death of his young daughter.[56] In addition, he had realized, rather too late, the awkward situation with Ursus into which he had inadvertently stepped. His laudatory letter, followed by the two copies of the *Mysterium* that he sent to Ursus—one of them intended for Tycho—was an (ill-conceived) attempt to obtain Ursus's patronal support. Evidently Maestlin had failed to inform Kepler about Ursus. From Graz, Kepler's main contacts with other practitioners circulated through Maestlin and Herwart von Hohenburg: he was not involved in the buzzing Tychonic correspondence networks. Thus, the *Mysterium,* although avidly endorsing Copernicus, now represented Kepler's best hope that Tycho would forgive him for a youthful indiscretion and perhaps bestow his support. "How happy I would be if Tycho would arrive at the same conclusion as Maestlin!" he wrote hopefully to Tycho at the end of 1597.[57]

Kepler received the invitation to join Tycho soon after the latter's arrival in Prague in June 1599. These propitious circumstances, with their far-reaching prospects for improving Kepler's security, raise the question of why Kepler's manifest Copernican sympathies proved to be no significant obstacle. One explanation is that throughout his career, Tycho had shown great respect for Copernicus's project, ultimately adopting the circumsolar orbits (for all planets but the Earth) and the equantless planetary models. Of course, that respect was mixed with genuine doubts. Tycho perceived in Kepler's *Mysterium* two difficulties that were essentially the same ones he had recently rejected in Rothmann's defense of Copernicus: the disproportionate (and hopelessly great) gap between Saturn and the fixed stars and the Earth's "unsuitability" to motion. In addition, Tycho had another worry: Kepler's willingness to take on board the Ptolemaic equant, a model more consistent with the dynamic, inversely active solar force than with Copernicus's compounded, uniform,

circular motions.[58] On the other hand, most strikingly, Tycho agreed with Ursus, Roeslin, and Rothmann in endorsing the criterion of *symmetria*. "There is no doubt," he wrote to Kepler, "that by divine inspiration everything in the universe is mutually related and ordained by a certain harmony and proportion; these [harmonies] are grasped succinctly as much in numbers as in figures, just as to some extent the Pythagoreans and Platonics foresaw a long time ago."[59] Once again, the issue was not the legitimacy or relevance of the standard but underdetermination of the manner of its application.

On that score, Tycho had two complaints. First, he rejected Kepler and Maestlin's a priori approach to determining the proportionalities of the relative distances. Second, if these relationships were to be determined at all, then it must be done directly from the sorts of observations that only he had been making "with great pains over many years," demonstrating proportionalities (*commensurationes*) that were "more exact" (*accuratiorem*) and that could yield a better determination of the true eccentricities of the individual planets.[60] Recourse to accurate observations as the ultimate standard of judgment was, of course, Tycho's watchword; but, unlike the Martian parallax observations on which the credibility of his planetary arrangement rested, nothing vital hung in the balance here. Claims based on *symmetria* were always reinterpreted to suit the requirements of alternative hypotheses.

The principal reason why Tycho sought to recruit Kepler to Prague was evidently his extraordinary astronomical skill. Tycho had a facility for recognizing and recruiting talent. Many other young men ended up bound to him by obligation. But there were two other important considerations: Tycho and Kepler were coreligionists, and in Kepler he also detected fealty and loyalty, the very qualities absent from the rebellious Ursus. Not only did Tycho expect Kepler to be the anti-Ursus, but, as a concrete demonstration of loyalty, he made it a condition that Kepler apply his skills to attacking and destroying him. Thanks to Ursus's misuse of the 1595 letter, Kepler had his own reasons for wanting to write a work that would allow him to give his side of the story. In the end, however, there was no concern about a rebuttal, as Kepler began writing shortly after Ursus died on August 15, 1600.[61]

Kepler's apologia for Tycho against Ursus, remarkable for its metalevel, epistemic reflections, became an occasion to work out his general ideas on the nature of hypotheses in astronomy and the logic of judging among alternatives.[62] Effectively, the disagreement with Ursus turned on different readings of *De Revolutionibus*. Ursus, who did not know Osiander's identity, believed the "Ad Lectorem" (which he called a "preface") to have been written by Copernicus and used it as the basis for a skeptical rendering of astronomical hypotheses. Kepler knew Osiander's identity both from an annotation on his copy of *De Revolutionibus* and from the two letters from Osiander contained therein (see chapter 4). From this evidence, he understood that Copernicus was making claims about the way the world really is. In turn, he read Osiander as "following the dictates of expediency" in seeking to conceal and thereby protect Copernicus's true opinions.[63] For Kepler, this kind of defense was unacceptable. He reiterated from the *Mysterium* that false premises could easily be detected and eliminated by reference to a further consideration: "Even if conclusions of two hypotheses coincide in the geometrical realm, each hypothesis will have its own peculiar corollary in the physical realm."[64] This view was entirely consistent with Clavius's arguments from 1581 against the ex falso verum objection.[65] Furthermore, from the 1596 *Episto-*

lae Astronomicae, Kepler had not failed to notice that physical criteria were crucial in the Brahe-Rothmann dispute: "Tycho does not think it absurd that the planets follow a moving sun wherever its goes as their center and yet do not depart from their own motions."[66] Although the *Apologia* deftly withheld any criticism of Tycho, it was precisely Tycho's physical theory of the sun dragging around the planets that Kepler—echoing Landgrave Wilhelm—privately regarded as deeply problematic. Prominently absent from the *Apologia* was any invocation of the *symmetria* criterion as determinative of a choice between two hypotheses coinciding in the geometrical realm. Kepler would not drop the search for harmonies, but it had become apparent that among the different world systems there was an emergent consensus between modernizers and traditionalists that the physical criterion was determinative. Both Kepler's contact with Galileo in 1597 and the appearance in 1600 of William Gilbert's *De Magnete* would fortify that impression. Although Maestlin would dissent, the problem was now to strengthen the Copernican hypothesis with specific physical arguments. And by 1605, physical considerations, robustly assimilated to the sort of accuracy for which Tycho had pleaded, guided Kepler to a radically new kind of planetary modeling.

PART V

Conflicted Modernizers at the Turn of the Century

13

The Third-Generation Copernicans

GALILEO AND KEPLER

The Kepler-Galileo relationship has a pronounced historiographical profile: two great figures, both members of the same age cohort, both followers of Copernicus, highly visible within and eventually across their own social networks, each productive of a remarkable trail of new claims and discoveries that wedded mathematics and natural philosophy—who yet scarcely communicated directly with one another. The failures of that relationship would have heavy consequences for seventeenth-century heavenly science and natural philosophy. Followers of Kepler and Galileo went in quite different directions. If viewed from the perspective of the 1580s, the Kepler-Galileo relationship seems no more than a special instance of the marked and often profound differences that characterized Copernican theorizers as dissimilar as Rothmann and Bruno, Maestlin and Digges. But by the following decade, Kepler and Galileo were already engaging one another within the discursive space created by the second-generation Copernicans and the practitioners of the contested *via media*. How those differences evolved in the decade prior to the telescopic discoveries and the elliptical astronomy is the concern of this chapter.

GALILEO AND THE SCIENCE OF THE STARS IN THE PISAN PERIOD

At the outset, it must be acknowledged that not much can be said from direct evidence about Galileo's earliest intellectual formation in the celestial subjects, astronomy and astrology. Many accounts jump to the telescope in 1609–10 or observe that Galileo taught the rudiments of astronomy and was (somehow) committed partially to "Copernicanism" by 1597 and fully by 1610.[1] How early or how late these commitments were formed carries interpretive consequences of no small significance. Was Galileo's new approach to the science of motion guided by a Copernican framework in thinking about the problem of falling bodies? Or did Galileo keep those problem domains separate? The unfortunate fact is that there are serious lacunae in Galileo's extant correspondence between 1585 and 1610.[2] What can be said with some confidence—and these are the usual claims—is that Galileo's training in mathematics derived from two different settings: his studies at the university in Pisa from 1580 to 1585, where he later held his first teaching position in mathematical subjects (1589–92); and his private studies of Euclid and Archimedes with Ostilio Ricci, a friend of Galileo's father, teacher of mathematics at the grand duke's court, and member of the Accademia dei Disegno.[3] That the information about Galileo's earliest encounter with the different parts of the science of the stars is so limited may explain the interpretive weight that many historians have placed on Galileo's middle (1610–32) and later periods (1633–42). But what can be said on the basis of circumstantial evidence?

The mathematics professors at Pisa followed a curriculum that, at least on the surface, resembled the teaching requirements elsewhere. Typically they lectured on one or two books of Euclid, Sacrobosco's *Sphere,* Peurbach's *New Theorics,* and "something from Ptolemy."[4] But, as elsewhere, the editions and commentaries used were tailored to local needs and settings. In this regard, it seems likely that Galileo was first introduced to the *Sphere* not through the editions of Melanchthon or Schreckenfuchs but through Clavius's and, hence, of course, to his classification of the science of the stars.[5] Furthermore, although "something from Ptolemy" could mean the *Geography* or the *Almagest,* as often as not it was the *Tetrabiblos*—and in this case Giuliano Ristori's 112 unpublished commentaries on that work. Ristori had delivered these commentaries for the first time as lectures in 1547 and for the last time in 1556, and they survived to furnish the text for the subsequent Pisan commentaries of Amerigo Ronzoni and the Camaldolese abbot Filippo Fantoni (d. 1591; lectured 1560–67, 1582–89).[6] Francesco Giuntini, who was Ristori's most prolific student, was undoubtedly influenced by Ristori in compiling his own edition of the *Tetrabiblos.* Dangerously following Pierre d'Ailly, Giuntini saw no conflict between astrology and theology—indeed, he believed that theology *derived* from astrology.[7] And among the many topics that Ristori and his commentators disputed was whether astrology was a true science; whether the first part (speculative astronomy) is more perfect than the second (prognostication of terrestrial effects); whether, as Pico had argued, astrologers were pseudoprophets; and whether the stars determine human actions.[8]

The Fantoni and Ristori lectures exemplify the tradition in which the mathematics lecturers were defending the astrology of the *Tetrabiblos* at Pisa in Galileo's student days.[9] Both organized their presentations in the style of a philosopher's commentary with invocation of authorities rather than mathematical arguments—something like Bellanti's reply to Pico. If the shadow of the nearly century-old debate of Pico and Savonarola with Bellanti still fell across views of the stars' influence at Florence and Pisa, it did not prevent the subject from being taught by a monk.[10] Like Ristori and Fantoni, Melanchthon himself lectured directly on the *Tetrabiblos,* but, as chapter 5 shows, the Wittenberg lecturers also approached astrology through practical, worked examples involving use of the *Prutenic Tables,* which were ulti-

mately amenable to absorption into a Lutheran apocalyptic framework.

This background is of relevance to Galileo's student days at Pisa because Fantoni was then the mathematics lecturer.[11] It raises confidence that Galileo had become acquainted with the *Tetrabiblos* at about the same time that he would have studied Peurbach's *New Theorics* and the *Almagest* in a context where Fantoni and Ristori explicitly defended astrological theory against Pico's arguments. That Galileo was also sympathetic to such anti-Piconian arguments seems likely in view of the fact that he had developed concrete expertise in casting nativities and, not more than twenty years later, had established a modestly successful astrological practice in Padua. Fantoni, however, is not known to have issued any prognostications, and, unlike Kepler, Galileo had nothing to say about astrological theory.

Galileo's avoidance of this domain of the science of the stars could be related to official Church proscriptions against certain kinds of divination. Here the chronology of events has some coherence. Just as Galileo ended his studies at Pisa, Clavius endorsed Pico's rebuttals in the 1585 edition of his *Sphere* commentary. In 1586, Pope Sixtus V's bull *Coeli et terrae* explicitly reserved foreknowledge for God, darkly warning of demons infecting superstitious "genethliacs" (nativity casters), "mathematicians," and "planetarians."[12] The Sistine bull was Piconian in spirit while leaving some room for the areas of astrology customarily considered safe (medicine, navigation, agriculture). It certainly did not encourage the sort of anti-Piconian physics that the Melanchthonian Georg Liebler imparted to Kepler at Tübingen. At the same time, Rome's new boundary did nothing to impede the warm contacts that commenced with Galileo sending to Christopher Clavius his work on the center of gravity of bodies in January 1588.[13]

What little can be said, therefore, of Galileo's early formation at Pisa in disciplines that made up the science of the stars suggests striking differences between his own astrological and physical ruminations in the 1580s and those of Kepler in the early 1590s. As a disciple of the leading academic Copernican in Europe, Kepler's early natural philosophizing was, from the start, organized by Copernican concerns; into these he injected Stoic and Neoplatonic resources gleaned from

Pico, making the sun into a dynamic, celestial motor. Whether or not his thinking was motivated directly by Brahe and Rothmann's negation of solid, planet-bearing spheres, Kepler's new, emanationist Sun constituted a response to the explanatory gap thereby created. Thus, the emphasis in Kepler's physics was truly on celestial order and light. The explanatory problem concerned why the planets retain an orderly regularity in their distances and relative velocities. Kepler had nothing to say about why and how heavy bodies accelerate toward a moving Earth.

Although *De Revolutionibus* was certainly known and even sympathetically mentioned in Pisa during Galileo's student years, references to it are somewhat scattered and unsystematic.[14] Certainly none of Galileo's teachers followed Copernicus's arrangement, and by the time he first encountered *De Revolutionibus* itself—probably between 1590 and 1592—the entailments on which he focused derived from his immersion in terrestrial mechanics. Thus the part of *De Revolutionibus* that he would have been most likely to notice was Copernicus's discussion of gravity, the emphasis on the primacy of circular motion—not just for the Earth but for the elements—and Copernicus's suggestive references to the cause of nonuniform, upward or downward rectilinear motion.[15] At Pisa, Galileo was already privy to vigorous local debates about why the elements, both heavy and light, accelerate as they approach their respective natural places. The Florentine Aristotelian Francesco Buonamici had defended Aristotle against the Archimedean notion that a medium composed of heavier bodies causes the upward acceleration of lighter ones, and he had also argued explicitly against the implication that all bodies are naturally heavy.[16] Fantoni was part of this discussion as well, and his philosophical interests further extended to the status of mathematics.[17] Galileo's earliest thinking about motion was thus organized within a literature quite different from the one in which Kepler found his compass. Galileo would be informed by the positions formulated and rejected by Buonamici.[18] In developing his own unique approach to the motion of the elements, Galileo would use the Archimedean balance and the related notion of hydrostatic thrust as powerfully influential heuristics.[19] These resources probably provided the interpretive grid through which Galileo first read Copernicus.

GALILEO AND THE WITTENBERG AND URANIBORG-KASSEL NETWORKS

Other considerations affected Galileo's reaction to Kepler's problematic. Galileo had very limited access to the resources of the Wittenberg network in the 1580s. Unlike Giordano Bruno, he had never crossed the Alps. Unlike Giovanni Antonio Magini, he did not have the typical publishing profile of an academic mathematician (ephemerides, spherics, theorics, prognostications). Further, he lacked local knowledge about the anonymous editorial work of Osiander and Rheticus's anger with him; he had no involvement with the Melanchthonian program of apocalyptic astrology and no familiarity with the Reinholdian annotations on Copernicus's planetary theory that had been so important in establishing the geoheliocentric problematic of Wittich and Tycho. What familiarity Galileo possessed with Wittenberg authors and themes, so far as we know, came solely through the mediation of printed works, such as Magini's and Reinhold's commentaries on Peurbach. Yet because all of Melanchthon's works were on the Index, whatever editions of Sacrobosco, Euclid, or Peurbach Galileo might have studied at Pisa would not have been the editions framed by Melanchthon's prefaces. If he knew anything about Maestlin at all, it was as Clavius's opponent in the calendar wars; he certainly had no private knowledge of Maestlin's or Kepler's difficulties with the Tübingen theologians.

But connections with Uraniborg are another matter. Massimo Bucciantini has recently discovered that, on his arrival at Padua in 1592, Galileo quickly gained access to the extraordinary library of the Genoese-born aristocrat Gian Vincenzo Pinelli (1535–1601).[20] Pinelli's library was rich not only in classic works of the ancients (Pappus, Euclid, Apollonius, Archimedes, Strabo, and Ptolemy) but in those of the moderns. A list of his collection allows us to say that he possessed many books of relevance to the present inquiry: *De Revolutionibus* (either 1543 or 1566), Brahe's *De Mundi Aetherei Recentioribus Phaenomenis* (1588), Ursus's *Fundamentum Astronomicum* (1588), Maestlin's *Ephemerides Nova* (1580) and *Epitome Astronomiae* (1582), Bruno's *Della Causa Principio et Uno* (1584) and *De Triplici Minimo et Mensura* (1591), and Digges's *Alae Seu Scalae* (1573).[21] The presence of Brahe's volume, which was not for sale, raises the further question of connections between Pinelli and Uraniborg. At least one intermediary was Gellius Sascerides (1562–1612), one of Tycho's principal disciples, the same person who had delivered to Maestlin a copy of *De Mundi Recentioribus Phaenomenis* in May 1588 and another to Magini at Bologna slightly later.[22] When Gellius came to Padua in 1589 to study medicine, he frequented Pinelli's house. Yet although Gellius assisted measurably in increasing Tycho's reputation in Italy, Pinelli obtained the copy of Tycho's book only in 1590, through the mediation of Joachim Camerarius II.[23] Much less certain is when and how he acquired a copy of Tycho's *Epistolae Astronomicae*—in all likelihood, not before 1599.[24]

GALILEO ON COPERNICUS
THE EXCHANGE WITH MAZZONI

Direct knowledge of Galileo's Copernican views comes to us from two well-known letters (May and August 1597), written as reactions to recently published books in which he found treatments of Copernicus: the *Mysterium Cosmographicum* and *In Universam Platonis et Aristotelis Philosophiam Praeludia* (Prelude to the Whole Philosophy of Plato and Aristotle) by Jacopo Mazzoni (1548–98), Galileo's onetime colleague at Pisa. These two books, while both giving prominence to Platonic themes, were quite different in character. Mazzoni shared the late-sixteenth-century humanist sensibility of seeking concordance among the ancients toward the goal of a "reconciliation of opposites in a harmonious ordering of all knowledge."[25] He was conversant with Giovanni Benedetti's discerning modifications of Aristotle's analysis of motion and was critical of what he took to be Aristotle's inadequate attention to mathematics in natural philosophy.[26] But Mazzoni was very far from accepting Kepler's union of Pythagorean-Platonic metaphysics with Copernican astronomy. Although he marshaled mathematical and optical arguments in situations where Aristotle had not, he did not thereby reject the conclusions of either Aristotle or Ptolemy. In this sense, one might regard his approach as in accord with the Clavius wing of the Jesuits.

Called to Pisa in 1588 by the grand duke Ferdinand de'Medici to lecture on Aristotle's *Physics,* in the following year Mazzoni was assigned the further duty of offering additional lectures on

Plato.[27] Mazzoni was one of a handful of Platonizing philosophers in the Italian universities of the late sixteenth century.[28] More than likely, Barozzi's edition of Proclus had contributed to a Pythagorean emphasis in this emerging group of Platonic sympathizers, much as it had stimulated such tendencies among John Dee, Thomas Digges, and their acolytes a decade earlier in England. Mazzoni's brief section concerning Copernicus appears in the midst of a much larger, eclectic work that fell into a long tradition explicitly aimed at comparing the philosophies of Plato and Aristotle, with the goal of explaining, adjudicating, and resolving their differences.[29] Galileo's letter to Mazzoni is especially critical because, among other things, it establishes that he had already developed a strong position about Copernicus prior to the arrival of Kepler's book. Equally important, Galileo used the phrase "much more probable" to describe how he held the Copernican-Pythagorean view: "To tell the truth, as much as I remain confidently in [agreement with your] other positions, so at first I remained confused and doubtful, seeing your Excellency so resolute and outspoken in clenching your fist against the opinion of the Pythagoreans and Copernicus about the motion and place of the Earth, which, being [now] held by me to be *much more probable* than that other view of Aristotle and Ptolemy, made me cock my ear to your argument; for I have some sentiments [*umore*] on this matter and other things that depend on it."[30] Galileo did not elaborate his position further, but he did trouble to point out an error his colleague had made in thinking about an alleged observational entailment of the Earth's motion. The difficulty was this: Mazzoni imagined that at the summit of a high mountain one ought to be able to see more than half of the celestial sphere. (The mountain was Mount Caucasus, not one that he had ever climbed; it is described by Aristotle in the *Meteorology*.) If the Earth revolves around the Sun, he argued, how much closer Earth dwellers would be to the starry sphere, and hence they ought to be able to see at least as much or more of it. But as they do not, Copernicus's proposition is "false and impossible."[31] That this entailment had never been brought against Copernicus in the sixteenth century suggests that Mazzoni had been stimulated to think about the problem either from his own reading of *De Revolutionibus*— or perhaps from an earlier discussion with Galileo himself—and had not simply lifted it from a teaching manual.

Galileo was the teacher and friendly critic in this exchange—a role in which he was generally most comfortable.[32] Mazzoni's book was not, however, principally concerned with "the opinion of the Pythagoreans and Copernicus." Although Mazzoni strongly believed in mathematics as a prelude to philosophizing about the natural world (for example, he attacked Aristotle's view that bodies move through a medium at speeds proportional to their sizes), the problems that attracted his interest did not arise principally from astronomical considerations. The reference to Copernicus, then, was slight.

What is interesting for us is that Galileo, who knew Mazzoni intimately from his three years at Pisa, chose to single out this lesser allusion for special comment. And this attention strongly suggests that Galileo was not merely calling notice to a philosophical error but that, of all the many topics addressed by Mazzoni in his lengthy book, *this* problem especially mattered to him. He ended his letter to Mazzoni with a characteristically Galilean rhetorical flourish: "But not to fatigue your excellency much longer, I do not wish to give you a long argument but only to beg you to tell me freely whether you judge that Copernicus can be saved in this manner."[33] Mazzoni is not known to have replied.

GALILEO AND KEPLER
THE 1597 EXCHANGE

In August 1597, a German emissary named Paul Homberger delivered to Galileo at Padua an unsolicited copy of the *Mysterium Cosmographicum*. Homberger, who taught music in the Graz *Stiftsschule* where Kepler was teaching mathematics, became the first intermediary between Kepler and Galileo. It seems clear that Kepler had instructed Homberger to distribute the book to unnamed "Italian mathematicians" because there were none he could name, whereas Homberger obviously had good contacts in Padua. He had studied briefly in the arts faculty of that university in 1595–96, and he might have attended one or more of Galileo's public lectures. Even if he had not, he would certainly have known that Galileo was the foremost teacher of mathematics at the university. There is no good evidence that either Homberger or Kepler knew Galileo to have

Copernican sympathies before Kepler's book arrived.[34] But because Homberger had studied at Padua, rather than at Bologna or Ferrara, it is not entirely accidental that Kepler's book was delivered into Galileo's hands. For some unknown reason, however, Homberger could not stay long. In the letter of thanks that Galileo wrote—or dashed off—he mentioned that he received the book "not days but only a few hours ago" and that Homberger was returning to Germany very soon.[35] It is certainly an encouraging letter—and also a tantalizing one.

Lacking direct evidence, we can use our earlier analysis of the *Mysterium* to imagine the scene of Galileo's initial reactions. Kepler's work blatantly proclaimed its Copernican-Pythagorico-Platonic intentions in the enticing polyhedral plate that begged even the disinclined to unfold it. Its author was a mathematician who had studied under Maestlin at Tübingen and who was now a teacher and prognosticator in Graz. If Galileo knew the name *Maestlin* from the calendar controversies, he had never heard of Kepler. As is already clear from his relationship to Mazzoni, however, Galileo had much more than a passing familiarity with Platonic philosophy, and he was already persuaded that Copernicus's reasons were "much more probable." In this respect, Galileo had evidently reached the same methodological difficulty that Copernicus, Rheticus, Digges, Rothmann, and Maestlin had faced: whether the heliocentric arrangement could be held as preferable to the traditional ordering even if it did not possess apodictic status.

Galileo had had little or no experience with *astronomers* who were such thoroughgoing disciples of Platonic philosophy. Mazzoni was a Platonist but not a practicing astronomer. Magini at Bologna and Benedetti at Pisa were mathematical practitioners, but they were not Platonists. At best, Galileo might have been reminded of Francesco Barozzi's translation of Proclus's commentary on Euclid and Clavius's fleetingly favorable references to Proclus.[36] Now, here was this Kepler, an obvious novice who frequently deferred to his teacher and openly referred to his student disputations, who spoke enthusiastically of his teaching mathematics in distant Graz, and who was deeply immersed in theological studies in a way that Galileo was surely not. Kepler also proclaimed that his work had nothing to offer to ordinary astrological practitioners. His goal was of the highest, transcendent sort. It dealt with astronomy as a domain of contemplative philosophy and theology. Perhaps too much for Galileo's taste? Kepler boldly claimed to have discovered, if you please, a new way to demonstrate Copernicus's opinion—with "physical or, if your prefer, metaphysical reasons."

Although Kepler was a mathematician, his aspirations were quite clearly demonstrative and theological. Furthermore, here was Maestlin—hitherto known to Galileo only as Clavius's opponent—presenting himself as the defender of Kepler and Rheticus (whose annotated and illustrated edition of the *Narratio* Galileo had never before seen); here also was Maestlin's thickly technical "Dimensions of the Heavenly Circles and Spheres" (which, for Galileo, may have evoked Magini's *Theorics*). Without any more than a cursory glance at this package of Copernicana—can we allow Galileo an hour or so with the preface, dedication, and main plate before Homberger's departure?—he must have been startled, and perhaps also anxious. The conventional wisdom among many historians is that Galileo read no further. Yet the entire letter is worthy of attention.

> It is not days but only a few hours ago that I received your book, which was brought to me by Paul Homberger *[Ombergero]*; as the same Paul mentioned that he was returning to Germany, I thought it would really be taken for ingratitude if I did not by this letter express my thanks for your gift *[munere]*. So please accept my thanks and, moreover, my gratitude for your having by this means graciously invited me to become your friend. So far I have only read the preface to your book, from which, however, I got a little insight into what you intend by it; really, I congratulate myself with all my heart to have such an ally *[socium]* in searching for the truth, such a friend *[amicum]* too of this same truth. For it is a sad thing that the students of truth *[studiosos veritatis]* are so rare that there are only a few who do not follow the corrupt way of philosophizing. However, this is not the place to deplore the miseries of our century; I should rather congratulate you on all the beautiful things you presented in support of the truth. I shall only add this and promise that I shall study your book patiently, being certain that I will find in it marvelous things. This I will do the more joyfully in that I have for many years past venerated the opinion of Copernicus *[quod in Copernici sententiam multis abhinc annis venerim]* and from it

the causes of many natural effects have been discovered by me, which without doubt are inexplicable by the ordinary hypothesis. I have written down many reasons *[rationes]*[37] as well as refutations of contrary arguments which, however, I thus far did not dare to make publicly known, being frightened *[perterritus]* by the fate of our teacher Copernicus who, though having gained immortal fame in the eyes of a few, has been mocked and driven off the stage *[explodendus]* by innumerable others (for so great is the number of fools). I should undoubtedly venture to disclose my opinions, if there were more men like you; since there are none, I shall refrain from any business of this sort.[38]

Galileo's position in this letter was stronger and more specific than in the previous one to Mazzoni. Rather than invoke the phrase "much more probable," he spoke of "the causes of many natural effects" and "many reasons as well as refutations of contrary arguments." The new language was clearly meant to hint strongly to Kepler that Galileo too had reached an apodictic proof. But Galileo did not produce such a proof in 1597 any more than in 1616, when, in his famous "Letter to the Grand Duchess Christina," he deployed a rhetoric of "experience and necessary demonstrations" without establishing any definitive arguments of apodictic strength.[39]

The 1597 exchange between Kepler and Galileo marks the emergence of a third-generation response, one in which physical questions pushed to the fore and clearly became dominant. But it was also a moment that exposed sharp differences between the two modernizers that are more difficult to explain. Commentators have not missed Galileo's competitive qualities in this letter, his seemingly inordinate need to dominate. Was this a feature of Galileo's personality? Or was Galileo a historical actor in a general system of social relations whose organization, quite apart from Galileo's unique temperament, structurally obligated that sort of behavior? In 1967, Willy Hartner opted for the first possibility. He interpreted the 1597 letter as "a moving document of human weakness." And further: "The wish not to lag behind when important novelties came to his knowledge seems to have been particularly strong during Galileo's adolescence and early manhood. At the same time, he also displayed a remarkable uneasiness on occasion about the open profession of opinions at variance with the ones commonly accepted."[40] Along the same lines, Francesco Barone has remarked on Galileo's "reaction of pride in comparison to his younger correspondent."[41]

Taking another approach, Massimo Bucciantini has proposed that Galileo's response was shaped by the dangerous political atmosphere resulting from the ever-tightening grip of the Inquisition and the Office of the Holy Index.[42] In this narrative, which foregrounds Counter-Reformation surveillance while moving it back before Bruno's trial, Galileo could not allow himself to engage more closely with Kepler because he was the student of Maestlin, a heretic of the first order, and well known from his recent polemic with Clavius over the calendar reform.[43] If correct, this interpretation would explain the otherwise strange allusion that Galileo made to Copernicus's fate in being "driven off the stage." Yet Bucciantini's attractive suggestion also raises new questions that counsel hesitation: if the *Mysterium* was seen to carry confessional import, why was it not immediately put on the Index? Indeed, had it actually been prohibited, the censors might simply have required that readers expunge Maestlin's name, thereby preserving the book's valuable core and essentially putting it into the category to which *De Revolutionibus* was later assigned: "prohibited until corrected."[44] And, finally, there is no evidence that fear of the Church's surveillance prevented Galileo from conducting further correspondence with the young cosmographer who thought that he had the keys to the order of the heavens. In short, there were ways to evade the censorship.[45]

The crucial issue, I suggest, was premised on a *pedagogical* model: the hierarchical teacher-student relationship. Throughout his life, Galileo, like Tycho Brahe, was most comfortable with disciples. Galileo presented himself as he had in the Mazzoni letter: as a teacher who appears always to know more than the student. He was something like the *praeses* who sets the questions and also provides the answers in a disputational performance. Much later on, when Galileo wrote skillfully in the genre of the dialogue, he reinscribed and caricatured the teaching relation in the form of what Tommaso Campanella called a "philosophical comedy," a conversation among Salviati (the teacher), Sagredo (a smart and open-minded student), and Simplicio (a dull, sometimes pedantic, and often resistant student).[46] The gift of Kepler's book had evidently put Galileo in an ambivalent state. Yet he used a language of amicability: Kepler was a potential "ally" and a "friend" who shared views that he himself was just then developing in a philosophical environment not especially well disposed to them. (Even the progressive Mazzoni, the sometime friend of Plato, was no particular friend of Copernicus). Did Galileo have nothing to teach Kepler? ("I shall study your book patiently. . . . I have written down many reasons.") Was Galileo afraid to share his views for fear that Kepler would appropriate them as his own? Or was it really so dangerous to express detailed Copernican convictions in a letter to a German prognosticator? For although it is quite clear from what we now know that Galileo harbored intentions similar to Kepler's, his caution and restraint at this crucial moment strongly recall Maestlin's at Tübingen. To Kepler, he must have seemed a sort of Italian Maestlin; to Galileo, Kepler appeared to be an irksome and uncertain candidate for discipleship, perhaps too much of an intellectual equal.

But to return to the question of Galileo's apparently adverse representation of the political and religious environment, was Galileo's situation at Padua, whose considerable fame rested on its medical and philosophical faculties, comparable to that of Maestlin's and Kepler's at theologically dominated Tübingen? There is no evidence that Padua at this time was in any way as hostile to open discussion as might be suggested by Galileo's "throwaway" remark to Kepler about the unfortunate fate of Copernicus: quite the opposite. Padua, where Copernicus had spent the last two years of his Italian studies, enjoyed considerable religious liberty, at least in the medieval sense of the legal assurance of certain rights, and it had the benefit of being the university city of the only Italian state that was really independent of imperial hegemony after 1530.[47] Moreover, in his own day, Galileo enjoyed excellent friendships with several important Paduan ecclesiastics, especially canons. At the cathedral in Padua, for example, was Antonio Querenghi, along with Paolo Gualdo (1548–1631), who, after 1596, became vicar general to the bishop of Padua, Marco Corner. At the parish church of San Lorenzo, where Galileo baptized his daughters Virginia and Livia, was his good friend Lorenzo Pignoria, the learned vicar known for his strong interest in Egyptian philology. At the parish church of San Martino, nearly adjacent to the university, was Martino San-delli, to whom Galileo would later entrust the Latin translation of his letters on sunspots.[48] And, of course, there was the welcoming atmosphere of the Villa Pinelli. Furthermore, although the Venetian Republic had not prevented the singular Giordano Bruno from being sent to Rome in 1592, it would protect its own faculty against Rome's interference, as Galileo would soon learn in his own case.[49] If Galileo did not yet appreciate this in 1597, that could help to explain his apparent caution.

It is more likely that Galileo's portrayal of a fearful Copernicus—"driven off the stage"—was not a sign of fear but quite the opposite. It was an explicit allusion to the ironical opening lines of the preface to De Revolutionibus where, echoing Horace, Copernicus spoke of himself as someone whose ideas *at first* appeared—even to himself— to be worthy of laughter and derision and would surely be repudiated (*explodendum*).[50] The use of Copernicus's word would surely communicate to Kepler that Galileo knew the text of De Revolutionibus and that, exactly like its author, he intended to publish proofs that would dispel the initial appearance of the theory's absurdity—but not just yet.[51] This allusion thus gave Galileo an excuse for delay and public silence on the epistemological status of the Copernican arrangement.

GALILEO AS A "MAESTLINIAN"

From the other side, I believe that Kepler's reply to Galileo can be read usefully against the background of his relationship to Maestlin. After seven years, he had grown accustomed to Maestlin's political and philosophical caution, his concealment of his real views, and his nevertheless extraordinarily supportive attitude toward his student once he saw the manuscript of the Mysterium Cosmographicum. Kepler's success in overcoming Maestlin's caution—the internal experience of pushing Maestlin into a public commitment to Copernicus—must have been very gratifying for the younger man. It was not unlike the role played by Rheticus in facilitating Copernicus's final decision to publish De Revolutionibus. Now, the two of them were in it together as Copernicans, and with Kepler's name as primary author: the Mysterium was "my little work (or rather, yours)."

It may be helpful to think of Kepler as trying to elicit from Galileo what he had successfully drawn from Maestlin. Overlooking Galileo's pre-sumed confessional identity as a Catholic, he was very happy to have a friendship with "an Italian" and he rejoiced in "our agreement on the Copernican cosmography" (*propter consensum nostrum in cosmographia Copernicana*). At the same time, he was eager to elicit Galileo's "judgment" about his book. By now, surely, Galileo had had enough time to study it: "For it is my nature to demand of all to whom I write for their uncorrupted judgment; and I want you to believe me, I much prefer even the sharpest criticism of a single wise man to all the unreasoned applause of the common crowd."[52]

Kepler naively believed that he could enlist Galileo in an enterprise much like the one that he had undertaken with Maestlin. His task was a difficult one. He knew nothing of Galileo personally and had none of his unpublished writings. Moreover, if Galileo's letter of August was tantalizing, it was also withholding; he had failed to specify the arguments in his alleged armamentarium of "the causes of many natural effects." And worse, he desisted from specific comments on the Mysterium. Why would such an "ally" need to be coaxed into joining a project of such presumably mutual advantage? Kepler set to work on the hesitant Galileo, flattering him by associations with the ancients, holding out the prospect of an alliance, and suggesting that perhaps the intellectual atmosphere since the time of Copernicus had changed for the better:

> Possessed of such an excellent intellect, would that you were of a different intent! Now although you warn discreetly and secretly, by your own example, that one should retreat before common ignorance, one should not rashly incite or oppose the madness of ordinary learned men—in which respect you follow our true masters, Plato and Pythagoras. Nevertheless, considering that in our era, a beginning to this prodigious labor was first begun by Copernicus himself and, after him, [continued by] a great many learned mathematicians, it is no longer novel to say that the Earth moves; thus, perhaps it would be better if, through our common efforts, with one impulse and without stopping, we would pull the chariot to the goal.[53]

Nowhere in this letter did Kepler speak of joining forces with Maestlin. Did he already sense that Galileo, unlike his own teacher, was more inclined to follow up on the search for efficient causes in Copernican cosmography? The answer is evidently affirmative. A year later, he speculated that Galileo had developed new arguments about the causes of the tides.[54]

Yet, in his effort to win over an ally presumed to be already sympathetic, Kepler avoided invoking specificities of his own position at the level with which we are familiar from his letters to Maestlin. He continued, instead, to flatter Galileo and to tempt him into a field of debate of his own construction. In this respect, Kepler's discursive and rhetorical categories more closely resemble the strategy of the preface of the *Mysterium,* which was directed to the general reader. He created latitude for himself by following the still-widespread practice of bypassing the use of proper names—understandably so, as the only disciples he could have named at that time were himself, Maestlin, Rheticus, and perhaps Digges.[55] He did not invoke any national advantage: rather, he said that in both Germany and Italy there were those who resisted Copernicus's ideas. Kepler approached Galileo as a member of a privileged mathematical elite. The recruitment was to occur within a generic coding of pairs of skill sets. There were those who were "ignorant of everything" as opposed to "those who are moderately learned, but not in mathematics";[56] and those who were "ignorant mathematicians" as opposed to those who were "learned or skilled mathematicians." Kepler's "moderately learned" types were the ones most amenable to conversion. They were so unskilled in how planetary ephemerides work that when they heard that some ephemerides were based on Copernicus's hypotheses, then they believed that "everyone who today composes ephemerides follows Copernicus; and if it is demanded of them that they grant that [these tables] can only be demonstrated by mathematical principles, [then they believe that] the phenomena cannot agree without the motion of the Earth. For though these postulates or pronouncements are not credible in themselves, still nonmathematicians should grant them; and since they are true, why should they not be passed off as irrefutables?"[57]

Operating at this general rhetorical level, Kepler encouraged Galileo to join him in an elite category of philosophical mathematicians. Kepler's construction left "ignorant mathematicians" as the real opponents, that is, the ordinary prognosticators represented in the preface to the *Mysterium*—a familiar casting that went back at least to the time of Regiomontanus and Paul of Middelburg. He treated the rest of this proposed enterprise as obvious—that Galileo would join Kepler in the sort of cosmographical project outlined in the book that he had just received.

The fatal flaw in Kepler's strategy was his apparent assumption that if he could win Galileo's collaboration in publicly endorsing the heliocentric order of the planets, then he would simultaneously gain assent to a common framework of physical principles governing the causes of planetary motion and astral influence, not to mention a coherent account of the behavior of bodies on a moving Earth. It did occur to Kepler that Galileo had a different argument in support of the Earth's motion (the tides), but he could not know at this point that their differences went much deeper. Galileo was working out a new set of physical principles that made time the crucial variable in understanding motion, rather than the efficient and final causality of pushes and purposes that still lay at the heart of Kepler's Pico-inspired solar astronomy.

Kepler's final proposals were based on the assumption that he and Galileo already had enough in common to launch something of a campaign to convince other like-minded mathematicians to come over to their view. The method that Kepler proposed was not a joint publication along the lines of the *Mysterium Cosmographicum* but a letter-writing project.

Only mathematicians remain, and with them greater work is required. Since they share the same title [as we do], these men do not grant postulates without a demonstration; among these, the more each one is unskilled, the more trouble he presents. Nevertheless, a remedy may be applied: isolation. In any one place there is just one mathematician; therefore, wherever he may be, he is the best man. So, if anywhere else he has a companion who shares his opinions, let him request letters of him. By this method, when he has shown the letters around (for which purpose yours is also profitable to me), it can excite in the souls of learned men the opinion that all professors of mathematics everywhere are in agreement. Truly, what deceit is necessary? Be confident, Galileo, and march on. If I have guessed correctly, few among the excellent mathematicians of Europe will wish to retreat from us—so great a force is truth.[58]

Mathematicians would be led to change their minds, along the lines that Copernicus had intoned in the famous passage from his preface, by the authority of the demonstrations of other mathematicians. And this persuasion would be achieved by epistolary rather than personal contact. The letter as a mode of humanist self-presentation was a well-established literary genre and one in which Kepler himself was already becoming well practiced. Printing a collection of letters in order to communicate philosophical ideas also had precedent, as with the mid-sixteenth-century edition of Ficino's letters.[59] Is it possible that Kepler was already aware that Tycho Brahe had published his *Epistolae Astronomicae* just a year before his epistolary proposal to Galileo?[60] Whatever the case, Kepler's humanist letter-writing proposal envisioned an approach that would bring together the best practicing mathematicians. It envisioned neither the mediation of princes nor the intervention of middle men nor the exchange of gifts. But who were the other Copernicans who might assist in this venture? Bruno was already in the dungeons; Tycho had suppressed Rothmann; Digges had disappeared from the scene of scholarly activity. Yet Kepler was clearly eager to establish a collaboration with Galileo on the basis of shared philosophical convictions, whether publicly or privately. "If Italy is less suitable to you for publishing and if you have encountered certain obstacles, perhaps Germany will grant this liberty to you. But enough about this. At least write often to me if you discover anything advantageous to Copernicus. If it is not agreeable for you to do so publicly, communicate privately."[61]

In continually referring to Galileo as an Italian, Kepler studiously ignored the fact that there might be confessional differences between the two of them. He chose to overlook the fact that the theological premises underlying his entire cosmography could be seen as risky propositions in Padua. Thus Kepler's invitation to Galileo to publish in Germany if the environment in Italy proved to be too hostile might be read as an invitation for Galileo to bypass his theologians, just as Kepler had done at Tübingen. Perhaps their common ground might be the Book of Nature—a kind of secularized or antidogmatic theology of the Creation.[62] Meanwhile, Kepler hoped that they could cooperate in the important observational work of trying to measure annual stellar parallax; for, as Kepler remarked at the end, he did not even have a quadrant in his possession.

Galileo did not reply to this letter for more than thirteen years. As early as 1597, it would appear that Galileo saw no hope of enlisting the younger man, whose philosophical framework was already so obviously entrenched. On the other hand, Kepler's failure to enroll Galileo in a collaborative venture should not be taken as evidence either that Galileo ignored the *Mysterium Cosmographicum* or that he had no deep interest in astronomy.[63] Galileo undoubtedly agreed with the logical and astronomical arguments in chapter 1.[64] Those arguments—the first systematic public explication of Copernicus's main theoretical postulates since 1543—were as compatible with the great deal that we know of Galileo's later convictions as with the little that we know of his earlier ones. There is no reason to doubt that Galileo would have found especially compelling Kepler's reply to Clavius's objection against Copernicus (that from a false premise, a true conclusion may be deduced); and there is even some evidence that he was sympathetic to Kepler's solar moving-power.[65] He also appears to have studied the numbers involved in Kepler's rule linking the planets' speeds with their mean distances from the Sun, although there is no basis for believing that he knew anything about the Piconian origins of the solar power.[66]

PADUAN SOCIABILITIES
THE PINELLI CIRCLE AND THE EDMUND BRUCE EPISODE, 1599–1605

The relationship between Kepler and Galileo, with its associated hopes and disappointments, did not altogether vanish in the years before 1610.[67] Although we lack evidence of any further correspondence between them, residual clues about their association remain in three puzzling letters to Kepler from an underappreciated Englishman named Edmund Bruce.[68]

Bruce appears near the end of a long tradition of English students who had made their way to Padua since the late fifteenth century. Padua had long boasted a rich tradition of transalpine students from Central Europe. Bruce had been elected representative (*consiliarius*) of the nation of English students from 1588 to 1594 (a post held by William Harvey in 1600).[69] He was a man of respectable learning who enjoyed the mixed company and conversation of university mathematicians, philosophers, and men of general letters like himself. Part of his interest in such associations, however, was quite specifically political. For many English students of the 1580s and '90s, educational travel had become an institution. As Jonathan Woolfson has shown, it was "a useful preparation for service to the state and especially diplomatic service, and a means by which the government could gather information on the disposition of foreign governments and on the activities of enemies to England abroad."[70] Among the chief recipients of such political intelligence was Anthony Bacon (1558–1601), the elder stepbrother of Francis and a close supporter of the powerful earl of Essex (1567–1601).[71]

With more than a decade's experience in Padua, Edmund Bruce was one of Anthony Bacon's main "intelligencers" in northern Italy.[72] Bacon himself had spent many years on the Continent and well understood what was involved in this kind of work. Among those with whom Bruce was associated was the prominent Paduan ecclesiastic Lorenzo Pignoria, who was a member of the famed circle of learned men who met regularly at the Paduan home of Gian Vincenzo Pinelli.[73] Bruce was also one of the self-described "friends" whose interests in mathematics, military affairs, and herbals Pinelli supported.

So, at least, we are told by Pinelli's biographer, Paolo Gualdo.[74] The image of Pinelli's circle as a self-conscious group, rather than an array of individual acquaintances, comes down to us from Gualdo's presentation, published six years after Pinelli's death. The purpose of writing the history of a human life at this time—much indebted to ancient Roman models—was to teach virtue by example.[75] Pinelli's life was exemplary for Gualdo not only for its embodiment of the virtues of learning, honesty, and modesty, but also because he valued certain conditions that made it possible for other men to realize these virtues as well. The chief means to this end was the great library of Latin and Greek manuscripts and recently printed works that he had collected in his house and which he made available to a wide range of scholars from all over Europe.[76] The house was located not far from the old city center, "at the crossroad of Saint Anthony's [Basilica] in the highest room facing the street."[77] Gualdo's fleeting description of the interior leaves a trace of Pinelli's vision: "He built and adorned the interior part of the house with large geographic maps and images of famous men. . . . He took over our

study room, and with a great expansion of the library, rooms and hall, he continued decorations of this sort. And as there were many delays in writing books for him, other ones were put in to enlarge the library so that those who wished would not be deprived of this library devoted to his image."[78]

Chapter 10 presents some examples of such humanistically inspired civil communities in the northern lands. As with the Dudith circle at Wrocław, mathematicians were sometimes involved, but rarely as a dominant presence. Among such groups, Tycho's Uraniborg represented an exceptional development: a privileged site, hierarchically structured around the figure of Brahe himself, organized with the primary objective of studying the stars and their influences. Uraniborg added significantly to the aura of noble authority attached to heavenly practitioners in the last quarter of the sixteenth century. Nonetheless, such groups were humanistic not merely in the limited sense that they respected the *studia humanitatis*, used ancient languages, and borrowed symbolic resources from Roman and Greek literature; more significant, they invigorated ancient ideals of sociability with the sumptuous material resources of their own age. Among various possible dimensions of friendship, the neo-Stoic ideal of constancy and inner discipline in the face of worldly adversity became an ideal for an affluent nobility that had the means to support it as a way of life.[79] It was materially bound by travel, correspondence, good conversation, and book and manuscript collecting, and it was represented at crucial nodal points with contact between learned aristocratic patrons (who could afford instruments and large libraries), ecclesiastics (with deep intellectual concerns), and academic mathematicians (who relished the books, conversation, and potential sources of patronage). In the last quarter of the sixteenth century, the most prominent and active of these groups included Tycho's Uraniborg on the isle of Hven; the Dudith circle; the wealthy patrician families of Augsburg; the cosmopolitan court circles in Rudolfine Prague centered on Thaddeus Hagecius, to which Brahe, Dee, and Kepler were drawn; the Northumberland network, which included Thomas Harriot, Nathaniel Torporley, Walter Warner, William Lower, and Nicholas Hill; and the circle gathered around the bibliophile Pinelli in Padua.[80]

Edmund Bruce, like Henry Savile, Paul Wittich, and Giordano Bruno, traveled along the margins of such sodalities, fertilizing them with his brief presence. And, like Kepler's aristocratic acquaintances and correspondents—Herwart von Hohenburg in Bavaria and Johannes Matthäus Wacker von Wackenfels in Prague—he was fascinated by the bold new philosophical thinking about the heavens that was heating up among the modernizers in these fin-de-siècle, late-humanist settings. Somehow he had established a connection with Kepler. The earliest contacts may have been mediated initially through the wealthy and large city of Augsburg, home of the great Fugger and Welser banking families. The Fuggers famously cultivated a network of news gatherers throughout Europe. From Augsburg, Hapsburg Vienna was within easy reach along the Danube. Vienna had good links to the Venetian Republic via a major overland trade route, especially significant for commerce in wines: the route passed just north of Graz and eventually down through Udine and Treviso.[81] The existence of correspondence between Marcus Welser and Wacker von Wackenfels in Prague suggests another important possible source of contact.[82] And there are further leads. In August 1602, Edmund Bruce recommended that Kepler communicate to him through the Augsburg Greek philologist David Hoeschel (an important collaborator of Marcus Welser), "through whom I send these letters to you and through whom they may come to me now without danger."[83] A year later, he urged Kepler to transmit his letters through Marcus Welser himself, whom he described as "most illustrious and my greatest friend."[84]

Welser's erudition and intense humanist interests made him a logical candidate to mediate the Bruce-Kepler connection—just as he would later mediate the encounter between Galileo and the Jesuit Christopher Scheiner (1575–1650) over the meaning of a group of dark patches said to be floating on or near the Sun. Tutored as a youth by Hieronymus Wolf, Welser was well versed in Greek, Latin, Italian, and French. His training with the anti-Piconian Wolf also makes it likely that his predilections included astrology—which would help to explain one dimension of his interest in Kepler (and perhaps in Galileo). However, another significant side of his late-humanist identity was the establishment of a printing business in 1594, where he produced a wide variety of ancient and modern works, including Christopher Scheiner's work on sunspots.[85] He was also fascinated by the original Roman settlements in Augsburg. Under his own name, he had published a well-illustrated, highly learned edition of the so-called Peutinger Map (Venice, 1591), based on Roman maps of Augsburg once in the possession of his relative Conrad Peutinger.[86] The only extant manuscript known to have been owned by Edmund Bruce is a copy of the Peutinger map, and it is listed in the inventory of Pinelli's (extant) library.[87] Thus the interest shared by Pinelli, Bruce, and Welser in Roman and sixteenth-century Augsburg highlights the importance of cartographic space for constructing the inner world of Pinelli's house and its symbolic evocation of the past.[88]

The full encounter between Bruce and Kepler—whether or not mediated through Welser—was clearly felt to be successful on both sides, as Bruce owned a copy of Kepler's *Mysterium* and represented himself as promoting it wherever possible.[89] Eventually, his network extended to contacts with Galileo and Magini. Through his involvement in the Pinelli circle, he must have been exposed to all sorts of philosophically heterodox views. Because he had close contacts with other English students as he circulated around Padua and Venice, he was also in a good position to hear about new ideas, while keeping a politically attuned ear to the ground. Until recently, the general view among historians of science has been that Bruce was merely a "gossip" and his reliability as a witness suspect because his report to Kepler, in August 1602, was based on hearsay and conjecture.[90]

Yet this judgment is not only too hasty but also misses the importance of Bruce as a figure representative of a certain kind of learned sociability that characterized the nascent civil communities that he traversed. Kepler's letters show that he treated Bruce as a trustworthy sounding board for his own ideas, much as he did the several learned noblemen with whom he enjoyed a correspondence. Clearly, he also assumed Bruce to be an adherent of the *Mysterium Cosmographicum*—a philosophical sympathizer, although not actively engaged in his own astronomical investigations. Kepler's long letter of 1599 revealed that he ascribed to Bruce a degree of familiarity with musical and astronomical theory sufficiently high to share with him his deepest, evolving thoughts concerning Copernican celestial harmonies.[91] And since Kepler had received no further communication from Galileo, he hoped that Bruce would send him "intelligence" on how the *Mysterium Cosmographicum* was being received in Italy—and especially by Galileo, whose active support he still hoped for.[92] Bruce probably expected to obtain both political information and further astronomical musings from Kepler. The 1599 letter also shows that while still in Graz, Kepler knew that his correspondent was acquainted with Galileo.[93] Furthermore, because Kepler had not heard from Galileo since 1597, he requested that Bruce directly pass on his letter to him—which he did.[94] Bruce's reply to Kepler is our main evidence for the subsequent development of the relationship with Galileo before 1610:

My most excellent Kepler, I hope that you received my letters sent from Padua; now I send you these from Florence, by which I assure you that it was my destiny to travel with Magini in the same coach from Padua to Bologna, and in whose home I was received graciously for a day and a night, and during the course of which time we spoke honorably about you. I showed him your *Prodromus*, and I said that you wonder very much that he never replied to his letters; but he swore to me that he had never before seen your *Prodromus*; yet he diligently expected its arrival every day and he faithfully promised me that, shortly, he wants to send his letters to you; and also, that not only does he like you but also he confessed that he also admires you for those things that you have found out [*inuentis*]. Galileo, however, told me that he wrote to you and that he had received your book, which, nevertheless, he denied to Magini; and I rebuked Galileo for praising you too softly, for I know for certain that he spreads as his own (to his pupils and others) the things that you have found out. I, however, acted and shall always act in a manner which redounds more to your honor than to his.[95]

Bruce's sharp eye for detail on behalf of Kepler betrays his sometime role as an intelligencer of the late Elizabethan age. The letter opens a window, all too briefly, onto a jumble of local interactions that historians of this period do not frequently encounter. Five years after the publication of the *Mysterium Cosmographicum*, so the letter suggests, there was continuing unease about Kepler's unsettling proposals that betokened an underlying anxiety among the two leading north Italian mathematical practitioners, Magini and

Galileo.[96] Bruce represented himself to Kepler as not caring much for Galileo—a sentiment that we have little reason to doubt was genuine. Bruce also confirms that Galileo had moved no closer to Kepler's cosmographical premises than he had been in 1597. His "many causes of natural effects" also remained unarticulated.

What, then, could Bruce have intended by an offhand remark intimating a suspicion of plagiarism on the part of Galileo? Instead of dismissing Bruce's report as "gossip," consider four alternative hypotheses. First, perhaps Galileo had conversed about Copernican arguments at Pinelli's house in a manner that sounded familiar to Bruce's ear simply because they resembled those in Kepler's first chapter.[97] Second, because Galileo had possession of Kepler's long letter of July 1599—which was mostly about musical proportionalities and planetary distances—Bruce might have heard that Galileo was presenting some such ideas to his students, without really knowing what judgment Galileo was putting on the different parts of the letter.[98] Third, Kepler told Bruce in 1599 that he "strongly desired" to know Galileo's opinion concerning the use of magnetic declination to establish the meridian line.[99] And finally, we may conjecture that Galileo was being cautious in not wishing his rival Magini to know that he had Kepler's book in his possession, let alone that he might be communicating its ideas as his own.[100]

Bruce reported nothing further about Galileo's activities. Yet it is now abundantly clear that Galileo was not pursuing a Keplerian physical cosmography during the early 1600s. He was engaged in a quite different sort of project, studying pendulums and motion along inclined planes. By October 1604, not too long after Bruce's report, he had found profound and unexpected regularities governing both the period of the pendulum and the relationship between speed and time (squared) in the case of objects in free fall.[101] Stillman Drake and others have done much important work to reconstruct Galileo's mechanical investigations in this period, but Drake doubts that Copernican considerations motivated these studies.[102] His position, still widely shared, is not entirely without merit, as there are no explicitly "Copernican" notes to be found among Galileo's extant papers. But Drake's interpretation also has a disabling, contrarian quality derived from his irritation with the view that Galileo was, in some manner, what he calls a "Copernican zealot." Drake, it should be noted, was zealous in his own conviction that "philosophers" could have had nothing to do with Galileo's most important discoveries. His real argument was with Alexandre Koyré. In effect, he wanted to overturn Koyré's image of "Galileo, philosopher," as a Platonized thought experimenter and replace it with his own representation of Galileo as a "real," experimental scientist.[103]

Yet, recalling Galileo's earlier stated support for Copernicus's theory, can it be true that, in the 1570s and '80s, figures like Digges, Bruno, Gilbert, Rothmann, Brahe, and Clavius all realized that the problem of falling bodies was relevant to a defense of the Earth's motion, but that in the 1590s Galileo did not? To doubt that Galileo saw such a connection seems unnecessary, even perverse. For all practical purposes, it would make the Galileo of this early period into a Simplicio character of just the sort that he later pilloried in the *Dialogue concerning the Two Chief World Systems*. One does not have to hold that Copernican considerations motivated all of Galileo's experimental investigations in this period in order to claim that he saw them as related.

As Massimo Bucciantini has shown, after his arrival in Padua Galileo had ample occasion to acquaint himself with aspects of the debates over planetary order that were taking place along the Uraniborg-Kassel axis.[104] Working without this material, but reaching conclusions compatible with it, Ron Naylor has proposed a Koyrean reconstruction of a Galileo who forged some kind of Copernican commitment as early as his unpublished treatise *De Motu*, dated 1590, but no later than his tidal theory of 1595—one that involved the notion that the same principles of circular motion might apply equally to terrestrial rotation and celestial revolution. In Naylor's account, by November 1602 Galileo had reached the radical insight that the isochronous motion of the pendulum in a circular arc was, like linear motion on an inclined plane, an instance of constrained free fall. Thus, "falling down" was an earthbound perception whose reality was the resultant of circular motions.[105] Such a reading would reinforce the case for Galileo's sincerity in the 1597 letter to Kepler. Naylor's account also might explain why Galileo's principled adherence to circularity would have made him unreceptive to Kepler's elliptical astronomy after 1609, but it leaves un-

explained why Galileo overlooked points of compatibility between his way of thinking about terrestrial physics and Kepler's way of thinking about celestial order between 1597 and 1610. Nor does it explain why Galileo broke off further contact. Much Galileo scholarship has focused on his conflicts with the traditionalists, but not with other modernizers.[106]

Yet, as opposed to conceptual reconstruction, direct evidence for or against Galileo's working explicitly within a Copernican frame after the 1597 letters to Mazzoni and Kepler and before the death of Pinelli in 1601 is slim at best. In his *Life of Pinelli*, Paolo Gualdo registered a reference to "The Commentary of Galileo Galilei, Mathematician of Florence and Paduan Professor, in favor of Copernicus against Jacopo Marroni."[107] "Marroni" is clearly a misspelling of "Mazzoni." The correction of the spelling error in Gualdo's printed errata sheet shows either that he had direct knowledge of this "commentary" or that he simply knew that the name was misspelled, or both. Whatever the case, Gualdo's language shows that within Pinelli's ambience, Galileo was already known to be "in favor of Copernicus." And it is possible that the Mazzoni commentary is the context in which Edmund Bruce heard Galileo "spread as his own the things that you [Kepler] have found out." And thus it is entirely plausible that Galileo continued to think about and pursue his own Copernican problematic after his exchanges with Mazzoni and Kepler. Yet in the early 1600s, he lacked a comprehensive account of the system of the world that could rival Kepler's cosmography or Aristotle's *Physics* in scope and demonstrative aspirations. Galileo had already conceived of such a work when he wrote the *Sidereus Nuncius* in 1610.[108] But until he could produce it, he evidently concluded that it was better to express himself covertly or to remain silent.[109]

1600: BRUNO'S EXECUTION

The network of Paduan intellectual friendships and covert mediations through which Galileo continued his association with Kepler provides one axis along which to read his involvement with the Copernican question. However, in 1600, an unexpected event dramatically complicated the space of political possibilities for this discussion: Giordano Bruno was executed on the Campo di Fiori in Rome. The location could not have been more public. The "Field of Flowers" was a popular region of the city, filled with a great concentration of businesses—grocers selling barley, other grains, and vegetables, and an animal market especially known for its horses. In addition, it contained a great number of bookstores and print shops. Together with the Ponte Sant'Angelo, it was a frequent site of executions, a "theater of capital punishment," as Eugenio Canone has aptly called it. The French ambassador occupied the Palazzo Orsini, which fronted on these proceedings, and he had already protested the "horrors" that took place in front of his palace—not because he objected to the execution of heretics but because he preferred that the actions take place at night, when they would not disturb his sound sleep, rather than early in the morning.[110]

The same Robert Bellarmine who had allowed scripture to direct his speculations about fluid heavens several years before Tycho Brahe was a consultor of the Holy Office. Bellarmine played a significant role in preparing a list of eight propositions that Bruno was asked to renounce as heretical. Did the list include Bruno's claims about the world's spatial infinitude and the plurality of suns? Frances Yates conjectured that the main charges against Bruno concerned his magical and Hermetic views and not his beliefs about the arrangement of the planets.[111] Unfortunately, it is not known if Bellarmine included any propositions about the Earth's motion. If he did not, it was hardly for any lack of familiarity with elementary Copernican propositions or because Bellarmine obdurately agreed with Aristotle on all matters of natural philosophy. He certainly was not an Averroist. In any event, just after he was appointed cardinal in 1599, Bellarmine became one of the judges who tried and convicted Bruno.[112] To the end, Bruno remained true to himself and to his life as a philosopher: "You perhaps pronounce a sentence against me with a fear greater than that with which I receive it," Bruno is famously reported to have said before he was burned alive.[113] His words were courageous and unrepentant—all the more so as his tongue was subsequently constrained by a vice.[114] Against the dogmatic, revealed theology of his judges, Bruno chose death over prudent dissimulation. His moral position, as Miguel Granada has suggested, was rationally consistent with an Averroist-inspired belief in man's perfectibility through philosophical contemplation of a divinity immanent in an infinite universe populated by innumerable Copernican worlds.[115] As for the true details of Bruno's conviction, they will always remain a matter of controversy, because after Napoleon ordered the trial records sent from Rome, they disappeared, probably to a pulp mill.[116]

Preoccupation with the tragic circumstances of the trial itself has tended to elide the question of the immediate impact of Bruno's death on learned opinion in Italy. Most recent Galileo scholars have all but ignored Bruno, perhaps in reaction to an earlier historiographical tendency to overspeculate, perhaps because Bruno's name is simply not to be found in any of Galileo's published or unpublished writings.[117] Regardless of why Bruno was convicted, however, nobody who received word of the trial—whether in liberal aristocratic circles or in the more conservative universities—could have doubted that it was dangerous to entertain any of Bruno's ideas publicly and perhaps even to speak of them in private.[118]

The Holy Index soon made the general perception specific. On August 7, 1603, by official decree of the Master of the Sacred Palace, Bruno's works were put into the most severe category of prohibition on the Index of Clement VIII: *opera omnia omnino prohibentur* (all works are completely prohibited).[119] They would remain so until 1900, which was more than enough time for Bruno to become an anticlerical symbol of the Risorgimento and later to earn an article portraying him as a martyr of science in the Soviet encyclopedia.[120] The works were described as being filled with "false, heretical, erroneous and scandalous doctrines, corruptive of good customs and of Christian piety."[121] The Index also specified that, in the city of Rome, all books in this category were not to be "printed, sold or discussed and handled in whatever manner."[122] Furthermore, it explicitly defined the places of possible transgression: "It is expected, expressly, that all booksellers in Rome and anyone else in that situation who has any of the above-mentioned books in their shop or study, should at once consign them to our Office; it warns these people that, besides the gravest offense that they commit in offending God and besides the ecclesiastical censure that they will incur, if it comes to our attention, they will be severely punished in accordance with the punishments threatened in the Sacred Canons, in the rules of the Index and in our edicts on material in books, published at another time, etc."[123] These punishments included "loss of books," a fine of "three hundred gold scudi," and the possibility that the Master of the Sacred Palace could inflict "corporal punishment at his will."[124] There was, in other words, no chance that Bruno's works could slip into the milder category of prohibition (*donec corrigatur*, prohibited until corrected), to which Copernicus's *De Revolutionibus* would be consigned in 1616.

The action against Bruno's writings occurred exactly one year after Edmund Bruce told Kepler that Galileo was spreading Keplerian ideas as though they were his own. It does not seem unreasonable to infer that in Italy, in the immediate years after 1600—and certainly no later than 1603—the Bruno episode rekindled old fears about the space for philosophizing. And when, in November of that year, Bruce asked Kepler about his "astronomical doubts," he did not mention Bruno's name at all.

Of course, one must avoid the misleading impression that absolutist political regimes and educational institutions in countries outside the Catholic Mediterranean world were significantly more open to change or dissent. Early modern knowledge makers were always in danger of boundary transgressions—and the more adventurous thinkers were always looking out for havens of philosophical heterodoxy. One has only to recall Kepler's maneuvers with the Tübingen theologians. The real question at stake was what sorts of intellectual opinions and practices regimes regarded as dangerous. For mathematical practitioners, predicting the death of a ruler was probably the riskiest enterprise of all. Yet what was considered a threat in Rome might not be so regarded in mercantile Amsterdam, Elizabethan London, or Rudolfine Prague. A salient example is the work of the London royal physician William Gilbert (1540–1603).

1600: WILLIAM GILBERT'S PROJECT FOR A MAGNETICAL PHILOSOPHY

In the same year that Bruno was put to death on the Campo di Fiori, Gilbert published a book, the bulk of which he seems to have completed as early as 1582 and the latter parts after 1588. It was grandly titled *Concerning the Magnet, Magnetic Bodies and this Great Magnet, the Earth; A New Philosophy, demonstrated by means of many arguments and experiments* (*De Magnete, Magneticisque Corporibus, et De Magno Magnete tellure; Physiologia Nova, plurimis & argumentis, & experimentis demonstrata*).[125] Like John Dee and Giordano Bruno, Gilbert presented himself as engaging in "a new kind of philosophizing" (*novo genere philosophandi*), of introducing "words [and doctrines] new and unheard-of."[126] He also presented himself boldly as deciding "to philosophize freely, as freely, as in the past, the Egyptians, Greeks, and Latins published their dogmas." And he engaged in a kind of rhetoric of antirhetoric, explicitly dissociating himself from rhetoric's "graces" and locating his work in opposition to the "veiled and pedantic terminology" of the alchemists and the learning of the ancients: "We do not at all quote the ancients and the Greeks as our supporters, for neither can paltry Greek argumentation demonstrate the truth more subtilly nor Greek terms more effectively, nor can both elucidate it better. Our doctrine of the loadstone is contradictory of most of the principles and axioms of the Greeks." In fact, Gilbert's anti-Hellenist rhetoric was not nearly as absolute as it might seem. Many of his references were to modernizing, contemporary natural philosophers, such as Nicholas of Cusa, Ficino, Cardano, Scaliger, and Giovanni Baptista della Porta—the very sorts of less traditional neo-Aristotelianizing or Neoplatonizing philosophers that Kepler read as a student. And Gilbert proclaimed that he "had no hesitation in setting forth in hypotheses that are provable, the things that we have through a long experience discovered."[127]

Gilbert modeled this new philosophy of the Earth as magnet on an explicit analogy between laying out meridian and equatorial circles on a celestial sphere and on a spherical lodestone. In this gesture at globe-making practice, the reader was instructed how to find the poles of a spherical lodestone by placing a needle or iron wire at various points on the surface, so that it was free to revolve; marking the direction of the needle pointer; and noting the convergence of the various, marked lines at a single point—"as is done by the astronomer in the heavens and on his spheres and by the geographer on the terrestrial globe."[128] Indeed, Gilbert's structured presentation cut through both the traditional genre of Aristotelian commentary and the loosely organized, aphoristic approach of Cardano, Scaliger, and Dee. The work began with a lengthy discourse on the lodestone that considered both the opinions of earlier writers (many of them Greek) and Gilbert's own experiences with magnets. This prelude built up to the main thesis about the character of the magnetic Earth and a typology of its motions. On the subject of the heavens, Gilbert showed himself abreast of second-generation trends, making references to Copernicus, Brahe, Magini, Offusius, Ursus, and Bruno. He mentioned all but the last two of these explicitly (and effortlessly) in *De Magnete;* but only in his posthumous *New Philosophy of Our Sublunar World* (1651) did he explicitly engage with Ursus and Bruno.[129] It seems all but certain that such a well-read English author knew the work of Digges and Dee, whom, for unknown reasons, he never mentions.

Gilbert could not have chosen a better weapon to deride both traditional Aristotelian school philosophy and its new-wave interpreters than the magnet's apparent ability to act at a distance. This phenomenon was a well-known anomaly for the Aristotelians. In *De Magnete,* Gilbert explicitly structured his presentation around the explanatory apparatus underlying magnetic effects. He ceaselessly castigated traditionalists' favorite explanatory categories: the four elements, "formal causes," "specific causes in mixed bodies," "second or prime forms," "propagating forms in generating bodies." "These," said Gilbert with Brunonian sarcasm, "we leave for roaches and moths to prey upon."[130] Against such Aristotelian-style causes, Gilbert opened up a new domain of philosophizing that, whatever its unstated debts to traditional or nontraditional natural philosophy, was organized around a single entity, the magnet. The central problem was to identify a causal agency that would account for the "coition of bodies that are separate from one another and that cohere naturally."[131]

The answer to the problem of cohesion was to be found in a reworking of Stoic explanatory resources, possibly inspired by Ficino: "magnetic potency," "inborn powers of attracting," "primary liveliness [*vigor*]," "corporeal and incorporeal effluvia," "magnetic coition." Gilbert remarked: "The Stoics attribute to the earth a soul, and hence they declare, amid the derision of the learned, that the earth is an animal. This magnetical form, be it liveliness [*vigor*] or be it soul, is astral." The new formulation gained authority partly from its opposition to traditionalist explanations of the magnet. "Let the learned lament and weep for that neither any of the better Peripatetics nor

as yet the ordinary philosophizers nor Johannes Costaeus, who looks down upon this sort of thing, were capable of appreciating this noble and excellent nature."[132] This magnetic soul worked by actively producing a stream of incorporeal emanations—an "orb of effluvia"—around the poles of the sphere of the lodestone. Within that domain, iron or some other kind of magnetic body would tend to move at either a right or an oblique angle to the pole. The closer the magnetic body came to the lodestone, the greater the force.[133] The magnetic soul purposively drew iron filings to itself and also made a needle revolve on a floating cork. In this rotation, it behaved, said Gilbert, like a "little Earth," or *terrella.* Each planet also had its own tendency to cohere "through continuity of substance," lunar bodies tending to the Moon, solar to the Sun, and so forth.[134]

What made Gilbert's magnetic philosophy all the more radical in its explanatory and unifying ambitions was its analogy with the heavens. Traditional natural philosophers of course followed Aristotle in keeping the heavenly and terrestrial domains ontologically distinct. Gilbert saw it as one of his tasks to expand the domain of the natural philosopher's purview, using the magnet to explain what astronomer-astrologers and ephemeridists had misunderstood. He must have had access to a fairly good collection of ephemerides, including Magini's of 1582, from which he cited in full the passage from Domenico Maria Novara's 1489 prognostication concerning the alleged changes in terrestrial latitudes.[135] For Gilbert, the "inexact" and "conjectural estimates" of Domenico Maria, Stadius, Reinhold, and Ptolemy were all examples of what was wrong with these "geographical" commentators: "These errors have crept into geography all the more easily because the magnetic force was quite unknown to authors." And besides, he added in an apparent reference to Tycho Brahe, "observations of latitudes cannot be made with exactitude save by skilled men, with the help of large instruments, and by taking account of refraction of lights."[136] Like Copernicus, whose text he was silently glossing, Gilbert argued with probability that a diurnally rotating Earth-magnet was a much simpler conception because it obviated the need for an outermost sphere hurtling about the center of the universe every twenty-four hours.[137] Furthermore—and most consequentially—the magnetic *vigor* or soul took over the explanatory function of Copernican gravity, substituting entirely natural, albeit incorporeal, effluvia for a divinely implanted "natural desire" of a globe's parts to cohere as a whole. But Gilbert could offer no "natural causes" for the complicated precessional motion, yielding the bow-like *intorta corolla* in *De Revolutionibus*, book 3, on the grounds that the motions themselves were "uncertain and unknown."[138]

Notwithstanding his explicit Copernican appropriations, Gilbert's picture went strangely fuzzy on the question of planetary order. He insistently chose evasive language to refer to the Sun's own mobility or immobility and to the Earth's annual revolution.[139] Similarly, while he praised the "symmetry and harmony" of the Earth's daily motion and the Moon's monthly motion, he failed to mention Copernicus's strikingly original claim for the *symmetria* rendered by the heliocentered revolutions.[140] This prevarication is all the more puzzling in view of his bold endorsement of an active Sun exerting an effect on the Earth and the planets.[141]

Gilbert's odd public stance cannot be explained functionally, as the consequence of an oppressive political climate; at least I have seen no evidence that would support such an interpretation. As a prominent member of the Royal College of Physicians and one who attended the queen herself, Gilbert was not subject to the constraints of the academic disciplinary hierarchy. This freedom is already evident in his bold willingness to philosophize in a new way. He had given a novel explanation for the Earth's daily rotation that he regarded "not with mere probability but with certainty." He boldly rejected the existence of a prime mover as a "fiction," a "hateful tyranny," and a "philosophic fable . . . beneath derision."[142] He embraced an aggressively anti-Aristotelian rhetoric throughout. He had clearly studied *De Revolutionibus* well enough to extract and comment on claims made in the technically difficult book 3.[143] He also knew Offusius's speculations about *symmetria* and planetary ordering. He had Ursus's *Fundamentum* and noted the contradiction of a "central" Earth in a universe deprived of spheres.[144] Furthermore, writing after 1588, he was also directly familiar with Tycho's geoheliocentric system and the objections to diurnal motion defended by the hypothetical "cannon experiment." And, most significant, Gilbert did not hesitate to rebut Tycho's argument:

The diurnal revolution of the Earth does not incite bodies along nor retard them: they neither outstrip the Earth's motion nor fall behind when shot violently, whether to the east or to the west. Let EFG be the Earth, A the center, LE the ascending effluvia. As the orb of the effluvia moves with the Earth, so the part of the sphere on the right line LE proceeds undisturbed in the general rolling around.[145] In LE the heavy body M falls perpendicularly to E, the shortest way toward the center; nor is this straight motion of the weight a composite motion or coacervated [i. e., mixed] with a circular motion, but rather simple and direct, never going out of the line LE. Indeed, a projectile shot with equal force from E toward F, and from E toward G, covers an equal distance in both directions, even as the Earth's daily whirling around proceeds; and even as a man who takes twenty steps toward the east or the west covers an equal distance. Hence the Earth's diurnal motion is not at all refuted by the illustrious Tycho Brahe through such arguments as these.[146]

Gilbert's argument directly inverted Tycho's interpretation of projectile motion on a moving Earth, going so far as to reject Copernicus's formulation that such aggregated motions were made up of rectilinear and circular components.

Gilbert was thus clearly aware of the main issues from the roiling debates of the 1580s, even if he had no direct access to the local struggles in the Uraniborg-Kassel network. More difficult to assess is his position on the annual motion: "I pass by the Earth's other movements, for here we treat only of the diurnal rotation, whereby it turns to the Sun and produces the natural day."[147]

Did Gilbert intend to address "the Earth's other movements" in another work? Or did he simply mean to refer to the precessional motion, treated in book 6, chapters 8 and 9? Logically, if the outermost sphere is deprived of any astronomical function, it is a short step to doing away with it altogether.[148] Unlike Copernicus, Gilbert was clearly prepared to make that step. Moreover, the daily motion alone raised the well-known problem of explaining how bodies in free fall are not left behind. Having gone that far, it hardly seems sensible to say that Gilbert believed that he could offer final, formal and efficient causes for the diurnal rotation (to frame the matter in Aristotelian categories) but deliberately refused to extend such explanations to the annual revolution. Rea-

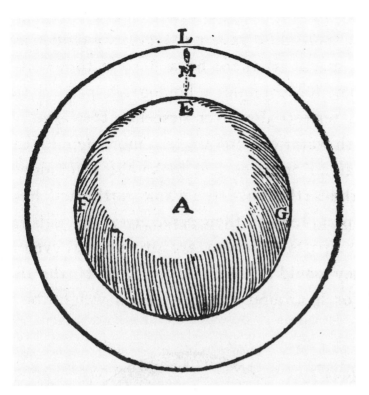

73. William Gilbert's rebuttal of
Tycho Brahe's imaginary cannon
experiment. Gilbert 1600, 341.
Image courtesy History of Science
Collections, University of Oklahoma
Libraries.

soning of this sort, joined with other considerations, has persuaded some interpreters that Gilbert must have been privately convinced of "the
heliocentric universe," possibly in Bruno's version.[149] But this reconstruction would imply that
Gilbert, like Bruno, decided to ignore just the sorts
of structural and ordering considerations that
had so attracted Rheticus, Maestlin, Rothmann,
and Kepler—or, at the very least, the Capellans
Offusius and Naibod.[150]

Here three different explanations may be suggested. First, Gilbert's problem may have been
entirely conceptual. Perhaps he simply did not
know how to make his magnetic physics extend
consistently to the annual revolution, much as he
had backed off proffering an explanation for the
precessional motion.[151] If each heavenly body (or
globe) had its own respective impulsion to cohere
and its own region of coition, then, how could circuits around the Sun be produced? How could
one magnetic soul produce three motions, one
rotational, another precessional, and the third a
revolution around another body? And why, if the
Sun was capable of inciting other bodies to motion, did it not move itself? Gilbert's waffling si-

lence on these questions marks an interesting
lapse in his otherwise aggressive, often Brunonian rhetoric and novel philosophizing.

A second possibility is that Gilbert simply lacked
the mathematical skill necessary to follow the in-
tricacies of Copernicus's precessional theory and
much else contained in *De Magnete*, book 6, and
that that section (like others) was written with the
skilled assistance of Edward Wright and Joseph
Jessop.[152] Here a third hypothesis may be proposed: that Gilbert, as a member of Elizabeth's
court, was very likely familiar at least with Bruno's Italian dialogues and possibly also the Latin
De Immenso. If so, Gilbert would have encountered Bruno's "Pythagorean" rendering of the
couplings of the Earth and Moon, and Mercury
and Venus. Bruno's emphasis on physical causes
rather than considerations of Copernican *symmetria* would have been quite consistent with Gilbert's own inclinations. Accordingly, Gilbert may
have found himself suspended in a state of uncertainty between two quite different readings and
uses of the Pythagoreans.[153] In all these hypotheses, Gilbert's dilemma highlights the constraints
of the genre in which he aspired to write: the nat-

ural philosophical demonstration—effectively, a theoric. Had he chosen the philosophical dialogue or the poem, as Bruno did, or the letter, as Tycho Brahe did, he would have had greater latitude to present mitigated claims to certainty without needing to "prove" every claim.

The legacy of uncertainty left by the world-system wars of the 1580s, further refracted through small print runs and the vagaries of circulation, may help to explain how a modernizing natural philosopher like Gilbert could land in such a quandary. When he published *De Magnete* in 1600, Gilbert clearly had access to the main works of the *via media* controversy of the late 1580s—Ursus's *Fundamentum Astronomicum*, whose defense of the Earth's diurnal rotation he shared; and Tycho's *De Mundi*, whose account of projectile motion he rejected. But there is no evidence that he had similar acquaintance with the literature of the mid-1590s: the Brahe-Rothmann exchange in the *Epistolae Astronomicae*, Ursus's virulent attack on Tycho in *De Astronomicis Hypothesibus*, Roeslin's selective appropriations from both, or Kepler's assertive and full advocacy of Copernicus in the *Mysterium*.[154] Gilbert was also caught on the Martian reef. Should he trust on Tycho's authority alone that the path of Mars intersects that of the Sun? Like Ursus and Roeslin, he apparently did not. Moreover, his attack on Tycho's cannonball thought experiment suggests again that he accepted something like Bruno's account of falling bodies in the *Cena*.[155] Although he could not adjudicate between Ursus and Tycho by producing Mars observations, he could associate himself with Ursus's diurnal motion on the basis of the magnetic philosophy. In that regard, Gilbert's practices, like those of Bruno, were ultimately a kind of negotiation with the conclusions of astronomers. Unlike Kepler, Gilbert could not initiate a "war on Mars"; but, like Descartes thirty years later, he could offer physical explanations for what the astronomers had already proposed and push them toward explanations not yet countenanced.

Initially, Gilbert's views attracted few followers in England.[156] One, however, was his friend Edward Wright, whose encomium to the book echoed Gilbert's position that it is "probable enough, on the ground of experiments and philosophical reasons not a few, that the earth, while it rests on its center as its basis and foundation, hath a spherical motion nevertheless."[157] More telling still, Wright, perhaps influenced by Bruno or Rothmann, freely invoked the accommodationist view of scripture: "It does not seem to have been the intention of Moses or the prophets to promulgate nice mathematical or physical distinctions: they rather accommodate themselves to the understanding of the common people and to the current fashions of speech."[158] Thomas Digges, who had speculated briefly on the magnet as a resource of Copernican physics, died before Gilbert's *De Magnete* appeared. The only sphere in which John Dee was interested by this time was the glass one that he placed on his window, the better to conduct conversations with angels. And there is no evidence that Tycho saw Gilbert's book before he died in 1601. But Gilbert's work—and especially Wright's remarks on the interpretation of scripture—would prove fertile ground for third-generation celestial theorizers.

Once again, it was Kepler who seized the opportunity, one afforded by the much wider circulation of *De Magnete* than of the *Mysterium*. Kepler had his own copy at least by November 1602, at precisely the moment that he was working to determine Mars's path, for the first time thinking of that circuit as an "orbit" and with a new physics at its foundation.[159] Gilbert's magnetical philosophy spoke to many of his immediate concerns, and it could not have escaped his attention that Gilbert provided an answer to Tycho's cannon thought experiment. For example, Gilbert answered the question that Kepler had put to Galileo concerning the magnet's role in affecting the motion of the terrestrial pole. But more than that, Gilbert's book also came packaged with compatible explanatory resources: final, efficient, and material causality. The magnet, a material body, produced effects on other bodies across a distance. It did all this without Gilbert's taking a step that must have been immediately obvious to the Copernicanizing Kepler. He took Gilbert's *terrella* and generalized its hints into a quantifiable model of motion for the Sun and the remaining planets. He also quickly integrated Gilbert's magnet into his work on the orbit of Mars, claiming that an incorporeal magnetic "species" issued from the central body. Gilbert's physical proposals fitted well with Kepler's Copernican cosmography; they also allowed him to refine and emend Pico's light physics with a force that could not only cross space but also pass through an intervening body.

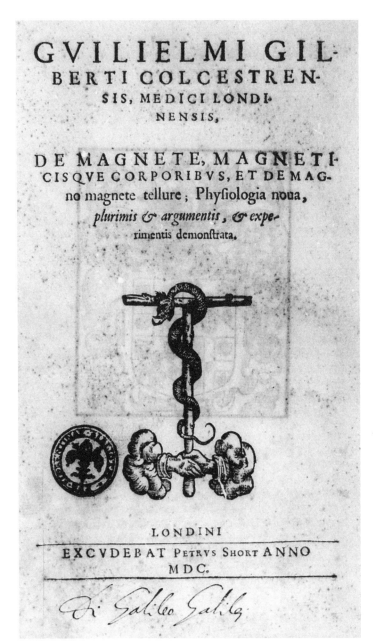

GVLIELMI GIL-
BERTI COLCESTREN-
SIS, MEDICI LONDI-
NENSIS,

DE MAGNETE, MAGNETI-
CISQVE CORPORIBVS, ET DE MAG-
no magnete tellure ; Phyſiologia noua,
plurimis & argumentis, & expe-
rimentis demonſtrata.

LONDINI
EXCVDEBAT Petrvs Short ANNO
MDC.

74. Title page of Gilbert, *De Magnete*. Galileo's provenance. Biblioteca Nazionale Centrale di Firenze B.R.121. By permission of Ministero per i Beni e le Attività Culturali della Repubblica Italiana / Biblioteca Nazionale Centrale di Firenze. Further reproduction by any means is prohibited.

Again, the differences with Galileo could not be more striking. Galileo obtained a copy of *De Magnete* at about the same time as Kepler, but he put it to no comparable use as a resource of heavenly motion.[160] He had, to be sure, genuine epistemic differences with the Englishman. Galileo valued Gilbert's observations—the effects that he described—and he would have been sympathetic to Gilbert's rebuttal of Tycho's cannon thought experiment, another reminder of how one might construct a physical basis for Copernicus's cosmography. But by this date, Galileo had no use for moving souls in natural philosophy, if ever he did. And that same disagreement about explanatory resources that had alienated him from Kepler's project also repelled him from any cosmographical uses for the Gilbertian magnet. As he said later in the *Dialogue*, "What I might have wished for in Gilbert would be a little more of the mathematician, and especially a thorough grounding in geometry, a discipline which would have rendered him less rash about accepting as rigorous proofs those reasons which he puts forward as *verae causae* for the correct conclusions he himself had observed. His reasons, candidly speaking, are not rigorous, and lack that force which must unquestionably be present in those adduced as necessary and eternal scientific conclusions."[161]

THE QUARREL AMONG
THE MODERNIZERS
NEW CONVERGENCES AT THE FIN DE SIÈCLE

The first decade of the seventeenth century was a time of remarkably rapid transition—not merely a continuation but an acceleration and complication of developments already under way in the late 1590s. In the earliest years of the 1600s, there was a convergence of new modes of natural philosophizing with theoretical astronomy such as had failed to occur in the years immediately following the publication of *De Revolutionibus*. Recall that in those much earlier times Tolosani had chastised Copernicus for threatening the order of knowledge but not for proposing an entirely new natural philosophy. Likewise, the prophetic concerns of the Wittenberg heavenly practitioners caused them to steer clear of new proposals in natural philosophy.

The situation at the fin de siècle could not have been more different. In their different ways Rothmann, Kepler, Bruno, Gilbert, Galileo, Harriot, Simon Stevin, and, to a lesser extent, Tycho Brahe reframed the Copernican problematic as an engagement between natural philosophy and planetary order. By no means was it the only consideration, as will be seen in succeeding chapters. After 1600, but before 1610, the hypothesis that the Sun was at rest near the center of planetary motion might be associated with either Gilbert's magnetical philosophy or Kepler's Platonic solids and solar motive force, with Bruno's infinite worlds or Thomas Harriot's fragmentary atomistic ruminations—or, as I will show in chapter 16, perhaps with some combination of the above.

Once again, Edmund Bruce has left a tantalizing bit of evidence that shows how this convergence appeared in the eyes of a lettered nonpractitioner. Brunonian philosophy, suspect among astronomers only fifteen years earlier, converged with astronomy to define one of the possible figurations of a new Copernican problematic. Bruce's final extant letter from Venice in November 1603 presented to Kepler a remarkable conundrum:

> I have many doubts in astronomy about which you alone [can] make me more certain. For I think that there are infinite worlds. Each one of these worlds is finite, as if the Sun's center is in the middle of the planets. And just as the Earth does not rest, so neither does the Sun. For it rotates most rapidly in its place around its axis, the motion of which the other planets follow. I think that the Earth is one among these [planets] but each [moves] more slowly the further it is from it [the Sun]. The stars are thus moved like the Sun, but they are not, as are the planets, driven around in a circle by the force of this body [of the Sun], since each one of these [stars] is a Sun—and not part of this, our smaller world of planets. I do not think that the Elemental world is peculiar and particular to us. For there is air between those bodies that we call stars; and consequently, there is also fire and water and earth. Moreover, I believe that this Earth that we trample under our feet is neither round nor spherical but that it comes closer to an oval figure. Also, I think that neither the light of the Sun nor the stars proceeds from matter but rather streams forth from their motion. In fact, the planets receive their light from the Sun because they are moved more slowly and they are hindered by their own proper motions. These and many other things appear to me to be probable; but now is neither the time nor the place for seeking proofs. It is enough for me if I can elicit your opinion about these ideas.[162]

This revealing letter sketches the elements of a new representation of the universe. It heralds efforts throughout the seventeenth century to join the Copernican or the Tychonic arrangement with new principles of matter, motion, space, and physical force. It also resonates with comparable interests among the learned friends of the ninth earl of Northumberland.[163] Strictly speaking, Bruce's sketch was neither Brunonian nor Keplerian but an attempt to bring together both—in a form to which neither Kepler nor Bruno themselves would have been comfortable granting full approval. Bruce created multiple, Sun-centered worlds, each with a moving power issuing from a sun that drives the planets. There was no privileged heavenly or elemental region but one unlimited space filled with elemental matter. In this account, Bruce seemed to sense only vaguely the problem of interactions among these individual worlds. His universe joined together physical elements with the barest hints of a quantitative approach, let alone the Keplerian musical harmonies that he had had several years to ponder. And, most remarkable, Bruce proposed that the Sun had its own rotational motion.

Had Bruce come in contact with Bruno's writings on infinite worlds before he left England around 1585, or perhaps after he became involved with the Pinelli Circle?[164] One inclines to the second alternative, as Kepler made no mention of Bruno in his copious letter to Bruce in 1599. Evidently it was not yet a topic of discussion for them. In either case, we do not know whether Bruce shared his bold world prospectus with Galileo or Magini, as he had with Kepler. There was plenty of opportunity to bring up the subject with Magini on the long carriage ride from Padua to Bologna, but Magini would hardly have been sympathetic to the Englishman's views. Similarly, we do not know whether Bruce spread other people's ideas around as he had accused Galileo of doing. As a political "intelligencer," it was his business to keep his ear to the ground. Did his political work also involve keeping an eye on the heavens?

Whatever the case may be, the very fact that a sometime member of the Pinelli circle was confiding such views to Kepler in late 1603 suggests that Giordano Bruno's ideas had received some hearing in Italy, even if now it had become much too dangerous to discuss them openly. But this ominously deteriorating intellectual situation in Italy was unlikely to have been the cause of the Englishman's decision to return to his homeland. More important, Bruce's political network was in upheaval. Anthony Bacon had died in May 1601. Two years later, in May 1603, James VI arrived to succeed Elizabeth on the English throne. After May 1601, Bruce seems to have acquired a new patron, reporting back now to Lord Burghley's secretary, the wealthy and influential Sir Michael Hicks (1543–1612).[165]

Unfortunately, after this letter, the brief trail left by Edmund Bruce dies out. Was he able to visit Kepler in Prague on his way back to England?[166] Indeed, did he ever return? There were many subjects that Kepler had mentioned to Bruce in 1603, among them the imminent appearance of a book on optics and a "new method" for approaching the motion of Mars. Kepler sent greetings to Magini and Galileo and dispatched his letter to Padua with Marcus Welser's brother, Matthaeus. If Bruce's return to England spelled the end of direct contacts with Kepler, it was only the beginning of Kepler's warm and friendly relations with the English.

GALILEO'S SILENCE ABOUT BRUNO

The Brunonian theme in the Kepler-Bruce exchange brings us back to the question of Galileo's silence over the news of Bruno's death. To cast any doubt on the idea that Bruno's execution was a topic of worried discussion in the last months of Pinelli's life, one is forced to make any number of improbable assumptions: for example, that word of the Nolan's demise had failed to reach Padua; that Galileo knew none of Bruno's works; that Galileo was too busy with his teaching and experiments to take any notice; and so on. There are also possibilities for which we lack direct evidence: could Galileo have been unaware that Bruno was held in a Venetian jail in 1592? Padua was in effect the Latin Quarter of Venice, and Galileo's numerous contacts with both clergymen and members of noble Venetian families included Alvise Mocenigo, a relation of Giovanni Mocenigo who had turned Bruno over to the Inquisition.[167] These men were all rich sources of political information in the Venetian Republic. Whatever specific ideas of Bruno might or might not have been discussed in the Pinelli circle, the lessons of his unfortunate ending cannot have been lost on the already cautious Galileo. The most obvious of these lessons would have been to take care about his public remarks, about references to the books he owned or was reading, and who knew about the topics in which he was interested. He might also have been wary about the kinds of explanations that would be most acceptable in natural philosophy. A great deal of Galileo's behavior after 1600 makes sense as the product of such deliberate circumspection, and some, at least, is probably to be attributed to his awareness of Bruno's fate. After the execution, any doubts he might have entertained about Kepler's cosmography and any personal insecurities he experienced in 1597 would be doubly reinforced by the unwelcome political atmosphere.

If Galileo indeed lacked exact details about how the Inquisition operated in Giordano Bruno's trial, he did not have to wait long to gain a firsthand acquaintance with some of its practices. Several documents discovered in 1992 by Antonino Poppi in the Venetian State Archives allow us a somewhat better vantage from which to understand Galileo's astrological activities and his earliest encounter with the Inquisition. Poppi's evidence shows that in April 1604, the Holy Office in Padua formally denounced Galileo's neighbor, the traditionalist Aristotelian philosopher Cesare Cremonini, and Galileo himself.[168] Cremonini was charged with teaching against the immortality of the soul; Galileo was charged with living in a state of heresy because he had allegedly maintained that the stars determine human fate. The Venetian government quickly and effectively defended both lecturers: it brushed aside the denunciation against Galileo as "extremely light and of no consequence" and that against Cremonini as motivated by "evil souls and self-interested persons."[169] Cremonini, it was affirmed, "had always lived 'catholically' and was a good Christian."[170] The immediate worry for the Venetian authorities was student unrest and the consequent damage to the university's reputation both at home and abroad. In the background lay smoldering tensions between the papacy and the Venetian Republic.

Galileo was denounced to the Holy Office by one Sylvester Pagnoni on April 21, 1604, just nine months after Bruno's works were placed on the Index. As Poppi surmises, Pagnoni was almost certainly the amanuensis who lived in Galileo's house for eighteen months between July 1602 and January 1604. This also happens to be the period to which Drake ascribes much of Galileo's work with pendulums and experiments with inclined planes. It also coincides with the period during which Edmund Bruce told Kepler that Galileo was spreading the latter's ideas as his own and the time when Galileo would have acquired Gilbert's *De Magnete* for his library. Nowhere in Galileo's correspondence for this period is there explicit evidence of this potent juxtaposition. But Pagnoni's report directly confirms what was at issue in 1604: neither Galileo's Copernican sympathies nor his studies of motion but his astrological and religious practices. "I have

seen him in his room making different nativities for different people upon which he [then] made his judgment. And he made these [judgments] one after another, and he said he had done so for some twenty years in order to make a living, and he maintained firmly and without question that his judgment[s] must come to pass."[171]

If the figure of twenty years is correct—and we have no further evidence to confirm it—then it would place Galileo's astrological activities as far back as his student years in Pisa (1583–84). And if one rejects this dating, one needs to explain why, suddenly and ex nihilo, Galileo began to engage in a new practice for which he had had no previous experience.[172] However, the serious accusation that Galileo believed that all his judgments "must come to pass" is obviously false: it was a standard charge against astrologers, with overtones of Arabic or Stoic fatalism. Pagnoni gave the further impression that Galileo's alleged conviction about his own astrological competence was not even merited. He reported that one of Galileo's clients, a German gentleman named Johannes Sweinitz, had complained that one of the nativities was not well done and made predictions quite contrary to what actually happened to him!

Pagnoni's testimony also sheds light on the ways family tensions could become a basis for inquisitorial mischief. He divulged to the inquisitor the disturbing revelation that Galileo's mother, Giulia Ammanati, had told him that her son never went to confession or took holy communion. But, in defense of his employer, Pagnoni said that he and Galileo had once gone to mass together to observe a holiday and that he had sometimes seen him attending with his mistress Marina Gamba, a Venetian woman, who lived nearby. While revealing such potentially incriminating information about Galileo's behavior as a Christian, Pagnoni said quite unambiguously that "concerning matters of faith, I have never heard him say anything bad . . . in matters of faith I believe that he believes."[173]

THE COPERNICAN PROBLEMATIC
AND ASTROLOGICAL THEORIZING
AFTER BRUNO'S TRIAL

If Galileo and Kepler were both engaged in astrological *practice* in the same period of their most deeply transformative work in mechanics and planetary theory, then one may ask, what was the

TABLE 5

Works published under Kepler's and Galileo's names, 1601-10

Year	Kepler	Galileo
1601	*De Fundamentis Astrologiae Certioribus*	
1602	*Calendarium und Prognosticum auf das Jahr 1603*	
1603	*Prognosticum auff das Jahr. . . 1604*	
1604	*Ad Vitellionem Paralipomena, quibus Astronomiae Pars Optica traditur.*	
	Gründtlicher Bericht von einem ungewohnlichen Newen Stern	
	Prognosticum auff das Jahr 1605	
1605	*De Solis Deliquio Epistola*	
	Calendarium und Prognosticum auf das Jahr 1606	
1606	*De Stella Nova*	*Le operazioni del compasso geometrico e militare*
1607		*Difesa contro alle calunnie ed imposture di Baldessar Capra*
1608	*Aussführlicher Bericht von dem newlich im Monat Septembri und Octobri diss 1607 Jahrs erschienenen HAARSTERN oder Cometen, und seinen Bedeutungen*	
1609	*Astronomia Nova Αιτιολογητοσ, seu Physica Coelestis, Tradita Commentarii de Motibus Stellae Martis*	
	Phaenomenon Singulare seu Mercurius in Sole	
	Antwort auff Röslini Discurs	
1610	*Tertius Interveniens, das is, Warnung an etliche Theologos, Medicos und Philosophos*	*Sidereus Nuncius*
	Dissertatio cum Nuncio Sidereo nuper ad mortales misso a Galilaeo Galilaeo	

theoretical grounding of such astrological exercises within a Copernican arrangement? We are speaking here of the years immediately after Bruno's death and just prior to 1609–10—the period that ended with Galileo's first telescope observations and their dramatic presentation in the *Sidereal Messenger;* the period culminating with Kepler's "war on Mars" and its publication as the *Astronomia Nova* in 1609. These well-known events tend to be lumped together within a common history of "Copernicanism," but that designation overlooks the obvious difficulty that they do not easily fit together. Why, after all, did Galileo altogether neglect elliptical astronomy when (like others in the seventeenth century) he might have dissociated the elliptical description of the orbit from Kepler's moving power? The chronology of our question deliberately avoids this commonly vexing question. Instead, it calls immediate attention to a neglected dimension of the strikingly different paths that Kepler and Galileo were already pursuing prior to the appearance of two of their best-known published works. Once the followers of Copernicus insisted on a physical grounding for the heliocentric ordering, once they insisted on that ordering as a true representation of the heavens, then a logically related question could be raised: What sort of theory of astral influence, if any, would be compatible with a moving Earth and a central Sun? This problem is the one that neither Copernicus nor Rheticus openly addressed, that midcentury star-theorizers like Dee and Offusius avoided by retaining a centrally resting Earth, and that the Copernicans Bruno and Rothmann rejected or avoided for other reasons.

From a contemporary perspective, the period 1601–1610 displays a remarkable contrast in Kepler's and Galileo's public productions. Kepler, the emperor's *mathematicus*, published nearly everything that he wrote, in spite of serious obstacles erected by the heirs of Tycho Brahe.[174] Many, indeed most, of these works concerned astrological theory and practice. By contrast, prior to the *Sidereal Messenger*, Galileo published under his own name only two very minor works: one a description of a versatile measuring and calculating instrument that he had invented and called the geometric and military compass, and the other a defense of his claim to priority in having invented this instrument.[175] Unlike Kepler, in the same period, Galileo never published any annual prognostications or any of his astrological judgments. The birth charts he prepared were intended only for private use by the individuals for whom they were cast. And there is no evidence that Galileo concerned himself at all about giving astrology "firmer foundations," let alone Copernican ones. He did not pursue any astrological entailments of the Copernican theory, and it is not obvious how such an investigation could have been reconciled with his Archimedean physics. Equally interesting are the generic domains in which he and Kepler chose to write.

Further explanation of the differences between these two publishing trajectories will be taken up in chapter 17. Here, it may be noted simply that throughout the decade, Kepler unflaggingly pursued his radical and comprehensive reform of the science of the stars, extending new physical or archetypal principles not only to the ordinary course of the planets but also to such extraordinary appearances as the new star of 1604 and the comet of 1607. Just as Galileo was drawing up horoscopes for private clients in Padua and Venice, Kepler was making public his annual prognostication for 1602 in Prague. However, unlike his limited theorizing in the Graz prognostications, all written in German for a general audience, the Prague forecast was composed in Latin for a learned audience. Here, Kepler followed Maestlin's earlier advice to limit the philosophizing in his *practicas*. Now, in 1602, he joined his forecast to an extensive theoretical section in a treatise titled *De Fundamentis Astrologiae Certioribus* (Concerning the More Certain Foundations of Astrology).[176]

KEPLER'S CONTINUING SEARCH FOR ASTROLOGY'S FOUNDATIONS

If the Earth moves, then either it is subject to the effects of astral influences or it is not. Copernicus remained silent on this question; but Kepler did not. He was the only Copernican to develop a heliostatic astrology. The seventy-five theses of Kepler's early, seminal work show him in full philosophical garb, working to establish his basic reform of the principles of astrological theory, or what he neologized as *Cosmotheory*. Following Ptolemy, he did not pretend to make astrology into a fully demonstrative science but only to give it "more certain foundations." His strategy resembled that of the *Mysterium* in its ruthless disparagement of ordinary prognosticators who, like ordinary astronomical theorizers, he viewed as failing to provide the reasons that grounded their practices. Those reasons, as first established in the *Mysterium,* were archetypal and physical. The physical approach was crucially motivated by Pico in its adoption of light as the singular heavenly condition for producing terrestrial effects.[177] However, Kepler was also driven by the ideal that if there are archetypal principles for astronomy, then such principles must also exist for astrology. This contention was Kepler's particular way of dealing with the problem that Copernicus had avoided. But the solution was not at all self-evident. Although Kepler's astrological theses obviously needed to be compatible with a moving Earth, he realized that he could not make the (continuous) solar moving power the efficient cause of both the Earth's annual motion *and* human and meteorological vicissitudes. Likewise, although his 1599 correspondence shows him playing with musical ratios and other archetypal principles, Kepler did not argue that the polyhedral distance and ordering determinations in the *Mysterium* were responsible for producing terrestrial effects. Did it make (astrological) sense that Saturn's influences, like the solar *motrices*, were consistently weaker because they were farther from Earth than Mars?

In the end, Kepler's reform would amount to a radical revision of Ptolemy's planetary order coupled with a moderate revision of his astrological theory. He decided that the efficacy of the planetary powers must lie in a revised version of the aspects, the distances of which were angular but not linear. Yet if this astrology was compatible with a moving Earth, it was not uniquely entailed by the Earth's motion. Even in the *Harmonice Mundi*, where Kepler famously announced a new specification of the period-distance relation (in the squares of the periods and the cubes of the distances), he did not use that discovery to argue for the necessity of his astrology. More to the point, the aspect-based astrology enabled him to break with the power-distance relation earlier favored by Jofrancus Offusius and John Dee. Nevertheless, such an aspectual astrology still presented its own difficulty: persuaded by Pico (and Maestlin) to reject the intrinsic power of the zodiacal divisions, Kepler needed to find a source of astral efficacy that would depend on a relationship between the aspectual angles independent of the twelve classical divisions of the zodiac.

If the solution to that problem did not lie in the zodiacal signs, then it had to lie somewhere in the angles formed by the rays between any two planets and the Earth (a triangle).[178] But because at any given moment *some* angular relationship between the planets and the Earth existed, how was one to know which ones were archetypically efficacious? For a geometrizing God who had already used the best of all possible geometric solids to space the orbs, the spacing of the rays had to lie in the best of all possible plane figures, "images" of the regular solids (thesis 37). This Platonizing gesture is particularly evident in the effort to derive astronomy and astrology from the same kinds of principles.

Within the bounds of traditional astrological theory, Kepler also moved to give the Sun a privileged position. Here his reforming move was bolder. He broke with the foundational notion that heating and humidifying, cooling and drying are separate functions of the Sun and Moon. Light alone was seen to be the common cause of heat (a quality proper to light) and of humidity (the result of reflected light), while cold and dryness were seen to be effects produced by the total absence of light.[179] This scheme naturally assigned a privileged place to the Sun: the Moon (no longer a planet) and the wandering planets received light from the Sun (in which case the light appeared cloudy) but may also have had their own intrinsic light (in which case it appeared bright).[180]

Kepler also proposed that the faculties of warming and humidifying should be reduced to another common category, the strengths of the planetary powers. In turn, these powers could be classified according to three degrees (excess, mean, and defect) in such a way that, in principle, fifteen different states were theoretically possible but only seven were realizable. The cause of this variation of the planetary forces was somehow to be connected to the way in which light was received, how it behaved as it reflected off different kinds of surfaces. "I am concerned with the kind of reflection that we see from a wall, even one with an uneven and rough surface, which reflects incident light from any one of its points over a complete hemisphere and imbues the reflected light with whatever color it may itself possess" (thesis 26). Reflections not just from walls but also from other kinds of surfaces— clouds (at sunset and sunrise, during and after a rain), the Moon and Sun in eclipse—constituted examples of visual effects that yielded variations in colors. In all these cases, the strength of the light ray diminished as it was bent—either by reflection or refraction—and produced a range of colors. From such information, one could infer the properties of the surface of a planet. For example, according to Kepler, a white face reflected in a black steel mirror appeared red; likewise, as Mars appears to us reddish and does not humidify greatly, its surface must be black. Kepler also attributed an intrinsic light to the planets, a power that is not on the surface but is transparent and "derived from the internal structure of its body" (as with a gemstone). He thought that this source of light could explain the planets' capacity to impart heat (thesis 29).

As a naturally occurring phenomenon of color production, the rainbow seemed to be a crucial analogy for explaining how the planets could yield different astrological effects. The issue is not whether Kepler properly understood how rainbows are produced (he did not) but that he associated the bending of light as it entered a different medium with the production of a sequence of colors, and colors were (already) linked to specific planets and their powers.[181] Can it be an accident that Kepler was then already at work on a new treatise concerning the foundations of optics at the very moment when he was trying to establish new foundations for astrology?

The behavior of light, however, was not the only area in need of reconsideration. The urgent question was what to do with the receptor of the astral rays, a Copernican Earth that was no longer the sole and unique destination for heavenly influences, as Tycho had wished, and which was now moved by incorporeal species emanating from the Sun. Kepler's proposal was to preserve the Earth's specialness while allowing its motion. How could this be done? Like Scaliger and Gilbert, Kepler turned over the causal agency to a soul, but not, in this case, a planetary soul. The Earth, he averred, has an "animal faculty" that practices geometry, in the image of God (of course), and is "roused to action by this celestial geometry or harmony of aspects" (theses 39 and 40). Effectively, the heavenly bodies were necessary but not sufficient to produce terrestrial consequences. The astral rays had to come together at certain angles, but the active, responding agency of the Earth's soul was necessary to read the harmonic instructions. For Kepler, this capacity was "instinctive" rather than ratiocinative. As such, the Earth possessed a kind of seventeenth-century genetic code: "Not human, nor properly speaking animal, nor like that of a plant, but of a particular kind which is defined from its activities, as are other kinds of animal faculty" (thesis 42). The plant cannot think, but nonetheless its "plastic faculty" can follow God's instructions (thesis 43) in producing five-part structures. Likewise, peasants do not philosophize (engage in formal reason), yet their ears respond to the harmony not because of the "soothing caress given to the ears" by the mixing together of musical notes (material necessity) but, as with plants, because "some geometrical relation connects the form to the musical consonances. This relation is familiar to everything else in the world, particularly to souls, and was indeed called harmony by some of the Ancients" (thesis 43). What was true for plants and peasants could now be extended to the Earth's capacity for geometry in its "vegetative animal force": "Although this force is always at work, it is stimulated most when it is fed with this kind of nourishment of aspects. Thus, just as the ear is aroused by a consonance, so that it listens carefully, and so hears that much more (seeking pleasure, which is the perfection of sensation), similarly the Earth is stimulated by the geometrical convergence of vegetative rays (for we have noted that they warm and humidify), so that it concentrates diligently, or to a greater extent, upon its vegetative work, and sweats out a large quantity of vapors."[182]

There now remained Pico's criticism that astrology could not account for specific differences among its effects. If the planetary-ray geometry affected all parts of the Earth equally, then the weather vapors ought to be the same everywhere, and the Earth should experience global rainstorms. In response to this objection, Kepler argued that because the parts of individual bodies have different dispositions, varying with time and place, different effects may follow from the same aspects.[183] For example, "when in Spring humours are abundant in the Northern hemisphere because of the increasing height of the Sun . . . then even the lightest aspect of any planets will spur that faculty in the Earth into action and it will sweat out a quantity of vapours to engender showers. At another time or place a far stronger aspect does indeed stimulate the Earth but since material is lacking it produces little result."[184] Other specific environmental conditions, internal predispositions of the Earth—such as an excessively dry or rainy year—might be longer-lasting than any particular planetary aspect. In such a case, "you may see that whenever even the lightest aspects occur they move quantities of rain or wind. Such a year was this 1601. In another year there is so much dryness that on a day of aspects nothing can be discerned except small clouds or smoke instead of vapors. Such was the year 1599."[185] And, of course, there were eclipses, those alignments of the Earth, Moon, and Sun given such astrological prominence by Ptolemy. For Kepler, the sudden interruption of light during an eclipse caused the Earth's animal faculty to experience "something like emotion." The Earth's humors, like those in the human body, waxed and waned, increasing and decreasing with the Moon's cycles. Likewise, "mariners say that the greatest tidal movements of the sea come back to the same days of the year after nineteen years; and the Moon, which governs humours, seems suited to play a part in this business, which involves an excess or defect in humours."[186]

In his prediction for 1602, Kepler marked out the domain of legitimate prognostication. His position represented an expansion of the standard formula that the stars cause only dispositions to action—"some excess in the inclination of souls" (thesis 69), while permitting a domain of freedom or unpredictable action. A better astrology was possible, as "Ptolemy's rules are ill-defined and not in close conformity with Nature" (thesis

63). For Kepler, however, the domain of freedom was much larger than for ordinary astrologers, "for Man is the image of God, not merely the offspring of Nature" (thesis 69). The size of harvests, for example, depended partly on "accidental causes" (thesis 64). A very bad wine crop? "Because the year was cold and damp." A good year suddenly interrupted by a frost, hailstorm, or flood? An unforeseen wind (thesis 65). A sudden earthquake in the Rhine Valley (in September 1601), an area not accustomed to tremors? "There is no star for earthquakes, but from observation of the world and of all ages" (thesis 70). And on the crucial subject of politics and war? "A matter for the judgment of those who are experienced in politics, for their power of prediction is no less than that of the astrologer. For the state has a will [*morem*], if I may call it such, no less [significant in this matter] than the influence of the heavens" (thesis 69).

In short, one could have a new kind of aspectual astrology, based on physical and archetypal causes, but, as with the current notion of the gene, such causes would still account only for general dispositions. Or, as Kepler put it in a postil to his *Harmonice Mundi*: "No individual results are due to the stars."[187] For just as individual light rays are bent as they enter a (variable) atmosphere, and then again as they enter the humors of a particular eye, so the farmer, the physician, and the statesman would be well advised to rely more on their own worldly experience than on the stars in making particular judgments. Likewise for Kepler himself. Could the stars account for his own astronomical discoveries?

An astrologer will seek in vain from the arrangement of the stars at my nativity the reasons for my discovery in the year 1596 of the proportions between the heavenly spheres; in the year 1604 of the way of seeing; in this year, 1618, of the reasons why to each planet has fallen an eccentricity of a particular size, neither smaller nor larger; in the intervening years, of my demonstration of the physics of the heavens, and of the ways in which the planets move and of their true motions; and lastly of the basis of the influentiality, which is metaphysical, of the heaven on this Earth below. Those things were not due to the influence of the characteristics of the heaven on that little flame of my vital faculty which had just been lit and brought into action; but in part they were hidden in the innermost essence of my soul, according to the Platonic theory of Proclus, and in part they were received within by another route, through my eyes of course. The one and only way in which the arrangement of the stars at my nativity operated was that it both polished up the little flames of talent and judgement and urged my mind on to untiring toil, as well as increasing my desire for knowledge. In short, it did not inspire my mind, or any of the faculties stated here, but roused it.[188]

The stars, it might be said, underdetermine specific differences. Here again was Kepler's selective middle way: with Pico, he rejected the total domination of the stars and sustained a confidence in human reason and capacity. Against Pico, he saw the stars as still necessary for inspiration, even if their influence was not sufficient to explain his particular discoveries. And much the same was true of his advice to rulers. Judgment based on the stars ought to be strictly managed so that it did not hinder what politically wise men knew directly from their own daily experience.[189] Politics, in short, should not be a monopoly of ordinary astrologers any more than a theology of the natural world should be the unique preserve of ordinary theologians. Or, as Kepler once quipped: "I act as do the Jesuits, who correct many [errors] that they may make men Catholic. More correctly, I do not act in that way, for those who defend all the nonsense are like the Jesuits. I am a Lutheran astrologer, who abandons the nonsense and keeps the kernel."[190]

14

The Naturalist Turn and Celestial Order

CONSTRUCTING THE NOVA OF 1604

THE PREDICTED CONJUNCTION OF THE THREE SUPERIOR PLANETS AND THE UNFORESEEN NOVA OF 1604

Sixteen hundred and four was a year of great astrological import for the prognosticators. It was supposed to mark the return of Saturn and Jupiter to conjunction in the Fiery Trigon (at 8° in the sign of Sagittarius) after eight centuries (see figure 75). One defined this astrologically significant zone by connecting the midpoints of every fourth zodiacal sign (each separated by 120°). Each trigon or triplicity thereby contained three signs with the same elemental and gendered nature. For example, Aries, Leo, and Sagittarius together formed the triangle of aggressive, masculine fire signs, while Taurus, Capricorn and Virgo constituted a grouping of receptive, feminine earth signs. Every twenty years, the two most distant planets, Jupiter and Saturn, entered conjunction within one of the signs of one of the four triplicities. After two centuries, the conjunction slipped into the next grouping of signs; and only after eight centuries did the same configuration recur in the original arrangement of signs. The year 1604 was also unusual in another way: Mars, the third superior planet, was expected to join the other two in Sagittarius on September 29. Such a conjunction was already a rare enough astrological event in the ordinary course of nature— a great opportunity for the ephemerists and an occasion to voice their usual opinions and disagreements about what it could mean. All the

more reason for astonishment, when, just over a week after this conjunction, various observers noticed a bright nova, never before seen, in the constellation of the Serpent.

A rush of publications ensued, although nothing that would begin to match the outpouring that had accompanied the 1572 nova. Kepler was among the first of those who issued a brief description and an astrological judgment, titled a *Report* (*Bericht*). Composed in the vernacular, it was essentially structured as a *practica*.[1] Besides glancing references to Leovitius ("Cyprian") and to Brahe's *Progymnasmata*, Kepler referred to no other astral writers or observations.[2] Most of his *Report* was devoted to the new star's import, which, he said, was of "more of a Martial nature as it shone in the place and on the day of the conjunction of Jupiter and Mars."[3] In his 1606 Latin treatise *On the New Star* (*De Stella Nova*), by contrast, Kepler composed a work for the learned, filled with a panoply of theoretical disputations that transformed the nova from an entity to be "reported" and "judged" into one to be explained— a topic for natural philosophy and the theoretical parts of the science of the stars.[4] Here he positioned himself as imperial mathematician in a wider sense—as the one who, like his Rudolfine predecessors Brahe and Hagecius, assembled, reported and, in a sense, presided over the observations of other practitioners.[5]

It is somewhat surprising, considering the remarkable conjoining of celestial events, that this occurrence precipitated nothing approaching the

mafculina. Aries igitur, & cum eo triangulantia, Leo & Sagittarius funt ignea. In primo quadrante principium Aries, in secundo me-

75. Signs and triplicities within which conjunctions of Saturn and Jupiter were predicted to occur, approximately every twenty years, from 1603 to 1763. Kepler 1606. Image courtesy History of Science Collections, University of Oklahoma Libraries.

cascade of writings that had occurred in association with the earlier nova. However, in 1602, the 1572 nova was effectively transformed into a new kind of event when Tycho's son-in-law, Franz Tengnagel, pushed through publication of the *Progymnasmata*. To his dismay, Kepler's own contribution to the preparation of the work was rendered invisible.[6] He had been responsible for an appendix at the end of the *Progymnasmata*, but to the reader his identity remained concealed.[7] Nevertheless, the appearance of the *Progymnasmata* had an unexpectedly significant impact. For the first time, the great majority of writings about the 1572 nova—hitherto scattered among many small, ephemeral publications—appeared together in a single massive representation. The whole work was comparable in intent to Tycho's *De Mundi Aetherei Recentioribus Phaenomenis* of 1588, but overall it included a much wider range of individual authors. By the time that it appeared, many of the 1572 nova writers, including Tycho, had passed from the scene. As a result, the *Progymnasmata* did not so much assist memory as sup-

plant and shape it. Now even those who had not lived through that earlier moment had available to them a body of specific materials—parallactic techniques and tabulation of the earlier nova's parameters—all framed by Tycho's arrangement of the sources. Furthermore, the assembly of writings in the *Progymnasmata* constructed for its readership an image of a European-wide community of observers—a representation of a collective enterprise.

The authority of the *Progymnasmata* was further enhanced by the publishing privileges granted by both the Emperor Rudolf II and King James VI of Scotland. Rudolf had a special relationship with Tycho, the emperor's mathematician; and James already had a special relationship with Denmark, having married Princess Anne (of Denmark) in Oslo in November 1589. A few months later, on March 29, 1590, James visited Uraniborg, where he received a grand tour of the island and was lavishly entertained.[8] In 1593, he contributed two poems to the *Progymnasmata*'s dedicatory apparatus; and, in the publishing privilege, which protected Tycho's work in Scotland for thirty years, he referred glowingly to his visit, witnessing "with our own eyes and ears."[9] One may well speculate about what the rising court official Francis Bacon might have made of James's presence at the front of the *Progymnasmata*, for Bacon made it his business to track the king's likes and dislikes; and just three years later he dedicated to the king his *Advancement of Knowledge*.[10] As will be seen in the next chapter, James's intervention did not go unnoticed by Kepler.

In another crucial way, however, the appearance of the much-delayed *Progymnasmata* established a new ground for many of the writings about the 1604 event. The nova of 1604 lacked the full burden of novelty of the earlier occurrence in the sense that the possibility of this kind of event was now established. Although it was widely believed that God could do anything, no one in 1572 had foreseen that he would choose to make a star appear where none had existed before. The 1604 nova, by contrast, appeared not as a solo event but as a kind of divine exclamation point for the conjunction that the prognosticators had anticipated. Perhaps this unanticipated union of ordinary and preternatural circumstances is one reason that the writings about the 1604 nova took a more urgently naturalistic turn. The early seventeenth-century controversialists (whatever positions they staked out) were as much preoccupied with the question of parallactic displacement as their predecessors, but many were no less concerned with defending physical explanations of the star's origins. The new star of 1604 pushed open the question of whether a deviant and unanticipated occurrence warranted an immanent, naturalistic explanation rather than one that was transcendental and miraculist.

GALILEO AND THE ITALIAN NOVA CONTROVERSIES

In Italy, the nova's appearance led quickly to an outbreak of local controversy, with universities at the center of contention. In early December 1604, Galileo's lectures at the university in Padua became the occasion for a series of exchanges.[11] Independently of that event, on December 23, a former student of Clavius's, the mathematician Odo van Maelcote, spoke at the Collegio Romano.[12] Between 1605 and 1606, Padua, Florence, Rome, and Venice were witness to no fewer than ten publications—some spaced as closely as a month apart. In contrast to contemporary discussion of the 1572 episode, these exchanges had a noticeably sharp edge. With Lodovico delle Colombe's reply to Alimberto Mauri's *Considerazioni* in 1608, the nova controversies petered out, although the underlying issues that they had articulated did not.[13]

The north Italian nova authors, or Novists, as I shall call them for analytic convenience, testify to abundant professional tensions. Most of these writers knew (or thought they knew) one another personally. Two areas of tension recurred. One concerned the relative social and intellectual status of writers in the genres of natural philosophy and mathematics. A second pertained to a sense of social marginality on the periphery of close-knit academic circles in Padua, Florence, and Rome. Many of these writers used the nova to proclaim the superiority of their disciplinary skills and discursive prerogatives—or to tear down those of their opponents. Even more than in the 1572 episode, the legitimacy of using parallax to make physical claims about the superlunary region came to be seen as emblematic of what mathematicians knew how to do and philosophers did not.

All sorts of authorial dodges and stratagems, many heavily personalized, contributed to the contentious atmosphere. Among various instances, Galileo's and Van Maelcote's lectures were delivered before public university audiences, but both withheld publication. The Milanese physician Baldassare Capra, who claimed to have attended Galileo's performances, quickly made his own representation of what happened—the only published reference to the event.[14] Meanwhile, at least two authors wrote under pseudonyms. Suspicions about their identities abounded (and, for different reasons, still do).[15] Lodovico delle Colombe, a gentleman and member of the Academy of Florence who published under his own name, stated his belief that Galileo was hiding behind the print name Cecco di Ronchitti. Subsequently, Colombe's own treatise on the new star was attacked in print by one Alimberto Mauri; neither contemporaries nor historians have been successful in identifying anyone by that name. And Galileo believed—with some reason, it seems—that Capra's teacher, the German prognosticator Simon Mayr (Simon Marius), had an important hand in motivating, if not actually writing, *Consideratione*

TABLE 6
The Italian nova controversies: Published treatises

January 15, 1605[a]	Antonio Lorenzini, *Discorso intorno alla nuova stella* (Padua)
February 16, 1605[a]	Baldassare Capra, *Consideratione astronomica circa la stella nova dell'anno 1604* (Padua)
End of February 1605[a]	Cecco di Ronchitti, *Dialogo . . . in perpuosito de la stella nova* (Padua)
December 1605	Raffael Gualterotti, *Sopra l'apparizione de la nuova stella* (Florence)
1605	Johann Van Heeck (Heckius), *De nova stella disputatio* (Rome)
1605	Astolfo Arnerio Marchiano, *Discorso sopra la stella nuova* (Padua)
23 December 1605/early 1606	Lodovico delle Colombe, *Discorso* (Florence)
June 1606	Alimberto Mauri, *Considerazioni . . . sopra alcuni luoghi del discorso di Lodovico delle Colombe* (Florence)
1607	Galileo Galilei, *Difesa contro alle calunnie e imposture di Baldessar Capra* (Venice)
1608	Lodovico delle Colombe, *Risposte . . . alle considerazioni di certa maschera saccente nominata Alimberto Mauri, fatte sopra alcuni luoghi del discorso dintorno alla stella apparita l'anno 1604* (Florence)

[a] Date submitted to printer.

Astronomica.[16] To add to the behind-the-scenes maneuvering, the Dutchman Johannes van Heeck (1579–1616) leveled a strident polemic from Prague against the "enemies" of Aristotle—"babbling New Philosophers," as he called them, who, in their "profane ignorance" and "stupid ostentation," were said to detract from the "sacred study of philosophy." Before Van Heeck's work could appear, however, Federigo Cesi, founder of the Accademia dei Lyncei in 1603 and the work's dedicatee, managed to moderate the tone of the manuscript by editing out such intemperate polemical language.[17]

This atmosphere of controlled tensions and explanatory caution suggests that Galileo was not alone in withholding public commitment to contentious issues, including the Copernican question. The net effect of the shadowy camouflaging of authorial identities and interventions was to keep discussion at a decidedly local level. At the same time, something interesting was happening at a deeper level of discussion. The nova was becoming a contested object at the interface of two crucial domains of natural events, the miraculous and the ordinary. If, like the 1572 nova, it was a miraculous, divinely created event in the stellar region, then it fell into the descriptive domain of the Nullists. But many of the more daring of the earlier writers (including Maestlin,

Hagecius, and Brahe) had not attempted any sort of physical explanation. In the end, they had shunted the explanatory task over to the domain of theology while criticizing traditionalist natural philosophers for maintaining as a fact that the heavens were immutable. By 1604, the problematic had shifted: many now believed that God had chosen to proceed through intermediate, recurring physical causes rather than directly through a single, miraculous ad hoc creation. Van Heeck, for example, argued that miracles are special: God does not "rashly" resort to them every day, and, hence, wherever possible one should look for natural causes of unusual phenomena. This put the nova squarely into the ordinary course of nature, where, as in the early considerations of Copernicus's theory, the struggle over disciplinary prerogative recurred.[18] Who had the best tools for explaining the *nova*, the mathematicians or the natural philosophers? If the former, then was the nova just a heavenly sign to be interpreted, or did it also have a causal influence on earthly affairs? If the latter, then which natural philosophers offered the most persuasive explanations?

Not unexpectedly, Galileo was in the antimiraculist camp but, because of the paucity of surviving materials, his role in this episode defies full elucidation. Compared to the surviving publications mentioned above, very little can be said about the structure and contents of Galileo's December 1604 lecture series. Although in 1607 he referred to "my three long lectures to more than a thousand auditors," we know little more than that he provided a report of his observations and that he claimed the nova to be above the Moon.[19] The extant exordium to one of the performances begins conventionally: that he saw Jupiter and Mars on October 9, 1604, at 5 P.M., in conjunction and about 8° from Saturn; and that on the following day, just after sunset, he saw "the new light" for the first time.[20] Because the planets were joined in 19° Sagittarius, he placed the new light at 18°, thus making a figure composed of four points of light. He said that some regarded the event as a "divine miracle"; others, moved by "vain superstition," saw it as a "portentous prodigy, messenger of a bad omen"; still others saw it as a true star existing in the heavens; and some thought it to be a "burning vapor" near the Earth.[21] Aside from such comments, there is no evidence that Galileo included an astrological judgment—which is not surprising in view of the fact that his brush with the Venetian Inquisition had occurred just a few months previously—and he made no attempt to link the nova to larger questions about planetary order or to the size of the universe.

The problem of situating Galileo's encounter with the nova in the immediate years before the telescope must now confront Stillman Drake's influential picture of an embattled Galileo, one who was already, in this early period, at war "against the philosophers" and who after 1605 "lost faith" in using the nova to confirm "Copernicanism" by "direct observation."[22] Drake's picture foreshadows Galileo's satirization of the Simplicio figure in the *Dialogue concerning the Two Chief World Systems.* Yet Galileo's initial exchange was not with "the philosophers." It was not Galileo's friend and colleague Cremonini or even the phantom philosopher Lorenzini who initially challenged Galileo by contesting his priority in observing the nova; it was rather Baldassare Capra, likely induced by his tutor, the sometime prognosticator Simon Mayr of Guntzenhausen (1573–1624).[23]

Capra's attack did not dispute the existence of the nova but rather the credit for having first observed it. He criticized Galileo's technical competence as an observer, but the real thrust of his attack was a narrative of moral transgression—moral in the sense that Capra accused Galileo of appropriating the initial observation of the nova from someone else, the Venetian nobleman Giacomo Cornaro. Cornaro was reported to have said that "he wished for the Excellent Galileo to see it."[24] But, in fact, so Capra relates, he himself had already seen the star on October 10—with Simon Mayr and "Signore [Camillo] Sasso, a gentleman from Calabria, unskilled in astrological matters"; and it was again Capra who allegedly told Cornaro about it a few days later and directed his attention to the star's coordinates. Galileo had allegedly failed to acknowledge his obligation to Cornaro and, by implication, to Capra. Capra apparently took the omission as an insult and used it to call into question Galileo's moral integrity in assigning credit, almost as though the observation was Capra's intellectual property. Unlike the servants who functioned as witnesses in the Tycho-versus-Ursus episode, Capra's witnessing narrative involved the presence of Venetian patricians. But for the time being, Galileo simply ignored the assault.

The curious character of Capra's initial attack on the leading mathematician of the Padua *studium* once more reminds us of the small scale of the venues within which early modern intellectual elites circulated. This was still true in the university towns of Europe a century after Copernicus's studies, and perhaps more so in the close quarters of the Italian *studia,* where the probability of people getting into one another's business was considerable.[25] Capra's attack suggests a desperate attempt to gain for himself some kind of status in university circles. For, in other respects, Capra and Mayr had no serious disagreements with Galileo about the preeminence of mathematicians and of their deployment of parallax. Indeed, it is easy to imagine that they could all have been united against a common enemy, as much of the *Consideratione* belittled Lorenzini as a philosopher and deemed him incompetent to make claims about the positions of objects in the heavens.

Capra's sometime mentor, Simon Mayr, was certainly not a traditional academic natural philosopher. The margrave of Brandenburg-Ansbach had sponsored Mayr's studies in mathematics and astronomy at Heilsbronn. Mayr became Ansbach court mathematician in 1599, and in the same year Kepler heard that he had just published a *New Table of Directions* for computing nativities.[26] With a letter of recommendation from the margrave, he went to Prague in May 1601, where he met Kepler and David Fabricius at Tycho's residence.[27] That he arrived in Padua in December 1601 to study medicine already suggests the direction of his interest in astrologically based medicine. Because of his Prague contacts it would not be surprising if he had acquired the *Progymnasmata* soon after its appearance.[28] Also, because he arrived with mathematical skills, it does not seem far-fetched that he might have met Galileo or attended some of his lectures. Sometime in 1602 he began to tutor Capra, whose interest in the science of the stars was also tied to medical ambitions.[29] Capra described him as "my dearest teacher" ("mio carissimo Maestro in questa profesione")—possibly as a *repetente*, as Mayr held no formal position on the faculty. In 1604–5, Mayr was elected *consiliarius* of the German nation, which makes it all the more likely that he was in attendance at Galileo's nova lectures.[30] That was evidently the initial site of provocation. In any event, Capra's debt to Mayr seems evident, as he invoked both Mayr's authority and that of Tycho Brahe ("most noble, learned, and ingenious") to support his contention that mathematicians deserved credit for demonstrating the nova's position in the starry heaven.[31] Like Van Heeck, Capra left to natural philosophy (rather than to miracles) the task of explaining the nova's generation.

The example of Capra and Mayr shows how parallactic observation was providing opportunities for prognosticators to engage in physical speculation to a far greater extent than in 1572. Having criticized Galileo's observational competence, Capra poked holes in Lorenzini's physical claims, in the course of which pursuit he created a little physical digression of his own. His critique concerned exhalations and light and raised arguments of the following sort: if the nova moved uniformly with the fixed stars, then it could not be an earthly exhalation. Likewise, the planets are not composed of exhalations, because they do not lose their brightness as they move circularly around the Earth. Furthermore, if the nova was generated by the union of light rays from different stars, then why aren't the stars constantly making novas? And how can something that is immovable (the nova) be caused by something that moves (the fixed stars)? Also, what appear to be spots (*machie*) on the moon are really nothing but vapors (*vapori*) dissipated by the

Moon's light; and it is unlikely that the unilluminated, opaque part of the Moon can disrupt vapors. Thus, because the nova was exposed to the Sun's rays for more than a month and did not dissipate, it could not have been an exhalation. In the same way, light from the stellar galaxy does not dissipate exhalations even though the stars are much greater. Light, exhalations, celestial mutations: "I don't know whether this is a way of philosophizing or jesting," Capra sighed. But there could be no doubt that the star had been generated in the heavens and that natural philosophers needed to find a way to explain its creation rather than obstinately (*ostinatamente*) persisting in believing that there could be no alteration at all. Another way had to be found to "save these accidents."[32]

On that note Capra left his physical speculations, but he briefly joined them to an astrological one. For Capra, as for Tycho in 1572, the nova was a "meaningful" sign (*significatione*) as evinced by the fact that the time between the appearance of the two novas (thirty-two years) was almost the age of Jesus Christ (thirty-three years). But Capra imputed no apocalyptic connotation to it, and he foresaw no "influences" on the "mysteries of religion." Unlike Tycho's interpretation of the nova in 1572, Capra's astrological treatment was trivially brief and restrained.

Galileo crisply and sarcastically annotated his copy of Capra's work but chose not to publish a response—or so it appears. In fact, before he had a chance to enter the fray, Cecco di Ronchitti's *Dialogue concerning the New Star* appeared. If this work was written by Galileo under a pseudonym, as some suspect, it would provide important evidence for understanding his views and strategies concerning the nova. And, in fact, the Cecco dialogue strongly evoked elements of the literary form and style later employed by Galileo himself. Antonio Favaro speculated that Ronchitti was indeed a nom de plume for both Galileo and his sometime acquaintance, the Benedictine monk Girolamo Spinelli.[33] The proposal is that Galileo provided the ideas and Spinelli composed the treatise.[34] Stillman Drake went further: he was convinced that the Cecco text "was certainly Galileo's."[35] Contrary to Drake and Favaro, however, Marisa Milani has argued persuasively that Ronchitti was a pseudonym not for Galileo but for Spinelli alone.[36] Although Spinelli was in some way part of Galileo's circle of acquaintances, this identification fails to explain why Spinelli wrote under a pseudonym; and if he was simply a surrogate for Galileo's views, why did Galileo need a stand-in to write under a concealed identity?

Identifying Cecco as Spinelli allows us to ask who was the opponent in the dialogue. Like Capra, Cecco di Ronchitti directed his attack mainly against Antonio Lorenzini. Stillman Drake suggests that Lorenzini was a pseudonym for Galileo's well-known friend, neighbor, and holder of the chair in natural philosophy, Cesare Cremonini—but that identification is also unconvincing.[37] Regardless of Lorenzini's true identity, there were larger issues at stake: Cecco's literary strategy of ridiculing Lorenzini involved an inversion of the positions of the mathematician and the philosopher in the social order. The *Dialogue* was an attempt at philosophical comedy, using popular characters and comic inversion to mock elements of the high culture of the university.

The Cecco work also reveals struggles over different kinds of disciplinary credibility in the circle of Paduan intellectual friendships during the years immediately after the death of Pinelli. Antonio Querenghi (1546–1633), a poet and one-time secretary to various Roman cardinals, was a close friend of Pinelli's, and after Pinelli's death, his home became the successor site for gatherings of the Paduan literati.[38] Cecco's dedication to the cleric Querenghi links his treatise to this Paduan learned civil society of ecclesiastics, professors, and aristocrats. The dedication immediately betrayed its learned authorship, using a Horatian rhetorical strategy to authorize the possibility of an absurdity: would it not be ridiculous if a common herdsman dressed up in an academic's gown conducted a dispute with "those Doctors of Padua"? "Wouldn't that make you laugh?" It soon turns out that the views of the herdsman about the nova are those attributed to Querenghi himself. Ronchitti says he would "put your [Querenghi's] gown on" and openly defend the superlunary location of the nova. If he did it well, Querenghi would get the credit; if not, then Querenghi would have to help him, "since I'm making you a present of it." Either way, Cecco purported to be representing Querenghi's views.[39]

Unlike Capra's formal, almost Latinate Italian, the language of Cecco's two interlocutors, Matteo and Natale, is an aggressively ribald and rustic Paduan dialect, reminiscent of Bruno's dialogues, the very use of which clearly signaled a break with academic convention. In their brief conversation, the two interlocutors greatly augmented Capra's line of attack. Both characters praised right-thinking, parallax-wielding mathematicians and heaped scorn on the figure of the bumbling philosopher who puts the nova beneath the Moon.

Matteo: What is this fellow that wrote the book? Is he a land surveyor?

Natale: No, he is a Philosopher.

Matteo: A Philosopher, is he? What has philosophy got to do with measuring? You know that a cobbler's helper can't figure out buckles. It's the Mathematicians you've got to believe. They are surveyors of empty air, just like I survey fields and can rightly tell you how long they are, and how wide. Just so can they.[40]

According to Natale, Lorenzini ascribed to mathematicians all sorts of ideas about generation and corruption of matter in the heavens.

The use of Brunonian rusticity permitted Cecco to go well beyond Capra's natural philosophical speculations and to debase the subject of heavenly matter, making it ordinary and even domestic: "If it was made of polenta, couldn't they still see it all right?" Later, meat, onions, milk, and omelets show up as resources in this domestication of academic natural philosophy. Having brought the discussion into the kitchen, Cecco di Ronchitti quickly suggested a series of further speculations about how this new star could have been created: from bits of air? From three or four very small, invisible stars now heaped together? Maybe it was not a star at all, but just a "bright spot"? And what of its consequences? It is said to "ruin the philosophy of those fellows [Aristotelians]." It might actually stop the heavens from moving—a view that the postil associates with Copernicus and which the text says is held by "lots of people and good ones too, who [already] believe that the sky doesn't move."[41]

The Cecco treatise, then, reveals several interesting developments. First, among the members of the Pinelli and Querenghi circles, the nova provided the occasion for experimentation with new literary and rhetorical possibilities for conducting natural philosophy outside the university. It flirted with Brunonian forms in its use of comedy, absurdity, dialogue, and the vernacular to create a new space for discussing an entity believed to be part of the natural order. Second, it considered the consequences of such an extraordinary event for the ordinary course of heavenly motion (e.g., the function of the sphere of the fixed stars). Third, Cecco made no mention of the conjunction of the superior planets and made no effort to render an astrological judgment. Finally, the author did not claim for his nova any astrological meaning at all.

Yet the question of the nova's possible astrological import could not be ignored. If it was a natural event—either generated from already-existing materials or itself already existing but for the first time visible to humans—then, like a fixed star, it should have had the potential to activate astral influences. And of course its position in the zodiac would be relevant to the meanings of its effects.

Already in 1606 these issues were aired. Lodovico delle Colombe, who soon became one of Galileo's great antagonists in Florence, believed the nova to be an already-existing entity, the visibility of which was blocked by patches of density in the starless crystalline sphere that lay beyond the fixed stars. As transparent regions of the sphere came between an observer and the previously hidden star, visibility was enabled in much the way that a pair of spectacles assists the nearsighted.[42] The last quarter of Colombe's book, however, was devoted to an all-out, Piconian-style attack on astrology and astrologers. "They [new stars] require that nativity casters be condemned for their wasteful *Almagests, Tetrabiblos[es], Ephemerides, Tables, Almanacs, Theorics, Spheres,* astrolabes, quadrants, and sextants as well as for rattling [our] brains and [causing] time to be lost."[43] Colombe's use of Piconian rhetoric enabled him to escape the need for any explanation of the nova's possible astrological effects or meanings. However, the pseudonymous Alimberto Mauri brushed aside this disparagement, generously misreading it as but a friendly warning to astronomers not to allow astrology's details to derail them from learning well the principles of astronomy.[44] Evidently, Mauri did not wish to rule out all forms of practical astrology. Later he cited Belanti in one of his postils.[45]

The heated and detailed Mauri-Colombe exchange does not raise confidence in Stillman Drake's conjecture that Mauri, like Cecco di Ronchitti, was a pseudonym for Galileo.[46] In fact, it may be helpful to try to piece together what can be said about Galileo's position in the nova controversies by *not* accepting Drake's identification of Mauri as Galileo. Following this route, one arrives at a rather different picture. As the debate raged through 1605 and 1606, Galileo chose not to insert himself into the public discussions: no *discorso, considerazione,* or *scherzo* came from his pen. Finally, in 1607, he could restrain himself no longer. When he finally unleashed his attack, the occasion was not specifically the nova but something of immediate, practical importance to him: a challenge to his proprietary claims concerning an instrument that he had described publicly the previous year and which he had dedicated to his noble pupil, the sixteen-year-old Cosimo de'Medici. This mattered directly to his livelihood as well as to his reputation.[47]

The 1606 work, entitled *On the Operation of the Geometric and Military Compass,* was a *practica*. It described a hinged brass compass that could be used for drawing both lines and circles, which was also calibrated for measuring and for use as a military sighting device. Galileo had been supplementing his university income by making and selling these compasses for many years. In his preface to the reader, Galileo named previous "satisfied customers"—all noblemen—for whom he had made these instruments over the past eight years.[48] Nevertheless, while providing guidelines for its operation, the book withheld details on how to construct the instrument—a craft in which Galileo had a commercial interest and for which he retained a skilled artisan who lived in his house. Although Galileo had hoped to publish many theoretical works, after Bruno's execution he restricted himself to this safer epistemic domain.[49]

HONOR AND CREDIBILITY IN THE CAPRA CONTROVERSY

Nevertheless, *practicas* could also function as flashpoints. In March 1607, Baldassare Capra published a work in Latin with an uncanny resemblance to Galileo's that promised to show not only how to operate but also how to build the compass.[50] It did not take Galileo long to dissect Capra's blatant plagiarisms, as is evident from the 151 systematic and detailed annotations on his personal copy. The following is typical: "Copied *to the word* from the first part of my *Operation.*"[51] Unlike Capra with his work on the nova, Galileo took Capra's appropriation of his compass as a direct threat to both his honor and a portion of his livelihood, not as a philosopher but as an instrument maker. Much of the *Defense against the Slanders and Deceits of Baldassare Capra* was, therefore, constructed as a legal brief in defense of his priority and his reputation—with explicit affidavits in the form of letters and sworn statements from various Venetian notables attesting to Capra's plagiarism. Before Galileo produced this evidence, however, he impugned Capra's credibility by addressing himself directly to his opponent's earlier charges about priority in observing the nova.

Galileo's main objection concerned the violation of scholarly etiquette. Neither Capra nor Mayr had behaved like responsible men desirous of learning *(studiosi)*. To describe their manner, Galileo used such words as "uncivil" *(incivile)*, "indiscreet" *(temeraria)*, "liar" *(falsidico)*, and "villainous" *(villanesca)*. They were also untrustworthy—like "spies" *(spie)*—so that one had to be circumspect in speaking in front of them.[52] Capra described himself as a "gentleman," and Galileo all but accused him of failing to act like one. Capra's failed gentility was certainly at stake in the charge of stealing Galileo's instrument.

Galileo also used the occasion to assail Capra's claims to scholarly credit for a heavenly discovery, and that episode casts an interesting light on how such credibility might be established. Galileo clearly believed that a prince—like Tycho Brahe or Wilhelm, landgrave of Hesse-Kassel— could act in a scholarly manner and lend his authority to a scholar's claims, but he never went so far as to assert that the ruler's social station bestowed on him the special capacity to authenticate those claims. Galileo constructed his standard of judgment on other grounds:

> I do not know in what school Capra has learned these brutish manners—certainly I do not believe that it was from his German teacher because, [Mayr] being a student of Tycho Brahe, he could have learned from him, and [he] would have shown his pupil which words to use if he wanted to publish—not only the things said by others but those things communicated and required in private writings. And both of them, the author himself and his teacher, would have learned modesty from him [Tycho] that they would have wished to insert into their writings, including some things written by their friend [Tycho] while he lived [*che ancor viveva*].[53]

Leaving aside the irony of holding up Tycho Brahe's turgid and at times aggressive prose as a standard of stylistic decorum, the interesting point here is that Galileo represented his own ideal of scholarly propriety as the one exemplified by Tycho Brahe himself and his *Progymnasmata*— a work of astronomical scholarship rather than a courtier's manual.

Galileo's concern was not just the right use of words but also the appropriate standard of precision in fixing the nova's time and place. There were two related points at issue: the actual time of initial observation and the standard that ought to be used to judge the degree of temporal precision. To the first, Galileo argued that what mattered was not when the nova was first seen but where it was located—above or below the Moon.[54] On the second point, Galileo maintained that establishing the time of first appearance was not a question of an individual report but of a collective one.

Galileo drew an important moral, already implicit in Tycho's assembly of author's reports in the *Progymnasmata:* he accentuated the correctness of *collective work as a practice.* But, unlike Francis Bacon in the collectivist scheme that he was hatching at the very same moment, Galileo used the *Progymnasmata* for his own specific polemical purposes, arguing that Capra's standard was too severe and that, in establishing the time and place of the 1572 nova, the writers whom Tycho cited were less exacting than was Capra.[55] With one exception, the examples that Galileo chose were, not by chance, members of the null-parallax group: Landgrave Wilhelm, Thaddeus Hagecius, Caspar Peucer, Paul Hainzel, Michael Maestlin, Cornelius Gemma, Jerónimo Muñoz, and Brahe himself.

The association of social status with the power to bestow credibility on natural knowledge has received much attention in the past twenty years. It is thus striking that Galileo did not invoke these authors' social station to warrant the credibility of their reports. His aim was to challenge what he regarded as Capra's excessively narrow and unnecessarily strict notion of precision—at a time, we might add, when there was no consensus about what such a standard should look like. Galileo instead emphasized wide variations among the descriptions contained in the *Progymnasmata.* Establishing the appearance of a new object involved comparing many reports from

CONFLICTED MODERNIZERS

different observers—a shared activity. For example, Wilhelm reported the new star on the third day of December, "while Venus was greater and clearer." Hagecius said that he had first seen the nova "around the birth of Our Lord." Peucer wrote on December 7 that this was the "fourth week" since he had observed the "new star." Paul Hainzel wrote that he had first seen "the light" on November 7 in the tenth house. Maestlin wrote that he had seen "a certain new star" in the "first week of November," while Cornelius Gemma put it on November 9 and Muñoz on November 2. And finally, Brahe himself would not confirm anything more exact than "near the end of the year 1572, near the beginning of the month of November or at least in its first third."[56]

While making no effort to tabulate or otherwise standardize these observational statements, Galileo extracted a significant moral corollary from the communal enterprise that Tycho Brahe had constructed in the *Progymnasmata:* true scholars are aware of their errors and understand the *corrigibility* provided by separate investigations. "The privileges and the abilities that time gives to all scholars—to be able to perceive errors, to correct them, to revise once, twice or a hundred times, to polish and to criticize their own writings—shall these be abolished and annulled by the censures of petulant and vigilant men?"[57] This lesson of communal corrigibility was the one that Galileo sought to teach Capra. And, in the end, he was successful both here and in the area that mattered most immediately to him: his priority in the invention and operation of the military-geometric compass. The Venetian senate found Capra guilty and banned him from the university. Galileo had gained a decisive legal and political victory. Along the way, he had constructed a new standard for defining the boundary between legitimate and illegitimate practice in observing the heavens and he had protected his livelihood as an instrument maker.

GALILEO AND KEPLER'S NOVA

If the Italians increasingly made the 1604 nova into an object to be explained by ordinary physical causes in the heavens, the local character of their controversies consumed the energies of the principals and kept their issues largely confined to the cities and geographically close correspondence networks in which they occurred. Kepler,

whose information about Italian (especially Paduan) activities seems to have come almost exclusively through the mediation of Edmund Bruce, was not at this time well connected to these webs of communication (and intrigue), as Bruce had either died or returned to England. Thus Kepler did not fully appreciate the extent to which the Italians were already advancing physicalist explanations of the nova. He did not know, for example, that Cecco and Capra had devoted treatises to exhaustive attacks on Lorenzini's *Discorso* or that Galileo and Maelcote had given lectures on the nova.

Kepler's understanding of the Italian response to the nova was skewed, as it was formed entirely by his reading of two works by Lorenzini: the *Discorso* of 1605 and a second (Latin) work that did not figure in the nova debates, *On the Number, Order and Motion of the Heavens, against the Moderns (De Numero, Ordine et Motu Coelorum,* 1606). Lorenzini exemplifies the growing tendency among traditionalist natural philosophers in the first decade of the seventeenth century to engage more openly and directly with at least limited representations of the claims of "more recent writers" (*recentiores*).[58] Kepler had encountered Lorenzini's works not through direct Italian contacts but through Herwart von Hohenburg, who, as one of his chief patrons and correspondents, often sent books that were difficult for Kepler to obtain.[59] But because Kepler was not part of the Pinelli and Querenghi circles that had produced the Cecco satire, he had no personal acquaintance with Lorenzini and knew of him only through his print identity. To Kepler, Lorenzini was just a philosopher who also happened to be a very bad astronomer and hence a perfect target. And Kepler was sufficiently appalled that he deposited a long footnote (the only one in the book) on the first page, in which he dismissed as "impossible" Lorenzini's assertion that the nova had appeared on October 8 in Sagittarius together with a conjunction of Mars and Jupiter, Saturn "not far absent," and the Moon "hurrying through Aries, opposite the Sun almost by a diameter." Lorenzini's undoing was his haste and lack of care: "If only he had used one word (such as, 'I think') or another (such as, 'I believe'), he could have saved himself the grief of saying what he affirmed in his excitement."[60] But worse was to come—and it provides another clue to Kepler's punctuated relationship with Galileo.

Kepler used Lorenzini, I believe, to challenge Galileo's continuing silence in their already-frayed relationship. Instead of addressing Galileo directly, he chided the entire community of Italian and French mathematicians. Why hadn't they taken care of the "Affaire Lorenzini" themselves? In a chapter on parallax and the size of the universe, Kepler deplored the "miserable condition of our times," in which the doctrine of parallax was misunderstood not by the common people but by "a philosopher, famous for little medical books, a most excellent man who came not from some uncivilized region but from Italy—and not just from some obscure part, but from Padua, a meeting place frequented as much by the most learned men as any other place in Europe." Now, here was this "famous philosopher" claiming that this star was below the moon and denying mathematicians' observations of a parallactic value of around $52\frac{1}{2}$ minutes, which would have placed it above the Moon. "Clavius, Ubaldo, Magini, Galilei, Ghetaldi, Rubeo and many others, What do you Italian mathematicians say to this? And what about [Bartolomeo] Cristini of Savoy?[61] What about the Frenchmen in whose country the excuse [for not reading] is that this is [just] another little book by an Italian, [even if] it has been translated into Latin? Why do you have such patience for this disgrace and shut your eyes to it? In fact, if as I suspect, you think that these are shameful frivolities, why do you not publicly rebut them? For clearly the method of this philosopher is a joke."[62] So far as I know, Galileo was the only person on this list with whom Kepler had directly corresponded. In bury-ing Galileo's name in the middle of a roster of Italian and French mathematicians, Kepler deftly avoided a frontal attack; at the same time, he preserved the possibility of a future alliance with Galileo while venting his frustration at the Florentine's public reticence to communicate.

He was doomed to remain disappointed. Galileo did not publicly criticize Lorenzini until his famous *Dialogue* of 1632, the year after Kepler died.[63] But this hardly means that he was ignoring Kepler. A single note in Galileo's hand shows that he had seen Kepler's *De Stella Nova* sometime after its publication in the autumn of 1606 and that he was actively trying to obtain a copy of the book in 1610.[64] Of greater significance than the content of Galileo's reference are its location and the dating. His paraphrase pertained to Kepler's discussion of the cause of the nova's sparkling or twinkling, which the latter had ascribed to the proper rotation of each of the fixed stars (but not, of course, to the sphere itself).[65] That passage from the *Stella Nova* occurs just a few pages beyond the reference to Galileo's own name, and it is difficult to imagine that Galileo could have overlooked it. Likewise, the positioning of Galileo's handwritten note is significant, as it occurs just before a series of other notes that he took from Tycho's *Progymnasmata* and used in his polemic against Capra.[66]

In addition to Galileo's silent engagement with Kepler, some of his notes on Tycho's *Progymnasmata* also provide hints that he was privately framing the nova in Copernican terms.[67] Elias Camerarius had described a diminishing daily parallax (10 minutes to 2 minutes) for the 1572 nova and hence a radically increasing linear distance from Earth—as much as twenty times the distance from its original position. In reporting Camerarius's claim, Tycho criticized his account as entailing a vast space between Saturn and the fixed stars (*nimiam mundi sensibilis asymmetriam*), much as he had earlier objected to Rothmann. Commenting, in turn, on Tycho's objection, Galileo wrote: "Camerarius's observation could be true, but the cause of the star's increase in distance from the meridian [*a vertice*] could be explained by the Earth's annual motion, which tended toward the south."[68] Galileo's note shows that he was using the 1572 nova to frame his thinking about the 1604 event and that he aligned himself with a Copernican rather than a Tychonic explanation for the nova's steady disappearance.[69]

Taken together, this evidence points to an important conclusion. Ten years after their original contact, Galileo was following Kepler's public arguments while also privately entertaining the possibility that the Earth's annual revolution explained the nova's apparent recession. But rather than form an alliance of mutual interest, Kepler and Galileo continued to circle one another much like characters in an unfolding drama— the former ever probing for an opportunity for contact, the latter ever avoiding but quietly watching the evolution of the other's ideas and taking his measure. In this setting, one can imagine Kepler's *De Stella Nova* briefly sitting on Galileo's desk in early 1607 as he composed his attack on Capra.[70]

CELESTIAL NATURAL PHILOSOPHY IN A NEW KEY
KEPLER'S DE STELLA NOVA AND THE MODERNIZERS

Kepler's *De Stella Nova* was a book unlike the Italian treatises with which Galileo was familiar or the works earlier assembled by Tycho in the *Progymnasmata*. It was not just another work about an unusual "marvel" that signified the end of the world. By the early seventeenth century, reports of comets, novas, unusual eclipses, and other preternatural events were the rule rather than the exception. Although Kepler believed that it was possible to assign an astrological meaning to the star and to claim that it stimulated effects in the sublunary region, he openly resisted the apocalyptic enthusiasms of other Lutheran contemporaries. He was in no rush to lo-

cate this nova either in the Elijah apocalyptic narrative or within what he called Lichtenberger's "hateful forebodings."[71] Time and again, Kepler positioned himself against popular prognosticators whose writings about heavenly marvels seemed to him hasty, tawdry, and devoid of the right explanatory ambitions.

Kepler claimed not merely to be doing natural philosophy in the heavens but to be doing it in a new register.[72] From the beginning of his studies, he had consistently used the generic image of everyday prognosticators and ordinary philosophers to foreground and legitimate the superiority of theoretical knowledge: he had done this for the Copernican arrangement (in the *Mysterium*), for astrology (in the *De Fundamentis Certioribus*), and for optics (in the *Paralipomena*). But now, for the first time, he confronted the question of how to explain the appearance and disappearance of a celestial entity in such a way that it would be consistent with all his other beliefs about God and the Creation. This approach, again, would differ from that of "ordinary people who avidly seek after knowledge of future specifics rather than philosophy."[73] In this sense, the *Stella Nova* was a book that, unlike earlier works, subsumed the question of the order of the planets within the broader question of the entire class of visible heavenly objects. Recall, for example, Maestlin's circumspect deployment of the 1577 comet as evidence for a reordering of Venus and Mercury and his highly tentative remarks—never further developed—about the 1572 nova and the size of the universe. Compare even Tycho's more determined and far-reaching use of vaguely specified observations of Mars and systematic representations of the path of the comet of 1577 to underwrite claims about the nonexistence of solid, impenetrable spheres. In drawing more general conclusions about the coherence of the heavens and the nature of its parts, the *Stella Nova* established the grounds of a new problematic. By contrast to his mentors, Kepler transformed the nova into an opportunity to argue a cluster of fundamental questions that moved the explanatory discourse out of the realm of the unusual and into that of the ordinary.

What would a Copernican, causal astrology look like? Could there be a causal connection between an unusual event in the ordinary course of nature (a conjunction) and an extraordinary event of (apparently) preternatural origins (the nova)?

What was the nature of the heavenly stuff? the size of the universe? the overall arrangement of its parts? And could such a set of causal and descriptive propositions be made to be consistent with the phenomenon of the nova itself—its varying distance, apparent size, color, and material constitution? Questions of this sort were resisted by the vast majority of mathematicians who wrote about the 1572 nova as inappropriate to mathematical subjects at best, illegitimate at worst, as few were willing to admit the possibility of any alteration above the Moon. Indeed, most traditional philosophers still doubted that mathematical practitioners could have anything new to say about natural philosophy. The example set by Copernicus and Rheticus was largely ignored. By contrast, although the decidedly untraditional Giordano Bruno had broken out of Aristotelian categories of space, matter, and motion, he had done so at the cost of writing off the science of the stars as conventionally practiced.

Kepler's *De Stella Nova* did not evade these dilemmas. Like Bruno, it classified novas in relation to natural philosophy and theology; but, unlike Bruno, Kepler's book included the science of the stars—with the title page explicitly representing his book as a group of "astronomical, physical, metaphysical, meteorological and astrological disputations." Historians, with few exceptions, have tended to cherry-pick their own favored themes from this somewhat awkwardly composed work, thereby failing to notice sufficiently the new and larger goal that Kepler had in mind.[74] And in developing such a perspective, Kepler also began to break away from the common sixteenth-century practice of arguing with generic opponents. Instead, he frequently named moderns (Pico, Tycho, Gilbert, Galileo, and Bruno), ancients (Aristotle, Ptolemy), and opponents like Lorenzini, whose names were not widely known within his social network. Characteristically, he also treated the nova as a forum for his wider Copernican project—an important point of contrast with the Italian treatises. But now, in 1606, he dared to use the nova even against Tycho Brahe in a way that, for good reason, he had avoided five years earlier in the *Apology for Tycho against Ursus* (*Apologia Tychonis contra Ursum*).

That point is worthy of note in light of Kepler's other possible choices of literary form. He might have written a treatise using the model of Tycho's recent *Progymnasmata* or Hagecius's *Inquiry*, collecting together reports from all over Europe as a step toward his own impending reform of theoretical astronomy. Alternatively, he might have followed another cautious route, producing an amplified *practica*, structured like his *Bericht* of 1604 but with further reports of his own and others' observations and parallactic measurements.

In choosing not to use those modes, *De Stella Nova* signified a new moment in the evolution of the Copernican question. It attempted a systematic engagement of planetary order, space, and astral causality within a world that was still for Kepler pregnant with divine purpose and meaning. His mix of scholarly authorities reflected this goal in its attention to the ancients but, as in his other major works, the balance always tipped to the moderns. However, the work was not "modern" in the (weaker) sense that one might use to characterize the humanist modality of a Mazzoni, where the ancient Plato sat as an equal at the table with Aristotle rather than as an infrequently invited guest, or of a Scaliger, where the disorderly sequence of propositions and authorities disrupted the orderly glosses on Aristotle's books without quite replacing Aristotle as the master authority.

Rather, what marked Kepler's approach in *De Stella Nova*—as with Gilbert, Bruno, and Francis Bacon—was a new kind of epistemic confidence, a critical edge in the twofold sense that there was hidden knowledge to be recovered from the ancients and, at the same time, that this knowledge could be fallible. He joined to this sensibility a willingness not merely to juxtapose previously ignored ancient opinion but actively to argue with both ancients and moderns using celestial distance estimates. Theoretical astronomy, in other words, became the active hermeneutic wedge, the resource of new questionings, within the space of natural philosophy in a way that was not apparent among other modernizing, anti-Aristotelian philosophers of the late sixteenth century, such as Giordano Bruno, Bernardino Telesio, Giovanni Battista della Porta, and Francesco Patrizi.[75] Kepler's remarkable scope and consistent audacity also showed in his engagement with philosophical positions that, for different reasons, Galileo regarded as difficult or dangerous: Pico della Mirandola's critique of astrology, William Gilbert's and Giordano Bruno's infinitist natural philosophies, Tycho Brahe's rejection of Copernicus's enlarged universe.

This brings us to the argumentational structure of the work as reflected in Kepler's organization of topics. Kepler was concerned with several major questions. Although the work opened with a rehearsal of astronomical observations (chapter 1), the succeeding chapters immediately moved into a discussion of the foundations of astrological theory (chapters 2–10) and eventually arrived at larger questions lying at the intersection of natural philosophy, astronomy, and theology. If the main issue for the 1572 writers was whether their star was above or below the Moon, Kepler added the new and provocative question of whether his star was beyond Saturn and the fixed stars. And here Kepler used the nova to argue against what might be called the minimal finitists, Ptolemy and Tycho, that the universe was much bigger than they would allow (chapters 15–16). Characteristically, he positioned himself in the middle when he argued against Bruno and Gilbert (but not Digges) that the universe, although immense, is not infinite (chapter 21). But he aligned himself with Tycho and Bruno in arguing against Aristotle for the alterability of the heavenly stuff (chapters 21–23).

If the heavens are subject to change, and hence devoid of solid, impenetrable planetary spheres, then the new question was how to explain such change: how and where was the star formed? Where did it go after it disappeared? What sort of matter was consistent with the motion of the nova? And was the agent of change a spirit or matter (chapter 24)? Did the occurrence of the anticipated great conjunction of Jupiter and Saturn in the Fiery Triangle (December 17, 1603) have any causal connection to the unexpected appearance of the *stella nova* in the constellation of the Serpent in October 1604 (chapter 26)? If it did, then was it physically possible for what was "below" (the planets) to have an effect on what is "above" (the nova) (chapter 27)? And, in the sublunar realm, were there meteorological consequences that followed either from the nova alone or from the unusual alignment of planetary and stellar events (chapter 28)? If, on the contrary, this incredible convergence of events came about by chance, then did it have any divine meaning or purpose at all, or was it the outcome of blind "material necessity" (chapter 27)? And thus he arrived at the question, what role did an omnipotent God play in this process? Perhaps, as many believed in 1572, it was just a miracle, a one-time divine act outside the regularities of nature. But,

if that was the case, did God act directly in the world or indirectly through intermediary agents, such as spirits?

To appreciate the scope and character of Kepler's argument—the diversity of problems that he presented to his readers—one needs to appreciate that the underlying issues were still simple: the universe was seen as finite, Sun-centered, and bound together by forces that moved both people and planets. Some of these problems, as will be seen, look like echoes of debates and conversations with various constituencies at Rudolf's court.

THE POSSIBILITY OF A REFORMED ASTROLOGICAL THEORIC

KEPLER FOR AND AGAINST PICO (AGAIN)

Let us begin by comparing Kepler's environment with that of Galileo. The Italians had already raised the question of the connection between the great conjunction and the appearance of the nova; but where Baldassare Capra had cautiously skirted the matter of the nova's astrological meaning, Colombe explicitly attacked any such import. In this respect, Colombe's views were clearly in line with post-Sistine orthodox opinion, as is clear from the censor's language on his book: "I, Brother Philip William, Dominican Lecturer in Theology, by order of the Most Illustrious and Reverend Archbishop of Florence, have reviewed the present discourse on the new star of Mr. Lodovico delle Colombe, [a book] which is very much in agreement with the true philosophy and the principles of Aristotle, and agrees with theology and contains many wonderful doctrines, explained with such clarity and ease that those who abhor the falsehood of judiciary astrology will find that it has much utility for them."[76] It is reasonable to assume that Galileo, so recently warned about his astrological activities in Padua, would have been aware, from Colombe's treatise, of the situation in Florence in 1606, just as he was surely aware of the severe prohibition against Bruno's works in 1603. By contrast, in Prague, the emperor's *mathematicus* faced a situation where, as part of his position, he was expected to issue regular prognostications, just as his predecessors Hagecius and Ursus had done.[77] And indeed it is likely that, in addition to the emperor, many of the (Paracelsian) alchemists held well-formed views about astral forces.[78]

Kepler, of course, had been concerned with the

problem of astrology's theoretical foundations since his student days. It was for him at once an epistemological and an ontological problem: whether the zodiac's signs could have any causal effect on anything. But in the discussion in the *Stella Nova*—unlike his discussion in the 1593 physics disputation, the *Mysterium*, and *De Fundamentis Astrologiae Certioribus*—Kepler opened up a full conversation, explicitly crediting Pico della Mirandola with the argument that the zodiacal names and pictorial representations were nothing but human constructions.[79] Names were not essences. The relationship between the words denoting the celestial animals of the zodiac and the meanings that they connoted were not fixed. Furthermore, because the names were arbitrary, many different ones could easily be assigned to the same disposition.[80] Likewise, astrologers made arbitrary associations of the signs with the four elements (see table 7). These were just the sorts of issues that Copernicus and Rheticus had failed to address. Kepler raised all sorts of difficulties: Why wasn't Aquarius, the water bearer, part of the group of three elemental signs related to water? Why weren't Taurus, the bull, and Cancer, the crab, included in the group of fire signs? Why were Taurus and Capricorn, the goat, associated with feminine instead of masculine qualities?[81] And what of incoherences in the ordering of the elemental groups? Fire was the first among the elements (hot, active, life giving, and masculine) and Aries the first sign in the zodiac; yet Kepler claimed that astrologers located the three earthy signs after the three fiery signs rather than after the triplicity of airy signs. All this, Kepler averred, was "pure fancy."[82]

The problems continued. Because it was a human invention, the zodiac also had a human history: it originated with the ancient astrologers of Babylon, India, Egypt, Arabia, Greece, and Rome; and it was (allegedly) opposed by philosophers at all times and places.[83] The twelvefold division of the zodiac into equal parts was also a human construction, arising from the fact that ancient rural farmers and sailors noticed that the Sun and the Moon were in opposition (although not in eclipse) twelve times annually; and each sign was divided into parts of thirty degrees because there was an average of about thirty days in the lunar cycle. Furthermore, because the longitudes of the constellations, with

TABLE 7
The four trigons or triplicities with associated zodiacal signs and their elemental and gendered natures

Water (F)	Fire (M)	Air (M)	Earth (F)[a]
Cancer	Aries	Gemini	Taurus
Scorpio	Leo	Libra	Virgo
Pisces	Sagittarius	Aquarius	Capricorn

SOURCE: Ptolemy, *Tetrabiblos*, bk. 1, chap. 18.

[a] F: feminine; M: masculine.

their animal images, no longer mapped precisely over the signs, this discrepancy further called into question the possibility of zodiacal causality.[84]

Kepler assigned credit for his constructivism directly to Pico: "All this was taught by Count Giovanni Pico della Mirandola more than 100 years before me, and I can justly be seen to concede his entire opinion concerning the vanity of astrology." But, unlike the Italians, Kepler held his ground on some matters: "There is also much that I do not [concede]."[85] Kepler's reform of astrology's general principles that had begun in the *More Certain Foundations* of 1602, just as he was about to become the emperor's mathematician, became in 1606 an open and systematic engagement with those elements of Pico's critique that had a direct bearing on the interpretation of the nova: great conjunctions and aspects. His stance cleared the way for a new astrology independent of the causal efficacy of the zodiacal signs and founded on the aspects alone. No longer a student defending a position midway between his Tübingen professors, Kepler openly argued the need for new foundations for theoretical astrology.

As usual, Kepler took what he liked and rejected the rest. Pico had stated that if one planet alone could not cause effects in the earthly realm, then a conjunction or another configuration of two or more planets could do no better, because what they possessed separately, they also possessed when conjoined.[86] In other words, no aspects of any sort, whether conjunction, trine, quartile, sextile, or another configuration, could cause terrestrial effects. Kepler agreed with Pico that conjunctions could not cause the rise and fall of religions and empires.[87] Against Pico, however, Kepler defended the (Ptolemaic) view that some aspects were efficacious. The questions were now three: How exactly were effects caused? Were conjunctions efficacious? And how could one pick out the efficacious aspects?

Kepler turned again to archetypal metaphysics. He believed that one should think of the aspects, first and foremost, as geometric relations (*relationes*) constituted by the angles formed by two planets and the Earth. A conjunction was simply an instance of a relation in which two planets lay at 0° along the same line. Yet, for purposes of calculation, it was common practice from medieval times to use mean rather than true conjunctions. And Pico denied that mean conjunctions could yield effects, as the planets were not truly aligned.[88] Moreover, there was also the physical question of what it was about such an alignment that was capable of causing effects. For example, every month the Sun and the Moon were lined up in this manner, yet without effect, "since the Moon moves very quickly, and hence the effect is not as violent or excessive."[89] And, as Messahalah argued, for slow-moving bodies like Saturn and Jupiter, the effect should be greater because the conjunction lasts longer; hence, slower speed should be associated with greater efficacy. Yet, Kepler denied that relative speed could count as the cause of astral effects: "Speed is noble in a race, but the king is at rest, firm and steady." And for all of his proper emphasis on heavenly light, Pico had missed the real significance of the Sun: "Pico, what do you say to Copernicus, who teaches that the Sun stands still because it is the most noble of the planets?"[90]

Kepler's answer to the question of astral causality built on the position already adumbrated in his 1602 *More Certain Foundations*. But, perhaps because the *Stella Nova* was constructed as a series of interlinked disputations, he openly triangulated his views with those of other authors rather than laying out a list of anonymous theses. And in his astrological disputation, beyond Pico, the main (named) opponent was Offusius. As part of his general critique, Pico had rejected all efficient heavenly causes, whether in the form of aspects or planetary influences. Offusius was, as Kepler said, "a sharp enemy of the aspects" who nonetheless retained the influences capable of impressing effects. Against both Offusius and Pico, Kepler preserved the aspects, but in so doing he accepted Pico's argument that the aspects cannot be efficient causes of terrestrial ef-

fects. The solution sketched five years earlier, but now further developed, was to locate the formal cause in the aspects and the efficient cause in the Earth's soul. In the *Stella Nova*, Kepler spelled out his position as follows: "It ought to be said that neither the stars nor their rays by themselves nor thereupon their configurations—since these are relations—cause effects except when under the power of an object [*sub ratione objecti*]. And [it ought to be said that] some faculty controls things that are capable of being acted upon (such as humors in the globe of the Earth), which can perceive and judge the shapes of radiations and therefore can rouse its body by some impetus either by moving—if it was a moving faculty [*facultas motrix*]—or by heating and sublimating [raising up] the humors."[91] The question of how an immaterial form (albeit extended in space) can bring about a material effect also continued to preoccupy Kepler in the *Harmonice Mundi*, where he spoke of the soul as an agent that produces motion by acting on itself when it finds similarities between sensible proportions and insensible archetypes.[92] The question was, which proportions were capable of moving spirits?

Kepler's answer, following Ptolemy, was that the number of efficacious aspects corresponded to the number of harmonic divisions. However, since his correspondence of 1599 with Herwart von Hohenburg (May) and Edmund Bruce (July), he had been steadily emending the ancients by increasing the number of allowable consonances beyond the perfect fourth, fifth, and octave of the Pythagoreans and the four aspects recognized by Ptolemy. Thus, beyond Ptolemy's opposition (180°, 1:1), trine (120°, 1:2), quadrature (90°, 1:3), and sextile (60°, 1:5), he now added the hard third (quintile, 72°, 1:4), hard sixth (biquintile, 144°, 2:3), and the hard sixth plus the hard third (sesquiquadrature or trioctile, 135°, 3:5); significantly, he also included conjunction (0°, undivided string).[93] If in the *Stella Nova* Kepler was still trying to explain the aspects from musical ratios, in the *Harmonice Mundi* he subsumed both the musical ratios and the astrological aspects under different properties of the polygons.[94] And once again Kepler framed his position as a response to Pico: "What reason does Pico bring forth to explain why geometry moves man through [musical] notes? May I say that it is the same reason why geometry affects sublunary nature through the rays of the stars?"[95]

As was typical of all Kepler's writings, he missed no opportunity to connect his subject with the Copernican ordering. The theme of planetary order sharply distinguishes his writing about the nova from those of his contemporaries—especially Galileo—and calls attention to Kepler's unique role (and freedom) in pushing it forward. The character of these arguments also calls into question the frequent claim that the inability of Copernicus and his followers to measure an annual parallactic angle for a distant star counted as serious empirical evidence against the existence of a real annual motion for the Earth.[96] This is, indeed, an objection against "the Copernican system," taken as a timeless representation; but it was not the argument made by Tycho Brahe. Tycho's objection, as Kepler knew from the *Progymnasmata* (and from the debate with Rothmann), was that in Copernicus's world, the distance of the fixed stars from the Sun at the center of the world was so great that it would destroy the proportionality of the distances among the parts: "If there is one sphere of the fixed stars, then because of its vastness, there would be contempt for the [relative] smallness of the moving spheres. For he [Tycho Brahe] says that in the human body there would be an incredible number of defects if the finger, nose, and other parts exceeded the mass of the remaining part of the body."[97] There would be a huge, "useless" gap between Saturn and the fixed stars. At stake, therefore, was not "the Copernican system," but Kepler's representation and defense of it grounded in the mystery of his cosmography as a divinely planned proportionality among the distances of the moving stars. What could be done about this apparently unharmonious hiatus in the Creator's plan? Put otherwise: if the same aesthetic standard underdetermined the claims of Brahe and Kepler, how could Kepler make his argument compelling?

The first consideration, taken up in chapter 15 of *De Stella Nova*, was to establish that a universe of much greater magnitude was more compatible with the phenomenon of the nova than Tycho's relatively smaller space. "If we open up the immense Copernican abysses, good God! How much higher will the star be elevated?"[98] Using a value of less than 2 minutes for the nova's parallax and assuming the distance between the Earth and the Sun to be 1,200 Earth radii (ER), Kepler estimated that the nova must be far beyond Saturn's distance of 720,000 ER, the most distant planet—indeed, 2,160,000 ER (3 × 720,000).[99] Given that the parallactic angle was invisibly small to the unaided eye, the underlying intuition was not unreasonable. When the nova was brighter than the fixed stars, it must have been closer to the Earth; as it grew dimmer and eventually disappeared, it must have been receding. The question was, how big did the universe need to be in order to accommodate this phenomenon?[100]

In chapter 16, Kepler calculated a value of 34,077,066 ⅔ Earth radii—about 2,434 times Tycho Brahe's value of 14,000 in the *Progymnasmata*.[101] Such an expansion would have further widened the gap between the outermost planet and the fixed stars. But now Kepler used the criterion of *symmetria* to change the terms of comparison. Following an argument championed by William Gilbert—but without mentioning him—he proposed that the right comparison was not between the sizes of the sphere of the fixed stars but between that sphere's velocity and its radius. Thus, the question was, how fast did Ptolemy's—or Tycho's—outermost sphere need to travel to produce the daily risings and settings of the celestial bodies? The answer? 2,625 ER/hour × 860 miles (= 1ER) × 24 hrs = 54,180,000 miles/day! "Whoever tries to grasp this incredible velocity will be equally exhausted and even more powerfully so than he who [tries to grasp] the Copernican immensity."[102] Gilbert, by comparison, had argued that the first mover would need to travel "three thousand great circles of the earth" in one hour, a speed so great that it would clearly require a substance "so strong, so tough, that it would not be wrecked and shattered to pieces by such mad and unimaginable velocity."[103] Surprisingly, Kepler did not develop the Gilbertian physical objection but instead showed himself to be more concerned with moving against Tycho Brahe. For Kepler, it was necessary to "reconcile the elegance of proportion to the world not as Brahe typically wished to do with size but according to beauty and reason. The perfection of the world is in motion, which has a kind of life to it. For motion, three things are required: a mover, a thing to be moved and a place. The mover is the Sun; the moved things extend from Mercury to Saturn and the place is the outermost sphere of the fixed stars . . . the moving things are the mean proportional between the mover and the place."[104] So Brahe was left with a horrendous asymmetry in the inordinate daily motion, compared with Kepler's claim that the interval between Saturn and the fixed stars was just the largest term in a mean proportional. As in the disagreement between Brahe and Rothmann, different applications of *symmetria* yielded different outcomes.

What space, then, was left for the followers of Tycho Brahe in which to locate the nova (chapter 17)? For Brahe, the distance between Saturn and the fixed stars was relatively small. Was it possible that the nova was to be found among the planets? There is no evidence that any Tychonic actually raised this possibility, but Kepler did not miss an opportunity to prosecute his case. If there were mutually touching, solid spheres "as the Aristotelians wish," and the nova was embedded within one of the orbs, then it would be carried around in a circle—which it clearly was not. On the other hand, if "there are no solid spheres, no contact, no drag, as Tycho wishes," then the nova, like the planets, would have a second inequality—that is, its motion would include an annual component. Hence, the Tychonics would have to expand their universe to the "immensity" required by Copernicus.[105]

MAKING ROOM
KEPLER BETWEEN WACKER VON WACKENFELS AND TYCHO BRAHE

A few chapters later, Kepler found himself arguing against Bruno and Gilbert. Alexandre Koyré's influential treatment of Kepler's arguments against an infinite universe—an encounter in which he pitted "The New Astronomy" (Kepler and Copernicus) against "The New Metaphysics" (Bruno and Gilbert)—was based almost entirely on his interesting reading of the difficult chapter 21 of the *Stella Nova* and forms the basis for all subsequent historical commentary.[106] Logically, Kepler sets up his rejection of both Bruno's and Gilbert's infinitism in an argument of the form *modus tollens*, used by astronomers since Ptolemy.

1. If the universe is infinite, then one ought to see a uniform distribution of fixed stars from any point in the visible universe.

2. One does not see such a uniform distribution.

Ergo: The universe is not infinite.

How could one demonstrate the second proposition? To begin with, for Kepler, the proposition (about what one ought to see) lay at the foundation of astronomy, because that discipline works from visible phenomena, fitting hypotheses to the appearances.[107] The gist of Kepler's attempted refutation was that the same heavenly objects would display a different pattern of distribution when seen at equal distances but from different locations. He developed the example of the three stars in Orion's belt, each with the same apparent diameter (2') and, as seen from Earth, separated by the same apparent distance (81'). Now imagine a change of perspective: "If somebody were placed in this belt of Orion, having our Sun and the center of the world above him, it would seem to him, first of all, as if immense stars were touching each other in an unbroken sea; and from there, the more he raised his eyes, the fewer stars would he see; moreover, the stars would no longer be in contact, but would gradually [appear to be] more and more scattered; and looking straight upward, he would see the same [stars] as we see, but twice as small and twice as near to each other."[108]

Kepler's example involved one of his favorite moves, reversal of perspective, the logic of which was intended to break down the necessity of one and only one point of view. The strategy resembled that of one of his student disputations, in which he had asked what the Earth would look like if we were Moon dwellers, or, as in the *Mysterium,* how the arcs of correction (prosthaphaeresis) differ in length when seen from the Sun, as opposed to from the Earth.[109] But now, instead of being directed against the ancients Aristotle and Ptolemy, Kepler aimed his well-tested weaponry against the modern Giordano Bruno. And the conclusion toward which he drove his argument was "an immense cavity, distinct and different in its proportions from the spaces that are between the fixed stars."[110] In other words, Kepler's Copernican arrangement, like Aristotle's Earth-centered one, must sit in a unique "place."

Logically, it is quite clear why Kepler needed to defend such a conclusion. His entire defense of the science of the stars—and not merely theoretical astronomy—was based on an archetypal metaphysics, a bounded universe well-proportioned according to the dimensions of the five regular solids. To concede an unbounded world would be to give up both the polyhedral defense of the Copernican ordering and the reformed, harmonic astrology.

But why do arguments about infinity show up at all in the *De Stella Nova?* This is an important question not raised by Koyré and his commentators. To defend the Copernican proposition that the universe was "similar to the infinite"—and hence capable of accommodating the 3,200,000 Earth radii required for the nova's maximum distance from the Sun—meant expanding the universe to 34,177,066 ER, well beyond the limits of Ptolemy and Tycho Brahe. None of the numerous nova writers had even raised the question of the infinite. Moreover, in Kepler's earlier writings, there was no trail of arguments against an infinite universe. The only other possibility is that Kepler was answering objections put to him either in correspondence or in person by followers of Bruno, for which there are only two possible sources: Edmund Bruce, and Kepler's good friend and Rudolf's court counselor, Johannes Matthias Wacker von Wackenfels.

Bruce is an unlikely candidate. His last letter to Kepler, as we saw in chapter 13, laid out a Brunonian vision—but without mentioning Bruno—and it predated the appearance of the nova. Wacker, however, is a very strong prospect. He had been one of Bruno's patrons when the latter visited Prague for a few months in 1588, and he possessed many of Bruno's works in his personal library, including his treatise on the infinity of the universe.[111] Gaspare Scioppio was a member of Wacker's circle of friends in the late 1590s and was an immediate witness to Bruno's execution in 1600.[112] Wacker had very close ties to Kepler, a fellow Swabian, even though he had converted to Catholicism in 1592. Kepler dedicated to Wacker his little treatise on the snowflake (1611). As will be seen in the final chapter , Wacker occupies an important place in Kepler's *Dissertatio cum Nuncio Sidereo.* In addition, Wacker had good friends among the Wrocław intellectual circle of Andreas Dudith, Nikolaus Rhediger, and Jacob Monau during the period when Wittich was in circulation.[113] And finally, Wacker was a close and loyal supporter of Rudolf and remained in his service until the end of his reign.[114]

This evidence helps to make sense of Kepler's chapter 21 as a public argument with Wacker. But Kepler represents his opponents generically, rather than by name, as a "sect of philosophizers" whom he compares to the theoretically undisciplined Pythagoreans, whose views Copernicus prudently and soberly revived and Aristotle un-

justly criticized. Of such kinds of thinkers as the Pythagoreans, Kepler says that they

> neither deduce their reasons from the senses nor accommodate the causes of things to experience but who immediately, and as if inspired (by some kind of enthusiasm) conceive and develop in the walls of their heads a certain opinion about the arrangement of the world. Once they have embraced it, they stick to it and they drag in by the hair [things] which occur and are experienced every day in order to accommodate them to their axioms. It pleases these philosophizers to want this new star and all others of its kind to descend little by little from the depths of nature, which, they assert, extend to an infinite altitude, until according to the rules of optics it becomes very large and attracts the eyes of men; then it goes back to an infinite altitude and every day [becomes] so much smaller as it moves higher.[115]

Elsewhere, Kepler refers to "the unfortunate Jord. Bruno . . . who made the world infinite so that [he posits] as many worlds as there are fixed stars. And he made this our region of the movable [planets] one of the innumerable worlds scarcely distinct from the others which surround it." And finally, he declares that "this sect misuses the authority of Copernicus as well as that of astronomy in general, which prove—particularly the Copernican one—that the fixed stars are at an incredible altitude."[116]

Kepler's inclusion of the argument with Wacker and Bruno offers a prime example of the kind of open, interconfessional sociability that enabled debate and engagement with heterodox philosophy among modernizers in the civil ambience of the Rudolfine court. The Catholic convert Wacker and the Lutheran Kepler openly disagreed with one another in a common discursive space where there was no obligation to reach agreement on what was proper to teach, what was orthodox to believe, or which philosophical opinions would please the emperor.

GENERATING THE NOVA
NEGOTIATING DIVINE ACTION
AND MATERIAL NECESSITY

If the universe was sufficiently large to contain the nova, then where did the new star come from? How was it formed? The argument with Bruno and Wacker continued, but now behind a screen of impersonal categories. Kepler used generic group-ings to set up different possible explanations: "as-trologers," "conjecturing physicists," "Epicurean physicists," and "theologians." Only among the first group did Kepler mention specific individuals, the prognosticators Johann Moller and Johann Krabbe. They had predicted that the conjunction would be the "generating cause" (*causa progene-trix*) of comets—"just as if," Kepler remarks, "ten months after I got married, a son would be born to me."[117] This sort of explanation implicated the zodiac; but if, following Pico, these stellar config-urations were no more than constructions, then a conjunction was to be seen as merely an acciden-tal relation visible from Earth. The "conjecturing physicists," another species of astrologer, con-tended that the new star appeared coincidentally with the great conjunction. This is the position that Kepler himself would eventually defend.[118]

Kepler's major difficulty was with the "Epicu-reans," who attributed the nova to the chance flowing together of atoms, like a throw of the dice.[119] The Epicureans were proxies for Bruno and Wacker.[120] However, the issue on this occa-sion was not the actual character of the atoms but of their chance combinations: how could one get a conjunction in the Fiery triplicity, unique not only in place but also in time? Given eternity and an infinite number of dice throws, innumerable new stars would appear, but would one of these novae coincide with just this particular great con-junction at this time? Another view considered was that of the Aristotelians, which posited two chains of causes, each completely independent of the other, which would result in the coincidence of the nova and the great conjunction.

Kepler confronted an interesting problem. He did not want a world of blind chance—a world without the fear and foreboding that, in his view, made possible an orderly, purposive life. Evil and free will must exist if the good is also to exist. But, because of his alignment with Pico, he could not accept that the nova had been produced by the conjunction of stars in the zodiac. His rejec-tion of the argument from blind chance, like the disagreement with Bruno in chapter 21, again re-calls discussions with Wacker von Wackenfels, and with Kepler's customarily inventive use of the personal as literary device of civil discussion:

I reported to my antagonists an opinion which is not mine but rather, my wife's. Now, having fol-lowed their reasoning, I had already conceded that order could result from chance; but she argued against it. My opponents ought to teach me to de-fend their opinion against such a redoubtable ad-versary. Yesterday, when I had grown tired of writ-ing and my mind was filled with dust motes from thinking about atoms, she called me to dinner and served me a salad. Whereupon, I said to her, if one were to throw into the air the pewter plates, let-tuce leaves, grains of salt, drops of oil, vinegar and water and the glorious eggs, and all these things were to remain there for eternity, then it would be a fact that this salad just fell together by chance. My beauty replied: "But it would be neither this presentation nor in this order."[121]

So Kepler's wife arrived at a classically Keple-rian conclusion: a coincidence both spatial and temporal must imply a unique result and, by im-plication, a unique cause—just as earlier Kepler had argued for the unique place of the universe. The cause of the coincidence is God—a conclu-sion that, momentarily, makes it appear that Kep-ler was turning back to the sort of miraculist ex-planation of the nova to which Maestlin and Tycho had reverted in 1572. But once again, Kep-ler may be seen as carving out a middle ground. Unlike the 1572 writers, Kepler wanted divine ac-tivity in the world to be specified physically and astronomically. It would not be enough for him to say, with his predecessors, that God just acts because he is all-powerful. Kepler's specification involved eliminating both the extremes of Bruno-nian blind chance and a God acting mysteriously and without the mediation of the natural world.

SUMMARY AND CONCLUSION

Both in Italy and the Empire, writers constructed the nova of 1604 in the small civil cultures of relatively autonomous, heterodox aristocrats or the courtly circles to which they belonged, freed from the hierarchically distributed authority of academic theologians and their traditional allies, the natural philosophers. In the nova episode, the realm of the extraordinary began to collapse into the ordinary course of nature. The outlines of a new problematic, not yet fully and consis-tently articulated, began to appear: God did not operate now in one way, now in another, but al-ways in one way. All celestial objects—whether visible and regular, like the planets, or invisible and unpredicted, like comets and novas—must somehow be explained by an immanent deity working through natural regularities either in a bounded or an unbounded space.

Yet there were important political differences between imperial Prague and the north Italian university towns of the early Seicento. Among all the contemporary writers on the nova, Kepler was the most systematic and inclusive, not merely in theorizing his positions but also in arguing against the claims of other modernizers. If the target of Kepler's opposition was among his noble friends at the intellectually heterogeneous and interconfessional Rudolfine court, he chose to conceal his local ambience by camouflaging the terms of his discussion, mixing the anonymous designators ("astronomers" and "philosophizers") with named but deceased adversaries (such as Pico and Bruno). The axis of conflict in Prague thus formed around the *novatores* and their followers: it was a *querelle des modernes*.

By contrast, the belittling and tendentious Italian treatises marked tensions colored by internal, university-dominated issues—with Rome always casting a shadow. In the Mauri and Cecco polemics, for example, the disputes still centered on the viability of parallax, much as in 1572. Galileo's controversy with Capra foregrounded issues of observational skill and standards of corrigibility. And Colombe's attack on astrology clearly denoted the boundaries that the Church would allow in defining the nova's influence. The Italian polemics thus prominently pitted traditionalist and modern voices, a *querelle des anciens et modernes*. But in all cases, the unstated background question was, which disciplinary groups got to speak for the nova and, by implication, for academic authority? Against this conflicted politics of representation, Galileo could not have missed the air of relative philosophical openness and possibility in Kepler's *De Stella Nova*. In Prague, not only could the imperial mathematician frame the problem of the nova in a Copernican universe, but he was also free to argue it publicly against Bruno's infinite worlds. In Padua, Venice, Florence, and Rome, Bruno's name could not even be mentioned.

15

How Kepler's New Star Traveled to England

KEPLER'S STAR OVER GERMANY AND ITALY

Although Kepler's *De Stella Nova* is a difficult, unruly, and, at times, indigestible book for modern readers (for whom no translation exists), it had broad appeal for different contemporary groups. As with the 1572 event, Kepler described a novelty that required no special technical skill to observe. Even the technical claim from parallax that it was a *stella* was far less controversial than in 1572. Lorenzini, the hapless opponent of parallax, was something of a soft target for Kepler—as he was for Capra, Mauri, and, much later, Galileo. Moreover, Kepler's systematic attack on Pico and in defense of a reformed, aspectual astrology linked the new star to the science of the stars. In these different ways, therefore, the nova moved out of the domain of the strictly miraculous and became part of the ordinary course of nature. At the same time, Kepler's discussions of heavenly alteration and the size of the universe brought him into explicit engagement with the natural philosophy of the modernizers—especially Bruno, Brahe, and Gilbert. In short, unlike the *Mysterium* and the technically forbidding *Astronomia Nova*, Kepler's book on the 1604 nova held appeal for diverse audiences.

As with the audience for the *Mysterium*, these readers included nobility of varying degrees of astronomic literacy; skilled heavenly practitioners in courts, universities, and aristocratic circles; a handful of academic theologians; and the king of England. To date, I have found no evidence of any university philosophers who read or commented on the book immediately following its appearance. Even in the absence of a "scientific community" as such, Kepler's work—somewhat like Tycho's *Progymnasmata*—structured a new space of possibilities for a diverse group of courtly and academic philosophizers.

What can be said about the manner in which the *Stella Nova* traveled between different sites, and how did it affect the Copernican question? Between November 1606 and the summer of 1607, copies circulated within Prague aristocratic circles and into many parts of Europe. The book was dedicated to Emperor Rudolf, following a pattern whereby Kepler associated his most substantial works with major rulers and lesser works with members of the upper nobility. Some copies were sent out with personal dedications. The most interesting was the one he sent to King James I.

More broadly, the distribution of copies shows the breadth of the learned, Latin-reading audience that Kepler was trying to reach in this first decade of the seventeenth century. Some were princely rulers from among the hundreds of scattered political territories of the Holy Roman Empire: Georg Friedrich, margrave of Baden; Baron Erasmus von Stahrremberg of Upper Austria; Maximilian, archduke of Tyrol; Christian II, duke of Saxony; and Archduke Ferdinand of Austria.[1] Others were members of the Bohemian nobility or advisers at the Rudolfine court, such as the emperor's confessor, Johannes Pistorius.[2] Some were regular correspondents—the chancellor of the electorate of Bavaria, J. G. Herwart von Hohenburg, and Kepler's close and prolific correspondent, the East Frisian mathematician and preacher David Fabricius. Still others were friends and teachers from his student days at Tübingen: Michael Maestlin; the theologian Matthias Hafenreffer and his son, Samuel; and his schoolmate Christopher Besold. A few were mathematicians in municipal posts, such as Johann Georg Brennger at the imperial city of Kaufbeuren, or at other German universities, like Ambrosius Rhodius at Wittenberg and Johann Reinhard Ziegler at Mainz.

Occasionally, specific reactions survive. Brennger's response illustrates how Kepler's engagement with Bruno and Gilbert opened a new space of legitimacy where the modernizers blended in as a seamless part of the discussion. The geocentrist Brennger agreed with Gilbert that the Earth and the planets possess magnetic natures that are in harmony with the Sun by virtue of their axes being inclined with respect to it; but he did not agree with Kepler that the Sun attracted and repelled the planets. He willingly accepted the Earth's sensitive faculty of Kepler's aspectual astrology but argued that that faculty should be seen as magnetic.[3] Reader response was configured partly by Kepler himself. Because he had organized his work as a group of disputations, it could serve quite different functions. Some might read it for its reformed astrology, others for its physical and metaphysical claims concerning space and matter, others for its contentions regarding the location and generation of the nova, and still others for Kepler's final speculations on its prophetic meaning.

Galileo was part of the group of skilled academic mathematicians who could have been expected to be interested in Kepler's arguments with Pico and the modernizers. Yet it is not known how Galileo came by a copy of the *Stella Nova* while he was working under the shadows of Bruno's execution, Pinelli's death, and his own difficulties with the Paduan Inquisition. There is no evidence in the surviving correspondence that Kepler had tried yet again to attract the Italian's notice. Had Galileo received the book directly from Kepler, presumably he would have sent an acknowledgment of a sort resembling the one that he had sent in 1597; but we know of no such response. One can only speculate that Galileo had encountered the book through one of his Paduan friends or perhaps through Edmund Bruce. It is possible that he borrowed it. In any case, he would have seen the book no earlier than the fall of 1606—certainly in time to be consulted before the attack on Capra. However it had come into his possession, Galileo continued by his silence to deny Kepler any public credit.

KEPLER'S ENGLISH CAMPAIGN
Converting King James

By contrast with the constrained environment in which Galileo worked in Italy, Kepler was beginning to attract significant notice outside the Empire among a handful of learned English and Welsh philosophizers, mostly educated at Oxford and subsequently associated with the household of Henry Percy, the ninth earl of Northumberland (known as the "Wizard Earl").[4] Many of these were aristocrats or clients of aristocrats disposed, like their Central European and Italian counterparts, to heterodox intellectual viewpoints. There is no evidence that Kepler's ideas provoked anything like the sharp retort that greeted Bruno at Oxford in 1583; indeed, there is no evidence that any of his ideas even became topics for disputations at Oxford and Cambridge.[5] The *Stella Nova* appears not to have been seen as a text for university practitioners.

As chapter 13 shows, Kepler was already something of a known quantity in England in the final years of Elizabeth's regime through the spy network that, unbeknownst to him, had links from 1599 to the earl of Essex via Anthony Bacon and, after 1602, to Sir Michael Hicks. His progress from obscure, confessionally beleaguered Graz prognosticator and teacher to the highly visible position of imperial mathematician left him, like his predecessors Ursus and Brahe, as a Protestant at the nominally Catholic Rudolfine court—a fact that might have been of political interest to English agents.

Kepler was also quite conscious of the power of print in his own time—undoubtedly influenced by the example of Tycho Brahe. After Tycho's untimely death in 1601, Kepler aggressively used print to spread his readings of the heavens. In the penultimate chapter of the *Stella Nova*, he explicitly departed from D'Ailly, Leovitius, Roeslin, and other astral prophesiers by attributing to the cumulative effect of planetary conjunctions not the usual rise and fall of kingdoms and religions but the unusual appearance of great secular achievements: new universities, voyages of discovery, weaponry, the critical study of texts, and the appearance of new systems of thought. Chief among these novelties was print.

> What shall I say of today's mechanical arts, countless in number and incomprehensible in subtlety? Do we not today bring to light by the art of printing every one of the extant ancient authors? Does not Cicero himself learn again how to speak Latin from our many critics? Every year, especially since 1563, the number of writings published in every field is greater than all those produced in the past thousand years. Through them there has today been created a new theology and a new jurisprudence; the Paracelsians have created medicine anew and the Copernicans have created astronomy anew. I really believe that at last the world is alive, indeed seething, and that the stimuli of these remarkable conjunctions did not act in vain.[6]

Kepler's representation of a historical shift toward humanity's control of its own destiny, of confidence in human learning and the power of print, still assumed a world subject to astral influences. But those influences had now been seriously circumscribed. Indeed, Kepler's sensibility in this passage exactly mirrors the advice that he gave to the emperor about astrology in 1611: study the stars, but trust more in the ordinary, worldly experience of your diplomatic advisers.[7]

The advice also applied to Kepler himself. With each passing year, he was turning out new works that, through book fairs, emissaries, and correspondence, increased the chances of his views becoming known by people whom he did not know personally. For example, Sir Christopher Heydon, an English nobleman with keen astrological interests, initially learned that Kepler had written a work on the "foundations of astrology" when he bought a copy of the *Optics* and then wrote to say that he could not obtain a copy of the astrological work from any booksellers.[8] Kepler replied that he was most thankful for the discovery of printing, which had enabled a most excellent man in England to become a reader of his book. As usual, Kepler then sought to conduct a conversation at a distance, to enroll a prospective friend by spilling out his ideas in one of his famously long letters, much as he had done in his earlier letters to Bruce and Galileo: his work on

Mars, his progress on musical harmonies, the disputations on the nova, and an extensive description of his aspectual astrology (and his differences with Offusius). The issue here was not, as it would be shortly with the telescope, the reliable replication of an observation, but the persuasiveness of Kepler's foundational principles for the science of the stars.[9]

Kepler also did not shrink from promulgating his views directly. Sometime after James came to the English throne in 1603, Kepler recognized a possibility for some kind of patronal relationship—in this case, endorsement of his reformed astrology. No doubt he had been encouraged in this view by James's public support of Brahe's *Progymnasmata*. But having failed abysmally, thus far, with Galileo, how could Kepler hope for the king's public endorsement? Could he even expect the king to read the *Stella Nova*? (Did patrons ever read any of the works dedicated to them?) And if so, did he expect James to plow through his arguments against Giordano Bruno and Pico della Mirandola? Indeed, what sort of prospective "convert" and patron was James I?

The case of King James is particularly interesting because, far more than earlier English monarchs, he established himself as an author. Indeed, his power as a ruler was bound closely to his use of language and his representation of himself as a writer. James's enhancement of the genre of monarchic writing made it a critical attribute of power.[10] As King James VI of Scotland, he had written a gloss on the book of Revelation.[11] A decade later, he set down his views on witches and astrologers in his *Daemonologie, in Forme of a Dialogue*, a work that Kepler knew in the 1604 Latin edition.[12] Among other things, James invoked the eschatology of Revelation to account for the prevalence of witchcraft in his own time: "The consummation of the worlde, and our deliverance drawing neare, makes Sathan to rage the more in his instruments, knowing his kingdome to be so neare the ende."[13] Just as some writers had taken the existence of deformities and monsters to be signs of the End, James saw witchcraft as a sign of the devil's distress with the inexorable progress of the eschatological narrative of the world. Most famously, James was also a prolific political writer. His *Basilikon Doron*—cast in the form of practical advice to his son Henry—emphasized the king's supremacy in affairs of the church while maintaining that a good king must never act against his subjects' well-being "by inuerting all good Lawes to serue onely for his unrulie priuate affections." Johann Sommerville characterizes James as a "moderate absolutist" who "combined absolutist principles with an emphasis upon the monarch's duty to rule according to law and in the public good."[14] James also sponsored a new translation of the Bible, culturally the indisputable centerpiece of sacred writing. The translators and composers of the King James Bible represented him as the "principal mover and author" of the whole undertaking through his "religious and learned discourse." In fact, James meant his Bible to speak in a single voice, eschewing the disputatious marginalia found in the Geneva edition.[15] Finally, the king used his writings to identify dangerous opponents. The devil had many faces: he might show up embodied in Puritans, Catholics, magicians, astrologers, or witches. As if to underline the reality of demonic dangers, a Catholic plot to blow up the English Parliament with gunpowder was foiled on November 5, 1605. Soon thereafter, Parliament passed legislation against Catholics, including a statute "for the better discovering and repressing of popish recusants."[16]

The king's political powers also seemed to possess a miraculous character whereby, as Stuart Clark persuasively argues, James represented himself as a divinely appointed judge engaged in contests with the spells and charms of the devil.[17] In his *Daemonologie,* he distinguished lawful from unlawful knowledge. The devil operated on both the learned and the unlearned, but the learned were most vulnerable to having "their curiositie wakened up and fedde by that which I call his schoole; this is the *Astrologie* judiciar."[18] Such men tried to gain for themselves "a greater name, by not onely knowing the course of things heauenly, but likewise to clim to the knowledge of things to come thereby." This was how one dangerous thing led to another: starting with what is "lawfull" and what "proceed of natural causes onelie," they were led "upon the slipperie and uncertaine scale of curiositie" until "lawfull artes or sciences failes, to satisfied their restlesse mindes, euen to seeke to that black and unlawfull science of *Magic.*"[19]

Not surprisingly, James helped himself to the standard, two-cell classification of the science of the stars. "The science of the Heauenly Creatures, the Planets, Starres, and such like: The one is

their course and ordinary motions, which for that cause is called *Astronomia* . . . that is to say, the law of the Starres. And this Arte indeede is one of the members of the *Mathematicques,* and not only lawfull, but most necessary and commendable. The other is called *Astrologia* . . . which is to say, the word and preaching of the starres." He then subdivided astrology into two parts, safe and dangerous:

> The first, by knowing thereby the powers of simples, and sicknesses, the course of the seasons and the weather, being ruled by their influence; which part depending upon the former [i.e., *Astronomia*], although it be not of it self a part of *Mathematicques:* yet it is not unlawful, being moderately used, suppose not so necessarie and commendable as the former. The second part is to trust so much to their influences, as thereby to fore-tell what common-weales shall flourish or decay: what persons shall be fortunate or unfortunate: what side shall winne in anie battell: what man shall obtain victorie at singular combate: what way, and of what age shall men die: what horse shall winne at match-running; and diuerse such like incredible things, wherein *Cardanus, Cornelius Agrippa,* and diuers others haue more curiouslie than profitablie written at large.[20]

The Jacobean divisions have the ring of familiarity. They are much the same lines of legitimation variously deployed by Calvin in Geneva, Andreae in Tübingen, and Clavius and Sixtus V in Rome, and they were probably passed down to the king through his Scottish humanist tutor, James Buchanan (1506–82).[21] James could also draw upon Sicke van Hemminga's devastating attack on the multiple uncertainties of nativities.[22] And when James granted the Company of Stationers a monopoly over print in 1603, their language presumed the distinctions of the *Daemonologie:* "All conjurers and framers of almanacs and prophecies exceeding the limits of *allowable astrology* shall be punished severely in their persons. And we forbid all printers and booksellers, under the same penalties, to print or expose for sale, any almanacs or prophecies which shall not first have been seen and revized by the archbishop, the bishop (or those who shall be expressly appointed for that purpose), and approved of by their certificates, and, in addition, shall have permission from us or from our ordinary judges."[23]

Kepler, of course, was not seeking the permis-

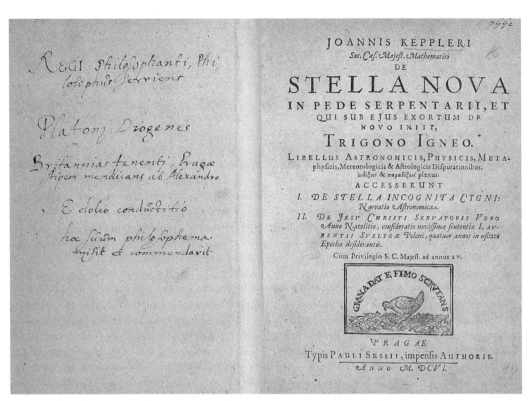

77. King James I's copy of Kepler's *De Stella Nova*, with poem inscribed by Kepler. © British Library Board. C.28.f.12.

sion of the stationers or, for that matter, of the archbishop. In his role as the Hapsburg emperor's official *mathematicus*, Kepler personally approached the king directly with a gift. This offering was neither a decorated material object meant for court display, like the polyhedral liquor dispenser he had proposed in 1596, nor a clever, flattering family emblem, but a copy of the *Stella Nova* adorned with a Keplerian poem. In this inscription, Kepler represented James as the "Philosophizing King" (*Rex Philosophantis*) and as unifier of England and Scotland—"Diogenes holding together the Britains"—while he himself was the "Philosopher serving Plato, begging money from Alexander of Prague"—an oblique reference to his chronically hopeless financial woes.[24]

In a telling letter that accompanied the book, Kepler communicated his respect for the king's learning by recommending specific chapters of the *Stella Nova* for His Majesty's study. Rather than promote a detailed reading of those chapters favoring Kepler's support of Pico's critique (with which the king would likely agree), he suggested that "for chapters 2, 3, 4, 5, 6 it suffices

just to read the titles." Then, after perusing those headings, the king should proceed directly to chapters 7, 8, 9, 10, 28, and 29, which would provide him with Kepler's arguments in favor of astrology and against Pico. These were especially important, as they were meant to bring James around to Kepler's own view of astrology—a view based on "lawful," harmonic, natural reasons. As for specifically astronomical and natural philosophical considerations, Kepler recommended that the king read chapter 1 in conjunction with the large fold-out plate of the nova in chapter 14; then, concerning "the very deepest questions," he should consult chapters 25, 26, and 27; and finally, he should look "most fully" at chapter 30, where he would find Kepler addressing the distinction between superstitious, demonic astrology and good, Christian astrology "practiced without magic."[25] In other words, Kepler was trying to win the king over to his kind of astrology by presuming to satisfy the Jacobean standard of an astrology grounded in "natural reasons," and with some hope, perhaps, that James would also pay attention to his more weighty and labyrinthine arguments about heavenly order. Cautiously excluded were the rather more challenging arguments against the universe's infinitude in chapter 21, later prized by Alexandre Koyré.

Because the dedication to the *Harmonice Mundi* continued themes adumbrated both in the *Stella Nova* and in his letter of 1607 to the king, it fills out what can be said of Kepler's general purpose. The dedication makes clear that Kepler hoped to convince the king to support both his reformed astrology and the entire project of archetypal principles. Might favorable reception of those principles in astrology lead to favorable consideration of Kepler's more adventuresome astronomical proposals, including the Copernican astronomy and the new rule connecting periods and distances? Kepler associated James's critical stance against astrology with his own reformed position: "James, on becoming a man, when he was at the helm of his kingdom, marked the excesses of astrology with public censure—which are in fact very clearly revealed in Book IV of this work, where the true bases of the effects of the stars are disclosed. . . . Thus nobody can have any doubt that you [James] will have complete understanding of the whole of this work and of *all its parts*."[26]

The dedication also boldly developed his earlier representation of the king as a friend of Plato; but now he introduced a calculated allusion to Tycho Brahe: "What more appropriate patron could I choose for a work on the harmony of the heavens, with its savor of Pythagoras and Plato, than that King who has borne witness to this study of Platonic learning by domestic tokens, which we know also from the public veneration of his subjects? Who when still a youth deemed the astronomy of Tycho Brahe, on which this works depends, worthy of the ornaments of his talent?"[27] Beyond evoking James's established association with Brahe, Kepler lauded the king's qualities as a philosopher in a way that mirrored and amplified the private dedication in the *Stella Nova*. Yet the dedication contains no hint that Kepler was angling for a position at the English court. If anything, what he clearly desired was public validation of his own philosophy of the heavens. This could come only from a king-philosopher who, unlike ordinary rulers, was supposed to be able to read and evaluate—and perhaps even to converse. And Kepler knew that James, as an author, was no ordinary ruler. Pre-

sumably, the king-philosopher would also recognize the worthy advice of a philosophical—as opposed to an ordinary—prognosticator. Thus, while Kepler recalls that in the *Stella Nova* he had made "a public prognostication . . . which burnt like a fiery coal (a verse well known in Scotland)," he then set himself a further, more transcendent goal "the more resolutely to chant the cosmic harmonies at some future time."[28] As in his earlier quests to win over the resistant Tübingen theologians and the reluctant hold-out Galileo, Kepler continued to see himself as actively interpreting and spreading God's Word as embodied in the heavens—a sort of secular theologian.[29]

To add to these appeals, Kepler strategically placed a digression just before the chapter on astrological principles, in which he laid out a harmonically grounded political philosophy directed against that of Jean Bodin (1530–96). Why else would the Bodin digression have been inserted than to show James the right sort of theoretical principles that ought to ground a philosophy of the polity?[30] The major theme of Bodin's political philosophy was the notion of an absolute sovereign power vested in the office rather than the person of the ruler. For Bodin, it was an attribute of sovereignty that the king had the right to make human laws but was subject only to divine and natural law—an idea that Kepler must have thought would appeal to King James. Bodin's musings about arithmetical progressions were of relatively minor importance in grounding his *Les six livres de la République* (*Six Books of the Commonwealth*).[31] Yet Kepler inflated this part of Bodin's philosophy to provide himself an occasion to publicize the superiority and relevance of his harmonic principles.

At the end of the *Harmonice*, Kepler defended his flank against a rival, prospective client of James, the polymath Rosicrucian physician Robert Fludd (1574–1637). Like the attack on Bodin, Kepler's detailed and devastating polemic against Fludd focused on (what he took to be) the mistaken and misdirected character of Fludd's metaphysics of macro- and microcosmical harmonies. Fludd represented these harmonies with elaborate symbol pictures, similar to Roeslin's, whereas Kepler held that his mathematical diagrams were not merely symbolic but corresponded to, and could be checked against, the physical world. James could not be allowed to think that Fludd's myriad pictures were in any way connected to the real world.[32] The heavens and their effects on human affairs—celestial and political philosophy—were united by a common body of archetypal doctrine. Thus, in composing the dedication to James, Kepler was explicit in the basis of his appeal for royal support: "The reasons for thinking of this patronage for my Harmony were supplied by the manifold dissonance in human affairs, which is indeed obvious, so that it cannot fail to offend, though it is compounded of melodic and distinct intervals." Indeed, "what else is a kingdom but a harmony?"[33] A Keplerian and Copernican harmony, of course.

Whether or not the learned James read and followed the imperial mathematician's advice is unknown, but the Jacobean court did not discourage Kepler. And, indeed, one may infer that the reception was quite positive, as in 1620 the learned and experienced diplomat Sir Henry Wotton (1568–1639) invited Kepler to England.[34] Wotton had just met Kepler in Linz and wrote of the encounter to another great promoter of natural knowledge, Lord Chancellor Francis Bacon, who had been seeking royal support for his schemes for many decades and whose *Novum Organon* had just appeared: "There I found Keplar," wrote Wotton, "a man famous in the Sciences, as your Lordship knowes, to whom I purpose to convey from hence one of your Books, that he may see we have some of our own that can honour our King, as well as he hath done with his *Harmonica*."[35] There is little doubt that the lord chancellor was less well-disposed to the celestial consonances of the "*Harmonica*" than was the king.[36]

Converting the Heavenly Practitioners

Kepler's strategy for enrolling the king in his program of safe, harmonic astrology carried on from 1607 into the debates with Robert Fludd in the 1620s. Yet as his astrological writings began to circulate in England, they landed in a political space that was more dangerous and conflicted than the situation in Prague. Hints of the astrologers' wars of the Continent were to be found some forty years earlier in William Fulke's *Antiprognosticon*. But now, in 1603, with the enthronement of a new ruler whose views on foretelling the future were, if anything, more pronounced and aggressively specified than those of Queen Elizabeth, the atmosphere for debate on this matter had begun to chill. English heavenly practitioners of these early years of the new century were thus highly conscious both of the king's voice in speaking for the future and of the possible uses to which Continental works might be put in their own writings. One consequence was that writers favorable to astrology tended to use Kepler's work in the selective way that marked the Wittenberg response to Copernicus.

The shift in the terms of discussion from the last years of Elizabeth to the early years of James's reign is well illustrated by a debate between John Chamber (1546–1604), prebendary of the royal chapel at Windsor and fellow of Eton College, and the astrological practitioner Sir Christopher Heydon. In 1601, Chamber initiated the controversy with his *Treatise against Iudicial Astrologie*, which echoed the diatribes of the godly divines of Geneva, Tübingen, Madrid, and Rome.[37] In the style typical of such theologically motivated denunciations, the treatise piled on the (negative) citations of ancient authorities and had no truck with technical mathematical detail. Yet, following Pico, Sextus Empiricus, Saint Augustine, and Sicke van Hemminga, Chamber augmented the familiar strategy of criticizing astrologers for their disagreements, uncertainties, and incoherencies. To the Augustinian objection of twins' experiencing different fates (underdetermined from birth!), Chamber added reports of monstrous births in which two boys joined at the hip died at different times; moreover, "by their birth they should have agreed wel, as hauing one constellation, but they did oft wrangle and fall out pittfully."[38] On the critical matter of the ordering of the planets' elemental qualities, Chamber criticized Ptolemy's assignments: "Why the Moone is moist, he ascribeth it to the vapours, which it draweth from the earth, then, how much more moist should the Sun be, which is known to draw infinitely more? then he saith that *Saturne* is cold because of his distance from the Sunne; and Mars hot by reason of his vicinity. To which may be said that of Mars hath his heat from the Sun, why is not the Sun as hot, or hotter then Mars? These things are ridiculous in philosophy & not worthy confutation."[39] Five years later, Kepler would mobilize similar Piconian objections against Ptolemy's arbitrary designations.

In 1603, just before James's ascension to the

throne, Christopher Heydon published a lengthy reply to Chamber.[40] Thus far, the debate was conducted in print—and in the vernacular—with Heydon clearly aligning himself with the modernizers and attacking his adversary for neither addressing "the whole nation of the learned" nor introducing new arguments.[41] A century after the Bellanti-Pico exchange, Heydon's subtitle displayed the full armory of skeptical resources deployed by Chamber and against which he felt it necessary to argue: "Reasons drawne out of Sixtus Empericus, Picus, Pererius, Sixtus ab Hemminga, and others, against this Arte, are particularly examined." His citations show that he was fully conversant with both English and Continental discussions, in which his sympathies lay with the likes of Copernicus, Rheticus, Maestlin, Magini, Ursus, Tycho Brahe, William Gilbert, Paracelsus, and Nicodemus Frischlin.[42] A student of the prominent Paracelsian physician Richard Forster—with whom he pursued a vigorous correspondence on astrological subjects—Heydon was influenced by the problematic made acute by the debates about celestial order in the 1580s and '90s as represented in the *Epistolae Astronomicae* and the *Progymnasmata:* Which system of the world should best underwrite the world of astral forces?

> For whether any of their opinions be true, or whether they be false, whether they be (as *Tycho* would have it) but one continued orbe, or many, or whether (as *Copernicus* saith) the Sun be the center of the world, and the earth be in the Sunnes place, between the sphere of *Mars* and *Venus,* the Astrologer careth not. For so that by any of these *Hypotheses,* he may come to the true place and motion of the Starres, this variety of opinions, whether such things be indeede, and in what order they be, is no impeachment to the principles of Arte.[43]

More than a century after Pico, Heydon was still preoccupied with securing the astronomical foundations of astrology against Piconian "Astrologiewhippers" and "opinionasters." Now, however, not merely the idiom but the occasions and choices for strengthening astrology had changed. Heydon took an agnostic position on planetary order—perhaps inspired by Ursus's blanket skepticism about astronomical hypotheses—an epistemic stance that would gradually become one of the important options for early seventeenth-century prognosticators. Yet, following Tycho and Roth-

mann—without deciding for either—he accepted that "our modern Mathematickes" had proved that there was "but one onely continued substance from the concaue superficies of the Moone, to the 8.Sphere."[44]

Chamber's reply to Heydon underlines just how politically freighted astrological practice was in England when James assumed the throne. The regime's opposition spread rapidly from the king to the Company of Stationers and now, with Chamber, to the Church of England. With the king's coat of arms displayed in brilliant colors opposite his frontispiece, Chamber directly echoed the king's language in his title: "A Confutation of Astrological Daemonologie, or the divells schole, in defence of a treatise intituled against Iudiciarie Astrologie, & oppugned in the name of Syr Christopher Heydon, Knight."[45] Further associating himself with the royal position, Chamber none too subtly intoned James's view that "Iudiciall Astrologie [is] an abuse, in my simple opinion, not inferiour to the superstition of witchcraft, and therefore very fitly censured by learned Epistemon in Your worthie work of Demonologie." Chamber also invoked the singularity of the king's learning and his special knowledge of the matter at hand: "Of all questions betwixt mine adversaire and mee, Ô King, I thinke mee happie, that I shall defende miselfe before your maiestie at this time; especially since You knowe the state of these questions, better than doeth either the Adversarie, or miselfe. Wherefore I am to beseech your Highnesse to reade mee with patience."[46] Chamber then pleaded that he had been an opponent of astrology even before the king's arrival but now, against their joint enemy—"the Demonologists"—he sought James as patron-protector: "The abuse of Astrologie I marked and misliked, before your Maiesties most happie comming to the State: neither did I onely obserue it, but detect it by writing, *nec tacui demens.* Whereupon I have procured to miselfe a bitter and sharpe adversarie supported with some other carying the countenance of learning. Being thus beset and baited of a number of Demonologists, in so iust a cause I had no way, but to flie to your sacred Maiestie for protection, who had already foreiudged the cause."[47]

A Jacobean shadow had fallen over astrological matters. The state was worried not about astrological theorizing per se but about the casting (and use) of horoscopes for the royal person. For controversialists like Chamber, all astrologers were suspect, a threat to the order of knowledge, to the state and the established church: he unflatteringly called them "figure-flingers and wizards," and Heydon returned the insult in kind, referring to his opponent as a "Urine-flinger."[48] The rhetorical resources for this controversy were almost as rich as those of the religious quarrels of the period.

Meanwhile, perhaps in the hope of advancing his cause against Chamber, Heydon initiated contact with Kepler, quickly acquiring the latter's *On the Firmer Foundations of Astrology*, the *Mysterium*, and the *Stella Nova*. He then drafted a reply to Chamber that was closely modeled on the first-mentioned of these works.[49] Here, then, was the prospect—unfortunately timed for Kepler—of his arguments for a reformed astrology being used against Chamber, a writer who blatantly sought to wrap himself in royal authority at just the moment that Kepler sought the king's ear and protection.

To add to Kepler's difficulties, Heydon then drafted a second treatise that appropriated elements of Keplerian astrology but explicitly refused to follow the Copernican arrangement. Although he completed it in 1608, Heydon withheld publication—probably because of the unwelcoming political atmosphere—and the manuscript did not appear until 1650, when its publication was financed by a group of newly invigorated astrologers of the Civil War period.[50] The obstacles to Kepler's campaign for a Copernican reform of the science of the stars could not be more strikingly exemplified than by this friendly, yet critically selective, reader.

Besides Heydon, Kepler had opened up other contacts with English mathematical practitioners through personal intermediaries. Two of Tycho Brahe's former assistants, Johannes Eriksen and the Dutch nobleman Franz Tengnagel, had traveled to England together. There, Eriksen met the polymath Thomas Harriot (1560–1621) in London and reported back to Kepler, who established a brief correspondence beginning in October 1606, largely focused on optical matters.[51] Eriksen told Kepler that Harriot had experienced certain difficulties (unspecified in the letter) associated with astrology and that he was "under suspicion" and "still circumscribed."[52] Kepler evidently assumed that Harriot was a practitioner and also a prisoner. He proposed help, informing the Englishman that for the past ten years he had rejected the common basis of astrology—the usual twelvefold division of the zodiac into houses, triplicities, and so forth—and had retained only the aspects "having led astrology across to the harmonic doctrine." The implication was that Kepler's astrology was both epistemologically more reliable and politically safer. If Harriot was further interested, he should consult the *Stella Nova,* and Kepler would welcome his opinions on it.[53] Kepler then added a sentence indicating to Harriot his awareness of the king's views: "Although I bow to no one in defending the truth for itself, nevertheless I think that King James will not be about to embrace it because he damns those things which I hold."[54]

How much Kepler actually knew from Eriksen of Harriot's true political circumstances and his views about the practice of astrology in the early years of the Jacobean regime is unclear. In Harriot's reply of December 2, 1606, he did not mention astrology at all but referred in general terms to "bad fortune which I now feel so that it is difficult for me either to write or to think and argue accurately about anything."[55] And in a letter of July 13, 1608, he confessed to Kepler that he could not "philosophize freely."[56] In fact, unlike Heydon, Harriot suffered real and considerable misfortunes. In 1594, a special commission mentioned him in its interrogatories while fishing for evidence of atheism against Sir Walter Raleigh, an early patron who was then rapidly falling from favor at Elizabeth's court.[57] This was bad enough for Harriot, but his difficulties continued in association with his next major patron, Henry Percy, the ninth earl of Northumberland. Harriot was jailed briefly in November 1605 on suspicion of association with the principals in the Catholic-led Gunpowder Plot. Thomas Percy, a distant cousin of Henry, was one of the leaders in the (unsuccessful) conspiracy to blow up the houses of Parliament with the king present. A tip-off led to Percy's capture and assassination. Meanwhile, the earl was tried, slapped with an impossible fine, and imprisoned in the Tower of London, where he became Raleigh's neighbor and remained until 1621.[58] Harriot, among others, was interrogated, but this time, unlike the earlier Raleigh investigation, the questions put to him were written in the king's own hand, and they provide a good indication of the sorts of worries that were entertained about him.

I. Herriote wolde be asked quhat purpose he hath haerde his lorde use anent my natiuitie or fortune.

II. if euer his lorde desyred him to caste it, or tell him my fortune.

III. if euer his lorde seemed to be discontented of the state.

IV. if euer he harde him talke, or aske him of my childrens fortunes.

V. if euer his lorde desyred to knowe quat should be his own fortune & ende.

VI. & if he did caste milorde, & his sonnes natiuities by his owin commande & knowledge.[59]

Although Harriot's answers have not survived, it is important to notice what was *not* at stake: the king made no reference to any experiments that Harriot might have performed or to any doctrines that he held in natural philosophy, including atomism, the Copernican planetary ordering, and the infinity of the universe. Indeed, the apprehensions about his alleged practices as an astrologer were not even the kinds of theological worries about astral determinism that had brought Galileo into conflict with the Venetian Inquisition a year earlier. The royal disquiet about Harriot concerned the single most sensitive issue touching astrological practice: using natural knowledge of the heavens' ordinary course to predict the death of a ruler or his family. This had been the main concern of Pope Sixtus V in the bull of 1586 and later for Pope Urban VIII in his reaffirmation of that bull in 1631; and, not surprisingly, it underlay King James's anxiety in 1605.[60]

Of course, just because Harriot was charged with casting the king's nativity does not mean that he did anything of the kind. In December 1605, imprisoned in the Gatehouse jail, he composed a plea to the Privy Council that John Shirley has called "probably the most personal record of the man which is left." The plea speaks of his "present misery," the details of "my sickness," and his commitment to "honest conversation and life."[61] However, his main defense was not the religious orthodoxy of his philosophical studies but his representation of himself as disinterested in political affairs. "I was neuer any busy medler in matters of state. I was neuer ambitious for pre-

ferments. But contented with a priuate life for the loue of learning that I might study freely. Wherein my labours & endeuours, if I may speak it without praesumption, haue ben paynfull & great. And I hoped & do yet hope by the grace of God & your Lordships fauour that the effectes shall so shew themselues shortly, to the good liking & allowance of the state and common weale."[62] Harriot did not address here the charge of casting the king's "natiuitie or fortune," but ultimately his answers to the Gunpowder Plot interrogation must have been satisfactory. Not long thereafter, he was released.

This outcome strongly suggests that no evidence was found of any astrological judgments touching the Crown. Two explanations may be suggested. First, it is possible that Harriot had already destroyed any potentially incriminating evidence.[63] Second, perhaps he was able to defend himself readily against the charge of casting nativities because he had already disavowed any firm basis for the practice in his natural philosophy. The second possibility opens up several important and difficult questions concerning the character of Harriot's general ambitions in natural philosophy. Had he been persuaded by Bruno's arguments before 1605 for an infinite, homogeneous space populated by innumerable bodies and made up of discontinuous minima separated by a void?[64] If so, had he rejected astrology for reasons of the sort that had led Bruno to cast aspersions on that discipline as conventionally practiced? And if Harriot accepted as true or probable the Earth's motion, as seems likely, was his position shaped decisively by Bruno's arguments rather than by those of Digges, Gilbert, or, perhaps, *De Revolutionibus* itself?[65]

To address these questions, we must take into account serious problems concerning the character and extent of Harriot's Brunonian and Copernican commitments as well as the meaning of what many scholars want to call the "Northumberland circle."[66] John Henry is surely right that at least one species of corpuscularianism held by Harriot—his view of mathematical points as indivisibles—is quite at odds with Bruno's indivisible minima and, if anything, closer to the views of the later Galileo.[67]

It seems clear that there was a group of men associated with the household of the ninth earl of Northumberland, from the early 1590s and that they exhibited a learned sociability characteristic of that moment—a mix of traditionalist and modernizing tendencies. The existence of an intellectual consensus among them, however, is much less certain. Their interests, talents and commitments were, if anything, quite varied, and it is difficult to find systematic linkages among Copernicus's theory, Democritus's atoms, Neoplatonism, Bruno's infinite space and innumerable worlds, and Kepler's planetary models. The problem is further exacerbated by the following difficulties: that the Harriot papers are fragmentary and frequently illegible; that there are precious few references to Bruno in these notes and even among the books in the Wizard Earl's library, let alone clear references to Harriot's Copernican views;[68] and that there is no evidence that the ideas embodied in the one published work of the circle (Nicholas Hill's *Epicurean, Democritean, and Theophrastic Philosophy, Proposed Simply rather than Taught as Doctrine* [Paris, 1601]) were held by anyone else in the group.[69]

Let us grant, with John Henry, that the high degree of consensus implied by the designation of a "Northumberland circle" is not warranted by the currently available evidence. Nevertheless, it is still clear that at least some of Bruno's ideas were well received—even if they came only indirectly from Bruno—by Nicholas Hill, Harriot's pupil Sir William Lower, and Harriot himself; and further, that at least some kind of sociability existed.

One kind of sociability, as is most obvious with Brahe, Kepler, and Galileo, was constituted by the medium of the scholarly letter. Next to the book, it was probably the most common form of communication among early modern heavenly practitioners, a resilient tool of thought and exchange. Moreover, this is often the only evidence we have. And it is through the letter that we learn that some members of Harriot's network of correspondents regarded Kepler's arguments against Bruno and Gilbert in *De Stella Nova* as coming up short. This is emphatically clear from the revealing letter written to Harriot by Sir William Lower from his residence in Trefenty, Carmarthenshire, South Wales, on June 21, 1610:

> Just at the instante that I receaved your letters wee Traventane Philosophers were a considering of Kepler's reasons by which he indeavours to overthrow Nolanus and Gilberts opinions concerning the immensities of the Spheare of the starres and

that opinion particularlie of Nolanus by which he affirmed that the eye beinge placed in anie parte of the universe the appearance would still be all one as unto us here. When I was sayinge that although Kepler had sayd somethinge to moste that mighte be urged for that opinion of Nolanus, yet be one principall thinge hee had not thought; for although it may be true that to the ey placed in anie starre of [Cancer] the starres in Capricorne will vanish, yet he hath not therfore so soundlie concluded (as he thinks) that therfore towards that parte of the world ther willbe a voidnesse or thin scattering of little starres wheras els round about ther will appeare huge starres close thrust togeather: for sayd I (havinge heard you say often as much) what if in that huge space between the starres and Saturne, ther remaine euer fixed infinite nombers which may supplie the apparence to the eye that shalbe placed in [Cancer], which by reason of ther lesser magnitudes doe flie our sighte, what if aboute Saturn, Jupiter, Mars, etc. ther moue other planets also which appeare not.[70]

Readerly selectivity can usefully point to vested commitments and interests. In this case, one may well ask why the "Traventane Philosophers" singled out chapter 21, the only place in the *Stella Nova* where Bruno's arguments on innumerable worlds in an infinite universe are explicitly mentioned. This is the same sort of problem earlier raised with respect to Galileo's focus on Mazzoni's critique of Copernicus in the *Comparison of Plato and Aristotle*. Evidently, Lower and his friends had at least two interests at stake. First, they found Kepler's argument against Bruno and Gilbert to be inadequate; at the very least, this objection raises the possibility that they had sympathies with Bruno that antedated their reading of *De Stella Nova*. It is also possible, however, that it was Kepler alone who had introduced them to Bruno's radical propositions about space, and hence that this was the extent of their knowledge of the Nolan's philosophy.[71] Second, prior to his knowledge of Galileo's telescopic discoveries, Lower knew that Harriot was already entertaining the possibility of invisible bodies circulating around the planets within the great space between the outermost visible moving star (Saturn) and the fixed visible stars. Harriot's pretelescopic speculations, therefore, were Brunonian in a sense quite similar to that of Edmund Bruce in his 1603 letter to Kepler. Thus in 1610, when Galileo reported new stars never before seen, these

novelties confirmed what were quite clearly Brunonian expectations.[72]

Like Kepler, Harriot could have asked himself
how it was possible to have a coherent astrology
either in a universe with no center and innumerable planetary worlds or in a unique, heliocentric
universe with a moving Earth. In the first case,
he had no alternative available but Bruno's total
rejection of conventional astrological theory or
its replacement with some sort of nonmathematical astral magic outside the structure of the science of the stars. In the second case, he had Kepler's strongly mathematical harmonic reform.
But, because there is no evidence that Harriot
was in any way inclined to Kepler's solution, one
must turn to the first possibility. And here, circumstantial evidence that Harriot rejected astrology because of the absence of hierarchy in the
unlimited Brunonian universe can be found only
if one is willing to associate him with Nicholas
Hill's *Epicurean, Democritean, and Theophrastic
Philosophy*—at best, an insecure influence by association. Even that evidence is not overwhelming
because, like Dee with his *Propadeumes,* Hill constructed his work as a series of discrete aphorisms.[73] Typical of this moment, it was, in Hill's
telling words, "a philosophy neither new nor old."[74]
Perhaps the best that one can say is that it included one aphorism condemning traditional astrology as the work of "cowardly philosophers"—
hardly a sufficient basis for excluding a new,
reformed astrology.[75]

Given what we know of Harriot's political situation after the unraveling of the "Gunpowder
Treason," such reasons may be sufficient to explain his unwillingness to take up Kepler's harmonic astrology. Even the favorably disposed
Christopher Heydon, who was in no way involved
in the plot, had encountered unremitting resistance to the publication of his *Astrological Discourse.*[76] But for Harriot there could have been
other reasons apart from those that had blocked
Heydon, considerations grounded in a different
context, where atoms were invoked not as part of
the infinite divisibility of lines but as an explanatory resource in optics. In his discussion with
Kepler about the transmission of light in a transparent body, Harriot explained the bending of
the ray from the surface physically as a series of
"internal reflections," the result of light meeting
resistance from "corporeal parts" and encountering no resistance from "incorporeal parts" or
vacuums between such indivisible bodies. Here,
in the context of his conversation with Kepler,
Harriot sounds like a proponent of a physical
rather than a mathematical atomism. If, at this
point, Harriot had read a few chapters beyond
chapter 21 in *De Stella Nova,* then he would have
known that Kepler had serious disagreements
with a picture of chance collisions in a universe
of Epicurean atoms. Perhaps for this reason he
chose not to engage Kepler's wider natural philosophy and astrology in a frontal attack when
instead he half-jokingly tweaked his correspondent: "Now I lead you through the doors into the
houses of nature where its secrets lie. If you cannot enter because of their narrowness, then abstract mathematically and contract yourself into
an atom and you will easily enter and later, after
you leave, tell me what marvels you have seen."[77]

Reading the Harriot archive leaves the impression that his deployment of atoms in this instance was restricted to piecemeal investigations
(and explanations), leaving open the question of
whether he held a genuinely systematic atomic
philosophy. Did Harriot ever ponder the question of how you could use atoms and the void to
explain one part of nature and not all others?
The problem relates to what may be thought of
as an "archival atomism"—islands of coherence
in the jumbled world of Harriot's surviving
manuscripts. Still again, indications of Harriot's
ideas about matter showed in his disagreement
with Kepler about *De Mundo,* an unpublished
manuscript by William Gilbert shown to him by
Gilbert's brother. *De Mundo* did not expound an
atomic natural philosophy, but it did propose
that space was a void inhabited by magnetic
planetary globes.[78] This work took an important
step beyond *De Magnete,* where, as Kepler already knew, Gilbert followed Copernicus in setting the immense sphere of the fixed stars at
rest. But whereas the Gilbert manuscript introduced both magnetic globes and a limitless void,
it resisted awarding planetary status to the
Earth.[79] Harriot, by then cognizant of Kepler's
objections against Bruno, pretended to accentuate their common cause in Gilbert's philosophy:
"I make mention of this [work of Gilbert] because, as I guess from your writings, this man's
philosophy is most agreeable to you. I have seen a
copy and I have read several chapters where I see
that *with us* he defends the vacuum against the
Peripatetics."[80] Kepler, of course, had no such

common cause with Harriot and Gilbert in the existence of a void. He had appropriated from Gilbert only what he needed—an incorporeal (magnetic) force to replace Pico's light physics.

Although Harriot's casual allusions to atoms and the vacuum in the 1608 letter to Kepler do not provide evidence for a systematic philosophy, they do point to real differences with Kepler while again exemplifying the persistent fault lines that divided the modernizers. And yet, per-haps because Harriot's views were not laid out systematically—either privately or in public—no one perceived in them any sort of hopeless in-commensurability with those of Kepler. And, in-deed, soon after the publication of Kepler's *Astronomia Nova* (1609), there is evidence that Harriot and Lower were quite convinced by Kepler's new planetary theory but not by his "magnetical natures."[81] In England, as in Prague, a *querelle des modernes* had begun.

PART VI

The Modernizers, Recurrent Novelties, and Celestial Order

16

The Struggle for Order

THE EMERGENT PROBLEMATIC OF THE *VIA MODERNA*

For celestial modernizers of the early seventeenth century, the problematic that had been emerging since the 1570s began to show signs of consensus: recurrent events (planets), the subject of the science of the stars, and nonrecurrent events (comets and new stars) somehow seemed to belong together in the realm of ordinary rather than extraordinary phenomena. Galileo's discoveries at the end of the first decade would further reinforce the sense that the heavens contained recurrent phenomena, marvels that, even if hidden, were still part of the natural order. But how did any of this pertain to the Copernican question? Was it necessary to assume the Earth's motion in order to explain why and when comets and novas appeared and disappeared or of what sort of stuff such novel entities must be made? Should the Earth be kept at rest, as various geoheliocentrists of the 1580s maintained? How large did the universe need to be to contain such a diverse array of appearances? As more questions of this sort accumulated and were seen to be part of a common problem of celestial order, there was an accompanying shift in the role of mathematically equipped heavenly practitioners. Yet engagement in philosophical practices, a disciplinary change in the astronomer's role, did not itself determine—let alone make it obvious—how to deduce from planetary order requisite principles of natural philosophy. Put otherwise,

both the new disciplinary practices in which astronomers were engaging, and the versions of planetary arrangements to which they committed themselves, *underdetermined*—and hence did not eliminate—the then-available alternatives in natural philosophy.[1]

Hints of the difficulty had already appeared in earlier practices of ignoring, such as characterized the Wittenberg interpretation. Maestlin's proposed Copernican model for the comet of 1577 simply ignored the possibility of a geoheliocentric alternative of the sort soon offered by Tycho Brahe, and neither model offered a natural explanation for the prediction that the comet would return in less than a year. In turn, Edmund Bruce's 1603 letter to Kepler shows both the novel suggestions and confusions of a nonpractitioner struggling to fit elements of Kepler's universe together with those of Bruno while ignoring (and hence dismissing) altogether the efforts of Galileo. A noteworthy cohort of natural philosophers who came of age in the early seventeenth century—Isaac Beeckman (b. 1588), Marin Mersenne (b. 1588), Thomas Hobbes (b. 1588), Pierre Gassendi (b. 1592), and René Descartes (b. 1596)—inherited these incomplete judgments. In turn, they would favor a different approach to the ordinary realm, constructed from new kinds of foundational principles of matter and motion. As natural philosophers, they rejected (or ignored) the Keplerian strategy of building a physics around the science of the stars; instead they tried to derive the Copernican ordering from their own physical principles.

These divergent strategies for managing problems of cosmic order and integration were prodigious enough even without attempting to make their principles consistent with predicting the future. If prognosticators assumed the Copernican ordering in order to solve the problems of comets and novas, they were confronted with the further question of how that arrangement would be compatible with a heliostatic astrology—unless, yet again, the issue was ignored. Among the modernizers, apart from Kepler with his new account of astral influence, most (including Galileo, Stevin, Rothmann, and Gilbert) chose to ignore the question of influence, separating it from the problems of scripture and physics, or renounced astrological prediction altogether (as did Maestlin). Kepler's controversies with Helisaeus Roeslin and Philip Feselius about the credibility of astrology did not directly implicate the order of the planets.

To complicate matters further, traditionalist academic writers did not fold their tents. For many reasons, Aristotelian natural philosophy proved to be sturdy and resilient—indeed, in manifold ways, it continued to provide scholars and students with the defining resources and methods for describing and explaining the natural world.[2] In part, as Edward Grant has observed, the long-standing medieval scholastic practice of treating topics in Aristotle as independent propositions allowed some writers to ignore problems that Aristotle himself had not raised or to resist unification of solutions among different questions that he had treated.[3] Thus, the analytic category *Aristotelianism* is as misleading as *Copernicanism* in suggesting more homogeneity—even more family resemblance—than actually existed in the interpretive practices of Aristotle's Renaissance commentators.[4] And in addressing the question of comets and novas, many traditionalist academic writers tinkered with the impermeable Aristotelian aether, readily making ontic adjustments and retrieving from Presocratic, Stoic, and early Christian writings a conception of the heavens as a perfect *fluid* substance in which new objects could appear, swim about, and disappear.[5]

The upshot is that just as turn-of-the-century heavenly modernizers were engaging in philosophical practices once considered off-limits, traditionalist natural philosophers were remaking and reinterpreting various elements of the Aristotelian corpus without giving up the overall structure of Aristotle's philosophy of nature. Depending on which group one focuses on, one will see either a sharp break or a moderate accommodation.[6] Either way, early-seventeenth-century thinkers were witness to a broader, more hybridized—and arguably messier—problematic than were celestial practitioners in the mid-sixteenth-century Wittenberg and Louvain networks. Those earlier groupings were centrally preoccupied with whether the planetary tables derived from the models of *De Revolutionibus* were superior to the ones based on the *Almagest*—a problem driven by the concerns of practical astrology. The new problematic was preoccupied with different questions: which ordering principle could predict and explain both old and new, recurrent and nonrecurrent phenomena while being consistent with holy scripture? The emerging criterion, in short, was that a theory of heavenly order must be consistent with both old and new phenomena believed to exist in the heavens. And this was a difficult standard to meet in a moment of rampant theoretical proliferation.

The revision of planetary order was, meanwhile, largely a tangential issue for the scattered group of anti-Aristotelian upstarts often called the "nature philosophers." Men like Paracelsus, Cardano, Telesio, Campanella, and Patrizi operated mostly without the formal authority of academic credentials and claimed the right to represent not just the heavens but the entire natural realm. They inventively mobilized all sorts of Presocratic, Platonic, and Stoic resources—bestowing on them Christian authority, sometimes countering Aristotle with positions that he himself had rejected, and often using the universities as a negative pole against which to mark their positions.[7] But the efforts of these naturalizing revisionists made no special claims grounded in the disciplinary authority of subjects mixed with mathematics. In fact, quite the opposite was the case: witness Bruno's biting parodies of the "mathematicians."

Cosmology is the analytic term regularly applied by historians to the enterprise of describing and explaining the universe according to general principles, and we would expect to find it in use by the likes of Kepler and Galileo as celestial order was emerging as an important preoccupation in the early 1600s. But cosmology was not a discipline, or a deposit of standard practices, or a regular category of reference. It did not yet define a

TABVLA MILITIAE
Scholasticæ.

AD IV.VEN.T.V.TEM.
En tibi præcipuos hostes, Studiosa Iuuentus,
Palladis ante Arcem, qui sua castra locant.
Hîc Ruditas: Metus: hîc Stupor: hîc ignaua Voluptas:
Hîc animum fractum, turba superba fugat.
Quos vbi constanti studio superaueris hostes:
Nec non septenos viceris arte gradûs.
Mox triplicis referat Turris penetralia Pallas:
Et te perpetuò GLORIA parta manet.
M. B. VV.

78. The student's army: late-sixteenth-century emblem, Altdorf Academy. Stopp 1974. To enter the castle of learning and the ring of the bacca-laureates, the masters assist students in climbing the steps of the liberal arts, which eventually lead to the three towers of the inner ring, which represent the upper faculties of medicine, law, and theology. The students must fight their way past the tents of the seven enemies of learning: Ignorance, Fear, Disinterest, Idleness, Pleasure, Arrogance, and Timidity.

space of actors' possibilities. Kepler, as will be recalled, had used the term *cosmography* in constructing his new systematics around the old categories of the science of the stars, framing his arguments within a balance of divine providence, mathematical archetypes, and natural causality. Even this term had caused confusion when applied to the heavens because of its conventional association with terrestrial geography. Galileo also used the term *cosmography* as a surrogate for the subject matter of an elementary course on spherics at Padua in 1602.[8] But when actively pointing to the goal of heavenly coherence in his 1613 work on sunspots (without systematic reference to divine planning and intervention)—which were addressed as letters to Marcus Welser, his old friend and sometime supporter from the days of the Pinelli gatherings—he wrote as follows: "The *true constitution of the universe* [is] the most important and most admirable problem that there is. For such a constitution exists; it is unique, true, real, and could not possibly be otherwise; and the greatness and nobility of this problem entitle it to be placed foremost among all questions capable of theoretical solution."[9]

Somewhat later, writing under the constraints

of Cardinal Bellarmine's 1616 injunction and unable to invoke Copernicus publicly, Galileo summoned up ancient Stoic (rather than Christian) authority to chide the modernizer Tycho Brahe and his Jesuit followers at the Collegio Romano: "Seneca recognized and wrote how important it was for the sure determination of these matters to have a firm and unquestionable knowledge of the order, arrangement, locations, and movements of the parts of the universe. In our age we still lack this; hence we must be content with what little we may conjecture here among shadows, until there shall be given to us the true constitution of the universe—inasmuch as that which Tycho promised us still remains imperfect."[10]

Ironically, the word *cosmology* seems to have been coined in 1605 by a successful Leipzig/Heidelberg traditionalist textbook writer named Clemens Timpler (1563/4–1624), a former student of Georg Liebler. As prolific as he was prolix, he was as blissfully unaware of Galileo's and Kepler's problematics as they were of his. Timpler's main goal was to present a compendium of traditional academic natural philosophy. He defined cosmology as "the physical doctrine which explains the world in general," adding that "the world is the most beautiful and ample corporeal structure, skillfully built by God of the heavens and elements for his glory and for man's use."[11] Ten years later, the Carmelite Paolo Antonio Foscarini published his *Treatise concerning Natural, Cosmological Divination* (*Trattato della divinatione naturale cosmologica ovvero de' pronostici e presagi naturali delle mutationi de TEMPI, &c.*).[12] Foscarini classified a domain of prediction restricted to natural environmental effects (e.g., wind, rain, storms, rainbows, and earthquakes) associated with natural signs (e.g., a halo around the moon, the shapes of clouds, and the color of the sky). His use of *cosmological* thus followed Sixtus V's bull of 1586 in limiting itself to one of the three safe areas of divination (medicine, navigation, and weather). Moreover, Foscarini made no attempt to connect planetary order to the prediction of such "cosmological" effects.[13]

Thus, while Foscarini's and Timpler's usages differed, both functioned something like the term *world system*, renaming distinctions that were already operative.[14] Indeed, Timpler's neologism still retained elements that were later shed from that term: Christian cosmogony, astrological influence, and the tenacious ontological distinction between the heavens and elementary region. Hence Timpler's *cosmology* belongs to a sixteenth-century network of meanings, closer to the theological mystery that Kepler believed he had revealed in his *Mysterium Cosmographicum*, or to the world described by William Cuningham in his *Cosmographical Glasse* (1559) and, a quarter century later, by John Blagrave in his *Mathematicall Jewel* (1585); or to Tycho Brahe's term *earthly astronomy*, which linked astral influence to alchemical processes in the Earth;[15] or, yet again, to John Dee's "whole and perfect description of the heauenly, and also elementall parte of the world, and their homologall application, and mutuall collation necessarie."[16] But although Timpler's heavens permitted astral influences in the celestial realm, like many another writer of academic physics textbooks of the very early seventeenth century, his universe reserved no "place" for novas and superlunary comets. Even when celestial novelties began to be accommodated to the heavens—usually without disturbing the traditional ordering—it was in the genre of the sphere.[17]

The problem can be approached from still another angle. Beginning in the mid-1580s, discussions about planetary ordering took on more of the character of a philosophical debate among astronomical practitioners.[18] As the modernizers moved Copernicus out of his customary position in the tables and the astronomical textbook literature, an interesting convergence occurred. New-style philosophizers (like Bruno and Gilbert), modernizing planetary theorizers of the Copernican persuasion (Kepler and Galileo), and modernizing traditionalists of the middle way (Brahe, Roeslin, and Ursus) adapted humanist literary forms and rhetorical resources to urge their celestial orderings. Copernicus and Rheticus had already pointed the way. Consequently, contemporaries found an increasing multiplicity of venues from which they could derive references and support for and against various claims for heavenly order. Indeed, the greater readers' familiarity with such works, the more there emerged an image of contestation and also uncertainty. It is no wonder that this proliferation of arrangements, together with the absence of a conclusive demonstration, induced a skeptical suspension of judgment in an astrological writer like Christopher Heydon or nurtured the Protestant mood of general epistemic unease betokening the decay of an orderly Nature, as reflected in the famous lines of the English poet John Donne:

And new Philosophy calls all in doubt,
The Element of fire is quite put out;
The Sun is lost, and th' earth, and no mans wit
Can well direct him where to looke for it.
And freely men confesse that this world's spent,
When in the planets and the Firmament
They seeke so many new; then see that this
Is crumbled out againe to his Atomies.
'Tis all in peeces, all cohaerence gone;
All just supply, and all Relation:
Prince, Subject, Father, Son, are things forgot.[19]

The new discursive space of the early seventeenth century also marked a shift in the authority of astronomy's earlier theoretical texts. Copernicus's *De Revolutionibus* was coming to be regarded as a different sort of book—a staple of the collections of the libraries of heavenly practitioners, princes and physicians, monasteries, Jesuit colleges and Protestant universities.[20] If, in the sixteenth century, mathematical practitioners and natural philosophers of the likes of Gemma Frisius, John Dee, Thomas Digges, Christopher Clavius, Giordano Bruno, Diego de Zuñiga, and Michael Maestlin owned and often annotated personal copies of *De Revolutionibus,* Thomas Hobbes could now easily consult the copies owned by the Cavendish family, whose children he tutored at Chatsworth House.[21] Some extant, well-studied copies of *De Revolutionibus* testify to its continuing utility among heavenly practitioners in the universities as a resource for learning the technical details of equantless planetary modeling and the complex precessional mechanism: Maestlin's eventual successor at Tübingen, Wilhelm Schickard (1592–1635), and the Leiden mathematician Willebrord Snell (1580–1626) both owned copies.[22] Yet if Peurbach's *New Theorics* and Ptolemy's *Almagest* were still hegemonic within the domain of astronomical theory, if works like Maestlin's *Epitome of Astronomy* and Clavius's *Commentary on the Sphere* continued to serve as authoritative texts of teaching in the universities, Copernicus's *De Revolutionibus* no longer functioned as the sole resource for contesting the traditional planetary arrangement displayed in pedagogical manuals of this sort. *De Revolutionibus* had become a residual text in a field of emergent theoretical possibilities—still read, studied, and plumbed, but now less vigorously and extensively annotated

than it had been between 1543 and 1600. It was gradually on its way to becoming a collectible for seventeenth-century antiquarians even as Kepler and Galileo were posing a new problematic, one of finding a unique set of principles of natural philosophy that could be brought into agreement with a single scheme of celestial order.[23] In short, even as literary conventions of compartmentalized exclusion persisted, the ideal of demonstrative knowledge in which competing alternatives would be decisively ruled out continued to carry great weight.

MANY ROADS FOR THE MODERNIZERS
THE SOCIAL DISUNITY OF COPERNICAN NATURAL PHILOSOPHY

In the first decade of the seventeenth century, the universities continued to be the traditional locus of philosophical authority. Copernicans who wished to air their views publicly as natural philosophy continued to be a tiny minority, with little or no power in that realm. Galileo and Maestlin—the only Copernicans in those institutions—held formal titles as mathematicians and not as natural philosophers. Mention of Copernicus's name and use of this or that bit of information from his writings were still common enough among academic textbook writers on the sphere or theorics, both on the Continent and in England; but by this period, such references were totally unspectacular. Copernicus's name was still positively associated with the *Prutenic Tables* and hence also with astrological prognostication. Proponents of one or another variant of Copernicus's views as philosophy were, for the most part, outside the universities. And within a few years of one another, an entire generation of Copernican proponents— Digges (d. 1595), Zuñiga (d. ca. 1600), Bruno (d. 1600), Gilbert (d. 1603), and Rothmann (d. 1608?)— had passed from the scene, while the cautious Galileo and Harriot were still little known outside their bounded friendship networks.

Maestlin, on the other hand, was now known to be an advocate of Copernicus to anyone (as, for example, Galileo) who had read the *Mysterium Cosmographicum*. Later seventeenth-century references regularly included his name. In 1640, for example, John Wilkins, knowing nothing of Maestlin's private evolution, represented him as "a Man very eminent for his singular skill in this Science [of astronomy]; who though at the first he were a follower of *Ptolomy*, yet upon his second and more exact thoughts, he concluded *Copernicus* to be in the right, and that the usual *Hypothesis, praescriptione potiùs quàm ratione valet,* do's prevail more by prescription then reason."[24] But after 1596, Maestlin chose not to involve himself in anything of a comparable nature; and although he had many students and many children, he would have no more Keplers at Tübingen. He may even have been an early casualty of too much academic committee work.[25]

The Copernicans' vanishing presence in the universities was somewhat offset by Kepler's unusual visibility and status as imperial mathematician and the wide circulation of his many publications. Yet this countervailing position belied his persistent failure to attract a significant following for his views—even though, because of their extensive development, his arguments might have been expected to be more widely discussed, if not more compelling. Besides Maestlin, who had his own reservations, only Kepler's correspondents Edmund Bruce and Herwart von Hohenburg showed any significantly favorable reception of his views. There was no getting around the fact that Kepler's heliostatist representation—in its 1596 version and even more so in its 1609 version—was philosophically idiosyncratic and impossibly demanding on traditionalist and, for that matter, even modernizing sensibilities. Kepler had decisively moved beyond *De Revolutionibus* and had broken with the Wittenberg interpretation. Who would be willing to follow him? Neither nouveau theorizers like Praetorius and Brahe nor even the early Copernicanizing Galileo were comfortable with Kepler's style of physical speculation about the cause of the motions, his peculiar appeal to archetypes for the spacing of the planets, and his apparently reactionary return to the Ptolemaic equant. If both Maestlin and Kepler invited readers of the *Mysterium* to see their work as improving on the arguments for planetary order in Rheticus's *Narratio Prima*, Maestlin himself never approved of Kepler's physical theorizing and would have been just as happy to stop with Rheticus. In breaking through the Tübingen Lutheran orthodoxy, the *Mysterium Cosmographicum*'s bold presentation may have gone too far in radicalizing the very reading of Copernicus that had been ignored by the Melanchthon circle. And where Kepler might have won further support, he did not mobilize his *Mysterium* in the service of popular prophecy. That early work, as a consequence, was as much out of temper with the widespread prophetic speculations of contemporaries as it was with everyday astrological prognosticators. What Kepler signaled as important through his Copernican grid was not Elijah's prediction of the end of the world but the story of its beginning, as laid out in Genesis. Thus the *Mysterium* was as profoundly unsuccessful in shifting any significant segment of courtly or academic opinion as Rheticus's *Narratio* had been in the middle decades of the sixteenth century. As late as 1619, Robert Fludd—another aspirant for King James's patronage—rejected Kepler's fully articulated, Pythagoreanizing exegesis of Genesis in favor of a more traditional use of world harmonies.[26] Perhaps the greatest achievement of the *Mysterium*—at least in the short term—was to help position its young author (with his irritating, anti-Tychonic convictions) as Brahe's successor in 1601.

Yet Kepler seemed to flourish intellectually even under the conditions of personal adversity fostered by Tycho Brahe's followers at Rudolf's court.[27] After 1602, Kepler faced serious resistance to his public use of Tycho's observational data, but unlike the opposition he had met from the Tübingen theologians, the conditions at the Prague court were interconfessionally welcoming. In Prague, Kepler retained an extraordinary inner focus and energy, which sustained his own relentless and remarkable evolution of Copernican positions throughout the philosophically tumultuous first decade of the seventeenth century. By 1606, he had assembled the elements of a new and expanded celestial philosophy against the followers of both Tycho Brahe and Giordano Bruno. In his characteristically imaginative way, he wedded parts of Gilbert's magnetical philosophy with Pico's critique of astrology and Copernicus's planetary ordering. Crucial to this intellectually rich period was Kepler's reform of the science of the stars through a Copernican-based, non-Melanchthonian, centrist astrology. He clearly hoped that this astrology would win over both practitioners and patrons, much as he hoped to use mathematical harmonies to ground moderate positions in political theory and theology. But, as chapter 15 shows, Kepler's status as a court Copernican did not automatically provide legitimation for his views.[28] He enjoyed, at best, uncertain gains in the Jacobean court and among aristocratic sodalities in Britain. And after his departure from Prague, he was acutely conscious of his own isolation.[29]

Elsewhere, Galileo had been following closely many of the above-mentioned developments from Padua, where Tychonic astronomy did not yet have a prominent following and where, in spite of modernizing developments among the Jesuits in natural philosophy, traditionalist Aristotelian natural philosophy still retained a powerful hold.[30] The university rolls show that he lectured regularly on Euclid, on the *Sphere* (using Clavius's commentary), on Peurbach's *New Theorics of the Planets,* and once on Ptolemy's *Almagest* in 1597.[31] These texts reflected the demands of a traditional university culture. Yet residual references to books that he had read or owned confirm that even before 1609, he was familiar with the writings of the modernizers. Save for the *Astronomia Nova,* he was acquainted with all the works mentioned above, along with Rheticus's *Narratio Prima,* Brahe's *Progymnasmata,* Gilbert's *De Magnete* and, probably, one or more of Bruno's works on the infinity of worlds. It is more than likely that he would have continued to follow these developments in silence had he not learned in May 1609 of an instrument made by a spectacle maker, Hans Lipperhey, coincidentally hailing from Middelburg, the same city as the late fifteenth-century prognosticator.

Perhaps one should not be surprised that Kepler, Galileo, Maestlin, and Harriot never formed any kind of alliance based even on such narrow agreement as the proper ordering of the planets. When Galileo composed a defense of Copernicus sometime in 1615, he constructed a list of ancients and moderns held together by the thin claim that this estimable group denied that the Earth's motion and the Sun's fixity were foolish (*stoltizia*). Among the ancients he listed Pythagoras, Philolaus, Plato, Heraclides of Pontus, Ecphantus, Aristarchus of Samos, Hicetus, Seleucus, and Seneca; among the moderns, Copernicus, Kepler, Gilbert, and Origanus.[32] In other words, Galileo's "group" merely mirrored Copernicus's rhetorical strategy in the preface to *De Revolutionibus,* where he represented himself as defending an apparent paradox.[33] The real problem was that even in 1615, Galileo could not name anyone in Italy who supported these propositions. He was precluded from mentioning them either for political reasons (as with Bruno and

Foscarini) or, worse, because few of them (except the unnamed Benedetto Castelli) really existed. Among potential allies beyond the Alps, he may have felt it dangerous to mention Maestlin; and there is no evidence that he knew of either Harriot or Simon Stevin. In short, his little tabulation was a wishful fantasy, at best a rhetoric of consensus. Yet this did not deter Galileo from writing that "there is no lack of other authors who have published their reasons on the matter. Furthermore, though they have not published anything, I could name very many followers of this doctrine living in Rome, Florence, Venice, Padua, Naples, Pisa, Parma and other places. This doctrine is not, therefore, ridiculous, having been accepted by great men; and, though their number is small compared to the followers of the common position, this is an indication of its being *difficult to understand rather than of its absurdity.*"[34] For Galileo, the last statement was the nub of the issue: Copernicus's technical arguments needed to be made more accessible for nonpractitioners, like theologians and traditional natural philosophers. In Galileo's view, Kepler's redigestion of Copernicus in the *Mysterium* hardly helped matters. And it was such considerations that may have moved Galileo to imagine writing a work like the *Dialogue.*

Yet despite all of these initiatives and wished-for groupings, the Copernicans simply did not constitute a coherent movement. There was no precedent for an astronomical hypothesis being used as the foundation of a new philosophy of nature—let alone a hypothesis whose main premises contradicted the evidence of uncorrected and unchallenged sensory experience. The handful of early-seventeenth-century adherents of the core Copernican propositions had neither the institutional tradition of a philosophical school (as, for example, the Averroist Aristotelians), nor the formal structure of a religious order (like the Jesuits), nor the compelling convictions of a political party (like the French Politiques), nor the explicit support of a single ruler, nor membership in a common humanistic circle, nor even the status of a literary assemblage of authors, grouped together in the manner of Tycho Brahe's *Progymnasmata.* Or, to take a later point of contrast, they lacked the social and political resources that made possible something like the late-seventeenth- and early-eighteenth-century reaction to Newton's natural philosophy—a coherent movement of disciples, public lecturers, and surrogates; spaces of public knowledge, such as the Royal Society; and regularly published vehicles, such as the *Journal des Sçavans,* that could display arguments, permit debate, and mobilize collective support.[35]

The mid-sixteenth- and early-seventeenth-century heliostatists had no such structured spaces of sociability. Typically, therefore, they embodied their undertakings in independent humanist narratives of self-discovery and persuasion—Rheticus's biographical narrative of self-revelation, Copernicus's story of rereading the ancients and rediscovering an old truth, Digges's trope of arguing Copernicus's case in a law court, Kepler's narrative in the *Mysterium* of hitting on the structure of the divine plan while teaching Euclid or his image of a long battle to win over Mars to the Copernican cause, Bruno's story of recovering ancient Egyptian mysteries. Collectively they may be seen to represent a *via moderna,* even if their rhetorics made no appeal for academic legitimation. Yet if the Copernicans of the early 1600s were socially fragmented and conceptually at odds over their physical premises, the period was marked by some remarkable proposals to eliminate lingering uncertainties about the Earth's motion. And although ultimately these efforts would fail to remove all objections to the Copernican theory, together they moved the debate over world systems to a new level of legitimacy and engagement. The generation that came to maturity between the 1620s and the 1640s would inherit far more consolidated representations of the new empirical and theoretical claims forged in the fin de siècle and the century's first decade, a period from which there would be no turning back.

ALONG THE *VIA MODERNA*

Simon Stevin

Separate from and uncoordinated with Galileo's unpublished efforts to work out a physics compatible with the motion of bodies on a moving Earth was the work of the polymath practitioner Simon Stevin (1548–1620). Stevin, who represented himself as neither an astronomer nor a theorist, held unambiguously that the Earth's motion is "what happens in nature."[36] His views, surprisingly, owed much to his engagement with Gilbert's *De Magnete.* He was one of a remarkable group of Dutch military engineers whose skills in fortification, harbor-drainage tech-

niques, and transportation methods were critical to the successful Dutch recovery of Spanish-occupied lands that began in the early 1590s. A native of Bruges (Brugge), Stevin had moved north after the fall of Antwerp in 1585, along with thousands of skilled craftsmen, wealthy merchants, printers, and publishers who fled Brabant and Flanders, in what Klaas van Berkel has called a "brain drain" to the north.[37] Eventually he became tutor and technical adviser to Maurice of Nassau, prince of Orange (1567–1625). As *Stadholder* (or leader) of Holland, the dominant province within the Dutch United Provinces, Maurice was a major military and political figure who possessed an unusual degree of literacy in the battlefield technologies that he deployed.[38]

Like Kepler, Stevin made vigorous use of print; but most of his production was in genres of "practice." Starting in the mid-1580s, he had put out a stream of practical mathematical works: a treatise on the "practice of weighing," another on statics (perhaps his best-known work today), another on hydrostatics, one teaching the use of decimal notation in arithmetic, still another on finding a port when at sea, and others on fortification and city planning.[39] In the Low Countries, this markedly practical character of Stevin's productions was not unusual, but the scope and depth of his writings were remarkable and left deep traces on the style of natural knowledge. That he also composed all his works in the vernacular Dutch—and even wrote a treatise on the singular virtues of that language—had much to do with their appeal in the Low Countries. Indeed, Stevin regarded Dutch as a special language of legitimation, the language of the "age of sages" (*wijsentijt*) that had preceded classical antiquity and from which a great wisdom had been lost. The idea of recovering an ancient, pristine wisdom held widespread attraction during the Renaissance, but no one before Stevin had interpreted this project as one to be conducted in the vernacular.[40] Yet, significantly, precedent for use of the vernacular in teaching had already begun in 1598 at the University of Franeker and was quickly institutionalized in the strikingly vernacular curriculum of Leiden University's self-contained school (*Duytsche mathematique*), built in 1600 with a focus on military engineering and surveying.[41]

But such familiar historiographical emphasis on "Dutch practicality" can easily be taken too far. The prognostication literature, as seen earlier, had long used the vernacular to spread its anticipations of the future. Yet Stevin is not known to have published any forecasts within the domain of practical astrology. It was not for lack of example. Surely Stevin knew the almanacs and prognostications of his close contemporary Nicholaus Mulerius (de Muliers), issued at least as early as 1604 and continuing until 1626.[42] Stevin also made considerable use of Stadius's *Ephemerides*, a work indisputably representative of Gemma Frisius's group in Louvain. And he mentions most of the leading ephemerides of the second half of the sixteenth century, which, it need hardly be said, were all explicitly devoted to the preparing of astrological prognostications: "Calculated ephemerides are now frequently printed, for example, those of *Johannes Stofflerus, Erasmus Reinhold, Leovitius, Stadius, Maginus, Martinus Everarti* [Everaerts], and the like."[43] Yet to my knowledge there are neither annual forecasts nor any (surviving) evidence of a concern with the theoretical parts of astrology, even though such areas might have been of interest to Prince Maurice.[44] Unlike Clavius, for example, Stevin had nothing to say about why one should not engage in different types of astrological practice. Were such omissions active? Was it because Stevin knew and accepted Pico's skeptical arguments? Or could he have been convinced by Sicke van Hemminga's brutal confrontation of rulers' horoscopes with their actual lives?

Whatever the explanation, the "practical" Stevin also chose to write in the domain of astronomical theory. Between 1605 and 1608, this work took the form of a study of heavenly motions (*De Hemelloop*) within a larger collection of mathematical works (*Wiscontige gedachtenissen*) that appeared also in the Latin translation of Willebrord Snell (1580–1626) and, by 1630, in French.[45] *De Hemelloop* was organized in a way that departed strikingly from the dominant textbook traditions of the sixteenth century. Stevin presented the mechanisms of the heavens first on the traditional assumption of a motionless Earth, building upon Stadius's ephemerides, and then following on Copernicus's assumption of the Earth as a moving planet.[46] Hence the real pedagogical innovation was that Copernicus's theory was taught by beginning with the problem of the almanac. If you could read and use an ephemerides, then you were prepared to move to an understanding of

the *Almagest*'s models and from there to *De Revolutionibus*. It is easier to say that his mode of presentation was unprecedented than to explain how he arrived at it.

Two considerations are paramount: one fairly obvious, the other not. The first is that Stevin composed the work as a didactic treatise for Maurice. Nominally it was meant to teach the ruler and was probably used in just that manner. But by publishing it Stevin was obviously trying to reach a wider readership, although there is no explicit evidence that he aspired to have his book incorporated into the university curriculum. Moreover, unlike the *Sidereus Nuncius*, which Galileo also dedicated to a ruler, this work was not a report or an announcement (because there were no celestial novelties to announce); nor, like the *Mysterium Cosmographicum*, did it present a "revelation" of the Creator's world plan. In fact, unlike Kepler, all of whose writings foregrounded Copernican themes or Digges's embedding of *De Revolutionibus*, book 1, in a prognostication, Stevin adopted a structure decidedly unlike other rhetorical formats used to present heliocentric ideas in the preceding century. It was simply a manual of pedagogy intended to teach the theoretical principles of the heavenly motions, and he may have intended it to be used in conjunction with *De Revolutionibus*.[47]

And this brings us to the second consideration. Stevin had not only studied *De Revolutionibus* carefully, but he had also read and absorbed Gilbert's recent *De Magnete*. That work must have held unusual interest for him, not least because Gilbert knew and explicitly approved of Stevin's proposal for solving the "longitude problem"—finding a ship's place at sea by means of measuring variations from due north on a magnetic compass ("easting"). But although Gilbert approved the idea in principle, he did not mince his criticism that Stevin's needle deviations failed to correspond to any predictable rule grounded in actual observations.[48] And when he took up the ordering of the planets, Stevin tactfully returned both the compliment and the criticism to Gilbert.[49]

Stevin readily applauded Gilbert's proposal of the Earth as a great magnet, but in other respects the two were clearly at variance. Stevin ignored any talk of a magnetic Earth-soul.[50] He was absolutely precise and clear that Mercury is closer to the Sun than Venus is because those planets, unlike Mars, Jupiter and Saturn, do not line up in opposition with the Sun. Similarly, because Mercury's limited elongations with respect to the Sun are smaller than those of Venus, it must lie within Venus's ambit. Acceptance of this Capellan ordering for Mercury and Venus already enjoyed considerable currency among Leiden academic humanists, at least partly because of its classical pedigree.[51] Stevin's arguments, however, in no way depended on ancient authority. In fact, Stevin was not only prepared to push beyond the Capellans, but he also invoked the crucial period-distance relation of *De Revolutionibus*, book 1, chapter 10, whose implications Gilbert himself had avoided: planets that have longer periods of revolution are farther away from the center.[52] Here, Stevin confronted the disparity pointed to by Copernicus (*De Revolutionibus*, book 1, chapters 7–8): as the heavens grow larger and larger, how can one justify a twenty-four-hour period for the largest heaven of all, the outermost sphere of the fixed stars? Gilbert had made much of this problem, denouncing this motion as a "superstition, a philosophic fable, now believed only by simpletons and the unlearned."[53] Stevin agreed with Gilbert but not in his aggressively Brunonian register: "It is more in accordance with natural reason to believe and to assume that this fastest motion of all is to be assigned to the smallest circle, to wit, the circle of the Earth in its place."[54]

Stevin's selective agreement and disagreement with the author of *De Magnete* has much to teach about both figures. Gilbert, whatever private beliefs he may have held, publicly sidestepped the question of the Earth's annual motion, leaving thereby a crucial ambiguity for those who followed him. For Stevin, the crux of the matter was how to explain physically that a sphere and the individual objects accompanying it could undergo different motions. To address these issues, Stevin produced analogies in a manner reminiscent of Galileo and far more richly developed than those of Rothmann. Compare the air surrounding and moving against and around buildings on the (fixed) Earth to a river flowing past a vertically standing stick. Next, imagine the same stick, still vertical, drawn at the same speed as the flowing river through a pool of standing water. "It must be admitted," says Stevin, "that the water will press equally against one stick and the other. And the same would evidently happen

with the air against the buildings or the buildings against the air." Thus, the earthly sphere and the sphere of air surrounding it "together form one sphere" and move together.[55] Next, considering the problem of how the Earth can undergo an annual and a diurnal motion simultaneously, Stevin produced a ship example reminiscent of Thomas Digges: "One motion is the daily rotation on its axis from West to East, but to explain this motion in its place somewhat more fully by means of an example, it might be said to be like a turning grindstone in a vessel sailing, which from the ship receives a motion from place to place, but its rotation on the axis meanwhile remains in the ship in the same place; and the same is the case with the Earth."[56] Gilbert was prepared to use the ship example to argue for the Earth's daily rotation,[57] and he even boldly speculated that "the space above the earth's exhalations is a vacuum," but, by contrast with Stevin, his avoidance of the annual motion left an obvious gap for any reader familiar with Copernicus's arguments.[58]

When Stevin turned to the Earth's third form of motion—asking how the axis maintains a fixed direction with respect to the fixed stars, and indeed why the axis remains always parallel to itself—he observed that Copernicus had described the problem in *De Revolutionibus*, book 1, chapter 11, with a diagram but had offered no "proof." At this point, Stevin directly credited Gilbert with offering a strong "natural cause," although not the full explanation. Taking Copernicus's model and Gilbert's causal mechanism together, Stevin proposed a way to think about the problem by imagining a magnetic needle free to rotate around a point in a box: as the box revolves to the right, the needle seems to revolve to the left. But because the net result is that the needle stands still, Stevin proposed to call the phenomenon "magnetic rest."

Subsequently he applied the same principle to his planetary theory. Here the question was how to maintain a system of spheres in mutual contact: if spheres carry the planets from west to east, and the motion of the outer spheres communicates through contact with the inner spheres, then why do not the inner spheres have the same velocity as the outer ones? Moreover, as the outermost sphere had the very short period of twenty-four hours, the contained spheres would need to rush around faster in the same period.

The problem was not a new one, as Aristotle's concentrically organized universe required complicated unrolling inner spheres. Stevin's spheres, however, were eccentric, and hence contact between adjacent spheres occurred only at the maximum distance or apogees of the lower spheres. Why then were the apogees not carried along by the higher spheres? Puzzled by this difficulty, Stevin considered that the planet might perhaps "fly through the air like birds about a tower, without the motion of the one causing any change in the motion of the other."[59] Had he been familiar with Brahe's *De Mundi Aetherei Recentioribus Phaenomenis*, Stevin might have considered the possibility of noninterfering fluid heavens. Moving incorporeal souls—Gilbertian or otherwise—had no appeal for him because he was in search of a materialist explanation. Thus he was obviously pleased to find a resolution in the concept of magnetic rest.

Stevin's particular use of Gilbertian resources to resolve objections against Copernicus's planetary arrangement added to the thickly diverse array of settlements between natural philosophy and the science of the stars that characterizes the century's first decade. It set Stevin apart not only from Copernican advocates in other countries but also from renowned Dutch academic mathematicians like Nicolaus Mulerius at Groningen, who, although thoroughly acquainted with *De Revolutionibus* (he published a third edition with commentary in 1617) still went no further than to endorse the daily motion and the Capellan ordering of Mercury and Venus. Mulerius's position, therefore, was essentially that of Ursus and, like the latter's, shows the persistent hesitation to move to a full adoption of the Copernican ordering even among those intimately familiar with *De Revolutionibus*. Rienk Vermij suggests that Mulerius saw no difficulties in reconciling the Earth's daily motion with scripture but, like Tycho earlier, found problematic the huge Copernican space between Saturn and the fixed stars. For Mulerius, however, the worry was not the emptiness of such a gap but the possibility that it could be filled with many suns, which would be "absurd and against Christian piety."[60]

Stevin, for his part, simply left out any consideration of scriptural concerns. A possible explanation—indeed altogether probable—lies again in his use of Gilbert's work. There Stevin could easily have found persuasive Edward Wright's remarks to the reader, placed prominently before Gilbert's own preface:

The passages quoted from Holy Writ do not appear to contradict very strongly the doctrine of the earth's [daily] mobility. It does not seem to have been the intention of Moses or the prophets to promulgate nice mathematical or physical distinctions: they rather adapt themselves [*sese accommodare*] to the understanding of the common people and to the current fashion of speech, as nurses do in dealing with babes; they do not attend to unessential minutiae. Thus, [in] Genesis i.16 and Psalm cxxxvi.7,9, the moon is called a great luminary, because it so appears to us, though, to those versed in astronomy, it is known that very many stars, fixed and planetary, are far larger. So too from Psalm civ.5, no argument of any weight can, I think, be drawn to contradict the earth's mobility, albeit it is said that God established the earth on her foundations to the end it should never be moved; for the earth may remain forevermore in its own place and in the selfsame place, in such manner that it shall not be moved away by any stray force of transference, nor carried beyond its abiding place where it was established in the beginning by the divine architect.[61]

Stevin's austere, heliostatic arrangement, with its use of Gilbert's magnetics, continued the pattern of selective appropriation and differential application that we have already noticed, time and again, in the Wittenberg tradition.[62]

Kepler's Radical Turn in Planetary Theory and the Elimination of Alternatives

Kepler's route along the *via moderna* was so singular that, as Bruce Stephenson suggests, his discoveries (unlike Newton's formulation of the principle of universal gravitation) might never have occurred had not Kepler made them.[63] From his Tübingen period onward, this unique development derived from a focus on the explanatory robustness of the Copernican hypothesis rather than worries about its predictive capacities.[64] Kepler, more than any other modernizer of this period, relentlessly pursued the standard of a strong demonstration. If one condition of a strict demonstration is not to ignore alternative explanations, then it may be said that a tenacious determination of just this sort united all of Kepler's intellectual efforts. This attitude was unprecedented in the Copernican debates. Such resolve is already strikingly apparent in Kepler's treatment of Gilbert, Galileo, Brahe, and Bruno in the *Stella Nova* of 1606 and continued unabated in the *Astronomia Nova* of 1609. More than any of Kepler's earlier work, the *Astronomia Nova* purported to determine the Earth's motion and the actual path of Mars in a three-dimensional orbit by a massively detailed, knock-down argument that involved nothing short of a complete rewriting of the foundations of theoretical astronomy itself. This work is known to most people now, above all, through the parts that Newton assimilated into his conception of natural philosophy: that Mars—and, by inference, all the other planets—move in a new kind of noncircular curve that turns out to be an ellipse, with the Sun at one focus; and that changes in the planet's velocity along that curve are modeled not by uniform circular motions, as long favored by tradition, but rather by a new parameter, the equal areas swept out by a radius vector extending from the Sun to the planet.

Because of the lingering Newtonian association, it was not clearly noticed until the recent work of William H. Donahue and James Voelkel that Kepler presented his conclusions as a concocted history of discovery meant to conceal not only its disagreements with Tycho's planetary theory and dependence on Tycho's observations but also elements of demonstrative uncertainty. The chief uncertainty concerned the problem of eliminating a Wittich-style derivation of Mars's orbit, a purely mathematical model involving extra epicycles that would require neither the Copernican ordering nor the solar moving power.[65] Voelkel suggests that the rhetorical form and composition of the *Astronomia Nova* were largely shaped by a combination of local political forces together with the above-mentioned objection. The resistance came from Tycho's heirs in Prague (chiefly his son-in-law Tengnagel) and a former member of the Uraniborg household (Christian Severin Longomontanus) as well as from David Fabricius, a Lutheran pastor in East Frisia who, apart from his Tychonic sympathies, produced several important regional maps and cast private horoscopes.[66] Whereas Kepler just wanted the use of Tycho's observations, the motivations of the Tychonic parties were diverse.

Tengnagel's uppermost concern was preserving Tycho's credit for the *Rudolfine Tables,* presumably for the attendant financial benefits that he imagined for himself. Besides striking a legal contract over the right to use Tycho's observations—highly disagreeable and constraining to Kepler—Tengnagel insisted on adding a (signed) letter to the *Astronomia Nova* that appeared after the author's dedicatory letter and several epigrams. Much like Osiander in his more famous "Ad Lectorem," Tengnagel tried to restrict the work's disciplinary domain: "I thought I should give you [the reader] just three words' warning, lest you be moved by anything of Kepler's, but especially his liberty in disagreeing with Brahe in physical arguments, which has groundlessly complicated the work on the *Rudolfine Tables.* But such liberties are habitual to all philosophers from the creation of the universe to the present."[67]

Tengnagel's letter differed from Osiander's in important ways.[68] The presence of his name left no possibility of the kind of deception engendered by anonymity: he was clearly speaking for himself and not the author. Unlike Osiander, Tengnagel did not go so far as to claim that astronomy was incapable of making true claims about the world. His warning had a distinctly legalistic tenor: Kepler was not "at liberty" to depart from Tycho by using "physical arguments," as such arguments would (in some untold way) "complicate" the preparation of the still-incomplete *Tables.* Of course, Kepler had known of Osiander's editorial mischief since his student days and commented on it with undisguised annoyance in the unpublished *Apology for Tycho against Ursus* (1601). But instead of using the full details of the story provided by the annotations on Maestlin's copy, Kepler used the briefer and less substantive annotations on his own copy to block the force of Tengnagel's letter. Employing Peter Ramus's plea for an "astronomy without hypotheses" as the occasion to advance his vision of theoretical astronomy based on true physical causes, Kepler dramatically exposed Osiander's identity:

It is a most absurd business, I admit, to demonstrate natural phenomena through false causes, but this is not what is happening in Copernicus. For he too considered his hypotheses true, no less than those whom you mention considered their old ones true, but he did not just consider them true, but demonstrates it; as evidence of which I offer this work.

But would you like to know who originated this tale, at which you wax so wroth? "Andreas Osiander" is written in my copy [of Copernicus's *De Revolutionibus*], in the hand of Hieronymus Schreiber of Nürnberg. This Andreas, when he was in charge of publishing Copernicus, thought this preface most prudent which you [Ramus] consider so absurd (as may be gathered from his letters to Copernicus), and placed it upon the frontispiece of the book, Copernicus himself being dead, or certainly unaware of this. Thus Copernicus does not mythologize, but seriously presents paradoxes; that is, he philosophizes. Which is what you wish of the astronomer.[69]

Kepler clearly assumed that readers would trust such personal information about the identity of the author of the anonymous letter. He hoped thereby to win a sort of legal credibility for his reading of Copernicus's intention while also allowing him to associate the search for true causes in the *Astronomia Nova* directly with *De Revolutionibus*. Kepler could thus represent Copernicus and more aptly himself as "philosophizing astronomers." Consequently, whatever Tengnagel's legal maneuvers to undermine and contain Kepler's anti-Tychonian projects and whatever success he may have had in pushing Kepler to structure his book as a contrived narrative, he had seriously failed to prevent the appearance of a massive and unprecedented revision of theoretical astronomy.[70]

Longomontanus fared no better. Unlike Tengnagel, he had no imperial lucre at stake. His investment seems to have come from resentment at losing his privileged position as Tycho's once highly favored assistant. Moreover, as a man of low social origins whose status had been raised by association with Tycho, he was strongly identified with Tycho's ideas, especially the Tychonic ordering scheme and his Copernican-style planetary theory. Although he paid lip service to Kepler's quest for physical causes in astronomy—and, indeed, later admitted the Earth's diurnal motion into his own world scheme—he was clearly envious of Kepler and had little sympathy for his project. More pointedly than Tengnagel, it would seem, he argued against the plausibility of searching for physical causes. Yet, in early 1605, Kepler communicated in no uncertain terms to Longomontanus that the physical consequences of Tycho's discoveries could not be ignored: "You Tychonic astronomers, having justly stripped physics of the solidity of the orbs, wrongfully left the flight of the planets whirling about in the greatest unbelievability and confusion. Why should

not I, on the other hand, assist them by inquiring into the physical forms of their motion through the transparent void? . . . Truly I think the sciences are intertwined in such a way that neither can be complete without the other. But I see that you are not particularly opposed to this."[71] Kepler threw back at Longomontanus not merely the need for a new physical theory but also a call for a new principle of the sciences as an interrelated web, a vision of inseparable and interdependent domains.

Most serious of all Kepler's opponents was Fabricius, who, although at odds with the key principles of Kepler's new sort of astronomy, served a crucial need as a sympathetic sounding board essential to Kepler in working out his ideas. Fabricius shared in the sixteenth-century passion for collecting nativities, and this was among the initial motives for his correspondence with Kepler. But, although astrological topics continued to show up in their extensive and intellectually intimate correspondence between 1602 and 1609, the dominant theme was astronomical theory. The main elements in what Kepler called his "war on Mars" are actually contained in these letters—the bisection of the eccentricity of the Earth's orbit (which he called "the key to astronomy"), the substitution of the true for the mean Sun, various provisional hypotheses (epicyclic, librational, oval), the character of the solar motive power, the ellipse, and, even more crucially, the area law. The work of explaining unfamiliar and difficult ideas—constancy of areas rather than angles, motion along inconstant rather constant arcs, variations in the planet's distance governed by the area law—involved much pedagogical effort and patience. Not surprisingly, Fabricius got stuck. In the course of the exchanges, he requested from Kepler worked examples; his cries for help and his counterproposals affected Kepler's theorizing, apparently pushing him to give up an early distance model that used an epicycle.[72]

As Voelkel argues, Kepler translated his experience with Fabricius directly into objections that he preemptively rebutted in the *Astronomia Nova*. Unlike Tengnagel, who produced legal obstacles but never any serious astronomical ones, Fabricius actually produced an elliptical model derived from Copernican-style double epicycles, without the Keplerian solar moving power and with the Earth at rest. To make the planet swing out from its circular path, Fabricius introduced a mobile line of apsides and a mechanism according to which the center of the eccentric librated along a line perpendicular to the apsidal line.[73] The model succeeded in generating an ellipse, but not according to the area rule. It was thus divorced from the physical intuitions that motivated Kepler's proposal. Fabricius's model was, rather, a performance worthy of Paul Wittich or Tycho Brahe. It was a post hoc, reactive "accommodation" from Kepler's novel physical theory into traditional form, but far more difficult to achieve than earlier such transformations and not without a degree of confusion.[74] Fabricius defended his efforts not by comparison with earlier attempts—of which we do not know him to have been aware—but by (re)stating a familiar principle: that the planets must move with uniform circular motions in accord with the heavens' spherical shape.

> I see that the motion of Mars in the sky agrees with your new hypotheses in every respect. But the procedure for calculating is intricate and difficult. Moreover, I would bring up something against your hypotheses generally. First, by means of your oval or ellipse you do away with uniform circular motion, which more than anything else seems to me unworthy of further consideration. As the sky is round, so it has circular motions, both regular and uniform around their centers to the greatest degree. The celestial bodies are perfectly round, as is manifest in the Sun and Moon. Therefore there is no doubt that their every motion comes about by means of a perfect circle, not an ellipse or a departure [from a circle]. And they are likewise moved uniformly about their centers. And since in your ellipse the center is not everywhere equally distant from the circumference, a uniform motion will certainly be nonuniform to the greatest degree about its proper center. Therefore if while retaining a perfect circle, you can justify the ellipse by means of another small circle, it would be more appropriate. It is not enough to be able to save the motion, but one must also put together the kind of hypotheses that differ in the least extent from natural principles. . . . It is absurd that the planet is really moved nonuniformly per se."[75]

Fabricius's proposal and the assumptions undergirding it threatened Kepler's larger demonstrative claim that the planetary path could be derived *only* from the new physical astronomy. If the Fabrician model did save the phenomena, as Kepler seemed willing to accept, then the allegedly absolute determinacy of Kepler's claim would fail to go through. The elliptical orbit would then underdetermine two different planetary arrangements, one geostatic, the other heliostatic. It is no wonder, then, that Kepler took such great pains to renew the logical objection first adumbrated in the *Mysterium*, where he refuted the application of arguments of the form "a true conclusion derived from false premises" (*ex falsa verum*). In the *Astronomia Nova*, he contended that "as false principles are fitted only to certain positions throughout the whole circle, it follows that they will not be entirely correct outside those positions." False principles work only accidentally. Or, as Kepler put it: "Sly Jezebel cannot gloat over the dragging of truth (a most chaste maiden) into her bordello. Any honest woman following this false predecessor would stay closely in her tracks owing to the narrowness of the streets and the press of the crowd, and the stupid, bleary-eyed professors of the subtleties of logic, who cannot tell a candid appearance from a shameless one, judge her to be the liar's maidservant."[76] Might Kepler have drafted the *Astronomia Nova* differently—less defensively, less cleverly—had his sounding board been a sympathetic Copernican rather than the Tychonic Fabricius? The fragmented state of the Copernicans foreclosed such a possibility, and in any case, as we have seen time and again, they differed over their physical theories.

Even with a more sympathetic audience, it seems likely that Kepler would have maintained his adherence to the same physical principles. But how was he to know that his own physical principles were not false? That seems to be the dilemma raised by the logic of underdetermination. In chapter 33 of the *Astronomia Nova*, Kepler posed the choice for the Tychonics as follows: "Only one of the following can be true: either the power residing in the sun, which moves all the planets, by the same action moves the earth as well; or the sun, together with the planets linked to it through its motive force, is borne about the earth by some power which is seated in the earth."[77]

The relevant point here is that at the moment when Kepler published the *Astronomia Nova* in 1609, there were simply no other rival physical theories. Thus, not unlike Copernicus in 1543, Kepler could set up the *immediate* alternative as the only one available. And, for him, this could only be Tycho Brahe:

Tycho himself destroyed the notion of real orbs, and I in turn have in this third part irrefutably demonstrated that there is an equant in the theory of the Sun or Earth. From this it follows that the motion of the Sun itself (if it is moved) is intensified and remitted according as it is nearer or farther from the Earth, and hence that the Sun is moved by the Earth. But if, on the other hand, the Earth is in motion, it too will be moved by the Sun with greater or lesser velocity according as it is nearer or farther from it, while the power in the body of the Sun remains perpetually constant. Between these two possibilities, therefore, there is no intermediate. I myself agree with Copernicus, and allow that the Earth is one of the planets.[78]

Kepler's eliminative reasoning here presumed that his physical theory (an evolution of the astrologizing physics of his student disputation) was the only possible theory that could (ever) be imagined, an assumption that would be roundly rejected as much by the Tychonics as by post-Keplerian Copernicans.

CONCLUSION

In the first decade of the seventeenth century, Kepler was not only one of the very few public adherents of the Copernican ordering in Europe, but, in his demonstrative goals, he was also by far its most ambitious and capable proponent. Among theorizing astronomers, he was the only one who attempted to rework the physical principles under-girding planetary order, planetary modeling, and astral influence. He was also the only one to try to shut down the claims of alternative celestial arrangements. His isolation, intensified by the failure to forge an alliance with Galileo, was but one instance of the general fragmentation among the Copernicans. Had a new technology for magnifying distant objects fallen into Kepler's hands in 1609, we can only imagine what he would done with it. We may speculate that he would have used it to enhance his reputation against the Rudolfine Tychonics and, surely, to further an intellectual vision that was already well established. But it is hard to imagine that he would have been any more successful in recruiting Galileo as an ally.

By chance, however, it was Galileo's opportunities that suddenly changed. The new magnifying technology opened unforeseen possibilities for mobilizing patronal support at the Medici court in Florence; it also created new conceptual possibilities for defending Copernicus's ordering, perhaps in its Keplerian or Brunonian versions. Looking back from the perspective of our own world, (over)saturated with expectations of technological innovation, it is hard to imagine a world that was not. Exactly what the promotion of new theoretical knowledge meant for the Florentine court and for the still-active culture of traditional prognosticators is the subject of the final two chapters.

17

Modernizing Theoretical Knowledge

PATRONAGE, REPUTATION, LEARNED SOCIABILITY, GENTLEMANLY VERACITY

The Copernican question is a subset of a larger problem: How did modernizers win credibility for new theoretical knowledge? The issue has already received considerable attention in earlier chapters. This chapter critically examines some recent, alternative proposals, with special focus on Galileo. There are two central issues. One concerns the nature and centrality of patronage as a kind of early modern sociability, the other the degree to which court sociabilities or aristocratic status in some way gave legitimacy to conditions of belief.

THEORETICAL KNOWLEDGE
AND SCHOLARLY REPUTATION

In the sixteenth and early seventeenth centuries, the typical celestial practitioner earned a living and official institutional status not by producing new theoretical knowledge but by repackaging, refining, and making more friendly what was already received, selectively incorporating bits of novelty that did not conflict with geocentric assumptions. The midcentury Wittenbergers exemplify this practice. And the thick successor quartos and folios of the late sixteenth century continued it, again differing not so much in their general treatment as in a reorganization and distillation of familiar topics with a sprinkling of novelty. This was the sense in which these compendia were considered to be new. The most successful of them still came from Italy and Central Europe: Clavius's commentaries on the *Sphere;* Maestlin's *Epitome of Astronomy;* the bulky commentaries on Peurbach's

New Theorics by Capuano, Reinhold, Schreckenfuchs, Wursteisen, and others; and the bulging *Ephemerides* of the likes of Magini and Moletti. The heyday of these works was roughly from the 1540s to the '80s. Late-sixteenth-century English textbooks, like Thomas Blundeville's *Exercises* and *Theoriques of the Seven Planets,* were heavily derived from such Continental works, "collected," as Blundeville readily confessed, "partly out of *Ptolomey,* and partly out of *Purbachius,* and of his Commentator *Reinholdus,* also out of *Copernicus,* but most out of *Mestelyn,* whom I haue cheefely followed, because his method and order of writing greatly contenteth my humor."[1]

Academic reputations were built on works of this sort. Writers who described themselves as mathematicians, astronomers, or astrologers became known for the quality and organizational style of their pedagogical manuals, for the utility and handiness of their tables, and, occasionally, for making a successful prediction. As late as 1617, when Kepler introduced the very first textbook built systematically on Copernican principles, he acknowledged the value of the by then substantial quantity of earlier spherics and theorics, including a good number of ancient writings. Copernican astronomy, as he recognized, could not be introduced without first drilling students in the geocentric rudiments of the sort in which he himself had been trained at Tübingen.

> Repetition of the Sphaerics ought not to be seen as useless or without value—either after the manner

of the ancients Euclid, Aratus, Cleomedes, Geminus, Proclus and Theon; or following the moderns, most of all Sacrobosco and his infinite number of commentators, among whom the most learned and plentiful are Christopher Clavius and Hartmann too, Virdung and Wursteisen, Peucer and Schreckenfuchs as well as Piccolomini, Brucaeus, Winsheim and Maestlin, and newest of all, the repetitions of Metius. There is no reason at all why these teaching compendia may not be used again with Peurbach's *Theorics* and [the commentaries of] Reinhold and Simi.[2]

But if new knowledge about the heavens was typically folded into conventional formats, how did it function in the academic reputations of its purveyors? When Magini and Galileo competed for the chair of mathematics at Bologna in 1588, a promoter of (the still very young) Galileo thought it enough to say that he was "well instructed in all the mathematical sciences"; but, in fact, it was Magini who, with his great ephemerides for 1581–1620 and his user-friendly rules for drawing up horoscopes (Venice, 1582), more closely fitted the typical profile of a university mathematician, and perhaps that is one reason why Magini got the chair and Galileo did not.[3] Tycho Brahe, on the other hand, although well traveled in German university circles in his youth and thereafter well connected to its professors, never wrote works in typical academic genres. Indeed, he even rejected the rectorship at Copenhagen in 1577 and generally regarded the university as unsuited to his projects.[4]

Reputation was thus largely based upon the display of the skills that went into improving the science of the stars. In a sense, it was the Kuhnian "normal science" of its day, although quite capable of being practiced un-Kuhnianly on the foundations of Copernican, Ptolemaic or Tychonic planetary theory.[5] Brahe, Kepler, and Galileo thus established wide reputations in their own lifetimes for their recognized mathematical or observational skills, but academic contemporaries did not automatically transfer their admiration to acceptance of theoretical reorderings of the heavens. Although Kepler, for example, wrote in all categories of the science of the stars, no university in Kepler's lifetime added the *Mysterium* or the *Astronomia Nova* to the official lists of texts to be taught by the professors of mathematics.[6] This is one reason why he was eventually motivated to condense

his ideas into the *Epitome*. Ironically, more students have read those books today as part of the *historiography* of science than when they were first introduced as part of the "science" of their own time.

The irony is not difficult to grasp: Brahe, Kepler, Bruno, and Galileo forged their new problematics in dialogue with the institutions that provided them with the tools and categories to set up their problems yet which simultaneously resisted any fundamental changes in their curricular arrangements. It is no accident, therefore, that these figures turned to more hospitable spaces and, in particular, to courts and aristocratic circles with receptive patrons. As noted earlier in this study, however, not every court was equally open to modernizing sensibilities: indeed, through bridge appointments, traditionalist academics often exercised a voice at court. And although a few noblemen had actively involved themselves in heavenly studies during the first nova episode, it was Tycho and the resources of his island castle that established a stunning new role model and new prestige for astronomical activity outside the universities.

Kepler's case also involved an early abandonment of the university as a site for his investigations after his initial training at Tübingen, even if, through correspondence, he continued his relationship with Maestlin and others. So pronounced was his aversion, or at least suspicion, of the academic environment that years later he even refused appointments at the universities in Bologna and Prague; he was considered and rejected at Wittenberg, and consideration of him as successor to Galileo at Padua also failed.[7] Kepler gained recognition through his publications, his voluminous correspondence, and his fortunate— if, at times, tortured—personal relationship with Tycho Brahe in the last two years of his life. After 1597, as prognosticators began to associate his name with Copernicus, Kepler became well known beyond Tübingen and Graz; and after arriving at the Rudolfine court, he continued to extend this reputation as an annual prognosticator. At the same time, like Galileo, he furthered his reputation through epistolary friendships with members of the nobility. Most famously, Kepler became involved as early as 1597 in a long and fruitful correspondence with the Bavarian chancellor Johann Georg Herwart von Hohenburg through the initial mediation of Clavius's student Christopher Grienberger.[8]

Galileo, in contrast to Brahe and Kepler, forged his early reputation as a university teacher. But unlike Reinhold, Magini, Maestlin, and Clavius, he never chose to publish any sort of textbook. Indeed, his use of print was closer to the likes of Thomas Harriot and the early Copernicus. Put bluntly, the early Galileo worked strictly through privately circulated exchanges and rarely used print to construct his renown as a mathematical practitioner and his personal identity as a man of learning.[9] Although *De Motu*, for example, was a seminally important writing in Galileo's private intellectual evolution, he never published it.[10] And when finally he did publish his little *practica* on the geometric-military compass, it immediately produced an unpleasant dispute whose consequences reverberated for years after. As late as 1623, he was still feeling the sting of Simon Mayr and Baldassare Capra's piracy of 1607: "I had many years previously shown it [the *Geometric and Military Compass*] and discussed it before a large number of gentlemen and had eventually publicized it in print. May I be pardoned this time if, against my nature, my habit, and my present intentions, I show resentment and cry out, perhaps with too much bitterness, about a thing which I have kept to myself these many years."[11]

Furthermore, even after he had used print to announce his telescopic discoveries, he soon turned back to scribal strategies. For example, he circulated his "Letter to the Grand Duchess Christina" (1615) in manuscript (it was not published until 1636), and he wrote out the "Letter on the Tides" (1616) at the request of Cardinal Alessandro Orsini, who also agreed with Galileo that "the extended discussion of such refutations [of the theories of other writers] for the general public [should] wait until I treat this subject at greater length in my *System of the World*."[12] An unspoken presentist tendency has obscured Galileo's methods of circulating his ideas, thereby permitting the unexamined assumption that, as is the case today, the printed book was the sine qua non for establishing and perpetuating intellectual fame in the sixteenth and seventeenth centuries. Galileo's career before the publication of the *Sidereus Nuncius* thus centered on university teaching and private ties to local aristocrats and ecclesiastics who were involved, one way or another, in diverse currents of learning and who also possessed influence with local rulers of the universities. How did these kinds of sociabilities shape the conduct of Galileo's work within the science of the stars?

PATRON-CENTERED HEAVENLY KNOWLEDGE

Court and aristocratic sociabilities afforded a markedly different space of possibilities for advocating and promulgating new theoretical knowledge from that of the dominant structure of pedagogical relations in the universities. Although many professors found ways to teach subjects that fell outside the regular curriculum, they were still constrained—as the examples of Galileo and Maestlin make clear—by what they could and did publish. The universities were built around older men teaching adolescent boys, who then displayed and honed their learning in fixed genres and ritualized disputations. The alternative cultural spaces involved powerful men who gained signs of homage and deference in exchange for protection and the bestowal of favors. The case of Galileo again differs in some interesting ways from those of Kepler and Brahe. Kepler had broken with the Tübingen theologians while still a student; Brahe had never sought an academic position. Galileo, however, had long made the university a way of life by the time that he decided to move. In his departure from the constraints that marked the university, the relevant questions concern the degree of constraint or freedom afforded by courtly practices. Specifically, what weight ought to be assigned to the demands and opportunities of patronage in shaping Galileo's judgments, his beliefs about the heavens, and the ways he should frame his ideas, as well as which he should push forward and which not?

Galileo's work as a conceptual innovator, problem solver, and self-described investigator of truth is complicated by its exceptional finale—the political disaster that ended with his condemnation by the Catholic Church in 1633.[13] Richard S. Westfall argued that the explanation for the gulf between the scientific and the political must be found *outside* Galileo's science but without relinquishing a place for truth as a motivating ideal.[14] Part of the explanation was to be found in Galileo's personality, which Westfall described as egotistical and insufferable; but the main part was to be found in "the patronage system," which purportedly fostered just those kinds of personality traits. Another important consideration was the acquisition and maintenance of social status.[15] And the two considerations—self-glorification and establishment of the truth—"tended to merge into one."[16] Westfall made the mitigated proposal that most of Galileo's major discoveries originated outside the framework of patronage but that some—especially the early telescope-based claims—were motivated by Galileo's overweening lust for the glory that it might bestow on him alone. As a prime example, Westfall alleged that competition for credit and patronage lay behind Galileo's failure to acknowledge his disciple Castelli for the original idea that observation of the phases of Venus confirmed that planet's circumsolar orbit.[17] Evidently Westfall assumed that the Medici would value the "cosmological" meaning of this discovery as much as Galileo did himself.

Mario Biagioli, on the other hand, presented a far more theoretically ambitious and thickly articulated model of patronage than that of Westfall, in which he subordinated the entire intellectual production (and reception) of Galileo's middle period to a well-integrated model of court social structures and dynamics. In this scheme, the well-known telescopic episode was not exceptional but rather paradigmatic of Galileo's entire career: it featured a "wannabe" courtier seeking preferment, who cleverly shaped his novel claims about the heavens to fit the political and social demands of the targeted patron.

On this account, patronage is "a key" for understanding the behavior of early modern actors in general and "the key" for understanding how early modern "scientists" behaved, communicated, disputed, argued, and, most important, established the legitimacy of their claims about the natural world. The social structure of patronage plays a major explanatory role, and truth essentially functions as a rhetorical strategy in the service of social climbing. This account has attracted much comment and carries wider implications for understanding patronage, but selective treatment has sometimes obscured its most salient claims and their logical relations.[18]

What makes this account especially interesting is its weaving together of a series of sociological and anthropological submodels. At its heart, status seeking and identity formation are bound together in a structural model not merely of patron-client relations but of such relations in court settings. Thus patrons and clients, following strict rules of etiquette, were said to operate in an economy of gift exchange in which clients got ahead by presenting gifts and incurring obligations to reciprocate. Galileo is said to have shaped his best-known claims about novelties in the heavens as symbolic vehicles to fit the political mythologies of his court patrons. In return, he gained "cognitive legitimacy" from the Medici.[19] He engaged in this "fitting" and "tuning" behavior not by arguing like a professor but by reinventing himself as a courtier, mastering displays of surface decorum and "fashioning" a new persona as court philosopher who used understanding of the natural world to win improvement in his social status.[20] Unlike Westfall's portrayal, which allows some room for Galileo's personality, Biagioli's Galileo is an epiphenomenon of the status contests in which he engages; he suffers no inner conflicts about his social identity.

More challenging for internalist accounts, Biagioli implicates Galileo's claims to truth about nature in relations of power. The reciprocity of the patronage relationship provides the entire dynamic for Galileo's efforts to achieve epistemic legitimation for his claims about nature: it mutually obligates patron and client to protect princely honor from "status pollution" by maintaining the right social distance. Hence it follows that patrons were (structurally) "noncommittal" in natural philosophical controversies and maintained a face of "instrumentalist" indifference. Likewise, just as the patron appeared not to be involved in controversy, so too did a high-visibility client like Galileo seek to craft a self-image of disinterestedness and objectivity.[21] Only when the patron saw some political gain to his image would he take a risk by endorsing a client's claims.[22] But a natural philosopher like Galileo, who was supposed to know how to behave like a courtier, could also advance his own social status by provoking and participating in lively controversies that entertained the prince but did not touch or endanger his honor. Controversies were thus a necessary part of the social structure of the career of a "scientific courtier." Moreover, because there were different levels of patrons, a client like Galileo who had once attracted the attention and benevolence of a "great patron" could never rest on his laurels because he was in a "high-risk, fast-track" relationship.[23] The logic of the patronage structure required him to increase the patron's honor by behaving something like a knight, continuing to make "spectacular" discoveries (like the telescopic ones), sponsoring allegedly controversial positions (like the Copernican theory), and winning debates staged, like jousts or duels, for the patron's amusement.[24]

This exchange model turns the patron and the court into both the primary audience and the primary source of legitimation for claims about the natural world, leaving little or no place for the universities.[25] As a consequence, it demotes natural philosophers and heavenly practitioners to the subsidiary role of socially anxious performers or entertainers, ever concerned to maintain social traction in the service of patron-rulers and court theatricality. And once having established the binary pair of courtier/other, it is a short step to claim further legitimacy for the model by dressing it up with an appeal to a revised Kuhnian account of incommensurability grounded in social—or, as Biagioli prefers, "anthropological"—boundaries.

Whatever its many apparent attractions, serious difficulties bedevil the court-centered model. The first is evidentiary. There are serious gaps between what Biagioli says he is doing—which can easily be read off the top—and the historical evidence as represented in his citation practices.[26] Another is the classic tension in structural-functionalist theory between the constraints exercised by general social structures and the relative freedom of individual agency.[27] Did Galileo's desire to join the Medici court actually determine or at least rigorously constrain the scientific questions he investigated? Were the positions that he adopted (such as his commitment to Copernicus's theory) shaped or caused by a perceived need to fit the structures of courtly games and the prospect of advancement through early modern "talent scouts" or "brokers"? Was his language constrained, if not determined, by the "codes of court culture"?[28] Although Biagioli frequently represents Galileo as a *bricoleur*—a creative fiddler and tinkerer—a good many passages leave the impression of a Galileo who is really a functionalist thriving in a heavily constrained structuralist idiom, all social relations and scientific positions deduced from means-end considerations.[29] Patronage, friendship, gift giving, self fashioning, status seeking, and honor preserving were just different kinds of instrumental sociabilities. Patronage provided the repertoire of motives; Galileo acted them out.

But whether or not Galileo was quite so heavily constrained by a "system" of patronage, the *historical generality* of its alleged structures must still be probed and secured rather than asserted. For the putative strength and value of the multiply layered scheme rests in the first instance on the generality—or, at best, the typicality—of its elements.[30] These features are the ones from which Galileo, the main social actor, was supposed to have constructed his bricolage. The model asserts that these attributes define early modern European court society and that they are all somehow connected to status seeking. The possibility does not seem to occur that these elements might exist independently of one another or that there might be significant variations from one court to another. Yet without demonstration of the generality or systematicity of the patronage structures, one is left with the claim that they applied only to the case of Galileo and only at the courts of the Medici and the Barberini.[31] For both Westfall and Biagioli, Galileo is, as it were, the prime exhibit. But here one must raise the more serious possibility that Biagioli's patronage model is the result of a rhetorical rather than a logical inference, one designed just to fit the singular case that he wished to explain.

A preliminary way to test this hunch is to ask whether the model covers some of the key episodes I study here. Because the reality of astral influence and the efficacy of astrological prediction were among the most heavily contested questions of natural knowledge in the sixteenth and early seventeenth centuries, one would like to know whether the arguments over the status of astrology fit the allegedly general structures for the conduct of court controversy.[32]

The answer is abundantly clear in one case after another: far from being neutral on that subject or treating it as a game "to be played 'after the table was cleared,'" rulers and patrons were anything but noncommittal and socially distantiated.[33] All the Hapsburg emperors, Pope Paul III, Duke Albrecht of Prussia, and Grand Duke Cosimo I de'Medici clearly believed in the reality of celestial influxes. For Cosimo I, the astrologers Ristori and Formiconi were serious advisers, not entertainers; at stake was how to get the best personal prognostications from the best astral number crunchers and interpreters.

Yet where there was debate, the ruler did not hide behind the mask of honor but stepped forward with his views. King James I—surely a "great patron" if ever there was one—openly published his grave doubts (in the *Daemonologie*) about practitioners of judicial astrology and ways to detect the devil's malign influence. In the Chamber-Heydon controversy, he did not stand neutrally in the background but allowed his publicly authored views to be represented through the Windsor chaplain John Chamber. Furthermore, although heavenly practitioners associated with courts were typically well trained in classical rhetoric, none (including Galileo) exhibited the "studied nonchalance" (*sprezzatura*) described in Castiglione's *Il cortegiano*. Far from maintaining a disciplined social distance or approaching a patron only through court brokers or even following the rules of courtiers' manuals—as the model predicts—Kepler wrote directly to the English king and held the expectation that James himself would actually read and judge specific sections of the *Stella Nova* (sent to him as a gift) and the *Harmonice Mundi*. James, however, was anything but noncommittal. Fifty years earlier, the same had been true of the relationship between Erasmus Reinhold and Duke Albrecht. Reinhold often wrote to the duke without benefit of "brokers"; although at one point the duke sent him a beer mug, Reinhold desperately (and candidly) expressed his need for money to publish his tables in the duke's name. And looking at the matter from the other side, the nobleman Tycho Brahe put his honor at stake when he vigorously struggled to maintain his priority claims against those of lower social status, like Christopher Rothmann and the swineherd Ursus, whom one might have expected him simply to ignore.

This failure to extend the background model's social-distance thesis increases the likelihood of a begged principle and clears the way for consideration of the specific question of Galileo's own commitment to Copernicus's theory. Always invoking the hegemony of patronage structures, Biagioli proposes that "Galileo's Copernicanism should be perceived as an *explanandum* rather than assumed as an *explanans*." He contends that only after 1611 did Galileo pursue and defend Copernicus's theory, because this was "the way a high-visibility client of a great patron *had to* maintain his status by keeping engaged in controversial, aggressive forms of intellectual production. And, quite literally, there was no more difficult—and therefore honourable—challenge for Galileo than to fight for Copernicanism."[34]

Once again, the structural demands of patronage and court-centered authorship drive the intellectual commitments, including Galileo's active support of the Copernican hypothesis.[35] This analysis leaves us to wonder whether Galileo's interest in tracking Kepler's explicitly Copernican program between 1597 and 1609 was also driven strictly by the desire for preferment at court. Or, indeed, whether Galileo was dissimulating in his earliest letter to Kepler when he said that he had long been a follower of Copernicus. Or whether the theoretical commitments of the church canon Copernicus and his disciple, the Wittenberg mathematician Rheticus, the cautious academic Maestlin, the landed gentleman Digges, the flamboyant "academic of no academy" Bruno, the military engineer Stevin, the Kassel court mathematician Rothmann, and Maestlin's famed student Kepler are to be attributed to office seeking and the urge for higher social status. If, as is clear, they were not, then it is difficult to warrant an appeal to what was socially typical and then to find Galileo to be an instance of such typified behavior.

Galileo's patronage issues were no more typical in his case than in that of other heliocentrists. But if Galileo's reluctance to use print or publicly to espouse his views on Copernicus's theory before 1610 was exceptional, then it is incumbent on us to take into account explanations other than patronage, including the effects of Bruno's trial in Italy, Galileo's encounter with the Venetian Inquisition, the absence of a necessary demonstration for the Earth's motion, and other such considerations that, in one way or another, were implicated in his sociabilities. Such difficulties with the model's generality as well as its specific application to the question of Galileo's engagement with the Copernican problematic are far from exhaustive. But in undermining confidence in the patron-centered model, they leave open the question of locating patronage in relation to the main problem of this chapter.

And on this point it can hardly be denied that patronage was a widespread and profoundly important institutional formation in early modern European culture.[36] In their different ways, Westfall and Biagioli rightly underscored the importance of patronage for understanding scientific culture. But now it must be recognized that it was just one *kind* of sociability, which overlapped other kinds of ties, such as kinship, friendship, citizenship, and confessional allegiance.[37] By fore-grounding and reducing all sociabilities to patron-client instrumentalities, we may easily miss the major character of Galileo's personal associations and other motivations for his intellectual projects.

PATRONAGE AT THE PERIPHERY
GALILEO AND THE ARISTOCRATIC SPHERE
OF LEARNED SOCIABILITY

I want to suggest another picture, one that does not eliminate patronage but resituates it. Galileo's primary forms of sociability, I suggest, lay in his friendships and in his pedagogical practices. Earlier studies of Galileo have not ignored friendship, but the historical character of his friendships ought not be taken for granted. Galileo surely must have encountered ideals of friendship much in favor in the learned circles with which he was associated. Constancy, reason, and personal restraint characterized the neo-Stoic philosophy of friendship found in Justus Lipsius's *De Constantia*, a work of great popularity in aristocratic circles from its initial publication in 1584.[38] Seneca's writings were a central resource for Lipsius and others: "Associate with those who will make a better man of you. Welcome those whom you yourself can improve. The process is mutual; for men learn while they teach." Seneca concluded with a distinction: a relationship that "regards convenience only and looks to the results is bargaining [*negotiatio*] and not a friendship [*amicitia*]."[39]

Galileo's actual friendships, at least as known through his correspondence practices (rather than deduced from moral ideals), reveal another dimension: the didactic role of the academic teacher. This is hardly surprising if one recalls that Galileo had spent twenty-one intellectually formative years at Pisa and Padua before moving to the Tuscan court. They often involved philosophical conversations, or shared descriptions of his experimental and observational work, in which Galileo was the pedagogue. If career aspirations were obviously present and at times urgent, they were also peripheral to his learning.

Many, if not most, of Galileo's important correspondents were members of a lettered sphere of nonpractitioners. Whether ecclesiastics, patricians, or aristocrats, men like Pinelli, Sarpi, Sagredo, del Monte, Gualdo, Querenghi, and Welser were friends of some considerable learning, knowledgeable in classical languages and the liberal arts, even substantial authors in their own right in other realms, and always remarkably capable of appreciating (although not necessarily agreeing with) the ideas that Galileo articulated in his correspondence. They were all engaged in scholarly activities of some sort, *but not because their living depended on it.* None had the obligation to teach. Though some had patrons, none were dependent on patronage in order to support their scholarly practices. Law courts were not yet the sites of support for scientific expertise that they would begin to become in the eighteenth century.[40]

Leaving aside ornate epistolary conventions of salutation that acknowledged social station, most of Galileo's letters, like Kepler's, were not cast in the idiom of court fashion for the amusement of polite society. A great many were effectively *written scholarly conversations* or arguments directed either to practitioners or to nonpracticing men of letters—frequently structured by academic problems, erudite references, epistemic standards, linguistic resources, and similar practices and concerns. In length, some of Kepler's letters verged on little treatises; by contrast, Galileo's were usually short. Somewhat later, Galileo's *Dialogue* exemplified argumentation cast in the form of a philosophical conversation. The move was not difficult: the dialogue and the letter were prominent forms of late humanist friendship practice that could easily be directed to didactic ends.[41] And in the days before scholarly journals, when the letter was a major form of learned communication, its rhetorical possibilities and relaxed, civil form of presentation made it one of the preferred literary models for Galileo, Kepler, and Brahe in their efforts to reformulate theoretical knowledge outside the traditional practices of the universities.[42]

A closer look at Galileo's early epistolary exchanges with various sorts of correspondents reveals the complexity of patronage themes and their involvement in both theoretical and practical knowledge. An important example is that of the marquis Guidobaldo del Monte (1545–1607), whose interest in materially helping Galileo grew organically out of common interests in Archimedes, problems of simple machines, and observing instruments. Guidobaldo had once been a student of Federico Commandino (1509–75) and shared elements of the Archimedean scholarly style in which Galileo sometimes wrote. Besides his acquaintance with Galileo, he had also cultivated contacts and exchanges with Mazzoni at the University of Pisa, and he introduced Galileo to Pinelli.[43] Yet, unlike Galileo, he used print early in his life to establish his views and interests. In 1579, for example, he published a book about the universal planisphere in which he located his account in relation to the Louvain tradition of Gemma Frisius and Johannes de Roias.[44] Guidobaldo explicitly classified his work as a theoric, presumably because it concerned purely geometric explorations of different cases of spherical projection into the plane of the instrument. He had nothing to say about any specific uses to which he had put his study. Del Monte also aspired to use the lever to provide theoretical foundations for the subject of mechanics and published a book on the subject.[45] Galileo clearly shared close intellectual interests with the marquis, even while differing with him over many issues concerning the science of motion; and he was thus comfortable in treating him as an epistolary partner.[46] In a 1602 letter to Guidobaldo from Padua, he excuses his urgency "in persisting with the wish to persuade [Guidobaldo] of the truth of the proposition that motion occurs in equal times in the same arc."[47] The letter was an effort to share and convince—with the expectation of the same from his correspondent.

There was also a significant practical or utilitarian side to their friendship. Earlier, in 1589, Guidobaldo had helped Galileo obtain his position at Pisa. In 1592, again through Guidobaldo, who had family connections in Venice, Galileo was introduced to the well-placed and immensely learned Pinelli, who was neither an author nor a practitioner but who also assisted him greatly in gaining his position at the university in Padua.[48] Even so, it would be misleading to say that their relationship was built principally or exclusively around what Guidobaldo could do for Galileo. The basis of their sociability resided in their common interest in learning and in its shared forms of expression.[49]

Still other exchanges with Galileo appear more focused on practical problems. For example, there are some letters from Giovanfrancesco Sagredo concerning the acquisition of magnets and the repair of mechanical devices—in which Sagredo had considerable interest and some ability—but with no indication of interest in the theory of the magnet.[50] In another letter, however, Sagredo wrote to Galileo thanking him for sending him some pieces of iron and informing him that he had personally given Paolo Sarpi a device for measuring declination—"which appears to me to be material on which to philosophize." The same letter also makes clear that Sagredo was one of Galileo's astrological clients and was familiar with others for whom Galileo had prepared nativities.[51]

Galileo also had amicable intellectual exchanges with politically powerful and learned men in Venice who did not directly help him to improve his salary or gain any preferment of office but who were, nevertheless, quite helpful to him in other ways—as he was to them. A notable example is that of Paolo Sarpi (1552–1623), the prominent state theologian, historian, legal adviser, and exceptional controversialist for the Venetian Republic. Sarpi became the great defender of civil authority and republican sovereignty in Venice's rupture with the papacy in 1606 and its subsequent placement under Church interdict.[52]

Like Guidobaldo del Monte and Sagredo, Sarpi was part of this sphere of learned nonpractitioners in Italy—men who did not make their living from studying and interpreting the heavens and other aspects of the natural world and yet who possessed some aptitude for such studies.[53] Some of their interests were quite theoretical. In September 1602, the versatile Sarpi wrote a detailed letter complaining of his difficulties in reading Gilbert's *De Magnete* and asking Galileo's help in understanding it.[54] As usual, Galileo was at his best in exchanges where he could be the teacher; and he, in turn, felt comfortable in communicating to Sarpi a new rule that he had discovered in which bodies accelerate uniformly by nature in free fall and also in violent projectile motion, in proportion to the time elapsed from rest (although from the incorrect assumption that speed is directly proportional to distance). He made clear his hope that Sarpi would give his "opinion" after "considering it a little bit," although clearly he did not expect Sarpi to perform an experiment.[55] Sarpi knew enough to recognize the importance of the news that he had received from Jacques Badovere in Paris about an instrument that could magnify distant objects by combining a convex and a concave lens in a tube. It was he who communicated this information to Galileo.[56] And to Giacomo Leschassier, another correspondent, Sarpi sent a copy of Galileo's *Sidereus Nuncius* with details on the sizes of the lenses and a brief description of "many other more remarkable things," such as the short periods of revolution of Jupiter's "stars."[57]

These letters indicate a shared culture of learned sociability—to use, once again, Peter Miller's apt and important expression. Reducing all friendships to a single, instrumental dimension—the power relations between patrons and clients—unnecessarily obscures and oversimplifies the character of Galileo's intellectual amities. Galileo had various sorts of relationships with members of the Venetian elite; the philosophically liberal, cautiously interconfessional Paduan sodalities of Pinelli and Querenghi; occasional foreign members, such as Nicolas-Claude Fabri de Peiresc and notable Englishmen like Edmund Bruce, Thomas Seggeth, and Henry Wotton; and the academic philosopher Cesare Cremonini. Deeply versed in neo-Aristotelian discussions of method, demonstration, and the classification of the sciences, Galileo was also drawn to anti-Aristotelian sensibilities wherever he found them.[58] He could maintain such connections because Venice and its university maintained a robust political independence even after the tragic case of Bruno marked the ascendancy of conservative voices in papal Rome. Although there were social differences between Galileo and his correspondents, multiple considerations ultimately bound him into this web of sociability: shared concerns in practical—and sometimes theoretical—domains of philosophy, dissatisfaction with conventional modes of academic discourse and a preference for late-humanist literary forms, an attraction to modernizing positions in natural philosophy, and support for liberalization of the rules for interpreting holy scripture. Galileo's correspondence with Venetian men of learning like Sarpi and Sagredo stands in distinct contrast with the single, prudently crafted letter to Kepler, the eager, would-be intellectual ally. Both modernizing practitioners like Magini and Kepler and traditionalists like Capra and Lodovico delle Colombe brought out the sharply combative (and competitive) side of Galileo's personality. And it was through the *philosophically unthreatening* patrician men of letters that Galileo benefited from occasional, important gestures of patronage that helped him move into more favorable institutional positions.

FLORENTINE COURT SOCIABILITIES

In the early seventeenth century, the Medici court in Florence did not display the same vigorous style of late-humanist learned sociability and active literary experimentation that flourished in Paduan and Venetian circles. It was also no longer the philosophically creative late Quattrocento Florence of Pico and Ficino. After his death in 1598, the prolific Platonizer Mazzoni was replaced at Pisa in 1600 by the more traditionalist Cosimo Boscaglia (155?–1621).[59] Traditionalist voices, like Colombe's, appear to have been ascendant. Nonetheless, the Medici—starting with Cosimo I in 1543—had initiated a strong tradition of support for their university in Pisa. As Charles Schmitt has emphasized, the Medici fertilized the *studio di Pisa* with new professorial appointments—including non-Italians—although their general intention was to conserve the institution as a professional training ground for their own citizenry, who, in turn, were forbidden to study at universities outside Tuscany.[60] Furthermore, evidence of the long and generous tradition of Medici patronage for art and court astrology remained strong, but in Paolo Galluzzi's judgment, Florence's taste for the "natural" in the early Seicento appeared mostly in special palace collections of exotic plants and animals.[61] And finally, there was also the powerful San Marco monastery, whose library Pico della Mirandola once used and in whose church he is buried, in whose convent Savonarola once lived and whose Dominicans would prove to be such a great nuisance for Galileo after he moved back to the city of his birth.

These observations raise a new and more focused question: what can be said of the environment for natural philosophy and the science of the stars in the orbit of the Medici court itself? Rather than generalize about "the Medici court" in the early seventeenth century—as though the views of its members were homogeneous—here it will prove helpful to assess the diversity of outlooks.

Antonio de'Medici (1576–1621) was the one member of the family who appears to have been especially interested in nontraditional or modernizing developments in natural philosophy. At his death in 1621, all sorts of medicinals were found at his residence in the Casino di San Marco: not only oils, powders, gums, and herbal leaves, but also cans of arsenic, animal fats, salts, and bottles of antimony—enough to make himself sick or, by chance, healthy. Perhaps not surprisingly, Antonio also had in his possession a manuscript of a work of the Paracelsian alchemist Gerhard Dorn, *Anatomy of Living Bodies,* as well as works by John Dee.[62] At the very least, it can be said of Antonio de'Medici that he had a special curiosity about the body. Galileo appreciated Antonio's predilections, as it was to him that Galileo directed a letter that detailed "some contemplations and different experiences" concerning various projects in mechanics and outlined both their theoretical elements and their practical applications.[63] This letter was not a special initiative on Galileo's part; it was Antonio himself who signaled his desire (through Galileo's cousin) to learn about "new things" in Galileo's "studies."[64] It was Antonio who approached Galileo when he heard of "the admirable proof and experience that you have made with the looking glass discovered by you and for which the Venetian Senate has rewarded you in conformity with your merit" and who was most eager to have Galileo build such an instrument for him.[65] It was also to Antonio that Galileo sent a detailed letter in which he summarized and visually illustrated the observations that would later be displayed more fully in the *Sidereus Nuncius:* the Moon, fixed stars ("that without it [the looking glass] could not be discerned"), and Jupiter "accompanied by three fixed stars, yet totally invisible by virtue of their smallness." These new stars, he told Antonio, appeared to be "most round and in the form of a little full moon, with a well-defined roundness and without any flaring."[66] It was the same Antonio who shot supportive glances across the breakfast table in the direction of Galileo's disciple Benedetto Castelli at the famous encounter that kicked off the Copernican controversies in Florence in December 1613.[67]

Besides Antonio, there was the powerful and sympathetic court secretary, Belisario Vinta (1542–1613), who also proved an influential and effective ally in encouraging Galileo's move back to Florence. Vinta was a continuous adviser to the Medici court from as far back as the time of Cosimo I (1567) and on into the early years of the reign of his grandson, Cosimo II. He thus represented a major link between those different eras of the court and had a significant role in shaping its domestic and foreign policy. Like his father, Vinta was something of a poet; his study of classical languages strongly suggests an assimilation of humanist values, as does his membership in the Accademia dei Filomati in Siena (1603). Professionally, he had been drawn to the study of law at Pisa (1561–66), some twenty years before Galileo's attendance.[68] Although we do not know his full course of study, there is evidence that he had more than a passing knowledge of astrology, as he owned a copy of Francesco Giuntini's compendious *Speculum Astrologiae.* Indeed, because both Vinta and Giuntini had studied at Pisa—albeit at different times—it is not impossible that Vinta could have known the Florentine astrologer personally or might have attended lectures based on Ristori's commentary on the *Tetrabiblos.*

The Vinta-Galileo correspondence that has survived is mostly from the period 1608–10. Many of the exchanges prior to the telescope episode show that Vinta was deeply respectful of Galileo's skills, but Galileo did not use Vinta as a sounding board for his mechanical investigations in the same way as he had done with Paolo Sarpi and Antonio de'Medici. Several letters reveal Galileo in the role of middleman in a business transaction that involved the sale of a large magnet owned by Sagredo. Vinta was eager to acquire the magnet for the court (at his price). Galileo eventually managed to find a price that satisfied both parties concerned; he also showed skill in designing a pithy saying (*simbolo*) to accompany the object: *Vim facit amor.*[69] On May 29, 1608, Vinta reported that Galileo's "diligenza" in the whole matter had left the Grand Duke Ferdinand and his wife "sopramodo sodisfatti et contentissimi."[70] At this moment, Galileo was clearly held in the highest esteem by the court, but thanks to the delay in receiving Vinta's letter from Florence, he was unaware of just how well he stood. In fact, quite the contrary: on the very next day, he received a letter from the Grand Duchess Christina whose formal tone upset him so much that he believed that he had somehow committed an unpardonable breach of etiquette. He was told to come to Florence in the summer by whatever means he could to continue his tutorial work with the young prince Cosimo. Then a second letter from an unnamed person threw doubt on the first and put Galileo into a very anxious state. He now feared that he had said something

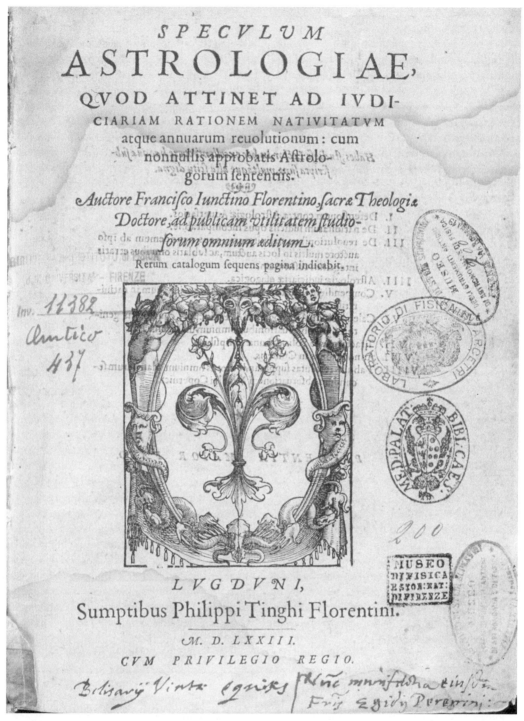

SPECVLVM
ASTROLOGIAE,
QVOD ATTINET AD IVDI-
CIARIAM RATIONEM NATIVITATVM
atque annuarum reuolutionum: cum
nonnullis approbatis Astrologorum sententiis.

Auctore Francisco Iunctino Florentino, sacræ Theologiæ
Doctore, ad publicam vtilitatem studio-
rum omnium æditum.

Rerum catalogum sequens pagina indicabit.

LVGDVNI,
Sumptibus Philippi Tinghi Florentini.

M. D. LXXIII.

CVM PRIVILEGIO REGIO.

79. Francesco Giuntini, *Speculum Astrologiae*, 1573. Provenance of Belisario Vinta. Courtesy Photographic Department, Istituto e Museo di Storia della Scienza, Florence.

wrong, something that had left the mistaken impression that he would not actually come to Florence. Moreover, because he had not yet heard that the magnet had actually arrived, he had no idea that his impressions were completely wrong.

The episode is illuminating because, in this moment of crisis over an apparent transgression of decorum in a social and economic contract, Galileo turned to Vinta as someone whom he could trust in political matters. And his rhetoric of supplication openly invoked Vinta's social status:

> It is necessary that I impart my difficulty to a trusted person to act to remove it, who in all respects ought to be none other than your eminence [Vinta]. And hence, I beg you, set aside your part as a courtier [*cortigiano*] and reserving only your knightly liberty and honest inbred disposition, tell me what I ought to do with the directness of the point of a short sword and do not keep me in the shadows with a pliable feather pen; because if you will only tell me: "Come, for whatever the Patrons wish to give," it will suffice for me; and to write to me otherwise will put me into a greater confusion than the one in which I find myself at present. . . . [Also] I beg of you not to put off further telling me something of the arrival or departure of the magnet because I have endured a fever for the past 25 days in which time I have also not had a free moment.[71]

Note that, amid all the gestures to Vinta's status at court, the issue at stake here was not Vinta's judgment about Galileo's credibility as a natural philosopher but only the security of Galileo's social-economic relationship to the Florentine rulers—about which he cared greatly. As a mathematical practitioner serving in the university of another political domain, the Venetian Republic, he would agree to perform a service for the court of a realm that he had once exclusively served. It was a variant of the pattern of familiar university-court bridge relationships. The immediate issue was Galileo's continuing willingness to tutor the young Cosimo in mathematics. Vinta was a politically trustworthy facilitator for Galileo at a moment when he feared that his relationship to the Florentine rulers—and his chances for a future, more stable office at that court—were imperiled. He appealed directly to Vinta's social status as a courtier and protector and his "inbred disposition" as a basis for using direct, honest language rather than a veiled, coded, decorous idiom. Vinta could be relied on to provide a reli-

able translation of what the rulers were "really thinking," and Galileo could feel confident in sincerely communicating his true needs and wishes.

Soon thereafter, Galileo received from Vinta the sort of reassurance that he wanted, in a letter dated June 11, 1608, which included a report of the grand duchess's sentiments in her own words:

> "Tell Galilei that he being the first and the most esteemed mathematician in Christendom and although it may be difficult for him, the Grand Duke and We desire that he come here this summer to provide exercises in mathematical subjects, which please our son the Prince so much; and that with the study that he will do with him this summer, he could then spare himself the time by not coming here and that we shall endeavor to make sure that he will not be sorry that he came." And so to you Sir this shows clearly how the matter stands and how much the sooner you come, the better.[72]

This exchange confirmed the great esteem in which Galileo was held at the highest levels of the court at the beginning of the summer of 1608. The main basis for this regard was his teaching and his astrological skill and, to some extent, his management of a business transaction—but not the telescope or gift giving. It also set an important precedent for a future moment when Galileo would again approach the politically reliable and experienced Vinta to assist in finding the right language in which to make his overtures to the Medici family.

By the end of the summer, having spent more time as tutor at close quarters with the family, Galileo composed a long letter directly to the grand duchess in which he proposed two mottoes for the magnet: *Vim facit amor* and *Magnus magnes cosmos*. Clearly, Galileo wished to make such associations local and personal. In the second motto, the cleverness involved punning both on the name of Cosimo and on the magnet itself as great (like Cosimo) and like the universe (*cosmos*).

This letter raises again the question of the state of natural knowledge in Medici court circles and how someone like Galileo might attempt to introduce some of the recent modernizing currents into them. Much has been made of this early letter to the Grand Duchess Christina as a prime instance of Galileo's pre-telescopic mastery of patronage techniques. In varying degrees, recent interpretations play on both political and natural philosophical attributions of meaning to the magnet. In 1985, Westfall framed its meaning in terms of patronage: "Obviously the lodestone represented the prince. The ancient symbol of the Medici embodied its shape. Moreover, the earth was known to be a great magnet, and the prince's name, Cosimo (or Cosmo) was a synonym for *Mondo*, or Earth. Hence it was possible, he concluded, 'through the most noble metaphor of a globe of lodestone to indicate our great Cosimo.' Such evidence reveals how the subtle alchemy of patronage transmuted an object of science into an *objet d'art* to amuse and flatter a prince."[73] In this passage, Westfall hints—without any significant evidence—that William Gilbert's equation of the Earth with a magnet was already generally known; that Ferdinand, Christina, and their son Cosimo were among those who knew (and understood) Gilbert's new philosophy of the magnet; and that Cosimo's name meant *Earth*.[74]

Five years later, building on Westfall's story—but then going well beyond it—Biagioli claimed that Galileo self-consciously used his emblem *Magnus magnes cosmos* to attach Gilbert's claim that "the world is a great lodestone" to the stability of Medici rule. This episode was said to be a good exemplification of Galileo's "tactics of socioprofessional self-fashioning" required by the structure of the patronage relations—and a telling anticipation of the telescopic moment: "It [Galileo's emblem] associated Gilbert's theory (one that could be used against the accepted Aristotelian cosmology) with that of the naturalness of Medici absolute rule. . . . Galileo's strategy was aimed at legitimizing scientific theories by including them in the representation of his patrons' power, thus securing both the patrons' involvement and endorsement. This was Galileo's attempt to get out of the deadlock caused by the patrons' noncommittal attitude."[75]

To date, no one has challenged these alleged connections, although they are anything but obvious. For Galileo, Gilbert's magnetic philosophy never occupied anything approaching the central position that Kepler or Stevin assigned to it. In fact Galileo remained critical of Gilbert as insufficiently mathematical in his natural-philosophical theorizing. Furthermore, although we know that Galileo happily explained aspects of Gilbert's ideas to his patrician friends in Venice, there is no evidence of such exchanges with members of the Medici court. Indeed, there is not a single letter to Vinta or Antonio de Medici—let alone the grand duchess or her son—indicating any Gilbertian associations for the celebratory magnet. And the generality of the *simbolo* could easily have fitted Giovanni Baptista della Porta's lengthy description of the magnet as a "wonder of nature."[76]

The relevant question, probed earlier with respect to Copernicus and Kepler, was how Galileo tried to gain a patron's support for a *theoretical* claim. Why would Galileo use an emblem and an inscription on a medal to communicate a new and profound philosophy, let alone one that involved descriptions of magnetic experiments? Biagioli suggests that the purpose was to "legitimate" Gilbert's philosophy in courtly society. Yet emblematic medals were in no way unique to the courtly milieu: they commonly served as commemorations for speech-day student orations and plays on moral or didactic topics.[77] Had Galileo struck his emblem as a means of introducing Gilbert's magnetic philosophy into the Medici court, one might have expected an accompanying oration interpreting the magnet's philosophical meaning. But no such evidence exists. For Galileo, as with Copernicus, Kepler, and Gilbert himself, theoretical claims of such an order were communicated within the proper rhetorical conventions: as arguments in books or in extensive letters.

This brings us finally to the adolescent prince Cosimo II. Unlike Vinta, he appears to have had no formal university education. But, of course, he had had an informal, if partial, one under Galileo's personal tutelage in some of his most impressionable years. And Galileo had had many years' experience giving private lessons to gentlemen in addition to his university lecturing. Thus it seems reasonable that he would have acquainted Cosimo with the uses of the military compass, which, after all, he had dedicated to him. Also, at some point—perhaps during the tutorial years—Galileo must have asked about the moment of his pupil's birth, with some thought to composing the heir apparent's horoscope.[78] Yet neither the title page nor the dedication of *Le operazioni del compasso geometrico e militare* contains any astrological references to the Medici family or to Cosimo. Perhaps more important was the financial advantage: this book offered an occasion for Galileo to publicize himself as an instrument maker with Medici associations. Galileo's angry reaction to Capra's plagiarism of the work on the military and geometric compass was clearly motivated by the threat that he perceived to his livelihood as a builder of these instruments and to his growing special relationship to the Medici family. In addition to teaching him the elementary mathematics required to understand the workings of the compass, Galileo might have introduced his pupil to Euclid's *Elements* and to Clavius's *Sphere* commentary.

It is also possible that Galileo gave the prince some instruction in the drawing up of an astrological scheme, or at the very least in how to read one. That was the sort of knowledge that his mother and father obviously valued, as Christina had requested a nativity for her husband when he fell gravely ill in 1609.[79] And she had maintained her faith in Galileo even after her husband died soon thereafter. If nothing else, the Medici, like other Italian princely families, trusted the mathematicians of their own local universities to render authoritative *individual* astrological judgments about their lives. Unlike annual prognostications, these judgments were not published. In Florence, this had been the practice since the time of Giuliano Ristori and Cosimo I; with Giovanni da Savoia, the custom had continued under the reign of Grand Duke Francesco. The persistence of such practices under Ferdinand and Christina shows the possibility of expressions of independence by the secular courts and the limited reach of the papal bull of 1586.

In short, some education in the practical realms of knowledge—including the science of the stars—was the sort of thing that a Medici heir would find important, much as other European rulers did. It was less common for a prince, like a Tycho Brahe or the landgrave of Hesse-Kassel, to develop skills in the heavenly sciences that were active and investigative.[80] The possibility that Galileo might have taken the occasion to introduce into the young Cosimo's head some of his own ideas or even to mention the works of Gilbert, Archimedes, Kepler, or Bruno cannot be excluded (or included) *tout court*. What can be said with considerable confidence is that, well before he had come by any knowledge of the telescope or had decided to use it as lure to be invited back to Florence, Galileo had succeeded in establishing a strong bond of personal trust with the heir apparent and his parents. Although the ties of sociability in Florence were perhaps less philo-

sophically compelling and stimulating than in Venice and Padua, they were not based, in the first instance, on gift giving. There were other considerations that drew Galileo back to the land of his birth.

GALILEO'S DECISION TO LEAVE PADUA FOR FLORENCE

It is well known that Galileo was trying to leave his teaching post at Padua as early as 1601, and that the construction and application of a new kind of observing instrument eventually provided the crucial occasion for that departure.[81] What we should make of this episode is a matter of debate. Galileo's decision involved a judgment about how to achieve certain goals. Retrospective understanding of that judgment colors interpretations of Galileo's social context and of what he thought he was doing. One part of the problem is assessing his objectives; another is how Galileo weighed and balanced them; still another is the methods that he used to achieve them. Furthermore, as in any decision with potentially great consequences, it was difficult to foresee all contingencies. And in leaving a place with many affective and material ties, Galileo faced a trade-off: something was gained, something given up. For all these reasons, Galileo turned for advice to trusted friends. It is interesting that although he gave astrological advice to clients facing important decisions, there is no evidence that he himself sought help from the stars in making the momentous decision to leave the university and his loyal circle of supporters in Padua, or even the day on which to make the journey.

Various secular considerations have been proposed as reasons for Galileo's decision to leave Padua for Florence: heavy financial obligations, low salary, lack of time for writing, resistance from the philosophical faculty, the honor and prestige of being a member of the Medici court, Florentine cultural pride and nostalgia for his homeland, and the religiously and politically charged situation in Venice. Not surprisingly, all these considerations are interrelated, and historians' interpretations turn on how they are weighted.

Between 1600 and 1606, with the out-of-wedlock births of his three children Virginia, Livia, and Vincenzio by his mistress Marina Gamba, Galileo discovered both the pleasures of fatherhood and its financial burdens.[82] The burdens are what concern us here because, as a university mathematician, Galileo was relatively impecunious. One measure of his situation is that his salary was always less than that of his philosophical colleague Cesare Cremonini, and it went up only slowly.[83] He earned no income from publishing until the appearance of his work on the proportional compass in 1606. His main sources of extramural income were four: building instruments; private tutoring; renting his house to lodgers; and casting horoscopes. He engaged in all of these income-generating activities while privately pursuing his experimental projects in natural philosophy and lecturing at the university. When the Inquisition investigated him in 1604, their action posed, as much as anything else, a threat to his revenues. Although he continued to cast nativities privately for friends and gentlemen, he never sought to extend his involvement with this kind of practice by publishing a work of astrological theory, as Kepler did, or a treatise on the astrological meanings of forehead furrows, as did Hagecius. In this regard, he more closely resembled his astrological predecessors at Florence, Ristori and Giovanni da Savoia.[84] When Baldassare Capra and Simon Mayr plagiarized his book on the military compass in 1607, the transgression also posed a threat to his livelihood as an instrument maker.

For patrons, a common motivation for bringing learned men from university to court—whether permanently or not—was to receive tutelage for themselves or their children. The Scottish humanist George Buchanan mentored the young James VI; likewise, the young noble scion Tycho Brahe was tutored by Anders Vedel before embarking on his tour of the German and Swiss universities; and Thomas Hobbes famously gave lessons to the children of the earl of Cavendish. The practice was no different in Italy. For much of the 1570s, Giuseppe Moletti tutored Vincenzo, the son of Guglielmo Gonzaga, duke of Mantua.[85] In 1599 and 1600, Vincenzo, now himself the duke, formally requested that the authorities release Magini from his late-spring lectures at Bologna to allow him to serve as a mathematical tutor for Vincenzo and his brother.[86] And so it was that in 1604 the duke approached Galileo with a request for such private instruction and an offer of three hundred ducats per annum, in addition to expenses for himself and a servant. Ga-

lileo replied (directly) that he could not accept less than five hundred ducats plus expenses.[87] He, like Magini, was not ready to give up his university post. But the next year, when the Medici court invited him to tutor the heir to the throne, Galileo saw long-term possibilities for improving many aspects of his situation in Padua—not least problematic of which was that the crush of activities in which he was engaged left him with precious little time to do everything else that he wanted to do. Unlike his prospects in Mantua, the position also offered the possibility of reestablishing his earlier connection with the university in Pisa, which carried its own prestige and opportunities as a public platform for his ideas. Money and time were obviously related, as the more he had of the former, the more he would have of the latter. Prestige alone, although clearly desirable, would buy no time or relief from financial burdens. Time to engage in extrapedagogical philosophical pursuits would be a prominent issue when he eventually opened full negotiations with the Medici court in early May 1610.[88]

STABILIZING THE TELESCOPIC NOVELTIES

And this brings us to some preliminary considerations of the telescopic episode. It is important to recognize that even before the publication of the *Sidereus Nuncius* on or around March 12, 1610, the *occhiale*—as Galileo often called it—and the reports based on it had already begun to initiate new discursive possibilities. Both in his letters and in the book that eventually appeared, Galileo did not cast his representations of new phenomena in the traditional language of epicycles, eccentrics, and equants, because he was no longer just addressing traditional moving points of light visible to the naked eye. Some of Galileo's novelties, like the Moon's surface, now appeared as three-dimensional bodies with hitherto unsuspected optical properties, made even more compelling by suggestive analogies to the Earth. Others, like the moving Jovian "planets," as he called them, or the large number of indiscernible fixed stars, were new in the sense that previously they had been hidden (*occulta*). Strikingly, the novel appearances that Galileo detected with his instrument and the astronomical meanings that he assigned to them were far more easily shared across the boundaries between heavenly practitioners, lettered nonpractitioners, and princes than earlier celestial novelties had been. With appropriate management and increasingly well-crafted instruments, the causal stimuli made themselves available to human receptor systems; and Galileo's favored practice of analogizing to familiar earthly phenomena, made even more vivid by his skilled engravings of the Moon, made his claims more accessible to nonmathematicians.[89] And thus, in spite of resistance—which was not inconsiderable, as chapter 18 shows—these representations would rapidly prove to be more "friendly" and hence open to agreement and debate.[90]

Historians do not usually think to contrast the early days of the Galilean telescope with the circumstances attending the earlier novas. Although many practitioners across Europe witnessed the nova of 1572, it disappeared within two years. After 1574, it had become a purely textual object; and it was not until 1602 that Tycho Brahe's fulsome representation of that event was publicly established. Controversies about the 1604 nova lasted for four years, and the publications concerning them were so scattered that in 1606 Kepler was unfamiliar with most of the Italian treatises, and Harriot and Stevin made no references to them in any of their extant writings. Moreover, Kepler's efforts to use the nova as a resource in the controversies about astrology and world systems met with limited success, even in England. And again, like the nova of 1572, the object disappeared. In 1609, anyone interested in observing novas would need to wait until the gods decided to produce another one.

The *occhiale*, however, was an engine for *re*producing novelties in the ordinary realm. This development was unprecedented. Rather than a group of mathematicians agreeing on the measurement of a tiny angle—the sine qua non for establishing the existence of an (unpredicted) *stella nova*—the prospect of an individual's using the instrument to seek out (predictable) samenesses in the surfaces, illuminations, circumferences, and relative positions of celestial objects changed the discursive possibilities for having an audience. As the instruments traveled, various representations that they permitted also traveled relatively quickly and easily, thereby opening a space as much for the objections of philosophers and the admiration of princes as for the once-exclusive considerations of mathematicians seeking the measurement of parallax.

To this wider audience, various eyepieces and Galileo's purported novelties would more readily become resources of debate at the interface of planetary order and the nature of the celestial region. And they were all over Europe within a few months—sometimes liked, sometimes disliked, often praised, at times envied, even for a time denied—and all this even though quite a few of the instruments were of such inferior quality that they yielded dubious and contestable images. But—and this is the crucial difference—neither the instrument nor the book nor the often blurry and disputed phenomena would go away. Unlike the ad hoc and mysterious circumstances that produced novas and comets, the occasions for discussing, debating, and studying the telescopic phenomena could eventually be made to come back. And although historians have rightly noted the serious limitations of the early telescopes—spherical and chromatic aberrations, uneven curvature of the lenses, limited field of view, and so forth—nevertheless, acquiring a *good-enough* viewing instrument soon proved to be much less problematic than waiting around for the next nova, comet, or two-headed monster.[91] In short, the possibilities for *stabilizing* agreement about the existence of the phenomena in the realm of ordinary, recurrent heavenly events were vastly better than producing consensus on the comet and Mars observations on which Tycho Brahe rested his principal claim to a unique world system, Maestlin's handful of observed positions for the short-lived comet of 1577, the unanticipated new stars of 1572 and 1604, or a major conjunction of the superior planets in the Fiery Triangle together with a nova. A related issue is how practitioners and patrons went about making judgments about the reality status of novelties without benefit of telescopes.

The Medici Court

It is no mystery why patrons and rulers generally coveted the lensed tubes: they satisfied the court taste for globes, clocks, maps, and other such celebratory collectibles; they could be displayed in palaces, presented as gifts, and also put to practical use on the battlefield or at sea. That they might be employed to view (impractical) novelties in the heavens was an unanticipated use. But in the very early days of the eyepieces, price and quality concerned the Medici court as they had with the lodestone. Shortly after Galileo's visit to Florence in the autumn of 1609 to demonstrate his instrument to Cosimo, Vinta began to conduct his own private investigations. In late September and early October 1609, he was already receiving reports from Giovanni Bartoli in Venice about the relative quality of various instruments and whether, in fact, Galileo really had the "secret" for making the best ones.[92] Bartoli himself had instructions from Vinta to purchase one or two of the "cannons for seeing far away" (*cannoni di veder lontano*) from a Frenchman in Venice, but he sent back the encouraging information that Galileo's instruments had the superior advantage of magnifying an image by a factor of ten.[93] Thus, well before the appearance of the *Nuncius,* Vinta was going about his customary business of assuring the court's financial interests.

Also before the appearance of the *Nuncius,* Galileo began to receive unambiguously positive signals from various quarters, including elements of the Tuscan court. The ever-curious and supportive Antonio de'Medici wrote directly to Galileo asking to have an eyepiece made for him. He had hearsay knowledge of "the marvelous proof and experience that you have achieved with the *occhiale,* discovered by you and for which therefore the Esteemed Senate of Venice has caused you to be remunerated in accord with your merit."[94] A week later, Galileo heard that the grand duke was "greatly admiring the marvelous effects of your most desirable *occhiale.*"[95] And from Enea Piccolomini, Cosimo's private secretary, he heard that the prince much desired the eyeglass as soon as possible.[96] In January 1610—still three months before the appearance of the *Nuncius*—Galileo sent Antonio a report with dramatic sketches of the Moon based on what he declared to be a twentyfold magnification, which magnified the surface by four hundred times and the lunar body by eight thousand times. He also felt it necessary to add—in case it might be doubted—that these phenomena could be observed only with the aid of the instrument and that he was the first ever to do so. But he was also conscious that the conditions for viewing needed to be carefully managed. Hence he advised care in anchoring the *occhiale* because of various motions that could affect the stability of the observed phenomena—for example, hand tremors caused by breathing and the motion of the blood, or other distortions caused by heat and airy vapors.[97]

The process of persuasion was thus already well under way before the end of January 1610, when Galileo told Vinta that the printing of the *Sidereus Nuncius* was proceeding apace in Venice. Attention to this prepublication development—conducted through epistolary *avvisi* or news reports from Galileo—is especially interesting because it helps to distinguish the role of the instrument from that of the printed book in bringing about a change in conviction about what was possible in the heavens. And indeed, there seems to have been no doubt that the grand duke accepted the existence of the early phenomena before the appearance of the *Nuncius* for, in spite of the variable quality of the early instruments, he and his old tutor had viewed the Moon together: "That the Moon is a body very similar to the Earth was already ascertained by me and shown in part to Our Most Serene Lord, although imperfectly since I did not yet have a spyglass of the excellence that I have now."[98] But now, with a note of excitement, Galileo built on the credibility he had already won by adding a second novelty to the first: "What exceeds all wonders, I have discovered four new planets and observed their proper and particular motions, different among themselves and from the motions of all the other stars; and these new planets move about another very large star like Venus and Mercury, and perchance the other known planets, move about the Sun. As soon as this tract, which I shall send to all the philosophers and mathematicians as an announcement [*avviso*], is finished, I shall send a copy to the Most Serene Grand Duke, together with an excellent spyglass, so that he can encounter [confirm for himself] all these truths."[99]

So here was Galileo's first disclosure of the Jovian phenomena, written as a private, news-bearing letter to a clearly receptive court just as his reports were already well on their way to being inscribed in a different form as a scholarly book for presentation to a wider circle of heavenly practitioners and philosophers. And it is important to underscore that the latest novelties were represented to Vinta and Cosimo as purely astronomical phenomena, with no associated astrological or court mythology—planets circling Jupiter just as Venus and Mercury circled the Sun. It is also significant that Galileo did not mention any other possible astronomical hypothesis, such as the idea that the new planets might revolve not around Jupiter but around a point lying along the line of sight from the Earth to Jupiter (as in the old problem of the ordering of Venus and Mercury with respect to the Sun). Vinta quickly replied: "The advisement [*avviso*] that you have given me of your new, wondrous and memorable observations strikes me as so marvelous and worthy of the ears of the Serene Patron that as soon as I received your letter, I read it to His Highness, who remained astonished [*stupefatte*] by this new proof of your almost supernatural nature."[100]

Vinta and Cosimo were thus excited by Galileo's personal reports of his observations—not by any hands-on ocular demonstrations, noble witnesses, stylized court flattery, or calculated gift giving. They accepted the report as true and let it be known that they were extremely eager to receive the book and the improved *occhiale* as soon as possible.[101] Their favorable response seems to have been simply a tacit inference from the cumulative history of positive firsthand experiences with Galileo as the prince's longtime tutor in mathematical matters, astrological adviser, skilled craftsman, and the prince's personal demonstrator of the eyepiece. These are the sorts of considerations that allowed credibility to be formed. Some might call it trust.

Only after receiving this unambiguous affirmation from Vinta and Cosimo did Galileo make his next move. He tried to use the court's acceptance of his astronomical claims to link his name with the Medici, and to the protection that he believed their name afforded with respect to his principal audience: "As to my new observations, I shall indeed send them as a news announcement [*avviso*] to all philosophers and mathematicians but not without the authority [*auspicii*] of our Most Serene Lord."[102] The court now had to consider how it preferred to be represented. In this case, we are fortunate to have evidence of how Galileo managed the negotiations concerning the most appropriate language, as early modern authors often composed their titles and drafted their dedicatory letters without prior approval from the prospective patron. Galileo provided two choices, one of which was a pun on Cosimo's name and the Greek word for universe (*Cosmica*), the other the name of the reigning family: "Since they are exactly four in number, [to] dedicate them to all four brothers with the name *Medicean Stars*."[103]

Galileo turned to Vinta, as he had in the past, for advice on such practical questions of court

protocol with further requests to keep the whole matter secret and, because of the printer's deadline, to reply as quickly as possible. Vinta immediately answered that Galileo's gesture was "generous and heroic and agrees with the other parts of your marvelous nature." He then explained that he had chosen Galileo's second proposal because the first involved a Greek word that "could be interpreted in different senses," whereas the second would serve "the glory of the Serene name of the House of Medici and of this nation and the city of Florence."[104] The naming was quite straightforward. It concerned only the immediate family, with whom Galileo had developed a long and solid relationship. It had nothing to do with the Medici dynasty as such.[105]

If Vinta and Cosimo had the advantage of a stream of *avvisi* from Galileo, bits of news about the novel phenomena—and what he made of them—were also circulating less systematically outside the Tuscan court. Galileo had not yet presented his findings to the university. But in the widening space of reception, we catch glimpses of how the new and unfamiliar was being processed within the frame of traditional ideas in flux.

Raffaelo Gualterotti (1544–1618)

From Raffaelo Gualterotti, 1604 Novist, poet, and court-festival author, the first reaction was skeptical but not immediately dismissive. In a letter to Alessandro Sertini, Gualterotti said that there could not be much new about Galileo's instrument (which he had not yet seen), as the "ancient astrologers" already had some sort of viewing device and, in any case, he himself had already described a *cerbottana* (a tube for shooting pellets or darts) in his book on the nova, whereby one would be able to see the stars by day. This *cerbottana*, of course, had no lenses. Furthermore, Gualterotti was not surprised that Galileo had found new things, "because we have known one another for 32 years and I have always known the excellence of his nature [*ingegno*]." He had heard of the lunar claims and he had his own explanation: vapors and exhalations from the Earth cast shadows on the Moon. He was also in possession of other extraordinary knowledge: he believed that he had once seen Venus eclipse the Moon.[106] Yet just after writing this letter, he must have received information directly from Galileo, as his posture abruptly changed: "I have read the letters about your *avvisi* and about your new observations. And concerning the first consideration, that the Sun is in the center, I have no doubt whatsoever. That there are more stars and more planets, I believe because many others have already believed it; but certainly I know nothing and I wish to understand better." It would be interesting to know how and in what form Gualterotti had received this information, as he seems not to have fully thought through this startling openness to a Sun-centered arrangement of the heavens. It did not, for example, preempt his earlier views concerning terrestrial and Venusian shadows on the Moon. On the other hand, he was clearly excited about the Jovian claims, whose reality he accepted on Galileo's authority: "Before I die, I wish to see this great star with the four planets that Your lordship has observed."[107]

Giovanni Battista Manso (1561–1645)

An even more remarkable pre-*Nuncius* respondent was the prolific literary writer, encyclopedist, and Neapolitan nobleman Giovanni Battista Manso, marquis de Villa.[108] News of Galileo's feats had reached him in Padua through Paolo Beni (1552/3–1625). Beni, like Manso came from a well-to-do family of noble ancestry. Although he held a knighthood, his career was both scholarly and religious. He held a doctorate in theology, wrote poetry, taught philosophy and rhetoric, composed a massive commentary on Plato's *Timaeus,* and played a prominent role in the debate about the proper usage of Italian prose (*questione della lingua*). He also led a troubled life: he did not get along with his brothers, his father virtually disinherited him, and he was expelled from the Jesuit order in 1593 for breaking rules concerning familial relations, property, and residence. In March 1600, he was back in Padua to teach humanities at the university (poetry, rhetoric, and history) and quickly joined the Accademia dei Recovrati, of which both Galileo and Cremonini were members.[109] The learned and humanistic Beni thus traveled in Galileo's circles, but he was no mathematician and clearly not a modernizer in natural philosophy.

The sense of wonder in Manso's expansive letter is palpable. He described reading over Beni's communication "many times with astonishment and the greatest delight" and immediately sharing the news with his friend della Porta and others whom he did not name. He reported a spectrum of responses. Most people were "terrified" (*atterrita*) by the novelties, but "the more learned did not judge it to be impossible." Manso himself went much further: "Moved by the authority of Your Excellency and Mr. Galileo I hold them [the novelties] to be not merely possible but most true" because "the observations made by two men of such unsurpassed learning and goodness, as those of Your Excellencies, ought not to conflict with things that exist (as I know these things do)."[110] Manso sensed that he lived in an unusual age: if Plato thanked the gods that he lived in the time of Socrates, Manso believed that he lived in "this happy century" when Galileo had brought the *occhiale* to perfection, revealing things that God had hidden.

Even without having the benefit of one of Galileo's own instruments, Manso understood something of what it promised. He believed that the *occhiale* "extends vision more than 60–80 times and renders the sight of the objects seen both near and great, as if not more than two miles away and brings the smallest things close." It was not the first time that he had heard of something like this—or so he said. His friend della Porta, with whom he would shortly initiate the Oziosi, one of those learned societies for which Italy was famous at this time, had already had an idea something like Galileo's—but not quite. Manso faithfully reported to Beni that "it provoked a not small jealousy in our Mr. Porta, who had thought to make an [eye]piece that could bring things in from the infinite (I say insofar as one could extend the visual line removed from all impediments), making proportionate the points of the concave and convex glass." Thus, Manso believed that Galileo had not so much invented the *occhiale* as improved on it.[111] In a sense, this was true.

In a similar vein, Manso took the measure of Galileo's other "marvels" (*meraviglie*) one by one. Already, he said, ancients like Ptolemy and Aristotle knew that there were stars yet unobserved— but now Galileo had actually observed them.[112] Likewise with the Milky Way. Averroës thought that there were "innumerable and very thick stars" that together were responsible for the "small and disturbed light." Ptolemy had described only the positions of the stars that he could see (*Almagest,* book 8, chapter 2). But what everyone agreed was that "the Milky Way is filled with many very small, scattered stars any one of which was so small that it could not be seen." Now, "Your Excellency and Sig. Galileo testify to having seen what many centuries believed with reason by means of good philosophy."[113]

Manso readily accepted the description of a lunar surface punctuated by valleys, mountains, hollows, shadows, and illuminations because he was able to see such things with his *occhiale* in Naples ("with which we can see a man as far away as three miles"). However, he was aware of his eyepiece's limitations, as he confessed that "because of the weakness of the instrument we cannot discern either the creases or the mountains and the folds in the surface that appear to you and Galileo." He then mentioned a different concern: "To tell the truth, I do not know where to assign this in philosophy, either according to the fifth essence imagined by Aristotle or according to the principles of Plato."[114] Here, at last, was an unprecedented novelty: Manso knew of no philosophical traditions in which to locate it. Rather, it struck him that there was a true similarity (*vera similitudine*) between the Earth and the Moon—not an analogy or mere resemblance but what he called a representation (*rappresentatione*). And this phenomenon left him to speculate on other matters: why were there spots (*macchie*) near the middle of the Moon but not on its edge? Perhaps the Moon could be thought of as a convex mirror.

Finally, Manso came to the marvel that "exceeded the others": "the observation of four or perhaps five new planets that Your Excellencies have seen."[115] He did not know that they had been named for the Medici. Yet he agreed that they could not be fixed stars, as they appeared to move with retrograde motion, and this caused him to wonder how the phenomena could be "accommodated": they seemed not to fit the "opinions" of any of the philosophers or astrologers, much less the epicycles of Ptolemy and Copernicus or the homocentrics of Fracastoro. He wanted to know from Beni whether the new planets were above or below the Sun. Their great speed suggested they revolved in a small circle below the Sun, but the relatively small space between the Moon and the Sun would not allow sufficient room for four or five new planets.

Manso then broached further questions and worries for Beni and Galileo to consider: Were the planets sometimes inside and sometimes outside the others' circuits? Were they really new, or were they among those already known? These issues led Manso to the question of the planetary hierarchy: according to Ptolemy, Venus and Mercury accompanied the Sun, "prince of the universe," like a guard (*satellitio*), "in a courtly manner as a sign of its seniority"; but by analogy, he worried that Jupiter, accompanied by four planets, would need to be bigger or equal to the Sun, or at least the largest of the five remaining ones. Saturn, "which is the biggest" (*sic*), did not deserve such an honor because of its wickedness and the slowness of its effects; even so, Jupiter, which was more deserving, must yield in size to Saturn. And finally, he reported "a sharp quarrel" among astrologers and many physicians in Naples who, provoked by knowledge of the new planets, feared the ruination of astrology and medicine ("rovinata l'astrologia e diroccata gran parte della medicina"). The issues were several: "The entire foundations will be destroyed because the following depend on the number of planets being seven: the distribution of the houses of the zodiac, the essential dignities of the signs, the quality of the natures of the fixed stars, the order of the rising stars [*cronicatori*], the governance of the ages of man, the months in the formation of the embryo, the causes of critical days, and both a hundred and a thousand other things."[116]

If Copernicus, Rheticus, and Kepler had worried about explaining the reduction from seven to six planets (when the Moon was removed), Manso was now concerned about the consequences of *adding* four new planets to the universe. And to cap off his concerns, he said that light was an instrument of astral influence. Might it be the case that stars with less light would have less effect on things below? If so, this might resolve the dilemma he had just outlined, as the weakness of the new planets' light would have no effect on earthly affairs. Such were his difficulties. But they were only so because he was trying to reason with the new, exciting but limited resources available to him. The addition of the new planets had upset the order of things as he knew them. He apologized to Beni for "the weakness of my nature and of my poor knowledge of the sciences," which he attributed to the crush of private and public occupations that often caused him to be away from the city and his study.[117]

CONCLUSION:
GENTLEMANLY TRUTH TELLERS?

These surviving letters raise general questions concerning the grounds for trust in a new sort of observation report and its theoretical import before the publication of the work in which Galileo announced his novelties. Why did Manso believe Galileo's report as transmitted (and translated) through Beni? Was Galileo's credibility linked directly to Beni's social status rather than to Galileo's scholarly reputation? Was gentlemanly class status a reliable marker of "perceptual competence" and "truth-telling," as Steven Shapin has inferred from the sixteenth- and seventeenth-century courtesy literature? In an honor culture, the courtesy manuals—the most influential of which were translations of Italian works—held up truth telling as an ideal moral quality ascribed to gentlemen. Like the structural-functionalist elements in Biagioli's story of patronage, Shapin's account implies a deduction: if X was a gentleman, then others would expect him to live up to the ethical norm of telling the truth; if not, then he would be relegated to the social category of a not-X—an unreliable servant or even a liar. The perceptual abilities of certain kinds of people could be trusted because of their inherited social class.[118] What this model of truth telling has to say about actual historical practice is another matter.[119]

Aristocratic or courtly sociability, I have argued, occupied a domain that was far less constrained by the literary, rhetorical, and disciplinary practices afforded by the universities. Consequently, the beliefs, judgments, and standards of proof held by gentlemen—and also of low-born, skilled court technicians—could be much freer and more varied, ranging from traditional to modernizing perspectives. The Manso and Gualterotti letters do not portray either a narrowly proscribed picture of gentlemen trusting uncritically in the testimonies of other gentlemen (just because they were gentlemen) or a representation of clients slavishly tuning their beliefs to fit court games. Instead, they reveal the conditions for a surprisingly open, learned, and critical intercourse available to a cultural sphere of lettered gentlemen and aristocrats, ecclesiastics, and modernizing professors from Padua to Naples and Florence—circumstances that were not essentially different from the cultural climate in the Hapsburg lands.[120] The report from Padua became an occasion for the Neapolitan Manso to reason and ruminate about what he had heard, to turn things over in his mind, to try to integrate the unfamiliar with what he already believed, and to weigh, sift, judge, and ask questions as they occurred within his frame of understanding, such as the implications of the four planets for the foundations of astrology. This openness is precisely what provided Galileo an entrée into these sociabilities. Perhaps high social class might have had some power to authorize factual claims in the political and legal spheres, but even in the seventeenth-century English courtroom, as Barbara Shapiro observes, "the mere status of gentleman could hardly be decisive . . . where one gentleman might well be contending against another and where witnesses of several classes might appear on both sides."[121]

Likewise, in matters of persuasion concerning knowledge of heavenly events, noble status per se carried no special testimonial weight, as may be recalled from the example of Tycho Brahe, who appealed to the witnessing of supposedly unreliable country folk concerning the existence of a new heavenly event. Similarly, Galileo invoked communal corrigibility rather than social station as the right standard for judging observational accuracy when he criticized Baldassare Capra and Simon Mayr for not following the norms of scholarly honesty allegedly exemplified by Tycho Brahe. And finally, when Galileo, Tycho, and Kepler broke out of established academic rhetorical and disciplinary conventions, they did not behave like ordinary courtiers; rather, they were moving the reinvigorated contest between the *via antiqua* and the *via moderna* out of the universities and into the financially uncertain but more flexible discursive space of the courts.

18

How Galileo's Recurrent Novelties Traveled

THE *SIDEREUS NUNCIUS*, THE NOVA CONTROVERSIES, AND GALILEO'S "COPERNICAN SILENCE"

Some influential interpreters of the *Sidereus Nuncius* have underscored Galileo's straightforward empiricist style in reporting observations and avoiding aggressive, systematic theorizing. An important function of this reading has been to dissociate Galileo from the Copernican convictions that he so clearly expressed in the 1597 letters to Mazzoni and Kepler. According to Drake, Galileo "lost faith in [Copernicanism] from 1605 until 1610." Galileo's silence on the Copernican question during those years allowed Drake to burnish an originary image of Galileo as the first modern scientific persona, at one with the temper of the nineteenth-century physicists James Clerk Maxwell and Heinrich Hertz and in sharp opposition to the "bookish philosophers" of his own time—the Ur-empiricist "forced to [conclusions] by systematic attempts to account for actual observations" and remaining "as nearly free from metaphysics as he could."[1] Like Drake, Ludovico Geymonat believed that Galileo did not use the *Sidereus Nuncius* to develop grander "revolutionary" claims. The telescopic discoveries were the result of "experience and not mathematics"; only later did Galileo quickly recognize their importance for the "system of the world."[2] Consistent with these views but seeking a different moral, Westfall and Biagioli proposed that Galileo's downplaying of cosmology involved a proxi-

mate political consideration; in Westfall's words, he "saw the telescope more as an instrument of patronage than as an instrument of astronomy."[3]

This curious alliance of internalist and externalist perspectives finds common cause in Galileo's silence on the Copernican question.[4] Yet Galileo was also silent on other concerns of potential relevance—including, as chapter 13 shows, his ongoing relationship with Kepler and his own astrological practices. Furthermore, Galileo did not explicitly frame his presentation using the tendentious nova controversies of the decade just ended. Indeed, because it was cast as a news announcement, the *Nuncius* avoided presenting itself officially as part of any controversy, although its material clearly lent itself to the kind of attack on the traditionalists for which his later work was famous. At the same time, Galileo did not entirely ignore the main issue raised by the Novists: from the beginning, the *Nuncius* spoke of stars "never before seen" (*antehac conspectas nunquam*). In this sense, Galileo's novas were not just the latest in a string of novelties to show up in the heavens. They were of a different kind. Galileo represented his novelties not by eliminating God's capacity and will to create entities never before seen but by representing himself as an indispensable intermediary. Although angels were conventionally represented as divine messengers, in this case it was a *human intermediary*, the *nuncius* Galileo Galilei, who claimed responsibility for rendering invisible objects visible "with the help of a *perspicillum*" or "looking glass."[5] Galileo's novas, unlike the earlier ones, which, so it was believed, God alone had caused to appear whenever he so desired, required human input of a new kind. It was this unprecedented, active meaning attached to the function of *nuncius*—a word that could mean both *message* and *messenger*—that marked the book and its instrument as unusual, exciting, and perhaps dangerous for contemporary audiences at court and elsewhere.

In another sense, however, Galileo quite resembled the 1604 Novists. He located all of his celestial novelties in the realm of ordinary phenomena—lunar peaks and valleys, new stars in the Pleiades, Orion, and the Milky Way and the four hitherto unknown Jovian "planets." Galileo, like the Novists, saw no need to invoke God as first cause. This is why he could put his entities squarely into the descriptive, explanatory and predictive domain of "philosophers and mathematicians" rather than the hyperphysical realm of the theologians.[6] Although his letters frequently refer to the book as *avviso astronomico*, his aim was not merely to "report" but also to give larger meaning to what he reported and, hence, to raise it to the level of a serious scholarly work. However, the Moon would now be "ordinary" in a sense hitherto forbidden by the traditionalists—"the great crowd of philosophers," as he enjoyed calling them—because Galileo's lunar surface was earthlike ("marked here and there with chains of mountains and depths of valleys").[7] The *Sidereus Nuncius* also made prominent on its title page the words "Medicean Stars," publicly cementing his private association with the Florence court and recalling the Reinholdian trope of permanent, celestial "monuments" rather than the temporary, human ones evoked in the dedication.[8]

That the book heralded the currently reigning Medici family in this manner did not mean, however, that it was written primarily for a court audience. The court was only one of Galileo's intended audiences—just as, for example, the papacy was only one of Copernicus's audiences.[9] Like the Novists, Galileo directed his treatise primarily and, in this case quite explicitly, to practitioners. He designated his novelties on the title page as "great and absolutely marvelous spectacles" (*magna, longeque admirabilia spectacula*)—accessible "to the gaze of everyone [*unicuique*], but especially [*praesertim*] philosophers and astronomers." Everyone, including the resident Medici family, was now invited to witness the new spectacles. But his private correspondence with the Medici court again contained clear reminders that he was directing his *avviso* to "all the philosophers and mathematicians."[10] Like the nova, the new phenomena were visually friendly, even if the optical and geometrical arguments that theoretically grounded them were intended for learned, Latinate readers, traditionalist and modern alike. Unlike the novas of 1572 and 1604 and the comets that appeared and disappeared between 1577 and 1585, Galileo's novelties did not vanish permanently: whoever possessed a good *perspicillum* could make them reappear. Thus they were neither deviant monsters, like two-headed monk-calves, nor unpredicted signs or portents of the end times beloved by the apocalypticians. And because Jupiter's "planets" were not extraordinary objects, this meant that they must have some "place" in the structure of the visible heavens, just like the already known planets. It also meant that in principle, their motions must be astronomically recurrent and predictable and hence potentially capable of bearing astrological meaning. But if the new Jovian "planets" had any astrological significance, Galileo said nothing about it in the *Nuncius*.[11]

Although Galileo chose not to use the *perspicillum* to foreground the world-systems debate, the instrument did not suddenly put the idea of "Copernicanism" onto his personal agenda. Rather, when news of the Flemish viewing device unexpectedly arrived, no one in Italy was more ready to exploit it than Galileo—both because he had considerable skills as an instrument builder and because he had been philosophizing about motion, including the Earth's—for more than twenty years. And, wherever possible, since 1597 he had been tracking Kepler's ideas about a new system of the universe. The telescope enabled Galileo to bring together ideas, predilections, and skills that had undergone a long fermentation. Court approbation would enable him to expand his view of the natural world to new audiences—not just the academic traditionalists and the small aristocratic circles of Padua, but the sort of wider reading public that Kepler commanded. In other words, the court would now become a political and economic lever that gave him greater freedom to publish and also to resist academic distinctions that relegated mathematics to a limited role in explaining natural phenomena. In this sense, Galileo was using the court association to continue his battle with the traditionalists at Padua and elsewhere. He had already forged an approach to natural philosophy through theoretical and practical mechanics, and he had begun framing a system of the theoretical domain of the science of the stars, but he had published none of it.

Passing references in the *Nuncius* indeed show that he wished not simply to "report" his observations but also to integrate them into a larger theoretical frame: "We will say more in our *System of the World*," he wrote, signaling a comprehensive treatise of natural philosophy delivered from an unmistakably Copernican platform: "We will demonstrate that [the Earth] is movable and surpasses the Moon in brightness, and that she is not the dump heap of the filth and dregs of the universe, and we will confirm this with innumerable arguments from nature."[12] It was a promissory note of sorts. Both in the dedication and later in the main body of his short text, he hinted at its possibilities when he gave a Copernican reading to the proposition that the four Medicean "planets" revolve around Jupiter, while that planet revolves around an immobile Sun. Yet, because Galileo surely realized that the Medicean novelties did not eliminate a Tychonic arrangement—they did not, after all, demonstrate that the Earth moved—he produced an objection that no one, including Tycho, had ever advanced. The Earth could not possibly be a planet, so his objection went, because if it were, it alone among the planets would have a moon, and the two together would travel around the Sun. But now the existence of a Jovian system of attendant planets showed that the Earth was not unique in having a moon. Because Galileo could not provide a necessary demonstration in the *Sidereus Nuncius*—once more pointing to the impasse of underdetermination—this was simply the best that he could do with his Jupiter material on the question of celestial order.

For all its boldness as an announcement, therefore, the book did not pretend to be an approach derived from principles like Kepler's *demonstratio*, which purported to unveil the whole mystery of the heavens. The *Nuncius* was not a *theorica*, a *principia*, or a *systema mundi*; yet, it was something more than a *practica*, because it was neither just a description of a new instrument like the military and proportional compass nor anything so ambitious as Tycho Brahe's instrument

book. As various commentators have noticed, it limited itself to the discursive expectations of the *avviso* to announce and report.[13] Hence the rhetorical form that Galileo (deliberately) chose underscored the limits of the book's horizon with respect to the Copernican theory. If its author's further general theoretical intentions were clear enough, he did not present his observations as part of a larger demonstration in which all alternative positions were refuted: hence he made no references to Bruno's countless worlds, Brahe's two-centered system, Kepler's elliptical orbits and polyhedral archetypes, Gilbert's magnetic philosophy, Maestlin's argument that the Moon has earthlike characteristics, or even the 1604 nova.[14] The intricate richness of the decade just passed was simply not acknowledged; and in erasing such complexity, the *Nuncius* inadvertently permitted recent commentators the latitude to find in it the image of the austere sensibility of a modern scientific report.

It is thus possible to emphasize quite different functions served by Galileo's instrument and first major printed book.[15] Surely both helped him to improve his financial situation; they opened a space for making his work public and relaxed the institutional obstacles to his cross-disciplinary goals. But, patron or no patron, neither the instrument nor the book could provide decisive evidence for the Earth's motion. This situation leaves open the questions of how the two, together or separately, succeeded first in stabilizing the representations of the heavens that Galileo promoted and how this outcome affected the Copernican question.

THROUGH A MACRO LENS
THE RECEPTION OF THE *SIDEREUS NUNCIUS* AND THE TELESCOPE, MID-MARCH TO EARLY MAY 1610

Albert Van Helden has rightly observed that the initial news contained in the *Sidereus Nuncius* traveled "surprisingly fast through diplomatic and commercial channels."[16] Van Helden's assessment is even more historically meaningful (and surprising) when calibrated against the stiff resistance accorded Kepler's 1596 *Mysterium Cosmographicum* and 1609 *Astronomia Nova* or the four-year, extended reaction to the 1604 nova. These sorts of comparisons reframe the questions of why, how, and where Galileo's claims broke through so rapidly—and among which audiences. Drawing in our focus on the week-by-week circulation of information—as the evidence makes possible in this unusual case—yields a remarkable pattern that allows reflections about the relative importance of the book and the instrument in the ways Galileo's claims circulated.

The book traveled more quickly and easily than did the telescopes made by Galileo himself. Many volumes moved independently of Galileo's personal instigation. In this sense, the *Sidereus Nuncius* traveled much like Kepler's *De Stella Nova*—a published description of new stars without benefit of actual, firsthand observation by readers of those works. Yet it would be unduly presentist to assume that contemporaries *expected* that a book about the heavens would require an accompanying instrument. To the contrary, the absence of the instrument seems not to have seriously hindered the credibility, let alone the excitement, generated by the book's reports. If anything, the book stimulated interest, heightened expectations, and raised demand for the *perspicillum*. And, ironically, Galileo's instrumental performances sometimes worked against the establishment of his claims.

The day after the *Nuncius* was published, the English gentleman-diplomat Henry Wotton had already read it and put a copy into the diplomatic pouch for King James, together with a relatively lengthy report to the earl of Salisbury that offers a rare glimpse of an early reaction outside Medicean channels. Ambassador to Venice since 1604, the lettered Wotton fed off the intelligence network of English students, spies, and tourists, among whom were such men as Edmund Bruce and the Scot Thomas Seggett, whom Wotton had obligingly helped out of a Venetian jail in 1604.[17] Wotton's remarkable letter to Salisbury shows, among other things, that he had no knowledge of the fact that his countryman Thomas Harriot had already been making celestial observations:

> Now touching the occurrents of the present, I send herewith unto his Majesty the strangest piece of news (as I may justly call it) that he hath ever yet received from any part of the world; which is the annexed book (come abroad this very day) of the Mathematical Professor at Padua, who by the help of an optical instrument (which both enlargeth and approximateth the object) invented first in Flanders, and bettered by himself, hath discovered four new planets rolling about the sphere of Jupiter, besides many other unknown fixed stars; likewise, the true cause of the *Via Lactea,* so long searched; and lastly, that the moon is not spherical, but endued with many prominences, and which is of all the strangest, illuminated with the solar light by reflection from the body of the earth, as he seemeth to say. So, as upon the whole subject he hath first overthrown all former astronomy—for we must have a new sphere to save the appearances—and next all astrology. For the virtue of these new planets must needs vary the judicial part, and why may there not yet be more? These things I have been bold thus to discourse unto your Lordship, whereof here all corners are full. And the author runneth a fortune to be either exceeding famous or exceeding ridiculous. By the next ship your Lordship shall receive from me one of the above-named instruments, as it is bettered by this man.[18]

Like Manso, Wotton expressed no serious doubts about the "strange news" that he had to report; in fact, he moved quickly from the new revelations to his understanding of their implications: "he hath first overthrown all former astronomy . . . and next all astrology." Also like Manso, Wotton immediately noticed the astrological implications of adding four new planets; but because he had the book (and Manso did not), he saw that as yet Galileo offered no assistance in that regard. Writing to another nobleman—and through him to the king himself—Wotton made no allusions to any Medicean dynastic mythologies, although, as a seasoned diplomat, he would have been acutely aware of such symbolism had it existed. In short, Wotton was conscious that Galileo's claims might eventually make their author look "ridiculous" (or, for that matter, "famous"), but his preliminary assessment of their credibility appears to have been based entirely on his own impressions—even without the aid of an "optical instrument." It was his job as a diplomat, after all, to sort out trustworthy from untrustworthy news. And if the king was interested, Wotton made it possible for His Majesty to read Galileo's book for himself; but whether he was successful in sending along one of the "bettered" optical instruments on the next ship is unknown.

Still another early act of book distribution—again without Galileo's direct intervention—occurred three days after Wotton's dispatch, when Paolo Sarpi sent the *Nuncius* to the Venetian ambassador in Paris, with instructions that it be shown to his friend Guillaume Leschassier. Sarpi's letter to Leschassier included a fairly exacting description of the *perspicillum* as an instrument already somewhat familiar—"what you call *lunettes*"—together with a brief accounting of the chief novelties "which you may read in the little book."[19] Yet the distribution of instruments had still not begun, and Sarpi made no mention of sending one—possibly because it was not seen as indispensable for securing belief. Only on March 19, a week after publication, was there evidence of a plan. Galileo notified Vinta that he had had some sixty telescopes "made with great trouble," of which he considered only ten to be good enough to send out. In an earlier message, he had also mentioned the figure of ten from a group of one hundred telescopes. At this point, it is evident that Galileo had a supply problem. He had more books than instruments, and the former were much easier to make than the latter, as no machine could reproduce the *perspicillum* in the same way that a printing press could duplicate an almanac or the *Nuncius*. Given the paucity of really high-quality instruments—presumably good enough to confirm what he had described in the book—Galileo evidently decided that the first consideration was to put this relative handful into the courts. And, of course, because he had chosen to make the Medici his protectors, the first recipients had to be other members of the family, such as his long-time supporter, Antonio. Beyond the immediate family, he also opted to send one to the elector of Cologne, at whose court Galileo's brother Michelangelo was a musician (but whose court mathematician, Johann Eutel Zugmesser, would turn out to be no friend of Galileo). He also named Cardinal del Monte, brother of his old patron Guidobaldo; and finally, without naming any individual recipients, he selected the courts of Spain, France, Poland, Austria, and Urbino. Conspicuous by its absence from the list was the court at Prague: after the long hiatus, it is possible that Galileo was worried about Kepler's response.

By March 27, in the third week after publication, there was still no mention of telescopes received. But the *Nuncius* continued to move. It was now in Florence, where it was read at a public gathering in the home of Francesco Nori. At the same time, the first signs of unfavorable reaction began to trickle in from university circles in Padua and from Bologna, where Magini's young secretary, Martin Horky, wrote to Kepler informing him that he had read the *Nuncius* and that its

novelties might be true . . . or false. There was no indication that these were Magini's own views, but that would soon change. And because of the temporal overlap in these developments, Galileo as yet had no knowledge about the negative reactions in Bologna academic circles. Nor did he know what was happening in Prague, where the emperor had (somehow) obtained a copy—probably in late March.

On April 3, at the beginning of the fourth week, activity increased at the court in Prague. More books were circulating. A second copy of the *Nuncius* arrived, carried to the Spanish ambassador by Matthaeus Welser, the sometime courier between Kepler and Edmund Bruce in 1603. And before the fourth week had drawn to a close, Giuliano de'Medici, the Tuscan ambassador, obtained a third copy of the *Nuncius* from Galileo through Wotton's man, Thomas Seggett, and had lent it to Kepler on April 8. However, Kepler had already seen the emperor's copy of the book, and thus, although Galileo had not sent him an instrument, it is safe to say that Kepler had already digested Galileo's claims no more than a month after publication. Again, because of travel time and normal delays, Galileo knew none of this until late April. From proximate Italian sources, he formed a much less sanguine view: according to rumors from Verona, some were saying that the *perspicillum* was the cause of the phenomena that he had described. Yet even as these sorts of doubts were beginning to grow, Benedetto Castelli in Brescia let Galileo know that he had obtained the book and had borrowed some sort of viewing instrument from a local priest. Thus an evolving discussion and assessment was well under way—for the most part, without the viewing tubes—in the first month after the publication of the *Nuncius*. And once again, Kepler and Galileo were linked through intermediaries rather than direct personal communication.[20]

In the fifth week, as more copies of the book began to circulate in Venice and Rome, activity in Prague began to intensify around Kepler. The Medici ambassador formally presented a request from Galileo for Kepler's opinion of the *Sidereus Nuncius*. Kepler himself is our source: "You made an appointment for me to meet you on April 13th. Then, when I appeared, you read me Galileo's request in his communication to you [a letter now lost], and you added your own exhortation. Upon hearing these words, I promised to prepare something in time for the scheduled departure of the couriers, and I kept my pledge."[21] Six days later, on April 19, Kepler indeed had a long letter ready for the courier to Italy. How he was able to respond so readily and in such depth will be discussed shortly. Suffice to say that for the first time since 1597, Kepler had broken the long silence and written directly to Galileo. Just two weeks later, on May 3, this letter appeared as a book titled *Dissertatio cum Nuncio Sidereo*. By the time Galileo actually received Kepler's long letter of April 19, the ground had been seeded by an informal report from a close Galilean ally in Prague, Martin Hasdale (in Czech, Hastal; in Italian, Asdalio). Hasdale had encountered Kepler in Prague at one of the ambassadorial homes (Saxony), where a great deal of court business was discussed, and he reported his impressions directly to Galileo. This letter is of considerable importance, as it forms a kind of private gloss on what Kepler wrote to Galileo:

> I asked him [Kepler] his opinion about the book and about his lordship [i.e., Galileo]. He replied to me that he corresponded many years ago with his lordship and that in truth he neither knew nor had ever known a more important person than your lordship in this profession; and that although Tycho was recognized as very important, he did not prevent your lordship from surpassing him by a lot. And as for the book, he says that you have shown the divinity of your spirit but that you have given reason for dissatisfaction not only to the German nation but also to your own in not having made any mention of those authors who have announced and given occasion to look for what you have nevertheless found. He names among these men Giordano Bruno, for Italy, and Copernicus and himself in claiming to have announced similar things (although without proof, like your lordship, and without demonstration); and he has brought along with him his book in order to show the place to the ambassador of Saxony.[22]

The Hasdale letter shows, among other things, that this was a world where practitioners already had their own *internal hierarchy of intellectual credit*, independent of court patronage and prestige. This is especially true among the small group of turn-of-the-century theoretical innovators in revolt against the universities, which would otherwise have been their natural source of authority. Kepler—not Emperor Rudolf—paid Galileo the high compliment of ranking him above

Tycho Brahe; but at the same time he criticized Galileo for failing to give adequate credit to the earlier work of Bruno and Copernicus. These intensely mixed judgments and the feelings that they reflected would inform Kepler's entire public exchange with the elusive and withholding Florentine.

KEPLER'S PHILOSOPHICAL CONVERSATION WITH GALILEO AND HIS BOOK

On April 13, Kepler was in pressed circumstances. The Medicean ambassador had formally communicated the request from Galileo (now lost) asking his opinion of the *Sidereus Nuncius*. Kepler had had access to a copy of the book for two weeks (if we assume that he saw the emperor's copy in late March), and the emperor had also asked to hear his views. Now he had a second copy. Yet, like most other readers, he did not have a *perspicillum*; nor was there anything else published by Galileo that might have provided some basis for reflection. Added to this, Galileo had flagrantly ignored him since 1597, when Kepler had extended his hand in friendship with the fervent hope of a collaboration. And Kepler's response gives no indication that Galileo had apologized in any way or that there was any mention of an instrument in the diplomatic pouch.

The background to this exchange was hardly auspicious. How would Kepler manage his own anger at Galileo's offhand treatment over many years? And what of the rumors in 1602 from Edmund Bruce that Galileo was spreading Kepler's ideas as though they were his own? And finally, what was Kepler to make of the ever more provocative letters from Martin Horky to the effect that he would have to correct all his work on "the ephemerides that you want to publish with Magini according to Tychonic principles?" In mitigation, Galileo's book appeared to lend unexpected support of a new kind to the Copernican planetary ordering that lay at the very heart of all of Kepler's own work. If Kepler was going to associate himself with Galileo publicly, it was going to require tremendous rhetorical delicacy and diplomatic discipline, not to mention epistemological acumen. It was also going to require him to decide whether and how to lend his support to claims based on the instrument when he himself could not claim to have been a personal witness to any observations or to know anyone, besides

Galileo himself, who was. And Kepler knew that there was skepticism among traditionalists at the Rudolfine court, based chiefly on the matter of the verisimilitude of the observational reports. Finally, all these issues had to be handled with great speed in order to meet the Tuscan ambassador's request.

What was to be done? First, Kepler had to choose the most effective rhetorical form for the occasion. As Isabelle Pantin has shown, he chose a word for his title that had some resonance with the public expounding of an academic thesis (*dissertatio*) yet did not fully evoke the ritualized oral contests that involved students defending propositions (*disputatio*). Also, Kepler avoided rhetorical possibilities that he had used on other occasions. For example, in his *Apologia* for Tycho Brahe (1601), he had had to defend Tycho's world system while avoiding arguments for Copernicus. Similarly, he eschewed the commentatorial form that would have required him to treat sequentially every single claim in the *Sidereus Nuncius*. And although he might have chosen the letter as an obvious form (e.g., *Letter to Galileo*), he instead chose a word that would permit him sufficient flexibility to engage in both serious philosophical argument and half-serious play. Pantin, in fact, insightfully urges us to notice the protective covering afforded by the choice of a ludic—as opposed to a demonstrative—framing of the reply to Galileo. Kepler's response would be a playful book filled with claims, counterclaims, paradoxes, human emotion, and many digressions. Kepler thus entered into a (nonlinear) *conversation* with Galileo's treatise—just the word, incidentally, that Edward Rosen aptly chose for the title of his English translation of Kepler's work. Even so, this was not going to be an idealized conversation between hereditary gentlemen of the sort described in the courtesy manuals but rather the occasion on which a long-brewing encounter would be thrust to the surface like the release of a prolonged buildup of seismic tensions.

There were four major reasons why Kepler was so remarkably successful in mobilizing a rapid reply to the Tuscan ambassador. First, he built the *Conversation* around his own writings (the *Mysterium*, the *Optica*, the *Stella Nova*, and to a much lesser extent the *Astronomia Nova*), aggressively yet deftly reconfiguring Galileo's discoveries within his own ambit so that effectively he controlled the categories of the "conversation."

Second, he positioned himself in a complementary, yet superior role to Galileo by privileging theorizing—an activity of the reasoning intellect that anticipated specific facts discovered through the senses. Third, just as he had done in the *Stella Nova*, he used his actual conversations (and disagreements) with Wacker von Wackenfels as a device to introduce the views of Giordano Bruno (with which Kepler strongly disagreed) and to represent Galileo in public dialogue with them. And fourth, he transformed the secret contacts mediated by Edmund Bruce into a public and, for Galileo, inconvenient fact. By choosing a flexible, humanistic format, at times revealing his own feelings, he was able to make considerable use of paradox and irony, moving around his target easily—now aligning, now backing off, jabbing with rhetorical questions, evoking uncomfortable precedents for what Galileo claimed to be unprecedented. Ultimately he represented himself as Galileo's Copernican ally, but only on his own terms. In short, Galileo's simple request brought him more than he had foreseen.

Kepler's "Notice to the Reader" immediately flagged an extraordinary juxtaposition of rhetorics: openness, philosophical independence and ambivalence. "I do not think that Galileo, an Italian, has treated me, a German, so well that in return I must flatter him, with injury to the truth or to my deepest convictions. Yet let no one assume that by my readiness to agree with Galileo I propose to deprive others of their right to disagree with him. What is more, I have undertaken herein to defend some of my own views also. . . . Yet I swear to reject them without reservation, as soon as any better informed person points out an error to me by a sound method."[23] Clearly, this was not to be an unqualified defense of Galileo. In the opening of the main text, Kepler immediately took a stab at Galileo for his long silence and his ignoring of the recent publication of his "new kind of astronomy or celestial physics." Self-consciously re-evoking the military imagery of the introduction to his *Astronomia Nova*, Kepler represented himself as a military general taking a respite from a long and strenuous campaign.[24] Yet, instead of receiving the anticipated acknowledgement and credit from Galileo, Kepler admitted being hit by a bombshell, a "surprise report" about the telescopic discovery of four previously unknown planets. It was a surprise both in the sense that what it revealed was unknown and in that the whole scene was completely unanticipated—an "unknown unknown."[25] Moreover, this report had come neither from Galileo personally nor through the published *Nuncius* but thirdhand, by word of mouth through a courier to Wacker, who, in turn, communicated the news to Kepler "from his carriage in front of my house." Thus opened a brief period of anxious anticipation for Kepler before he was able to see the text of the *Nuncius*. This anxiety nominally concerned Kepler's fear (and Wacker's hope) that Galileo had confirmed Bruno's theory of countless worlds in an infinite space and that irreparable harm would be done to the finitist, archetypal world of the *Mysterium:* "Our emotions were strongly aroused. . . . He was so overcome with joy by the news, I with shame, both of us with laughter, that he scarcely managed to talk, and I to listen."[26]

While there seems little doubt that something like this episode did occur, one must remember that Kepler had the *Nuncius* by the time he wrote this account, and hence we should read it less as a faithful report than as a lens that Kepler used to focus the attention of Galileo and other readers on a serious political issue. By invoking Wacker and identifying him as a Rudolfine court official, Kepler was speaking to modernizing elements in Prague and elsewhere, including Catholics who might be open to Bruno; at the same time, Kepler was publicizing names and philosophical doctrines that Galileo had *excluded* from the *Nuncius* and about which he could write only at his peril. In fact, their omission is just what contributes to the impression that Galileo's work was like a "modern scientific report" and devoid of "cosmology." By contrast, Kepler was brutally frank in kindling matters that Galileo extinguished by his silence: "If four planets have hitherto been concealed up there, what stops us from believing that countless others will be hereafter discovered in the same region, now that this start has been made? Therefore, either this world is itself infinite, as Melissus thought and also the Englishman William Gilbert, the author of the magnetic philosophy; or, as Democritus and Leucippus taught, and among the moderns, Bruno and Bruce, who is your friend, Galileo, as well as mine, [and who taught that] there is an infinite number of other worlds (or earths, as Bruno puts it), similar to ours."[27]

Consider the political implications of this passage. Kepler was exposing in print the hitherto invisible social connections that had bound him to Galileo in the early 1600s, and he was doing so in a way that dangerously linked Galileo to Bruno through their mutual friend, the otherwise unidentified Edmund Bruce. Kepler constructed Galileo's discovery of hitherto unknown bodies in the heavens as a step toward the Brunonian thesis of innumerable worlds. But, of course, the Holy Index of 1603 proscribed Bruno's writings altogether. And because Kepler knew the fate of Bruno and his works through the Catholic convert Wacker, Kepler's references to Bruno should be read as even more politically aggressive toward Galileo than one might at first imagine.[28] In fact, Kepler left no ambiguity in the matter, returning to the association between Bruce and Bruno and producing the taunting impression that Galileo was directly engaged with them in critical philosophical dialogue:

You [Galileo] correct and, in part [*partim*], render doubtful our Bruce's doctrine, borrowed from Bruno. These men thought that other celestial bodies have their own moons revolving around them, like our earth with its moon. But you prove that they were talking in generalities. Moreover, they supposed it was the fixed stars that are so accompanied. Bruno even expounded the reason why this must necessarily be so. The fixed stars, forsooth, have the quality of sun and fire, but the planets, of water. By an inviolable law of nature these opposites combine. The sun cannot be deprived of the planets; the fire, of its water, nor in turn the water, of the fire. Now the weakness of his reasoning is exposed by your observations. In the first place, suppose that each and every fixed star is a sun. No moons have yet been seen revolving around them. Hence this will remain an open question until this phenomenon too is detected by someone equipped for marvelously refined observations. At any rate, this is what your success threatens us with, in the judgment of certain persons. On the other hand, Jupiter is one of the planets, which Bruno describes as earths. And behold, there are four other planets around Jupiter. Yet Bruno's argument made this claim not for the earths but for the suns.[29]

Looking at the other side of the coin, Kepler let it be known that he was relieved that Galileo had not advanced Brunonian claims: "In the first place, I rejoice that I am to some extent restored to life by your work. If you had discovered any planets revolving around one of the fixed stars, there would now be waiting for me chains and a prison amid Bruno's innumerabilities—I should rather say, exile to his infinite space."[30]

So Kepler now aligned himself with Galileo in his disagreements with the Brunonian Wacker, thereby continuing the discussion begun four years earlier in the *Stella Nova*. But immediately he took back some of his praise by invoking the superiority of knowledge of causes over knowledge derived from the senses: "What Galileo recently saw with his own eyes, he [Bruno] had many years before not only proposed as a surmise, but thoroughly established by reasoning. It is doubtless with perfect justice that those men attain fame whose intellect anticipates the senses in closely related branches of philosophy."[31] And Kepler now instantiated his claim by grouping the a posteriori Copernicus of the *Mysterium* with Galileo: "[Copernicus] brought to light only the *tou hoti* [the 'what is']."[32] Meanwhile, Kepler located himself in the tradition of Plato, Euclid, and the Pythagoreans who, although they had hit a priori on the divinity of the five regular solids, failed to see the application of those forms to the organization of the heavens. Where the ancients had failed, however, Kepler believed he had inferred correctly: "From the visual sight of the Copernican system he [Kepler] rises, as it were, from the facts [*ek tou hoti*] to the causes [*tou dioti*], and to the same explanation as Plato had set forth *a priori* and deductively so many centuries before. He shows that the rational principle of the five Platonic solids is impressed in the Copernican system of the world. . . . Certainly those who grasp the causes of things with the mind before those things are revealed more resemble the Architect than those who, after having seen these things, reflect on their causes."[33]

Kepler's gesture, evoking Rheticus's *Narratio*, used the *Mysterium* rather than the elliptical astronomy of the *Astronomia Nova* to crown himself with the mantle of heavenly theoretical knowledge. (Wishing to paint Galileo as fact-bound and himself as a philosopher, Kepler evidently found it in his interest at this moment to downplay the crucial importance of Brahe's observations in the *Astronomia Nova*.) At the same time, Kepler implicitly realigned himself with the apriorist Bruno while relegating Galileo to the decidedly lesser status of a mere beholder of new facts.

From there, Kepler proposed to defend Galileo as a Copernican ally against a difficulty that he had failed to address in the *Nuncius:* the implications of the Jovian planets for Copernican astrology. As a practicing astrologer, Galileo could not have overlooked this problem. It was immediately obvious to contemporaries. The gentleman-amateur Manso had called attention to it even before reading the *Nuncius,* and the diplomat Henry Wotton had also noticed the difficulty immediately on reading the book. Remarkably, Kepler's proposal was not so different from Manso's conjectured solution: because the four new planets never diverged more than fourteen minutes from Jupiter, the orbit of the outermost body made the entire area less than the apparent diameter of the Sun or Moon. This meant that Jupiter could be treated as a slightly larger body whose (apparent) diameter, augmented by the four Jovian planets, would not exceed that of the Sun. "In this way astrology maintains its standing," wrote Kepler, while playfully dropping in a bit of teleology that Galileo could never tolerate: "It becomes evident that these four new planets were ordained not primarily for us who live on the earth, but undoubtedly for the Jovian beings who dwell around Jupiter."[34] At this point, Kepler sealed the alliance according to his conditions: "This conclusion is more obvious to those who, together with you, Galileo, and with me, accept Copernicus's system of the universe. . . . Our moon exists for us on the earth, not for the other globes. Those four little moons exist for Jupiter, not for us. Each planet in turn, together with its occupants, is served by its own circulators. From this line of reasoning we deduce with the highest degree of probability that Jupiter is inhabited. Tycho Brahe likewise drew the same inference, based exclusively on a consideration of the hugeness of those globes."[35]

Paradoxically, Kepler was now back with Galileo as a "Copernican," but their proximity to Bruno was hardly the neighborhood in which Galileo would have wished to reside. Kepler's speculations about inhabited planets, his aspectualist astrology, elliptical astronomy, and archetypal hypothesizing grounded in teleological explanations might have fascinated the emperor and Wacker (who had finally accepted Kepler's magnetic principles in his Brunonian universe), but such manner of philosophizing was as disagreeable to Galileo's Archimedean sensibilities and threatening to his political situation as it was outrageously out of step with stodgy elements in the Tuscan court. In Florence, there were no such debates about Bruno of which we are aware.

On the contrary, at the famous after-breakfast court debate in December 1613, the standard kick-off point for Galileo's troubles with Rome, the discussion centered on whether Copernicus's theory was compatible with the Bible—a question raised by the traditionalist Cosimo Boscaglia but which had stirred no special interest in Prague when Kepler published his apposite understanding of that issue in 1609.[36] Whereas Galileo had to contend with the Pisan traditionalists and their allies in the Dominican monastery of San Marco, which would eventually lead to censures from Rome, Prague was still alive with talk of inhabited worlds and magnetic forces even as Rudolf slid ignominiously off his throne in 1612, bringing down a brilliant cultural era. This contrast between Prague and Florence contained many of the same elements that had marked the nova controversies but a few years earlier.

Under the circumstances, how did Galileo plan to deal with Kepler's complicated reply to the ambassador's simple request? As early as March 1610, Galileo already had great plans for a second edition of the *Nuncius:* he was starting to translate it into Italian, he was collecting observations of the periods of Jupiter's planets, he was planning to add more numerous and beautiful copper engravings and introductory poems that would glorify the grand duke and himself.[37] But how could Galileo publish such an edition without mentioning Kepler's *Conversation,* without engaging Bruno and Gilbert, without taking a stand on astrology—the practice of which was under surveillance in Venice, Florence, and Rome—and without acknowledging Kepler's degrading of his claims to originality in building and applying the telescope? A book treating such topics would have been an obvious choice for a putatively "high-risk, fast-track" court client eager to show off his mettle; but the much-promised second edition never appeared.

In fact, in Italy's politically restrictive atmosphere, especially after the Holy Index's suppression of Bruno's writings in 1603, it had become dangerous to be seen to be too close to modernist currents in natural philosophy.[38] As the nova controversies played out within and between different cultural spaces (courts, universities, and religious orders), there was a strategic muffling of positions and camouflaging of identities in the ongoing struggle between traditionalists and modernizers, and among the modernizers themselves. For example, Galileo's next major work on the heavens, the *Letters on Sunspots* (1613), continued the *Nuncius*'s strategy of avoiding the names Bruno and Kepler even as it put into the ordinary realm of Nature eastward-moving, alterable spots on the Sun and suggestively compared them to the heliocentered planets.[39]

Galileo was not alone in maneuvering with political caution. His learned and independent-minded opponent, the Jesuit Christopher Scheiner, published under the pseudonym "Apelles" and worked within and around his order's tightening rules of obedience.[40] Yet Scheiner's 1612 *Tres Epistolae de Maculis Solaribus (Three Letters on Sunspots)* also shows that he had made the naturalist turn—and then some. As in Galileo's work, a modernizing tone is apparent, even a tone of constrained excitement: his observations are couched in a language of astonishment and surprise (*nova et incredibilia*) tempered by care and circumspection. Going beyond Galileo's solo approach, Scheiner appealed to unnamed witnesses and to the use of multiple, differently powered optical tubes in order to deflect charges of instrumental deception and to support the consistency of the observations of the new entities' place, arrangement, and number.[41] Scheiner's entities were neither ad hoc nor preternatural events; although they were also natural phenomena, they were neither comets nor clouds. They were natural, opaque, solid, and shadow producing—perhaps to be explained as the denser parts of a Sun-centered celestial sphere, perhaps independent, planetlike bodies circulating like Jupiter's satellites—but always detached, contrary to Galileo's view, from what he assumed to be a solid, inalterable solar surface.[42]

This position represented as much a new direction of theorizing within the *via media* as it represented a shift toward a new ground of argumentation that was inextricably bound to the instrument. Because Scheiner's "solar stars" (*sidera heliaca*) were not *on* the Sun, they could not be used to infer a Keplerian or Galilean solar moving power; but because he believed them to be revolving around the Sun (although tracking their recurrence was uncertain), they could be invoked in support of either a Capellan or Tychonic arrangement.[43] Thus, as public disagreement flared theoretical movement and consolidation of positions continued—even under surveillance.

GALILEO'S NEGOTIATIONS WITH THE TUSCAN COURT, MAY 1610

It is now apt to revisit the question of Galileo's motivations and means for returning to Florence. The timing reveals what he imagined was necessary from the court to achieve his goals. Among these considerations, there was no role for Medici dynastic mythology; but numerous other conditions obtained. One of these conditions was Medici acceptance of Galileo's claim that the phenomena that he described really existed in the heavens. Indeed, full and specific negotiations with Florence began only after Galileo had achieved stabilization of his astronomical novelties with key members of the court, after publication of the *Sidereus Nuncius,* after he had presented his discoveries in public lectures at the university, after he had distributed his instrument to numerous princes, and after Kepler had published his *Conversation.* Clearly, Galileo's credibility resided in multiple sources, both inside and outside the court.

The idea that a one-size-fits-all (scientific) patronage system always required clients to work through brokers to preserve the patron's honor is not coherent with the evidence.[44] If Galileo communicated through Belisario Vinta, the court's chief secretary, that was simply because his relationship to that particular court had evolved differently from, say, the earlier direct exchanges with the duke of Mantua. Through his tutorial work, as we have seen, he already had a well-developed relationship with Vinta and a close personal acquaintance with Cosimo. Sometimes intermediaries were involved and sometimes not: in this case, Vinta managed the practical negotiations because there was reliable precedent, and the long letter of May 7, 1610, from Galileo, following a visit to Florence, laid down his specific requests to the court. Written almost two months after the publication of the *Nuncius* and about two weeks after Kepler's personal reply to it, its importance should not be underestimated.[45]

The opening strategy of the letter shows the sorts of appeals that Galileo believed would be convincing to the Medicean court. It left aside most of the arguments already found in the *Nuncius*—from optics, the technology of the instrument, or the circumstances of observation—and turned rather to his public performance at the university. The court already had the book; hence the letter would bring to bear new evidence that Galileo thought would matter. The university, as the conventional site of disputation, was the first space of authority that he believed the court would find compelling. Much as he had done earlier with the nova of 1604, he gave lectures, representing them to Vinta as follows: "The whole university turned out, and I so convinced and satisfied everyone that in the end those very leaders who at first were my sharpest critics and the most stubborn opponents of the things that I had written, seeing their case to be desperate and in fact lost, publicly stated that they are not only persuaded but are ready to defend and support my doctrines against any philosopher who dares to attack them."[46]

Although Galileo soon had ill-wishers in Padua, his message on this occasion was clear: the traditionalists had changed their minds in a public setting. He says nothing here about entertainment and jousting. There was a kind of academic self-fashioning consistent with Galileo's private persona as invincible professor: "Things have turned out just the opposite [for my opponents]; and indeed it was necessary that truth should remain on top."[47] Of course, as we now know, Galileo's report was exaggerated, but if Vinta had mistrusted it, he could have conducted his own investigation of the matter, just as he had made sure to get the best price for the *occhiale.* Yet (we may infer) he did trust Galileo's report, because eventually he acceded to all his requests.

Galileo's second authority was Kepler's letter of April 19, "written in approbation of every detail contained in my book without the slightest doubt or contradiction of anything."[48] Here was an obvious exaggeration, not to mention an unpleasant irony. Was it not shameless (or just ironic?) that the authority of a man whom Galileo had studiously ignored for twelve years should now be brought forward in support of his attempts to secure advantage? It was not principally Kepler's court title that was significant in this case, but Kepler's approval of Galileo's arguments and the long geographical distance that that approbation had traveled. Galileo conveniently omitted reference to Kepler's multiple and variegated ambivalences. Hence it was now up to Galileo's "noble patrons" to join with Kepler and the Paduan academics in "render[ing] it [the *Sidereus Nuncius*] the esteem deserved by such distinguished novelties."[49]

Galileo next took up the matter of accessibility and claim to the instrument. Although he was receiving requests from various quarters, he responded almost exclusively to Roman cardinals and the Florentine court.[50] This strangely narrow strategy suggests that he judged the Italian courts to be his major sources of protection against the resistance that he was encountering in the universities whose approbation he valued (Pisa, Padua, and soon Bologna). Foreign patronage and approbation were of significance to him only in the context of his local struggles within the Italian universities. Thus, Galileo esteemed Kepler's recent expressions of approval because they could be cited to Vinta. Likewise, he declared that he would only share his method of making viewing devices with "a subject of the Grand Duke"— once again using the occasion to affirm his link to the authority and protection of the Medici.[51] Galileo's indifference in sending one of his viewing instruments to Prague appears especially odd in view of Kepler's clear support and the general value that surely would have accrued from Rudolf's public approbation. But just as the emperor's wishes were not to be satisfied at all by Galileo, so even a request from Queen Marie de Medici to Vinta in early July went unfulfilled until mid-September because Galileo ranked Cardinal Odoardo Farnese ahead of her.[52]

Of course, acquiring some kind of instrument was not difficult: viewing glasses were known to be easily obtainable in Paris, Venice, and Naples. Rudolf, an avid reader of Giovanni Baptista della Porta's *Natural Magic*, already had one in 1609.[53] But, as news of Galileo's discoveries spread, both princes and heavenly practitioners wanted better-quality devices. The issue at stake, then, was not retaining a monopoly over telescopes and future discoveries, but to protect what he had already achieved. Behind this motive lay the unpleasant plagiarism of 1607: after having been "burned" by Capra and Mayr, Galileo was extremely sensitive to the possibility that another of his books based on an instrument would be pirated.[54]

Fear of piracy in fact seems to explain much of Galileo's haste both in rushing the *Nuncius* to press and in hesitating to distribute instruments more quickly.[55] As we have seen earlier, the literary model that he chose owed something to handwritten newsletters, widely circulated yet uncertain purveyors of all manner of political and commercial information.[56] Yet unlike the typically anonymous writers of these political *avvisi*,[57] Galileo was eager not only to announce his discoveries quickly but also to proclaim himself the eponymous "messenger," the one who brought news of new stars. And the care with which he sought to tie himself closely to the Medici had much to do with the desire for the protection that he imagined to be associated with their name. Many of the fears of early modern heavenly practitioners, as this study has shown, were well founded, and their insecurities were much the most common motivation for seeking patronage. In this case, being publicly associated with a powerful family might protect Galileo from competing claimants at a time when no systematic legal notion of intellectual property existed and where such legal remedies as he had just employed against Capra had left him exhausted and bitter. For although Galileo had many needs, he understood that future possibilities for advancing a philosophical program concerned with the "true constitution of the universe" rested on identifying himself securely with the *perspicillum* and the novelties that it alone could reveal.

Although Galileo's dedication to the *Sidereus Nuncius* is sometimes read as meant only for the Medici, his intended audience was far broader. The rhetoric of the dedication advertised to anyone who read the book that Galileo had a special relationship to the Medici family. For this reason, its language emphatically departed from the fairly ordinary gestures found in his book on the proportional compass. To achieve credibility for his close association, he had to offer convincing personal evidence that was not generally known. Thus, for example, he let it be known to his larger readership what the grand duke already knew that he had "instructed Your Highness in the mathematical disciplines, which task I fulfilled during the past four years, at that time of the year when it is custom to rest from more severe studies." He also informed the general reader that he had privileged information about Cosimo II's nativity, in which (so he claimed) Jupiter occupied an important place in the midheaven and "looked down upon Your most fortunate birth from the sublime throne and poured out all his splendor and grandeur into the most pure air."[58] This is the only published reference that Galileo ever made to an activity in which he had been engaged for years. Moreover, as Isabelle Pantin has shown, he rounded off the edges of his case, omitting evidence, for example, that might complicate or obscure: he failed to mention the positions of the other planets, and it is very likely that the entire exercise was not carefully based on the prince's actual time of birth.[59] And finally, Galileo unabashedly burnished his own role as a chosen messenger. Just as "divine inspiration" had supposedly influenced him to instruct the young Cosimo, so it was also "under Your auspices Most Serene Cosimo, I discovered these stars unknown to all previous astronomers." Hence, Galileo constructed himself as having the "right" to name the new entities the "Medicean stars"—although, of course, the real decision had been made by Belisario Vinta.[60] The crucial message to the general reader of the dedication, therefore, was that any Capras or Mayrs who might try to take away Galileo's glory would have to deal not just with him but also with his protectors.

The matter of public protection, then, was handled by the conventions of the dedication. It was not, however, the main topic of Galileo's private negotiations with the court. Hence, in the remaining part of the letter of May 7, 1610, we find no mention at all of the grand duke's horoscope. Rather, Galileo laid out his personal requests in remarkably direct and frank language. Two elements of this request are often emphasized: he asked for no more than the new, higher salary offered to him by the Venetians (one thousand scudi per annum), and he asked for a new title that would include the word *philosopher* (*Filosofo*) in addition to the title that he already held at the university in Padua (*Matematico*). This would give him the kind of disciplinary authority that he could not claim in his post at Padua. Why he wanted the new title is plainly tied to the third item of his wish list: "If I am to return to my native land, I desire that the primary intention of His Highness shall be to give me leave and leisure to draw my works to a conclusion without my being occupied in teaching."[61]

Galileo then explained at length. First, he had been giving private lessons and taking scholars into his home for many years, and while these involvements still left him some time for his non-

university studies, he declared that these commitments "constitute something of an obstacle to me and impede my studies [so that] I should like to live completely free from the one and largely free of the other." Second, he complained that in his "public lessons" at the university "I can teach only those rudiments for which the majority of people are prepared, and such teaching is merely a hindrance and no help in completing my works." Of course, he hastened to add, he did not mind teaching such elementary subjects to princes (of whom Cosimo was of course the only example). Third, he now wished to replace the income that he received from teaching with income from writing books and making inventions. In other words, now that he anticipated Medici protection ("books dedicated always to my lord"), he hoped effectively to change his longtime practices of circulating knowledge only through manuscripts.[62]

To make his proposal concrete, he supplied Vinta with a list of works that he hoped to put into print. This critically revealing enumeration resembled the sort of inventory of promised works that Reinhold and Maestlin had proposed when seeking publishing privileges some forty to fifty years earlier. Full quotation is desirable because the list shows clearly what Galileo hoped to do with his "free time" at court:

> The works which I must bring to conclusion are principally these: two books *Concerning the System or Constitution of the Universe,* an immense conception full of philosophy, astronomy and geometry. [Also] three books *Concerning Local Motion*—an entirely new science in which no one else, ancient or modern, has discovered any of the most remarkable things shared in common which I demonstrate to exist in both natural and violent movement; hence I may call this a new science and one discovered by me, if not from its first principles. [Also] three books on mechanics, two relating to demonstrations of its principles and foundations, and one concerning its problems; and although other men have written on this subject, what has been done is not one-quarter of what I write, either in quantity or otherwise. I have also various little works on topics concerning nature, such as *On Sound and the Voice, On Vision and Colors, On the Sea Tides, On the Nature of Continuous Quantities, On the Motions of Animals,* and yet other works. I have also in mind the writing of some books about military matters, framing them not merely as ideas but showing by very exquisite rules every-

thing in that science which pertains to the understanding of, and depends upon, mathematics, such as the knowledge of pitching camp, fortification, ordnance, assaults, sieges, estimation of distances, understanding of artillery, use of various instruments, and so on. I must also reprint [the book] for the use of my Geometric Compass, dedicated to His Highness, as no more copies are to be found, and this instrument has been so widely embraced everywhere that other instruments of this kind are no longer made, while I have made thousands.[63]

It is apparent from this substantial ensemble of titles that Galileo had in mind an ambitious publishing program. He was proposing to publish books in two general categories, theoretical and practical. Effectively, he wanted the court's sanction to carry forward a public philosophical program that, despite its support among the Venetian aristocracy, was blocked by traditionalists at the university. In addition to these ambitious *theoricas*, he larded the list with some military *practicas* that would appeal to the traditional court appetite for such writings. Thus, if he were to return to Florence, it would not be to a teaching position at Pisa, although he still relished the honor of being associated with his old institution. It would be a bridge position with none of the disadvantages that he had experienced in Padua.

On the other hand, Galileo clearly had no intention of fashioning himself—or his writings—as a conventional courtier, let alone a Medici court jester. Aspiring courtiers did not typically present lists of books that they hoped to publish.[64] His social identity as a teacher and his philosophical identity as a modernizer were clearly rooted in the academic practices and struggles in which he had spent most of his adult life. What he wanted from the Medici was what he imagined Kepler to possess: freedom to philosophize in new ways and protection for the ideas expressed in his publications. (Of course, he knew nothing of Kepler's own difficulties in Prague). For Galileo, as for Kepler, the court was an institution that seemed to offer better possibilities than the universities for philosophizing in the *via moderna*. The Medici court would become a setting in which academic struggles were played out. Galileo understood well the importance of the university's intellectual authority. And eventually, unlike Kepler, he was successful in using his court position to influence the appointment,

at Pisa, of a disciple among the traditionalists: the mathematically talented modernizer and Copernican Benedetto Castelli.[65] This move established an early-seventeenth-century tradition of Galileans in Italian universities, whereas Kepler had no academic followers until the 1620s.[66]

VIRTUAL WITNESSING, PRINT, AND THE GREAT RESISTANCE

In just over four months, between April and the end of August 1610, Galileo's telescopic claims—especially his Jovian observations—came under heavy attack. They were variously portrayed as deceptions, tricks, and fictions. But none of the opponents went beyond these sorts of denunciations to broader claims about the status of the Copernican arrangement. This is curious and even surprising, because although the *Nuncius* had made highly suggestive gestures to Copernicus's ordering, after May 1610 Kepler's *Dissertatio* explicitly linked the sidereal novelties to the Copernican arrangement and—even worse—to Bruno's innumerable worlds. The question now was, what would Galileo and his opponents do with Kepler's metabolization of the *Nuncius*?

The principal resistance emanated from Magini's sphere of influence in academic Bologna, but it rapidly spread to court circles in Prague, where both Galileo's supporters and his opponents struggled for Kepler's favor. The details are complex enough,[67] but one should not lose sight of a larger pattern of struggle. The conflict was not, as it is sometimes portrayed, just a two-sided stand-off between Galileo and hard-line academic traditionalists. It also involved his ongoing differences with the modernizer Kepler and the mitigated traditionalist Magini—not to mention the entanglements among the three of them. In any case, three- and sometimes four-sided struggles were a feature of the wild-growth social ecology of natural philosophy in this period: among modernizers (like Kepler and Bruno), between modernizers and traditionalists (Galileo and Cremonini), between modernizers and mitigated traditionalists (Galileo and Magini), and between mitigated and progressive traditionalists of the *via media* (Magini and Origanus). The two-sided contest that Galileo used to frame his *Dialogue* simply buried these differences by largely erasing them.

Proximate, personal connections are important for understanding the character of the resistance. Nominally, the episode centered on Magini's secretary, Martin Horky (ca. 1590–ca. 1650). Horky was a young man of some learning and promise who had come from Lobkoviče, southwest of Prague, and had been living with Magini in Bologna for a year while also tutoring his son.[68] It may have been useful to Magini to have a man with Hapsburg connections living in his household, as he was then engaged in a bitter conflict concerning the superiority of his ephemerides for 1608–30 over those of David Origanus, the ordinary professor of mathematics at Frankfurt.[69] Horky had connections with both the university and the court in Prague.[70] Horky also had Paduan associations, having studied there around 1605, where it is quite possible that he encountered Baldassare Capra and Simon Mayr.[71] By the very end of March 1610, Horky had the *Nuncius* but no telescope. The absence of an instrument, however, did not prevent him from approaching Kepler on April 6, proclaiming his intention to publish against Galileo's "four fictitious planets."[72] It seems very likely that Horky was reflecting Magini's own views, as he also alluded at length to a proposed collaboration between Kepler and Magini on a new ephemerides, but in a more combative register. Whatever motives underlay Horky's aggressive language and secretive approach, his letter shows that different readings of the *Nuncius* (with or without the telescope) could support both positive evaluations of its claims and equally negative ones.

A few days later, Magini himself vigorously courted Kepler on the matter of the ephemerides in an effort to draw him into his struggle with Origanus. His idea was a collaboration: it would be based upon Tycho's numbers, it would improve calculations for judiciary astrology, and, even more important, it would surpass in accuracy Origanus's just-published *Brandenburg Ephemerides*. Magini was nothing if not a great tabulator of numbers, and his overture marks a mentality that still operated squarely within the classic categories of the science of the stars; yet at that very moment the intellectual ground was (literally) moving under his feet. In seeking the alliance, Magini protected himself by a well-defended tunnel vision. He ignored Kepler's Copernican commitments and he simply disregarded Origanus's remarkable initiatives into celestial natural philosophy: a diurnally rotating Earth

rooted in the Gilbertian magnet, a geoheliocentric ordering of the planets, and explicit rejection of the physical materiality of the heavenly orbs.[73] Worse yet, on Magini's other flank was Galileo. At the end of his letter to Kepler, Magini casually asked his opinion of "the four new planets of Galileo."[74] Magini's curiosity (and anxiety) about Kepler's opinion seems to have been connected with the worry that Galileo's four additional planets would undermine the whole enterprise of the traditional, seven-planet ephemerides. Horky had said as much to Kepler: "If Galileo's story is taken to be true, the ephemerides based on Tychonic foundations that you want to publish with Magini would require eleven planets."[75]

As Magini's and Horky's concerns show, what mattered in these (still) early days of the *perspicillum* was the increase in the number of planets rather than how they were ordered. Certain astrological consequences of Galileo's discoveries were noticed quickly both by mathematically skilled practitioners and by learned men of no special mathematical talent (like Manso and Wotton). Although these were not Piconian worries, as soon as practitioners noticed the repercussions for astral prognostication and the principles underlying it, tremors and anxieties ensued. In 1609, the forward-looking ephemeridist Origanus saw no such threat to the accuracy and value of his tables. But by the spring of 1610, Magini and Horky were openly worried. Moreover, Galileo's naming of the new planets for the Medici and his allusions to Jupiter in the grand duke's horoscope did nothing to allay concerns. Of what value would a seven-planet ephemerides be if there were eleven planets in the heavens? Kepler had already intervened on Galileo's behalf by proposing the Copernican-style solution of a moving Earth-soul, resonating with the angles of planetary rays. A few days later, on April 24, it suddenly looked as though the matter might be resolved when Galileo himself showed up in person in Bologna on his return from Pisa and brought with him one of his good telescopes.

This episode is one of the interesting moments in early-seventeenth-century witnessed observations. As soon as Galileo departed, Horky sent Kepler a detailed accounting. On the evening of his arrival, Horky recounted that he engaged in a deception: "I never slept, but I tested Galileo's instrument in innumerable ways as much on what is below as on what is above. On things below, it rendered marvels; in the heavens, it failed because what appeared to be other stars were fixed stars doubly enlarged."[76] Horky also confessed that he secretly made for himself a wax impression of the lenses, bragging that he could make an even better instrument.[77] The representation of the scene nicely evokes Tycho's recounting of Ursus's nocturnal sniffings around the diagrams lying about in his library—except, in this instance, the report came from the offender himself.[78] Also, like Edmund Bruce a few years earlier, Horky presented himself as an agent for Kepler against Galileo.

On the following evening, April 25, Galileo personally conducted a viewing before a public gathering. Unfortunately, the surviving testimonials of that event come only from Horky and Magini, the most negatively disposed parties. Those men—not yet knowing that Kepler's *Dissertatio* was already at the press—shaped their representation with the intent of winning Kepler over to their viewpoint. Horky referred to "many witnesses, most excellent men and most noble doctors," among whom he named only "Antonio Roffeni, a most learned mathematician at the Academy of Bologna." According to Horky, the results were uniformly disappointing: "All confessed that the instrument deceived."[79] A month later, Magini described the same scene to Kepler in slightly more precise language: "On the nights of April 24 and 25, he stayed overnight with me at my house with his *perspicillum* daring to [try to] show these new Jovial circulators, but he achieved nothing. For more than twenty most learned men attended, but no one saw the new planets perfectly."[80] Magini also mentioned among these "learned" the same Giovanni Antonio Roffeni (ca. 1580–1643), perhaps to impress Kepler with the presence of a mathematically skilled witness. This Roffeni, otherwise unidentified by Magini and Horky, was a sometime student and amanuensis of Magini's from a noble Bologna family, awarded a doctorate in philosophy and medicine in 1607 and, between 1609 and 1644, a prolific author of almanacs and prognostications.[81] The performance took place at the home of a nobleman, Massimiano Cavrara. The other members of the audience are not known.

The position of both Magini and Horky in the early spring of 1610 is interesting. They did not deny that the instrument worked, in the sense that it enlarged already known objects on Earth and in the heavens, but they did deny that it revealed any new, hitherto unknown entities. Of course, the question of "newness" also depended on what Galileo had said, what expectations he had raised before the illustrious gathering on the evening of April 25. Unfortunately, there is no independent report from him. Yet, it is possible to make some reasonable inferences.

Galileo's personal log shows that on April 24, he observed two stars to the west of Jupiter and, on the following night, one to the east and three to the west.[82] Neither this distribution nor its sequel were predictable patterns. For example, in the *Nuncius,* a comparable set of configurations is described on, respectively, February 19 and January 22. Of course, even without predicting the changes, if Galileo had arranged two viewing performances, at least he could have predicted and shown some redistribution; but he had only one such audience in Bologna.

Within these limitations, surely Galileo could not and did not advance an even more ambitious prediction, such as: "You will see four little points of light revolving around Jupiter." To have made this prognostication would have gone counter to the almost day-by-day *descriptive* history of "freeze-frame" sequences described in the *Nuncius.*[83] Galileo's strategy in the *Nuncius* was a retrospective one—drawing in the reader with a history of first attempts, surprises, changes of interpretation, and sometimes dramatic visual representations—in its use of humanist, personal narrativity similar to strategies adopted by Copernicus in *De Revolutionibus,* Maestlin in the *Observatio et Demonstratio Cometae,* and Kepler in the *Mysterium* and the *Astronomia Nova.* But compared with Kepler's *Astronomia,* Galileo's representations were much easier to follow, even for those disinclined to trust them. Thus, on January 7, he informed readers of "three little stars positioned near [Jupiter]—small, but yet very bright." Immediately after, he gave the reader his initial interpretation: "I believed them to be among the number of fixed stars." (The printer used asterisks to represent the stars, and in the first of his diagrams, there were two to the east and one to the west.) Then, still carrying the reader along with the history of his observations and the meanings that he attached to them, he declared that "I was not in the least concerned with their distances from Jupiter, for, as we said above, I first believed them to be fixed stars."

Clearly, if Galileo had first believed that he was seeing fixed stars, he would have known that others could be similarly confused. Two days later, however, on January 8, Galileo introduced the surprise: "Guided by I know not what fate, I found a very different arrangement." The new ordering put the three "little stars" to the West. Why had this occurred? Galileo again presented himself in the person of the open-minded reader, led along by the unexpected: "I had by no means turned my thought to the mutual motions of these stars, yet I was aroused by the question of how Jupiter could be to the east of all the said fixed stars when the day before he had been to the west of two of them." Perhaps Jupiter had moved, and the little stars were really fixed. Once again, instead of jumping to the next observation, Galileo evoked the personal when he wrote of his "disappointment" on finding that "the sky was everywhere covered with clouds." And, on January 10, the *Nuncius* delivered another surprise when two stars appeared to the east and one of the three disappeared from sight. Galileo now advanced two bolder proposals: first, that the third star was "hidden behind Jupiter"; second, that the perceived changes in position were to be attributed to the stars rather than to Jupiter, since they remained in the same alignment along the ecliptic but shifted their positions relative to Jupiter in a way that Jupiter could not do with respect to them.

Now, five days into his narrative, Galileo named the eleventh as the day on which he advanced a still bolder hypothesis—"entirely beyond doubt"—in which he put forth a daring analogy to the motions of the Jovian stars: "That in the heavens there are three stars wandering around Jupiter like Venus and Mercury around the Sun." He had now moved methodically from "little stars" to "wandering stars" and hence to planets. And having introduced planets, he began to raise questions of order. He avoided any Ptolemaic talk of Mercury and Venus traveling on epicycles around empty centers; in fact, he went so far as to claim that "no one can doubt that they complete their revolutions about him [Jupiter] while, in the meantime, all together they complete a 12-year period about the center of the world."[84] Galileo then concluded this discussion—and the book itself—with an even more general interpretation of his observations: "We have moreover an excellent and splendid argument for taking away the scruples of those who,

while patiently accepting the revolution of the planets around the sun in the Copernican system, are so disturbed by the attendance of one Moon around the Earth while the two together complete the annual orb around the Sun that they conclude that this constitution of the universe must be overthrown as impossible."[85]

Some commentators have read this passage as proposing "a visible model of Copernicus's solar system."[86] And certainly, by comparing Jupiter's moons and the Earth's moon, it can be allowed that Galileo was beginning to develop a stronger sense of systematicity than that found in Tycho Brahe's use of *systema*. But, as Wade Robison has observed, the meaning of *system* was not that of a dynamic, physical order—one that explained why neither Earth nor Jupiter loses its moons as it revolves—but, at most, a descriptive, architectonic structure.[87] In this sense, Galileo cautiously avoided the assertive physicalist posture on which his later reputation would rest and did not even gesture at Copernicus's "natural appetite" to account for the coherence of the Jovian bodies. Furthermore, the syntax of Galileo's allusion to "those who patiently accept the revolution of the planets around the Sun in the Copernican system" strongly suggests that this passage was aimed at the (still unnamed) Tychonics.[88] For first-time traditionalist witnesses at Magini's house in Bologna, however, it is hard to imagine that Galileo would have attempted any such aggressive claims. At most, he might have attempted to convince such observers that "there are three new stars to the west and one to the east of Jupiter; and there are no fixed stars known in this vicinity." Perhaps Horky and Magini had deceptively primed the audience's perceptions beforehand to "see revolutions" of Jupiter's moons.

In any case, for Horky and Magini—both of whom had read the *Nuncius*—there were no new revelations, and hence there was nothing left to discuss. There was no message except Galileo's performative failure. Unlike Magini's relatively straightforward telling, however, Horky's inversive rhetoric left no middle ground: Galileo was to be seen as a "fable-telling celestial merchant" ("fabulosum mercatorem caelestem"):[89]

His hair falling out; in his weak replies, his skin covered with the pimples of the French disease; his skull attacked, his ravings finding lodging in his brain; his optical nerves worn out because he has observed the minutes and seconds surrounding Jupiter with too much curiosity and presumption; his sight, his hearing, his taste and his touch destroyed; his hands tied up in knots by the gout because he has stolen the mathematical and philosophical treasure; his heart throbbing because he has sold to everyone a celestial fable; his guts producing an unnatural tumor because he exercises no more charm by contrast with learned men and notables; his feet cry out with gout because he is everywhere mistaken in all four cardinal points [i.e., in every direction]. Lucky the physician, three and four times fortunate, who would bring back to health this crippled messenger. With the illness cured, I return to you the stars, you small clear gems, you dear small gems.[90]

Horky's elemental, bodily idiom of emotion gave voice to the fear of novelty—of looking too closely. Here, once again, was the familiar, if unusually vivid, trope of the disorderly monster, recalling the elevated temperature of Hagecius's vitriolic polemic with Raimondi in 1572 and Ursus's attack on Tycho in 1597 (see chapter 8 above). At the end of Galileo's abortive act of representation, Horky declared, he remained silent and left sadly the following morning. Pantin plausibly suggests that by allowing Horky free rein, Magini engaged in a kind of covert aggression against Galileo.[91]

Ultimately, the question of how Kepler would appraise the veracity of the Bologna dispatches is directly related to how he evaluated the *Sidereus Nuncius*. Horky, Magini and the twenty "learned men" of Bologna had allegedly looked through a good-quality (perhaps 20× or 30× power) *perspicillum* with Galileo present. Kepler had already endorsed Galileo's findings, both privately and publicly, based only on his reading of the *Nuncius*. In Steven Shapin and Simon Schaffer's currency, Kepler was a "virtual witness," someone who trusts in the narrative representation of an observational scene at which he or she was not personally present.[92] How, then, did the nonwitness Kepler defend Galileo's claims? Kepler acknowledged that his own judgment might appear rash, as it was not based on personal experience; but in the *Dissertatio*, he provided not one but eight reasons for trusting in Galileo's report: (1) the sure quality of his style; (2) the lack of a motive for practicing deception ("Is there any reason why the author should have thought of misleading the world with regard to only four planets?"); (3) a commitment to truth regardless of opposi-

tion from common opinion;[93] (4) Galileo's status as both a "gentleman of Florence" and a "learned mathematician"; (5) his "keen sight" as compared to Kepler's "poor vision"; (6) his open invitation to others to view the same sights; (7) his offer to provide "his own instrument in order to gain support on the strength of observations"; and (8) his risk in "mock[ing] the family of the Grand Dukes of Tuscany, and attach[ing] the name of the Medici to figments of his imagination, while he promises real planets."[94]

In short, Kepler had devised an original approach to a problem that neither he nor anyone else had previously faced. In the absence of direct observational experience, Kepler inferred Galileo's honesty not from some moral essence about his character or his class but from the logic of his contingent social circumstances. Galileo was stylistically open, he was taking various kinds of risks to his reputation, he was opposing received opinion, and he was inviting others to look; if he was lying and deceiving, then he would suffer serious consequences. Kepler could find no reason why Galileo would not want to be sincere.[95] Moreover, for Kepler, although Galileo had not provided a *dioti* demonstration, there were still good-enough reasons, based on practical considerations, to trust his *tou hoti* claims. And Kepler would go some way toward providing a *dioti* demonstration himself by offering speculations in his own philosophical idiom. Yet, among the reasons offered up by Galileo, Kepler mentioned neither the remarkable pictures of the Moon, nor the naming of the Jovian satellites, nor the significance of Jupiter in the grand duke's horoscope.

For all that, Kepler's remarkable "conversation" with Galileo did not end the matter. The Horky-Magini affair involved him in what can only be described as a political mess, an imbroglio that further tested his relationship with Galileo and involved him in delicate diplomatic exchanges with Magini and Horky. The complications that evolved provide further insight into how the *Nuncius* traveled and how its claims were judged after the appearance of Kepler's *Dissertatio*.

Kepler's ambivalent presentation in the *Dissertatio* was significant, as it opened opportunities for others to make different readings. And, in fact, after the book arrived in Bologna on May 20, Horky and Magini quickly constructed Kepler as an ally against Galileo. They systematically ignored Kepler's Copernican framework of interpretation, together with his reasons for accepting Galileo's novelties as true, and read his book as though it completely undermined Galileo. For example, in those passages where Kepler gave some credit to Giovanni Baptista della Porta for devising a looking glass before Galileo, Magini told Kepler: "Your method pleases me. I believe that it will not be agreeable to Galileo because you have judged him by his principles, decently and amicably. It remains only to eliminate and destroy the four new servants of Jupiter. He will hardly succeed."[96]

While Horky and Magini were working hard to enroll Kepler in their cause, Magini had opened another front. Through Hasdale, Galileo learned that Magini had written to Zugmesser, the elector of Cologne's *mathematicus,* as well as to "all the mathematicians of Germany, France, Flanders, Poland, England, etc." Information about Magini's initiatives was already widespread; Hasdale said that he had heard about them from various agents and informers associated with the court. Magini had fertile ground to plow. Zugmesser's animosity toward Galileo was rooted in the earlier dispute about the military-geometric compass, and Galileo, with his occasional tendency toward misstatement and lack of tact, had swung a heavy sledgehammer. In the *Difesa* against Capra and Mayr, he had erroneously identified Zugmesser as Flemish rather than as a German from Speyer. Moreover, at Cornaro's house, Galileo had charged Zugmesser with receiving the idea of his instrument from Tycho Brahe, but Zugmesser said that he had never met Tycho. Zugmesser also told Hasdale that Galileo had admitted the inferiority of his instrument.[97] To add to the miscalculation, Galileo had put the elector of Cologne (and hence Zugmesser) near the top of his list of select recipients of a choice *perspicillum.*[98] But possession and use of the instrument in no way moved Zugmesser into Galileo's camp. The circumstances surrounding the military-geometric instrument continued to cast a long shadow over the *perspicillum* and its claim to produce novelties.

Efforts in Prague to quell the rising storm came to naught. In early June, Hasdale hoped to defuse the growing conflict between Zugmesser and Galileo by communicating Zugmesser's grievances to Galileo in the (vain) hope that the latter would engage in appropriate diplomacy. At exactly the same time, Kepler wrote a remarkably patient letter to Horky and tried to explain why it was perfectly possible that he, Magini, and even Galileo had had trouble seeing the new stars in Bologna. He told Horky: "I strongly doubt that Galileo has the eye of a lynx, and you who deny recognizing the same things, you are myopic."[99] In the *Dissertatio* Kepler had proposed that Galileo increase the number of lenses, a solution that Kepler would modify several months later, after he had had access to a telescope.[100] But tinkering with the instrument was not what Horky had in mind. In the middle of June the machinations of the previous two months took a sudden turn that again underscored the unusual power of print.

Horky took off for Modena with a short but acerbic manuscript titled *A Most Brief Peregrination.* It was immodestly dedicated to all the professors of the philosophy and medical faculties at Bologna, his hoped-for protectors. On June 18, he was granted an imprimatur by the inquisitor in Modena, and on June 21, the book came off the presses.[101] The pamphlet continued the main theme of Horky's letters: the Jovian planets were "fictions." But now a yawning gap opened between Horky's emphatic, sometimes carnivalesque rhetoric and his optical arguments. Horky admitted that he had, in fact, seen four "spots" [*maculas*] "in the heavens," but he ascribed these appearances to the refraction of Jupiter's rays in the intervening fog and mist of the air.[102] Effectively, Horky had devised an objection constructed along the lines of stock parallax arguments against the superlunary location of comets and novas: Galileo's "spots" were celestial deceptions but atmospheric realities. To add to the growing philosophical mayhem, Horky helped himself to a hastily conceived hypothesis in Kepler's *Dissertatio* intended to explain the variations in the appearances of the four "planets" by reference to differential exposure of their surfaces to Jupiter's light. Entirely contrary to Kepler's intentions, Horky interpreted this optical speculation as an endorsement of a visual hallucination, thereby inserting the imperial mathematician's authority into the *Peregrination.*[103] By now, the situation was spinning out of control.

Magini's opportunities for damage control were less numerous than those of Tycho Brahe in his dealings with Ursus and Wittich some fifteen years earlier. Nonetheless, he was not without resources: the next day, Magini's former student Roffeni began the process. He sent an embarrassed letter to Galileo narrating an extraordinary and woeful tale of mischief: "His [Magini's] servant had caused to be printed a work against Galileo." Magini had tried to prevent it, but to no avail. When Horky returned to Bologna, Magini blew up and evicted the hapless "servant."[104] Unlike Roffeni, however, Magini located his mistrust neither in Horky's social class nor in his lack of mathematical skill but in his linguistic and national identity. On the same day that Roffeni composed his letter, Magini himself wrote to a friend of Galileo's, heaping scorn on "Mr. Martin Horky, a German," referring to him as "uncivil and inconsiderate" and ending with the charge that "all Germans are enemies of us high Italians."[105] This was the same kind of language that Magini had used in his attack on Origanus. However, neither Magini's cursing nor his efforts to track down his former secretary met with success. When last seen by Magini, Horky was on his way back to Modena to fetch several hundred copies of his book.

Distribution began at once. Within a week after publication, copies of the *Peregrinatio* were on their way to key locations: Prague (to Kepler), Venice (to Sarpi), and Florence (to Francesco Sizzi, ca. 1585–1618). And in the succeeding two months copies are known to have made their way to Roffeni (in Bologna), Matthaeus Welser (brother of Marcus, in Prague), Michael Maestlin (in Tübingen), and Lodovico delle Colombe (in Florence).[106] Many other copies were surely in circulation. Hostility to Galileo found receptivity in different places, some of them quite unlikely. Maestlin, in whom one surely would have predicted Copernican sympathies comparable to those expressed by Kepler, was unexpectedly quite approving: "This Martin really freed me from great anxiety. Martin's writing greatly pleases me." And further: "You [Kepler] in your work, have deplumed Galileo," Maestlin chortled. At stake was Maestlin's belief that he had discovered in the ancient writings what Galileo claimed to be the first to know with the *perspicillum:* the Moon's surface as unpolished and spotted, and the existence of more stars in the heavens than the ancients believed. "I thank you again for the most honest mention that you make of me in this dissertation."[107] Kepler's ambivalence in the *Conversation* thus easily played into the hands of anyone who had mixed feelings about Galileo and his discoveries. Maestlin's response confirms, once again, the weak ties

that bound the Copernicans while also corroborating that practitioners rather than patrons regulated the flow of intellectual credit.

Francesco Sizzi came from an old and distinguished Florentine family mentioned even by Dante in the *Divine Comedy*.[108] Unlike the pious, schoolmasterly Maestlin, he was neither an academic nor a courtier; his sociability was learned but strongly inclined to traditional premises. Somehow the young Sizzi had become connected with the Horky-Magini network. He appears briefly in Horky's *Peregrinatio* among those who deny the reality of the Jovian planets ("Let Galileo listen to the young, very learned Florentine patrician Francesco Sizzi").[109] In late June, Horky confided in him his difficulties with Magini and quoted Kepler's letter of June 7 in detail. Horky emphasized that Galileo's four new planets were fictions, the result of a visual hallucination.[110] By early August, Sizzi was moved to compose a *disputatio* critical of the "rumor" of Galileo's four new planets, whose existence—were it to be demonstrated—would undermine the foundations of astrology. The treatise was decorated with a bouquet of learned references that masked its rather simple thesis: both holy scripture and natural reason support the proposition that there are only seven planets. The argument for the sevenness of civil and natural order recalls Rheticus's arguments for the sixness of the Copernican system. Sizzi, however, did not find his views transparently reflected in holy scripture without the aid of modern interpreters. Pico della Mirandola's *Heptaplus* and its Jewish sources were at hand to assist him in interpreting the seven lamps of the menorah (Exodus 25:37; Zechariah 4:2) as allegorically referring to the seven planets.[111] And this interpretation sufficed to launch a vigorous investigation into the various ways in which the number seven, rather than eleven, brought order to the macro- and microcosmos. Along the way, Sizzi dismissed the "school of astronomers" that put all the planets in motion around a stationary Sun, thereby (said he) eliminating the tedious hodgepodge of equants, deferents, and epicycles but simultaneously failing to support the "ancient arrangement of the [astrological] houses and the dignities of the signs distributed among each of the planets." Consider that Venus and Mercury do not digress far from the Sun but nevertheless have "their own proper houses, exaltations and trigonocracies." "By what comparable reason," asked Sizzi, "may one conclude that these fictitious planets choose their dignities with Jupiter?"[112] With Horky and Magini, Sizzi regarded the 7/11 problem as a threat to the traditional numbering of planets.

As elements of opposition coalesced around Horky and his self-interested uses of Kepler, Magini and Roffeni were on the offensive to win back Galileo's support. And what better way than to inform him that Horky was in league with his old enemies? On June 29, Roffeni told Galileo that Horky had been in touch with Baldassare Capra in Pavia while lodging with the Jesuits in the College of Nobles.[113] A few days later he communicated that "[Horky's] books were left at the home of Baldassare Capra, with whom he had stayed for a few days in Pavia."[114]

The Capra link underscores two persistent themes in this study: the small scale of social spaces in which heavenly knowledge was conducted, and the uses of print both to mobilize and to undermine credit. Nodes of resistance to the *Nuncius* in Italy were organized around the same people and issues as during the nova controversies in Padua just a few years earlier. Galileo had used print and his relationship to the Medici to identify himself publicly with the military-geometrical compass in 1606. In 1610, he was seeking to do the same—even more boldly— with the *perspicillum* and its novelties. But here there is an unexpectedly new theme: even with Medici protection of Galileo, men of low academic status were able to use print to challenge and subvert his special claims. Print turned out to be a highly flexible resource of representation that took considerable social energy to control, just as it had in the Tycho-Ursus affair and as it would in Galileo's later affair with Rome of 1632–33.

On the other side, the *perspicillum* by itself did not establish the claims for a new world. Private correspondence networks—sometimes with, sometimes without, the *Nuncius*—successfully carried and maintained the credibility of Galileo's novelties in the first half of 1610. Moreover, personally managed performances offered no guarantee of winning adherents. They were vulnerable, as Galileo's failed Bologna demonstration showed; and conversely, the absence of the Galilean *perspicillum* did not prevent Kepler from using the representational power of print to offer cogent support. Kepler's narrative freedom also

permitted him to situate Galileo's discoveries in a Copernican world of his own design.

On the other hand, whatever the successes of these different scenes for the establishment of credibility, one must not underestimate the sense in which the *perspicillum* had changed the conditions of evidence and debate. Although the very early observing instruments could yield contestable results, one must again compare that situation to that of the handful of Nullists in the 1572 nova controversies. Against that backdrop, one can say that it was really not so very long before the *perspicillum* began to make a difference. Magini himself acquired an instrument sometime in the latter part of June 1610, at the very moment when the Horky affair blew up. Toward Galileo he rapidly changed his tune. Magini now acknowledged that he had seen spots on the Moon "very well," that they appeared like "drops of oil on the surface of the water," and that the Moon itself looked like "a ball of snow—not well formed but gross, which makes for obscurity and inequality in certain places." He juxtaposed these remarks with defensive and apologetic words to Galileo: he had no part of the great deception [*coglionaria*] in which his "German" had embroiled him. He was "ashamed" of Horky's work, and he had now heard that Horky had gone to Milan four days earlier to the house of Capra, "already an enemy of Signore Galilei."[115] He was evidently smart enough to realize that he had chosen the wrong side. And in the tangle of considerations—print, patronage, old hostilities, envy, and academic and court authority—the telescope proved to be a different kind of resource. It was imperfect by modern standards, but, unlike the earlier novas, it produced representations that would not go away; and hence it had a new sort of capability to force beliefs to change.

This consideration has a direct bearing on the relevance of Medici patronage. On July 10, 1610, Galileo received the letter from Cosimo II for which he had long been waiting. All his requests were met—the salary, the court title, and the post of "First Mathematician of our Studium in Pisa"—"without obligation to live in Pisa, nor obligation to read there, except when it may please you as an honor." It ended with the desirable stipulation that "ordinarily residing in Florence and pursuing the perfection of your studies and your labor, but with obligation to come to us whenever we may call you, even if you are outside Florence."[116] Although the Horky affair had spiked opposition for a brief moment, it made no difference to the court appointment, just as Galileo's prospective connection to the Medici, already announced in the *Sidereus Nuncius,* had made little or no difference to the forces arrayed against him in Bologna, Prague, or anywhere else. Medici patronage gave Galileo an unusual measure of philosophical and pedagogical freedom to develop his ideas outside academic constraints; but court associations with published heavenly knowledge claims, from Copernicus to Galileo, proved to be a surprisingly weak form of protection.

MAGINI'S STRATEGIC RETREAT AND THE 7/11 PROBLEM

As the summer of 1610 wore on, the Magini group continued to reverse direction, Galileo continued to proclaim new discoveries, and the relationship between Kepler and Galileo briefly reached a point of forthright reciprocity. Each of these developments illuminates different dimensions of how the Galilean novelties traveled as they engaged the resistance.

In late July, Roffeni discovered that Horky had implicated him as a primary witness against Galileo and immediately offered to compose a refutation.[117] But here a new problem arose: how would he and Magini manage the retraction of their previous, vehement opposition to the *Nuncius?* Roffeni chose the form of an "apologetic letter"—a polemic addressed to Galileo which, in its brief and undemanding compass, enabled sequential replies to extensive citations culled from Horky's four "problems."[118] One strategy for undermining Horky was to portray his contentions as "childish" (*puerilem doctrinam*) and filled with inconsistencies rather than to designate him as completely "unskilled" (which he was not). This approach would exonerate Magini's judgment in taking Horky into his house in the first place, as his errors could then be ascribed to the passions of youth. One manifestation of Horky's immaturity was taken to be his choice of language, a criterion that Roffeni easily obtained from his reading of Kepler's *Dissertatio.* Just as Kepler had ranked good style as an important standard for judging the reliability of the *Nuncius,* Roffeni criticized Horky's inflated and bumptious language as a basis for mistrusting his objections. Indeed, Roffeni pointed out that some of Horky's language

was not even original, as it came "word for word" from Ursus's 1597 attack on Tycho! To add further insult, Horky contradicted himself when he alleged on Tycho's authority that the Dane had found "a thousand of the very smallest little stars"—implying that Galileo had merely found some new fixed stars.[119] Roffeni said that he had no idea where Horky had read this in Tycho's writings: at most, Tycho had discovered thirty-one stars previously unknown to the ancients.

Roffeni next disputed Horky's description of the scene of witnessing. Against the assertion that the Bologna gathering had completely failed to see anything at all, Roffeni claimed that he had faith in "those men at Magini's house who affirmed that they saw those planets in some way rather than those who Martin brought forth as witnesses from I know not which letters of testimony." He did not explicitly mention Magini, but instead reported that "Antonio Santini, a nobleman from Lucca and most skilled in mathematical matters, has often seen these planets in Venice; and in that city, it was confirmed by that same man [Santini] in the presence of some Bolognese patricians who journeyed to that very place at the same time."[120] Perhaps Roffeni's decision to locate the scene of positive sightings at Magini's house, without mentioning Magini as a directly affirming witness, was an attempt to further protect his former teacher by allowing him to enjoy the testimony of place without the testimony of his word.

The next argument cycle shows how well Roffeni had absorbed his reading of Kepler's *Dissertatio.* Ostensibly, Horky's objection was that the alleged entities were not planets because they lacked a necessary (albeit unusual) astrological property—emitting a scent. Roffeni replied by transforming the discussion of the Jovian planets' "nature" into an astronomical question of distances. Because the new planets always remained extremely close to Jupiter, "Was it not pleasing to the Creator of all things to bestow this dignity on Jupiter beyond all others? Or rather, as Kepler says, could the other planets also have their 'circulators' which cannot be seen by us on account of their small size and great distance from us? In fact, if Venus and Mercury go around the Sun—as is the opinion of Copernicus—why could there not be four other planets going around Jupiter with a common circuit of more or less twelve years? In what way do Venus and

Mercury complete their course in one year with the Sun?"[121] By putting recurrent—rather than singular—novelties into play, this intriguing passage shows how Kepler's reading of Galileo's telescopic discoveries was pushing the previously resistant Roffeni in new directions. Venus and Mercury were circumsolar on *Copernicus's* authority rather than on Tycho's. Instead of arguing strictly from the grounds of witnessed observations, Roffeni now argued from the grounds of plausible analogy and possibility. But in casting his discussion in the rhetorical form of an apology, he also avoided making definitive judgments about the probable or demonstrative status of these claims.

Roffeni then turned to what may be called the 7/11 problem. Here, for the first time, Horky's competence as an astrologer was brought into question. He did not know the proper way to cast a celestial chart. He claimed audaciously to be able to make predictions for an entire year. And he wrote much impudent nonsense to Kepler when he was living at Magini's house that only showed his ineptitude in mathematical disciplines. Clearly, Horky had not allowed for the fact that the new planets were extremely small, and hence their effects could only be "weak and stunted." And suppose there were indeed eleven planets? Would this really undermine the standard view, upheld by Ptolemy, Cardano, and other mathematicians? Would it be necessary to ascribe houses and exaltations to each of the new planets? Why would this undermine the principles of astrology "constantly confirmed throughout the centuries by repeated observations"? Roffeni's reply was essentially the same as that privately advanced by Manso several months earlier.

Finally, Roffeni criticized Horky for rushing into print with his views about visual deceptions. First he should have assembled men of great learning and mathematical skill so that he, "a rash German," would not have advanced such doubts about Galileo, "a man most highly regarded privately and publicly in mathematical matters." Roffeni ended his *Epistola Apologetica* by affirming to the reader that the discoveries were "true and solid." He claimed that he spoke for unnamed "noble patricians and learned men," unlike Horky, who was alone in his doubts. Roffeni declared to Galileo that when such men "gather together with me in order to discuss this new discovery of Astrology, I am able to vindicate you from the charges of your adversaries because you were the first to publish the principles of the instrument" (*theoricam Organi*).[122] And the *Epistola* ended with a passage from the *Nuncius* that describes the construction of the double-lensed instrument.

Roffeni's defense of Galileo shows at a level of fine detail how a group that had been aggressively resistant could shift its position into reverse, combining both vague appeals to "the noble and learned" with more specific appeals to new arguments from Kepler. The Roffenis, Heydons, Origanuses, and Maginis—all practicing astrologers—were beginning to engage issues of the Copernican planetary order in a way that had not happened in the period of the novas and comets. Planetary order and astrology could be seen to be connected because the Galileo novelties were recurrent.

GALILEO AND KEPLER
THE DENOUEMENT

In late July 1610, the tension between the two Copernican advocates was rapidly approaching a moment of culmination. Several considerations were converging, not the least of which was that Galileo had neither acknowledged Kepler's support for his cause nor sent him a telescope. Beyond this glaring contentiousness were further unknowns. First was the future yield of the instrument. Galileo had no way to be sure that he would be able to produce new discoveries within months of the publication of the *Nuncius*. He might find nothing more for the rest of his life. Yet on July 30, Galileo was lucky: he informed Vinta that, indeed, he had found "another most extravagant marvel": Saturn was not one star but a composite of three.[123] This was certainly a dramatic astronomical observation in its own right and presumably good material for a supposedly fast-track client; but Galileo made no effort to use this latest novelty to gain public credit—either as a naming opportunity for the Medici family in Florence or even for the regent Marie de'Medici, wife of the recently assassinated Henri IV (d. May 14, 1610). In fact, Galileo was well aware that the French king had wished to have a "beautiful star" named after him.[124] But he told Vinta to keep the discovery secret until he had a chance to reprint the *Nuncius*. This reluctance to publish fits the pattern of Galileo's practices before 1610 but stands in contrast to his haste to put his first telescopic discoveries into the public realm.

The explanation for this newfound caution was tied to a recent development: Galileo was beginning to acknowledge to himself the dangers of the Horky-Magini-Capra-Zugmesser network that he had previously underestimated. Furthermore, as Galileo learned from Martin Hasdale, his chief court confidant in Prague, Matthaeus Welser was now circulating Horky's *Peregrinatio*. And Hasdale, or Asdalio, now framed the difficulties even more ominously: "For reasons of state, the Spaniards judge it to be necessary to suppress your Highness's book since [they claim that] it harms religion and under its mantle all kinds of villainies are permitted to occur to the monarchy. Only the servants and partisans of this league plot against Your Highness." Horky's little pamphlet was beginning to take on unexpected political meanings in Prague: it served Welser's propapal interests as a member of the Spanish party at the court, while Asdalio, an ally of Sarpi, was seen to be associated with the antipapal "Venetians."[125]

Whether or not he liked it, therefore, Galileo's political image in Prague had a Venetian cast. The title page of the *Nuncius* reinforced this representation of him as a professor at Padua, even if the reality was that the Venetians themselves were unhappy with Galileo for leaving the republic. And Galileo now was developing an acute sense of how easily his observations and the meanings that he ascribed to them could be distorted, politicized, and pirated. These circumstances beyond his control may help to explain his hesitation in issuing an *avviso* for Saturn; instead, he concealed it within an anagram ("Smaismrmilmepoetaleumibunenugttaurias"), which he sent to Prague sometime in mid-August 1610.[126] Although this gesture was understandable, it was also a tactless maneuver since, once again, it failed to signal his full trust in Kepler.

Even before receiving the anagram, Kepler had decided to try again to move Galileo. On August 9 he expressed a number of frank concerns.[127] Foremost was the need for one of Galileo's instruments. The viewing devices (*ocularia*) in Prague were weak—at best they magnified by two or three times. He was just able to make out some stars in the Milky Way and some spots on the Moon. But, having only just received a copy of

Horky's *Peregrinatio* through Matthaeus Welser ("as I write this letter"), Kepler underscored that the absence of an adequate instrument compromised Galileo's position in the Horky matter and Kepler's consequent ability to defend him.[128] Horky's low credibility was not in question: he was obviously an "impetuous adolescent," an actor given to excessive stage display. But Kepler implied that Galileo had done little to deflect the Bohemian's influence. With uncharacteristic sharpness, he ascribed the Horky matter to envious and aggressive Italians "who directed this stranger's work, in a manner to exact vengeance on my German *Dissertatio*." Later, he again evoked the national theme in a veiled reference to Lorenzini and Galileo: "Is it not astonishing that the professors of the universities in that country [Italy] indiscriminately oppose themselves to the discovery of novelty? this country where the thing that is most witnessed and best known among all astronomers—I mean parallaxes—find adversaries who occupy very eminent positions and enjoy the highest scientific reputation? For I do not wish to conceal from you that there have arrived in Prague the letters of several Italians who deny having seen these planets with the looking glass [*perspicillo*]."[129]

In Kepler's view, Galileo had permitted the question of establishing the existence of his celestial novelties to become a "struggle of virtue and vice"—in Kepler's language, a "juridical" problem and not a "philosophical" one.[130] In the *Dissertatio*, Kepler had defended Galileo's discoveries on the moral and social grounds that he had no reason to deceive. And he now reminded Galileo that "there are so many people who would prefer to believe in your dishonesty rather than in the discovery of a new thing."[131] Meanwhile, there was the danger that, by tossing off a few ill-conceived remarks about optical reflections, Horky could sway ordinary people with little knowledge of such matters. The solution, to Kepler, was evident: "I beg of you, Galileo, to produce several witnesses as soon as you can. For from the letters that you have sent to different people, I understand that you do not lack witnesses; but, to defend the reputation of my *Letter* [i.e., the *Dissertatio*], I cannot produce anyone, except you who proclaim [that you have witnesses]. All the authority of observation remains with you alone."[132] Although the testimony of witnesses would presumably transform a moral proof ("he has no reason to deceive") into one that was at least freer of such human qualities, Kepler went no further than to recommend an increase in the *number* of testimonials.

Kepler's letter finally had its desired effect. On August 19, 1610, Galileo broke his thirteen-year silence. How would he explain himself? How would he deal with Kepler's Copernican arguments, comparisons with Bruno and Bruce, the failure to send a telescope, the question of witnesses, and so on? Kepler did not have to read far. Galileo acknowledged receipt only of Kepler's last two letters (late April and early August) but complimented Kepler for being "the first and almost the only one" to support him. That was the extent of Galileo's expressed gratitude. Indeed, he avoided singling out for praise any of Kepler's specific arguments, instead paying general homage to his moral and intellectual superiority and promising to reply soon in a second edition of the *Nuncius* (which, of course, never appeared). With considerable plausibility, Isabelle Pantin judges that Galileo acted as though the *Dissertatio* was a kind of "heavenly favor," an unexpected gift that was "largely merited and which he had the right to exploit to his liking but which had not cost him great effort and which ought not to inspire from him any excessive gratitude."[133] In short, it was an unrequited gift.

Explaining why Galileo felt no apparent obligation to reciprocate—even where his self-interest was so evidently at stake—is a difficult and intriguing problem. But perhaps the most crucial observation to be made is that although Galileo's environment provided more than enough good reasons for a high degree of personal suspicion, his leeriness of Kepler was part of a longstanding pattern of unwarranted suspicion. Nothing that Kepler did to support and reassure his brother in arms was ever enough; Galileo always felt wary and, apparently, fear of being dominated. In reality, although there were profound differences in their approach to the science of the stars, they also shared highly significant areas of intellectual agreement. One of these was Kepler's need for a good telescope and for the testimony of viable witnesses in order to provide stronger "philosophical" confirmation of Galileo's discoveries.

In the letter of August 19, Galileo acknowledged Kepler's wish for one of his telescopes, but he quickly provided further excuses for delay. He described the manufacture of the instruments as exceedingly painstaking (*ualdè laboriosa*). He also explained that he did not wish to make lens-grinding and polishing machines in Padua, as it would be impossible to transport them to Florence; but once he was settled there, "I shall send them to my friends."[134]

However, on the matter of how Kepler might further anchor Galileo's credit, there was more than just a vague promise. Here the issue was not specificity but rather what should count as an adequate and persuasive resource of credit.

> My very dear Kepler, you ask for other witnesses. I produce the grand duke of Tuscany who, having frequently observed the Medicean Planets with me at Pisa for a few months, gave me on my departure a gift of more than one thousand ducats, and who is about to call me to my country with the same annual salary of one thousand ducats and with the title of Philosopher and Mathematician to His Highness without imposing on me any burden; and moreover, in offering me the most tranquil leisure so that I might finish my works on mechanics, the system of the world, as well as local motion—natural as well as violent—which I demonstrate geometrically from very many unknown and admirable laws [*sinthemata*].[135]

Galileo seemed to believe that Kepler would be impressed by this display of court largesse—the grand duke's witnessing, the salary, the title, and the support for publishing time. And no doubt Kepler was affected. The grand duke of Tuscany was certainly a worthy reference. But is his testimonial sufficient evidence of a widespread *structure* of noble accreditation—or, better, a clear display of state power—that Kepler would have understood? Certainly it was not the approach taken by Kepler himself in the *Dissertatio*, where the emperor's authority in matters of adjudicating heavenly knowledge was never accorded any unusual status. Nor was Galileo's exposition of courtly rewards a display of *sprezzatura*, the studied nonchalance of one courtier engaging with another, or an instance of the patron's honor being kept at a distance from a client's assertions of truth. If anything, it was a display of braggadocio and domination—the kind of relationship in which Galileo felt most secure and comfortable—indeed, exactly the narrow terms which he could tolerate with his German colleague. Galileo was letting Kepler know, in the most vulgarly explicit, material terms, what he had received from the Medici, all of it private material that came directly from his letter to Vinta of May 7, 1610.

Was Kepler, who was already convinced of Galileo's claims, supposed to use this information publicly to persuade others, like Horky, Magini, and Zugmesser? These men had already read the *Nuncius* with deep skepticism and well understood the association between the Medici and the Jovian planets. Was knowledge of Galileo's salary and title supposed to compel them to change their minds? To Kepler's disappointment, Galileo could produce as additional witnesses only Giulio de'Medici, brother of the ambassador to Prague, and unnamed others: "My dear Kepler, at Pisa, at Florence, at Bologna, at Venice, at Padua there are very many people who have seen, [but] all hesitate and keep quiet; in fact, most of them do not know [enough] to recognize either Jupiter or Venus as a planet, and it is with difficulty that they [manage to reach] the Moon."[136] On this paltry list—reminiscent of the list of "Copernicans" to which he alluded in 1615—Galileo had not even mentioned the several Roman cardinals to whom he had sent his instrument.

What was at stake for Kepler were not Galileo's social status and salary but the public adequacy of his claims about the heavens. The Galilean representations fitted Kepler's cosmographical project: he could continue to defend them along the lines of the *Dissertatio*. The anti-Galilean network, however, appeared to demand a different sort of response. And because Galileo had not produced a telescope in timely fashion, Kepler now took it upon himself to find a good one from another source. At the end of August, just as Kepler received the letter of August 19, the elector of Cologne passed through Prague and lent Kepler the very instrument earlier sent to him by Galileo. Consequently, in just over one week (from August 30 to September 8), Kepler was able to observe what he now called for the first time the "satellites" of Jupiter, and he was careful to do so with the testimony of various named and carefully described witnesses. Presumably, these were the kinds of testimonials that Kepler had expected from Galileo. The first was Benjamin Ursinus (Behr, 1587–1633), "a diligent student of astronomy who, from the start, because he loves this art and has decided to practice philosophizing in it, never dreams of ruining the credit necessary to a future astronomer by false wit-

ness." But there was more to Ursinus's reliability than concern for his future reputation. Kepler explained: "We adopted the following method: with a piece of chalk and out of sight of the other, each of us drew on a wall what he had been able to observe; afterward, each of us went at the same time to see the other's picture to see if it was in agreement. This [method] is also to be understood for the following [observations]."[137] Perhaps the procedure owed something to Tycho Brahe's method of having two assistants observe the same star simultaneously through different slits in the alidade. In any case, this method of anchoring the empirical by controlling for error is evidently what Kepler had meant earlier by a "philosophical" approach.

From August 30 to September 5, Benjamin Ursinus was Kepler's principal co-witness. For the remaining three days, Kepler mentioned three other witnesses: Thomas Seggett, "an Englishman already known by his books and correspondence with famous men and who therefore takes care of his reputation"; Franz Tengnagel, secret councillor to Archduke Leopold; and Tobias Scultetus, imperial councillor in Silesia.[138] If reputation—whether earned or ascribed by title—was supposed to guarantee against deception, it was the independently drawn chalk pictures that decisively assisted epistemic conviction. And because the phenomena were allegedly recurrent, other observers with comparably good instruments could seek to emulate the scene of observation that Kepler had described.

In typical Keplerian fashion, he was ready for the press within days. On September 11, 1610, his Frankfurt publisher Zacharias Palthenius had in hand his *Report on the Observations of the Four Satellites of Jupiter*, which appeared in mid-October of the same year, adjoined to the *Dissertatio*, and with a 1611 imprint.[139] All these details were described in the book. The use of the word *satellites*, a Keplerian neologism, had the astrological advantage of addressing—if not effectively wiping out—Magini's worry that the Jovian circulators would upset his ephemerides. Thus, by late October, Kepler had decisively moved his defense of Galileo's novelties from the status of juridical to astronomically witnessed objects. Galileo's novelties now traveled in tandem with Kepler's representation of a new scene of witnessing. Once again, however, it was hardly enough for Galileo. He soon heard of Kepler's latest efforts but still hoped most eagerly for him to demolish Horky in print.[140]

SCOTTISH SCIENTIFIC DIPLOMACY
JOHN WEDDERBURN'S *CONFUTATIO*

Just as Kepler's latest work was appearing in mid-October, another counterattack against Horky was launched. Surprisingly, it involved a different sort of perturbation, as it occurred outside the private correspondence networks running through Bologna, Florence, Padua, and Prague. The new intervention involved a Scottish student in Padua, John Wedderburn (Joannes Wedderburnius, Scottobritannicus). Wedderburn's encounter with Horky came in a dispute between two foreign students who had both benefited from the intellectual and national diversity of Padua. His choice of format—a "refutation" (*confutatio*) of Horky's "four problems against the *Nuncius Sidereus*"—shows that he was allowing Horky's organization to configure his own approach. At another level, it mimicked the range of tensions and considerations that marked the divide between Galileo and the traditionalists. Yet although Wedderburn's little treatise was structured by Horky's "problems," his answers were decidedly conditioned by Kepler's *Dissertatio.* A crucial reason is that, like the Kepler of May 1610, Wedderburn did not have a telescope. Hence he explicitly used Kepler's eight criteria for advocating trust in Galileo's observations without an instrument.[141]

For this reason, Wedderburn cannot just be considered a "follower" of Galileo. Kepler provided the grid through which he read. For example, he showed himself to be no sympathizer of Giordano Bruno and Edmund Bruce. Thus Wedderburn was something of a budding modernizer who signaled his clear allegiance to both Kepler and Galileo and fittingly dedicated his *Refutation* to King James's Venetian ambassador, Henry Wotton. Wedderburn's associations with Wotton must have been close, as his language precisely echoed the latter's early reaction to the *Nuncius* in portraying that work as "designed to overthrow the first principles and the skilled practice of astrology."[142]

Once again, print enabled secondary figures who had not initiated major ideas to perturb the debate in unexpected ways. Not only did Wedderburn introduce his arguments outside the compass of Magini's strategic summer withdrawal, but, equally important, because he was unaware of Kepler and Galileo's personal tensions as well as Kepler's *Narratio,* he confidently and unknowingly brought onstage the earlier arguments of 1610, joined in a common struggle against Horky. Moreover, he stated that he knew of no one at all who had defended Galileo, least of all "most famous mathematicians": "Clavius was silent, Magini abstained, others deferred."[143] As far as Wedderburn knew, Horky and Kepler were the only ones to react to the *Nuncius.*

Wedderburn's selective approach also reveals another dimension of the manner in which the *Nuncius* traveled: it echoed the Wittenberg interpretation. Practical astronomy could always be treated apart from theoretical issues. Wedderburn tried to develop Kepler's pretelescopic arguments while still managing to keep separate all Keplerian efforts to link the existence of the Jovian planets to the Copernican arrangement. Against Horky, he claimed that the existence of the four Medicean planets would not disrupt the accuracy of calculations based on the seven-planet ephemerides, as astronomy could infer true conclusions even from the most absurd suppositions.[144] He defended Galileo's priority in discovering the "new planets" not by confirming their existence through his own observations but by denying that Galileo could have deduced their existence from the "old wives' tales" of Giordano Bruno and Edmund Bruce.[145] Galileo could not have been entirely satisfied by this Keplerian defense of his discoveries. Nor could he have been amused when Kepler remarked to him: "I have seen Wodderborn's *Confutatio*: It is gratifying."[146] After the *Narratio de Satellitibus,* Kepler gradually gave up his long efforts to establish an alliance with Galileo.

GALILEO'S NOVELTIES AND THE JESUITS

Against this jumble of contested and often uncoordinated engagements with Galileo's celestial representations throughout 1610, it is difficult to say how much was known (and exactly when) by Clavius and his disciples at the Collegio Romano.[147] Although Clavius enjoyed a considerable correspondence with Magini, there is no direct evidence that he was aware of the Horky controversy. His relationship with Galileo went back more than twenty years, to 1588, but their correspondence was sporadic. The source of Clavius's earliest information about Galileo's discoveries came from Augsburg, through the humanist printer and wealthy banker Marcus Welser; and it was sent at virtually the moment the *Nuncius* was published in Florence.[148] This development shows that Welser maintained excellent contacts with Italy (going back to the days of the Pinelli circle) and that, like some other learned nonpractitioners, he knew of Galileo's claims prior to publication. But he may have drawn away from Galileo politically after the Venetian Interdict of 1607. Rather than write directly to Galileo, Welser asked for Clavius's judgment about the existence of the alleged celestial novelties. Therefore, one may be confident that Clavius had obtained the *Nuncius* sometime thereafter. Six months later, a letter of mid-September 1610 from Rome mentions that he had acquired a copy of Kepler's *Dissertatio*. Because of Clavius's authoritative position, this means that other Jesuits, like Cardinal Bellarmine, could have been made aware of the Galileo-Bruno associations that Kepler had drawn.[149] Just a few days later, Clavius suddenly heard from Galileo himself, apologizing for the long break in their correspondence and informing him of his recent appointment by the grand duke but, notably, without the details of salary and other benefits that he had thrust at Kepler a month earlier.[150]

It is interesting that Galileo chose this moment to break the long silence with Clavius. After Kepler's letter of early August, it is likely that he was motivated to garner witnesses. Also, Galileo knew that there had been earlier Jesuit efforts to observe. He specifically mentioned hearing that Clavius "and one of your brothers" had an *occhiale* with which they had been trying unsuccessfully to view Jupiter's stars. Galileo did not promise a new *occhiale* but acknowledged that such observing difficulties were "no great surprise to me" and proffered practical advice on mounting and deploying the instrument. He also mentioned that he had not done much work on Jupiter since publishing his first observations, although he had continued to perfect the instrument. Unlike his more heavy-handed approach to Kepler, his tone with Clavius was measured; he mentioned no witnesses and referred only to his own observing experience. All this points to a cautious effort to gain support from Clavius and his

circle. Soon after, however, he heard a rumor from the artist Ludovico Cigoli in Rome that must have caused him to worry that the Horky faction had gotten to Clavius.[151] But Galileo's letter, pointing out some possible causes of observing difficulties, could well have provided a counterbalance to those concerns.

This convergence of circumstances leading up to and including Galileo's letter of mid-September 1610 helps to explain why Clavius and his disciples did not veer off into the polemics that were consuming the energies of the highly capable Italian and Hapsburg practitioners. Moreover, the Jesuit rejection of practical astrology may well have reduced anxieties, like those suffered by Magini, about a possible threat to the science of the stars. Indeed, if the Jesuit mathematicians had encountered Kepler's witnessed observations of mid-October, these could only have encouraged them to persevere. For by late November, the *Clavisi* had begun a renewed—and ultimately successful—campaign to plot the positions of Jupiter's moons.[152] In January 1611, Clavius acknowledged to Marcus Welser that he no longer believed the Jovian observations to be illusions created by the instrument. Recurrent observations of what he now took to be the same objects had convinced him to change his mind. In fact, he was prepared to go further: "I believe that gradually other remarkable things about the planets will be discovered."[153]

By the time that Galileo arrived in Rome in the spring of 1611 for a highly gratifying *festa* with the mathematicians of the Collegio Romano, the Jesuits had also accepted the existence of phases of Venus and that Saturn's appearance was an "oblong and ovate" cluster of three contiguous bodies.[154] Odo Van Maelcote's lengthy and detailed eulogy adopted the language of Galileo's own conceit of the "messenger" (*nuncius*), following quite closely the text of the *Sidereus Nuncius* and acknowledging, but without direct comment, Galileo's conviction in the heliocentric ordering.[155] And, sometime during this same visit, Galileo further stimulated the *Clavisi* by showing Van Maelcote "spots that changed position and order" with respect to [*sub*] the surface of the sun.[156] As the new year got under way, these recurring novelties were firmly joining the evanescent novas and comets of 1572–1604 in the realm of ordinary phenomena.

As long as the question of novelties could be

classified as a bounded observational problem, confined entirely to the existence of the phenomena, it could be quarantined away from the modernizers' concern with the true constitution of the universe. Even John Wedderburn had (barely) managed to keep those issues separate. But once the Jesuit mathematicians admitted singular and recurrent novelties into the domain of the ordinary, they found it more and more difficult to resist acknowledging the question of how the universe made sense as an ordered whole.

They were pressed on different flanks. On one side were traditionalists in natural philosophy, like Cardinal Bellarmine, who was willing to allow holy scripture—but not astronomers' novelties—to modify the claims of natural philosophy.[157] In a sense, the question for Bellarmine was the same as for Horky and Wedderburn: were the phenomena real or not? If they were, then Bellarmine would be pleased to confine them to the domain of traditional practical astronomy, as he sharply reminded Brother Foscarini in his famous letter of 1615.[158]

On the other side were mathematical modernizers like Galileo and Kepler, as well as proponents of the *via media,* such as Brahe, who, in their different ways, believed that accommodation of novelties could be achieved only by reordering the planets and hence by making them part of celestial natural philosophy. For Clavius, accommodation meant admitting new entities not just into the world but also into the thirteenth-century text to which he was deeply committed as the basis for teaching spherics. The belief that the same text could always be used to teach the right constitution of the world—save for a few "digressions"—had its obvious, albeit imperfect, parallel with biblical commentary. Yet, in a famous passage from the very last edition of his *Sphere,* Clavius included a short but astonishing reference to Galileo's discoveries—the Moon's uneven surface, Venus's sometimes-crescent shape, a star joined to either side of Saturn, four stars roving along with Jupiter—while also explicitly crediting Galileo for having described these phenomena "carefully and accurately" in the *Nuncius.* He then concluded: "Because these things are so, let astronomers consider how the celestial orbs may be arranged in order to save the phenomena."[159]

Because Clavius died soon after writing these suggestive words and because he situated them on a separate line immediately following his endorsement of Galileo's descriptions, they were easy to notice and obviously carried great weight with his successors. But what did he intend? The formulation seems to contain the characteristic Jesuit combination of tradition and novelty. On the one hand, one should be open to what has been discovered with the telescope, and one should think about how to rearrange the planets to accommodate them; but the purpose of such rearranging should be restricted to saving the phenomena. That is, Clavius appeared to advocate not a change in what ought to count as the true arrangement of the universe but only the hypothetical arrangement that would best conserve traditional natural philosophy and the Ptolemaic practices contained in his last commentary on Sacrobosco.[160] In this sense, his position resembled the one stated just four years later by Cardinal Bellarmine.

By ignoring or underplaying the phrase "in order to save the phenomena," subsequent readers found a way to strengthen Clavius's meaning. Perhaps Clavius had deliberately left open this possibility. In the "digression" on how to decide between Copernicus and Ptolemy—still included in the 1611 edition of his *Sphere*—Clavius had urged that astronomers turn to natural philosophy and scripture when two hypotheses provide geometrically equivalent predictions of the same observations. Would that advice not apply as well to the Galilean novelties? Ironically, because Jesuit astronomy was not driven by astrological concerns, it was not resistant to the Galilean novelties in the way that had consumed Magini and his circle. Yet, in 1611, the space of possibilities had evolved: which natural philosophy was the right one? And if one read the introduction to Kepler's *Astronomia Nova,* then which standard was the right one for accommodating scripture to the claims of astronomers? Could Galileo's novelties be entertained apart from such questions?

When Galileo was honored with great fanfare by the mathematicians of the Collegio Romano in April 1611, Clavius was personally present, but, as far as we know, there was absolutely no mention of the consequences for celestial order. Yet how could one admit the existence of recurrent celestial novelties, like the phases of Venus, while retaining loyalty to Aristotle in natural philosophy and Ptolemy in theoretical astronomy? After

1611, these questions traveled together. Jesuit mathematicians did not avoid the new problematic created by Galileo's novelties. At the same time, most were not compelled by his increasingly public advocacy of a Copernican solution. Like the earlier Wittenbergers, the Jesuits selectively accommodated into their representations of the heavens whatever elements did not threaten to undermine traditionalist arguments for the Earth's centrality and stability. It is therefore not surprising that many eventually saw in Tycho Brahe's arrangement—or some modification of it—a viable alternative that Clavius had found insufficiently attractive in 1600.[161]

The Great Controversy

At the end of the seventeenth century, the European social order was still a world of privilege, faith, and tradition. The distribution of land, positions, wealth, and status favored churches, monarchies, princes, aristocrats, and pockets of rich merchants and bankers. Universities fitted comfortably into this order; they were hierarchically structured, exclusively male foundations supported by princely and ecclesiastical benefactors. The university, the monarchical or princely state, and the Church constituted the three pillars of cultural authority—the principal structures of these still-traditional societies.[1] Within these arrangements, representations of celestial order and the future were advanced by men who identified themselves variously as university teachers of mathematical subjects; court, municipal, or academic prognosticators; natural philosophers; almanac makers; physicians; learned aristocrats; high ecclesiastics; and doctors of divinity—but not yet scientists. As they pursued their prognosticatory objectives and representational constructions, these multifarious social types circulated mostly within but sometimes between their local social networks. They sought to persuade in a mix of religious and secular vocabularies. They maneuvered their arguments carefully between the astral influences of the pagan gods and the grave words of holy scripture, Church councils, and papal decrees; and they negotiated their positions within resistantly hierarchical disciplinary domains. Gradually, a small group of modernizers and modernizing

traditionalists of the middle way constructed the problem of alternative planetary arrangements as a new kind of controversy that involved weighing, balancing, and adjudicating among different sorts of evidence. Yet all such evidence was vulnerable to the objection that no single piece could decisively determine a final choice, and in some cases the same evidence supported opposing claims. In this sense, the Copernican question introduced a problem that became a notable feature of the physical and social sciences of later centuries. On the other hand, if many terms and categories in which the Copernican controversy was conducted embodied unmistakable elements of a scientific world that, in its instrumentalities, theoretical commitments and provisionality was beginning to look discernibly modern, other terms seem strange—or, not to lose focus, early modern—for all of the practitioners in this study, even those who broke most profoundly with the ancients, still regarded their projects as engagements with antiquity and divinity.[2] In sum, these tangled developments may be thought of as marking distinctive elements of the first phase of an early modern scientific movement, stretching from the late fifteenth to the early seventeenth centuries—in truth, a long sixteenth century.

This conclusion makes another pass through the narrative, bringing into focus larger themes and patterns, and then points forward, suggesting lines of organization for a periodization of the Copernican question in its later phases.

Copernicus's problematic involved several major areas of concern—planetary modeling, the ordering of the planets, the consequences of such ordering for natural philosophy, and the prediction of future configurations of the heavens and their influences. It is customary to inflect the first three; in this book, I have placed an uncustomary emphasis on the last. That is exactly where the Wittenberg reading cast its own emphasis, selectively constructing the author of *De Revolutionibus* as a "second Ptolemy" and using elements of Copernican planetary theory compatible with its own interests in astral prognostication while ignoring the eloquently framed humanist appeals to *symmetria* advanced both by Rheticus and his admired teacher. The practice of ignoring contravening arguments of the sort that Rheticus and Copernicus provided in the *Narratio Prima* continued to operate as an important practice for bounding off difficulties; it often functioned to maintain existing lines of disciplinary authority and was easily reinforced by the genera of the heavenly literature.

If the Wittenberg consensus marginalized planetary ordering in the persistent debate over prognosticatory credibility, it is striking that Copernicus and a handful of his followers succeeded at all in establishing that their version of the "Pythagorean opinion" was a legitimate alternative, one that could not be ignored even when most considered it to be false. Against such persistently overwhelming judgment, practitioners who opted to associate themselves with the central Copernican claims believed those claims not only to be predictively useful but to yield superior understanding by virtue of the unity and greater scope of their explanations. For the Copernicans, this expanded domain of understanding begat a more secure confidence in the theory's truth (and not merely its likelihood) and motivated the pursuit for explanations of singular novelties from the 1570s onward. That was as much the case for Digges and Rothmann as for the publicly cautious Maestlin and his remarkable student, Kepler.

Even in its initial stages, this nascent early modern scientific movement took a different course from the medieval academic practice of considering and rejecting hypothetical alternatives. Unlike theologically motivated fourteenth-century flirtations with the possibility of the Earth's daily rotation, Copernicus's revision of the planetary arrangement occurred at a juncture with the emergent fifteenth-century cultures of print and prognostication: the mobilization of print in the service of both the theoretical and the practical literature of forecast, the creation of new conditions of prognosticatory authorship, the appropriation of resources of humanist rhetoric and dialectic, and the upsurge in apocalyptic expectation. Different types of evidence also find a place in this accounting—values newly derived from Copernicus's models, without adopting his ordering (Reinhold's planetary tables); observations of unanticipated and nonrecurrent events (the celestial novelties of the 1570s onward); specifically targeted observations requiring very large instruments (Tycho Brahe's Mars parallax campaign); and observations of unanticipated, recurrent phenomena aided by novel instrumentation (Galileo's telescopic novelties). These various kinds of evidence persistently failed to shut down all the uncertainties among the competing theoretical alternatives. Indeed, two centrally defining characteristics of the Copernican question were the weakening authority of traditional stratagems of closure and the failure of even the most novel evidence and argumentation to determine decisively the choice between competing celestial representations. Coincidentally or not, this sense of fallibility about heavenly knowledge occurred at just that moment in European history when the fundamental criteria of religious faith had also been thrown into doubt, fragmenting the unity of Christendom into multiple, adversarial churches.[3]

The much-told story of an astronomical revolution that starts with Copernicus, autonomous and triumphant in its development, marginalized astrology as an inconvenient presence. This book has tried to reverse that picture: astrology and astronomy made up an interlinked complex of theoretical and practical subjects that connected the heavens to the social realm. Considering the four-cell classification of the science of the stars that had evolved by the end of the fifteenth century enables us to pick out some important distinctions and connections that are otherwise puzzling or invisible. For one thing, works like Koyré's *Astronomical Revolution* and Kuhn's *Copernican Revolution* can now be read as informed by a historiographical decision to restrict attention not just to astronomy but to the domain of *theoretical* astronomy.[4] For another, the unprecedented outpouring of printed annual prognostications in the final three decades of the fifteenth century helps to distinguish Renaissance astrology as wider in scope and more public than its medieval antecedents. The observation that Kepler and Galileo practiced astrology strictly as a means of earning a living has often seemed useful in defending an image of their scientific purity.[5] But this standard throwaway gesture ignores the prevailing classification of knowledge and writes off a crucial boundary dispute: the handful of practitioners who were trying to reform astronomical theory did so in the service of what they regarded as a better astrology. Kepler, for example, regarded the great majority of run-of-the-mill prognosticators as idiots—from his perspective, they were the ones who were in it only for the money. Their forecasts were failures because he regarded their theoretical principles—unlike his own—as incorrect. Explicit where Copernicus was not, Kepler defined his great projects as the reform of the theoretical principles underwriting practical astrology; indeed, Kepler's ambitious aim was to reform every part of the science of the stars—including theoretical astronomy, the part on which his current fame rests. While Kepler constructed his "new astronomy" as part of physics, he used convention to treat theoretical and practical astrology in separate genres.

Galileo, by contrast, seems to have held quite traditional views about astrology's theoretical foundations, and he showed no Keplerian inclination to reform that subject any more than he did planetary theory. Thus, although both Kepler and Galileo engaged in practical astrology, they pursued different kinds. After Kepler left Tübingen for Graz and Prague, he published annual prognostications generally resembling those of a hundred years before, whereas all of Galileo's surviving manuscripts show that he was casting nativities or horoscopes for private individuals—some of them for the money, but some clearly because he believed that he could learn something, as when he produced nativities for his two daughters, Livia and Virginia.[6] Of course, Galileo had no motivation to publish the individual horoscopes that he constructed, as they were intended solely for the private use of the individual in question.

But an even more telling consideration for Galileo was the political situation in Counter-Reformation Italy, especially after the bull of 1586, which was unfavorable to the kinds of public astrological practices in which Kepler regularly engaged in Prague. Galileo, unlike Kepler, never made forecasts for a city or a region; nor, did he publish annual nativities for the local ruler, like Pietro Avogario of Ferrara; and he never made any astrologically motivated judgments about the Medici dynasty. The one and only time that he ever made published reference to an individual nativity was in the famous preface to the *Sidereus Nuncius,* when he called attention to the important place of Jupiter in the horoscope of his sometime pupil Cosimo II, thereby publicizing to a general audience his personal association with the immediate Medici family.

COPERNICANS AND MASTER-DISCIPLE RELATIONS

The striking diversity of Copernicans during the period of the Wittenberg consensus has also occupied considerable attention in this book. That heterogeneity alone (*pace* the category *Copernicanism*) should be enough to complicate loose Kuhnian claims about radical meaning change and "bandwagon effects" between incommensurable scientific paradigms. But some general comment is necessary concerning the contexts in which practitioners decided for the Copernican arrangement. The available evidence, although incomplete, points to the space of the master-pupil relationship—perhaps not surprisingly. These relationships had their origin in the paternalistic structures of the family and the all-male cultures of the universities. As the cases of Wittenberg and Bologna show, it was mutually advantageous for students to room in professors' houses. Relations between older and younger men served both as a means for transmitting, conserving, and developing tradition and as an occasion for resisting and changing it. The *via antiqua* and the *via moderna* arose from a common source, like a pair of staircases ascending a building in opposite directions from a common starting point.

In this study, the model of master-student relationships seems to provide the essential setting for breaks and stresses with the *via antiqua* of various kinds and degrees. Such pairings are associated with numerous critical junctures in the

Copernican question: the productive revision of Ptolemy's models (Peurbach and Regiomontanus); the radical break with Ptolemy's planetary ordering (Copernicus and Domenico Maria Novara); Copernicus's collaboration with his first disciple, Rheticus; the break with the emerging Wittenberg consensus (Rheticus and Melanchthon); the failure to attract new disciples (Rheticus and Cardano); Thomas Digges's replacement of his father's traditional geocentric scheme with his infinitist interpretation of Copernicus's *De Revolutionibus;* the effect of Paul Wittich's influential reading of *De Revolutionibus* on Tycho Brahe; Tycho's complicated and ultimately unsuccessful attempt to control Kepler's theorizing; the cooperative relationship that enabled Kepler to publish the *Mysterium* and Maestlin to associate himself publicly with Copernicus; and the love-hate relationship between Galileo and Kepler.

If Kepler's vision was collaborative, Galileo's, like Tycho Brahe's, was hierarchical: he was more comfortable with disciples. The man who actually did the teaching at Pisa when Galileo acquired the title of chief mathematician and philosopher to the grand duke of Tuscany was his one-time student and Copernican sympathizer, the Benedictine Benedetto Castelli. Discipleship is the role structure of the later *Dialogue:* a conversation between a "teacher" (Salviati-Galileo), and two "students": Sagredo, smart, receptive, and aristocratic; Simplicio, dull, sometimes pedantic, often resistant, and academic. When Galileo put Pope Urban VIII's views on divine omnipotence and the fallibility of human knowledge (the so-called medicine of the end) into the mouth of Simplicio, was he acting impulsively? Or, as seems more likely, was he incautiously reenacting a longstanding pattern of dominance and control that, for the most part, had worked well for him in the past?

SEVENTEENTH-CENTURY THOUGHTS ABOUT BELIEF CHANGE

In the Kuhnian era, commitment and resistance to theories, conceptual schemes, research programs or paradigms, by whatever name, became a well-established topic of investigation, normalized as a subject for doctoral examinations and conference sessions. By contrast, the period of concern here had a literature of method and logic built largely on Aristotelian foundations, but certainly no self-conscious literature of scientific change. Yet, from time to time, actors voiced unsystematic, localized reflections that provide a glimpse into their metalevel understandings. Their explanations were less purely logical than social and psychological, and in that sense they curiously anticipate Kuhn's more rigorous, historically sourced formulations. For example, in explaining his own willingness in 1615 to entertain the newly revived and unfamiliar "Pythagorean opinion," the Carmelite theologian Paolo Antonio Foscarini (1565–1616) adduced what he called variously the force of custom, usage or habit: "Once a custom [*consuetudine*] is established and men are hardened into opinions which are trite and plausible, and which are part of everyone's common sense, then both the educated and the uneducated embrace them and are hardly able to be dislodged from them. The force of habit [*abito*] is so great that it is said to be another nature. Thus it happens that something with which one is familiar, even if it be an evil, becomes more loved and desired than a good which is unfamiliar." When opinions harden into a disposition of the mind, according to Foscarini, they become authoritative in their own right; effectively, authority is just the habitual made invisible: "As soon as they [opinions] have firmly established their roots in the mind [*animo*], any opinion which is different from the customary one seems as a result to be like disharmony to hearing, as darkness to vision, as a stench to smell, as bitterness to taste, and as roughness to touch. For ordinarily we do not weigh and judge a thing according to what it is itself, but according to the decree of an authority which remains unmentioned."[7]

Bold and incisive words from the Carmelite. Indeed, they could easily have been applied to an explanation of religious belief. To escape the weight of authority requires us only to see it for what it is: "Since this authority is only human, it should not be held in such importance that it causes us to condemn, to reject, or to put aside what is evidently true to the contrary, whether this be shown accidentally by some better proof not previously noticed or occasionally by sensation itself. The road to the future should not be closed so that our descendants are neither able nor venturesome enough to discover more and better things than the ancients handed on to us."[8]

If Aristotle and the other ancients were just human, then the moderns were at least their equals and, if truth be told, their betters: "Could it not be said . . . that the experiences [trials, *isperienze*] of the moderns have on some particular issues closed the venerable mouths of the ancients and have established that some of their most important and solemn teachings are empty and false?"[9] It is not difficult to see how Cardinal Bellarmine could have read as threatening such a challenge which, in vigorously contesting the authority of tradition, made not even a single reference to the Church fathers; equally disturbing were the specific arguments purporting to reconcile scripture with the Copernican theory.[10] More difficult to understand is Pope John Paul II's complete neglect of Foscarini in his 1992 address to the Pontifical Academy of Sciences, which aspired but failed to bring closure to the Galileo affair.[11]

Another notable reflection, sixty years after Foscarini, is to be found in a public lecture by Robert Hooke, the famed curator of the Royal Society, Gresham College lecturer, and experimental assistant to Robert Boyle.[12] Hooke acknowledged that the Copernican argument from the symmetry of the universe might not seem sufficient to persuade geocentrists. How could one explain such resistance, even among learned geometers, astronomers, and philosophers?

> Most of those, when young, have been imbued with principles as gross and rude as those of the Vulgar, especially as to the frame and fabrick of the World, which leave so deep an impression on fancy that they are not without great pain and trouble obliterated: Others as a further confirmation in their childish opinion, have been instructed in the *Ptolomaick* or *Tichonick* System, and by the authority of their Tutors, over-awed into a belief, if not a veneration thereof: Whence for the most part such persons will not indure to hear Arguments against it, and if they do, 'tis only to find answers to confute them.

Hooke, like Foscarini, ascribed the formation of early beliefs to personal temperament and authority, but now also to the space of the pedagogical relationship. He noticed that some managed to resist what they were taught even though they did not possess absolute demonstrations: "On the other side, some out of a contradicting nature to their Tutors; others by a great prejudice of institution; and some few others upon better reasoned grounds, from the proportion and harmony of the World, cannot but imbrace the *Copernican*

Arguments, as demonstrations that the Earth moves, and that the Sun and Stars stand still."[13]

These illuminating contemporary observations resonate well with the theme of the master-disciple pairing as the locus of intellectual formation and rupture. Temperament and reason were conjointly implicated in a narrative of belief change. Personal identification with the teacher and his convictions set up the conditions for subsequent approval or rejection. Shortly, we shall see that Hooke had his own card to play in the Copernican question.

THE END OF THE LONG SIXTEENTH CENTURY

If we organize a long sixteenth century around the Copernican question, then it began in the 1490s with Copernicus's response to the crisis triggered by Pico's attack on the science of the stars, a body of knowledge whose practitioners bridged the Church, princely courts, and universities; and it appears to have ended somewhat abruptly in the mid-1610s, as higher ecclesiastical elements in Rome and their allies in the universities came to regard modernizers, both in the Church and the courts, as a significant threat to their authority to control the representation of celestial order. The issue of protecting the boundary between safe (divine) and dangerous (natural) foreknowledge, as embodied in the papal bull of 1586, did not disappear—indeed Urban VIII's bull *Inscrutabilis* (1631) reaffirmed the prohibition against predicting the death of a pope—but the question of celestial order was quickly emerging as a threat of comparable concern.[14] In both cases—astrological foreknowledge and celestial order—skepticism proved to be the theologians' epistemic stance of first resort; secret censures, bulls, injunctions, book prohibitions and corrections, incarcerations, and the occasional trial were among the preferred means of enforcing discipline.

Signs of the emergent shift were already evident in the prohibitions that the Index slapped on all of Bruno's writings in 1603. Another was the rehabilitation of Tolosani's mid-sixteenth-century criticisms of Copernicus by the Dominican Tommaso Caccini in a sermon at Santa Maria Novella in Florence in 1614.[15] And perhaps the best-known sign of difficulty is the argument over scripture and the Copernican arrangement that began at the Tuscan court in 1613.[16] It was this event that prompted Galileo to write down his preliminary thoughts on the status of the Bible's utterances concerning the Earth's motion, later expanded and published as the famous *Letter to the Grand Duchess Christina*.[17] Cardinal Bellarmine's now equally famous letter of April 12, 1615, to Foscarini succinctly formulated the emergent position:

> To say that the assumption that the Earth moves and the Sun stands still saves all the appearances better than do eccentrics and epicycles is to speak well, and *contains nothing dangerous [non ha pericolo nessuno]*. But to wish to assert that the Sun is really [*realmente*] located in the center of the world and revolves only on itself without moving from east to west, and that the Earth is located in the third heaven and revolves with great speed around the Sun, *is a very dangerous thing [è molto pericoloso]*, not only because it irritates all the philosophers and scholastic theologians, but also because it is damaging to the holy faith by making the holy scriptures false.[18]

Talk of supposition was not threatening, as it bespoke astronomy's limited epistemic capability; but talk of "what really exists" was very dangerous because it intruded on the authority of theology and natural philosophy. It is interesting that Bellarmine, ever the scourge of Protestants, did not locate the danger in any Wittenberg writings. (The Melanchthonians, after all, had not postulated a moving Earth in order to save the appearances).[19] The worry in Rome was, at first, about dissent or deviation within the Church. Cardinal Bellarmine took his stance in reaction to a reading of Clavius's final (and much-noticed) statement in the 1611 edition of his *Sphere* commentary.[20] It was this statement that Foscarini freely paraphrased (and modified) to his own purposes, using it to justify an investigation that breezily referenced Galileo and Kepler while omitting any reference to the Church fathers: "Father Clavius, a most learned man . . . concedes that, in order to alleviate the many difficulties which the common system does not fully resolve, astronomers are forced to try to provide some other system, which he exhorts them to do with strong encouragement."[21] Bellarmine, for his part, did not think that Clavius had meant to encourage so aggressively the sort of enterprise which Foscarini incautiously described in the fol-

lowing terms: "The opinion of Pythagoras and Copernicus is not contrary to the principles of astronomy and cosmography, but rather it has no small degree of probability and likelihood. It is much better than the many other opinions which challenge the common system but are only delirious searchings. . . . The Pythagorean opinion surpasses all of these as easier, more accommodated to all the phenomena, and more useful in calculating the motions of the celestial bodies with a fixed rule and without any epicycles, eccentrics, deferents, or swift motions."[22]

According to Foscarini, this revived ancient opinion derived its "degree of probability" not only from its consistency with theoretical principles but also from a consensus among the moderns Galileo, Kepler, and the members of the Accademia dei Lincei. Such an opinion was not merely worthy of the Church's consideration, but it was Foscarini's purpose, "speaking for the profession of which I am a member," to show that it was not inconsistent with a considerable number of scriptural passages.[23] And even after Bellarmine's warning to treat the heliocentric theory hypothetically (*ex suppositione*), Foscarini did not drop his efforts to convince his own church of the Earth's diurnal motion from the existence of easterly trade winds, "that perhaps will have not a little force of demonstration and of necessary argument."[24]

Foscarini's work was, of course, hardly the first to accommodate scriptural authority to the Copernican ordering, but unlike those of his predecessors Bruno, Rothmann, Wright, Zuñiga, and Kepler, his presentation was systematic and thorough. Bellarmine, who certainly could not have forgotten that Bruno's infinitist-Copernican writings had been prohibited thirteen years earlier, precisely echoed the Brunonian language in representing Foscarini's position as an opposition between the mathematician's "suppositions" and the physicist's "demonstrations."[25] There was to be no talk of probabilities in relation to the certitudes of scripture. Around 1615–16, the mood in Rome had changed quickly and in a way that had not happened elsewhere in Europe. Bellarmine, strongly committed to a literalist reading and a literalist standard of interpretation (although well aware of criteria of exegetical latitude), now showed his readiness to make the Earth's motion not only a "matter of faith" but also one that evoked the Council of Trent, "based upon the au-

CONCLUSION

thority of the speaker [*ex parte dicentis*]": "Anyone who would say that Abraham did not have two sons and Jacob twelve would be just as much of a heretic as someone who would say that Christ was not born of a virgin, for the Holy Spirit has said both of these things through the mouths of the Prophets and the Apostles."[26]

Bellarmine's carefully considered but unmistakably heavy response to Foscarini, a fellow theologian and member of a minor order, was just one of several gestures that issued from the Church's highest levels in 1616. The "Consultants' Report" of February 24, 1616, cast its judgment in a form and a language that one would expect from a juridico-theological body of that moment—beholden in its qualitative categories and assumptions to traditional natural philosophy and canon law rather than to the conditional logic and comparative probability of the major claim in *De Revolutionibus*.

Propositions to be assessed:

(1) The Sun is the center of the world and completely devoid of local motion.

Assessment: All said that this proposition is foolish and absurd in philosophy, and formally heretical since it explicitly contradicts in many places the sense of Holy Scripture, according to the literal meaning of the words and according to the common interpretations and understanding of the Holy Fathers and the doctors of theology.

(2) The Earth is not the center of the world, nor motionless, but it moves as a whole and also with diurnal motion.

Assessment: All said that this proposition receives the same judgment in philosophy and that in regard to theological truth it is at least erroneous in faith.[27]

Like Tolosani's judgment of 1546, the consultants moved the question of celestial order out of the hands of enthusiasts for new discoveries in the mixed mathematical sciences and into their own zone of disciplinary comfort: traditional natural philosophy and theology. Their formulation left no room for compromise.[28] Rome took two further actions in 1616: one, a formal warning to Galileo, ordered by Pope Paul V and transmitted by Cardinal Bellarmine, to desist altogether from defending or holding the Copernican opinion;[29] the other, acts of censorship in March by the Holy

Congregation of the Index, completely prohibiting Foscarini's *Lettera* and putting Diego de Zuñiga's *Commentary on Job* and Copernicus's *De Revolutionibus* into the milder category of "prohibited until corrected."[30] Unlike the censorship, however, the warning to Galileo was secret, unknown even to his important friend Cardinal Maffeo Barberini, the future Pope Urban VIII.[31]

The Copernican question was now clearly taken to be of significance at the very highest levels of the Church. The Index published its list of specific corrections of *De Revolutionibus* only in 1620; yet, because it was not merely the author who was to be censored, but the work itself, the Index had to decide which passages to edit. In other words, someone actually had to read the book—but not with the favorable consequences for which Galileo had hoped. It was also necessary to specify differences among at least the first and second editions—not to mention Nicholas Mulerius's just-published third (1617). The 1620 Index even overlooked that the second edition included the *Narratio Prima*, with its comment that the "monks" might well declare the work to be "heretical."[32] In the end, the actual censoring was left to the moral and legal obedience of individual readers. Of some six hundred extant copies in Owen Gingerich's unprecedented survey, about 8 percent bear the marks of the 1620 censorship.[33] However, because readers themselves were supposed to carry out the acts of correction on their own books, their relationship to the text allowed a certain degree of control, permitting them first to read (and think about) the offending passages and then, by an act of studied obligation, to cross them out. Or not.

The tremendous attention that has been focused on the Galileo affair overshadows the fact that the Church's move to put Copernicus on the Index came relatively late: seventy-three years after the first edition of *De Revolutionibus*, fifty years after the second, and twenty years after Tycho Brahe's debate with Christopher Rothmann in the *Epistolae Astronomicae*. It represented a shift from what had been regarded throughout much of the sixteenth century as a legitimate resource for obtaining foreknowledge to a threatening kind of philosophy that purported to refer to the actual structure of the heavens. Thus, within a relatively short period, Copernicus's hypothesis was transformed from a resource of prognostication and a matter of philosophical debate into a question of uniformity and obedience.

Looking back just a few years before these ominous developments, it is notable that Church traditionalists did not overreact to the invocation of natural explanations for the nova of 1604; nor did the Protestant Kepler's *Astronomia Nova* cause anything faintly resembling the stir caused by Galileo's announcement of his telescopic discoveries and his move to the Tuscan court. Furthermore, as long as the claims based upon the telescopic observations could be kept separate from the question of celestial order, as they were in the Jesuit celebration of Galileo in 1611, they posed no evident difficulties. But Galileo's claim concerning the existence of phases of Venus was difficult to ignore, because at the very least it lent itself to modest, heliocentered, Capellan, or Tychonic interpretations.[34] Yet as long as theologians, natural philosophers, and prognosticators endorsed the stringent Aristotelian standard of necessary demonstration, the telescope did no more to uniquely determine one account of celestial order than it did to make astrological prognostication more certain. Matters changed when local sensitivities about the disciplinary authority of theologians and their traditionalist philosophical allies in the universities were perceived to be at stake.

THE ERA OF CONSOLIDATION
WORLD SYSTEMS AND COMPARATIVE PROBABILITY

Appearing slightly more than a decade apart, Kepler's *Epitome of Copernican Astronomy* (1618–21) and Galileo's *Dialogue Concerning the Two Chief World Systems* (1632) consolidated the main resources of controversy developed between the 1580s and the telescopic revelations of 1610–12. In different but complementary ways, these works now reframed the earlier controversy explicitly as a battle between competing *world systems*. The term, as will be recalled, was already in circulation on the caption to Brahe's scheme of the planets in the 1588 *De Mundi,* a work reissued in 1603 and 1610.[35] Yet the full scope of the new claims and arguments was not yet thoroughly and widely appreciated, in part because, as late as 1618, all of Kepler's major theoretical achievements and Galileo's main telescopic discoveries were still available only in scattered form. Nicholas Mulerius's edition of Copernicus's work, for example, contained many annotations to the text

but no reference to Osiander's identity as the author of the "Ad Lectorem," although the information had been made public by Kepler in 1609.[36] Bellarmine's 1616 injunction to Galileo made no acknowledgment that Keplerian or Tychonian arguments existed. Uncertainty regarding the standards for assessing the burgeoning corpus of literature was already manifest in the exchange with Foscarini. Foscarini's appeal to probability was itself a kind of rhetorical gesture in the sense that it merely signaled the existence of "superior" arguments without reviewing their content. In different ways, the *Epitome* and the *Dialogue* significantly changed this situation. Forcefully consolidating all the major arguments then available, these works framed the problematic as one of weighing a considerable body of probable arguments for and against the diurnal and annual motions of the Earth. The new problematic was how to assess the cumulative and comparative weight of this corpus.

Here, the contrast between the *Epitome* and Kepler's earlier writings is worth noting. Kepler's modeling practices in the *Astronomia Nova* pushed Ptolemaic and Tychonic tolerances to the breaking point. The difficulties that David Fabricius expressed in his correspondence of 1603–8 proved to be paradigmatic of many later reactions: even the most skilled practitioners found Kepler's work troublesome. Many of the diagrams were not visually friendly, the calculations were complex and long, and the narrative was composed under circumstances that limited its appeal to other audiences. Consequently, the project would not travel easily in its original form. If accepting the elliptical orbit also meant adopting the solar-magnetic force, then anyone who rejected the Sun as mover might easily reject the noncircular orbit unless the two could be separated. In 1610, for example, Sir William Lower told Thomas Harriot that Kepler's elliptical hypothesis "overthrowes the circular astronomie" but that he could not "phansie those magnetical natures."[37] Another notable example is that of the Gdańsk astronomer and prognosticator Peter Crüger (1580–1639). In 1620, Crüger wrote to Philipp Müller in Leipzig that Kepler's "Work on Mars" required a man to spend not a day but an entire year to understand it. And further: "I have carefully read through the Kepplerian *Harmony of the World.* It appears that this work is as equally obscure as the Martian [book]." To Kepler he confessed: "I like the diagram of your lunar hypothesis . . . but because I dislike the ellipses, everything is obscure to me." By 1629, however, the *Epitome* and the *Rudolfine Tables* were available. Crüger then made a startling revelation to Müller:

> For myself, so far as other less liberal occupations allow, I am wholly occupied with trying to understand the foundations upon which the Rudolfine rules and tables are based, and I am using for this purpose the *Epitome of* [*Copernican*] *Astronomy* previously published as an introduction to the tables. This *Epitome* which previously I had read so many times and so little understood and so many times thrown aside, I now take up again and study with rather more success seeing that it was intended for use with the tables and is itself clarified by them. . . . I am no longer repelled by the elliptical form of the planetary orbits; Keppler's proofs in his *Commentaries on Mars* have persuaded me.[38]

Note that what helped Crüger was earlier absent for Fabricius: the possibility of using the elliptical model, as described in the *Epitome* and the *Astronomia Nova,* in conjunction with the *Rudolfines*. Yet even Crüger's eventual willingness to use the ellipse for calculation did not mean that he accepted Kepler's solar moving power:

> While Keppler labors to demonstrate Copernicus's hypotheses with physical reasons, he introduces remarkable speculations pertaining not so much to astronomy as to physics, such as magnetic fibers of the planets. . . . In order to defend the Earth's annual motion, he reforms almost all of philosophy and introduces a new one of his own; he also invents new astronomical terms, such as "focus," "sun-seeking and sun-fleeing fibers," "diacenters" [etc.]. . . . These things are pleasing, but quite obscure. . . . This being the case, not a few may be enticed by [Kepler's] speculations to his celestial physics and to Copernican astronomy; but many also will be deterred, especially when they have seen the publication of that other work [Longomontanus's *Astronomia Danica* (1622)], which reforms all of astronomy according to the Tychonic hypotheses [and observations].[39]

Crüger's circumspect reactions suggest that Fabricius's sidestepping of Kepler's physics, twenty years earlier, was not unusual; nor were the difficulties that Kepler posed for the Tychonics in Prague. For one thing, the use of Kepler's elliptical hypothesis illustrates the same kind of selective appropriation found already in the assimilation of Copernicus's planetary models in the sixteenth century. Not everyone who accepted Keplerian ellipses in the seventeenth century also adopted the terrestrial motions; nor was anyone converted by the new planetary theory in the sense of conversion described in Kuhn's *Structure*.[40] The astrologer and French royal professor of mathematics, Jean-Baptiste Morin (Morinus; 1583–1653), for example, claimed to accommodate Keplerian ellipses to a Tychonic scheme following a strategy similar to that of the mid-sixteenth-century Wittenberg practitioners who accommodated Copernicus's equantless devices to a geostatic frame of reference.[41]

As presented in the *Astronomia Nova,* therefore, Kepler's version of the Copernican system faced obstacles to successful circulation not only because of the technical novelty and difficulty of its astronomical models but also because of its complicated story of detours and dead ends. Kepler anticipated the difficulty. He explicitly invited professors of natural philosophy to read his lengthy summary in the introduction, acknowledging that such readers would be quite annoyed with him and Copernicus for "having shaken the foundations of the sciences with the motion of the earth." But he then offered them a choice "either of reading through and understanding the proofs themselves with much exertion, or of trusting me, a professional mathematician, concerning the sound and geometrical method presented."[42]

With the publication of Kepler's *Epitome,* the first Copernican academic textbook, the elements of the heliocentric theory were clearly and unambiguously laid down as a "world system" (*systema mundana*). The *Epitome* would become the single most important theoretical resource for seventeenth-century Copernicans: astronomer-astrologers like Crüger would find it a valuable propaedeutic to the *Rudolfine Tables* and hence to prognostication. Modernizing natural philosophers like Galileo, René Descartes, Pierre Gassendi, and John Wilkins would find in it an explicit and convenient summary of Kepler's physical arguments against the Tychonic model. It would appeal to both court and university audiences. But because it was cast in the question-and-answer form of the universities, it took aim at the heart of philosophical and theological authority. Kepler was explicit about challenging traditional disciplinary authority in his dedication: at stake, he wrote, are the "rules of the Academies" (*leges Academiarum*), the "honor of Academics" (*honor Academicorum*), and the "boundaries of Academic Philosophy" (*Academicae Philosophicae limites*). Patrons were obliged to protect these boundaries, Kepler acknowledged, but a wise prince "knows that the boundary posts of true speculation are the same as those of the fabric of the world" and not those "set up in the narrow minds of a few men." Kepler understood the resistance that he could expect from the universities: "They are established in order to regulate the studies of the pupils and are concerned not to have the rules of teaching change very often: in such places, because it is a question of the progress of the students, it frequently happens that the things which have to be chosen are not those which are most true but those which are most easy."[43] Later authors, like Ismaël Boulliau, Thomas Hobbes, and Descartes, would find that they could reconstitute Kepler's astronomical claims while ignoring his new physics and his Neoplatonic and Lutheran deity.

Although Rome quickly prohibited the *Epitome*, Galileo was looking for a copy of it in Florence in the summer of 1619, soon after the first of its three parts appeared.[44] Certainly the work must have emboldened him, suggesting possibilities for presenting the Copernican problem to a broader kind of audience, much as he had already succeeded in doing with his little telescopic reports. However, in the same year, the *Epitome* was put on the Index, thereby marking the Copernican question not merely as a matter of internal Catholic dissent, as in the case of Foscarini and Zuñiga, but also as one of Protestant disobedience and heresy. Thus any passing thought that Galileo might have had of making substantial reference to Kepler was now out of the question. In any case, beyond such political considerations, the later Galileo still had no sympathy for Kepler's archetypal arguments, his solar moving power, his explanation of the tides or, even more blatantly, the elliptical astronomy and the harmonic relation of the periods (squared) to the distances (cubed). And, most consequentially, when Galileo finally managed to publish the *Dialogue* in 1632, it had nothing to offer astrologers.[45]

The very literary form that made the *Epitome* so apt a vehicle for doing battle with the tradi-

alism of the universities—and especially for theoretically equipping prognosticators—made it somewhat less suited for the aristocratic and ecclesiastical audiences in which Galileo had long circulated. Thus whereas the *Epitome* and the *Dialogue* both attacked "the schools"—Kepler typically gave direct quotations from Aristotle—Galileo's work skillfully dramatized the cosmic alternatives using the generic form of the conversation, where the academic voice was represented by the pedant Simplicio, clearly subordinated to the aristocrats Salviati and Sagredo. Simplicio's speeches did not lack verisimilitude, nor were they made to be unreasonable—thus, all the more devastating when his arguments were systematically demolished. The Dominican Tommaso Campanella (1568–1639) quipped that the work was a "philosophical comedy" (*questa comedia filosofica*):

> Everyone plays his part marvelously: Simplicio as the laughing stock of this philosophical comedy, who, at the same time, shows the foolishness of his sect—the manner of speaking, the insecurity, the stubbornness and what not. Clearly we need not envy Plato. Salviati is a great Socrates who causes things to be born that are not yet born, and Sagredo is a free intellect who, not corrupted by the schools, judges all with great wisdom. . . . You have done what I wished when I wrote to you from Naples [many years ago], to wit, that you ought to put your teachings into [the form of] a dialogue in order to assure reaching all, etc.[46]

Where the *Epitome* (as a work of theoretical astronomy) was completely focused on the heavens, the *Dialogue* used novel celestial phenomena (such as novas, sunspots, and Jupiter's moons) to undermine the Aristotelian claim of heavenly nonalterability, and it used the motions of sublunar objects to infer the Earth's motion (bodies in vertical free fall from a tower, a moving ship, or motion along an inclined plane). But by the beginning of the *Dialogue's* Fourth Day, devoted entirely to the tidal motions, Galileo was willing to say only that the earlier arguments showed that "all terrestrial events except the ocean tides are impartial as to the earth's motion or rest."[47] Had the *Dialogue* ended with the Third Day, there might have been no trial.

Yet, in placing it near the end, Galileo obviously signaled that the tidal discussion was a serious part of his overall argument; indeed, it consummated the earlier arguments for the Earth's diurnal and annual motion, as both were needed to explain the regular tidal effects. What complicates the interpretation of its meaning is the mix of demonstrative and mitigated dialectical language with which Galileo tagged his discussion.[48] Thus, in the heavily strategic language of the preface, he said of the tidal argument: "Now, so that no foreigner can ever appear who, strengthened by our own weapons, would blame us for our insufficient attention to such an important phenomenon, I decided to disclose these probable arguments which would render it persuasive [*persuasibile*], given that the Earth were in motion."[49] But, like Kepler's approach in the *Astronomia Nova*, the Fourth Day moves along an eliminationist track, sketching out and dispensing with one alternative after another until finally Simplicio is made to announce the standard explicitly: "I know that the primary and true cause of an effect is only one, and so I understand very well and am sure that at most one can be true, and I know that all the rest are fictitious and false; and perhaps the true one is not even among those which have been produced so far."[50]

The only "foreigner" whom Galileo associated with an alternative tidal theory was Kepler; the rest, although unnamed, were Italians. Moreover, Galileo caricatured Kepler's views on the tides and, following his earlier pattern of withholding credit, brushing him aside as having "lent his ear and g[iven] his assent to the dominion of the moon over the water, to occult properties, and to similar childish ideas."[51] In the end, Galileo famously planted the omnipotence argument in the mouth of Simplicio.[52] Thus, for all his rhetorical agility, Galileo managed to leave the air dangerously thick with ambiguity. In the Fourth Day his tidal theory was the only one left standing against the other named alternatives; yet in the preface, it was described as probable and persuasive, and in the conclusion as uncertain and possibly false. Learned readers quickly spotted empirical difficulties, such as the daily delay in the turning of the tides; and, as Carla Rita Palmerino has shown, Pierre Gassendi's "tardy attempt to reconcile Galileo's explanation of the tides with Kepler's model of planetary motion ended up in failure."[53] As for the pope, he had only to have the *Dialogue*'s preface and conclusion read to him at dinner to stoke his notoriously short temper.

FROM PHILOSOPHIZING ASTRONOMER-ASTROLOGERS TO NEW-STYLE NATURAL PHILOSOPHERS

This consolidation of a critical mass of sophisticated printed arguments, diagrams, and references in support of the Earth's motions made possible a new sort of multifaceted, robust public debate from the 1620s to the 1640s such as had not occurred even in the last two decades of the long sixteenth century. Moreover, shortly after the trial of Galileo, in 1636, when his *Letter to the Grand Duchess Christina* was finally published in the Netherlands (frequently bound with the Latin translation of the *Dialogue*), the Galilean arguments about the Earth's motion began to circulate bundled together with the theological arguments based on the principle of accommodation.[54] In short, Rome's efforts to block the circulation of Galileo's attractive defense of the Copernican side in the world-systems debate were variously, and often ingeniously, circumvented. The evolving discussion was led largely by modernizing natural philosophers who, with few exceptions, were not astronomical practitioners and were explicitly opposed—sometimes vigorously—to astrological prognostication. The new generation encountered the Copernican question framed by works that no longer recirculated the ambiguities, terse phrasing, and diagrammatic limitations of *De Revolutionibus*. Ironically, just at the moment that the Church singled out for censorship the author Copernicus and the book *De Revolutionibus,* the latter had long since ceased to serve its original function. This phase of the debate was no longer dominated by prognosticators like Crüger, Magini, Origanus, and Fabricius. The emergent voices were those of a new breed of natural philosophers, the likes of Descartes, Gassendi, Marin Mersenne, Hobbes and Wilkins. In ways that can only be hinted at, these thinkers relied quite readily on the writings of Kepler, Galileo, and the Dutch astronomer Philip Lansbergen for both their evidentiary materials and their theoretical options; and they quickly subordinated the exclusively astronomical and astrological issues of the earlier period to questions of agreement with their own physical principles and issues of biblical compatibility. The period from the 1620s to the

1640s was, in short, the moment when modernizing natural philosophers captured the Copernican question.

This period was marked by residual disciplinary practices, even as important shifts in disciplinary goals emerged. Astronomers, even those who regarded physical objectives as desirable for theoretical astronomy, continued to regard prognostication as their primary objective. Modernizing philosophers who opposed the practice of astrological forecast, like Gassendi, typically grounded their opposition to astrology in the same Piconian arguments that had circulated in the sixteenth century. One of these was the argument from proximate and remote causes: the same effect, say a rich corn crop, could be better explained by a proximate, sublunary cause (the farmer's application of good fertilizers) than a general, remote cause, the influences of the Sun and stars.[55] But the proliferation of new arguments pushed these thinkers to engage more fully with questions of comets, novas, and planetary arrangement that might have been disregarded or treated superficially in the previous century.

Marin Mersenne (1588–1648), the polymathic Minim monk, is a salient example of a churchman who intensively engaged with modernizing developments within the framework of traditionalist concerns. In 1623, that framework was his massive commentary on the book of Genesis, where he attacked those whom he designated generically as "atheists," "magicians," "deists," "pagans," "heretics," and other detractors from the Catholic faith; but he also included an extensive discussion of "authors of new philosophy," including Campanella, Telesio, Kepler, Galileo, Gilbert, "and others."[56] Nothing like this had happened in the sixteenth century, the heyday of Genesis commentary.[57] Bellarmine had not brought up the Copernican ordering in his Louvain lectures of 1570–72; nor did Benito Pereira in his daunting four-volume Genesis commentary of 1591–99. Why, then, did Mersenne devote a section of his commentary to planetary ordering, citing some twenty-eight arguments and using materials from the Copernican modernizers Kepler and Lansbergen, together with an odd assortment of other, less obvious writers (perhaps conveniently on hand in the library of his order)?[58] Was he already aware of the sharp polemic between Kepler and Robert Fludd (with

whom he would soon have his own polemic) over the right use of Platonic resources for interpreting Genesis?[59] On the Copernican question, Mersenne was flexible: he neither endorsed the Earth's motion nor regarded the Church's judgment as irreversible.[60] On other matters, however, Mersenne showed an unusual passion for engagement with the ideas of the modernizers. In fact, his famously proactive stance in circulating the books and ideas of the modernizers—while carefully ignoring their more contentious and potentially dangerous positions—probably contributed more to the legitimation of Copernican and Galilean ideas than if he had staked out an affirmation based on a list of Scholastic-style arguments.[61]

Familiarity and articulated discussion thus bred accommodation even in a highly traditionalist format of presentation. Still, for every Zuñiga, Foscarini, Campanella, Gassendi, and Mersenne, there were a good many conservative Church theologians closed to the new currents. The trial of Galileo in 1633 marked the high-water line of the traditionalist backlash. It reframed the Copernican question as a matter of obedience to Church legal and scientific authority, an episode whose cultural meaning would transcend the moment of its immediate occurrence.[62] Yet, while the infamous episode dampened expressions of full public commitment in Catholic circles, it failed to shut down various degrees of positive expression. Partly, this was because the trial's implications were seen to be a matter of moral prudence and obedience rather than a matter of faith or papal infallibility; partly, it was because the Roman Inquisition had no practical power outside Italy.[63]

Consequently, if Catholic heavenly practitioners were somewhat reticent in their public declarations about the Earth's motion, nevertheless, they did not cease to argue the Copernican question or to pursue investigations within the domain of practical astronomy.[64] In fact, to expand the frame of reference, there was more discussion, more debate, and more disagreement about issues pertaining to natural philosophy and celestial order from the 1620s onward than in the first forty years after the publication of *De Revolutionibus*. If the 1580s and '90s had witnessed a proliferation of alternative astronomical orderings, the debate about celestial order shifted with Descartes to the context of alternative natural philosophies. Descartes's presentation of the first comprehensive, theoretical alternative to the natural philosophy of Aristotle in 1644 became a different kind of occasion within which to situate the Copernican question.[65]

Unlike Tycho Brahe or even Copernicus himself, Descartes did not merely add auxiliary physical hypotheses to a nontraditional celestial ordering. He was rethinking the fundamental physical principles themselves. And all this occurred at a time when the heavens were yielding up recurrent, novel entities. Descartes, like Mersenne, was a member of the generation for whom Galileo's telescopic discoveries—the moons of Jupiter, sunspots, the phases of Venus—were formative events of their youth.[66] All these now became phenomena to be deduced from mechanical principles. But Descartes also encountered Kepler's theorizing in the later 1620s through his friendship with the Dutch schoolmaster Isaac Beeckman (another master-disciple relationship). Beeckman's private journals reveal that he had thought carefully (and critically) about Kepler's physics in the *Astronomia Nova* and the *Epitome*. Beeckman rejected Kepler's solar moving power and formulated a remarkable inertial conception of motion in which, once placed in motion, a body required no further application of force to maintain it in that state.[67]

Descartes's relation to Kepler's writings is complicated (for us) by his generic citation practices (for example, ascribing views to "les astronomes"). Leibniz believed that Descartes used Kepler's ideas "brilliantly, although as is his custom, he concealed their author."[68] There are clear indications of Descartes using terminology of unambiguously Keplerian coinage: he uses the term *vortex* to describe the medium in which the planets moved, along with terms like *aphelion, perihelion,* and *natural inertia;* and there are further indications that Descartes's vortices resemble Keplerian orbits in their being flattened at the sides, although he explained this phenomenon by the mechanical pressure of contiguous heavenly whirlpools; and Descartes's placement of the Sun at the intersection of the planets' orbital planes is strikingly Keplerian.[69] In sum, Cartesian astronomy selectively appropriated residues of Keplerian physical theory, expunged incorporeal magnetic movers, and completely ignored the mathematical apparatus of Keplerian planetary modeling.[70]

To these observations, one may add a suggestive parallelism in the threefold uses of light as the differentiating feature of the two world schemes: for the trinitarian Kepler, the sun (visible symbol of God the Father) emits light, the planets (symbol of Christ the Son) receive it, and the intervening aether (the Holy Spirit) transmits it; for Descartes, the sun and fixed stars (very rapid, indefinitely small particles) emit light, the Earth, Moon, comets, and other planets (bulky particles, less suited to movement) reflect it, and the heavens (very small, spherical particles) transmit it.[71] Perhaps the use of these Keplerian resources should not surprise us, as Descartes was in close contact with men who were actively engaged with, although not uniformly favorable to, Keplerian planetary theory: Beeckman, Morin, Boulliau, Hortensius, Gilles Personne de Roberval, and Mersenne.[72]

The crucial point is not merely the existence of such appropriations but how the Copernican question was absorbed and transformed under the Cartesian impress. Like Kepler, Descartes readily inhaled Jupiter's moons and sunspots into his system.[73] Yet after Galileo's trial in 1633—although not necessarily because of it—Descartes was seeking to maintain the Earth's motion in a new sense—at rest with respect to the particles surrounding it, yet carried along by the fluidlike properties of its own whirlpool.[74] Like Kepler in the introduction to the *Astronomia Nova*, Descartes positioned himself explicitly against Tycho's scheme: "I deny the motion of the Earth more carefully than Copernicus and more truthfully than Tycho," as he dexterously phrased it in the *Principles of Philosophy*.[75] The Cartesian approach to Tycho's system was more direct—more Keplerian—than that of Galileo: from the planetary vortices, Descartes deduced the physical impossibility of the Sun's carrying with it all the planets except the Earth.[76] At the same time, in a bold stroke, he rejected the incorporeal souls of Keplerian and Gilbertian magnetism. Thus, even if one did not accept Descartes's moving yet resting Earth, one might still be convinced by his arguments *against* Tycho, the alternative to which most traditionalists were now flocking. Finally, Descartes, in the even more radical spirit of Bruno, broke fundamentally with the astronomer's traditional sphere, even the much-enlarged version of Copernicus and Kepler.[77] Cartesian space was indefinitely extended, and thus accommodating the distances traversed by comets and novas was no longer a problem. Not only were these non-recurrent phenomena part of the ordinary course of nature, but there was now an explanation for their existence that had been overlooked by Tycho:

> We must not be influenced by the fact that Tycho and the other Astronomers who carefully searched for the parallaxes of Comets said only that they were situated beyond the Moon, toward the sphere of Venus or Mercury but not beyond Saturn; for they could no less satisfactorily have deduced from their calculations that the Comets were beyond Saturn. But because they were disputing the views of the ancients, who included Comets among the meteors [formed in the air] below the Moon, they contented themselves with showing that they are in the heaven, and did not dare to attribute to them the altitude which their calculations were revealing, for fear of making their proposition less believable.[78]

As passages of this sort attest, Descartes left little standing in his path. Rather than celebrating Tycho's break with Aristotle, Descartes simply crushed the *via media*, portraying Tycho as excessively tied to the ancients. The only thing that Descartes could not do was to *predict* the occurrence of comets and novas. He did not claim to be an astronomer; nor did he know of any who could make such predictions. Indeed, as a natural philosopher, it was good enough for him to produce conclusions that were true from hypotheses that might be false, but, as he said, were probably not.[79]

By the end of the 1630s, a whiff of historical temporality began to appear in arguments over the credibility of the Copernican proposal: not only the arguments of authorities but also social variability and time were now invoked as part of a larger claim about the probability and relative superiority of Copernicus's opinion. As John Wilkins put it in his *Discourse Concerning a New Planet, Tending to Prove that 'tis probable our* EARTH *is one of the* PLANETS: "All Men have not the same way of apprehending things; but according to the variety of their Temper, Custom and Abilities, their Understandings are severally fashioned to different Assents." And further: "*Copernicus,* who was a Man very exact and diligent in these studies [of astronomy] for above 30 years together, from the year 1500. to 1530. and upwards: And since him, most of the best Astronomers have been on his side. So that now, there is scarce any

of note and skill, who are not *Copernicus* his followers, and if we should go to most voices, this Opinion would carry it from any other."[80] Among the "best" and most skilled proponents he named were Rheticus, Rothmann, Maestlin, Reinhold, Gilbert, Kepler, and Galileo—the last three "with sundry others, who have much beautified and confirmed this *Hypothesis,* with their new inventions."[81] Wilkins's own errors (on Reinhold), hasty judgments (on Gilbert), and omissions (of Zuñiga, Foscarini, Stevin, Crüger, and Wilkins's own countrymen Digges and Harriot) are less interesting than the persuasive uses to which he sought to apply his list. One such claim—without mention of specific names—was that the Copernicans were more open-minded than the traditionalists: "Amongst the followers of *Copernicus,* there are scarce any, who were not formerly against him; and such, as at first, had been throughly [*sic*] seasoned with the Principles of *Aristotle.* . . . Whereas on the contrary, there are very few to be found amongst the followers of *Aristotle* and *Ptolomy,* that have read any thing in *Copernicus,* or do fully understand the Ground of his Opinion; and I think, not any, who having been once setled [*sic*] with any strong assent on this side, that have afterwards revolted from it."[82] Wilkins's dialectical argument, unprecedented in this matter, introduced temporality and change of mind into the assessment of beliefs, to which he urged the moral that "in all probability, this is the righter side."[83]

Another side of the dialectical argument was, in the absence of decisive affirmative arguments, to point out the greater deficiencies of the prevailing alternative—a standard move since Copernicus. Like Foscarini and Hooke, Wilkins adduced a psychological explanation: it was not habit that was at issue but self-love—"an over-fond and partial conceit of their proper Inventions." This was Wilkins's explanation for Tycho Brahe's resistance to Copernicus: "Every Man is naturally more affected to his own Brood, than to that of which another is the Author; though perhaps it may be more agreeable to reason."[84] To overweening self-estimation he added the fear of opposing ancient authority (especially scriptural)—adding, correctly, that Copernicus's opinion was not condemned by the Council of Trent but only "in later times hath been so strictly forbidden, and punished."[85]

In the mid-seventeenth century, the Copernican question was significantly reframed by the Jesuit Giovanni Battista Riccioli (1598–1671): an astronomer of considerable skill, philosophically a Scholastic of the baroque era, an encyclopedist in the final days of encyclopedism, a controversialist unusually literate in the modernizers; a man of conscience and consummate learning, yet also carefully attendant to church decrees and the authority of scripture.[86] The frontispiece of the *Almagestum Novum*, a wondrous example of Jesuit emblematic skill, captured both the spirit and the argument of the book in the image of the scales of justice, the balance. It embodied preservation of tradition along with openness to novelty.

But which way did the scales tip? In 1615, Foscarini had invoked the authority of Clavius to warrant balancing probability in favor of Copernicus and the moderns. In 1638, John Wilkins appropriated the passage for similar ends, but embellished it in his translation with a concocted deathbed scene: "'Tis reported of *Clavius*, that when lying upon his Death-bed, he heard the first News of those Discoveries which were made by *Gallileus* his Glass, he brake forth into these words. . . . That it did behove Astronomers, to consider of some other *Hypothesis*, beside that of *Ptolomy*, whereby they might salve all those new appearances. Intimating that this old one, which formerly he had defended, would not now serve the turn: And doubtless, if it had been informed how congruous all these might have been unto the Opinion of *Copernicus*, he would quickly have turned on that side."[87]

In 1651, Riccioli again invoked the Clavius passage while reversing the balance of meaning to favor the *via media*: "Upon considering the new celestial phenomena detected by Galileo with the Belgian looking-glass and disclosed in the *Sidereal Messenger*, the old man exclaims at the end of his life: 'Let the astronomers consider how the celestial orbs ought to be arranged so that these phaenomena may be saved.'"[88] Riccioli represented the telescope on his frontispiece, held not by Galileo but by the mythical Argus, man of a hundred eyes. His knee genuflected reverentially, Argus points his finger in the direction of the di-

vine hand with which God has created the natural world according to "Number, Measure and Weight," the oft-cited words of the book of Wisdom, chapter 11. Also shown on the frontispiece (upper right) are Galileo's discoveries—lunar craters, Saturn's anses, and Jupiter's moons—but all held by flying angels. Similarly, angels have replaced Keplerian celestial magnets and Cartesian vortices, as Mercury, Venus, Mars, and the Sun are all shown in the hands of cherubs (upper left). Reason and the corporeal senses are acknowledged, but only if guided by divine illumination mediated by angelic agency. Finally, the balance itself, held by the virgin Astraea, shows Riccioli's world system outweighing the Copernican. Old, chastened Ptolemy rests on his shield while firmly grasping the heraldic shield of Riccioli's patron, the House of Grimaldi: "I am raised that I may be corrected," he is made to utter. The *Almagest* is subject to improvement, but without entirely throwing away its most critical foundation. Hence Riccioli called his own system (which he says that he taught at Parma) the "semi-Tychonic" because Jupiter and Saturn have the Earth as the center of their motions even as Mars, Venus, and Mercury Tychonically circumnavigate a Sun that revolves around the Earth.[89]

Behind the emblematic of the balance lay a massively detailed apparatus of argumentation and also a functional bibliography. (In his citation practices, Riccioli proved himself to be more modern than Descartes or Galileo, typically giving authors' names, titles of works, editions, and page numbers.) A reader with access to the books cited in the *Almagestum Novum* could follow virtually any argument or claim directly back to its original source, thereby opening up Riccioli's position to further reflection and criticism and loosening up still more the author's control over his conclusions. In addition, one can gauge Riccioli's representations precisely by noting that he classified (without comment) Protestants as well as Catholics as followers of the "Copernican System": Copernicus, Rheticus, Kepler, Maestlin, Rothmann, Galileo, Gilbert ("who nevertheless only asserts the daily motion of the Earth"), Foscarini, Zuñiga, the author of *Aristarchus Redivivus* ("the author's name suppressed"),[90] Boulliau, Jacob Lansbergen, Pierre Herigone, Gassendi, Stevin, Wilhelm Schickard, Giordano Bruno, and Descartes. Notably absent were the Englishmen Digges, Har-

80. Frontispiece, Riccioli, *Almagestum Novum*, 1653 (reissue of 1651). Mandeville Special Collections Library, University of California, San Diego.

riot, Henry More, and Wilkins, and the Dutchman Isaac Beeckman; questionable was Celio Calcagnini, who in 1544 endorsed a daily but not an annual motion, and Gassendi, who, at least outwardly, displayed adherence to Tycho Brahe's arrangement.[91] From these eighteen or so authors,

Riccioli constructed forty-nine arguments in favor of the daily and annual motions—a disproportionate number derived from the simplicity and symmetry claims in Copernicus, Kepler and Galileo. Against these, he listed thirty-three authors: a good many were *Sphere* and *De caelo* commenta-

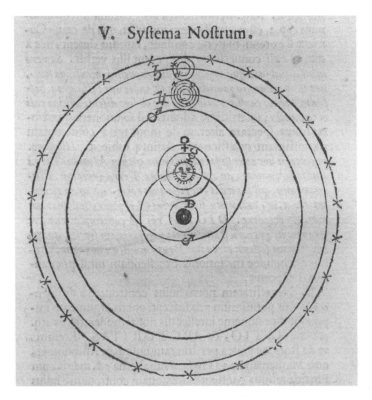

V. Syſtema Noſtrum.

81. "Our System," according to Riccioli 1653 [1651]. Note that Jupiter and Saturn incorporate Galileo's discoveries, but both planets retain the Earth as their center of motion. Mandeville Special Collections Library, University of California, San Diego.

tors, but they also included recent authors, like Mersenne and Riccioli's own onetime teacher, Giuseppe Biancani, as well as other important members of the Jesuit order, Clavius, Christopher Scheiner, and Melchior Inchofer (the latter, of course, having served on the commission that judged Galileo). From these authors, Riccioli derived seventy-seven arguments against the Earth's motion, many of them reducible to standard Aristotelian accounts of projectiles and falling bodies, and decisively secured, in his view, by his own extensive "physico-mathematical argument" against Galileo's tower experiment.[92] Yet Riccioli also honestly admitted that in the realm of probabilities there was a class of arguments that favored both hypotheses: "They are so many of such a kind that with some ability, the intellect can be inclined toward one or the other hypothesis."[93]

Riccioli sought to present his case "purely from physics and astronomy;"[94] but in the end, he decisively eliminated whatever uncertainties of underdetermination might remain "when both sacred authority and Divine Scriptures are taken into account."[95] By "sacred authority," he meant, in the first instance, the Church fathers and the authority of the fourth session of the Council of Trent on the interpretation of the Bible.[96] Thus he left no doubt about who had the authority to set the standards of interpretation; but he went further, claiming that in the Copernican case, in particular, biblical passages bearing on the Earth's resting and the Sun's standing still were to be read "in a literal and proper sense" rather than in a figural or moral sense.[97] Capping his arguments with the appearance of juridical finality, he added a detailed reading of the relevant scriptural passages, the texts of the 1616 decree against Foscarini, Zuñiga and Copernicus, the specific passages of De Revolutionibus to be corrected, and the full text of the sentence against Galileo and of his abjuration. Riccioli's rebalancing would thereafter truly define the Copernican question as a historic controversy—"long-famous," as he said, "especially in this century."[98]

THE COPERNICAN QUESTION AFTER MIDCENTURY

Riccioli's New Almagest cleared a new space for discussion, becoming the preeminent critical text, an encyclopedic compendium of sources for writers on the Copernican question after 1651.

With the availability of such a resource, it may be no accident that, after midcentury, it is difficult to find a modernizer who did not also associate his philosophizing with Copernicus. Yet, much as in the fin-de-siècle period, significant lines of divergence continued to divide the character of Copernican representations. One of these joined the thirteenth-century principle of divine omnipotence to Galileo's telescopic discoveries: (1) God could have created innumerable worlds; (2) Giordano Bruno claimed that God must have used his infinite power to so do ("all the stars *are* systems"); (3) Galileo had inferred a moon with Earthlike characteristics. Just five years after Galileo's condemnation, John Wilkins connected the dots: an infinite, active god must have used his power to create other beings to occupy other worlds. He broached in his *Discovery of a World in the Moone* the probability of an Earthlike moon with lunar inhabitants, the dark areas interpreted as seas, the entire body surrounded by a vaporous atmosphere. The existence of Lunatics was further testimony to the divine wisdom. Scriptural silence on the plurality of worlds was merely an indication that the Holy Ghost did not use the Bible "to reveale any thing unto us concerning the secrets of Philosophy."[99] A few years later, in 1646, Henry More (1614–87), early under the spell of Descartes, postulated an infinite number of inhabited planets.[100] In 1686, Bernard le Bovier de Fontenelle (1657–1757) placed a male Cartesian philosopher and a noble lady in an imaginary garden where the former proceeded to teach, through Socratic "conversations," a universe with an infinite number of Earthlike planets circulating in vortices around Sun-centered worlds: "If the fix'd Stars are so many Suns, and our Sun the centre of a Vortex that turns round him, why may not every fix'd Star be the center of a Vortex that turns round the fix'd Star? Our Sun enlightens the Planets; why may not every fix'd Star have Planets to which they give light?"[101]

Such imaginative expansions, exploiting the almost limitless inferential resources of analogy, quickly opened up not only new heavenly prospects but also new kinds of audiences for the Copernican question, early signs of an emergent civil sociability.[102] Wilkins's *Discovery* enjoyed two printings in 1638, further additions in 1640, and new issues in 1684, 1707, and 1802.[103] Fontenelle's attractive little *Entretiens sur la pluralité des mondes* went through thirty-three French editions by 1757, with five separate English and two German translations; and Joseph Glanville's 1688 translation appeared in Philadelphia as late as 1803.[104] Readers with little technical expertise could thus easily sidestep the subtleties of the erudite, baroque Riccioli and the brilliant but difficult Kepler and jump directly to Enlightenment with an easy conviction in the Copernican ordering derived from the analogies and probabilities of the world pluralists.

The broader readership attracted to Wilkins, Fontenelle, and the growing gaggle of pluralist writers was also drawn to the almanac literature which, in England, according to Bernard Capp's estimate, had grown to sales of 300,000 to 400,000 per annum by the 1660s.[105] Just as in the late fifteenth and early sixteenth centuries, almanacs were typically produced with accompanying astrological prognostications. But after the collapse of state censorship in 1641, there was "an explosion of new publications, on every conceivable subject and in every possible vein. Before 1640, there were no printed newspapers; by 1645, there were several hundreds."[106] Contributors to this wider literature wrote in the vernacular and, by their efforts, brought a new vigor to the science of the stars. They sometimes translated earlier Continental prophecies; they collected and sometimes brought to press astrological writings that still existed in manuscript, like Christopher Heydon's *An Astrological Discourse*, and lost writings, such as Rheticus's reconciliation of Scripture with Copernicus's hypothesis; they inundated the expanding reading public with their own prognostications; they produced their own compendia of theoretical astronomy and astrology, and they published practical ephemerides. They tried to reform astrology, often along lines first mapped in the sixteenth century.[107] And like Riccioli, they were self-conscious of their place in their subject's history—if not in its historical bibliography—as is evident from the "Catalogue of Astrological Authors Now Extant, Where Printed, and in what Yeer," located at the end of William Lilly's *Christian Astrology*.[108] Patrick Curry aptly calls the period of the Interregnum (spanning the Civil War, the Commonwealth, and the Protectorate) the "Halcyon Days" of English astrology.[109] One might add that such publications opened a new kind of space for public discussion alongside the aristocratic circles of the century previous.

The most prominent English astrologers of the

Interregnum who were Royalist in politics were also Keplerian Copernicans on the question of planetary order, accommodationist in their reading of scripture, anti-Aristotelian in natural philosophy, yet diverse in the alternative forms that such leanings might take.[110] Indeed, philosophizing astronomers of the century's middle decades had much more to choose from in natural philosophy than those of a century earlier. Seth Ward (1617–89), for example, noted that at Oxford "there is scarce any Hypothesis . . . but hath here its strenuous Assertours, as the Atomicall and Magneticall in Philosophy, the Copernican in Astronomy &c."[111] Controversies among astrologers were as sharp and personal as any in this period; yet they differed not principally over planetary ordering—a matter on which, like Heydon, they could always plead agnosticism—but over the accuracy of the tables on which the credibility of their almanacs and prognostications ultimately rested.

Vincent Wing (1619–68), one of the most prolific and eloquent practitioners of this moment, made no special arguments for a motionless Earth in his first publication, a coauthored compendium of practical astronomy based on the tables of Tycho, Argoli, and Lansbergen, "the best approved Uraniscopers now extant."[112] Within two years, however, he had quickly associated himself with the modernizers in theory, producing the Keplerian-sounding *Harmonicon Coeleste or, The Coelestiall Harmony of the Visible World,* in which he clearly declared, on the authority of Boulliau and Galileo's *Dialogue,* for "the Copernican Systeme . . . proved by apodicticall Mediums." By "apodictic," he intended the much sought-after, full demonstration that would eliminate other alternatives; but he did not even dignify the Tychonic system as worthy of refutation, and merely held of the Copernican that as philosophy, "I know nothing of worth opposeth it." Wing knew that this dismissal alone did not yield an apodictic demonstration.

No? saith my adversary, what think you of the Scripture? Why, I answer, that whatsoever is there spoken of the Earth's Rest or the Sun's Motion . . . is to be understood . . . as the Philosophers say; according to our apprehension and vulgar manner of speaking, and not according to the nature of the things. . . . we must study Philosophy, and not make the holy Scripture the reconciler of those doubts, which are infinite. But if any be so refractory, he cannot assent to the Physical truth, let him construe it as a meer *Hypothesis,* and I will return to my task in hand.[113]

Such accommodationist and agnostic sentiments, appearing in exactly the same year as Riccioli's careful balancing act (of which Wing was not yet aware), capture telling differences in the political and religious possibilities open to mid-century astral practitioners who, apart from their convictions about the order of the heavens, were fighting difficulties in shared domains of modeling and calculational practice. In planetary theory, for example, Wing, like Riccioli, now associated himself with Kepler's elliptical astronomy in the version of "the painfull and learned *Bullialdus* [Boulliau]," whereby he reduced the ellipse to uniform circular motions.[114] Yet, as a practitioner independent of both Rome and the universities, Wing was not encumbered by academic and ecclesiastical authority as Riccioli was. Signaling his autonomy by the print identity "Lover of Mathematics" (Philomathematicus), Wing readily dispatched the traditionalists not with Ricciolian arguments, but with disdain: "These and many other enormious Tenents of the Peripatetick, the pure Astronomer laughs at as meer figments of Mans brain, or *Entia rationis,* as they call them."[115] By 1669, Wing had absorbed Riccioli, and, with more respect than he had leveled against the Aristotelians, he disposed of Tycho's "hypotyposis" as "more ingenuous than true."[116] Against Riccioli, with a nod to Kepler's *Epitome,* Wing now offered an amendment to his earlier Copernican diagram—a sidebar comment, on the explicit authority of Descartes, that the planets circulate in vortices of "celestial matter" around the rotating Sun, the closer parts moving more rapidly than those further out.[117]

Wing's rival, Thomas Streete (1621?–89), another important English astrologer of the Interregnum, was also a Keplerian Copernican, a prolific ephemeridist, Royalist, and conservative Anglican. Unlike Wing, however, he did not trouble to argue against alternative planetary orderings. Moreover, he was silent on scripture's bearing on the Earth's motion; in natural philosophy, he declared neo-Paracelsian principles that could have appealed easily enough to Tycho Brahe.[118] A clerk in the Excise Office, Streete greeted the restoration of Charles II to the throne in 1660 with his *Astronomia Carolina* (1661).[119] In other respects,

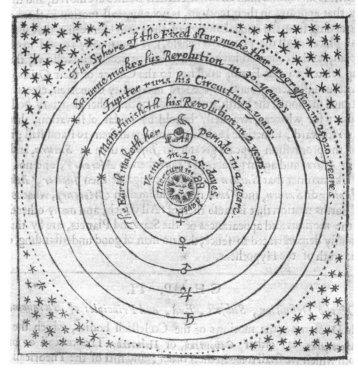

however, Streete's approach could not have been more different from Wing's. He made no secret of his nontraditional intentions: "Wee intend not here to insist on the great Utility, Antiquity and Excellency of Astronomy."[120] There would be standard gestures to the Creator, but, unlike Wing, Streete offered no special discussion of scripture's status. Where Wing was historical, expansive, allusive to controversy, and filled with erudite, Latinate asides, Streete was spare, economical in his prose, and ceaselessly droning in his precisely worked exercises. For him there was simply no controversy about the order of the heavens. Past astronomers, in his account, functioned less as theorizers than as long-gone witnesses, their names mere appendages to the times, places, and positions reported.[121] Streete's minimalist style of empiricism did not derive from a social grouping that yielded credibility by dint of shared status; rather he represented evidence in the form of an eponymous sequence, an observational trail whose ending he marked with a dated, witnessed, and located observation.[122] Grounding his practical astronomy, Streete provided the restored king

and his followers with a nonmechanical, alchemical philosophy of the heavens: "The Visible World and every part thereof consisteth of three Principles, *Sulphur, Salt and Mercury.* The *Sulphur* or Soul of the World, from whence proceedeth heat and light, is most manifest in the bodies of the Sun and Fixt-Stars. The Salt or Corporeity of all things is the chief consistence of the Planets *Saturn* and *Jupiter,* with their Attendants, *Mars,* our Earth with the *Moon, Venus,* and *Mercury.* The *Mercury* or Spirit of the Universe operates through the *Aether* and fluid *Medium,* wherein all visible bodies have their place and motion."[123] Like Kepler and Descartes, Streete proposed that the world's principles were three; but for reasons entirely unspecified, he left readers only with a hint that they were "chymical" and failed to specify what work they performed.

ROBERT HOOKE, ISAAC NEWTON, AND THE CRUCIAL EXPERIMENT

Both Robert Hooke (1635–1702) and Isaac Newton (1642–1727) were members of a generation that en-

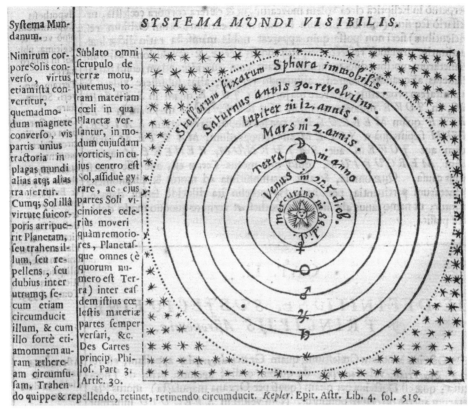

83. The visible world embedded in a Cartesian vortex, from Wing 1669. The sidebar juxtaposes Descartes's *Principles of Philosophy*, pt. 3, article 30, and Kepler's *Epitome of Copernican Astronomy*. Reproduced by permission of The Huntington Library, San Marino, California.

countered the Copernican question not through the originary texts of *De Revolutionibus* and the *Narratio Prima* but as a controversy already matured and refracted through the reinvigorated midcentury literature of the heavens and natural philosophy. Substantially dominated by a broad spectrum of modernizers like Kepler, Wilkins, Wing, Streete, Descartes, and others, this literature provided theoretical resources for many independent English prognosticators even as it still bore important structural similarities to the classification of knowledge of the Wittenberg era. But, thanks to the gradually improved accessibility of Tycho Brahe's data through Kepler's *Rudolfine Tables*, the predictions yielded by practical astronomy increased in accuracy during the seventeenth century.[124] By the 1650s, the new ephemerides were all considerably more accurate than those based on Reinhold's *Prutenics* a century earlier; and although not all midcentury planetary theorists were Copernicans, the great majority of ephemerides from Kepler onward bore the unmistakable authority of Copernican authors.[125] Many of the most prolific authors of practical astrology responsible for the revival of prognostication in the 1640s and '50s associated the accuracy of their ephemerides with the Copernican ordering in theoretical astronomy.

Hooke and Newton encountered this new space of possibilities in their youth. Newton's early reading notes and page-marking practices show that he was already acquainted with the Copernican ordering through the astronomer-astrologers Wing and Streete. Streete provided Newton with an uncontested representation of the Copernican ordering and of Kepler's elliptical astronomy.[126] Newton also owned and annotated both Wing's *Harmonicon Coeleste* and his *Astronomia Britannica*.[127] Galileo's *Dialogo* and Foscarini's *Lettera* were available to him through Thomas Salusbury's 1661 English translation.[128] Through Descartes's *Principia* he had an image of the Earth, at rest, floating with respect to its own whirlpool, yet moving with respect to the solar vortex. And through Walter Charleton's *Physiologia Epicuro-Gassendo-Charletoniana, or A Fabrick of Science Natural upon the Hypothesis of Atoms,* he would have found Copernicus classified as one of the "Renovators who dig for Truth in the rubbish of the Greek Patriarchs" and "who hath rescued from the jawes of oblivion, the almost extinct Astrology of *Samius Aristarchus.*"[129] Nowhere in Newton's writings is there evidence that he saw planetary ordering as contested or that alternative ordering systems were in need of refutation. He did not pose the problem in the manner of Wilkins, for example, as a balance between probable arguments, or in the manner of his own famous crucial experiments with prisms to determine the nature of light.[130] Thus, in the literature of the 1650s and 1660s of which we know him to have been aware, Newton was effectively faced with a consensus in which different physical theories were consistent with the Copernican arrangement but not with one another.

In the early 1660s, it was clearly the Cartesian framework that fixed the fundamental coordinates of natural philosophy and the Copernican question for Newton and Hooke, just as it had been at the Dutch universities in the previous decade.[131] Newton aptly used the title "Philosophical Questions" for his student "Waste Book." In these musings, Newton called attention to all sorts of inconsistencies in the Cartesian system. Whatever its difficulties, however, Newton accepted that Descartes offered a genuine alternative to Aristotle—a unified system that purported to reduce the properties of all bodies, both heavenly and earthly, to the same sort of reality, the so-called *res extensa* or extended stuff.

On the question of an *astronomical* proof of the Earth's motion, however, Hooke differed importantly from Newton. Captivated, like Newton, by Cartesian principles, he nonetheless acknowledged that celestial order still lay open to some doubt. For him, the accuracy of Streete's or Wing's ephemerides did not count decisively in favor of their Copernican convictions, and he did not share the sixteenth-century aspiration to improve astrological predictions by solving theoretical astronomy's difficulties. In a sense, part 1 of that earlier problematic (improving predictive accuracy for planetary positions) had been solved. Yet, as a young man, Hooke had come into close con-

tact with John Wilkins and other members of the Oxford circle, including Seth Ward and Christopher Wren; and through these associations Hooke had undoubtedly imbibed the masterfully assembled probability arguments of Wilkins's *Discourse,* tipped in favor of the Copernican ordering.[132] And now, sometime in the early 1660s, Hooke had come upon and studied Riccioli's equally impressive balance of probabilities in favor of the semi-Tychonic ordering.[133] This encounter must have left him with a conundrum: If the Holy Ghost had not used the Bible to teach natural philosophy, as Galileo had argued, and following him Wilkins and Wing, then Riccioli's appeal to the literal reading of scripture as the final arbiter must be disqualified as simply irrelevant.

In 1668, a striking event put Riccioli's physico-mathematical experiment at the center of attention in the third volume of the *Philosophical Transactions* of the newly founded Royal Society of London. James Gregory presented Riccioli's argument against Galileo's tower experiment, concerning falling bodies on a moving Earth, as a claim already highly contested in Padua.[134] The circulation of English and Scottish students between Britain and Italy now assisted in foregrounding the Riccioli-Galileo disagreement at a significant node of English scientific life. Riccioli had employed various strategies in the *Almagestum Novum* for persuading readers that bodies of different weights, dropped simultaneously from one of Bologna's highest towers, had landed at different times: he cited witnesses, referred to "frequently repeated" experiments, and narrated experimental events as time-bound.[135] Yet Riccioli's observational rhetoric failed to block criticism of his representation of Galileo's solution to the falling-bodies problem as a compounding of circular motions. Gregory's report of 1668 was probably the occasion for Robert Hooke's 1674 Cutlerian Lecture, in which he proposed an astronomical solution to the question of the Earth's motion as a prelude to a dynamic analysis of the falling-bodies problem. This revealing episode exposes a critical aspect of high-end English theorizing about the Copernican question in the Restoration era.

Robert Hooke, as Michael Hunter and Simon Schaffer urge, "can be taken as representative of the enterprise of Restoration natural philosophy—the commitment to progress through instrumentation, the link between mechanism and design,

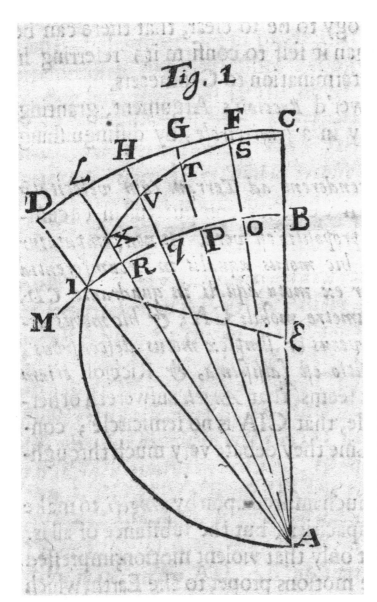

Fig. 1

84. Galileo's tower experiment, Gregory 1668. As the tower moves along the parallel circular arcs defined by its top, CFGHLD, and its base, BOPQRl, the body follows circular arc STVXl. An observer at the base of the tower does not perceive the circular motion shared with the tower but sees only the body as it appears to fall along the straight lines FS, GT, HV, etc. Reproduced by permission of The Huntington Library, San Marino, California.

the careful juxtaposition of hypothesis and empiricism."[136] It was Hooke who famously built Robert Boyle's air pump. More than that, he saw instrumentation as a kind of royal road to the disciplining of the passions, a means, as he proclaimed in his *Micrographia,* to "rectifying the operations of the *Sense,* the *Memory,* and *Reason,*" a "Universal cure of the Mind."[137] And now, just a few years beyond the *Micrographia,* it was again he who claimed that the great "Controversie" between world systems could be settled decisively with another instrument: a zenith telescope that he built and poked out the window of his rooms

at Gresham College.[138] The purpose? Like the ink dot he made famous with his microscope—he compared that optically-magnified mark to "a splatch of London dirt"—Hooke proposed to resolve the Copernican question by measuring a single, tiny angle: the angle of annual stellar parallax.

Like Tycho Brahe nearly a century earlier, Hooke knew that this measurement was difficult. In the *Micrographia* he had already diagnosed the problem of refraction: "The atmosphere is a transparent Globe, or at least a transparent shell, encompasing an opacous Globe, which, being more dense then the medium encompassing it, refracts

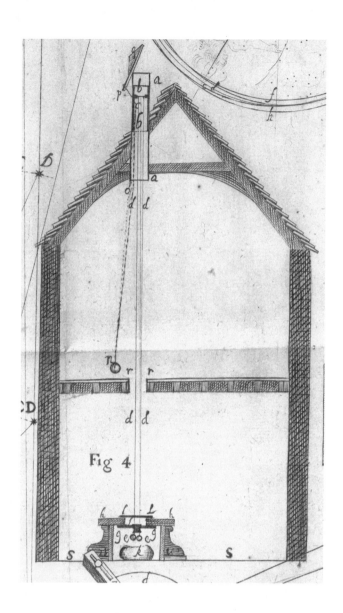

85. Diagram of Robert Hooke's zenith telescope, protruding through the roof of his room in Gresham College, London. From Hooke 1674. Reproduced by permission of The Huntington Library, San Marino, California.

or inflects all the entring parallel Rays into a point or focus."[139] But, in the *Attempt to Prove*, he confidently laid out the problem as amenable to solution: "'Tis not here my business to instruct them [the unlearned] in the first principles of Astronomy, there being already Introductions enough for that purpose: But rather to furnish the Learned with an *experimentum crucis* to determine between the *Tychonick* and *Copernican* Hypotheses. That which hath hitherto continued the dispute hath been the plausibleness of some Arguments alledged by the one and the other party, with such who have been by nature or education prejudiced to this or that way."[140] Passions and probabilities: every argument involved some subjectivity. For

Hooke, the "crucial experiment" was thus needed because other criteria were not decisive to "any determination of the Controversie":

I confess that there is somewhat of reason on both sides, but there is also something of prejudice even on that side that seems the most rational. For by way of objection, what way of demonstration have we that the frame and constitution of the World is so harmonious according to our notion of its harmony, as we suppose? Is there not a possibility that the things may be otherwise? nay, is there not something of probability? may not the Sun move as *Ticho* supposes, and the Planets make their Revolutions about it whilst the Earth stands still, and by its magnetism attracts the Sun, and so keeps him moving about it, whilst at the same time Mercury and Venus move about the Sun, after the same manner as Saturn and Jupiter move about the Sun whilst the Satellites move about them? . . . Is there not much reason for the Hypothesis of *Ticho* at least, when he with all the accurateness that he arrived to with his vast Instruments, or *Riccioli*, who pretends much to out-strip him, were not able to find any sensible Parallax of the Earths Orb among the fixt Stars, especially if the observations upon which they ground their assertions, were made to the accurateness of some few Seconds?

No fuller characterization can be found of the Copernican question at this historical moment. The problem for Hookean "Observators" was that the naked eye cannot distinguish an angle of less than a minute. Tycho and Riccioli thus claimed more than they could deliver. "The Controversie therefore notwithstanding, all that hath been said either by the one or by the other Party, remains yet undetermined, Whether the Earth move about the Sun, or the Sun about the Earth; and all the Arguments alledged either on this or that side, are but probabilities at best, and admit not of a necessary and positive conclusion." Were sensible parallax ever to be detected, wrote Hooke, it would constitute "a most undenyable Arguement of the truth of the *Copernican Systeme.*"[141]

In their encounter with the Copernican question, a striking difference separated Hooke from Newton. For Newton, as for Streete, there was no controversy; for Hooke, on the other hand, it was Riccioli who organized the problematic to which he offered his "crucial experiment": "The Inquisitive Jesuit *Riccioli* has taken great pains by 77 arguments to overthrow the *Copernican* Hypothesis, and is therein so earnest and zealous, that though otherwise a very learned man and good Astronomer, he seems to believe his own Arguments; but all his other 76 Arguments might have been spared as to most men, if upon making observations, as I have done, he could have proved there had been no sensible Parallax this way discoverable, as I believe this one Discovery will answer them, and 77 more, if so many can be thought of and produced against it."[142] What unwieldy probabilities could not decide, the instrument would cleanly determine. To the "grand objection of the *Anti-copernicans* . . . [which] kept me from declaring absolutely for the *Copernican* Hypothesis," the solution was to build a telescope

better than any of the instruments that Tycho and Riccioli had used—a veritable "Archimedean Engine that was to move the Earth."[143]

With a fanfare appropriate to his installation as the Royal Society's curator of experiments, Hooke had already announced in his *Micrographia* an optimism in the power of optical instrumentation so great as to overwhelm any and all infirmities of the senses. That work is filled with an unrestrained modernist confidence in what he called "the certainty of Particulars," "the plainness and soundness of Observations on material and obvious things," and the use of instruments or "Artificial Organs." Stunning fold-outs showed everything from flies and nettles to lunar features and instrumental designs, all yielding an abundance of striking visual representations of the microscopic world and a few of the far-distant heavens by which Hooke hoped to persuade. Hooke displayed a new sort of skill in managing these visualities. "These drawings," as Catherine Wilson has observed, "were composites, not records of individual observations."[144] Hooke thus enhanced the credibility of his representations both by cleaning up the idiosyncrasies of individual specimens and by using pictures of the instruments themselves to trumpet their precision. He claimed superiority for his well-articulated pictures, taken together, over the traditionalists' visually deficient "infallible Deductions, or certainty of Axioms." Hooke was clearly indebted to Francis Bacon's strictures that gathering and cataloguing particulars was the necessary prelude to explanation. He was also careful to reserve for a different work the redeployment of his description of particulars "into Philosophical Tables, as may make them most useful for the raising of Axioms and Theories." In sum, the *Micrographia*'s caution about inferring too far beyond particulars was unique to its problematic and disciplinary subject matter; it was directly comparable to Robert Boyle's reticence concerning sure explanations for the phenomena produced by his air pump.[145]

By contrast, the issue at stake in this phase of the Copernican question was the standard taken to be appropriate to a different problem and disciplinary practice. Hooke's *Attempt to Prove the Motion of the Earth* was concerned not with knowledge claims based on the amassing of particulars or their deployment and assessment in gentlemanly conversation.[146] His project on that occasion was with a quite different aspect of Baconian logic: the so-called "Fingerpost or Crucial Instance," whereby of two hypotheses, if one is refuted, the remaining one is established.[147] To Hooke, that sort of logic—a once-and-for-all demonstration—seemed applicable in the controversy between opposing world systems but *not* in experimental philosophy, where, as Hobbes reminded Boyle, one could not be sure that all causes had been countenanced. It was becoming apparent that *different proof structures* might be needed in different areas of natural knowledge. With the Bible now excluded as an arbiter of natural knowledge, Hooke confined himself entirely to the ordinary realm of natural explanation. In the domain of natural history, Hooke would show himself to be confident in his descriptive goals and reserved—even deflationary—in his inferences to explanations; whereas in the world-systems controversy, his pronounced demonstrativist ambitions would prove, in due course, to have been premature and overconfident.[148]

THE COPERNICAN QUESTION
CLOSURE AND PROOF

Twenty years after Galileo's first encounters with Rome, his trial dramatically transformed the Copernican question from a hypothesis about planetary ordering, known as a famous alternative to traditional astronomical schemes, into a European-wide controversy freighted with divisive confessional and disciplinary import. After 1633, identifying with the Copernicans was clearly associated with having decided for the modern way not just in astronomy but also in natural philosophy. To contemporaries, traditionalists and moderns alike, it seemed a very great controversy of longstanding in an age marked by a relentless stream of religious controversies—at a time when *scientific revolution* had not yet become the preferred trope of change that it would be for twentieth-century historians of science.[149]

Yet in spite of the consolidation of arguments, the diversity already evident at the beginning of the century persisted among Copernicus's mid-century followers. To identify oneself publicly with the Copernican arrangement or to declare its truth did not entail allegiance to the uniform set of commitments in natural philosophy evoked by the nineteenth- and twentieth-century term *Copernicanism*. The Copernican question achieved closure—an end to questioning and criticism from competing alternatives—in different ways among different audiences. These endings occurred through no single proof and with audiences as variously overlapping as almanac readers, practicing astrologers, planetary table makers, extraterrestrializers, itinerant scientific lecturers, and, of course, philosophizing astronomers and high-end, new-style natural philosophers.

It is a "standard move," as Peter Dear cogently remarks, to end accounts of seventeenth-century natural philosophy with Newton; but what kind of closure did Newton provide to the Copernican question?[150] For Thomas Kuhn, "Copernicanism" between Kepler and Newton functioned as a positive heuristic, a theory-laden engine that motivated new and fruitful questions, like "What moves the planets?" or revitalized old doctrines, like the infinite universe and atomism, and recast them in a form amenable to "scientific" quantification. Undoubtedly there is some truth in this formulation—as there is in so much of Kuhn's writing—although he overestimated both the inevitability of the process and the extent to which Copernicans as a group were uniquely responsible for these developments.[151] Moreover, because of the original exclusion of astrology from the conceptual scheme that Copernicus was said to have replaced, the relationship between planetary order and astral influence had no place in that narrative and, hence, as I have been at pains to emphasize, was not seen as part of the problem of closure. "The construction of Newton's corpuscular world machine completes the conceptual revolution that Copernicus had initiated a century and a half earlier. Within this new universe the questions raised by Copernicus's astronomical innovation were at last resolved and Copernican astronomy became *for the first time physically and cosmologically plausible.* . . . Only as Copernicanism became credible, through the dissemination and acceptance of this new conceptual framework, did the last significant opposition to the conception of a planetary earth disappear."[152]

There is a further historical difficulty in Kuhn's concluding reflections that merits comment. Well before Newton, a good number of practitioners regarded Copernicus's ordering of the planets as not merely plausible but probable or the best of the available alternatives or even divinely revealed. Riccioli, the Church's most respected opponent of Copernicus after Galileo's trial, even assigned enough probability to the heliocentric scheme, on Astraea's scales, to create (in his terms) a credible, plausible counterweight to his own world scheme.[153] He did not dismiss Copernicus's system as an absurdity, as was common among traditionalists in the previous century. On the other side of the balance, Robert Hooke accorded enough probability to Riccioli's assessment that he thought it necessary to address and overrule it with a crucial astronomical measurement. Meanwhile, after 1644, proponents of Cartesian corpuscular principles rapidly broke into the philosophy curricula across Europe in a way that Kepler and Galileo could never have imagined about twenty or thirty years earlier. At Leiden and Utrecht, where students from Germany and Central Europe often came to study, Descartes's curious, but certainly "credible" version of the Copernican arrangement—ill-suited, to be sure, for purposes of prognostication— first began to take hold.[154] Across the channel, the astronomer-astrologers Wing and Streete regarded Copernicus as being of decided utility to prognostication; but in Cambridge, the Neoplatonist Ralph Cudworth (1617–88), who, like Newton, regarded Descartes's corpuscular views as atheistic, accepted Copernicus's theory as established but harbored no prognosticatory ambitions.[155] If the Copernican question is seen to involve the securing of a range of commitments to a heliostatic ordering of the planets, then it had multiple kinds of closures, and Newton's was but one of them.

As we have seen time and again, the Copernican question was no simple story of astronomers and natural philosophers battling against traditionalists to demonstrate the truth of one and only one system. Astronomy's chief raison d'être in the seventeenth century was still the making of successful astral prognostications. Systematic doubts about astrology had traveled well—in fact, extremely well. For a brief time at midcentury, it appeared that the momentary flourishing of English Keplerian-Copernican astrologers had finally brought closure to Piconian doubts concerning the accuracy of practical astronomy and, at least by implication, the premises of theoretical astronomy. But the securing of such quantitative foundations failed to improve the astrologers' forecasts: although more accurate planetary tables and Keplerian planetary theory were mutually supportive, the astrologers still disagreed among themselves.[156] Worse still, the astrologers could be spectacularly wrong for reasons having nothing whatsoever to do with the accuracy of theoretical or practical astronomy, as in the infamous solar eclipse of March 29, 1652 ("Black Monday"), which did indeed temporarily darken the skies but still failed embarrassingly to produce the horrendous, eternal, millennial darkness that had been forecast.[157] Moreover, the heyday of the philosophizing astronomer-astrologers was rapidly being superseded by new-style natural philosophers like Descartes, Hobbes, and Gassendi, for whom the accuracy of the tables was less important than how celestial order fitted into the compass of their general epistemic principles, theologies, and political concerns. Further, most of these new philosophical systematizers maintained a distinct hostility to astrology. And there was an audience for the skeptics as much as for the prognosticators: Gassendi's Piconian attack on the "Superstitious Credulity" of the "Star-Prophets" easily found an English translator (an anonymous "Person of Quality") and a London printer willing to publish it in 1659.

Newton shared in the Restoration's general skepticism about astrology, which was already widespread among leading, modernizing natural philosophers—although not, it would seem, because he had read Pico but because astrology offended his theological convictions as much as those he held in natural philosophy. It is ironic that at just the moment that Newton withdrew the support of his system of the world from astrology, he was able to provide a new kind of closure to the problematic of theoretical astronomy raised by the Copernican question, supplying Keplerian astronomy with unprecedented physical foundations. The matter is complicated, as with many earlier thinkers, by what was put into print and what remained in manuscript. Recent commentators are in more solid agreement about Newton's public grounding of the Copernican system than about its relation to his private ruminations. The Principia publicly derived the Earth's motion around the Sun neither from the Kepler-Gilbert magnetic solar force nor from Kepler's trinitarian celestial theology, but rather from a novel congress of active and passive principles. Newton's gravity acted centripetally toward the Sun across a space devoid of all Cartesian matter, its power varying inversely as the distance was found to be squared. When acted upon by this inverse-square force, the Earth, passively resistant to changes of state, described Keplerian equal areas in equal times with respect to the Sun and was deflected into a stable elliptical orbit, that stability being effectively the outcome of a balance of gravitational and inertial forces.

Newton's presentation differed from Hooke's otherwise remarkably similar model in that gravity was, for the first time, accessible—indeed, operationally defined—as a quantity, measurable only through its effects.[158] The much-celebrated achievement was unprecedented both in its extraordinary unification of diverse phenomena and also in its generality: the same laws unified the planets, the tides, subvisible particles, and comets while also predicting the interactions between bodies still undiscovered. As with Descartes, the planetary ordering was seen to be simply a deductive outcome of the corpus of laws that informed the whole project. "The Copernican system is proved *a priori*," wrote Newton; "for if the common center of gravity is calculated for any position of the planets it either falls in the body of the Sun or will always be very close to it."[159] As for comets, like Kepler and Descartes, Newton took them out of the extraordinary realm; but, unlike them, he produced a model that for the first time made one class of these entities predictable and recurrent; and for good measure he produced an explanation of why they were necessary to the general economy of the universe.[160] This was quite the reverse of what had occurred in the previous century, when prognosticators of all kinds had simply failed to predict the appearances of comets. Newton's new analysis of comets allowed him to collapse heavenly influence and catastrophic import into a very small class of predictable, periodic events. Comets could cause physical effects, now including collisions with the Earth: for Newton's followers Edmund Halley and William Whiston, they carried much the kind of millennialist import ascribed in the previous century to planetary conjunctions.[161] But importantly, in this new accounting, the planets were entirely stripped of both astrological influence and apocalyptic meaning.

Yet, if the remarkable unification of diverse phenomena was compelling to Newton's dedicated followers, it did not end debate about alternatives in natural philosophy, because the account notoriously still left vulnerable to Cartesian criticism the nature of gravity's cause. It was evi-

dently this intractable problem that led Kuhn to assess Newton's conceptual achievement as (descriptively) credible rather than (demonstratively) proved: "Though the achievements of Copernicus and Newton are permanent, the concepts that made those achievements possible are not. Only the list of explicable phenomena grows; there is no similar cumulative process for the explanations themselves."[162] In this moderately continuist, Duhemian passage from *The Copernican Revolution*, the full bite of Kuhn's revolutionary paradigm shift is not yet evident.

But for contemporaries, the controversy went on in its own local registers. In seventeenth-century natural philosophy, to leave unaddressed a cause as fundamental as gravity was, at worst, to invite rejection under the charge of occult unintelligibility and, at best, to invite dispute.[163] Not only was Newton aware of the difficulty, but, as J. E. McGuire and P. M. Rattansi showed in a classic paper, he engaged in a prolonged and intensely private effort to find a complementary grounding for his natural philosophy in ancient historical and theological sources.[164] The question of what Newton chose to make public and what he withheld has since occasioned much discussion among commentators. In Rob Iliffe's apt formulation, "Newton initially envisaged a treatise quite different from the *Principia* as it eventually appeared."[165] Important indications of what he intended for a subsequent edition leaked out through a small circle of trusted friends and disciples. Essentially, Newton saw himself as recovering an ancient, prelapsarian wisdom that the moderns later defiled, distorted, and rendered idolatrous: the Egyptians already knew and taught the Copernican system in its Newtonian version, and it was passed down through Pythagoras, who concealed it in the form of mystical allegories. For Newton, the idolatry began when men started to worship the stars, to assign them human characteristics and powers: "To make this hypothesis the more plausible they feigned that ye stars by virtue of these souls were endued with the qualities of ye men ['& according to these qualities governed the world . . . ']. And by means of these fictions ye soules of ye dead men grew into veneration wth ye stars and by as many as received this kind of theology were taken for ye Gods wch governed the world."[166]

From Ricciolian angels to Cartesian matter to planetary influences, Newton would have no truck with entities mediating between God and the universe. Throughout his writings, he worked to make divine activity in nature preeminent. "This most beautiful system of the sun, planets, and comets, could only proceed from the counsel and dominion of an intelligent and powerful Being," he famously wrote in the General Scholium to the 1713 edition of the *Principia*, where, more explicitly than in the first edition, he tried to manage that work's natural-theological aims.[167] In Newton's private rendering of the ancient wisdom, this celestial order was explicitly divorced from the idolatrous order of astral influences. In its public rendering, he achieved by silence the erasure of traditionalist astrology as well as of the reformed astrology of Kepler and his many English acolytes. Some late-seventeenth-century moderns who followed Newton in natural philosophy produced refutations of astrology's premises that differed little from those articulated by Pico in 1496.[168] Long before it became philosophically tenable, following Duhem and Quine, to justify salvaging refuted scientific hypotheses by making pragmatic adjustments to them, practicing astronomer-astrologers diluted, or altogether evaded, Piconian refutations of their predictions and theories by seeking to improve their planetary tables and change their astronomical theories.[169] Unlike their predecessors in the long sixteenth century, however, late-seventeenth-century high-end reformers of natural philosophy simply stopped trying to fix the astrological division of the science of the stars. And when the moderns ceased to provide theoretical explanations for astrology's chronic predictive difficulties, they deprived astrology of new resources and credibility to compete with emergent projects, such as social mathematics, that privileged proximate causes and calculable probabilities in the study of human uncertainty about the future.[170]

NOTES

INTRODUCTION

1. Many such questions were also integrated into commentaries on John of Sacrobosco's *Sphere* (see, for example, Thorndike 1949, 30–31). See further Edward Grant's valuable "Catalog of [400] Questions on Medieval Cosmology, 1200–1687," in Grant 1994, 681–741.

2. "In addition, they invent another earth, lying opposite to our own, which they call by the name of 'counter-earth' " (Aristotle 1939, bk. 2, chap. 13, 217; cf. bk. 2, chap. 14, 241–45). There are some differences in how sixteenth-century translators rendered the Greek, e.g., "Pythagorici autem habitantes Italiam contradicunt illis, et dicunt quod ignis est positus in medio, et quod terra est stellarum una, et revolvitur circulariter, et ex motu eius circulari fit nox, et dies, et faciunt aliam terram, quam vocant antugamonani" (Aristotle 1962a, V, fol. 146K–L); "Pythagorei, dicunt in medio enim ignem esse inquiunt: terram autem astrorum unum existentem circulariter latam circa medium, noctem, & diem facere. Amplius autem oppositam aliam huic conficiunt terram, quam antichthona nomine vocant" (Aristotle 1597, T. 72, 643–44).

3. See Robin Bruce Barnes's excellent treatment of this problem (1988, 1–59, 73–75).

4. See Smoller 1994, 3–4; Smoller 1998.

5. Columbus to a member of the royal court (1500) on his return in chains from his third voyage to the "New World Indies," quoted by Watts 1985, 73.

6. See Osiander 1527, well discussed in Scribner 1981, 142–47.

7. See Kuhn 1957, 93–94. Kuhn's exclusion of astrology was entirely typical of other "big-picture" narratives of his historiographical moment: Koyré 1992; Blumenberg 1987, 1965; Zinner 1988. See Westman 1997.

8. North 1975; LeMay 1978. North credits Giacon (1943) as the earliest modern historian to maintain that "Copernicus was an astrologer." Charles Webster was not able to connect Copernicus himself to the wider themes of astrology and Christian eschatology, but he was looking for the right kinds of connections (Webster 1982, 15–47). See also my brief preliminary study anticipating the present work (Westman 1993).

9. Whewell 1857, 271–331.

10. Quoted from Kuhn's classic *Copernican Revolution* (Kuhn 1957, 230). By now it is well established that Copernicus's book was widely consulted (see Gingerich 2002). For recent appreciations, see Westman 1994; Swerdlow 2004a.

11. In reading Kuhn's *Structure of Scientific Revolutions,* Ernan McMullin offers an illuminating distinction between "deep" and "shallow" revolutions, suggesting that the Copernican episode exemplified "a revolution of a much more fundamental sort because it involved a change in what counted as a good theory, in the procedures of justification themselves. . . . And what made it revolutionary was . . . the very idea of what constitutes valid evidence for a claim about the natural world, as well as in people's beliefs about how that world is ordered at the most fundamental level" (McMullin 1998, 123).

12. The classic statement of this view is Geertz 1983, 55–70.

13. On the perils of extreme localism, see Dear 1995, 4, 245–46; Schuster and Watchirs 1990, 38–39.

14. Regiomontanus 1496, in Regiomontanus 1972, fifth conclusion, 68.

15. Here it must be emphasized that practitioners were neither making their own observations nor in some sense "testing" earlier theories against new evidence but utilizing what Bernard Goldstein has aptly characterized as "a literary tradition of scientific treatises" (Goldstein 1994, 189).

16. How Apollonius and Hipparchus interpreted the *choice* between these different hypotheses is a separate matter. Duhem (1996) held that the Greeks, in general, saw the choice as entirely one between theories used as calculational instruments with no claim to truth; but Lloyd (1978) has raised serious questions about Duhem's readings of the Greek sources.

17. Ptolemy 1998, bk. 3, chap. 4, 153.

18. See Swerdlow 1973, 472: "It is even possible that, had Regiomontanus not written his detailed description of the eccentric model, Copernicus would never have developed the heliocentric theory."

19. Copernicus 1543, bk. 2, introduction, fol. 27v; see also bk. 3, chaps. 15, 20, 25. For less literal translations than mine, cf. Copernicus 1978, 51; Copernicus 1976, 79.

20. Copernicus 1543, bk. 1, chap. 8/Copernicus 1978, 16; Virgil, *Aeneid*, bk. 3, 72.

21. Kuhn 1957, 23–24, develops the analogy of a merry-go-round ticket collector to assist understanding of the Sun's daily and annual motions.

22. Copernicus 1543, bk. 1, chap. 9, 7r–v; book 5, 133v–134/Copernicus 1978, 227–29: "Primum non iniuria motum commutationis dicere placuit. . . . Nam motus commutationis nihil aliud esse dicimus. " See also Walters 1997.

23. Aiton 1987, 23. Albert of Brudzewo lectured on Peurbach at Krakow; in 1494, it became the earliest printed commentary on that work. Aiton suggests that Peurbach's comment may have played an important role in suggesting to Copernicus the primacy of the Sun (9).

24. Copernicus 1978, bk. 4, 173: "The moon, taken by itself, gives no indication that the earth moves, except perhaps in its daily rotation."

25. On the phases of Venus, see Thomason 2000. On the gaps between the spheres, see Van Helden 1985, 46–47. On the ontology of the spheres, see Aiton 1981; Jardine 1982; Grant 1994, 346; Lerner 1996–97, 1:121–38, 2: 67–73; Goddu 2004.

26. Rheticus 1971, 137; Rheticus 1982, 107.

27. See Van Helden 1985, 50–52.

28. Ptolemy 1998, bk. 1, chap. 7, 45. Ptolemy also considers and rejects the possibility that the air surrounding the earth carries objects around, for either they would be left behind by the more rapidly moving air or, if "fused" to it, would never appear to change position.

29. See Lloyd 1979, 25–28, 71, 73–74, 76–78, 205–6.

30. Copernicus 1978, bk. 1, chap. 8, 16–17; for intelligibility, see Dear 2006, 8–14.

31. Copernicus 1978, bk. 1, chap. 9, 18; for an exhaustive study of the possible sources of this theory, see Knox 2005, esp. 203–11.

32. See Wallis in Copernicus 1952, 528–29; Toulmin 1975, 384–91; McMullin 1998, 134–35. For entail-

ments of the heliocentric theory that Copernicus might have—perhaps should have—foregrounded in *De Revolutionibus,* see Swerdlow 2004a.

33. Of course, even if false, potential explanations may be very appealing (see P. Lipton 1991, 56–74).

34. Copernicus 1978, preface, 5; bk. 5, 227.

35. For what Duhem did and did not hold, see Ariew 1984.

36. Duhem 1894, 85.

37. See esp. Laudan 1990, 320–53.

38. See the clear and helpful discussion of Klee 1997, 65–67.

39. Duhem 1894, 87.

40. Philip Kitcher maintains that the question of underdetermination is now a philosophical commonplace (1993, 247–56); for its widespread influence in science studies, see Kuhn 1957, 36–41, 75; Kuhn 1970, 4; Dietrich 1993; Zammito 2004.

41. Duhem believed that eventually physical theories do reach ultimate truth, but he was wary of making untimely metaphysical claims. See Roger Ariew and Peter Barker in Duhem 1996, xi.

42. For the seventeenth-century theme of the use of divine powers, see Funkenstein 1986.

43. Duhem 1908, 150–51. Duhem's attribution to Bellarmine of the omnipotence argument is hasty and unwarranted.

44. Duhem 1996, 150–83; Finocchiaro 2005, 266–69. Little is known of Urban's beliefs about the natural world, although it is clear that, apart from his traditionalism in natural philosophy, he subscribed to a belief in astral forces (D. Walker 1958, 205–12).

45. Lloyd 1978.

46. Barker and Goldstein 1998.

47. Clavelin 2006, 16–17.

48. John Marino grounds the periodization of the Italian states ca. 1450–1650 in economic history (Marino 1994); for M. S. Anderson, it is armed struggle among the European states from the French invasion of Italy to the beginning of the Thirty Years' War (1998); in contrast, Eric Hobsbawm has argued for a "short twentieth century" (1994).

49. See Westman 1975b.

50. Schilling 1981, 1986; Headley 2004, xvii–xxv; Brady 2004.

51. In some of my earlier writings, I worked in the historiographical framework that organized the period 1543–96 around the Copernican proposals. See, for example, Westman 1975b.

52. Voelkel 2001, 130–210.

53. For Victorian representations, see Barton 2003. Rudwick argues that the undifferentiated term *savant* or *learned* was the predominant designator (2005, 22–23).

54. For an introduction to different ideas of progress, see Ginsberg 1973.

55. See Oreskes 1999.

56. Kuhn 1970, 1–9; Duhem 1996, 79.

57. See, for example, I. Cohen 1985, 161–75 ("The Newtonian Revolution"). For a recent popular distillation, see Gleick 2003. I do not mean to suggest that the precise sense of Newton's contribution is without philosophical interest.

58. S. Weinberg 2001, 194–95; S. Weinberg 2008. Weinberg means here two things: that Newton's geometric style is no longer the language in which physicists understand Newton's laws; and that although theological considerations were personally important to Newton, they played no role in the formulation of his laws of motion as they were later accepted.

59. Jardine 1991, 160–67.

60. See Hall 1980; Bertoloni Meli 1993; Dear 2001, 164–67.

61. Lawrence and Molland 1970 (italics in original). For an interesting historicization of this passage, see Friesen 2003, 179–81.

62. Cf. Kuhn's and Duhem's understanding that explanations good at one moment drop away in another while fueling the continuous advance of descriptive laws of nature. For discussion of these issues, see Westman 1994, 83–85.

63. S. Weinberg 2001, 158.

64. Ibid., 136–37.

65. Weinberg says that "the languages in which we describe rocks or in which we state physical laws are certainly created socially, so I am making an implicit assumption . . . that our statements about the laws of physics are in a one-to-one correspondence with reality" (ibid., 150). For Descartes' position on eternal truths, see Funkenstein 1975a.

66. See Grant 1994, 350.

67. Much worse monsters—like nuclear weapons and the human immunodeficiency virus—have taken their place.

68. Baade and Zwicky 1934.

69. See Marschall 1994, 101–6. The *Oxford English Dictionary* gives this example from *Chain Store Age*, 1933: "The 'One-stop-drive-in super-market' provides free parking and every kind of food under one roof." The invincible flying "superman" who helps the weak was not introduced until the summer of 1938 in *Action Comics*; prior to that time, the term seems to have had primarily a Nietzschean meaning, connoting a superior kind of man.

70. Geertz 1983, 55–70; Jardine 1991.

71. However, Ludwik Fleck's "harmony of illusions" has the advantage of being more flexible than Kuhn's "paradigms" (Fleck 1979).

72. As Ernan McMullin has observed, Kuhn later departed from this radical view, holding that a small group of criteria involved in theory choice persist unaffected across theoretical divides (McMullin 1993, 125–26).

73. Alan Richardson argues that Kuhn allowed himself to be bamboozled by philosophers into accepting a notion of incommensurability as a strictly semantic notion (and hence one of intertranslatability), whereas his original text shows that scientists in the world of the new paradigm do not merely believe differently but also live and work differently (Richardson 2002; also Hacking 1993); Steven Weinberg rejects Kuhn's incommensurability as incompatible with anything that corresponds to his experience as a physicist and rightly calls attention to the fact that "Kuhn himself in his earlier book on the Copernican revolution told how parts of scientific theories survive in the more successful theories that supplant them, and seemed to have no trouble with the idea." But in a later essay, he says that Kuhn's account of the shift from Aristotelian to Newtonian physics was indeed a "paradigm shift" (Weinberg 2001, 187–206, 269).

74. For an analytical usage, see Gingerich and Westman 1988.

75. Certeau 1984, xi–xiv, 29–39.

76. See Lerner 2005.

77. On the new social dynamics made possible by the steam press in the early nineteenth century, see Secord 2000. As Bernard Lightman remarks, the invention of "-ist" labels was an accepted neologizing practice among late-nineteenth-century members of the Metaphysical Society (Lightman 2002).

78. De Morgan 1855, 5–25. On the importance of historical legitimation to mathematics, see Richards 1987.

79. De Morgan 1855, 5–6.

80. Ibid., 16, 18.

81. Masson 1859–94, 6:525–45.

82. See Jay 1984, 21–80. For a sophisticated defense of the use of present analytic categories, see Jardine 2003.

1. THE LITERATURE OF THE HEAVENS AND THE SCIENCE OF THE STARS

1. Albumasar 1994. This stripped-down version of the *Great Introduction* was translated into Latin by Adelard of Bath (ca. 1080–1152) early in the twelfth century. See also Lemay 1962; Lipton 1978.

2. For example, a cold diet might be recommended to counteract a hot disease (French 1994, 30–59, 39–42). For a good example and fine color reproduction of a 1399 zodiac man, see Page 2002, 56, fig. 46.

3. Arrizabalaga 1994, 237–88, 245. On planetary conjunctions, see North 1980; Smoller 1994; Hayton 2004.

4. Pedersen 1978a, 304; Kuhn 1957, 94: "Astrology provided the principal motive for wrestling with the problem of the planets, so that astrology became a particularly important determinant of the astronomical imagination." On the problem of writing the history of a disease entity, like syphilis, see Fleck 1979.

5. Boudet calls this "the most important astrological library possessed by an individual in Europe at the end of the fifteenth century" (1994, xi).

6. Ptolemy 1940, bk. 2, chap. 1, 117–21.

7. On estimating numbers of copies, see Köhler 1986; Niccoli 1990, 1–12; Wagner 1975. Richard Kremer has been studying the literature of astrological prognostication for some years, and when published, his work will improve significantly on Zinner's statistics.

8. The essential bibliographical source for these claims is Zinner 1941, supplemented by Grassi 1989; Hain 1826–28; and Houzeau and Lancaster 1882–89.

9. Zinner 1990, 110–17; Lowood and Rider 1994, 4–8.

10. On the origins and geography of printing, see Febvre and Martin 1984; Eisenstein 1979; Hirsch 1967; Krafft and Wuttke 1977; Chrisman 1982; Lowry 1979, 1991; Tyson and Wagonheim 1986; Chartier 1989; Johns 1998.

11. See Westman 1980a, n. 68. I presented additional evidence at the 16th International Congress of the History of Science, Bucharest, September 1, 1981.

12. Two scholars deserve credit for raising and pushing the matter as far as then possible: North (1975, 169–84) and Lemay (1978). See also Thorndike 1923–58, 5:419 ff.

13. Rosen 1984b, 111; see also Kuhn 1957, 94: "It may even be significant that Copernicus, the author of the theory that ultimately deprived the heavens of special power, belonged to the minority group of Renaissance astronomers who did not cast horoscopes."

14. Copernicus 1978, 344. Note that Rosen does not distinguish between different types of astrology.

15. See Webster 1982, 15: "Copernicanism did not directly confront judicial astrology, but there can be no doubt that the Copernicans of the seventeenth century led the trend against judicial astrology, so finally emancipating astronomy from its bondage under medicine." In other respects, I find myself in close agreement with many of Webster's prescient conclusions, which are not inconsistent with the views I develop in the present work.

16. Hutchison 1987.

17. Dreyer 1953, 332–33. Willy Hartner and Owen Gingerich also share Dreyer's view (see Gingerich 2004, 186–89).

18. Grendler 1989, 222.

19. For notions of ideal form, see Colie 1973; Minnis 1984, 118–59.

20. Turner 1974, 495–531, esp. 510–11; Schaffer 1997.

21. See Weber 1919, 129–56; Farrar 1975, esp. 191.

22. On the development of epistemic hierarchy in the Middle Ages, see McInerny 1983; on the ways it was applied to astronomy, see McMenomy 1984.

23. See McClure 2004.

24. Regiomontanus 1537, 51.

25. This *laus* appears in the suppressed introduction to *De Revolutionibus* (Copernicus 1978, 7).

26. Capuano 1518, fol. 25v, col. b: "Ex hoc patet quod cum subiectum astrologie sit naturale, et modus demonstrandi mathematicus, quod participat nobilitatem scientie naturalis: et certitudine mathematice"; McMenomy 1984, 418.

27. Capuano 1518, fol. 26v b; McMenomy 1984, 239, 419.

28. See Clark 2006, 1989.

29. Among medieval Arabic authors in Latin translation, *scientia astrorum* appears at least in Al-Kindi and *scientia stellarum* in Alcabitius Abdylaziz (Carmody 1958, 84, 149). Charles Burnett found the term *astronodia* in a twelfth-century manuscript; it covered both *astronomia*, the study of the motions of the heavens with instruments, and *astrologia*, study of the heavens without instruments (1987a, 137–38).

30. On *scientia*, see Raymond Williams (1976, 276–80). For the local mathematics training of nineteenth-century Cambridge students , see Warwick 2003; on the post–World War II dispersion of Feynman diagrams, see Kaiser 2005; and on the unique politico-ethical struggles of individual scientists working within the technobureaucratic culture and institutional arrangements of the mid-twentieth-century United States, see Thorpe 2006.

31. Ross 1962.

32. See now Dear 1995. Butterfield 1957 already uses *natural philosophy* (50, 139), but it is not used as a consistent and self-conscious replacement for *science* until much later (e.g., Schaffer 1985; Henry 1997, 4).

33. Copernicus 1978, 5: lines 44–45.

34. Experimental mathematics uses computers to test for consistency with cases that usually have nothing to do with the physical world.

35. Avogadro 1521, 1523; Zacuth 1518; Biblioteca Medicea Laurenziana: James of Spain 1479, fol. 22r.

36. See Zinner 1941, 60.

37. Giuntini 1573, fol. +2v, : "Docteur en Theologie et Aumosnier de nostre trescher et tresamé frere le duc d'Aniou."

38. The full bundle of works does not appear until the editions of Giuntini 1581 and 1583.

39. Giuntini 1583, 540; in the text, however, Giuntini describes him as "magnus Platonicus et doctus philosophus."

40. Ibid: "Fuit orator et Poeta et philosophus celeberrimus et elegantissimus et multos edidit libros elegantissimos, et unum volumen contra Astrologos."

41. Ibid., 550–51.

42. Giuntini 1573, fol. 290b.

43. See entries in Zinner 1941.

44. Goldstein (1967) was the first to attach Ptolemy's authorship to the missing text of the *Planetary Hypotheses*.

45. Aristotle 1963, bk. 2, chap. 2.

46. McKirahan 1978, 199–201.

47. Aristotle 1975, 78b 35–79a 17.

48. Ibid., 79a 9–13.

49. Ptolemy 1998, bk. 1, chap. 7, 45.

50. Taub 1993, 52–58.

51. Stephenson 1987, 28.

52. See Goldstein 1967.

53. Reinhold 1542, fol. C5v.

54. On subalternation, see McKirahan 1978; Mc-Menomy 1984; Livesey 1982, 1985. See also Weisheipl 1965, 1978, 47–49; Lindberg 1982; Wallace 1984b, 99–148. On Osiander, see Westman 1980a, 107–9.

55. Salio 1493, preface, unpag.

56. Schreckenfuchs 1569, 2, my italics:

Unde Astronomicae Disciplinae principia petenda sint. Astrorum Scientiam in duas diuidi partes. Haec scientia despescitur in Astronomiam et Astrologiam. Astronomia est doctrina, quae mediantibus Geometria, et Arithmetica, inquirit, ac demonstrat motus uarios, magnitudines, et distantias corporum coelestium, ut paucis multa dicam, ipsa omnes diuersitates, et mutationes apparentiarum, tam in planetis quam in reliquis stellis, saluat. Astrologia autem est doctrina, quae ex stellarum motu ac uirtute, naturae atque situ diuersos qualitatum et quantitatum motus in corporibus, praedicit. De Astrologia in hoc libello nihil agitur, cum requirat propriam, et specialem tractationem, quae intricatior est, atque latius patet, quàm ut breuibus enarri possit. Titulus libri est de Sphaera, propterea quod continet tractationem de Sphaera, hoc est, de corpore globoso seu rotundo, quod constat ex diuersis circulis, qui ex materiali sphaera, per imaginationem ad coelestem sphaera à discente transferri debent. Quod sit subiectum huius libri, ex praedictis satis superque constat, nempe aliud quàm primum mobile.

57. As does Schreckenfuchs 1556.

58. Ptolemy 1940, bk. 1, chap. 1, 3.

59. Ibid.

60. Benjamin and Toomer 1971, 139. I have made slight changes in the translation.

61. Ibid., 141, my emphasis.

62. Ibid.

63. Poulle 1975; Poulle 1980, 1:42 ff.

64. Benjamin and Toomer 1971, 30–33.

65. One is reminded here of the "equatorie of the planetis" ascribed to Geoffrey Chaucer: "Take therefore a metal plate, or else a board planed smooth, tested with a level, and polished evenly; and when it has been made into a perfect circle by your compasses, it shall be 72 large inches or six feet in diameter. The circumference of this circular board should be bound with an iron rim, just like a cart wheel, so it does not warp or become crooked. If you wish, the board can be varnished, or parchment can be glued over it to give a good surface" (Price and Wilson 1955, 47).

66. Poulle 1975, 100.

67. This amalgamation of different elements of the science of the stars into a single, weighty volume is especially evident in the massive ephemerides produced after the appearance of Reinhold's planetary tables (1551): Carelli 1557; Leovitius 1556–57; Magini 1582, 1585; Origanus 1609.

68. Pedersen 1978b, 160; for the Erfurt theoric, see Poulle 1975, 101, 108. The Adler Planetarium in Chicago holds a beautiful series of such hands-on, didactic models from early-sixteenth-century Italy. A student could manipulate the thick paper tabs attached to the orbs to help visualize various kinds of relationships, such as between true and mean motions.

69. Consider Christian Wursteisen's question: "Is not Mercury's equant a special kind of orb?" which he answers thus: "It is not an orb but a circle with a certain eccentricity, as with Venus and the three superiors, that the deferent orb describes in the imagination." (Wursteisen 1573, 214). See also Frischlin 1601, bk. 4, chap. 5, 221, 228; and Kepler 1937–, 7:293, ll. 27–28: "Aristoteles, solidis orbibus coelum refertum credens (licet aequivocae materiae) et philosophi posteriores, quos secuti esse videntur Arabes, et post eos Purbachius Theoriarum scriptor"); M.-P. Lerner states that Peurbach suspended judgment on the ontology of the particular orbs and treated them purely geometrically (Lerner 1996–97, 1:128–29); on Copernicus, see Goddu 2010, 370–80.

70. See Bennett 2003, 142–43; Bennett 1986; Van Helden 1994; Warner 1994.

71. See Aristotle 1961–62, 1, 2.1, 87; Aristotle 1977, xvii. Eustachius a Sancto Paolo, a major seventeenth-century commentator on Aristotle, offers this gloss: "Hujus pars duplex; altera nempe theoretica seu contemplatrix, altera practica seu operatrix: quod colligitur ex Aristotele, 2.Metaph.c.1. quo loco *finem theoreticae* ait *esse veritatem; practicae verò, opus:* Unde illa definitur, Quae in sui subjecti sola contemplatione conquiescit, haec verò, Quae sui subjecti contemplationem ad praxin seu opus refert" (Sancto Paulo 1648 [1609], 1).

72. As Edward Grant points out, the classification scheme of Domingo Gundisalvo (*De Divisione Philosophiae*, ca. 1150) already reflects a complexity occasioned by the arrival of new Greek and Arabic texts (see Grant 1974, 53, 59–76).

73. This omission may account for its absence from recent studies of medieval divisions of the sciences, such as Weisheipl 1978 and McInerny 1983.

74. Ptolemy 1493, fols. 10v–11: *andatores* or *circuitores*.

75. Aristotle 1962b, 339a 31–32; see North 1986b, 46.

76. See Federici-Vescovini 1996.

77. In what sense were Peurbach's theorics physical embodiments of Ptolemy's mathematical models? Olaf Pedersen regarded them as "pseudo-physical spheres . . . a kind of lip-service to the Aristotelians but devoid of real astronomical importance" (Pedersen 1978, 165; see also Lerner 1996–97, 1:128–29); but cf. Shank 2007, 2; Swerdlow 1973, 437; Swerdlow 1996, 188; Jardine 1982.

78. See Pedersen 1978a, 314–22.

79. See Pedersen 1975. Pedersen's neologism for these collections of works is *corpus astronomicum*.

80. To secure this usage, further study of the term in McMenomy's more extensive survey of such collections would be required (1984, 481–516).

81. See Grant 1974, 67–68, 71.

82. Densmore 1995, 1.

83. Contrast this with Peter Galison's account of the situation in late-twentieth-century physics: "The intellectual and social world of the experimenter is different from that of the theorist. Arguments take place in different physical and social settings, with different standards of demonstration" (Galison 1997, 9).

84. Cf. Biagioli 1989.

85. In 1888, Oliver Heaviside wrote a letter to *The Electrician* titled "Practice vs. Theory: Electro-Magnetic Waves," in which he stated: "The duty of the theorist [is] to try to keep the engineer . . . straight, if the engineer should plainly show that he is behind the age, and has got shunted onto a siding. The engineer should be amenable to criticism." Quoted in Hunt 1983, 353; Kline 1995.

86. See Morell and Thackray 1981, 19–20, 96. As the authors point out, the immediate occasion for Whewell's term was the 1833 meeting of the Cambridge Meeting of the British Association, at which Samuel Coleridge pushed for the formation of a national church of intellect, a "scientific clerisy" modeled after "the Reformation ideal of a clergyman and a schoolmaster in every parish."

87. In the America of the nineteenth-century Gilded Age, the word *professional* carried clear commercial connotations that provoked cultural anxieties about the term *professional scientist*. See Lucier 2009, 723–32; Ross 1962.

88. For the role of Avicenna as a basis for the medical curriculum, see Siraisi 1987, 10–12.

89. See Aristotle 1936, 215–17: *De sensu* 436a18–b2; quoted in Schmitt 1985, 8; Siraisi 1987, 97.

90. On the development of the humanist tradition in Renaissance medicine, see, inter alia, Schmitt 1985; Nutton 1985; Wear 1985.

91. "Ad lecturam tertij Avicenne de egritudinibus acapite usque ad pedes," as we read in the rolls for 1480–81 (Dallari 1888, 1:112). See Park 1985, 60, 245–48; Siraisi 1990, 152; Siraisi 1987, 55–56.

92. See Wear 1985, 123.

93. Marquardi 1589. Marquardi was ordinary professor of medicine at Bologna.

94. Thus, a typical requirement is that of 1482–83: "Ad Astronomiam de mane diebus continuis et ordinarijs. D. M. Hieronymus de Manfredis, cum hoc quod faciat iudicium et tacuinum" (Dallari 1888, 1:118).

95. See Thorndike 1923–58, 4:232–42. The judgment begins: "Iohannis Pauli de Fundis Tacuinus astronomico-medicus." Undoubtedly de Fundis composed other judgments, but this is the only one of whose existence I am aware.

96. Mendoça 1596.

97. Gaffurio 1967; Gaffurio 1979. See also Gaffurio 1993, 1969; Moyer 1992, 66–77.

98. Swetz 1987, 33. See also Van Egmond 1980; Strong 1936; Schrader 1968.

99. See Sanford 1939; Swetz 1987, 29.

100. Swetz points out that " 'Do' is a characteristic feature either implicitly or explicitly given in the text. Latin arithmeticians used to write 'Fac ita,' 'do it thus,' and the Germans, 'Thu ihm also,' 'do it as before.' " (Swetz 1987, 195).

101. Ibid., 40.

102. Vieri 1568, 114–15; cited in Crombie 1977, 75.

103. Papia 1482: *Damnum, Error, Fictum, Gradus, Hereticus, Ignorancia, Judeus, Mulier, Notarius, Officium, Sciencia, Socius.*

104. Frisius 1548, 11: "Delibatis theorices huius artis principiis, ad praxim accingimur, quam et si variis multisque instrumentis possemus docere, cum tamen inter omnia nullum tam perfectum, tam generale sit, quod tantum praestare possit. "

105. Lomazzo 1584, 17: "Pittura è Arte laquale con linee proportionate, et con colori simili à la natura de le cose, seguitando il lume perspettiuo imita talmente la natura de le cose corporee, che non solo rappresenta nel piano la grossezza, & il rilieuo de' corpi, ma anco il moto, e visibilmente dimostra à gl'occhi nostri molti affetti, et passioni de l'animo."

106. Ibid., bk. 6, 279: "Della virtù della prattica." Lomazzo also believed that theory was a property of the soul and immortality and that it could counteract the decline of the corruptible body on which artistic practice depended (see Campbell 2002).

107. Rothmann 1595.

108. Fandi 1514.

109. "The mingled influences of the stars can be understood by no one who has not previously acquired knowledge of the combinations and varieties existing in nature" (Ptolemy 1822, 153).

110. Curiously, astrology is not usually seen as one of the areas characterized by humanist practices (see, for example, Mann 2004).

111. Pedersen's study of such collections has been extended by McMenomy (1984, 481–528).

112. The situation is analogous to the one encountered by Charles Schmitt in the early 1970s (Schmitt 1973).

113. Ptolemy 1484a.

114. For example: "Judgment must be regulated by thyself, as well as by science; for it is not possible that particular forms of events should be declared by any person . . . since the understanding conceives only a certain general idea of some sensible event, and not its particular form. It is, therefore, necessary, for him who practices herein to adopt inference [*conjectura*]. They only who are inspired by the deity [*numine*] can predict particulars. . . . Love and hatred prohibit the true accomplishment of judgments; and, inasmuch

as they lessen the most important, so likewise they magnify the most trivial things" (Ptolemy 1822, 225).

115. For a brief but useful summary, see Sudhoff 1902, 8–9.

116. The title of Johannes Stadius's translation clearly specifies the joining of medicine and mathematics: "Hermetis Trismegisti Iatromathematica (hoc est medicinae cum mathematica coniunctio) ad Ammonem Aegyptium conscripta" (Sudhoff 1902, 10–13). There is no reference to this writing in Yates 1964.

117. For the House of Scotto, see Bernstein 1998, 29–54.

118. Ptolemy 1493.

119. Herzog August Bibliothek Wolfenbüttel copies: shelfmark 12 Astron. 2; 9.4 Astron.2°; 20.3. Astron. 2°.

120. Ptolemy 1493, repr. Venice, February 1519. Copy used: British Lib. 8610.f.22.

121. Ptolemy 1519.

122. "Reuolutionem annorum mundi uel natiuitatis alicuius," etc. (Alfonso X 1483, fol. a4); see also, Lilly 1647, 738–41.

123. Ptolemy 1533, fol. a1v: "Ornatissimo Simul ac Doctissimo Viro Othoni Brunfelsio, Medicinae Doctori, Nicolaus Prucknerus."

124. Richard Harvey, however, complained in his copy (now in the British Library) about the editing: "Pruckner, you would have done well if you had better corrected your Otto's little book" (Ptolemy 1533, fol. a2). On Brunfels's role in the Strasbourg humanist community, see Chrisman 1982, 51.

125. Tamsyn Barton (1994) questions whether it had any real practical value.

126. Following the *Tetrabiblos* itself came Giovanni Pontano's Latin translation of the *One Hundred Aphorisms*, Nicolaus Leonicus's translation of Ptolemy's *Inerrantium stellarum singularum significationes*, and a section containing seven works "from the Arabs and Chaldeans": *Hermetis uetustissimi Astrologi centum Aphoris. Lib. I; Bethem Centiloquium; Eiusdem de Horis Planetarum Liber alius; Almansoris Astrologi propositiones ad Saracenorum regem; Zahelis Arabis de Electionibus Lib. I; Messahalah de ratione Circuli & Stellarum, & qualiter in hoc seculo operentur, Lib. I; Omar de Natiuitatibus Lib. III.* The volume concluded with Manilius's astrological poem *Astronomicon,* an unfinished work rediscovered by Poggio Bracciolini in 1416 and published by Regiomontanus in 1472.

127. Ptolemy 1535. Apart from Giovanni Pontano's by-now standard translation of the *Centiloquium,* it included a handful of Camerarius's "little annotations" (*annotatiunculae*) on the first two books, Matteo Guarimberto's "Little Work on the Rays and Aspects of the Planets" (*Opusculum de Radiis & Aspectibus Planetarum*) and Ludwig of Riga's "Astrological Aphorisms" (*Aphorismi Astrologici*).

128. Ptolemy 1535, fol. aa.

129. Ptolemy 1541, fol. a2r. Copy used: Herzog August Bibliothek N.46.20. Helmst.(1).

130. Ibid.: "Quas sculptas in primo huius operis limine posuimus, quoniam magnam lucem ui debantur allaturæ rebus sua natura obscurioribus. . . . Item omnium constellationum figuras graphicè, propter singulare studiosorum commodum, depiximus . . . quae est omnium, quae in Almagesto demonstrantur, epitome & compendium, quod ad reminiscentim conducet plurimum, Georgio Valla Placentino interprete."

131. Abenragel 1551, fols. A3–b: Epistola Nuncupatoria.

132. Ptolemy 1548.

133. Ptolemy 1553. Melanchthon translated the title thus: "Concerning Astronomical Predictions, to which the Greeks and Latins gave the title 'Four Books' " (*De praedicationibus astronomicis, cui titulum fecerunt Quadripartitum*). He also rendered the opening phrase of book I as *praedictiones astrologicae* ("Of the means of astrological predictions "). Isaac Casaubon annotated British Library 718.b.4.(1,2); Andreas Lemmel owned Herzog August Bibliothek 642. Astron.).

134. Ptolemy 1554.

135. Ptolemy 1554, 2: "Haly Heben Rodoan Arabem . . . qui prodierit in lucem tanto authore dignus: Is uero si ueram mentem Ptolemaei uerborum translatione explicatam habuisset, forsan nos hoc labore liberasset. Nunc uero cum neque per se clarus sit liber hic ob breuitatem, neque aliorum expositio quae in lucem nondum prodierit utilis sit, nec quae prodierit Haly ut dixi perfect sit, cogor utilitatis publicae causarum Ptolemaei gloriae ad hunc nouum laborem descendere."

136. Ptolemy 1578.

137. See, for example, Fabri de Budweis 1490, unpag., 2v.

138. Schmitt 1983, 49–51, 121–24.

139. See Bentley 1983, 70–111.

140. See Pomian 1986; Zambelli 1987, esp. 103–8; Gregory 1983.

141. See Thorndike 1923–58, 4:114–31.

142. See Zambelli 1987, 106–7; Caroti 1987.

143. Zambelli 1987, 107; Schöner 1545, fol. XCV.

144. *Sphaerae Mundi Compendium* 1490, fol. B2v. Cf. Thorndike 1949, 119: "Around the elementary region revolves with continuous circular motion the ethereal, which is lucid and immune from all variation in its immutable essence."

145. Thorndike 1949, 120.

146. Regiomontanus 1482, 511–30; see Pedersen 1978b, 168–85.

147. See, for example, Peurbach omnibus edition, 1491.

148. An excellent example of the latter is Herzog August Bibliothek 59 Astron 1–3, whose owner bundled three different editions of Peurbach, each with

its own individual assets: Reinhold's commentary (1580) and Christianus Wursteisen's *Quaestiones Novae* on Peurbach (Basel: 1569, 1573). For another reader's composite, see "Ratdolt–British Library Bundled Copy."

149. Peurbach 1472 (Nuremberg).

150. Aiton 1987, 23.

151. Ibid., "On the Three Superior Planets," 19.

152. Peurbach 1472, "On Mercury," unpag.: "Ex his igitur et dictis superius manifestum est, singulos sex planetas, in motibus eorum aliquid cum Sole communicare, motumque illius quasi commune speculum & naturae regulam esse, motibus illlorum illud." *Aliquid* can also be translated as an adverb (e.g., "somewhat" or "to some extent"); I follow Aiton in rendering it as a pronoun (1987, 23). On the annual component, Dennis Duke's computer animation, titled "Ptolemy's Cosmology," is most helpful (www.csit.fsu.edu/~dduke/models ["Ptolemy.exe"]).

153. Peurbach 1485, fol. 29v: "Omnes planetae mensuramque proportionalem sine dubitatione habent ad Solem; ideo Sol est tanquam dux, princeps et moderator omnis utique." Copy used: Zinner Collection, San Diego State University (QB41.S3 1485).

154. Capuano de Manfredonia 1515, fol. xxxiii. Elsewhere, Capuano maintained that Venus and Mercury receive the sun's light but are never seen to eclipse the Sun because of their small diameters; furthermore, they move with the Sun's mean motion, even though the Sun is the mean between the three higher and three lower planets and the Sun's sphere is larger, and hence it ought to move more slowly than planets below it (Capuano de Manfredonia 1518, fols. 32vb–33ra).

155. Peurbach 1542, fols. N6v–O: "Theorica mercurii, scholion." Using language strikingly resonant with that of both Copernicus and Kepler, Reinhold underscored "the harmony and ratio of each planet to the Sun's motion" and observed that "this universal wheel of things does not exist by chance but is divinely conserved by and arising from some wise, architectural mind."

156. Rheticus 1982, chap. 10, 60, ll. 74–75: "Communis orbium planetarum inter se dimensio."

157. Aiton 1987, 9; Zinner (1988, 97) drew attention to this important passage.

158. Ptolemy 1988, 419.

159. G. J. Toomer points out that Ptolemy well understood the observational problem (ibid., 419 n.).

160. For example, after presenting Ptolemy's optical-astronomical arguments for the Earth's centrality and stability, Regiomontanus briefly introduced the physical argument from the observation of heavy, falling bodies: "We can confirm the same thing by direct argument" (Regiomontanus 1496, bk. 1, chap. 3, 67).

161. Ibid., bk. 9, chap. 1, 192–93. I take up this question more thoroughly in chapter 3.

162. Capuano de Manfredonia 1518, fols. 32va–33ra.

163. Ptolemy 1940, bk. 1, chap. 4, 34–39: "Of the Power of the Planets"; bk. 1, chap. 17, 79–83: "Of the Houses of the Several Planets; see also Simonetta Feraboli's commentary (Ptolemy 1985, 369–70).

164. Ptolemy 1940, bk, 1, chaps. 5–6, 39–41; cf. Magini 1582, pt. 1, chap. 3, fol. E3, titled "De planetarum vi, atque potestate iuxta primas qualitates": "Minime vero putandum est, has qualitates eis vere inesse, sed potius virtutes harum qualitatum effectrices."

165. Ibid., bk. 1, chaps. 17–19, 79–91.

166. On Regiomontanus's printing project, see Zinner 1990, 110–17; Lowood and Rider 1994.

167. The forename is sometimes cited by historians as Adalbertus or Woijciech, the surname as Blar, Brudzevius or Brudzewski. I follow the title-page usage from the 1495 edition.

168. Brudzewo 1900; see Brudzewo 1495. For convenience, all citations are to the Birkenmajer edition (1900).

169. See especially Goddu 2010, 31–37; Brudzewo 1900, p. L; Birkenmajer 1924, 85–96; Birkenmajer 1972a, 622; Jardine 1982, 189–90 n.

170. Brudzewo 1900, 13. Albertus Magnus had a well-developed appreciation of astrology, a subject that he defended against standard Augustinian objections (see Zambelli 1992, 259–61; Zinner 1990, 73).

171. As with the *Tetrabiblos*, the typical practice of later publishers was to bundle Peurbach with auxiliary works, further crafting it as a different product. For example, Heinrich Petri bundled Christian Wursteisen's *New Questions* on Peurbach with Regiomontanus's *Disputations* and Johann Essler's *Useful Treatise . . . The Mirror of the Astrologers, in which Astrologers' Errors are Shown from having neglected the Equation of Time*.

172. Brudzewo 1900, 16–17.

173. Ibid., 17: "Sed practice a diversis diversimodo tradita est, ab aliquibus per instrumenta varia, ab aliis autem per tabulas diversas." This is confusing because Brudzewo has equated "practical astronomy" with astrology, whereas Campanus made instruments and tables the subject of practical astronomy. In his marginal diagramming of this passage, Georg Tanstetter opted for the Campanus version (Brudzewo 1495: Columbia University copy).

174. Ibid., 16: "Prima [i.e., astronomia] vocatur theorica seu speculativa, secunda vocatur practica, quam segregato nomine astrologiam dicimus."

175. Valla 1501.

176. Rosen 1981, 450.

177. Valla made these brief allusions—for that is what they are—in his section on physics rather than the one on astronomy (Valla 1501, XXI, 24; Rosen 1981, 451).

178. Czartoryski 1978, 355–96. Of course, owning such early editions shows only the possibility, not proof, that Copernicus acquired them at or near the time of publication.

179. Aristotle 1939, bk. 2, chap. 13, 217 ff.

180. Prowe 1883–84, 2:187–88; Swerdlow 1973, 439–40; Bilińksi 1977, 56–57. Cf. *De Revolutionibus* (bk. 1, chap. 5, fol. 3v): "It is said that Philolaus the Pythagorean, no ordinary *mathematicus,* thought that the Earth rotates, wanders with several motions and is one of the stars."

181. Valla 1501, chap. 43: "De terra positione."

182. The copy of Capella (No. 84844) held in the Burndy Collection at the Huntington Library is bound with the first compendium edition of the *Sphere* commentaries of Cecco d'Ascoli, Capuano de Manfredonia, and Jacque Lefebvre d'Etaples (1499). Although the present cloth binding probably dates from the eighteenth or early nineteenth century, this bundling suggests that at least one early owner saw these works as belonging together.

183. Regiomontanus 1496, bk. 9, chap. 1: "De reliquis autem tribus controuersia fuit." Michael H. Shank called attention to a passing comment in Regiomontanus's massive, unpublished polemic "Defense of Theon against George of Trebizond" (now in the Academy of Sciences, St. Petersburg, Russia, fol. 153v), wherein Regiomontanus sarcastically dismisses the ordering criteria as "rhetorical": the superior planets are grouped together by a shared "epicycle of the sun," whereas the inferior planets are grouped by their "longitudinal motion" (Shank 2005a).

184. Regiomontanus 1496, bk. 9, chap. 1, 192: "Fiet igitur ut distantia inter duo luminaria sibi quam vicinissmie approximata: semidiametrum terre 1006 fere vicibus contineat. Hoc autem spatium natura non sinit vacuum: necessario igitur quoddam celeste corpus ipsum occupabit. Sed id corpus de integritate erit orbium Solis et Lune; frustra enim tanta moles in celo permitteretur"; for important comment on this passage, see Shank 1998, 164 n. 6; Shank 2007, 2.

185. Goldstein 2002, 219.

186. For an excellent critical treatment of the current state of the question, see Goddu 2006.

187. Wilson 1975, 17–39; Swerdlow 1973; for passages concerning nonuniform circular motion, 434–35; Copernicus 1978, 4.2, 5.25, 5.2; cf. Clutton-Brock 2005, 210.

188. On the ontology of the orbs and spheres, see Aiton 1981; Westman 1980a; Jardine 1982; Lerner 1996–97.

189. Regiomontanus 1496, 12.12; Swerdlow 1973, 471–78; see Dennis Duke's animation (www.csit.fsu .edu/~dduke/models ["Venhelio2.exe"]).

190. Curtis Wilson 1975, 34 n., was careful to note that his figure "cannot be easily adapted to the case of the inferior planets."

191. To see how the Sun's motion is still mirrored in the planets even in a geoheliocentric arrangement, see Duke's animation (www.csit.fsu.edu/~dduke/models ["Tycho.exe"]).

192. It may be questioned whether the proposition to which Copernicus appeals is, in fact, something as strong as a principle which is "necessarily true." An epistemically weaker reading would be "assuming the consideration." For "assumpta ratione," Rosen translates "their principle assumes" (Copernicus 1978, bk. 10, chap. 1, 18:27; Goldstein offers "assuming the principle" (Goldstein 2002, 222).

193. Swerdlow 1973, 440.

194. Goldstein 2002, 220.

195. In the *Commentariolus,* Copernicus reported Venus's period as "in the ninth month"; it is also given as nine months, instead of 225 days or 7½ months, in *De Revolutionibus* (see Goldstein 2002, 221, 229–31; Swerdlow 1973, 440). Capella 1499, bk. 8, fol. r5: "Quinque uero sydera nesciunt obumbrari. Tria item ex his cum Sole Lunaque orbem Terrae circumeunt. Venus uero ac Mercurius non ambiunt Terram. Quod Tellus non sit centrum omnibus planetis." ("Three of these planets [i.e., Saturn, Jupiter and Mars], together with the Sun and the Moon, go around the Earth's globe, but Venus and Mercury do not go around the Earth. That the Earth is not the center of all the planets"); Vitruvius 1496, 9.1.6: "The stars of Mercury and Venus, making their paths in the form of a crown around the rays of the Sun as around a center, perform back and forth motions, and retardations."

196. These are the features of Copernicus's system commonly noted; but, for an especially useful (and very lengthy) list of such entailments, see Swerdlow 2004a, 64–120.

197. Goldstein, 2002, 221–22. Goldstein also discounts the importance that Swerdlow attaches to the intersection of the orbs of Mars and the Sun: "There is no evidence that Copernicus was concerned with this intersection of orbs, and I think it unnecessary to ascribe such a view to him." This position is strongly confirmed by Goddu's recent study of Copernicus in the context of fifteenth-century Krakovian natural philosophy (Goddu 2004, 83–90).

198. Goldstein does not consider Aristotelian commentary at Krakow; but for important preliminary investigations, see Goddu 2004.

199. Quoted and trans. in Goldstein 2002, 225: Achillini 1498; Achillini 1545, fols. 34v col. b–35r col. a.

200. Achillini's values for the geocentric periods of Mercury (335 days), Venus (344 days), and the Sun (365 days) confirm a period-distance relation, although not a proportional one (Achillini 1498, fol. 15r). See further chap. 3, this volume.

201. Brudzewo 1900, 117, my emphasis: "Ponit autem Magister correlarie, quomodo omnes planetae in motibus suis habent communicationem cum motu

Solis. Hoc ideo, quia cum eo habent naturalem connexionem, sicut cum luminoso, sicut testaur Ptolemaeus primo Quadripartiti, et ergo participant cum ipsius motu influxu et operatione."

202. Ficino 1989, chaps. 4, 14.

203. For an alternative reconstruction, see Shank 2009. Unlike Kepler, Copernicus did not attribute any sort of motive power to the Sun in *De Revolutionibus;* but in the *Narratio Prima,* Rheticus and Copernicus refer to the Sun as the "principle of motion and of light" (Rheticus 1982, 113, 169; cf. Granada 2004a).

204. See Swerdlow and Neugebauer 1984, 20–21, 26–27; Clutton-Brock 2005, 19; Goddu 2006, 37.

205. In his unpublished *Defensio Theonis,* Regiomontanus used a period-distance relation to criticize the system of al-Bitruji (Shank 2007, 10), but he failed to make the same criticism of Ptolemy in his *Epytoma almagesti.*

206. Although Regiomontanus endorsed Ptolemy's ordering in both the *Defensio Theonis* and the *Epitome,* he was trenchantly critical of George of Trebizond's reasons for holding the same conclusion (Shank 2007, 11–12).

207. Regiomontanus 1537, 38. The Earth's rotation would impede walking; birds could not fly well toward the east because to lift them from the ground, their feathers would have to be aligned against the wind; buildings would be destroyed by the "violent impetus"; a projectile thrown into the air would not return to the same place; and so forth.

2. CONSTRUCTING THE FUTURE

1. For the ancient period, see Evans 2004; Rochberg 2004; Swerdlow 1998.

2. Hale 1971, 3.

3. Luther 1967, 97.

4. Cornwallis 1601, sig. H1.

5. Kepler 1992, dedicatory letter, 31.

6. D'Amico 1983, 8–9.

7. Regiomontanus 1496, fol. a2v.

8. See, for example, Regiomontanus 1481.

9. Boudet 1994; Carmody 1958.

10. The four major collections that I have used are in the British Library; the Herzog August Bibliothek, Wolfenbüttel; the Biblioteca Universitaria, Bologna; and the Biblioteca Colombina y Capitular, Seviile

11. From the nineteenth century, there are the valuable writings of Gabotto 1891 and Bertolotti 1878, on which Thorndike (1923–58) relied. Among recent sources, the most important are Garin 1983; Zambelli 1986, 1992; Niccoli 1990; Vasoli 1980c; Grafton 1999; Hayton 2004, 2007; Vanden Broecke 2003; Rutkin 2005; and Azzolini 2009.

12. Pietramellara 1500, n.p.

13. Novara, 1484–1504.

14. About the former, see Pomata 1998.

15. On Wittgenstein and the problem of rule following, see Winch 1958, 21–39; Lynch 1992a,b; Bloor 1992.

16. Ptolemy 1940, 167 n. 4.

17. Ibid., bk. 2, chap. 8, 189.

18. Ibid., bk. 1, chap. 2, 7. This aetherial power or *dunamis* affected first the elements of fire and air, and these, in turn, affected earth and water, plants and animals.

19. See Garin's note in Pico della Mirandola 1946–52, 2:539.

20. Boudet 1994; Kibre 1966.

21. In two early fifteenth-century judgments, both authors describe themselves simply as "master": Blasius of Parma 1405, and Melletus de Russis de Forlivio 1405, 23r).

22. Niccoli 1990, 6. This extraordinary collection is today located in the Biblioteca Colombina y Capitular, Seville (see Wagner 1975).

23. *Profetia over Pronosticatione trovata in Roma in una piramide in versi latini tradotti in vulgare;* Niccoli 1990, 7.

24. Niccoli 1990, 143.

25. "Listen, mortals, to the horrible signs that announce great trials to our age; I wish to bring an end to my song, excellent listener, . . . this noble prophecy is ended, to your honor; O merry listener, here I make an end to my discourse." Quoted and translated in ibid., 18.

26. Ibid., 19.

27. *El se movera un gato* (n.p.: Angelo Ugoletti, 1495), fol. 2v, quoted and translated in Niccoli 1990, 20; Daston and Park 1998, 177–78.

28. Kurze 1960, 8.

29. Lichtenberger 1488, bk. 2, chap. 13; Reeves 1969, 349.

30. Lichtenberger 1488, bk 1. chap. 3; Reeves 1969, 349–50.

31. Literally, "the king of chaste appearance"; Lichtenberger 1488, bk. 2, chap. 4; Reeves 1969, 350.

32. Kurze 1960, 69.

33. Ibid., 185.

34. Warburg 1999, 623, 627; Kurze 1986, 183.

35. Hammerstein 1986, 132.

36. Middelburg 1492.

37. In this regard, it is interesting that Paul of Middelburg was described by his pupil J. C. Scaliger as "omnium sui temporis mathematicorum facile princeps" (Scaliger 1582, 807; Marzi 1896, 47).

38. Middelburg 1492, unpag., fol. 1r–v.

39. Ibid.

40. See Pedersen 1978b, 173.

41. See Swerdlow 1993; Rose 1975a, 95–98.

42. At the universities, this group included the following. *Bologna:* Baldino di Baldini, Girolamo Manfredi, Marco Scribanario, Domenico Maria Novara, Giacomo Pietramellara, Giacomo Benazzi, Lattanzio Benazzi. *Ferrara:* Pietro Buono Avogario, Antonio Arquato. *Padua:* Luca Gaurico. *Leipzig:* Wenceslaus

Faber von Budweis, Martin Pollich of Mellerstadt (first rector of Wittenberg). *Ingolstadt:* Johann Engel, Johann Stab, Lucas Eindorfer. *Krakow:* Matthias Schinagel, Johannes Schultetus, Marcin Bylica, Johann of Glogau, Michael Falkner of Vratislavia, Albertus of Brudzewo, Bernard of Krakow. *Heidelberg:* Johannes Virdung of Hassfurt. *Vienna:* Andreas Stiborius, Johannes Stabius, Johann Muntz, Georg Tannstetter, Andreas Perlach.

Court associations included the following. *Wiener Neustadt:* Emperor Frederick III (1440–93): Johann Nihil Bohemus, Georg Peurbach, Johannes Lichtenberger. *Vienna:* Maximilian I (1493–1519): Joseph Grünpeck, Johannes Stabius, Andreas Stiborius. *Urbino:* Paul of Middelburg. *Ferrara:* Giovanni Bianchini, Pietro Buono Avogario. *Mantua:* Bartolomeo Manfredi.

43. See Zinner 1941, nos. 531, 514, 735.

44. For a list of the prognostications from which these authorial identifications are derived, see Pèrcopo 1894, 210–16.

45. For Krakow, see Birkenmajer 1972b, 474–82; Markowski 1974, 83–89. For Vienna, see Hayton 2004.

46. See Knoll 1975, 137–40.

47. See Birkenmajer 1972b, 479–81.

48. Hasfurt 1492.

49. Thorndike 1943, 291. This information strongly suggests that Brudzewo's astrologically inflected commentary on Peurbach was associated with his own activities in practical astrology.

50. Ibid.

51. Glogau 1480, unpag.: "Ego magister Johannes de Glogovia maiori stilo ad leui ad honorem dei famamque incliti studii Cracouiensis scribere institui. Hoc autem pronosticum meum in tres distinxi differencias in quibus status elementorum mutationes aeree singulis diebus euenientes in primis legentur. Rerum tunc elementorum disposicio in altera parte videbitur. In fine vero dies electos ad balneandi et aduentosandi et actus humanos dirigendos per conuiuia hominem utilitate iuxta stellarum testimonia subiungas."

52. See Hannemann 1975; Scribner 1981, 49–69.

53. Quoted in Strauss 1966, 4.

54. On Ferrara as a cultural center, see Gundersheimer 1973; on the astrological murals of the Palazzo Schifanoia, see Warburg 1912; *Lo zodiaco del principe* 1992.

55. See Swerdlow 1993; Chabás and Goldstein 2009.

56. Vasoli 1980c, 129–58.

57. Giorgio Valla may have been his pupil (see Rose 1975a, 48, 71 n.).

58. "Petrus Bonus Advogarius Ferrariensis. Medicus. Insignis. Astrologus. Insignior," *Dizionario biographico degli Italiani.* (Other spellings: Advogario, Avogario, dell'Avogaro.)

59. Advogarius [1495], unpag., fol. 2v: "Lo illustrissimo & excellentissimo duca di Ferrara & singulare mio signore lanno presente perche el segno della vita nella reuolutione possiede la somma del cielo la luna a Venere aggiunta di piu felice fortuna godera: non pero senza pericoli sara sua excellentia. Tutte queste cose si confermano per la riuolutione di sua genitura: guardisi pero sua excellentia del mese dagosto che non sia da qualche infermita vexato per moto o per altra causa/accioche questo principe excellentissimo possa lastabilita del suo regno possedere."

60. Advogarius [1496], unpag., fol. 3r: "Lo illustrissimo & excellentissimo duca di Ferrara signor mio singularissimo per lo influxo delle benivole stelle questo anno sallegrera per felice sorte: nientedimeno non passera senza grandi spese: sara molto intento allacquisitara: alla stabilita del imperio sara punto: guardasi pero la sua excellentia da uiaggi per acqua piu presto che per terra: el luogo del suo imperio & la.x.casa celeste gli pormette felicita & del imperio stabilita indubitamente."

61. Gabotto 1891, 25–26:

Ill.me ac invictissime Dux Domine, Domine mi singularissime. Salutem perpetuam ac de inimicis victoriam et triumphum etc. Io al presente ho compito el juditio de lo anno proximo che vene, et perche temp e de publicarlo, como e usanza, prima lo mando a V.S.ria azò che quella prima el veda che niuno altro, ut moris est. El iudito è assai terribile, como vederà V.S.: at[t]amen summus rex, cuius habenis tota mundi machina gubernatur, hec omnia mutare, variare et ut sue voluntati placet disponere potest, qui in omnibus laudatus sit et benedictus. El iuditio mando a V.S. ligato cum la presente cum li di de l'anno boni per assaltar li inimici, quando bisognasse, per havere victoria, et anche li mando li di infortunati de tuto lo anno, ne li quali non se deve pigliare bataglia ne assaltare inimici perchè seria pericolo grandissimo a chi comenzasse. Io me arecomando infinite volte a V.S. la quale Dio conservi, imo augumenti in stato felicissimo. Feliciter voleat. Ex.tia V. In Ferraria, di ultimo Februarij 1479.

62. Ibid., 27: "Credo che V. Ill.ma Signoria serà contenta ch'io lo habij a publicare como io soglio fare ne li anni passati. Io voluntera lo mando a Ducal V.S., perchè no gi e influentia trista, per quello ch'io ho veduto, in la revolutione de V.E. Questo iudicio haverà ad esser mandato per tuta Italia e fora de Italia, et dara pur nome a questa nostra felice patria, ma prius se lezerà el titolo del presente iuditio che è a laude et gloria de V. Ill.ma S., la quale Dio conservi in stato felicissimo" (Ferrara, February 14, 1490).

63. Ibid., 27; this is clearly shown in Azzolini 2009.

64. Gabotto 1891, 26: "Dico che chi as saltasse li inimici prima, se bene havesse piu zente d'arme in decuplo, seria forza che perdesse e seria rotto cum tutte le sue zente per lo maraveglioso influxo celeste che tunc corre" (June 20, 1484).

65. Ibid.:

La V.S. have de mi l'altro heri la ellectione pro itinere per duy dì, zoè 21 et 22. Se possibile fusse che V.S.

andasse a di 26 de zugno, zoè sabato proximo che vene, V.S. haverla optima ellectione ad expugnandum inimicos et ad ottinere ogni victoria, et V.S. haveria optimo fine ne le sue facende, perche tunc la luna abraza Jupiter et Venus de aspecti beati, et ipsa luna erit lumine crescens; et iddo V.S. ogni modo et omnibus remotis pigli li predicti 26 dì et sera bon per lei, auxilliante deo. Fatilo, fatilo, fatilo. Io me arecomando mille volte a V. Ill. S., la quale Dio conservi in stato felicissimo.

3. COPERNICUS AND THE CRISIS OF THE BOLOGNA PROGNOSTICATORS, 1496–1500

1. Spellings of proper names were quite varied in this period. Novara was also known as Dominicus Maria de Novaria Ferrariensis or Domenico Maria di Novara. And for various prognostications, we find, inter alia, Domenego Maria da Nouara (Italian, 1497), Dominicho fer. da noara (Italian, 1492), Dominicus Marie de Ferraria (Lat., 1490), and Dominici Marie fer. de noaria (Latin, 1492). See also Rosen 1995a, 129.

2. Rosen 1995a has made much of this motion of the pole; for useful references to the nineteenth- and early-twentieth-century literature, see Rosen 1974.

3. The existence of such an institution as a private humanist school, a gathering of learned men of letters, or even a physical structure, such as a house or villa, has been called into serious question by Hankins 1991.

4. See esp. Kuhn 1957, 129; Burtt 1932, 54–55; Stimson 1917, 25.

5. Richard Lemay situates Copernicus in a Krakovian astrological context. He suggests that Copernicus's freedom from the need for patronage allowed him the unusual luxury of being able to attend directly to theoretical astronomy: "To be able to indulge in the detached and independent pursuit of astronomical studies goaded by nothing else than the love of truth, and furthermore to contemplate with serenity a lifetime dedication to the realization of a single major idea, was not given to the ordinary astronomer-astrologer of Copernicus' time" (Lemay 1978, 354).

6. John lectured from 1468 to 1507, Albert from 1474 to 1495 (Birkenmajer 1900).

7. See chapter 1. For a useful inventory of Krakow prognostications, see Markowski 1992.

8. Abenragel 1485. Swerdlow and Neugebauer call this "one of the most comprehensive and influential Arabic astrological treatises to be translated into Latin" (Swerdlow and Neugebauer 1984, 4). In addition, Copernicus owned the 1492 Venice edition of the Alfonsine tables and the 1490 Augsburg edition of Regiomontanus's Tabulae Directionum. The Euclid edition was published in Venice in 1482, with the commentary of Campanus of Novara (see Czartoryski 1978, 365–66).

9. Prowe 1883–84, 2: pt. 2, 516–17; trans. Rosen 1984b, 144. I have modified Rosen's translation in order to bring out more precisely the legal terms: "Statuimus, quod quilibet Canonicus de novo intrans, nisi in Sacra pagina Magister vel Baccalaureus formatus, aut in Decretis, vel in iure civili, aut in medicina seu physica Doctor aut Licenciatus exstiterit, post residentiam primi anni, si Capitulo visum et expediens fuerit, teneatur ad triennium ad minus in aliquo studio privilegiato in una dictarum facultatum studere." On the distinction between *medicine* and *physic*, see Cook 1990, 398–99.

10. See Monfasani 1993, esp. 252–57.

11. See Knoll 1975, 137–56.

12. Prowe 1883–84, 1:73–82.

13. Rheticus 1971, 111.

14. See Ady 1937, chaps. 6–7.

15. The result was a *governo misto*: "It pleases the lord Pope that the statutes dealing with the authority, jurisdiction and powers of all the magistracies of the said city shall be observed, and that none of the said magistracies shall have power to determine anything without the consent of the [papal] Legate or governor, and that, in like manner, the Legate governor shall not have power to determine anything without the consent of the magistracies deputed to rule the city" (ibid., 39–40).

16. Ibid., 90.

17. Ibid., 132.

18. Ibid., 132–33.

19. See Watson 1993, 5.

20. Christoph Scheurl, "Ad Sixtum Tucherum" (November 22, 1506), in Knaake and Von Soden 1962, 1:39; quoted and trans. in Watson 1993, 157. The earthquake occurred in 1505.

21. For brief but useful details on the pontificate of the Borgia pope Alexander VI (1492–1503), see Kelly 1986, 252–54.

22. "Professor mathematum" is the somewhat ambiguous phrase used by Rheticus to report what Copernicus told him in 1539 (*Narratio Prima* in Prowe 1883–84, 2: pt. 1, 297). Edward Rosen's "lectured in mathematics" (Rheticus 1971, 111) and "lectured publicly in Rome" (Rosen 1984b, 71) are helpful renditions. But note that Rheticus did not say that Copernicus "disputed in Rome," language that he might have used to signal the format of a philosophical disputation. Further, he does not specify which of the several mathematical subjects Copernicus might have been explaining. Rather, he seems to have lectured to a mixed audience that could have included prognosticators, painters, and instrument makers, students at the Roman Sapienza, and learned members of the papal court.

23. This judgment is based on Bonney 1991, 88.

24. Ibid. For the wider historiographical issues in the international system of alliances of which Italy was a part, see Marino 1994.

25. Guicciardini 1969, 48–49.

26. Ibid.

27. Phillips 1977, 122.

28. See Abulafia 1995, introduction, 48.

29. Guicciardini 1623, fol. 22r. Although *science* is a possible translation, I avoid it. Guicciardini is clearly distinguishing between those who make their predictions employing certain acquired, intellectual skills and those for whom knowledge of the future is somehow revealed.

30. Guicciardini 1969, 43–45.

31. See esp. Niccoli 1990. I discuss this distinction in chapter 1.

32. See Skinner 1978, 1:116–18.

33. Niccoli 1990, 163–66. *Cuius, cuia, coioni,* is a nonsensical declension with probable sexual connotations: *cuia = cuglia,* a pinnacle or spire of a steeple; *coioni = coglioni,* testicles.

34. A good example may be found in Giovanni Garzoni's "Laus Astrologie" (Biblioteca Universitaria di Bologna: Garzoni 1500, fol. 207v).

35. See Bühler 1958, 45. Hectoris's production may have exceeded Bühler's estimates, however, as may be seen by consulting the books listed under his name in the catalogue of printers of sixteenth-century books held by the Biblioteca Estense, Modena.

36. Discussed in chapter 2. On the Este astrological culture, see Biondi 1986; *Lo Zodiaco del Principe* 1992.

37. Savonarola 1497, 370, 339–40; quoted and trans. in Niccoli 1990, 165. Apparently without realizing it, Carl Sagan furthered the objection of Pico and Savonarola: "Despite the efforts of ancient astronomers and astrologers to put pictures in the skies, a constellation is nothing more than an arbitrary grouping of stars, composed of intrinsically dim stars that seem to us bright because they are nearby, and intrinsically brighter stars that are somewhat more distant" (Sagan 1980, 196–97).

38. See Weinstein 1970, 286–88.

39. The 1502 edition (Venice: Bernardino Vitali) eliminated many of the wearisome abbreviations of the first edition. The 1554 edition adjusted Bellanti's print identity to read "Mathematicus et Physicus;" all page references are to the 1554 edition.

40. Bellanti 1554, "Ad Lectorem," fol. A1v.

41. Here Bellanti was indebted to Marsilio Ficino ("amico meo"; ibid., 171), whose *De Triplici Vita* (1489) was already causing a stir.

42. Basel: Hervagius, 1553, 1554; Cologne, 1578, 1580. See Vasoli 1965, 598.

43. He was, in fact, thirty-one at his death, although his epitaph at San Marco in Florence gives his age as thirty-two: "Johannes iacet hic Mirandula. / Caetera norunt et Tagus et Ganges forsan et antipodes / Ob. an. sal. 1494. Vix. an. 32" (Rocca 1964, 19); cf. Cardano 1547b, 162: "Vixit igitur annis XXXIII. cum eius obitum Astrologus eodem anno praedixisset, qui etiam aduersus illum scripsit."

44. Walker 1958, 54–59.

45. "Veniamus ad neotericos. Nicolaus Oresmius, et philosophus acutissimus et peritissimus mathematicus, astrologicam superstitionem peculiari commenatrio indignabundus etiam insectatur" (Pico della Mirandola 1946–52, 1:58, bk. 1).

46. See Popkin 2003, 1993, 1996; Schmitt 1972a. In private conversations, Richard Popkin urged me to consider the possible influence of Sextus Empiricus on Pico; cf. Garin 1983, 87.

47. Pico della Mirandola 1946–52, 1:27.

48. Ibid., 29; see especially the helpful paraphrases of Craven 1981, 131–55, and Parel 1992, 19. Other useful, although not always reliable, summaries of Pico are Allen 1966, 19–34; Tester 1987, 207–13; and Thorndike 1923–58, 4:531–39.

49. Pico della Mirandola 1946–52, 1:29.

50. Ibid., 1:196, bk. 3, chap. 4. Cf. Ficino 1989, bk. 3, chap. 3, 257: "Spirit is a very tenuous body, as if now it were soul and not body, and now body and not soul. In its power there is very little of the earthy nature, but more of the watery, more likewise of the airy, and again the greatest proportion of the stellar fire. The very quantities of the stars and elements have come into being according to the measures of these degrees. This spirit assuredly lives in all as the proximate cause of all generation and motion, concerning which the poet said, 'A Spirit nourishes within' [*Spiritus intus alit*]." Walker (1958, 57), argues for a very close connection between Pico and Ficino: "The chapters in the *Adversus Astrologiam* on celestial and human spirit are so close to Ficino's thought that it seems highly probable they derive from him."

51. Pico della Mirandola 1946–52, 1:190, bk. 3, chap.3.

52. Ibid., 1:516–18, bk. 4, chap. 16. The opponent was Pierre d'Ailly.

53. The best account is, in fact, Eugenio Garin's excellent critical apparatus to his edition of the *Disputationes;* but the otherwise thorough and admirably reliable D. P. Walker gives this section no consideration.

54. On the problem of domification, see North 1986a, 27–30; Vanden Broecke 2003. These two methods, like the others, involved dividing a circle chosen from one of the three major astronomical reference frames—equatorial, ecliptic, and horizon—into arcs or "houses"—not to be confused with the zodiacal signs. The method ascribed to Regiomontanus involved dividing the equatorial circle into twelve parts, starting from the intersection of the equator and horizon; the Campanus method divided the prime vertical, a great circle that joined the observer's zenith to the equinoctial points, and then projected these points into the ecliptic.

55. On this point, Pico confused the tropical with the sidereal year, and the original printed text mistakenly reads "12 degrees" instead of "12 seconds."

Pico della Mirandola 1496, fol. Giii: "Thebit annum constare dixit cccclxv. diebus horis sex. minutis ix. gra[dus] xii." Although the published value may contain a typographical error, Garin, who made a critical comparison with the original manuscript, makes no comment here.

56. Pico della Mirandola 1946–52, 2:330, bk. 9, chap. 9.

57. Ibid., 2:322, bk. 9, chap. 8.

58. See Shapin and Schaffer 1985, esp. chaps. 2, 4, and 5.

59. Pico della Mirandola 1946–52, 2:334, bk. 9, chap. 10.

60. Ibid., 2:354, bk. 9, chap. 12.

61. Ibid., 2:370, bk. 10, chap. 4.

62. Ibid., 2:372, bk. 10, chap. 4: "Quam quod mediorum situs et ordo penitus in ambiguo."

63. Ibid., 2:374, bk. 10, chap. 4.

64. See Rocca 1964, 5, 7.

65. Pico della Mirandola 1946–52, 2:374, bk. 10, chap. 4. A manuscript titled "Almagestus Auerois" appears on the inventory of Pico's library (Kibre 1966, no. 626, 203–4).

66. Pico della Mirandola 1946–52, 2:374, bk. 10, chap. 4: "Quomodo vero tres aliae se habeant, Sol, Venus, Mercurius, incertum."

67. "Vixerat cum Dominico Maria Bononiensi, cuius rationes plane cognoverat, et observationes adiuverat" (Rheticus 1550a in Prowe 1883–84, 2:390).

68. Tabarroni 1987, 177; Biliński 1989, 38–39. At the time, the via Galliera was one of the most important and prestigious streets in the city; leaving the city, it pointed toward Ferrara, Domenico's place of birth.

69. Sighinolfi 1920. Much of this information is based, in fact, on Malagola 1878; see also Birkenmajer 1900, chap. 19, s.v. "Dominicus Maria Novara," 424–48; Birkenmajer 1975, 738–96. In the 1890s, Birkenmajer attempted to follow up on the fate of Domenico's library. He wrote to Antonio Favaro, who on February 6, 1898, informed him, "Having examined in the two libraries—the University and the Communal [in Bologna]—the books that could have belonged to Domenico Maria, no signs of ownership were found in any of the respective copies" (Birkenmajer 1975, 762–63). On November 29, 1994, I discovered the provenance of Domenico Maria Nouaria Ferrariensis in the 1493 edition of the *Tetrabiblos* at the Biblioteca Universitaria, Bologna.

70. Sighinolfi 1920, 235: "Item solvit ser Laurentio de Benatijs pro pensione domus duorum annorum libras 100."

71. He held the position perhaps until 1528. See Zambelli 1966a, 180–81; Mazzetti 1988, 47. Mazzetti's source was Fantuzzi 1781–84, 2: 62.

72. Benatius 1502 (Copy used: British Library): "Exquisitissimus praeceptor noster Dominicus Maria Novari[a]."

73. Zambelli 1966b, 181.

74. Cattini-Marzio and Romani 1982, 60, 62; for an illustration of the castle, see Portoghesi n.d., 8; cf. Prowe 1883–84, 1:236.

75. See Ridolfi 1989; Frati 1908; cf. Nussdorfer 1993, 103–18.

76. Archivio di Stato di Bologna: Archivio del notaio Lorenzo Benazzi, 1459–1508.

77. This situation is not unlike the problematic of contemporary social scientists, as described by Clifford Geertz: "Social events do have causes and social institutions effects; but it just may be that the road to discovering what we assert in asserting this lies less through postulating forces and measuring them than through noting expressions and inspecting them" (Geertz 1983, 34).

78. Novara 1492. There are some slight differences between the Italian (*Iudicio*) and the Latin (*Pronosticon*).

79. See Niccoli 1990, 10. To my knowledge, the question of notarial uses of astrological prognostication has simply not been investigated. In addition to the Bologna prognosticators who came from notarial families, we might mention the great naturalist Ulisse Aldrovandi (1522–65), whose father, Teseo, was both a notary and the secretary of the Bologna senate (see Franchini et. al. 1979, 10).

80. Malagola 1878, 572–73.

81. Biblioteca Universitaria di Bologna: de Fundis 1435, opening: "Altissimi dei nostri Ihesus Christi virtute chooperante primo in hoc meo iudiciolo." The colophon reads: "Datum Bonon. die septima febr. 1435 per doctorem artium Iohannem paulum de fundis actu legentem in astronomia et in medicina nostris studentibus et necnon inclite et excelse com(mun) itatis Bonon. astrologum bene meritum."

82. For example, "Per mi Hieronimo di Manfredi doctore de le arte & medicina nel studio famoso de bologna madre di studij, 1479" (Biblioteca Universitaria di Bologna).

83. Novara 1500, fol. 96v (copy used: Biblioteca Universitaria di Bologna).

84. For example, there are small differences for Novara 1492.

85. The 1484 prognostication was published in Venice by Bernardino Benali: "Per me magistrum Domenicum mariam Ferrariensem. iudicium editum in almo ac inclito studio Bon. anno Domini m. cccc.lxxxiiii" (copy used: Herzog August Bibliothek, Wolfenbüttel). The 1497 prognostication was published in Rome (in Italian) by Stephan Plannck (copy used: Biblioteca Colombina y Capitular, Seville).

86. See Sorbelli 1938, 114; Thorndike 1923–58, 5:234–51.

87. Pietramellara 1500. It was issued on January 18, two days before Novara's (copy used: Biblioteca Universitaria di Bologna).

88. Novara 1489 (copy used: Biblioteca Colombina

y Capitular, Seville). Magini 1585, 29–30. Magini was a rival of Galileo, an astrologer and mathematician at Bologna, and, effectively, a much later successor to Domenico Maria. William Gilbert, although critical of Domenico's judgment, lifted the same passage from Magini without attribution (Gilbert 1958, bk. 4, chap. 2, 315–16).

89. The only prognostication that I have been unable to find for Copernicus's Bologna period is that of 1498.

90. Novara 1499 (copy used: Archiginnasio, Bologna):

We regard those judges as unjust who presume to judge something about which they know nothing. For only the good man is a [true] judge among those who do know. How many of these unfair judges there are in our time [who classify] the science of the stars among the other disciplines of the liberal arts. This is not a surprise. For it is the customary role, especially among ignorant men, to criticize and revile because they know nothing. Others believe this science of the stars [scientia astrorum] to be deceitful and worthless and of no civil use. Still others, wearing the skull caps of dark ignorance, declaim in their arrogant orations that astronomers argue for necessity in human affairs. Another group, on the other hand, argues against the latter, appearing to dispute about everything. They compete in agonistic disputations and imitate certain astronomical words, names and rhetorical styles. Entirely forsaking the office of wise men, however, they prefer to be seen as wise men rather than to be [wise] and not to be seen. For, as Aristotle observes, the wise man's work is the first of the two pearls: it concerns the one who knows that he does not deceive. . . . Those accusing the astronomers do not understand astronomical matters. . . . they are only imitators of the words compared to the beholding astronomer. The art of imitation, however, deceives many. As you know, the imitators stray far from the truth and express with words and names a pretense [to understand] the individual arts when they understand nothing at all about these arts. So, when they contemplate the words, at least let them be imitated in such a way that they appear to be well spoken and so that these imitators may stroke the ears sweetly in some natural way.

91. Ptolemy 1493, comment on Tetrabiblos, "Prohemium," 3: "Et qui dixit in virtute fuit, quia iste demonstrationes firmiores et fortiores sunt illis quae sunt in arte iudiciorum et quas de geometria et arismetica [sic] sunt accepte."

92. Ptolemy 1493, Centiloquium, "First Saying," 107.

93. Scientifica, in the sense of satisfying Aristotle's requirements for apodictic knowledge.

94. The section begins: "In fact, he who thinks that astronomers reckon things by necessity is lost in ignorance about the astronomical discipline. For what astronomers say is that from a fixed position of the stars a fixed and necessary inclination follows" (Novara 1499).

95. Aristotle 1966, bk. 1, chap. 2, 70b 20f., 31.

96. Novara 1499.

97. For the years preceding the 1499 forecast, I have seen those for 1484, 1487, 1489, 1490, 1492, 1496, and 1497.

98. Benatius 1502.

99. In the Italian universities, the connection between medicine and natural philosophy was well established (Siraisi 1987, 221–23; Schmitt 1985); but the connection between astronomy and medicine has been less well appreciated.

100. See Kibre 1967; Lind 1993, 9. According to Nancy Siraisi, Avicenna's canon concerned "parts of the body with their anatomy, physiology, and diseases, arranged from head to toe. Judging from the content of the sections specified, the first year was devoted to the head and brain; the second to the lungs, heart, and thoracic cavity; the third to the liver, stomach, and intestines; and the fourth to the urinary and reproductive systems" (Siraisi 1987, 55–56).

101. See Ady 1937, 144–45; Raimondi 1950, 69–70; Kibre 1967, 506.

102. Garzoni (Opusculum de Dignitate Urbis Bononiae), cited in Raimondi 1950, 71:

Ho sempre pensato che non vi sia alcuna scienza che possa essere messa a pari con l'astrologia perché questa porta agli uomini un bene sommamente utile e onorevole. Coloro che ne sono esperti annunciano la morte dei principi e le mutazioni degli stati; predicono le guerre, le pestilenze, le carestie; insegnano ciò che bisogna fuggire o seguire. Quante sciagure si sarebbero potute evitare se si fosse ascoltato il consiglio degli astrologi! Io credo che chi ignora l'astrologia, non possa riuscire buon filosofo, medico e poeta. D'altra parte è quasi impossibile trovare un geografo che non possieda nozioni astrologiche, come attestano Claudio Tolomeo, Strabone, Gnosio e tutti gli altri. Che dire poi della scienza militare, dell'agricoltura, della navigazione, alle quali l'ausilio della astrologia è più che necessario?

103. Quoted in Raimondi 1950, 65 (novella 65). The recommendation that Gabriel should read both theorica and practica underlines the complementarity of the two genres.

104. Ady 1937, 162; Raimondi 1950, 54; Sighinolfi 1914.

105. Ady 1937, 144.

106. For the display of Latinity as a sign of superiority and its use in academic games of dominance, see Grafton and Jardine 1986, 92–94.

107. Zaccagnini 1930, 125; Ady 1937, 144.

108. Raimondi 1950, 58.

109. Cayado 1501. Of two copies of the 1501 edition in the Biblioteca Estense, Modena (shelfmark a.H.7.15), one has extensive hand illuminations, suggesting that it was intended for presentation.

110. For Cayado's connection to Szdłowiecki, Beroaldo, and Copernicus, see Gorski 1978, 397–401.

111. Pico della Mirandola 1496a; on the back appear Hectoris's symbol and the date, March 6, 1496.

112. I translate the Latin titles thus: A Little Erudite Work wherein Is Contained a Declamation on the Excel-

lence of the Philosopher's, Physician's and Orator's Disputations; And a Little Book Concerning the Best State and Prince (December 1497); *On Happiness* (April 1499). These works exemplify Beroaldo's classicizing spirit and his celebration of worldly values, such as friendship. Beroaldo dedicates the *Declamation* to Paul Szdłowiecki, described as a "Polish scholar" (*Scholasticum Polonum*) and "archigrammates, auricularius illustratissimi principis nuncuparis / Cancellarium uulgo nouitant."

113. Ady 1937, 161.

114. "Quo mihi homo nemo neque amicior neque carior neque coniunctior fuit": Beroaldus n.d., sigs. b 4r–c 2r; Malagola 1878, 275; "Mine mi eruditorum nobilissime: nobilium eruditissime" (Beroaldus 1488, epistolary dedication).

115. Cayado 1931, 86.

116. Quoted in Lind 1993,1992.

117. Malagola 1878, 275–76. All this information based upon Archivio di Stato di Bologna: Liber Partitorum magnificorum dominorum Sedicem, vols. 10–12.

118. Beroaldus 1500, bk. 5, fols. 100v–101r. See Rhodes 1982b, 14–17. A copy of this edition was owned by the Varmian canon Johann Langhenk; although Copernicus's hand does not appear on this item, its presence in Varmia shows that Hectoris's productions were making their way into that region (see Czartoryski 1978, no. 23, 374).

119. See Cayado 1931, 13.

120. Pico della Mirandola 1496b (March), fol. YYiir: Beroaldus to Pico, April 10, 1486; see also Rhodes 1982b, 14–17; Garin 1942, 588. Because these letters were publicly available, both Copernicus and Novara could have read them. Beroaldus's reference to dinner with Pico and Mino Rossi also signals Pico's importance in Bolognese noble circles.

121. Rhodes 1982a, 17: "On 31 July 1502 Beroaldus had another book printed by Benedictus Hectoris, *Orationes et carmina Beroaldi,* in which he addressed an epistle and some verses to Roscius, and on S3 recto he included a poem about a supper party given by Roscius to Prince Bentivoglio."

122. *Dubium:* a proposition of uncertain truth and suited for debate, examples of which are often to be found listed in the practical medical manuals with which Novara was well acquainted (e.g., Savonarola 1502, Tabula, capitulum 3, "De pupilla": "Whether dryness causes the pupil of the eye to narrow").

123. Because the daily business of the Sedici was documented in standard, notarial Latin, Domenico must have intended the compliment as a sign that Rossi's Latin was as learned as that used at the university (see Zaccagnini 1930, 124; Malagola 1878, 275–76).

124. The only prognostications directed to Rossi that I have found were authored by Novara.

125. Rhodes 1982c, 229–31.

126. The Venetian publisher and bookseller Ottaviano Scotto sold the 1471 and the 1496 editions of Pietro d'Abano's *Conciliator* as well as the 1493 edition of the *Tetrabiblos*. As Martin Lowry (1991, 187) has observed, it was customary for scholars at the universities of Ferrara, Padua, Pavia, and Bologna to obtain their larger textbooks from Venice: thus it is not surprising that books like the *Conciliator* and the *Tetrabiblos* would also be found in this market.

127. Salio was the "corrector" of Beroaldus 1488.

128. Two transpositions are necessary to reach the proposed reading: The *A* and the *n* must first be exchanged; then the *a* must be converted to an *o*. See Ptolemy 1493, dedication: "Hieronymus salius fauentinus* artium et medicine doctor:dnico marie de anuaria ferrariensi artium et medicie doctori astrologoque excellentissimo de nobilitate astrologie." Cf. Birkenmajer 1975, 756–62. The several copies of this rare edition that I have seen contain the same uncorrected error.

129. "Mei Dominici Marie de Nouari Ferr[ariensis]." The provenance appears after the colophon of the last-bound work, Albubater 1492. I conjecture that Novara bundled together Eschenden 1489 and Albubater 1492 soon after he acquired Ptolemy 1493, with its personal dedication. In a note beneath the provenance, partially legible with the aid of ultraviolet light, Novara refers to the erroneous spelling of his name in the published dedication and forgives the editor for his human weakness ("Dedicatio uero mei Ptolomei Faventino Bono: et [cur?] suo [rum?] error[um?] vir humanum [?]"). The two additional works are noted (by hand) in the table of contents, also on the spine (perhaps by a librarian) and on the top and outside of the volume. The entire collection is bound in an early vellum binding. When the collection was put together, the margins were cut down, partially slicing off some annotations. The name "Ul[isse] Aldr[ovandi]" is penciled in on the dorso of the front board, but Paula Findlen confirmed (personal communication) that there are no internal markings that correspond to Aldrovandi's cataloguing practices, that is, provenance and book number. Moreover, although the 1493 *Tetrabiblos* and Eschenden's *Summa* do appear in Aldrovandi's catalogue, the work by Albubater does not. This absence suggests that an early librarian arbitrarily entered Aldrovandi's initials into the book.

130. Österreichische Nationalbibliothek: Novara, n.d., fol. 199r. As Ernst Zinner points out, this manuscript was kept from 1519 onward in the library of Johannes Schöner in Nuremberg. The amanuensis, one Johannes Michael Budorensis, had contacts with Ratdolt's publishing house in Venice and perhaps also with Novara himself. Budorensis may have acquired papers of both Regiomontanus and Novara on the latter's death and transmitted them to Nuremberg (Zinner 1990, 153–54).

131. Novara probably meant "my teacher" in a literary rather than a literal sense, that is, from having read his books. Although it is just possible that Novara and Regiomontanus had met personally, I regard such an encounter as unlikely.

132. Copernicus 1978, 218.

133. Swerdlow and Neugebauer 1984, 66.

134. See Dobrzycki and Kremer 1996.

135. Copernicus 1978, bk. 4, chap. 14, 200.

136. Swerdlow (1973, 456–63) and Swerdlow and Neugebauer (1984, 47–48) argue for the virtual identity of Copernicus's lunar model with that of the "Maragha school" astronomer Ibn ash-Shatir.

137. Rheticus 1971, 133.

138. Tabarroni 1987, 184.

139. Achillini 1498, fol. 15r a: "tamen ipse 3° almagesti capitulo primo concedit Mercurium et Venerem cadere super eadem linea inter solem et oculum nostri. et demonstrat quae necesse est sic esse et Geber ibidem. et sic videtur contradictio in dictis ptolomei in hoc an venus et mercurius cadant in eodem epipodo [periodo?] cum sole etc." Because the *Almagest* has no such discussion of Mercury and Venus in book 3, chapter 1, Achillini might have been using Geber's *Correction of the Almagest* (*Islah al-Majisti*).

140. Achillini 1498 (7 August), fol. 15r b.

141. Bellanti 1498, bk. 10, 213: "Pariter de Mercurij et Veneris situ sub sole ignorat demonstrationem apertam, quae ex epicyclorum quantitate tantaque vel tanta apparentia maioritatis corporis ipsorum elicitur postquam sequaces hoc ignorant. Quae vero dicit Auerrois nullius sunt momenti, asserit enim XII Metaph. erraticarum earum apparentias in eodem orbe diuersis polis posse saluari, epicyclos negans ineptissime, qui etiam quandoque in philosophia deceptus est, dolet tamen ob senium ne possit astrologiam discere, quam antiquorum & sui temporis ignorabat."

142. Because Bellanti's book appeared in May and Achillini's in August 1498, Copernicus would have had to draw the connections himself. For Achillini's gloss on Averroës' commentary on Aristotle's *Metaphysics* discussion, see Goldstein 2002, 225, and chapter 1, this volume.

143. See Goddu 2004, 71 ff.; Hatfield 1990, 93–166.

144. See Dobrzycki 2001.

145. See Swerdlow 1973; Copernicus 1985, 3:75–126; Dobrzycki 1973.

146. Copernicus 1884, 186: "Si nobis aliquae petitiones, quas axiomata vocant, concedantur." The date of composition of this work, like so much else about this period of Copernicus's life, is uncertain. Swerdlow (1973, 431) believes that "there is insufficient evidence to determine how long before 1514 Copernicus developed his new planetary theory," whereas Rosen opts for 1508–15 (Copernicus 1985, 79–80).

147. Aristotle 1966, bk. 1, chap.10, 76b, 14–15. Some further light might be thrown on this passage if it could be established which edition and commentary Copernicus was using.

148. "Omnes orbes ambire Solem, tanquam in medio omnium existentem, ideoque circa Solem esse centrum mundi" (Copernicus 1884, 186; cf. Swerdlow 1973, 436; Copernicus 1985, 81).

149. "Quicquid nobis ex motibus circa Solem apparet, non esse occasione ipsius, sed telluris et nostri orbis, cum quo circa Solem volvimur ceu aliquot alio sidere, sicque terram pluribus motibus ferri." Copernicus 1884, 186; cf. Swerdlow 1973, 436; Copernicus 1985, 81–82.

150. This important point, which allowed Copernicus to sidestep the Aristotelian standard of demonstration, was later emphasized by Rheticus (1982, 58): "Cum autem tum in physicis, tum in astronomicis ab effectibus et observationibus ut plurimum ad principia sit processus, ego quidem statuo Aristotelem, auditis novarum hypothesium rationibus, ut disputationes de gravi, levi, circulari latione, motu et quiete terrae diligentissime excussit, ita dubio procul candide confessurum, quid a se in his demonstratum sit, et quid tanquam principium since demonstratione assumptum."

151. "Proinde ne quis temere mobilitatem telluris asseverasse cum Pythagoricis nos arbitretur, magnum quoque et hic argumentum accipiet in circulorum declaratione. Etenim quibus Physiologi stabilitatem eius astruere potissime conantur, apparentiis plerumque innituntur; quae omnia hic in primis corrunt, cum etiam propter apparentiam versemus eandem" (Copernicus 1884, 187–88). Copernicus's reference to the Pythagoreans as "natural philosophers" shows that he associated them with Aristotle rather than Ptolemy. Thus Copernicus's reconsideration of Aristotle's rejection of the Pythagorean position played an important part in Copernicus's explanation of the shared-motion problem. Cf. Swerdlow 1973, 439–40; Copernicus 1971a, 82; Bilińksi, 1977, 56–57.

152. Rheticus 1982, 55: "Quare, cum hoc unico terrae motus infinitis quasi apparentiis satisfieri videremus"—echoing Copernicus 1543, bk. 1, chap. 10, fol. 10: "Quae omnia ex eadem causa procedunt, quae in telluris est motu."

153. Rheticus 1982, 61:3–8. The passage explicitly attributes to Aristotle the recognition that if one motion was ascribed to the Earth, then other motions could equally well be assigned to it (cf. Aristotle 1939, 243; bk. 2, chap. 14, 296b, 1–5).

154. Swerdlow 1973, 440.

155. Copernicus produces values of nine months (Venus) and three months (Mercury): Copernicus 1884, 2:188.

156. Rheticus explicitly denies that a higher sphere could cause any inequality in the motion of a lower sphere: "Quilibet planetae orbis suo a natura sibi at-

tributo motu uniformiter incedens suam periodum conficit et nullam a superiori orbe vim patitur, ut in diversum rapiatur" (Rheticus 1982, 60; Rheticus 1971, 146).

157. Swerdlow 1973, 437; cf. Rosen 1985, 92; Dobrzycki 2001.

158. For analysis of such textualities, see Hallyn 1990, 60–61.

159. See Matsen 1977, 1994.

160. See Westman 1980a.

161. Zinner 1988, 186, basing his argument on Birkenmajer 1924, 199–224, notes that Miechow had many "astronomical" works in his library; but using "the science of the stars" as our classification, we can easily see that the *sexternus* found its place amid kindred books—for example, copies of Ptolemy's *Tetrabiblos* and Stöffler's *Almanach*. Rosen's translation of Miechów's entry is problematic (Copernicus 1985, 75).

162. The designation *sexternus* undoubtedly comes from Matthew of Miechów, as the term refers to the size of the item and hence to a catalogue entry rather than to the subject matter. However, the term *theorica* is more problematic. In general, Matthew's entries reflect accurate condensations of actual titles; hence, following this practice, he might have been using a title that was part of the manuscript itself. On the other hand, he already had two items with the title *theorica* in his library, and he might simply have decided to assign this word of his own account.

163. Copernicus 1884, 2:187.

164. Aristotle 1966, bk. 1, chap. 1, 273.

165. See Goddu 2010, 275–300. Swerdlow also wrestled with this question: "It could . . . be intelligently argued that because Copernicus calls these statements postulates (*petitiones*), he is therefore not asserting that they are necessarily true. Yet, if he had any doubts about the truth of the heliocentric theory, he probably would not have advanced it in the first place" (1973, 437 n.). Swerdlow's first statement seems to me to be exactly right. Perhaps we might say that by leaving the *Commentariolus* as a *sexternus*, Copernicus was not yet "advancing" it. In fact, Copernicus's argument with the natural philosophers was that their claims "rest for the most part on appearances," which they do not fully "save"; on the other hand, Copernicus believed that from his postulates "the motions can be saved in a systematic way." Cf. Swerdlow and Neugebauer 1984, 9: "The heliocentric theory and the motion of the earth were presented as a series of postulates, although there is no doubt that Copernicus considered them true. This was not really objectionable, and was in fact entirely reasonable, because Copernicus knew that at the time he had no way of proving that the earth in fact moves." Cf. Rosen's footnotes, Copernicus 1985, 38, 56, 66, 83, 192.

166. Rheticus 1971, 126–27.

167. Another source from which Copernicus could have drawn this information is Regiomontanus 1496, book 9, prop. 1, fol. klv. Regiomontanus knew al-Bitruji's views directly from *De Motibus Celorum*, of which he owned a copy (see Shank 1992, 17).

168. To the best of my knowledge, Ludwik Birkenmajer was the first to point out that Copernicus had found the passage in Pico 1984a (1900, 94). Ernst Zinner claimed that Copernicus had mistaken Averroës for Aven Rodan, based on a misreading of Pico (Zinner 1943, 510 n.: "Tatsächlich handelte es sich um Aven Rodan ['Ali ben Ridwān]; Copernicus hatte die Stelle wohl dem Werke des Pico della Mirandola wider die Sterndeutung entnommen und den Namen Aven Rodan in Averroes verschrieben"). Goldstein (1969, 58 n.) rightly called attention to Zinner's error.

169. See Rosen's note in Copernicus 1978, 356–57. Beginning with Erasmus Reinhold, many sixteenth-century readers of this passage noted that the same observation could also be found in another source: "Idem est in historia Carolj Magnj" (The same is [to be found] in the *History of Charlemagne*); cf. Gingerich 2002, Edinburgh 1, 268–78 (1543, fol. 8); Prague 3, 23–28 (1566, fol. 8). The source is the ninth-century Abbot Einhard, and the earliest printed edition that I have found is Einhard 1532. This volume also contains *Vita et Gesta Caroli Cognomento Magni, Francorum Regis Fortissimi, et Germaniae suae illustratoris, autorisque optime meriti, per Eginhartum, illius quandoque alumnum atque scribam adiuratum, Germanum conscripta*. On page 122, Einhard reports that Charlemagne had died in 814 but that afterward it was said that the event had been presaged as follows: "Appropinquantis finis complura fuere praesagia, ut non solum alij, sed et ipse hoc minitari sentiret. Per tres continuos, uitaque termino proximos annos et solis et lunae creberrima defectio, ac in sole macula quaedam atri coloris septem dierum spatio uisa." It is clear that this description scarcely resembles the passage from Copernicus on Averroës. Moreover, even if Copernicus had known this book, it is clearly not his original source, as he does not cite it.

170. Copernicus 1978, bk. 1, chap. 10, 19; Copernicus 1543, fol. 8: "Quamuis & Averroes in Ptolemaica paraphrasi, nigricans quiddam se uidisse meminit, quando Solis & Mercurij copulam numeris inueniebat expositam: & ita decernunt haec duo sydera sub solari circulo moueri."

171. Ibid.

172. See Swerdlow 1976, 122. This language and Swerdlow's diagrammatic reconstructions make it quite plausible that Copernicus was allowing here for the existence of solid orbs, in the geometric sense, but without pronouncing in any way on their materiality or impenetrability.

173. Copernicus 1978, bk. 1, chap. 10, 22; cf. Giovanni Gioviano Pontano (1429–1503) in Pontano 1512,

bk. 1, sig. A2. For further discussion, see also Trinkaus 1985, 450–51.

174. Rheticus 1971, 109–10, my italics.

175. Copernicus 1972, fol. 13, my italics: "Assumpsimus extra quibusdam revolutionibus mobilem esse tellurem quibus tamque primario lapidi totam astrorum scientiam instruere initiam"; Copernicus 1978, 26; again, in the *Letter against Werner* (1524), he says that "the science of the stars is one of those subjects which we learn in the order opposite to the natural order" (Copernicus 1971c, 98; Copernicus 1985, 146).

176. For Copernicus's reconfiguring of gravity and the elements, see Knox 2002, 2005.

177. The manuscript of *De Revolutionibus* contains a suppressed passage from Lysis's letter to Hipparchus, available to Copernicus both in Bessarion's *In Calumniatorem Platonis*, fols. 2v–3r, and in *Epistolae Diversorum Philosophorum* (Venice, 1499). Introducing the text of the letter, Copernicus mentions that "Philolaus believed in the earth's motion" and that "Aristarchus of Samos too held the same view according to some people"; Copernicus also explains that these views are not widely known because of the Pythagoreans' practice of not committing "the secrets of philosophy to writing" (see Rosen's discussion in Copernicus 1978, 25–26, 361–63; Prowe 1883–84, 2:128–31; Africa 1961).

4. BETWEEN WITTENBERG AND ROME

1. See Barnes 1988.

2. See Hammerstein 1986; Köhler 1986; Reeves 1969, v–vi.

3. Quoted in Rupp 1983, 257, and Bonney 1991, 21. Luther completed his translation of the Bible in 1534.

4. See Dobrzycki and Szczucki 1989.

5. The *Commentariolus* was not published until the nineteenth century (Copernicus 1884, 2:184–202); it lacked an explicit public strategy of persuasion and, therefore played a somewhat different role in promoting Copernicus's work.

6. He was known to his classmates as Hosen Enderle (see Swerdlow and Neugebauer 1984, 1:13.).

7. On Schöner, see Wrightsman 1970, 120. Copernicus is not known to have had a mistress, but he did have a female housekeeper, whose presence in his house made him the subject of fairly strong censures by the Varmia bishop (see Rosen 1984b, 149–57).

8. Giese attributed Copernicus's failure to mention Rheticus to a kind of absentmindedness about anything that was not "philosophical": "incommodi, quo in praefatione operis praeceptor tuus tui mentionem omisit. quod ego non tui neglectu, sed lentitudine et incuria quaedam (ut erat ad monia quae philosophical non essent, minus attentus), praesertim iam languescenti evenisse interpretor, non ignarus, quanti facere solitus fuerit tuam in se adiuvando operam et facilitatem" (Giese to Rheticus in Leipzig, 26

July 1543; Prowe 1883–84, 2:420). However, Hooykaas endorsed the view of Bruce Wrightsman that "Copernicus shrewdly declined to name his Lutheran disciple, Rheticus, in his letter of dedication to the Pope, as one of those whose assistance and encouragement persuaded him to have the work published. What other possible reason could there be for such a significant omission?" (Hooykaas 1984, 38; Wrightsman 1975, 234). By the same token, Rheticus nowhere mentions the pope in the *Narratio Prima*.

9. Edward Grant concludes: "Astrologers and natural philosophers may have shared the Aristotelian conviction that celestial bodies were the ultimate causes of terrestrial effects, but natural philosophers largely excluded the prognosticative aspects of astrology from their deliberations. Except for the attribution of certain qualities to certain planets, astrological details and concepts are virtually ignored in questions on Aristotle's natural books, especially on *De caelo*. The properties, positions, and relationships of the planets used for astrological prognostication were of little significance for the scholastic tradition in natural philosophy" (Grant 1994, 36 n. 66). Of course, the general exclusion of astrology from natural philosophy did not mean that specific churchmen were averse to engaging in practical astrology.

10. Luther 1969; Ludolphy 1986; Barnes 1988, 46–53.

11. Caroti 1986, 120; Kusukawa 1991, 1995.

12. Barnes 1988, 97; Bretschneider et al. 1834–, 20:677–85.

13. See, for example, Hammer 1951, 313: "At the sight of these beautiful luminaries, they may meditate upon the entire arrangement of the year and upon the reason why God, the Author of all, created differences in seasons and annual cycles. And finally, that at such contemplation they may acknowledge God as the Creator and praise His wisdom and goodness shining forth from the infinite variety of blessings by which He shows His care for mankind. May they also realize that the wise and just Creator has shed the rays of His light upon us, namely, in order to distinguish between the concepts of good and evil."

14. Barnes 1988, 96–99; Caroti 1986, 109–21.

15. See Bretschneider et al. 1834–, 8:63, no. 5362; Caroti 1986, 120. Of course, although Stöffler's 1499 *Almanach* had been a major resource in the flood predictions, he himself had thrown cold water on the rising expectations as the time grew close (see Stöffler 1523).

16. Quoted and trans. in Warburg 1999, 656–57.

17. Ludolphy 1986, 106.

18. For an excellent analysis of the meaning of *superstition* in this period, see Clark 1991, 233–35.

19. See Caroti 1986, 118. On D'Ailly's astrology, see Smoller 1994.

20. Cited by Barnes 1988, 97; the example comes

from Melanchthon's preface to Johannes Funck, the son-in-law of Andreas Osiander (Funck 1559).

21. Ibid.

22. Peucer 1591.

23. Ibid., 389–91v.

24. "Johannes Schonerus dicebat se vidisse antiquissimum librum apud episcopum Bambergensem manu scriptum ex quo Joannes iste Picus omnia descripsit, impudenter sibi ea vindicatus, quibus contra astrologos arbitratur. Liber autem ille ignoti auctoris erat." Discovered by Aleksander Charles Gorfunkel in a copy of Pico's *Disputationes* (Pico della Mirandola 1504). Eugenio Garin cites this reference without further identification of location (Garin 1983, 85). It is not clear who wrote this note.

25. According to Garin 1983, 86: "The diffusion of the knowledge of this annotation was attributed to 'George Joachim Rheticus the famous mathematician and doctor,' who said that he had heard it in person." Because Rheticus could only have heard this remark from Schöner on the occasion of his visit in 1539, he would have been able to pass it on to Copernicus in the same year and thereupon to Melanchthon on his return to Wittenberg. "Est in manibus hominum farrago criminationum a Pico non scripta, sed excerpta ex vetustioribus commentariis, qui ad huius divinatricis reprehensionem multo ante collecti fuerunt" (preface to Schöner 1545, fol. β2r; also in Bretschneider et al. 1834–, 5:819).

26. Bretschneider et al. 1834–, 3:119, no. 1455; Caroti 1986, 114 n.

27. See also Garin 1983, chap. 4.

28. Bretschneider et al. 1834–, 5:818. Cf. Cicero 1959, bk. I, i, 1, 223: "There is an ancient belief, handed down to us even from mythical times and firmly established by the general agreement of the Roman people and of all nations, that divination of some kind exists among men; this the Greeks call *mantike*—that is, the foresight and knowledge of future events."

29. "Hos refutarunt docti viri, Bellantius et alii quidam; et multae leves et ieiunae cavillationes obiiciuntur, quarum repetitio longa, et refutatio non necessaria est" (Bretschneider et al. 1834–, 5:819).

30. Preface to Schöner's *De Iudiciis*, Bretschneider et al. 1834–, 5:823: "At saepe fallimur, saepius etiam quam in ceteris artibus. Fateor hoc quoque. Nec tamen ars nulla est. Quid enim familiarius homini, quam hallucinari ac errare? Sed manent tamen aliquae verae notitiae, quas alii magis, alii minus dextre ad ea, de quibus iudicant, accommodant; et de futuris rebus etiam pauca prospicere et utile et magnume est." Pico had already noticed Ptolemy's admission that the art of astrology is uncertain (Pico della Mirandola 1496, bk. 2, chap. 6) and that "only those inspired by the divine predict particulars" (bk. 2, chap. 1). See Belluci 1988, 619.

31. Bretschneider et al. 1834–, 13:537: "Ars est ordo certarum propositionum, exercitatione cognitarum

ad finem utilem in vita." Cited in Bellucci 1988, 615–616.

32. Bretschneider et al. 1834–, 13:185.

33. Pico della Mirandola 1496, bk. 11, chap. 1; see Belluci 1988, 616–17.

34. "Cum natura uno et eodem modo agat, postquam multa exempla congruere compertum est, recete inde extruitur universalis. Hoc modo et Medicus suas universales constituit. Non colligi omnes singulares experientias de Cichorio, cuius magnus usus est in febribus, possunt, et saepe effectio eius impeditur, sed tamen consensus multorum exemplorum, quia natura uno modo agit, vim speciei ostendit. ita de astris, recte dicimus universales experientias esse, quas recitavimus de Solis et Lunae effectionibus: item de insignibus coniunctionibus, quia compertum est, similes esse effectiones plerumque" (Bretschneider et al. 1834–, 13:333).

35. Bretschneider et al. 1834–, 13:223–91, 335–36; see 261 for the equant.

36. Hooykaas 1984.

37. Thorndike 1923–58, 5:354–69.

38. The press was located first at Bamberg and later at Kirchehrenbach (see Zinner 1941, 57, e.g., nos. 1151 and 1266); for Apianus's press, see Günther 1882, 11.

39. Zinner 1990, 115.

40. See Rosen 1971b, 393.

41. Zinner 1941, nos. 1038, 1080, 1099, 1100, 1151, 1186, 1217, 1266, 1303, 1304, 1394–96, 1459, 1463–64, 1503–4, 1573, 1575–77, 1677, 1702, 1728, 1790–91, 1837, 1857, 1884, 1892, 1920–21.

42. Burmeister 1967–68, 3:50.

43. Buonincontro 1540. It is noteworthy that the Basel publisher of this work, Robert Winter, issued Rheticus's *Narratio Prima* in the following year.

44. On April 8, 1535, Johannes Apelt, the former chancellor to Duke Albrecht of Brandenburg, sent Albrecht from Nuremberg a nativity prepared by Joachim Camerarius with the suggestion that if he could not find someone to explain it to him, he should consult "an old canon from Frauenburg" (Prowe 1883–84, 1:401 n.; Biskup 1973, 155).

45. Hooykaas 1984, 14; Prowe 1883–84, 1:516; Burmeister 1967–68, 1:19; Rheticus 1982, 209–22; Swerdlow 1992.

46. Gaurico 1552.

47. Trans. Swerdlow 1992, 274; however, I render *pars* as "part" rather than "branch."

48. Albubater 1540, preface: "Quare ne illius difficultate deterreatur, pedissequam illius Astrologiam, ceu fructum ac mercedem quendam illi adiungendam esse putamus, quae & ipsa multas affert utilitates. In qua cum hoc tempore aliquid typis excudere uellemus, commodum ad manus nostras prouenit, Albubatris Liber genethliacus, siue De natiuitatibus inscriptus: quem non solum propter rerum copiam & authoris diligentiam, caeteris praeferendum putamus, uerum etiam propter iucundam ordinis nouita-

tem. Ita enim rerum per stellas significatarum ordinem secutus est, ut tamen ordinem Domorum non inuerterit."

49. Swerdlow 1992, 272.

50. Ludwik Birkenmajer pointed out that the horoscope must have been made while Copernicus was alive, as it made no sense to prepare a forecast for someone who was not living. Furthermore, he believed that Rheticus was the source of the information and that the Wittenbergers were interested to judge the worth of Copernicus's doctrine based upon the horoscope of its creator (see Birkenmajer 1975, 726–27, 728–33). Swerdlow and Neugebauer have analyzed Cod. lat. Monac. 27003, fol. 33 (see figure 32; also reproduced in Biskup 1973, plate 22), and conclude that the horoscope is in somewhat closer agreement with the Alfonsine-based *Tabulae Resolutae* than the numbers predicted by Copernicus's theory (Swerdlow and Neugebauer 1984, 454–57).

51. See Thorndike 1923–58, 5:367. From my inspection of the copy held by the Bibliothèque Nationale, I can find no evidence for Thorndike's comment that "Schoner maintained that the Copernican system was not unfavorable to astrology" (Schöner 1545). In the description of a copy of this work for sale by Jonathan Hill in 1995 (catalogue no. 88, item no. 89, 37), Thorndike's comment is endorsed. However, when I checked with the dealer, he too was unable to find any references to Copernicus.

52. For the hypothesis that Copernicus functioned as a powerful father figure to Rheticus, see Westman 1975b.

53. See Burmeister 1970, 1:46–105.

54. For a listing of these works, see ibid., 2:29–37.

55. Ibid., no. 21, 2:43.

56. Ibid., 1:69–70; 2:39–40. The Latin works are all called *Prognosticum Astrologicum;* the German works are called *Gemeine Anzeigung* (for 1545) and *Practica* (for 1547). The prognostication for 1546 appeared also in English (London: Richard Grafton, 1545).

57. Copernicus 1543, Vatican copy, title page; for illustration and description, see Gingerich 2002, Vatican 2, 108–10.

58. Achilles Gasser in Feldkirch to Georg Vögeli in Konstanz, 1540, in Burmeister 1967–68, 3:15–19. Burmeister published the original text with German translation; there is also an excellent French translation in Rheticus 1982, 197–99.

59. Burmeister 1967–68, 3:15: "Videtur tamen novae et verissimae astronomiae restitutionem, immo τὴν παλιγγευυησίυ haud dubie prae se ferre, praesertim cum de eiusmodi propositionibus evidentissima decreta iactitet, super quibus a doctissimis non modo mathematicis, sed philosophis maximis etiam non citra sudorem."

60. Rheticus 1971, 109–10.

61. Ibid., 110.

62. Later, however, Rheticus does return to the question of the *primus motus,* making it appear that he has indeed been working his way through the manuscript. This passage creates the impression that either Rheticus was studying and writing in great haste (without revising) or that the order of treatment was meant to make it appear as if Rheticus was faithfully reporting his own study of the manuscript.

63. J. L. E. Dreyer believed this to be the case: "Nothing of this theory of monarchies is mentioned by Copernicus himself but we cannot doubt that Rheticus would not have inserted it in his account if he had not had it from his 'D.Doctor Praeceptor,' as he always calls him" (Dreyer 1953, 333). Alexandre Koyré was more cautious: "It is difficult to say if Copernicus shared the views of his young friend, or was merely indifferent to them" (Koyré 1992, 32–33). See also North 1994, 289–90.

64. Rheticus 1982, 11. The title of the Basel edition is somewhat different: instead of putting Schöner's name first, the title now begins "Concerning the Books of Revolutions." It then continues unchanged with Copernicus's full title: "of that most erudite man and most excellent Mathematician Nicolaus Copernicus, Canon of Varmia." However, where the Gdańsk edition then identifies only "a certain young man studious of Mathematics," the Basel edition names Georg Joachim Rheticus.

65. This is the view of the translators of the French edition: "Ce passage astrologique de Rheticus interrompt l'exposé sur la variation de l'excentricité et sur le mouvement de l'apogée du soleil, commencé au chapitre IV. Aussitôt après cette digression, d'ailleurs, Rheticus poursuit la description de la théorie copernicienne du soleil. On voit bien sur cet exemple que les indications portées en manchettes par H. Zell dans l'édition de Gdansk visent à attirer l'attention du lecteur sans toujours remplir la fonction de titres de chapitres" (Rheticus 1982, 155).

66. Here I differ with the brilliant French team of editors and translators, who interpret this section as a digression from the astronomical material (Rheticus 1982, 155, n. 47).

67. "Addam et vaticinium aliquod" (Rheticus 1982, 47; Rheticus 1550b).

68. "De dignitate astrologiae" (Burmeister 1967–68, 3:25 n., 88, 90, 1:27).

69. Rheticus 1971, 121–22; Rheticus 1982, 47–48. Figure 33 is a reconstruction not found in the original work (Rheticus 1982, 153–54).

70. Derived not from the Bible but from the Babylonian Talmud. *Elias* is the Greek form and *Helias* the Latin of "Elijah." See further Warburg 1999, 693–95; Rheticus 1971, 122 n.; Rheticus 1982, 155 n. Barnes 1988, 78, 104–5, 107–8, 113, 279 n.; Granada 2000a, 109.

71. Reeves 1969, 309. For the Elijah prophecy, besides Rheticus 1982, 155 n. 46; Barnes 1988; Granada 1997b.

72. Carion 1550, *Biv; Melanchthon and Peucer 1624, preface, 120: "Sex millia annorum mundus, & deinde conflagratio. Duo millia Inane. Duo millia Lex. Duo millia dies Messiae. Et propter peccata nostra, quae multa & magna sunt, deerunt anni, qui deerunt."

73. See Burmeister 1970, 1:85–91.

74. Barnes 1988, 107.

75. Melanchthon 1532 in Bretschneider et al. 1834–, 12:708; Melanchthon and Peucer 1624, 27: "Nomen Chronici Carionis retinui, quod mutare illud autor primu sanctae beataeque memoriae Philipp. Melanthon socer meus noluit. Occasio nominis huius inde extitit, quod cum Ioannes Carion Mathematicus ante annos XL. coepisset contexere Chronicon, & recognoscendum illud atque emendandum, priusquam prelo subiiceretur, misisset ad Philippum Melanthonem, hic, quod parum probaretur, totum aboleuit [aboluit?] una litura, alio conscripto, cui tamen Carionis nomen praefixit. Sed et hoc cum retexuisset, amici nomen et memoriam, à cuius primoridijs [Gr.: aformi] prima Chronici contexendi natà atque profecta esset, titulo posteritati commendare voluit." Peucer prepared the tabular index to the book ("Tabella ostendens quo origine legenda et cognoscenda sit series historiarum mundi; 49); cf. Barnes 1988, 106–8.

76. The passage ends exactly like Carion's: "Et si qui deerunt, deerunt propter nostra peccata quae multa sunt." Bretschneider et al. 1834–, 12:46 f. Melanchthon says in a letter to Carion on the comet of 1531 that Paul of Santa Maria is the source of the Elijah prophecy cited by Carion himself in his *Chronica*. On the Elijah prophecy in the Renaissance, see Secret 1964, 11; for these references, see Rheticus 1982, 155 n. 46.

77. Barnes 1988, 50–52; Headley 1963, 108–10.

78. Pico della Mirandola 1969, exp. vii, chap. 4, 53–55; Pico della Mirandola 1965, 159–62.

79. See Hooykaas 1984, 56, 87.

80. Pico della Mirandola 1965, 160.

81. Ibid., 161.

82. Ibid., 159.

83. Ibid., 159.

84. Cf. Rosen (1943, 468), who recognized the possible astrological associations of the first part of the work's title but avoided any further discussion of it.

85. Rheticus 1971, 122. I have modified Rosen's translation.

86. "Letter to Emperor Ferdinand," preface to Johannes Werner, *De Triangulis Sphaericis* and *De Meteoroscopiis* (Krakow, 1557), in Rheticus 1982, 233; see also Rheticus 1971, 123 n.

87. "Preface to Werner" in Rheticus 1982, 233.

88. Rheticus 1971, 126–27.

89. "As for the fact that the planets are each year observed as direct, stationary, retrograde, near to and remote from the earth, etc., my teacher shows that this can be due to a regular motion of the terrestrial globe. . . . This [terrestrial] movement is such that the Sun occupies the middle of the universe while the earth, in place of the Sun, revolves on an eccentric that my teacher has decided to call the Great Orb" (Rheticus 1982, 54, 106; cf. Rheticus 1971, 135–36).

90. Ibid.

91. Granada (1996a, 794–97) also regards Copernicus as playing a silently supportive role for Rheticus's views about scripture and the Earth's motion.

92. Rheticus 1971, 136; Rheticus 1982, 106. Cf. Copernicus 1543, bk. 3, chaps. 1, 3.

93. Ravetz 1965, 1966; Curtis Wilson 1975. Working at a different historiographical moment, neither Ravetz nor Wilson was concerned with the rhetoric of Rheticus's arguments.

94. Rheticus 1971, 137; Rheticus 1982, 107.

95. See Copernicus 1543, bk. 1, chap. 4. The immediately preceding sentence reads: "Terrae igitur, ad Martis et aliorum planetarum motus restituendos, alium locum deputandum esse patet" (Rheticus 1982, 55).

96. In a highly suggestive and influential interpretation of this *petitio*, Noel Swerdlow argued that Copernicus meant the spheres to be taken as material, impenetrably solid entities whose motions would be incompatible with the nondiametral axes of the equant (Swerdlow 1973, 424–25, 432, 438–40). Later, in outlining the arrangement of the universe, Rheticus wrote in such a way as to suggest that the planets are situated in eccentric orbs (*orbes*): "Intra concavam superficiem orbis Martis et convexam Veneris, cum satis amplum relictum sit spatium, globum telluris cum adiacentibus elementis orbe Lunari circumdatum circumferri." In my opinion, although talk of "concave and convex surfaces" warrants talk of "thickness," it does not logically imply impenetrability.

97. Swerdlow (1973) has suggested that Copernicus's starting point was not the Earth's motion but his dissatisfaction with the equant, already evident in the first *petitio* of the *Commentariolus*.

98. Rheticus 1982, 186–187; Kepler 1937–, 1:119–120; the letter was dated September 13, 1588 (Brahe 1913–29, 7:129); when the letter came into Maestlin's possession is not known.

99. Rheticus 1971, no. 5, 137–38; Rheticus 1982, 107: "Since we see that this one motion of the earth satisfies an almost infinite number of appearances, should we not attribute to God, the creator of nature, that skill which we observe in the common makers of clocks? For they carefully avoid inserting in the mechanism any superfluous wheel or any whose function could be served better by another with a slight change of position."

100. Ibid.

101. For an excellent discussion of dialectical topics, see esp. Goddu 2010, 275–300; Moss 1993, 7–9.

102. For further analysis of Copernicus's logical resources, see Goddu 1996; Goddu 2010, 300–24.

103. Rheticus 1971, 165.

104. Ibid., 145.

105. Ibid., 138.

106. Rheticus 1971, 139; Rheticus 1982, 108.

107. Rheticus 1971, 139.

108. Rheticus 1971, 140; Rheticus 1982, 57, 109.

109. Ibid.

110. Rheticus 1971, 140–41; Rheticus 1982, 110; Aristotle 1960, bk. 2, chap. 5, 287b34–288a1.

111. "Ego quidem statuo Aristotelem, auditis novarum hypothesium rationibus, ut disputationes de gravi, levi, circulari latione, motu et quiete terrae diligentissime excussit, ita dubio procul candide confessurum, quid a se in his demonstratum sit, et quid tanquam principium sine demonstratione assumptum " (Rheticus 1982, 58, 110; Rheticus 1971, 142).

112. Rheticus 1971, 141; Rheticus 1982, 110.

113. Aristotle 1961–62, bk. 2, chap. 1, 993b 26–27; Rheticus 1971, 142–43; Rheticus 1982, 111. See Goddu's important discussion (2010, 321–23). According to Rosen, Rheticus generally quoted from Greek authors, but in this instance, the passage that Rheticus produced came from Cardinal Bessarion's Latin translation of Aristotle's *Metaphysics*—which means that that work could have been in the Varmia library.

114. Rheticus 1971, 146; I agree with the translation in Rheticus 1982, 113, which takes "orb" for *orbis*, rather than Rosen's "sphere."

115. See Achillini 1498; chap. 1 above.

116. Rheticus 1971, 146; Rheticus 1982, 60, 113: *principium motus et lucis*. The editors of Rheticus (1982, 169) comment here that "for Copernicus, the sun is simply a light that illuminates the world . . . the thesis of the sun as a principle of motion does not appear in Copernicus." Kepler later found inspiration in this particular passage, which he developed in ways anticipated by neither Copernicus nor Rheticus (see chap. 11, this volume; Kepler 1937–, 1:70, l. 34).

117. Rheticus 1971, 147; Rheticus 1982, 113.

118. As Rosen observes, Rheticus sometimes does not follow carefully the lettering of the diagram that he is reporting from Copernicus's manuscript (see Rheticus 1971, 155 n.).

119. For an example of a diagram added by Maestlin, see Kepler 1937–, 1:111; Rheticus 1982, 175–76; Maestlin 1596b, 134 ff. See also Grafton 1973.

120. Rheticus 1971, 186; Rheticus 1982, 138.

121. For a representation of the equant, compared with Kepler's ellipse, see Dennis Duke's animation (http://people.scs.fsu.edu/~dduke/Kepler.html; accessed July 19, 2008).

122. As Edward Grant maintains: "Nothing that Copernicus said or implied in *De Revolutionibus* enables us to decide with any confidence whether he assumed hard or fluid spheres. Copernicus fits the pattern of the Middle Ages, when explicit opinions about the rigidity or fluidity of the orbs were rarely presented" (Grant 1994, 346; cf. Lerner 1996–97, 1:131–38; Westman 1980a, 107–16; Goddu 1996, 28–32).

123. See especially chap. 12: "On Librations" (Rheticus 1971, 153–62; Rheticus1982, 118–22, 172–75. This is the so-called Tūsi couple. See Hartner 1973, 420–22; Ragep 2007. It is noteworthy that Rheticus makes no ascription to Arabic authority.

124. Rheticus 1971, 148; Rheticus 1982, 114.

125. Rheticus 1971, 194; Rheticus 1982, 144.

126. Ibid.

127. Rheticus 1982, 189, 141. Melanchthon used the story of traces on the Rhodian shore in his preface to a 1537 edition of Euclid's *Elements* (see Moore 1959, 147).

128. "Judicabat Alfonsinos potius quam Ptolemaeum imitandum et tabulas cum diligentibus canonibus sine demonstrationibus proponendas; sic futurum, ut nullam inter philosophos moveret turbanm: vulgares mathematici correctum haberent motuum calculum, veros autem artifices, quos aequioribus oculis respexisset Iupiter, ex numeris propositis facile perventuros ad principia et fontes, unde deducta essent omnia . . . atque illud Pythagoreorum observaretur, ita philosophandum, ut doctis et mathematicae initiatis philosophiae penetralia reserantur, etc." (Rheticus 1982, 85, 143; Rheticus 1971, 192).

129. The word *artifex* means, literally, "author." Rosen translates it as "scholar"; the French team uses "savant." Had Rheticus used *homo doctus, eruditus,* or perhaps even *scholasticus,* the translation would have been straightforward. Clearly, Rheticus wants to contrast more than just *learned* and *unlearned,* as "ordinary mathematicians" are not unlearned. I suggest that the distinction that Rheticus is urging is between *theorica* and *practica:* only a few are capable of grasping the theoretical assumptions from which the tables are derived. A few paragraphs later, Rheticus offers clarification when he cites Aulus Gellius's *Noctes Atticae* (bk. 1, chap. 9, no. 8): "As for the unlearned, whom the Greeks call 'people incapable of speculation, people who are strangers to the muses, to philosophy, and to geometry,' their shouts ought to be ignored" (see Rheticus 1982, 86, 144; Rheticus 1971, 195).

130. Rheticus 1971, 193; Rheticus 1982, 143.

131. *De caelo,* bk. 2, chap. 14. Rheticus quotes the full passage in Greek. This confirms what already seems obvious: Rheticus availed himself of books in the library of Copernicus and Giese.

132. For Osiander's life in relation to his views on natural knowledge, see Wrightsman 1975, esp. 215–21; Wrightsman 1970.

133. Osiander 1532; Wrightsman 1970, 46.

134. Quoted and trans. in Seebass 1972, 36.

135. Bretschneider et al. 1834–, 3:115.

136. Rosen 1971b, 403.

137. See Shipman 1967.

138. Kepler 1858–71, 1:246; Prowe 1883–84, 1: pt. 2, 523 n.; Burmeister 1967–68, 3:25–26; Rheticus 1982, 208–9.

139. Prowe 1883–84 1: pt. 2, 522.

140. "Peripathetici et theologi facile placabuntur, si audierint, eiusdem apparentis motus varias esse posse hypotheses, nec eas afferri, quod certo ita sint, sed quod calculum apparentis et compositi motus quam commodissime gubernent, et fieri posse, et alius quis alias hypotheses exogitet, et imagines hic aptas, ille aptiores, eandem tamen motus apparentiam causantes, ac esse unicuique liberum, immo gratificaturum, si commodiores excogitet. Ita a vindicandi severitate ad exquirendi illecebras avocati ac provocati primum aequiores, tum frustra quaerentes pedibus in auctoris sententiam ibunt" (Prowe 1883–84, 1: pt. 2, 523; Burmeister 1967–68, 3:25; Rheticus 1982, 208; Hooykaas 1984, 36–37).

141. See Wrightsman 1975; Williams 1992, 249–54, 483–88, 998–1001.

142. My italics. For English translations of the text of the "Ad Lectorem," see Rosen 1971a, 24–25; Copernicus 1978, xvi.

143. Ibid.: "Is there anyone who is not aware that from this assumption it necessarily follows that the diameter of the planet in the perigee should appear more than four times, and the body of the planet more than sixteen times as great as in the apogee, a result contradicted by the experience of every age?"

144. Rheticus 1971, 146; Rheticus 1982, 113.

145. Rheticus crossed out the offending Osiander letter with a red crayon before he sent it off to Wittenberg (see Gingerich 1992a, 72–73).

146. See Barnes 1988, p. 129.

147. This point is correctly stressed by Wrightsman 1975, 222.

148. Osiander 1548 (trans. George Joye).

149. Wrightsman cites a letter from Osiander to Luther's chaplain, Justus Jonas, following the comet of 1538 (Wrightsman 1970, 229): "I do not wish to tell Germany's future on the basis of the stars; but on the basis of theology, I announce to Germany the wrath of God."

150. This interpretation appears to be supported by a fragment of an Osiander letter written from Nuremberg, 13 March 1540 (List 1978, 455–56).

151. Wrightsman 1970, 161 ff.

152. "Quia optem etiam praemitti vitam auctoris quem a te eleganter scriptam olim legi. . . . Vellem adnecti quoque opusculum tuum, quo a sacrarum scripturarum dissidentia aptissime vindicasti telluris motum" (Giese from Lubawa to Rheticus in Leipzig, July 16, 1543; Prowe 1883–84, 1: (2), 537–39; 2:419–421; Burmeister 1967–68, 3:54–55).

153. Hooykaas 1984, 144.

154. Hooykaas offers no speculations or evidence

on this matter, although in October 1541, Melanchthon wrote to Mithobius criticizing "that Sarmatian astronomer" (ibid., 145).

155. Hooykaas 1984, 82 n. (referring to original pp. 1, 11, 16, 32, 33, 59, 63).

156. On the principle of accommodation, see Funkenstein 1986, 11–12, 222–70; Scholder 1966, 56–78; Westman 1986, 89–93.

157. Hooykaas 1984, 8 (45, 68): "Quemadmodum Scriptura genus sermonis, consuetudinem loquendi, et rationem docendi a populo et vulgo sumit." Subsequent references to this text in Hooykaas 1984 provide page numbers for the original Latin followed in parentheses by those for the modern Latin text and English translation.

158. Ibid., 10 (46, 70).

159. Ibid., 35 (54, 84).

160. Ibid., 39 (56, 87): "Manifestum est propter eandem causam, excepto Sole et Luna nihil de reliquis Planetis ibidem dici, utcunque Picus in suo Heptaplo eos conetur inde eruere, ut et alia taceam, quae ibidem praetermittuntur."

161. Pico della Mirandola 1965, 69.

162. Ibid., 69–70.

163. Ibid., 95–96.

164. Ibid., 97–98.

165. Ibid., 100–101.

166. Ibid.

167. Ibid.

168. Ibid., 101. Hooykaas (1984, 33), who partly paraphrases and partly quotes these passages, ends with his own irritable judgment: "Evidently some allegorical exegetes did not shrink back from the most tortuous reasonings and the most gratuitous assumptions in order to reach their goal."

169. Hamilton 2004, 108, 115.

170. On this episode, see Prowe 1883–84, 1: pt. 2, 274; Müller 1908, 25; Rose 1975a, 131; also Striedl 1953, 96–120; Rosen 1971b, 387–88.

171. For Edward Rosen, however, only the pope's personal knowledge and approval of De Revolutionibus could have indicated positive sentiment in Rome: "Had Copernicus actually received Paul III's permission to print the De Revolutionibus, what earthly reason would have deterred him from making a public proclamation to that effect at some prominent point in his Preface?" (Copernicus 1978, 337; see also Rosen 1975b). However, Rosen ignores the serious possibility that there was a negative shift among the Roman authorities only after the arrival, in July 1542, of a new master of the sacred palace, Bartolomeo Spina (see Kempfi 1980, 252).

172. Copernicus 1978, xvii. I have emended Rosen's translation. On Schönberg's life, see Walz 1930; Rose 1975a, 131.

173. See Burmeister 1967–68; Gingerich 1993c.

174. See Faculty of Law, University of Cambridge,

"Primary Sources on Copyright, 1450–1900," www
.copyrighthistory.org/htdocs/index.html, accessed September 9, 2010. For an important account of the transitional moment in the conception of literary property, see Rose 1994.

175. See Frugoni 1950; Thorndike 1923–58, 5:252–74.

176. D'Amico 1983, 47.

177. Gauricus 1541. I have not been able to consult this early edition (1504).

178. Granada and Tessicini (2005) point out important rhetorical and linguistic parallels between Copernicus's and Fracastoro's prefaces while suggesting that "at least part of [Copernicus's dedicatory letter] was intended to neutralize and oppose the Fracastorian reform in order to win support for his own system" (472).

179. See Thorndike 1923–58, 5:256–59; Pèrcopo 1894, 123–69.

180. See, for example, the *Liber de revolutionibus et nativitatibus* of the Jewish astrologer and Talmudic scholar Abraham ben Meir ibn Ezra (1092–1167), also known as Abraham Judaeus and Abraham Abenare. Pico cited him frequently as "Avenazram" (e.g., Pico della Mirandola 1496, bk. 1. chap. 1, 106–7). The first published edition appeared in 1485 (Venice: Erhard Ratdolt) with the title *De Nativitatibus* and advertising itself as "utilissimus in ea parte astrologie qui de nativitatabus tractat: cum figuris exemplaribus singulis domibus antepositis." See also Petrus Pitatus, *Tractatus de Revolutionibus Mundi atque Natiuitatum:* "Reuolutionem annorum mundi, uidelicet Introitum Solis in primum Arietis, uel etiam in quodcunque aliud Zodiaci punctum, utputa natiuitatum, uel aedificiorum inuenire." As usual, Edward Rosen tried to dissociate Copernicus from any taint of astrology (Rosen 1943, esp. 468).

181. Cf. the interesting remark written by a sixteenth-century commentator—probably from the Wittenberg orbit—on the title page of a copy of *De Revolutionibus*. The commentator guessed that "Copernicus derived the title of his volume from that passage in the *Astronomical Hypotheses* of Proclus [*Hypotyposis*, IV, 98] where he mentions 'Sosigenes the peripatetic and his work περὶ τῶν ἀνελιττουσῶν, that is, *de revolutionibus;* it was not Copernicus who added *orbium caelestium*, but someone else. In these six books Copernicus embraced the whole of astronomy, stating and proving individual propositions mathematically and by the geometrical method in imitation of Ptolemy" (Prowe 1883–84, 1: pt. 2, 541–42; Rosen 1943, 468–70; Gingerich 2002, Wolfenbüttel 1, 96–98). The last statement comes almost verbatim from Rheticus's *Narratio Prima*. Rosen argues that Copernicus was probably unfamiliar with Proclus's text, contrary to what the commentator believed; but Copernicus certainly was aware, from Pico's *Disputationes*, of the more familiar association between revolutions and nativities.

182. All quotes from suppressed introduction to bk. 1, Copernicus 1978, 7–8.

183. *Chrys.:* I think I've never read anything in pagan writers more proper to a true Christian than what Socrates spoke to Crito shortly before drinking the hemlock: "Whether God will approve of my works," he said, "I know not; certainly I have tried hard to please him. Yet I have good hope that he will accept my efforts."

Neph.: An admirable spirit, surely, in one who had not known Christ and the Sacred Scriptures. And so, when I read such things of such men, I can hardly help exclaiming, "Saint Socrates, pray for us!" (Erasmus 1965, 67–8).

184. "Qui apud me pressus non in nonum annum solum, sed iam in quartum nouennium, latitasset" (Copernicus 1543, fol. iii, ll. 13–14).

185. "Pleraque tamen interim admiserunt, quae primis principis, de motus aequalitate, uidentur contrauenire" (ibid., fol. iii b, ll. 12–13). This refers to the equant's violation of the principle of uniform, circular motion. Cf. "Prima petitio" in the *Commentariolus:* "Omnium orbium caelestium sive sphaerarum unum centrum non esse": "There is no one center of all the celestial orbs or spheres" (Copernicus 1884, 2:186).

186. "Mundi formam, ac partium eius certam symmetriam non potuerunt inuenire, vel ex illis colligere" (Copernicus 1543, fol. iii b, ll. 14–15). I translate *forma* as "arrangement" in order to convey *symmetria* (the due proportion of each part to another with respect to the whole). The sixteenth-century editor of Vitruvius, Guillaume Philandrier (1505–65), pointed out that there is no specific Latin word for the Greek *symmetria* and that Vitruvius appears to favor the noun *commensum*, from the verb *commetior*. Philandrier's notes are cited in Laet 1649, 38.

187. "Sed accidit eis perinde, ac si quis a diuersis locis, manus, pedes, caput, aliaque membra optime quidem, sed non unius corporis comparatione, depicta sumeret, nullatenus inuicem sibi respondentibus, ut monstrum potius quam homo ex illis componeretur" (Copernicus 1543, fol. iii b, ll. 15–19). Cf. Pierre Gassendi's paraphrase of the same passage, where, as in my translation, the analogy to painting is stressed: "Sed iis perinde evenire, ac si quis Pictor manus, caput, pedes, membra caetera, optime illa quidem, sed non unius corporis comparationis depicta adunaret, sicque ex illis monstrum potius, quam hominem compingeret" (Gassendi 1655, 296).

188. Horace 1926, 451, ll. 1–13.

189. See Weinberg 1961, 1:74.

190. Landino 1482, clvii v.

191. Fracastoro's difficulty was at the level of the assumed premises (eccentric vs. concentric spheres), whereas Copernicus's objection was at the level of the consequence, the *mundi formam:* neither the one nor

the other kind of spheres produced the right entailment. (For further discussion of Fracastoro and Copernicus, see Granada and Tessicini 2005, 462–63.)

192. "Sed & syderum atque orbium omnium ordines, magnitudines, & coelum ipsum ita connectat, ut in nulla sui parte possit transponi aliquid, sine reliquarum partium, ac totius uniuersitatis confusione" (Copernicus 1543, fol. iii j, ll. 22–25). Cf. the translations in Kuhn 1957, 142; Copernicus 1978, 5; 1976, 26. For discussion of the terms *ordo* and *symmetria* in *De Revolutionibus*, see Rose 1975b, 153–58.

193. For Copernicus's knowledge of the *Posterior Analytics*, see Birkenmajer 1972a, 615. On demonstration, see Bennett 1943; see also Wallace's illuminating discussion of the later career of this ideal of knowledge in Galileo's period (Wallace 1984b, 99–148).

194. For discussion of the dialectical intrinsic topos from an integral whole, see Goddu 2010, 64–65, 67, 69, 83–84, 182, 283–84; Goddu 1996, 41, 50; Moss 1993, 44.

195. According to Lucio Bellanti, Paul of Middelburg excelled "in both parts of astrology," that is, astronomy and "true astrology" (Bellanti 1554, 218).

196. Cf. Hallyn 1990, 73–103.

197. Nardi 1971, 99–120, and, following Nardi, Bettini 1975. A description of Copernicus's alleged self-portrait is known only through Biliński 1983, 276: "[One sees] a scar on Copernicus's face, at first not visible, and furthermore, what is more surprising, in his pupil one sees a reflection of the bell tower of a Gothic church." In response to my queries, Jerzy Dobrzycki reported that this painting is undoubtedly the one hanging today in the municipal high school of Toruń. However, although it is very like an original portrait, it seems to have been painted not by Copernicus but by a professional artist of northern European origin. Bettini and Nardi also suggest that Giorgione's *The Three Philosophers* portrays young Copernicus, al-Battani, and Ptolemy as the figures, but this hypothesis is just as conjectural as the interpretations of other art historians who have suggested three Aristotelians or three magi: cf. Wind 1969, 4–7, 25–26. For a more sober treatment of this painting, see Meller 1981, 227–47.

198. Klein 1961, 215–16.

199. Gauricus 1969, 92–93: "Mensuram igitur, hoc enim nomine Symmetriam Intelligamus, cum in caeteris omnibus quas natura progenuit rebus, tum uero in homine ipso admirabilissimam et contemplari et amare debebimus, Ita enim undique exactissime dimetatis partibus compositum est nostrum hoc corpus, ut nihil plane aliud quam Harmonicum quoddam omnibus absolutissimum numeris instrumentum esse uideatur." Cf. Vitruvius: "Symmetria est ex ipsius operis membris conueniens consensus": Vitruvius 1496, bk. 1, chap. 2; bk. 3, chap. 1; bk. 6, chap. 2. Vitruvius, in turn, had drawn the notion of *symmetria* from ancient rhetoric.

200. Copernicus 1543, fols. iii b, iv b. Copernicus's emphasis on God's ordained power, or *potentia ordinata* ("ab optimo et regularissimo omnium opifice"), rather than his absolute power (*potentia absoluta*), fits well with the association of papal authority with natural order.

201. Maffei 1518, fol. 141: "Imprimis urbs tua [papa] curanda interpolandaque quin frustra (ut inquit apostolus) aliis praesideat, qui domum propriam neglexerit. Ante omnia tuis contraria moribus auaritia purganda ac pristinae libertati protinus restituenda, quum natura id expetat ut membra capiti congrua, ciues principi ac greges pastori similes in hac parte reddantur" (quoted in D'Amico 1980, 182; my translation). Maffei's measures included reform of the collection of revenues, curbing of lawyers' fees, policing of crime near the Curia, preservation of a regular and steady food supply, personal papal involvement in acts of charity, and the establishment of seminaries where the *artes liberales* might be taught (D'Amico 1980, 183).

202. D'Amico 1983, 223.

203. For a superb treatment of this theme, see Scribner 1981, 100–104, 165, 232–34; for illustrations, see Westman 1990, 190–91.

204. Luther 1883–, 10:2, 458; quoted and trans. in Scribner 1981, 245.

205. By contrast: "If the hypotheses assumed by [traditional astronomers] were not false, everything which follows from their hypotheses would be confirmed beyond any doubt" (Copernicus 1543, fol. iii b). Cf. Gauricus: "As for what is said about poets and painters, that they may do what they please, this is valid to the extent that they do not depart from nature" (Gauricus 1541, fol. Aiii). Here again, Copernicus's humanism is profoundly evident. As Paul Oskar Kristeller reminds us, "Moral teaching is often contained in literary genres cultivated by the humanists where a modern reader might not expect to find it. . . . The humanists also followed ancient and medieval theory and practice in their belief that the orator and prose writer is a moral teacher and ought to adorn his compositions with pithy sentences quoted from the poets or coined by himself" (Kristeller 1961b, 1:295).

206. See Kempfi 1972.

207. Rheticus 1982, 84–85, 142–43; see Drewnowski 1978.

208. See Copernicus 1978, 342.

209. In 1525, Tiedemann Giese wrote one of the earliest anti-Lutheran polemics (*Anthelogikon*), in which he advocated tolerant persuasion and mutual compromise. See Borawska 1984, 303–43; Kempfi 1972, 397–406, esp. 400; Hooykaas 1984, 20–27; Hipler 1868.

210. Alexandre Birkenmajer (1965, 15), has observed that on the four occasions when Copernicus refers to a deity, he nowhere uses the word *God* but rather employs such terms as *Opifex omnium, Opifex universorum,* and *Opifex Maximus.*

211. From Piccolomini 1551, 964, verse 32. See also Drewnowski 1973; Prowe 1883–84, 2:278–80; Hipler 1875, 21.

212. As J. G. A. Pocock has observed about the languages of political thought, "We wish to study the languages in which utterances were performed, rather than the utterances which were performed in them. . . . When we speak of 'languages,' therefore, we mean for the most part sub-languages: idioms, rhetorics, ways of talking about politics, distinguishable language games of which each may have its own vocabulary, rules, preconditions and implications, tone and style" (Pocock 1987, 21).

213. Laudan 1990, 323.

214. Dear 1995, 12.

5. THE WITTENBERG INTERPRETATION OF COPERNICUS'S THEORY

1. It was hardly "the book that nobody read": see Gingerich 2002, 2004.

2. Clavius 1594, 68.

3. Rothmann to Brahe, April 18, 1590, Brahe 1913–29, 6:217, ll. 6–7, 20–22.

4. "Nam ipsos Copernici libros Reuolutionum legere non omnibus vacat" (Kepler 1984, 31; Kepler 1937–, 1: chap. 1, p. 15).

5. "Considerations on the Copernican Opinion" (1615), in Finocchiario 1989, 71.

6. Hortensius made these remarks in the dedication to a work by Wilhelm Blaeu (1571–1638), once an assistant of Tycho Brahe, which provided just such illustrations of Copernican and Ptolemaic globes as were called for: "Candido ac Benevolo Lectori M. Hortensius," in Blaeu 1690, unpag. The passage is referred to also by Thorndike 1923–58, 6:7, and F. Johnson 1953, 286.

7. In 1551 Mercator made an elegant but still traditional celestial sphere and an astrological disc (see Vanden Broecke 2001).

8. In recent historiography, Copernicus's beliefs regarding the ontology of the celestial spheres and the logic of his main claim are good examples of areas of philological difficulty. See, for example, Swerdlow 1973, 432, 437–39, 477–78; Rosen 1971a, 13–21; Aiton 1981, 96–98; Jardine 1982; Westman 1980a, 112–16.

9. The classic work on the *Prutenics* and their subsequent use is Gingerich 1973c.

10. See Hartfelder 1889, 419–36, 491–500; see also Woodward 1924.

11. Hartfelder 1889, 210–22.

12. For the application of Melanchthonian principles at the Nuremberg *Gymnasium*, see Strauss 1966, 236 ff.

13. Hartfelder 1889, 489–538; Kusukawa 1995, 185–88; cf. Eulenburg 1904.

14. Quoted and translated by Thorndike 1923–58, 5:378; also quoted in Kusukawa 1995, 186–87.

15. Euclid 1537; Moore 1959, 150.

16. Bretschneider et al., 1834–, 9:261–66.

17. Melanchthon 1536, 1537.

18. Zinner 1941, nos. 1602, 1647 (also Bretschneider et al., 1834–, 3:115), 1701, 1802, 1833, 1881, 2025. See also Pantin 1987, 85–101.

19. Zinner 1941, no. 1969.

20. Ibid., nos. 2027, 2047 (Augsburg 1551).

21. As I did in Westman 1975b, 165–93. Although breaking with internalism in its heyday, the article failed to give any place to astrology. See now Kusukawa 1995; Brosseder 2005.

22. My emphasis. See Hammer 1951.

23. Bretschneider et al., 1834–, 11:265–66.

24. Peucer 1553, preface.

25. G. Kepler 1931, 2:137–38; Jarrell 1971, 36.

26. Höss 1972; Burmeister 1967–68, 3:8.

27. See Schilling 1981, 1986.

28. Williams 1992, 610; Swerdlow and Neugebauer 1984, 1:10, 2: figs. 1, 2, pp. 564–65.

29. Williams 1992, 613.

30. Rheticus 1971, 190.

31. Dantiscus, in his *Jonas propheta* or *Prophecy of the Destruction of the Free City of Danzig* (1538), warned the Danzigers of, among other things, Lutheran "impiety" (see Williams 1992, 615–16).

32. See Hartmann to Duke Albrecht, April 27, 1543, in Voigt 1841, 283.

33. See Voigt 1841, 111–12.

34. Burmeister 1967–68, 1:47.

35. Ptolemy 1991, 25.

36. See Barton 1994, 179–81, 206, 209, 212.

37. Burmeister 1967–68, 3:28–29.

38. Ibid., 28–38. Presumably, Rheticus had in mind that part of a particular prediction that concerns the region of the Earth affected by eclipses (see Ptolemy 1940, bk. 2, chaps. 4–5).

39. Burmeister 1967–68, 1:65–69.

40. See especially Moran 1978, 190–96; Moran 1977, 1991b; Feingold 1984; Zinner 1956, 604–5; Clulee 1988, 32–33, 192–93; Hill 1998; Hayton 2004. Consider also the great market for ivory sundials (Gouk 1988).

41. As I argued in Westman 1980a, 117–18.

42. Burmeister 1967–68, 1:67.

43. Ibid., 1:69.

44. Ibid., 1:72.

45. For descriptions of these copies and their owners, all from 1543, see Gingerich 2002: Peucer (Paris 12, p. 40); Heller (Rostock, p. 90); Schreiber, later owned by Kepler (Leipzig, pp. 76–80); Stoius (Copenhagen 1, p. 32); Homelius, later owned by Praetorius (New Haven 1, CT, Yale, 306–13); Reinhold (Edinburgh, Crawford Library, 268–78).

46. See Gingerich 2002, Vatican 2, 108–10.

47. Blumenberg 1965, 109.

48. Bretschneider et al., 1834–, 4:847.

49. New evidence, based on the records of Iserin's trial, undermines an earlier claim that the main charge

was that of sorcery (Burmeister 1967–68, 1:14–17; West-man 1975b, 187. See now Burmeister 1977; Tschai-kner 1989; Danielson 2006, 15–17).

50. See Rosen 1969.

51. Lorraine Daston has articulated this feeling most cogently: "Whatever and however vehement their other confessional differences, historians, sociologists, and philosophers of science share a certain horror of the psychological, properly so-called, and I confess I am no exception to this general hostility" (Daston 1995, 4). As will be apparent, I do not share this horror.

52. Rheticus had written: "In my teacher's revival of astronomy I see, as the saying is, with both eyes and as though a fog had lifted and the sky were now clear, the force of that wise statement of Socrates in the *Phaedrus:* 'If I think any other man is able to see things that can naturally be collected into one and divided into many, him I follow after and "walk in his footsteps as if he were a god" ' " (Rheticus 1971, 167–68).

53. On the nature of psychoanalytic validation, see Wisdom 1974; Glymour 1974, 332–48, 285–304.

54. Manuel 1968, 5.

55. Rheticus mentions that he had met Paracelsus in 1532 (when Rheticus was eighteen) and was im-pressed by him. It is interesting that he did not then decide on a career in medicine, a vocational choice that would have identified him directly with his fa-ther. But in 1554, at the age of forty and after his trav-els had in effect ended, Rheticus became a practicing physician in Krakow, and his interest in Paracelsus revived. See Burmeister 1967–68, 1:35, 152–55.

56. Rheticus 1971, 109.

57. Ibid., 108. A facsimile of the title page is re-printed in Burmeister 1967–68, 2:58–59.

58. Rheticus 1971, 186.

59. Ibid., 186–87.

60. Burmeister 1967–68, 1:175.

61. Prowe 1883–84, 1: pt. 2, 388: "Profectus itaque in Ungariam, ubi tum agebat Rheticus, humanis-sime ab eo sum exceptus. Vix autem pauco sermone ultro citroque habito, cum meae ad se profectionis causam accepisset, in has voces erupit: 'profecto', in-quit, 'in eadem aetate ad me venis, qua ego ad Coper-nicum veni. Nisi ego illum adiissem, opus ipsius omnino lucem non vidisset.' " Also quoted and trans-lated in Koestler 1959, 189–90.

62. In 1557 Rheticus wrote of Copernicus, "whom I cherished not only as a teacher, but as a father" (quem non solum tanquam praeceptorem, sed ut patrem colui). This letter (Rheticus to King Ferdinand I in Burmeister 1967–68, 3:139), which shows no animos-ity toward Copernicus, weighs against Arthur Koes-tler's thesis that Rheticus felt betrayed by Copernicus when the latter failed to mention him in *De Revolu-tionibus.* This omission is perhaps to be understood as part of a deliberate strategy to direct the two works to different confessional audiences.

63. Presumably same-sex and represented as an Italian vice (another term being *florenzen,* "to florence"). For an excellent treatment of sodomy's practice and diverse meanings, see Puff 2003. We learn the de-tails about Rheticus from Jacob Kröger (d. 1582), a Hamburg pastor who, like many readers, kept per-sonal notes in his copy of Johannes Stadius's *Ephe-merides, 1554–1576* (Stadius 1560): "Excellens Mathe: qui vixit et docuit Lipsiae aliquandiu, post vero circa annum 1550 ea urbe aufugit propter Sodomitica et Italica peccata. Ego hominem novi" (cited by Voss 1931, 179–84, 182–83; also Zinner 1988, 259). I have not been able to locate the original, formerly located at the Hamburger Staats- und Universitätsbibliothek; it may have been destroyed in World War II.

64. Konrad Schellig, *In pustulas malas morbum quem malum de francia vulgus appellat,* in Stevenson 1886, no. 1270, 56.

65. See Gingerich 1970; Voigt 1841, 514–46; Rheti-cus to Duke Albrecht, 29 August 1541, in Burmeister 1967–68, 3:38–39.

66. Reinhold 1542, fol. C2v.

67. Ibid., fol. C7r.

68. Rheticus 1971, 134; Rheticus 1982, 54, 105.

69. Rheticus 1971, 135.

70. Reinhold's copy was first identified by Owen Gingerich (1993b, 176–77; Gingerich 2002, Edin-burgh 1, 268–78).

71. Copernicus 1978, bk. 5, chap. 2, 240 (alluding to Cicero, *Republic,* VI).

72. Copernicus 1543, bk. 5, chap. 2: "principia artis"; Copernicus 1978, bk. 5, chap. 4, 242.

73. Copernicus 1543 (copy in Crawford Library, Edinburgh), bk. 5, chap. 4, fol. 142r: "Primus Modus per Eccentrepicyclum;" fol. 142v: "Secundus Modus per Homocentrepicyclos"; Ibid., bk. 5, chap. 25, fol. 164v.

74. Ibid., bk. 3, chap. 3, fol. 67v.

75. Ibid., bk. 1, chap. 6, fol. 4v.

76. Cf. Ptolemy 1998, bk. 3, chap. 1, 132: "The only points which we can consider proper starting points for the sun's revolution are those defined by the equi-noxes and solstices on that circle."

77. Reinhold 1551, "Praeceptum xxi," fol. 34v. Rein-hold 1571, held by the Schweinfurt Stadtbibliothek, contains Maestlin's extensive annotations. I have used this copy.

78. Reinhold to Albrecht, May 12, 1542, in Voigt 1841, 516: "Wiewohl ich E.F.G. unbekannt bin."

79. Ibid.: "Bin ich doch aus der Ursache, dass E.F.G. vor andern Fürsten Tugend und löbliche Kün-ste und besonders die Astronomie und Cosmographie lieben, ehren und förden, bewogen worden."

80. Ibid.: "Dieses Büchlein Schulmaterie ist"; "Und nicht ein grosses Gepränge macht, wie wenn man viele Instrumente malet u.s.w."

81. Ibid.: "So ist es doch der Grund der rechten Kunst, woraus die Instrumente kommen, und ein

Schlüssel dieser Künste, dem ohne eisen Anfang kann man den Ptolemäus und die Tafeln nicht verstehen oder brauchen."

82. Albrecht to Reinhold, August 8, 1542, ibid., 517: "Wir haben euer Schrieben sammt dem Büchlein, welches ihr uns zugeschrieben, empfangen, gelesen und wohl vernommen."

83. Ibid.: "Dass ihr uns den Grund derselben löblichen Kunst zugeschrieben habt und damit unsern Namen rühmen thun."

84. Reinhold to Albrecht, October 8, 1542, ibid., 518.

85. Albrecht to Reinhold, November 27, 1542, ibid., 518–19. The duke's language closely resembles phrasing that he had used a month earlier in acknowledging thanks for a gift to the Nuremberg instrument maker Georg Hartmann: "Your high thanks, however, were not necessary because what I have done for you in this case is done out of grace for you and to show our affection to your person and the praiseworthy arts" (Albrecht to Hartmann, 5 October 1542, ibid., 279).

86. Ibid.: "Wollten wir uns immer als der gnädige Herr in allem Ziemlichen finden lassen"; "In Gnaden bittend, ihr wollet uns bisweilen *ex astris* euer Judicium."

87. Ibid., Reinhold to Albrecht, January 8, 1544, 519–20.

88. Melanchthon to Albrecht, July 16, 1544, ibid., 520–21, my italics. Shortly thereafter, Melanchthon's son-in-law Sabinus, who was working to help establish the duke's university at Königsberg, put in a further endorsement of Reinhold (p. 521).

89. Albrecht to Melanchthon, August 2, 1544, ibid., 521.

90. Melanchthon to Albrecht, October 18, 1544, ibid., 523; Bretschneider et al., 1834–, 5:510–11.

91. Reinhold to Albrecht, October 14,1544, in Voigt 1841, 522.

92. Zinner 1988, 503.

93. Albrecht to Reinhold, December 11, 1545, in Voigt 1841, 524.

94. Reinhold to Albrecht, December 13, 1545, ibid.

95. See Grimm 1973, 205–8.

96. Albrecht to Reinhold, April 20, 1547, in Voigt 1841, 525: "Your apology that you have not been able to complete everything, given the present tumultuous times, was not necessary. Since we have already pledged to support your studies with a gracious assistance, you should have complete faith in us; because we are always inclined to show you our goodwill." Albrecht to Melanchthon, June 1547, ibid.: "We are sorry from the bottom of our hearts that Erasmus and other pious and very learned people, because of the unsettledness of the current times, have been impeded from their work. But we are very pleased that you continue to praise him as a highly useful man and, on account of your recommendation, we want to show him that we are not going to cut his funding."

97. Albrecht to Melanchthon, October 18, 1547, ibid., 526.

98. Grimm 1973, 183–84.

99. Melanchthon to Albrecht, April 29, 1548, in Voigt 1841, 526–27.

100. Melanchthon to Albrecht, November 1548, ibid., 527. Isinder was first dean of the philosophy faculty and was recommended for the job by Camerarius (Voigt 1841, 117–18).

101. Reinhold to Albrecht, May 2, 1549, ibid., 527–28. My italics.

102. Melanchthon to Schöner, November 13, 1544. no. 3073, Bretschneider et al., 1834–, 9:526–27: "Sciunt typographi, scholasticos libellos, qui artium praecepta continent, excudi foeliciter. Ideo si authoritas accesserit tua, obtineri res poterit. Solvi pretium pro labore describendi iustum est. Quaeso igitur scribe, vel per Ioachimum Leucopetr. significa, quid de editione earum tabularum Erasmo sperandum sit. Ioachimo Leucopetreo pro muneribus missis gratias ago."

103. Reinhold 1551, fols. αɪv–α2v: "Diploma Caesareum Concessum Erasmo Reinholt Salveldensi" (June 24, 1549).

104. Reinhold to Staphylus, September 8, 1549, in Voigt 1841, 529.

105. "Ich habe nun aber viele Gründe, warum ich die Tafeln *Tabulae Prutenicae* nennen und dem erlauchten Fürsten Herzog Albrecht von Preussen dediciren möchte; und zwar ist der vornehmste der, dass ich die meisten Beobachtungen, von welchen als den Principien und Fundamenten ausgehend ich dies Tafeln entworfen und ausgeführt, von dem hochberühmtesten Nicolaus Copernicus, einem Preussen, entliehen habe" (ibid., 528–32, esp. 530).

106. Ibid., 531.

107. Staphylus to Albrecht, September 1549, ibid., 532.

108. Albrecht to Staphylus, November 29, 1549, ibid., 533.

109. Reinhold to Staphylus, Day of the Innocents, 1550, ibid., 534.

110. Ibid., 534.

111. These excellent examples can be found in Strauss 1966, 206–7; see also Strauss's helpful discussion of Nuremberg monetary units and values, 203–8.

112. We do not know how much Petreius intended to contribute to the publication expenses.

113. See Voigt 1841, 537. Luther, for example, received from his printers only a few copies of his writings.

114. Reinhold 1551, fol. α3v.

115. Ibid., author's preface, quoted in Gingerich 1973c, 48.

116. Albrecht to Reinhold, July 27, 1550, in Voigt 1841, 540; for monetary values, see Strauss 1966, 203–8.

117. Reinhold to Albrecht, October 14, 1551, ibid., 542.

118. Albrecht to Reinhold, March 21, 1552, ibid., 543.

119. Albrecht was particularly interested in the nativity of the King of France (Albrecht to Reinhold, April 12, 1552, 543).

120. Voigt 1841, 543.

121. Ibid., 543–46.

122. The first printed edition occurred in the heyday of the early publication of Arabic astrological works: Alfonso X 1483. See the useful commentary in Thoren 1974.

123. Reinhold 1551, author's preface, quoted in Gingerich 1973c.

124. Gingerich 1973c, 49–50.

125. Reinhold 1551, author's preface, facing fol. α3: "Causas verò & rationem singularum compositionum exposui in commentarijs nostris, quos scripsi in opus reuolutionum Copernici." For Reinhold's commentary, see Henderson 1975; Birkenmajer 1972c, 765.

126. For detailed description of the family of interrelated Reinhold copies and their subsequent lineage, see Gingerich and Westman 1988, 27–41; see further Gingerich 2002.

127. Reinhold 1551, "Praefatio," β2v: "Vetus nomen est Astrologiae, qua intelligebant olim doctrinam non solum de viribus seu effectibus: verumetiam de motibus syderum ac corporum coelestium. Posterior autem aetas eam doctrinam, quae rationem motus stellarum contemplatur ac numeris persequitur. Astronomiam consueuit dicere, & Astrologiae nomen accomodauit ad solas praedictiones de euentibus, qui astrorum motibus & positu efficiuntur, aut significantur in hac inferiori natura. Verum de hac diuinatrice parte alias dicitur."

128. Ibid., fols. 4r, 15r, 18v, 25v, 29r.

129. Ibid., fol. 21.

130. Heller 1549. See also Heller's dedicatory poem to Schöner 1545: "In Astrologicum opus Clarissimi Mathematici D. Ioannis Schoneri, Carmen Ioachimi Helleri Leucopetraei Ludimagistri Noribergensis."

131. Ibid.; for Heller's copy, see Gingerich 2002 (Rostock), 90.

132. Heller 1551, fol. Aii r. We cannot assume that Heller refers here to De Revolutionibus, as he might have had prepublication access to Reinhold 1551 or Reinhold 1550.

133. Gingerich 2002 (Rostock), 90: "Clarissimo & doctiss: artium philosophicarum ac medicinae Doctori D. Christop[horo] Stathmioni amico cariss[im]o: suo d[ono] d[edit]. Joach[im] Hellerus Leucopetreus." A note directly adjacent to Heller's, but in a different hand, contains some natal information about Copernicus identical to that found in the nativity shown earlier (see fig. 32, this volume): "Natus est 1473. 19 Febr. hor. 4. min. 48 mer. Thorunij in Borussia / Nicolaus Copernicus Mathematicorum nostro seculo

praeceptor obijt 1543. aetat. 70." Further research is needed to establish whether this note was added by Stathmion.

134. The Imperial Privilege printed with the *Prutenic Tables* (Reinhold 1551) lists a "Commentarius in Opus Revolutionum Copernici" (see Gingerich 1973c, 58–59; Henderson 1975). It seems reasonable to infer that knowledge of Reinhold's unpublished commentary must have been common among close members of the Melanchthon circle.

135. For biographical details, see *Nouvelle biographie générale*, 767–70; Blount 1710, 735–37; Voigt 1841, 500–508.

136. Bretschneider et al., 1834–, 13: cols. 179–411.

137. Ibid., 216–17.

138. Ibid., 244, 225, 241, 262.

139. The change in tone from the 1549 edition to the 1550 edition was first pointed out by Wohlwill (1904, 260–67), and was then widely adopted by later German historians such as Maurer (1962, 223), Müller (1963), and Blumenberg (1965, 101–21, 174; the texts are printed in parallel columns). Thorndike (1923–58, 5:385), made use of Wohlwill's discovery, but it was overlooked by Kuhn (1957, 191–92, 196), and Boas (1962, 126). Wohlwill's position is found in more recent American studies: see Christianson 1973; Wrightsman 1975, 347–48; Moran 1973.

140. Peucer 1553, fols. E3v, P2.

141. Ibid., fols. E4, G2v.

142. Gingerich 1992a, 81.

143. Theodoricus 1570, 116–19.

144. The Strigelius/Maestlin copy is now located at the Stadtbibliothek Schaffhausen.

145. Ramus 1569, 66: "Homilius etiam Carolo imperatori & Augusto Saxoniae electori mathematum magister fuit, & ab utroque ampla praemia consecutus." Notes derived from Homelius appear in the following extant copies of *De Revolutionibus*: 1543 New Haven 1; 1543 Schweinfurt; 1543 Gotha 1 (see Gingerich 2002).

146. Scultetus kept a notebook in which he recorded the annotations from one of Homelius's copies of *De Revolutionibus* (Bamberg, Staatliche Bibliothek, shelfmark J H Msc astr 3). See further Gingerich and Westman 1988, 30–31.

147. Neander 1561. Neander was promoted to Master of Arts in 1550 under the deanship of Reinhold (*Album Academicae Vitebergensis* 1841 1:239).

148. Poggendorff 1863, 1:757; Smith 1958, 135. Flach is not known to have published anything.

149. Witekind 1574, 63, 79–83 (on Copernican values), 108–11 (on solar and lunar distances), 122–27 (on the equinoxes), 224–38 (on Reinhold). Copy used: Schweinfurt Stadtbibliothek. I have not been able to inspect Witekind's *Oratio de Doctrina et Studio Astronomiae* (Neapoli Nemetum, 1581). Witekind also wrote several polemical works against the Jesuits.

150. Hildericus 1568, 1590.

151. Poggendorff 1863, 762–63. In 1550, he assisted in the production of an edition of Ptolemy's *Almagest* (Zinner 1941, no. 1997).

152. Bauer 1999, 357–60, 417–24; Poggendorff 1863, 2:833; Gundlach 1927–2001, 1:365.

153. The work is curious because *De Revolutionibus*, bk. 2, is essentially a work of trigonometry. See Dibvadius 1569; Moesgaard 1972b, 117–18.

154. See also Siderocrates 1563; Thorndike 1923–58, 6:123.

155. Apianus studied in his hometown of Ingolstadt, received a medical degree from Bologna in 1564, and then taught geometry and astronomy at Tübingen from 1568 (Poggendorff 1863, 1:52; Schaff 1912, 54).

156. See Rosen 1967, 33 n. 14.

157. Reinhold 1542, "Prefatio," fols. C4v–C5; Aristotle 1975, bk. 1, chap. 13, 19–21.

158. Universitätsbibliothek Erlangen-Nürnberg: Straub 1575, fol. 2r; Schadt 1577, ibid., fol. 72r.

159. In his brief *Vita et Opera Procli*, the Freiburg *mathematicus* Erasmus Oswaldus Schreckenfuchsius lists "Hypotyposes Astronomicorum" among the works of Proclus Diadochus (Proclus Diadochus Lycii 1561, fol. A1v). Proclus's title may have been the inspiration for the work of Reinhold and Peucer and the title that Tycho Brahe used to designate his planetary arrangement.

160. Peucer 1568.

161. For a discussion of the different versions of this work, see Zinner 1988, 273–74.

162. Peucer 1568, 312: "Est autem hoc anno 1559."

163. Ibid., 7, 12, 113, 333, 403.

164. Rheticus 1971, 139.

165. Peucer 1568, 299–301.

166. Ibid., 485–91.

167. Ibid., 516–17.

168. Cf. Copernicus 1543, bk. 3, chap. 3, and chap. 4, fols. 66v–67r; Peucer 1568, 523–25, 526.

169. Gonville and Caius College: University of Wittenberg lecture notes, 1564–70.

170. Quoted in Thorndike 1949, 29.

171. Ibid., 39; Smoller 1994, 44–54.

172. The note taker was apparently Laurentius Rankghe of Colberg, who put his name and the date (beginning April 10, 1564) at the head of the commentary, but the handwriting is not consistent throughout.

173. See Cambridge University, Gonville and Caius College MS. 387.

174. Schönborn 1570–72.

175. Messahalah 1549.

176. Garcaeus 1556.

177. Garcaeus 1569; Barnes 1988, 62–63.

178. Garcaeus 1576. See Thorndike 1923–58, 6:104, 595–98.

179. Garcaeus 1576, fol. A5v: "De diuinatrice parte nihil dico, nisi hoc. Etamsi nullae essent tempestatum aut temperamentorum significationes in positu stellarum, tamen hance mantiki, quae uerissima et longe potior est, magnifaciendam esse, quod uidelicet pulchritudo corporum et ordo motuum, illustria testimonia sunt de Deo, et de prouidentia. . . . Sed in alijs scriptis saepe uiri docti et Deum timentes dizerunt de dignitate et usu harum artium. Quare hanc meam commemorationem uolui breuiorem esse. Totus autem hic meus labor tantum illustrat doctrinam de motibus. Adiuuo discentium laborem in computanda tota anni ratione."

180. Here he made considerable use of precepts 23, 24, and 33 in the *Prutenics* and, again, Copernicus's values for the sidereal and tropical years in *De Revolutionibus*, bk. 3 (Garcaeus 1576, prop. 21).

181. See Biblioteka Uniwersytecka Wrocławiu, "Brevis Repetitio." The lecture is anonymous, but the hand resembles that of Peucer.

182. Ramus 1569, 64–75. For a good treatment, see Hooykaas 1958, 75–90.

183. Ramus 1970b: "Germaniam ut mathematum altricem animo complector.".

184. Ramus 1569, 65. Fust was involved with Gutenberg, but this edition of Cicero was not "the first book ever printed."

185. Ibid., 66.

186. Ibid.: "Literae erant in eo latinae & graecae semonis ea facilitas, ut Melanchthonis discipulum facile posses agnoscere: mathesis & matheseos diligentia tanta, quae bibliothecas omnium mathematicorum, si superfuisset aetas, mathematicis cuiusquemodi libris expletur uideretur."

187. Ibid., 50, 66. Previous commentators, including myself, have translated this phrase as "*astronomy* without hypotheses," failing to notice that the Latin consistently employs *astrologia*. From the context, it is clear that Ramus intended to equate the two terms, perhaps following Reinhold's *Prutenics*. See Jardine 1987, esp. 95; Hooykaas 1984, 157–64.

188. Ramus 1569, 50: "Atque utinam Copernicus in istam Astrologiae absque hypothesibus constituendae cogitationem potius incubuisset, longé enim facilius ei fuisset astrologiam astrorum suorum veritati respondentem describere, quam gigantei cujusdam laboris instar terram movere, ut ad terrae motum quietas stellas specularemur."

189. Ibid.: "At in posteris fabula est longé absurdissima, naturalium rerum veritatem per falsas causas demonstrare."

190. Ibid.: "Rheticus etiam Cracoviam mathematis illustravit & literis nostris ad studium liberandae hypothesibus Astrologiae spem quoque illustrandae parisiensis academiae dederat."

191. Ramus 1569, copy held at University of Aberdeen, shelf no. pi 5102 LaR s1.

192. Brahe to Rothmann, January 21, 1587, in Brahe 1913–29, 6:88–89.

193. Kepler 1992, 28; Kepler 1937–, 3:6.

194. See Dobrzycki and Kremer 1996.

6. VARIETIES OF ASTROLOGICAL CREDIBILITY

1. Seznec 1953.

2. On this matter, see Walker 1958, 45–53.

3. The problem of separating legitimate from illegitimate prognostication was analogous to determining the criteria for detecting witches. See S. Clark 1997.

4. Calvin 1962, 7: "Mais les affronteurs qui ont voulu, sous ombre de l'art, passer plus outre, en ont controuvé une autre espèce qu'ils ont nommée judiciaire, laquelle gît en deux articles principaux: c'est de savoir non-seulement la nature et complexion des hommes, mais aussi toutes leurs aventures, qu'on appelle, et tout ce qu'ils doivent ou faire ou souffrir en leur vie; secondement, quelles issues doivent avoir les entreprises qu'ils font, trafiquant les uns avec les autres; et en général de tout l'état du monde."

5. Ibid., 9: "S'il falloit faire comparaison, il est plus que certain que la semence du père et de la mère ont une influence cent fois plus vertueuse que n'ont pas tous les astres, et ce nonobstant on voit qu'elle défaut souvent, et aussi la disposition peut être diverse." Perhaps because Calvin cast his work as an "advertissement" for the unlearned, he did not see fit to cite learned authorities, such as Pico.

6. Cardano 1967, 5:490.

7. Giuntini 1573, fol. 6.

8. See Cox-Rearick 1993, 22 ff.; Hemminga 1583, 105–16.

9. Gaurico 1552, 9: "Haec coelestis figura fuit supputata per Fratrem Iulianum ordinis Carmelitanorum, et fundata Iussu Ducis Alexandri Medices, qui circa mediam noctem in cubili suo fuit iugulatus a suo Consobrino Anno Seruatoris 1537.vertente, uti colligitur ex Sole et Saturno partiliter alligatis. Labentibus 1583. mutabit sceptra. Anno autem 1627. eradicabitur et Solo aequabitur Arx illa infausto sydere fundata." The assailant was Lorenzino (1514–48), son of Pierfrancesco the Younger (1487–1525), a member of the collateral branch of the Medici family (see Cox-Rearick 1984, 4, 49).

10. Hemminga 1583, 105–16.

11. For well-developed cases of rectification, see Quinlan-McGrath 2001; for Luther, see Grafton 1999, 74–75.

12. According to Kuhn (1970, 80), the object of normal science "is to solve a puzzle for whose very existence the validity of the paradigm must be assumed. Failure to achieve a solution discredits only the scientist and not the theory. Here, even more than above, the proverb applies: 'It is a poor carpenter who blames his tools.'"

13. Goclenius the Younger 1618, 49; Giuntini 1581–83, tome I, 126, 621; cf. Thorndike 1923–58, 5:326. Goclenius (Goeckel) held chairs in physics (1608–11), medicine (1611–12), and mathematics (1612–21) at Marburg.

14. Biblioteca Medicea Laurenziana, Florence: Giuliano 1537. Castagnola (1989) has transcribed and edited the full text with brief introduction. I have made comparisons with the original.

15. Castagnola 1989, 133.

16. Ibid., 128.

17. Castagnola places the appointment at Florence (ibid., 130), whereas Cox-Rearick (1984, 256 n.) and Charles Schmitt (1972b, 259) put it at Pisa. It is certain that Ristori was lecturing at Pisa in 1548.

18. The astrological foundations of Medici dynastic imagery is the subject of Cox-Rearick's important study (1984, 206 ff.). Allegri and Cecchi (1980) have produced a useful guide to the palace, but they failed to note the astrological theme running throughout the motifs.

19. Cox-Rearick 1984, 257.

20. Ibid.

21. Already in the 1520s, Cosimo's association with Saturn-in-Capricorn themes shows up in Jacopo da Pontormo's *Vertumnus and Pomona*, a lunette for the Salone at the Medici villa in Poggio a Caiano (Cox-Rearick 1984, 117–42, 212–20).

22. Ibid., 256; Cox-Rearick 1964, 1:303–4 and n. 32.

23. See Rousseau 1983, 476–83.

24. Rousseau suggests both that propagandistic considerations factored into the duke's deliberately delayed military engagement and that he felt confident of victory because of the astrologers' predictions (ibid., 479).

25. However, Agostino Nifo, who wrote on the prognostication of 1524, did use Castiglione in a book that he wrote on the courtier (see Burke 1996, 48).

26. Giuntini (1581, 1: 14–15) located his own defense against Pico and Averroës; see further Ernst 1991, esp. 254–58.

27. See Righini-Bonelli and Settle 1979.

28. For Savoia's geniture of Cosimo I, see Biblioteca Nazionale Centrale di Firenze: da Savoia 1537, XX.10; for Guidi's geniture of Francesco, see Biblioteca Nazionale Centrale di Firenze, Guidi 1561; Guidi 1566.

29. See Grafton 1999, 64–65.

30. Ibid., 65.

31. This is Thorndike's hypothesis (1923–58, 6:100).

32. Carey 1992.

33. See Ashworth 1990, esp. 305–16.

34. See Grafton 1999, 109–26.

35. Examples include the horoscope collection of Nicolaus Gugler at Wittenberg (see ibid., 74–75).

36. See, for example, the copy of Garcaeus 1576 (British Library 718.k.32.), or the loose insertions of horoscopes in Giuntini 1581 (Bologna University Library [IV.K.I.68]). Such compilations of astrological particulars also motivated comprehensive indexing and tabulation of information.

37. Thorndike 1923–58, 6: 105. Thorndike does not give the full title, but the copy is in the Bibliothèque Nationale, Paris (Rés. V.1300).

38. Grafton 1999, 65–70.

39. Ibid., 82.

40. Grafton (ibid., 94–96) has provided an excellent account.

41. Ibid., 92.

42. Cardano 1547a; quoted and trans. in Grafton 1999, 94.

43. Grafton 1999, 94–95.

44. Cardano, *Liber de exemplis centum geniturarum* in: *Opera Omnia*, vol. 5, geniture 67, 491; quoted and trans. in ibid., 92.

45. See the evidence cited in ibid., 96.

46. Gaurico 1552.

47. Grafton (1999, 75–76, 96–108) cites important annotations from two English readers, Gabriel Harvey and Thomas Smith, showing that they appreciated Gaurico's hostility to Cardano.

48. Garcaeus 1576. The dedicatory letter is dated 1570. Herzog August Bibliothek 6.1.Astron.2° (2) is bound with Bellanti 1553 and extensively annotated; cf. Thorndike 1923–58, 6:42.

49. Rantzov 1585.

50. For Zabarella on the method of resolution and composition, see Jardine 1988, 690–91.

51. Cardano 1967, 5:94 ("Prooemium Expositoris").

52. On Cardano's handling of his errors in the geniture of King Edward, see Grafton 1999, 121–23; for applied predictions as a futures market, see Lindorff 2003.

53. Cardano, 1543, epistolary dedication, fol. Aiij v–r; quoted and trans. in Grafton 1999, 80.

54. See Vanden Broecke 2005.

55. Garcaeus 1576, preface, fol. A3v.

56. Because of his strong Nuremberg contacts, it would be surprising if Cardano had not received a copy of *De Revolutionibus* from Rheticus, Osiander, or Petreius; although no such identification appears in Gingerich 2002, Gingerich recently reported that the copy stolen from the Biblioteca Palatina, Parma—and since recovered—once belonged to Cardano (Gingerich 2008).

57. He remarks only that "those who fashion celestial spheres are accustomed to join Venus to the Sun in one sphere" (Cardano 1967, vol. 5, text xxx, 122–25, p. 124b).

58. Quoted in Vanden Broecke 2003, 103; on de Scepper, see 97–111.

59. Waterbolk 1974, 233–34; Lammens 2002, 61–62.

60. For preliminary comments, see chap. 8 below and Westman 1980a, 119–20.

61. For Gemma, see Lammens 2002; Vanden Broecke 2003; Hallyn 2004; Waterbolk 1974.

62. For Cornelius, see Vanden Broecke 2003, 186–90.

63. Vanden Broecke has suggested that they constituted some sort of *familia*: ibid., 149, 160–63, 177, 181.

64. Dee 1975, fol. b. iiij.

65. Nicholas Clulee (1988, 304) rightly calls attention to the fact that John Dee's so-called diary was really a group of discrete entries in various ephemerides.

66. Lammens 2002; Gingerich 2002, Provinciale Bibliotheek van Friesland, Leeuwarden, Netherlands, 146–50.

67. Gemma to Dantiscus, July 20, 1541 (Lammens 2002, 1:35, 213). If Gemma actually had access to the *Commentariolus*, which seems dubious, there are certainly no telltale references to its language.

68. Quoted in Lammens 2002, 1:60–65.

69. Ibid., 1:108–9.

70. Ibid., vols. 2–3.

71. The same attitude prevails in Gemma's writings on instruments (see Goldstein 1987, 167–80).

72. Ptolemy 1548. It has been suggested that this work first appeared in 1543, but Gogava's preface is dated August 1548, and the appended letter from Gemma Frisius is dated October 1548.

73. Ibid., Gemma Frisius, "Letter to the Reader": "Lacunis Arabum ad nos deductas, foedas sane illas et ineptas, ut reprehensione et crimine carere tempora nostra uix possent, si renatis artibus ac literis oblini nos fecibus et barbarie ista pateremur"; "Cedant sanè uetera, dummodo meliora succedant."

74. Ptolemy 1548.

75. Vanden Broecke 2003, 177–78.

76. Bretschneider et al. 1834–, 13:267: "Nam quas vires in adficiendis his corporibus infimis hos Planetas habere experientia docuit, has necesse est illos exercere languidius, quando in summis epicyclorum apsidibus collocati, longissime a terra abscesserunt: sed longe efficacius et potentius, quando in imis partibus suorum epicyclorum constituti, aliquot millibus diametrorum terrae, propriores nobis facti sunt. / Habet autem Saturnus vim frigefaciendi, et leniter exiccandi. Mars vero vehementer exiccat et urit. Sed Iupiter inter hos medius temperatam naturam habet. Calefacit enim simul et humectat, et foecundos generationique rerum aptos spiritus excitat et fovet."

77. Ibid., 276: "Animadversum est, horum trium Planetarum, Solis, Veneris et Mercurii, eandem esse lineam medii seu aequalis motus, hoc est, centra epicyclorum semper circumferri cum linea aequalis motus Solis. Unde etiam dicitur horum trium Planetarum esse perpetua coniunctio, secundum medium seu aequalem ipsorum motum."; 285: "Tres superiores Planetae, qui non ita ad Solis viciniam alligati sunt, sed libero incessu per totum Zodiacum vagantur (nisi quod in epicyclo suum cursum accommodant ad Solis motum)."

78. Ibid., 286: "Perpetuo igitur quasi satellites, qui ministerio et custodiae corporis regii praefecti sunt."

79. Ibid., 276: "Sed nos retinemus antiquissimorum Astrologorum sententiam, quam Cicero quoque, Ptolemaeus et alii recentes Mathematici magno consensu sunt secuti. / Dicimus igitur Venerem proxime

infra Solem, et sub hac Mercurium supra Lunae sphaeram collocari."

80. Gemma Frisius in Stadius 1560, sig. b3–b3v.

81. Ibid.

82. On *dioti/tou hoti*, see chap. 5, this volume; Hallyn 2004; Lammens 2002, 1: 114–15.

83. Copernicus 1543, bk. 1, chap. 10, fol. 8v; Copernicus 1978, 15:20–26. Copernicus says nothing at this point about the Ptolemaic alternative; moreover, his discussion is preceded by an argument for a reordering of the inferior planets that Gemma Frisius entirely ignores.

84. This feature can be inspected just by consulting the simplified Copernican arrangement in *De Revolutionibus*, bk. 1, chap. 10. For a much more helpful approach, see the visual animations on Dennis Duke's website: http://people.scs.fsu.edu/~dduke/models.htm.

85. Hallyn 2004, 83.

86. Noel Swerdlow (2004a, 88–90) lists some twenty-seven nonarbitrary (i.e., *dioti*) entailments of the Copernican theory.

87. Gemma Frisius in Stadius 1560, sig. b3–b3v. In making my own translations, I have benefited from Cindy Lammens's helpful readings (2002, 1:110–17).

88. Deborah Harkness (1999, 129–30) has called attention to the importance of eirenic tendencies that may have attracted Dee—especially the ideas of the group known as the Family of Love, which was involved in promoting toleration in the frame of a universal religion. Among its members were Gerard Mercator, Gemma Frisius, and the major Antwerp publisher Plantin.

89. For what little can be reconstructed of Dee's early studies at Cambridge and Louvain, see Nicholas Clulee's judicious remarks (1988, 22–29); for more general background, see Feingold 1984.

90. Conversation with Steven Vanden Broecke, History of Science Society meeting, Pittsburgh, November 1999. For further details on Gemini (Lambert, Lambrit, or Lamnbrechts), see Turner 1994, 347; O'Malley 1972.

91. Turner 1994, 348.

92. In Roberts and Watson 1990: Albohaly's *De Judicijs Nativitatum* (1546; #693), Joachim Heller's edition of Messahala (1549, #509), and a collection of Messahala's works (1532; #510). Hereafter references are to the Roberts-Watson numbering of Dee's catalogues of 1557 (designated "B") and 1583 (designated "#").

93. Camerarius (B15; #375, #526), Gogava (Ptolemy 1548; B17, #462), Melanchthon (1553; B115), and Cardano (B202).

94. Roberts and Watson 1990; Clulee 1988, 251 n. 1.

95. Dee 1975, fols. b. iijv–b. iiij. See Clulee's excellent discussion of this question (1988, 60–64).

96. See Maclean 1984.

97. Dee 1978, 113.

98. Especially useful are Bowden 1974, 62–78; Clulee 1988, 19–73; Heilbron in Dee 1978, 204–41.

99. Dee owned a copy of Bellanti's critique of Pico (Bellanti 1554; see Roberts and Watson 1990, B92, #106); on Dee and Pico, see Vanden Broecke 2003, 170–78; Bowden 1974, 63–78.

100. In February 1583, Dee (1583 in Dee 1968b, 19) was also using Aristotelian demonstrative ideals in his work on the correction of the Julian calendar. His personal inscription of this work to Lord Burghley, "Lorde Threasorer of Englande," shows that he also took for granted that his patrons would understand his reform in such terms:

το ὅτι and το διοτι,
I shew the thing and reason why;
At large, in breif, in middle wise,
I humbly give a playne advise;
For want of tyme, the tyme untrew
Yf I have muyst, commaund anew
Your honor may so shall you see
That love of truth doth govern me.

101. "By so much as the passage of a star above the horizon takes longer, by so much it is better fitted to make a stronger impression of its virtue by means of its direct rays" (Dee 1978, aphorism 51, p. 175).

102. Bodleian Library: Dee 1582, art. 7, fols. 38, 65.

103. Vanden Broecke (2001) has found suggestive evidence concerning Mercator's role in forming Dee's project for an astrological physics in Louvain.

104. What is known of Mercator's views dates to much later in his life (see Bowden 1974, 90 n. 6).

105. Dee's 1557 lists "Physica Melanchthonis" (B118) and the 1583 catalogue "Philippi Melanth Physices Epitome 8° Oporin, 1550" (#827), both of which I take to refer to the *Initia Doctrinae Physicae*.

106. Dee 1978, aphorism 54, pp. 149, 151.

107. Ibid., aphorism 89, p. 175: "Planets situated at their greatest distances from the earth, near their apogees, exercise their powers more strongly and splendidly in matters of which they would then be proper significators than they do in the same matters when they are borne close to the earth, near their perigees. In contrast, they act more vigorously and effectively in other matters subjected to them at their greatest nearness to the earth than they can when they are as distant as possible from the earth. . . . [L]et the positions of greatest and least distance from the earth first be known to you for each planet individually." See further Heilbron's commentary in the same volume, 218–19, 233–34.

108. Offusius 1570. For discussions of Offusius, see Bowden 1974, 78–107; Stephenson 1994, 47–74.

109. Gingerich and Dobrzycki (1993, 239) checked Louvain, Cologne, Wittenberg, Leipzig, Cambridge, Oxford, and Basel.

110. A reference to December 1557 (Offusius 1570, fol. 28v) suggests that the text was nearly finished by the end of the year in which the author completed his ephemerides (Offusius 1557, published at the end of January).

111. Francisca de Feines to Elizabeth, Queen of France (Offusius 1570, fol. aij: "Viri mei deffuncti in patrem tuum"); for further clues to Offusius's Parisian years, see Sanders 1990, 212–17.

112. "D[omi]no Lazaro Schonero paedagogiarchae Marpurgensi F Risnerus Lutetia misit 10. Calend. Martii 1577" (Columbia University copy). On Risner, see Lindberg 1976, 185.

113. Offusius 1570: "Franciscus Rassius Noëns Chirurgiens Parisiensis. 1572" (Biblioteca Nazionale di Firenze). Noens had already acquired *De Revolutionibus* in 1559 (Gingerich 2002, copy in Bibliothèque Municipale, Évreux, France, 53).

114. *Progymnasmata Astronomiae* (1602), Brahe 1913–29, 2:421–22.

115. Bodleian: Savile T 3.

116. Offusius's book continued to be of interest to mid-seventeenth-century astrologers (William Lilly bound his copy with Kepler's *Ephemerides* of 1617: Bodleian: Ashm. 470 [2.]).

117. Pico della Mirandola 1946–52, 1: 42: "Videas umbram aut quasi *larvam,* et sub aperta luce fallaciam tenebrarum abomineris."

118. Offusius 1570, fol. eiij r: "Opusculum perpusillum est, ac magna in eo latere cognoscet studiosus rimator, quod etsi in laruatam Astrologiam nominem, non abs re hoc à nobis factum est, nec omnino temerè. Picus Mirandulus eam impugnauit meo iudicio, quae hactenus ineruditorum auriculas pulcherrimè detexit, qui proposita omnia sine ulla ratione, doctrina vel consideratione, tanquam vera assumpsere, déque futuris iucundissimé cum Philosophiae quadam iniuria, fabulati sunt ex illa."

119. Ptolemy 1940, bk. 1, chap. 4, 37: "Jupiter has a temperate active force because his movement takes place between the cooling influence of Saturn and the burning power of Mars. He both heats and humidifies; and because his heating power is the greater by reason of the underlying spheres, he produces fertilizing winds."

120. Offusius 1570, Dedication to Maximillian (1556), fol. aiiii: "Ut videre est illius libri I. cap. 3. ubi Lunam humectare asserit: quia, inquit, prope terram fertur, unde humidae exhalationes exeunt. Item ait Saturnum exiccare, quia à terrae humiditate longissimè distat. Certe aeternam pati à corruptibili, non est sapientis dictum: Imò si eius ibidem rationes robur in se haberent, sequeretur Iouem esse exiccantem, quia inter duos arefacientes fertur, cùm dicat temperatum esse, qui inter frigidificam et aestuosam, Stellas vehitur. Sunt et alia apud illum, homine Philosopho indigna." Cf. discussion of bk. 1, chap. 4 in chap. 3 above.

121. Ibid., fol. e: "In qua partim Physicè, partim Mathematicè procedemus, in dubijs verisimilia amplexantes, speramúsque ut etiam scientiae nomine digna in posterum fiat"; "Ab inspectione operis diuini, longaque experientia collecta norma et regula."

122. Ibid., general letter to the reader, fol. eij v: "Quibusdam nostras obtuli circiter MMD.CC. obseruationes, è coelo ipso scrupulosè sumptas inter peregrinandum, sperans ab ijs talionem, hoc est, quid obseruatum recipere ad hanc artem stabiliendam, sed me hercle nihil minus."

123. For the so-called nesting hypothesis, see Van Helden 1985; Stephenson 1994.

124. Offusius 1570, chap. 1, fol. 1r.

125. Ibid., fol. 3v: "Non solùm praefatam distantiam (ex qua Sol tanquam naturalis caloris fons, vitam nobis infundit) aperiunt regularium corporum Trigoni per numerum illum 576, veteribus quoque *Animam mundi* dictum, verùm per eorundem corporum inter se proportionem quantitatis (eodem orbe contentorum et circumscriptorum) se manifestat, quod hic unicè desideramus, nempe in qua harmonia sit qualitatum quantitas (scilicet intelligendo eas quantas, quantitate intelligibili) quae nobis à coelo causatur."

126. For a clear discussion of Offusius's scheme, see Stephenson 1994, 48–51.

127. Ibid., fols. 6–6v:

Quod autem hic Venerem sub Mercurio locarim, ne propterea damnetur symmetria. Martianus Capella qui Encyclopaediam edidit, putauit quod hae stellae circuncurrunt solem. Huius credulitatis (sola causam reuolutionis excogitans in singulis) et ego diu fui. Timaeus Platonis credidit ambas supra Solem positas. Alpetragius Mercurium sub Sole, Venerem verò suprà, Ptol. et post eum tota ferè recentiorum Mathematicorum Schola, ponit Venerem intere Solem et Mercurium: tamen sub illo tandem Copernicus, vir reliquis non inferior, imò meliorem quàm illi fortunam in hoc nactus (omnium obseruationibus se solus ac recentiorum quoque iuuit) demonstrat amborum siderum (sic et aliorum quoque) errores saluabiles, si Solem quiescentem circuncurrant, nec aliud sequitur ex eius Theoria (singulis consideratis) quàm Venus ipsa aliquando longè proprior sit terrae quàm Mercurius. Haec igitur libertas mea ne ob hoc reprehendatur, nam ubilibet has stellas locari quis velit, praescribam illi Theoriam obseruatis excursionibus apparentibusque diametris omnino conuenientem, quid ergo superest quàm concedere?

128. Gingerich and Dobrzycki 1993, 239, 245.

129. Quoted and translated (ibid., 240) from Offusius's notes to *De Revolutionibus,* bk. 1, chap. 5, referring to the preface. The entire Latin passage is given in 252 n. The authors render the crucial adverb *omnino* as "arbitrarily"; I have emended their translation.

130. Ibid., 241: "When the arguments from geometry and physics are inadequate, we do not doubt, following the testimony of the Holy Scripture, that the Earth is at rest and that the Sun moves. For the Psalmist clearly confirms the Sun's motion: 'He set the tabernacle for the Sun, which is as a bridegroom coming out of his chamber, and rejoiceth as a strong man to run a race. His going forth is from the end of the heaven, and his circuit unto the ends of it' [Psalms

19:4–6]. Another Psalm tells of the Earth, 'which He hath established for ever' [Psalms 104:5]. And Ecclesiastes in the first chapter states: 'But the Earth abideth for ever. The Sun also ariseth, and the Sun goes down and hastens to his place where he arose' [Ecclesiastes 1:4–5]." These are exactly the verses cited by Melanchthon in his *Initia Doctrina Physicae* (Bretschneider et al. 1834–, 13: col. 217) and later echoed by his followers (cf. chapter 5 above).

131. Ibid., 245–47.

132. Ficino 1989, bk. 1, chap. 3, 113.

133. Ibid., bk. 1, chap. 4, 115.

134. Walker 1958, 36.

135. See Clulee 1988, 12–13.

136. Ficino 1989, 160; Folger Shakespeare Library, call no. BF1501 J2 Copy 2 Cage, signed "Joannes Dee": "Similem ego lapidem vidi et eiusdem qualitatis. anno 1552 vel 1553. Aderant Cardanus Mediolanensis, Joannes Franciscus et Monsie[u]r Beaudulphius Legatus Regis Gallici in aedibus Legati in Sowthwerk." Cited in Roberts and Watson 1990, 85 n. 256; Gingerich and Dobrzycki 1993, 252 n. On the Offusius-Dee contact, see Clulee 1988, 36; Heilbron (in Dee 1978, 54, 59, 60). Grafton (1999, 112 and 112 n. 18) read "Joannes Franciscus" as John Francis Cheke.

137. See Thomas 1971, 289. Thomas cites Morley 1854, 2: chap. 6.

138. See, for example, Dee 1968b, 15: "May 23, 1582, Robert Gardener declared unto me hora 4½ a certeyn great philosophical secret, as he had termed it, of a spirituall creatuer, and was this day willed to come to me and declare it, which was solemnly done, and with common prayer."

139. Ficino 1989, bk. 3 chap. 15, 317; on music spirit, bk. 3, chap. 21. For further discussion, see Walker 1958, 16 ff.

140. Ficino 1989, 160; Folger Shakespeare Library, call no. BF1501 J2 Copy 2 Cage.

141. Clulee's evidence (1988, 36) is based on Dee's later recollection (*A Necessary Advertisement,* by an Unknown Friend . . . , from Dee 1968a, 58–59).

142. Dee 1851, 58; cited by Heilbron (in Dee 1978, 54).

143. Leovitius 1558, fol. E2v ("De Fortitudine et Debilitate Planetarum"). See also Clulee 1988, 35.

144. Wolf's dialogue first appeared in Leovitius's *Ephemerides* (1556–57, fols. y5v–z4v). I quote below from Turner's English translation (1665).

145. Turner 1667, 163; Leovitius 1556–57, fol. z2r.

146. Turner 1667, 165; Leovitius 1556–57, fol. z2v. Among those mentioned by Wolf was Lucio Bellanti.

147. Dee 1975, bi v.

148. Clulee 1988, 65.

149. Dee 1975, fol. B.iiij. *Bayard* refers to one who is blindly self-confident, an ignoramus.

150. Clulee 1988, 34, 35, 122, 147, 161, 169, 193, 225.

151. See Harkness 1999, 127.

152. Ibid., 20–21.

153. Ibid., 22.

7. FOREKNOWLEDGE, SKEPTICISM, AND CELESTIAL ORDER IN ROME

1. See Hsia 1998, 10–41; Bireley 1999, 25–69.

2. See esp. Partner 1976, 25–41.

3. See Delumeau 1977, 126.

4. This statement is based on a study of the indexes of Ehses 1961–76.

5. This section is a revised version of a section from Westman 1986, 87–89, a highly compressed article written for a general audience.

6. This statement based on a study of the indexes of Ehses 1961–76.

7. Biblioteca Nazionale Centrale di Firenze: Tolosani 1546–47, with the provenance of the important monastery of San Marco in Florence.

8. See Marzi 1896.

9. Scudder 1620, sig. A3 (cited in Bennett 1970, 26).

10. See, for example, Sir Richard Barckley, *The Felicite of Man* (1631): "All bookes should bee protected by such noble patrones whose dispositions and indowments have a sympathy & correspondence with the arguments on which they intreate" (cited in Bennett 1970, 31).

11. If this copy has survived, it lacks Tolosani's provenance.

12. Tolosani 1546–47, fols. 339r–343r: "De coelo supremo immobile et terra infima stabili ceterisque coelis et elementis intermediis mobilibus." The entire text of this little work is transcribed by Garin 1975. All citations are to the Garin transcription, although I have also noted a few discrepancies with the original. See further Granada 1997a; Lerner 2002; Rosen 1975b.

13. Andrzej Kempfi (1980, 252) believes that there was a hardening of the papal attitude between the last years of Clement VII, who, as chap. 4 shows, was introduced to Copernicus's ideas through Johann Widmanstetter in 1533, and the time of Bartolomeo Spina, Master of the Sacred Palace from July 1542.

14. Garin 1975, 38: "Ut facilius lectores cognoscant Nicolaum Copernicum non legisse nec agnovisse rationes Aristotelis philosophi ac Ptholomaei astronomi."

15. Ibid.: "Haec ille ignotus author."

16. Ibid.: "Unde author ille cuius nomen ibi non annotatur, qui ante libri eius exordium loquitur 'ad lectorem de hypothesibus eiusdem operis,' licet in priori parte Copernico blandiatur, in calce tamen verborum, recte considerata rei veritate absque assentatione sic inquit: 'Neque quisquam (quod ad hypotheses attinet) quicquam certi ab Astronomia expectet, cum ipsa nihil tale praestare queat, ne si in alium usum conficta[m] pro veris arripiat, stultior ab hac disciplina discedat quam accesserit.' Haec ille ignotus author."

17. Ibid.: "Ex quibus verbis authoris eiusdem libri taxatur insipientia, quod stulto labore conatus fuerit Pictagoricam confictam opinionem iam diu merito

extinctam denuo suscitare, cum expresse contraria sit rationi humanae atque sacris adversa literis, ex qua facile possent oriri dissensiones inter divinae scripturae catholicos expositores et eos qui huic falsae opinioni pertinaci animo adhaerere vellent."

18. Ibid., 35, 39–41: "Aristoteles vero in 2. de coelo et mundo, textu commenti 72, proponit rationes Pictagoricum et inde solvit eas. Tandem ponit et exprimit opinionem sua iuxta rei veritatem, textu commenti 97, de loco et quiete terrae et per naturales rationes et per signa astrologica. Mittimus ergo lectorem ad librum secundum Aristotelis de coelo et mundo, et ad commentarios divi Thomae super eum, lectione 20, 21, et 26, ubi plena veritas manifestatur" (Aristotle, *De Coelo*, bk. 2, chap. 14, 296a 24–297a 8; Thomas Aquinas, *In Aristoteli Libros de Coelo et Mundo Expositio*).

19. For important discussions of the doctrine of subalternation, see McKirahan 1978; Livesey 1982, 1985.

20. Garin 1975, 42.

21. Spina presented articles concerning baptism and justification (Ehses 1961–76, 12:676, 725); he was succeeded on his death by the Bolognese Dominican Egidius Fuschararus (ibid., 5:728 n.).

22. Caccini's deposition, March 20, 1615 (Finocchiaro 1989, 136–41).

23. Partner 1976, 45–46.

24. Bujanda 1994, 22.

25. See Blackwell 1991, 13.

26. Bujanda 1994, 22.

27. Thorndike (1923–58, 6:147) makes a similar point about "the occult."

28. "Index Paulus IV . . . January 1559," in Reusch 1961, 176–208.

29. Nonetheless, Reinhold's *Commentary on Peurbach* is listed on the Rome Index in 1590 (Bujanda 1994, 082E), and his name has been crossed out, probably by the 1572 owner, on a copy once held by the Biblioteca Nazionale di Roma (now Huntington #701416), which contains many books from the Collegio Romano. Albert of Brandenburg appears on the Munich 1581 Index (Reusch 1961, 344).

30. Schöner 1545 ("Ne extra hanc Bibliothecam efferatur. Ex obedientia"). The same label is found in the same location on the frontispiece of G. B. Riccioli's *Almagestum Novum* (Bologna, 1651), held by the Stanford University Library, reproduced in Westman 1994.

31. Reusch 1961, 196–97. This declaration was anticipated by the less refined 1554 Indexes of Milan and Venice, wherein the inquisitors used a generic listing of the form "[All] Books of ——," without making any exclusions (ibid., 175).

32. See, for example, the 1593 Rome Index, "Rule 9" (Bujanda 1994, 857).

33. Pico della Mirandola 1496, bk. 3, chap. 19.

34. See Franz 1977.

35. Bujanda 1994, 35–36: "Noms déformés, titres

bouleversés, prohibitions qui se répétaient ou se chevauchaient, précisément parce que les censeurs n'arrivaient pas à dominer la matière. Toutes ces erreurs, ces imprécisions, ces incompréhensions montrent d'abord, et d'une manière évidente, les limites culturelles et intellectuelles . . . des fonctionnaires qui, à divers niveaux, recueillaient et transmettaient les informations."

36. Reusch 1961, 99. The Venetian Inquisition had begun to confiscate and burn books in 1547; in July 1548, the copy of Osiander's book owned by the printer Antonio Brucioli was one of several burned (see P. Grendler 1977, 82).

37. Reusch 1961, 194.

38. Ibid., 321–22. Also included were Jacobus Zicolerus, Joachim Vadianus, Martin Crusius, Martin Borrhaeus, and Michael Saselius. Not until the Munich Index of 1581 do we find Johannes Garcaeus listed (p. 347).

39. Ibid., 586–87.

40. Ibid., 381.

41. *The Catholic Encyclopedia*, 10:144. Medina was posthumously exonerated.

42. Medina 1564, bk. 2, chap. 1, fol. 9v: "Prophetica euentuum praenunciatio; de prophetico euentuum uaticinio."

43. Ibid., fol. 15v.

44. Ibid., fols. 10r, 20r–v, 24r, 28r.

45. Pico della Mirandola 1946–52, bk. 3, chap. 19, 357–63: "Cur nautae, medici, agricolae, vera saepius praedicant quam astrologi."

46. Here is a good example of the way that the interchangeability of the terms *astrology* and *astronomy*, noted earlier by Reinhold and Peucer, can lead to confusion among historians.

47. Ibid., fol. 16v.

48. Aristotle 1962b, bk. 1, chap. 2, 339a 11 22 ff.: "This [terrestrial] region must be continuous with the motions of the heavens, which therefore regulate its whole capacity for movement: for the celestial element as source of all motion must be regarded as first cause" (etc.). See North 1986b, 46.

49. Medina 1564, fol. 17.

50. Ibid., fol. 17v.

51. Ibid., fol. 18r.

52. Ibid.

53. Medina used the Scholastic term *difformiter*, referring to a body's nonuniform motion.

54. Medina 1564, fol. 18–18v.

55. Ibid., fol. 18v.

56. Proclus Diadochus Lycii 1560.

57. Kessler 1995, 289.

58. The Zamberti translation was completed in 1539 (P. Rose 1975a, 52).

59. Kessler 1995, 292.

60. Pantin 1987.

61. Giacobbe 1972a, 1972b, 1977; Mancosu 1996, 8–19.

62. Kessler 1995, 295–308.

63. For biographical details, see, inter alia, Palmesi 1899.

64. Righini-Bonelli and Settle 1979.

65. Settle 1990, 27.

66. Ibid., 30.

67. Danti 1572, dedicated to the "most excellent and most illustrious Signora Donna Isabella Medici Orsini, Duchessa di Bracciano."

68. Danti 1577. The work had evidently been under way for many years before Danti arrived in Bologna, as he dedicated the section on astronomy to the Duchess Isabella Medici, with a date of November 18, 1571:

> Se bene ho fino à qui tenuto appresso di me questo trattato della Sfera, il quale è già quattr'anni, che per mio passatempo ridussi in tauole, senza mai lasciarmi da preghi altrui persuadee, che ne chiedeuon copia, ho volsuto hor nondimeno farne humilmente dono a Vosta Signoria Illustrissima, à fin che, se mai gli conuerrà uscire in publico, non esca pe altre mani, ne sotto altro nome che'l suo, dal quale harà non piccolo fauore poi ch'ella (fra molt'altre) tanto di questa nobilissima, et piaceuole scienza diletta. Io mi sono ingegnato nel raccor queste tauole di non lassare à dietro cosa alcuna, che alla intelligenza di tal facultà sia necessaria, è di non offendere ancora chi legge con cose troppo sottili, ò superflue come vedrà bene ella, la quale conquella reeuerentia ch'à me si conuiene supplico ch'al puro, et sincero affetto dell'animo mio riguardando accetti il piccol presente è l'ardentissimo desiderio ch'ho di seruirla in qual si voglia occasion maggiore. Di Firenze alli 18 di. Novembre 1571.

69. Ibid., 4: "E però (come Proclo afferma) deuono essere collocate quanto al subietto nel mezzo fra le scienze naturali, & le Metafisicali. Atteso che essendo il subietto della Metafisica separato da ogni materia, & quanto all'essistenza, & ancho quanto all cogitatione, seguirà, che il subietto delle Matematiche sia nel mezzo fra quello delle due sopradette, poi che esso essendo materiale viene considerato senza materia alcuna sensibile."

70. Ibid., 8:

> PRATICA, la quale và misurando le cose secondo ciascuna delle tre misure per longo, largo, et profondo, appllicandoui per tutto i numeri. dimostra la quantità delle linee, delle superficie, et de'corpi. SPECULATIVA, che considera i principij, cioè i punti, linee, superficie, et corpi, comparando le linee alle linee, le superficie, alle supeficie, et i corpi à i corpi secondo la uguale, et ineguale ragione, che hanno fra di loro, senza l'applicatione de'numeri, et di essa i primi principij sono. MISTA (per dir così) laquale considera i principij come fa la speculatiua, ma ci aggionge i numeri, et le misure nella materia sensata come fa l'Architettura, et altre simili mechanice.

71. Ibid., 15. "Se bene l'Astronomia è una delle scientie soggette et subalterne (parlando come i Filosofi) nondimeno essendo ella nobile, et degna, meritamente deue essere posta fra qual si voglia scienza principiae, si perche tratta del Cielo et di quei risplen-denti corpi celesti da noi chiamati stelle, si ancora perche le Matematiche hanno per soggetto, et quasi sempre sopra certissime cose discorrono" (Danti 1569, dedicated to Ferdinando de Medici. The work was issued again in 1578 at Florence after Danti had announced himself as "Publico Lettore delle Mathematiche nello Studio di Bologna" and dedicated to "Francesco de'Medici, Secondo Gran Duca di Toscana.")

72. Danti 1577, 23. Cf. Proclus Diadochus Lycii 1561.

73. Danti 1577, 24:

> Et se in alcuna parte delle Matematiche sono necessarie le figure in questa parte della Astronomia, che s'aspetta al trattato della Sfera, et delle Teoriche de'Pianteti, sono necessarissime. La onde potremmo dubitare di recuere non piciolbiasimo, hauendo stampate queste tauole senza le solite figure, se non sapessimo, che così fatti compendij sono per quelli, che di già hauendo apprese cotali scienze, possono con essi ridursele a memoria, et che quelli che in esse sono meno esperti so potranno seruire delle figure, specialmente delle Teoriche de'Pianete con le annotationi del Reinoldo stampate à Parigi, con le quali dallo Autore sono state ordinate le presenti tauole, et scusar noi, che non poteuamo che bene stesse, adattare le figure in esse tauole senza guastare i ordine loro; Oltre la difficultà che habbiamo delli Intagliatori; poiche questa pestilente contagione ci ha di maniera serrati i passi da ogni intorno, che non si può hauer copia de'periti di cotal mestiero."

This is especially interesting because it shows that Danti regarded Reinhold's commentary as providing a *tou hoti* preparation for the *Almagest*.

74. For example, ibid., 35: "Il centro del corpo solare quando è nell'Auge, ò nel suo opposto, hoggi è più vicino all terra, che non era al tempo di Tolomeo tret'uno semidiametri della terra, si come egregiamente è dimostrato dal Copernico."

75. Dallari 1888, 2:231.

76. O'Malley 1993, 202–3.

77. Ibid., 149.

78. Ibid., 206, esp. 226.

79. Ibid., 211–12.

80. Ibid., 200.

81. See Giard 1995b, liii–lxiv.

82. As usual, because of the ready interchangeability of the terms *astrology* and *astronomy*, one needs to pay close attention to context.

83. Clavius 1992, 1: pt. 1, 63; as John O'Malley writes: "More than any other individual, he [Nadal] instilled in the first two generations their *esprit de corps* and taught them what it meant to be a Jesuit" (1993, 12).

84. Pereira 1609.

85. Pereira based his distinctions on two Aristotelian commentators—Simplicius, reporting on the views of Geminus, and Averroës: Simplicius, 2 *Physics, Text 17*. Averroës, 1 *Metaphysics, Commentary 19*; *De coelo* 2, commentary 59 (ibid., bk. 2, chap. 3, 83).

86. Ibid.: "An differat Astrologus a Physico" (in response to the tenth doubt).

87. Pereira 1661. According to Thorndike (1923–58, 6:410), the work first appeared at Ingolstadt and Venice in 1591, followed by subsequent printings in 1592 (Lyons and Venice), 1598 (Cologne), 1602 and 1603 (Lyons), 1612 (Cologne), and 1616 (Paris). An English translation of the book against astrological divination (only) was prepared by Percy Enderbie in London (1661) and reprinted in 1674 as *The Astrologer Anatomiz'd Or, the Vanity of Star-Gazing Art.* All citations are to the 1661 edition and preserve the seventeenth-century English spelling.

88. For example, Pereira 1609, bk. 9. chap. 5, 535: "Idemque firmatur auctoritate Ptolemaei, qui *in opere suo quadripartito* dicit, effectus qui ex superioribus corporibus procedunt, non evenire inevitabiliter. In *Centiloquiuo* etiam dicit, signa coelestia quae sunt indicia rerum sublunarium, esse media inter id quod est necessarium et possibile" (my italics). He then follows with a passage from Thomas Aquinas on variations of effects from celestial causes.

89. Pereira 1661, fol. A4: "*Johannes Picus Mirandulanus* hath written very largely and learnedly upon this subject, but his prolixity retards his Readers from perusing his whole works; therefore in this Tract (though the field be large and spacious) such method shall be used, that nothing shall be brought upon the stage (yet many curious Questions offer themselves) but onely such as shall be succinct, and most conducing to the purpose." Elsewhere, he refers the reader directly to Pico's *Disputationes* (bk. 2, chap. 5, and bk. 5, chap. 11) for evidence of astrology's threat to Christianity (fol. C3).

90. Pereira 1661, 23–24 (referring to *De caelo* bk. 2, chaps. 2, 5, 9).

91. Ibid., chap. 3, 69.

92. Ibid., 38.

93. Ibid., reason 3, p. 44.

94. Ibid., 57–58.

95. Ibid., 55.

96. Ibid., 59, 67.

97. Ibid., 64, 60.

98. See esp. Giard 1995b, liii–lxvi; Baldini 1992b; Crombie 1977; Brizzi 1995; Hellyer 2005, 119–22.

99. For Clavius's life, see the excellent annotated chronology provided by Ugo Baldini and P. D. Napolitani (Clavius 1992, 1:33–58).

100. Ibid., 38–39. We have this on the authority of Clavius's student Christopher Grienberger: "Audiebatur quidem in scholis, et in Posterioribus ab Aristotele saepius repetebatur geometricum illud, quo asseritur *Tres angulos cuiuscunque Trianguli aequales esse duobus rectis,* et fortassis nomen triangli adhuc agnoscebatur: sed quid esset, tres angulos esse aequales duobus rectis, vix erat qui explicaret, et fortassis meno qui demonstraret. Eaedem voces feriere quoque non semel aures Clavii, atque ad Geometrim

iam olim a Natura factas etiam vulneravere: non enim sonos, sed verborum sensum, atque sententiam percipere cupiebant. Quare iam diu multumque solicitum tandem P. Petrus Fonseca, quo tunc Conimbricae utebatur Magistro in Philosophia."

101. Ibid., 43, 63, based on Ms. Vatican, Urb. Lat. 1303 and 1304. Exactly which elements of astrology Clavius treats is unclear without examination of the evidence.

102. For Jesuit worries about Averroist tendencies, see Hellyer 2005, 16–19.

103. For extensive analysis of Clavius's *Sphere* commentary, see Lattis 1994.

104. Clavius 1570, second dedication, unpag.: "Quid, quod ex una eorum obseruatione, uaria et maxima ad hominum usus in omni genere commoda promanarunt? quomodo enim uel agricultura, uel nauigandi medendiue ars, uel ad omnes uitae functiones, ususque necessaria illa anni in menses diesque descriptio, aut principio excogitari, aut deinceps retineri, sine Solis ac Lunae ductu, et quadam quasi institutione, potuissent? mitto supputationes Ecclesiasticas, quae *a Tridentina sacrosancta oecumenica synodo tantopere commendantur,* Deique ac diuorum cultum, stataque ac solennia sacrificia; quorum omnis ratio atque constantia, quin ex eodem profluant fonte nemini dubium est."

105. Clavius 1591, 5. Unless otherwise noted, all citations are to this edition.

106. Ibid., 5. Later, when describing the division of the zodiac, Clavius noted that "many things ought to be said here about the various properties and names of the signs but because they pertain more to judiciary astrologers, should be omitted here" (ibid., 244).

107. Mirandulanus 1562, 397–432.

108. Clavius 1591, 5–6.

109. To the 1581 edition, Clavius merely added a postil at the bottom right, which reads: "Astrologia Iudiciaria res est superstitiosa." There is a partly underlined copy of the 1496 Hectoris edition of Pico's collected works in the extant library of the Collegio Romano, but there is no indication of when the book was acquired (Biblioteca Nazionale di Roma, shelf no. 70.2.B.9,2).

110. See, for example, Blancanus 1620, 399.

111. Clavius 1591, 463: "Rationes autem, quibus haec omnia inuestigari possint, et examinari, (Distantias enim centrorum, et magnitudines semidiametrorum examinare per tempus hic non licuit, sed eas ex alijs auctoribus, ut scriptae sunt, accepimus) in nostris theoricis explicabuntur." Following this final sentence of the commentary, Clavius provides a prelude to "our *Theorics,*" a schematic tabulation of the basic terminology of the orbs and spheres (464–83).

112. Ibid., 5.

113. Ibid., 64; see also Lattis's discussion of this passage, Lattis 1994, 117–18.

114. Copernicus 1543, bk. 1, chap. 11, Biblioteca Nazionale di Roma, shelf no. 201 39 I 26 (Provenance: "Collegij Rom Societ[atis] Jesu/Cat[alog]o Inscripsit C CL"): "Fallitur hic Copernicus" (fol. 20); "Lapsus est hic Copernicus. Qua de re vide scholium nostrum propositionis 21 de triangulis sphaericis" (fol. 3); "Hallucinatur hic Copernicus. Lege scholium nostrum propos[itionis] 24 de triangulis sphaericis" (f. 25); "Labitur hic Copernicus Consule scholium nostrum propos[itionis] 23 (f. 25). These references to Clavius's *Theodosii Tripolitae Sphaericorum Libri III* (Rome, 1586) show that he had studied parts of the text quite carefully.

115. Clavius 1591, 64.

116. Ibid.

117. Ibid., 64. The postil reads: "Verior sententia de ordine caelorum." Wallace (1984a, 33) first pointed out the shift from *verum* (1570) to *veriorem* (1581). This point is well discussed by Lattis 1994, 76–77.

118. Clavius 1591, 63–69.

119. Ibid., 68–69. Clavius is clearly referring to Peurbach's shared-motions passage ("ut in Theoricis planetarum explicatur").

120. Ibid., 65: "Atqui Luna maximam deprehensa est pati aspectus diuersitatem." Lattis (1994, 74) speculates that Clavius had obtained observations of daily parallax from his teacher Maurolyco; but if he did, he does not cite them, and this lack of citation still leaves open the question of how Clavius conceived of evidence.

121. Clavius 1591, 67.

122. Ibid., 68.

123. Ibid., 67:: "Ut autem plenior cognitio huius ordnis habeatur, non abs re facturum me arbitror, si rationes alias Astronomorum in medium adducam, ex quibus conuenientia huiusce ordinis elucescet."

124. Ibid., 69–70. Clavius does not refer here to his own promised work, which he always designates as "our *Theorics*," but to the description of the Mercury model in Peurbach 1472, "On Mercury," unpag.

125. See chapter 4 above.

126. Clavius 1591, 69: "Estenim Sol omnium rex; Saturnus autem, ob senectutem, eius consiliarius; Iuppiter, ob magnanimitate, iudex omnium; Mars dux militiae; Venus, dispensatrix omnium bonorudm, instar matrisfamilias; Mercurius eius scriba, ac cancellarius; Luna denique nuntij officio fungitur."

127. See Seznec 1953; Cox-Rearick 1984; Rousseau 1983.

128. Clavius 1591, 69–70.

129. Ibid., 71: "Praeterea secundum Albategnium & Tebith, & alios Astronomos, diameter uisualis Solis, ad diametrum visualem Veneris (sunt autem visuales diametri illorum circulorum, qui nobis apparent in astris) proportionem habet decuplam."

130. Ibid.

131. Ibid., 72: "Extra hunc uero mundum, seu extra coelum Empyreum, nullum prorsus corpus existit, sed est spatium quoddam infinitum, (si ita loqui fas sit) in quo etiam toto Deus existit sua essentia, in quo infinitos alios mundos, perfectiores etiam hoc, fabricare posset, si uellet, ut Theologi asserunt."

132. The full text and translation of the relevant passages is given conveniently in Crombie 1977, 65–66.

133. Clavius 1591, 455: "Auerroem, et Auerroistas, quos uerius hac in parte Erroistas dixeris."

134. Clavius 1591, 69.

135. Melanchthon in Bretschneider et al. 1834–, 13: col. 232: "Hic vero perversitas et petulantia Averrois, et multorum aliorum detestanda est, qui hanc doctrinam magna arte extructam derident, propterea quod non possit adfirmari in coelo re ipsa tales ubique machinas esse. Hac calumnia non deterreri se sinant studiosi, quo minus cognoscant motus, quorum leges ut aliquomodo ostendi possent, haec erudite tradita sunt, etiamsi non necesse est talem in coelo sculpturam esse orbium."

136. See Swerdlow 1972; Di Bono 1990; Baldini 1991, esp. 51–53.

137. Amico 1536; Fracastoro 1538. The San Diego State University copies are bound together.

138. Copernicus 1971b, fol. iii v.

139. Clavius's arguments about the status of eccentrics and epicycles have been discussed extensively (Jardine 1979; Lattis 1994, 113–15, 126–44).

140. But cf. Barker 1999, esp. 354–55.

141. For discussion see Swerdlow 1973; Aiton 1981; Jardine 1982; Barker 1999; Goldstein 2002; Barker 2003.

142. Clavius 1581: "Disputationem perutilem de orbibus Eccentricis, et Epicyclis contra nonnullos philosophos." This disputation remained in all subsequent editions.

143. Clavius 1591, 450: "Sicut in philosophiae naturali per effectus deuenimus in cognitione causarum, ita etiam in Astronomia, quae de corporibus coelestibus a nobis remotissimis agit, necesse est, ut in cognitionem ipsorum, coordinatione, constitutionemque perueniamus ex effectibus, hoc est, ex motibus stellarum per sensus nostros perceptis."

144. Ibid., 456.

145. Ibid., 450.

146. Ibid., 451: "Vera causa illarum apparentiarum: quemadmodum etiam ex falso verum colligere licet, ut ex Dialectica Aristotelis constat."

147. Ibid., 452; translated and quoted in Lattis 1994, 135.

148. Clavius 1591, 452–53.

149. Lattis (1994, 138) suggests plausibly that Clavius was "not rigidly dogmatic." Clavius's copy of *De Revolutionibus* did not include the *Narratio Prima*.

150. Clavius 1591, 452. Clavius accuses Copernicus of having proceeded "just as in many syllogisms we can prove some already known conclusions even from false premises" (Lattis 1994, 137).

151. Clavius 1591, 453; Lattis 1994, 139 (trans. slightly emended).

152. Clavius 1591, 453; Lattis 1994, 140 (trans. slightly emended).

153. Regarding utility, Clavius argued that astronomy is useful for theology, metaphysics, natural philosophy, medicine, poetry, navigation, and cosmography and also valuable to ecclesiastics, kings, and emperors. Next to its value in fostering admiration for the divine handiwork, Clavius devoted the most space to the popes' longstanding interest in reforming the calendar, an occupation that finally reached fruition in 1583. Clavius alluded in passing to his own role: "I was occupied in this matter for not a few years at the command of the highest pope and not enough in my studies and works" (Clavius 1591, 7–10).

154. Clavius 1591, 7: "Astronomy is called by most 'natural theology' [*Theologia naturalis*] because it examines the most superior bodies."

155. Ibid.: "Caeli enarrant gloriam Dei, & opera manuum eius ennunciat firmamentum,"and "Quoniam uidebo caelos tuos, opera digitorum tuorum, Lunam & stellas, quae tu fundasti." For detailed analysis of Riccioli's frontispiece, see Westman 1994, 80–81.

156. See Blackwell 1991, 36 ff.

157. See Baldini and Coyne 1984. For the dating of the manuscript, see 5.

158. Besides the important notes of Baldini and Coyne (1984) on Bellarmine's lectures, see Lattis (1994, 94–102) on Clavius's reply to Bellarmine.

159. See esp. Baldini 1984; Blackwell 1991, 40–45.

160. For example, there is only one brief allusion to Genesis in Thorndike 1949. For Genesis commentary, see A. Williams 1948; Steneck 1976.

161. Grant 1978.

162. Baldini and Coyne 1984, 10.

163. Bellarmine's reference is erroneous (ibid., 32 n. 33).

164. Cf. Blackwell 1991, 41–42: "It is purely a historical accident that he had already taken such an anti-Aristotelian stance only a few months before the observation of a nova in November 1572, which was the beginning of the decline of the Aristotelian notion of the immutability of the heavens."

165. Baldini and Coyne 1984, 20, 21: "Respondeo primum ad theologum non spectat hoc diligenter investigare. Et idcirco dum inter astrologos durat lis, sicut vero adhuc durat de modo explicandi huiusmodi apparentias. Nam alii explicant per motum terrae, et quietem omnium stellarum, alii per quaedam figmenta epyciclorum, et eccentricorum; alii per motum syderum a se ipsis: possumus nos eligere id constiterit, stellas moveri ad motum coeli, non a se, hoc videndum erit, quod recte intelligantur scripturae, ut cum ea perspecta veritate non pugnent. Certum enim est verum sensum scripturae cum nulla alia veritate sive philosophica, sive astrologica pugnare."

8. PLANETARY ORDER, ASTRONOMICAL REFORM, AND THE EXTRAORDINARY COURSE OF NATURE

1. Charles Webster (1982, 16 ff.) noticed and extrapolated on the important image of God as monarch in his Eddington Memorial Lectures at Cambridge, November 1980.

2. See again Barnes 1988, 82–93, 118, 170 ff.

3. Pauli 1856; Barnes 1988, 162.

4. For an excellent account of this evolving philosophical strain, see Lerner 1996–97, 2:3–15.

5. See Caspar 1993, 41–50; Jarrell 1971; Methuen 1998.

6. I surmise Muñoz's date of birth by subtracting twenty years from the date on which he obtained his bachelor of arts degree from the university in Valencia, the city of his birth. For further biographical details, see Víctor Navarro Brotóns, "La obra astronómica de Jerónimo Muñoz," in Muñoz 1981, 18–22.

7. R. Evans 1973, 146 ff., 255. This is still the best study of Rudolf's court.

8. For the actual reception of the most famous among courtiers' manuals, Castiglione's *Il cortegiano*, see Burke 1996.

9. See esp. Lerner 1996–97, 2:9; Granada 1997b, 415–17.

10. For a clear statement of this distinction, see Timpler 1605, 27, problema 5.

11. Kuhn 1957, 208.

12. Kuhn 1970, 116–17.

13. *Julius Caesar*, I.ii. On Machiavelli, astrology, and Fortuna, see Parel 1992, 63–85; but for critical reservations about Parel, see Najemy 1995.

14. Erastus 1557, fol. Aiv: "Der inhalt dieses Buchs. Erstlich ist ein büchlein in drei theil oder tractat getheilt / welches ungefehrlich vor lx jharen / durch Jeronimum Sauonarolam in welscher sprach geschriben worden / und im truck ausgangen ist / zu einter betrefftigung des / so her Joannes Picus ein herr zu Mirandulen wieder der Astrologos dem gemeinen mann zur warnung geschrieben."

15. Savonarola 1581; D. P. Walker 1958, 57.

16. Erastus 1569 (preface, Heidelberg, 1568).

17. Erastus 1571–73, 140–65.

18. Fulco 1560.

19. See Scepperius 1548.

20. Hemminga 1583, preface, fol. *6r.

21. Ibid. Hemminga did not explain his omission of Giuntini and Garcaeus, who might have served his purposes equally well. In the spirit of Hemminga, Kepler expressed disdain for the compendia of Schöner, Garcaeus, Leovitius, and Giuntini, which he regarded as constructed on "frivolous foundations" (Kepler to Herwart von Hohenburg, May 30, 1599, Kepler 1937–, 13:354, ll. 607–10).

22. Hemminga 1583, 117; see also Clarke 1985, 263.

23. Hemminga 1583, 118–19.

24. Ibid., 113.

25. Ibid., 225–86.

26. Frischlin 1586. (Copy used: San Diego State University, 1601.)

27. Ibid.

28. On Leovitius, see Birkenmajer 1972d.

29. Brahe 1913–29, 3:221–22.

30. Leovitius 1564, fols. H4v–K1v.

31. "Prognosticon ab anno domini 1564 usque in viginti annos sequentes, desumptum ex Coniuntionibus et Oppositionibus superiorum planetarum, Solis eclipsibus, alijsque Stellarum configurationibus, quae intra id tempus accident" (ibid., fols. L2, N2v).

32. Ibid., fols. N2v–N3; cf. Granada 1997b, 398–401.

33. Geveren appeared in many editions: 1577, 1578, 1580, 1582, 1583, and 1589. A note to the 1589 edition advises that the author did not intend his predictions as "demonstrations" but "as probable thinges so long to be embraced, till we learne more certaine: in these and like things" (fol. A2). Harvey 1583, 2–3: "The slight arguments of Picus Mirandula, Cornelius Agrippa and diuers other to yet contrary, haue been thoroughly answered by Balantius, Schonerus, Melancton, Cardane, and sundry other, but specially of late by Iunctinus, who in his confutation proceedeth compendiously, and directly from argument to argument, leauing in a manner nothing untouched that hath beene, or can been objected in disgrace of this knowledge." See also Granada 1997b, 402–4.

34. Brahe 1913–29, 3:224. By the time he wrote these words, Brahe was aware of Van Hemminga's attack on Leovitius: "A quibus non saltem Sixtus ab Hemminga nuper contra Astrologiam, non tam ratis, quam plausibilibus ratiocinijs & experimentationibus facile conuellendis, scribens, sibi haud temperauit."

35. Translated in Raeder, Strömgren, and Strömgren 1946.

36. Descartes 1998, pt. 1, 5: "As soon as age permitted . . . I completely abandoned the study of letters . . . resolving to search for no knowledge other than what could be found within myself, or else in the great book of the world."

37. See Burke 1996.

38. Brahe 1598, 107.

39. See Gingerich 1973c, 52–53.

40. Thorndike 1923–58, 6:68.

41. Barnes 1988, 141–81; Niccoli 1990, 140–67; Zambelli 1986a.

42. See Barnes 1989, 170 ff.

43. Magnitude −9.5, brighter than the quarter Moon and visible in the daytime. See Clark and Stephenson 1977, chap. 7; Marschall 1994, 56–59.

44. Clark and Stephenson (1977, 114) remark that "the intellectual climate in Europe which prevailed before the Renaissance was anything but favourably disposed towards the recording of a new star . . . [yet] not until AD 1572 do we find another new star which is so well documented."

45. P. Fabricius 1556, fol. Aij: "Ich habe in meiner Pratica / welche ich ampts halben auff das 56. Jar habe machen unnd vor zehen monat aufgehen lassen mussen / im ersten Capitel under andern gemet / das diss jar ohn Cometen nicht ergenehn werde / wie denn die all sehen unnd lesen werden / dieselb meine Practica haben"; Heller 1557.

46. Zambelli 1986a, 239–63.

47. See Van Helden 1985, 6–8, 16–19.

48. Regiomontanus 1531. Another edition appeared at Nuremberg in 1544. Jervis (1985) provides a translation, commentary, and facsimile edition of this later edition; for further discussion, see Barker and Goldstein 1988, 303–7, 312–13.

49. Jervis 1985.

50. Donahue 1975, 254–55; Christianson 1979, 121–22, 133. Hellman 1944, 92–93; Lerner 1996–97, 2:7–15; Granada 2006, 128–30.

51. As Barker and Goldstein observe, it was still possible to argue against the terrestrial location of comets, as Jean Pena did, based on prior physical or optical claims about their constitution as spherical lenses (Barker and Goldstein 1988, 316–17).

52. The notion of a *Denkkollektiv* was first introduced in 1935 by Ludwik Fleck (1979).

53. Ralph Cudworth's usage gives a clue: "A *Progymnasma* or Praelusory attempt, towards the proving of a God from his Idea as including necessary existence" (Cudworth, *Intellectual System*, I.v.,724; *Oxford English Dictionary*, 2nd ed., s.v. "Progymnasma"). *Praelusory* denotes that which is a precursor or preparation to something that comes next. In an excerpt translated into English from the *Progymnasmata*, Tycho's son-in-law Franz Tengnagel lamented that the book "deserveth a more famous title, than to be called Astronomicall Exercises" (Brahe 1632, fol. Av).

54. Unfortunately, few, if any, of these books survive. For Tycho's library, see references in Gingerich and Westman 1988, 5 n.

55. Brahe 1913–29, 3:307–8. Both Dreyer (1963, 38) and Thoren (1990, 55) describe the event well in their own words, but the former offers no source and the latter's citations are, uncharacteristically, inaccurate; the passage is cited correctly in Clark and Stephenson 1977, 174.

56. This episode does not fit the representation of servants as unreliable witnesses that is found in contemporary courtesy books (cf. Shapin 1994, 91–92).

57. Dreyer 1963, 42; Brahe 1913–29, 1:35–44; ibid., 44: "From our study at Herreward [Abbey], December 1572."

58. Brahe 1913–29, 1:36, my italics.

59. See, for example, Dreyer 1963, 52. Cf. Brahe 1913–29, 3:93: "Diarium illud una cum ijs, quae de Noua Stella adiunxeram, inspiciendum dedi."

60. Brahe 1913–29, 1:9, l. 21; 3:93.

61. Brahe 1913–29, 1:18, ll. 33–37; Segonds 1993, 370–75.

62. Dreyer 1963, 87; Tycho was also granted other holdings by the crown (108–13).

63. Brahe 1913–29, 5:143; Brahe 1598, 130. On the stone-laying ceremony, see Christianson 2000, 53–57. On interpretations of Uraniborg, see Hannaway 1986; Shackelford 1993.

64. On the use of *symmetria* as an architectural principle at Uraniborg, see Thoren 1990, 106–10.

65. On the accuracy of Tycho's observations, see Dreyer 1963, 356–60; Wesley 1978; Thoren 1990, 188–91.

66. Dreyer 1963, 115.

67. Mosely 2007, 125; and for an excellent discussion of Tycho's entire publishing program, 119–26.

68. R. Evans 1973, 204.

69. He sometimes signed himself "Nemicus," sometimes "Hayko." He was known in French as "Hagèce" and in Czech as Hajek, and Galileo called him "Agecio."

70. Dee 1659, 212; Evans 1973, 204 n.; Harkness 1999, 28–29.

71. For a brief entrée to this subject, see the illustrations accompanying Švejda 1997, 618–26; Slouka 1952. Stolle's life dates are unknown.

72. Cf. R. Evans 1973, 245.

73. Robert Evans' judgment here seems to me to be correct: "We cannot say for certain how far they formed a conscious, closed circle, but it seems probable that there was a tendency in that direction" (ibid., 203).

74. Ibid., 87–88.

75. Ibid., 196–242.

76. In a particularly long and extremely learned letter to Ferdinand, Rheticus referred frequently to the Turkish wars and also to Copernicus as "not only my teacher but also whom I revered as a father" (Rheticus in Krakow to Ferdinand in Vienna, 1557, Burmeister 1967–68, 3: no. 34, pp. 132–40).

77. Evans 1973, 203–4 n.

78. Hagecius was living in Vienna as late as January 1576 and in Prague no later than February 1578. He "had been the personal physician of Rudolf's father and grandfather" (Fučíkova 1997b, 26). Mattioli was best known for his important commentaries on Dioscorides; he published *Epistolarum Medicinalium Libri Quinque* in 1561, several years after arriving in Prague.

79. On the Prague Castle, see Fučíkova 1997b, 2–71.

80. Carelli 1557. The printer acknowledged that it was "difficult to print astronomical tables" and that "ephemerides were the most difficult of all to adjust accurately" (fol. 92v). On Hagecius, see Kořán 1959; Evans 1973, 152.

81. Horský and Urbánková 1975, 12–13. I have seen only the *Practica teutsch auff das 1554 jar zu Wienn* (Vienna: Syngriener, 1553).

82. For example: "Latitudo frontis incipit à radice nasi ubi terminantur cilia, versus commissuram coronalem. Longitudo frontis intelligatur per latitudinem corporis, ut villi ac nerui incedunt" (Hagecius 1584, 31).

83. Hagecius 1562. One of the copies at the British Library is partially hand watercolored, which suggests the strong possibility that this book was used for identifying plants as well as being considered a work worthy of presentation to the emperor. Melantrich was the official imperial printer. At Vienna, Hagecius is known to have studied with Andreas Perlach; he may also have worked with Joachim or Elias Camerarius at Frankfurt.

84. Mattioli's contacts with Aldrovandi continued after his arrival at the court (see Findlen 1994, 178, 270, 356).

85. Hagecius 1574. Hagecius announced himself on the title page as "Aulae Caesareae Maiestatis Medicum." On Mattioli, see Evans 1973, 118.

86. See Burke 1996, 19–20.

87. These were Paul Fabricius, *Stellae nouae vel nothae potius, in coelo nuper exportae, & adhuc lucentis, Phaenomenon descriptum & explicatum* (The appearance of a new, or rather illegitimate/bastard star recently risen and still shining in the heavens); Cornelius Gemma, *Stellae peregrinae iam primum exortae, & coelo constanter haerentis, Phaenomenon vel obseruatum, diuinae Prouidentiae vim, & gloriae Maiestatem abundeconcelebrans* (The appearance of a strange star, risen now for the first time and constantly adhering to the heaven, abundantly celebrating the power of divine providence and the majesty of his glory).

88. Hagecius 1574, 6: "Quod ex eodem veritatis fonte deprompta essent, cùm & temporibus diuersis & locis magno interuallo disiuncti, uterque eidem rei explorandae incumberemus."

89. Ibid., 7: "Nam consensus in doctrina veritatis, argumentum est probabile, non tamen necessarium."

90. Ibid., 108: "Ut frustra timere videatur hic Aristot. nequid maculae aspergeretur coelestibus, aut ea collabi aliquando & interire necesse sit, si quid elementaris naturae illuc deferatur: aut vicissim coelesti naturae indignum incompeténsue, in domicilio caducarum rerum aliquandiu hospitari."

91. Ibid., 60–62.

92. Ibid.: "Dubium non est, ut omnium miraculorum, ita huius quoque, Deum supremam efficientem causam esse, nec illi ullam aliam cooperari. Materia hîc penitus ferè à sensibus & intellectu abstractus est. Nam quali materia Deus usus sit in efformando illo prodigio, dici haud potest: cui aequè promptum & facile est, ex nihilo, solo verbo, vel etiam qualicunque assumpta materia, quiduis facere, & quae maximè discordantis naturae inter se videntur."

93. Hagecius 1576. Two years later, Hagecius (1578) appended to his treatise on the comet of 1577–78 a new attack on Raimondi. A dedication copy in the Prague Klementinum is to Wenceslaus Wyesovitz.

94. On the predominance of social class, see Shapin 1991; Biagioli 1993, 115–16, 288.

95. Hagecius 1576, fol. B4v.

96. Dreyer (1963, 73) says that the lectures took place at the behest of some aristocratic students at the university; but it is just as likely that Pratensis had been involved, as one might infer from his earlier plea for Tycho to publish his nova tract (May 3, 1573; Brahe 1913–29, 1:6–8).

97. Brahe 1913–29, 1:145: "Clarissimi viri, vosque studiosi adolescentes, rogatus sum, non solum a quibusdam vestrum, amicis meis, sed ab ipso etiam Serenissimo Rege nostro, ut nonnulla in Mathematicis disciplinis publice proponerem. Id muneris, etsi a meis conditionibus, et ingenij ac exercitationis tenuitate, admodum sit alienum: tamen Regiae Majestatis petitioni resistere non licuit, vestrae non placuit, et meapte sponte ab ineunte aetate eo propensus fui."

98. See Bailly 1779, 429–42; Dreyer 1963, 73–78; Moesgaard 1972a, 32; Westman 1975c, 307–8; Christianson 1979; Westman 1980a, 123; Thoren 1990, 80–86; Jardine 1984, 263–64.

99. Brahe to Pratensis, February 14, 1576, Brahe 1913–29, 7:25–26; Christianson 1979, 111.

100. Ibid., 7:39, ll. 25–26; Brahe to Severinus, September 3, 1576, Christianson 1979, 111.

101. Brahe 1913–29, 1:146.

102. Ibid., 1:146–49, 1:166–67, 1:172.

103. Dreyer 1963, 34.

104. Brahe 1913–29 , 1:149: "Ex his duobus artificibus, Ptolomaeo et Copernico, omnia illa, quae nostra aetate in astrorum reuolutionibus perspecta et cognita habemus, constituta ac tradita sunt."

105. See Kaufmann 1993, 136–50.

106. Brahe 1913–29, 1:149.

107. Ibid. Quoted and trans. in Moesgaard 1972a, 32; quoted with modifications in Westman 1975c, 307.

108. Brahe 1913–29, 1:172–73; Moesgaard 1972a, 32. The reference is clearly to Peucer 1568 (perhaps under the title *Hypotheses Astronomicae*; see chap. 5).

109. Brahe 1913–29, 1:152: "Non dubium est enim, hunc inferiorem mundum a superiori regi et impregnari: 'O quam mira et magna potentia coeli est, Quo sine nil pareret tellus, nil gigneret aequor.' Hinc nata est alia occultior et a sensibus externis magis separata doctrina, quam Astrologiam appellarunt. Haec enim de effectibus et influentia siderum in elementarem mundum et corpora, quae ex elementis constant, judicium profert."

110. Ibid., 1:152–53:

De qua quidem non libenter hic verba facerem, siquidem non ita demonstrationi indubitatae pateat atque ea, de quibus prius diximus: tamen, quoniam plures inveniantur, qui hac mantica et coniecturali potius, quam demonstratiua cognitione delectentur, plusque hac, quam reliquis antedictis afficiantur, lubet etiam in eorum gratiam, siquidem in Astrologiae mentionem incidimus, nonnulla disserere, praesertim cum

haec partim Mathematica, ob eam quam cum Astronomia, de qua antea diximus, habeat cognationem, partim Physica sit, nec satis demonstratiua: cumque insuper multi sint, qui cum alias Mathematum partes ita suis demonstrationibus esse fulcita viderent, ut eas in dubium vocare non possent—quis enim a Geometra edoctus, omnes trianguli tres angulos simul sumptos esse aequales duobus rectis, non fatebitur?

111. Ibid., 1:172: "Nam & ipse Danzaeus tacite Astrologicis praedictionibus, praesertim Genethliacis, minus fauebat, utut is in adolescentia hoc etiam studium excoluerat, atque mangum in eo profectum fecerat, ita ut multa uero euentu hinc in priuatis personis praedicere potuerat. . . . Putabat uero etiam is, Astrologicas praedictiones Euangelicae doctrinae refragari."

112. "Haec post orationem subiungenda" (Brahe 1913–29, 1:170); Christianson 1979, 113–15.

113. Brahe 1913–29, 1:70: "Is uero, quo conscius esset, se eiuscemodi argumenta ijsdem studiosis contra Astrologiam dictasse, & postmodum Commentarijs suis ad Paulinas epistolas inserta publicasse, quae tamen pro maiori parte e Caluini libello contra Astrologos desumta videntur."

114. Ibid.: "Existimo autem ipsius & aliorum argumentis in hac oratiuncula satis obuiatum esse, et quantum ad Caluini eruditum alias libellum attinet, quem contra Astrologos uibrauit, is non tam contra eos, quam pro illis facit, uel ipso autore, alias satis perspicaci & ingeniosi, ignorante."

115. Ibid., 1:166–67: "Mirum tamen est, nonnullos, inter quos famosus ille Erastus, quicum Medicinam, quae physica quaedam est cognitio, et ex naturae inferioris investigatione dependet, coelestia, unde haec vires et mutationes suas sortitur, inconsiderate negligere."

116. Ibid., 1:168.

117. Ibid., 1:36, ll. 39–42.

118. Ibid,1:168, my italics: "Quin et eo licentiae devenit dictus Comes Mirandulanus, ut non solum Astrologica (quae cum Physica & Mantica sint, atque prosubiectae materiae fluxibilitate varie alterari queant, facile in controuersiam veniunt), sed et Astronomica, adeoque mutationem maximae obliquitatis Eccliptiicae a veterum temporibus hucusque factam in dubium vocare non sit veritus."

119. See J. Evans 1998, 31–32, 54–55.

120. Ibid., 1:168–69:

Aiunt enim pro certo Italici quidam scriptores, inter quos est Lucas Gauricus, Episcopus Geoponensis appellatus, et ob Astrologiae professionem clarus, eidem Pico tres praestantes in Italia Astrologos annum aetatis 33 fatalem, ex directione Horoscopi ipsius Aphetae ad corpus Martis anaretae: quod et in Genethliaco ipsius Themate (modo id quod circumfertur, verum sit) satis quadrat. Et quamuis idem Picus hanc praedictionem amoliri, atque irritam reddere, quantum in ipso erat laborarit, adeo ut se in coenobium quoddam circa idem aetatis tempus abdidisse dicatur, nihilominus eodem anno, quo praedictum fuerat, satis concessit, ut sic in proprio corpore, adeoque vitâ ipsâ Astrologiae certitudinem expertus fuerit, quam ingenio et calamo labefactare nitebatur.

121. Brahe 1913–29, 5:117:

In ASTROLOGICIS quoque effectus siderum scrutantibus non contemnendam locavimus operam, ut & haec, a mendis & superstitionibus vindicata, experientiae, cui innituntur utplurimum consona sint. Nam exactissimam in iis adinvenire rationem, quae Geometricae & Astronomicae veritati par sit, minus duco possibile. Cum vero huic Prognosticae Astronomiae parti, quae mantica & Stochastica est, in adolescentiâ impensius addictus fuissem, posteaque ob motus Siderum, quibus fundatur, non satis perspectos eam seposuissem, donec huic incommodo subveniretur; compertis demum exactius Siderum viis, eam subinde in manus resumendo, majorem subesse certitudinem huic cognitioni, utut vana & frustranea non solum vulgo, sed & plerisque Doctis, adeoque nonnullis inter eos Mathematicis habeatur, comperi, quam quis facile existimârit: Idque tam in influentiis & praedictionibus meteorologicis, quam Genethliacis.

I have made modifications to Brahe 1598, 117.

122. Dreyer 1953, 367.

123. Capella 1499, bk. 8, fol. r5; Simplicius, 519.9–11 in Cohen and Drabkin 1966, 107; Eastwood 2001; Grant 1994, 312–13.

124. See Goldstein 2002, 229.

125. Brahe 1913–29, 1:172: "Sequenti uero die . . . praelectionem inchoauj . . . iuxta Copernici mentem et numeros, reducendo tamen omnia ad stabilitatem terrae, quam is triplici cieri motu finxerat, idque circa Fixas stellas, et duo mundi luminaria."

126. Brahe 1913–29, 2:428.

127. Christianson (2000, 102) cites a figure of "well over three thousand books."

128. Naibod 1573; his print identity was "Physicus et Astronomus."

129. Ibid., fols. 39v–42r.

130. Whoever labeled the diagram left it ambiguous as to whether the (unlabeled) circle on which the Sun is riding defines its orb or, as I have suggested, the orb is given by the surfaces of the Moon and Mars. If the former is the case, then there would be a problem concerning the two points where Mercury and Venus "cut" the solar circle (see further Lerner 1996–97, 2:49).

131. Westman 1975a, 324 n. 91.

132. Hellman 1944, 318–430.

133. Maestlin 1578, 14; cf. Hagecius 1578, 15–16: "Nostra etiam aetate *plurimi viri docti et pij* eandem opinionem habuisse videntur [Cometas ex occultis naturae causis prouenire]: ut Iacobus Zieglerus, Ioannes Vogelinus, praeclarus olim Viennae Astronomus . . . nostro Gemma. . . . Ego haec mea eius, & Tychonis Brahe Dani, viri nobilissimi doctissimique item Ioannis Praetorij Norimbergensis Mathematici, ac etiam Hieronymi Munnos Hispani, Hebreae linguae & Mathematum professoris, in Academia Val-

entiniana, eximij, & *aliorum doctorum virorum* censurae and iudicio lubens subijcio" (my italics). Hagecius 1576, fol. B4v: "Nomina autorum qui de stella scripserunt, ac mihi cognita sunt, haec sunt: Cornelius Gemma Louaniensis, Hieronymus Munnos Hebreae linguae et Mathematices professor in Academia Valentiana, idiomate Hispanico, Thomas Diggesseus in Anglia, & Tycho Brahe in Dania, sermone latino: *viri in Mathematicis exercitatissimi, & verè inter artifices numerandi*" (my italics).

134. Maestlin 1578, 17. Likewise, he uses the term *optici* ("opticians" or "opticists") to refer to two well-known authorities: "At contrà Vitellio & Alhazen Optici demonstrant" (18).

135. See Granada 1997b, 401.

136. Christianson 1979, 139: "It seems to me that the new star *anno '72* was a harbinger of the maximum conjunction, for it was united across the poles of the world with the beginning of Aries, in which location this aforenamed maximum conjunction will be celebrated and held."

137. Ibid., 139–40.

138. Thoren (1990, 128–32) makes the interesting suggestion that Tycho's position reflects a local conflict with Jørgen Dybvad, the Copenhagen professor of theology and mathematics, for royal favor; but Dybvad's apocalyptic views about the comet were by no means unusual, and the evidence for the whole account is thin.

139. See Diesner 1938.

140. On observational practices in early-seventeenth-century astronomy, see Dear 1995, 25, 66, 93–123; the notion of the collectively witnessed observation lies at the heart of the story of late-seventeenth-century experimental science, famously narrated by Shapin and Schaffer (1985, 336).

141. For this important distinction, see Dear 1995, 6–7.

142. Christianson 1979, 134.

143. See Donahue 1981. Grant emphasizes the juxtaposition of heterogeneous and, at times, inconsistent positions within the larger compass of Aristotelian natural philosophy (1994, 676–79); see also Granada 1997b; Lerner 1996–97.

144. On the last point, see Lerner 1996–97, 2:54.

145. Brahe 1913–29, 4:249–50; for Gemma's Dutch treatise on the nova, see Van Nouhuys 1998, 150–56.

146. Gemma 1578, 57. The postil reads: "Cometa praesens in caelo Mercurii."

147. He came to Alsace around 1572 and was in Hagenau for more than twenty years. For details, see Diesner 1938.

148. Roeslin 1578, fols. E–Ev: "Ut ex veris fundamentis parallaxeos notavit & demonstravit Cornelius Gemma in Mercurij Sphaeram illum collocans, quem sanè virum, utpote in isto genere studiorum, quasi hereditate paterna exercitatissimum, longè maiorem facimus quam istos Astrorum malos observatores omnes."

149. Hagecius 1574, 60: "Quòd igitur ad aetheream non ad elementarem regionem accensenda sit haec stella, duo sunt firmissima quae id confirmant: quorum alterum est aequabilis & perfecta ipsius cum motu proprio conuersio. Quae enim in elementari consistunt regione, non possunt ea aequabilitate conuerti: Alterum carentia parallaxeos."

150. Roeslin 1578, fol. Ev: "Dicere mihi nunc Physicus velit, quomodo tam aequalis constans & proportionalis motus cadere possit in elementarem regionem aëris vel ignis? cum elementorum partibus innatum sit vagari & huc illucque incertis sedibus agitari? Aut si ductu & tractu materiae viseosae flamma serpsit, & quasi peculiare iter pabulo allecta fecit, ut Aristoteles velle videtur: Quomodo quaeso materia fuit ita aequaliter disposita, ut eandem proportionem flamma prorependo semper servaverit, & iter suum ad locum Stellae novae direxerit? Quare hunc Cometam & aethereum & in aethereum regionem collocandum iudicamus."

151. Ibid., fols. C2–v.

152. Ibid., fol. E2v: "Quemadmodum enim Mercurius in annuo spacio, quatuor Zodiaci quadrantes perficiens, quater etiam ferè est retrogradus, quater velox & directus, octies verò stationarius. Sic Cometa noster exactè unum circuli sui quadrantem perficiens in unius anni quadrante, semel fuit retrogradus, semel directus & velox, bis verò stationarius, in longitudinibus Epicycli medijs, ut vel hoc nomine constet in Mercurij Sphaera rectè observatum fuisse Cometam & collocatum." At the end of chap. 6, Roeslin speaks of the comet as being located "in the region of Mercury" (E3v).

153. These connections are nicely described by Granada 1997b, 444–52.

154. Roeslin 1578, fol. G4v: "Illi critici autem debent nobis eò magis commendati esse, quod ijs aptè correspondeat Analogia veteris & novi Testamenti: coincidant prophetiae omnes & oracula, in ijsque finiantur & claudantur: accedant insuper naturae ipsius testimonia, desumpta ex magno Libro Mundi & creaturarum."

155. Granada 1997b, 450 n.

156. For a fuller account, see Westman 1972a.

157. Brahe 1913–29, 4:266.

158. Ibid., 4:190; Barker and Goldstein 2001, 94.

159. Christianson 1979, 129, my italics. Variant 3 reads: "Mercury has its orb around the sun and Venus around Mercury."

160. Maestlin 1580; see Granada 1997b, 446 n.

161. For the importance of the distinction between God's absolute and ordained power, see Funkenstein 1986, 121–52.

162. Montaigne 1958, 29, italics in original: "Of Prognostications."

163. On Montaigne's skepticism, see Popkin 2003, 44–63.

164. Montaigne 1958, 429: "Apology for Raymond Sebonde."

9. THE SECOND-GENERATION COPERNICANS: MAESTLIN AND DIGGES

1. Note that this actors' formulation avoids the excessively global analytic categories *instrumentalism* and *realism*, without relinquishing the possibility that a particular agent could be, say, "realist" or "instrumentalist" with respect to different parts of the natural world.

2. See Maestlin 1597; Maestlin 1624.

3. Maestlin 1586.

4. Maestlin 1588.

5. Dated September 5, 1571 (Zinner 1941, no. 2553). Maestlin's copy, today located at the Schweinfurt Stadtarchiv, contains a number of his own corrections and additions as well as those of its subsequent owner, the Altdorf mathematician Johannes Praetorius.

6. For biographical details, see Betsch and Hamel 2002; Jarrell 1971, 10–44.

7. On heavenly poetry, see Pantin 1995.

8. Frischlin 1573, 1–26; Maestlin's work follows on 27–32. Whatever else Frischlin may have imbibed from the young Maestlin, he did not so much as mention the name Copernicus in his poem.

9. This was the same year in which the physician and theologian Thomas Erastus left Heidelberg for Basel. The two shared an opposition to astrology; but it is not known whether Erastus influenced Maestlin's appointment or whether they ever met.

10. Apianus's copy of *De Revolutionibus* contains a few notes by Maestlin (Gingerich 2002, Stuttgart 1, 93–94).

11. Maestlin 1576; reissued in 1580.

12. Kepler 1858–91, 1:56–58.

13. *Demonstratio Astronomica Loci Stellae Novae:* "Quod nullo modo fieret, si Orbi alicuius Planetae Affixa esset, nam ut uidere licet 5. Lib. Coper. commutationibus motus expers non esset" (Brahe 1913–29, 3:60). Tycho omitted Maestlin's postils from his edition. Yet it is useful to know that references to Copernicus occur four times in the postils, so that a reader such as Frischlin could not have missed them.

14. Ibid.: "Quoniam immensa est Altitudo Orbis stelliferi, quae quousque se extendat, non constat, ad quam, quae inter Solem et Terram est distantia, concerni nequit (ut testatur Copernicus, Astronomorum post Ptolemaeum Princeps, qui omnium Orbium Planetarum certas distantias a Centro Mundi demonstrans, in Orbe Stellato subsistit) ideoque impossibile veram huius Stellae, uel magnitudinem uel Altitud. a Centro Mundi dimetiri, certium tamen est." In this passage, Maestlin avoided identifying the Sun with the center of the world, although without that assumption the statement makes no sense.

15. Ibid.: "Ex dictis patet noui huius luminis apparitionem, non a naturali causa dependere, qualem sane supra enumerati plaerique reddere conati sunt, nec Cometam, sed potius Stella Nouam dicendam

esse: nisi Cometas non tantum in Elementari Regione, sed etiam in Orbe stellato, qui secundum Copernicum est Coelum extremum, seipsum & omnia continens, generari posse, adeoque Coelum generationis & corruptionis, contra Aristotelem omnesque Physicos & Astronomicos, non expers esse, dicere uelimus."

16. In his treatise on the comet of 1580, Maestlin again referred to this passage, using the Earth-Sun distance as a comparison reference to the insensible distance from the center of the world to the starry sphere: "Sic nollem inficiari Stellam nouam anni 1572 similiter caudatam fuisse, sed quoniam distantia eius, sicut & totius orbis stelliferi, in quo versata est, tanta fuit, ut ad eam integra Solis & terrae distantia non sentiatur, sicut Copernicus asserit, nobis certe in terris eius cauda sursum porrecta non apparuit" (Maestlin 1581, xiiii).

17. Brahe 1913–29, 3:60.

18. He appears to assume here something like Descartes's scholastic premise that the cause of a being must have at least as much perfection as the being itself.

19. Brahe 1913–29, 3:60.

20. "Jar nach der zeit Messiae stehn soll / dann an den letzten zweytausent Jaren sind nu nit vil mehr uber 400 jar uberig. Es komme nun diser Spruch von Elia oder nicht / dann in Heyliger Schrifft wird er nicht gefunden: So ists doch gewiss / dass wir von der Welt end nit ferrn sind. Alle Propheceynungen der Schrifft lauffen aus. Paulus 2.Thess.2.wil in weit zil stecken / den Thessalonichern damit die Gedancken zunemmen." (Maestlin 1583, 37; cf. Barnes 1988, 113).

21. Kepler 1937–, 1:93, ll. 12 ff.

22. Brahe 1913–29, 3:62: "Quid vero Nova haec Stella portendat, aliis disputandum relinquemus; nobis autem tantum illa, quae Astronomus Veritatis amans, de ea pronunciaret, conscribere placuit"; Granada 1997b, 415. Granada has recently found evidence that Maestlin initially included eschatological sentiments in his treatise but suppressed them in the published version (2007b, 109).

23. Maestlin 1578, fol. A4r–v:

Veruntamen mihi hîc confitendum est, me multorum expectationi, in quorum manus hac mea incident, non omnino satisfecisse: nam licet quae Astronomus de hoc Cometa dicere potest, compilauerim, quae tamen is portendat, ego coniecturas tantùm, non ex Astrologiae fontibus promanantes, sed aliunde deriuatas, notaui. Spero autem, me eius rei causam venia non indignam afferre. Etsi enim hactenus Mathematicam abstractam & concretam mihi nonnihil familiarem fecerim, in concreta tamen, cui motuum coelestium considerationes subiacent, ego Astronomiae potius, quàm Astrologiae incubui. Cùm enim ex multiplicibus aliorum eruditorum virorum querelis, & etiam proprijs experimentis intellexissem, in motuum tabulis & calculo aliquid desiderari, quanquam motuum rationes siue hypotheses ab Artificum diuina solertia probè inuentae & demonstratae sint, quòd ipse calcu-

lus tamen faciem coeli nonnihil vel excedat, vel ab eo deficiat: Ideo illi me dedere coepi, ut obseruationes in coelo complures ego ipse notarem si forsan ex earum collatione cum antiquissimorum Hipparchi, Ptolemaei, Albategni, & recentiorum Regiomontani, Peurbachij, Copernici & aliorum obseruationbus, possem breui (si modo Deus vitam & vires mihi largiatur) calculum ad absolutam & diu expectatim integritatem reducere. Hinc factum est, ut Astronomiam Astrologiae perpetuò praeposuerim. Quare iudicium Astrologicum mihi hîc arrogare nec possum nec volo, sed id alijs relinquo, quorum multos video admodum esse solicitos, ut audacter (siquidem hoc facile est) diuinent. Hanc igitur ob causam quatenus Astronomicae scientiae Cometa subditur, à me explicatus est, ut Dei Opt. Max. sapientia & omnipotentia hîc, ut & in alijs, conspiciatur.

24. Wurttembergische Landesbibliothek Stuttgart, 4°15b no. 55; quoted and trans. in Jarrell 1971, 139.

25. Maestlin 1580, fols.):(2–2v:

Compendium Astronomiae. Explicationem uberiorem, siue commentarium in doctrinam sphaericam, seu priorem Astronomiae partem, de primo motu. Theorias Planetarum, siue commentarium in alteram Astronomiae partem. Arithmeticam vulgarem, perspicuam. Doctrinam Triangulorum planorum et Sphaericorum absolutissimam. Commentarium in Cleomedem. Commentarium item eruditum, et demonstrationes propositionum Theodosij librorum de Sphaera. Varia scioterica et suspensilia noua, quibus per umbram Solis, vel altitudinem eius, aut stellarum, hora diurna vel nocturna inuestigatur: Item alia instrumenta, ad obseruationes Phaenomenon coelestium, et ad dimensiones planimetricas et stereometricas utilia. Reuolutiones orbium coelestium, ad imitationem Almagesti Ptolemaei, et Reuolutionum Nicolai Copernici. Tabulas motuum orbium coelestium, ex istis Reuolutionibus nouis deriuatas, ad imitationem Tabularum Alphonsinarum et Prutenicarum. Tabulas item resolutas, ad imitationem Tabularum Blanchni [sic]. Nec non Ephemerides nouas, ex nouis his Tabulis computatas.

26. See Wolf in Leovitius 1556–57, fols. y5v–z4v; Wolf (in Maestlin 1580), refers to the ephemerides of both Leovitius and Stadius (1560).

27. Maestlin 1597.

28. The occasion was the appearance of a comet in 1618. Maestlin to Johann Faulhaber, January 18, 1619, MSS. University of Tübingen, Mi XII.27b: "I am not an astrologer" (quoted by Jarrell 1971, 176).

29. See Methuen 1999, 105.

30. For Heerbrand, see Hübner 1975; Methuen 1998, 132–37; Hellman 1944, 262–65.

31. Andreae 1567, fol. Aa iiir: "Das es aber nach solchem mutinassen allwegen und zu aller zeit gewisslich und nicht anderst geschehen solt / wann gleich des Himmels Lauff getroffen / und umb ein Minuten die Rechnung nicht fehlete / das würdt kein verstendiger Mathematicus / noch vil weniger ein Christ sagen / wie ich dann von dem berhümtesten Mathematico (so meines wissens auff disen Tag in Teutschland lebet) dergleichen vil und offt gehört / welcher die

Weissagungen / so auss des Himmels Lauff gemacht / da sie auff besondere Personen und Landschaffen gezogen / vergleicht einem / der mit Würffel spilet / da gantz ungewiss ist / ob er alle Sei / oder alle Es werffen werde." See also fol. Aa iiir–v and Methuen 1996, 125–29.

32. Frischlin's work was based on astronomical lectures delivered between 1569 and 1572 while the regular lecturer, Philip Apianus, was away. See Hofmann 1982, 247 n. 58.

33. Frischlin 1601, 420–21:

> Anyone who dares to adduce proofs against the astrologers from among those about whom we have spoken or reckoned thus far, let him read the Sacred Bible, Basil, Chrysostum, Nazianzus, Theodoretus, Augustine, Ambrose, Lactantius, Eusebius, Girolamo Savonarola; from the ancient philosophers [let him read] Plato, Aristotle, Hippocrates, Galen, Celsus; from the moderns, Celius Rhodingus, Pico Mirandola, Angelo Poliziano, Luis Vives, Mainardi, Fuchs, Valleriola, Lang, Schegck, Thomas Erastus, each one cited by me in places; and also in books written publicly against the vanity of astrologers. Luther also taught that the astrological art is diabolical, and Calvin published a singular book against the same thing. No one can object to me, therefore, either on the basis of how long astrology has endured or on the [universal] consent of men.

34. Earlier in his treatise he even felt it necessary to defend himself against the charge that he had endorsed Melanchthon's ethical, logical, grammatical, and physical writings. Ibid., fol. 5: "Quasi verò Tubingenses non iamdudum, Melanthonis utramque Grammaticam, cum Rhetorica et Dialectica ex Academia sua eiecerint: aut quasi ullus ibi reperiatur artium Studiosus, qui initia doctrinae Physicae atq; Ethicae, conscripta à Philippo in manibus habeat: aut quasi non iam olim Philippus cum suis locis communibus et toto corpore doctrinae à Tubingensibus, publico concilio, sit ad orcum damnatus. Et tamen isti homines affirmare audent, non sordere ipsis Philippi scripta. Sed quia de hac re exit iam peculiaris à me Dialogus, in quo mea demonstratur innocentia, & noxa aduersariorum, iccirco pluribus pro me dicendis nunc supersedeo."

35. "Hic itaque Frischlini liber . . . deprehenditur habere Methodum quidem talem, quae cum modo tradendarum scientiarum non admodum congruit: Res autem quae negocio haud sufficienter satisfaciunt. Plaeraque enim ibi compraehensa a rectitudine non leuiter recedunt: multa item satis intricatè et imperfectè traduntur. Ex quibus non obscurè colligitur, Autorem scientiae Mathematicae esse oblitum." (Hauptstaatsarchiv Stuttgart: Maestlin 1586; see also Methuen 1998, 101–6, 129–32).

36. Maestlin to Kepler, May 2, 1598, no. 97, Kepler 1937–, 13:210.

37. See Westman 1972; Hellman 1944, 137–59.

38. Brahe 1913–29, 3: 62, ll. 39–45; 63, ll. 1–15.

39. "Rationi omnino consentaneum est. Talem esse constitutionem machinae totius immensis, quod firmiores admittit demonstrationes: quod ita totum universum conuertit, ut nihil sine istius confusione transponi possit: per quas omnia motuum phaenomena exactisime demonstrari possunt: et in quod nullum in progressu occurit inconueniens." (Gingerich 2002, Schaffhausen, 219–27 [1543, fol. iiij]).

40. Ibid., fol. iv. At the top of fol. ij r, Maestlin wrote that he had learned of Osiander's identity from a letter in a book that he had purchased from the widow of his old teacher, Phillip Apianus (d. 1589). Many years later, in 1613, Maestlin told Kepler that Apianus's widow, "as a result of certain occult insinuations (undeserved and against me), sold every copy that had survived and allowed me to inspect what books that remained" (Kepler 1937–, 13:58).

41. "Magnum [est] certe argumentum omnia tam phenomena, quam ordinem et magnitudem orbium, in terra mobilitatem conspirare" (Copernicus [Schaffhausen] 1543, fol. iiij r).

42. Schreckenfuchs 1569, 36: "Caeterum cùm mobilitas terrae uarie disputari potest, ut uidere est apud Nicolaum Copernicum, virum incomparabilis ingenij, quem meritò possem dicere mundi miraculum, ni uererer quosdam uiros, ueterum philosophorum sanctionum tenacissmos, & non immeritò, offendi. Et ne in re dubia multa adducam argumenta, quae longissima egeant explicatione, de hac re in Commentarijs nostris in Copernicum, si fata sinent, prolixius et manifestiùs dicemus."

43. Copernicus (Schaffhausen) 1543, fol. iiij: for full Latin text, see Westman 1975d, 62–63.

44. Although Maestlin did not directly annotate the "symmetria" passage, he gave a full citation from the *Ars Poetica* opposite the sentence where Copernicus says how long it took him to complete his work. This annotation shows that he was completely familiar with Horace.

45. See Schreckenfuchs 1569, 36.

46. Copernicus (Schaffhausen) 1543, fol. iiv.

47. André Goddu opens a new perspective by suggesting that Copernicus had in mind here a criterion of relevance that he could have learned at Krakow from Peter of Spain's topical logic: "If hypotheses must be relevant to the observations, then we must be able to see how all of the observations follow from the hypotheses and in that sense are confirmed beyond a doubt" (see Goddu 1996, 51; Goddu 2010, 285–300; 321–3). Goddu also identifies a statement in Aristotle's *Nicomachean Ethics* (bk. 1, ch. 8, 1098b11–12) that Copernicus could have known: "With a true view all the data harmonize, but with a false one the facts soon clash" (Goddu 2010, 390).

48. Copernicus (Schaffhausen copy) 1543, fol. iiv: "Verum vero consonat et ex vero non nisi verum sequitur. Et si in processu ex dogmate vel hypothesibus aliquod falsum et impossibile sequitur, necesse est in hypothesibus latere vitium. Si ergo hypothesis de ter-

rae immobilitate vera esset, vera etiam essent quae inde sequuntur. At sequuntur in Astronomia plurima inconvenientia et absurda tam orbium constitutionis quam orbium motus planetae. Ergo in ipsa hypothesi vitium erit. Minor patet in motu Solis: in anni tropici magnitudine; item trium superiorum planetarum motu; maxime autem Venere et orbe stellato." Here and in n. 51, I accept the critical improvements made by Alain Segonds to my original transcription and translation (Kepler 1984, 261–62; cf. Westman 1975d, 60).

49. This objection fails to appear in other *Sphere* commentaries, including that of so sophisticated a commentator as Schreckenfuchs.

50. This is close to the notion of Gérard Genette's *paratext*, as applied by Fernand Hallyn to Oresme and Buridan (Hallyn 1990, 61).

51. Copernicus (Schaffhausen) 1543, fol. 141r: "Aliud argumentum quod terrae mobilitatem confirmat. In hypothesi de terrae immobilitate, planè absque revolutione dicitur, orbes Solis, Veneris et Mercurii, qui sunt à se inuicem distincti, sunt contigui, non continui: uno et eodem motu medio cieri, cum tamen nunquam alias usu veniat ut duorum vel plurium planetarum orbes unum medium motum habeant. Huius vera revolutio egregiè patet in hypothesi de mobilitate terrae, ut videre est in textu."

52. "Est ergo motus quem prisci epicycli dixerunt, nihil aliud, quam differentia, qua terra motum planetae velocitate superat, ut in 3 superioribus, vel qua terra velocitate superatur, un in duobus inferioribus. Et hic motus Copernico commutationis dicitur" (ibid.). These passages incidentally help to date the annotations to the period 1570–73, as Maestlin had referred to *De Revolutionibus*, bk. 5, chap. 3 in his nova tract.

53. See Westman 1975c, 332–33.

54. Maestlin to Kepler, March 9, 1597, Kepler 1937–, 13:111.

55. T. Digges 1573. Cf. "Praefatio authoris": "Thomas Digges, a Learned Man, to Ingenious Seekers of Heavenly and Astronomical Wisdom."

56. Johnson 1952.

57. T. Digges 1573, fol. A2v: "Licet etenim alij de Parallaxibus Pheonomenon scripsere, fuerunt tamen (ut veritatem eloqui non pertimescam) demonstrationes eorum omnes vere Dedalicae Alae, quibus aut infima haec sublunari Regione volitare cogerentur, aut si altius contenderent cum Icaro in errorum pelagus liquefactis pennis praecipites agi necesse fuerit."

58. Johnson (1952) inclines toward placing his death in the year 1559, but Leonard may have died as late as 1563 (see Patterson 1951).

59. Johnson 1952.

60. Two other important examples of this period are Cuningham 1559 and Recorde 1556.

61. L. Digges 1562. Although the date 1556 is clearly marked on the frontispiece, the imprint is London: Thomas Gemini, 1562. And the printer says that he is

"there ready exactly to make all the Instruments apperteynynge to this Booke."

62. L. Digges 1571.

63. Ibid., 97–103 (following pagination of the 1591 ed.). It is likely that Digges was using a post-1560 edition of Euclid, as the extant 1551 edition, which contains his provenance (dated 1558), lacked the proofs showing the construction of the five regular solids in book 13 (Euclid, *Euclidis Elementorum Liber Decimus*, Petro Montaureo interprete. Paris: M. Vasconsan, 1551. See Jonathan A. Hill, *Catalogue 150: 25th Anniversary*, n.d. item 27, 37–38).

64. See Sanders 1990; Johnston 1994, 69–70.

65. T. Digges 1573, sig. A2r.

66. Dee 1573, sig. A2v.

67. Johnson 1937, 157.

68. I take his explicit dissociation from "mysticall appliances" to be a coded reference to Offusius's views, proffered the year before. See also Johnson 1937, 31–32.

69. See item no. 68 in Dee's 1583 catalogue, signed "Thomas Diggius 1559" (Roberts and Watson 1990, 43, 82 n.). One of the earliest books owned by Digges was a copy of Euclid's *Elements* (Paris: M. Vascosan, 1551). Signed by Thomas Digges with the date 1558, this item appeared on the market in 2003 (see Hill, *Catalogue 150*, item 27).

70. Dee's copy: Roberts and Watson 1990, item no. 2109. There is a dedication copy to Henry Savile in the Bodleian, signed "H.S. Ex dono Th. Diggesej auctoris."

71. T. Digges 1573, fol. A1v: "Plato said that men are given eyes in order to do astronomy."

72. T. Digges 1572; transcription by F. R. Johnson, now in my possession. The judgment itself is no longer extant.

73. T. Digges 1573, fol. A2: "At qui Platonicis seu ut veriùs loquar Mathematicis istis instructus Alis, sursum in Aetherea contendat, Elementaribusque prorsùs Regionibus traiectis, longè remotiorem Cometarum locis esse perspexerit."

74. Ibid., sig. A2; "De stella admiranda in Cassiopeiae Asterismo, coelitus demissa ad orbem usque Veneris, iterumque in Coeli penetralia perpendiculariter retracta. Lib. 3. A. 1573," cited in Dee 1851b, 25; Johnson 1937, 156 n. Surprisingly, there is no reference to the nova in the diary that Dee kept in his copy of Stadius's *Ephemerides*. See also Clulee 1988, 177.

75. For example, copies at the Paris Observatoire (shelf no. 21144); Bibliothèque Nationale (shelf nos. V.7738, V.6556); Bibliothèque Mazarine (15828); Oxford, Bodleian (Ashmole 133 (6)): bound with Dee's *Propaedeumata* (1568), two copies of the *Nucleus* and several late-seventeenth-century works; Cambridge, Peterhouse College (Pet.3.18); Cambridge, Wren Library, (V.I.9.110^2; V.I.1.114^3); Biblioteca Nazionale di Firenze (1108.12, destroyed in the 1956 flood); Biblioteca Nazionale, Rome (69.5.B.17.1).

76. T. Digges 1573, sig. L2.

77. Ibid., sig. L3r: "portentosi Syderis a *Potentissimo terricolis* exhibiti . . . CHRISTI DEI adventum MAGIS dennutiantis [sic] oppositum . . . stupendum DEI miraculum." On this point, see Granada 1994, 12.

78. T. Digges 1573, A4v: "Licèt Saturni, Iouis, et Martis, Parallaxeis adèo sint exiguè ut sensuum imbecillitate vix discerni possint."

79. Ibid., sig. A3: "Mutilum et mancum potiùs quoddam, ex repugnantibus et mutuò collidentibus eccentricis Orbibus, et Epiciclis irregularitèr super propriis centris currentibus."

80. Ibid.: "Illisque perinde accidebat ac si quis ex diuersis hominum picturis, Manus, Pedes, Caput, aliaque membra, elgantèr equidem sed non unius hominis consideratione depicta assumeret, atque inuicem coniuncta, hominis picturam perfectam sese exhibere putaret. Haec autem si veras Hypotheses assumpsisent, illis accidere nullo modo possent." Digges's insertion of the word *picture* strongly suggests that he knew the Horatian subtext of Copernicus's image.

81. Ibid.: "Hoc saltèm admonere statui ansam oblatam esse, et occasionem maximè oportunam experiendi an Terrae motus in Copernici Theoricis suppositus, sola causa fiet cur haec stella magnitudine apparente minuatur."

82. Ibid., A4v–B1r: "Promitto, quibus (non probabilibus solummodò argumentis, sed firmissimis fortassè Apodixibus) demonstrabitur, verissimam esse Copernici hactenus explosum de Terrae motu Paradoxum."

83. Ibid., A3r–A3v:

For, if it were thus, always decreasing towards the spring Equinox, it would be observed to be very small in its own magnitude. If, afterwards, increasing little by little towards the following June, it shall have continued in existence, it will scarcely be of the same brightness as when it first appeared, but in the autumn Equinox it will be seen of unusual magnitude and splendor. However, no cause of diversity of apparent quantities of this sort can be assigned other than that of its elongations from the earth, since not only would it be contrary to the basic principles of Physics that a star should increase or diminish in the Sky, but by the clear measures which have been set in this art, it will be perceived to be otherwise. Therefore, I have thought not only that a treatment of this subject is necessary, but also that Mathematics has rules for measuring the location, distance and magnitude of this stupendous star, and for manifesting the wonderful work of God to the whole race of mortals (who strive to understand something celestial and lie not wholly buried in the earth); also for examining theorics and establishing the true system of the universe, as well as for measuring most accurately the parallaxes of celestial phenomena.

Quoted and trans. in Johnson 1937, 158–59.

84. L. Digges 1576. The running page headings that commence with Thomas Digges's diagram read "The Addition."

85. Johnson and Larkey 1934, 76.

86. Thomas Digges in L. Digges 1576.

87. Johnson and Larkey 1934, 83.

88. Dee claimed that he had provided the main principles of the "power of celestial bodies" (*de Caelestium corporum virtute*), sufficient for others, proceeding demonstratively" (*Apodicticè procedendi*) to find further principles (1978, 112–13).

89. Johnson and Larkey 1934, 82.

90. L. Digges 1576, fol. 15v: "Mercury is next under Venus, somwhat shyning, but not very brighte: neuer aboue 29. degrees from the Sunne, his course, is like to Venus, or the Sunnes motion."

91. Johnson and Larkey 1934, 79. This passage shows that Digges was aware that Copernicus himself had not inserted the "Ad Lectorem," but he did not know Osiander's identity because he was not associated with the German network that linked Rheticus, Apianus, Praetorius, and Maestlin.

92. Ibid., 79–80.

93. Ibid., 80.

94. Ibid.: "If therefore the Earth be situate immoueable in the Center of the worlde, why finde we not Theorickes upon that grounde to produce effects as true and certain as these of Copernicus? Why cast we not away those Circulos Aequantes and motions irregulare, seeing our owne Philosopher Aristotle him selfe the light of our Universities hath taught us: Simplicis corporis simplicem oportet esse motum."

95. Ibid., 95, 80

96. See Feingold 1984, 186: "Thomas Digges virtually abandoned his theoretical studies once he entered the service of the Earl of Leicester."

97. Johnson and Larkey 1934, 91: "Whether the worlde haue his boundes or bee in deede infinite and without boundes, let us leaue that to be discussed of Philosophers, sure we are yt the Earthe is not infinite but hath a circumference lymitted, seinge therefore all Philosophers consent that lymitted bodyes maye haue Motion, and infinyte cannot haue anye."

98. De Morgan 1839, 455a; De Morgan 1855; both cited in Johnson and Larkey (1934, 74), who rightly conjectured that the copy consulted by De Morgan was missing the diagram of the infinitized Copernican universe. A small piece of the title page of the second copy has been damaged and repaired (see figure 51). The torn-off part has been incorrectly relabeled "1594"; the actual date of the edition is 1596.

99. For example, it appeared on the cover of the first paperback edition of Kuhn 1957 and in Koyré 1957, 37; forty years later, its popularity was again attested to by its reproduction in Shapin 1997, 22, and Jacob 1997, 29.

100. L. Digges 1555, xiii. Digges's first citation is to the thirteenth-century Bolognese astrologer Guido Bonatti, "where he writeth 'against those who say that the science of the stars cannot be known by anyone; against those who say that the science of the stars is damnable rather than useful, etc. and against those who contradict the judgment of astronomy and

who rebuke it while not knowing its worth insofar as it is not lucrative.'"

101. Ibid., xiv.

102. Ibid., xv: "He proueth it one of the chief sciences *Mathematical,* by the autoritie of the best learned, and by *Aristotele* in hys *Posteriorum.* Howe commeth it to passe louinge Reader, seynge it is a noble science, *et scientia est notitia vera conclusionum, quibus propter demonstrationem firmiter assentimur,* that it is counted vayne, and of so smal strengthe."

103. The diagram appears on fols. 4v and 16 in both the 1576 and 1583 editions.

104. For convenience, all quotations are from Johnson and Larkey 1934, 79. The Huntington Library copy was acquired at the Anderson Galleries sale in May 1919. Johnson was a research fellow at the Huntington, 1933–35. I have checked all passages against the Huntington copy.

105. Miguel Angel Granada (1994, 16–20), in particular, has emphasized the hierarchical gradations still present in Thomas Digges's universe.

106. Marcellus Palingenius (1947, 183). The work may well have been inspired locally by the extraordinary astrological murals of the Schifanoia Palace; see *Lo zodiaco del principe* 1992.

107. Johnson and Larkey 1934, 88–89.

108. Copernicus 1978, bk. 1, chap. 8, 16. Later, this passage undoubtedly served as inspiration for Galileo's famous passage in the *Dialogue* where Salviati describes various motions within the reference frame of a ship at rest and in motion (Galilei 1967, Second Day, 126, 140–44).

109. Ibid. Digges rendered this passage thus: "And of thinges ascending and descending in respect of the worlde we must confesse them to haue a mixt motion of right and circular, albeit it seeme to us right & streight" (Johnson and Larkey 1934, 92–93).

110. Ibid., 93.

111. Johnson 1937, 164.

112. See Dear 1995 for an excellent analysis of the shift in the notion of experience from a generalized report to a historically specific description.

113. Johnson and Larkey 1934, 81. The passage continues: "This ball euery 24. houres by naturall, uniforme and wonderfull slie & smoth motion rouleth rounde, making with his Periode our naturall daye, whereby it seemes to us that the huge infinite immoueable Globe should sway and tourne about."

10. A PROLIFERATION OF READINGS

1. For a valuable history of usages, see Lerner 2005.

2. Vermij 2002, 34–42. For the Capellan ordering in the ninth century, see Eastwood 2001.

3. Contrary to early misidentifications by Zdenek Horský ("Copernicus' Writings on the Revolutions of the Celestial Spheres with Marginal notes by Tycho Brahe" [Horský 1971, 12–13; accompanied by a facsim-

ile of *De Revolutionibus*]), Gingerich 1973b, Gingerich 1974, and Westman 1975c, but subsequently corrected in Gingerich and Westman 1988.

4. See Christianson 2000, 67, fig. 12.

5. Quoted and trans. in Christianson 1979, 129.

6. Brahe 1913–29, 7:40, ll. 14–20; trans. and quoted in Mosely 2007, 271.

7. Four of these are discussed in Gingerich and Westman 1988; subsequently, Gingerich identified a fifth copy with only one Wittich annotation (Gingerich 2002, xx).

8. Gingerich 2002, 12.

9. Dudith became acquainted with Rheticus in the late 1560s; Praetorius stayed at Dudith's home from 1569 to 1571 and, like Wittich, tutored his host in elements of the science of the stars (see Dobrzycki and Szczucki 1989, 26).

10. Christianson (2000, 74–77) suggests a number of possible social models for what he wants to characterize as Tycho's *familia*—the ecclesiastical or monastic household, the Italian humanist academy, the Renaissance court, and the professorial household. According to the social historian David Herlihy (1991), the *familia* denoted in antiquity an aggregation of slaves, in the Middle Ages an organized and stable community, and throughout both periods and into the Renaissance a household bound by ties of affection. Christianson's important suggestions point to the need for further, detailed work on the kinds of ties that bound together Tycho's island.

11. For an example of this method, see Gingerich 2002, 12.

12. See Swerdlow 1975.

13. Transcriptions with commentary on the complete set of diagrams, all of which come from the Vatican copy, may be found in Gingerich and Westman 1988, 77–140.

14. Copernicus (Vatican) 1543, bk. 1, chap. 9, fol. 7r.

15. See P. Lipton 1991.

16. Copernicus 1971b, fol. 202v. Wittich cross-referenced this passage on fol. iij r of both the Prague and Vatican copies.

17. Ibid., fol. 10. I read Wittich's use of the word *confirmare* in the dialectical sense of "to fortify" rather than the apodictic "to demonstrate."

18. Copernicus (Vatican) 1543, fol. 9v. For Venus's mean sidereal period, Wittich corrects Copernicus's nine months to eight, and for Mercury's, eighty days to eighty-eight.

19. See Gingerich and Westman 1988, 138–40.

20. See the evocative account in Christianson 2000, 77–79.

21. For Tycho's efforts to obtain Wittich's library, see Gingerich and Westman 1988, 120–23.

22. See again Peter Lipton's helpful discussion (Lipton 1991, 56–74).

23. Tycho to Brucaeus, 1584, Brahe 1913–29, 7:80, ll. 8–16: "Expertus sum in fine anni elapsi 82 et prin-

cipio 83 ex parallaxibus Martis tum temporis achronychij, et ob id Terrae in minori distantia vicini, quam Sol a Terra removetur, idque iuxta COPERNICI placita ad ⅓ quasi partem, unde maiores parallaxes, quam Sol ipse inducere debuisset, cum tamen longe minores fuisse"; Brahe to Landgrave Wilhelm, January 18, 1587, Brahe 1913–29, 6:70, ll. 34–35: "Nam tertia fere parte per hanc Mars in oppositionem Solis terris redditur propior quam ipse Sol."

24. See Gingerich and Voelkel 1998, 3.

25. Copernicus 1971b, fol. 202; see also Rheticus 1982, 107, 186–87 n. 227.

26. Brahe to Landgrave Wilhelm, January 18, 1587, Brahe 1913–29, 6:70, ll. 38–40.

27. Miguel Granada (2006, 137) rightly observed that the question at stake here was "something more than choosing via this crucial experiment between two conflicting cosmological systems."

28. These instruments included the great quadrant, the zodiacal armillary, the bifurcated sextant, the trigonal sextant, the largest steel quadrant, and the famous mural quadrant. For an excellent, detailed account of these early efforts to find Mars's parallax, see Gingerich and Voelkel 1998, 5–9.

29. Ibid., 16; Mosely 2007, 67.

30. Brahe to Landgrave Wilhelm, Brahe 1913–29, 6:70, ll. 29–42.

31. Rothmann 1619, chap. 5; cited and trans. in Rosen 1985, 28–29; Moran 1982.

32. Brahe 1913–29, 6:388; trans. Christianson 1979, 136; discussed in Granada 2006, 135.

33. Christianson 1979, 133; cited in Gingerich and Westman 1988, 73.

34. Brahe to Rothmann, January 20, 1587, Brahe 1913–29, 6:88, ll. 4–15. The complete passage is rightly stressed by Mosely 2007, 70. Cf. Rosen 1985, 27; Gingerich and Westman 1988, 75.

35. Brahe 1913–29, 6:88, ll. 9–12; Gingerich and Westman 1988, 75.

36. Rothmann to Brahe, October 11, 1587, Brahe 1913–29, 6:111, ll. 24–26; Gingerich and Westman 1988, 75.

37. Brahe to Peucer, September 13, 1588, Brahe 1913–29, 7:130, ll. 9–11; Gingerich and Westman, 1988, 74 n.

38. Bellanti 1554, 40–41: "Quaestio Tertia De Natura Partium Coeli: An caelum sit substantiae fluxibilis."

39. Ibid.

40. Mosely 2007, 76, citing Aiton 1981, 99–100; for further discussion, see Westman 1980a, 113, 139 n. 45.

41. Gingerich and Voelkel 1998, 11–16.

42. Bellanti 1554, 41: "Ad secundum dicitur negando soliditatem exigere maiorem quantitatem materiae quam aliquod fluxibile ut dictum est: negatur etiam quod long distantia in maxime diaphanis radios impediat, refrangat, reflectat &c."

43. For a brief but reliable biographical sketch, see Rothmann 2003, 10–14.

44. Assuming that this was Rothmann's first year;

given a typical age of fifteen at initial entry, this would put his birthdate in 1560. If it were known that Rothmann had been in Wittenberg just a year earlier than 1575, it would allow for the possibility that he had personally met Peucer.

45. Straub, Erlangen MS., fol. 2r; Schadt, Erlangen MS., fol. 72r.

46. Evidently a reissue of [Peucer] 1568.

47. Gingerich 2002, Schweinfurt 1543, 91: "Andr. Osiandri (ut aiunt)."

48. Gingerich 2002, Yale University, Beinecke Library 1543, 308.

49. For further discussion, see Schofield 1981, 27–34; Barker 2004; Granada 1996b, 61–66; Granada 2007a.

50. Rothmann to Brahe, September 21, 1587, Brahe 1913–29, 6:116, ll. 17–18.

51. Landgrave Wilhelm to Brahe, October 20, 1585, ibid., 6:31–32.

52. See Mosely 2007, 257–65.

53. Rothmann to Brahe, October 13, 1588, Brahe 1913–29, 6:157, ll. 8–16, 158, ll. 21–26; Mosely 2007, 282–83. Bürgi was always described in the correspondence as self-educated, perhaps to stress his lack of a university education and his status as a craftsman.

54. Barker and Goldstein 1995, 390–91; Granada 2002a, 115–36; Granada 2004b; Mosely 2007, 74–75; Thorndike 1923–58, 6:19–20, 71–72, 83–84.

55. Rothmann 1619, 104–5; Mosely 2007, 74 n.: "Si refractio ista esset ab orbibus coelestibus, non tantum usque ad 15 aut 20 ab horizonte gradus, verum (quemadmodum Alhazen et Vitellio in dictis locis demonstrare conantur) usque ad verticem duraret, adeoque omnium observationum certitudo turbaretur necesse esset." He reiterated the point on November 14, 1587 (Rothmann to Brahe, Brahe 1913–29, 6:121): "Ita vides, hoc unico argumento, quod nimirum Refractiones non durent usque ad verticem, firmissime demonstrari, non esse diuersa Aetheris & Aëris Diaphana. Nec enim ipsi Optici negare possunt, quin praesupposito diuerso Aetheris & Aëris Diaphano necessarium sit, ut Refractio duret usque ad verticem, ut ex Alhazeno lib 7 & Vitellione lib. 10 P. 51 manifestum est." In 1604, Kepler commented extensively on the Rothmann-Tycho skirmish (Kepler 1937–, 2:78–80; Kepler 2000, 93–96).

56. Indeed, Tycho claimed that it *was* his own and that Ursus had stolen it from his study. The Ursus episode has now received extensive treatment by several scholars, and, largely for reasons of limited space, I have nothing much to say about it in this book. See Jardine 1984; Rosen 1986; Gingerich and Westman 1988, 50–69; Granada 1996b, 77–107; Mosely 2007, 78, 177, 185; Jardine and Segonds, 2008.

57. Bruce Moran felicitously characterized Wilhelm as a "prince-practitioner" (Moran 1981, 1982).

58. Important new evidence cited by Granada 1996b, 119–20, from correspondence in the Württember-

gische Landesbibliothek, Cod. Math. 4° 14b, fol. 19r: "Librum [*De mundi recentioribus phaenomenis*] et litteras a Tabellario bene accepi. Et habeo imprimis gratias pro duro iudicio de libro Raymari. Et quantum ad Systema Mundi attinet, iudico Raymarum sua habere a Tychone (cuius discipulus fuit) et terram ille mobilem statuit, ne videatur cum Tychone consentire."

59. Adam Mosely (2007, 298–306) has compiled a valuable list of "known and presumed owners of Tycho's work prior to 1602."

60. Rothmann to Brahe, October 13, 1588, Brahe 1913–29, 6:150.

61. Ibid., 6:151, ll. 16–21.

62. Ibid., 6:156, ll. 29–30; 6:152, ll. 41–42. At the horizon, refraction suddenly decreases ("Quod autem circa Horizontem tam subito decrescunt Refractiones, id a meris vaporibus est").

63. Ibid., 6:149, ll. 35–39.

64. Ibid., ll. 16–28.

65. Brahe to Rothmann, November 24, 1589, ibid., 6:196, ll. 11–14.

66. Brahe to Rothmann, February 21, 1589, ibid., 6:168, ll. 14 ff.

67. In medicine, Rothmann liked Tycho's Paracelsian notion that the spirit is the link between the body and the soul (Rothmann to Brahe, October 13, 1588, ibid., 6:154, ll. 12–35).

68. Brahe to Rothmann, 24 November 1589, ibid., 6:187, ll. 11–19.

69. Ibid., 6:195, l. 39–196, l. 12. These usages appear interchangeable and occur in close proximity to one another.

70. Segonds 1993; Hannaway (1986, 63), suggests that Tycho's aims in his *laboratorium* were essentially contemplative; but this view overlooks the activist strand of astronomy that was ultimately intended to improve the foundations of practical astrology, alchemy, and medicine.

71. I discuss this matter more generally in chapter 17.

72. Rothmann to Brahe, April 18, 1590, Brahe 1913–29, 6: 214, ll. 13–18: "Profundissimae & subtilissimae Mathematicae Disputationes."

73. Rothmann to Brahe, October 13, 1588, ibid., 6:149. At various places in the letter, Rothmann carefully distinguished the landgrave's own opinions from his own (pp. 155, 157, 158, 161).

74. Rothmann to Brahe, August 22, 1589, ibid., 6:182, l. 40–183, l. 2: "Iudicabunt postea Doctissimi Mathematici, cuius sententia vera sit. Nec enim nos ipsi in hac materia & Actoris & Rei & Iudicis personam sustinere possumus, sed requiritur ad sententiam pronunciandam persona tertia, *philalethes* & nullo prorsus praeiudicio fascinata." He also appealed to the aphorism "Truth is the daughter of time."

75. Brahe to Rothmann, November 24, 1589, ibid., 6:193, l. 24–194, l. 3.

76. Ibid., 6:187, ll. 5–9.

77. Ibid., 6:191, ll. 13–30. At the time that Tycho wrote this letter, Rothmann had recently told him that he knew, through his brother Johannes, "that you [Tycho] have sent our disputations to Master Peucer and that Master Peucer mentioned that he did not wish to express public favor for either me or you" (Rothmann to Brahe, August 6, 1589, ibid., 6:201, ll. 39–202).

78. For a careful exploration of Tycho's editorial probity in the *Epistolae Astronomicae*, see Jardine, Mosely, and Tybjerg 2003, 421–51; and on the question of Tycho's astronomical letters in the context of Renaissance epistolary culture, see Mosely 2007, 31–115.

79. Rothmann to Brahe, 13 October 1588, Brahe 1913–29, 6:157, ll. 5–9. If Wittich had shown Rothmann his copies of *De Revolutionibus*, it is likely that he would have pointed out the annotations that had been copied from Reinhold.

80. Ibid., 6:157, ll. 11–16.

81. This sort of materialist thinking would later show up in Kepler's insistence that a point cannot cause a physical effect.

82. Rothmann to Brahe, October 13, 1588, Brahe 1913–29, 6:158, ll. 18–28.

83. Ibid., 6:159, ll. 1–3: "Aliud inuenire non possum, quam nullam praeter unicam Copernici Hypothesin veram esse."

84. Ibid., 6:158, ll. 29–38. It is interesting that Rothmann does not attribute to Copernicus himself a belief in solid spheres.

85. Ibid., 6:157, ll. 23–26. This was, of course, the upshot of Wittich's transformation diagrams.

86. Ibid., 6:160, ll. 13–22. Even Maestlin, Rothmann averred, had not properly understood Copernicus's account of the library motion of the ecliptic's obliquity.

87. Ibid., 6:159, ll. 14–18.

88. Rothmann could not have seen Rheticus's as yet unpublished treatise. Had he heard of its argument through the oral tradition enabled by Praetorius?

89. On Rothmann's exegetical practices, see Howell 2002, 93–94, 100–101.

90. Rothmann to Brahe, October 13, 1588, Brahe 1913–29, 6:159, l. 41–160, l. 1: "Paulus quoque cum Roman:1. ait, Deum ex visibilibus hisce agnosci, non obscure arguit, longe maiorem sapientiam Dei latere in Natura, quam in sacris literis sit reuelata."

91. Ibid., 6:160, ll. 26–29.

92. Brahe to Rothmann, February 21, 1589, ibid., 6:176, ll. 39–40: "Has nostras Hypotheses Apparentijs caelestibus ad amussim satisfacere, & tam Ptolemaicas, quam Copernianas longe antecellere, ipsique veritate magis correspondere."

93. Ibid., 6:176, l. 41–177, l. 9.

94. Ibid., 6:178, ll. 1–4: "Verum cum animaduertissem subtili & accurata Obseruatione, praesertim Anno 82 habita, Martem Acronychum Terris propriorem fieri ipso Sole, & ob id Ptolemaicas diu receptas Hypotheses constare non posse."

95. Ibid., 6:178, l. 40–179, l. 4.

96. See Granada 1996a. Granada 2002a, 106–7, suggests that the Lutheran Rothmann's exegesis followed that of Calvin; see Howell 2002, 92–106.

97. Brahe to Rothmann, November 24, 1589, Brahe 1913–29, 6:187, ll. 5–9. Peucer's letter is not extant, and Brahe did not mention to Rothmann which passages Peucer had glossed; but he was obviously extremely pleased by Peucer's endorsement (Brahe to Peucer, September 13, 1588, ibid., 7:133, ll. 23–26); see also Howell 2002, 101.

98. Brahe to Rothmann, November 24, 1589, Brahe 1913–29, 6:186.

99. Ibid., 6:197, l. 7–198, l. 5. These arguments were developments of positions already adumbrated in *De Mundi Recentioribus Phaenomenis*.

100. On Rothmann's illness, see Barker 2004.

101. Rothman to Brahe, Brahe 1913–29, 6:215, ll. 36–38.

102. Ibid., 6:215–16; see also Granada 2007a, 103–5.

103. Ibid., 6:222, ll. 36–47.

104. Ibid., 6:221, ll. 24–30: "Sicque Terra, tanquam patiens & quiescens, Caeli agentis & reuoluti uires, ac influxus ad Centrum tendentes commodius recipit, atque altera Mundi pars, utut minima, non immerito simul existit: cum tot tantaque praeter animantia ipsi coelo analoga contineat. Ideoque scriptum est, Creauit Deus Coelum & Terram, ubi Terra altera, & Coelo quasi conferenda Mundi pars censetur, & praedicatur: Nec instar minimi cuiusdam, imo obscuri, & abjecti Astri (prout fert Hypotyposis Copernicea) abijcitur aut negligitur."

105. Ibid., 6:223, ll. 4–8.

106. Bruno 2000. Other possible translations of *fastidito*, following the entry for *fastidioso* in the dictionary of Bruno's close friend John Florio (Florio 1611), include someone who is "Yrkesome," "Wearysome," or "Lothsome to the Minde."

107. Court figures whom Bruno would have encountered through the embassy include Sir Philip Sidney; Robert Dudley, earl of Leicester; some members of the Catholic party; the Howards; the earl of Oxford, and perhaps through one of these contacts, the printer Charlewood (see Providera 2002, 174).

108. Canone and Spruit 2007.

109. For example, while in Toulouse between 1579 and 1581, Bruno is known to have lectured for at least six months on Sacrobosco's *Sphere* (Canone 2000, cxxxvi); it would be surprising if Bruno had not already been acquainted with Clavius's *Sphere* in either or both the 1570 and the 1581 editions (ibid., 62–63).

110. Canone and Spruit 2007, 376.

111. See Snyder 1989, 96–102.

112. Miguel Angel Granada, "Introduction," in Bruno 1995, xxi–xxx.

113. Kořán 1969; Horský 1975, 65; Westman 1980b, 97; Sturlese 1985.

114. Granada (1990, 358–59) suggests plausibly that Bruno possessed the second edition of *De Revolutionibus;* if so, then he would have had available to him Rheticus's *Narratio Prima*. A second edition of *De Revolutionibus* found by Owen Gingerich in the Biblioteca Casanatense has the provenance "Brunus Fr[ater] D[ominicanus]" (Gingerich 2002, 115). It is uncertain whether the provenance is in Bruno's hand, although this uncertainty would not exclude his having owned this copy.

115. In *De l'infinito,* Bruno invokes the relation between period and distance to explain that as planets move in circles with greater radii, they move more slowly but are still able to receive some of the sun's vital heat (Bruno 1995, 189; Singer 1950, 305). However, Bruno neither associates this position with Copernicus nor uses it to criticize Ptolemy.

116. See, for example, Bruno 1586, 1609; Michel 1973, 180.

117. Bruno 1977, dialogue 4, 183–84.

118. Ciliberto 1979.

119. Bruno 1977, dialogue 1, 86–87.

120. Ibid., dialogue 4, 190.

121. Ibid., 192 "Secondo il senso del Copernico" is translated as "according to Copernicus's meaning" rather than "according to Copernicus's theory."

122. Ibid., 193.

123. Possibly a reference to Copernicus that Bruno found in one of the editions of Pontus de Tyard's *L'univers,* "in which the Earth and all the elementary region, with the orb of the Moon, are contained, as if by an epicycle" (1557, 99); Yates 1947, 102–3; well discussed in McMullin 1987, 57–58.

124. For discussion of these issues, see McMullin 1987, 68–74.

125. See Gatti 1999, 65 ff.

126. Tessicini 2001; Tessicini 2007, 15–58.

127. The matter is carefully discussed by Sturlese (1985, 324–25), who, against Kořán and Horský's proposal of Hagecius as a candidate for this intermediary, points to unreliable communication between Uraniborg and Prague and the existence of only one letter from Hagecius in 1588. Perhaps the intermediary was the same person who made available to Bruno the first Uraniborg publication, the *Diarium* of Elias Olsen Morsing.

128. For example, through court connections, might Digges, Dee, or William Gilbert have encountered Bruno's London dialogues? (For a highly suggestive reconstruction of a "Gilbert Circle," see Gatti 1999, 86–98.)

129. The *Oratio Valedictoria* appeared in March 1588 (Bruno 1588a, 1–52); Bruno 1588b, 55–190. Singer (1950, 140) proposes as a translation "The Abruptly Ended Discourse"; for further discussion, see Granada 1996b, 15–30.

130. Granada 1996b, 15–17; Sturlese 1985, 325–29.

131. Tessicini 2007, 159–69.

132. Bruno 1962, vol. 1, part 1, bk. 1, chap. 5, 221; for discussion, see Tessicini 2007, 160.

133. Bruno 1962, vol. 1, part 1, bk. 1, chap. 5, 219: "Ista fuere mihi physica ratione reperta / Pluribus abhinc lustris, sensu interiore probata, / Sed tandem et docti accipio firmata Tichonis / Servatis Dani, ingenio qui multa sagaci / Invenit, atque aperit conformia sensibus hisce." See Tessicini 2007, 159.

134. Ibid., vol. 1, part 2, bk. 4, chap. 9, 53. See Tessicini 2007, 163.

135. For the kind of experience typically deployed, see Dear 1995.

136. Cited in Spruit 2002, 244.

137. Ibid., 245–46. Morsing's *Diarium* contained a section on the astrological meaning of the comet of 1585, but Bruno simply ignored it (Brahe 1913–29, 6:408–14).

138. See Spruit 2002, 247–49. Bruno 1995, 42–43: "Make then your forecasts, Mr. Astrologers, with your slavish physicians, with the help of those circles with which you describe those nine, moving, imaginary spheres and by means of which you imprison your brains, so that as you appear to me to be like parrots in a cage as I watch you jumping up and down, twirling around and hopping within those circles."

11. THE EMERGENCE OF KEPLER'S COPERNICAN REPRESENTATION

1. On Zuñiga, see the important article by Navarro-Brotóns 1995; Westman 1986, 92–93.

2. See Kepler 1981, chap. 14, 155–56; Lerner 1996–97, 2: 68–69; Van Helden 1985, 46.

3. See for example Kuhn 1957; Copernicus 1952, 527, 529; Price 1959, 202; Crowe 1990, 90–99; Hanson 1973; Jacobsen 1999.

4. For Liddel, see Westman 1975c, 320. Liddel's notes derive from associations with Wittich (see Gingerich 2002, Aberdeen 4, 266–67).

5. See Schofield 1981; Jardine 1984; Granada 1996b; Mosely 2007.

6. Richard Jarrell (1981, 15) has written: "Of all the Copernicans in Europe near the end of the sixteenth century, Maestlin was virtually the only one holding a teaching post in a university." Jarrell's important observation is evidently a general impression, as he offers no specific evidence for the claim.

7. See Methuen 1996.

8. Navarro-Brotóns 1995.

9. See Bunzl 2004.

10. See Wallace 1984b, 257–60.

11. Benedetto Castelli (1578–1643), generally taken to be an endorser of the Copernican ordering, is believed to have studied with Galileo in Padua between 1604 and 1606 (Drake 1978, 121). I have encountered no explanation for Castelli's reasoning in supporting the Copernican order other than the fact of his association with Galileo.

12. Isabelle Pantin (Galilei 1992, 54 n.) shows that in calculating the horoscope of Cosimo II, Galileo's values best fit those of Magini's *Ephemerides* (Venice, 1582). For Galileo's astrology, see Ernst 1984; Kollerstrom 2001; Swerdlow 2004b; Rutkin 2005.

13. Zinner 1943, 424.

14. Gingerich 2002, Schweinfurt, 91–93.

15. Tycho Brahe to Thaddeus Hagecius, November 1, 1589, Brahe 1913–29, 7:206–7.

16. For further details, see Westman 1975c, 290–96.

17. See Universitätsbibliothek Erlangen-Nürnberg: Praetorius 1594, fol. 94v.

18. A dedication copy of Tycho's *De Mundi Aetherei Recentioribus Phaenomenis* to Praetorius is extant in the Wrocław University Library (see Norlind 1970, 124).

19. Universitätsbibliothek Erlangen-Nürnberg: Praetorius 1594, fol. 98v; Westman 1975c, 299–301.

20. Universitätsbibliothek Erlangen-Nürnberg: Praetorius 1594, fols. 97r, 99v; Westman 1975c, 301.

21. Methuen represents Maestlin as a skeptic, teaching Copernicus in the classroom but without committing himself (Methuen 1996).

22. Kepler 1984, 21; Kepler 1981, 63, my italics.

23. Kepler makes first reference to a personal copy of *De Revolutionibus* after having left Tübingen: "Nam exemplar meum libro 5. Revol: Cap:4. quo loco nodus quaestionis haeret, aut mendosum est, aut ego caecus" (Kepler to Maestlin, October 3, 1595, Kepler 1937–, 13: no. 23, p. 45).

24. Kepler reports the existence of these two disputations in his "Preface to the Reader" in Kepler 1981, 63.

25. Again, we cannot be certain that Maestlin would have drawn such diagrams in his classroom lectures. Tredwell (2004) rightly underscores the importance of Maestlin's edition of the *Narratio Prima* as preparing readers to understand Kepler's *Mysterium Cosmographicum*.

26. Koyré 1992, 129; Toulmin 1975.

27. Kepler 1984, chap. 1, 31, my italics.

28. Koyré 1992, 130.

29. Ibid., 129; Kuhn 1957, 39–41, 75–77. See also Lakatos and Zahar 1975; Toulmin 1975; Thomason 2000.

30. Kepler ignored or did not recognize that the comet was underdeterminative: it could be viewed as evidence in favor of either Tycho's or Copernicus's overall arrangements, as it appeared to confirm the view they shared of the ordering of Mercury and Venus (see Westman 1972a, 26–30; Westman 1975d).

31. See this volume, chap. 9, n. 39

32. "Preface to the Reader," Kepler 1981, 63.

33. "Preface to the Reader," Kepler 1984, 21.

34. Kepler 1981, note to 1621 ed., 51:

There exist in Germany cosmographies by Münster and others, in which indeed the beginning is about the whole universe and the heavenly regions, but they are finished off in a few pages. The main bulk of this book, however, comprises descriptions of territories and cities. Thus, the word cosmography is commonly used to mean geography; and that title, though it is drawn from *universe*, has induced bookshops and those who compose catalogues of books, to include my little book under geography. Nevertheless, I have taken the mystery as a secret [*pro Arcano*], and marketed this discovery as such: and indeed I had never read anything of the sort in any philosopher's book.

35. Because of these contemporary resonances, I prefer to retain the term *cosmographic* from the Latin title *Mysterium Cosmographicum* rather than to introduce the word *universe*, for which Latin equivalents might be *mundanum* or *caelum*.

36. See Methuen 1999.

37. For Liebler, see Methuen 1998, 193–97, 203, 221–22.

38. Alain Segonds points out that Maestlin defined efficient and final causes as foreign to astronomy: "Efficientis et finalis causae tanquam ab Astronomia alienae nulla fit mentio" (Kepler 1984, 232 n.). Yet of course Maestlin followed the generally accepted notion that astronomy embraced both mathematical and physical parts.

39. Burtt 1932, 58–59. For a cogent critique of Burtt's general approach, see Hatfield 1990, 93–166; for Burtt's place in the historiography of the scientific revolution, see Cohen 1994, 88–97.

40. Kuhn 1970, 152–53.

41. Burtt's "diffused throughout an orbit" does not stay close enough to the Latin, where the word *orbit* does not appear: "Cum igitur primum motorem non deceat orbiculariter esse diffusam" (Kepler 1937–, 20: pt. 1, p. 148, ll. 37–38). The adverb *orbiculariter* (orbicularly) can be rendered as "in a circular or spherical form." The word does not appear in *De Revolutionibus*, but Pico della Mirandola uses *orbicularis* (1946–52, 1:194, bk. 3, chap. 4).).

42. Burtt 1932, 59; Kepler 1937–, 20: pt. 1, 148, ll. 19–32.

43. In the *Epitome of Copernican Astronomy*, bk. 4, chap. 2 (Kepler 1937–, 7:263, ll. 3–7; Kepler 1939, 859), Kepler clearly spelled out this matter: "When we ask in what place in the world the Sun is situated, Copernicus, as being skilled in the knowledge of the heavens, shows us that the Sun is in the midpart. The others who exhibit its place as elsewhere are not forced to do this by astronomical arguments but by certain others of a metaphysical character drawn from the consideration of the Earth and its place." See also Westman 1977, 15–18.

44. In his disputation, however, Kepler wrote as though Copernicus already attributed a motive power to the Sun: "Tantis igitur mactum honoribus, tantis onustum Solem muneribus putat Copernicus se obtinere posse, ut in medium mundi collocet primum, *ut motor ipse*, sicut per se immobilis necessariô, ita

etiam in immobili domicilio haereat" (Kepler 1937–, 20: pt. 1, p. 148, ll. 43–45, my italics).

45. Copernicus 1978, 22; Granada (2004a) has recently argued for the presence of an "incipient solar dynamics" in Copernicus and Rheticus.

46. This is one of a handful of instances where Kepler informs us of when and where he first studied a particular book: "After I came to the study of philosophy, in my eighteenth year, the year of Christ 1589, the *Exercitationes exotericae* of Julius C. Scaliger were passing through the hands of the younger generation" (Kepler 1981, 1621 ed., 51 n. 1).

47. Kepler 1937–, 7:294; Kepler 1939, 891.

48. See Wolfson 1962.

49. Brahe 1588; see fig. 61.

50. Kepler 1939, 891–92; Kepler 1937–, 7:294.

51. Kepler's later note adds credibility to this interpretation: "For once I believed that the cause which moves the planets was precisely a soul [as I was of course imbued with the doctrines of J. C. Scaliger on moving intelligences]. But when I pondered that this moving cause grows weaker with distance, and that the Sun's light also grows thinner with distance from the Sun, from that I concluded, that this force is something corporeal, that is, a species which a body emits, but an immaterial one" (Kepler 1981, 1621 ed., 203 n. 3; bracketed portion omitted by Valcke 1996, 293).

52. Two years later, Kepler remarked to Maestlin that he already held this idea as an "axiom" at Tübingen (Kepler to Maestlin, October 3, 1595, Kepler 1937–, 13: no. 23, p. 35).

53. Kepler 1937–, 20: pt. 1, p. 149.

54. As Stephenson (1987, 26) aptly notes: "The solid-sphere models had long coexisted with mathematical astronomy. They were compatible with the geometrical models for the very direct reason that they too were geometrical models."

55. Methuen 1998, 203–4.

56. Lindberg 1986.

57. I tried to determine these sources in my doctoral dissertation (Westman 1971; see also Westman 1972b).

58. Valcke 1996; for the Keplerian passages, Valcke draws liberally from Simon 1979. See also Rabin 1987.

59. Pico della Mirandola 1946–52, 2:236: bk. 3, chap. 9.

60. Ibid., 2:242, 244: bk.3, chap. 10.

61. Ibid., 2:196: bk. 4, chap. 4; Valcke 1996, 291–92.

62. Pico's source in this passage is, I believe, Ficino rather than, as Valcke argues, Aristotle (1996, 291 n.).

63. Kepler 1937–, 2:34–36, prop. 32; Kepler 2000, 39–41; Simon 1979, 197–98; Valcke 1996, 292 n. 50.

64. Kepler 1937–, 3:240; Kepler 1992, 379; cf. Kepler 1937–, 4:168, ll. 15–19, thesis 22; Rabin 1987, 152 n. 39.

65. Rabin believes that the 1599 reference was also the first, "so it is possible that Kepler had not even read Pico's treatise when he began revising his own ideas about astrology, although he may have heard of it" (1997, 762).

66. Kepler 1858–71, 7: 753: "J. Picus Mirandulae comes, Italus, ante 100 annos scripsit contra astrologos, cumque quodam operis sui libro demonstraturus esset, falsum esse, quod astrologi dicerent, ad mutationem trigonorum coelestium mutari imperia et posse ex doctrina astrologica corrigi vitiosam rationem temporum, si sc. memorabilia eventa ad memorabiles constellationes accommodentur, hoc inquam refutaturus ille seriem aetatis mundanae ex suo ingenio constituit."

67. Kepler 1997, 384; Kepler 1937–, 6:285.

68. Quoted and trans. in Rabin 1997, 754; Kepler 1937–, 4:161. I have slightly emended Rabin's translation.

69. Rosen 1984a, 253–56.

70. Kepler to Roeslin, quoted and trans. in ibid., 255; Kepler 1937–, 14:328, ll. 426–28.

71. Roeslin to Kepler, October 17, 1592, quoted and trans. in ibid., 255; Kepler 1937–, 19:320–21.

72. Kepler's preoccupation was sufficiently intense that he began to build up an extensive horoscope collection (as will be documented in a final volume of Kepler 1937–).

73. In the extant correspondence, Kepler's first reference to Roeslin was made, in passing, in a letter to Maestlin dated October 3, 1595 (Kepler 1937–, 13: no. 23, p. 39, ll. 237–38).

74. In the Narratio Prima, Rheticus maintained that "Pico would have had no opportunity, in his eighth and ninth books, of impugning not merely astrology but also astronomy" if he had known Copernicus's teachings. To this passage Maestlin affixed a simple reader's postil showing that he knew the full name intended: "Picus Mirandola" (Kepler 1937–, 1:94).

75. Kepler 1984, 78.

76. Curiously, among earlier sixteenth-century writers, this objection of Pico and Savonarola had but a limited following.

77. See Paulsen 1906, 24–25. On the development of the institution of the disputation, see W. Clark 1989, 115, 145; W. Clark 2006, 68–92.

78. Liebler 1589. Liebler mentioned J. C. Scaliger's work in a postil to the 1589 edition: "Scaligerus exercitat 23. Ad Cardanum." Liebler presented a copy of this edition to the Tübingen philosophical faculty on April 9, 1594 (Universitätsbibliothek Tübingen, shelf no. Aa 834).

79. Methuen makes an important contribution by correcting Max Caspar's claim that Veit Müller was the principal philosophy professor during Kepler's time in Tübingen (cf. Caspar 1993, 44; Methuen 1998, 193–97, 203, 221–22, 226).

80. Liebler, 1589, 234–36; the full Latin passage, under the question "Of what substance are the stars [made]?" is given by Methuen 1998, 195 n. 101.

81. Koyré 1992, 379.

82. Stadius followed Hieronymus Lauterbach (1561–77), who had studied at Vienna, where he succeeded Paul Fabricius to the mathematics professorship in 1558 (see Sutter 1975, 257–75; see also Boner 2009).

83. For Kepler's extant calendar prognostications, see Kepler 1937–, 11: pt. 2; Kremer 2009.

84. Kepler 1937–, 11: pt. 2, pp. 14, 16, 498; on Caesius, see Sutter 1975, 255.

85. Sutter 1975, 293; Kremer 2009.

86. Kepler 1937–, 11: pt. 2, p. 9, ll. 25–29; Sutter discovered two copies of this hitherto unknown practica in Ljubljana (Sutter 1964, 254, no. 655; Sutter 1975, 292).

87. Kepler 1937–, 7:301, l. 24; see Stephenson 1987, 142.

88. Kepler 1937–, 11: pt. 2, p. 9, ll. 34–41.

89. Ibid., 48; cf. Field 1984a, 251–54, theses 34–45.

90. See Alain Segonds's excellent commentary (Kepler 1984, 231–33).

91. For example, in his Trattato della sfera, Galileo says that "il soggetto della cosmografia essere il mondo, o vogliamo dire l'universo, . . . è la speculazione intorno al numero e distribuzione delle parti d'esso mondo, intorno alla figura, grandezza e distanza d'esse e, più che nel resto, intorno ai moti loro; lasciando la considerazione della sostanza et delle qualità delle medesime parti al filosofo naturale" (quoted in ibid., 231).

92. See further Barker and Goldstein 2001.

93. See this volume, chap. 7. "Here we are concerned with the book of Nature, so greatly celebrated in sacred writings. It is in this that Paul proposes to the Gentiles that they should contemplate God like the Sun in water or in a mirror" ("Original Dedication," Kepler 1981, 53).

94. Maestlin to Kepler, March 9, 1597, Kepler 1937–, 13: no. 63, pp. 108–12; "Non aspernor hanc de anima et virtute motrice speculationem. Verum metuo ne nimis subtilis sit, si nimium extendatur."

95. "In the year 1595, on the 9/19th of July . . . I discovered this secret; and turning at once to the study of it, in the following October [1595], in the dedication of the prognostication of that year, which I had to compose as part of my office, I promised to publish a small work to announce publicly how tedious it was for a lover of philosophy like myself to make these conjectures [about the future]" (Kepler 1984, 17).

96. Ibid., 12, my italics.

97. Ibid., 13; Kepler 1981, 55.

98. Kepler 1981, 57. The allusion had also an immediate local referent, as Segonds points out: the Graz Stiftsschule where Kepler taught mathematics to the sons of the Protestant nobility (Kepler 1984, 26 n.).

99. Field 1984a, thesis 1, 232.

100. Ibid., theses 2–3, 232.

101. Ibid., thesis 15.

102. Kepler 1937–, 1: chap. 20. For good accounts, see Koyré 1992; Voelkel 2001, 52–59.

103. Kepler 1937–, 1: chap. 16; Kepler 1981, 165; Kepler 1965, 29–31.

104. Kepler 1937–, 1: chap. 16, p. 56; Westman 1971, 118–19.

105. Ibid., ch. 12; see further this volume, chap. 14.

106. "All my books are Copernican" (Kepler to Johannes Quietanus Remus, August 4, 1619, Kepler 1937–, 17: no. 846, p. 364).

107. "Original Preface to Reader," Kepler 1937–, vol. 1; Kepler 1984, 22.

108. Rhonda Martens (2000, 70–71) makes the interesting suggestion that Kepler had in mind a convergence of different kinds of arguments from different disciplines rather than a unification of principles common to all.

109. Kepler 1984, 4, 236–37.

110. See also Kepler 1937–, 7:261–62; on the *prisca* tradition, see D. P. Walker 1958.

111. Kepler 1997, 301. The passage is a postil to a long quotation from Proclus's *Commentary on the First Book of Euclid's Elements*. Judith V. Field has emphasized this theme (1988, 50–51).

112. See Williams 1948.

113. Kepler 1984, chap. 2, 48–54; see also Segonds's important note, ibid., 275 n. 25.

114. On Kepler's humanist scholarship, see Grafton 1991b.

115. Ibid., 195–97.

116. Kepler 1981, chap. 23, p. 223.

117. Lerner (1996–97, 2:70–73) rightly underscores this point.

118. Kepler 1981, "Original Preface," 65: "For in discussing the foundation of the universe itself, one ought not to draw explanations from those numbers which have acquired some special significance from things which follow after the creation of the universe."

119. I follow Segonds's reading here, which is based on the *Lexicon* of Rudolf Goclenius: "Corpus, quod est quantitas, est tres dimensiones. Itaque non potest intelligi et definiri sine his" (Kepler 1984, 272 n. 1).

120. For recent accounts, see Koyré 1992, 140–55; Field 1988, 35–60; Stephenson 1994, 75–89; Martens 2000, 39–56 ; Voelkel 2001, 32–41; Barker and Goldstein 2001, 99–103.

121. Maestlin to Kepler, 27 February 1596, Kepler 1937–, 13: no. 29, pp. 1–10. See Voelkel 2001, 67–69. Barker and Goldstein go so far as to claim that "rather than an exercise in astronomy or a defense of Copernicanism as a novel cosmology, Kepler's first book must be read as essentially theological" (2001, 99).

122. See this volume, chap. 9; L. Digges 1571.

123. See Field 1984b, 273–96.

124. See Alain Segonds's note (Kepler 1984, 296–97 n. 13); Westman 1972b.

125. Kepler mentions Offusius, in passing, for the first time in a letter to Christopher Heydon (October 1605, Kepler 1937–, 15: no. 357, p. 234). I have been unable to locate Kepler's copy of this exceedingly rare book.

126. For Offusius's astrology, see Bowden 1974, 78–107; Stephenson 1994, 47–63; Sanders 1990, 204–49.

127. Kepler to Maestlin, August, 2, 1595, Kepler 1937–, 13: no. 21.

128. Around 1616, some twenty years after the appearance of the *Mysterium Cosmographicum*, Maestlin tried unsuccessfully to produce a new, annotated edition of *De Revolutionibus* through the Basel publisher Heinrich Petri (Kepler 1858–91, 56–58).

129. "Ego verò studeo, ut haec ad Dej gloriam, qui vult ex libro Naturae agnoscj, quam maturrimè vulgentur: quo plus alij inde extruxerint, hoc magis gaudebo: nullj invidebo. Sic vovj Deo, sic stat sententia. Theologus esse volebam: diu angebar: Deus ecce meâ operâ etiam in astronomiâ celebratur" (Kepler to Maestlin, October 3, 1595, Kepler 1937–, 13: no. 23, p. 40).

130. Kepler 1981, chap. 1, 85 n. 1.

131. The question of what Maestlin felt able to teach in the classroom has been the subject of some dispute. Did he actually defend Copernicus's main propositions in his regular lectures, or only in special private classes or separate tutorials reserved for superior students like Kepler? (See Rosen 1975a; Methuen 1996.)

132. Kepler 1984, xvi–xviii.

133. In a letter to Kepler of April 12, 1598, Hafenreffer recalled having recommended "not only in my own name but also in the name of my colleagues, the omission from your treatise of the chapter (I think it was number five) which dealt with this agreement [between Copernicus and the Bible], lest [theological] disputes arise therefrom" (Kepler 1937–, 13: no. 93, p. 203). Later Kepler's arguments were published in the *Astronomia Nova*. In the *Mysterium Cosmographicum*, he wrote: "Although it is proper to consider right from the start of this dissertation on Nature whether anything contrary to Holy Scripture is being said, nevertheless I judge that it is (1) premature to enter into a dispute on that point now, before I am criticized. I promise generally that I shall say nothing which would be an affront to Holy Scripture and that if Copernicus is convicted of anything along with me I shall dismiss him [Copernicus] as worthless. That has always been my intention, since I first made the acquaintance of Copernicus's *On the Revolutions*" (trans. in Kepler 1981, 75).

134. Kepler to Maestlin, October 3, 1595, Kepler 1937–, 13: no. 23, pp. 34, 45 ff.

135. Kepler 1992, chap. 6, 155–80.

136. Kepler could have become acquainted with

Clavius's *Sphaera* through Christopher Grienberger at Graz (see Caspar 1993, 80–81).

137. Kepler 1984, 253–54.

138. Kepler thought highly of Reinhold's *Commentary on Peurbach* (1553 ed.), and according to Edward Rosen, it is likely that Kepler used that work to prepare his first classes at Graz in 1594 (Rosen 1967, 33).

139. Kepler's son Ludwig published Kepler's *Dream* in 1634, an influential little work in which Kepler imagined an inhabited world on the Moon (see Rosen 1967).

140. Kepler later realized the omission: see Westman 1972a, 26–29.

141. Dreyer 1953, 373.

142. "Among the Germans, there are the *Cosmographies* of Münster and others, where one begins by treating the whole universe and the parts of the heavens; but these subjects are examined in a few, brief pages, the principal mass of the book being constituted by the description of regions and towns. This is because the common use of the word 'cosmography' is in the sense of geography. And this word, since it is derived from 'cosmos', is imposed by booksellers and those who make book catalogues so that they place my book among works of geography" ("Note to the Title," Kepler 1984, 11); "My name suffered a hard fate because the printers miscopied it as 'Repleus' rather than 'Keplerus'" ("Author's Note to the Old Dedication," ibid., 17).

12. KEPLER'S EARLY AUDIENCES, 1596–1600

1. In the notes to the 1621 edition, Kepler was more critical of his own astrological arguments than of any other part of the *Mysterium* (see Kepler 1981, 125 n. 1).

2. In relation to English natural philosophy, see esp. Shapin 1994, 65–125.

3. Bellarmine 1586.

4. See, for example, Milward 1978b, 177–86.

5. As an example of one of the replies to Bellarmine, see Willet 1593. For an excellent survey of these clashes in England, see Milward 1978a.

6. "We cannot be blamed for the other material that was added, especially the foreword by Maestlin [to Rheticus's *Narratio*] since none of these later additions were seen by us before they were sent to the printer" (Hafenreffer to Kepler, April 12, 1598, Kepler 1937–, 13:203; quoted and trans. in Rosen 1975a, 327).

7. Michael Maestlin, "Preface to the Reader," in Rheticus 1596, 83. Nevertheless, even in later editions of his *Epitome*, Maestlin continued to maintain that the pursuit of efficient and final causes was alien to astronomy (Maestlin 1624, 30: "Efficientis autem et finalis causa tanquam ad Astronomia alienae, nulla fit mentio").

8. Maestlin to Kepler, October 30, 1597, Kepler 1937–, 13:151; quoted and trans. in Rosen 1975a, 326.

9. Kepler to Duke Frederick, February 29, 1596, Kepler 1937–, 13: no. 30, p. 66.

10. Maestlin to Duke Frederick, March 12, 1596, Kepler 1937–, 13: no. 31, pp. 67–69.

11. For example, brandy (Sun-Mercury), mead (Mercury-Venus), ice water (Venus–Earth and Moon), strong red vermouth (Earth-Mars), and a costly new white wine (Mars-Jupiter): Kepler to Duke Frederick, February 17, 1596, Kepler 1937–, 13: no. 28, pp. 50–54; see also Jarrell 1971, 160.

12. Kepler underestimated the difficulties of making a working model of a Copernican planetarium, some of whose metal components would require cogwheels with as many as 324 teeth (see Prager 1971, 385–92).

13. "Plate III. Showing the Dimensions and Distances of the Orbs of the Planets by means of the Five Regular Geometrical Bodies, Dedicated to the Most Illustrious Prince and Lord, Lord Friedrich, Duke of Württemberg and of Teck, Count of Mömpelgard, etc."

14. Thomas Kuhn's illustration of the Keplerian polyhedra is typical of later representations: it is redrawn from the original and fails to show the base-stand or the dedication (1957, 218).

15. "Idem (quod tamen tibi hîc concreditum velim) nostros Theologos etiam nonnihil offendit, authorite tamen Principis nostri, cui principale Schema dedicatum est, moti, in medio relinquunt" (Maestlin to Kepler, October 30, 1597, Kepler 1937–, 13: no. 80, p. 151).

16. Zinner (1941, no. 3190) lists the prognostication as Erfurt, 1585; although we know little about Limnaeus, he spent his entire career at Jena (see Jöcher 1784–1897, 3:1836; Rosen 1986, 104, 345 n.; Voelkel 2001, 88).

17. Georg Limnaeus to Kepler, April 24, 1598, Kepler 1937–, 13: no. 96, 207–8.

18. See Rosen 1986, 94–101.

19. See Westman 1975c, 304 n.

20. Praetorius to Herwart von Hohenburg, April 23, 1598, Kepler 1937–, 13: no. 95, pp. 205–6.

21. Ibid.

22. Stadtarchiv und Stadtbibliothek Schweinfurt: Praetorius 1605, fol. 2v.

23. Kepler 1937–, 13: no. 95, p. 206. *Speculatio* seems best translated as "belonging to theory," as in *philosophia speculativa*; its closest cognate is *contemplativa*.

24. Ibid.: "Et videtur ipsa phaenomena corrigi velle, ut corporum speculatio subsistere possit, quod quale sit alij videant, me haec non intelligere fateor." Praetorius also mentions that he had found nothing blameworthy in Maestlin's Appendix.

25. Fabian 1972–2001, 5:370, 372, 373; Mosely 2007, 193.

26. For a list of those known to have received copies, see Mosely 2007, 298–306.

27. For details, see Jardine 1984; Rosen 1986; Schofield 1981.

28. Beyond his activist engagement with precision

instrumentation and the science of the stars, the landgrave had interests typical of the culture of certain late-sixteenth-century German courts, including strong Paracelsian sympathies and interests in plant collecting and cultivation (see Schimkat 2007, 77–90; cf. Watanabe-O'Kelly 2002, 71–129).

29. Brahe 1588 (British Library, C.61.c.6; see fig. 59). Maestlin's underlinings and occasional notes show that he had studied key passages concerning Tycho's world system (see esp. 186–87, 190).

30. For details see Granada 1996b, 114–24, based on hitherto unstudied archival material at the Württembergische Landesbibliothek, Stuttgart.

31. Ibid., 116.

32. Ibid., 120 n.: "Si Mars tam prope accedit ad terram, propius scil. quam Sol secundum Copernici et Tychonis placita: enim necesse erit illum cum terrae proximus est habere maiorem quam solem parallaxin. Sed hoc nunquam a quicumque astronomorum intelligere potui."

33. Ibid., 124.

34. "Ego non sum Astronomus . . . hoc saltem cupio dass ich generaliter wissen möchte wie es doch in Mundo und mitt den orbibus geschaffen. Generalem rationem scire cupio. Specialia deinde ego ab artificibus accipio" (Roeslin to Maestlin, December 15, 1588, Württembergische Landesbibliothek, Stuttgart, Cod. Math. 4° 14b, fol. 20r; cited in Granada 1996b, 147 n.).

35. At Tübingen, Heerbrand's emphasis on the *liber naturae* was important for both Roeslin and Kepler's projects (see Heerbrand 1571, 32–33; cf. Methuen 1998, 137ff).

36. Roeslin (2000, prop. 109, p. 37), claimed that 2 Kings 20 supported the astrological doctrine of directions ("[King] Hezekiah being sick, is told by Isaiah that he shall die; but praying to God, he obtaineth longer life, and in confirmation thereof receiveth a sign by the sun's returning back"); but cf. Jacob Andreae, who argued that this passage only referred to God's occasional use of the heavens (see Methuen 1998, 128).

37. On Bongars and for a fine summary and analysis of the whole work, see Granada 2000b, vii–xv.

38. Roeslin 2000, 51 (image I).

39. Ibid., 45, 55 (image V).

40. Ibid., 45–48, 54 (image III); Granada 1996b, 140–44. In fact, Ursus was circumspect and left open the question of the infinitude of the universe; in 1597, he shifted his position toward that of Tycho and accepted the intersection of the paths of Mars and the Sun. To add to the confusion, Ursus's system is the one labeled "according to the hypotheses of Nicolaus Copernicus."

41. Roeslin 2000, 52, 54, 56.

42. Perhaps he had learned of Herwart's interests through his contacts with Ernst of Bavaria, archbishop of Cologne, the dedicatee of *De Opere Dei* (Roeslin 2000, 3–7). For translation and discussion of key passages, see Voelkel 2001, 80–82.

43. Roeslin to Herwart von Hohenburg, May 4/14, 1597, Kepler 1937–, 13: 123–24; quoted and trans. in Voelkel 2001, 80.

44. Ibid.

45. Ibid., Voelkel 2001, 82.

46. Jardine 1984, 53–54; Ursus 1597, fols. D1–Dv.

47. Ibid., 54.

48. Ursus 1597, fol. D: "Veriorem enim scio Hypothesibus Roeslini."

49. Jardine 1984, 49.

50. Ibid., 55–56; Mosely 2007, 190.

51. See Mosely 2007, 190–93.

52. Ibid., 189.

53. See Christianson 2000, 197–236.

54. As Mosely points out, the presentation of this book with the manuscript of the valuable star catalogue actually followed Rudolf's decision to appoint Tycho (Mosely 2007, 136; Brahe 1913–29, 8:163–66; Thoren 1990, 410–13).

55. See Gingerich and Westman 1988, 20–23.

56. Caspar 1993, 96–99, 108–15.

57. Kepler to Tycho, December 13, 1597, Kepler 1937–, 13:154; trans. Rosen 1986, 90; cited in Voelkel 2001, 83–84.

58. Kepler 1981, chap. 22, 215–19: "Why a Planet Moves Uniformly about the Center of the Equant"; Tycho to Maestlin, April 21, 1598, Kepler 1937–, 13:205, ll. 27–36.

59. Tycho to Kepler, April 1, 1598, Kepler 1937–, 13:99, ll. 90–95.

60. Ibid., 13:197, ll. 13–21.

61. Rosen 1986, 322.

62. The work was composed between October 1600 and April 1601, during which time Kepler suffered from a chronic fever (ibid., 322–23). The full text, together with translation and commentary, appears in Jardine 1984.

63. Jardine 1984, 152.

64. Ibid., 141–42.

65. Jardine 1979.

66. Jardine 1984, 146. Shortly afterward, Kepler mentioned that "William Gilbert the Englishman appears to have made good what was lacking in my arguments on Copernicus's behalf" (ibid.).

13. THE THIRD-GENERATION COPERNICANS: GALILEO AND KEPLER

1. See Drake 1978, 110; Wallace 1984a, 37; Wallace 1984b, 260; Drake 1987.

2. On efforts to reconstruct the Galilean chronology in his Pisan and Paduan periods, see the aptly sobering remarks of Reeves 1997, 25, and Bucciantini 2003, 29; cf. Drake 1978, 6–156; Schmitt 1972b, 243–71; Wallace 1998, 27–52.

3. Ricci lived ca. 1530–1600; see further Schmitt 1972b, esp. 246.

4. The university statues specified: "Astronomi

primo anno legant Auctorem Spherae, secundo Euclidem interpretent, tertio quaedam Ptolomaei" (quoted in Schmitt 1972b, 257).

5. Michele Camerota's argument for dating Galileo's *Juvenalia* to the Pisan period supports the likelihood of Galileo's use of Clavius (see Camerota 2004, 42).

6. Biblioteca Nazionale Centrale di Firenze: Ristori 1547–58, 98, 259 n. References to observations in 1547 suggest that the lectures were prepared and delivered for the first time in that year. Rutkin (2010, 141) believes that Ristori was using the 1548 Camerarius-Gogava edition. Ristori's original lectures are in Biblioteca Nazionale Centrale di Firenze: Ristori, n.d. Fantoni wrote his own name over that of Ristori at the bottom of the final page of the manuscript (430). Another hand has made numerous interventions in the text, which are usually expansions of abbreviations. This manuscript (shelfmark B.7.479) is a copy of the previous manuscript, prepared either by Fantoni himself or by an amanuensis. It is to this copy that Fantoni has added his own textual observations. Ronzoni appears to have made another copy directly from Ristori, who is also credited explicitly in the title: "Lectura super Ptolomei Quadripartitum . . . ac eximij magistri Iuliani Ristorij Pratensis, per me Amerigum roncionibus dum eum publice legeret in almo Pisarum gimnasio currenti calamo collecta" (Biblioteca Riccardiana: Ristori, n.d.). However, the Ronzoni copy is missing books 3–4.

7. See Ernst's important discussion (1991, 255–58); Giuntini 1581. Giuntini was heavily attacked by the Jesuit Antonio Possevino.

8. On Pico, see Biblioteca Nazionale Centrale di Firenze: Ristori 1547–48, lectio 52, fols. 196v–197; on the status of astrology as a science, see MS. Riccardiana 157, fol. 8.

9. As far as I know, Charles Schmitt was the first to call attention to these important manuscripts. Writing in the framework of the historiography of the early 1970s, when he was trying to make the study of universities an important part of the history of science, he regarded it as important that they showed "a strong occult element in the teaching of mathematics and astronomy at Pisa from the reopening of the studio in 1543 until the time of Galileo" (Schmitt 1972b, 259).

10. In 1581 the Florentine Dominican theologian Tommaso Buoninsegni issued a Latin translation of Savonarola's attack on astrology (Buoninsegni 1581).

11. Fantoni also referred to Copernicus as *vir peritissimus* in an unpublished commentary on Peurbach's *Theoricae Novae* (see Camerota 1989, 91).

12. On the bull, see Ernst 1991, 249–51, 254–55. Wallace believes that Galileo used the 1581 edition of Clavius's *Sphaera* for his *Tractatio de Caelo*, but this would not rule out his knowledge of subsequent editions (Wallace 1984b, 257–59; Wallace 1984a, 33–34).

13. Camerota 2004, 52.

14. See Helbing 1997.

15. Copernicus 1543, bk. 1, chaps. 8–9; Copernicus 1978, 24–28.

16. Camerota and Helbing 2000, 361–62. Buonamici taught at Pisa from 1565 to 1603.

17. Schmitt 1972b, 260 n.: *An Demonstrationes Mathematicae Sint Certissimae; Absolutissima Quaestio de Motu Gravium et Levium ex Praelegentis Doctoris Ore Excerpta in Accademia Pisana* (Biblioteca Nazionale di Firenze, Conventi Soppressi B.10.480).

18. See Funkenstein 1975b.

19. See Camerota and Helbing 2000; Camerota 2004, 43–50.

20. Bucciantini 2003, 33–48.

21. Ibid., 33–34.

22. Ibid., 36–37; for Maestlin's copy of *De Mundi*, see figure 59, this volume; Westman 1972a; Gingerich and Westman 1988.

23. Brahe to Camerarius, October 21, 1590, Brahe 1913–29, 7:276. Pinelli was also very interested in the globes belonging to Tycho and the landgrave (Bucciantini 2003, 40–41).

24. Bucciantini believes that it is "quite probable" that Pinelli (and Galileo) had the book very soon after its publication in 1596 (Bucciantini 2003, 57); and much of how one interprets Galileo's crucial letter to Kepler of October 1597 hangs on this dating. Adam Mosely, who has inventoried extant copies as well as Tycho's references to dispatched copies of the *Epistolae Astronomicae*, is disinclined to date Galileo's acquisition of the *Epistolae* before 1599 (Mosely 2007, 300).

25. The phrase is borrowed from R. J. W. Evans, who points to the *Concordia Platonis cum Aristotele* aspired to by the Bohemian Johann Jessenius (Evans 1979, 32).

26. Bertoloni Meli 1992, 9.

27. Mazzoni's print identity announced that he was "ordinary" professor of Aristotle, but on Plato "extraordinary" (Mazzoni 1597).

28. See Purnell 1972.

29. "Quod Terra sit Centrum Mundi. Et quod non moueatur, reijctur commentum Pythagoraeorum, Aristarchi Samij, & Nicolai Copernici" (Mazzoni 1597, 129). On the *comparatio* tradition, see Purnell 1971, 31–92.

30. Galilei 1890–1909, 2:198; cf. Drake's translation (1978, 40).

31. Mazzoni 1597, 132–33; Mazzoni's argument is well explicated in Shea 1972, 111–13.

32. Early in the letter, Galileo expresses "the greatest satisfaction and relief" that Mazzoni now "inclines to that part which was judged by me to be true and by you the contrary"; he also refers to how "in the first years of our friendship we disputed together with such joy" (Galilei 1890–1909, 2:197).

33. Galilei 1890–1909, 2:202. There is reason to

believe that Galileo had written much more about this problem than he expressed in the letter to Mazzoni.

34. Martinelli 2004.

35. Galilei 1890–1909, 10:67–68, quoted and translated by Hartner 1967, 181, and Koestler 1959, 356. I have made a number of substantive changes in Hartner's translation.

36. Or Egnazio Danti's disciplinary scheme (Danti 1577). I am not suggesting that there is any secure evidence that Galileo knew Barozzi's work at this time, but simply that his own views about the pre-eminence of mathematics would have been quite compatible with those of the Barozzi Proclus.

37. "Multas conscripsi et rationes et argumentorum in contrarium eversiones" (Galilei 1890–1909, 10:68). Hartner's translation suggests more than is warranted by Galileo's language: "I have worked out proofs, as well as computations of contrary arguments" (Hartner 1967, 181). Galileo's use of the term *rationes* (reasons, considerations, causes, reckonings) is rather vague compared to words that he might have used, such as *causa, demonstratio,* or even *theoria.*

38. Galilei 1890–1909, 10:67–68.

39. Moss 1993, 187–88, 198–200. Nonetheless, it is clear from Galileo's sketch of a theory of the tides at around this time that he believed that he was within reach of a necessary demonstration (Galileo to Cardinal Orsini, January 8, 1616, Galilei 1890–1909 5:377–95; translated in Finocchiaro 1989, 119–33).

40. Hartner 1967, 180–81.

41. Barone 1995, 370.

42. Bucciantini 2003, esp. 74–81.

43. Bucciantini's proposal is an important one, but it is mitigated by the fact that virtually every Protestant author of any reputation—including most of the members of the Melanchthon circle and the major publishers of works about the science of the stars—was also on the Roman Indexes of 1590, 1593, and 1596, among them Melanchthon, Rheticus, Neander, Ramus, Garcaeus, Schöner, E. O. Schreckenfuchs, Peucer, Peurbach (Reinhold's edition), and the publishers of *De Revolutionibus,* Johannes Petreius and Heinrich Petri. (For details, see Bujanda 1994, 979–1074.)

44. Two Jesuit copies of the 1543 *De Revolutionibus,* including that of Clavius, have censored the name of the publisher Petreius, and two 1566 editions have censored the name of Rheticus (Gingerich 2002, Rome, copies 1,2,3,4, pp. 112–14). Rheticus's name was already on the Index prior to Maestlin's edition of the *Narratio,* which accompanied the *Mysterium.* The inscription on Rome 4, 114, suggests the locus of the inquisition's worries: "Vidit P. Rd Inquisitor inde Corrigatur si qua erant astronomiae judiciariae die 2 apr[ri]l[is] 1597."

45. As Paul Grendler has observed, "A great deal of freedom of enquiry existed so long as speculation did not touch essential religious doctrine, or the scholar did not publish his views" (Grendler 1988, 51).

46. Campanella to Galileo, August 5, 1632, Galilei 1890–1909, 14:366.

47. Camerota 2004, 75–82; Woolfson 1998, 5. An early-seventeenth-century quip leaves different impressions of the four leading Italian universities: "Bologna innamorati, Padova scolari, Pavia soldati, Pisa frati" (Bologna [for] lovers, Padua scholars, Pavia soldiers, Pisa brothers): cited by Charles Schmitt from a section of a manuscript by Girolamo da Sommaia, titled "Delli studii e de dottori" (Biblioteca Nazionale di Firenze, Magliabechiana VIII, 75, fol. 70r in Schmitt 1972b, 248 n.).

48. Mattiazzo 1992, 289–305; Bellinati 1992, 257–65.

49. See Poppi 1992.

50. See chap. 4, this volume; Westman 1990, 179; Gingerich 2004, 135.

51. The word also appears in Mazzoni's description of Copernicus's objections to Ptolemy in *De Revolutionibus* bk. 1, chap. 7: "Illud itaque in primis notamus non esse adeo ridiculam Ptolomaei rationem, ut Copernicus existimat, quando nempè ad *explodendum* terrae motum dixit, quod quae repentina vertigine concitantur, videntur ad collectionem prorsus inepta, magisque unita dispergi, nisi cohaerenti aliqua firmitate contineantur" (Mazzoni 1597, 130).

52. Kepler to Galileo, October 13, 1597, Kepler 1937–, 13: no. 76, pp. 144–46.

53. Ibid., 13:144–45.

54. Kepler to Herwart von Hohenburg, March 26, 1598, Kepler 1937-, 13: no. 91, ll. 162–68: "Now with regard to your thinking that arguments for the motion of the earth can also be taken from reasons of the winds and the seas' motions, I too certainly have several thoughts about these matters. Recently, when the Paduan mathematician Galileo testified in a letter to me that he had most correctly deduced the causes of very many natural things from Copernicus's hypotheses which others could not render from the conventional [hypotheses], although he did not relate any specifically, I suspected this [cause] of the tides" (quoted and translated in Voelkel 2001, 71–72).

55. I am assuming that at this time Kepler was unaware of Zuñiga and Bruno.

56. Perhaps Kepler was referring to someone like the Styrian physician Johannes Oberndörfer, to whom Kepler sent a copy of the *Mysterium,* dated June 16, 1597 (for an illustration, see Beer and Beer 1975, 98).

57. Kepler to Galileo, October 13, 1597, Kepler 1937–, 13: no. 76, ll. 35–42; for further discussion of this passage, see Voelkel 2001, 70–71.

58. Kepler to Galileo, October 13, 1597, Kepler 1937–, 13: no. 76, ll. 43–51.

59. Ficino 1546–48. Dedicated to Cosimo I.

60. Tycho sent a copy to Kepler in December 1599 (Kepler 1937–, 14: no. 145, ll. 170–12); prior to 1600,

there is no evidence from Kepler's correspondence that he had been able to obtain a copy (see Mosely 2007, 300–301).

61. Kepler to Galileo, October 13, 1597, Kepler 1937–, 13: no. 76, ll. 53–56.

62. See Westman 2008. Could Kepler have inspired Galileo's well-known use of the Book of Nature trope in *The Assayer*?

63. Stillman Drake maintains that Galileo was not interested in carrying out observations in support of Copernicus: "Galileo's letters and papers are devoid of astronomical observations before 1604, and after 1605 again until 1610." But, shortly thereafter, Drake informs us that hardly any astronomical papers related to the *Dialogue* still exist because "they were removed by friends (and probably destroyed) while Galileo was on trial in Rome in 1633, lest they be found to incriminate him" (Drake 1973, 184–85). The same argument could be used to explain why there is so little information for Galileo's views about the heavens from the earlier part of his life.

64. Cf. Drake 1973, 176: "The first nineteen chapters of the *Prodromus* proceeded along lines quite alien to Galileo's outlook. But Chapter 20 dealt with motions and distances, which had been Galileo's chief interests from his earliest days. The idea of an exact relation between planetary periods and orbital distances caught his fancy, even though Kepler's reasoning failed to convince him." Even if Drake's reconstruction of Galileo's "pseudo-Platonic cosmogony" is accepted, he provides no persuasive evidence for dating the manuscripts on which his conjecture is based, and hence no secure basis for linking this putative cosmogony to the Bruce episode (see below).

65. Galileo to Castelli, December 21, 1613, Galilei 1890–1909, 5:288: "Di più molto probabile e ragionevole che il Sole, come strumento e ministro massimo della natura, quasi cuor del mondo, dia non solamente, com'egli chiaramente dà, luce, ma il moto ancora a tutti i pianeti che intorno se gli raggirano" (trans. in Finocchiaro 1989, 54).

66. Stillman Drake conjectured that Galileo used a table of Copernican mean solar distances from chapters 20–21 in Kepler's *Mysterium* to construct a kinematic account of the origins of the planetary motions. Such a story, if true, would be quite at variance with Kepler's archetypal account of the structure of the heavens (Drake 1973; Drake 1978, 63–65; but see Bucciantini 2003, 108–10).

67. My views on this question, worked out independently between 1998 and 2000, are largely supported by Bucciantini's excellent investigations (2003, 93–116).

68. Kepler's manner of address is revealing: "Nobilissimo viro D. Edmundo Brutio Anglo, amico meo, Patavij nunc agentj reddantur" (Kepler to Edmund Bruce, September 4, 1603, Kepler 1937–, 14: no. 268, ll. 41–43). Arthur Koestler's characterization of Bruce is an example of a novelist's excessive license: "Among Kepler's admirers was a certain Edmund Bruce, a sentimental English traveller in Italy, amateur philosopher and science snob, who loved to rub shoulders with scholars and to spread gossip about them" (Koestler 1959, 360–61). Stillman Drake refers to "Bruce, a Scot then residing in Padua," perhaps confusing him with the Scottish hero Robert Bruce (Drake 1978, 46), but later corrects himself when he describes "an English gentleman well versed in mathematics, military matters, and botany [who] appears to have known Kepler at Graz before moving to Padua in 1597" (ibid., 442).

69. Woolfson 1998, 131.

70. Ibid., 18, although Harvey was elected in the law university.

71. Ibid., 215. On Anthony Bacon, see Stephen and Lee 1891, vol. 2; Zagorin 1998, 4. On Francis Bacon's activities as an intelligencer, see Martin 1992, 50.

72. Woolfson 1998, 133.

73. Bellinati 1992, 341.

74. Gualdo 1607, 43: "Edmundum Brutium in his nobilem Anglum, disciplinarum Mathematicarum, rerumque militaris, & herbariae apprimè scientem, cuius ille commentationes non semel suspexit, cuius se quandoque imparem curiositati est ingenuè professus."

75. On this point, see Miller 2000, 4.

76. Ibid., 71.

77. "In contrata Crosariae divi Antonii, in camera superiori, versus viam" (according to an unpublished ms. dated July 27, 1601; cited by Bellinati 1992, 337 n.).

78. Gualdo 1607, 72: "Domum qua parte se in conspectum interiorem dabat, instruxerat ornaratque Geographicis grandioribus tabulis, iconibusque illustrium virorum. . . . Arripuit studium hic noster, grandique accessione [increase] bibliothecam, conclauia, atrium, eiusmodi ornatu hoestauit. Ut enim plerisque moris fuit libros illi suos inscribere, aliisue inseriptos elargiri, ita non defuerunt, qui imaginem suam bibliothecae eius dicatam vellent."

79. Miller 2000, 110–20.

80. Drake 1978, 459. On Pinelli's library and the Index, see Grendler 1977, 288–89; Stella 1992. Further research is needed to establish the character of the friendships and social connections of the earl of Northumberland (Kargon 1966).

81. See Magocsi 1993, 35, map 11.

82. R. J. W. Evans (1984, 270 n. 19) refers to fragments of correspondence at the Österreichische Nationalbibliotek (MS 9734, fols. 21, 24).

83. Edmund Bruce in Florence to Kepler in Prague, August 15, 1602, Kepler 1937–, 14: no. 222. David Hoeschel taught Greek at a *Gymnasium* in Augsburg (Favaro 1884, 7).

84. Edmund Bruce in Padua to Kepler in Prague, August 21, 1603, Kepler 1937–, 14: no. 265, ll. 6–9.

85. Scheiner 1612. For Scheiner and Welser, see Galilei 2010.

86. Welser 1591. On Welser and his ambience, see also Evans's important treatment (1984).

87. It is not known when Edmund Bruce brought this work to Pinelli's attention (Biblioteca Ambrosiana, Milan: Library Inventory of Gian Vincenzo Pinelli, fols. 237–39). In November 1603 Bruce used a different courier, traveling from Venice to Prague. "Da tuas simul cum D. Van Tau literis: ei, a quo has accipies" (Edmund Bruce in Venice to Kepler in Prague, November 5, 1603, Kepler 1937–, 14: no. 272, l. 35). I have been unable to find any information about the courier, Van Tau.

88. In 1604, as the collection was being transported by sea to Naples, pirates attacked the ship and, disappointed by its contents, dumped the cargo overboard. Fortunately, twenty-two chests were retrieved, but eight containing books and manuscripts—as well as portraits and mathematical instruments—were lost. On the fate of the library, see M. Grendler 1980, 388–89.

89. "Tuumque Prodromum multis monstraui quem omnes laudant" (Bruce to Kepler, August 21, 1603, Kepler 1937–, 14: no. 265).

90. The word most commonly used to describe (and dismiss) him is *gossip*. Bruce's charge that Galileo had plagiarized some of Kepler's views is also generally dismissed. Stillman Drake believed that Galileo could not possibly be plagiarizing from Kepler, as he did not agree with anything in the *Cosmographic Mystery* (1978, 63). Cf. Bucciantini 2003, 93–116.

91. Kepler to Bruce, July 18, 1599, Kepler 1937–, 14: no. 128, ll. 11–15: "Jam de coelorum Harmonia dicamus, qua materia praecipuè Pythagoraei celebres sunt. Invenio in hypothesibus Copernici harmoniam talem. . . . Harmonia sive proportio Geometrica non est (ut nos opinamur ex judicio aurium) in ipso materiali vocum, sed potiùs in formali." Kepler noted that he had spent all day composing this letter: "Totum diem scribendo consumpsj" (ibid., l. 351).

92. Ibid., ll. 262–63: "Itaque et Italorum judicia habere pervelim, ut cum Germanis conferre possim."

93. Ibid., ll. 5–6: "D. Galilaeum praecipuè hoc nomine saluta, à quo miror me responsum nullum accipere." To whom other than Bruce could this request apply?

94. Biblioteca Nazionale di Firenze: Kepler 1599, fols. 35–40. The first page of the letter exists only in the form of a copy that uses one side of the page (fol. 35r only); the original addressee's name is lacking. Antonio Favaro published very short excerpts from the letter only in those places where Galileo is mentioned by name, but he was unable to identify the addressee (Galilei 1890–1909, 10:75–76). An annotation by "E. A." on Favaro's index to the manuscript volume speculates that the letter might have been sent to Paul Homberger, the same diplomat who carried Kepler's book to Italy two years earlier (Biblioteca Nazionale di Firenze MS. Galileiana 88, fol. 3v). On the other hand, Max Caspar gives the full text (Kepler 1937–, 14: no. 128), and he argues persuasively that the addressee could only have been Edmund Bruce, partly on the basis of comparisons with the contents of a later letter from him to Kepler (ibid., 14: no. 265). Heeding Kepler's request to transmit information to Galileo, Bruce then must have passed the letter over to Galileo, whence it ended up in the Galileiana collection in Florence (ibid., 14:459), immediately following Kepler's letter to Galileo of October 1597 (Biblioteca Nazionale di Firenze MS. Galileiana 88, fols. 33r–34r).

95. Bruce to Kepler, August 15, 1602, Kepler 1937–, 14: no. 222.

96. Because the letter never reached Kepler, who had left Graz for Prague, Bruce repeated in abbreviated form much of the same information a year later, while adding that Magini had only just received the *Mysterium:* "Inter omnes litteratos totius Italiae, de te loquutus sim; diceres me non solum tui amatorem sed Amicum fore: dixi illis de tua mirabili inuentione in arte musica; de observationibus Martis: tuumque Prodromum multis monstraui quem omnes laudant; reliquosque tuos libros avidèque expectant: Maginus ultra septimanam hic fuit tuumque Prodromum a quodam nobili Veneto pro dono nuperrimè accepit: Galeleus tuum librum habet tuaque inuenta tanquam sua suis auditoribus proponit: multa alia tibi scriberem si mihi tempus daretur" (Bruce to Kepler, August 21, 1603, Kepler 1937–, 14: no. 265). This repetition suggests that the initial information was accurate.

97. Paolo Gualdo, whose book on Pinelli is our main source about his circle, knew somehow that Galileo had argued against Mazzoni; yet, the only source of information was either verbal or direct knowledge about the contents of something that Galileo wrote, titled "Commentarius Galilaei Galilaei, florentini mathematici, Patavini Professoris pro Copernico adversus Iacobum Marronium [corrected afterward to *Mazzonium*]" (Gualdo 1607, 29; cited in Bellinati 1992, 352).

98. In the letter of August 21, 1603 (Kepler 1937–, 14: no. 265, l. 15), Bruce had referred to "tua mirabili inuentione in arte musica" (your remarkable discovery in music). In the 1599 letter, Kepler had written: "Velim tamen ex aliquo excellenti Musico quibus abundaıt Italia, discere artificiosam et Geometricam tensionem totius clavichordij, aut si solo aurium judicio feruntur, quaero ex ipsis, an non alicubj in Organis et instrumentis duplex F, duplex A etc. fiat" (ibid., 14: no. 128, ll. 265–68).

99. "Tertio vehementer cuperem a Galilaeo post exactè constitutam lineam meridianam, observarj declinationem magnetis ab illa linea meridiana: sic

ut magnetica lingula libere in quadrato vase ad perpendiculum erecto et latere ad meridianam applicato natet [etc.]" (ibid., 14: no. 128, ll. 338–41). Kepler then remarked that, using this magnetic hypothesis, one could explain the variation in the altitude of the terrestrial pole that Domenico Maria Novara had noticed.

100. An ungenerous reading is that Magini wanted to leave Bruce with the impression that Galileo had lied about not receiving Kepler's book. But there is no evidence that Magini behaved in this manner here or in other circumstances; and further, as he himself had not seen the *Mysterium* at this point, he would not have known what Galileo had seen in the volume. It is more likely that Galileo wished to dampen any speculations about a possible connection with Kepler.

101. This claim is based on Drake's dating and analysis of a large collection of notes in Galileo's hand (Drake 1978, 74–104).

102. It is one of Drake's persistent themes, now fairly widely accepted in the literature, that Galileo's studies of motion in the years after he received Kepler's book and until 1609 were not connected to any larger claim about the order of the planets (Drake 1976, 142–43). Yet Drake also believes that, for some reason, in August 1602 Galileo was thinking about the ratios of the planetary distances and speeds presented by Kepler in chapter 20 of the *Mysterium Cosmographicum* (Drake 1978, 63–65, 478 n.). Moreover, Drake also interprets the absence of references to Copernicus in Galileo's extant correspondence between 1605 and 1609 (except for a brief moment in 1605 when Galileo allegedly thought that he could "confirm" Copernicus's theory by means of measurements of the parallax of a new star) to mean that Galileo had "lost faith in it [Copernicanism]" until his telescopic observations (ibid., 110, 483). By 1990, he had shifted to the view that Galileo was a "semi-Copernican" until he had "physical evidence from the tides"; the telescope provided evidence only "against the Ptolemaic system" rather than in support of Copernicus (Drake 1990, 131–32). In contrast to these diverse and often forced ad hoc explanations, Drake makes only two minor allusions to Giordano Bruno (ibid., 159, 440). On other grounds, Hans Blumenberg also thought that Galileo saw no connection between his work on the physical problems of free fall and projectile motion and "the Copernican system's need for proof" (Blumenberg 1987, 393).

103. Michael Sharratt (1994, 75) sensibly suggests a middle ground: Galileo did imagine some idealized, counterfactual experiments while also conducting some actual trials.

104. Bucciantini 2003, 74–81.

105. Naylor 2003. After 1595, Naylor argues that "motion in a circular arc had moved from a position of no obvious significance to one of surprising theoretical importance. Within a decade it had evidently, rapidly assumed theoretical prominence. It would not be an exaggeration to say that this kind of notable change in direction and emphasis in the study of motion appears unprecedented. Moreover, the only visible source capable of prompting such a remarkable transformation is Copernicanism. In fact no others seem available. Certainly the attempt to find an alternative origin for this radical change has up till now proved unsuccessful" (ibid., 177). Clavelin has shown that Koyré's position ("good physics is made *a priori*"), which located Copernicus's theory at the origins of Galilean dynamics, was anticipated by Paul Tannery's reading of Galileo in 1901 (Clavelin 2006, 15).

106. Important exceptions are Bucciantini 2003 and Camerota 2004.

107. Gualdo 1607, 29.

108. Galilei 1989a, 57: "We will say more in our *System of the World*, where with very many arguments and experiments a very strong reflection of solar light from the Earth is demonstrated to those who claim that the Earth is to be excluded from the dance of the stars, especially because she is devoid of motion and light. For we will demonstrate that she is movable and surpasses the Moon in brightness, and that she is not the dump heap of the filth and dregs of the universe, and we will confirm this with innumerable arguments from nature."

109. When describing what he hoped to write if he were invited to the Medici court, Galileo mentioned a *systema mundi*: "Two books on the system and constitution of the universe—an immense conception full of philosophy, astronomy, and geometry; three books on local motion, an entirely new science, no one else, ancient or modern, having discovered some of the very many admirable properties that I demonstrate to exist in natural and forced motions, whence I may reasonably call this a new science discovered by me from its first principles" (translated and quoted in Drake 1978, 160).

110. Canone 1995, 46–49, 59.

111. Yates 1964, 354–55; cf. Finocchiaro 2002.

112. See Le Bachelet 1923; Blackwell 1991, 45–48.

113. The witness was Gaspar Schoppius. See Spampanato 1933, Documenti Spampanato 1933, 202, no. 30; Blackwell 1991, 48.

114. The tongue vice was commonly used on heretical impenitents so that they could not utter further blasphemies before being burned (Canone 1995, 54 n.).

115. See Granada 1999b.

116. See Finocchiaro 2002.

117. Among recent writers, Hilary Gatti (1997) is one of the few to point to significant, detailed parallels between Bruno and Galileo. Richard J. Blackwell (1991, 47–48) comes closest to recognizing the importance of the Bruno question for Bellarmine.

118. A good deal of speculation has focused on what impact the trial of Bruno might have had on Bellar-

mine's attitude toward Galileo in 1616 and at Galileo's own trial in 1633 (see again Blackwell 1991, 48 ff.). Little or no attention has focused on the period 1600–1610.

119. Giovanni Maria Guanzelli da Brisighella, a Dominican, issued the decree, by which about forty authors and seventy titles were pronounced "suspect and prohibited" (see Canone 1995, 44–61).

120. Ibid., 59–60; Bruno 1969–78.

121. *Indicis Librorum Expurgandorum In studiosorum gratiam confecti Tomus Primus, In quo quinquaginta Auctorum Libri praecaeteris desiderati emendatur* (Rome, 1608; [August 7, 1603]), 600. Cited in Ricci 1990, 239–40.

122. Canone 1995, 45.

123. Ibid.

124. Ibid.

125. Gilbert 1958. Edward Wright's prefatory address noted that "that work held back not for nine years only, according to Horace's Counsel, but for almost [an]other nine [i.e. about 1582]" (xliv). Throughout, I have emended Mottelay's generally serviceable late-nineteenth-century translation with word choices closer to Gilbert's own text. (For Mottelay's translation principles, see pp. vii–viii.)

126. Ibid., xlix. Gilbert actually included a glossary of new terms at the start of the book ("Verborum Quorundam Interpretatio").

127. Ibid., l–li.

128. Ibid., bk., 1, chap. 3, 24.

129. Gilbert 1965, 192–93, 196–205.

130. Gilbert 1958, bk. 2, chap. 3, 104; chap. 4, 105.

131. Ibid., bk. 3, chap. 6, 121.

132. Ibid., bk. 6, chap. 5, 339.

133. Ibid., bk. 3, chap. 6, 121.

134. Ibid., bk. 6, chap. 5, 340.

135. Ibid., bk. 6, chap. 2, 315–17. Gilbert makes only passing reference to stellar influence and altogether neglects planetary astrology (see bk. 6, chap. 7, 349–50).

136. Ibid.

137. "Non probabilis modò, sed manifesta videtur terra diurna circumuolutio, cum natura semper agit per pauciora magis, quàm plura" (Gilbert 1600, bk. 6, chap. 3, 220; cf. Copernicus 1543, bk. 1, chaps. 5, 8). Gilbert (1958) argues for the diurnal motion in bk. 6, chaps. 3–4, 317–35.

138. Gilbert 1958, bk. 6, chap. 9, 358. Pumfrey goes further in suggesting that Gilbert regarded astronomy as incapable of achieving certainty, along the lines of Osiander's letter to the reader (2002, 166).

139. Gilbert 1958, bk. 6, chap. 3, 321; chap. 6, 344. For the periods of revolution of Venus and Mercury, Gilbert used values taken directly from *De Revolutionibus*, bk. 1, chap. 10.

140. Gilbert 1958, bk. 6, chap. 6, 343–44.

141. Ibid., bk. 6, chap. 4, 333: "The sun (chief inciter of action in nature), as he causes the planets to advance in their courses, so, too, doth bring about this revolution of the globe by sending forth the virtues of his spheres—his light being effused"; see also bk. 6, chap. 6, 344.

142. Ibid., bk. 6, chap. 3, 322.

143. Ibid., bk. 6, chaps. 7–9.

144. Gilbert 1965, bk. 2, chap. 10, 151; see further Lerner 1996–97, 2:282.

145. Gilbert's term is *volutatio* rather than *revolutio, motus,* or *rotatio*.

146. Gilbert 1600, bk. 6, chap. 5, 228, my translation. Cf. Gilbert 1958, bk. 6, chap. 5, 340–41.

147. Ibid., bk. 6, chap. 5, 327. In *De Mundo*, Gilbert explicitly stated that Tycho accepted "Copernicus's reckoning" with respect to the order of Mercury and Venus: "Illustrissimus Tycho Brahe Solem vult centrum esse secundorum mobilium, sive planetarum, terram vero constituit centrum universi; Mercurium & Venerem Copernici ratione circa Solem cieri" (Gilbert 1965, 192).

148. For discussion of this point, see Lerner 1996–97, 2:151–52.

149. See esp. Freudenthal 1983, 34–35; Gatti 1999, 97–98.

150. In *De Mundo*, Gilbert briefly invokes the principle that "what is closer [to the Sun] is moved more rapidly," but in the paragraph that follows, he does not use this principle to affirm the Copernican arrangement (Gilbert 1965, 194).

151. Freudenthal 1983, 33: "Rather than admit explicitly his failure to supply magnetic foundations for all celestial motions, Gilbert chose not to discuss the annual revolution at all."

152. Pumfrey 2002, 175–81; Gatti 1999, 86–98.

153. In *De Mundo*, he cited only the "mathematical" reading and represented Copernicus as following it: "Nonnulli Pythagorei, qui ex vetustioribus Graecis mathematicae auxiliis philosophiae fundamenta posuerunt, terram non in centro aliquo quiescere, sed in obliquo circulo volvi existimabant; ut Philolaus apud Plutarchum. Respuit hanc opinionem reliqua antiquitas. Copernicus; ut absurdiorem circulorum numerum, & implicitas vias evitaret, motum etiam telluris supponit in obliquo circulo" (Gilbert 1965, 192; but cf. his treatment in 1958, chap. 6, bk. 3, 317–18).

154. The works of Tycho, Roeslin, and Kepler—but not Ursus—had all appeared together at the Frankfurt Book Fair in the spring of 1597 (Mosely 2007, 193).

155. For Bruno on falling bodies, see Westman 1977. If Gilbert did know Tycho's *Epistolae Astronomicae*, he was also not sufficiently persuaded by Rothmann's Copernican arguments.

156. See esp. Pumfrey 1987.

157. Gilbert 1958, "Address by Edward Wright," xliii. In Gilbert 1600, Wright's eulogy follows the author's own preface.

158. Ibid., xlii; the entire passage is quoted below in chapter 16.

159. First mentioned in Herwart von Hohenburg to Kepler (November 21, 1602, Kepler 1937–, 14: no. 235, ll. 36–39), but already cited by Kepler from memory in early December (Kepler to Fabricius, December 2, 1602, ibid., 14: no. 239, ll. 437–40).

160. "And perhaps Gilbert's book would never have come into my hands if a famous Peripatetic philosopher had not made me a present of it, I think in order to protect his library from its contagion" (Galilei 1967, 400). Drake (1978, 67) dates Galileo's acquisition to the period 1600–1602.

161. Galileo 1967, Third Day, 406.

162. Bruce to Kepler, November 5, 1603, Kepler 1937–, 14: no. 272, ll. 9–28.

163. See, inter alia, Jacquot 1974; Kargon 1966, 5–17; Gatti 1993.

164. According to Woolfson (1998, 215, 236), Bruce was in Padua as early as 1585–86, as reported by the widely traveled Oxford fellow Samuel Foxe (1560–1630).

165. I infer this claim from the anomalous presence of Kepler's last surviving letter to Bruce in a collection of Hicks's correspondence (British Library: Kepler 1603, Lansdowne 89, fol. 26 [Kepler 1937–, 14: no. 268]). The British Museum Library acquired the marquis of Lansdowne's collection in 1807 (see Harris 1998, 33). Kepler addressed the letter to "Edmund Bruce, my friend and most noble man *nunc agentj reddantur.*" Hicks entertained James I at his estate in Ruckholt, Essex, on June 16, 1604, and lent money to Francis Bacon between 1593 and 1608 (Stephen and Lee 1891, 26:350).

166. Bruce to Kepler, August 21, 1603, Kepler 1937–, 14: no. 265, ll. 10–12: "Etiam forsan egomet ipse ad te volarem antequam ad meam patriam ruerto; nam nullus est in toto hoc mundo cum quo libentius conloquar."

167. In September 1592, when Pinelli was actively trying to bring Galileo to the university in Padua, he informed Galileo of a promising meeting with "un gentilissimo Mocenigo." At the very least, this suggests the existence of a network through which Galileo could have learned about Bruno (Pinelli to Galileo, September 25, 1592, Galilei 1890–1909, 10:49–50).

168. Poppi 1992. The following discussion is based on my review of this volume (Westman 1996).

169. By this phrase they meant Camillo Belloni, the "concurrent" lecturer and rival of Cremonini in natural philosophy.

170. Poppi 1992, 63, 74.

171. Ibid., document 5, 56. Poppi provides a facsimile of the original document, pp. 51–54.

172. Several of Galileo's nativities from this period survive. See Biblioteca Nazionale di Firenze: Galilei,

"Astrologica nonnulla"; Campion and Kollerstrom 2003, 147–67).

173. Poppi 1992, 60.

174. See Voelkel 1999.

175. For an excellent history and description of Galileo's instrument, see the video demonstration on the Museo Galileo Web site, http://brunelleschi .imss.fi.it/esplora/compasso/dswmedia/storia/estoria1 .html.

176. On this point, see the apt observations of Gérard Simon (1979, 36).

177. Later on in the treatise, Kepler again rejects the reality of the zodiac and its divisions, as he had done in the *Mysterium Cosmographicum,* but on this occasion, he expressly says: "This silly part of Astrology has already been refuted indirectly, on physical grounds, by the Astrologer Stöffler (to avoid appealing to the testimony of the hostile Pico della Mirandola)" (Field 1984a, 257; hereafter all references to specific theses are from the Field translation). In subsequent works, Kepler explicitly associated himself with Pico.

178. Thesis 24 refers to three superior and two inferior planets (Field 1984a, 241); in thesis 25, Kepler refers to five planets, leaving out the Earth, Sun, and Moon (243); in thesis 37, he says that "the positions, the spacing and the bulk of the bodies should bear to one another the proportions that arise from the regular solid figures—as I proved in my *Mysterium Cosmographicum*" (250). Such phrasings show that he was deliberately avoiding a specific engagement with the Copernican arrangement in this treatise.

179. Field 1984a, thesis 19, 237; thesis 21, 238.

180. Ibid., thesis 25.

181. Ptolemy 1940, book 2, chap. 9. Ptolemy discusses the colors of eclipses and comets but does not associate them with the rainbow. Besides Aristotle's *Meteorologica,* Kepler would have found resources for his discussion in Cardan's *De Subtilitate* and Scaliger's *Exotericarum Exercitationum* (see Boyer 1959, 151–53).

182. Field 1984a, thesis 43, 253; Kepler further developed his notion of the Earth's soul in the *Harmonice Mundi* (Kepler 1997, 362–76).

183. *Aspect* refers to the angle separating two planets in the zodiac. Major aspects are conjunction (0°), opposition (180°), quadrature (90°), trine (120°), and sextile (60°).

184. Field 1984a, thesis 44, 253.

185. Ibid., thesis 45, 254.

186. Ibid., thesis 47, 255.

187. Kepler 1997, 377.

188. Ibid., bk. 4, chap. 7, 377–78.

189. See Simon 1979; Rabin 1997.

190. Kepler to Maestlin, March 15, 1598, Kepler 1937–, 13: no. 89, ll. 175–8.

1. Kepler 1604; Kepler 1937–, 1:393–99.

2. Ibid., 396.

3. Ibid., 398.

4. The subtitle of this work reads: "A Little Book Filled with Astronomical, Physical, Metaphysical, Meteorological and Astrological Disputations, Paradoxes, and Common Opinions" (Kepler 1937–, 1:149).

5. Ibid., 159. The others included Magini, Roeslin, Fabricius, Maestlin, Bartholomeus Crestinus, and Jost Byrgi.

6. See Caspar 1993, 139.

7. Brahe 1913–29, 3:320–23. Kepler made known his identity in a letter to Magini many years later (February 1, 1610, Kepler 1937–, 16:279–80).

8. Christianson 2000, 279, 140–41.

9. Brahe 1913–29, 2:5–12. James visited Hven on March 20, 1590, for a few hours (Brahe 1913–29, 7:224). Tycho read this correctly as a sign of future patronage; subsequently, he worked through James's former tutor, Peter Junius, to arrange for James to issue a privilege (ibid., 7:331). He judged the poems to be "quite good," although a modern reader might reach a different judgment (ibid., 7:282–83, 6:307–9).

10. But no royal poetry accompanied that treatise; on Bacon's relationship with James, see Gaukroger 2001, 73–74, 161–65; Zagorin 1998, 18–24, 58–59, 168–69; Martin 1992.

11. Antonio Favaro published the only fragment that survives from one of these lectures, in Galileo's hand—exactly one page (MS. Florence, Biblioteca Nazionale Centrale, Galileiana 47, fol. 4); there is also a copy in another hand (ibid., fols. 5–7r). In addition, Favaro assembled and organized chronologically a further group of fragments of notes and observations that include a handful of Galileo's reading notes, in Latin, of Kepler's *De Stella Nova* and Tycho Brahe's *Progymnasmata* (Galilei 1890–1909, 2:277–84; MS. Galileiana 47, fols. 4r–13r). A further group of notes, in Italian, taken from Tycho's *Progymnasmata*, were evidently made later and were used in the composition of the Third Day of the *Dialogue concerning the Two Chief World Systems*.

12. Baldini 1981.

13. Drake 1976, xv–xvi.

14. Galilei 1890–1909, 2:291: "Havendo veduto che l'Eccellentissimo Sig. Galileo, nelle sue dotte lettioni, che di questa Stella alli giorni passati publicamente fece."

15. Stillman Drake (1976, 12) argued that both Cecco di Ronchitti and Alimberto Mauri were pseudonyms for Galileo and that parts of Lorenzini's treatise were written by Cremonini.

16. Favaro found a note that Mayr wrote on a copy of his *Prognosticon Astrologicum* (1623): "Dieveil auf vorge-dachte grosse Vereinigung / Saturn und Jupiter / im Schützen folgents 1604.Jahr im Herbst der herrliche Neue Stern im Schützen erschienen ist. Davon viel schreibens gewesen, ich auch zu Padua in Welschland meinem in Mathematicis discipulo Balthasar Capra, einen Meyländischen vom Adel, einen Tractat in die Feder dictirt, welchen er auch unter seinem Namen, mir zum besten, in welscher sprach hat trucken lassen, dieweil ich in solchen einen vornehmen Professorem Philosophiae daselbsten, welcher gantz ungeschickte sachen wider die observationes Astronomorum hatte in truck publicirt, nach nohtturfft widerleget habe" (Herzog August Bibliothek, Wolfenbüttel, fols. A2, 2; cited in Favaro 1983, 2:630–31).

17. Ricci 1988, 126–127.

18. Heckius 1605, 16: "Sed quia Deus non operatur temere, nec ullus aparet finis propter quem creauerit has stellas, non est ita facile recurrendum ad miracula, praeterea causam nos quaerimus naturalem, quando possumus, harum stellarum assignari potest causa naturalis."

19. Galilei 1890–1909 2:520, 523: "E non necessarie all'intento delle mie lezioni, che fu di provare solamente come la Stella nuova era fuori della sfera elementare."

20. Ibid., 277.

21. Ibid., 278.

22. See Drake 1977, 110, 483 n. 26.

23. Mayr was a member of the German nation in the faculty of arts at Padua (see Favaro 1966, 1:137 n.).

24. Galilei 1890–1909, 2:294.

25. For example, the average enrollment for twelve German universities in the period 1516–1520 was 1,177, with a peak of 3,157 for Vienna (see Overfield 1984). For astute observations on the state of universities in the sixteenth century, see Giard 1991, 19–25.

26. Herwart von Hohenburg had read this book and reported that it dealt with domification and the theory of aspects according to Ptolemy and the ancients (Herwart to Kepler, March 18, 1600, Kepler 1937–, 14: no. 158, p. 111).

27. See Christianson 2000, 320.

28. For helpful biographical details, see ibid., 319–21.

29. If one may so interpret Capra's print identity on the *Consideratione:* "Baldesar Capra Gentil'homo Milanese studioso d'Astronomia, & Medicina."

30. Drake (1976, 456) provides helpful biographical details on Mayr but does not speculate on his possible relationship with Galileo at the university.

31. Galilei 1890–1909, 2:294.

32. Ibid., 2:299–300, 304.

33. Spinelli was a member of the convent of San Giustina, along with the better-known Benedetto Castelli, who was, like him, a sometime student of Galileo; moreover, he was also on familiar terms with

Galileo's friend Giacomo Alvise Cornaro, a Venetian patrician (Milani 1993, 72).

34. Favaro 1966, 1:231–33; see also Tomba 1990, 92.

35. Drake 1976, 25–27.

36. Milani 1993. Favaro proposed that the Cecco *Dialogue* was a joint production, with Galileo the "scientist" providing the ideas and Spinelli the "man of letters" translating the Galilean work into Paduan dialect (Galilei 1890–1909, 2:272). Ludovico Maschietto (1992, esp. 432) follows this view without providing further evidence. Stillman Drake (1976, 1977) wishfully believed that Galileo was the sole author because the satire against philosophers fitted his own preconceptions about Galileo's unyielding hostility to that group. In fact, Drake's arguments are forced, and he cites little evidence for his claims. Because Drake published his 1976 work himself, under the auspices of the famous Los Angeles antiquarian book dealer Jake Zeitlin, it is likely that it was not critically refereed.

37. For example, it would be necessary to show that Lorenzini was not an actual figure and that Cremonini authored Lorenzini 1606. Drake's attempt to prove that Galileo wrote the Cecco treatise encouraged him to regard Lorenzini as a pseudonym for Cremonini, Galileo's sometime philosophical opponent (Drake 1976, 5–7, 9).

38. Of the five dedicatory poems to Gualdo's *Vita* (1607), Querenghi's is the first, followed by those of three Jesuits and Lorenzo Pignoria; see also Drake 1976, 36 n.

39. Ibid.

40. Ronchitti 1605; Drake 1976, 38.

41. Drake 1976, 38–40.

42. Ibid., 59.

43. Colombe 1606, 70; cf. 49, reference to Pico.

44. Mauri 1606; Drake 1976, 80.

45. Drake 1976, 110. However, the postil concerns the nature of the celestial substance ("Quaestio Tertia De Natura Partium Coeli: An Caelum Sit Substantiae Fluxibilis") rather than the attack on Pico.

46. Drake makes an admirable attempt to claim Mauri's identity for Galileo, but I do not find either his internal or his external arguments to be convincing; moreover, to the extent that his claim rests on the identification of Cecco with Galileo, it is further weakened (ibid., 55–71).

47. Galilei 1890–1909, 2:367–68. Galileo described the instrument to Cosimo as suited for "mathematical play in your first youthful studies."

48. Ibid., 370. They included Friedrich, prince of Holsazia (1598); Ferdinand, archduke of Austria; Philip, landgrave of Hesse (1601); and the duke of Mantua (1604).

49. For a different explanation of Galileo's decision not to publish his theoretical writings, see Biagioli 2006, 3–13.

50. Capra 1607.

51. Ibid., 453, emphasis in original. Favaro provides all the annotations.

52. *Difesa contro alle Calunnie ed Imposture di Baldessar Capra* (Galilei 1890–1909, 2:521).

53. Ibid., 521–22: "Io non so in quali scuole abbia il Capra imparato questa bruttissima creanza: dal suo maestro alemanno non credo certo, perchè, facendosi egli scolare di Tico Brae, aveva da quello potuto imparare, ed al suo discepolo mostrare, quali termini usare si devino nel publicare non solamente le cose dette da altri, ma le già communicate e mandate attorno con scritture private; ed ambidue, come studiosi del medesimo autore, potevano avere appresa la modestia da quello, il quale, volendo inserir ne'suoi scritti alcune cose di un amico suo, che ancor viveva, e pure in materia della nuova Stella di Cassiopea."

54. "The first words of my first lecture were these: 'A certain strange light was observed for the first time on the 10th day of October in the highest [heaven]'" (Galilei 1890–1909, 2:524).

55. Ibid.: "Ma se si deve esser così severo critico in queste precisioni, perchè non si è posto il Capra a riprendere in Tico Brae, prima il medesimo Ticone, e poi tanti autori segnalati, le scritture de i quali sono da lui registrate nei *Proginnasmati*, le quali sono così poco scrupulosi nell'assegnare il luogo ed il tempo dell'apparizione della Stella di Cassiopea?"

56. Ibid. Galileo cites the exact page numbers in the *Progymnasmata* from which he has extracted this information.

57. Ibid., 521.

58. Lorenzini 1606, 32: "Copernici, Magini, et Clavii opinio refutatur." Cf. Drake 1976, 81.

59. Caspar 1993, 85; Kepler 1937–, 1:477. The note is to the Latin work; I infer that Kepler also had his copy of the Italian treatise from Herwart.

60. Kepler 1937–, 1:158.

61. Bartolomeo Cristini, mathematician to Carlo Emanuele I, duke of Savoy (See Favaro 1886, 51–52).

62. Kepler 1937–, 1:229.

63. *Dialogue concerning the Two Chief World Systems*, Galilei 1890–1909, 7:303. Even almost thirty years after Lorenzini's treatise first appeared, Galileo did not identify Cremonini as the author.

64. Galileo to Giuliani de'Medici, October 1, 1610, no. 402, Galilei 1890–1909, 10:441: "Io prego V.S. Ill. ma a favorirmi di mandarmi l'*Optica* del S. Keplero e il trattato sopra la *Stella Nuova*, perchè nè in Venezia nè qua gli ho potuti trovare."

65. Galilei 1890–1909, 2:280 (MSS. Galileiana 47, car. 11r): "Kepplerus, De stella nova, car. 95, de scintillatione ait, fieri posse ex rotatione fixarum; et licet ad ipsas ☿ insensibilis omnino sit, ita ut a nobis, eo constitutis, nulla ratione videri possit, tamen non evanescit ipsi naturae, etc. Consideretur, quod multo citius evanescit illuminatio corporis lucidi, quam conspectus eiusdem: et die a longissima distantia vide-

mus facem ardentem, quae tamen corpora nobis adi-
acentia non illustrat." See further Bucciantini 2003,
140–41. Cf. Kepler 1937–, 1:243–44.

66. Galilei 1890–1909, 2:280–84. Because this
note on Kepler appears on a single scrap of paper, it is
likely that other such scraps have simply been lost.

67. Bucciantini 1997, 244–45.

68. Ibid., 281–83; cited and discussed in Bucciantini 1997, 243–44.

69. Bucciantini argues from scattered references
that Galileo's position was consistent over his whole
life (ibid., 245–48).

70. Bucciantini (2003, 140 n. 77) surmises that
Galileo could have borrowed the book. The copy of
the *Stella Nova* at the Biblioteca Universitaria in Padua
was once held in the Benedictine monastery of Santa
Giustina where Galileo's disciple Benedetto Castelli
lived.

71. Kepler 1937–, 1: chap. 28, p. 324: "Haec est Philosophia famosissimi illius Liechtembergii; quam
verissimam exemplis compluribus, si non essent odi-
osa, comprobrare possem."

72. Ibid., 1:320: "De naturae arcanis hoc ipso libro,
quem scribo, quemque hic evulgo, tam multa com-
mentus non essem: nisi ex Naturae arcanis nova haec
Stella prodijsset: Itaque si quibusdam Philosophorum
absurda videtur esse mea haec nova philosophia."

73. Ibid., 1:314.

74. See esp., Koyré 1957, 58–87; Simon 1979 is
somewhat the exception in this regard.

75. Kristeller 1964; Copenhaver and Schmitt 1992,
303–28.

76. Colombe 1606, 71: "Io Fra Filippo Guidi Dome-
nicano Lettore di Teologia, per ordi dell'Illustris &
Reuerendiss. Monsignore l'Arciuescovo di Fiorenza,
ho riuisto il presente discorso sopra la nuoua stella
del S. Lodouico delle Colombe, ilquale è molto conforme alla vera Filosofia & a i principi d'Aristotile,
e concorda con la Teologia e contiene molte belle
dottrine, spiegare con molta chiarezza, e facilità dal
quale potranno trarre utilità quelli, che abboriscano
la falsità dell'Astrologia iudiciaria."

77. As chapter 8 shows, Kepler's position was close
to that of Tycho Brahe, but evidently Tycho had no occasion to issue any forecasts prior to his untimely
death. There is one known astrological forecast by
Ursus for 1593 (Launert 1999, 239–42).

78. A good place to start on this question is R.
Evans 1973, 199–218.

79. Kepler 1937–, 1: ch. 7, pp. 172–81. "Esto et tertia
causa, primae permixta à Pico etiam commemorata,
quae effecit, ut constellationes nonnullae humanam
repraesentantes effigiem, quorundam individuorum
nomina meruerint, historiae nempe seu verae seu fabu-
losae" (175). Evidently, until he saw the right opportunity to establish his own position in relation to Pico's
claims, Kepler avoided mentioning them in print. Cf.
Field 1984a, 257: "This silly part of astrology has al-

ready been refuted indirectly, on physical grounds, by
the astrologer Stöffler (to avoid appealing to the testimony of the hostile Pico della Mirandola)."

80. Kepler 1937–, 1:176: "Non enim omnia nomina
sunt à dispositione stellarum: contrà saepius diversa
nomina ab eadem dispositione sunt orta."

81. Ibid., 1:178.

82. Ibid., 1:180: "Signa zodiaci ab elementis deno-
minata mero inventorum arbitrio."

83. Ibid., 1: chap. 3, p. 168.

84. Ibid., 1:170–71.

85. Ibid., 1:184.

86. Pico della Mirandola 1496, bk. 5, chap. 5.

87. Kepler 1937–, 1:188–89: "Fateor, nec hoc tantum,
sed totam hanc artem, religionum et imperiorum
periodos ex conjunctionibus determinandi, ego quoque cum Pico, ineptiarum et superstitionis damno."

88. Pico della Mirandola 1496, bk. 5, chap. 6.

89. Kepler 1937–, 1:188.

90. Ibid.

91. Ibid., 1:190.

92. Kepler 1997, 304–27 (*Harmonice Mundi*, bk. 4,
chaps. 2–5).

93. This point has been shown nicely by Judith V.
Field (1984a, 201–7).

94. As Field observes: "Musical ratios are derived
from those polygons whose sides are most closely related to the diameter of the circle in which they are
inscribed, while the astrological ratios are derived
from those polygons which will fit together to form
tesselations or polyhedra" (ibid., 207).

95. Kepler 1937–, 1:194. Concerning the historical
origins of the idea of the aspects, Kepler credited Pico
with the "ingenious guess" that early astrologers had
derived them from the four lunar phases.

96. For example, Francis Johnson (1959, 220)
claimed: "The fact that should be emphasized and re-
emphasized is that there were no means whereby the
validity of the Copernican planetary system could be
verified by observation until instruments were developed, nearly three centuries later, capable of measuring the parallax of the nearest fixed star. For that
length of time the truth or falsity of the Copernican
hypothesis had to remain an open question in science."
Against Johnson and Karl Popper, Imre Lakatos and
Elie Zahar argued that such a "crucial experiment"
would have made the abandonment of geocentric astronomy rational only after Bessel's observations of
1838 (Lakatos and Zahar 1975, 360).

97. Kepler 1937–, 1:232.

98. Ibid., 1:231.

99. Kepler writes "vicies semel centena et sexa-
ginta millia" for the nova's distance. I am most grateful to Bill Donahue for sorting out my confusion with
this number and with the subsequent calculation.

100. Kepler 1937–, 7: bk. 4, chap. 2; Kepler 1939, 887.

101. Kepler 1937–, 1:235; Brahe 1913–29, 2:428–30.
See Van Helden 1985, 50–51, 62–63.

102. Kepler 1937–, 1:234.

103. Gilbert 1958, bk. 6, chap. 3, p. 324.

104. Kepler 1937–, 1:234.

105. Ibid., 1:239.

106. Koyré 1957, 58–87. Koyré translates the most crucial passages.

107. Ibid., 62. Koyré remarks: "Astronomy therefore is closely related to sight, that is, to optics. It cannot admit things that contradict optical laws." Although Koyré's point is clear enough, it is not sufficient to say that the problem concerns the part of astronomy called optics, because what Kepler really wanted from that mixed science was its geometrical resources. In other words, nothing about the physical or metaphysical nature of light was at stake in this reasoning because Kepler was dealing strictly with how things ought to appear rather than with invisible forces.

108. Ibid., 63; Kepler 1937–, 1:254. I have emended Koyré's translation.

109. See chap. 11, this volume, figs 66 and 67.

110. Koyré 1957, 62; Kepler 1937–, 1:253.

111. R. Evans 1973, 232.

112. D'Addio 1962.

113. R. Evans 1973, 154–55.

114. Ibid., 155.

115. Koyré 1957, 59 (I have emended Koyré's translation); Kepler 1937–, 1:251–52.

116. Ibid., 61; Kepler 1937–, 1:253. On Kepler's association with Wacker, see the important study by Granada (2009).

117. Kepler 1937–, 1:275.

118. Kepler was aware that the matter of "coincidence" was itself subject to difficulties: even if the nova did appear in the same "place" as the conjunction, it did not appear on the precise day of its occurrence.

119. Ibid., 1:276: "Stella ex atomis confluxit." For especially useful discussions of this section, see Simon 1979, 60–64; Boner 2007.

120. Bruno often cites Epicurus in *De l'infinito* (Bruno 1996, 25, 31, 43, 115, 141, 191).

121. Ibid., 285; Simon 1979, 63.

15. HOW KEPLER'S NEW STAR TRAVELED TO ENGLAND

1. The archducal copy, gold-tooled with the two-headed Hapsburg eagle, is held by the Österreichische Nationalbibliothek (*48.H.1). Underlinings in chapters 28 and 30 suggest that the reader was especially interested in the book's prophecies.

2. There is a Kepler dedication copy in the Strahov Library, Prague (A G III 89) which, to judge by what I could read, was probably sent to a member of the nobility; unfortunately, the name of the dedicatee has been all but rubbed out.

3. Johann Georg Brennger to Kepler, September 1, 1607, Kepler 1937–, 16: no. 441, pp. 34–41; for Bruno, ibid., p. 39, ll.24–26.

4. Among them was Thomas Harriot (see Feingold 1984, 104, 136–37, 207).

5. My claim is based on the evidence in Feingold 1984.

6. Kepler 1937–, 1:330–32; Rosen 1967, 141–58; Jardine 1984, 277–78.

7. Evans 1973, 84.

8. Christopher Heydon to Kepler, February 4, 1605, Kepler 1937–, 15: no. 327, p. 150.

9. Kepler to Christopher Heydon, [October 1605], Kepler 1937–, 15: no. 357, pp. 231–39; against Offusius, p. 234, ll.120–23 ff.: "Nihil hic tribuo reflexioni, nam aspectus est in mera incidentia seu concursu radiorum. Nec sequitur operatio, quia alterius radius à Tellure in alterius corpus reflectitur (quae philosophia, puto est Jofranci Offuciij)."

10. See Fischlin and Fortier 2002a.

11. James VI and I 1588.

12. James VI and I 1603; James VI and I 1604.

13. James VI and I 1603, 81.

14. Sommerville 1994, xv.

15. See Peck 1991a, 5–6.

16. Sommerville 1994, xx.

17. Clark 1997, 631–32.

18. James VI and I 1603, 10.

19. Ibid.

20. Ibid., 12–13. James also added to the second, unlawful category "the knowledge of natiuities; the *Chiromancie, Geomantie, Hydromanti, Arithmanti, Physiognomie,* & a thousand others. . . . And this last part of *Astrologie* whereof I have spoken, which is the root of their branches, was called by them *pars fortunae*. This parte now is utterlie unlawful to be trusted in, or practized amongst christians, as leauing to no ground of naturall reason" (ibid., 14).

21. Buchanan's didactic, neo-Latin poem, the *Sphaera,* included a section on the dangers of astrological prophecy and magic and drew on a typical medley of midcentury astronomical manuals (MacFarlane 1981, 369; Naiden 1952; Pantin 1995).

22. James VI and I 1603, "To the Reader," fol. A4.

23. Nicholson 1939, quoted in Curry 1989, 20, my italics.

24. Kepler 1937–, 19:344, item 7 (51). The inscription appears in the British Library copy of the *Stella Nova:* C.28.f.12. See figure 77, this volume.

25. Kepler to James I, 1607, Kepler 1937–, 16: no. 470, pp. 103–4. For the plate of the nova, see Kepler 1937–, vol. 1, between pp. 226 and 227; on the condemnation of magic, see ibid., 1:336.

26. Kepler 1997, 2–3, my italics.

27. Ibid.

28. Ibid., 4.

29. Westman 2008.

30. Kepler knew from the *Daemonologie* that James had critically read Bodin's treatise against witches: "Who likes to be curious in these things [particular

rites and secrets of these unlawful arts], he may reade, if he will, here of their practises, *Bodinus Daemonomanie,* collected with greater diligence, then written with judgment." (James VI and I 1603, fol. A4).

31. See esp. Bodin 1576, bk. 1, chap. 8; on Bodin's natural philosophy, see Blair 1997.

32. W. Pauli 1952; Westman 1984.

33. Kepler 1997, 3–4.

34. Kepler to Matthias Bernegger, August 29, 1620, Kepler 1937–, 18: no. 891, p. 41. The invitation was extended by Sir Henry Wotton in Linz: "Ill. D. Wotonii non minor erga me humanitas in visitando fuit; doluit praeproperus eius transitus. Hortatur ut in Angliam transeam. Mihi tamen haec altera mea patria propter ignominiam istam, quam sustinet, deserenda non est ultrò, nisi velim ingratus haberi." And a few months later, he remarked of Wotton's invitation that it would be tempting to leave the civil wars of Germany, but "an igitur mare transibo, quo me vocat Wotonus? Ego Germanus? Continentis amans, insulae angustias horrens? Periculorum eius praesagus? Uxorculam trahens et gregem liberorum?" (ibid., 5/15 February 1621, 18, no. 909:63).

35. Henry Wotton to Francis Bacon, summer 1620, Kepler 1937–, 18: no. 892, p. 42.

36. On Bacon, see Lerner 1996–97, 2:137–41.

37. Chamber 1601.

38. Ibid., 53–54.

39. Ibid., 67–68.

40. Heydon 1603.

41. Ibid., fol. P3: "His whole tractate is nothing, but a rhapsody of other mens fragments, and fancies. Wherein as he hath brought nothing of his own, besides superfluous digressions, and much intemperancie."

42. Bowden (1974, 135–36) remarks on his "confusing eclecticism" and his "peculiar inconsistencies" in using the work of Brahe and Kepler.

43. Heydon 1603, 371.

44. Ibid., 370:

But this opinion he confirmeth by those motions, which haue of late been deuised by our modern Mathematickes, which they say, their predecessors *never knew.* Yet their deuise proueth not that multiplicite of reall orbs, which they have imagined. For these are but inuentions, to make us conceive the *Theoricks,* whereas in trueth our late and most exact Astrologers hold, that there are no such eccentricks, epicicles, concentrickes, and circles of equation, as are mentioned by them, and as both *Tycho Brahe* and *Rothmann,* doe at large prouue: and therefore in his *Progymnasmata* deuiseth newe *Hypotheses,* quite differing from the olde. In the meane time constituting but one onely continued substance from the concaue superficies of the Moone, to the 8. Sphere, with whome in this point *Rheticus, Ramus, Scultetus, Frischlinus, Ursus, Aslacus* and *Fracastorius* doe concurre.

45. Bodleian Library: Chamber 1603. On the title page, Chamber identified himself as "prebend of Windesor [Chappell] & fellowe of Eton."

46. Ibid., unpaginated dedication: "Many worthy kings and emperours haue troden and traced this path before You, yet for abilitie and gifts to discerne of the cause, being all farre behinde You, whom God hath annointed with the oyle of wisedome, *prae consortibus tuis,* above all Your fellowe kings and princes."

47. Ibid.

48. Heydon 1603, 123–24.

49. Bowden 1974, 129–40.

50. William Lilly, foreword to Heydon 1650:

This exquisite Treatise having been near 40. years detained in private hands, is now by the good hand of God made publike; it being the One and only Copy of this Subject extant in the World: Pen'd it was by the incomparably learned Sir Christopher Heydon *Knight,* Whose able Pen hath so strenuously vindicated Judicial Astrologie; as to this day not any Antagonist durst encounter with his unanswerable Arguments. In this Tractate that very thing which all Antagonists cry out for, *viz.* Where's the demonstration of the Art? is hear in this Book by understandable Mathematical Demonstrations so judiciously proved, that the most scrupulous may received full satisfaction.

Elias Ashmole funded "the charges of cutting the Diagrams in brass, that so the work might appear in its greater lustre" [fol. A4v].

51. Kepler to Thomas Harriot, October 2, 1606, Kepler 1937–, 15: no. 394, pp. 348–49. Evidently, Eriksen told Kepler of Harriot's optical work, as Kepler's initial letter foregrounds his own difficulties in measuring the angle of refraction and requests Harriot's assistance: "I hear that your experiences disagree by two or three degrees from those of Witelo, whom I have followed."

52. Ibid., 15:349: "Audio tibi malum ex Astrologia conflatum. Obsecro an tu putas dignam esse, cuius causa talia sint ferenda" (I hear that evil flowed to you from [the practice of] astrology. I beg to know, do you think that a [subject] is worthy from which such causes are endured?). It seems very unlikely that Eriksen met Harriot while he was still in prison.

53. Ibid., 15:350.

54. Ibid.

55. Thomas Harriot to Kepler, December 2, 1606, Kepler 1937–, 15: no. 403, p. 365. Nonetheless, the letter immediately provided a table of refraction values for different liquid media.

56. Harriot to Kepler, July 13, 1608, Kepler 1937–, 16: no. 497, p. 172, ll. 32–34: "Ita se res habent apud nos, ut non liceat mihi adhuc libere philosophari. Haerimus adhuc in luto. Spero Deum optimum maximum his brevi daturum finem." However, a few lines further on, he mentioned having read some chapters of the manuscript of Gilbert (1651), "where I see that, with us, he defends the vacuum against the Peripatetics" (ibid.: 16:173, ll. 43–48).

57. The charges all came from local ministers in Dorset. Although the testimonies did not stick, it is

interesting to see that their character was entirely theological. For example, John Jessop, minister at Gillingham, reported that he "hath harde that one Herryott of Sr Walter Rawleigh his howse hath brought the godhedd in question, and the whole course of the scriptures, but of whome he soe harde it he doth not remember," and another minister said "that he harde of one Herryott of Sr Walter Rawleigh his house to be suspected of Atheisme" (Shirley 1974b, 24–25).

58. Fuller accounts of the episode are to be found in ibid., 16–35; Kargon 1966, 15–17.

59. Shirley 1974b, 28.

60. For Urban, see D. Walker 1958, 205–12; Headley 1997; Shank 2005b.

61. His illness "was more then three weeks old before; being great windenes in my stomack and fumings into my head rising from my spleen, besides other infirmityes, as my Doctor knoweth & some effectes my keeper can witness" (cited in Shirley 1974b, 29).

62. Ibid.

63. As Shirley has established (ibid., 33 n.), investigations of Harriot's papers in 1603 and 1605 turned up nothing incriminating: "I have made as diligent search of Mr Herriotts Lodging and studie at Sion as the time would permitt; . . . Letters there were few, and almost none at all; and such as are, carrie an olde date; scarcely one written of late" (Sir Thomas Smith to the Earl of Salisbury, n.d., Hatfield House MSS: 113:43). It is possible that at one time Harriot did practice astrology but later gave it up. Hilary Gatti suggests tantalizingly that British Library MS. Add. 6789, fol. 183v, might contain evidence that Harriot cast a horoscope on November 16, 1596, predicting the death of Elizabeth I in 1617 (Gatti 1993, 12). Unfortunately, I find no horoscope on fol. 183v.

64. See Jacquot 1974.

65. This is the implication of Kargon's reading of the manuscripts of Harriot's notes (British Library: Harriot, MS. Birch, 4458, fols. 6–8; MS. Add. 6782, fol. 374 [Kargon 1966, 11]). Unfortunately, these pages contain only some jottings on Zeno's paradox and offer nothing about the order of the planets.

66. On these matters, see the important critique of John Henry (1982).

67. Ibid., 280–89. Henry persuasively contends that Harriot's notes on atomism are incomplete and do not form a complete system. He believes that Harriot's problematic derived from Aristotle's "mathematical" argument about the infinite divisibility of a line. Of course, this account does not definitively rule out the possibility that Bruno's version of atomism provoked Harriot to turn back to Aristotle's contention that infinitely small points cannot heap up to form a finite entity.

68. There are just two references: a letter from William Lower to Harriot, in which Bruno is mentioned, and an abbreviated reference to the title of one of Bruno's works, "Nolanus de universo et mundis" (British Library: Harriot, Add. MS 6788, fol. 67v). Stephen Clucas argues that this admittedly slim evidence has been used to bolster unreasonable claims about the influence of Bruno on Harriot (Clucas 2000); cf. Jacquot 1974.

69. Kargon (1966, 5–17); Jacquot 1974, 110 ff.

70. William Lower in Trefenty, Wales, to Harriot in London (British Library: Lower 1610, fols. 425–26).

71. I strongly doubt that Kepler was the first to introduce Harriot to certain of Bruno's "cosmological" arguments. On this matter, John Henry does not give any ground. When Harriot referred to "Nolanus, de immenso et mundi" in a list of books (British Library Add. MS. 6788, fol. 67v), Henry argues that because "there is no book by Bruno with quite that title," it must represent, at worst, a conflation of the titles of two different works by Bruno, and hence a "confusion" that suggests that Harriot had not yet read anything by Bruno. Yet the entry is merely an abbreviation for the title De Innumerabilibus Immenso Infigurabili; Seu de Universo et Mundis Libri Octo (Frankfurt, 1591), and Henry does not compare this entry with others on Harriot's list of books to see if there is a pattern of such abbreviations.

72. John Henry (1982, 275) rightly argues that Lower's letter alone does not prove that Harriot knew Bruno's arguments from a firsthand acquaintance with any of Bruno's works; but Henry does not acknowledge that Harriot and Lower—however they came upon the problem of stellar distribution—take the side of Bruno and Gilbert against Kepler.

73. To my knowledge, the association with Dee has not been investigated. As Jacquot (1974, 113) aptly remarks, however, the form of exposition is deliberate, as it would have allowed Hill to escape the potential charge of advocating a doctrine. Without contradicting this view, Saverio Ricci characterizes Hill's work as "a collection of reading notes, marginal postils to a library of libertine authors in which the Hermetica and Lucretius, the Neoplatonists and Cardano accompany Bruno and Gilbert" (Ricci 1990, 63). Hugh Trevor-Roper regards Hill's work as "a series of terse philosophical propositions, like the theses which challenging philosophers or theologians undertook to defend against all comers. There are over five hundred of these propositions, and they set out, in a disorderly and sometimes obscure apophthegmatic form, a comprehensive picture of the universe. Essentially it is the universe of Giordano Bruno" (Trevor-Roper 1987b, 31). Robert Kargon is the most critical, characterizing Hill's work as "a confused, self-contradictory mélange of the views of many thinkers. Particularly, it is a blend of the thought of the atomists, Aristotle, Nicholas of Cusa, the fabled Hermes Trismegistus, Bruno, Gilbert, and Copernicus. The work is chiefly of interest in that it illuminates the various streams

which fed into the group around Percy. Hill, a thinker of minor ability, could only imperfectly reproduce the thought of Harriot, Warner, Percy, and the others" (Kargon 1966, 15). Further analysis of Hill's book is clearly desirable.

74. Hill 1619, 7.

75. Ibid., 131–32: "Domus autem, exultatio, facies, triplicitas, terminus, decanus, graduum foeminitas putealitas luciditas & caetera illius farinae contemnendissima sunt, & pusillanimorum inuenta philosophi *makrofilia* [?], id est, magnanimitate nullatenus digna."

76. As reported by Nicholas Fiske in his letter to the reader: "I have many times endeavored its impression, but without success; for until of late years such was the error or rather malice of the Clergy, who only had priviledg of licensing Books of this nature, that they wilfully refused the publication" (Heydon 1650, fols. A4r–v).

77. Harriot to Kepler, December 2, 1606, Kepler 1937–, 15: no. 403, pp. 367–68; Jacquot 1974, 115; Henry 1982, 287–88. Henry does not show how the "physical" atomism of this letter coheres with the "mathematical" atomism of Harriot's manuscripts. In any case, his important point here would be that neither is coherent with Bruno's atoms.

78. Harriot to Kepler, July 13, 1608, Kepler 1937–, 16: no. 497, p. 173. The posthumous work to which Harriot referred was Gilbert 1965, bk.1, chap. 22, 196–205.

79. Lerner 1996–97, 2:146–51, my italics.

80. Harriot to Kepler, July 13, 1608, Kepler 1937–, 16: no. 497, p. 173.

81. Lower to Harriot, February 6, 1610, British Library MS. Add. 6789, fols. 427–29. For a partial transcript, see Lohne 1973, 208–9.

16. THE STRUGGLE FOR ORDER

1. I argue this point in Westman 1980a, 134.

2. See the essays in Di Liscia et al. 1997.

3. On this important point, see Grant 1978, esp. 105 n. 13.

4. Ibid. The diversity of "Aristotelianisms" was first recognized by Schmitt (1973).

5. The view that the heavens were made of an incorruptible fluid stuff was already prevalent before the recovery of the full corpus of Aristotle's writings in the thirteenth-century Latin West, and it never quite died out. See Grant 1994, 350; Donahue 1981; Lerner 1996–97.

6. Alexandre Koyré and Thomas Kuhn emphasized rupture, whereas writers like Charles Schmitt and Edward Grant have demonstrated important elements of Scholastic accommodation.

7. For a general overview and characterization of this group, see Ingegno 1988; Kristeller 1964; Brickman 1941.

8. Galilei 1890–1909, 2:211–12; see Drake 1978, 52.

9. Galilei 1957, 97, my italics.

10. Guiducci 1960, 57.

11. Timpler 1605, chap. 2., questions 1 and 2, pp. 17–19. The principal authority on Timpler is Freedman 1988. The term also appears in John Florio's 1611 Italian dictionary. The self-described physician William Cuningham uses the term *cosmography* in much the same sense as Timpler uses *cosmology*: "Cosmographie teacheth the discription of the universal world, and not of th'earth only: and Geographie of th'earth, and of none other part" (*The Cosmographical Glasse, conteinying the pleasant Principles of Cosmograhie, Geographie, Hydrographie, or Nauigation* [London: Ioan. Daij, 1559], fol. 6). Cf. John Blagrave: "Cosmographie is as much to say, the description of the world: as well his Aethereall part as, as Elementall, and in this differeth from Geographie , bicause it distinguisheth the earth by the celestiall circles only and not by Hilles, Riuers and such like" (1585, bk. 1, chap. 1, 6).

12. Foscarini 2001.

13. Foscarini claimed to treat only material or "purely natural" causes (e.g., elemental vapors) and excluded both supernatural and astrological causes from his discussion, although he alluded to a larger work (*Institutioni di tutte le dottrine*, tome 2, bk. 4, treatise 4) in which he treated those topics (Foscarini 2001, 50–52).

14. No one equated the terms *world system* and *cosmology*; they recapitulated disciplinary distinctions. The former functioned as a synonym for planetary arrangement within the discourse of the science of the stars, whereas the latter functioned as a synonym for natural philosophy or one of its subsidiary parts. Foscarini would later refer to the "Pythagorean opinion" rather than to a world system or cosmology.

15. Brahe 1913–29, 5:117–18.

16. Dee 1975, fol. biii; Dee classified the study of the effects of astral influences on the lower world as *astronomia inferior* (see Clulee 2001, 174); see also Thomas Blundeville (1597, 134): "What is Cosmographie? . . . These foure, Astronomie, Astrologie, Geographie, and Chorographie."

17. For France, Roger Ariew (1999, 103–15) shows that as early as 1623, the Aristotelian Jacques du Chevreul had incorporated into his geostatic account the concept of sunspots as denser parts of celestial spheres. The interesting question here is how early and with what consequences traditionalists in natural philosophy began to accommodate celestial novelties to the ordinary realm.

18. Perhaps it is not surprising that as the Copernican question took on the character of a debate, rhetorical elements became more prominent (see Moss 1993). On Kepler's humanist practices, see Grafton 1997, 185–224.

19. Donne 1611, fol. B. See Nicholson 1935, 457–58; Johnson 1937, 243–44.

20. See Gingerich 2002.

21. Ibid. The two copies are Chatsworth 1, Derbyshire (1543); Chatsworth 2, Derbyshire (1566).

22. Ibid., Basel 1 (1543); Glasgow 1 (1543).

23. Ibid.

24. Wilkins 1684, 13.

25. See Wischnath 2002.

26. See Pauli 1955; Westman 1984, 177–229.

27. Resistance by the Tychonics has been discussed most fully by Voelkel (2001, 142–69).

28. Cf. Biagioli 1992, 17.

29. See Applebaum 1996, 475, 499; Kepler 1937–, 7: bk. 4, pref., 249; Kepler to Bianchi, February 17, 1619, Kepler 1937–, 17: no. 827, pp. 321–28.

30. Charles Schmitt (1973) was the first to call special attention to the emergence of alternative versions of Aristotelian natural philosophy in the late sixteenth century. See also Wallace 1988; Jardine 1988. I do not know which edition of the *Almagest* Galileo used.

31. Favaro 1966, 2:113–15.

32. Galilei 1890–1909, 5:351–2; Galilei 1989b, 70–71. And when we allow for the failure of either Gilbert or Origanus to endorse an annual motion for the Earth, the list reduces to Copernicus and Kepler!

33. See Westman 1990; and chapter 4 above.

34. Galilei 1989b, 71, my italics.

35. See Jacob 1976, chap. 2; Dobbs and Jacob 1995, chap. 2; Stewart 1992; Heilbron 1983; Hall 1991; Dear 2001, 164–67.

36. But Stevin also held that because of the huge size of the universe, even Saturn could be at its center; although Copernicus had placed the Sun like a lamp in the middle of a beautiful church, Stevin believed that one could only justify the Sun's centrality as a matter of "convenience" (Stevin 1961, 3: bk. 3, 138–39).

37. Klaas van Berkel is careful to say that Stevin's precise motives for moving north are not known (Berkel 1999, 12–36, esp. 14–16).

38. See Israel 1995, 242–53, 273.

39. Berkel 1999, 17. The work on decimals appeared in Antwerp, the others in Leiden.

40. See Vermij 2002, 59; Walker 1972.

41. Vermij 2002, 18, 21.

42. On Mulerius, see ibid., 45–52.

43. Stevin 1961, 3:45.

44. Dijksterhuis 1943. Thanks to Floris Cohen for confirming the absence of any references to astrology in this work.

45. *Hypomnemata Mathematica, hoc est eruditus ille pulvis, in quo exercuit . . . Mauritius, princeps Auraicus* (Lugduni Batavorum: I. Patius, 1605–8), tome 1, *De Cosmographia* (1608). The three parts of cosmography are the doctrine of triangles, geography, and astronomy.

46. Gingerich (2002) found no copy of *De Revolutionibus* owned by Stevin.

47. In Stevin 1961, 3: bk. 3, chap, 1, prop. 2, p. 129: for example, he refers to "a drawing in the 11th chapter of his first book," as if he assumed that the reader should be able readily to consult this diagram.

48. Gilbert 1958, bk. 4, chap. 9, pp. 252–54. Gilbert was careful to indicate that his knowledge came from a passage cited by Hugo Grotius rather than directly from Stevin 1599. The question of magnetic "dip" was already a well-known problem in treatises on magnetism of the period (see the important collection in Hellmann 1898).

49. Gilbert 1958, bk. 4, chap. 9, p. 253: "The grounds of variation in the southern regions of the earth, which Stevinus searches into in the same way, are utterly vain and absurd; they have been put forth by some Portuguese mariners, but they do not agree with investigations: equally absurd are sundry observations wrongly accepted as correct."

50. Stevin 1961, 3:129.

51. On the prevalence of Capellan sympathies, see Vermij 2002, 32–42.

52. Gilbert clearly knew the principle from *De Revolutionibus*, but his statement of it is curiously incomplete and ambiguous: "Saturn, having a greater course to run, revolves in a longer time, while Venus revolves in nine months, and Mercury in 80 days, according to Copernicus; and the moon makes the circuit of the earth in 29 days 12 hours 44 minutes" (Gilbert 1958, bk. 6, chap. 6, p. 344).

53. Ibid., bk. 6, chap. 3, pp. 321–22.

54. Stevin 1961, 3:125.

55. Ibid., 127.

56. Ibid.

57. Gilbert 1958, bk. 6, chap. 3, p. 323.

58. Ibid.: "I pass by the earth's other movements, for here we treat only of the diurnal rotation" (327); "And if there be but the one diurnal motion of the earth round its poles . . . there may be another movement for which we are not contending" (336).

59. Stevin 1961, 3:133.

60. Mulerius 1616, preface; quoted in Vermij 2002, 51.

61. Gilbert 1958, xlii–xliii.

62. On differential uses of Gilbert's magnetic philosophy, see Pumfrey 1989, 45–53.

63. This is Bruce Stephenson's interesting observation (1987, 203).

64. Ultimately, Kepler's physical commitments forced him to struggle with the considerable predictive power and interpretive elasticity of deferent-epicycle models; see Gearhart 1985.

65. Donahue 1988; Voelkel 2001, 170–210. Building on Donahue's groundbreaking work, Voelkel's is the first attempt to offer sustained local explanations for the structure of the *Astronomia Nova*.

66. Christianson 2000, 273–76.

67. Kepler 1992, 43; Voelkel 2001, 168.

68. Voelkel (2001, 218–19) is clearly right to sug-

gest that Kepler's exposé of Osiander was designed to counterbalance Tengnagel's own letter.

69. Kepler 1992, 28; Gingerich 2002, Schaffhausen 1543, 218–21.

70. After vigorously and persuasively arguing for Tengnagel's interference as "censor," Voelkel concludes that "Kepler gave him little time to prepare it [his preface], and also that he was distracted by his activities at court, so we can conclude that Tengnagel had perhaps lost interest in the matter" (Voelkel 2001, 227; also 167–69).

71. Kepler to Longomontanus, early 1605, Kepler 1937–, 15: no. 323, ll. 101–9; quoted and translated in Voelkel 2001, 161.

72. Voelkel 2001, 186.

73. Ibid., 207–10. See Fabricius to Kepler, March 12, 1609, Kepler 1937–, 16: no. 524, ll. 330–429. The diagram appears on p. 235.

74. Voelkel reiterates Fabricius's confusion about Kepler's project as well as his own theory designed to replace it (Voelkel 2001,182–210). But it is sometimes hard to discern the line between confusion and genuinely plausible disagreement over what counted as a "natural principle."

75. Fabricius to Kepler, January 20, 1607, Kepler 1937–, 15: no. 408, ll. 15–30, 110–11. Quoted and translated in Voelkel 2001, 200–201.

76. Kepler 1992, chap. 21 ("Why, and to what extent, may a false hypothesis yield the truth?"), 298, 300.

77. Ibid., chap. 33, 379.

78. Ibid.

17. MODERNIZING THEORETICAL KNOWLEDGE

1. Blundeville 1594, 1597, 1605, 1613, 1621, 1636, 1638; Blundeville 1602. Excluding Robert Recorde's *Castle of Knowledge* (1556) and various works on instruments, comets, and novas, Blundeville's two works seem to have been the only indigenous textbooks of astronomy to come out of England between 1560 and 1640. The English translated or republished a considerable number of Continental works (see Johnson 1937, 301–35; Feingold 1984, 215).

2. Kepler 1937–, 7:7.

3. Giovanni dall'Armi, a well-placed Bolognese senator, recommended Galileo as "a noble Florentine, a young man of about 26 [sic] and well instructed in all the mathematical sciences." Although dall'Armi was sometimes mistaken in what he said and seemed to be unaware of Galileo's studies at Pisa, he emphasized that Galileo had been trained at the Florentine court by Ostilio Ricci, "huomo segnalatissimo e provvisionato dal Gran Duca Francesco" (Malagola 1881, 7–23).

4. See Westman 1980a, 123.

5. Normal science, in the strong sense adumbrated by the early Kuhn, implies puzzle solving governed and limited by tacitly held, core, paradigmatic assumptions (Kuhn 1970, 23–42).

6. There is evidence that some of Kepler's books had made their way into the personal libraries of some of the students and faculty at Oxford and Cambridge, although it is not clear how these works were read and used (see Feingold 1984, 52, 66, 100, 110, 113, 118, 139, 140).

7. For Bologna, see Kepler to Roffeni, April 17, 1617, Kepler 1937–, 17: no. 761, pp. 222–24; for Prague, Johannes Jessenius to Kepler, November 30, 1617, ibid., 17: no. 776, p. 243; for Wittenberg, ibid., 19:349–50; for Padua, Galileo to Giuliano de'Medici, October 1, 1610, ibid., 16: no. 593, p. 335; R. Evans 1973, 134–36.

8. See Caspar 1993, 80–81.

9. See Galilei 1992, xii–xiii.

10. As indeed he did not publish his little works on the balance (1586), "Cosmography" (1596), and mechanics (1593–1600) at the time of initial composition (see Drake and Drabkin 1969, 402–3).

11. Galilei 1960, 164.

12. Galilei 1989c, 119. This work later became the fourth book of the *Dialogue*.

13. Richard S. Westfall presented the "scientific" Galileo in his volume for the Cambridge History of Science Series (1971, 16–24). But he had nothing to say there about Galileo's trial.

14. "Galileo published the *Dialogue* because he believed it stated the truth, and any account that leaves out so basic a consideration must surely be defective. . . . I ask then the following question: can the internal dynamics of the system of patronage help us to understand—I say 'help us to understand,' not 'explain'—the two related decisions, by Galileo and by the Church, that allowed the publication of the *Dialogue*?" (Westfall 1989, 63). These essays were published separately between 1987 and 1989.

15. "Why did Galileo decide to complete the *Dialogue* and to publish it? Not least because he could not remain the most admired intellectual of the age and not do so" (ibid., 68).

16. Ibid., 69–70.

17. Westfall's patronage-based explanation was immediately disputed by reconstructions of how Venus would have appeared in the fall of 1610: Gingerich 1984 ("fuzzy and uninteresting"), 1992b; Drake 1984; Peters 1984.

18. For Biagioli's claim concerning the generality of his analysis of "the patronage system," see Biagioli, 1993, 353.

19. Ibid., 84. Although anthropologists and early modern historians have long recognized the significance of gift giving as a major form of social exchange (see Mauss 1954; Bourdieu 1990; Kettering 1986; Davis 2000), the burden of Biagioli's case rests primarily on the exchange model's subtheses.

20. Biagioli appropriates Stephen Greenblatt's notion of "self-fashioning," itself the English Renais-

sance version of Clifford Geertz's (1983) constructiv-ist account of humans as "cultural artifacts" (Green-blatt 1980, 3–4; Biagioli 1993, 2–3; for important dif-ferences, cf. Greenblatt 1980, 8–9).

21. Biagioli 1993, 87. Here Biagioli trades on sug-gestive analogies with the modern French university system, as represented by Pierre Bourdieu and Jean-Claude Passeron (1970, 82).

22. Biagioli 1993, 73–84. The "noncommittal pa-tron" is one of Biagioli's original and most interest-ing theses, and much of the force of his argument for his particular historical claims hinges on it rather than on the more general claims about gift giving.

23. Viala 1985, 51–84. Biagioli sees no problem in generalizing Viala's distinctions, without argument, from the realm of seventeenth-century French liter-ary authors to that of Galileo's Florence, where he as-serts that the Medici counted as "great patrons" or *mecenats* (1993, 84–90).

24. Biagioli 1993, 85, 90.

25. In the spirit of Westfall, Biagioli initially spoke of "Galileo's system of patronage" (1990c) and later, following Greenblatt, "Galileo's self-fashioning" (1993, chap. 1).

26. In the only substantively critical review of *Gali-leo Courtier* and in a subsequent vigorous exchange, Michael H. Shank showed convincingly that Biagioli had overlooked or misread historical sources both crucial and inconvenient to his argument. Most tell-ingly, Shank demonstrated that Jupiter was neither astrologically nor mythologically central to the motifs in the plumb-aligned upper and lower rooms of the Palazzo Vecchio; hence, the client Galileo's formative moment, knowingly fashioning his Jupiter discover-ies of 1609–10 to match an alleged sixteenth-century Medici dynastic mythology, was shown to be entirely without foundation. Shank showed that Biagioli had even overlooked evidence in Cox-Rearick 1984. In the latter, persuasive evidence of astrological themes is to be found in the pictures rather than in the rela-tionship between the iconography of the rooms up-stairs and downstairs (Shank 1994; Biagioli 1996; Shank 1996). Some later commentators have quietly dropped the "dynastic" part of the thesis, thereby failing to acknowledge the loss of some of the most dramatic art-historical and political meanings and leaving only the unobjectionable and conventional proposition that Galileo was seeking to attach his re-cent discoveries to the *reigning* Medici family.

27. See Hollis 1977, 1–21; and, as applied to disci-plinary categories, Westman 1980a, 133–34.

28. See, for example, Biagioli 1993, 162–69.

29. A further difficulty is the question of in what sense Galileo can be considered a courtier. Some-times Biagioli presents Galileo as an "honorary" but not an ordinary courtier in a "fairly unarticulated so-cioprofessional space"; in his own eyes, according to Biagioli, "he tried to represent himself as a noble,"

but "in the court taxonomies he was only a gentle-man." Leaving aside Biagioli's uncertain handling of this central consideration, it is important to note how much this picture is at odds with that of the free-wheeling, opportunistic *bricoleur*: "Galileo had little control over the questions that were asked him. Nev-ertheless, he *had to* answer them somehow, and in a witty manner fitting the codes of court culture" (ibid., 160–63, my italics).

30. In the sweeping generalizations of the final chapter, there is talk of "homologies" and "definitely comparable" protocols within European court society and "quite consistent features of the patronage sys-tem throughout Europe (and across many decades)" (ibid., 353–54).

31. Biagioli's "system of patronage" is not much different in its most general structural format from that of Westfall's theoretically modest, homespun sociology—a reciprocal relation built almost entirely around the singular character of Galileo himself and his most important court patrons. Westfall came to such externalism haltingly (and with apologies) only late in his career (Westfall 1985, 130–32). Biagioli crit-icized Westfall's externalism as epistemically weak (1990c, 2–3, 42–43); but in *Galileo Courtier*, Biagioli attributed a position to Westfall that more closely re-sembled his own than the one that Westfall actually held: "Westfall has argued, quite correctly, that the Medici rewarded Galileo's discoveries not because of their technological usefulness or scientific importance, but because they prized them as spectacles, as exotic marvels" (1993, 105).

32. "Disputes were a common dimension of the life of the court and academies" (Biagioli 1993, 164).

33. "Games to be played 'after the table was cleared' were so common as to be discussed in classic text-books of polite and courtly behavior like Stefano Guazzo's *La civil conversazione*" (ibid., 165).

34. Biagioli 1990c, 44–45, my italics. In this early statement of his views, Biagioli is more aggressively forthcoming in his argument against Westfall, whose position he describes as "the magico-ethical compul-sion inevitably associated with the intellectual recog-nition of the 'truth' of a theory," and with "the catego-ries of Kuhnian historiography," that he characterizes as "substantialistico-idealistic."

35. Biagioli's formulations of this matter do not permit an easy reading of his position. At times, he wavers as to "whether Galileo's commitment to Co-pernicanism was a cause or an effect of his move to court." He finds himself undecided over the primacy of the "externalist" chicken ("Obviously, I am not suggesting that Galileo decided to become a Coperni-can in order to move up in the social scale") or the "internalist" egg (as when he speaks of "the increas-ing [but still not decisive] evidence he had in favor of heliocentrism"), a dilemma he wants to resolve as fol-lows: "It was only by being *both* a social and cognitive

resource that it [Copernicanism] attracted a few astronomers and mobilized them to articulate it further" (1993, 226–27, 218–25; for similar considerations, see 90–101). Cf. Westfall on the "social turn": "All during this period [summer 1610] Galileo seems to have used his telescope to further his advancement *rather than* Copernicanism" (1985, 26, my italics).

36. A particularly useful volume on this prolific subject is Lytle and Orgel 1982.

37. See Lytle 1987; Weissman 1987, 25–45; Pumfrey and Dawbarn 2004.

38. Indeed, it would be most valuable to know more about Galileo's own ethos of friendship. On the general question of neo-Stoicism as an ideology of aristocratic sociability, see Miller 1996; Miller 2000, 49–75; Oestreich 1982.

39. Miller 2000, 51 nn. 8, 10, quoting Seneca, *Epistulae morales,* bk. 7, chap. 8, p. 35; bk 9, chap. 11.1, p. 49: "Ista quam tu describis, negotiatio est, non amicitia, quae ad commodum accedit, quae quid consecutura sit spectat."

40. See Golan 2004, 123–25.

41. See Chatelain 2001.

42. On early modern epistolary culture, see Hatch 1982; Lux and Cook 1998; Jardine, Mosely, and Tybjerg 2003, 422–25.

43. Bertoloni Meli 2003, 29–30; Camerota 2004, 80.

44. Del Monte 1579, 3: "Primùm itaque universalis planispherii à Gemma Frisio aediti (cuius alii quoquè mentionem fecere) cùm sit altero simplicius, contemplationem aggrediamur."

45. See Henninger-Voss 2000, 251–52.

46. For the differences between del Monte and Galileo, see Bertoloni Meli 1992, 21–26.

47. Galileo to Guidobaldo del Monte, November 29, 1602, Galilei 1890–1909, 10: no. 88, pp. 97–100.

48. Favaro 1966, 1:37–38; del Monte to Galileo, September 3, 1593, Galilei 1890–1909, 10: no. 51, p. 62; Rose 1975a, 225–28. In turn, Pinelli maintained friendships with Antonio Possevino, Ludovico Gagliardi, and Antonio Barisoni, important members of the Jesuit College in Padua (Cozzi 1979, 149).

49. Likewise, Commandino was born of a noble family, and his grandfather had been in the service of the duke of Urbino (see Rose 1975a, 185–88).

50. Giovanfrancesco Sagredo to Galileo, January 17, 1602, Galilei 1890–1909, 10: no. 75, p. 86; Sagredo to Galileo, August 8, 1602, ibid.,10: no. 80, p. 89; Sagredo to Galileo, August 2, 1602, ibid., 10: no. 82, pp. 90–91.

51. Sagredo to Galileo, October 18, 1602, ibid., 10: no. 87, pp. 96–97.

52. The interdict and Sarpi's role in it are extensively treated in Bouwsma 1968, 339–628. Although Bellarmine was ambivalent about the papacy's aggressive stance toward Venice, Sarpi would become the cardinal's inveterate enemy (Godman 2000, 198–99).

53. Sarpi became acquainted with Peiresc and Galileo through Pinelli's circle (Miller 2000, 86–87).

54. Paolo Sarpi to Galileo, September 2, 1602, Galilei 1890–1909, 10: no. 83, pp. 91–93.

55. See Shea 1972, 8; Galileo to Sarpi, October 16, 1604, Galilei 1890–1909, 10: no. 105, p. 115.

56. For the reconstruction of this episode, see Galilei 1989a, 4–5.

57. Galilei 1890–1909, March 16, 1610, 10: no. 272, pp. 290.

58. See Wallace 1984b; Jardine 1988, 708–11.

59. According to Drake (1978, 441), who gives no references, Boscaglia was appointed to the faculty in 1600, taught logic and philosophy, wrote poetry, and "was known as an expert in Plato's philosophy"; however, Favaro (1983, 1:181) notes that he also appears as a doctoral examiner with the title "Florentino, Phil. et med. Doctore."

60. See Schmitt 1972b, 248–49.

61. Galluzzi 1980. Interest in natural history was not restricted to Florence (see Findlen 1994; Freedberg 2002, 151–345).

62. Galluzzi 1980, 209–10; Galluzzi 1982.

63. Galileo to Antonio de'Medici, February 11, 1609, Galilei 1890–1909, 10: no.207, pp. 228–30.

64. Ibid.: "Mi ordina in oltre mio cognato, che io deva scrivere a V.E. qualche cosa di nuovo intorno a i miei studii, sendo tale il suo desiderio; il che ricevo a grandissimo favore, et mi è stimolo a speculare più del mio ordinario." Galileo's cousin was Benedetto Landucci, and it was Antonio who intervened on his behalf at the court.

65. Antonio de'Medici to Galileo, September 12, 1609, ibid., 10: no. 238, p. 257; see Biagioli (1993, 42–43), who reads the letter as exemplifying instrumental friendship within a reciprocal system of gifts and counter-gifts.

66. Galileo to Antonio de'Medici, January 7, 1610, Galilei 1890–1909, 10: no. 259, p. 277.

67. Castelli to Galileo, December 14, 1613, ibid., 11: no. 956, pp. 605–6; Finocchiaro 1989, 47–48.

68. His degree was in "both laws," that is, civil and canon (Fusai 1975, 12–13).

69. Galileo to Vinta, May 3, 1608, Galilei 1890–1909, 10: no. 187, pp. 205–9.

70. Vinta to Galileo, May 20, 1608, ibid., 10: no. 189, p. 210.

71. Galileo to Vinta, May 30, 1608, ibid., 10, no. 190, pp. 210–13.

72. Vinta to Galileo, June 11, 1608, ibid., 10: no. 192, pp. 214–15.

73. "We might search some time . . . to find a better description of the mores of patronage" (Westfall 1985, 117).

74. John Florio's dictionary (1611) gives the following meaning for *mondo:* "The world, the universe. Also a Mound or Globe, as Princes hold in their hands. Also cleane, cleansed, pure, neate, spotlesse,

purged. Also pared, pilled. Also winnowed, etc. Also, as we say, a world, a multitude, or great quantitie." Better for Westfall's purpose would have been the word *terra*: "The element called earth. Also our generall mother the earth. Used also for the whole world."

75. Biagioli 1993, 125. Like Westfall, Biagioli fails to comment on Gilbert's analogy between the Earth (*terra*) as a great magnet and the magnet as a little Earth (*terrella*).

76. See Porta 1658, bk. 7, chap. 5: "Of the Wonders of the Loadstone."

77. Frederick Stopp (1974, 196–97) has constructed an inventory of a large number of such medals struck in the sixteenth century at the university in Altdorf (where Praetorius taught until 1616). For example, one bears the inscription *Princeps amat hanc colat illum* ("A prince favors justice but pursues war"). The medal shows "Justice, with a sword and balance, hand[ing] a crown to Mars, armed and with a torch." Another bears the motto *Sciolus Persaepius Errat* ("The charlatan is mostly wrong").

78. The exact date of the composition of Cosimo II's horoscope is unknown; but Galileo certainly had the information readily available in 1610, when he was rushing the *Sidereus* through its final stages of publication, as he sketched the natal scheme using one side of one of his lunar wash diagrams (see Gingerich 1975, 88; for the six-week rush to print, see Gingerich and Van Helden 2003).

79. Galileo to Christina, January 16, 1609, Galilei 1890–1909, 10: no. 204, pp. 226–27. Galileo here mentions that he has spent much time on the grand duke's nativity because he is not sure that he has the correct birthdate and has had to correct the *Prutenic Tables* with Tycho Brahe's solar theory.

80. In his *De Stella Cygni* (1606), Kepler says of the landgrave that he was very diligent and studious in the science of the stars, more so than one expects in a prince: "Olim in ministerio Illustrissimi Landgravij Hassiae Gulielmi (cujus in siderali scientia studium et diligentia, major quàm in Principe requireres, inventaque praeclarissima, Tychonem Brahe ad aemulationem extimulârunt, ut passim in ejus viri operibus, maximé in libro Epistolarum videre est), is quo de ago, Byrgius automaton coeleste apparans, globum coelestem ex argento adjecerat" (Kepler 1937–, 1:307).

81. See, for example, Drake 1957, 59–72; Cochrane 1973, 167–80; Westfall 1985; Van Helden in Galilei 1989a.

82. Dava Sobel (1999) foregrounds what is known of Galileo's family life in her popular treatment. For some other popular accounts of Galileo, including information about his daughter Sister Maria Celeste (b. Virginia), see Levinger 1952; Reston 1994.

83. On Galileo's university income, see Drake 1978, 51, 141, 160–61; Westfall 1985, 119; Favaro 1:57–95; Biagioli 2006, 8–9.

84. But he was rather unlike the fascinating but undisciplined medical astrologer and autodidact Simon Forman (see esp. Kassell 2005; Traister 2001).

85. See Laird 2000, 18–19.

86. Malagola 1881, 21–22, documents 5 and 6.

87. Galilei 1890–1909, Galileo to [Vincenzo Gonzaga], May 22, 1604, 10:107: "Venni, pensai, parlai et tornai; et dissi al S. Giulio Cesare [Caietano] che rispondesse all'A.V.S., che havendo io esaminate le mie necessità et lo stato mio, non potevo per li ducati 300 et spesa per me et per un servitore offertami partirmi di qua, et che però mi scusasse apresso V.A.S. etc., soggiungendoli che caso che V.A.S. li havesse domandato quali fussero state le mie pretensioni, li dicesse ducati 500 et 3 spese." Note that Galileo here negotiated directly with the duke (cf. Biagioli 1993, 81, referencing Goffman 1956, 481).

88. Biagioli acknowledges that tutoring a prince was a common route to patronage, but he then presumes that Galileo's main objective in seeking to move to court was "the social status one could obtain by serving a single princely patron" and asserts that "Galileo was seeking much more than free time at the Medici court" (Biagioli 1993, 29, 30; Biagioli 2006, 11). Here and elsewhere, he is at pains to downplay Galileo's continuing relationship with the university and his use of the court to promote his philosophical program. Yet in his negotiations with the Medici, Galileo was quite clear that he wanted to be freed of obligations to give private lessons to students who lived with him: "Desiderei che la prima intenzione di S.A.S. fusse di darmi odio et comodità di potere tirare a fine le mie opere, senza occuparme in leggere" (Galileo to Vinta, May 7, 1610, Galilei 1890–1909, 10: no. 307, p. 350).

89. See Winkler and Van Helden 1992; Bredekamp 2000.

90. See David Gooding's (1986) outstanding treatment of this problem.

91. Galilei 1989a, 13–14; Van Helden (1994, esp. 9–16) has also called attention to the uncertainties surrounding the initial use of an untried instrument: for example, "how and where to look for the image." For a useful discussion of further difficulties, see Zik 2001.

92. Bartoli to Vinta, September 26, 1609 Galilei 1890–1909, 10: no. 241, p. 259; Bartoli to Vinta, ibid., October 3, 1609, 10: no. 242, p. 260.

93. Bartoli to Vinta, October 24, 1609, ibid., 10: no. 245, p. 261.

94. Antonio de'Medici to Galileo, September 12, 1609, ibid., 10: no. 238, p. 257.

95. Giovanni Battista Strozzi to Galileo, September 19, 1609, ibid., 10: no. 239, p. 258.

96. Piccolomini to Galileo, September 19, 1609, ibid., 10: no. 240, pp. 258–59.

97. Galileo to Antonio de'Medici, January 7, 1610, ibid., 10: no. 245, pp. 273–78.

98. Galileo to Vinta, January 30, 1610, ibid., 10: no. 262, p. 280; quoted and translated in Galilei 1989a, 17.

99. Ibid.

100. Vinta to Galileo, February 6, 1610, Galilei 1890–1909, 10: no. 263, p. 281.

101. Ibid.

102. Galileo to Vinta, February 13, 1610, ibid., 10: no. 265, pp. 282–83; quoted and translated in Galilei 1989a, 18. I read the word *auspicii* as Latin for "authority" or "protection "rather than "favor."

103. Ibid.; Galilei 1989a, p. 19.

104. Vinta to Galileo, February 20, 1610, Galilei 1890–1909, 10: no. 266, pp. 284–85.

105. At this crucial juncture, where one would expect exact advice, Vinta made no mention of either the horoscope of Cosimo II or the mythologies represented in the paintings of the Palazzo Vecchio. The primacy of astronomical rather than astrological material again shows up in the title that Galileo used to designate his work in a letter to Cosimo of March 19, 1610: "I send to your Highness my *Avviso Astronomico*, dedicated to your blessed name. That which is contained in it and the occasion of writing it to you you will find in the *dedicatoria* of the work, to which I refer you in order not to trouble you twice" (ibid., 10: no. 276, p. 297; cf. Biagioli 1993, 128–29; Shank 1996).

106. Gualterotti to Alessandro Sertini, March 1, 1610, Galilei 1890–1909, 10: no. 267, pp. 285–86.

107. Gualterotti to Galileo. March 6, 1610, ibid., 10: no. 268, pp. 286–87. In a follow-up letter to Galileo, Gualterotti proposed that "to me is due that praise which you give to a Dutchman [*Belga*], and which you can give to your country" (April 24, 1610, ibid., 10: no. 300, p. 342; trans. in Galilei 1989a, 45–46).

108. He wrote at least a dozen works. See Manfredi 1919; De Filippis 1937.

109. Evidence about the complexities of Beni's life is carefully collected in Diffley 1988.

110. Manso to Paolo Beni, March 1610, Galilei 1890–1909, 10: no. 274, p. 292.

111. Ibid.

112. Ibid. Manso cites, without reference, Ptolemy in the *Almagest,* Aristotle's *Meteorology,* and "Alfagranus."

113. Ibid., 10:293.

114. Ibid.

115. Ibid., 10:294.

116. Ibid., 10:295.

117. Ibid.

118. See Shapin 1994, esp. chap. 3, "A Social History of Truth-Telling: Knowledge, Social Practice, and the Credibility of Gentlemen." For all its talk of practice, Shapin's account is about norms and aspirations rather than actual behavior, as "the historian has no privileged knowledge of how much or how little the early modern period was marked by genuine truth-telling" (101, xxvii–xxviii).

119. See P. Lipton 1998.

120. On the Central European scene, see R. Evans 1979, 38–40; on courtly patronage as fostering a realist agenda for astronomy, see Jardine 1998.

121. B. Shapiro 2000, 118.

18. HOW GALILEO'S RECURRENT NOVELTIES TRAVELED

1. Drake 1978, 110, 152 (Drake believed that Galileo was not yet "wedded" to the Copernican system in 1610); Drake 1990, 233–34. Wallace (1984, 259–260) follows Drake on this matter.

2. Geymonat 1965, 39–40.

3. Westfall 1985, 128; Biagioli 1993, 94–96.

4. On the question of Galileo's silence, see esp. Clavelin 2004, 17–28; Bucciantini 2003, xxii–xxiii.

5. Literally, a glass through which one looks. The neologism is derived from *specillum* (*speculum* or glass), best rendered in French as *lunette* or *lentille*. The Italian *occhiale*, derived from *occhio* (eye), the term Galileo often used in his letters, has the connotation of an eyepiece—hence, a glass through which one looks with the eye. It was variously Englished in the seventeenth century as "Mathematicians perspicil," "perplexive glasse," "optick magnifying Glasse," "trunk-spectacle," and "optick tube"(see Nicholson 1935, 442). For a recent critical summary of the literature on the telescope and optics, see Pantin's fine discussion in Galilei 1992, lxviii–lxxxviii; I have also benefited from the excellent English translation and commentary by Albert Van Helden (Galilei 1989a) and his important discussion of the invention of the telescope (1977).

6. In an early letter to Vinta, Galileo characterized his work in progress as follows: "As soon as this tract, which I shall send to all the philosophers and mathematicians as an announcement [*avviso*], is finished, I shall send a copy to the Most Serene Grand Duke, together with an excellent *occhiale*, so that he can reencounter [*riscontrare,* "meet with,"] all these truths" (Galileo to Vinta, January 30, 1610, Galilei 1890–1909, 10: no. 262, pp. 280–82; quoted and translated in Galilei 1989a, 18, by Van Helden, who uses the more precise *verify*).

7. Galilei 1989a, 40.

8. Cf. Reinhold 1551, fol. α3v: "Etsi autem honorificum est relinquere nominis & virtutum memoriam in scriptis, historijs, in tropheis in aedificijs, tamen multò splendidius est, & gratius habere monumenta in his pulcherrimis, & perpetuis corporibus, coelo & stellis quasi flixa, quas quoties adspiciunt homines docti, & benè morati excitantur, primùm ut celebrent Deum conditorem huius mirandi operis, deinde ut gratias agant, quòd monstrauit motus, postea etiam de beneficijs magnorum Principum, & scriptorum cogitant, quorum laboribus haec sapientia conseruata & propagata est."

9. Ibid., 20–21.

10. In his reply to the January 30 letter, Vinta echoed Galileo's language in referring to the *"avviso* sent to all the philosophers and mathematicians" (Vinta to Galileo, February 6, 1610, Galilei 1890–1909, 10: no. 263, p. 281). Cf. Biagioli 1993, 120: "He needed an absolute prince because his marvels could best gain value and grant him social legitimation if they were made to fit the dynastic discourse of such a ruler. . . . [H]e correctly realized that Venice was not the best marketplace for his marvels."

11. Galilei 1890–1909, 11:105–16; see Kollerstrom 2001, 425.

12. Galilei 1890–1909, 3:75; Galilei 1989a, 57. It is difficult to reconcile this passage with Stillman Drake's remark that "he did not, however, declare in the *Starry Messenger* that the earth was a planet" (1978, 157).

13. And was so called by Galileo and his correspondents (e.g., Castelli to Galileo, September 27, 1610, Galilei 1890–1909, 10: no. 399, p. 436). See Galilei 1992, xxxiii; Galilei 1989a, ix; see also Dooley 1999.

14. Through Kepler, Galileo had known since 1596 that "Maestlin proves by many inferences, of which I have not a few, that it [the Moon] has also got many of the features of the terrestrial globe, such as continents, seas, mountains, and air, or what somehow corresponds to them"(Kepler 1981, 164–65). In 1621, Kepler claimed that such was the "consensus of many philosophers" and that "Galileo has at last thoroughly confirmed this belief with the Belgian telescope" (ibid., 168–69).

15. Galilei 1992, lii.

16. Galilei 1989a, 87.

17. Wotton had taken his B.A. at Merton College, Oxford, in 1588 and later visited Kepler (Feingold 1984, 83).

18. Henry Wotton to the earl of Salisbury, March 13, 1610, Smith 1907, 1: no. 181, pp. 486–87.

19. Paolo Sarpi to [Guillaume Leschassier], March 16, 1610, Galilei 1890–1909, 10: no. 272, p. 290.

20. Biagioli reads this episode as an official, state-level diplomatic exchange where patronage governed scientific communication: "While Galileo was communicating with Kepler, Cosimo II was asking Rudolph II to confirm the existence of the Medicean planets. From such a confirmation, the Medici would improve their international image, but Rudolph would also have his very high status confirmed since he (through his mathematician) had been given the status of a judge of the matter. Kepler and Galileo communicated as clients (and therefore representatives) of the Emperor and the Grand Duke and not as scientists" (1990c, 27–28). If this was in fact an official communication, it is curious that Giuliano de'Medici's copy of the *Sidereus Nuncius* arrived through Thomas Seggett unaccompanied by a formal letter from Vinta or Cosimo II or by a telescope. On April 19, the Tuscan ambassador asked Galileo for a telescope on behalf of the emperor, suggesting that it be sent from Venice by Asdrubale da Montauto (Galilei 1890–1909, 10: no. 291, p. 319); but, as far as can be determined, an instrument was never sent.

21. Kepler 1965, 3.

22. Martin Hasdale to Galileo, April 15, 1610, Galilei 1890–1909, 10: no. 291, pp. 314–15.

23. Kepler 1965, 7.

24. Ibid., 6: "For the military analogy, which I jokingly used in that public book, has been continued, with no less propriety in this introduction to a private letter."

25. Or "unk-unk," in the felicious phrase of Duncan Agnew. Personal comment, Science Studies Colloquium, University of California, San Diego, March 12, 2003.

26. Kepler 1965, 10.

27. Ibid., 11; Kepler 1993, 7. I have made minor adjustments to Rosen's translation.

28. Kepler to Johann Georg Brengger, April 5, 1608, Kepler 1937–, 16: no. 488, p. 142: "Brunum Romae crematum ex D. Wackherio didici, ait constanter supplicum tulisse. Religionum omnium uanitatem asseruit, Deum in Mundum in circulos in puncta conuertit." Wacker had an eyewitness report from Gaspare Scioppio of Bruno's execution (see R. Evans 1979, 58).

29. Kepler 1965, 38–39; Kepler 1993, 26. I have slightly adjusted Rosen's translation.

30. Kepler 1965, 36–37; Kepler 1993, 24. Rosen observes that in the April 19 letter to Galileo, Kepler referred to Wacker as the one who would put him in chains and prison (Kepler 1965, 133 n.). But this may also be a subtle reference to Bruno's own fate.

31. Kepler 1965, 37 (modifying Rosen's trans.); Kepler 1993, 25.

32. Kepler 1965, 38; Kepler 1993, 25. Rosen renders *tou hoti* less convincingly as "bare facts."

33. Kepler 1965, 38; Kepler 1993, 25–26. I follow Pantin's translation.

34. Kepler 1965, 41; Kepler 1993, 28–29.

35. Kepler 1965, 42; Kepler 1993, 27–28.

36. Castelli to Galileo, December 14, 1613, in Finocchiaro 1989, 47–48. Biagioli asserts that the 1613 episode was a typical form of court entertainment, a game that enabled Galileo to improve his status and credibility by impressing the prince in "present[ing] himself not as a lowly mathematician but as a true philosopher." Evidently, Galileo did not regard a debate with Kepler or Bruno to be a game that would improve his status and credibility, even though parts of Kepler's *Dissertatio* could be read as public challenges (see Biagioli 1993, 164–69). By contrast, there is no evidence that Kepler's status with the emperor was raised, although his work improved his credibility among self-interested traditionalists in both Prague and Italy.

37. See Galilei 1992, liii–liv.

38. The atmosphere was different in the Low Coun-

tries. Two years before Galileo, David Fabricius's son Johannes publicly referred to Bruno and Kepler's thesis that the Sun rotates on its own axis and also claimed to have observed spots adhering to the body of the Sun (Fabricius 1611, fols. D1v–D2v).

39. "The spots' motion with respect to [*rispetto al*] the Sun appears to be similar to those of Venus and Mercury and also to those of the other planets around the same Sun, which is from west to east" (Galilei 1890–1909, 5:96).

40. Scheiner's *Tres Epistolae* was published without permission of the censors (see Galilei 2010, 57, 174; for censorship within the order, see Hellyer 2005, 36–38).

41. Galilei 1890–1909, 5: 25, ll. 3–4, p. 26.

42. Ibid., 5:306–13: "Reliquum ergo, ut sint vel partes alicuius caeli densiores, et sic erunt, secundum philosophos, stellae; aut sint corpora per se existentia, solida et opaca, et hoc ipso erunt stellae, non minus atque Luna et Venus, quae ex aversa a Sola parte nigrae apparent"; Galilei 2010, 56–57, 72–73.

43. Galilei 2010, 180, 229.

44. Sharon Kettering's 1986 study of French patronage makes considerable use of the category of broker, but she has nothing to say about writers or intellectuals; hence it is at least questionable whether her categories should be applied without further discussion to the case of Galileo. Similarly, it is far from clear that brokers or learned advisers, let alone patrons, simply laid down the iconographical conventions or themes that an artist was supposed to follow in a commission (see esp. Hope 1982, 293–343).

45. Galileo to Vinta, May 7, 1610, Galilei 1890–1909, 10: no. 307, pp. 348–53; available in Drake's English translation (1957, 60–65). Biagioli makes highly selective use of this letter throughout his chapter on Galileo's self-fashioning to underwrite his model of patronage (1993, 11, 29 n., 57 n., 97 n.; 1990c, 13).

46. Quoted in Drake 1957, 60.

47. Ibid.

48. Ibid.

49. "And you may believe that this is the way leading men of letters in Italy would have spoken from the beginning if I had been in Germany or somewhere far away" (ibid.).

50. See Kepler 1993, xxxii–xxxiv.

51. Galileo to Vinta, May 7, 1610, Galileo 1890–1909, 10: no. 307, p. 350.

52. Matteo Botti to Belisario Vinta, July 6, 1610, Galilei 1890–1909, 10: no. 353, p. 392; Andrea Cioli to Belisario Vinta, September 13, 1610, ibid., 10: no. 389, p. 430.

53. Giovanni Bartoli to Belisario Vinta, September 26, 1609, ibid., 10: no. 241, p. 259; for further discussion, see Kepler 1993, 54 n. 30, 63–64 n. 57.

54. On piratical aspects of book culture in late-seventeenth- and early-eighteenth-century England, see Johns 1998, chap. 7.

55. Biagioli (2000) has recently argued for the "monopoly" view, an interesting position that seems to be meant to sustain the structural-functionalist assumption in his patronage model that clients were continually required to produce spectacular discoveries.

56. See Dooley 1999, 9–44.

57. As aptly noted by Isabelle Pantin, Galileo spoke of his treatise as an *avviso* intended for philosophers and mathematicians even before he drafted the dedication to Cosimo II (Galilei 1992, 49 n.).

58. Galilei 1989a, 31; Galilei 1992, 3.

59. Ibid., 54 n.

60. Galilei 1989a, 32.

61. Galileo to Vinta, May 7, 1610, Galilei 1890–1909, 10: no. 307, pp. 351–52. I have introduced a few changes to Stillman Drake's translation (1957, 60–65; and again, citing the proposed publication list, Drake 1978, 160).

62. Galileo to Vinta, May 7, 1610, Galilei 1890–1909, 10: no. 307, p. 351.

63. Drake 1957, 62. Cf. Biagioli: "A more complex reading of patronage dynamics shows that Galileo was seeking much more than free time at the Medici court" (1993, 30).

64. For example, in his otherwise remarkable *Court Society*, Norbert Elias (1983) does not even consider the social category of the natural philosopher or heavenly practitioner and has very little to say about universities.

65. In his *Letters on Sunspots*, Galileo described Castelli as "a monk of Cassino . . . of a noble family of Brescia—a man of excellent mind, and free (as one must be) in philosophizing" (Drake 1957, 115).

66. The early Keplerians included Peter Crüger (Danzig), Philipp Mueller (Leipzig), Henry Briggs (Gresham), and John Bainbridge (Oxford).

67. See Pantin's lucid treatment (Galilei 1993, xxviii–lii).

68. Horky to Kepler, April 6, 1610, Kepler 1937–, 16: no. 563, p. 299, ll. 24–26.

69. Magini's *Ephemerides coelestium motuum . . . ab anno Domini 1608 usque ad annum 1630* appeared at Frankfurt in 1608; the next year the work was again published in Venice. Origanus published Magini's letter ("Apology of Giovanni Antonio Magini of Padua for his Ephemerides against David Origanus") and his own reply in Origanus 1609, fols. (c)3v–(d)2.

70. Horky refers to the rector of the Prague Karolinum, Martin Bacháček, who was a friend of Kepler; and to two patrons, Ladislaus Zeydlič à Schönfeldt, councillor to the emperor, and Christopher Wratislaus à Mitrowic, burgrave to the Crown in Lochovice (Kepler 1937–, 16: no. 563, p. 300; also Horky to Kepler, April 27, 1610, ibid., 10: no. 507, p. 307).

71. Favaro 1966, 1:138; Kepler 1993, xlvi. Because Horky would have been a member of the German nation at the university, it is likely that he would have encountered Simon Mayr.

72. Horky to Kepler, April 6, 1610, Kepler 1937–, 16: no. 563, p. 300. By writing in a mixture of German and Latin, Horky believed that he was protecting the confidentiality of his letter.

73. Origanus 1609, fols. (a)3–(c)3; 121–22.

74. Magini to Kepler, April 20, 1610, Kepler 1937–, 16: no. 569, pp. 304–5; Favaro provides only the last line of this letter (Galilei 1890–1909, 10: no. 298, p. 341).

75. Horky to Kepler, April 16, 1610, Galilei 1890–1909, 10: no. 565, p. 301. Horky's arithmetic (7 + 4) suggests that he did not understand—or did not take into account—that, following the Copernican model, Kepler's universe would have had only ten planets.

76. Horky to Kepler, 27 April 1610, ibid., 10: no. 571, p. 308.

77. Ibid.: "Ich hab das Perspicillum als in Wachs abgestochen das niemandt weiss undt wen mir Gott wider zue hauss hilft will ich fiel ein pessers Perspicillum machen als der Galileus."

78. Gingerich and Westman 1988, 55.

79. Horky to Kepler, April 27, 1610, Kepler 1937–, 16: no. 570, p. 308.

80. Horky to Kepler, May 24, 1610, ibid., 16: no. 575, pp. 311–12; Magini to Kepler, May 26, 1610, ibid., 16: no. 576, p. 313.

81. Kepler 1993, xxxix n. See Roffeni 1614.

82. Galilei 1890–1909, 3: pt. 2, p. 436; see Galilei 1989a, 93–94; Biagioli 2006, 113–14.

83. Galilei 1989a, 64–66. All quotations follow in sequence on these pages.

84. Ibid., 84.

85. Ibid. I have made minor adjustments to the translation.

86. Kuhn 1957, 222.

87. Robison 1974, 167.

88. Ibid. This view would be consistent with Galileo's having read and assimilated Brahe's controversy with Rothmann.

89. Horky to Kepler, April 27, 1610, Kepler 1937–, 16: no. 571, p. 306.

90. Ibid., 16:306–8.

91. Kepler 1993, xv.

92. Shapin and Schaffer 1985, 60–65.

93. Kepler to Magini, May 10, 1610, Galilei 1890–1909, 10: no. 308, p. 353: "We are both Copernicans: like praise like."

94. Kepler 1965, 12–13; Kepler 1993, 8–9.

95. Kepler's attribution of sincerity to Galileo resonates with John Martin's claim that the Renaissance witnessed the emergence of a new moral imperative, especially strong among the Protestant reformers, that placed a high value on the honest expression of one's feelings and convictions (1997, esp. 1326–42).

96. Magini to Kepler, May 26, 1610, Kepler 1937–, 16: no. 576, p. 313. Pantin has argued that Zugmesser was innocent of the charges of knowingly stealing Galileo's compass (Kepler 1993, lx–lxi).

97. Galileo, Difesa contro Capra, Galilei 1890–1909, 2:545; Hasdale to Galileo, June 7, 1610, Galilei 1890–1909, 10: no. 328, p. 370; Kepler 1993, lxi. Although he verbally charged Zugmesser with obtaining information from Tycho, Galileo said nothing about this in the Difesa.

98. Galileo to Vinta, March 19, 1610, Galilei 1890–1909, 10: no. 277, p. 298.

99. Kepler to Horky, June 7, 1610, Kepler 1937–, 16: no. 580, p. 315.

100. Kepler 1965, 19:86 n. 141.

101. Galilei 1890–1909, 3:131–45.

102. Ibid., 3:145: "Cur quatuor ficti planetae circa corpus Iovis sint, superius in altero problemate rationem dixi eam, quia bis ac ter quod pulchrum est, hic repeto; et dico, illos esse in Caelo circa corpus Iovis quia intermedium caliginosum, puta aërem et refractionem Iovis, cum radios perfecte egerere potest, illas quatuor maculas omnes ostendit."

103. Ibid., 141. Horky's argument seems to have been that the four "spots" were optical effects produced by the refraction of Jupiter's rays in the airy mist.

104. Roffeni to Galileo, June 22, 1610, Galilei 1890–1909, 10: no. 334, pp. 375–76; Alessandro Santini to Galileo, June 24, 1610, ibid., 10: no. 337, pp. 377–78.

105. Magini to Alessandro Santini, June 22, 1610, ibid., 10: no. 335, p. 377.

106. The copy that Horky sent to Maestlin is still extant (in the Schaffhausen Stadtbibliothek) with the provenance "Clarissimo viro Michaeli Maestlino Mathematum in celeberri Tubingensi Academia Professori Ordinario Domino patrono suorum collendissimo levidente hoc exemplar dono dedit Martinus Horky. 18 Jun. 1610."

107. Maestlin to Kepler, September 7, 1610, Kepler 1937–, 16: no. 592, p. 333. The copy of Dissertatio cum Nuncio Sidereo that Kepler sent to Maestlin is extant (in the Schaffhausen Stadtsbibliothek).

108. For excellent biographical information on Sizzi, see Kepler 1993, p. l, n. 74.

109. Horky á Lochovič 1610, 138.

110. "Tota hallucinatio" ([Horky to Sizzi], [June 1610], Galilei 1890–1909, 10: no. 347, pp. 386–87.

111. For Pico's exegetical practices, see Black 2006.

112. Sizzi 1611, 217.

113. Roffeni to Galileo, June 29, 1610, Galilei 1890–1909, 10: no. 344, pp. 384–85.

114. Roffeni to Galileo, July 6, 1610, ibid., 10: no. 352, pp. 391–92.

115. Magini to Alessandro Santini, [June 1610], ibid., 10: no. 338, pp. 378–79.

116. Cosimo II to Galileo, July 10, 1610, ibid., 10: no. 359, pp. 400–401.

117. Roffeni to Galileo, July 27, 1610, ibid., 10: no. 368, p. 408.

118. Quickly completed in an Italian version, the brief work did not appear until translated into Latin in early 1611 as the Apologetic Letter against the Blind

Foreign Travels of the Same Deranged Martin who, under the family name Horky, published against the Nuncius Sidereus concerning the Four New Planets of Galileo Galilei, formerly Public Mathematician in the Padua Gymnasium (Roffeni 1611).

119. Ibid., 197; Horky á Lochovič 1610, 137–38.

120. Roffeni 1611, 198.

121. Horky á Lochovič 1610, 138: "Nihil vidi, quod naturam veri planetae redoleat" (quoted by Roffeni 1611, 198).

122. Roffeni 1611, 200.

123. Galileo to Vinta, July 30, 1610, Galilei 1890–1909, 10: no. 370, p. 410; Galileo to Castelli, December 30, 1610, ibid., 10: no. 447, p. 504; Galileo to Clavius, December 30, 1610, ibid., 10: no. 446, p. 500.

124. Galileo to Vincenzio Giugni, June 25, 1610, ibid., 10: no. 339, pp. 379–82; Kepler 1993, xiii–xv.

125. Hasdale to Galileo, August 9, 1610, Galilei 1890–1909, 10: no. 375, pp. 417–18: "Velsero Augustano, tutto spagnuolo et poco amico de' Venetiani." See Kepler 1993, lxiii–lxiv.

126. The letter is not extant but is referred to in replies to Galileo from Giuliano de'Medici and Martin Hasdale; Kepler attempted unsuccessfully to decipher the anagram in his treatise on Jupiter's satellites, and Galileo finally revealed the solution to Giuliano de'Medici on November 13, 1610 (Galilei 1890–1909, 10: no. 427, p. 474 ("Altissimum planetam tergeminum observavi": "I have observed that the highest planet is threefold"); Kepler 1937–, 4:345–46; for discussion of this episode, see Kepler 1993, p. lv, n. 12).

127. Kepler to Galileo, August 9, 1610, Kepler 1937–, 16: no. 584, pp. 319–23.

128. Kepler to Horky, August 9, 1610, ibid., 16: no. 585, p. 323. Kepler did not mention to Galileo that he had received Horky's book through Welser.

129. Kepler to Galileo, August 9, 1610, ibid., 16: no. 584, pp. 320, 322. Kepler assumed that Galileo would make the connection to Lorenzini and perhaps also to the *Stella Nova*.

130. Ibid., 16:321: "Certamen hoc virtutis est cum vitio. . . . Et verò, non problema philosophicum, sed quaestio juridica facti est, an studio Galilaeus orbem deluserit."

131. Ibid.

132. Ibid., 16:322: "In te uno recumbit tota observationis authoritas."

133. Kepler 1993, xxxiv.

134. Galileo to Kepler, August 19, 1610, Kepler 1937–, 16: no. 587, p. 328.

135. Ibid.

136. Ibid.

137. Kepler 1993, 37–38.

138. Ibid., 39.

139. For the printing history, see Kepler 1993, cxx–cxxi.

140. Kepler to Galileo, January 9, 1611, Galilei 1890–1909, 11: no. 455, p. 17.

141. Kepler to Galileo, August 9, 1610, Kepler 1937–, 16: no. 584, p. 156, ll. 4–13. This looks like a passage from Wedderburn.

142. Wedderburn 1610, 153.

143. Ibid.

144. Ibid., 167: "Nam qui exactiores fuerunt in supputando caelestium orbium motus, absurdissimis interdum nitebantur suppositionibus: nec mirum; quoniam supposito falso, sequi potest verum, quamvis non e contra."

145. Ibid., 162: "Profecto Brutii et Bruni aniles fabulae."

146. Kepler to Galileo, [December 1610], Kepler 1937–, 16: no. 603, p. 355.

147. On the Jesuit encounter with Galileo and the telescope, see Lattis 1994, 180–216; Galilei 2010.

148. Welser to Clavius, March 12, 1610, Galilei 1890–1909, 10: no. 270, p. 291.

149. Francesco Stelluti to Giovanni Battista Stelluti, September 15,1610, ibid., 10: no. 390, p. 430.

150. Galileo to Christopher Clavius, September 17, 1610, ibid., 10: no. 391, pp. 431–32.

151. Lodovico Cardi de Cigoli to Galileo, October 1, 1610, ibid., 10: no. 403, pp. 441–42: "These followers of Clavius, all of them, believe nothing. Clavius, among others, the head of them all, said to a friend of mine concerning the four stars that he [Clavius] was laughing about them and that one would first have to build a spyglass that makes [*faccia*] them, and [only] then would it show them. And he [Clavius] said that Galileo should keep his own opinion and he [Clavius] would keep his." Quoted and trans. in Lattis 1994, 184 (with slight modification, as noted).

152. There is a Jesuit observation log in Galileo's hand for November 28,1610–April 11,1611 (Galilei 1890–1909, 3: pt. 2, pp. 863–64; reproduced in Lattis 1994, 189).

153. Clavius to Welser, January 29, 1611, Clavius 1992, 6: no. 324, p. 168.

154. Collegio Romano mathematicians to Bellarmine, April 24, 1611, Galilei 1890–1909, 10: no. 520, pp. 92–93. Odo van Maelcote's praise of Galileo before the Collegio Romano (Galilei 1890–1909, 3: pt. 1, pp. 293–98) described him as "Galilaeus Patritius Florentinus, inter astronomos nostri temporis et celeberrimos et foelicissimos merito numerandus.."

155. Galilei 1890–1909, 3:297: "En tibi iam certum, Venerem moveri circa Solem (et idem, procul dubio, dicendum de Mercurio) tanquam centrum maximarum revolutionum omnium planetarum. Sed et illud indubitatum, Planetas non nisi mutuato a Sole lumine illustratos splendescere: quod tamen non existimo verum esse in stellis fixis."

156. Odo van Maelcote to Kepler, December 11, 1612, 3: pt. 2, no. 810, p. 445. Van Maelcote, who, by this date, had also read Kepler's *Dioptrice*, the *Astronomia Nova*, and the *Conversation with Galileo's "Sidereal Messenger,"* cautiously did not describe the spots as *in Sole*.

157. On April 19, 1611, Bellarmine wrote as follows to the mathematicians of his order, submitting a list of the new observations made by Galileo: "I hear various opinions spoken about these matters and Your Reverences, versed as you are in the mathematical sciences, will easily be able to tell me if these new discoveries are well founded, or if they are rather appearances and not real" (Bellarmine to the mathematicians of the Collegio Romano, Galilei 1890–1909, 11: no. 515, pp. 87–88; quoted and trans. in Lattis 1994, 190).

158. See my conclusion, this volume.

159. Clavius 1611, 3:75:

Inter alia, quae hoc instrumento visuntur, hoc non postremum, locum obtinet, nimirum Venerem recipere lumen à Sole instar Lunae, ita ut corniculata nunc magis, nunc minus, pro distantia eius à Sole, appareat. id quod non semel cum alijs hic Romae obseruaui. Saturnus quoque habet coniunctas duas stellas ipso minores unam versus Orientem, & versus Occidentem alteram. Iuppiter denique habet quatuor stellas erraticas, quae mirum in modum situm & inter se, & cum Ioue variant, ut diligenter & accurate Galilaeus Galilaei describit.

Quae cum ita sint, videant Astronomi, quo pacto orbes coelestes constituendi sint, ut haec phaenomena possint saluari.

160. For further discussion, see Lattis 1994, 198–202.

161. See Grant 1984; Ariew 1999, 97–119; Schofield 1981, 1989; Dinis 1989; Romano 1999; Galilei 2010, 44.

CONCLUSION

1. These were also still the main structures underlying what Jonathan Israel (2001, 714–15) calls "the Radical Enlightenment."

2. By contrast, late-nineteenth- and twentieth-century modernism saw itself as ever more independent of the past (Schorske 1980, xvii).

3. On the crisis of religious certitude, see Richard H. Popkin's classic history of skepticism (2003); Popkin 1996.

4. Less attention has been paid to practical astronomy, but see the pioneering work of James Evans (1998); Gingerich 1993d.

5. Accompanying "A Statement by 192 Leading Scientists," which "caution[ed] the public against the unquestioning acceptance of the predictions and advice given privately and publicly by astrologers," the astrophysicist Bart Bok acknowledged that that up to the time of Newton, "there were good reasons for exploring astrology"; but in a supporting companion article, the science writer Lawrence E. Jerome explained that Kepler "used his position as an astrologer strictly as a means of earning a living and supporting his astronomical observations" (Bok and Jerome 1975, 23, 48).

6. On Galileo's astrological activities, see Campion and Kollerstrom 2003 (including "Galileo's Horoscopes

for His Daughters," 101–5); Rutkin 2005; Bucciantini and Camerota 2005; Swerdlow 2004b; Kollerstrom 2001; Ernst 1984; Righini 1976; Favaro 1881.

7. Foscarini 1615, 218–19. I have followed Blackwell's translation except where indicated in parentheses.

8. Ibid.

9. Ibid., 219.

10. In the judgment of Ernan McMullin, the specific arguments of Foscarini's *Letter* and his defense of it against an unnamed critic, together with Galileo's earlier letter to Castelli, constituted "what might seem a fairly strong case for not proceeding against Copernicanism" (McMullin 2005c, 104–5).

11. John Paul II 1992. See Blackwell 1998.

12. Hooke 1674: "I have begun with a Discourse composed and read in Gresham Colledge in the Year 1670. when I designed to have printed it, but was diverted by the advice of some Friends to stay the repeating the Observation, rather then publish it upon the Experience of one Year only."

13. Ibid., 2–3.

14. On the importance of Urban's bull in precipitating Galileo's downfall, see Shank 2005b; Walker 1958, 205–12; Ernst 1984, 1991; Shea in Lindberg and Numbers 1986, 128–29; Dooley 2002.

15. See Garin 1975, 31–32.

16. Castelli to Galileo, December 14, 1613, in Finocchiaro 1989, 47–48.

17. Galileo to Castelli, December 21, 1613, in ibid., 49–54.

18. Bellarmine to Foscarini, April 12, 1615, ibid., 166; Galilei 1890–1909, 12: no. 1110, pp. 171–72, my italics.

19. In fact, Bellarmine, evidently following Foscarini's formulation, mistakenly believed that supposing the Earth's annual motion might better save the appearances than using eccentrics and epicycles!

20. See Clavius 1611, 3:75; Lattis 1994, 181.

21. Blackwell 1991, 222.

22. Ibid., 221.

23. Ibid., 223. Foscarini was careful not to invoke the name of Giordano Bruno.

24. See Kelter 1992.

25. "Dove non solo fa ufficio di matematico che *suppone:* ma anco di fisico che *dimostra* il moto de la terra" (Bruno 1955, dialogue 3, p. 149, ll. 12–14, my emphasis). This passage would seem to strengthen Blackwell's suggestion (1991, 48): "It seems to be highly likely that Bellarmine would have clearly remembered, and would have been personally influenced by, his experiences in the difficult Bruno affair when he came to deal with Copernicanism and Galileo in 1615–16." To be sure, Bellarmine's and Bruno's formulations echo Osiander's language in his letter to the reader: "*Causas . . .* seu hypotheses, cum *ueras* assequi nulla ratione possit, qualescunque excogitare & confingere, quibus *suppositis*" (Copernicus 1566, fol. 1v).

26. Bellarmine to Foscarini, April 12, 1615, Finoc-

chiaro 1989, 68. On the influence of Bellarmine's scripturalist reading of the heavens in this matter, see McMullin 2005b.

27. Consultant's Report, Finocchiaro 1989, 146–47.

28. Cardinal Paul Poupard, who in 1990 coordinated the results of the commission to study the Galileo affair, read Bellarmine's response as embodying a degree of flexibility that might permit future shifts in the Church's judgment on the proof of the Earth's motion. But Annibale Fantoli (2002, 6–9) has seriously called Poupard's reading into question, and Ernan McMullin (2005b) has asked incisively why Bellarmine and the theologians did not avail themselves of mitigated alternatives, such as the Tychonic arrangement or Urban VIII's theologically grounded, voluntarist exclusion of the possibility of a necessarily true demonstration or, indeed, why the Church went beyond a simple ban on publication.

29. Cardinal Bellarmine's certificate (May 26, 1616), in Finocchiaro 1989, 153.

30. Decree of the Index (March 5, 1616), in ibid., 148–50.

31. These events, much raked over, are finally becoming clearer on the basis of new archival information and persuasive synthesis (see Fantoli 2005, 117–49; Artigas, Martínez, and Shea 2005, 213–33).

32. Lerner 2004, 70–71.

33. Lerner points out that some sixteenth-century readers carried out their own independent acts of censorship by crossing out the names of Lutherans (ibid., 72).

34. For differing assessments of the prehistory of Galileo's claim, see Ariew 1987; Thomason 2000; Goldstein 1969.

35. Brahe 1913–29, 4:158: "Nova Mundani Systematis Hypotyposis."

36. Mulerius 1617.

37. Lower to Harriot, February 6, 1610, in Rigaud 1833, 42–43.

38. Both letters to Müller are quoted and translated in Russell 1964, 8; see also Peter Crüger to Johannes Kepler, July 15, 1624, Kepler 1937–, 18: no. 990, p. 191.

39. Crüger to Philipp Müller, July 1, 1622, Kepler 1937–, 18: no. 933, p. 92.

40. In *The Structure of Scientific Revolutions*, chap. 12, Kuhn says that in a scientific revolution, "there can be no proof" and that the issue is one of "persuasion"; further, this persuasion occurs "for all sorts of reasons and usually for several at once" (1970, 152–53). Paul Hoyningen-Huene reminds us that Kuhn was not ruling out the persuasiveness of good reasons (1993, 252–53).

41. Morinus 1641, 17–18. In this treatise, Morin shows only the example of Mercury. A rigorous demonstration of equivalence would have involved more than the elliptical epicycle that Morin showed, including proper calibration of the eccentricities, apsidal lines maintaining fixed directions, epicycles revolving in a

direction opposite to their motion around the Earth, and, most difficult to conceive, conformity to Kepler's area law (on this problem, see Swerdlow 2004a,97).

42. Kepler 1992, 46–47; Kepler 1937–, 3:19.

43. Kepler 1937–, 7:253; Kepler 1939, 847–48.

44. Johannes Remus Quietanus wrote to Kepler on July 23, 1619, that Galileo wanted "your Copernican book," but although it was prohibited in Florence, he thought that Leopold (of Austria) could easily get it for him ("Desiderat Galilaeus habere librum tuum Copernicanum quia est prohibitus et Florentiae non haberi potest, unde petijt á Serenissimo nostro eundem librum, se enim facilè habiturum licentiam asserit": Kepler 1937–,17, no. 845, p. 362; also Galilei 1890–1909, 12, no. 1403, p. 469). Kepler replied on August 4 that "all of my books are Copernican" but that he suspected the book in question to be the *Epitome* ("Omnes enim mei sunt copernicani. . . . Suspicor igitur, de Epitoma Astronomiae Copernicanae tibi sermonem esse": Kepler 1937–, 17: no. 846, p. 364).

45. Galilei 1997, Second Day, 122: "Salviati: And where do you leave the predictions of astrologers, which after the event can be so clearly seen in the horoscope, or should we say in the configuration of the heavens?"

46. Campanella to Galileo, August 5, 1632, Galilei 1890–1909, 14: no. 2284, p. 366.

47. Galilei 1967, 416 (postil).

48. Pointing to a note that Galileo wrote on a flyleaf of the *Dialogue,* Jean D. Moss suggests that Galileo believed privately that his argument was "dialectically convincing but not completely demonstrated" (1993, 297 ff.).

49. Galilei 1997, 80. Finocchiaro translates *persuasibile* as *plausible,* whereas I follow John Florio's "that may be perswaded" (1611). See Finocchiaro's extensive gloss on the epistemic strength of Galileo's term and on the meaning of "defense" (ibid., 80 n. 11).

50. Ibid., 284. Slightly later, Salviati endorses an Aristotelian demonstration, advancing the thesis that the tidal effects must follow from the Earth's motions "with necessity, so that it is impossible for them to happen otherwise" (288; also 302 n. 44).

51. Ibid., 304.

52. Ibid., 307: "Whether God with His infinite power and wisdom could give to the element water the back and forth motion we see in it by some means other than by moving the containing basin; I say you will answer that He would have the power and the knowledge to do this in many ways, some of them even inconceivable by our intellect." See Finocchiaro's important gloss on this passage.

53. Palmerino 2004, 234.

54. See Westman 1983, 329–72.

55. Gassendi 1659, 25: "Forasmuch as the Stars are General Causes only, in respect of sublunary things; we may well demand a reason, why any singular effect may not be ascribed to some singular Cause here

below, where are such multitudes of natural and convenient Actives and Passives, rather than to those remote ones, the Stars." Gassendi apologized for "so tedious a list of whimzies" (ibid., 37).

56. Mersenne 1623.

57. See Williams 1948.

58. Hine (1973) provides a helpful accounting; Mersenne 1623, question 9 ("The Earth"), article 4 ("Whether or not the Earth may be moved, reasons for its mobility are affirmed"). Among the authors to whom Mersenne attributed arguments were Andreas Libavius, William Gilbert, Celio Calcagnini, David Origanus, Tycho Brahe, and Michael Maestlin.

59. Kepler and Fludd disagreed on the use of pictures and symbols to interpret the birth and layout of the universe (see Pauli 1955, 147–240; Westman 1984).

60. Mersenne 1623, question 9, article 4, col. 894.

61. Garber 2004. Mersenne displays a simplified diagram of the Copernican system to which he applies the objection that from the central Sun, "heavy bodies ascend absolutely and naturally; Christ ascends to the lower heavens and descends to the upper heavens" (Mersenne 1623, question 9, article 5, col. 897).

62. On the history of the Galileo affair after the trial, see Finocchiaro 2001; Finocchiaro 2005; Segre 1998; Galluzzi 1998; Fantoli 2003, 345–74.

63. See Russell 1989, esp. 367–68; Heilbron 2005; Sarasohn 1988.

64. See Heilbron 1999.

65. Rienk Vermij, for example, argues this position for the Low Countries, where Descartes secured a beachhead at Utrecht and Leiden in the 1640s (2002, 156–87).

66. As a student at La Flèche in June 1611, Descartes participated in a memorial celebration on the death of Henri IV, the school's patron, at which one of the poems recited was entitled "Concerning the Death of King Henry the Great and on the Discovery of Some New Planets or Wandering Stars around Jupiter, Noted the Previous Year by Galileo, Famous Mathematician of the Grand Duke of Florence" (quoted in Ariew 1999, 100).

67. Beeckman's diary references Kepler's *Astronomia Nova*, chap. 35, and adds his own opinion, as follows: "*Id quod semel movetur, semper moveri*. Manent igitur planetae post alium latitantes in eo motu, in quo erant; imo propter refractionem nonnihil lucis ad eum venit. Ergo nullius est momenti retardatio, quae aliqua potest esse ob absentiam partis alicujus virtutis moventis. Est tamen aliqua, quae possit esse causa motus apheliÿ" (Beeckman 1939–53, 3:101). The main study of Beeckman is Berkel 1983, but Schuster (1977) first called my attention to important connections between Beeckman and Descartes. See also Gaukroger 1995, 68–103; Vermij 2002, 113–19.

68. "Tentamen de Motuum Coelestium Causis,"

Acta Eruditorum, February 1689; quoted by I. B. Cohen 1972, 205.

69. Descartes 1983, pt. 3, prop. 35, p. 98; see Applebaum 1996, 454.

70. For Descartes's vortices and planetary theory, see Aiton 1989; Aiton 1972, 30–64; Gaukroger 2002, 144–46.

71. Descartes 1983, pt. 3, prop. 52, p. 110; Kepler 1939, bk. 4, chap. 2, 855.

72. See, in particular, letters from Descartes to Beeckman, August 22, 1634, Descartes 1897–1913, 1: no. 57, p. 307, and to Golius, May, 19, 1635, ibid., 1: no. 60: 324. In the latter Descartes refers to Morinus 1634, which attacked Jacob Landsbergen, an ardent follower of Kepler.

73. Descartes 1983, pt. 3, props. 32–33, pp. 97–98.

74. The standard account, already articulated by Henry More in the 1660s, is that, in reaction to Galileo's condemnation, Descartes shifted his position from explicit endorsement of Copernicus's theory in *Le monde* (1633) to a relativist conception of the Earth's motion in the *Principia Philosophiae* (see, for example, Koyré 1966, 333). Contrary to this view, see Michael Mahoney's introduction to Descartes 1979, xviii. Daniel Garber argues that Descartes' position was both "political and prudential" and cognitively sincere (1992, 181–88). Stephen Gaukroger endorses the first explanation in his (1995, 185, 290–92, 304, 408); but in Gaukroger 2002 (142–46), he moves away from the earlier externalist explanation, associating himself now with the explanation favored by Garber, grounded in the internal coherence of Descartes's principles of matter, space, and motion.

75. Descartes 1983, pt. 3, prop. 19, p. 91.

76. Ibid., pt. 3, props. 38–39, pp. 101–2.

77. See Lerner 1996–97, 2:177, 181, 187.

78. Descartes 1983, pt. 3, prop. 41, pp. 103–4.

79. Ibid., pt. 3, props. 43–45, pp. 104–6.

80. Wilkins 1684, "To the Reader," fol. A3r; 13.

81. Ibid., 14.

82. Ibid., 16.

83. Jean Moss emphasizes the centrality of Aristotelian dialectical reasoning in the Copernican episode and for Wilkins in particular (1993, 301–29).

84. Wilkins 1684, 17.

85. Ibid., 18.

86. See Dinis 2003, 195–224; Dinis 1989.

87. Wilkins 1684, 16.

88. Riccioli 1651, pt. 1, xviii. The world-systems controversy is treated in bk. 9, sections iii–iv.

89. Diagrams and descriptions of the Copernican/Aristarchan and Tychonic systems precede the diagram of Riccioli's own system (ibid., bk. 3, 102–3). Dinis suggests that Riccioli centers Jupiter and Saturn on the Earth in order to avoid an excessive privileging of the Sun (1989, 133). Six hundred and twenty pages later, in an appendix to book 3, Riccioli qualifies his own system as "merely probable . . . provided that the

eccentricities of the planets . . . are so constituted that the phenomena of the motions may be saved without evident error" (731).

90. Riccioli, who derived some of his sources from Mersenne, probably intended here Roberval 1644; Mersenne published a second edition of Roberval's treatise in his *Novarum Observationum* (1647, 3:1–64).

91. Riccioli claimed that although Gassendi was aware of the decree against Galileo, he did not feel that it should restrain his opinions; later, Riccioli included Gassendi among the Copernicans. For Gassendi's position, see Palmerino 2004, 234.

92. The details of this argument are too extensive to treat here. Adrien Auzout and Christiaan Huygens regarded it as suspiciously beneath Riccioli's capabilities (Auzout 1664–65, 58–59; Huygens 1888–1950, 21:824–25). Dinis claims that the supposed proof is based on a single sentence in Galileo's *Dialogue:* "The true and real motion of the stone is never accelerated at all, but is always equable and uniform" (Galilei 1967, Second Day, 166) against which Riccioli argued that the motion was, indeed, accelerated! (Dinis 2003, 208 n. 71). Alexandre Koyré (1955) emphasized the sheer conceptual difficulties faced not only by Riccioli but even by those who, like Giovanni Alfonso Borelli, understood Galileo's problem better. With considerable dexterity, Peter Dear (1995, 76–85) has shown how Riccioli sought to establish his expertise in the manner by which he represented his observations and experiences with falling bodies and the inferences that he drew from those events.

93. Riccioli 1651, pt. 2, bk. 9, sect. 4, chap. 35, conclusion 4, p. 478; quoted in Dinis 2003, 208–9.

94. Riccioli associated his careful separation between "pure" or "free" philosophical reasons and sacred scripture with Kepler's discussion of holy scripture in the introduction to the *Astronomia Nova* (1651, bk. 9, sect. 4, chap. 35, conclusion 7, p. 479).

95. The question of Riccioli's intellectual sincerity in light of the Church's condemnation of Galileo continues to vex commentators. Jean-Baptiste Delambre confidently expressed the view that "without his robe, he would have been a Copernican" (1821, 2:279). Present opinion leans toward consistency between Riccioli's private and public views (Schofield 1981; Grant 1984, 13–15; Dinis 2003, 208–9; Riccioli, 1651, bk. 9, sect. 4, ch. 35, pp. 478–79). Dinis presents evidence that Riccioli's position hardened considerably after the *Almagestum Novum.* See also Martin 1997.

96. For the decree of the council's fourth session, April 8, 1546, see Blackwell 1991, 181–84.

97. Riccioli 1651, bk. 9, sect. 4, ch. 35, p. 493. For excellent discussion, see Dinis 2003, 210.

98. Riccioli 1651, bk. 9, sect. 4, chap. 1, p. 290: "Iam tandem controuersiam aggredimur, inter Astronomicas, hoc praesertim saeculo, longè celeberrima." Riccioli defines *systema mundi* as "nothing other than

the coordination or composition of the great parts of the world, i.e., the elements and the heavens, the matter and number of which has as a whole and between its parts, form, order and place relative to the center of the Universe" (bk. 9, sect. 3, chap. 2, p. 276).

99. Dick 1982, 97–105. According to Barbara Shapiro, John Wilkins "became the chief English exponent of the doctrine of the plurality of worlds" (1969, 33). For Bruno's influence on Wilkins, see Ricci 1990, 110–14.

100. Dick 1982, 117; More 1646.

101. Quoted in Dick (1982, 126) from the 1688 London trans. of Joseph Glanville, *A Plurality of Worlds.*

102. On the utility of Habermas's notion of civil society, see Broman 1998; Broman 2002.

103. Dick 1982, 97–98.

104. Ibid., table 2, 136–38.

105. Capp 1982, 280.

106. Curry 1989, 19. Curry's claim is grounded in the work of Christopher Hill, Bernard Capp, and Keith Thomas.

107. John Gadbury, for example, put together a large collection of genitures, and John Goad followed a Baconian course in accumulating weather information (ibid., 74–78). Bowden 1974, 176–95.

108. Lilly 1647, unpaginated, following 832.

109. Ibid., 19.

110. Curry associates political and religious affiliation with astronomical and astrological positionings (1987, 245–59).

111. Ward 1654, 2.

112. Wing and Leybourne 1649.

113. Wing 1651, "To the Reader," unpag.

114. Ibid; Wing 1656, 37: "The learned and painful Bullialdus (to make the operation more easie) shews how to effect the same by and Epicycle, whose motion is supposed to be double to the motion of the Planet in his Orbe, and so . . . it may be found with more ease, which way in my judgment is the most rationall and absolute of all other." For Boulliau's derivation of the elliptical orbit from uniform circular motion, see Wilson 1989, 172–76; Hatch 1982, xxvii–xxxiv.

115. Wing 1651, fol. Aa3v.

116. Wing 1656, 33; Wing 1669, 115. A long postil to this passage makes obvious Wing's rejection of Riccioli.

117. Wing 1669, 116.

118. D. T. Whiteside judges him to have been "a careful observer of celestial phenomena with a good knowledge of current computational techniques, but not a man strongly endowed with mathematical ability" (1970, 7).

119. Streete 1661, fol. A2: "In the sad and doleful Night of Your Sacred Majesties absence from your People, was this small Astronomical Work begun, and carefully continued for some years."

120. Ibid., 6.

121. On the importance of the historically witnessed experimental event, see Dear 1995, 63–92.

122. As an example of such a trail ending, consider: "*Anno 1661. April* the 23th being the day of the Coronation of our most Gracious Soveraign King *Charles* the Second; That ingenious Gent.*Christianus Hugenius*, of *Zulichem*, Mr. Reeves, with other Mathematical friends and my self, being together at Long-Acre, by help of a good Telescope, with red glasses for saving our eyes, saw Mercury from a little past one until two of Clock, appearing in the Sun as a round black spot, below and to the right hand, so that in the Heavens he was above, and to the left from the Suns Center, and entred on the Sun much about one of Clock" (Streete 1661, 118).

123. Ibid., 7.

124. Curtis Wilson 1989a, 161–63. Wilson regards the striking improvements in observing instruments (the filar micrometer, pendulum clock, and telescopic sights) between the 1640s and the 1660s as a revolution in its own right: it produced better solar and lunar parallax values and raised new anomalies, but it did not lead to "the sharp improvements in the accuracy of the tables that one might have expected."

125. For example, Philip Lansbergen's *Tabulae Perpetuae* (1632), the tables in Ismael Boulliau's *Astronomia Philolaica* (1645), John Newton's *Astronomica Britannica* (1657), Riccioli's *Astronomia Reformata* (1665), and Jeremy Shakerley's *Tabulae Britannicae* (1653). Isaac Newton could cite Kepler's and Boulliau's mean solar distances as authoritative in confirming Kepler's third law on the grounds that they had "with great care determined the distances of the planets from the sun; and hence it is that their tables agree best with the heavens" (quoted in Wilson 1989b, 241).

126. Whiteside 1970, 7, 16 n.; Westfall 1980, 94.

127. These copies are located, respectively, in the libraries of Columbia University and Trinity College, Cambridge. See McGuire and Tamny 1983, 300–301; Westfall 1980, 155 n. 44.

128. Whiteside (1970, 16) references a section of Newton's notebook titled "Systema Mundanum Secundu[m] Copernicum" (Pierpont Morgan Library, New York).

129. London: Thomas Newcomb, 1654, bk. 1, sect. 1, 4.

130. For Newton's crucial experiment in optics, see A. Shapiro 1996.

131. For the Dutch context, see Vermij 2002, 139–76.

132. Hooke 1665, Preface, fol. g2r–v; Simpson 1989, 34.

133. References to some of Riccioli's observations are already found in Hooke 1665, 230, 238.

134. Gregory 1668, 693–98; reprinted in Koyré 1955, 354–58.

135. Riccioli's mobilization of experience is nicely captured by Dear 1995, 83–85.

136. Hunter and Schaffer 1990, introduction, 19.

137. Johns 1998, 428; Dennis 1989.

138. See J. Bennett 1990, 21–32.

139. Hooke 1665, 230.

140. Hooke 1674, 2.

141. Ibid., 4.

142. Ibid., 5–6.

143. Ibid., 8–9. "Though *Riccioli* and his ingenious and accurate Companion *Grimaldi* affirm it possible to make observations by their way, with the naked edge to the accurateness of five seconds; Yet *Kepler* did affirm and that justly, that 'twas impossible to be sure to a less Angle then 12 seconds: And I from my own experience do find it exceeding difficult by any of the common sights yet used to be sure to a minute" (10).

144. See Catherine Wilson 1995, 87–88; on the political meanings and uses of Hooke's representations, see Dennis 1989.

145. On the probabilistic views of English experimentalists, see Shapin and Schaffer 1985, 23–24.

146. Cf. Shapin 1994, 121: "The practice which emerged with the Interregnum work of Boyle and his Oxford associates, and which was institutionalized at the Restoration in the Royal Society of London, was strongly marked by its *rejection* of the quest for absolutely certain knowledge, by its suspicion of logical methods and demonstrative models for natural science, and by its tolerant posture towards the character of scientific truth." See also ibid., 287, 309.

147. *Novum Organon*, Bacon 1857–1859, 2: xxxvi; Urbach 1987, 169–71.

148. Cf. Shapin 1994, 307–9. It was later determined that Hooke's attempt to detect annual parallax had actually failed, although Flamsteed believed that he had confirmed it, and Newton also accepted the result (see Curtis Wilson 1989c, 240).

149. For religious controversies, see Milward 1978a, 1978b. Professionalizing historians of science of the 1940s and '50s, led by Koyré, Herbert Butterfield, A. Rupert Hall, and Kuhn, appropriated the meaning of *revolution* as punctuated discontinuity but displaced the Marxist trope with an image of science shorn of any association to social class (see I. B. Cohen 1985, 389–404; R. S. Porter 1992; H. F. Cohen 1994). When social class regained historiographical legitimacy, it had lost its connection to scientific revolution (see esp. Shapin and Schaffer 1985).

150. Dear himself offers a nonstandard reading of Newton (1995, 248–49). As early as 1839, Augustus De Morgan maintained: "The controversy ceases to have any interest after the publication of the *Principia* of Newton. Even to this day, we believe there are some who deny the earth's motion, on the authority of the Scriptures, and every now and then a work appears producing mathematical reasons for that denial; these works, as fast as published, after making each two converts and a half in a country town, are heard of no more until fifty years afterwards, when they are discovered by bibliomaniacs bound up in volumes of tracts with dissertations on squaring the circle, and perpetual motion, and pamphlets predicting national bankruptcy" (De Morgan 1839).

151. Kuhn's treatment of this period was indebted in no small way to Koyré 1955; but see now Lerner 1996–97, 2: chap. 6, 137–89.

152. Kuhn 1957, 261, my italics. I do not wish to understate the full scope of Kuhn's interpretation, which, beyond asserting that Newton provided "an economical derivation and plausible explanation of Kepler's laws," also claims for him "a new way of looking at nature, man, and God."

153. Jon Dorling (n.d.) attempts a Bayesian interpretation of Riccioli's compendium of probabilities.

154. See Vermij 2002, 139–237; Moesgaard 1972b; Sandblad 1972; Zemplén 1972.

155. Cudworth 1678, preface: "The True Intellectual System of the Universe; in such a sense, as Atheism may be called, a False System thereof: The Word Intellectual being added, to distinguish it from the other, Vulgarly so called, Systems of the World, (that is, the Visible and Corporeal World) the Ptolemaick, Tychonick, and Copernican; the two Former of which, are now commonly accounted to be False, the Latter True." See further McGuire 1977.

156. See Curry 1987, 254–55.

157. See Burns 2000.

158. See Pugliese 1990; cf. Kuhn 1957, 249–52.

159. Newton, *De Motu*; quoted in Curtis Wilson 1970, 160–61.

160. See Schaffer 1987, 1993; Genuth 1997, 133–55.

161. On Whiston, see Force 1985; on Halley, Genuth 1997, 156–77; Schaffer 1993.

162. Kuhn 1957, 265. I read this position as compatible with Duhem's (Westman 1994, 83–85).

163. Newton (1962, 547) anticipates the objection in his General Scholium to the *Principia*: "But hitherto I have not been able to discover the cause of those properties of gravity from phenomena, and I frame no hypotheses; for whatever is not deduced from the phenomena is to be called an hypothesis; and hypotheses, whether metaphysical or physical, whether of occult qualities or mechanical, have no place in experimental philosophy." On the tension between explanatory hypotheses and mathematical description in Newton's work, see McMullin 1990, 67–76.

164. McGuire and Rattansi 1966, 125.

165. Iliffe 1995, 164.

166. Iliffe 1995, 168 (citing Yahuda MS 41, fol. 9v). The bracketed text was added by Newton.

167. See Snobelen 2001, 174–75.

168. See Hunter 1987; Bowden 1974, 218–24.

169. On Quine, see Klee 1997, 63–69; Gillies 1998.

170. As Lorraine Daston contends, the emergent eighteenth-century calculus of risk "regarded expectation as a mathematical rendering of pragmatic rationality" (1988, 49–111, 182–87). On sources of credibility for social theory, see T. Porter 2003.

BIBLIOGRAPHY

Manuscript citations are listed alphabetically in the first section of the bibliography according to the name and location of the repository, the title of the collection, folio reference and date. They are cited in the endnotes according to the name of the repository, the author's name or the manuscript title, and the date (as applicable). Primary and secondary sources are integrated into a single list and cited in the endnotes by the conventional short form of author's name and publication date.

In this bibliography I refer to two kinds of bundled volumes. The first is an omnibus or compendium edition, a group of works issued as a single volume by a publisher and usually paginated continuously. An example of an omnibus bundle is the 1566 edition of De Revolutionibus published at Basel by Heinrich Petri. The second is a singular volume in which an owner has bound together several works of his or her own choosing. An example is the 1493 omnibus edition of Ptolemy's Tetrabiblos, published at Venice by Ottaviano Scotto and bound with two other, separately published works; it is held by the Biblioteca Universitaria di Bologna.

MANUSCRIPTS

Archivio di Stato di Bologna
Archivio del notaio Lorenzo Benazzi, 1459–1508. Accessible at http://patrimonio.archiviodistatobologna.it.
Liber Partitorum magnificorum dominorum Sedicem, 1480–, vols. 10–12. Acts of the chief magistracy of Bologna, the Sedici Riformatori.

Biblioteca Ambrosiana, Milan
Library Inventory of Gian Vincenzo Pinelli. MS R104 Sup., fols. 237–39.

Biblioteca Medicea Laurenziana, Florence
James of Spain. 1479. MS Plutei Principali 34, Sup. 22.
Ristori, Giuliano. 1537. "Prognostic upon the Geniture of the Most Illustrious Duke Cosimo de Medici." MS Plutei Principali 89, Sup. 34.

Biblioteca Nazionale Centrale di Firenze
da Savoia, Giovanni. 1537. "Judicium de Commutationibus Saturni et Martis et eius Saturni cum Joue fato." MS Magliabechiano XX.10.
Galilei, Galileo. n.d. Lecture on the nova of 1604; various reading notes. MS Galileana 47, fols. 4r–13r.
———. n.d. "Astrologica Nonnulla." MS Galileana 81.
Guidi, Giovanni Battista. 1561. "Natività di Francesco I." MS Magliabechiano XX.19.
———. 1566. "Natività di Francesco I." MS Magliabechiano XX.38.
Kepler, Johannes. July 18, 1599. Letter to Edmund Bruce. MS Galileana 88, fols. 35–40.
Ristori, Giuliano. 1547–58. [Provenance of Filippo Fantoni.] Lectures on Ptolemy's *Quadripartitum.* MS Conventi Soppressi F.IX.478.
———. n.d. [Commentaries of Filippo Fantoni]. Copy of Giuliano Ristori's lectures on Ptolemy's *Quadripartitum.* MS Conventi Soppressi B.VII.479.
Tolosani, Giovanni Maria. 1546–47. "De Veritate S. Scripturae." MS Conventi Soppressi J.I.25.

Biblioteca Riccardiana, Florence
Ristori, Giuliano. n.d. [1547]. "Lectura super Ptolomei Quadripartitum . . . ac eximij magistri Iuliani Ristorij Pratensis, per me Amerigum roncionibus dum eum publice legeret in almo Pisarum gimnasio currenti calamo collecta." MS. 157.

Biblioteca Universitaria di Bologna
de Fundis, Johannes Paulus. 1435. "Tacuinus astronomico-medicus." MS L.iv, fols. 1–10r.
Garzoni, Giovanni. 1500. "Laus astrologie." MS 1391
(2648), fols. 207v–208.

Biblioteka Uniwersytecka Wrocławiu
"Brevis Repetitio Doctrinae de Erigendis Coeli Figu-
ris." June 13-October, 1570. University of Wittenberg.
MS. M. 1565.
Schönborn, Barth[olomeus]. 1570–72. University of
Wittenberg. MS. M. 1330. Contains various lectu-
res on mathematics, geography, and astronomy:
"Annotationes in libellum sphaericum Casp. Peu-
cer" (June 19–October 22, 1570); "Tractatus de
nativitatibus" ([October 1570?]– January 17, 1571);
"In Arithmeticen Gemmae Frisii Annotationes
Traditae Witebergae à M. Barth. Schönborn"
(January 29–April 26, 1571); "In Theoricas Plan-
etarum Georgij Purbachij" (October 9, 1571–
February 6, 1572); "In Logisticen Astronomicen
Sebastiani Theodorici" (October 25–November
25, 1571); "In Euclides Elementa" (February 5–
March 31, 1572); and "Initia Doctrinae Geographi-
cae Tradita Publicè in Academia Witebergensis
à Clariss. Viro Dn. M. Barth. Schönborn" (Febru-
ary 7–March 19, 1571).

Bibliothèque Nationale de France, Paris
Blasius of Parma. 1405. "Iudicium revolutionis anni
1405." MS Lat. 7443.
Melletus de Russis de Forlivio. 1405. "Iudicium super
anno 1405." MS Lat. 7443.

Bodleian Library, Oxford
Chamber, John. 1603. "A Confutation of Astrological
Daemonologie, or the divells schole, in defence of
a treatise intituled against Iudiciarie Astrologie,
and oppugned in the name of Syr Christopher
Heydon, Knight." MS Savile 42.
Dee, John. March 25, 1582. "A Playne discourse . . .
concerning ye needful reformation of ye vulgar
kallender." MS Ashmole, 179.

British Library, London
Harriot, Thomas. Notes. MS Birch, 4458, fols. 6–8.
———. Notes. MS Add. 6782, fol. 374.
———. Notes. MS Add. 6788, fol. 67v.
Kepler, Johannes. September 4, 1603. Letter to Ed-
mund Bruce. MS Lansdowne, 89, fol. 26.
Lower, William. June 21, 1610. Letter to Thomas Har-
riot. MS Add. 6789, fols. 427–29.

Gonville and Caius College, University of Cambridge
University of Wittenberg. Lecture notes, 1564–70.
MS 387.

Hauptstaatsarchiv Stuttgart
Maestlin, Michael. January 15, 1586. Letter to Ludwig,
duke of Württemberg, MS. A274 Bü 46.

Österreichische Nationalbibliothek, Vienna
Novara, Domenico Maria. n.d. "De Mora Nati." MS
Vin 5303, fols. 196r–199r.

Stadtarchiv und Stadtbibliothek Schweinfurt
Praetorius, Johannes. 1605. Planetarum Theoriae Incho-
atae. MS H73.

Universitätsbibliothek Erlangen-Nürnberg
Praetorius, Johannes. 1594. "Compendiosa Enarratio
Hypothesium Nic. Copernici, Earundem insuper
alia dispositio super Ptolemaica principia." MS 814.
Schadt, Andreas. 1577. "In Theorias Planetarum Pur-
bachij Annotationes Vitebergae Privatim Traditae."
MS 840.
Straub, Caspar. 1575. "Annotata in Theorias Planeta-
rum Georgii Purbachi." MS 840.

PRINTED SOURCES,
PRIMARY AND SECONDARY

Abenragel, Haly (Haly ibn Ragel, Ibn Abi'l-Ridjāl,
Abu 'l-Ḥasan 'Alī al-Shaybānī al-Kātib al-Maghribī
al-Ḳayrawānī). 1485. In Judiciis Astrorum. Venice:
Erhard Ratdolt.
———. 1551. Libri de Iudiciis Astrorum. Basel: Henricus
Petri.
Abenrodan, Haly (Haly Abenrudian, Ibn Riḍwān,
Abu 'l-Ḥasan 'Alī b. Riḍwān b. 'Alī b. Dja'far al-Miṣrī).
1484. See Ptolemy 1484a.
Abulafia, David. 1995. "Introduction." In The French
Descent into Renaissance Italy, 1494–95: Antecedents
and Effects, ed David Abulafia. Aldershot: Ashgate.
Achillini, A. August 7, 1498. De Orbibus Libri 4. Bolo-
gna: Benedictus Hectoris.
———. 1545. Opera Omnia. Venice.
Advogarius, Petrus Bonus (Avogario, Pietro Buono).
[1495]. Pronostico dell anno MCCCCLXXXXVI. Ferrara: n.p.
———. [1496]. Pronostico dell anno MCCCCLXXXXVII.
Ferrara: n.p.
Ady, Cecilia M. 1937. The Bentivoglio of Bologna: A Study
in Despotism. Oxford: Oxford University Press.
Africa, Thomas W. 1961. "Copernicus' Relation to Aristarchus and Pythagoras." Isis 52:403–9.
Aiton, Eric J. 1972. The Vortex Theory of Planetary
Motions. London: MacDonald.
———. 1981. "Celestial Spheres and Circles." History
of Science 19:75–114.
———. 1987. "Peurbach's 'Theoricae Novae Planetarum': A Translation with Commentary." Osiris 3:
4–43.

———. 1989. "The Cartesian Vortex Theory." In Taton and Wilson 1989, 207–21.

Albubater (Ibn al-Khasīb, Abū Bakr al-Hasan b. al-Khasīb). 1492. *De Nativitatibus*. Venice. June.

———. 1540. *Albubatris Astrologi Diligentissimi, Liber Genethliacus siue De natiuitatibus, Non Solum Ingenti Rerum Scitu Dignarum Copia, Verum Etiam Iucundissimo Illarum Ordine Conspicuus*. Nuremberg: Johannes Petreius.

Album Academicae Vitebergensis. 1841. Ed. C. E. Foerstermann. 3 vols. Leipzig.

Albumasar (Abū Maʿshar Jaʿfar b. Muḥammad b. ʿUmar al-Balkhī). 1994. *The Abbreviation of the Introduction to Astrology*. Ed. and trans. Charles Burnett, Keiji Yamamoto, and Michio Yano. New York: Brill.

Alchabitius (al-Kabīsī, ʿAbd al-ʿAzīz b. ʿUthmān b. ʿAlī, Abu ʾl-Sakr). 1485. *Libellus Isagogicus Abdilasi, id est, Servi Gloriosi Dei: Qui dicitur Alchabitius ad Magisterium Iuditiorum Astrorum: Interpretatus a Ioanne Hispalensi. Scriptumque in eundem a Iohanne Saxonie editum utili serie connexum incipient*. Venice: Erhard Ratdolt. (In Ratdolt–British Library Bundled Copy, q.v.)

Alfonso X, King of Castile and Leon. 1483. *Alfontij regis castelle illustrissimi celestium motuum tabule necnon stellarum fixarum longitudines ac latitudines Alfontij tempore ad motus veritatem mira diligentia reducte. Ac primo Joannis Saxoniensis in tabulas Alfontij canones ordinati incipiunt faustissime*. Venice: Erhard Ratdolt. (In Ratdolt–British Library Bundled Copy, q.v.)

Allegri, Ettore, and Alessandro Cecchi. 1980. *Palazzo Vecchio e i Medici. Guida Storica*. Florence: Studio per Ed. Scelte.

Allen, Don Cameron. 1966 [1941]. *The Star-Crossed Renaissance: The Quarrel about Astrology and Its Influence in England*. New York: Octagon Books.

Allen, Michael J. B., Valery Rees, and Martin Davies, eds. 2002. *Marsilio Ficino: His Theology, His Philosophy, His Legacy*. Leiden: Brill.

Alliaco, Petrus de [Pierre d'Ailly]. 1490. *Concordantia astronomie cum theologia. Concordantia astronomie cum hystorica narratione. Et elucidarium duorum precedentium*. Augsburg: Erhard Ratdolt.

Amico, Giovanni Battista. 1536. *De Motibus Corporum Coelestium Iuxta Principia Peripatetica sine Eccentricis Epicyclis*. Venice: I. Patavino and V. Roffinello.

Anderson, Matthew Smith. 1998. *The Origins of the Modern European State System, 1494–1618*. London and New York: Longman.

Andreae, Jacob. 1567. *Christliche/notwendige und ernstliche Erinnerung/Nach dem Lauff der irdischen Planeten gestelt/Darauss ein jeder einfeltiger Christ zusehen/was für glück oder unglück/Teutschland diser zeit zugewarten. Auss der vermanung Christi Luc 21 in fünf Predigen verfasset*. Tübingen.

Applebaum, Wilbur. 1996. "Keplerian Astronomy after Kepler: Researches and Problems." *History of Science* 24:451–504.

Aquinas, Thomas. 1952. *The Summa Theologica of Saint Thomas Aquinas*. Trans. E. D. Province, Chicago: Encyclopaedia Britannica.

Ariew, Roger. 1984. "The Duhem Thesis." *British Journal for the Philosophy of Science* 35:313–25.

———. 1987. "The Phases of Venus before 1610." *Studies in History and Philosophy of Science* 18:81–92.

———. 1999. *Descartes and the Last Scholastics*. Ithaca, NY: Cornell University Press.

Ariew, Roger, and Peter Barker. 1996. "Pierre Duhem: Life and Works." In Duhem 1996.

Aristotle . 1597. *Operum Aristotelis Stagiritae Philosophorum Omnium Longè Principis Noua Editio, Graecè & Latinè*. 2 vols. Trans. Julius Pacius. [Geneva]: Gulielmus Laemarius.

———. 1936. *On the Soul, Parva Naturalia, On Breath*. Trans. W. S. Hett. Cambridge, MA: Harvard University Press.

———. 1960. *On the Heavens*. Trans. W. K. C. Guthrie. London: Heinemann.

———. 1961–62. *The Metaphysics*. 2 vols. Trans. H. Tredennick. Cambridge, MA: Harvard University Press.

———. 1962a [1562–74]. *Aristotelis Opera cum Averrois Commentariis*. 15 vols. Facsimile ed. [Venice: Iunctas.] Frankfurt: Minerva Verlag.

———. 1962b. *Meteorologica*. Trans. H. D. P. Lee. Cambridge, MA: Harvard University Press.

———. 1963 [1929]. *Physics*. 2 vols. Trans. P. H. Wicksteed and F. M. Cornford. Cambridge, MA: Harvard University Press.

———. 1966. *Posterior Analytics*. Trans. H. Tredennick. *The Topics*. Trans. E. S. Forster. Cambridge, MA: Harvard University Press.

———. 1975. *Posterior Analytics*. Trans. J. Barnes. Oxford: Clarendon Press.

———. 1977. *Politics*. Trans. H. Rackham. Cambridge, MA: Harvard University Press.

Arrizabalaga, Jon. 1994. "Facing the Black Death: Perceptions and Reactions of University Medical Practitioners." In García-Ballester et al. 1994, 237–88.

Artigas, Mariano, Rafael Martínez, and William R. Shea. 2005. "New Light on the Galileo Affair?" In McMullin 2005a, 213–33.

Ashworth, Willam B., Jr. 1990. "Natural History and the Emblematic World View." In Lindberg and Westman 1990, 303–32.

Aulotte, Robert, ed. 1987. *Divination et controverse religieuse en France au XVIe siècle*. Paris: Belles Lettres.

Auzout, Adrien. 1664–65. *Lettres . . . sur les grandes lunettes*. Amsterdam.

Avogadro, Sigismondo. 1521. *Pronostico dell'anno 1521*. N.p.

———. 1523. *Pronostico dell'anno 1523*. N.p.

Avogario, Pietro Buono. *See* Advogarius, Petrus Bonus.

Azzolini, Monica. 2009. "The Politics of Prognostication: Astrology, Political Conspiracy and Murder
in Fifteenth-Century Milan." *History of Universities*
23: 4–34.

Baade, Walter, and Fritz Zwicky. 1934. "Supernovae
and Cosmic Rays." *Physical Review* 45:138.

Bacon, Francis. 1859–64. *The Works of Francis Bacon.*
7 vols. Ed. James Spedding, Robert Leslie Ellis, and
Douglas Denon Heath. London.

Bailly, Jean Sylvain. 1779–82. *Histoire de l'astronomie
moderne depuis la fondation de l'école d'Alexandrie
jusqu'à l'epoque de M.D.CC.XXX.* 3 vols. Paris: Frères
de Bure.

Baldini, Ugo. 1981. "La Nova del 1604 e i matematici
e filosofi del Collegio Romano: Note sur un testo
inedito." *Annali dell'Istituto e Museo di Storia della
Scienza di Firenze* 6, no. 2: 63–97.

———. 1984. "L'astronomia del Cardinale Bellar-
mino." In Galluzzi 1984, 293–305.

———. 1988. "La conoscenza dell'astronomia copernicana nell'Italia meridionale anteriormente al
Sidereus Nuncius." In Nastasi 1988, 127–68.

———. 1991. "La teoria astronomica in Italia durante
gli anni della formazione di Galileo: 1560–1610."
In Casini 1991, 39–67.

———. 1992a. *"Legem Impone Subactis": Studi su filosofia e scienza dei Gesuiti in Italia.* Rome: Bulzoni.

———. 1992b. *"Legem Impone Subactis:* Teologia, filosofia e scienze mathematiche nella didattica e nella
dottrina della Compagnia di Gesù." In Baldini 1992a,
19–73.

Baldini, Ugo, and George V. Coyne. 1984. *The Louvain
Lectures (Lectiones Lovanienses) of Bellarmine and
the Autograph Copy of his 1616 Declaration to Galileo.* Studi Galileiani Special Series, vol. 1, no. 2. Vatican City: Specola Vaticana.

Barker, Peter. 1999. "Copernicus and the Critics of
Ptolemy." *Journal for the History of Astronomy* 30:
343–58.

———. 2003. "Constructing Copernicus." *Perspectives
on Science* 10:208–27.

———. 2004. "How Rothmann Changed His Mind."
Centaurus 46:41–57.

Barker, Peter, and Bernard R. Goldstein. 1988.
"The Role of Comets in the Copernican Revolution." *Studies in History and Philosophy of Science*
19:299–319.

———. 1995. "The Role of Rothmann in the Dissolution of the Celestial Spheres." *British Journal for
the History of Science* 28:385–403.

———. 1998. "Realism and Instrumentalism in Sixteenth Century Astronomy: A Reappraisal." *Perspectives on Science* 6:232–58.

———. 2001. "Theological Foundations of Kepler's
Astronomy." *Osiris* 16 88–113.

Barnes, Robin Bruce. 1988. *Prophecy and Gnosis: Apo-
calypticism in the Wake of the German Reformation.*
Stanford, CA: Stanford University Press.

Barone, Francesco. 1995. "Galileo e Copernico." In
Galileo Galilei e la cultura veneziana, 363–79.

Barton, Ruth. 2003. " 'Men of Science': Language, Identity and Professionalization in the Mid-Victorian
Scientific Community." *History of Science* 41:73–119.

Barton, Tamsyn. 1994. *Ancient Astrology.* London:
Routledge.

Bauer, Barbara, ed. 1999. "Naturphilosophie, Astron-
omie, Astrologie." In Bauer, *Melanchthon und die
Marburger Professoren (1527–1627),* 345–439. 2 vols.
Marburg: Universitätsbibliothek; Völker & Ritter.

Beeckman, Isaac. 1939–53. *Journal tenu par Isaac Beeckman de 1604 à 1634.* 4 vols. Ed. Cornelius de Waard.
La Haye: M. Nijhoff.

Beer, Arthur, and Peter Beer, eds. 1975. *Kepler: Four
Hundred Years.* Vistas in Astronomy, 18. Oxford:
Pergamon Press.

Bellanti, Lucio. 1498. *Lucii Bellantii Senensis Physici
Liber de Astrologica Ueritate; Et, In Disputationes Ioan-
nis Pici aduersus Astrologos Responsiones.* Florence:
Gherardus de Haerlem.

———. 1502. *Liber de Astrologica Veritate.* Venice: Bernardino Vitali.

———. 1553. *Liber de Astrologica Veritate.* Basel: Jacobus
Parcus.

———. 1554. *Lucii Bellantii Senensis Mathematici et
Physici Liber de Astrologica Veritate.* Basel: Hervagius.

———. 1578 [1580]. *Liber de Astrologica Veritate.* Cologne.

Bellarmine, Robert. 1586. *De Controversiis Christianae
Fidei, adversus huis Temporis Haereticos.* 3 vols. Rome.

Bellinati, Claudio. 1992. "Galileo e il Sodalizio con Ec-
clesiastici Padovani." In Santinello 1992, 257–65.

Belluci, D. 1988. "Mélanchthon et la Défense de l'Astrologie." *Bibliothèque d'Humanisme et Renaissance*
50:587–622.

Benatius, Jacobus. 1502. *Pronosticon.* Bologna, n.p.

Benjamin, Francis C., Jr., and G. J. Toomer, eds. 1971.
*Campanus of Novara and Medieval Planetary The-
ory: "Theorica Planetarum."* Madison: University of
Wisconsin Press.

Bennett, Henry Stanley. 1970. *English Books and
Readers, 1603 to 1640.* Cambridge: Cambridge University Press.

Bennett, Jim A. 1986. "The Mechanics' Philosophy
and the Mechanical Philosophy." *History of Science*
24:1–28.

———. 1990. "Hooke's Instruments for Astronomy
and Navigation." In Hunter and Schaffer 1990,
21–32.

———. 2003. "Presidential Address: Knowing and
Doing in the Sixteenth Century; What Were Instruments For?" *British Journal for the History of Science* 36:129–50.

Bennett, Owen. 1943. *The Nature of Demonstrative
Proof according to the Principles of Aristotle and St.
Thomas Aquinas.* Washington, DC: Catholic Uni-
versity of America.

Bentley, Jerry H. 1983. *Humanists and Holy Writ: New*

Testament Scholarship in the Renaissance. Princeton, NJ: Princeton University Press.

Berggren, J. L., and Bernard R. Goldstein, eds. 1987. *From Ancient Omens to Statistical Mechanics: Essays on the Exact Sciences Presented to Asger Aaboe*. Copenhagen: University Library.

Berkel, Klaas van. 1983. *Isaac Beeckman (1588–1637) en de mechanisierung van het wereldbeeld*. Amsterdam.

———. 1999. "Stevin and the Mathematical Practitioners, 1580–1620." In Berkel, Van Helden, and Palm 1999, 12–36.

Berkel, Klaas van, Albert Van Helden, and Lodewijk Palm, eds. 1999. *A History of Science in the Netherlands: Survey, Themes and Reference*. Leiden: Brill.

Bernstein, Jane A. 1998. *Music Printing in Renaissance Venice: The Scotto Press (1539–1572)*. Oxford: Oxford University Press.

Beroaldus, Philippus (Filippo Beroaldo). n.d. *Symbola Pythagore*. Bologna.

———. 1488. *Annotationes Centum*. Bologna: Franciscus de Benedictis for Benedictis Hectoris.

———. 1500. *Commentarii a Philippo Beroaldo Conditi in Asinum Aureum Lucii Apuleii: Mox in Reliqua Opuscula eiusdem Annotationes Imprimentur*. Bologna: Benedictus Hectoris.

Bertoloni Meli, Domenico. 1992. "Guidobaldo del Monte and the Archimedean Revival." *Nuncius* 7:3–34.

———. 1993. *Equivalence and Priority: Newton versus Leibniz; Including Leibniz's Unpublished Manuscripts on the Principia*. Oxford: Oxford University Press.

———. 2006. *Thinking with Objects: The Transformation of Mechanics in the Seventeenth Century*. Baltimore: Johns Hopkins University Press.

Bertolotti, A. 1878. "Giornalisti, astrologi e negromanti in Roma nel secolo XVII." *Rivista Europea* 5:466–514.

Beste, August Friedrich Wilhelm. 1856. *Die bedeutendsten Kanzelredner der älteren lutherschen Kirche*, 3 vols. Leipzig: Gustav Mayer.

Betsch, Gerhard, and Jürgen Hamel, eds. 2002. *Zwischen Copernicus und Kepler: M. Michael Maestlinus Mathematicus Goeppingensis 1550–1631*. Frankfurt: Harri Deutsch.

Bettini, Sergio. 1975. "Copernico e la pittura Veneta." *Notizie dal Palazzo Albani* 4, no. 2: 22–30.

Biagioli, Mario. 1989. "The Social Status of Italian Mathematicians, 1450–1600." *History of Science* 27:41–95.

———. 1990a. "The Anthropology of Incommensurability." *Studies in History and Philosophy of Science* 21, no. 2:183–209.

———. 1990b. "Galileo the Emblem-Maker." *Isis* 81: 230–58.

———. 1990c. "Galileo's System of Patronage." *History of Science* 79:1–62.

———. 1992. "Scientific Revolution, Social Bricolage and Etiquette." In Porter and Teich 1992, 11–54.

———. 1993. *Galileo Courtier*. Chicago: University of Chicago Press.

———. 1996. "Playing with the Evidence." *Early Science and Medicine* 1:70–105.

———. 2000. "Replication or Monopoly? The Economies of Invention and Discovery in Galileo's Observations of 1610." *Science in Context* 11:547–90.

———. 2006. *Galileo's Instruments of Credit: Telescopes, Images, Secrecy*. Chicago: University of Chicago Press.

Biliński, Bronisław. 1977. *Il Pitagorismo di Niccolò Copernico*. Wrocław: Polskiej Akademii Nauk.

———. 1983. "Il periodo Padovano di Niccolò Copernico (1501–1503)." In Poppi 1983.

———. 1989. *Messaggio e itinerari Copernicani*. (Celebrazioni italiane del V centenario della nascità di Niccolò Copernico, 1473–1973.) Warsaw: Polskiej Akademii Nauk.

Biondi, Grazia. 1986. "Minima astrologica: Gli astrologi e la guida della vita quotidiana." *Schifanoia* 2:41–48.

Bireley, Robert. 1999. *The Refashioning of Catholicism, 1450–1700*. Washington, DC: Catholic University of America Press.

Birkenmajer, Alexandre. 1965. "Copernic comme philosophe." In *Le soleil à la Renaissance: Sciences et mythes*, 9–17. Brussels: Presses Universitaires de Bruxelles.

———. 1972a. "Copernic philosophe." In *Études d'histoire des sciences en Pologne* (Studia Copernicana 4), 563–78.

———. 1972b. "L'astrologie cracovienne à son apogée." In *Études d'histoire des sciences en Pologne* (Studia Copernicana 4), 474–82.

———. 1972c. "Le commentaire inédit d'Erasmus Reinhold sur le *De revolutionibus* de Nicholas Copernic." In *Études d'histoire des sciences en Pologne* (Studia Copernicana 4), 761–66.

———. 1972d. "Leovitius etait-il un adversaire de Copernic?" In *Études d'histoire des sciences en Pologne* (Studia Copernicana 4), 767–78.

Birkenmajer, Ludwik. 1900. *Mikołaj Kopernik*. Krakow: Polska Akademia Umiejętności.

———. 1924. *Stromata Copernicana*. Krakow: Polska Akademia Umiejętności.

———. 1975. *Nicolas Copernicus, Part One: Studies on the Works of Copernicus and Biographical Materials*. 2 parts. Trans. Jerzy Dobrzycki, Zofia Piekarec, Zofia Potkowska, and Michal Rozbicki; ed. Owen Gingerich. Ann Arbor, MI: University Microfilms.

Biskup, Marian. 1973. *Regesta Copernicana (Calendar of Copernicus's Papers)*. (Studia Copernicana 8.) Wrocław: Polskiej Akademii Nauk.

Black, Crofton. 2006. *Pico's "Heptaplus" and Biblical Exegesis*. Studies in Medieval and Reformation Traditions, 116. Leiden: Brill.

Blackwell, Richard J. 1991. *Galileo, Bellarmine and the Bible*. Notre Dame, IN: University of Notre Dame Press.

———. 1998. "Could There Be Another Galileo Case?" In Machamer 1998, 348–66.

Blaeu, William. 1690. *Institutio Astronomica, De usu Globorum et Sphaerarum Caelestium ac Terrestrium: Duabus Partibus Adornata, Una, secundum hypothesin Ptolemaei, per Terram Quiescentem; Altera, juxta mentem N. Copernici, per Terram Mobilem.* Amsterdam: Joannis Wolters.

Blagrave, John. 1585. *The Mathematical Iewel, Shewing the Making, and Most Excellent Vse of a Singuler Instrument So Called.* London: Walter Venge.

Blair, Ann. 1997. *The Theater of Nature: Jean Bodin and Renaissance Science.* Princeton, NJ: Princeton University Press.

Blancanus, Joseph. 1620. *Sphaera Mundi seu Cosmographia.* Bologna: Hieron. Tamburini.

Bloor, David. 1992. "Left and Right Wittgensteinians." In Pickering 1992, 266–82.

Blount, Thomas Pope. 1710. *Censura Celebriorum Authorum.* Geneva.

Blumenberg, Hans. 1965. *Die kopernikanische Wende.* Frankfurt: Suhrkamp.

———. 1987. *The Genesis of the Copernican World.* Trans. R. M. Wallace. Cambridge, MA: MIT Press.

Blundeville, Thomas. 1594 [further editions 1597, 1605, 1613, 1621, 1636, 1638]. *M. Blundevile His Exercises, Containing Six Treatises.* London: John Windet.

———. 1602. *The Theoriques of the Seven Planets.* London: Adam Islip.

Boas, Marie. 1962. *The Scientific Renaissance, 1450–1630.* New York: Harper.

Bodin, Jean. 1576. *Les six livres de la République.* Paris: Jacques du Puy.

Bok, Bart J., and Lawrence E. Jerome. 1975. *Objections to Astrology.* New York: Prometheus Books.

Boner, Patrick. 2007. "Kepler *v.* the Epicureans: Causality, Coincidence and the Origins of the New Star of 1604." *Journal for the History of Astronomy* 38:207–21.

———. 2009. "Finding Favour in the Heavens and the Earth: Stadius, Kepler and Astrological Calendars in Early Modern Graz." In Kremer and Włodarczyk 2009, 159–78.

Bonney, Richard. 1991. *The European Dynastic States, 1494–1660.* Oxford: Oxford University Press.

Borawska, Teresa. 1984. *Tiedemann Giese (1480–1550).* Olstyn: Pojezierze.

Bouazzati, Bennacer el, ed. 2004. *Les éléments paradigmatiques, thématiques et stylistiques dans la pensée scientifique.* Najah el Jadida: Publications de la Faculté des Lettres, Rabat.

Boudet, Jean-Patrice. 1994. *Lire dans le ciel: La bibliothèque de Simon de Phares astrologue du XVe siècle.* Les Publications de Scriptorium, 10. Brussels: Centre d'Études des Manuscrits.

Bourdieu, Pierre. 1990. *The Logic of Practice.* Stanford, CA: Stanford University Press.

Bourdieu, Pierre, and Jean-Claude Passeron. 1970. *La réproduction: Elements pour une théorie du système d'enseignement.* Paris: Minuit.

Bouwsma, William. 1968. *Venice and the Defense of Republican Liberty: Renaissance Values in the Age of the Counter Reformation.* Berkeley: University of California Press.

Bowden, Mary Ellen. 1974. "The Scientific Revolution in Astrology: The English Reformers, 1558–1686." PhD diss., Yale University.

Boyer, Carl. 1959. *The Rainbow: From Myth to Mathematics.* New York: Sagamore Press.

Brady, Thomas A., Jr. 2004. "Confessionalization: The Career of a Concept." In Headley, Hillerbrand, and Papalas 2004, 1–20.

Brady, Thomas A., Jr., Heiko A. Oberman, and James D. Tracy, eds. 1994–95. *Handbook of European History, 1400–1650.* Leiden: Brill.

Brahe, Tycho. 1588. *De Mundi Aetherei Recentioribus Phaenomenis.* Uraniborg.

———. 1598. *Astronomiae Instauratae Mechanica.* See Raeder, Strömgren, and Strömgren 1946.

———. 1632. *Learned: Ticho Brahae his Astronomicall Coniectur of the New and much admired Starre Which Appered in the year 1572.* London: By BA and TF for Michaell and Samuell Nialand.

———. 1913–29. *Opera Omnia.* 15 vols. Ed. J. L. E. Dreyer. Copenhagen: Axel Simmelkaer.

Brecht, Martin, ed. 1977. *Theologen und Theologie an der Universität Tübingen: Beiträge zur Geschichte der evangelisch-theologischen Fakultät.* Tübingen: Mohr.

Bredekamp, Horst. 2000. "Gazing Hands and Blind Spots: Galileo as Draftsman." *Science in Context* 13:423–62.

Bretschneider, Carolus Gottliebus, et al., eds. 1834–. *Corpus Reformatorum.* Halle [1834–60], Brunswick [1863–1900], Berlin [1905–]: C. A. Schwetschke.

Brickman, Benjamin. 1941. *An Introduction to Francesco Patrizi's "Nova de universis philosophia."* New York: Columbia University Press.

Brizzi, Gian Paolo. 1995. "Les jésuites et l'école en Italie (XVIe–XVIIIe siècles)." In Giard 1995a, 35–53.

Broman, Thomas H. 1998. "The Habermasian Public Sphere and 'Science in the Enlightenment.'" *History of Science* 36:123–49.

———. 2002. "Introduction: Some Preliminary Considerations on Science and Civil Society." In *Osiris* 17, "Science and Civil Society," ed. Lynn K. Nyhardt and Thomas H. Broman, 1–21.

Brooks, Peter Newman, ed. 1983. *Seven-Headed Luther: Essays in Commemoration of a Quincentenary, 1483–1983.* Oxford: Oxford University Press.

Brosseder, Claudia. 2005. "The Writing in the Wittenberg Sky: Astrology in Sixteenth-Century Germany." *Journal of the History of Ideas* 66:557–76.

Brudzewo, Albertus de. 1495. *Commentaria Utilissima in Theoricis Planetarum.* Milan: Uldericus Scinzenzaler.

———. 1900. *Commentariolum super Theoricas Novas*

Planetarum Georgii Purbachii in Studio Generale Cracoviensis per Magistrum Albertum de Brudzewo Diligenter Corrogatum, A.D. 1482. Trans. and ed. L. Birkenmajer. Krakow: Joseph Filipowski.

Bruno, Giordano. 1586. *Figuratio Aristotelici Physici Auditus ad eiusdem Intelligentiam atque Retentionem per Quindecim Imagines Explicanda.* Paris.

———. 1588a. *Oratio Valedictoria.* In Bruno 1962, 1: pt. 1, 1–52.

———. 1588b. *Camoeracensis Acrotismus seu Rationes Articulorum Physicorum adversus Peripateticos Parisiis Propositorum.* Wittenberg: Zacharias Krafft. In Bruno 1962, 1: pt. 1, 55–190.

———. 1609. *Summa Terminorum Metaphysicorum.* Marburg: Rodolphus Hutwelcker.

———. 1955 [1584]. *La cena de le ceneri.* Giovanni Aquilecchia, ed. Turin: Einaudi.

———. 1962 [1879–1891]. *Opere Latine.* 3 vols. in 8 parts. Ed. F. Fiorentino et al. Facsimile ed. Stuttgart–Bad Canstatt: Friedrich Froman Verlag.

———. 1977 [1584]. *The Ash Wednesday Supper: La cena de le ceneri.* Trans. and ed. Edward A. Gosselin and Lawrence S. Lerner. New York: Archon Books.

———. 1995. *Oeuvres complètes de Giordano Bruno: Oeuvres italiennes.* Vol. 4. *Del'Infini, de l'univers et des mondes,* ed. Giovanni Acquilecchia. Paris: Les Belles Lettres.

———. 2000 [1582]. *The Candlebearer.* Ed. Gino Moliterno. Ottawa: Dovehouse Editions.

"Bruno, Giordano." 1969–78. In *Bolshaia Sovietskaya Entsiklopedia.* 30 vols. Moscow.

Bucciantini, Massimo. 1997. "Galileo e la *Nova* del 1604." In Bucciantini and Torrini 1997, 237–48.

———. 2003. *Galileo e Keplero: Filosofia, cosmologia et teologia nell'età della Controriforma.* Torino: Giulio Einaudi.

Bucciantini, Massimo, and Michele Camerota. 2005. "Once More about Galileo and Astrology: A Neglected Testimony." *Galilaeana* 2:229–32.

Bucciantini, Massimo, Michele Camerota, and Sophie Roux, eds. 2007. *Mechanics and Cosmology in the Early Modern Period.* Florence: Leo S. Olschki, 2007.

Bucciantini, Massimo, and Maurizio Torrini, eds. 1997. *La diffusione del copernicanesimo in Italia, 1543–1610.* Florence: Leo S. Olschki.

Buck, Lawrence P., and Jonathan W. Zophy, eds. 1972. *The Social History of the Reformation.* Columbus: Ohio University Press.

Budweis, Wenceslaus de. [1490?]. *Judicium Liptzense.* N.p.

Bühler, Kurt. 1958. *The University and the Press in Fifteenth-Century Bologna.* Notre Dame, IN: University of Notre Dame.

Bujanda, Jesús Martínez de, et al., eds. 1994. *Index de Rome 1590, 1593, 1596: Avec étude des index de Parme 1580 et Munich 1582.* Index des livres interdits, 9. Sherbrooke, Quebec: Libraries Droz.

Bunzl, Martin. 2004. "Counterfactual History: A User's Guide." *American Historical Review* 109:845–58.

Buonincontro, Lorenzo. 1540. *Rerum Naturalium & Divinarum sive de rebus Coelestibus . . . Eclipsium Solis & Lunae Annis Iam Aliquot Uisarum usque ad Postrema Huius Anni MD XXXX Descriptiones per Philippvm Melanchthonem & Alios.* Basel: Robert Winter.

Buoninsegni, Tommaso. 1581. *Hieron. Savonarolæ . . . opus eximium adversus divinatricem astronomiam in confirmationem confutationis ejusdem astronomicæ prædictionis J. Pici Mirandulæ Comitis, ex Italico in Latinum translatum, interprete . . . T. Boninsignio, . . . ab eodem scholiis, adnotationibus illustratum. Accedit ejusdem interpretis apologeticus adversus hujus operis vituperatores.* Florence: Georgius Marescotus.

Burke, Peter. 1996. *The Fortunes of the Courtier: The European Reception of Castiglione's "Cortegiano."* University Park: Pennsylvania State University Press.

Burmeister, Karl H. 1967–68. *Georg Joachim Rheticus, 1514–1574: Eine Bio-Bibliographie.* 3 vols. Wiesbaden: Pressler.

———. 1970. *Achilles Pirmin Gasser, 1505–1577: Arzt und Naturforscher, Historiker und Humanist.* 3 vols. Wiesbaden: Guido Pressler Verlag.

———. 1977. "Neue Forschung über Georg Joachim Rhetickus." In *Jahrbuch des Vorarlberger Landesmuseumsvereins, 1974–75,* 37–47. Bregenz: Freunde der Landeskunde, 1977.

Burnett, Charles. 1987a. "Adelard, Ergaphalau and the Science of the Stars." In Burnett 1987b, 133–45.

———, ed. 1987b. *Adelard of Bath: An English Scientist and Arabist of the Early Twelfth Century.* London: Warburg Institute.

Burns, William E. 2000. "'The Terriblest Eclipse that Hath Been Seen in Our Days': Black Monday and the Debate on Astrology during the Interregnum." In Osler 2000, 137–52.

Burtt, Edwin Arthur. 1932. *The Metaphysical Foundations of Modern Science.* 2nd rev. ed. New York: Doubleday Anchor.

Butterfield, Herbert. 1957. *The Origins of Modern Science, 1300–1800.* Rev. ed. New York: Free Press.

Calvin, Jean. 1962. *Traité ou avertissement contre l'astrologie qu'on appele judiciaire et autres curiosités qui règnent aujourd'hui au monde.* Paris: Armand Colin.

Camerota, Michele. 1989. "Un breve scritto attinente alla 'Quaestio de certitudine mathematicarum' tra le carte di Filippo Fantoni, predecessore di Galileo alla cattedra di matematica dell'Università di Pisa." *Annali della Facoltà di Magistero dell'Università di Cagliari* 13:91–155.

———. 2004. *Galileo Galilei e la cultura scientifica nell'età della Controriforma.* Rome: Salerno Editrice.

Camerota, Michele, and Mario Helbing. 2000. "Galileo and Pisan Aristotelianism: Galileo's 'De Motu Antiquiora' and the 'Questiones De Motu Elementorum' of the Pisan Professors." *Early Science and Medicine* 5:319–65.

Campbell, Erin J. 2002. "The Art of Aging Gracefully: The Elderly Artist as Courtier in Early Modern Art Theory and Criticism." *The Sixteenth Century Journal* 33:321–31.

Campion, Nicholas, and Nick Kollerstrom, eds. 2003. "Galileo's Astrology." Special issue of *Culture and Cosmos* 7, no. 1.

Canone, Eugenio. 1995. "L'Editto di Proibizione delle Opere di Bruno e Campanella." *Bruniana & Campanelliana* 1: 43–61.

———, ed. 2000. *Giordano Bruno: 1548–1600 ; Mostra storico documentaria*. Biblioteca di Bibliografia Italiana, 164. Rome: Biblioteca Casanatense.

Canone, Eugenio, and Leen Spruit. 2007. "Rhetorical and Philosophical Discourse in Giordano Bruno's Italian Dialogues." *Poetics Today* 28, no. 3:363–91.

Capella, Martianus. December 16, 1499. *De Nuptiis Philologia et Mercurii*. Ed. Franciscus Vitalis Bodianus. Vicenza: Rigo gi ca Zeno.

Capitani, Ovidio, ed. 1987. *L'Università a Bologna: Personaggi, momenti e luoghi dalle origini a XVI secolo*. Bologna: Amilcare Pizzi.

Capp, Bernard. 1982. "The Status and Role of Astrology in Seventeenth-Century England: The Evidence of the Almanac." In Zambelli 1982, 279–90.

Capra, Baldassare. 1607. *Usus et Fabrica Circini Cuiusdam Proportionis*. [Padua: Peter Paul Tozzio] In Galilei 1890–1909, 2:427–511.1.

Capuano de Manfredonia, Franciscus. 1515 [1495]. *Theorice noue planetarum Georgij Purbachij astronomi celebratissimi. Ac in eas Eximij Artium et medicine doctoris Dominum Francisci Capuani de Manfredonia: in studio Patauino astronomiam publice legentis: sublimis expositio et luculentissimum scriptum*. Paris: Parvus.

———. 1518 [19 January]. *Sphaera cum Commentis . . . Expositio*. Venice: Heirs of Ottaviano Scotto.

Cardano, Girolamo. 1543. *Libelli duo*. Nuremberg: J. Petreius.

———. 1547a. *Aphorismorum Astronomicorum*. In Cardano 1663.

———. 1547b. *De Iudiciis Geniturarum*. Nuremberg: Johannes Petreius.

———. 1554. See Ptolemy 1554.

———. 1557. *Ephemerides . . . ad Annos XIX Incipientes ab Anno Christi MDLCCVII usque ad Annum MDLXXXV*. Venice: Vincentius Valgrisius.

———. 1578. See Ptolemy 1578.

———. 1967 [1663]. *Opera Omnia*. Ed. C. Spon. 10 vols. Facsimile ed. Ed. August Buck. [Louvain: I. A. Huguetan and M. A. Ravaud.] New York: Johnson Reprint.

Carelli, Giovanni Battista. 1557. *Ephemerides . . . ad Annos XIX Incipientes ab Anno Christi MDLCCVII usque ad Annum MDLXXXV*. Venice: Vincentius Valgrisius.

Carey, Hilary. 1992. *Courting Disaster: Astrology at the English Court and University in the Later Middle Ages*. New York: St. Martin's Press.

Carion, John. 1550. *The Thre bokes of Cronicles, which John Carion (a man syngularly well sene in the Mathematicall sciences) Gathered Wyth great diligence of the beste Authours that have written in Hebrue, Greke or Latine*. London: Walter Lynne.

Carmody, Francis J. 1958. *The Arabic Corpus of Greek Astronomers and Mathematicians*. Bologna: Arti Grafiche Tamani.

Caroti, Stefano. 1986. "Melanchthon's Astrology." In Zambelli 1986b, 109–21.

———. 1987. "Nicole Oresme's Polemic Against Astrology in his 'Quodlibeta.'" In Curry 1987, 75–93.

Casini, Paolo, ed. 1991. *Lezioni Galileiane, I: Alle origini della rivoluzione scientifica*, Rome: Istituto della Enciclopedia Italiana.

Caspar, Max. 1993 [1948]. *Kepler*. Trans. C. D. Hellman; notes by O. Gingerich and A. Segonds. New York: Dover.

Castagnola, Raffaella. 1989. "Un oroscopo per Cosimo I." *Rinascimento* 29: 125–89.

The Catholic Encyclopedia. 15 vols. 1907–12. New York: Appleton.

Cattini-Marzio, Marco, and Marzio A. Romani. 1982. "Le corti parallele: Per una tipologia delle corti padane dal XIII ad XVI secolo." In Papagno and Quondam 1982.

Cayado, Hermico. 1501 [1496]. *Aeclogae Epigrammata Sylvae*. Bologna: Benedictus Hectoris.

———. 1931. *The Eclogues of Henrique Cayado*. Ed. Wilfred P. Mustard. Baltimore: Johns Hopkins University Press;.

Certeau, Michel de. 1984. *The Practice of Everyday Life*. Trans. Steven Rendall. Berkeley: University of California Press.

Chabás, José, and Bernard Goldstein. 2009. *The Astronomical Tables of Giovanni Bianchini*. Leiden: Brill.

Chamber, John. 1601. *A Treatise against Iudicial Astrologie*. London: John Harrison.

Chartier, Roger, ed. 1989. *The Culture of Print: Power and the Uses of Print in Early Modern Europe*. Trans. Lydia G. Cochrane. Princeton, NJ: Princeton University Press.

Chatelain, Jean-Marc. 2001. "Polymathie et science antiquaire à la Renaissance." In Giard and Jacob 2001, 443–60.

Chrisman, Miriam Usher. 1982. *Lay Culture, Learned Culture: Books and Social Change in Strasbourg, 1480–1599*. New Haven: Yale University Press.

Christianson, John Robert. 1973. "Copernicus and the Lutherans." *The Sixteenth Century Journal* 4: 1–10.

———. 1979. "Tycho Brahe's German Treatise on the Comet of 1577: A Study in Science and Politics." *Isis* 70: 110–40.

———. 2000. *On Tycho's Island: Tycho Brahe and His Assistants, 1570–1601*. Cambridge: Cambridge University Press.

Cicero. 1959. *De divinatione*. Trans. W. A. Falconer. Cambridge, MA: Harvard University Press.

Ciliberto, Michele. 1979. *Lessico di Giordano Bruno.* 2 vols. Rome: Ateneo & Bizzarri.

Clagett, Marshall, ed. 1959. *Critical Problems in the History of Science.* Madison: University of Wisconsin Press.

Clark, David, and F. Richard Stephenson. 1977. *The Historical Supernovae.* Elmsford, NY: Pergamon Press.

Clark, Stuart. 1991. "The Rational Witchfinder: Conscience, Demonological Naturalism and Popular Superstition." In Pumfrey, Rossi, and Slawinski 1991, 222–48.

———. 1997. *Thinking with Demons: The Idea of Witchcraft in Early Modern Europe.* Oxford: Oxford University Press.

Clark, William. 1989. "On the Dialectical Origins of the Research Seminar." *History of Science* 17 111–54.

———. 2006. *Academic Charisma and the Origins of the Research University.* Chicago: University of Chicago Press.

Clarke, Angus. 1985. "Giovanni Antonio Magini (1555–1617) and Late Renaissance Astrology." PhD diss., University of London.

Clavelin, Maurice. 2004. *Galilée copernicien.* Paris: Albin Michel.

———. 2006. "Duhem et Tannery, lecteurs de Galilée." *Galilaeana* 3:3–17.

Clavius, Christoph. 1570. *In Sphaeram Joannis de Sacrobosco Commentarius.* Rome: Victorium Helianum.

———. 1581. In *Sphaeram Ioannis de Sacro Bosco commentarius: Nunc iterum ab ipso auctore recognitus et multis ac variis locis locupletatus.* Rome: Dominici Basae.

———. 1591. *In Sphaeram Joannis de Sacrobosco Commentarius.* 3rd ed. Venice: Ioannes Baptista Cioti. Reprint of Rome 1585 ed.:"Nunc tertio ab ipso auctore recognitus, et plerisque in locis locupletatus."

———. 1594. *In Sphaeram Joannis de Sacrobosco Commentarius.* 4th ed. Louvain.

———. 1611. *Opera Mathematica.* 5 vols. Moguntia: Antonii Hierat

———. 1992. *Corrispondenza.* Ed. Ugo Baldini and P. D. Napolitani. 7 vols. Pisa: Università di Pisa, Dipartimento di Matematica, Sezione di Didattica e Storia della Matematica.

Clucas, Stephen. 2000. "Thomas Harriot and the Field of Knowledge in the English Renaissance" in Fox 2000, 93–136.

Clulee, Nicholas H. 1988. *John Dee's Natural Philosophy: Between Science and Religion.* London: Routledge.

———. 2001. "Astronomia Inferior: Legacies of Johannes Trithemius and John Dee." In Newman and Grafton 2001, 173–233.

Clutton-Brock, Martin. 2005. "Copernicus's Path to His Cosmology: An Attempted Reconstruction." *Journal for the History of Astronomy* 36:197–216.

Cochrane, Eric. 1973. *Florence in the Forgotten Centuries, 1527–1800.* Chicago: University of Chicago Press.

Cohen, H. Floris. 1994. *The Scientific Revolution: A Historiographical Inquiry.* Chicago: University of Chicago Press.

Cohen, I. Bernard. 1972. "Newton and Keplerian Inertia: An Echo of Newton's Controversy with Leibniz." In Debus 1972, 192–211.

———. 1985. *Revolution in Science.* Cambridge, MA: Harvard University Press.

Cohen, Morris Raphael, and Israel Drabkin, eds. 1966. *A Source Book in Greek Science.* Cambridge, MA: Harvard University Press.

Colie, Rosalie. 1973. *The Resources of Kind: Genre-Theory in the Renaissance.* Berkeley: University of California Press.

Colombe, Lodovico delle. 1606. *Discorso nel quale si dimostra, che la nuova stella apparita l'ottobre passato 1604. nel Sagittario non è cometa, ne stella generata, ò creata di nuovo, ne apparente ; ma una di quelle che furono da principio nel cielo; e ciò esser conforme alla vera filosofia, teologia, e astronomiche demostrazioni ; con alquanto di esagerazione contro a' giudiciari astrologi.* Florence: Giunti.

———. 1608. *Risposte piacevoli e curiose alle Considerazioni di certa maschera saccente nominata Alimberto Mauri, fatte sopra alcuni luoghi del discorso del medesimo Lodovico dintorno alla stella apparita l'ano 1604.* Florence: Caneo.

Contopoulos, George, ed. 1974. *Highlights of Astronomy.* vol. 3. Dordrecht: Reidel.

Cook, Harold J. 1990. "The New Philosophy and Medicine in Seventeenth-Century England." In Lindberg and Westman 1990, 397–436.

Copenhaver, Brian P., and Charles B. Schmitt. 1992. *Renaissance Philosophy.* Oxford: Oxford University Press.

Copernicus, Nicholas. 1543. *De Revolutionibus Orbium Coelestium.* Nuremberg: Iohannes Petreius.

———. 1566. *De Revolutionibus Orbium Coelestum.* Basel: Heinrich Petri.

———. 1884. "De Hypothesibus Motuum Caelestium a Se Constitutis Commentariolus." In Prowe 1883–84, 2:184–202.

———. 1952. *Revolutions of the Heavenly Spheres.* In *Great Books of the Western World,* vol. 16. Trans. Charles Glenn Wallis. Chicago: Encyclopaedia Britannica.

———. 1971a. "Commentariolus." Trans. Edward Rosen. In Rosen 1971a.

———. 1971b [1566]. *De Revolutionibus Orbium Coelestium Libri Sex (Editio Basileensis) cum Commentariis Manu Scriptis Tychonis Brahe.* Facsimile ed. Ed. Zdeněk Horský. Prague: Pragopress.

———. 1971c. "Letter against Werner." Trans. Edward Rosen. In Rosen 1971a.

———. 1972. *Nicholas Copernicus Complete Works.* Vol. 1. *De Revolutionibus Orbium Coelestium.* Facsimile ed. Wrocław: Polskiej Akademii Nauk.

———. 1976. *Copernicus: On the Revolutions of the Heavenly Spheres.* Trans. A. M. Duncan. New York: Barnes and Noble.

———. 1978. *Nicholas Copernicus: Complete Works.* Vol. 2. *On the Revolutions of Heavenly Spheres.* Trans. Edward Rosen; ed. J. Dobrzycki. Wrocław: Polskiej Akademii Nauk.

———. 1985. *Nicholas Copernicus: Complete Works* Vol. 3. *Minor Works.* Trans. Edward Rosen; Ed. Paul Czartoryski. Wrocław: Polskiej Akademii Nauk.

Cornwallis, William. 1601. *Discourse upon Seneca the Tragedian.* London: Edmund Mattes.

Corsi, Pietro, and Paul Weindling, eds. 1983. *Information Sources in the History of Science and Medicine.* London: Butterworth Scientific.

Cox-Rearick, Janet. 1964. *The Drawings of Pontormo.* Cambridge, MA: Harvard University Press.

———. 1984. *Dynasty and Destiny in Medici Art: Pontormo, Leo X, and the Two Cosimos.* Princeton, NJ: Princeton University Press.

———. 1993. *Bronzino's Chapel of Eleanora in the Palazzo Vecchio.* Berkeley: University of California Press.

Cozzi, Gaetano. 1979. *Paolo Sarpi tra Venezia e l'Europa.* Turin: Einaudi.

Craven, William G. 1981. *Giovanni Pico della Mirandola, Symbol of His Age: Modern Interpretations of a Renaissance Philosopher.* Geneva: Librairie Droz.

Crombie, Alastair C. 1977. "Mathematics and Platonism in the Sixteenth-Century Italian Universities and in Jesuit Educational Policy." In Maeyama and Saltzer 1977.

Crosland, Maurice, ed.. 1975. *The Emergence of Science in Western Europe.* New York: Science History Pub.

Crowe, Michael J. 1990. *Theories of the World from Antiquity to the Copernican Revolution.* New York: Dover.

Cudworth, Ralph. 1678. *The True Intellectual System of the Universe.* London: Printed for Richard Royston.

Cuningham, William. 1559. *The Cosmographical Glasse, Conteinyng the Pleasant Principles of Cosmographie, Geographie, Hydrographie, or Nauigation.* London: John Daye.

Cunningham, Andrew, and Perry Williams. 1993. "Decentring the 'Big Picture': *The Origins of Modern Science* and the Modern Origins of Science." *British Journal for the History of Science* 26:407–32.

Curd, Martin, and Jan A. Cover, eds. 1998. *Philosophy of Science: The Central Issues.* New York: Norton.

Curry, Patrick, ed. 1987. *Astrology, Science and Society: Historical Essays.* Woodbridge, UK: Boydell Press.

———. 1987a. "Saving Astrology in Restoration England: 'Whig' and 'Tory' Reforms." In Curry 1987, 245–59.

———. 1989. *Prophecy and Power: Astrology in Early Modern England.* Princeton, NJ: Princeton University Press.

Czartoryski, Paweł. 1978. "The Library of Copernicus." *Studia Copernicana* 16:355–96.

D'Addio, Mario. 1962. *Il pensiero politico di Gasparo Scioppio.* Milan.

Dallari, Umberto. 1888. *I rotuli dei lettori legisiti e artisti dello studio bolognese dal 1384 al 1799.* 3 vols. Bologna: Merlani.

D'Amico, John F. 1980. "Papal History and Curial Reform in the Renaissance." *Archivum Historiae Pontificiae* 18.

———. 1983. *Renaissance Humanism in Papal Rome: Humanists and Churchmen on the Eve of the Reformation.* Baltimore: Johns Hopkins University Press.

Danielson, Dennis. 2006. *The First Copernican: Georg Joachim Rheticus and the Rise of the Copernican Revolution.* New York: Walker and Co.

Danti, Egnazio. 1569. *Trattato dell'uso et della fabbrica dell'astrolabio.* Florence: Giunti.

———. 1572. *La sfera del mondo ridotta in cinque tavole.* Florence.

———. 1577. *Le scienze matematiche ridotte in tavole.* Bologna: Appresso la Compagnia della Stampa.

Daston, Lorraine. 1988. *Classical Probability in the Enlightenment.* Princeton, NJ: Princeton University Press.

———. 1995. "The Moral Economy of Science." *Osiris* 10:2–24.

Daston, Lorraine, and Katharine Park. 1998. *Wonders and the Order of Nature, 1150–1750.* New York: Zone Books.

Davis, Natalie Zemon 2000. *The Gift in Sixteenth-Century France.* Madison: University of Wisconsin Press.

Deane, William. 1738. *The Description of the Copernican System, with the Theory of the Planets . . . Being an Introduction to the Description and Use of the Grand Orrery.* London.

Dear, Peter. 1985. "Totius in Verba: Rhetoric and Authority in the Early Royal Society." *Isis* 78, no. 2:144–61.

———. 1995. *Discipline and Experience: The Mathematical Way in the Scientific Revolution.* Chicago: University of Chicago Press.

———. 2001. *Revolutionizing the Sciences: European Knowledge and Its Ambitions, 1500–1700.* Princeton, NJ: Princeton University Press.

———. 2006. *The Intelligibility of Nature: How Science Makes Sense of the World.* Chicago: University of Chicago Press.

Debus, Allen G., ed. 1972. *Medicine and Society in the Renaissance: Essays to Honor Walter Pagel.* New York: Neale Watson.

Dee, John. 1558. *Propaedeumata aphoristica.* In Leovitius 1558.

———. 1573. *Parallacticae Commentationes Praxeosque Nucleus Quidam.* London: John Day.

———. 1583. *To the Right Honorable and my singular good Lorde, the Lorde Burghley, Lorde Threasorer of*

Englande, A playne Discourse and humble Advise for our Gratious Queene Elizabeth, her most Excellent Majestie to peruse and consider, as concerning the needful Reformation of the Vulgar Kalender for the civile yeres and daies accompting, or verifyeng, according to the tyme truely spent. In Dee 1842.

———. 1659. *A True and Faithful Relation of what passed for many yeers between Dr. John Dee . . . and some spirits tending (had it succeeded) to a General Alteration of most States and kingdoms in the World . . . Out of the original copy . . . with a Preface by Meric Casaubon.* London: D. Maxwell.

———. 1851. *Autobiographical Tracts of Dr. John Dee.* Ed. James Crossley London: Chetham Society.

———. 1968a [1577]. *General and Rare Memorials Pertayning to the Perfect Art of Navigation.* [London: John Day.] Amsterdam: Da Capo.

———. 1968b [1842]. *The Private Diary of Dr. John Dee and the Catalogue of His Library of Manuscripts.* Ed. James O. Halliwell. [London: Camden Society.] Repr. New York: AMS Press.

———. 1975. *The Mathematicall Preface to the Elements of Geometrie of Euclid of Megara (1570).* New York: Science History Publications.

———. 1978 [1558, 1568]. *John Dee on Astronomy: Propaedeumata Aphoristica.* Trans. and ed. W. Shumaker; intro. John L. Heilbron. Berkeley: University of California Press.

De Filippis, Michele. 1937. *G. B. Manso's "Enciclopedia."* Berkeley: University of California Press.

Delambre, Jean-Baptiste. 1821. *Histoire de l'astronomie moderne.* 2 vols. Paris: Courcier.

del Monte, Guidobaldo. 1579. *Planisphaeriorum Universalium Theorica.* Pisa: Hieronymus Concordia.

Delorme, Suzanne, ed. 1975. *Avant, avec, après Copernic: La représentation de l'univers et ses conséquences épistémologiques.* Paris: Blanchard.

Delumeau, Jean. 1977. *Catholicism between Luther and Voltaire: A New View of the Counter-Reformation.* Trans. Jeremy Moiser. London: Burns and Oates.

De Morgan, Augustus. 1839. "Motion of the Earth." In *The Penny Cyclopaedia,* 15:454–58.

———. 1855. "The Progress of the Doctrine of the Earth's Motion between the Times of Copernicus and Galileo; being Notes on the Antegalilean Copernicans." In *Companion to the Almanac,* 5–25. London: Knight & Co.

Dennis, Michael. 1989. "Graphic Understanding: Instruments and Interpretation in Robert Hooke's *Micrographia.*" *Science in Context* 3:309–64.

Densmore, Dana. 1995. *Newton's "Principia": The Central Argument.* Trans. W. H. Donahue. Santa Fe, NM: Green Lion Press.

De Santillana, Giorgio. 1955. *The Crime of Galileo.* Chicago: University of Chicago Press.

Descartes, René. 1897–1913. *Oeuvres de Descartes.* 13 vols. Ed. Charles Adam and Paul Tannery. Paris: L. Cerf.

———. 1979. *Le Monde.* Trans. M. Mahoney. New York: Abaris Books.

———. 1983. *Principles of Philosophy.* Trans. V. R. Miller and R. P. Miller. Dordrecht: Reidel.

———. 1998. *Discourse on the Method for Conducting One's Reason Well and for Seeking Truth in the Sciences.* 3rd ed. Trans. Donald A. Cress. Indianapolis: Hackett.

Di Bono, Mario. 1990. *Le Sfere Omocentriche di Giovan Battista Amico nell'Astronomia dell'Cinquecento con il testo "De motibus corporum coelestium. . . . "* Genoa: Centro di Studio sulla Storia della Tecnica.

Dibvadius, Georgius Christophorus. 1569. *Commentarii breves in secundum librum Copernici, in quibus argumentis infallibilibus demonstratur veritas doctrinae de primo motu, et ostenditur Tabularum compositio.* Wittenberg: Clemens Schleich.

Dick, Steven J. 1982. *Plurality of Worlds: The Origins of the Extraterrestrial Life Debate from Democritus to Kant.* Cambridge: Cambridge University Press.

Dick, Wolfgang R., and Jürgen Hamel. 1999. *Beiträge zur Astronomiegeschichte.* Acta historica astronomiae 2. Thun: Harri Deutsch.

Dictionary of Scientific Biography. 1970–84. 16 vols. New York: Scribner.

Dictionary of the History of Ideas. 1973. 5 vols. New York: Scribner.

Diefendorf, Barbara B. and Carla Hesse, eds. 1993. *Essays in Honor of Natalie Davis.* Ann Arbor: University of Michigan Press.

Diesner, Paul. 1938. "Der elsässische Arzt Dr. Helisaeus Roeslin als Forscher und Publizist am Vorabend des dreissigjährigen Krieges." *Jahrbuch der Elsaß-Lothringischen Wissenschaftlichen Gesellschaft zu Strassburg* 11:192–215.

Dietrich, Michael. 1993. "Underdetermination and the Limits of Interpretative Flexibility." *Perspectives on Science* 1:109–26.

Diffley, Paul Brian. 1988. *Paolo Beni: A Biographical and Critical Study.* Oxford: Oxford University Press.

Digges, Leonard. 1555. *A Prognostication of Right Good Effect, fructfully augmented contayninge playne, briefe, pleasant, chosen rules, to judge the wether for euer, by the Sunne, Moone, Sterres, Cometes, Raynbowe, Thunder, Cloudes, with other Extraordinarie tokens, not omitting the Aspectes of Planetes, with a brefe Iudgement for euer, of Plentie, Lacke, Sickenes, Death, Warres etc. Openinge also many naturall causes, woorthy to be knowen, to these and others, notw at the last are adioyned, diuers generall pleasunte Tables: for euer manyfolde wayes profitable, to al maner men of any understanding.* London: Thomas Gemini.

———. 1562. *A Boke named Tectonicon briefely shewinge the exacte measurynge, and speady reckenynge all maner Lande, squared Timber, Stone, Steaples, Pyllers, Globes, etc. Further, declaringe the perfecte makinge and large use of the Carpenters Ruler, conteyninge a Quadrant Geometricall: comprehendinge also the rare use of the Squire. And in thend a lyttle treatise adioyned, openinge the composicion and appliancie of an Instrument called the profitable Staffe. With other thinges pleasaunt and necessary, most conducible for Surveyers, Landemeaters, Joyners, Carpenters, and Masons.* London: Thomas Gemini.

———. 1571. *A Geometrical Practise, named* PANTO-METRIA, *diuided into three Bookes, Longimetra, Planimetra, and Stereometria, containing Rules manifolde for mensuration of all lines, Superficies and Solides: With sundry straunge conclusions both by instrument and without, and also by Perspective glasses, to set forth the true description or exact plat of an whole Region: framed by Leonard Digges Gentleman, lately finished by Thomas Digges his sonne.* London: Henrie Bynneman.

———. 1576. *A Prognostication euerlastinge of right good effecte . . . Lately corrected and augmented by Thomas Digges.* London: Thomas Marsh. (Same as T. Digges 1576.)

Digges, Thomas. 1572. "Thomas Digges to Lord Burghley, December 11, 1572." *Calendar of State Papers* (Domestic), 1547–80, 454.

———. 1573. *Alæ seu Scalæ Mathematicæ, quibus visibilium remotissima Cœlorum Theatra consocendi, & Planetarum omnium itinera nouis & inauditis Methodis explorari: tu`m huius portentosi Syderis in Mundi Boreali plaga insolito fulgore coruscantis. Distantia, & Magnitudo immensa, Situsq' protinu`s tremendus indagari, Deiq' stupendum ostentum, Terricolis expositum cognosci liquidissime` prossit.* London: Thomas Marsh.

———. 1576. *A Perfit Description of the Caelestiall Orbes according to the most aunciente doctrine of the Pythagoreans, latelye reuiued by Copernicus and by Geometricall Demonstrations approued.* London: Thomas Marsh. (Same as L. Digges 1576.)

Dijksterhuis, Eduard Jan. 1943. *Simon Stevin.* The Hague: Nijhoff.

Di Liscia, Daniel A., Eckhard Kessler, and Charlotte Methuen, eds. 1997. *Method and Order in Renaissance Philosophy of Nature.* Aldershot: Ashgate.

Dinis, Alfredo de Oliveira. 1989. "The Cosmology of Giovanni Battista Riccioli (1598–1671)." PhD diss., Cambridge University.

———. 2003. "Giovanni Battista Riccioli and the Science of His Time." In Feingold 2003, 195–224.

Dizionario biografico degli Italiani. 1960–. 70 vols. Rome: Istituto della Enciclopedia Italiana.

Dobbs, Betty Jo, and Margaret Jacob. 1995. *Newton and the Culture of Newtonianism.* Atlantic Highlands, NJ: Humanities Press.

Dobrzycki, Jerzy, ed. 1972. *The Reception of Copernicus' Heliocentric Theory.* Wrocław: Polskiej Akademii Nauki.

———. 1973. "The Aberdeen Copy of Copernicus's *Commentariolus.*" *Journal for the History of Astronomy* 4:124–27.

———. 2001. "Notes on Copernicus's Early Helio-

centrism." *Journal for the History of Astronomy* 32:223–25.

Dobrzycki, Jerzy, and Richard L. Kremer. 1996. "Peurbach and Maragha Astronomy? The Ephemerides of Johannes Angelus and Their Implications." *Journal for the History of Astronomy* 27:187–238.

Dobrzycki, Jerzy, and Lech Szczucki. 1989. "The Transmission of Copernicus's *Commentariolus* in the Sixteenth Century." *Journal for the History of Astronomy* 20:25–28.

Donahue, William H. 1972. "The Dissolution of the Celestial Spheres, 1595–1650." PhD diss., University of Cambridge.

———. 1975. "The Solid Planetary Spheres in Post-Copernican Natural Philosophy." In Westman 1975a, 244–75.

———. 1981. *The Dissolution of the Celestial Spheres, 1595–1650.* New York: Arno Press.

———. 1988. "Kepler's Fabricated Figures: Covering up the Mess in the *Astronomia nova.*" *Journal for the History of Astronomy,* 19:217–37.

Donne, John. 1611. *An Anatomy of the World: Wherein by the occasion of the untimely death of Mistris Elizabeth Drury the frailty and the decay of this whole world is represented.* London: Samuel Macham.

Dooley, Brendan. 1999. *The Social History of Skepticism: Experience and Doubt in Early Modern Culture.* Baltimore: Johns Hopkins University Press.

———. 2002. *Morandi's Last Prophecy and the End of Renaissance Politics.* Princeton, NJ: Princeton University Press.

Dorling, Jon. n.d. "Mid-Seventeenth Century Arguments for and against Copernicanism: A Probabilistic Appeal." Unpublished paper.

Dottorati dal 1609 al 1614. Archivio della Curia Arcivescovile di Pisa, car. 74t. Pisa.

Drake, Stillman. 1957. *Discoveries and Opinions of Galileo.* New York: Anchor Books.

———. 1973. "Galileo's 'Platonic' Cosmogony and Kepler's Prodromus." *Journal for the History of Astronomy* 4:174–91.

———. 1976. *Galileo against the Philosophers.* Los Angeles: Zeitlin and Ver Brugge.

———. 1977. "Galileo and the Career of Philosophy." *Journal of the History of Ideas* 38:19–32.

———. 1978. *Galileo at Work.* Chicago: University of Chicago Press.

———. 1984. "Galileo, Kepler, and Phases of Venus." *Journal for the History of Astronomy* 15:198–208.

———. 1987. "Galileo's Steps to Full Copernicanism and Back." *Studies in History and Philosophy of Science* 17:93–105.

———. 1990. *Galileo: Pioneer Scientist.* Toronto: University of Toronto Press.

Drake, Stillman, and Israel Drabkin, eds. 1969. *Mechanics in Sixteenth-Century Italy: Selections from Tartaglia, Benedetti, Guido Ubaldo, and Galileo.* Madison: University of Wisconsin Press.

Drake, Stillman, and Charles Donald O'Malley, eds. 1960. *The Controversy on the Comets of 1618*. Philadelphia: University of Pennsylvania Press.

Drewnowski, Jerzy 1973. "Rzekomy Portret Epitafijny Mikołaj Kopernika, Ojca Astronoma." *Kwartalnik Historii Nauki i Techniki* 18:511–26.

———. 1978. "Autour de la parution de *De Revolutionibus* (Essai d'une nouvelle interprétation du témoinage de Rheticus dans la correspondance de Copernic avec Giese)." *Organon* 14:253–61.

Dreyer, John Louis Emil. 1953 [1905]. *A History of Astronomy from Thales to Kepler*. Ed. W. H. Stahl. 2nd ed. New York: Dover.

———. 1963 [1890]. *Tycho Brahe: A Picture of Scientific Life and Work in the Sixteenth Century*. New York: Dover.

Duhem, Pierre 1894. "Quelques réflexions au sujet de la physique expérimentale." *Revue des questions scientifiques* 36:179–229.

———. 1908. *Sozein ta Phainomena: Essai sur la notion de théorie physique de Platon à Galilée* Paris: Hermann.

———. 1969. *To Save the Phenomena: An Essay on the Idea of Physical Theory from Plato to Galileo*. Trans. Edmund Doland and Chaninah Maschler. Chicago: University of Chicago Press.

———. 1996. *Pierre Duhem: Essays in the History and Philosophy of Science*. Trans. and ed. R. Ariew and P. Barker. Indianapolis: Hackett.

Eastwood, Bruce. 2001. "Johannes Scotus Eriugena, Sun-Centred Planets, and Carolingian Astronomy." *Journal for the History of Astronomy* 32:281–324.

Ehses, Stephanus et al., eds. 1961–76. *Concilium Tridentinum: Diariorum, Actorum, Epistularum*. Freiburg im Breisgau: Herder.

Einhard, Abbot. 1532, March. *Wittichindi Saxonis rerum ab Henrico et Ottone I Imp. Gestarum Libri III, unà cum alijs quibsdam raris et antehac non lectis diuersorum autorum historijs, ab Anno salutis D.CCC. usque ad praesentem aetatem: quorum catalogus proxima patebit pagina*. Basel: J. Hervagius.

Eisenstein, Elizabeth. 1979. *The Printing Press as an Agent of Change: Communications and Cultural Transformations in Early-Modern Europe*. 2 vols. Cambridge: Cambridge University Press.

Elias, Norbert. 1983 [1969]. *The Court Society*. Trans. Edmund Jephcott. New York: Pantheon.

Erasmus, Desiderius. 1965. *The Colloquies of Erasmus*. Trans. C. R. Thompson. Chicago: University of Chicago Press.

Erastus, Thomas. 1557. *Astrologia Confutata: Ein warhafte gegründte unwidersprechliche Confutation der falschen Astrologei . . . von neuen ins deutsch gebracht*. Schleusingen: Hamsing.

———. 1569. *Defensio libelli Hieronymi Savonarolae de astrologia divinatrice adversus Christophorum Stathmionem medicum Coburgensem*. [Geneva?]: Johannes Le Preux and Johannes Paruum.

———. 1571–73. *Disputationum de medicina nova P. Paracelsi pars prima in qua quae de remedis superstitiosis et magicis curationibus ille prodidit praecipue examinantur*. Basel: Peter Perna.

Ernst, Germana. 1984. "Aspetti dell'astrologia e della profezia in Galileo e Campanella." In Galluzzi 1984, 255–66.

———. 1991. "Astrology, Religion and Politics in Counter-Reformation Rome." In Pumfrey, Rossi, and Slawinski 1991, 249–73.

Eschenden, John of. 1489. *Summa Iudicialis de Accidentibus Mundi*. Venice.

Euclid. 1537. *Elementorum geometricorum libri XV*. Basel: J. Hervagius.

———. 1551. *Elementorum liber decimus.*. Paris: M. Vascosan.

Eulenburg, Franz. 1904. *Die Frequenz der deutschen Universitäten von ihrer Gründung bis zur Gegenwart*. Leipzig: B. Teubner.

Evans, James. 1998. *The History and Practice of Ancient Astronomy*. New York: Oxford University Press.

———. 2004. "The Astrologer's Apparatus: A Picture of Professional Practice in Greco-Roman Egypt." *Journal for the History of Astronomy* 35:1–44.

Evans, R. J. W. 1973. *Rudolf II and His World: A Study in Intellectual History, 1576–1612*. Oxford: Oxford University Press.

———. 1979. *The Making of the Habsburg Monarchy, 1550–1700: An Interpretation*. Oxford: Clarendon Press.

———. 1984. "Rantzau and Welser: Aspects of Later German Humanism." *History of European Ideas* 5:257–72.

Fabian, B., ed. 1972–2001. *Die Messkataloge Georg Willers*. Hildesheim: Georg Olms Verlag. 5 vols.

Fabricius, Johannes. 1611. *De Maculis in Sole Observatis et Apparente earum cum Sole Conversione Narratio*. Wittenberg: Johan Borener Senioris & Elias Rehifledius.

Fabricius, Paul. 1556. *Der Comet im Mertzen des LVI. Jhars zu Wien in Osterreych erschinen*. Nuremberg: Merckel.

Fabri de Budweis, Wenceslaus. 1490. *Judicium Liptzense*. N.p.

Fandi, Sigismondo. 1514. *Teorica et Practica perspicacissimi Sigismundi de Fantis Ferrariensis in Artem Mathematice Professoris de Modo Scribendi Fabricandique Omnes Litterarum Species*. Venice: Ioannem Rubeum Vercellensem.

Fantoli, Annibale. 2002. "Galileo and the Catholic Church: A Critique of the 'Closure' of the Galileo Commission's Work." *Vatican Observatory Publications*, Special Series, Studi Galileiani 4, no. 1. Vatican City: Specola Vaticana.

———. 2003. *Galileo: For Copernicanism and for the Church*. 3rd ed. Trans. George Coyne. Studi Galileiani, 6. Notre Dame, IN: University of Notre Dame Press.

———. 2005. "The Disputed Injunction and Its Role in Galileo's Trial." In McMullin 2005a, 117–49.

Fantuzzi, G. 1781–84. *Notizie degli scrittori bolognesi*. 9 vols. Bologna.

Farrar, W. V. 1975. "Science and the German University System, 1790–1850." In Crosland 1975, 179–92.

Favaro, Antonio 1881. "Galileo Astrologo." *Mente e Cuore* 8:99–108.

———. 1884. "Sulla morte di Marco Velsero e sopra alcuni particolari della vita di Galileo." *Estratto dal Bulletino di Bibliografia e di Storia delle Scienze Matematiche e Fisiche* 17:1–21.

———. 1886. *Carteggio inedito di Ticone Brahe, Giovanni Keplero e di altri celebri astronomi e matematici dei secoli XVI e XVII*. Bologna: Nicola Zanichelli.

———. 1966. *Galileo Galilei e lo studio di Padova*. 2 vols. Padua: Editrice Antenore.

———. 1983. *Amici e corrispondenti di Galileo*. 3 vols. Florence: Salimbeni.

Febvre, Lucien, and Henri-Jean Martin. 1984 [1958]. *The Coming of the Book: The Impact of Printing, 1450–1800*. Trans. David Gerard. London: Verso.

Federici-Vescovini, Graziella. 1996. "Michel Scot et la 'Theorica Planetarum Gerardi.'" *Early Science and Medicine* 1:272–82.

Feingold, Mordechai. 1984. *The Mathematicians' Apprenticeship: Science, Universities and Society in England, 1560–1640*. Cambridge: Cambridge University Press.

———, ed. 2003. *Jesuit Science and the Republic of Letters*. Cambridge, MA: MIT Press.

Feldhay, Rivka. 1995. *Galileo and the Church: Political Inquisition or Critical Dialogue?* Cambridge: Cambridge University Press.

Ferreiro, Alberto, ed. 1998. *The Devil, Heresy, and Witchcraft in the Middle Ages*. Leiden: Brill.

Ficino, Marsilio. 1546–48. *Le divine lettere del gran Marsilio Ficino tradotte in lingua toscana per Felice Figliucci*. 2 vols. Venice: G. Giolito di Ferrara.

———. 1989 [1489]. *Three Books on Life: A Critical Edition and Translation* Trans. and ed. C. V. Kaske and J. R. Clark. Medieval and Renaissance Texts and Studies, 57. Binghamton, NY: Renaissance Society of America.

Field, Judith V. 1984a. "A Lutheran Astrologer: Johannes Kepler." *Archive for History of Exact Sciences* 31:190–268.

———. 1984b. "Kepler's Rejection of Numerology." In Vickers 1984, 273–96.

———. 1988. *Kepler's Geometrical Cosmology*. Chicago: University of Chicago Press.

Findlen, Paula. 1994. *Possessing Nature: Museums, Collecting, and Scientific Culture in Early Modern Italy*. Berkeley: University of California Press.

Finocchiaro, Maurice A., ed. and trans. 1989. *The Galileo Affair: A Documentary History*. Berkeley: University of California Press.

———. 2001. "Science, Religion, and the Historiography of the Galileo Affair: On the Undesirability of Oversimplification." *Osiris* 16:114–32.

———. 2002. "Philosophy versus Religion and Science versus Religion: The Trials of Bruno and Galileo." In Gatti 2002, 51–96.

———. 2005. *Retrying Galileo, 1633–1992*. Berkeley: University of California Press.

Fischlin, Daniel, and Mark Fortier. 2002a. "'Enregistrate Speech': Stratagems of Monarchic Writing in the Work of James VI and I." In Fischlin and Fortier 2002b, 37–58.

———. 2002b. *Royal Subjects: Essays on the Writings of James VI and I*. Detroit, MI: Wayne State University Press.

Fleck, Ludwik. 1979 [1935]. *Genesis and Development of a Scientific Fact*. Trans. F. Bradley and T. Trenn; ed. T. Trenn and R. K. Merton; foreword by T. S. Kuhn. Chicago: University of Chicago Press.

Florio, John. 1611. *Queen Anna's New World of Words, or Dictionarie of the Italian and English tongues*. London: Melchior Bradwood.

Force, James. 1985. *William Whiston, Honest Newtonian*. Cambridge: Cambridge University Press.

Foscarini, Paolo Antonio. 1615. *A Letter . . . Concerning the Opinion of the Pythagoreans and Copernicus About the Mobility of the Earth and the Stability of the Sun and the New Pythagorean System of the World*. Naples: Lazaro Scoriggio. In Blackwell 1991.

———. 2001 [1615]. *Trattato della divinatione naturale cosmologica ovvero de' pronostici e presagi naturali delle mutationi de TEMPI, &c.* Facsimile ed. Ed. Luciano Romeo. [Naples: Lazaro Scoriggio] Cosenza: Progetto 2001.

Fox, Robert, ed. *Thomas Harriot: An Elizabethan Man of Science*. Aldershot: Ashgate.

Fracastoro, Girolamo. 1538. *Homocentrica*. Venice: Joannes Patauino et Venturino Roffinello.

Franchini, Dario A., et al. 1979. *La scienza a corte: Collezionismo eclettico, natura e immagine a Mantova fra Rinascimento e Manierismo*. Rome: Bulzoni Editore.

Franz, Günther. 1977. "Bücherzensur und Irenik: Die theologische Zensur im Herzogtum Württemberg in den Konkurrenz von Universität und Regierung." In Brecht 1977, 123–94.

Frati, Lodovico. 1908. "Ricordanze Domestiche di Notai Bolognesi." *Archivio Storico Italiano* (series 5, tome 41): 3–15.

Freedberg, David. 2002. *The Eye of the Lynx: Galileo, His Friends, and the Beginnings of Modern Natural History*. Chicago: University of Chicago Press.

Freedman, Joseph S. 1984. *Deutsche Schulphilosophie im Reformationszeitalter (1500–1650): Ein Handbuch für den Hochschulunterricht*. Münster: MAKS Publikationen.

———. 1988. *European Academic Philosophy in the Late Sixteenth and Early Seventeenth Centuries: The*

Life, Significance and Philosophy of Clemens Timpler (1563/4–1624). 2 vols. Hildesheim: Olms.

Freeland, Guy, and Anthony Corones, eds. 2000. *1543 and All That: Image and Word, Change and Continuity in the Proto-scientific Revolution.* Boston, MA: Kluwer.

French, Roger. 1994. "Astrology in Medical Practice." In García-Ballester et al. 1994, 30–49.

Freudenthal, Gad. 1983. "Theory of Matter and Cosmology in William Gilbert's *De Magnete.*" *Isis* 74:22–37.

Friesen, John. 2003. "Archibald Pitcairne, David Gregory and the Scottish Origins of English Tory Newtonianism, 1688–1715." *History of Science* 41:163–91.

Frischlin, Nicodemus. 1573. *Consideratio novae stellae, quae mense Novembri, anno salutis MDLXXII in Signo Cassiopeae populis Septentrionalibus longè apparuit.* Tübingen: n.p.

———. 1601 [1586]. *De Astronomicae Artis cum Doctrina Coelesti et Naturali Philosophia, Congruentia, Ex Optimis quibusque Graecis Latinisque scriptoribus, Theologis, Medicis, Mathematicis, Philosophis et Poëtis collecta: Libri Quinque. Passim inserta est huic operi solida diuinationum Astrologicarum confutatio, repetita ex optimis quibusque Auctoribus, tàm recentibus quàm veteribus, quorum nomina post praefationem inuenies.* Frankfurt: Typis Wolffgangi Richteri.

Frugoni, Arsenio, ed. 1950. *Carteggio umanistico di Alessandro Farnese.* Florence: Olschki.

Fučíková, Eliška, ed. 1997a. *Rudolf II and Prague: The Court and the City.* Prague: Thames and Hudson.

———. 1997b. "Prague Castle under Rudolf II, His Predecessors and Successors." In Fučíková 1997a.

Fulco, Gulielmus [William Fulke]. 1560. *Antiprognosticon Contra Inutiles Astrologorum Praedictiones Nostradami, Cuninghami, Loui, Hilli, Vaghami, & Reliquorum Omnium.* London: Henry Sutton.

Funck, Johann. 1559. *Apocalypsis: Der Offenbarung Künfftiger Geschicht Johannis, . . . bis an der welt ende, Auslegung . . . Mit einer Vorrede Philip. Melanth.* Schleusingen: Hamsing.

Funkenstein, Amos. 1975a. "Descartes, Eternal Truths, and the Divine Omnipotence." *Studies in History and Philosophy of Science* 6:185–99.

———. 1975b. "The Dialectical Preparation for Scientific Revolutions: On the Role of Hypothetical Reasoning in the Emergence of Copernican Astronomy and Galilean Mechanics." In Westman 1975a, 165–203.

———. 1986. *Theology and the Scientific Imagination from the Middle Ages to the Seventeenth Century.* Princeton, NJ: Princeton University Press.

Fusai, Giuseppe. 1975 [1905]. *Belisario Vinta: Ministro e consigliere di stato dei Granduchi Ferdinando I e Cosimo II de'Medici (1542–1613).* Florence: Gozzini.

Gabotto, Ferdinando. 1891. *Nuove ricerche e documenti sull'astrologia all corte degli Estensi e degli Sforza.* Turin: La Letteratura.

Gaffurio, Franchino. 1967 [1492]. *Theorica Musice.* Facsimile ed. [Milan.] New York: Broude Bros.

———. 1969. *The "Practica Musicae" of Franchinus Gafurius.* Trans. I. Young. Madison: University of Wisconsin Press.

———. 1979 [1496]. *Practica Musice.* [Milan.] New York: Broude Bros.

———. 1993. *The Theory of Music.* Trans. W. K. Kreyszig. New Haven: Yale University Press.

Galilei, Galileo. 1890–1909. *Le opere di Galileo Galilei: Edizione nazionale sotto gli auspicii di Sua Maestà il re d'Italia.* 20 vols. in 21. Florence: Tip. di G. Barbèra.

———. 1957. *Letters on Sunspots.* In Drake 1957.

———. 1960. *The Assayer.* Trans. S. Drake and C. D. O'Malley. In Drake and O'Malley 1960.

———. 1967. *Dialogue Concerning the Two Chief World Systems.* Trans. S. Drake. Berkeley: University of California Press.

———. 1989a. *Sidereus Nuncius, or The Sidereal Messenger.* Trans. Albert Van Helden. Chicago: University of Chicago Press.

———. 1989b [1615]. "Considerations on the Copernican Opinion." In Finocchiaro 1989, 70–86.

———. 1989c. "Galileo's Discourse on the Tides." In Finocchiaro 1989, 119–33.

———. 1992. *Sidereus Nuncius: Le Messager Celeste.* Trans. and ed. Isabelle Pantin. Paris: Les Belles Lettres.

———. 1997. *Galileo on the World Systems: A New Abridged Translation and Guide.* Trans. M. A. Finocchiaro. Berkeley: University of California Press.

———. 2010. *Galileo Galilei and Christoph Scheiner, "On Sunspots."* Trans. and intro. Eileen Reeves and Albert Van Helden. Chicago: University of Chicago Press.

Galileo Galilei e la cultura Veneziana. 1995. Venice: Istituto Veneto di Scienze, Lettere ed Arti.

Galison, Peter. 1997. *Image and Logic: A Material Culture of Microphysics.* Chicago: University of Chicago Press.

Galluzzi, Paolo. 1980. "Il mecenatismo mediceo e le scienze." In Vasoli 1980a, 189–215.

———. 1982. "Motivi paracelsiani nella Toscana di Cosimo II e di Don Antonio dei Medici: Alchimia, medicina 'chimica' e riforma del sapere." In Zambelli 1982, 31–62.

———, ed. 1984. *Novità celesti e crisi del sapere.* Atti del Convegno Internazionale di Studi Galileiani. Florence: Giunti Barbèra.

———. 1998. "The Sepulchers of Galileo: The 'Living' Remains of a Hero of Science." In Machamer 1997, 417–47.

Garber, Daniel. 1992. *Descartes' Metaphysical Physics.* Chicago: University of Chicago Press.

———. 2004. "On the Frontlines of the Scientific Revolution: How Mersenne Learned to Love Galileo." *Perspectives on Science* 12: 135–63.

Garcaeus, Johannes, Jr. 1556. *Tractatus Brevis et Utilis, de Erigendis Figuris Coeli, Verificationibus, Revolutionibus et Directionibus. Ad illustrissimum Principem ac Dominum Pomeraniae Ducem.* Wittenberg: Heirs of Georg Rhau.

———. 1569. *Eine Christliche kurze Widerholung der warhafftigen Lere und bekentnis unsers Glaubens von der Zukunfft des Herrn Christi zum Gericht.* Wittenberg.

———. 1576. *Astrologiae Methodus in qua Secundum Doctrinam Ptolemaei Genituras Qualescunque Iudicandi Ratio Traditur.* Basel: Henricus Petri.

García-Ballester, Luis, et al. 1994. *Practical Medicine from Salerno to the Black Death.* Cambridge: Cambridge University Press.

Garin, Eugenio. 1942. "Il Carteggio di Pico della Mirandola." *La Rinascita* 5:567–91.

———. 1975. "Alle origini della polemica anticopernicana." *Studia Copernicana* 6 (Colloquia Copernicana 2): 31–42.

———. 1983. *Astrology in the Renaissance: The Zodiac of Life.* Rev. trans. E. Garin and C. Robertson. London: Routledge and Kegan Paul.

Gassendi, Pierre. 1655. *Tychonis Brahei, Equitis Dani, astronomorum coryphaei, vita . . . Accessit Nicolai Copernici, Georgii Peurbachii, et Joannis Regiomontani Astronomorum celebrium vita.* Paris: Apud Viduam Mathurini Dupuis.

———. 1659. *The Vanity of Judiciary Astrology, or Divination by the Stars.* Trans. A Person of Quality. London: Printed for Humphrey Moseley.

Gatti, Hilary. 1993. *The Natural Philosophy of Thomas Harriot.* Oxford: Oxford University Press.

———. 1997. "Giordano Bruno's *Ash Wednesday Supper* and Galileo's *Dialogue of the Two Major World Systems.*" *Bruniana and Campanelliana* 3:283–300.

———. 1999. *Giordano Bruno and Renaissance Science.* Ithaca, NY: Cornell University Press.

———, ed. 2002. *Giordano Bruno: Philosopher of the Renaissance.* Aldershot: Ashgate.

Gaukroger, Stephen. 1995. *Descartes: An Intellectual Biography.* Oxford: Clarendon Press.

———. 2001. *Francis Bacon and the Transformation of Early-Modern Philosophy.* Cambridge: Cambridge University Press.

———. 2002. *Descartes' System of Natural Philosophy.* Cambridge: Cambridge University Press.

Gaulke, Karsten, ed. 2007. *Der Ptolemäus von Kassel: Landgraf Wilhelm IV von Hessen-Kassel und die Astronomie.* Hessen: Museumlandschaft Hessen-Kassel.

Gaurico, Luca. 1552. *Tractatus Astrologicus, In quo agitur de praeteritis multorum hominum accidentibus per proprias eorum genituras ad unguem examinatis. Quorum exemplis consimilibus unusquisque de medio genethliacus vaticinari poterit de futuris, Quippe qui per varios casus artem experientia fecit, Exemplo monstrante uiam.* Venice: Curius Troianus Nauo.

Gauricus, Pomponius. 1541. *Super Arte Poetica Horatii.* Rome.

———. 1969 [1504]. *De Sculptura.* Trans. A. Chastel and R. Klein. Geneva: Droz.

Gearhart, C. A. 1985. "Epicycles, Eccentrics, and Ellipses: The Predictive Capabilities of Copernican Planetary Models." *Archive for the History of Exact Sciences* 32:207–23.

Geertz, Clifford. 1983. *Local Knowledge: Further Essays in Interpretive Anthropology.* New York: Basic Books.

Geminus. 1590. *Elementa Astronomiae.* Altdorf: Christoph Lochner.

Gemma, Cornelius. 1578. *De Prodigiosa Specie, Naturaque Cometae, qui Nobis Effulsit Altior Lunae Sedibus, Insolita Prorsus Figura, ac Magnitudine, anno 1577 plus Soprimanis 10 Apodeixis tum Physica tum Mathematica.* Antwerp: Christopher Plantin.

Gemma Frisius, Reiner. 1548. *De Principiis Astronomiae et Cosmographiae, deque usu Globi Cosmographici.* Antwerp: Joannes Steels.

Gentili, Augusto, and Claudia Cieri Via, eds. 1981. *Giorgione e la cultura veneta tra '400 e '500: Mito, allegoria, analisi iconologica.* Rome: De Luca.

Genuth, Sara Schechner. 1997. *Comets, Popular Culture, and the Birth of Modern Cosmology.* Princeton, NJ: Princeton University Press.

Gerth, Hans Heinrich, and C. Wright Mills, eds. 2000. *From Max Weber: Essays in Sociology.* London: Routledge and Kegan Paul.

Geymonat, Ludovico. 1965. *Galileo Galilei: A Biography and Inquiry into His Philosophy of Science.* New York: McGraw-Hill.

Giacobbe, G. C. 1972a. "Il Commentarium de Certitudine Mathematicarum Disciplinarum di Alessandro Piccolomini." *Physis* 14: 162–93.

———. 1972b. "Francesco Barozzi e la 'Quaestio de Certitudine Mathematicarum.' " *Physis* 14: 357–94.

———. 1977. "Un gesuita progressista nella 'Quaestio de Certitudine Mathematicarum' Rinascimentale." *Physis* 19: 51–86.

Giacon, Carlo. 1943. "Copernico, la filosofia e la teologia." *Civiltà Cattolica* 94:281–90, 367–74.

Giard, Luce. 1991. "Remapping Knowledge, Reshaping Institutions." In Pumfrey, Rossi, and Slawinski 1991, 19–47.

———, ed. 1995a. *Les jésuites à la Renaissance: Système éducatif et production du savoir.* Paris: Presses Universitaires.

———. 1995b. "Le devoir d'intelligence ou l'insertion des jésuites dans le monde du savoir." In Giard 1995a, xi–lxxix.

Giard, Luce, and Christian Jacob, eds. 2001. *Des Alexandries, I: Du livre au texte.* Paris: Bibliothèque nationale.

Gilbert, William. 1600. *De Magnete.* London: Peter Short.

———. 1965 [1651]. *De Mundo Nostro Sublunari Philosophia Nova.* [Amsterdam: Elzevir.] Amsterdam: Hertzberger.

———. 1958 [1893]. *Concerning the Magnet, Magnetic Bodies and about this Great Magnet, the Earth; A New Philosophy, demonstrated with many arguments and experiments.* Trans. P. Fleury Mottelay. New York: Dover.

Gillies, Donald. 1998. "The Duhem Thesis and the Quine Thesis." In Curd and Cover 1998, 302–19.

Gingerich, Owen. 1970. "Erasmus Reinhold." In *Dictionary of Scientific Biography* 1970–84.

———. 1973a. "From Copernicus to Kepler: Heliocentrism as Model and as Reality." *Proceedings of the American Philosophical Society* 117:513–22.

———. 1973b. "Copernicus and Tycho." *Scientific American* 229:86–101.

———. 1973c. "The Role of Erasmus Reinhold and the Prutenic Tables in the Dissemination of the Copernican Theory." *Studia Copernicana* 6 (Colloquia Copernicana 2): 43–62, 123–25.

———. 1974. "The Astronomy and Cosmology of Copernicus." In Contopoulos 1974, 67–85.

———. 1975. "Dissertatio cum Professore Righini et Sidereo Nuncio." In Righini-Bonelli and Shea 1975, 77–88.

———. 1984. "Phases of Venus in 1610." *Journal for the History of Astronomy* 15:209–10.

———. 1992a. *The Great Copernicus Chase and Other Adventures in Astronomical History.* Cambridge: Cambridge University Press.

———. 1992b. "Galileo and the Phases of Venus." In Gingerich 1992a, 98–104.

———. 1993a. *The Eye of Heaven: Ptolemy, Copernicus, Kepler.* New York: American Institute of Physics.

———. 1993b. "The Astronomy and Cosmology of Copernicus." In Gingerich 1993a.

———. 1993c. "*De Revolutionibus:* An Example of Renaissance Scientific Printing." In Gingerich 1993a, 252–68.

———. 1993d. "Early Copernican Ephemerides." In Gingerich 1993a, 205–20.

———. 2002. *An Annotated Census of Copernicus' "De Revolutionibus" (Nuremberg, 1543 and Basel, 1566).* Leiden: Brill.

———. 2004. *The Book Nobody Read: Chasing the Revolutions of Nicolaus Copernicus.* New York: Walker and Co.

———. 2008. Lecture delivered at the conference "Kepler 2008: From Tübingen to Żagań," University of Zielona Góra, June 22.

Gingerich, Owen, and Jerzy Dobrzycki. 1993. "The Master of the 1550 Radices: Jofrancus Offusius." *Journal for the History of Astronomy* 24:235–53.

Gingerich, Owen, and Albert Van Helden. 2003. "From *Occhiale* to Printed Page: The Making of Galileo's *Sidereus Nuncius.*" *Journal for the History of Astronomy* 34:251–67.

Gingerich, Owen, and James Voelkel. 1998. "Tycho Brahe's Copernican Campaign." *Journal for the History of Astronomy* 29:1–34.

Gingerich, Owen, and Robert S. Westman. 1988. *The Wittich Connection: Conflict and Priority in Sixteenth-Century Cosmology.* Philadelphia: Transactions of the American Philosophical Society, 78, pt. 7

Ginsberg, Morris. 1973. "Progress in the Modern Era." In *Dictionary of the History of Ideas,* 3:633–50.

Giordano Bruno, 1548–1600: Mostra storico documentaria, 2000. Florence: Leo S. Olschiki.

Giuntini, Francesco. 1573. *Speculum Astrologiae, quod Attinet ad Iudiciariam Rationem Natiuitatum atque Annuarum Reuolutionum: Cum Nonnullis Approbatis Astrologorum Sententiis.* Louvain: Philippus Tinghi.

———. 1581. *Speculum Astrologiae, Universam Mathematicam Scientiam in Certas Classes Digestam Complectens. Accesserunt Etiam Commentaria . . . in Duos Posteriores Quadripartiti Ptolemaei Libros, etc.* 2 vols. Lyons: Philippus Tinghi.

Gleick, James. 2003. *Isaac Newton.* New York: Pantheon.

Glogau, John of. 1480. *Prognosticum:* n.p.

Glymour, Clark. 1974. "Freud, Kepler, and the Clinical Evidence." In Wollheim 1974, 285–304.

Goclenius, Rudolf, the Younger. 1618. *Acroteleuticon Astrologicum.* Marburg.

Goddu, André. 1996. "The Logic of Copernicus's Arguments and His Education in Logic at Krakow." *Early Science and Medicine* 1:28–68.

———. 2004. "Hypotheses, Spheres and Equants in Copernicus's *De Revolutionibus.*" In Bouazzati 2004, 71–95.

———. 2006. "Reflections on the Origin of Copernicus's Cosmology." *Journal for the History of Astronomy* 37:37–53.

———. 2010. *Copernicus and the Aristotelian Tradition: Education, Reading, and Philosophy in Copernicus's Path to Heliocentrism.* Leiden: Brill.

Godman, Peter. 2000. *The Saint as Censor: Robert Bellarmine between Inquisition and Index.* Leiden: Brill.

Goffman, Erving. 1956. "The Nature of Deference and Demeanor." *American Anthropologist* 58:475–99.

Golan, Tal. 2004. *Laws of Men, Laws of Nature.* Cambridge, MA: Harvard University Press.

Goldstein, Bernard R. 1967. *The Arabic Version of Ptolemy's Planetary Hypotheses.* Philadelphia: Transactions of the American Philosophical Society 57.

———. 1969. "Some Medieval Reports of Venus and Mercury Transits." *Centaurus* 14:49–59.

———. 1987. "Remarks on Gemma Frisius's *De Radio Astronomico et Geometrico.*" In Berggren and Goldstein 1987, 167–80.

———. 1994. "Historical Perspectives on Copernicus's Account of Precession." *Journal for the History of Astronomy* 25:189–97.

———. 2002. "Copernicus and the Origins of His Heliocentric System." *Journal for the History of Astronomy* 33:219–35.

Gooding, David. 1986. "How Do Scientists Reach Agreement about Novel Observations?" *Studies in History and Philosophy of Science* 17:205–30.

Gorski, Karol. 1978. "Copernicus and Cayado." *Studia Copernicana* 16:397–401.

Gouk, Penelope. 1988. *The Ivory Sundials of Nuremberg, 1500–1700.* Cambridge: Whipple Museum.

Grafton, Anthony. 1973. "Michael Maestlin's Account of Copernican Planetary Theory." *Proceedings of the American Philosophical Society* 117:523–50.

———, ed. 1991a. *Defenders of the Text: The Traditions of Scholarship in an Age of Science, 1450–1800.* Cambridge, MA: Harvard University Press.

———. 1991b. "Humanism and Science in Rudolphine Prague: Kepler in Context." In Grafton 1991a, 178–203.

———. 1997. *Commerce with the Classics: Ancient Books and Renaissance Readers.* Ann Arbor: University of Michigan Press.

———. 1999. *Cardano's Cosmos: The Worlds and Works of a Renaissance Astrologer.* Cambridge, MA: Harvard University Press.

Grafton, Anthony, and Lisa Jardine, eds. 1986. *From Humanism to the Humanities.* London: Duckworth.

Granada, Miguel A. 1990. "L'Interpretazione Bruniana di Copernico e la 'Narratio Prima' di Rheticus." *Rinascimento* 30:343–65.

———. 1994. "Thomas Digges, Giordano Bruno y el Desarrollo del Copernicanismo en Inglaterra." *Éndoxa: Series Filosóficas* 4:7–42.

———. 1995. "Introduction." In Bruno 1995, xxi–xxx.

———. 1996a. "Il problema astronomico-cosmlogico e le sacre scritture dopo Copernico: Christoph Rothmann e la 'Teoria dell'Accomodazione.'" *Rivista di Storia della Filosofia* 4:789–828.

———. 1996b. *El debate cosmológico en 1588: Bruno, Brahe, Rothmann, Ursus, Röslin.* Naples: Bibliopolis, 1996.

———. 1997a. "Giovanni Maria Tolosani e la prima reazione romana di fronte al 'De revolutionibus': La critica di Copernico nell'opuscolo 'De coelo et elementis.'" In Bucciantini and Torrini 1997, 11–35.

———. 1997b. "Cálculos cronológicos, novedades cosmológicas y expectativas escatológicas en la Europa del siglo XVI." In Granada 2000a, 379–478.

———. 1999a. "Christoph Rothmann und die Auflösung der himmlischen Sphären. Die Briefe an den Landgrafen von Hessen-Kassel 1585." In Dick and Hamel 1999, 34–57.

———. 1999b. "'Esser Spogliato dall'Umana Perfezione e Giustizia': Nueva evidencia de la presencia de Averroes en la obra y en el proceso de Giordano Bruno." *Bruniana e Campanelliana: Ricerche filosofiche e materiali storico-testuali* 5: 305–31.

———. 2000a. *El Umbral de la Modernidad: Estudios sobre filosofía, religión y ciencia entre Petrarca y Descartes.* Barcelona: Herder.

———. 2000b. "Prologue." In Roeslin 2000, vii–xv.

———, ed. 2001. *Cosmología, teología y religión en la obra y en el proceso de Giordano Bruno.* Barcelona: Publicacions Universitat Barcelona.

———. 2002a. *Sfere solide e cielo fluide: Momenti del dibattito cosmologico nella seconda metà del Cinquecento.* Naples: Angelo Guerini.

———. 2002b. *Giordano Bruno: Universo infinito, unión con Dios, perfección del hombre.* Barcelona: Herder, 2002.

———. 2004a. "Aristotle, Copernicus, Bruno: Centrality, the Principle of Movement, and the Extension of the Universe." *Studies in History and Philosophy of Science* 35:91–114.

———. 2004b. "Astronomy and Cosmology in Kassel: The Contributions of Christoph Rothmann and His Relationship to Tycho Brahe and Jean Pena." In Zamrzlová 2004, 237–48.

———. 2006. "Did Tycho Eliminate the Celestial Spheres before 1586?" *Journal for the History of Astronomy* 37:125–45.

———. 2007a. "The Defence of the Movement of the Earth in Rothmann, Maestlin and Kepler: From Heavenly Geometry to Celestial Physics." In Bucciantini, Camerota, and Roux 2007, 95–119.

———. 2007b. "Michael Maestlin and the New Star of 1572." *Journal for the History of Astronomy* 38:99–124.

———. 2009. "Kepler and Bruno on the Infinity of the Universe and of Solar Systems." In Kremer and Włodarczyk 2009, 131–58.

Granada, Miguel A., and Dario Tessicini. 2005. "Copernicus and Fracastoro: The Dedicatory Letters to Pope Paul III, the History of Astronomy, and the Quest for Patronage." *Studies in History and Philosophy of Science* 36:431–76.

Grant, Edward. 1974. *A Source Book in Medieval Science.* Cambridge, MA: Harvard University Press.

———. 1978. "Aristotelianism and the Longevity of the Medieval World View." *History of Science* 16:93–106.

———. 1984. "In Defense of the Earth's Centrality and Immobility: Scholastic Reaction to Copernicanism in the Seventeenth Century." *Transactions of the American Philosophical Society* 74:1–69.

———. 1994. *Planets, Orbs and Spheres: The Medieval Cosmos, 1280–1687.* Cambridge: Cambridge University Press.

———. 1996. *The Foundations of Modern Science in the Middle Ages: Their Religious, Institutional and Intellectual Contexts.* Cambridge: Cambridge University Press.

Grassi, Giovanna. 1989. *Union Catalogue of Printed Books of 15th, 16th and 17th Centuries in European Astronomical Observatories.* Rome: Vecchiarelli.

Greenblatt, Stephen. 1980. *Renaissance Self-Fashioning: From More to Shakespeare.* Chicago: University of Chicago Press.

Gregory, James. 1668. "An Account of a Controversy

betwixt Stephano de Angelis, Professor of the Mathematics in Padua, and Joh. Baptista Riccioli Jesuite; as it was communicated out of their lately Printed Books, by that Learned Mathematician Mr. Jacob Gregory, a Fellow of the R. Society." *Philosophical Transactions of the Royal Society* 3:693–98.

Gregory, Tullio. 1983. "Temps astrologique et temps chrétien." In Leroux 1984, 557–73.

Grendler, Marcella. 1980. "A Greek Collection in Padua: The Library of Gian Vincenzo Pinelli (1535–1601)." *Renaissance Quarterly* 33:386–16.

Grendler, Paul F. 1977. *The Roman Inquisition and the Venetian Press, 1540–1605*. Princeton, NJ: Princeton University Press.

———. 1988. "Printing and Censorship." In Schmitt and Skinner 1988, 25–53.

———. 1989. *Schooling in Renaissance Italy: Literacy and Learning, 1300–1600*. Baltimore: Johns Hopkins University Press.

Grimm, Harold. 1973. *The Reformation Era, 1500–1650*. 2nd ed. New York: MacMillan.

Gualdo, Paolo. 1607. *Vita Ioannis Vincentii Pinelli, Patricii Genuensis*. Augsburg.

Guicciardini, Francesco. 1623. *Storia d'Italia*. Venice: Agostin Pasini.

———. 1969 [1561]. *The History of Italy*. Trans. S. Alexander. Princeton, NJ: Princeton University Press.

Guiducci, Mario. 1960 [1619]. *Discourse on the Comets*. Trans. S. Drake and C. D. O'Malley. Philadelphia: University of Pennsylvania Press.

Gundersheimer, Werner. 1973. *Ferrara: The Style of a Renaissance Despotism*. Princeton, NJ: Princeton University Press.

Gundlach, Franz. 1927–2001. *Die akademischen Lehrer der Philipps-Universität in Marburg von 1527 bis 1910*. 3 vols. Marburg: N. G. Elwert'sche Verlagsbuchhandlung, G. Braun.

Günther, Siegmund. 1882. *Peter und Philipp Apian, zwei deutsche Mathematiker u. Kartographen: Ein Beitrag zur Gelehrten-Geschichte des XVI. Jahrhunderts*. Prague: Königl. Böhm. Gesellschaft der Wissenschaften.

Hacking, Ian. 1993. "Working in a New World: The Taxonomic Solution." In Horwich 1993, 275–310.

Hagecius, Thaddaeus. 1553. *Practica teutsch auff das 1554 jar zu Wienn*. Vienna: Syngriener.

———. 1562. *Herbář, jinak bilinář* [Herbarium, or A Most Useful Herbal]. Prague: Melantrich.

———. 1574. *Dialexis de Novae et Prius Incognitae Stellae Apparitione*. Frankfurt.

———. 1576. *Responsio ad Virulentum . . . H. Raymundi . . . Scriptum: quo iterum confirmare nititur, stellam, quæ anno 72 & 73 supra sesquimillesimum fulsit, non novam, sed veterem fuisse, etc.* Prague: George Nigrin.

———. 1578. *Descriptio Cometae, qui apparuit Anno Domini MDLXXVII à ix die Nouembris usque ad xiii diem Ianuarij, Anni etc. LXXVIII; Adiecta est Spongia contra rimosas et fatuas Cucurbitulas Hannibalis Raymundi, Veronae sub monte Baldo nati, in larua Zanini Petoloti à monte Tonali*. Prague: George Melantrich.

———. 1584 [1562]. *Aphorismorum Metoposcopicorum Libellus Unus*. Frankfurt: Wechel.

Hain, Ludwig. 1826–28. *Reportorium Bibliographicum*. Stuttgart: Cottae.

Hale, John Rigby. 1971. "Sixteenth-Century Explanations of War and Violence." *Past and Present* 51:3–26.

Hall, A. Rupert. 1980. *Philosophers at War: The Quarrel between Newton and Leibniz*. Cambridge: Cambridge University Press.

Hall, Marie Boas. 1991. *Promoting Experimental Learning: Experiment and the Royal Society, 1660–1727*. Cambridge: Cambridge University Press.

Hallyn, Fernand. 1990. *The Poetic Structure of the World: Copernicus and Kepler*. Trans. Donald Leslie. New York: Zone Books.

———. 2004. "Gemma Frisius: A Convinced Copernican in 1555." *Filozofski vestnik* 25:69–83.

Hamesse, Jürgen, ed. 1994. *Manuel, PROGRAMMES de cours et techniques d'enseignement dans les universités médiévales*. Louvain-la-Neuve: Institut d'Etudes Médiévales.

Hamilton, Alastair. 2004. "Humanists and the Bible." In Kraye 2004, 100–117.

Hammer, William. 1951. "Melanchthon, Inspirer of the Study of Astronomy, with a Translation of His Oration in Praise of Astronomy (*De Orione*, 1553)." *Popular Astronomy* 51:308–19.

Hammerstein, Helga Robinson. 1986. "The Battle of the Booklets: Prognostic Tradition and Proclamation of the Word in Early Sixteenth-Century Germany." In Zambelli 1986b, 129–51.

Hankins, James. 1991. "The Myth of the Platonic Academy of Florence." *Renaissance Quarterly* 44:429–75.

Hannaway, Owen. 1986. "Laboratory Design and the Aim of Science: Andreas Libavius versus Tycho Brahe." *Isis* 77:585–610.

Hannemann, Manfred. 1975. *The Diffusion of the Reformation in Southwestern Germany, 1518–1534*. Chicago: University of Chicago Press.

Hanson, Norwood Russell. 1973. *Constellations and Conjectures*. Ed. Willard C. Humphreys. Dordrecht: Reidel.

Harkness, Deborah. 1999. *John Dee's Conversations with Angels: Cabala, Alchemy and the End of Nature*. Cambridge: Cambridge University Press.

Harris, Philip Rowland. 1998. *A History of the British Museum Library*. London: British Museum.

Hartfelder, Karl. 1889. *Philipp Melanchthon als Praeceptor Germaniae*. Berlin: A. Hofmann.

Hartner, Willy. 1967. "Galileo's Contribution to Astronomy." In McMullin 1967, 178–94.

———. 1973. "Copernicus, the Man and the Work." *Proceedings of the American Philosophical Society* 117:420–22.

Hasfurt, John of. 1492. *Judicium Baccalarij Johannis Cracoviensis de Hasfurt*. Leipzig: Konrad Kachelofen.

Hatch, Robert. 1982. *The Collection Boulliau (BN, FF.13019–13059): An Inventory*. Philadelphia: American Philosophical Society.

Hatfield, Gary. 1990. "Metaphysics and the New Science." In Lindberg and Westman 1990, 93–166.

Hayton, Darin. 2004. "Astrologers and Astrology in Vienna during the Era of Emperor Maximilian I (1493–1519)." PhD diss., University of Notre Dame.

———. 2007. "Astrology as Political Propaganda: Humanist Responses to the Turkish Threat in Early-Sixteenth Century Vienna." *Austrian History Yearbook* 38:61–91.

Headley, John M. 1963. *Luther's View of Church History*. New Haven: Yale University Press.

———. 1997. *Tommaso Campanella and the Transformation of the World*. Princeton, NJ: Princeton University Press.

———. 2004. "Introduction." In Headley, Hillerbrand, and Papalas 2004, xvii–xxv.

Headley, John M., Hans J. Hillerbrand, and Anthony J. Papalas, eds. 2004. *Confessionalization in Europe, 1555–1700*. Aldershot: Ashgate.

Heckius [van Heeck], Johannes. 1605. *De Nova Stella Disputatio*. Rome: A. Zanetti.

Heerbrand, Jacob. 1571. *Compendium Theologiae Methodi Quaestionibus Tractatum*. Tübingen: Georg Gruppenbach.

Heilbron, John L. 1983. *Physics at the Royal Society during Newton's Presidency*. Los Angeles: William Andrews Clark Memorial Library.

———. 1999. *The Sun in the Church*. Cambridge, MA: Harvard University Press.

———. 2005. "Censorship of Astronomy in Italy after Galileo." In McMullin 2005a, 279–322.

Helbing, Mario Otto. 1997. "Mobilità della terra e riferimenti a Copernico nelle opere dei professori dello Studio di Pisa." In Bucciantini and Torrini 1997, 57–66.

Heller, Joachim. 1549. *Practica auff das MDXLIX Jar/ Gestelt durch Joachim Heller/der Astronomey verordenten leser zu Nürmberg*. Nuremberg: Johann vom Berg und Ulrich Neuber.

———. 1551. *Practica auff 1551*. Nuremberg: Johann vom Berg und Ulrich Neuber.

———. 1557. *Practica auf das MDLVII Jar sampt Anzeygung unnd erclerung/Was die erscheinung/unnd bewegung/ des vergangenen unnd quuor angezeygten Cometen Im sechs und funfftstigsten Jar gewesen und bedeutet habe. Aus warem grundt der Astronomey von denem Practicirt und gestellet durch*. Nuremberg: J. Heller.

Hellman, C. Doris. 1944. *The Comet of 1577: Its Place in the History of Astronomy*. New York: Octagon.

Hellmann, Gustav, ed. 1898. *Rara Magnetica, 1296–1599*. Berlin: A. Asher and Co.

———. 1924. *Versuch einer Geschichte der Wetter-Vorhersage im XVI Jahrhundert*. Abhandlungen der Preussischen Akademie der Wissenschaften, Physikalische-Mathematische Klasse, 1. Berlin: Akademie der Wissenschaften.

Hellyer, Marcus. 2005. *Catholic Physics: Jesuit Early Modern Philosophy in Early Modern Germany*. Notre Dame, IN: University of Notre Dame Press.

Hemminga, Sixtus ab. 1583. *Astrologiae Ratione et Experientia Refutate Liber: Continens breuem quandam Apodixin de incertitudine & vanitate Astrologica, & particularium praedictionum exempla triginta: nunc primùm in lucem editus contra Astrologos Cyprianum Leouitium, Hieronymum Cardanum; & Lucam Gauricum*. Antwerp: Christopher Plantin.

Henderson, Janice. 1975. "Erasmus Reinhold's Determination of the Distance of the Sun from the Earth." In Westman 1975a, 108–30.

Henninger-Voss, Mary J. 2000. "Working Machines and Noble Mechanics: Guidobaldo del Monte and the Translation of Knowledge." *Isis* 91:233–59.

Henry, John. 1982. "Thomas Harriot and Atomism: A Reappraisal." *History of Science* 20:267–303.

———. 1997. *The Scientific Revolution and the Origins of Modern Science*. New York: St. Martin's Press.

Henry, John, and Sarah Hutton, eds. 1990. *New Perspectives on Renaissance Thought: Essays in the History of Science, Education and Philosophy in Memory of Charles B. Schmitt*. London: Duckworth.

Herlihy, David. 1991. "Family." *American Historical Review* 96:1–16.

Heydon, Christopher. 1603. *A Defense of Iudiciall Astrologie, in Answer to a Treatise Lately Published by M. Iohn Chamber*. Cambridge: Iohn Legat.

———. 1650. *An Astrological Discourse with Mathematical Demonstrations, Prouing the Powerful and Harmonical Influence of the Planets and Fixed Stars upon Elementary Bodies in Justification of the Validity of Astrology: Together with an Astrological Judgment upon the Great Conjunction of Saturn and Jupiter 1603*. London: Nicholas Fiske.

Hildericus, Theodorico Edo. 1568. *Logistice Astronomica*. Wittenberg.

———. 1590. *Gemini Elementa Astronomiae, Graece et Latine*. Altdorf.

Hill, Katherine. 1998. " 'Juglers or Schollers?': Negotiating the Role of a Mathematical Practitioner." *British Journal for the History of Science* 31:253–74.

Hill, Nicholas. 1619 [1601]. *Philosophia Epicurea, Democritiani, Theophrastica proposita simpliciter, non edocta*. Geneva: Fabriana.

Hine, William L. 1973. "Mersenne and Copernicanism." *Isis* 64:18–32.

Hipler, Franz. 1868. *Nikolaus Kopernikus und Martin Luther: Nach Ermländischen Archivalien*. Braunsberg: Eduard Peter.

———. 1875. "Die Portraits des Nikolaus Kopernikus." *Mitteilungen des Ermländischen Kunstvereins* 3.

Hirsch, Rudolf. 1967. *Printing, Selling, and Reading, 1450–1550*. Wiesbaden: Otto Harrassowitz.

Hobsbawm, Eric. 1994. *The Age of Extremes: A History of the World, 1914–1991*. New York: Vintage.

Hofmann, Norbert. 1982. *Die Artistenfakultät an der Universität Tübingen, 1534–1601*. Tübingen: Franz Steiner Verlag.

Hollis, Martin. 1977. *Models of Man: Philosophical Thoughts on Social Action*. Cambridge: Cambridge University Press.

Hooke, Robert. 1665. *Micrographia: Or Some Physiological Descriptions of Minute Bodies made by Magnifying Glasses*. London: Martyn and Allestry.

———. 1674. *An Attempt to Prove the Motion of the Earth from Observations*. London: John Martyn.

Hooykaas, Reijer. 1958. *Humanisme, science et réforme: Pierre de la Ramée (1515–1572)*. Leiden: Brill.

———. 1984. *G. J. Rheticus' Treatise on Holy Scripture and the Motion of the Earth*. Amsterdam: North Holland Publishing.

Hope, Charles. 1982. "Artists, Patrons and Advisers in the Italian Renaissance." In Lytle and Orgel 1982, 293–343.

Horace. 1926. *Satires, Epistles, and Ars Poetica*. Trans. H. R. Fairclough. Cambridge, MA: Harvard University Press.

Horky á Lochovič, Martin. 1610. *Brevissima Peregrinatio contra Nuncium Sidereum nuper ad Omnes Philosophos et Mathematicos Emissum*. [Mutinae (Modena): Julianus Cassianus]. In Galilei 1890–1909, 3:129–44.

Horský, Zdeněk. 1975. "Bohemia and Moravia and Copernicus." In *The 500th Anniversary of the Birth of Nicholas Copernicus*, 46–100. Prague: Czechoslovak Astronomical Society.

Horský, Zdeněk, and Emma Urbánková. 1975. *Tadeáš Hájek z Hájku (1525–1600) a jeho doba*. Prague: Statní Knihovna.

Horwich, Paul. 1993. *World Changes: Thomas Kuhn and the Nature of Science*. Cambridge, MA: MIT Press.

Höss, Irmgard. 1972. "The Lutheran Church of the Reformation: Problems of Its Formation and Organization in the Middle and North German Territories." In Buck and Zophy 1972, 317–39.

Houzeau, Jean-Charles, and Albert Lancaster. 1882–89. *Bibliographie générale de l'astronomie*. 2 vols. Brussels: F. Hayez.

Howell, Kenneth J. 2002. *God's Two Books: Copernican Cosmology and Biblical Interpretation in Early Modern Science*. Notre Dame, IN: University of Notre Dame Press.

Hoyningen-Huene, Paul 1993. *Reconstructing Scientific Revolutions: Thomas S. Kuhn's Philosophy of Science*. Trans. A. T. Levine. Chicago: University of Chicago Press.

Hsia, R. Po-Chia. 1998. *The World of Catholic Renewal, 1540–1770*. Cambridge: Cambridge University Press.

Hübner, Jürgen. 1975. *Die Theologie Johannes Keplers: Zwischen Orthodoxie und Naturwissenschaft*. Tübingen.

Hunt, Bruce. 1983. "'Practice vs. Theory': The British Electrical Debate, 1888–1891." *Isis* 74:341–55.

Hunter, Michael. 1987. "Science and Astrology in Seventeenth-Century England: An Unpublished Polemic by John Flamsteed." In Curry 1987, 261–300.

Hunter, Michael, and Simon Schaffer, eds. 1990. *Robert Hooke: New Studies*. London: Boydell and Brewer.

Hutchison, Keith. 1987. "Towards a Political Iconology of the Copernican Revolution." In Curry 1987, 95–142.

Huygens, Christiaan. 1888–1950. *Oeuvres complètes*. 22 vols. The Hague: M. Nijhoff.

Iliffe, Robert. 1995. "'Is He Like Other Men?' The Meaning of the *Principia Mathematica* and the Author as Idol." In Maclean 1995, 159–76.

"Index Paulus IV . . . January 1559." In Reusch 1961, 176–208.

Ingegno, Alfonso. 1988. "The New Philosophy of Nature." In Schmitt and Skinner 1988, 236–63.

Israel, Jonathan. 1995. *The Dutch Republic: Its Rise, Greatness, and Fall, 1477–1806*. Oxford: Clarendon.

———. 2001. *Radical Enlightenment: Philosophy and the Making of Modernity, 1650–1750*. Oxford: Oxford University Press.

Jacob, James R. 1997. *The Scientific Revolution: Aspirations and Achievements, 1500–1700*. Atlantic Highlands, NJ: Humanities Press.

Jacob, Margaret C. 1976. *The Newtonians and the English Revolution, 1689–1720*. Ithaca, NY: Cornell University Press.

Jacobsen, Theodor S. 1999. *Planetary Systems from the Ancient Greeks to Kepler*. Seattle, WA: University of Washington Press.

Jacquot, Jean. 1974. "Harriot, Hill, Warner and the New Philosophy." In Shirley 1974a, 107–28.

James VI and I. 1588. *A fruitefull meditation, containing a plaine and easie exposition. . . . of the 7.8.9. and 10. verses of the 20. Chap. Of the Revelation*. Edinburgh: Henry Charteris.

———. 1603. *Daemonologie, in Forme of a Dialogue*. London: Willam Cotton and William Aspley.

———. 1604. *Daemonologia: Hoc est, Adversus Incantationem siue Magiam Institutio Forma Dialogi Concepta, & in Libros III. Distincta*. Hannover: Guilielmus Antonius.

———. 1994. *Political Writings*. Cambridge: Cambridge University Press.

Jardine, Nicholas 1979. "The Forging of Modern Realism: Clavius and Kepler against the Sceptics." *Studies in History and Philosophy of Science* 10:141–13.

———. 1982. "The Significance of the Copernican Orbs." *Journal for the History of Astronomy* 13:168–94.

———. 1984. *The Birth of History and Philosophy of*

Science: Kepler's "A Defense of Tycho against Ursus," *with Essays on Its Provenance and Significance.* Cambridge: Cambridge University Press.

———. 1987. "Scepticism in Renaissance Astronomy: A Preliminary Study." In Popkin and Schmitt 1987, 83–102.

———. 1988. "Epistemology of the Sciences." In Schmitt and Skinner 1988, 685–711.

———. 1991. *Scenes of Inquiry: On the Reality of Questions in the Sciences.* Oxford: Clarendon.

———. 1998. "The Places of Astronomy in Early-Modern Culture." *Journal for the History of Astronomy* 29:49–62.

———. 2003. "Whigs and Stories: Herbert Butterfield and the Historiography of Science." *History of Science* 41:125–40.

Jardine, Nicholas, and Alain Segonds. 1987. "A Challenge to the Reader: Petrus Ramus on *Astrologia* without Hypotheses." In Popkin and Schmitt 1987, 83–102.

———. 2008. *La guerre des astronomes: La querelle au sujet de l'origine du système géo-héliocentrique à la fin du XVIe siècle.* 2 vols. Paris: Les Belles Lettres.

Jardine, Nicholas, Adam Mosley, and Karin Tybjerg. 2003. "Epistolary Culture, Editorial Practices, and the Propriety of Tycho's *Astronomical Letters.*" *Journal for the History of Astronomy* 34:421–51.

Jarrell, Richard. 1971. "The Life and Scientific Work of the Tübingen Astronomer, Michael Maestlin, 1550–1631." PhD diss., University of Toronto.

———. 1981. "Astronomy at the University of Tübingen: The Work of Michael Mästlin." In Seck 1981, 9–19.

Jaszi, Peter, and Martha Woodmansee, eds. 1994. *The Construction of Authorship: Textual Appropriation in Law and Literature.* Durham, NC: Duke University Press.

Jay, Martin. 1984. *Marxism and Totality: The Adventures of a Concept from Lukács to Habermas.* Berkeley: University of California Press.

Jervis, Jane. 1985. *Cometary Theory in Fifteenth-Century Europe.* Warsaw: Polskiej Akademii Nauk.

Jöcher, Christian Gottlieb. 1784–1897. *Gelehrtenlexikon.* 4 vols. Leipzig.

John Paul II. 1992. "Lessons of the Galileo Case: Discourse to the Pontifical Academy of Sciences." *Origins* 22:370–74.

Johns, Adrian. 1998. *The Nature of the Book: Print and Knowledge in the Making.* Chicago: University of Chicago Press.

Johnson, Francis R. 1937. *Astronomical Thought in Renaissance England: A Study of the English Scientific Writings from 1500 to 1645.* Baltimore: Johns Hopkins University Press.

———. 1952. "The History of Science in Elizabethan England: The Life and Times of Thomas and Leonard Digges." Unpublished lecture, History of Science Society.

———. 1953. "Astronomical Textbooks in the Sixteenth Century." In Underwood 1953, 1:285–302.

———. 1959. "Commentary on Derek J. deSolla Price." In Clagett 1959, 219–21.

Johnson, Francis R., and S. V. Larkey. 1934. "Thomas Digges, the Copernican System, and the Idea of the Infinity of the Universe in 1576." *Huntington Library Bulletin* 5:69–117.

Johnston, Stephen Andrew. 1994. "Making Mathematical Practice: Gentlemen, Practitioners and Artisans in Elizabethan England." PhD diss., University of Cambridge.

Jones, Alexander, ed. 2010. *Ptolemy in Perspective: Use and Criticism of his Work from Antiquity to the Nineteenth Century.* New York: Springer.

Jung, Carl G., and Wolfgang Pauli. 1955. *The Interpretation of Nature and the Psyche.* Trans. P. Sitz. London: Routledge and Kegan Paul.

Kaiser, David. 2005. *Drawing Theories Apart: The Dispersion of Feynman Diagrams in Postwar Physics.* Chicago: University of Chicago Press.

Kargon, Robert. 1966. *Atomism in England from Hariot to Newton.* Oxford: Oxford University Press.

Kassell, Lauren. 2005. *Medicine and Magic in Elizabethan London. Simon Forman: Astrologer, Alchemist and Physician.* Oxford: Clarendon.

Kaufmann, Thomas DaCosta. 1993. *The Mastery of Nature: Aspects of Art, Science, and Humanism in the Renaissance.* Princeton, NJ: Princeton University Press, 1993.

Kelly, John Norman Davidson. 1986. *The Oxford Dictionary of Popes.* Oxford: Oxford University Press.

Kelter, Irving A. 1992. "Paolo Foscarini's Letter to Galileo: The Search for Proofs of the Earth's Motion." *Modern Schoolman* 70:31–44.

Kempfi, Andrzej. 1969. "Erasme et la vie intellectuelle en Warmie au temps de Nicolas Copernic." In Margolin 1972, 397–406.

———. 1972. "Tydeman Giese jako uczen i korespondent Erazmu z Rotterdamu miedzy Fromborkiem a Bazylea." *Kommentarze Fromborskie* 4:26–44.

———. 1980. "Tolosani versus Copernicus." *Organon* 16–17:239–54.

Kent, F. W., Patricia Simmons, and J. C. Eade, eds. 1987. *Patronage, Art, and Society in Renaissance Italy.* Oxford: Clarendon Press.

Kepler, Gustav. 1931. *Familiengeschichte Keppler.* 2 vols. Görlitz: C. A. Starke.

Kepler, Johannes. 1604. *Gründtlicher Bericht von einem ungewöhnlichen newen Stern.* Prague: Schumans Druckerei.

———. 1606. *De Stella Nova.* Prague: Paulus Sessius.

———. 1858–71. *Joannis Kepleri Astronomi Opera Omnia.* Ed. Christian Frisch. 8 vols. Frankfurt: Heyder and Zimmer.

———. 1937–. *Gesammelte Werke.* Ed. Max Caspar et al. 22 vols. Munich: C. H. Beck

———. 1939. *Epitome of Copernican Astronomy: Books IV*

and V. Great Books of the Western World, vol. 16. Trans. C. G. Wallis. Chicago: Encyclopaedia Britannica.

———. 1965. *Kepler's Conversation with Galileo's Sidereal Messenger.* Trans. Edward Rosen. New York: Johnson Reprint.

———. 1981. *Mysterium Cosmographicum: The Secret of the Universe.* Trans. A. M. Duncan; intro. and commentary by E. J. Aiton. New York: Abaris Books.

———. 1984. *Le secret du monde.* Ed. and trans. Alain Segonds. Paris: Les Belles Lettres.

———. 1992. *New Astronomy.* Trans. W. H. Donahue. Cambridge: Cambridge University Press.

———. 1993. *Discussion avec le messager céleste; Rapport sur l'observation des satellites de Jupiter.* Trans. Isabelle Pantin. Paris: Belles Lettres.

———. 1997. *The Harmony of the World.* Trans. E. J. Aiton, A. M. Duncan, and J. V. Field. Philadelphia: American Philosophical Society.

———. 2000. *Optics: Paralipomena to Witelo and the Optical Part of Astronomy.* Trans. William H. Donahue. Santa Fe, NM: Green Lion Press.

Kessler, Eckhard. 1995. "Clavius entre Proclus et Descartes." In Giard 1995a, 295–308.

Kettering, Sharon. 1986. *Patrons, Brokers and Clients in Seventeenth-Century France.* Oxford: Oxford University Press.

Kibre, Pearl. 1966 [1936]. *The Library of Pico della Mirandola.* New York: AMS Press.

———. 1967. "Giovanni Garzoni of Bologna (1419–1505), Professor of Medicine and Defender of Astrology." *Isis* 58:504–14.

Kitcher, Philip. 1993. *The Advancement of Science: Science without Legend, Objectivity without Illusions.* Oxford: Oxford University Press.

Kittleson, James M., and Pamela J. Transue, eds. 1984. *Rebirth, Reform and Resilience: Universities in Transition, 1300–1700.* Columbus: Ohio State University Press.

Klee, Robert. 1997. *Introduction to the Philosophy of Science: Cutting Nature at Its Seams.* New York: Oxford University Press.

Klein, Robert. 1961. "Pomponius Gauricus on Perspective." *Art Bulletin* 43:211–30.

Kline, Ronald. 1995. "Construing 'Technology' as 'Applied Science': Public Rhetoric of Scientists and Engineers in the United States, 1880–1945." *Isis* 86:194–221.

Knaake, J. F. K., and Franz von Soden, eds. 1962. *Christoph Scheurls Briefbuch, ein Beitrag zur Geschichte der Reformation und ihrer Zeit.* Aalen: Zeller.

Knoll, Paul W. 1975. "The Arts Faculty at the University of Cracow at the End of the Fifteenth Century." In Westman 1975a, 137–56.

Knox, Dilwyn. 2002. "Ficino and Copernicus." In *Marsilio Ficino: His Theology, His Philosophy, His Legacy,* ed. Michael J. B. Allen and Valery Rees with Martin Davies, 399–418. Leiden: Brill.

———. 2005. "Copernicus's Doctrine of Gravity and the Natural Circular Motion of the Elements." *Journal of the Warburg and Courtauld Institutes* 58:157–211.

Koestler, Arthur. 1959. *The Sleepwalkers: A History of Man's Changing Vision of the Universe.* New York: Macmillan.

Köhler, Hans-Joachim. 1986. "The *Flugschriften* and Their Importance in Religious Debate: A Quantitative Approach." In Zambelli 1986b, 153–75.

Kollerstrom, Nicholas. 2001. "Galileo's Astrology." In Montesinos and Solís 2001, 421–31.

Kořán, Ivo. 1959. "Kniha Efemerid z biblioteky Tadeáše Hájka z Hájku" [Books of ephemerides in the library of Thaddeus Hayek of Hayek]. *Sborník pro dějiny přírodních věd a techniky* 6: 221–27.

———. 1969. "Praski krąg humanistów wokół Giordana Bruna" [The circle of Prague humanists around Giordano Bruno]. *Euhemer* 71–72: 81–93.

Koyré, Alexandre. 1955. "A Documentary History of the Problem of Fall from Kepler to Newton: De Motu Gravium Naturaliter Cadentium in Hypothesi Terrae Motae." *Transactions of the American Philosophical Society* 45: 329–95.

———. 1957. *From the Closed World to the Infinite Universe.* New York: Harper.

———. 1966 [1939]. *Études galiléennes.* Paris: Hermann.

———. 1992 [1961]. *The Astronomical Revolution: Copernicus-Kepler-Borelli.* Trans. R. E. W. Maddison. New York: Dover.

Krafft, Fritz, Karl Meyer, and Bernhard Sticker, eds. 1973. *Internationales Kepler-Symposium, Weil der Stadt 1971.* Hildesheim: Gerstenberg.

Krafft, Fritz, and Dieter Wuttke, eds. 1977. *Das Verhältnis der Humanisten zum Buch.* Boppard: Boldt.

Kraye, Jill, ed. 2004. *The Cambridge Companion to Renaissance Humanism.* Cambridge: Cambridge University Press.

Kremer, Richard L. 2009. "Kepler and the Graz Calendar Makers: Computational Foundations for Astrological Prognostication." In Kremer and Włodarczyk 2009, 77–100.

Kremer, Richard L., and Jarosław Włodarczyk, eds. 2009. *Johannes Kepler from Tübingen to Żagań.* Studia Copernicana 42. Warsaw: Instytut Historii Nauki PAN.

Kristeller, Paul Oskar. 1961a. *Chapters in Western Civilization.* 2 vols. 3rd ed. New York: Columbia University Press.

———. 1961b. "The Moral Thought of the Renaissance." In Kristeller 1961a.

———. 1964. *Eight Philosophers of the Italian Renaissance.* Stanford, CA: Stanford University Press.

Kuhn, Thomas S. 1957. *The Copernican Revolution.* Cambridge, MA: Harvard University Press.

———. 1970. *The Structure of Scientific Revolutions.* 2nd ed. Chicago: University of Chicago Press.

Kurze, Dietrich. 1960. *Johannes Lichtenberger: Eine*

Studie zur Geschichte der Prophetie und Astrologie. Lübeck: Mattisen.

———. 1986. "Popular Astrology and Prophecy in the Fifteenth and Sixteenth Centuries: Johannes Lichtenberger." In Zambelli 1986b, 177–93.

Kusukawa, Sachiko. 1991. "Providence Made Visible: The Creation and Establishment of Lutheran Natural Philosophy." PhD diss., University of Cambridge.

———. 1995. *The Transformation of Natural Philosophy: The Case of Philip Melanchthon.* Cambridge: Cambridge University Press.

Laet, Johannes de. 1649. *De Architectura.* Amsterdam: Elsevier.

Laird, Walter Roy. 2000. *The Unfinished Mechanics of Giuseppe Moletti: An Edition and English Translation of His Dialogue on Mechanics, 1576.* Toronto: University of Toronto Press.

Lakatos, Imre, and Elie Zahar. 1975. "Why Did Copernicus' Research Program Supersede Ptolemy's?" In Westman 1975a, 354–83.

Lammens, Cindy. 2002. *"Sic Patet Iter ad Astra:* A Critical Examination of Gemma Frisius' Annotations in Copernicus' *De Revolutionibus* and His Qualified Appraisal of the Copernican Theory." PhD diss., University of Ghent.

Landino, Christoforo. 1482. *Q. Horatii Flacci Opera Omnia.* Florence: Antonio di Bartolomeo Miscomini.

Lattis, James M. 1994. *Between Copernicus and Galileo: Christoph Clavius and the Collapse of Ptolemaic Cosmology.* Chicago: University of Chicago Press.

Laudan, Larry. 1990. "Demystifying Underdetermination." In Curd and Cover 1998, 320–54.

Launert, Dieter. 1999. *Nicolaus Reimers (Raimarus Ursus): Günstling Rantzaus—Brahes Feind—Leben und Werk.* Munich: Institut für Geschichte der Naturwissenschaften München.

Lawrence, P. D., and A. G. Molland. 1970. "David Gregory's Inaugural Lecture at Oxford." *Notes and Records of the Royal Society of London* 25:143–78.

Le Bachelet, Xavier-Marie. 1923. "Bellarmin et Giordano Bruno." *Gregorianum* 4:193–210.

LeGrand, H. E., ed. 1990. *Experimental Inquiries: Historical, Philosophical and Social Studies of Experimentation in Science.* Dordrecht: Kluwer.

Lehmann-Haupt, Hellmut, ed. 1967. *Homage to a Bookman: Essays on Manuscripts, Books and Printing, Written for Hans P. Kraus on His 60th Birthday.* Berlin: Gebr. Mann Verlag.

Lemay, Richard. 1962. *Abu Ma'sar and Latin Aristotelianism in the Twelfth Century: The Recovery of Aristotle's Natural Philosophy through Arabic Astrology.* Beirut: American University Press.

———. 1978. "The Late Medieval Astrological School at Cracow and the Copernican System." *Studia Copernicana* 16:337–54.

Leovitius, Cyprianus. 1556–57. *Ephemeridum Novum atque Insigne Opus ab Anno Domini 1556 usque in*

1606 *Accuratissimè Supputatum.* Augsburg: Phillippus Ulhardus.

———. 1558. *Brevis et Perspicua Ratio Iudicandi Genituras, ex Physicis Causis et vera experientia extructa: & ea Methodo tradita, ut quiuis facilè, in genere, omnium Thematum iuditia inde colligere possit:* CYPRIANO *Leouitio à Leonicia, excellente Mathematico, Autore. Praefixa est Admonitio de vero & licito Astrologiae usu: per Hieronymum VVolfium, virum in omni humaniore literatura, linguarum, artiumque Mathematicarum cognitione praestantem, in Dialogo conscripta. Adiectus est praeterea libellus de Praestantioribus quibusdam Naturae virtutibus: Ioanne Dee Londiniense Authore.* London: Henry Sutton.

———. 1564. *De conjunctionibus Magnis Insignioribus Superiorum planetarum, Solis defectionibus, et Cometis, in quarta Monarchia, cum eorundem effectuum historica expositione; his ad calcem accessit Prognosticon ab anno Domini 1564 in Viginti sequentes annos.* Lauingen: Emanuel Salczer.

Lerner, Michel-Pierre. 1996–97. *Le monde des sphères.* 2 vols. Paris: Les Belles Lettres.

———. 2002. "Aux origines de la polémique anticoperniciene (I)." *Revue des sciences philosophiques et théologiques.* 86:681–721.

———. 2004. "Copernic suspendu et corrigé: Sur deux décrets de la congregation romain de l'Index (1616–1620)." *Galilaeana* 1:21–89.

———. 2005. "The Origin and Meaning of 'World System.'" *Journal for the History of Astronomy* 36: 407–41.

Leroux, Jean-Marie, ed. 1984. *Le temps chrétien de la fin de l'antiquité au moyen âge: III–XIIIe siècles.* Paris: CNRS.

Levinger, Elma Ehrlich. 1952. *Galileo: First Observer of Marvelous Things.* New York: Julian Messner.

Lewis, Archibald R., ed. 1967. *Aspects of the Renaissance.* Austin: University of Texas Press.

Lichtenberger, Johannes. 1488. *Pronosticatio in Latino.* Heidelberg: Heinrich Knoblochtzer.

Liebler, Georg. 1589 [1561, 1566, 1573, 1575, 1576, 1584, 1581, 1587, 1593, 1596]. *Epitome Philosophiae Naturalis.* Basel: J. Oporinus.

Lightman, Bernard, ed. 1997. *Victorian Science in Context.* Chicago: University of Chicago Press.

———. 2002. "Huxley and Scientific Agnosticism: The Strange History of a Failed Rhetorical Strategy." *British Journal for the History of Science* 35:271–89.

Lilly, William. 1647. *Christian Astrology Modestly Treated of in Three Books.* London: John Partridge and Humphrey Blunden.

Lind, L. R., ed. 1992. *The Letters of Giovanni Garzoni, Bolognese Humanist and Physician, 1419–1505.* Atlanta: Scholars Press.

———. 1993. "Giovanni Garzoni 1419–1505: Bolognese Humanist and Physician." *Classical and Modern Literature* 14:7–24.

Lindberg, David C. 1976. *Theories of Vision from al-Kindi to Kepler.* Chicago: University of Chicago Press.

———, ed. 1978. *Science in the Middle Ages.* Chicago: University of Chicago Press.

———. 1982. "On the Applicability of Mathematics to Nature: Roger Bacon and His Predecessors." *British Journal for the History of Science* 15:3–25.

———. 1986. "The Genesis of Kepler's Theory of Light: Light Metaphysics from Plotinus to Kepler." *Osiris* 2:5–42.

Lindberg, David C., and Ronald L. Numbers, eds. 1986. *God and Nature: Historical Essays on the Encounter between Christianity and Science.* Berkeley: University of California Press.

Lindberg, David C., and Robert S. Westman, eds. 1990. *Reappraisals of the Scientific Revolution.* Cambridge: Cambridge University Press.

Lindorff, David. 2003. "Poindexter the Terror Bookie: Why Stop with an Assassination Market?" *Counterpunch.* July 30.

Lipton, Joshua. 1978. "The Rational Evaluation of Astrology in the Period of Arabo-Latin Translation, ca. 1126–1187." PhD diss., University of California, Los Angeles.

Lipton, Peter. 1991. *Inference to the Best Explanation.* London: Routledge.

———. 1998. "The Epistemology of Testimony." *Studies in History and Philosophy of Science* 29:1–31.

List, Martha. 1978. "Marginalien zum Handexemplar Keplers von Copernicus: *De Revolutionibus Orbium Coelestium,* Nürnberg, 1543." *Studia Copernicana* 16:443–40.

Livesey, Steven J. 1982. "*Metabasis:* The Interrelationship of the Sciences in Antiquity and the Middle Ages." PhD diss., University of California, Los Angeles.

———. 1985. "William of Ockham, the Subalternate Sciences, and Aristotle's Theory of Metabasis." *British Journal for the History of Science* 59:128–45.

Lloyd, Geoffrey E. R. 1978. "Saving the Appearances." *Classical Quarterly* n.s. 28:202–22.

———. 1979. *Magic, Reason and Experience: Studies in the Origins and Development of Greek Science.* Cambridge: Cambridge University Press.

———. 1996. *Adversaries and Authorities: Investigations into Ancient Greek and Chinese Science.* Cambridge: Cambridge University Press.

Lo zodiaco del principe: I decani di schifanoia di Maurizio Bonora. 1992. Ferrara: Maurizio Tosi.

Lohne, Johannes. 1973. "Kepler und Harriot, ihre wege zum Brechungsgesetz." In Krafft, Meyer, and Sticker 1973, 187–214.

Lomazzo, Giovanni Paolo. 1584. *Trattato dell'arte de la pittura.* Milan: Paolo Gottardo Pontio.

Lorenzini, Antonio. 1606. *De Numero, Ordine et Motu Coelorum.* Paris: D. Hilaire.

Lowood, Henry E., and Robin E. Rider. 1994. "Literary Technology and Typographic Culture: The Instrument of Print in Early Modern Science." *Perspectives on Science* 2:1–37.

Lowry, Martin. 1979. *The World of Aldus Manutius: Business and Scholarship in Renaissance Venice.* Ithaca, NY: Cornell University Press.

———. 1991. *Nicholas Jenson and the Rise of Venetian Publishing in Renaissance Europe.* Oxford: Basil Blackwell.

Lucier, Paul. 2009. "The Professional and the Scientist in Nineteenth-Century America." *Isis* 100:699–732.

Ludolphy, Ingetraut. 1986. "Luther und die Astrologie." In Zambelli 1986b, 101–7.

Luther, Martin. 1883–. *D. Martin Luthers Werke: Kritische Gesamtausgabe.* 71 vols. Weimar: Verlag Hermann Böhlaus Nochfolger.

———. 1967. *Whether Soldiers, Too, Can be Saved* (1526). In *Luther's Works,* 46:155–206. St. Louis, MO: Concordia Press.

———. 1969. "Vorrhede Martini Luthers: Auff die Weissagung des Johannis Lichtenbergers." In Warburg 1969, 545–50.

Lux, David S., and Harold J. Cook. 1998. "Closed Circles or Open Networks?: Communicating at a Distance during the Scientific Revolution." *History of Science* 36:179–211.

Lvovický de Lvoviče, Cyprián Karásek. *See* Leovitius, Cyprianus.

Lynch, Michael. 1992a. "Extending Wittgenstein: The Pivotal Move from Epistemology to the Sociology of Science." In Pickering 1992, 215–65.

———. 1992b. "From the 'Will to Theory' to the Discursive Collage: A Reply to Bloor's 'Left and Right Wittgensteinians.' " In Pickering 1992, 283–300.

Lytle, Guy Fitch. 1987. "Friendship and Patronage in Renaissance Europe." In Kent, Simmons, and Eade 1987, 47–62.

Lytle, Guy Fitch, and Stephen Orgel, eds. 1982. *Patronage in the Renaissance.* Princeton, NJ: Princeton University Press.

MacFarlane, Alan D. J. 1970. *Witchcraft in Tudor and Stuart England.* New York: Harper and Row.

MacFarlane, Ian Dalrymple. 1981. *Buchanan.* London: Duckworth.

Machamer, Peter, ed. 1998. *The Cambridge Companion to Galileo.* Cambridge: Cambridge University Press.

Maclean, Gerald, ed. 1995. *Culture and Society in the Stuart Restoration: Literature, Drama, History.* Cambridge: Cambridge University Press.

Maclean, Ian. 1984. "The Interpretation of Natural Signs: Cardano's *De subtilitate* versus Scaliger's *Exercitationes.*" In Vickers 1984, 231–52.

Maestlin, Michael. 1576. *Ephemeris Nova Anni 1577: Seqvens vltimam hactenvs a Ioanne Stadio Leonou-*

thesio editarum Ephemeridum, supputata ex tabulis Prutenicis. Tübingen: Georg Gruppenbach.

———. 1578. *Obseruatio et Demonstratio Cometae Aetherei, qui Anno 1577 et 1578, Constitutus in Sphaera Veneris, Apparuit.* Tübingen: Georg Gruppenbach.

———. 1580. *Ephemerides novae ab anno salutiferae incarnationis 1577 ad annum 1590. Supputatae ex Tabulis Prutenicis. Ad Horizontem Tubingensem, cuius longitudo est 29.grad. 45.Scru. Latitudo verò 48. grad. 24.scru.* Tübingen: Georg Gruppenbach.

———. 1583. *Aussführlicher und Gründtlicher Bericht von der Allgemainen/und Nunmehr bey sechtzehen Hundert jaren/von dem ersten Keyser Julio/biss auff setzige unsere Zeit/im gantzen H. Römischen Reich gebrauchter Jarrechnung oder Kalender.* Heidelberg: Jacob Müller.

———. 1581. *Consideratio et Observatio Cometae Aetherei Astronomica, qui anno MDLXXX in alto Aethere apparuit.* Heidelberg: Mylius.

———. 1586. *Alterum Examen Novi Pontificalis Gregoriani Kalendarii, quo ex ipsis fontibus demonstratur, quod novum Kalendarium omnibus suis partibus, Quibus quam rectissimè reformatum vel est, vel esse putatur multis mouis mendosum, et in ipsis fundamentis vitiosum sit.* Tübingen: Georg Gruppenbach.

———. 1588. *Defensio Alterius sui Examinis, quo ex ipsis fundamentis demonstraverat, quod Gregorianum Nouum Kalendarium omnibus suis partibus, quibus quàm rectissimè reformatum vel esse debebat, vel esse putatur totum sit vitiosum, Adversus, Cuiusdam Antonii Posseuini Iesuitae ineptissimas elusiones, quibus ipse dum Examine illud extenuat, et calumnijs carpit, non solum imperitiam et vanitatem suam prodit, verum etiam. Licet inuitus, et non cogitans Nouam Gregorianam Kalendarij emendationem magis confundit, et funditus euertit.* Tübingen: Georg Gruppenbach.

———. 1596a. "Preface to the Reader." In Rheticus 1596.

———. 1596b. *De Dimensionibus Orbium et Sphaerarum Coelestium iuxta Tabulas Prutenicas, ex Sententia Nicolai Copernici.* Tübingen: Georg Gruppenbach. In Kepler 1937–, 1:132–45.

———. 1597. *Epitome Astronomiae, qua brevi explicatione omnia, tam ad Sphaericam quam Theoricam eius partem pertinentia, ex ipsius scientiae fontibus deducta, perspicuè per quaestiones traduntur, Conscripta per Michaelem Maestlinum Goeppingensem, Matheseos in Academica Tubingensi Professorem.* Tübingen: Gruppenbach.

———. 1624. *Epitome Astronomiae.* Tübingen: Gruppenbach.

Maeyama, Yasukatsu, and Walter G. Saltzer, eds. 1977. *Prismata: Naturwissenschaftsgeschichtliche Studien; Festschrift für Willy Hartner.* Wiesbaden: F. Steiner.

Maffei, Raffaele. 1518. *De Institutione Christiana.* Rome: Mazochium.

Magini, Giovanni Antonio. 1582. *Ephemerides.* Venice: Zenarius.

———. 1585. *Tabulae Secundorum Mobilium Coelestium.* Venice: Zenarius.

Magocsi, Paul Robert. 1993. *Historical Atlas of East Central Europe.* Toronto: University of Toronto Press.

Malagola, Carlo. 1878. *Della vita e delle opere di Antonio Urceo Detto Codro: Studi e ricerche.* Bologna: Fava e Garagnani.

———. 1881. *Galileo e l'Università di Bologna.* Florence: M. Cellini.

Mancosu, Paolo. 1996. *Philosophy of Mathematics and Mathematical Practice in the Seventeenth Century.* New York: Oxford University Press.

Manfredi, Michele. 1919. *Gio. Battista Manso: Nella vita e nelle opere.* Naples: N. Jovene.

Mann, Nicholas. 2004. "The Origins of Humanism." In Kraye 2004, 1–19.

Manuel, Frank. 1968. *A Portrait of Isaac Newton.* Cambridge, MA: Harvard University Press.

Margolin, Jean-Claude, ed. 1972. *Colloquia Erasmiana Turonensia.* Toronto: University of Toronto Press.

Margolin, Jean-Claude, and Sylvain Matton, eds. 1993. *Alchimie et philosophie à la Renaissance.* Paris: J. Vrin.

Marino, John. 1994. "The Italian States in the 'Long Sixteenth Century.' " In Brady, Oberman, and Tracy 1994–95, 331–67.

Markowski, Mieczysław. 1974. "Die Astrologie an der Krakauer Universität in den Jahren 1450–1550." In Szczucki 1974, 83–89.

———. 1992. "Repertorium Bio-bibliographicum Astronomorum Cracoviensium Medii Aevi." *Studi Mediewistyczne* 28:91–155.

Marquardi, Ioannis. 1589. *Practica Theorica Empirica Morborum Interiorum, a Capite ad Calcem usque.* Spira: Typis Bernardi Albini.

Marschall, Laurence A. 1994. *The Supernova Story.* Princeton, NJ: Princeton University Press.

Martens, Rhonda. 2000. *Kepler's Philosophy: The Conceptual Foundations of the New Astronomy.* Princeton, NJ: Princeton University Press.

Martin, John. 1997. "Inventing Sincerity, Refashioning Prudence: The Discovery of the Individual in Renaissance Europe." *American Historical Review* 102:1309–42.

Martin, Julian. 1992. *Francis Bacon, the State, and the Reform of Natural Philosophy.* Cambridge: Cambridge University Press.

Martinelli, Roberto Biancarelli. 2004. "Paul Homberger: Il primo intermediario tra Galileo e Keplero." *Galilaeana: Journal of Galilean Studies* 1:171–82.

Marzi, Demetrio. 1896. *La questione della riforma del calendario nel Quinto Concilio Lateranense (1512–1517).* Florence: G. Carnesecchi.

Maschietto, Ludovico. 1992. "Girolamo Spinelli e Benedetto Castelli Benedettini di Sta. Giustina, discepoli e amici di Galileo Galilei." In Santinello 1992, 431–44.

Masson, David. 1859–94. *The Life of John Milton, Nar-*

rated in Connexion with the Political, Ecclesiastical, and Literary History of His Time. 7 vols. London: Macmillan.

Matsen, Herbert S. 1977. "Students' 'Arts' Disputations at Bologna around 1500, Illustrated from the Career of Alessandro Achillini (1463–1512)." History of Education 6:169–81.

———. 1994. "Students' 'Arts' Disputations at Bologna around 1500." Renaissance Quarterly 47:533–55.

Mattiazzo, Antonio. 1992. "La diocesi di Padova nel periodo dell'insegnamento di Galileo (1592–1610)." In Santinello 1992, 289–305.

Maurer, Wilhelm. 1962. "Melanchthon und die Naturwissenschaft seiner Zeit." Archiv für Kulturgeschichte 44:199–226.

Mauri, Alimberto. 1606. Considerazioni . . . sopra alcuni luoghi del discorso di Lodovico delle Colombe intorno alla stella apparita 1604. Florence: G. A. Caneo.

Mauss, Marcel. 1954 [1924]. The Gift. London: Cohen & West.

Mazzetti, Serafino. 1988 [1848]. Repertorio dei professori dell'Università e dell'Istituto delle Scienze di Bologna. Bologna: S. Tommaso d'Aquino.

Mazzoni, Jacopo. 1597. In Universam Platonis et Aristotelis Philosophiam Praeludia, Sive de comparatione Platonis et Aristotelis Liber Primus. Venice: Iohannes Guerilius.

McClure, George. 2004. The Culture of Profession in Late Renaissance Italy. Toronto: University of Toronto Press.

McGuire, James E. 1977. "Neoplatonism and Active Principles: Newton and the Corpus Hermeticum." In McGuire and Westman 1977, 95–142.

McGuire, James E., and P. M. Rattansi. 1966. "Newton and the 'Pipes of Pan.'" Notes and Records of the Royal Society of London 21:108–43.

McGuire, James E., and Martin Tamny. 1983. Certain Philosophical Questions: Newton's Trinity Notebook. Cambridge: Cambridge University Press.

McGuire, James E., and Robert S. Westman. 1977. Hermeticism and the Scientific Revolution. Los Angeles: William Andrews Clark Memorial Library.

McInerny, Ralph. 1983. "Beyond the Liberal Arts." In Wagner 1983, 248–72.

McKirahan, Richard. 1978. "Aristotle's Subordinate Sciences." British Journal for the History of Science 11:197–220.

McMenomy, Christie. 1984. "The Discipline of Astronomy in the Middle Ages." PhD diss., University of California, Los Angeles.

McMullin, Ernan, ed. 1967. Galileo, Man of Science. New York: Basic Books.

———. 1987. "Bruno and Copernicus." Isis 78:55–74.

———. 1990. "Conceptions of Science in the Scientific Revolution." In Lindberg and Westman 1990, 27–92.

———. 1998. "Rationality and Paradigm Change in Science." In Curd and Cover 1998, 119–38.

———, ed. 2005a. The Church and Galileo. Notre Dame, IN: University of Notre Dame Press.

———. 2005b. "The Church's Ban on Copernicanism, 1616." In McMullin 2005a, 150–90.

———. 2005c. "Galileo's Theological Venture." In McMullin 2005a, 88–116.

Medina, Miguel de. 1564. Christianae Paraenesis siue De Recta in Deum Fidei. Venice: Giordano Ziletti e Gio. Griffio.

Melanchthon, Philipp. 1532. Chronica durch Magistrum Johann Carion fleissig zusammengezogen, menigklich nützlich zu lesen. Wittenberg.

———. 1536. Mathematicarum Disciplinarum tum Etiam Astrologiae Encomia. (Strasbourg.) In Bretschneider et al. 1834–, 9:292–98.

———. 1537. Rudimenta Astronomica Alfragani. Item Albategnius astronomus peritissimus de motu stellarum, ex observationibus tum propriis, tum Ptolemaei, omnia cum demonstrationibus Geometricis et Additionibus Joannis de Regiomonte. Item Oratio introductoria in omnes scientias Mathematicas Joannis de Regiomonte, Patauij habita, cum Alfraganum publice praelgeret. Eiusdem utilissima introductio in elementa Euclidis. Nuremberg: Johannes Petreius.

———. 1834–60. Opera Quae Supersunt Omnia. Ed. C. G. Bretschneider. 28 vols. Halle [1834–52] and Brunswick [1853–60]: C. A. Schwetschke. See Bretschneider et al. 1834–.

Melanchthon, Philipp, and Peucer, Caspar. 1624. Chronicon Carionis Expositum et Auctum multis et veteribus et recentibus historiis, in descriptionibus regnorum et gentium antiquarum, et narrationibus rerum Ecclesiasticarum, et Politicarum Graecarum, Romanarum, Germanicarum et aliarum, ab exordio Mundi usque ad Carolum V Imperatorem. 2 vols. Frankfurt am Main: Godefrid Tampachius.

Meller, Peter. 1981. "I 'Tre Filosofi' di Giorgione." In Gentili and Via 1981, 227–47.

Mendoça, Bernardino de. 1596. Theorica y practica de guerra. Antwerp: Emprenta Plantiniana.

Mersenne, Marin. 1623. Quaestiones Celeberrimae in Genesim. Paris: Sebastian Cramoisy.

———. 1647. Novarum Observationum Physico-mathematicarum. Paris: Antonius Bertier.

Messahalah (Masha' Allāh). 1549. De Revolutione Annorum Mundi, de Significatione Planetarum Nativatibus, de Receptionibus, ed. Joachimus Hellerus Leucopetreus. Nuremberg: J. Montanus and U. Neuber.

Methuen, Charlotte. 1996. "Maestlin's Teaching of Copernicus: The Evidence of His University Textbook and Disputations." Isis 87:230–47.

———. 1998. Kepler's Tübingen: Stimulus to a Theological Mathematics. Brookfield, VT: Ashgate.

———. 1999. "Special Providence and Sixteenth-Century Astronomical Observation: Some Preliminary Reflections." Early Science and Medicine 4:99–113.

Michel, Paul-Henri. 1973. The Cosmology of Giordano

Bruno. Trans. R. E. W. Maddison. Ithaca, NY: Cornell University Press.

Middelburg, Paul of. 1492. *Inuectiva magistri Pauli de Myddelburgo vatis profecto celeberrimi in supersticiosum quendam astrologum et sortilegum una quoque et decem venustas vel astronomicas questiones.* Venice: E. Ratdolt.

Milani, Marisa. 1993. "Il 'Dialogo in Perpuosito de la Stella Nuova' di Cecco di Ronchitti da Brugine." *Giornale Storico della Letteratura Italiana* 170:66–86.

Miller, Peter N. 1996. "Citizenship and Culture in Early Modern Europe." *Journal of the History of Ideas* 57:725–42.

———. 2000. *Peiresc's Europe: Learning and Virtue in the Seventeenth Century.* New Haven: Yale University Press.

Milward, Peter. 1978a. *Religious Controversies of the Elizabethan Age.* London: Scolar Press.

———. 1978b. *Religious Controversies of the Jacobean Age.* London: Scolar Press.

Minnis, Alastair J. 1984. *Medieval Theory of Authorship: Scholastic Literary Attitudes in the Later Middle Ages.* London: Scolar Press.

Mirandulanus, Antonius Bernardus. 1562. *Antonii Bernardi Mirandulani, episcopi Casertanio, Disputationes in quibus primum ex professo monomachia (quam Singulare certamen Latini, recentiores Duellum uocant) philosophicis rationibus astruitur, & mox diuina authoritate labefactata penitùs euertitur: mones quoque iniuriarum species declarantur, easque conciliandi et è medio tollendi certissimae rationes traduntur. Deinde verò omnes utriusque philosophiae, tam contemplatiuae quàm actiuae, Loci obscuriores, & ambiguae Quaestiones (praesertim de Animae immortalitate, & Astrologiae iudiciariae diuinationibus) Aristotelica methodo luculentissimè examinantur & explicantur.* Basel: Henricus Petri.

Moesgaard, Kristian P. 1972a. "Copernican Influence on Tycho Brahe." *Studia Copernicana* 5 (Colloquia Copernicana 1), 31–55.

———. 1972b. "How Copernicanism Took Root in Denmark and Norway." *Studia Copernicana* 5 (Colloquia Copernicana 1), 117–52.

Monfasani, John. 1993. "Aristotelians, Platonists, and the Missing Ockhamists: Philosophical Liberty in Pre-Reformation Italy." *Renaissance Quarterly* 46:247–76.

Montaigne, Michel de. 1958. *The Complete Essays of Montaigne.* Trans. D. M. Frame. Stanford, CA: Stanford University Press.

Montesinos, José, and Carlos Solís, eds. 2001. *Largo campo di filosofare: Eurosymposium Galilei, 2001.* Orotava: Fundacion Canaria Orotava de Historia de la Ciencia.

Monumenta Paedagogica Societatis Jesu Quae Primum Rationem Studiorum, Anno 1586 Editam Praeces-

sere. 1901. Ed. C. G. Rodeles et al. Madrid: Augustino Avrial.

Moore, Marian A.. 1959. "A Letter of Philip Melanchthon to the Reader." *Isis* 50, no. 2: 145–150.

Moran, Bruce T. 1973. "The Universe of Philip Melanchthon: Criticism and the Use of the Copernican Theory." *Comitatus* 4:1–23.

———. 1977. "Princes, Machines and the Valuation of Precision in the Sixteenth Century." *Sudhoffs Archiv* 61:209–28.

———. 1978. "Science at the Court of Hesse-Kassel." PhD diss., University of California, Los Angeles.

———. 1981. "German Prince-Practitioners: Aspects in the Development of Courtly Science, Technology, and Procedures in the Renaissance." *Technology and Culture* 22:253–74.

———. 1982. "Christoph Rothmann, the Copernican Theory, and Institutional and Technical Influences on the Criticism of Aristotelian Cosmology." *The Sixteenth Century Journal* 13:85–108.

———, ed. 1991a. *Patronage and Institutions: Science, Technology, and Medicine at the European Court (1500–1750).* Woodbridge, UK: Boydell.

———. 1991b. "Patronage and Institutions: Courts, Universities, and Academies in Germany; an Overview: 1550–1750." In Moran 1991a, 169–84.

More, Henry. 1646. *Democritus Platonissans, or, An Essay upon the Infinity of Worlds out of Platonick Principles.* Cambridge: Roger Daniel.

Morell, Jack, and Arnold Thackray. 1981. *Gentlemen of Science: Early Years of the British Association for the Advancement of Science.* Oxford: Clarendon Press.

Morinus, Ioannis Baptista. 1634. *Responsio pro Telluris Motu Quiete ad Jacobi Lansbergii Apologiam pro Telluris Motu.* Paris: Johannes Libert.

———. 1641. *Coronis astronomiae iam a fundamentis integre et exacte restitutae: Qua respondetur ad introductionem in thearum astronomicum, clarissimi viri Christiani Longomontani; Hafniae in Dania Regij Mathematum Professoris.* Paris: Apud Authorem.

Morley, Henry. 1854. *Jerome Cardan.* London: Chapman and Hall.

Mosely, Adam. 2007. *Bearing the Heavens: Tycho Brahe and the Astronomical Community of the Late Sixteenth Century.* Cambridge: Cambridge University Press.

Moss, Jean Dietz. 1993. *Novelties in the Heavens.* Chicago: University of Chicago Press.

Moyer, Ann E. 1992. *Musica Scientia: Musical Scholarship in the Italian Renaissance.* Ithaca, NY: Cornell University Press.

Mulerius, Nicholas. 1616. *Institutionum Astronomicarum Libri Duo.* Groningen: Sassius.

———. 1617. *Astronomia Instaurata, Libris Sex Comprehensa, Qui de Revolutionibus Orbium Coelestium Inscribuntur.* Amsterdam: Willem Janszoon Blaeu.

Müller, Konrad. 1963. "Ph. Melanchthon und das kopernikanische Weltsystem." *Centaurus* 9:16–28.

Müller, Max. 1908. *Johann Albrecht von Widmanstetter (1506–1577): Sein Leben und Wirken.* Bamberg: Handels-Druckerei.

Muñoz, Jerónimo. 1981. *Libro del nuevo cometa.* Valencia: Gráficas Soler.

Naibod, Valentine. 1573. *Primarum de Coelo et Terra Institutionum Quotidianarumque Mundi Revolutionum Libri Tres.* Venice: n.p.

Naiden, James R. 1952. *The Sphaera of George Buchanan (1506–1582).* n.p.

Najemy, John J. 1995. Review of Anthony Parel, *The Machiavellian Cosmos. Journal of Modern History,* 67:676–80.

Nardi, Bruno. 1971. *Saggi sulla cultura Veneta del quattro e cinquecento.* (Medioevo e Umanesimo, no. 12). Padua: Antenore.

Nastasi, Pietro, ed. 1988. *Atti del Convegno "Il Meridione e le Scienze": Seculi XVI–XIX.* Palermo: University of Palermo.

Navarro-Brotóns, Victor. 1995. "The Reception of Copernicus in Sixteenth-Century Spain: The Case of Diego de Zuñiga." *Isis* 86:52–78.

Naylor, Ron. 2003. "Galileo, Copernicanism and the Origins of the New Science of Motion." *British Journal for the History of Science* 36:151–81.

Neander, Michael. 1561. *Elementa Sphaericae Doctrinae seu de Primo Motu.* Basel: Johannes Oporinus.

Newman, William R., and Anthony Grafton, eds. 2001. *Secrets of Nature: Astrology and Alchemy in Early Modern Europe.* Cambridge, MA: MIT Press.

Newton, Isaac. 1962 [1st English ed. 1728]. *Sir Isaac Newton's Mathematical Principles of Natural Philosophy and His System of the World.* 2 vols. Trans. A. Motte. Berkeley: University of California Press.

Newton, John. 1657. *Astronomica Britannica.* London: Leybourn.

Niccoli, Ottavia. 1990. *Prophecy and People in Renaissance Italy.* Trans. L. G. Cochrane. Princeton, NJ: Princeton University Press.

Nicholson, Marjorie Hope. 1935. "The 'New Astronomy' and English Literary Imagination." *Studies in Philology* 32:428–62.

———. 1939. "English Almanacks and the New Astronomy." *Annals of Science* 4:1–33.

Norlind, Wilhelm. 1970. *En levnadsteckning med nya bidrag belysande hans liv och verk.* Lund: Gleerup.

North, John D. 1975. "The Reluctant Revolutionaries: Astronomy after Copernicus." *Studia Copernicana* 13:169–84.

———. 1980. "Astrology and the Fortunes of the Churches." *Centaurus* 24:181–211.

———. 1986a. *Horoscopes and History.* London: Warburg Institute.

———. 1986b. "Celestial Influence: The Major Premiss of Astrology." In Zambelli 1986b, 45–100.

———. 1994. *The Norton History of Astronomy and Cosmology.* New York: W. W. Norton.

Novara, Domenico Maria. 1484–1504. *Pronosticon.* [Latin prognostications, issued annually.] Bologna.

Nussdorfer, Laurie. 1993. "Writing and the Power of Speech: Notaries and Artisans in Baroque Rome." In Diefendorf and Hesse 1993, 103–18.

Nutton, Vivian. 1985. "Humanist Surgery." In Wear, French, and Lonie 1985, 75–99.

Oestreich, Gerhard. 1982. *Neostoicism and the Early Modern State.* Trans. D. McLintock. Cambridge: Cambridge University Press.

Oestmann, Günther, H. Darrel Rutkin, and Kocku von Stuckrad, eds. 2005. *Horoscopes and Public Spheres: Essays on the History of Astrology.* Berlin: W. de Gruyter.

Offusius, Jofrancus. 1557. *Tabula Cardinalis Galliae Medio Accommodata.* Paris: Ex officina Ioannis Royerij typographi Regij.

———. 1570. *De Divina Astrorum Facultate, In Laruatam Astrologiam.* Paris: Ex officina Ioannis Royerij typographi Regij.

O'Malley, Charles Donald. 1972. "Andreas Vesalius." In *Dictionary of Scientific Biography* 1970–84, 5:347–49.

O'Malley, John W. 1993. *The First Jesuits.* Cambridge, MA: Harvard University Press.

Oreskes, Naomi. 1999. *The Rejection of Continental Drift: Theory and Method in American Earth Science.* New York: Oxford University Press.

Origanus, David. 1609. *Novae Motuum Coelestium Ephemerides Brandenburgicae.* Frankfurt (Oder): J. Eichhorn.

Osiander, Andreas. 1527. *Eyn Wunderliche Weyssagung von dem Bapstum* [Wondrous Prophecy of the Papacy]. Nuremberg: Güldenmundt.

———. 1532. *Gutachten über die Scheidung der Ehe Heinrich VIII von England mit Katharina von Aragon.* In Strype 1848, 1:19 ff.

———. 1544. *Coniecturae de Ultimis Temporibus, ac de Fine Mundi ex Sacris Literis.* Nuremberg: Johannes Petreius.

———. 1548. *The Conjectures of the Ende of the Worlde* (gathered out of Scripture by A. Oseander). Trans. George Joye. [Antwerp: S. Mierdman].

Osler, Margaret, ed. 2000. *Rethinking the Scientific Revolution.* Cambridge: Cambridge University Press.

Ottaviano Scotto omnibus edition. 1490. Contains Sacrobosco 1490, Regiomontanus 1490, and Peurbach 1490. Venice: Ottaviano Scotto.

Overfield, James H. 1984. "University Studies and the Clergy in Pre-Reformation Germany." In Kittleson and Transue 1984, 254–92.

Pagden, Anthony, ed. 1987. *The Languages of Political Theory in Early-Modern Europe.* Cambridge: Cambridge University Press.

Page, Sophie. 2002. *Astrology in Medieval Manuscripts.* Toronto: University of Toronto Press.

Palingenius, Marcellus. 1947. *The Zodiake of Life.* Trans.

Barnabe Googe. New York: Scholars' Facsimiles and Reprints.

Palmerino, Carla Rita. 2004. "Gassendi's Reinterpretation of the Galilean Theory of the Tides." *Perspectives on Science* 12:212–37.

Palmesi, Vincenzo. 1899. "Ignazio Danti." *Bolletino della R. Deputazione di Storia Patria per l'Umbria* 5, no. 1:81–125.

Pantin, Isabelle. 1987. "La Lettre de Melanchthon à S. Grynaeus: Les avatars d'une apologie de l'astrologie." In Aulotte 1987, 85–101.

———. 1995. *La poésie du ciel en France dans la seconde moitié du seizième siècle.* Geneva: Droz.

Papagno, Giuseppe, and Amadeo Quondam, eds. 1982. *La corte e lo Spazio: Ferrara Estense.* Rome: Bulzoni Editore.

Papia, Petrus de. 1482. *Practica Nova Iudicialis.* Nuremberg: Anton Koburger.

Parel, Anthony. 1992. *The Machiavellian Cosmos.* New Haven: Yale University Press.

Park, Katharine. 1985. *Doctors and Medicine in Early Renaissance Florence.* Princeton, NJ: Princeton University Press.

Partner, Peter. 1976. *Renaissance Rome, 1500–1559.* Berkeley: University of California Press.

Patterson, Louise Diehl. 1951. "Leonard and Thomas Digges: Biographical Notes." *Isis* 42:120–21.

Pauli, Simon. 1856 [1574]. *Postilla.* In Beste 1856, 2:272–87.

Pauli, Wolfgang. 1955. "The Influence of Archetypal Ideas on the Scientific Theories of Kepler." In Jung and Pauli 1955, 147–240.

Paulsen, Friedrich. 1906. *The German Universities and University Study.* Trans. F. Thilly and W. Elwang. New York: Charles Scribner's Sons.

Peck, Linda Levy, ed. 1991a. *The Mental World of the Jacobean Court.* Cambridge: Cambridge University Press.

———. 1991b. "The Mental World of the Jacobean Court: An Introduction." In Peck 1991a.

Pedersen, Olaf. 1975. "The *Corpus Astronomicum* and the Traditions of Mediaeval Latin Astronomy." *Studia Copernicana* 13:57–96.

———. 1978a. "Astronomy." In Lindberg 1978, 303–37.

———. 1978b. "The Decline and Fall of the Theorica Planetarum: Renaissance Astronomy and the Art of Printing." *Studia Copernicana* 16: 157–85.

The Penny Cyclopaedia of the Society for the Diffusion of Useful Knowledge. 20 vols. 1833–43. London: C. Knight.

Pèrcopo, Erasmo. 1894. *Pomponio Gàurico: Umanista Napoletano.* Napoli: Luigi Pierro.

Pereira, Benito. 1609 [1562]. *De Communibus Omnium Rerum Naturalium Principiis et Affectionibus.* Rome: Impensis Venturini Tramezini, Apud Franciscum Zanettum et Barthol. Tosium socios.

———. 1591. *Adversus Fallaces et Superstitiosas Artes,*
Id Est, de Magia, de Observatione Somniorum, et de Divinatione Astrologica, Libri Tres. Venice: Ciottus.

———. 1661. *The Astrologer Anatomiz'd Or, the Vanity of the Star-Gazing Art.* Trans. Percy Enderby. London: Benj. Needham.

Peters, William T. 1984. "The Appearance of Venus and Mars in 1610." *Journal for the History of Astronomy* 15:211–14.

Peucer, Caspar. 1553. *Elementa Doctrinae de Circulis Coelestibus, et Primo Motu, Recognita et Correcta.* Wittenberg: Crato.

[Peucer, Caspar]. 1568. *Hypotyposes orbium Coelestium, quas appellant Theoricas Planetarum: congruentes cum Tabulis Alphonsinis et Copernici, seu etiam tabulis Prutenicis: in usum Scholarum publicatae.* Argentorati: Theodosius Rihelius.

———. 1570. *Brevis Repetitio Doctrinae de Erigendis Coeli Figuris.* June–October.

———. 1591. *Commentarius de praecipuis divinationum generibus divinationum, in quo a prophetiis autoritate divina traditis et a physicis coniecturis discernuntur artes et imposturae diabolicae atque observationes natae ex superstitione et cum hac coniunctae.* Wittenberg: J. Crato.

Peurbach, Georg. 1472. *Theoricae Novae Planetarum.* Nürnberg: Johann Müller.

———. 1482. *Theoricae Novae Planetarum.* In Ratdolt omnibus edition 1482.

———. 1485. *Theoricae Novae Planetarum.* In Ratdolt omnibus edition 1485.

———. 1490. *Theoricae Novae Planetarum.* In Ottaviano Scotto omnibus edition.

———. 1491. *Theorica Novae Planetarum.* In Peurbach omnibus edition 1491 and in Ratdolt omnibus edition 1491.

———. 1515 [1495]. *See* Capuano de Manfredonia 1515.

———. 1535. *Theoricae Novae Planetarum.* Wittenberg: Joseph Klug.

———. 1542. *See* Reinhold 1542.

———. 1543. *Theoricae Novae Planetarum.* Paris: Christian Wechel.

Peurbach omnibus edition, 1491. *Sphaerae mundi Compendium feliciter inchoat . . . Iohannis de Sacro Busto sphaericum opusculum una cum additionibus nonnullis . . . Contra Cremonensia in planeta theoricas delyramenta Ioannis de Monteregio disputationes . . . nec non Georgii Purbachii in eorundem motus planetarum accuratiss. theoricae . . .* Venice: Monteferrato. Also includes Sacrobosco 1491, Regiomontanus 1491, and Peurbach 1491.

Phillips, Mark. 1977. *Francesco Guicciardini: The Historian's Craft.* Toronto: University of Toronto Press.

Piccolomini, Aeneas Sylvius. 1551. *Opera Omnia.* Basel: Henricus Petri.

Pickering, Andrew, ed.. 1992. *Science as Practice and Culture.* Chicago: University of Chicago Press.

Pico della Mirandola, Giovanni. 1496. *Disputationes*

adversus Astrologiam Divinatricem. Bologna: Benedictus Hectoris.

———. 1496a. *Opuscula haec Ioannis Pici Mirandulae Concordiae Comitis. Diligenter impressit Benedictus Hectoris Bononien. adhibita per uiribus solertia & diligentia ne ab archetypo aberraret: Bononiae anno Salutis Mcccclxxxxvi. die uero xx. Martii*. Bologna: Benedictus Hectoris.

———. 1496b. *Opuscula e Disputationes*. Bologna: Benedictus Hectoris.

———. 1504. *Disputationes*. Strasbourg: Johann Prüss.

———. 1946–52. *Disputationes adversus Astrologiam Divinatricem*. 2 vols. Ed. E. Garin. Florence: Vallechi.

———. 1965. *The Heptaplus, or the Sevenfold Narration of the Six Days of Genesis*. Trans. D. Carmichael. Indianapolis: Bobbs-Merrill.

———. 1969. *Opera Omnia (1557–1573)*. 2 vols. Facsimile ed. Hildesheim: Georg Olms Verlag.

Pietramellara, Giacomo. 1500. *Iuditio*. Bologna: Giustiniano da Rubiera.

Pocock, John Greville Agard. 1987. "The Concept of a Language and the Métier d'Historien: Some Considerations on Practice." In Pagden 1987, 19–38.

Poggendorff, Johann Christian. 1863. *Biographisch-literarisches Handwörterbuch zur Geschichte der exacten Wissenschaften*. Vols. 1–2. Leipzig: Johann Ambrosius Barth.

Pomata, Gianna. 1998. *Contracting a Cure: Patients, Healers, and the Law in Early Modern Bologna*. Baltimore: Johns Hopkins University Press.

Pomian, Krzystof. 1986. "Astrology as a Naturalistic Theology of History" in Zambelli 1986b, 29–43.

Pontano, Giovanni Gioviano. 1512. *De Rebus Coelestibus*. Naples: Sigismund Mayr.

Pontoppidan, Erich. 1760. *Origines hafnienses eller Den kongelige residentzstad Kiøbenhavn*. Copenhagen: Andreas Hartvig.

Popkin, Richard H. 1979. *The History of Scepticism from Erasmus to Spinoza*. Berkeley: University of California Press.

———. 1993. "The Role of Scepticism in Modern Philosophy Reconsidered." *Journal of the History of Philosophy* 31:501–17.

———. 1996. "Prophecy and Scepticism in the Sixteenth and Seventeenth Centuries." *British Journal for the History of Philosophy* 4:1–20.

———. 2003. *The History of Scepticism from Savonarola to Bayle*. Rev. ed.. Oxford: Oxford University Press.

Popkin, Richard H., and Charles B. Schmitt, eds. 1987. *Scepticism from the Renaissance to the Enlightenment*. Wiesbaden: O. Harrassowitz.

Poppi, Antonino, ed. 1983. *Scienza e filosofia all'Università di Padova nel Quattrocento*. Padua: Edizioni Lint.

———. 1992. *Cremonini e Galilei inquisiti a Padova nel 1604: Nuovi documenti d'archivio*. Padua: Editrice Antenore.

Porta, Giovanni Battista della. 1658 [1558]. *Natural Magick*. London: John Wright.

Porter, Roy S. 1986. "The Scientific Revolution: A Spoke in the Wheel?" In Porter and Teich 1986, 290–316.

Porter, Roy, and Mikuláš Teich, eds. 1986. *Revolution in History*. Cambridge: Cambridge University Press.

———. 1992. *The Scientific Revolution in National Context*. Cambridge: Cambridge University Press.

Porter, Theodore M. 2003. "Genres and Objects of Social Inquiry, from the Enlightenment to 1890." In *The Cambridge History of Science: The Modern Social Sciences*, 7:13–39. Cambridge: Cambridge University Press.

Portoghesi, Paolo. n.d. *Ferrara, the Estense City*. Bologna: Italcards.

Poulle, Emmanuel. 1975. "Les équatoires, instruments de la théorie des planètes au moyen âge." *Studia Copernicana* 13 (Colloquia Copernicana 3): 97–112.

———. 1980. *Les instruments de la théorie des planètes selon Ptolémée: Équatoires et horlogerie planétaire du XIIIe au XVIe siècle*. 2 vols. Geneva: Droz.

Prager, Frank D. 1971. "Kepler als Erfinder." In Krafft, Meyer, and Sticker 1973, 385–92.

Price, Derek J. de Solla. 1959. "Contra-Copernicus: A Critical Re-estimation of the Mathematical Planetary Theory of Ptolemy, Copernicus and Kepler." In Clagett 1959, 197–218.

Price, Derek J. de Solla, and R. M. Wilson. 1955. *The Equatorie of the Planetis, edited from Peterhouse MS. 75.I*. Cambridge: Cambridge University Press.

Proclus Diadochus Lycii. 1560. *In Primum Euclidis Elementorum Librum Commentariorum ad Universam Mathematicum Disciplinam Principium Eruditionis Tradentium Libri IIII*. Trans. Francesco Barozzi. Padua: Gratiosus Perchacius.

———. 1561. *Procli Lycii de Sphaera, hoc est, De Circulis coelestibus, Liber Unus*. With "Vita et opera Procli" and commentaries by Erasmus Oswaldus Schreckenfuchsius. Basel: Henricus Petri.

Providera, Tiziana. 2002. "John Charlewood, Printer of Giordano Bruno's Italian Dialogues, and his Book Production." In Gatti 2002, 167–86.

Prowe, Leopold. 1883–84. *Nicolaus Coppernicus*. 2 vols. Berlin: Weidmannsche Buchhandlung.

Ptolemy, Claudius. 1484a, January 15. *Liber Quadripartiti Ptolomaei id est quattuor tractatuum: In radicanti discretione per stellas de futuris et in hoc mundo constructionis et destructionis contingentibus. Liber Ptholomei quattuor tractatuum: cum Centiloquio eiusdem Ptholomei: et commento Haly*. Venice: Erhard Ratdolt. (In Ratdolt-British Library Bundled Copy.)

———. 1484b. *Quadripartitum*. Venice: Erhard Ratdolt. (in Ratdolt-British Library Bundled Copy)

———. 1493. *Liber quadripartiti Ptholomei. Centiloquium eiusdem. Centiloquium hermetis. Eiusdem de stellis beibenijs. Centiloquium bethem et de horis plane-*

tarum. Eiusdem de significatione triplicitatum ortus. Centus quinquagenta propositiones Almansoris. Zahel de interrogationibus. Eiusdem de electionibus. Eiusdem de temporum significationibus in iudiciis. Messahallach de receptionibus planetarum. Eiusdem de interrogationibus. Epistola eiusdem cum duodecim capitulis. Eiusdem de reuolutionibus annorum mundi. Venice: Boneto Locatelli.

———. 1519. *Quadripartitum iudiciorum opus Claudij Ptolemei Pheludiensis Joanne Sieurreo brittuliano Bellouacensi perbelle recognitum.* Paris: Joannis de Porta.

———. 1533. *Quadripartitum.* Basel: Johannes Hervagius.

———. 1535. *Libri Quatuor Compositi Syro Fratri.* Nuremberg: Johannes Petreius.

———. 1541. *Claudii Ptolemaei Pelusiensis Alexandrini Omnia, Quae Extant, Opera, Geographia Excepta, quam seorsim quoque hac forma impressimus.* Basel: Henricus Petri.

———. 1548. *Cl. Ptolemaei Pelusiensis Mathematici Operis Quadripartiti, in Latinum Sermonem Traductio: Adiectis Libris Posterioribus, Antonio Gogava Graviens Interprete. A Clarissimum Principem Maximilianum Comite Burens. Item De Sectione Conica Orthogona, quae parabola dicitur: Deque Speculo Ustorio, Libelli duo, Hactenus desiderati: restituti ab Antonio Gogava Graviensis. Cum praefatione D. Gemmae Frisii Medici et Mathematici clariss.* Louvain: Petrus Phalesius and Martinus Rotarius.

———. 1553. *Claudii Ptolemaei de praedictionibus astronomicis, cui titulum fecerunt Quadripartitum Grecè et Latinè, libri III. P. Melanthone interprete. Eiusdem fructus librorum suorum, sive centum dicta, ex conversione J. Pontani.* Basel: Ioannes Oporinus.

———. 1554. *De Astrorum Iudiciis, aut ut vulgò vocant quadripartitae Constructionis libros commentaria.* Basel: Henricus Petri.

———. 1578. *Hieronymi Cardani in Cl. Ptolemaei de Astrorum Judiciis, aut . . . quadripartitæ constructionis lib. IIII. commentaria ab autore castigata: his accesserunt ejusdem Cardani de septem erraticarum stellarum qualitatibus at viribus liber posthumus, geniturarum item XII. . . . exempla. item C. Dasypodii . . . scholia et resolutiones seu tabulæ in lib. IIII. apotelesmaticos. Cl. Ptolemæi una cum aphorismis eorundem librorum.* Basel: Henricus Petri.

———. 1822. *Ptolemy's Tetrabiblos or Quadripartite Being Four Books of the Influence of the Stars . . . Centiloquy.* Trans. J. M. Ashmand. London: Davis and Dickson.

———. 1940. *Tetrabiblos.* Trans. F. E. Robbins. Cambridge: Harvard University Press.

———. 1985. *Le previsioni astrologiche (Tetrabiblos).* Trans. S. Feraboli. Milan: Arnoldo Mondadori.

———. 1991 [1932]. *The Geography.* Trans. Edward Luther Stevenson. New York: Dover.

———. 1998. *Almagest.* Trans. G. J. Toomer. Princeton, NJ: Princeton University Press.

Puff, Helmut. 2003. *Sodomy in Reformation Germany and Switzerland, 1400–1600.* Chicago: University of Chicago Press.

Pugliese, Patri J. 1990. "Robert Hooke and the Dynamics of Motion in a Curved Path." In Hunter and Schaffer 1990, 181–205.

Pumfrey, Stephen. 1987. "Mechanizing Magnetism in Restoration England: The Decline of Magnetic Philosophy." *Annals of Science* 44:1–22.

———. 1989. "Magnetical Philosophy and Astronomy, 1600–1650." In Taton and Wilson 1989, 45–53.

———. 2002. *Latitude and the Magnetic Earth: The First Story of Queen Elizabeth's Most Distinguished Man of Science.* Cambridge: Icon Books.

Pumfrey, Stephen, and Frances Dawbarn. 2004. "Science and Patronage in England, 1570–1625: A Preliminary Study." *History of Science* 42:137–88.

Pumfrey, Stephen, Paolo L. Rossi, and Maurice Slawinski, eds. 1991. *Science, Culture and Popular Belief in Renaissance Europe.* Manchester: Manchester University Press.

Purnell, Frederick, Jr. 1971. "Jacopo Mazzoni and His Comparison of Plato and Aristotle." PhD diss., Columbia University.

———. 1972. "Jacopo Mazzoni and Galileo." *Physis* 14: 273–94.

Quinlan-McGrath, Mary. 2001. "The Foundation Horoscope(s) for St. Peter's Basilica, Rome, 1506." *Isis* 92: 716–74.

Rabin, Sheila. 1987. "Two Renaissance Views of Astrology: Pico and Kepler." PhD diss., City University of New York.

———. 1997. "Kepler's Attitude toward Pico and the Anti-Astrology Polemic." *Renaissance Quarterly* 50: 750–70.

Raeder, Hans, Elis Strömgren, and Bengt Strömgren, eds. and trans. 1946. *Tycho Brahe's Description of His Instruments and Scientific Work.* Copenhagen: Ejnar Munksgaard.

Ragep, F. Jamil. 2007. "Copernicus and His Islamic Predecessors: Some Historical Resources." *History of Science* 45: 65–81.

Raimondi, Ezio. 1950. *Codro e l'Umanesimo a Bologna.* Bologna: Cesare Zuffi.

Ramus, Petrus. 1569. *Scholarum Mathematicarum, Libri Unus et Triginta.* Basel: Eusebius Episcopius.

———. 1970a [1569]. *Scholae in Liberales Artes.* Facsimile ed. Intro. Walter J. Ong. Hildesheim: Georg Olms Verlag.

———. 1970b. *Scholae Physicae Praefatio.* In Ramus 1970a, unpag., following p. 616.

Rantzov, Heinrich (Ranzovius). 1585. *Exempla quibus astrologicae scientiae certitudo astruitur. Item de annis climatericis et periodis imperiorum, cum pluribus aliis artem astrologicam illustrantibus.* Cologne: M. Cholin.

Ratdolt omnibus edition 1482. Contains Sacrobosco

1482, Regiomontanus 1482, and Peurbach 1482. Venice: Erhard Ratdolt.

Ratdolt omnibus edition 1485. Contains Sacrobosco 1485, Regiomontanus 1485, and Peurbach 1485. Venice: Erhard Ratdolt.

Ratdolt omnibus edition 1491. Contains Sacrobosco 1491, Regiomontanus 1491, and Peurbach 1491. Venice: Erhard Ratdolt.

Ratdolt-British Library Bundled Copy. Contains Ratdolt omnibus edition 1482, Ptolemy 1484a, Ptolemy 1484b, Alchabitius 1485, and Alfonso X 1483.

Ravetz, Jerome R. 1965. *Astronomy and Cosmology in the Achievement of Nicolaus Copernicus.* Wrocław: Ossolineum.

————. 1966, October. "The Origins of the Copernican Revolution." *Scientific American,* 88–98.

Recorde, Robert. 1556. *Castle of Knowledge Containing the Explication of the Sphere Both Celestiall and Materiall.* London: Reginald Wolfe.

Reeves, Eileen. 1997. *Painting the Heavens: Art and Science in the Age of Galileo.* Princeton: Princeton University Press.

Reeves, Marjorie. 1969. *The Influence of Prophecy in the Later Middle Ages: A Study in Joachimism.* Oxford: Clarendon.

————. 1992. *Prophetic Rome in the High Renaissance Period.* Oxford: Oxford University Press.

Regiomontanus, Johannes. 1481. *Ephemerides* [1482–1506]: *Ioannis de Monte Regio: Germanorum Decoris: Aetatis nostrae atronomorum principis Ephemerides.* Venice: Ratdolt.

————. 1482. *Dialogus inter Viennensem et Cracoviensem adversus Gerardum Cremonensem in Planetarum Theoricas Deliramenta.* In Ratdolt omnibus edition 1482 and Regiomontanus 1972.

————. 1485. *Dialogus . . . adversus Gerardum Cremonensem in Planetarum Theoricas Deliramenta.* In Ratdolt omnibus edition 1485.

————. 1490. *Dialogus . . . adversus Gerardum Cremonensem in Planetarum Theoricas Deliramenta.* In Ottaviano Scotto omnibus edition.

————. 1491. *Dialogus . . . adversus Gerardum Cremonensem in Planetarum Theoricas Deliramenta.* In Ratdolt omnibus edition 1491 and Peurbach omnibus edition 1491.

————. 1496. *Epytoma Almagesti.* [Venice]. In Regiomontanus 1972.

————. 1531. *De Cometae Magnitudine Longitudineque ac de Loco Ejus Vero Problemata XVI.* Nuremberg: Fridericus Pepyus.

————. 1533. *An Terra Moveatur an Quiescat Disputatio.* In Schöner 1553 and in Regiomontanus 1972.

————. 1537. *Oratio Johannis de Monteregio, Habita Patavij in Praelectione Alfragani.* (Nuremberg: Petreius.) In Regiomontanus 1972.

————. 1553. *An Terra Moveatur an Quiescat Disputatio.* In Johannes Schöner, *Opusculum Geographicum.* [Nuremberg.] In Regiomontanus 1972, 37–39.

————. 1972 [1949]. *Opera Collectanea.* Ed. Felix Schmeidler. Osnabrück: Zeller.

Reinhold, Erasmus. 1542. *Theoricae novae planetarum Georgii Purbachii Germani ab Erasmo Reinholdo Salueldensi pluribus figuris auctae et illustratae scholijs, quibus studiosi, praeparentur, ac inuitentur ad lectionem ipsius Ptolemaei . . . Inserta item methodica tractatio de illuminatione Lunae. Typus Eclipsis solis futurae Anno 1544.* Wittenberg: Hans Lufft.

————. 1550. *Ephemerides duorum annorum 50. et 51. supputatae ex novis tabulis astronomicis Erasmum Reinholdum Saluedensum ad meridianum Wittebergensem.* Tübingen: Ulrich Morhard.

————. 1551. *Prutenicae Tabulae Coelestium Motuum.* Tübingen: Ulrich Morhard.

————. 1571. *Prutenicae Tabulae Coelestium Motuum.* Ed. Michael Maestlin. Tübingen: Oswald and Georg Gruppenbach.

Reston, James, Jr. 1994. *Galileo: A Life.* New York: Harper-Collins.

Reusch, Franz Heinrich. 1883. *Ein Beitrag zur Kirchen- und Literaturgeschichte.* 2 vols. Bonn; repr. Aalen: Scientia Verlag.

————, ed. 1961 [Tübingen, 1886]. *Die Indices Librorum Prohibitorum des Sechzehnten Jahrhunderts.* Nieuwkoop: B. de Graaf.

Rheticus, Georg Joachim. 1540. *Narratio Prima.* Gdańsk: Rhodvs.

————. 1541. *Narratio Prima.* Basel: Winter.

————. 1550a. *Ephemerides Novae seu Expositio Positus Diurni Siderum.* Leipzig: Wolfphgang Gunter.

————. 1550b. *Prognosticon oder Practica Deutsch.* Leipzig: Valentin Bapst.

————. 1566. *Narratio Prima.* Basel: Henricus Petri.

————. 1596. *Narratio Prima.* Tübingen: Gruppenbach.

————. 1971 [1939]. *Narratio Prima.* Trans. Edward Rosen. In Rosen 1971a. New York: Octagon.

————. 1982. *Narratio Prima* (Studia Copernicana 20). Ed. and trans. (into French) Henri Hugonnard-Roche, Jean-Pierre Verdet, Michel-Pierre Lerner, and Alain Segonds. Wrocław: Polskiej Akademii Nauk.

Rhodes, Dennis. 1982a. *Studies in Early Italian Printing.* London: Pindar Press.

————. 1982b. "Philippus Beroaldus, Minus Roscius and an Undated Book." In Rhodes 1982a, 14–17.

————. 1982c. "Benedictus Hectoris of Bologna and His Complaints against Typographical Pirates." In Rhodes 1982a, 229–31.

Ricci, Saverio. 1988. "Federico Cesi e la Nova del 1604: La Teoria della Fluidità del Cielo e un Opuscolo Dimenticato di Johannes van Heeck." *Atti della Accademia Nazionale del Lincei Rendiconti* 43: 111–33.

————. 1990. *La fortuna del pensiero di Giordano Bruno, 1600–1750.* Florence: Le Lettere.

Riccioli, Giovanni Battista. 1651 [reissued Frankfurt, 1653]. *Almagestum Novum.* Bologna: Victor Benatij.

———. 1665. *Astronomia Reformata*. Bologna: Benatius.

Richards, Joan L. 1987. "Augustus de Morgan and the History of Mathematics." *Isis* 78: 7–30.

Richardson, Alan. 2002. "Narrating the History of Reason Itself: Friedman, Kuhn, and a Constitutive A Priori for the Twenty-First Century." *Perspectives on Science* 10: 253–274.

Ridolfi, Angelo Calisto. 1989. *Indice dei Notai Bolognesi dal XIII al XIX Secolo*. (Graziella Grandi Venturi, ed.; con premesse di Mario Fanti e Diana Tura.) Bologna: Estratto da L'Archiginnasio, 1989.

Rigaud, Stephen P. 1833. *"Account of Harriot's Astronomical Papers."* In Shirley 1981.

Righini, Guglielmo. 1976. "L'oroscopo galileiano di Cosimo II de' Medici." *Annali dell'Istituto e Museo di Storia della Scienza di Firenze* 1:29–36.

Righini-Bonelli, Maria Luisa, and Thomas B. Settle. 1979. "Egnatio Danti's Great Astronomical Quadrant." *Annali dell'Istituto e Museo di Storia della Scienza di Firenze* 4, no. 2: 1–13.

Righini-Bonelli, Maria Luisa, and William R. Shea, eds. 1975. *Reason, Experiment, and Mysticism in the Scientific Revolution*. New York: Science History Publications.

Roberts, Julian, and Andrew G. Watson, eds. 1990. *John Dee's Library Catalogue*. London: Bibliographical Society.

Roberval, Gilles Personne de. 1644. *Aristarchus Samus de Mundi Systemate, partibus et motibus ejusdem libellus*. Paris: G. Baudry.

Robison, Wade L. 1974. "Galileo on the Moons of Jupiter," *Annals of Science* 31:165–69.

Rocca, Paolo. 1964. *Giovanni Pico della Mirandola nei Suoi Rapporti di Amicizia con Gerolamo Savonarola*. (Quaderni di Storia della Scienze e della Medicina, III) Ferrara: University degli Studi di Ferrara.

Rochberg, Francesca. 2004. *The Heavenly Writing*. Cambridge: Cambridge University Press.

Roeslin, Helisaeus. 1578. *Theoria Nova Coelestium Metwepωn, in qua ex plurium cometarum phoenomenis Epilogisticώs quaedam afferuntur, de novis tertiae cuiusdam Miraculorum Sphaerae Circulis, Polis et Axi: Super quibus Cometa Anni MDLXXVII nouo motum et regularissimo ad superioribus annis conspectam Stellam; tanquam ad Cynosuram progressus, Harmoniam singularem undique ad Mundi Cardines habuit, maximè verò medium Europae, et exactè Germaniae Horizontem non sine numine certo respexit*. Argentorati: Bernhardus Iobinus.

———. 2000 [1597]. *De Opere Dei Creationis*. Facsimile ed. Lecce, Italy: Conte.

Roffeni, Giovanni Antonio. 1611. *Epistola Apologetica contra Caecam Peregrinationem*. (Bologna, Johannes Rossi). In Galilei 1890–1909, 3:193–200.

———. 1614. *Pronosticon ad annum Dom. 1614: Additis Laudibus, & Responsionibus Aduersus Verae Astrologiae Calumniatores*. Bologna: Bartolomeo Cochi.

Romano, Antonella. 1999. *La contra-réforme mathématique: Constitution et diffusion d'une culture mathématique jésuite à la Renaissance*. Paris: École Française de Rome.

Ronchitti, Cecco di. 1605. *Dialogo de Cecco di Ronchitti da Bruzene in Perpuosito de la Stella Nuova*. Padua: Pietro Paulo Tozzi.

Rose, Mark. 1994. "The Author in Court: Pope v. Curll (1741)." In Jaszi and Woodmansee 1994, 212–29.

Rose, Paul Lawrence. 1975a. *The Italian Renaissance of Mathematics: Studies on Humanists and Mathematicians from Petrarch to Galileo*. Geneva: Droz.

———. 1975b. "Universal Harmony in Regiomontanus and Copernicus." In Delorme 1975, 153–58.

Rosen, Edward. 1943. "The Authentic Title of Copernicus' Major Work." *Journal of the History of Ideas* 4:457–74.

———. 1958. "Galileo's Misstatements about Copernicus." *Isis* 49:319–30.

———. 1967. "In Defense of Kepler." In Lewis 1967, 141–58.

———. 1969. Review of Burmeister 1967–8, *Isis* 59:231.

———. 1971a [1939]. *Three Copernican Treatises*. New York: Octagon.

———. 1971b. "Biography of Copernicus." In Rosen 1971a.

———. 1974. "Domenico Maria Novara." In *Dictionary of Scientific Biography*, 10:153–55.

———. 1975a. "Kepler and the Lutheran Attitude towards Copernicanism in the Context of the Struggle Between Science and Religion." In Beer and Beer 1975, 317–37.

———. 1975b. "Was Copernicus' *Revolutions* Approved by the Pope?" *Journal of the History of Ideas* 36: 531–42.

———. 1981. "Nicholas Copernicus and Giorgio Valla." *Physis* 23:449–57.

———. 1984a. "Kepler's Attitude Toward Astrology and Mysticism." In Vickers 1984, 253–72.

———. 1984b. *Copernicus and the Scientific Revolution*. Malabar, FL: Robert E. Krieger.

———. 1985. "The Dissolution of the Solid Celestial Spheres." *Journal for the History of Ideas* 45:13–31.

———. 1986. *Three Imperial Mathematicians: Kepler Trapped between Tycho Brahe and Ursus*. New York: Abaris Books.

———. 1995a. "Copernicus and His Relation to Italian Science." In Rosen 1995b, 127–37.

———. 1995b. *Copernicus and His Predecessors*. London and Rio Grande: Hambledon.

Ross, Sydney. 1962. "'Scientist': The Story of a Word." *Annals of Science* 18: 65–86.

Rothmann, Johann. 1595. *Chiromantiae Theorica Practica Concordantia Genethliaca, Vetustis Novitate Addita*. Erfurt: Ioannes Pistorius.

Rothmann, Christoph. 1619. *See* Snellius 1619.

———. 2003. *Christoph Rothmanns Handbuch der Astronomie von 1589.* Ed. Miguel A. Granada, Jürgen Hamel, and Ludolph von Mackensen. Acta historica astronomiae, 19. Frankfurt: Harri Deutsch.

Rousseau, Claudia. 1983. "Cosimo de Medici and Astrology: The Symbolism of Prophecy." PhD diss., Columbia University.

Rudwick, Martin, J. S. 2005. *Bursting the Limits of Time: The Reconstruction of Geohistory in the Age of Revolution.* Chicago: University of Chicago Press.

Rupp, E. Gordon. 1983. "Luther against the Turk, the Pope and the Devil." In Brooks 1983, 255–73.

Russell, John L. 1964. "Kepler's Laws of Planetary Motion, 1609–1666." *British Journal for the History of Science* 2:1–24.

———. 1989. "Catholic Astronomers and the Copernican System after the Condemnation of Galileo." *Annals of Science* 46:365–86.

Rutkin, H. Darrel. 2005. "Galileo Astrologer: Astrology and Mathematical Practice in the Late-Sixteenth and Early-Seventeenth Centuries." *Galilaeana* 2:107–43.

———. 2010. "The Use and Abuse of Ptolemy's *Tetrabiblos* in Renaissance and Early Modern Europe (Giovanni Pico della Mirandola and Filippo Fantoni)." In Jones 2010, 135–49.

Sacrobosco, Johannes. 1478. *Sphaera Mundi.* Venice: Adam von Rottweil.

———. 1482. *Sphaericum Opusculum.* In Ratdolt omnibus edition 1482.

———. 1485. *Sphaericum Opusculum.* In Ratdolt omnibus edition 1485.

———. 1490. *Sphaericum Opusculum.* In Ottaviano Scotto omnibus edition 1490.

———. 1491a. *Sphaera Mundi.* In Peurbach omnibus edition 1491.

———. 1491b. *Sphaericum Opusculum.* In Peurbach omnibus edition 1491 and Ratdolt omnibus edition 1491.

———. 1527. *Textus de sphaera Joannis de Sacrobosco . . . ad utilitatem studentiu(m) philosophiae Parisiensis academiæ illustratus. . . .* Paris: Simon Colinaeus.

Sagan, Carl. 1980. *Cosmos.* New York: Random House.

Salio, Girolamo [of Faventino], ed. 1493. *Tetrabiblos.* See Ptolemy 1493.

Sancto Paolo, Eustachius à. 1648 [1609]. *Summa Philosophiae Quadripartita, de Rebus Dialecticis, Ethicis, Physicis, et Metaphysicis.* Cambridge: Roger Daniels.

Sandblad, Henrik. 1972. "The Reception of the Copernican System in Sweden." In Dobrzycki 1972, 241–70.

Sanders, Philip Morris. 1990. "The Regular Polyhedra in Renaissance Science and Philosophy." PhD diss., University of London (Warburg Institute).

Sanford, Vera. 1939. "The Art of Reckoning." *The Mathematics Teacher* 32:243–48.

Santinello, Giovanni, ed. 1992. *Galileo e la cultura padovana.* Padua: CEDAM.

Sarasohn, Lisa T. 1988. "French Reaction to the Condemnation of Galileo, 1632–1642." *Catholic Historical Review* 74: 34–54.

Savonarola, Girolamo. 1497. *Tractato contra li astrologi* [Florence: Bartolomeo de'Libri]. In Savonarola 1982.

———. 1557. *Astrologia Confutata: Ein warhafte gegründte unwidersprechliche Confutation der falschen Astrologei . . . von neuen ins deutsch gebracht.* T. Erastus. Schleusingen: Hamsing.

———. 1581. *Opus Eximium adversus Divinatricem Astronomiam . . . Interprete F. Thomasso Boninsignio.* Florence.

———. 1982. *Scritti filosofici.* Ed. Giancarlo Garfagnini and Eugenio Garin. Rome: A. Belardetti.

Savonarola, Johannis Michaelis. 1497. *Practica Medicinae, sive De Aegritudinibus.* Venice: Bonetus Locatellus.

———. 1502. *Practica.* Venice: Bernardinus Vercellensis.

Scaliger, Julius Caesar. 1582. *Exotericae Exercitationes ad Cardanum.* Frankfurt.

Scepperius, Cornelius Duplicius. 1548 [1523]. *Adversus Falsos Quorundam Astrologorum Augurationes Assertio.* Cologne: Birckmann.

Schaff, Josef. 1912. *Geschichte der Physik an der Universität Ingolstadt.* Erlangen, 1912. Diss. Phil. Erlangen.

Schaffer, Simon. 1983. "History of Physical Science." In Corsi and Weindling 1983, 285–314.

———. 1987. "Newton's Comets and the Transformation of Astrology." In Curry 1987, 219–43.

———. 1993. "Comets and Idols: Newton's Cosmology and Political Theology." In Seef and Theerman 1993, 206–31.

———. 1997. "Metrology, Metrication, and Victorian Values." In Lightman 1997, 438–74.

Scheiner, Christopher. 1612. *Tres Epistolae de Maculis Solaribus . . . Accuratior Disquisitio.* Augsburg: Ad insigne pinus.

Scheurl, Christoph. 1962. "Ad Sixtum Tucherum (November 22, 1506)." In Knaake and Soden 1962.

Schilling, Heinz. 1981. *Konfessionskonflikt und Staatsbildung: Eine Fallstudie über das Verhältnis von religiösem und sozialem Wandel in der Frühneuzeit am Beispiel der Grafschaft Lippe.* Quellen und Forschungen zur Reformationsgeschichte, 48. Gütersloh: Mohn.

———. 1986. "The Reformation and the Rise of the Early Modern State." In Tracy 1986, 21–30.

Schimkat, Peter. 2007. "Wilhelm IV als Naturforscher, Ökonom und Landesherr." In Gaulke 2007, 77–90.

Schmitt, Charles B. 1972a. *Cicero Scepticus: A Study of*

the Influence of the Academica in the Renaissance. The Hague: Nijhoff.

———. 1972b. "The Faculty of Arts at Pisa at the Time of Galileo." *Physis* 14: 243–72.

———. 1973. "Towards a Reassessment of Renaissance Aristotelianism." *History of Science* 11: 159–93.

———. 1981. *Studies in Renaissance Philosophy and Science*. London: Aldershot.

———. 1983. *Aristotle and the Renaissance*. Cambridge, MA: Harvard University Press.

———. 1985. "Aristotle among the Physicians." In Wear, French, and Lonie 1985, 1–15.

Schmitt, Charles B., and Quentin Skinner, eds. 1988. *The Cambridge History of Renaissance Philosophy*. Cambridge: Cambridge University Press.

Schmitz, Rudolf, and Fritz Krafft, eds. 1980. *Humanismus und Naturwissenschaften*. Beiträge zur Humanismusforschung, 4. Boppard: Harald Boldt.

Schofield, Christine Jones. 1964. "Tychonic and Semi-Tychonic World Systems." PhD diss., University of Cambridge.

———. 1981. *Tychonic and Semi-Tychonic World Systems*. New York: Abaris.

———. 1989. "The Tychonic and Semi-Tychonic World Systems." In Taton and Wilson 1989, 33–44.

Scholder, Klaus. 1966. *Ursprunge und Probleme der Bibelkritik im 17. Jahrhundert*. Munich: Kaiser.

Schöner, Johannes. 1545. *De iudiciis nativitatum Libri Tres. Scripti a Ioanne Schonero Carolostadio, Professore Publico Mathematum, in celebri Germaniae Norimberga. Item Praefatio D. Philippi Melanthonis in hos de Iudicijs Natiuitatum Ioannis Schoner libros.* Nuremberg: Ioannis Montani et Ulrici Neuber.

———. 1553 [1533]. *Opusculum Geographicum*. Nuremberg.

Schorske, Carl. 1980. *Fin-de-Siècle Vienna*. New York: Knopf.

Schrader, Dorothy V. 1968. "De Arithmetica, Book I of Boethius." *Mathematics Teacher* 61: 615–28.

Schreckenfuchs, Erasmus Oswald. 1556. *Commentaria in Novas Theoricas Planetarum Georgii Purbachii*. Basel: Henricus Petri.

———. 1569. *Commentaria in Sphaeram Ioannis de Sacrobusto*. Basel: Henricus Petreius.

Schuster, John. 1977. "Descartes and the Scientific Revolution, 1618–1634." 2 vols. PhD diss., Princeton University.

Schuster, John, and Graeme Watchirs. 1990. "Natural Philosophy, Experiment and Discourse in the 18th Century." In LeGrand 1990, 1–47.

Scribner, Robert W. 1981. *For the Sake of Simple Folk: Popular Propaganda for the German Reformation*. Cambridge: Cambridge University Press.

Scudder, Henry. 1620. *A Key of Heaven*. London: R. Field.

Seck, Friedrich, ed. 1981. *Wissenschaftsgeschichte um Wilhelm Schickard*. Contubernium: Beiträge zur Geschichte der Eberhard-Karls-Universität Tübingen, 26. Tübingen: J. C. B. Mohr.

Secord, James A. 2000. *Victorian Sensation: The Extraordinary Publication, Reception, and Secret Authorship of "Vestiges of the Natural History of Creation."* Chicago: University of Chicago Press.

Secret, François. 1964. *Les kabbalistes chrétiens à la Renaissance*. Paris.

Seebass, Gottfried. 1972. "The Reformation in Nürnberg." In Buck and Zophy 1972, 17–41.

Seef, Adele F. and Paul Theerman, eds. 1993. *Action and Reaction: Proceedings of a Symposium to Commemorate the Tercentenary of Newton's Principia*. Newark: University of Delaware Press.

Segonds, Alain Philippe. 1993. "Tycho Brahe et l'Alchimie." In Margolin and Matton 1993, 365–78.

Segre, Michael. 1998. "The Never Ending Galileo Story." In Machamer 1998, 388–416.

Settle, Thomas B. 1990. "Egnazio Danti and Mathematical Education in Late Sixteenth-Century Florence." In Henry and Hutton 1990, 24–37.

Seznec, Jean. 1953. *The Survival of the Pagan Gods: The Mythological Tradition and Its Place in Renaissance Humanism and Art*. Trans. Barbara Sessions. New York: Pantheon.

Shackelford, Jole. 1993. "Tycho Brahe, Laboratory Design, and the Aim of Science: Reading Plans in Context." *Isis* 84: 211–30.

Shakerley, Jeremy. 1653. *Tabulae Britannicae*. London: R. and W. Leybourn.

Shank, Michael H. 1992. "The 'Notes on Al-Bitruji' Attributed to Regiomontanus: Second Thoughts." *Journal for the History of Astronomy* 23: 15–30.

———. 1994. "Galileo's Day in Court." *Journal for the History of Astronomy* 25: 236–42.

———. 1996. "How Shall We Practice History?" *Early Science and Medicine* 1:106–50.

———. 1998. "Regiomontanus and Homocentric Astronomy." *Journal for the History of Astronomy* 29: 157–66.

———. 2005a. "Before the Revolution: Fifteenth-Century European Astronomy in Context." Paper delivered at Max Planck Institute for the History of Science Conference, "Before the Revolution: The Forgotten Fifteenth Century." January 13–15.

———. 2005b. "Setting the Stage: Galileo in Tuscany, the Veneto, and Rome." In McMullin 2005a, 57–87.

———. 2007. "Regiomontanus as a Physical Astronomer: Samplings from *The Defence of Theon Against George of Trebizond*." *Journal for the History of Astronomy* 38: 325–49.

———. 2009. "Setting up Copernicus? Astronomy and Natural Philosophy in Giambattista Capuano da Manfredonia's *Expositio* on the *Sphere*." *Early Science and Medicine* 14: 290–315.

Shapin, Steven. 1991. " 'A Scholar and a Gentleman':

The Problematic Identity of the Scientific Pracitioner in Early Modern England." *History of Science* 29: 279–327.

———. 1994. *A Social History of Truth: Civility and Science in Seventeenth-Century England*. Chicago: University of Chicago Press.

———. 1997. *The Scientific Revolution*. Chicago: University of Chicago Press.

Shapin, Steven, and Simon Schaffer. 1985. *Leviathan and the Air Pump: Hobbes, Boyle and the Experimental Life*. Princeton, NJ: Princeton University Press.

Shapiro, Alan. 1996. "The Gradual Acceptance of Newton's Theory of Light and Color, 1672–1727." *Perspectives on Science* 4:59–140.

Shapiro, Barbara. 1969. *John Wilkins, 1614–1672: An Intellectual Biography*. Berkeley: University of California Press.

———. 2000. *A Culture of Fact: England, 1550–1720*. Ithaca, NY: Cornell University Press.

Sharratt, Michael. 1994. *Galileo: Decisive Innovator*. Cambridge: Cambridge University Press.

Shea, William. 1972. *Galileo's Intellectual Revolution: The Middle Period, 1610–1632*. New York: Neale Watson.

Shipman, Joseph C. 1967. "Johannes Petreius, Nuremberg Publisher of Scientific Works, 1524–1550." In Lehmann-Haupt 1967, 147–62.

Shirley, John W., ed. 1974a. *Thomas Harriot, Renaissance Scientist*. Oxford: Clarendon Press.

———. 1974b. "Sir Walter Ralegh and Thomas Harriot." In Shirley 1974a, 36–53.

———, ed. 1981. *A Source Book for the Study of Thomas Harriot*. New York: Arno Press.

Siderocrates (Eisenmenger), Samuel. 1563. *De Usu Partium Coeli Oratio*. Tübingen: Morhard.

Sighinolfi, Lino. 1914. "Francesco Puteolano e le origini della Stampa in Bologna e in Parma," " *La Bibliofilia* 15: 383–92.

———. 1920. "Domenico Maria Novaria e Niccolò Copernico allo Studio di Bologna." *Studi e Memorie per la Storia dell'Università di Bologna* 5:207–236.

Simon, Gérard. 1979. *Kepler, Astronome-Astrologue*. Paris: Gallimard.

Simpson, A. D. C. 1989. "Robert Hooke and Practical Optics: Technical Support at a Scientific Frontier." In Hunter and Schaffer 1989, 33–61.

Singer, Dorothy Waley. 1950. *Giordano Bruno: His Life and Thought*. With Annotated Translation of His Work *On the Infinite Universe and Worlds*. New York: Henry Schuman.

Siraisi, Nancy G. 1987. *Avicenna in Renaissance Italy: The "Canon" and Medical Teaching in Italian Universities after 1500*. Princeton, NJ: Princeton University Press.

———. 1990. *Medieval and Early Renaissance Medicine: An Introduction to Knowledge and Practice*. Chicago: University of Chicago Press.

Sizzi, Francesco. 1611. *Dianoia Astronomica, Optica, Physica* [Venice: Petrus Marius Bertinus.] In Galilei 1890–1909, 3:201–50.

Skinner, Quentin. 1978. *The Foundations of Modern Political Thought*. 2 vols. Cambridge: Cambridge University Press.

Slouka, Hubert. 1952. *Astronomie v Československu od dob Nejstarších do Dneška*. Prague: Osvěta.

Smith, David Eugene. 1958. *A History of Mathematics*. 2 vols. New York: Dover.

Smith, Logan Pearsall. 1907. *The Life and Letters of Sir Henry Wotton*. 2 vols. Oxford: Clarendon.

Smoller, Laura Ackerman. 1994. *History, Prophecy and the Stars: The Christian Astrology of Pierre D'Ailly, 1350–1420*. Princeton, NJ: Princeton University Press.

———. 1998. "The Alfonsine Tables and the End of the World: Astrology and Apocalyptic Calculation in the Later Middle Ages." In Ferreiro 1998, 211–39.

Snellius, Willebrord. 1619. *Descriptio Cometae Qui Anno 1618 Mense Novembri Primum Effulsit. Huic accessit Christophori Rhotmanni Ill. Princ. Wilhelmi Hassiae Lantgravii Mathematici descriptio accurata cometae anni 1585*. Leiden: Officina Elzviriana.

Snobelen, Stephen D. 2001. " 'God of gods, and Lord of Lords': The Theology of Isaac Newton's General Scholium to the Principia." *Osiris* 16: 169–208.

Snyder, John. 1989. *Writing the Scene of Speaking: Theories of Dialogue in the Late Renaissance*. Stanford, CA: Stanford University Press.

Sobel, Dava. 1999. *Galileo's Daughter: A Historical Memoir of Science, Faith, and Love*. New York: Walker.

Sommerville, Johann P. 1994. "Introduction." In James VI and I 1994.

Sorbelli, Albano. 1938. "Il 'Tacuinus' dell'Università di Bologna e le sue prime edizioni." *Gutenberg-Jahrbuch* 33: 109–14.

Sphaerae Mundi Compendium. 1490. Venice: Octavianus Scotus.

Spampanato, Vincenzo. 1933. *Documenti della vita di Giordano Bruno*. Florence: Olschki.

Spruit, Leen. 2002. "Giordano Bruno and Astrology." In Gatti 2002, 229–50.

Stadius, Johannes. 1560. *Ephemerides Novae et Auctae, 1554–1576*. Coloniae Agrippinae: Birckmann.

Stella, Aldo. 1992. "Galileo, il Circolo Culturale di Gian Vincenzo Pinelli e la 'Patavina Libertas.' " In Santinello 1992, 307–25.

Steneck, Nicholas H. 1976. *Science and Creation in the Middle Ages: Henry of Langenstein (d. 1397) on Genesis*. Notre Dame, IN: University of Notre Dame Press.

Stephen, Leslie, and Sidney Lee, eds. 1891. *Dictionary of National Biography*. Vol. 26. London: Smith, Elder and Co.

Stephenson, Bruce. 1987. *Kepler's Physical Astronomy*. Princeton, NJ: Princeton University Press.

———. 1994. *The Music of the Heavens: Kepler's Har-*

monic Astronomy. Chicago: University of Chicago Press.

Stevenson, Enrico. 1886. Inventario dei Libri Stampati Palatino Vaticani. Rome.

Stevin, Simon. 1599. Portuum Investigandorum Ratio. Leiden.

———. 1605–8. Hypomnemata Mathematica, hoc est eruditus ille pulvis, in quo exercuit . . . Mauritius, princeps Auraicus. Leiden: I. Patius.

———. 1961. The Principal Works of Simon Stevin. 5 vols. Ed. E. Crone, A. Pannekoek et al. Amsterdam: C. V. Swets and Zeitlinger.

Stewart, Larry. 1992. The Rise of Public Science: Rhetoric, Technology, and Natural Philosophy in Newtonian Britain, 1660–1750. Cambridge: Cambridge University Press.

Stierius, Johannes. 1671 [1647]. Praecepta Logicae, Ethicae, Physicae, Metaphysicae, Sphaericaeque Brevibus Tabelis Compacta: Una cum Questionibus Physicae Controversis. 7th ed. London: J. Redmayne.

Stimson, Dorothy. 1917. The Gradual Acceptance of the Copernican Theory of the Universe. Hanover, NH: n.p.

Stöffler, Johannes. 1523. Expurgatio aduersus diuinationum XXIIII anni suspitiones. Tübingen: U. Morhard.

Stone, Lawrence, ed. 1974. The University in Society. Princeton, NJ: Princeton University Press.

Stopp, Frederick John. 1974. The Emblems of the Altdorf Academy: Medals and Medal Orations, 1577–1626. London: Modern Humanities Research Association.

Strauss, Gerald. 1966. Nuremberg in the Sixteenth Century: City Politics and Life between Middle Ages and Modern Times. Bloomington: Indiana University Press.

Streete, Thomas. 1661. Astronomia Carolina, A New Theorie of the Coelestiall Motions. London: Lodowick Lloyd.

Striedl, Hans. 1953. "Der Humanist Johann Albrecht Widmanstetter (1506–1577) als klassischer Philologe." In Festgabe der bayerischen Staatsbibliothek, 96–120. Wiesbaden: Harrassowitz.

Strong, Edward William. 1936. Procedures and Metaphysics: A Study in the Philosophy of Mathematical-Physical Science in the Sixteenth and Seventeenth Centuries. Berkeley: University of California Press.

Strype, John. 1848. Memorials of the Most Reverend Father in God Thomas Cranmer, Sometime Lord Archbishop of Canterbury. Oxford.

Sturlese, Maria Rita Pagnoni. 1985. "Su Bruno e Tycho Brahe." Rinascimento 25: 309–33.

Sudhoff, Karl. 1902. Iatromathematiker vornehmlich im 15. und 16. Jahrhundert. Abhandlungen zur Geschichte der Medicin. Wrocław: Max Müller.

Sutter, Berthold. 1964. Graz als Residenz: Innerösterreich, 1564–1619. Graz: Katalog der Ausstellung.

———. 1975. "Johannes Keplers Stellung innerhalb der Grazer Kalendertradition des 16. Jahrhunderts." In Sutter and Urban 1975, 209–373.

Sutter, Berthold, and Urban, Paul, eds. 1975. Johannes Kepler 1571–1971: Gedenkschrift der Universität Graz. Graz: Leykam.

Švejda, Antonin. 1997. "Science and Instruments." In Fučíková 1997a, 618–26.

Swerdlow, Noel M. 1972. "Aristotelian Planetary Theory in the Renaissance: Giovanni Battista Amico's Planetary Spheres." Journal for the History of Astronomy 3:36–48.

———. 1973. "The Derivation and First Draft of Copernicus's Planetary Theory: A Translation of the Commentariolus with Commentary." Proceedings of the American Philosophical Society 117: 423–512.

———. 1975. "Copernicus's Four Models of Mercury." Studia Copernicana 13 (Colloquia Copernicana 3). Wrocław: Polskiej Akademii Nauk, 141–60.

———. 1976. "Pseudodoxia Copernicana: Or, Enquiries into Very Many Received Tenets and Commonly Presumed Truths, Mostly Concerning Spheres." Archives internationales d'histoire des sciences 26: 108–58.

———. 1992. "Annals of Scientific Publishing: Johannes Petreius's Letter to Rheticus." Isis 83: 270–74.

———. 1993. "Science and Humanism in the Renaissance: Regiomontanus's Oration on the Dignity and Utility of the Mathematical Sciences." In Horwich 1993, 133–68.

———. 1996. "Astronomy in the Renaissance." In Walker 1996, 187–230.

———. 1998. The Babylonian Theory of the Planets. Princeton, NJ: Princeton University Press.

———. 2004a. "An Essay on Thomas Kuhn's First Scientific Revolution, The Copernican Revolution." Proceedings of the American Philosophical Society 148, no. 1: 64–120.

———. 2004b. "Galileo's Horoscopes." Journal for the History of Astronomy 35: 135–41.

Swerdlow, Noel M., and Otto Neugebauer. 1984. Mathematical Astronomy in Copernicus's "De Revolutionibus." New York and Berlin: Springer Verlag.

Swetz, Frank J. 1987. Capitalism and Arithmetic: The New Math of the 15th Century; Including the Full Text of the "Treviso Arithmetic" of 1478. Trans. D. E. Smith. LaSalle, IL: Open Court.

Szczucki, Lech, ed. 1974. Astrologia e religione nel Rinascimento. Wrocław: Zakład.

Tabarroni, Giorgio. 1987. "Copernico e gli Aristotelici Bolognesi." In Capitani 1987.

Taton, René, and Curtis A. Wilson, eds. 1989. Planetary Astronomy from the Renaissance to the Rise of Astrophysics. Cambridge: Cambridge University Press.

Taub, Liba Chaia. 1993. Ptolemy's Universe: The Natural Philosophical and Ethical Foundations of Ptolemy's Astronomy. Chicago: Open Court.

Tessicini, Dario. 2001. " 'Pianeti consorti': la Terra e la Luna nel diagramma eliocentrico di Giordano Bruno." In Granada 2001, 159–88.

———. 2007. *I dintorni dell'infinito: Giordano Bruno e l'astronomia del Cinquecento. Bruniana & Campanelliana*. Supplementi, xx. Studi, 9. Pisa: Fabrizio Serra.

Tester, S. Jim. 1987. *A History of Western Astrology*. Woodbridge, UK: Boydell Press.

Theodoricus Winshemius, Sebastianus. 1570. *Novae Quaestiones Sphaerae, hoc est De Circulis Coelestiubus et Primo Mobili, in gratiam studiosi iuuentutis scriptae*. Wittenberg: J. Crato.

Thomas, Keith. 1971. *Religion and the Decline of Magic*. New York: Scribner.

Thomason, Neil. 2000. "1543—The Year that Copernicus Didn't Predict the Phases of Venus." In Freeland and Corones 2000, 291–332.

Thoren, Victor. 1974. "Extracts from the Alfonsine Tables and Rules for Their Use." In Grant 1974, 465–87.

Thoren, Victor (with contributions from John R. Christianson). 1990. *The Lord of Uraniborg*. Cambridge: Cambridge University Press.

Thorndike, Lynn. 1923–58. *A History of Magic and Experimental Science*. 8 vols. New York: Columbia University Press.

———. 1943. "Another Virdung Manuscript." *Isis* 34: 291–93.

———. 1949. *The Sphere of Sacrobosco and Its Commentators*. Chicago: University of Chicago Press.

Thorpe, Charles. 2006. *Oppenheimer: The Tragic Intellect*. Chicago: University of Chicago Press.

Timpler, Clemens. 1605. *Physicae seu Philosophiae Naturalis Systema Methodicum, in tres partes digestum: in quo tamquam in speculo seu theatro universa Natura, per Theoremata et Problemata breuiter et perspicuè explicata et disceptata, contemplanda proponitur, Pars Prima; complectens Physicam Generalem. Auctore Clemente Timplero Stopensi Misnico*. Hannover: Apud Guilielmum Antonium.

Tomba, Tullio. 1990. "L'osservazione della stella nuova del 1604 nell'ambito filosofico e scientifico padovano." *Cesare Cremonini (1550–1631). Il suo pensiero e il suo tempo* (Convegno di Studi Cento, 7 April 1984). Ferrara: Cento.

Toulmin, Stephen E. 1975. "Commentary." In Westman 1975a, 384–91.

Tracy, James D., ed. 1986. *Luther and the Modern State in Germany*. Sixteenth Century Essays and Studies, 7. Kirksville, MO: Sixteenth Century Journal.

Traister, Barbara Howard. 2001. *The Notorious Astrological Physician of London: Works and Days of Simon Forman*. Chicago: University of Chicago Press.

Tredwell, Katherine A. 2004. "Michael Maestlin and the Fate of the *Narratio Prima*." *Journal for the History of Astronomy* 35:305–25.

Trevor-Roper, Hugh. 1987a. *Catholics, Anglicans and Puritans: Seventeenth Century Essays*. Chicago: University of Chicago Press.

———. 1987b. "Nicholas Hill, the English Atomist." In Trevor-Roper 1987a, 1–39.

Trinkaus, Charles. 1985. "The Astrological Cosmos and Rhetorical Culture of Giovanni Gioviano Pontano." *Renaissance Quarterly* 38:446–72.

Tschaikner, Manfred. 1989. "Der verzauberte Dr. Iserin." *Kulturinformationen Vorarlberger Oberland* 2:147–51.

Turner, Gerard L'Estrange. 1994. "The Three Astrolabes of Gerard Mercator." *Annals of Science* 51, no. 4: 329–53.

Turner, Robert. 1657. *Ars Notoria: the Notory Art of Solomon, shewing the cabalistical key of magical operations, the liberal sciences, divine revelation, and the art of memory. Whereunto is added an Astrological Catechism, fully demonstrating the art of Judicial Astrology. Together with a rare Natural secret, necessary to be learn'd by all persons; especially Sea-men, Merchants, and Travellers. Written originally in Latine [by Apollonius, Leovitius, and others. Collected] and now Englished by R. Turner, Φιλομαθης*. London: J. Cottrel.

Turner, R. Steven. 1974. "University Reformers and Professorial Scholarship in Germany, 1760–1806." In Stone 1974, 2:495–531.

Tyard, Pontus de. 1557. *L'univers, ou Discours des parties de la nature du monde*. Lyon: Ian de Toures and Guillaume Gazeau.

Tyson, Gerald P., and Sylvia S. Wagonheim, eds. 1986. *Print and Culture in the Renaissance: Essays on the Advent of Printing in Europe*. Newark: University of Delaware.

Underwood, E. Ashworth. 1953. *Science, Medicine, and History*. London: Oxford University Press.

Urbach, Peter. 1987. *Francis Bacon's Philosophy of Science*. La Salle, IL: Open Court.

Ursus, Nicolaus Reimarus. 1588. *Fundamentum Astronomicum*. Strasbourg.

———. 1597. *De Hypothesibus Astronomicis: seu Systemate Mundano, Tractatus Astronomicus & Cosmographicus. Item Astronomicarum Hypothesium a se inventarum, oblatarum, & editarum, contra quosdam eas sibi temerario seu potius nefario ausu arrogantes, Vendicatio et Defensio, Eque Sacris Demonstratio*. Prague: Apud Autorem.

Valcke, Louis. 1996. "Jean Pic de la Mirandole et Johannes Kepler: De la Mathématique à la Physique." *Rinascimento* 36: 275–96.

Valla, Giorgio. 1501. *De Expetendis et Fugiendis Rebus Opus*. Venice: Aldus Manutius.

Vanden Broecke, Steven. 2001. "Dee, Mercator, and Louvain Instrument Making: an Undescribed Astrological Disc by Gerard Mercator (1551)." *Annals of Science* 58:219–40.

———. 2003. *The Limits of Influence: Pico, Louvain, and the Crisis of Renaissance Astrology*. Leiden: Brill.

———. 2005. "Evidence and Conjecture in Cardano's Horoscope Collection." In Oestmann, Rutkin, and von Stuckrad 2005, 207–23.

Van Egmond, Warren. 1980. *Practical Mathematics in the Italian Renaissance: A Catalogue of Italian Abacus Manuscripts and Printed Books to 1600.* Florence: Istituto e Museo di Storia della Scienza di Firenze.

Van Helden, Albert. 1977. *The Invention of the Telescope.* Transactions of the American Philosophical Society 67, no. 4.

———. 1985. *Measuring the Universe: Cosmic Dimensions from Aristarchus to Halley.* Chicago: University of Chicago Press.

———. 1994. "Telescopes and Authority from Galileo to Cassini." *Osiris* 9:9–29.

Van Nouhuys, Tabitta. 1998. *The Age of Two-Faced Janus: The Comets of 1577 and 1618 and the Decline of the Aristotelian World View in the Netherlands.* Leiden: Brill.

Vasoli, Cesare. 1965. "Lucio Bellanti." In *Dizionario biografico degli Italiani.*

———, ed. 1980a. *Idee, istituzioni, scienza ed arti nella Firenze dei Medici.* Florence: Giunti-Mart.

———, ed. 1980b. *La cultura delle corti.* Bologna: Capelli.

———. 1980c. "Gli astri e la corte: L'astrologia a Ferrara nell'età ariostesca." In Vasoli 1980b, 129–58.

Vermij, Rienk. 2002. *The Calvinist Copernicans: The Reception of the New Astronomy in the Dutch Republic, 1575–1750.* Amsterdam: Royal Netherlands Academy of Arts and Sciences.

Viala, Alain. 1985. *Naissance de l'écrivain: Sociologie de la littérature à l'âge classique.* Paris: Les éditions de Minuit.

Vickers, Brian, ed. 1984. *Occult and Scientific Mentalities in the Renaissance.* Cambridge: Cambridge University Press.

Vieri, Francesco de. 1568. *Discorso di M. Francesco di Vieri, cognominato il Secondo Verino, del soggeto, del numero, dell'uso, et della dignità et ordine degl'habiti dell'animo, cio dell'arti, dottrine morali, scienze specolative, e facoltà strumentali.* Florence.

Vitruvius Pollio, Marcus. 1496. *De Architectura.* Venice.

Voelkel, J. R. 1999. "Publish or Perish: Legal Contingencies and the Publication of Kepler's *Astronomia Nova.*" *Science in Context* 12: 33–59.

———. 2001. *The Composition of Kepler's "Astronomia Nova."* Princeton and Oxford: Princeton University Press.

Voigt, Johannes Kaspar. 1841. *Briefwechsel der berühmtesten Gelehrten des Zeitalters der Reformation mit Herzog Albrecht von Preussen.* Königsberg: Gebrüder Bornträge.

Voss, Wilhelm. 1931. "Handschriftliche Bemerkungen in alten Büchern." *Die Sterne,* 179–84.

Wagner, David L., ed. 1983. *The Seven Liberal Arts in the Middle Ages.* Bloomington: Indiana University Press.

Wagner, Klaus. 1975. "Judicia Astrologica Colombiniana. Bibliographisches Verzeichnis einer Sammlung von Praktiken des 15. und 16. Jahrhunderts der Biblioteca Colombina (Sevilla)." *Archiv für Geschichte des Buchwesens* 15: 1–98.

Walker, Christopher, ed. 1996. *Astronomy before the Telescope.* New York: St. Martin's Press.

Walker, D. P. 1958. *Spiritual and Demonic Magic from Ficino to Campanella.* London: Warburg Institute.

———. 1972. *The Ancient Theology: Studies in Christian Platonism from the Fifteenth to the Eighteenth Century.* Ithaca, NY: Cornell University Press.

Wallace, William A. 1984a. "Galileo's Early Arguments for Geocentrism and His Later Rejection of Them." In Galluzzi 1984, 31–40.

———. 1984b. *Galileo and His Sources: The Heritage of the Collegio Romano in Galileo's Science.* Princeton, NJ: Princeton University Press.

———. 1988. "Traditional Natural Philosophy." In Schmitt and Skinner 1988, 201–35.

———. 1998. "Galileo's Pisan Studies in Science and Philosophy." In Machamer 1998, 27–52.

Walters, Alice N. 1997. "Conversation Pieces: Science and Politeness in Eighteenth-Century England." *History of Science* 35: 121–54.

Walz, Angelus Maria. 1930. "Zur Lebensgeschichte des Kardinals Nikolaus von Schönberg, O.P." In *Mélanges Mandonnet* (2) (Bibliothèque Thomiste, no. 14). Paris: Vrin, 371–87.

Warburg, Aby. 1912. "Italian Art and International Astrology in the Palazzo Schifanoia, Ferrara." In Warburg 1999, 732–57.

———. 1999. *The Renewal of Pagan Antiquity: Contributions to the Cultural History of the European Renaissance.* Ed. Kurt W. Forster; trans. David Britt. Los Angeles: Getty Research Institute for the History of Art and the Humanities.

Ward, Seth. 1654. *Vindiciae Academiarum.* Oxford.

Warner, Deborah. 1994. "Terrestrial Magnetism: For the Glory of God and the Benefit of Mankind." *Osiris* 9:67–84.

Warwick, Andrew. 2003. *Masters of Theory: Cambridge and the Rise of Mathematical Physics.* Chicago: University of Chicago Press.

Watanabe-O'Kelly, Helen. 2002. *Court Culture in Dresden: From Renaissance to Baroque.* Basingstoke: Palgrave.

Waterbolk, Edzo H. 1974. "The 'Reception' of Copernicus's Teachings by Gemma Frisius." *Lias: Sources and Documents Relating to the Early Modern History of Ideas* 1:225–42.

Watson, Elizabeth See. 1993. *Achille Bocchi and the Emblem Book as Symbolic Form.* Cambridge: Cambridge University Press.

Watts, Pauline Moffitt. 1985. "Prophecy and Discovery: On the Spiritual Origins of Christopher Columbus's 'Enterprise of the Indies.'" *The American Historical Review* 90:73–102.

Wear, Andrew. 1985. "Explorations in Renaissance Writings on the Practice of Medicine." In Wear, French, and Lonie 1985, 118–45.

Wear, Andrew, Roger French, and I. M. Lonie, eds. 1985. *The Medical Renaissance of the Sixteenth Century.* Cambridge: Cambridge University Press.

Weber, Max. 1919. "Science as a Vocation." English translation in Gerth and Mills 2000, 129–56.

Webster, Charles. 1982. *From Paracelsus to Newton: Magic and the Making of Modern Science.* (Eddington Memorial Lectures, Cambridge University, November 1980.) Cambridge: Cambridge University Press.

Wedderburn, John. 1610. *Quatuor Problematum Quae Martinus Horky Contra Nuntium Sidereum de Quatuor Planetis Novis Disputanda Proposuit: Confutatio.* (Padua: Petrus Marinelli.) In Galilei 1890–1909, 3:153–78.

Weinberg, Bernard. 1961. *A History of Literary Criticism in the Italian Renaissance.* 2 vols. Chicago: University of Chicago Press.

Weinberg, Steven. 2001. *Facing Up: Science and Its Cultural Adversaries.* Cambridge: Harvard University Press.

———. 2008. "Without God." *New York Review of Books,* September 25.

Weinstein, Donald. 1970. *Savonarola and Florence: Prophecy and Patriotism in the Renaissance.* Princeton, NJ: Princeton University Press.

Weisheipl, James. 1965. "Classification of the Sciences in Medieval Thought." *Medieval Studies* 27: 54–90.

———. 1978. "The Nature, Scope and Classification of the Sciences." In Lindberg 1978, 461–82.

Weissman, Ronald. 1987. "Taking Patronage Seriously: Mediterranean Values and Renaissance Society." In Kent, Simmons, and Eade 1987, 25–45.

Welser, Marcus. 1591. *Fragmenta tabulae antiquae in quis aliquot per Romanas provincias itinera. Ex Peutingorum bibliotheca. Edente, et explicante. Marco Velsero Matthaei F. Aug. Vind.* Venice: Aldus Manutius.

Wesley, Walter. 1978. "The Accuracy of Tycho Brahe's Instruments." *Journal for the History of Astronomy* 9: 42–53.

Westfall, Richard S. 1971. *The Construction of Modern Science: Mechanisms and Mechanics.* Cambridge: Cambridge University Press.

———. 1980. *Never at Rest: A Biography of Isaac Newton.* Cambridge: Cambridge University Press.

———. 1985. "Science and Patronage: Galileo and the Telescope." *Isis* 76: 11–30.

———. 1989. *Essays on the Trial of Galileo.* Notre Dame, IN: University of Notre Dame Press.

Westman, Robert S. 1971. "Johannes Kepler's Adoption of the Copernican Hypothesis." PhD diss., University of Michigan.

———. 1972a. "The Comet and the Cosmos: Kepler, Mästlin and the Copernican Hypothesis." In Dobrzycki 1972, 7–30.

———. 1972b. "Kepler's Theory of Hypothesis and the 'Realist Dilemma.'" *Studies in History and Philosophy of Science* 3: 233–64.

———, ed. 1975a. *The Copernican Achievement.* Berkeley: University of California Press.

———. 1975b. "The Melanchthon Circle, Rheticus, and the Wittenberg Interpretation of the Copernican Theory." *Isis* 66: 165–93.

———. 1975c. "Three Responses to the Copernican Theory: Johannes Praetorius, Tycho Brahe and Michael Maestlin." In Westman 1975a, 285–345.

———. 1975d. "Michael Mästlin's Adoption of the Copernican Theory." *Studia Copernicana* 14: 53–63.

———. 1977. "Magical Reform and Astronomical Reform: The Yates Thesis Reconsidered." In McGuire and Westman, 1977, 2–91.

———. 1980a. "The Astronomer's Role in the Sixteenth Century: A Preliminary Study." *History of Science* 18: 105–47.

———. 1980b. "Humanism and Scientific Roles in the Sixteenth Century." In Schmitz and Krafft 1980, 83–99.

———. 1983. "The Reception of Galileo's *Dialogue* in the Seventeenth Century: A Partial World Census of Extant Copies." In Galluzzi 1983, 329–72.

———. 1984. "Nature, Art, and Psyche: Jung, Pauli, and the Kepler-Fludd Polemic." In Vickers 1984, 177–229.

———. 1986. "The Copernicans and the Churches." In Lindberg and Numbers 1986, 76–113.

———. 1990. "Proof, Poetics and Patronage: Copernicus's Preface to *De revolutionibus.*" In Lindberg and Westman 1990, 167–206.

———. 1993. "Copernicus and the Prognosticators: The Bologna Period, 1496–1500." *Universitas* 5 (December): 1–5.

———. 1994. "Two Cultures or One? A Second Look at Kuhn's *The Copernican Revolution.*" *Isis* 85: 79–115.

———. 1996. Review of Antonino Poppi, *Cremonini e Galilei. Isis* 87: 166–67.

———. 1997. "Zinner, Copernicus and the Nazis." *Journal for the History of Astronomy* 28: 259–70.

———. 2008. "Was Kepler a Secular Theologian?" In Westman and Biale 2008, 24–52.

Westman, Robert S., and David Biale, eds. 2008. *Thinking Impossibilities: The Legacy of Amos Funkenstein.* Toronto: University of Toronto Press.

Whewell, William. 1857 [1837]. *History of the Inductive Sciences.* 3rd ed. London: John W. Parker and Son.

Whiteside, D[erek] T[homas]. 1970. "Before the *Principia:* The Maturing of Newton's Thoughts on

Dynamical Astronomy, 1664–1684." *Journal for the History of Astronomy* 1: 5–19.

Wilkins, John. 1684 [1638]. *A Discourse concerning a New Planet, Tending to prove That 'Tis Probable our Earth is One of the Planets.* London: John Gellibrand.

Willet, Andrew. 1593. *Tetrastylon Papisticum, That is, The Foure Principal Pillers of Papistrie, the first conteyning their raylings, slanders, forgeries, untruthes: the second their blasphemies, flat contradictions to scripture, heresies, absurdities: the third their loose arguments, weake solutions, subtill distinctions: the fourth and last the repugnant opinions of New Papisters With the old; of the newe one with another; of the same writers with themselves: yea of Popish religion with and in it selfe.* N.p.

Williams, Arnold. 1948. *The Common Expositor: An Account of the Commentaries on Genesis, 1527–1633.* Chapel Hill: University of North Carolina Press.

Williams, George Huntston. 1992. *The Radical Reformation.* 3rd ed. (Sixteenth Century Essays and Studies, 16.) Kirksville, MO: Sixteenth Century Journal Publishers.

Williams, Raymond. 1976. *Keywords: A Vocabulary of Culture and Society.* New York: Oxford University Press.

Wilson, Catherine. 1995. *The Invisible World: Early Modern Philosophy and the Invention of the Microscope.* Princeton, NJ: Princeton University Press.

Wilson, Curtis A. 1970. "From Kepler's Laws, So-called, to Universal Gravitation: Empirical Factors." *Archive for History of Exact Sciences* 6: 89–170.

———. 1975. "Rheticus, Ravetz, and the 'Necessity' of Copernicus's Innovation." In Westman 1975a, 17–39.

———. 1978. "Horrocks, Harmonies, and the Exactitude of Kepler's Third Law." In Wilson 1989b, 236–59.

———. 1989a. "Predictive Astronomy in the Century after Kepler." In Taton and Wilson 1989, 161–206.

———. 1989b. *Astronomy from Kepler to Newton.* London: Variorum Reprints.

———. 1989c. "The Newtonian Achievement in Astronomy." In Taton and Wilson 1989, 233–74.

Winch, Peter. 1958. *The Idea of a Social Science and Its Relation to Philosophy.* London and Henley: Routledge and Kegan Paul.

Wind, Edgar. 1969. *Giorgione's 'Tempesta': With Comments on Giorgione's Poetic Allegories.* Oxford: Oxford University Press.

Wing, Vincent. 1651. *Harmonicon Coeleste or, The Coelestiall Harmony of the Visible World.* London: Robert Leybourn.

———. 1656. *Astronomia Instaurata: or, A New and Compendious Restauration of Astronomie.* London: R. and W. Leybourn.

———. 1669. *Astronomia Britannica.* London: Sawbridge.

Wing, Vincent, and William Leybourne. 1649. *Ura-nia Practica, or, Practical Astronomy in Vi Parts.* London: R. Leybourn.

Winkler, Mary G., and Albert Van Helden. 1992. "Representing the Heavens: Galileo and Visual Astronomy." *Isis* 83: 195–217.

Wischnath, Johannes Michael. 2002. "Michael Mästlin als Tübinger Professor: akademischer Alltag an der Schwelle zum 17. Jahrhundert." In Betsch and Hamel 2002, 195–231.

Wisdom, J. O. 1974. "Testing an Interpretation within a Session." In Wollheim 1974, 332–48.

Witekind, Hermann. 1574. *De Sphaera Mundi.* Heidelberg.

Wohlwill, Emil. 1904. "Melanchthon und Copernicus." *Mitteilungen zur Geschichte der Medizin under der Naturwissenschaft* 3: 260–67.

Wolf, Hieronymus. 1558. See Leovitius 1558.

———. 1657. See Turner 1657.

Wolfson, Harry A. 1962. "The Souls of the Spheres from the Byzantine Commentaries on Aristotle through the Arabs and St. Thomas to Kepler." *Dumbarton Oaks Papers* 16: 65–94.

Wollheim, Richard, ed. 1974. *Freud: A Collection of Critical Essays.* New York: Doubleday Anchor.

Woodward, William Harrison. 1924. *Studies in Education during the Age of the Renaissance, 1400–1600.* Cambridge: Cambridge University Press.

Woolfson, Jonathan. 1998. *Padua and the Tudors: English Students in Italy, 1485–1603.* Toronto: University of Toronto Press.

Wrightsman, A. Bruce. 1970. "Andreas Osiander and Lutheran Contributions to the Scientific Revolution." PhD diss., University of Wisconsin.

———. 1975. "Andreas Osiander's Contribution to the Copernican Achievement." In Westman 1975a, 213–43.

Wursteisen, Christanus. 1573. *Quaestiones Novae in Theoricas Planetarum.* Basel: Henricus Petri.

Yates, Frances A. 1947. *The French Academies of the Sixteenth Century.* London: Warburg Institute.

———. 1964. *Giordano Bruno and the Hermetic Tradition.* Chicago: University of Chicago Press.

Zaccagnini. 1930. *Storia dello Studio di Bologna durante il Rinascimento.* (Biblioteca dell' "Archivum Romanicum.") Geneva: Leo S. Olschki S.A.

Zacuth, Abraham ben Samuel de Ferrara. 1518. *Pronostico dell'anno 1519.* Bologna?: n.p.

Zagorin, Perez. 1998. *Francis Bacon.* Princeton, NJ: Princeton University Press.

Zambelli, Paola. 1966a. "Giacomo Benazzi." In *Dizionario biografico degli Italiani,* vol. 8, 180–81.

———. 1966b. "Lattanzio Benazzi." In *Dizionario biografico degli Italiani,* vol. 8, 181.

———, ed. 1982. *Scienza, credenze, occulte, livelli di cultura.* Florence: Olschki.

———. 1986a. "Many Ends for the World: Luca Gaurico, Instigator of the Debate in Italy and in Germany." In Zambelli 1986b.

————, ed. 1986b. *Astrologi Hallucinati: Stars and the End of the World in Luther's Time.* Berlin: W. de Gruyter.

————. 1987. "Teorie su Astrologia, Magia e Alchimia (1348–1586) nelle Interpretazioni Recenti." *Rinascimento* 26: 95–119.

————. 1992. *The Speculum Astronomiae and Its Enigma: Astrology, Theology and Science in Albertus Magnus and His Contemporaries.* Dordrecht: Kluwer.

Zammito, John H. 2004. *A Nice Derangement of Epistemes: Post-Positivism in the Study of Science from Quine to Latour.* Chicago: University of Chicago Press.

Zamrzlová, Jitka, ed. 2004. *Science in Contact at the Beginning of the Scientific Revolution.* Prague: National Technical Museum.

Zemplén, Jolan. 1972. "The Reception of Copernicanism in Hungary (A Contribution to the History of Natural Philosophy and Physics in the 17th and 18th Centuries)." In Dobrzycki 1972, 311–56.

Zik, Yaakov. 2001. "Science and Instruments: The Telescope as a Scientific Instrument at the Beginning of the Seventeenth Century." *Perspectives on Science* 9:259–84.

Zinner, Ernst. 1941. *Geschichte und Bibliographie der astronomischen Literatur in Deutschland zur Zeit der Renaissance.* Leipzig: Karl W. Hiersemann.

————. 1956. *Deutsche und niederländische Instrumente des 11.–18. Jahrhunderts.* Munich.

————. 1988 [1943]. *Entstehung und Ausbreitung der copernicanischen Lehre.* 2nd ed. Ed. H. M. Nobis and F. Schmeidler. Munich: C. H. Beck.

————. 1990. *Regiomontanus: His Life and Work.* E. Brown. Amsterdam: North Holland Publishers.

INDEX

Italic page references indicate illustrations.

Antwerp: astrological texts, 28; Inquisition, 198; Plantin, 548n88

Apianus, Petrus, 39, 114; *Astronomicum Caesareum*, 160, 283

Apianus, Philip, 164, 168, 224, 545n55, 562n32; Maestlin and, 260, 261, 265, 560n10, 562n40

Apocalypse of St. John, 2

apocalyptics, 13–14, 109, 263, 486; Bellarmine, 218; Galileo, 355; Garcaeus, 167; Kepler and, 14, 329, 393, 424; Leovitius, 47, 228; Lichtenberger, 14, 68–70, 263, 327, 393; Lutherans, 109, 125, 223, 309, 354, 390, 393; Maestlin and, 14, 262; Melanchthon, 167, 230, 355; Newton, 512; Osiander, 130; Roeslin, 254, 255; Tycho and, 14, 228, 252–53; witchcraft, 405. *See also* Elijah prophecy

apodictic standard of demonstration, 101, 123, 130, 169–70, 184, 267, 271; Wing and, 503. *See also* Aristotle, standard of demonstration.

Apollonius, 5, 180, 348, 516n16; *Conic Elements*, 180

Apuleius, *Golden Ass*, 94

Arabic: Alhazen, 180; astrology, 28, 47, 115, 123; *Tetrabiblos*, 43, 45, 46. *See also* al-Battani

Aratus/Aratea, 153; *Phaenomena*, 282

Archimedes, Galileo and, 353, 355, 378, 440–41, 463–64

Arienti, Sabadino degli, *Le porretane*, 93

Ariew, Roger, 516n35, 588n17

Aristarchus, 55, 161, 209–10, 231, 348, 425, 533n177

Aristarchus Redivivus, 499

Aristotle, 5, 47; aether, 225, 420; Alexander's patronage, 158; anti-Aristotelian stances, 19, 223–24, 254, 276, 305, 370, 394, 420, 442, 503, 555n164, 588n6; authority, 39–40, 41, 90, 123–25, 126, 169–70, 177, 195–96, 206, 341, 488; Bellanti and, 289–90; Bellarmine and, 367; Benedetti critique, 356; Bruno and, 301, 303, 304, 305, 394; Clavius and, 207, 209, 214, 215; comets and novas and, 225, 230, 254, 257, 258; concentric spheres, 99, 429; *De caelo/On the Heavens*, 55–56, 61, 124, 128, 209, 223–24, 304, 325, 500–501, 533n9; *De mundo/On the World*, 124, 209; *De sensu et sensato*, 41; dialectical reasoning, 601n83; Digges and, 19, 273, 279–80; disciplinary classifications, 32–35, 39–40; diversity of "Aristotelianisms," 420, 588n4; Earth's motion/doctrine of simple motion, 4, 161, 279, 531n153; elements theory, 104, 295; falling bodies, 310; fluid heavens, 19, 420, 588n5; four causes, 318; Frischlin and, 260–61; Galileo and, 355, 442; Gilbert and, 369, 370; gravity, 7; Harriot and, 587n67; Hill and, 587n73; immutability of the heavens, 19, 230, 242, 258, 261, 295, 555n164; intuitions, 4, 6; Italian authors and, 13, 385; Kepler and, 14, 317–20, 332–33, 394, 395, 494, 581n181; *Logic*, 215; Maestlin and, 19, 264, 279–80; Mazzoni and, 575n27; Melanchthon critique, 113, 199; *Metaphysics*, 537n113; *Meteorologica*, 40, 63, 200, 209, 356, 551n48, 581n181; *Narratio Prima* and, 123–25, 126, 128, 215, 537n113; natural philosophy, 41, 110, 126, 195, 205, 356, 367, 420, 425, 497, 531n151, 559n143, 588n17, 589n30; Newton and, 506; *Nicomachean Ethics*, 562n47; Novara and, 529n90; novas and, 225, 230, 401, 555n164; *Physics*, 32, 63, 113, 205, 223–24, 273, 279, 356; planetary order, 99, 209, 213; *Posterior Analytics*, 33, 100, 123,

124, 136, 207; and Pythagoreans, 1, 400, 515n2, 531n151; Riccioli and, 501; Scaliger and, 319; standard of demonstration, 5, 8, 123, 242, 272, 274, 279–80, 298, 492, 529n93, 531n115, 548n100, 600n50; Sun's position, 86, 123–24, 213; "theoretical philosophy," 34–35, 39; *Topics*, 103; world's eternity, 218

Arquato, Antonio, 91, 524n42

arts: astrology among, 177, 178; "liberal," 42–43; mathematics classified in, 147; Melanchthon's definition of, 113

Arzachel/al-Zarqali, 86, 200, 201

Ashmole, Elias, 586n50

aspectual astrology, 582n26; Kepler, 379, 381, 403–5, 463, 581n183, 584n95

astral-elemental qualities, 54, 57; Chamber's critique of, 409; Clavius, 212; Copernicus, 102, 105; Jupiter, 52–54, 132, 133, 181, 187, 212, 549n119; Kepler's critique of, 396; Mars, 52, 132, 181, 212, 409, 549n119; ordering of, 57; Pereira's critique of, 205–6; Pico's critique of, 86, 87, 105, 133; Ptolemy's *Tetrabiblos*, 52–54, 57, 65, 184–88, 192, 321, 409; "quantity of the qualities," 184, 188, *188*, 189; Saturn, 52–54, 86, 132, 181, 187, 189, 206, 409, 549n119; Sun, 52, 53, 54, 61, 132, 133, 144, 189, 200, 212, 318–21, 324, 379

astrolabes, 86

Astrologia, Urania as, *48*

astrology, 517n4; academic opposition to, 244; Albertus Magnus defending, 522n170; Arabic, 28, 47, 115, 123; Aristotle as authority, 39–40, 206; Augustine vs., 172, 206, 208, 522n170; Catholic Church vs., 12–13, 83, 110, 172, 202, 205–8, 354; causal claims of, 25–29, 69, 91, 96, 113, 178, 200, 205, 206, 393–94, 396–97; classifications of, 30–40, *36*, 43; conjectural, 38, 91–93, 177, 206, 208, 245, 246, 248, 401; Copernicus praising, 30; Copernicus's exceptionalism, 3, 10–11, 28–29, 104–5, 515n8, 518n13, 539n180; *De Revolutionibus* title and, 134–35; Digges on, 269–70, 275; doctrine of directions, 574n36; facial lines, *241*; Florentine tradition of, 174, 354; and geography, 146; Gilbert and, 580n135; good/bad, 31, 84, 187, 190–92; "Halcyon Days" of English, 502; harmonic, 397, 399, 409, 411, 414; Harriot and, 414, 587n63; heliostatic arrangement and, 20, 378, 420, 424, 426; Islamic, 40; legitimacy for Christians, 11–13, 115, 119–21, 170; manuals of, 12, 192; mathematical, 166, 242, 245; between mathematics and physics, 323; Medici court and, 12, 172–74, 227, 447, 546; Medina's orthodoxy and, 199–202, 217; Melanchthon on, 11, 110–13, 143–44, 181, 202–3, 213–14, 227, 245–46, 252–53, 259, 323, 355; natural philosophers and, 110, 511, 533n9; new Piconians vs., 226–27; Newton and, 511–13; optics, 12, 183–85, 190; physical, 13, 147, 205, 214, 245, 246, 289, 320–24, 336, 372, 378–81, 431; post-Tridentine, 202–4, 207; Protestant reformers and, 11–12, 109–12, 120; reform of, 183–85, 310, 395–99, 403, 414, 487, 502; safe/dangerous, 406, 585nn20, 21; Savonarola vs., 82–83, 226, 322, 527n37, 575n10; as superstitious, 76–78, 82, 111–12, 115, 171, 193, 199; term use, 10, 30; "true," 540n195; Tycho and, 28, 234, 245–47, 252–

and, 184, 323; English Keplerian-Copernican astrologers and, 511; Frischlin drawing on, 227; Garin edition, 527n53; Gassendi and, 511; on heat and light, 85, 320–21; Hemminga and, 227, 322–23; Heydon and, 410; Inquisition and, 198; Kepler and, 14, 16, 320–30, 332, 372, 380–81, 394, 396–97, 401, 403, 407, 415, 571n65, 581n177, 584nn79,95; Maestlin and, 259, 321, 323; Medina and, 199, 200, 322; Melanchthon and, 112–13, 130, 181, 534n25; modernizers and, 496, 513; *Narratio Prima* vs., 103, 121, 148, 571n74; Novara and, 91–93; Offusius vs., 185–90, 323, 397; Osiander and, 130; Pereira and, 206, 553n89; Pisa curriculum and, 354; plagiarism, 112, 113; planetary order, 11, 12, 86–87, 92, 99–100, 103–5, 113, 169, 209; *Prutenic Tables* (Reinhold) and, 170; publications, 10–11, 82, 96, 97, 100, 226; revolutions and nativities, 539n181; Sagan and, 527n37; scripture ahead of natural divination, 132, 133; Strasbourg edition (1504), 112; Tycho and, 245–50; Wolf vs., 192–93

divination: Calvin on, 171–72; Cicero's *On Divination*, 82, 96, 199, 534n28; classes of, 199; Inquisition vs., 198; naturalistic, 110–13, 144, 159, 170, 202, 422. *See also* foreknowledge; prognostication

Dobrzycki, Jerzy, 98, 102, 189, 540n197

domification, 167, 527n54, 582n26

Donahue, William H., 430, 584n99, 589n65

Donne, John, 423

Dorling, Jon, 604n153

Dorn, Gerhard, *Anatomy of Living Bodies*, 443

Dousa, Janus, "Delineation of the Orbs of Venus and Mercury," 282

Drake, Stillman, 365, 376, 386–89, 455, 577–83, 592n59, 594n1, 595n12

Dreyer, J. L. E., 29, 234, 244, 248, 335, 535n63, 556n55, 558n96; *History of Astronomy from Thales to Kepler*, 29

Dudith, Andreas, 282–83, 565n9; circle, 282, 363, 400

Duhem, Pierre, 180, 512, 516nn16,40, 517n62; *To Save the Phenomena*, 8–9

Duhem-Quine thesis, 8, 85, 140, 513

Duke, Dennis, website of planetary animations, 522n152, 548n84

Dürer, Albrecht, 157, 168, 175, 299

Dutch, 426–29. *See also* Leiden

Dybvadius, Georgius Christophorus (Joergen/Jørgen Christoffersen), 313, 559n138; *Short Comments on Copernicus's Second Book*, 164

Earth: magnetic, 368–74, 427–29, 446, 497; soul, 380, 397, 427, 581n182; stellar parallax, 6. *See also* Earth's motion; geocentric ordering

Earth's motion, 1, 426, 495, 496, 589n32; Aristotle, 4, 161, 279, 531n153; Bellarmine, 490–91, 599n19, 600n28; Bible and, 4, 130–31, 298, 429, 536n91, 603n150; Boulliau, 602n114; Clavius, 210, 216, 554n119; *Commentariolus*, 56, 101–2, 122, 532n165, 536n97; De Morgan, 21; *De Revolutionibus*, 5–6, 56–57, 169, 189–90, 265, 279, 282–85, 285, 316, 428–29, 495, 516n24, 589n32; Descartes, 497; Digges, 271; Galileo, 361, 366, 425, 439, 456, 457, 490–92, 495, 599n19, 600nn28,50,52; Gemma, 183; Gilbert, 370;

428, 589nn32,58; Harriot and, 412; Hooke and, 506, 509–10; Kepler, 316, 361, 379, 430, 492, 589n32; Maestlin, 265, 266; Mersenne, 496; *Narratio Prima*, 126, 536n99; Newton, 511; Offusius, 188, 189; Pythagorean opinion, 1, 33, 56, 523n180, 533n177; Regiomontanus, 61, 524n207; Reinhold and, 152, 189, 285; Riccioli, 501; shared-motions conundrum, 50–51, 53, 57, 61, 101, 169, 211–12, 267, 523n183, 531n151, 554n119; Stevin, 426–29; Tycho, 57, 248, 282, 370; Wittich, 282–85

eccentrics, 50, 63, 135, 309; Achillini, 99; Alexandrian Greeks, 5; Averroës, 205; Bellarmine and, 599n19; Clavius, 215–16; *De Revolutionibus*, 135, 151–52, 169, 215–16, 283, 539–40n181; Peurbach, 33–34, 35, 53; Ptolemy, 5, 61, 99; Regiomontanus, 5, 56, 58, 156n18; Rheticus, 29, 118, 119, 125; Roeslin, 347; Stevin, 429

eclipses: conjunctionist astrology, 47; Leovitius, 228; lunar, 98–99, 234–35; Ptolemy, 65; Rheticus, 99, 541n38; solar, 511; Tycho, 234–35

economic support: of Copernicus, 78, 526n5; of Galileo, 389, 448, 467; of Tycho, 237. *See also* patronage

Ecphantus, 425

Edward VI, King of England, 191, 227

Egenolphus, Christanus, 198

Einhard, Abbot, 532n169

The Electrician, 520n85

elements, 35, 36, 61; Aristotelian theory, 104, 295; Chamber on, 409; Euclid's *Elements*, 78, 143, 202–3; planetary order, 49, 51, 86, 100, 102; prognostication, 71; Schreckenfuchs, 37. *See also* astral-elemental qualities

Elias, Norbert, *Court Society*, 596n64

Elijah prophecy: Carion, 119–20, 536n76; comet and, 255, 536n76; Gasser and, 150; Kepler and, 393; Leovitius and, 228; Maestlin and, 262; Melanchthon on, 232, 536n76; Osiander and, 130; Rheticus and, 118, 125, 130, 150, 232, 309; Wittenbergers and, 262

Elizabeth, Queen of England, 268, 371, 375, 404, 409, 411, 587n63

elliptical astronomy, 16–17, 320, 325, 353, 366, 377, 431–32, 492–93, 503, 505

El se movera un gato, 67

emblems, 421, 446, 593n77

encyclopedism, 499

Enderbie, Percy, 553n87

end of the world. *See* apocalyptics

Engelhardt, Valentine, 168

England: apocalypticians, 228, 252; Bruce, 375, 391; Bruno, 300, 404, 568nn107,128; Cardano, 191, 227; Dee, 185, 356; Digges, 224, 268–76, 356; Gilbert, 16, 372; "intelligencers" for, 363, 404, 458; Kepler, 16, 375, 403–15, 449; King Charles II, 503–4; King Edward VI, 191, 227; King Henry VIII, 128, 227; Lord Burghley, 269–70, 375, 548n100; professionalization, 29; prognostication, 409, 411–12, 502–5; Queen Elizabeth, 268, 371, 375, 404, 409, 411, 587n63; and Riccioli-Galileo disagreement, 506; textbook publishing, 423, 434, 590n1; works in vernacular, 268. *See also* James VI of Scotland/James I of England; London; Oxford

ephemerides, 427, 434, 505, 519n67; Carelli, 240; Leovitius, 228, 263, 427; London, 229; Maestlin, 257, 263;

Medicinalium Libri Quinque, 557n78; *Herbarium*, 240

Mauri, Alimberto, 389, 402, 582n15, 583n46; *Considera-zioni*, 384

Maurice of Nassau, prince of Orange, 426–27

Maximilian, archduke of Tyrol, 403

Maximilian courts, 238–39; Maximilian I, 82, 238, 525n42; Maximilian II, 185, 240

Maxwell, James Clerk, 19, 455

Mayr, Simon, 384–85, 386–87, 390, 436, 448, 454; German nation at Italian university, 582n23, 596n71; *New Table of Directions*, 386; *Prognosticon Astrologi-cum*, 582n16

Mazzoni, Jacopo, 575n27; and Galileo, 356–59, 366, 413, 575–76nn32,33, 578n97; Guidobaldo and, 441; *In Universam Platonis et Aristotelis Philosophiam Praeludia/Comparison of Plato and Aristotle*, 356–57, 413; Pisa, 441, 442

McGuire, J. E., 512

McKirahan, Richard, 33

McMullin, Ernan, 515n11, 517n72, 599n10, 600n28

Medici court, 17, 79, 172–74, 442–51; Alessandro, 12, 172–73, 227; Antonio, 442–43, 450; astrology at, 12, 172–74, 227, 447, 546nn18,21; Christina, 358, 436, 443–46, 447, 490, 495; Cosimo I, 172, 173–74, 203, 438, 443, 447, 546n21; Cosimo II, 389, 443–47, 450–51, 465–69, 472, 487, 569n12, 583n47, 593nn78,79, 594n105, 595n20, 596n57; court sensibilities, 442–47; Ferdinand, 356, 443, 447, 552n71; Francesco, 174, 447; Galileo and, 17, 358, 433, 436–38, 443–51, 456–59, 465–68, 472, 475–80, 487, 579n109, 583n47, 591nn23,26, 593nn87,88, 595n20, 596n57, 598n126; Giuliano, 459, 595n20, 598n126; Giulio, 480; Isa-bella, 552n68; Marie, 466, 477; university in Pisa, 442, 448, 465

medicine: astrology/astronomy and, 64–66, 83, 91, 93, 96, 113, 199, 244–45, 354, 387, 422, 529n99; Coper-nicus studying, 137; Copernicus's Varmia practice, 104; Medici court, 442–43; and scholarly melancholy, 191–92; Tycho's Paracelsian notion of, 567n67; in universities, 41, 78, 93, 147, 162, 224, 238, 529n99. *See also* disease

Medigo, Elia del, 87

Medina, Miguel, 13, 199–202, 206, 217, 322, 551nn41,53; *A Christian Exhortation, or Concerning the Right Faith in God/De Recta Fidei*, 199, 205, 208

melancholy, scholarly, 191

Melanchthon, Philipp, 11–12, 15, 110–13, 114, 141–71, 193; and Albrecht, 145, 154–55, 214, 543n196; apocalyptics, 167, 230, 355; Aristotle critique, 113, 199; astrology, 11, 110–13, 143–44, 181, 202–3, 213–14, 227, 245–46, 252–53, 259, 323, 355; astronomy textbooks, 164–68, 213; and Clavius, 213–14; conjunctionist astrology, 47; Daniel commentary, 130; daughter Magdalena, 144; and *De Revolutionibus*, 11–12, 141–71, 182, 190; Digges and, 275; Elijah prophecy, 232, 536n76; Frischlin and, 227, 264, 562n34; Gasser, 116; Gemma and, 181–83; and Grynaeus, 51–52, 202; Heerbrand student of, 263; Hemmingsen and, 245–46; horoscope by Giun-tini, 32; Index, 197, 311, 355, 576n43; *Initia Doctrinae Physicae*, 113, 161, 165, 170, 181, 182, 190, 205, 550n130;

Liebler and, 323, 336, 354; and mathematical disci-plines, 143, 168, 202–3; Offusius and, 187; *Oratio de Orione*, 120; "Oration on the Dignity of Astrology," 143, 144; Osiander and, 128, 130; perpendicularity, 185; *Praeceptor Germaniae*, 110; Rheticus and, 11–12, 109, 114–15, 127–30, 144, 147–50, 162, 170, 181; *schola privata*, 143; Schöner, 110, 112, 534n25; Socrates of, 150; sons-in-law, 112, 144, 171, 543n88; stacking prin-ciple, 187; *Tetrabiblos* commentaries and lectures, 46, 354; Tübingen, 120, 264; university system, 143–70, 214; Wolf friend of, 192

Melanchthon circle, 11–12, 144–64, 260, 424, 544n134; Reinhold, 11–12, 144–60, 543n96, 544n134; Rheticus, 11–12, 109, 114–15, 127–30, 144, 147–50, 162, 170, 181; Wittenberg orbit, 163. *See also* Wittenberg response to Copernicus

Melantrich, 557n83

Mercator, Gerard, 12, 548n104; and Dee, 179, 183, 184, 548n103; and Family of Love, 548n88; globe maker, 141, 541n7

Mersenne, Marin, 419, 495–97, 501, 601nn58,61, 602n90; *Novarum Observationum*, 602n90

Messahalah, 32, 90, 397; *De Revolutionibus Annorum*, 45; *Three Books*, 166

meteorology, 234

meteors, 223, 242

Methuen, Charlotte, 323, 569n21

Metrodorus, 213

Michelangelo, 79

microscope, 39; Hooke, 507

Miechów, Matthew of, 102–3, 532nn161,162

Milan: Index of, 551n31; publishers, 53, 96

Milani, Marisa, 387

military capacity, German mathematicians, 168

military coalitions, Venice, 79

military events: prognostication focus, 92. *See also* wars

military and political concerns, Digges, 274–75, 564n96

Milius, Crato, 198

Miller, Peter, 442, 592n38

Mirandola, Antonio Bernardo, *Monomachia*, 208

Mocenigo, Alvise, 375

Mocenigo, Giovanni, 375

modernizers, 15, 228, 351–415, 417–85, 495–513, 599n2; and Bruno, 301, 394; Descartes, 305, 493; emergent problematic of the, 419–23; Galileo, 366, 419, 425, 434–54, 468, 483, 495; Harriot, 16, 411–13; Heydon, 16, 410; quarrel among, 374–75, 402, 415; Regiomon-tanus, 63; Rudolfine court, 16, 17, 395, 400, 462; *via media* and, 286; world systems and physical crite-rion, 350. *See also* Kepler, Johannes

modus tollens, 6–8, 61, 162, 262, 291, 298, 399

Moletti, Giuseppe, 311–12, 434, 448

Moller, Johann, 401

Monau, Jacob, 400

Montaigne, Michel de, vi, 258

Montulmo, Antonius de, *De iudiciis nativitatum (Con-cerning the Judgments of Nativities)*, 115, 166

Moon: Copernicus, 97–99, 125, 151, 161, 516n24, 531n136; eclipse of, 98–99, 234–35; Galileo and, 449, 450, 453,

Moon *(continued)*
456, 483, 502; Kepler and, 328, 380; motion of, 324;
parallaxes of, 97–98, 603n124; Wilkins and, 502
Moran, Bruce, 566n57
More, Henry, 500, 502, 601n74
Morin, Jean-Baptiste, 493, 497, 600n41, 601n72
Morsing, Elias Olsen, *Diarium*, 568n127, 569n137
Mosely, Adam, 566n34, 567n59, 574n54, 575n24
Moss, Jean D., 600n48, 601n83
motion: doctrine of simple, 161, 279; laws of, 113, 517n58;
Moon, 324; science of, 9, 353, 366, 441, 579n105;
shared-motions conundrum, 50–51, 53, 57, 61, 101,
169, 211–12, 267, 523n183, 531n151, 554n119; Sun, 1, 5,
6, 50–52, 92, 101, 123–24, 318–19, 375, 425, 516n21,
524n203, 537n116, 549–50n130, 570n44; tables, 5,
262. *See also* Earth's motion
Mottelay, Fleury, 580n125
Mulerius, Nicholaus, 427, 429, 491, 492
Müller, Philipp, 492–94
multiples, emergent fashion of, 175
Munich Indexes, 551nn29,38
Muñoz, Jerónimo, 224, 240–41, 242, 311, 390–91, 555n6
Münster, Sebastian, 114, 168, 198, 569–70n34, 573n142
Mylichius, Jacobus, 197
Mysterium Cosmographicum (Kepler), 378, 426, 432,
488, 570n35, 573n1, 576n56, 581n177; audiences
for, 403–4; Bruce and, 364–65; as cosmographical,
316, 323–50; and *De Revolutionibus*, 141, 266, 315,
316, 323–50, 393; Galileo and, 356–62, 425, 462, 463,
577nn64,66, 578n90, 579n102; Gilbert and, 372; Heydon and, 411; Maestlin and, 266, 315, 316, 317, 423–
24, 569n25; Magini and, 578n96, 579n100; preface
of, 361, 569n24, 572n118; resistance to, 335, 457;
reversal of perspective, 399; and scripture, 572n133;
universities not adopting as official text, 435

Nadal, Jerónimo, 204, 205, 552n83
Naibod, Valentine, 168, 248–49, 558n128; planetary
order, 248–49, 250, 282, 285, 301; *Three Books of Primary Instruction concerning the Heavens and Earth
and the Daily Revolutions of the World*, 248, 282
Napoleon, 260, 367
Narratio Prima, 114–30, 139–41, 170, 309–10, 486,
535–37; Aristotle, 123–25, 126, 128, 215, 537n113; audiences for, 109–10, 133; authorial responsibility for,
11, 114; Basel edition, 161, 534n43, 535n64; Bruno
and, 568n114; Clavius and, 209, 215, 554n149; *Commentariolus* compared with, 100; Copernicus with
Novara, 78; Copernicus as "professor mathematum,"
526n22; dedicated to Schöner, 109, 110, 114–24, 149;
De Revolutionibus bundled with, 161, 282, 287, 491;
De Revolutionibus not mentioning, 110, 135, 137, 139,
533n8, 542n62; Earth's motion, 126, 536n99; eclipse
theory, 99; *Encomium Prussiae*, 126, 127, 129, 138–39,
145; Galileo and, 425, 463; Gasser prefatory letter,
109; Gemma reading, 179–80; heliocentric arrangement first proposed, 11, 121–22; Kepler and, 325–26,
330, 337, 340, 424, 463, 569n25; Maestlin and, 122,
125–26, 141, 261, 262, 265, 315, 325–26, 326, 358, 424,
569n25, 573n6, 576n44; necessity insisted upon,

121–26, 139, 165; no group of followers formed about,
147, 170; vs. Pico, 103, 121, 148, 571n74; planetary
order, 103, 125, 139–40, 179; pope excluded, 110; Praetorius
copy, 312; psychodynamic hypothesis and, 149, 150;
Reinhold and, 150, 178; Rothmann and, 297–98,
567n88; Socrates in, 150; Sun's motion, 524n203;
title, 120; Tolosani and, 196; Tycho and, 122, 287;
Wheel of Fortune, *118;* Wittich, 285, 287; world-
historical prophecies, 29, 118–21, 535n63
nativities. *See* horoscopes/nativities/genitures
natural philosophy, 374, 382–402, 464, 510; anti-
Aristotelian, 420, 503, 588n6; Aristotle, 41, 110, 126,
195, 205, 356, 367, 420, 425, 497, 531n151, 559n143,
588n17, 589n30; and astrology, 110, 511, 533n9; Bellarmine, 367; Bible and, 4–5, 9, 19, 110, 130, 131, 196, 213,
217–19, 226, 489–91, 496, 506, 510, 572n133; Bruno's
infinitist, 281, 304–6, 317, 394, 490; celestial signs
and, 257; *Commentariolus* and, 532n165; Copernican,
100, 101, 102, 374, 423–26; cosmography and, 325;
Descartes, 496–98; divination based on, 110–13, 144,
159, 170, 202, 422; English Restoration, 506–7; Galileo and, 353, 373, 423, 437, 457, 493, 495, 497; Gilbert
and, 371–72; hyperphysical, 225; Jesuits and, 205, 213–
19; Kepler and, 14, 320–29, 341, 353, 393–95, 403, 493,
495–97; Maestlin and, 267–68, 423; Medici court,
442–47; medicine and, 529n99; Melanchthon, 110–
13, 144, 161, 170, 202, 213, 336; *Narratio Prima*, 124,
282; new-style, 495–513; Newton, 31, 425–26, 430,
510–13; nova (1604) and, 385, 387, 388, 393–95; Nullists and, 230, 232; Origanus, 469; patronage, 146–
47, 436–40; Peripatetics, 232, 309; Pico and, 85;
planetary order, 16, 420–22, 496–98; practices of
ignoring, 419–20; Pythagoreans, 101, 531n151; social
status, 596n64; term, 17, 518n32; Tolosani, 195–96,
202, 374; Tycho, 234; Urban VIII, 516n44. *See also*
astronomy; modernizers; Paracelsus; traditionalists
natural theology, Clavius, 217
Nature, 19; Book of, 226, 263, 332, 362, 571n93, 577n62
Naylor, Ron, 366, 579n105
Neander, Michael, 163, 544n147, 576n43
necessity, 91, 121–26, 139, 165, 167
Neoplatonism: Cudworth, 511; in development of early
modern science, 317; Ficino, 190–91, 192; Florentine,
135; Kepler and, 317–18, 320, 321; and Novara, 10, 76,
135, 317; Tycho and, 234
Neo-Stoic philosophy of friendship, 363, 440, 592n38
nested spheres, 6, *50,* 314
Neugebauer, Otto, 97, 526n8, 531n136
newspapers, 502
New Theorics of the Planets (Peurbach), 49–55, 208, 423,
434; Brudzewo commentary on, 53–56, *55,* 70, 77–
78, 87, 212, 516n23, 525n49; Capuano commentary
on, 45, 51, 96, 434; in classroom, 164, 165, 166; Clavius and, 211–12, 554nn119,124; *De Revolutionibus*
title and, 135; Fantoni unpublished commentary on,
575n11; Galileo and, 354, 425; Kepler and, 164, 311,
334; Melanchthon additions, 143; "partial sphere"/
"total sphere," 49; Ptolemy's mathematical models,
519n77; Ratdolt's omnibus edition of (1491), 326;
Regiomontanus printing of, 27–28, 50, 53, 57; Rein-

Peurbach, Georg: authorial designation, 32; Bruno and, 301; epicycles, 50, 285; Ferrara, 71; geniture sent to Cardano, 175; Herzog August Bibliothek copies of, 44–45, 521–22n148; *Narratio Prima* and, 124; patron Emperor Frederick III, 71, 525n42; and Ptolemy's *Almagest*, 4, 51, 52, 164; and Regiomontanus, 63, 488; "theoric of orbs," 33–34, 35, 101, 519n69; Wittenberg curriculum, 290–91; work included in publisher's bundle, 522n171. See also *New Theorics of the Planets*

Peutinger, Conrad/Peutinger Map, 364

Phares, Simon de, 25–26, 63, 66

Philip II, King of Spain, 199

Philolaus, 55, 56, 425, 523n180, 533n177

physics: Aristotle's *Physics*, 32, 63, 113, 205, 223–24, 273, 279; astrology and, 13, 147, 205, 214, 245, 246, 289, 320–24, 336, 372, 378–81, 431; astronomy and, 13, 30–34, 39, 55, 152, 158–59, 161–62, 169, 205, 215, 289, 328, 340–41, 342, 431–33, 487; crucial experiments in, 8; division of roles in twentieth century, 520n83; epistemic status, in Ptolemy, 33; epistemic status, in Weinberg, 517n65; Kepler, 315, 317–28, 336, 372, 431, 497; location in science of stars, 30–36, 36; Melanchthon, 161, 214; *Narratio Prima*, 124; Riccioli, 501; seventeenth-century textbooks of, 422

Piccolomini, Aeneas Sylvius/Pope Pius II, 63, 139

Piccolomini, Enea, 450

Pico, Gian Francesco, 82, 84, 199, 208

Pico della Mirandola, Giovanni, vi, *83;* and astronomy, 85, 87, 103, 135, 332, 571n74; authorial designation, 32; Bolognese friends, 94, 95, *95,* 530n120; and Christian Kabbalah, 130; *Commentariolus* and, 103; death, 84, 246, 527n43; Ficino and, 84, 85, 527n50; geniture, 175, 246; *Heptaplus*, 120, 132–33; new Piconians, 226–27; Sextus Empiricus influence, 527n46. See also *Disputations against Divinatory Astrology*

Pietramellara, Giacomo de, 64, 70, 89, 91, 524n42

Pignoria, Lorenzo, 363, 583n38

Pindar's Olympian ode, 127

Pinelli, Gian Vincenzo, 355–56, 391, 575n23, 578nn87,88; circle, 362–66, 375, 421, 482, 578n97, 592nn48,53; death (1601), 366, 388; Galileo and, 366, 421, 440, 441, 442, 581n167

pirated literature, 46–47, 82, 96, 165, 466

Pisa: Benedetti, 356, 357; Castelli, 468, 488; Galileo, 300, 311–12, 353–57, 376, 441–43, 448, 464–65, 475, 488, 575n5; Mazzoni, 441, 442; traditionalists, 442, 464, 465; university, 442, 448, 465, 488, 575nn9,16, 576n47

Pistorius, Johannes, 403

Pius II (Aeneas Sylvius Piccolomini), Pope, 63, 139

plagiarism: Capra (1607), 389, 390, 447, 466; Galileo accused of, 365, 460, 578n90; Offusius, 269; Pico accused of, 112, 113

plague, Black Death/bubonic plague, 25

planetarium, 339, 519n68, 573n12

planetary conjunctions. *See* conjunctionist astrology

planetary equatorium, 38–39

planetary modeling, 169, 257, 309, 423, 486; Copernicus, 3, 12, 59; Goldstein and, 56–59, *60;* Keplerian, 297, 350, 433, 493, 497

planetary order, 1, 3–7, 14, 19, 29, 48–61, 76; comets and, 254, 257–58; as deductive outcome of common

center of gravity, 512; Galileo, recurrent novelties and, 477; Gilbert and, 370; Kuhn, 225; Mersenne and, 496; orreries, 6, 141, *142;* Peurbach's independent representation of planets, 49–50; Pico's criticisms of, 11, 12, 86–87, 92, 99–100, 103–5, 113, 169, 209; and political disorder, 81; Ptolemy and Copernicus compared, 57, 104, 301–3, *302,* 310, 335; scriptural compatibility with, 4, 130–33, 506; second-generation Copernicans and, 226, 259–81; uncontested, for Newton, 506; Wilkins and, 506; world pluralists and, 502. *See also* Capella, Martianus; Clavius, Christopher; *Commentariolus; De Revolutionibus Orbium Coelestium;* Descartes, René; heliocentric arrangement; Heydon, Christopher; Kepler, Johannes; Maestlin, Michael; Naibod, Valentine; *Narratio Prima;* natural philosophy; period-distance principle; Ptolemy, Claudius; Regiomontanus; Roeslin, Helisaeus; Stevin, Simon; *symmetria;* Tycho Brahe; underdetermination; Venus/Mercury ordering; Wittenberg response to Copernicus

planetary refraction, 290, 473, 597n103

Plato: Clavius, 209, 213, 214; *Commentariolus* on, 101; *De Revolutionibus* and, 318; Digges, 268, 269, 280, 563n71; doctrine of forms, 32–33; in evangelical curriculum, 143; five regular solids, 188, 269, 273, 463; Galileo, 357, 425; Kepler, 188, 317–18, 320, 321, 329, 330, 336, 343, 356, 408, 463; Limnaeus, 340; Mazzoni, 356–57, 413, 575n27; Mersenne, 496; *Philebus,* 42; Sun's position, 86, 181, 213; *Timaeus,* 188, 189, 200, 209, 329, 330, 342, 452. *See also* Neoplatonism

Platonic Academy, 76, 202, 526n3

Pliny, 122, 200

Plotinus, 85, 320

Plutarch, 135

Pocock, J. G. A., 541n212

Polanco, Juan Alfonso de, 20–24

Poland: cold war boundaries, 145. *See also* Frombork; Krakow; Varmia

politics: Gregorian calendar, 260; and cometary prognostication, 252; English Royalist, 503; "intelligencers," 363, 375, 404, 458; Kepler and court, 430, 462; languages of, 541n212; Maestlin's experience in, 262, 266; patron-client, 437, 455; political theory independent of the stars, 226. *See also* Inquisition; military events; modernizers; patronage; rulers

polyhedra, 584n94; Digges's, 269, 270, 273, 330, 563n63; Kepler's, 188, 329–41, *338,* 347–48, 357–58, 399, 463, 573n14; Plato's, 188, 269, 273, 463

Pomian, Krzysztow, 47

Pontano, Giovanni, 47; *Centiloquium* translation, 46, 521n127; *De Rebus Coelestis,* 104; *One Hundred Aphorisms* translation, 521n126

Pontormo, Jacopo da, *Vertumnus and Pomona,* 546n21

popes: Alexander VI, 79; Clement VII, 63, 133–34, 194; Clement VIII, 367; Gregory XIII, 260, 262; John Paul II, 489; Julius II, 79; Leo X, 63, 137; Nicholas V, 63; Paul IV, 197, 198, 207; Paul V, 491; Pius II (Aeneas Sylvius Piccolomini), 63, 139; Sixtus V, 202, 207, 354, 406, 412, 422; Urban VIII (Maffeo Barberini), 9, 412, 488, 489, 491, 516n44, 600n28. *See also* papal bulls; Paul III

prophecies. *See* divination; prognostication; world-historical prophecies

prosthaphaeresis, defined, 283

Protestants, 11, 110, 145, 194, 337; authorial classification, 32; Bellarmine as scourge of, 490; Calvinist, 171–72, 260, 290; Church of England, 410; Copernicans, 499–500; and Gregorian calendar, 260; on Index, 198, 494, 576n43; legitimacy of astrology for, 11–13; and Rudolf II, 239; Schmalkaldic League, 155. *See also* Bible; Christianity; Lutherans; Reformation/reformers, Protestant

Prowe, Leopold, 76

Pruckner, Nicolaus, 45

Prutenic Tables (Reinhold), 12, 141, 152–62, 182, 228, 309, 423, 486; Albrecht patronage and, 12, 128, 152–60, 170; and Clavius calendar, 260; ephemerides (general) using, 167, 505, 519n67; Galileo and, 593n79; Garcaeus and, 291, 545n180; Index, 197; Kepler and, 311, 332; Maestlin and, 257, 260, 261, 263, 560n5; Offusius and, 190; and safe prognosticatory practice, 171; Stadius's *Ephemerides* based on, 179, 181, 190, 261, 427, 542n63, 563n74; Tycho and, 234, 248; and Wittenberg curriculum, 166, 167, 169, 354

psychodynamic hypothesis, on Rheticus, 147–50, 542

Ptolemy, King, 153

Ptolemy, Claudius: aether, 33, 65–66, 524n18; astrology vs. astronomy, 47, 165, 173, 177, 245; *Astronomia* on her throne, 27; authority of, 63, 113, 214, 329; Bruno and, 301–3, 302, 568n115; chorography vs. geography, 146; classifications by, 26, 32, 33, 34–35, 37–38, 39, 40, 43; Clavius and, 208–17, 218; Copernicus compared, 57, 104, 182, 301–3, 302, 310, 335; Copernicus imitating, 1, 11, 102, 104, 117, 127, 539n181; Copernicus as second Ptolemy, 102, 219, 244, 265, 486; eccentrics and epicycles, 5, 61, 99, 335; equants, 126, 349, 424; *Geography*, 46, 91, 529n88; Kepler and, 329, 332, 335, 380–81, 395, 396–97, 409, 581n181; Kuhn and, 225–26; length of tropical year, 85–86; lunar parallax, 97–98; and *modus tollens* reasoning, use of, 6; *Narratio Prima* and, 124, 127–28; nested spheres, 314; Novara's critique of, 76; *Planetary Hypotheses*, 32, 33, 40, 289, 518n44; Regiomontanus and, 63, 488, 524nn205,206; rejects air moving with Earth, 516n28; stacking principle, 187, 188; Sun's position, 86, 105, 181, 211; "theoretical philosophy," 34–35; in Tycho's Copenhagen Oration, 244–45; *Urania* and, 27, 48; "world system" of, 345. See also *Almagest; Centiloquium; One Hundred Aphorisms;* planetary order; *Tetrabiblos*

publishers: Augsburg, 28, 364; Barozzi, 202–3; Company of Stationers, 406–7, 410; *De Revolutionibus*, 128–30, 133, 134, 141, 161, 177, 282, 287, 325, 576nn43,44; first printed book claimed, 168; heavenly literature, 1480s–1550s, 28; Hectoris, 82, 94–100, 527n35, 530nn118,121; on Index, 198, 576n43; Italy, 43, 45, 49, 53, 82, 94–100, 530nn126,130; London, 183, 184, 226, 252, 300, 406–7; Maestlin, 263; Nuremberg, 3, 45–46, 53, 96, 109, 114, 115, 116, 530n130; pirating by, 46–47, 82, 96, 165, 466; Royer, 185; *Tetrabiblos*, 1530s–1550s, 45–47; Tübingen, 166, 260, 263–64, 332, 337; Tycho, 237–38, 296, 383; Utrecht, 131; Wittenberg, 28, 51–52, 167. *See also* Basel publishers; litera-

ture; Petreius, Johannes; print technology; Ratdolt, Erhard; Salio, Girolamo

Pumfrey, Stephen, 580n138, 589n62

Puteolano, Francesco, 94

Pyrnesius, Melchior, 139

Pythagorean opinion: Aristotle and, 1, 400, 515n2, 531n151; Bruno and, 301–5, 371; celestial harmonies, 255, 282; Copernicus and, 486, 523n180, 533n177; Earth's motion, 1, 33, 56, 523n180, 533n177; Foscarini on, 488, 490, 588n14; Galileo and, 356, 425; Kepler and, 329, 330, 336, 337, 356, 397, 400, 408, 463; Mazzoni, 356–57; *Narratio Prima*, 125, 127–28, 330; natural philosophy, 101, 531n151; Newton and, 512; planetary arrangement, 304; Roeslin and, 343, 347; Tolosani vs., 196–97; Tycho's castle architecture, 237

Querenghi, Antonio, 359, 388, 391, 442, 583n38

Quietanus, Johannes Remus, 600n44

Quine, W. V. O., 74; Duhem-Quine thesis, 8, 85, 140, 513

Rabelaisian language, 242, 281

Rabin, Sheila, 321, 571n65

Raimondo, Annibale, 242, 557n93

rainbows, 379, 581n181

Raleigh, Walter, 411, 587n57

Ramus, Peter, 168–69, 185, 198, 430–31, 576n43; *Scholarum Mathematicarum*, 169, 203, 244, 545n187

Rankghe, Laurentius, 545n172

Rantzov, Heinrich: *Exempla quibus Astrologicae Scientiae Certitudo Astruitur*, 176; Tycho friend, 176, 348, 349

Rassius de Noens, Franciscus, 185

Ratdolt, Erhard (publisher), 43, 44, 49, 55; *New Theorics* omnibus edition (1491), 326

Rattansi, P. M., 512

Ravetz, Jerry, 122

realism, 8–9, 560n1

rectification, 173

reform: astrology, 183–85, 310, 395–99, 414, 487, 502; astronomy, 121, 230, 233–34, 244, 247–48, 310, 430–31, 487; calendar, 137, 138, 145, 194, 548n100, 555n153; science of the stars, 121, 169, 411, 424. *See also* Reformation/reformers, Protestant; revolutions

Reformation/reformers, Protestant, 71, 109–10, 520n86; Counter-Reformation, 359, 487; *De Revolutionibus* and, 137–39; Melanchthon's curriculum, 143; Nuremberg, 110, 128; popular propaganda, 137, 138; schools, 143; sincerity valued, 597n95; Wittenberg, 11–12, 15, 109–13, 120, 171, 207

refraction, 290–92, 295, 299, 369, 379, 473, 507, 567n62, 586nn51,55, 597n103

Regiomontanus: astrology excluded, 104; authorial designation, 32; and al-Bitruji, 52, 524n205, 532n167; Bruno and, 301; and Clavius, 209, 213; and comets, 231, 241, 270; *Defensio Theonis*, 524nn205,206; *Disputations against . . . Cremona's Theorics of the Planets (Deliramenta)*, 49, 70; against Earth's motion, 61, 524n207; and eighth sphere, 200–201; epicycle-eccentric transformation, 5, 58; Ferrara, 71; Manilius poem, 521n126; *New Theorics*, printing of, 4, 27–28, 50, 53, 57; and Novara, 97, 531n131; Nuremberg printing project, 96; *Paduan Oration* and praise of math-

Regiomontanus (continued)

ematics, 30, 143, 202, 219, 244; parallax, 231, 241, 270; period-distance relation, 7, 61, 524n205; and planetary order, 52, 56, 57, 58, 61, 99, 213, 523n183, 524n206; prognosticatory authority, 63, 97, 167; Reinhold's oration to, 143–44; Schöner and, 114, 116, 231–32; "second Ptolemy," 63, 488; and shared motions, 523n183; *Sixteen Problems on ˙ . . . Location of Comets*, 231–32; Sun's position, 87, 211; *Tabulae Directionum*, 55, 114, 144, 526n8; *Tradelist*, 63, 66, 115. See also *Epitome of the Almagest (Epytoma almagesti)*

Regius, Petrus, 46

Reinhold, Erasmus: Albertine patronage, 12, 128, 152–60, 170, 174, 439, 543nn88,96; astronomical axiom, 283; brother Johannes, 164; Bruno and, 301; Clavius and, 208, 218–19; *Commentary on New Theorics*, 33–34, 52, 53, 150–53, 164, 182, 204, 208, 285, 290–91, 355, 434, 522n155, 551n29, 573n138, 576n43; death, 158, 160, 166; Dee and, 184; and *De Revolutionibus*, 11–12, 147–48, 150–61, 151, 167, 169–70, 178, 245, 280, 283, 355, 567n79; and Earth's motion, 152, 189, 285; ephemerides, 427, 505, 519n67; Galileo and, 355, 456, 467; Gilbert and, 369; *Hypotyposes Orbium Coelestium (Hypotheses Astronomicae)*, 164–65, 245, 291; Index, 197, 551n29, 576n43; Kepler and, 320, 434–35; Maestlin and, 260, 263; and Melanchthon circle, 11–12, 144–60, 181, 214, 543n96, 544n134; Neander under deanship of, 544n147; and Offusius, 190; oration to Regiomontanus, 143–44; and Osiander dedication, 128; Praetorius and, 341; Proclus title, 545n159; Ptolemy's *Almagest*, 182, 552n73; Ramus on, 168; Rothmann and, 296–97; shared motion with Sun, 53, 522n155; Stevin and, 427; Tycho and, 248; Wilkins and, 498; Wittich and, 283, 285. See also *Prutenic Tables*

Reisacher, Bartholomew, 241

religions: agnosticism, 21, 410, 503; dissenters, 145; Family of Love and, 548n88; war, 155. *See also* Christianity; God; theology

Renaissance: Aristotle as authority, 39–40; astrology's scope, 487; Copernicus's exceptionalism, 518n13; Dutch as special language, 426; grammar schools, 29; Horace commentators, 136–37; humanist New Testament scholarship, 47; *Narratio Prima* and, 124–25; princely patronage in, 146; rhetorical fashion for praising or satirizing professions, 30; sensibility of order, 56; sincerity valued, 597n95

reputation, 12, 15, 73, 75, 82, 112, 114–16, 134, 141, 158, 172–74, 217, 239, 293, 311, 337, 348, 356, 389–90, 433, 434–36, 454, 471–72, 478, 480

resistance: to Copernicus, 266, 489, 498; as discussed in Kuhnian era, 488; to Galileo, 16, 17, 447, 466, 468–81; to Heydon, 414; to Kepler, 14–15, 335, 424, 449, 457, 494; Maestlin understanding of, 266; meaning within traditional society, 15. *See also* traditionalists

Restoration era, 506–7, 511, 603n146

revolutions: astronomical, 486–87, 539n181; celestial, 119–21; "deep" and "shallow," 515n11; *De Revolutionibus* title, 45, 135; Kuhn and, 3–4, 225–26, 259, 510; Marxism and social class, 603n149. See also *De Revolutionibus Orbium Coelestium*; Scientific Revolution

Revolutions of the Heavenly Spheres. See *De Revolutionibus Orbium Coelestium* (Copernicus)

Rhau, Georg, 167

Rhediger, Nikolaus, 400

Rheticus, Georg Joachim, 114–33, 145–47, 152, 172; aging, 150; "another Copernicus," 168; and astrology, 28; biographical narrative, 121, 426; Bruno and, 300, 301, 568n114; and Cardano, 175–76; *Chorographia*, 128, 146; Copernicus in Italy, 78, 80, 87–88; Copernicus's disciple, 6, 11–12, 78, 87, 103, 104, 114–18, 121, 131, 139, 145–50, 488, 542nn52,62, 557n76; and court diversity, 239; death (1574), 290; *De Revolutionibus* reading, 13, 110, 121–26, 131–36, 146–48, 169, 170, 178, 280, 330; and the Elijah prophecy, 118, 125, 130, 150, 232, 309; failure to attract new disciples, 488; father Georg Iserin, 148, 149, 150, 193, 271, 541–42n49; in Frombork with Copernicus, 78, 103, 114–15, 131, 139, 145–48; Gasser dedication, 116; and Giese, 117, 126–31, 138, 145, 146, 533n8; on importance of eclipses, 99, 541n38; on Index, 197, 311, 576nn43,44; Kepler and, 325–26, 329, 330, 335, 337, 340; Leipzig, 128, 147, 150, 153, 162; lost writing, 502; Lutheran, 11, 125, 131, 138–39, 195, 533n8; Maestlin and, 122, 125–26, 261, 262, 265, 266, 315, 325–26, 326, 358, 424; Mars parallax, 6, 287; and Melanchthon, 11–12, 109, 114–15, 127–30, 144, 147–50, 162, 170, 181; name, 148; Nuremberg, 114–15, 128; *Opusculum*, 130–33; and Osiander, 128, 291, 340, 355; and Pico's alleged plagiarism, 112, 113, 534n25; and planetary order, 101, 125, 139–40, 179, 209, 232, 422; Praetorius and, 150, 312; psychodynamic hypothesis on, 147–50, 542; Ramus and, 169; Riccioli and, 499; sodomy charge, 150, 542n63; standard of demonstration, 18; Sun's motion, 52; Wilkins and, 498; world system, 125, 281. See also *Narratio Prima*

Rhodius, Ambrosius, 404

Ricci, Agostino, 201

Ricci, Ostilio, 353, 590n3

Ricci, Saverio, 587n73

Riccioli, Giovanni Battista, 499–502; Gregory and, 506; Hooke and, 506, 509, 603nn133,143; *New Almagest (Almagestum Novum)*, 135, 217, 499–502, 500, 506, 551n30, 602nn90,95; "Our System," 288, 501; and probabilities, 501, 506, 510–11, 604n153; and Wing, 503; world system, 499, 510–11, 601–2nn89,98

Richardson, Alan, 517n73

Risner, Friedrich, 185, 186

Risorgimento, 367

Ristori, Giuliano, 12, 32, 172–74, 238, 438, 447; *Tetrabiblos* commentaries, 173, 354, 443, 575n6

Roberval, Gilles Personne de, 497, 602n90

Robison, Wade, 471

Roeslin, Helisaeus: astrological doctrine of directions, 574n36; Bruno and, 305; career, 224; comets and nova, 253, 254–57, 256, 342; *De Opere Dei Seu de Mundo Hypotheses*, 341, 574n42; and Fludd, compared, 408; Frankfurt Book Fair, 580n154; Gilbert and, 372; Heerbrand and, 574n35; Kepler and, 322, 339–49, 404, 420, 571n73; Maestlin lends book to, 293; planetary orderings, 343–48; as Schwenckfeldian, 256; Signaculum Mundi/"World's Seal," 343, *343*;

Swerdlow, Noel, 523nn 196,197, 548n86; Copernicus's *Commentariolus*, 56–57, 59, 97, 101, 102, 104, 516n18, 530n50, 531nn,136,146,151, 532nn165,172, 536nn96,97, 548n86

Swetz, Frank J., 520n100

symmetria: Copernicus, 7, 104, 125, 135–37, 187–90, 248, 265, 282, 299, 486, 539n186, 540n192; Digges, 271–72; Dürer, 299; Gilbert, 370–71; Kepler, 190, 330, 343, 349, 350, 398–99; Maestlin, 265, 343, 562n44; Offusius, 187–90, 248, 370; Praetorius, 313; Roeslin, 343, 347; Tycho, 237, 248, 299, 349, 350, 393, 399, 557n64; Vitruvius, 236–37, 248, 539n186, 540n199, Wittich, 283–85

syphilis/"French disease," 25–26, 62, 81

Syrenius, Julius, *On Fate*, 208

systema mundi. See world system/*systema mundi*

Szdłowiecki, Pawel, 94, 530n112

Tannery, Paul, 579n105

Tebaldi, Aegidius de, 43, 46

telescope, 39; *cerbottana* and, 451; Galileo, 15–17, 353, 377, 413, 433, 437, 441–42, 447–82, 486, 492, 497, 502, 579n102, 594n5, 595nn14,20; Hooke, 507, 508, 509; Riccioli, 499

Telesio, Bernardino, 13, 394, 420, 496

Tengnagel, Franz, 233, 383, 411, 430–31, 480, 556n53, 590nn68,70

Tessicini, Dario, 303, 304, 539n178

Tetrabiblos (Ptolemy), 3, 12, 29, 43–49, 289, 549n119; Abenrodan's commentary, 39–40, 43–46, 54, 90, 183, 230; Arabic, 43, 45, 46; astral-elemental qualities and, 52–54, 57, 65, 184–88, 192, 321, 409; astronomy in relation to astrology, 32, 37, 43, 173, 177; Camerarius translations, 45, 46, 113, 145, 166, 180, 575n6; Cardano's commentary, 177, 178; classification of, *36*; Dee and, 183–84, 190; *De Revolutionibus* and, 135, 170; Galileo and, 354; Gemma's "Letter to the Reader," 180; Giuntini and, 31, 173, 354, 575n7; Gogava translation/Louvain edition, 45–47, 179–81, 184, 547n72, 575n6; Greek, 43, 45, 47, 180; Latin, 43, 45, 46–47, 179, 180; Locatelli edition, 44, 183; Medina and, 199; Melanchthon commentaries and lectures, 46, 354; Miechów library, 532n161; Novara and, 96–97, 97, 530n129; Offusius and, 185, 187, 188; *One Hundred Aphorisms/Centiloquium*, 43; Peucer's textbook and, 165; planetary order, 56, 61, 99, 190, 310; prognostication with, 63, 65–66, 67, 90; Rheticus, 105; Ristori commentaries, 173, 354, 443, 575n6; Salio edition (1493), 35, 44, 45, 56, 96–97, 97, 98, 530n126; Strasbourg edition, 45; translators, 43–47, 113, 145; University of Bologna Library, 96–97

textbooks, 309, 422, 423, 427; Blundeville, 434, 590n1; Clavius, 13, 208–13, 423; English, 434, 590n1; first Copernican, 493–94; Frischlin, 263–64; Kepler, 434–35, 493–94; Maestlin, 260, 263–64, 311, 434; scholarly reputation and, 436; Stevin, 427; Strigelius, 260; Timpler, 422; Tübingen publishers, 166, 260, 263–64; Venetian publishers, 530n126; Wittenberg curriculum, 143–44, 164–68, 214

Thabit ibn-Qurrah, 85–86, 152, 200, 213

Theodoric, Sebastian, 164, 166, 198; *New Questions on the Sphere*, 161–62, 166

theology: astrology derived from, 174, 354; astronomer-astrologer seeking concordance with, 2; astronomy as natural theology, 217, 555n154; authority, 490; celestial sign interpretation, 230, 263, 385; Epicurean, 112; Kepler, 19, 217, 325, 327–32, 408, 511, 572n121; Maestlin, 224, 260, 263; Newton, 511; Osiander, 129; prognostication vs. divine foreknowledge, 12, 171–72, 263; Tübingen faculty and, 15, 224, 263, 329, 331–32, 336–39, 355, 362, 424, 436. *See also* Bible; God; religions

Theon, 9, 86

theoretical astrology, 36, 43–47, 44, 199, 202, 320; Brudzewo, 54; dilemma of underdetermination, 74, 202; Gemma, 180–81; Giuntini, 31; Kepler, 14, 264, 320–29, 336, 372, 376–81, 395–97, 403–15, 424, 463, 487, 573n1, 581n177, 584n79; Maestlin, 262–64, 280, 321–24, 420, 486, 561n28; Magini, 204; Melanchthon, 113, 181; Nadal, 205; Pico vs., 84, 101, 105, 113; planetary order and, 170; prognostication and, 43–47, 63, 65–66, 502; Ptolemy, 52, 56, 74, 96–97, 173, 199; Ristori, 173; Schöner, 115. *See also* astrology

theoretical astronomy, 13, 26, 29, 36, 38–39, 126, 172–73, 486–87, 511; Aristotle's distinctions applied, 164; Bellanti, 84, 93; Benazzi, 88–89, 92–93; Brudzewo, 54; Bruno and, 305; Clavius, 14, 207–8; *Commentariolus*, 103–4, 135; Copernicus's competences in, 78, 93; Copernicus's freedom from patronage and, 526n5; *De Revolutionibus* and, 109, 113, 135–38, 141, 150–58, 181, 208, 215, 282, 539n181; Digges, 280; dilemma of underdetermination, 202; equantless, 126–27, 152, 215, 248, 349, 493; Galileo, 15, 311–12, 353, 357, 362, 377, 450–51, 577n63; Gemma, 180–81; Kepler, 14, 316–35, 340–41, 349, 357–61, 394–95, 399, 430–31, 487, 585n107; Maestlin and, 14, 256–57, 260, 262–68, 311, 317, 323, 570n38, 573n7; Magini, 204; Medina, 199; *Narratio Prima*, 119, 121–27; natural philosophy and, 230, 232, 374; Newton, 511; nova (1572) and rise of theoretical astronomer, 230–34; Novara, 91–93, 97, 529nn90,94; Nullists, 230, 232, 250, 253, 281, 385, 390, 475; Paul of Middelburg, 137; Pereira, 205; Peurbach's *New Theorics*, 53–55, 208, 519n77; planetary tables, 93, 97–98, 128, 420; practice of, 267–68; prognostication and, 12, 14, 26, 28, 38–43, 280, 496, 502–11; Ptolemy, 29, 54, 208, 214; Regiomontanus, 54, 97; Rothmann, 294–95, 296; Sacrobosco's *Sphere*, 55; Schreckenfuchs, 37, 207; Stevin, 427–29, 439; Tycho, 13, 14, 152, 228–35, 244–48, 267, 294–96, 310, 327, 349, 422, 430–31, 567n70; warfare, 62; Wedderburn, 481. *See also* astronomy; planetary order; textbooks

theoretical holism, 8–9, 21

theorica: Campanus usage, 39–40; Miechów usage, 532n162. *See also* theoretical astrology; theoretical astronomy

theorica/practica distinction, 38–43, 63, 168, 520n106, 537n129. See also *practica; theorica*

Thirty Years' War, 239, 516n48

Thomas Aquinas/Thomism, 83, 84, 195–96, 205, 553n88; "Treatise on the Work of the Six Days," 218

Thoren, Victor, 556n55, 558n138

Text:	9.25/11.75 Scala Pro
Display:	Scala Pro
Compositor:	Integrated Composition Systems
Indexer:	Barbara Roos
Illustrator:	B. Harun Küçük
Printer and Binder:	Thomson-Shore, Inc.